Brooke Holmes, Klaus-Dietrich Fischer (Eds.)
The Frontiers of Ancient Science

Beiträge zur Altertumskunde

Herausgegeben von Michael Erler, Dorothee Gall,
Ludwig Koenen und Clemens Zintzen

Band 338

The Frontiers
of Ancient Science

―

Essays in Honor of Heinrich von Staden

Edited by
Brooke Holmes
and
Klaus-Dietrich Fischer

with the assistance of
Emilio Capettini

DE GRUYTER

ISBN 978-3-11-055922-4
e-ISBN (PDF) 978-3-11-033633-7
e-ISBN (EPUB) 978-3-11-038930-2
ISSN 1616-0452

Library of Congress Cataloging-in-Publication Data
A CIP catalog record for this book has been applied for at the Library of Congress.

Bibliografische Information der Deutschen Nationalbibliothek
Die Deutsche Nationalbibliothek verzeichnet diese Publikation in der Deutschen Nationalbibliografie; detaillierte bibliografische Daten sind im Internet über http://dnb.dnb.de abrufbar.

© 2017 Walter de Gruyter GmbH, Berlin/München/Boston
Dieser Band ist text- und seitenidentisch mit der 2015 erschienenen gebundenen Ausgabe.
Druck und Bindung: Hubert & Co. GmbH & Co. KG, Göttingen
∞ Gedruckt auf säurefreiem Papier
Printed in Germany

www.degruyter.com

Preface

It would indeed mean sending owls to Athens, or coals to Newcastle, were one to explain the importance of Heinrich von Staden for the study of ancient science and medicine to his colleagues in the field, to say nothing of his impact on the history of philosophy, classics, and literary theory. In a career spanning over four decades, Heinrich has transformed the study of ancient Graeco-Roman medicine by pioneering a scholarly approach that is equally attentive to the specificities of language, culture, history, and individual authors and large-scale philosophical and methodological questions.[1] Theoretical sophistication and philological precision, vast erudition and conceptual flair—these are the hallmarks of the inimitable von Staden style. Heinrich's ability to move effortlessly between different scholarly traditions and cultures has enabled him to bring contemporary problems in the history and philosophy of science to bear on classical antiquity while at the same time challenging outmoded paradigms of "premodern" science and medicine and integrating ancient texts into transhistorical conversations. Born and raised in South Africa, Heinrich also has a remarkable talent for moving between cultures and languages. With a publication record in multiple languages, he embodies the twenty-first-century ideals of globalism and multiculturalism as much as the traditionally polyglot cosmopolitanism of classics as a discipline. Heinrich's insatiable curiosity about the world is rivaled only by his generosity towards his far-flung friends, young and old. Even in the midst of a whirlwind of international engagements, he will always manage to find time for one-on-one conversation (almost certainly in the mother tongue of his interlocutor).

Heinrich's support of scholars working in premodern science and medicine has nowhere been more evident than in his creation of a robust, international research community in the history of science during his twelve-year tenure as Professor of Classics and History of Science in the School of Historical Studies at the Institute for Advanced Study in Princeton (1998–2010), a post he took up after teaching at Yale University for thirty years (and as William Lampson Professor of Classics and Comparative Literature from 1996–1998). It was at the Institute that the idea for a volume honoring Heinrich was first generated, when Arsenio Ferraces Rodríguez and Cloudy Fischer were working there during the summer of 2009; Brooke Holmes joined the editorial team the following year. Sadly, Arsenio found himself unable to continue with his collaboration soon

---

1 See the bibliography on pp. 707–12.

after the first call for papers had gone out in 2011, and the project had to proceed without him, on both sides of the Atlantic.

The present volume gathers contributions from twenty-nine historians of ancient and early modern medicine, science, religion, and philosophy who were fortunate to be Visitors or Members of the Institute during Heinrich's time there. Ranging from mechanics and mathematics to medicine and magic, from Bronze Age archaeology to modern receptions of Hippocratic texts, from peacocks to badgers, they speak to the richness and breadth of Heinrich's own expertise. As is fitting for a tribute to a true citizen of the world, the volume includes contributions in four languages (all languages in which Heinrich is fluent) and represents a range of national traditions and styles: as Heinrich himself might say, "E pluribus unum" and "Vive la différence!". The editors hope that this volume will not only be seen as a tribute to Heinrich's exquisite scholarship long overdue, his generous support of scholars young and old, and his kindness and charm, but that it will also invite those who open it to explore again Heinrich's own rich corpus of scholarship.

The editors would like to thank David Kaufman and especially Emilio Capettini for their assistance with the preparation of this volume; special thanks is due to Emilio for the preparation of the general index. We are also grateful to Terrie Bramley for her help with various aspects of the publication; and to Katharina Legutke, Florian Ruppenstein, and Mirko Vonderstein at de Gruyter for their support of the project. BH is also grateful to Caroline Bynum for sage guidance during the editing and publication process. Finally, acknowledgment is due to the David Magie Research Fund in the Department of Classics at Princeton University for very generous support of the volume's preparation and publication.

Table of Contents

Preface —— V

Abbreviations —— XI

Isabella Andorlini
Egypt and the Medicinal Use of Papyrus According to Soranus and Other Physicians —— 1

Markus Asper
Medical Acculturation?: Early Greek Texts and the Question of Near Eastern Influence —— 19

Han Baltussen
"Hippocratic" Oaths?: A Cross-Cultural Exploration of Medical Ethics in the Ancient World —— 47

Alan C. Bowen
Simplicius in Thirteenth-Century Paris: A Question —— 67

Andrea Falcon
Aristotle and the Study of Animals and Plants —— 75

Christopher A. Faraone
A Case of Cultural (Mis)translation?: Egyptian Eyes on Two Greek Amulets for Ophthalmia —— 93

Klaus-Dietrich Fischer
Gesund durchs Jahr mit Dr. Hippokrates – Monat für Monat! —— 111

Allan Gotthelf
Teleology and Embryogenesis in Aristotle's *Generation of Animals* 2.6 —— 139

Danielle Gourevitch, Philippe Charlier
Un ex-voto ophtalmologique inscrit à Rougiers (Var) : pour une étude des maladies des yeux dans le monde gallo-romain —— 175

Brooke Holmes
Medicine and Misfortune: *Symptōma* in Greek Medical Writing —— 191

Carl A. Huffman
Mathematics in Plato's *Republic* —— 211

Katerina Ierodiakonou
Hellenistic Philosophers on the Phenomenon of Changing Colors —— 227

Jacques Jouanna
Erotian, Reader of Hippocrates' *Prognostic:* A New Discovery —— 251

Joshua T. Katz
Aristotle's Badger —— 267

W. R. Laird
Heron of Alexandria and the Principles of Mechanics —— 289

Helen Lang
Plato on Divine Art and the Production of Body —— 307

Roberto Lo Presti
"For sleep, in some way, is an epileptic seizure"
(*somn. vig.* 3, 457a9–10) —— 339

Arnaldo Marcone
Il *Numen Augusti* nel *senatus consultum de Cn. Pisone patre* —— 397

Stephen Menn
How Archytas Doubled the Cube —— 407

Ian S. Moyer
A Revised Astronomical Dating of Thessalus' *De virtutibus herbarum* —— 437

Vivian Nutton
What's in a *Nomen*?: Vlatadon 14 and an Old Theory Resurrected —— 451

D. T. Potts
An Archaeological Meditation on Trepanation —— 463

Christine Proust
Des listes pour apprendre, résoudre, classer, archiver, explorer ou inventer —— 493

Francesca Rochberg
Conceiving the History of Science Forward —— 515

Amneris Roselli
Galeno sull'autenticità del *Prorretico I* —— 533

Thomas Rütten
Hippokrateskommentare im 16. Jahrhundert: Peter Memms Eidkommentar als Paradigma eines gegenwartsbezogenen Genres —— 561

Mark J. Schiefsky
***Technē* and Method in Ancient Artillery Construction: The *Belopoeica* of Philo of Byzantium —— 615**

Heinrich Schlange-Schöningen
Herrschaftskritik bei Galen —— 655

Philip van der Eijk
Galen on the Assessment of Bodily Mixtures —— 675

Contributors —— 699

Heinrich von Staden: Bibliography 1975–2012 —— 707

Index locorum qui e scriptoribus antiquis et medii quod dicunt aevi Graecis atque Latinis citantur —— 713

General Index —— 737

Abbreviations

ABL	Harper, R. F. *Assyrian and Babylonian Letters* (Chicago: The University of Chicago Press, 1892–1914)
AE	*L'année épigraphique: revue des publications épigraphiques relatives à l'Antiquité romaine* (Paris: Presses Universitaires de France, 1889–)
Ael.	Aelianus
nat. anim.	*de natura animalium*
Aesop.	Aesopus
Aesch.	Aeschylus
sept.	*septem contra Thebas*
Aët.	Aëtius philosophus
Aëtius Amid.	Aëtius Amidenus
lib. med.	*libri medicinales*
Afric.	Iulius Africanus
Alex. Aphr.	Alexander Aphrodisiensis
de an.	*de anima*
in de sens.	*in librum de sensu commentarium*
mantissa	*mantissa*
quaest.	*quaestiones*
Alex. Trall.	Alexander Trallianus
Ps.-Alex. Trall.	Pseudo-Alexander Trallianus
puls. ur.	*de pulsibus et urinis*
Anax.	Anaxagoras
Anon. Lond.	Anonymus Londinensis
Anon. Paris.	Anonymus Parisinus
de morb. acut. et chron.	*de morbis acutis et chroniis*
Antyll.	Antyllus
AO	Cuneiform tablets in the collections of the Musée du Louvre, Paris
Apollon. Perg.	Apollonius Pergaeus
con.	*conica*
Apul.	Apuleius
apol.	*apologia*
Archig.	Archigenes
Archim.	Archimedes
con. sph.	*conoidea et sphaeroidea*
aequil.	*de planorum aequilibriis*

sph. cyl.	*de sphaera et cylindro*
Archyt.	Archytas Tarentinus
Aret.	Aretaeus
Aristoph.	Aristophanes
Acharn.	*Acharnenses*
ran.	*ranae*
Arist.	Aristoteles
de an.	*de anima*
an. post.	*analytica posteriora*
cael.	*de caelo*
cat.	*categoriae*
div. somn.	*de divinatione per somnum*
eth. Nic.	*ethica Nicomachea*
gen. anim.	*de generatione animalium*
hist. anim.	*historia animalium*
incess. anim.	*de incessu animalium*
insomn.	*de insomniis*
iuv.	*de iuventute*
long. vit.	*de longitudine vitae*
mem.	*de memoria*
metaph.	*metaphysica*
meteorol.	*meteorologica*
part. anim.	*de partibus animalium*
phys.	*physica*
poet.	*poetica*
pol.	*politica*
pr.	*problemata*
sens.	*de sensu*
somn. vig.	*de somno et vigilia*
top.	*topica*
Ps.-Arist.	Pseudo-Aristoteles
col.	*de coloribus*
mech.	*mechanica*
Artem.	Artemidorus Daldianus
Aug.	Augustus
Beda	Beda
temp. rat.	*de temporum ratione*
BM	British Museum, London, museum siglum

BRM 4	Clay, A. T. *Babylonian Records in the Library of J. Pierpont Morgan*, vol. 4 (New Haven: Yale University Press, 1923)
CAD	Biggs, R. D., et al., eds. *The Assyrian Dictionary of the Oriental Institute of the University of Chicago* (Chicago: The Oriental Institute, 1956–)
Cael. Aur.	Caelius Aurelianus
acut. pass.	*celeres vel acutae passiones*
diaet. pass.	*de speciali significatione diaeticarum passionum fragmentum*
tard. pass.	*tardae passiones*
CAG	*Commentaria in Aristotelem Graeca*
Cass. Fel.	Cassius Felix
Cels.	Celsus
Chalc.	Chalcidius
in Tim.	*in Platonis Timaeum*
Cic.	Cicero
acad.	*academica*
fin.	*de finibus*
Phil.	*Philippicae*
Tusc.	*Tusculanae disputationes*
CIL	Mommsen, Theodor, et al., eds. *Corpus Inscriptionum Latinarum* (Berlin: G. Reimer/De Gruyter, 1863–)
Colum.	Columella
CT	British Museum, Department of Egyptian and Assyrian Antiquities. *Cuneiform Texts from Babylonian Tablets in the British Museum* (London: The Trustees of the British Museum, 1896–1990)
Democr.	Democritus
Ps.-Democr.	Pseudo-Democritus
med.	*liber medicinalis*
Diocl.	Diocles Carystius
Diodor. Sic.	Diodorus Siculus
Diog. Laërt.	Diogenes Laërtius
Diosc.	Dioscorides
mat. med.	*de materia medica*
simpl.	*de simplicibus*
Ps.-Diosc.	Pseudo-Dioscorides
alex.	*alexipharmaca*

Elias	Elias
in cat.	*in Aristotelis categorias commentaria*
Emped.	Empedocles
Epicur.	Epicurus
ep. Hdt.	*epistula ad Herodotum*
Erasistr.	Erasistratus
Erm	Hermitage (Ermitage, Eremitage) Museum, St. Petersburg, museum siglum
Erot.	Erotianus
Euc.	Euclides
elem.	*elementa*
Eur.	Euripides
Herc. fur.	*Hercules furens*
Eutoc.	Eutocius
in de sph. cyl.	*commentarii in libros de sphaera et cylindro*
in con.	*commentaria in conica*
Fest.	Festus
Gal.	Galenus
de alim. facult.	*de alimentorum facultatibus*
de antidot.	*de antidotis libri II*
ars med.	*ars medica*
de atra bile	*de atra bile*
de bon. mal. suc.	*de bonis malisque sucis*
de caus. puls.	*de causis pulsuum*
de com. sec. Hipp.	*de comate secundum Hippocratem*
de comp. med. per gen.	*de compositione medicamentorum per genera*
de comp. med. sec. loc.	*de compositione medicamentorum secundum locos*
de const. art. med.	*de constitutione artis medicae ad Patrophilum*
de cris.	*de crisibus*
de diebus decr.	*de diebus decretoriis*
de diff. resp.	*de difficultate respirationis*
de exp. med.	*de experientia medica*
ad Glauc. meth. med.	*ad Glauconem de methodo medendi*
gloss.	*linguarum seu dictionum exoletarum Hippocratis explicatio*
in Hp. aph. comm.	*in Hippocratis aphorismos commentaria*
in Hp. art. comm.	*in Hippocratis de articulis commentaria*
in Hp. epid. I comm.	*in Hippocratis epidemiarum librum primum commentaria III*

in Hp. epid. II comm.	in Hippocratis epidemiarum librum secundum commentaria V
in Hp. epid. III comm.	in Hippocratis epidemiarum librum tertium commentaria III
in Hp. epid. VI comm.	in Hippocratis epidemiarum librum sextum commentaria I–VIII
in Hp. fract. comm.	in Hippocratis de fracturis librum commentaria III
in Hp. nat. hom. comm.	in Hippocratis de natura hominis librum commentaria II
in Hp. off. comm.	in Hippocratis de officina medici librum commentaria III
in Hp. prog. comm.	in Hippocratis prognosticum commentaria III
in Hp. prorrh. I comm.	in Hippocratis prorrheticum I commentaria III
in Hp. acut. comm.	in Hippocratis de victu acutorum librum commentaria IV
indol.	de indolentia
de libr. propr.	de libris propriis
de loc. aff.	de locis affectis
de meth. med.	de methodo medendi
de nomin. med.	de nominibus medicis
de opt. med. cogn.	de optimo medico cognoscendo
de ord. libr. suor.	de ordine librorum suorum ad Eugenianum
de plac. Hipp. et Plat.	de placitis Hippocratis et Platonis
de praecogn.	de praecognitione ad Epigenem
de an. aff. dign. et cur.	de propriorum animi cuiuslibet affectuum dignotione et curatione
protrept.	adhortatio ad artes addiscendas
quod opt. med.	quod optimus medicus sit quoque philosophus
qualit. incorp.	quod qualitates incorporeae sint
de san. tuenda	de sanitate tuenda
de sectis	de sectis ad eos qui introducuntur
de simpl. med. temp. ac fac.	de simplicium medicamentorum temperamentis et/ac facultatibus
subf. emp.	subfiguratio empirica
de sympt. caus.	de symptomatum causis
de sympt. diff.	de symptomatum differentiis
de temper.	de temperamentis
de ther. ad Pis.	de theriaca ad Pisonem
de usu part.	de usu partium
de venae sect. adv. Erasistrateos	de venae sectione adversus Erasistrateos Romae degentes

Ps.-Gal.	Pseudo-Galenus
def. med.	*definitiones medicae*
de hist. philos.	*de historia philosopha*
introd. s. medic.	*introductio sive medicus*
remed. parab.	*de remediis parabilibus*
succed.	*de succedaneis*
Gell.	Aulus Gellius
noct. Att.	*noctes Atticae*
Hdt.	Herodotus
Hero Alex.	Hero Alexandrinus
bel.	*belopoeica*
dioptra	*commentatio dioptrica*
mech.	*mechanica*
pneum.	*pneumatica*
Herodianus	Herodianus
soloec.	*de soloecismo et barbarismo*
Herophil.	Herophilus
Hipp.	Hippocrates
acut.	*de victu acutorum*
acut. (sp.)	*de victu acutorum (spuria)*
aff.	*de affectionibus*
aph.	*aphorismi*
art.	*de arte*
Coac.	*Coacae praenotiones*
decent.	*de decenti habitu*
epid. I–VII	*epidemiarum* I–VII
fist.	*de fistulis*
flat.	*de flatibus*
foet. exsect.	*de foetus exsectione*
int.	*de internis affectionibus*
iusi.	*iusiurandum*
loc. hom.	*de locis in homine*
medic.	*de medico*
morb. I–IV	*de morbis* I–IV
morb. sacr.	*de morbo sacro*
mul. I–II	*de morbis mulierum* I–II
nat. hom.	*de natura hominis*
prog.	*prognosticum*
prorrh. I	*prorrheticus* I

superf.	*de superfetatione*
ulc.	*de ulceribus*
vet. med.	*de vetere medicina*
vict.	*de victu*
virg.	*de virginum morbis*
vuln. cap.	*de vulneribus in capite*
Ps.-Hipp.	Pseudo-Hippocrates
epist. ad Maecen.	*epistula ad Maecenatem*
epist. ad Antig.	*epistula ad Antigonum (Antiochum)*
hipp. Berol.	*hippiatrica Berolinensia*
hipp. Cant.	*hippiatrica Cantabrigiensia*
hipp. Lond.	*hippiatrica Londinensia*
hipp. Lugd.	*hippiatrica Lugdunensia*
hipp. Par.	*hippiatrica Parisina*
Hist. Aug.	*Historia Augusta* (see SHA)
Hom.	Homerus
Od.	*Odyssea*
Hor.	Horatius
carm.	*carmina*
ep.	*epistulae*
Hsch.	Hesychius
ILS	Dessau, Hermann. *Inscriptiones Latinae Selectae* (Berlin: Weidmann, 1892–1916)
Inscr. It.	*Inscriptiones Italiae* (Rome: Libreria dello Stato, 1931–) [Vol. XIII 2 = Degrassi, Attilio. *Fasti et elogia: Fasti anni Numani et Iuliani* (Rome: Istituto poligrafico dello Stato, 1963)]
Ios.	Iosephus
ant. Iud.	*antiquitates Iudaicae*
IRT	Reynolds, J. M., and J. B. Ward-Perkins. *The Inscriptions of Roman Tripolitania* (Rome: British School at Rome, 1952)
Isid.	Isidorus Hispalensis
orig.	*origines*
Ku.	Kuyunjik, British Museum cuneiform catalogue siglum
K.	Kühn, C. G. *Claudii Galeni opera omnia*, 20 vols. (Leipzig: C. Cnobloch, 1821–1833) [repr. Hildesheim: Olms, 1964–1965]

LAS	Parpola, Simo. *Letters from Assyrian Scholars to the Kings Esarhaddon and Assurbanipal*, vol. 1, Alter Orient und Altes Testament 5.1 (Kevelaer: Butzon & Bercker, 1970)
L.	Littré, É. *Œuvres complètes d'Hippocrate*, 10 vols. (Paris: J. B. Baillière, 1839–1861)
Leo medic.	Leo medicus
de nat. hom. syn.	*de natura hominum synopsis*
Lib.	Libanius
or.	*orationes*
Lucr.	Lucretius
Marcell.	Marcellus
med.	*de medicamentis*
Marcellin.	Marcellinus
puls.	*de pulsibus*
Men.	Menander
Nemesius Emesenus	Nemesius Emesenus
nat. hom.	*de natura hominis*
Olymp.	Olympiodorus
in cat.	*in categorias commentarium*
Opp.	Oppianus Apamensis
cyn.	*cynegetica*
Orib.	Oribasius
coll. med.	*collectionum medicarum reliquiae*
ecl. med.	*eclogae medicamentorum*
eupor.	*libri ad Eunapium*
syn.	*synopsis ad Eustathium filium*
Ov.	Ovidius
met.	*metamorphoses*
trist.	*tristia*
Paneg.	*Panegyrici Latini*
Papp.	Pappus Alexandrinus
Parm.	Parmenides
Paul. Aeg.	Paulus Aegineta
Pelagon.	Pelagonius
Philo Alex.	Philo Alexandrinus
ebr.	*de ebrietate*
Philod.	Philodemus
sign.	*de signis*

Philop.	Philoponus
in de an.	*in Aristotelis de anima libros commentaria*
in de gen. et corr.	*in Aristotelis libros de generatione et corruptione commentaria*
Philumen.	Philumenus
ven.	*de venenatis animalibus*
Pl.	Plato
Gorg.	*Gorgias*
leg.	*leges*
Phd.	*Phaedo*
Phdr.	*Phaedrus*
Phlb.	*Philebus*
pol.	*politicus*
resp.	*respublica*
soph.	*sophista*
Tim.	*Timaeus*
Ps.-Pl.	Pseudo-Plato
Ax.	*Axiochus*
Plin.	Plinius
nat.	*naturalis historia*
Plut.	Plutarchus
adv. Colot.	*adversus Colotem*
comm. not.	*de communibus notitiis adversus Stoicos*
prim. frig.	*de primo frigido*
de prof. virt.	*de profectu in virtute*
Stoic. repug.	*de Stoicorum repugnantiis*
Polyb.	Polybius
Posidon. Phil.	Posidonius
Procl.	Proclus
in Euc.	*in primum Euclidis elementorum librum commentarii*
Q. Ser.	Quintus Serenus
Ruf.	Rufus Ephesius
onom.	*de corporis humani appellationibus*
quaest. med.	*quaestiones medicinales*
satyr. gon.	*de satyriasi et gonorrhoea*
Ps.-Ruf.	Pseudo-Rufus
anat.	*de anatomia partium corporis humani*

SAA 10	Parpola, Simo. *Letters from Assyrian and Babylonian Scholars*, State Archives of Assyria 10 (Helsinki: Helsinki University Press, 1993)
Scrib. Larg.	Scribonius Largus
comp.	*compositiones*
Sen.	Seneca
ep.	*epistulae*
nat. quaest.	*naturales quaestiones*
Sever.	Severus
clyst.	*de clysteribus*
Sext. Emp.	Sextus Empiricus
adv. math.	*adversus mathematicos*
Pyrrh. subfig.	*Pyrrhoneae hypotyposes*
SHA	Scriptores Historiae Augustae
Hadr.	*Hadrianus*
Simp.	Simplicius
in de cael.	*in Aristotelis de caelo commentaria*
in cat.	*in Aristotelis categorias commentarium*
in phys.	*in Aristotelis physica commentaria*
Sor.	Soranus
gyn.	*gynaecia*
Ps.-Sor.	Pseudo-Soranus
isag.	*isagoge*
Speus.	Speusippus
SpTU	Hunger, Hermann. *Spätbabylonische Texte aus Uruk*, vol. 1, Ausgrabungen der Deutschen Forschungsgemeinschaft in Uruk-Warka 9 (Berlin: G. Mann, 1976)
Stob.	Stobaeus
Strat.	Strato Lampsacenus
STT	Gurney, O., and J. J. Finkelstein. *The Sultantepe Tablets*, 2 vols. (London: British Institute of Archaeology at Ankara, 1957–1964)
Suet.	Suetonius
Aug.	*de vita Caesarum lib. II: divus Augustus*
Theodorus Priscianus	Theodorus Priscianus
eupor.	*euporiston libri III*
Theodosius	Theodosius
sph.	*sphaerica*

Theophr.	Theophrastus
hist. plant.	*historia plantarum*
odor.	*de odoribus*
sens.	*de sensibus*
Thuc.	Thucydides
Var.	Varro
ling. Lat.	*de lingua Latina*
Veg.	Vegetius
mulom.	*digesta artis mulomedicinalis*
Verg.	Vergilius
georg.	*georgica*
Vitr.	Vitruvius
arch.	*de architectura*
Xen.	Xenophon
an.	*anabasis*
hist. Graec.	*historia Graeca (Hellenica)*
YBC	Yale Babylonian Collection

Isabella Andorlini
Egypt and the Medicinal Use of Papyrus According to Soranus and Other Physicians

Abstract: In his account of the manufacture of papyrus in *nat.* 13.72, Pliny the Elder makes no mention of its medicinal application among the miscellaneous uses popular in the Egyptian *chōra*. He does, however, refer in a number of other places to the reputation among physicians of the ash that is obtained from burning papyrus. Ancient doctors, too, recognized the therapeutic value of both the plant and the paper made from it. The present contribution focuses on the therapeutic uses attested in the medical writers; it considers, too, the additional information supplied by the *Gynecology* of Soranus, the distinguished Roman physician, who studied in Alexandria in the first and second centuries A.D. Soranus' original comparison of the layers of the uterus with the arrangement of fibers in layers of papyrus will be illustrated together with similar analogies employed by the Byzantine writers Meletius and Leo. It will be shown how physicians visiting Alexandria and Egypt were likely to have gained firsthand experience both in the medical schools and in the headquarters of the papyrus industry, where they became acquainted with the usefulness of papyrus in treatment and healing.

> πάπυρος γνώριμος πᾶσιν, ἀφ' ἧς ὁ χάρτης κατασκευάζεται,
> εὔχρηστος δὲ εἰς τὴν ἰατρικὴν χρῆσιν
>
> Papyrus, from which papyrus roll is made, is familiar to all and highly useful in medical practice.
> Dioscorides, *De materia medica* 1.86.1 (trans. Beck)

Those who look closely at the literary and documentary sources will find in them much evidence for the ancient awareness that papyrus could serve medicinal purposes. This evidence mainly concerns the specific cases of the application of the plant or the paper made from it. A comprehensive survey of the data on

---

This paper grew out of my presentation at the 25th International Congress of Papyrology (Ann Arbor, July 29–August 4, 2007). Unless otherwise stated, all translations are my own. I am most grateful to David Leith and John Lundon for revising my English text and offering invaluable advice.

the use of papyrus in a therapeutic context can contribute to the history of this practice, which was very popular throughout Egypt and beyond.[1]

The aim of the following investigation is to address three related questions.

(i) Our information on the medicinal employment of the plant spans the period from the Egyptian Ebers papyrus, written in the second millennium B.C., through a single Hippocratic citation, to the medical writers of Roman date, such as Celsus, Dioscorides, and Pliny the Elder in the first century A.D., to Galen in the second century A.D. More specifically, papyrus ash served as an ingredient of medical recipes while the paper product functioned as a bandage or a blistering plaster. Moreover, Byzantine medical writers merely repeated the uses of papyrus already known, so that there is no further evidence beyond what can be gained from their predecessors.

(ii) Aside from the knowledge displayed by the medical writers, evidence for the use of papyrus in a therapeutic context is scant. Non-medical sources, however, demonstrate that *Cyperus papyrus* served in everyday life in Egypt as a foodstuff and as a fragrance and substitute for incense.[2]

(iii) Although the medical tradition extending from Hippocrates to Paul of Aegina is conservative and the therapeutic applications recorded by Dioscorides outlived classical antiquity, surviving into the Coptic and Arabic periods, the evidence found in Soranus' *Gynecology* is original and deserves attention. The comparison of the layers of the uterus with the arrangement of fibers in papyrus layers is not referred to elsewhere,[3] and perhaps reveals a close familiarity with the papyrus products with which the distinguished physician became acquainted in the Alexandrian *milieu*.

When we turn to the medicinal employment of the papyrus plant and sheets of writing material, we have to reckon with the terms *papyros* or *byblos*, which can refer to the plant or the artificial product made from it. The words *chartēs*, *chartion*, or *chartarion*, in turn, commonly denoted papyrus rolls or pieces of

---

[1] The sources concerned with the use of papyrus as a drug are conveniently assembled by Naphtali Lewis, *Papyrus in Classical Antiquity* (Oxford: Clarendon Press, 1974), 31, 97, who draws on Egyptian, Greek, and Arabic evidence. For the Latin references, see *Thesaurus linguae Latinae*, vol. X, 1. *papyrus*. I. *de herba*, B.3, and II. *de charta*, B: 259–60 [Paśkiewicz].

[2] See Theophr. *odor*. 28: τὸ δὲ χρίσμα τὸ Ἐρετρικὸν ἐκ τοῦ κυπείρου (87 Eigler-Wöhrle-Herzhoff) ("the Eretrian unguent is made from the root of *kypeiron*"); Diosc. *mat. med.* 1.4.1: ῥίζαι ... εὐώδεις ("the roots ... have a pleasant smell"). The stalks of Cyperacae are said to burn with a pleasant smell: see Bernard P. Grenfell, Arthur S. Hunt, and David G. Hogarth, *Fayûm Towns and Their Papyri* (London: Egypt Exploration Fund, 1900), 17.

[3] This passage of Sor. *gyn.* 1.13.1 (10.1–2 Ilberg) is copied by Orib. *coll. med.* 24.31.21–22 (44.3–7 Raeder), as part of a long section taken over from Soranus.

them.[4] Despite the pervasiveness of these words in our Greek and Latin sources, confirmed by the roughly two thousand citations identified through computer-assisted searches in the corpora of both literary and documentary texts, the evidence for the medicinal use of papyrus is strictly confined to technical literature. As far as we know, no non-technical source ever refers to *byblos*, *papyros*, or *chartēs* being employed medicinally.

The results of an extensive study of the evidence can be grouped into the following four categories: (a) no relevant reference to the medicinal use of papyrus in non-medical Greek or Latin literature; (b) around 180 relevant citations of *chartēs* and around 40 of *papyros* in the corpus of Greek medical sources (Hippocrates, Dioscorides, Galen, Severus Iatrosophista, Oribasius, Aëtius of Amida, Alexander of Tralles, and Paul of Aegina);[5] (c) around 40 relevant citations in the

---

4 Unlike *chartēs*, the foreign origins of both *papyros* and *byblos* have been the subjects of scholarly debate and still remain an open question. While *byblos/biblos* might have a Semitic origin and derive from the Phoenician port Byblos, the word *papyros* is said to come from the Egyptian *pa-p-ouro*, denoting "the material of the Pharaoh"; cf. Paul Chantraine, *Dictionnaire étymologique de la langue grecque*, 4 vols. (Paris: Klincksieck, 1968–1980), 856; Lewis, *Papyrus*, 4, with n. 2; Françoise Skoda, "De quelques phytonymes empruntés," *LAMA* 4 (1979): 306–308; Johannes Kramer, *Von der Papyrologie zur Romanistik*, Archiv für Papyrusforschung, Beih. 30 (Berlin: De Gruyter, 2011), §6 ("*Papyrus* in den antiken und modernen Sprachen"), 91.
5 I refer to the relevant instances selectively in what follows. (i) χάρτης κεκαυμένος *vel* πάπυρος (ashes): Diosc. *simpl.* 1.75.2; 78.2; 79; 176; 190.1; 2.54.1; Gal. *de simpl. med. temp. ac fac.* 12.94.13 K.: ἡ τέφρα τοῦ κεκαυμένου χάρτου ("ashes of burnt papyrus sheet"); *de comp. med. sec. loc.* 13.295.17 K.; 13.296.7, 14–15 K.; 13.297.1, 4–5, 9, 13 K.; 13.298.3, 10 K.; 13.299.9, 13 K.; 13.300.5, 16 K.; 13.304.9 K.; 13.305.12 K.; 13.315.18 K.; *de comp. med. per gen.* 13.841.7, 10, 14 K.; 13.852.7, 9, 11, 15 K.; 13.853.2, 8, 12 K.; 13.854.1, 4, 7, 10–11, 13 K.; ps.-Gal. *remed. parab.* 14.324.10 K.; 14.381.3–4 K.; *succed.* 19.728.7 K.; 19.729.5 K.: ἀντὶ ἐλλεβόρου μέλανος ... ῥίζα παπύρου ("instead of black hellebore ... use the root of a papyrus plant"); 19.739.18 K.; Orib. *syn.* 1.19.18; 3.97, 113; *eupor.* 2.5.3; 3.13.4; 4.74.1; *coll. med.* 7.1.5; 8.25.15–16, 19; 10.24.7; 14.23.3; 15.16.3; 50.52.4; *ecl. med.* 54.6–10; *hipp. Berol.* 55.5; *hipp. Par.* 290; *hipp. Lond.* 19; *hipp. Cant.* 100.7; Aëtius Amid. *lib. med.* 6.50; 7.61, 80, 85; 8.25; 9.42; 15.11; Alex. Trall. 2.427.17; Paul. Aeg. 7.3.16 s.v. πάπυρος, 12.1, 24–27, 38; 13.1, 14; 17.36; Paul. Nic. 65.18. (ii) τὸ διὰ χάρτου *vel sim.* (a remedy containing papyrus sheet): Sor. *gyn.* 3.41.8; Gal. *de meth. med.* 10.382.5–6 K.; *ad Glauc. meth. med.* 11.125.8 K.; *de comp. med. sec. loc.* 12.465.16; 466.1, 5, 8; 611.8; 880.1; 13.500.18; 554.3; 853.4 K.; Sever. *clyst.* 39.6–7 Dietz: καὶ τὸ διὰ τῶν χαρτῶν δὲ ἄριστόν ἐστι βοήθημα ("the remedy made from burnt papyrus sheets is the best"); Orib. *syn.* 3.113; 9.34.1; *eupor.* 4.12.11; 101.1; 129; *coll. med.* 8.24.55; 44.12.2; *ecl. med.* 63.7; 83.3–4; 147.14; Aëtius Amid. *lib. med.* 6.68; 9.42; 11.29; 16.62, 119; Paul. Aeg. 3.3.4, 42.4, 45.7, 59.1, 66.3, 75.1; 4.44.5, 48.2; 7.3.16 s.v. πάπυρος, 12.24–25, 13.14, 17.36. (iii) χαρτίον *vel sim.*, or παπύριον, as a bandage (or wrapping material, or instrument): Diosc. *mat. med.* 1.8.1: προυποκειμένου χαρτίου ("putting first a small sheet of papyrus underneath"), and Orib. *coll. med.* 12 v 2; Diosc. *mat. med.* 2.76.16: καινῷ ἀποδήσας χάρτῃ ἀποτίθεσο ("then wrap [the fat] in a fresh sheet of papyrus and store it"); *simpl.* 1.183.1: ἐπὶ τῶν περὶ τὸν δακτύλιον συρίγγων σὺν παπυρίῳ ἐντιθεμένη ("it is also useful for perianal fistulas introduced with a small piece of papyrus [i.e., as a

corpus of Latin literary texts (Celsus, Pliny, Columella, Scribonius Largus, Chiron, Q. Serenus, Vegetius, Caelius Aurelianus, Marcellus Empiricus, Pelagonius, Cassius Felix);[6] (d) no evidence of any medicinal use among the approximately 300 occurrences in Greek documentary papyri.

Even the Roman encyclopedist Pliny the Elder, in his account of the usefulness of the papyrus plant in the *Natural History* (13.72),[7] offers no indication of

---

tampon]"); 1.197.3; Gal. *de meth. med.* 10.1000.12–13 K.: ἵνα χάρτου μαλακὴν καὶ εὔτονον ἐν κύκλῳ περιελίττων ("wrapping a soft and elastic strip of papyrus sheet around"), copied by Orib. *coll. med.* 50.1.1; Gal. *de meth. med.* 10.1001.7–8, 10–11 K.: τὸ χαρτίον ἐν κύκλῳ περιελιττόμενον ..., τοῦ χάρτου σύμμετρον ἐλίττων ἐνθεῖναι ("if not much is missing, it is sufficient, as was said before, to place a small strip of papyrus sheet around"); *de comp. med. sec. loc.* 12.881.2 K.: ἔνδησον εἰς χάρτην ("wrap them in a sheet of papyrus"); 13.339.13–14 K.: καὶ ἄνωθεν ἐπίρριπτε χάρτην καὶ ἔα μέχρις ἀφ' ἑαυτοῦ ἀποστῇ ("put a sheet of papyrus on top and leave it there until it falls off by itself"); ps.-Gal. *remed. parab.* 14.358.1 K.; 14.419.8 K.; 14.444.11 K.; 14.479.16 K.; 14.525.6 K.; Orib. *syn.* 1.15.4; *eupor.* 3.13.4; *coll. med.* 7.21.9: ἔπειτα χαρτίον ὄξει βεβρεγμένον ἐπιθετέον καὶ ἐπιδετέον ("then one should apply and fasten on top a small sheet of papyrus soaked in vinegar"); 10.23.8; 12 σ 48; 44.21.7: ὅταν βρέξας τις ἔτι αὐτῷ πάπυρον ἢ σπόγγον ("if you steep a piece of papyrus or a dried sponge in it [i.e., the caustic]"); 46.30.3: χαρτίῳ σκεπάζων αὐτὸ καὶ οὐκ ὀθονίῳ, ἵνα μὴ διὰ τῶν ἀραιωμάτων ἐκρεύσῃ τὸ φάρμακον ("cover the part with a small sheet of papyrus and not with linen, so that the remedy cannot escape through the holes"); 50.1.1, 4; 5.7–8; *ecl. med.* 74.5; 141.1; *hipp. Berol.* 52.18; 130.129; *hipp. Lugd.* 30; 180; Aëtius Amid. *lib. med.* 3.22; 12.1; 15.11, 15 (ἡ διὰ ψυλλίου, "a plaster bandage with *plantago*"); 16.20, 62, 124; Paul. Aeg. 3.77.4; 6.55.2. (iv) κόλλα, cellulose gum, juice, glue: Aëtius Amid. *lib. med.* 12.53: κόλλης τῶν χαρτῶν τοῦτ' ἔστι γύρεως ἡψημένης ("the glue of papyrus rolls, i.e., fine flour, boiled"). (v) σφαιρίον, a pill: Sever. *clyst.* 41.11–13 Dietz: λαμβάνοντες οὖν τὴν πάπυρον καὶ οἱονεὶ τῇ συναγωγῇ μικρὸν σφαιρίον ποιήσαντες ("we take just the papyrus plant and roll it into a kind of small ball") (add 41.15, 19, 22 Dietz).

6 The references are cited in n. 22 below.

7 Pliny does, however, refer to the medicinal ash obtained from burning papyrus and to its usefulness as a bandage in a number of other passages. Cf. *nat.* 24.88: *Cogn⟨a⟩ta in Aegypto res est harundini papyrum, praecipuae utilitatis, cum inaruit, ad laxandas siccandasque fistulas et intumescendo ad introitum medicamentorum aperiendas. charta, quae fit ex eo, cremata inter caustica est. cinis eius ex vino potus somnum facit. ipsa ex aqua inposita callum sanat* ("of a kindred nature with the reed is the papyrus of Egypt; a plant that is remarkably useful, in a dried state, for dilating and drying up fistulas, and, by its expansive powers, opening an entrance for the necessary medicaments; the ashes of paper prepared from the papyrus are reckoned among the caustics: those of the plant, taken in wine, have a narcotic effect; the plant, applied topically in water, removes callosities of the skin"); 28.61: *extremitates corporis velleribus perstringi contra horrores sanguinemve narium inmodicum, [- - -] lino vel papyro principia genitalium* ("for excessive nose-bleeds, the extremities of the body should be well rubbed, [- - -] the extremities of the generative organs should be tied with a thread of linen or papyrus"); 28.168: *optime e⟨l⟩lychnio papyraceo oleoque sesamino fuligine in novum vas pinnis detersa, efficacissime tamen evolsos ibi pilos coercet* ("the best of all being that made from a wick of papyrus mixed with oil of sesame; the soot removed with feathers into a new vessel; this will prevent the growth of hair

any medicinal application, although he does mention the multiple uses popular in the Egyptian *chōra*. These ranged from the manufacture of such articles as sandals, ropes, crowns, and baskets to the construction of river craft.[8] In the relevant passage of his *Enquiry into Plants*, repeated in part by Pliny, Theophrastus observes succinctly that papyrus served "very many uses" (αὐτὸς δὲ ὁ πάπυρος πρὸς πλεῖστα χρήσιμος, *hist. plant.* 4.8.4). Nevertheless, in the following paragraph he stresses its principal utility as a foodstuff, enumerating the ways in which people could be nourished by its various parts.

> μάλιστα δὲ καὶ πλείστη βοήθεια πρὸς τὴν τροφὴν ἀπ' αὐτοῦ γίνεται. Μασῶνται γὰρ ἅπαντες οἱ ἐν τῇ χώρᾳ τὸν πάπυρον καὶ ὠμὸν καὶ ἑφθὸν καὶ ὀπτόν· καὶ τὸν μὲν χυλὸν καταπίνουσι, τὸ δὲ μάσημα ἐκβάλλουσιν. (Theophr. *hist. plant.* 4.8.4)
>
> But above all the plant also is of very great use in the way of food. For all the natives chew the papyrus both raw, boiled, and roasted; they swallow the juice and spit out the quid. (trans. Hort)

From the absence of any specific evidence in non-technical sources, on the one hand, and the presence of roasted papyrus as an ingredient in the Ebers papyrus, on the other, one might reasonably conclude that any awareness of the plant's therapeutic utility required a significant amount of technical knowledge and professional competence. Thus, although native to Egyptian culture and widely consumed in the countryside, *Cyperus papyrus* never became one of the healing tools of folk medicine but was closely affiliated with practices of professional distinction.

The modern visitors to Egypt who have borne witness to the uses of the plant acknowledged by the Egyptian doctors of the time, such as the famous botanist Prosper Alpinus, who travelled to Egypt in 1580, were intellectual tourists, too.[9]

---

that was removed there"); 28.214: *vitia vero, quae in eadem parte serpunt, iocur eorum combustum ... cum charta et arrhenico sanat* ("and for serpiginous affections of those parts, the liver of those animals is used burnt ... and mixed with papyrus and arsenic"); 29.106: *Alopecias ... inlinunt cum cinere chartae* ("where the hair has been lost through alopecia ... apply the ashes of papyrus sheets"); 34.170: *cinis autem usti ad serpentia ulcera aut sordida, eademque quae chartis ratio profectus* ("the ashes of calcined lead are used for serpiginous or sordid ulcers, these producing the same advantageous effects as the ashes of burnt papyrus sheets").

8 For the articles made from *Cyperus papyrus*, see Bridget Leach and John Tait, "Papyrus," in Paul T. Nicholson and Ian Shaw (eds.), *Ancient Egyptian Materials and Technology* (Cambridge: Cambridge University Press, 2000), 227–53.

9 Prosper Alpinus is the first person to provide us with a drawing of the papyrus, which the Egyptians call *berdi:* cf. *Plantes d'Egypte par Prosper Alpin (1581–1584)*. Traduit du latin, présenté et annoté par R. de Fenoyl (Cairo: IFAO, 1980), 110–11. The famous botanist and physician mentions a number of medicinal uses made of *Cyperus papyrus* in his own time: "Les chi-

Referring to the medicinal ash obtained by burning papyrus paper and effective for wounds and eye disorders, Alpinus apparently draws upon sophisticated ancient sources such as Dioscorides and Galen, as will be seen below.

One important clue to the plant's health-promoting potential, however, was the fact that people in Egypt enjoyed eating papyrus prepared in many ways. Another was that some ancient authorities recognized its nutritional value, especially that of its stalk, juices, and roots, whether roasted or not. Herodotus remarks that the lower extremity of the plant was a delicacy when baked on the fire, while Pliny, drawing on Theophrastus, emphasizes the quality of the juice contained in its stalk.[10] Highly instructive, too, is a joke from *The Frogs* by Aristophanes, where the comic poet plays on the terms for papyrus and book by alluding to Euripides as "giving her [i. e., the art of tragedy] the juice of chatterings, pressing it from the books" (χυλὸν διδοὺς στωμυλμάτων, ἀπὸ βιβλίων ἀπηθῶν, 943). Although this custom was regarded as typically Egyptian by Herodotus, Theophrastus, Diodorus of Sicily, and Pliny, the Greek settlers themselves progressively introduced papyrus into their diet.[11]

---

rurgiens égyptiens utilisent la *moelle* pour élargir les lèvres des ulcères. La *cendre* faite avec le *rouleau* leur sert à guérir les ulcères récents et aussi à empêcher les ulcères pernicieux de s'étendre (si on les en saupoudre fréquemment). Avec les *rouleaux frais*, on fait un distillat très efficace contre la cataracte, l'obscurcissement et l'affaiblissement de la vue" (110). The enduring appreciation of the medicinal value of papyrus is confirmed by a thirteenth-century Arabic author, the botanist Abû-l-ʿAbbâs an-Nabâtî, who remarks that "man verwendet ihn [i.e., den Papyrus] bei der ärztlichen Behandlung" (trans. Adolf Grohmann, *Allgemeine Einführung in die arabischen Papyri* [Vienna: F. Zöllner, 1924], 36).

10 The most important passages are Hdt. 2.92.5: τὴν δὲ βύβλον τὴν ἐπέτειον γινομένην ἐπεὰν ἀνασπάσωσι ἐκ τῶν ἑλέων, τὰ μὲν ἄνω αὐτῆς ἀποτάμνοντες ἐς ἄλλο τι τρέπουσι, τὸ δὲ κάτω λελειμμένον ὅσον τε ἐπὶ πῆχυν τρώγουσι καὶ πωλέουσι. οἳ δὲ ἂν καὶ κάρτα βούλωνται χρηστῇ τῇ βύβλῳ χρᾶσθαι, ἐν κλιβάνῳ διαφανέι πνίξαντες οὕτω τρώγουσι ("they also use the papyrus which grows annually: it is gathered from the marshes, the top of it cut off and put to other uses, and the lower part, about twenty inches long, eaten or sold; those who wish to use the papyrus at its very best, roast it before eating in a red-hot oven"); Theophr. *hist. plant.* 4.8.2–4 (cited above); Diodor. Sic. 1.80.5–6: καὶ τῶν ἐκ τῆς βύβλου πυθμένων τοὺς δυναμένους εἰς τὸ πῦρ ἐγκρύβεσθαι, καὶ τῶν ῥιζῶν καὶ τῶν καυλῶν τῶν ἑλείων τὰ μὲν ὠμά, τὰ δ' ἕψοντες, τὰ δ' ὀπτῶντες, διδόασιν ("they give their children such stalks of the papyrus plant as can be roasted in the coals, and the roots and stems of marsh plants, either raw or boiled or baked"); Plin. *nat.* 13.72: *mandunt quoque crudum decoctumque, sucum tantum devorantes* ("they chew it also, both raw and boiled, though they swallow the juice only").

11 See, for example, Ulrich Wilcken, *Urkunden der Ptolemäerzeit (ältere Funde)*, vol. 1 (Berlin: De Gruyter, 1927), 409 n. 8, and his comments on documents 91–93, 96, where "Papyrusstengel" are recommended for food. The ancient evidence dealing with the importance of papyrus as a foodstuff has been collected by Georg Wöhrle, "Papyrophagie," in Raimar Eberhard et al. (eds.), *"... vor dem Papyrus sind alle gleich!" Papyrologische Beiträge zu Ehren von Bärbel Kramer (P.*

As for specifically medicinal applications of both the papyrus plant and the paper made from it, our evidence goes back to the Egyptian pharmacopoeia of the Ebers papyrus, written about 1500 B.C.

"Roasted papyrus" figures prominently in a few prescriptions dealing with external remedies. "Papyrus ash" was applied effectively with other drugs in a bandage for stiff limbs (*P. Ebers* 669) and in an eye compress (*P. Ebers* 340). "Cooked unwritten papyrus," furthermore, mixed together with "wax, oil, and *wah*-legume" appears to be active in the fourth day of a cure to relieve the pain of a burn (*P. Ebers* 482):

> The beginning of remedy against burn (i.e., *combustio*). ...What is done the fourth day: wax, grease of ox, papyrus are burnt with manna, mixed together, and (it) is bandaged therewith. (trans. Ebbell)[12]

In particular, the last of these Egyptian recipes can serve as a link to other pieces of evidence for a Greek tradition in the therapeutic use of papyrus sheets.

The value assigned by the Egyptians to the ash obtained by burning papyrus is subsequently confirmed in a Greek context by a single citation surviving in Hippocrates' *Diseases of Women*. Here we come across a plant remedy native to Egypt in a gynecological text probably going back to the fifth century B.C.[13] Together with squill, ashes, white lead, and myrrh, the "third part of the ash resulting from a burnt papyrus sheet" is recommended in a poultice good for diseases of the eye.[14]

Kramer), Archiv für Papyrusforschung, Beih. 27 (Berlin: De Gruyter, 2009), 243–47. Lewis, *Papyrus*, 22–23 remains a useful analysis.

12 A German translation is given by Wolfhart Westendorf, *Handbuch der altägyptischen Medizin*, 2 vols., Handbuch der Orientalistik I 36.1–2 (Leiden: Brill, 1999), 2:632. The Copts also appreciated the powder of the burnt plant and of the burnt sheets, which figures in the recipes of the Chassinat papyrus (*Ch.* 121, 165, 177, 178): cf. Walter C. Till, *Die Arzneikunde der Kopten* (Berlin: Akademie Verlag, 1951), 83, 122 ("Ein Papyruspulver gegen Geschwüre"), and 125 ("Asche von hieratischem Papyrus"; "verbrannter neuer Papyrus"; "Ein Papyruspulver für die Zähne"); Lisa Manniche, *An Ancient Egyptian Herbal* (Austin, Tx.: University of Texas Press & British Museum Publications, 1989), 99–100 (*Cyperus papyrus* L.).

13 Ingredients native to Egypt appear in a number of Hippocratic gynecological prescriptions. See Laurence M. V. Totelin, *Hippocratic Recipes: Oral and Written Transmission of Pharmacological Knowledge in Fifth- and Fourth-Century Greece*, Studies in Ancient Medicine 34 (Leiden: Brill, 2009), 179–84, on the relation of the Hippocratic recipes to Egyptian medicine.

14 On this dry poultice, see Dietlinde Goltz, *Studien zur altorientalischen und griechischen Heilkunde. Therapie – Arzneibereitung – Rezeptstruktur*, Sudhoffs Archiv, Beih. 16 (Wiesbaden: Steiner Verlag, 1974), 221.

Παράπαστον· μόλιβος κεκαυμένος καὶ σποδὸς ἴσα, σμύρνης δέκατον μέρος, ὁποῦ μήκωνος σμικρόν, οἶνος παλαιός· ξηρὰ τρίψας χρῶ. Σκίλλα, καὶ σποδοῦ τρίτον μέρος, καὶ ψιμυθίου, τρίτον μέρος χάρτου κεκαυμένου, μέρος δέκατον σμύρνης. (Hipp. *mul.* I 105, 8.228.20–23 L.)

A dry poultice. The same amount of burnt lead and of lead oxide, the tenth part of myrrh, a bit of poppy juice, old wine; grind together the dry ingredients and use. Squill, the third part of lead oxide and of white lead, the third part of a burnt papyrus sheet, the tenth part of myrrh.

Among the Greek medical writers of the Roman period, Dioscorides is the first to report accurately that the Egyptians ate the papyrus root and swallowed the juice. Dioscorides' entry on *papyros* in his *De materia medica* is concise but exhaustive, apparently providing the basic information for descriptions compiled later or expanded in turn by Galen, Oribasius, Aëtius of Amida, Alexander of Tralles, and Paul of Aegina.[15]

πάπυρος γνώριμος πᾶσιν, ἀφ᾽ ἧς ὁ χάρτης κατασκευάζεται, εὔχρηστος δὲ εἰς τὴν ἰατρικὴν χρῆσιν, πρὸς ἀναστόμωσιν συρίγγων σκευασθεῖσα διάβροχος περιειλουμένου λίνου ἄχρι ξηρασίας· στεγνουμένη γὰρ καὶ καθιεμένη ἐμπίπλαται ὑγρασίας καὶ ἐξοιδοῦσα διανοίγει τὰς σύριγγας. ἔχει δέ τι ἡ ῥίζα αὐτῆς καὶ τρόφιμον· διαμασώμενοι γοῦν αὐτὴν οἱ ἐν Αἰγύπτῳ ἀποχυλίζουσιν ἐκπτύοντες τὸ διαμάσημα, χρῶνται δὲ καὶ ἀντὶ ξύλων αὐταῖς. ἡ δὲ κεκαυμένη πάπυρος ἄχρι τεφρώσεως δύναται νομὰς ἐπέχειν τὰς ἐν στόματι καὶ παντὶ μέρει· βέλτιον δὲ ὁ χάρτης καεὶς δρᾷ τὸ τοιοῦτον. (Diosc. *mat. med.* 1.86 = 1:81.18–82.5 Wellmann)

Papyrus, from which papyrus roll is made, is familiar to all and highly useful in medical practice for opening fistulas: it is prepared, after it has been soaked, by wrapping it with a linen thread, until it dries. For as it is inserted compressed, it becomes filled with moisture and, as it swells, it opens the fistulas. Its root, moreover, has something that is even nutritive: the people in Egypt, after chewing it, extract its juice and spit out the chewed matter; they also use these reeds for timber. Papyrus that is burned to ashes keeps in check sores in the mouth and everywhere else; but papyrus roll that was set on fire does this kind of thing better. (trans. Beck)

While remarking on the general reputation of papyrus as a writing material, Dioscorides focuses on the following therapeutic purposes. (a) The substances of the papyrus plant exhibit cicatrizing properties. If applied moistened as a lotion, papyrus helps to cure ulcers; when prepared as a dry compress for open wounds, it helps to keep them dry. (b) Ulcers of the mouth or in other areas benefit from the local use of papyrus ash. (c) Finally, Dioscorides points out that the ash obtained from direct combustion of papyrus sheets was regarded as having a more

---

[15] See above n. 5.

potent therapeutic effect.¹⁶ This variation in the cicatrizing properties ascribed to different products of the plant, perceptively observed by the ancient pharmacologist, is probably due to the mineral elements present in the plant, which increase their drying effects in the paper-making process.¹⁷

As it bends without breaking and is extremely light, papyrus paper competed with linen as a means of bandaging the affected part of the body in combination with various poultices. Strips of papyrus served on occasion as bandages, but far more frequent was the use of papyrus sheets of different sizes as a sort of band-aid to hold the poultice in place on the affected part of the body.¹⁸

Not surprisingly, evidence for papyrus in local therapeutic practices is provided by the mention of these specific applications in medical papyri of the early Roman period excavated in the temple context of Tebtunis in Egypt (*PSI* X 1180 and *P. Tebt.* II 273).¹⁹

Indeed, papyrus was a favorite healing aid within such a context. In the recipes surviving in the Greek receptarium, ashes of burnt papyrus soaked in water are the component of a lotion used specifically to treat leprosy, while a small

---

16 This statement is adopted by Galen, who stresses the weakness of the ash produced from the burning of the plant (ἐπειδὰν δὲ καυθῇ, φάρμακον ἤδη γίνεται ξηραντικόν, ὥσπερ καὶ ἡ τέφρα τοῦ κεκαυμένου χάρτου, πλὴν ὅσον ἀσθενεστέρα ἐστὶν ἡ τῆς παπύρου, *de simpl. med. temp. ac fac.* 12.94.12–14 K.; "if it is burnt, it already turns into a drying remedy, the same as the ashes of burnt [manufactured] papyrus, with the only difference that the ash of the plant is less powerful"), and later by Orib. *eupor.* 2.5.3; *coll. med.* 10.23.8; 15.16.3. The same passage recurs in Paul. Aeg. 7.3.16 s.v. πάπυρος.

17 The healing properties of the papyrus plant are apparently due to the approximately 60 % cellulosic material in the stems, and to the high mineral concentrations (potassium, sodium, calcium, magnesium, iron, and manganese). As papyrus swamps present considerable surfaces for the absorption of substances, large amounts of nutrient elements are incorporated into the plant. Modern analysis of *Cyperus papyrus* L. has indicated that the amount of nutrients accumulated by papyrus is higher than that of most other macrophytes. Cf. John J. Gaudet, "Mineral Concentrations in Papyrus in Various African Swamps," *Journal of Ecology* 63 (1975): 483–91 (with earlier bibliography).

18 Evidence for these applications is further supplied by Diosc. *simpl.* 1.183.1 (3:221.12–13 Wellmann): ποιεῖ καὶ ἐπὶ τῶν περὶ τὸν δακτύλιον συρίγγων σὺν παπυρίῳ ἐντιθεμένη ("it also works for perianal fistulas introduced into them with a small piece of papyrus") and *simpl.* 1.197.3 (3:226.11–12 Wellmann): δεῖ δὲ προαναστομοῦν τὰς σύριγγας σπογγίῳ ἢ παπύρῳ ἐσκελετευμένοις ("it is necessary to open up the fistulas first with a dried sponge or piece of papyrus").

19 Full editions in Isabella Andorlini, "Un ricettario da Tebtynis: parti inedite di PSI 1180," in Isabella Andorlini (ed.), *Testi medici su papiro; atti del Seminario di Studio (Firenze, 3–4 giugno 2002)* (Florence: Istituto Papirologico "G. Vitelli," 2004), 81–118, and in Ann E. Hanson, "A Receptarium from Tebtynis," in Isabella Andorlini (ed.), *Greek Medical Papyri* II (Florence: Istituto Papirologico "G. Vitelli," 2009), 71–103 (no. 5), respectively.

sheet of medicated paper (i.e., *chartarion*) is applied locally for lichen.[20] This extensive collection of recipes for treating dermatological conditions was compiled in the late first or early second century A.D. and increases our evidence for the adaptation of recipes to an Egyptian environment by revealing the penetration of Egyptian elements into a Greek text produced in a culturally indigenous *milieu*.[21] Furthermore, a small quantity of *chartēs* appears in another receptarium from Tebtunis dated to the late second century A.D., where an eye-salve is prescribed (*P. Tebt.* II 273, col. VI, 9 χάρτου (δρ.) α).

Even a cursory glance at the Latin evidence on the subject reveals that this tradition is neither independent nor original with respect to the Greek one.[22] Cel-

---

[20] *PSI* X 1180, fr. A, col. II, 10–11: πρὸς λέπρας, ἐὰν ἐκ|δέρης αὐτάς, βάμμα παπύρου κεκαυ-μ(ένης) ("against leprosy: when you have scraped off these lesions, prepare an ointment with burnt papyrus"); fr. A, col. III, 5–7: τὸν λιχῆνα προεζμησάμενον κα|τάχριε καὶ ἔξωθεν γῦριν· ἐπάνω δὲ | το[ῦ] φαρμάκου χαρτάριον ἐπίθες ("scrape the area affected by lichen first, smear with the finest meal externally; then cover the application with a bandage made from a papyrus sheet"). Burnt papyrus also figures in a pill prescribed in *P. Ant.* III 127, fr. 2, 5, where the ash is considered to be effective against dysentery.
[21] Cf. Andorlini, "Un ricettario da Tebtynis," 91.
[22] The relevant references concern (i) the ash, *charta combusta*, or *chartae combustae cinis*: Cels. 5.22.2b, 5; 6.4.3, 15.1, 19.2; Scrib. Larg. *comp.* 114, 237; Q. Ser. 139: *cerussam et chartam, quam gens Aegyptia mittit* ("white lead and papyrus are materials that come from Egypt"); Chiron 88: *si fistula facta fuerit, curabis, ... papyro* ("if a fistula results, treat it with a piece of papyrus"); 92; Veg. *mulom.* 2.13.5; 23.2; 27.3: *papyri iniectione* ("by inserting a piece of papyrus"); 96: *fistulae curantur papyro* ("fistulas are treated with papyrus"); Cael. Aur. *chron.* 4.8.117: *chartae exustae* ("[a dose] of burnt papyrus sheet"); Marcell. *med.* 34.101: *ad uarices ... lanuginem de papyro adpone ... (id est illam lanuginem, quae uiridi papyro in summitate est quasi paniculae eminentis)* ("apply the soft tufts of papyrus to varicose veins, i.e., those soft tufts on top of the green papyrus plant that stand out"); Pelagon. 134: *chartam puram combures et bibere dabis cum vino veteri* ("burn a piece of clean papyrus and give it to drink with old wine"); 344; and (ii) the bandage, strip (or instrument): Cels. 5.28.12K: *facile tamen est callum quibuslibet adurentibus medicamentis erodere: satis est vel papyrum intortum vel aliquid ex penicillo in modum collyri adstrictum eo inlini* ("it is easy to eat away the callus with any of the caustic medicaments; it is enough to smear one of them on rolled papyrus, or upon a pledget of wool twisted into the shape of a tent"); Colum. 6.6.4: [i.e., *sanguis*] *inhibetur papyri ligamine* ("[the blood] is stopped by tying [the tail] with a strip of papyrus"); Veg. *mulom.* 2.57.1; 4.4.4: *papyro ligata cauda restringitur* ("the blood is stopped with a strip of papyrus tied round the tail"); Marcell. *med.* 10.43; 58: *papyrus ... involuta naribus inseratur* ("by wrapping up a piece of papyrus, insert it through the nose"); Cass. Fel. 20.3: *angustas cavernulas ... papyro patefacies. quod papyrum sic praeparabis. papyrum vitriariorum eliges carnosum, id est quod non fuerit fragile vel flaccidum ..., et iterum alio papyro paulo robustiore mutabis* ("for opening the narrow hollows (of the fistulas) use a piece of papyrus; prepare the papyrus in this way: choose a fleshy papyrus such as used in the manufacture of glass, i.e., that is not fragile or flaccid ... and then, replace it with another and more robust papyrus").

sus, Pliny, Scribonius, and Caelius Aurelianus all include papyrus as a component (i.e., *charta combusta*) of twenty or so prescriptions for diseases of the skin.

We can thus summarize the results of the foregoing analysis by organizing the range of the medical applications of papyrus under a few main headings. First, given its mildly caustic and desiccating properties, papyrus ash served as a medicinal ingredient; also, the ash of the roasted papyrus plant or sheets was valued as an antiseptic and drying agent mixed into various external remedies for wounds, ulcers, and surgical incisions. Second, by far the most common use of papyrus documented in our Greek medical sources is its application in an often cited plaster named after papyrus as its characteristic ingredient (τὸ διὰ χάρτου).[23] Third, papyrus paper was popularly employed instead of linen as a means of keeping the poultice in place on the affected parts of the body, as attested in a Greek papyrus as early as the late first century A.D.[24]

Besides the points raised earlier, two other topics of interest concern the widespread presence of papyrus within the sophisticated *milieu* of the frontier capital, Alexandria, often considered the cradle of advanced medical education.

The evidence of Soranus of Ephesus, the distinguished Methodist physician active in the late first century A.D. and early second century A.D., requires brief discussion here. According to the biographical data provided by our sources, Soranus spent the first part of his career at Alexandria in Egypt, where he probably studied anatomy and gained firsthand experience in everyday anatomical practice.[25] Soranus' *Gynecology* contains many instructions for the treatment and care of women. The author uses the terms *papyros* and *chartēs* in only three passages, one of which refers, however, to the traditional employment of papyrus in vaginal suppositories as an astringent agent.[26] Discussing the anatomy of the uterus in the first book of his *Gynecology*, Soranus describes the tunics (χιτῶνες) by analogy with layers of papyrus (i.e., ἶνες, a term used for the fibrous

---

23 See above n. 5 (ii).
24 See above n. 20.
25 Cf. Sudae *Lexicon*, Σ 851 and 852 (4.407.20–27 Adler). For Soranus (*fl.* 98–138 A.D.), see now Monica Green and Ann E. Hanson, "Soranus of Ephesus: *Methodicorum princeps*," in Wolfgang Haase (ed.), *Aufstieg und Niedergang der römischen Welt*, II.37.2 (Berlin: De Gruyter, 1994), 968–1075.
26 Cf. Sor. *gyn.* 3.41.8 (121.1–2 Ilberg): εἰ δὲ πρὸς ἀνάβρωσις εἴη, καὶ τῷ διὰ χάρτου μέλανι μετ' ὄξους ἤ τινι τῶν πρὸς τοὺς δυσεντερικοὺς ἀναγραφομένων τροχίσκων ("if, besides, there is an erosion, one should also use the 'black remedy' made of papyrus, together with vinegar, or any of the troches that are prescribed for dysentery" [trans. Temkin]).

tissues of the human body).²⁷ He describes the features of the membranes as follows:

> ἡ δὲ ὅλη μήτρα συνέστηκεν ἐκ δυοῖν χιτώνων ἐναντίως ἑαυτοῖς ἐσχηματισμένων ἐμφερῶς ταῖς τῶν χαρτῶν ἰσίν. ὁ μὲν οὖν ἔξωθεν νευρωδέστερός ἐστι καὶ λειότερος καὶ σκληρότερος καὶ λευκότερος, ὁ δὲ ἔσωθεν σαρκωδέστερος καὶ δασύτερος καὶ ἁπαλώτερος καὶ ἐνερευθέστερος, δι' ὅλου μὲν καταπεπλεγμένος ἀγγείοις, πλείοσιν δὲ καὶ ἀξιολόγοις κατὰ τὸν πυθμένα καὶ τοῦ σπέρματος ἐκεῖ κολλωμένου καὶ τῆς καθάρσεως ἐκεῖθεν φερομένης. οἱ μέντοι δύο χιτῶνες οὗτοι συνέχονται πρὸς ἀλλήλους ὑμέσι λαγαροῖς καὶ νεύροις, ὥστε πολλάκις ἐπεκτεινομένων αὐτῶν προπίπτειν τὴν ὑστέραν, τοῦ μὲν νευρώδους χιτῶνος κατὰ χώραν μένοντος, τοῦ δὲ ἔσωθεν κατ' ἐκτροπὴν προπίπτοντος. (Sor. *gyn.* 1.13.1–2 = 10.1–12 Ilberg)
>
> The whole uterus is composed of two layers which are arranged crosswise, similarly to the strips of papyrus. The outer layer is relatively sinewy, smooth, hard, and white whereas the inner layer is fleshy, rough, soft, and reddish. The latter is interwoven throughout with vessels, which, however, are more numerous and noteworthy in the region of the fundus, since it is here that the seed adheres and since from here the menses are produced. Now these two layers are interconnected by flexible membranes and nerves and if these are often stretched, the uterus may prolapse, the sinewy layer remaining in its place, whereas the inner layer prolapses by eversion. (trans. Temkin)

This explicit comparison is used by way of illustration and has the ring of authenticity. Soranus' detailed knowledge of the uterine tunics is apparently the result of his own investigation of the female organs, likely carried out at Alexandria within an anatomical tradition going back to Herophilus' remarkable investigations.²⁸ The image of two separate overlapping layers of tissue opposite

---

27 Galen adopts the term *is* for a tape consisting of a single papyrus strip, which has to be soft and resistant, in a surgical context where he offers bandaging directions: cf. Gal. *de meth. med.* 10.1000.14–16 K.: καὶ τὸ τῆς ἰνὸς πέρας ἐπικολλᾶν χρὴ διὰ κόμμεως τῷ ὑποβεβλημένῳ ἄνω μέρει τῆς ἰνός· ἐν τάχει τε γὰρ ξηραίνεται καὶ ἀλύπως σφίγγει ... ὃ καὶ μετὰ τὸ κολλῆσαι τὴν ἶνα ῥᾳδίως ἐξαιρήσεις ("it is also necessary to smear the upper edge of the strip with gum by placing it underneath, for then it dries quickly, and binds painlessly ... something you will remove easily after gluing the strip of papyrus" [trans. Johnston-Horsley]); and above, n. 5 (iii). In a philological context, furthermore, he uses *is* for a strip of a papyrus roll containing writing which has become detached and lost (*in Hp. epid. VI comm.* prooem., 17A.794.17–795.1–2 K. [= 4.12–13 Wenkebach]): δυνατὸν γὰρ δὴ οὕτως καὶ λεπτῆς ἰνὸς ἀπολωλυίας συναπολέσθαι τὴν γραμμὴν ταύτην ("thus it is possible that, having lost a thin strip, one has lost the corresponding letter").
28 Cf. Heinrich von Staden, *Herophilus: The Art of Medicine in Early Alexandria* (Cambridge: Cambridge University Press, 1989), 139–53. The tradition of continuous skeletal anatomy or dissection at Alexandria beyond the time of Herophilus and Erasistratus, however, is a controversial issue, cf. von Staden, *Herophilus*, 142, 146, and Vivian Nutton, "Galen in Egypt," in Jutta Kollesch and Diethard Nickel (eds.), *Galen und das hellenistische Erbe*, Sudhoffs Archiv, Beih. 32 (Stuttgart: Steiner, 1993), 11–32, at 15–17.

one another is compatible both with female anatomy and with the layers of papyrus strips laid across one another at right angles.²⁹

In a passage from the second book dealing with obstetric practice, Soranus describes the task of the midwife at the final stage of childbirth.

> λοιπὸν δὲ ἡ μαῖα δι' ἑαυτῆς ἀποδεχέσθω τὸ ἔμβρυον, προϋποβεβλημένου ῥάκους κατὰ τῶν χειρῶν ἤ, ὡς αἱ ἐν Αἰγύπτῳ ποιοῦσιν, λεπτῆς παπύρου ξεσμάτων πρὸς τὸ μήτε ἀπολισθάνειν αὐτὸ μήτε θλίβεσθαι, τρυφερῶς δὲ ἐφεδράσθαι. (Sor. gyn. 2.6.4 = 55.5–8 Ilberg)
>
> Finally the midwife herself should receive the infant, having first covered her hands with pieces of cloth or, as those in Egypt do, with scraps of thin papyrus, so that it may neither slip off nor be squeezed, but rest softly. (trans. Temkin)

The obstetric practice to which Soranus refers here is the use of the woody root of the mature papyrus plant, attesting to the enduring popularity of an Egyptian custom.³⁰

Both the comparisons, originating with Soranus and not repeated elsewhere, exhibit sophistication and reveal the author's predilection for integrating narrative discussion with etymologies, analogies, and learned digressions. Furthermore, the appropriate distinction between the term *chartēs*, referring to papyrus paper in the former citation, and the word *papyros*, denoting the plant in the latter, seems to reflect scholarly readings and firsthand experience gained in various local contexts.

Although Soranus' comparison between the uterine tunics and papyrus layers does not occur elsewhere, this stylistic device is encapsulated in Meletius' description of the ocular tunics in his *On the Constitution of Man* 2 (63.7–10 Cramer), a passage repeated verbatim in Leo the Physician's *Epitome On the Nature of Man*, or *Synopsis* 35 (30.20–22 Renehan). The dates of the two Byzantine writers remain controversial, ranging from the ninth to the late twelfth or early thirteenth century.³¹

---

29 See, for example, the drawing of the uterine layers in Enzo Brizzi et al., *Anatomia topografica* (Milan: Edi. Ermes, 1978), 392–95.

30 The ξέσματα are the shavings produced by peeling either the stem or the root of the papyrus plant. This procedure echoes Theophr. *hist. plant.* 4.8.4, where he remarks that "the Egyptians use the roots instead of wood, not only for burning but also for making all kinds of utensils" (χρῶνται δὲ ταῖς μὲν ῥίζαις ἀντὶ ξύλων οὐ μόνον τῷ κάειν ἀλλὰ καὶ τῷ σκεύη ἄλλα ποιεῖν ἐξ αὐτῶν παντοδαπά).

31 It is customary to place Meletius the Monk and Leo the Physician, whose *Epitome* is a series of excerpts from the similar work of Meletius, tentatively in a ninth- and tenth-century context. For other views on their chronology, see Robert Renehan, "Meletius' Chapter on the Eyes: An Unidentified Source," in *Symposium on Byzantine Medicine, Dumbarton Oaks Papers* 38 (1984),

Meletius' exposition of the number and character of the ocular tunics furnishes the nomenclature for a four-tunic system which corresponds to our retina (ὁ ἀμφιβληστροειδής), uvea (ὁ ῥαγοειδής), cornea (ὁ κερατοειδής), and conjunctiva (ὁ ἐπιπεφυκώς) 2 (68.3–70.3 Cramer). In another informative paragraph, whose source is not named, Meletius not only gives the various words for the tunics in Greek but also provides the etymologies of these words. To describe the nervous structure of the retina, Meletius uses the comparison with papyrus and makes the etymology explicit by stating that the internal tunic was so called because of its similarity to a net:

> Διασχίζεται δὲ τὰ νεῦρα εἰς τὰς θαλάμους, ὥσπερ εἴ τις λαβὼν πάπυρον, ταύτην εἰς λεπτὰ κατατεμὼν διασχίζει, ἀναπλέκει τε πάλιν, καὶ ποιεῖ χιτῶνα τὸν λεγόμενον ἀμφιβληστροειδῆ, ὅμοιον ἀμφιβλήστρῳ· ὄργανον δὲ τοῦτο θηρευτικὸν ἰχθύων. 2 (63.7–11 Cramer)[32]
>
> And the nerves are split apart in the thalami as if someone, taking a papyrus and splitting it into fine pieces, entwines it again and makes the so-called net-like tunic in similar fashion to a net; this is an instrument for catching fish.

The correlation between fibrous tunics and strips of papyrus, apparently the result of personal examination, might reflect a mannerism of the original writer on whom Meletius heavily depends. It could be argued that the explicit comparison of papyrus with the uterine membranes introduced by Soranus and with the ocular tunics by Meletius must have been easily understood by readers familiar with both the fibrous nature of papyrus and the construction of the sheet. That this stylistic feature is redeployed in another context by Meletius leads us to believe that he had access to a medical work of some importance or to an abridgement of an otherwise lost treatise. The points of resemblance in the accounts of Soranus, Meletius, and Leo, all of whom adopt the same comparative clause ("just as ..., so also"), provides us with an excellent clue for narrowing down the possible sources of Meletius' chapter on the eye. Although little is known of Soranus' work on the names and etymologies of the parts of the

---

159–68 (esp. 159 n. 5), and Anna Maria Ieraci Bio, "Leone medico," in Antonio Garzya et al. (eds.), *Medici bizantini* (Turin: UTET, 2006), 787–99. Meletius' work was printed by John Anthony Cramer, *Anecdota Græca e codicibus manuscriptis bibliothecarum Oxoniensium*, vol. 3 (Oxonii: Typogr. Academ., 1836; repr. Amsterdam: Hakkert, 1963), 1–157.

[32] Meletius' account, with minor variations, is repeated in Leo medic. *de nat. hom. syn.* 35 (30.20–23 Renehan): διασχίζονται δὲ ἐν τοῖς θαλάμοις ὥσπερ τις, λαβὼν πάπυρον, ταύτην εἰς λεπτὰ κατατεμὼν ἀναπλέκει τε πάλιν καὶ ποιεῖ χιτῶνα τὸν λεγόμενον ἀμφιβληστροειδῆ ὁμοίως ἀμφιβλήστρῳ· ὄργανον δὲ τοῦτο θηρευταῖς ἰχθύων ("[and the nerves] are split apart in the thalami as if someone, taking a papyrus and splitting it into fine pieces, were to entwine it again and make the so-called net-like tunic similar to a net; this is an instrument for hunters of fish").

body, at least Meletius' section on the optic nerves and the four tunics of the eye has been shown to go back to Soranus.³³ Despite the predilection of Soranus for explaining his points by means of analogy (e. g., ὡς, ὥσπερ, ... καί, gyn. 1.16, or ὥσπερ καί ... οὕτως καί, gyn. 1.40, or καθάπερ γάρ, ... οὕτω καί, gyn. 1.35), it is not inconceivable that Soranus himself relied on Hellenistic models.³⁴ As we have seen, Soranus' adult career began with studies in Egypt, where he became familiar with the dissections of the uterus undertaken by Herophilus in Alexandria and with specifically obstetric practices.³⁵

* * *

The selection of topics and ideas that I have put forward creates a picture of a multicultural environment where book-learning and practical training could interact and where Greek physicians managed to pick up therapies or drugs validated by direct experience. Medical studies and the papyrus industry flourished in Alexandria, for a long time the city where Greek doctors travelling to Egypt could achieve fame and fortune and become familiar with the wide variety of

---

33 Cf. Meletius' *On the Constitution of Man*, 2 (Περὶ ὀφθαλμῶν) (61–72 Cramer). For a discussion of methods of detecting the "anonymous treatise embedded in the pages of Meletius," see Renehan, "Meletius' Chapter on the Eyes," 166–68.
34 Herophilus (ca. 330/20–260/50 B.C.), for example, was credited with works on the anatomy of both the reproductive organs and the eye (see von Staden, *Herophilus*, T61, T87–89 with commentary; T193–96, and p. 300). It has been persuasively argued that the "four-coat" scheme of the anatomy of the eye originated with Herophilus, who also compared the ῥαγοειδής membrane of the eye to a grape skin and the third coat (i. e., the retina) to a casting net (cf. Ruf. *onom.* 153 [154.9–10 Daremberg-Ruelle]: ἐπειδὴ δὲ Ἡρόφιλος εἰκάζει αὐτὸν ἀμφιβλήστρῳ ἀνασπωμένῳ, ἔνιοι καὶ ἀμφιβληστροειδῆ καλοῦσιν, "since Herophilus, however, compares it [the third coat] to a casting-net that is drawn up, some also call it net-like"; and ps.-Ruf. *anat.* 15 [171–72 Daremberg-Ruelle]). Nonetheless, any comparison with papyrus is absent from testimonia to Herophilus.
35 Cf. Sor. *gyn.* 1.10 (8 Ilberg), 3.2–3 (94–95 Ilberg) (with regard to Herophilus' anatomy), or 2.6 (54–55 Ilberg) (referring to Egyptian midwives), and the fuller text of the papyrus fragment of *gyn.* 3.2–3 (*PSI* II 117, fourth c. A.D.), which preserves a passage that is missing from the *Par. Gr.* 2153, where the similarity between the membrane (ὁ χιτών) in the female and that in the male is explicitly described (cf. *recto*, 14–15: ὁ μὲν ἔνδοθεν ‖[αὐτῆς] χιτὼν σαρκωδέστερός ἐστιν ("the inner membrane of the uterus is fleshier"); *recto*, 16–18: ὁ ‖[δὲ ἔξωθεν πε]ριτενὴς καὶ λεῖος ὁμοιούμενος‖[? ("while the outer membrane is taut and thin like it"). On this direct testimony, see Isabella Andorlini, "Riconsiderazione di PSI II 117: Sorani *Gynaecia*," in Véronique Boudon-Millot, Alessia Guardasole, and Caroline Magdelaine (eds.), *La science médicale antique: nouveaux regards: études réunies en l'honneur de Jacques Jouanna* (Paris: Beauchesne, 2007), 41–71 (esp. 56).

uses of *Cyperus papyrus*, at that time the principal resource for transmission of the written word.[36]

That Soranus could have combined his intellectual interests there with those of a practicing anatomist and clinician is assured by the familiarity he displays not only with the anatomy of the human body, but also with the "anatomy of the papyrus roll."[37]

Bibliography

Alpin, Prosper. *Plantes d'Egypte par Prosper Alpin (1581–1584)*. Traduit du latin, présenté et annoté par R. de Fenoyl (Cairo: IFAO, 1980).

Andorlini, Isabella. "Riconsiderazione di PSI II 117: Sorani *Gynaecia*," in Véronique Boudon-Millot, Alessia Guardasole, and Caroline Magdelaine (eds.), *La science médicale antique: nouveaux regards: études réunies en l'honneur de Jacques Jouanna* (Paris: Beauchesne, 2007), 41–71.

Andorlini, Isabella. "Un ricettario da Tebtynis: parti inedite di PSI 1180," in Isabella Andorlini (ed.), *Testi medici su papiro; atti del Seminario di Studio (Firenze, 3–4 giugno 2002)* (Florence: Istituto Papirologico "G. Vitelli," 2004), 81–118.

Beck, Lily Y., trans. *Pedanius Dioscorides of Anazarbus, De materia medica*, Altertumswissenschaftliche Texte und Studien 38 (Hildesheim: Olms-Weidmann, 2005).

Boudon-Millot, Véronique, and Jacques Jouanna. *Galien, Ne pas se chagriner*, t. IV (Paris: Les Belles Lettres, 2010).

Brizzi, Enzo, et al. *Anatomia topografica* (Milan: Edi. Ermes, 1978).

Chantraine, Paul. *Dictionnaire étymologique de la langue grecque*, 4 vols. (Paris: Klincksieck, 1968–1980).

Cramer, John Anthony. *Anecdota Græca e codicibus manuscriptis bibliothecarum Oxoniensium*, vol. 3 (Oxonii: Typogr. Academ., 1836; repr. Amsterdam: Hakkert, 1963).

---

[36] The bookish nature of a doctor's intellectual life is mirrored in Galen's private collection of books, as emerges from the newly rediscovered *On Consolation from Grief*. See *indol.* 7–8, in Véronique Boudon-Millot and Jacques Jouanna, *Galien, Ne pas se chagriner*, t. IV (Paris: Les Belles Lettres, 2010), 4, 41–44. On this topic, see Vivian Nutton's article "Galen's Library," in Christopher Gill, Tim Whitmarsh, and John Wilkins (eds.), *Galen and the World of Knowledge: Greek Culture in the Roman World* (Cambridge: Cambridge University Press, 2009), 19–34, and Amneris Roselli, "Libri e biblioteche a Roma al tempo di Galeno: la testimonianza del *de indolentia*," *Galenos* 4 (2010), 127–48 (with earlier bibliography).

[37] This anatomically oriented approach goes back to Eric G. Turner, *Recto and Verso: The Anatomy of the Papyrus Roll*, Pap. Brux. 16 (Brussels: Fond. Égypt. Reine Élisabeth, 1978), who writes that "these ribands [of the papyrus stem] are laid side by side; ... above them a second layer is placed with equal care at right angles" (14). In his works concerned with the manufacture of papyrus, Turner was apparently unaware of the comparison made by ancient medical writers between papyrus layers and tunics of the uterus and the eye.

Daremberg, Charles, and Charles-Émile Ruelle. *Œuvres de Rufus d'Éphèse* (Paris: Imprimerie nationale, 1879; repr. Amsterdam: Hakkert, 1963).
Dietz, Friedrich R. *Severi Iatrosophistae De clysteribus liber* (Regimontii Prussorum: Borntraeger, 1836).
Ebbell, Bendix. *The Papyrus Ebers: The Greatest Egyptian Medical Document* (Copenhagen: Levin and Munksgaard, 1937).
Eigler, Ulrich, Georg Wöhrle, and Bernhard Herzhoff. *Theophrast, De odoribus*, Beiträge zur Altertumskunde 37 (Stuttgart: Teubner, 1993).
Fischer, Klaus-Dietrich. *Pelagonii Ars Veterinaria* (Leipzig: Teubner, 1980).
Gaudet, John J. "Mineral Concentrations in Papyrus in Various African Swamps," *Journal of Ecology* 63 (1975): 483–91.
Goltz, Dietlinde. *Studien zur altorientalischen und griechischen Heilkunde. Therapie – Arzneibereitung – Rezeptstruktur*, Sudhoffs Archiv, Beih. 16 (Wiesbaden: Steiner, 1974).
Green, Monica, and Ann E. Hanson. "Soranus of Ephesus: *Methodicorum princeps*," in Wolfgang Haase (ed.), *Aufstieg und Niedergang der römischen Welt*, II.37.2 (Berlin: De Gruyter, 1994), 968–1075.
Grenfell, Bernard P., Arthur S. Hunt, and David G. Hogarth. *Fayûm Towns and Their Papyri* (London: Egypt Exploration Fund, 1900).
Grohmann, Adolf. *Allgemeine Einführung in die arabischen Papyri* (Vienna: F. Zöllner, 1924).
Hanson, Ann E. "A Receptarium from Tebtunis," in Isabella Andorlini (ed.), *Greek Medical Papyri* II (Florence: Istituto Papirologico "G. Vitelli," 2009), 71–103 (no. 5).
Hort, Arthur, trans. *Theophrastus, Enquiry into Plants*, vol. 1 (Cambridge, Mass.: Harvard University Press, 1916).
Ieraci Bio, Anna Maria. "*Leone medico*," in Antonio Garzya et al. (eds.), *Medici bizantini* (Turin: UTET, 2006), 787–99.
Ieraci Bio, Anna Maria. *Paolo di Nicea, Manuale medico* (Naples: Bibliopolis, 1996).
Ilberg, Johannes. *Sorani gynaeciorum libri IV, De signis fracturarum; De fasciis; Vita Hippocratis secundum Soranum*, CMG IV (Leipzig: Teubner, 1927).
Kramer, Johannes. *Von der Papyrologie zur Romanistik*, Archiv für Papyrusforschung, Beih. 30 (Berlin: De Gruyter, 2011).
Leach, Bridget, and John Tait. "Papyrus," in Paul T. Nicholson and Ian Shaw (eds.), *Ancient Egyptian Materials and Technology* (Cambridge: Cambridge University Press, 2000), 227–53.
Lewis, Naphtali. *Papyrus in Classical Antiquity* (Oxford: Clarendon Press, 1974).
Manniche, Lisa. *An Ancient Egyptian Herbal* (Austin, Tx.: University of Texas Press & British Museum Publications, 1989).
Nutton, Vivian. "Galen in Egypt," in Jutta Kollesch and Diethard Nickel (eds.), *Galen und das hellenistische Erbe*, Sudhoffs Archiv, Beih. 32 (Stuttgart: Steiner, 1993), 11–31.
Nutton, Vivian. "Galen's Library," in Christopher Gill, Tim Whitmarsh, and John Wilkins (eds.), *Galen and the World of Knowledge: Greek Culture in the Roman World* (Cambridge: Cambridge University Press, 2009), 19–34.
Raeder, Ioannes. *Oribasii Collectionum medicarum reliquiae, libri XXIV–XXV; XLIII–XLVIII*, CMG VI 2,1 (Leipzig: Teubner, 1931).
Renehan, Robert. *Leonis medici De natura hominum synopsis*, CMG X 4 (Berlin: Akademie-Verlag, 1969)
Renehan, Robert. "Meletius' Chapter on the Eyes: An Unidentified Source," in *Symposium on Byzantine Medicine*, Dumbarton Oaks Papers 38 (1984), 159–68.

Roselli, Amneris. "Libri e biblioteche a Roma al tempo di Galeno: la testimonianza del *de indolentia*," *Galenos* 4 (2010): 127–48.
Skoda, Françoise. "De quelques phytonymes empruntés," *LAMA* (Centre de Recherches Comparatives sur Les Langues de La Méditerranée Ancienne) 4 (1979): 306–308.
Temkin, Oswei (with the assistance of Nicholson J. Eastman, Ludwig Edelstein, and Alan F. Guttmacher), trans. *Soranus' Gynecology* (Baltimore: The Johns Hopkins University Press, 1956; repr., 1991).
Till, Walter C. *Die Arzneikunde der Kopten* (Berlin: Akademie-Verlag, 1951).
Totelin, Laurence M. V. *Hippocratic Recipes: Oral and Written Transmission of Pharmacological Knowledge in Fifth- and Fourth-Century Greece*, Studies in Ancient Medicine 34 (Leiden: Brill, 2009).
Turner, Eric G. *Recto and Verso: The Anatomy of the Papyrus Roll*, Pap. Brux. 16 (Brussels: Fond. Égypt. Reine Élisabeth, 1978).
von Staden, Heinrich. *Herophilus: The Art of Medicine in Early Alexandria* (Cambridge: Cambridge University Press, 1989).
Wellmann, Max. *Pedanii Dioscuridis Anazarbei De materia medica libri quinque*, 3 vols. (Berlin: Weidmann, 1907–1914).
Westendorf, Wolfhart. *Handbuch der altägyptischen Medizin*, 2 vols., Handbuch der Orientalistik I 36.1-2 (Leiden: Brill, 1999).
Wilcken, Ulrich. *Urkunden der Ptolemäerzeit (ältere Funde)*, vol. 1 (Berlin: De Gruyter, 1927).
Wöhrle, Georg. "Papyrophagie," in Raimar Eberhard et al. (eds.), "... *vor dem Papyrus sind alle gleich!*" *Papyrologische Beiträge zu Ehren von Bärbel Kramer (P. Kramer)*, Archiv für Papyrusforschung, Beih. 27 (Berlin: De Gruyter, 2009), 243–47.

Markus Asper
Medical Acculturation?: Early Greek Texts and the Question of Near Eastern Influence

Abstract: This paper looks at early Greek medicine as a case-study of acculturation. First, it tries to establish that there was medical acculturation by giving a survey of non-coincidental points of concurrence between Hippocratic and first-millennium Mesopotamian medicine. The paper focuses on textual conventions as the kind of concurrence that can lead to hypotheses about functions and social backgrounds of such knowledge. Linguistics suggests that the mode of transmission seems to have been one of slowly migrating, rather closed groups of practitioners. Several writings in the Hippocratic Corpus (*De victu acutorum*, *De insomniis* = *De victu* IV, *De morbo sacro*) attack such groups. The "rational" strands of Hippocratic medicine seem to differentiate themselves from those acculturated practitioners, not the least by turning to texts for new usages. The paper assumes that these groups of practitioners are both the starting-point and the contrast against which early Greek theoretical medicine unfolded, that is, the recent result of a development both tiny and local when compared with the "global" traditions of practitioners. Thus, instead of the two dominant narratives about the relations between Greece and her neighboring cultures that offer either seamless continuity or radical break, by way of a case-study, a third one is suggested that locates the significant break within Greek culture.

In sixth- to fourth-century B.C. Greece, certain peculiar forms of discourse emerged, concerned with medicine and zoology, but also with astronomy and mathematics. These (often fragmentary) texts all share a modern-looking obsession with truth and a modern-looking interest in method, argument, explanation, sometimes even proof, and refutation. When studying these forms of discourse, it is difficult, even for the modern reader who is aware of the dangers of anachronism, to avoid terminology and notions like "theory," "science," and "rationality." From a contemporary, ahistorical, and Eurocentric point of view, therefore, these forms of discourse do not look peculiar; rather, one is tempted to take them as clear harbingers of modern science. From the perspec-

---

Thanks to Maria Börno, Klaus-Dietrich Fischer, Kerstin P. Hofmann, Brooke Holmes, Alexandra von Lieven, Steve Kidd, Dave Lunt, and Gonzalo Rubio for their generous help at various stages of this paper. It is my great pleasure to dedicate it to Heinrich von Staden who has changed my academic life more than once, all for the better.

tive of a comparative historian, however, who looks at how neighboring cultures in the Near East and in Egypt engaged with the same questions and problems at roughly the same time, it is evident just how peculiar the Greek way of dealing with these phenomena actually was. In order to make sense of the facts, one has essentially a choice between two narratives: one describes the Greek scenario as a brilliant "explosion" against a Near Eastern-Egyptian background painted in dull colors; the other one prefers a model of continuous exchange and gradual acculturation,[1] which regards Greek forms of rational practice[2] as offshoots of a Near Eastern/Egyptian substrate. The former stresses how singular "Greek" discourse is; the latter shows how many elements it shares with ancient Near Eastern and Egyptian culture. While the former has been the orthodox position for a long time, recent research and the debate about orientalism have shattered both the evidence and the underlying assumptions for what once looked like a clear cultural divide.[3]

In this paper, I will look at medicine as providing material for a case study that one can (and should) supplement with parallel studies, focusing on, e.g., mathematics,[4] astronomy, omen-texts, and perhaps practices of jurisdiction as well. I am convinced that all such studies would point towards the same result,

---

[1] For recent discussion of the loaded term "acculturation," see U. Gotter, "'Akkulturation' als Methodenproblem der historischen Wissenschaften," in S. Altekamp, M. R. Hofter, and M. Krumme (eds.), *Posthumanistische Klassische Archäologie. Historizität und Wissenschaftlichkeit von Interessen und Methoden* (Munich: Hirmer, 2001), 255–80; and, in my context, the especially relevant C. Ulf, "Rethinking Cultural Contacts," *Ancient West & East* 8 (2009): 81–132; for a survey of the debate among ethnologists, A. M. Ervin, "A Review of the Acculturation Approach in Anthropology with Special Reference to Recent Change in Native Alaska," *Journal of Anthropological Research* 36 (1980): 49–70 is still useful.
[2] For "rational practice" as an umbrella concept useful for avoiding the anachronistic notions of "science," see J. Ritter, "Translating Rational-Practice Texts," in A. Imhausen and T. Pommerening (eds.), *Writings of Early Scholars in the Ancient Near East, Egypt, Rome, and Greece* (Berlin: De Gruyter, 2010), 349–83.
[3] For example, W. Burkert, *Orientalizing Revolution: Near Eastern Influence on Greek Culture in the Early Archaic Age* (Cambridge, Mass.: Harvard University Press, 1992); M. L. West, *The East Face of Helicon: West Asiatic Elements in Greek Poetry and Myth* (Oxford: Oxford University Press, 1997). See also the reactions to Bernal's Black Athena (e.g., as charted in M. R. Lefkowitz and G. MacLean Rogers, eds., *"Black Athena" Revisited* [Chapel Hill: University of North Carolina Press, 1996]). From this perspective, R. Palter, "Black Athena, Afrocentrism, and the History of Science," in Lefkowitz and Rogers, *"Black Athena" Revisited*, 209–66, offers a useful discussion of, especially, mathematical and medical continuities (he arrives at a negative conclusion, especially for Egyptian traditions).
[4] See my essay on Greek mathematics, arguing for a more or less parallel case: M. Asper, "The Two Cultures of Mathematics in Ancient Greece," in E. Robson and J. Stedall (eds.), *The Oxford Handbook of the History of Mathematics* (Oxford: Oxford University Press, 2009), 107–32.

namely that in trying to historically contextualize the phenomenon of Greek "scientific" discourse within the Eastern Mediterranean *koinē*, rather than choosing one of the two narratives mentioned, one arrives at a third. From my point of view, that which is usually presented as a difference between Greek and non-Greek discourse is located within the Greek cultures of rational practice themselves. In the course of my argument, the respective marginality of theoretical and non-theoretical approaches to medicine will be reversed, at least partly. For lack of other evidence, my argument relies almost exclusively on texts, but "writing science" is only one of several practices that comes with "doing science." Thus, what is true for the former also throws a light upon the latter, which means that texts will be used here almost like "index fossils."

In ancient Mesopotamia and Egypt, professional medicine and its literature have a history that reaches back into the third millennium B.C. Given the many cultural contacts between Greece and the ancient Near East, particularly Egypt and Mesopotamia and its surrounding cultures, from at least Mycenaean times down to Seleucid culture in the third century B.C., one would expect to find some signs of contact and acculturation in the realm of medicine, just as, for example, they exist in the fields of writing, metallurgy, and time-reckoning. Sometimes, close connections between early Greek medicine and its Near Eastern neighbors are simply taken for granted.[5] If by "signs" one means certain, identifiable pieces of knowledge, however, the results are scarce.

I Early Greek Medicine and Eastern Mediterranean Traditions: Contacts

Among the writings in the so-called Hippocratic Corpus there is a sub-group, once believed to be "Cnidian" (as opposed to the "Coan" school of Hippocrates and his disciples).[6] These treatises are *De internis affectionibus*, *De morbis* II, *De*

---

[5] E.g., J. Laskaris, *The Art Is Long: "On the Sacred Disease" and the Scientific Tradition* (Leiden: Brill, 2002), 52; compare, however, V. Nutton, *Ancient Medicine* (London: Routledge, 2004), 40–44. Brief discussion of points of contact in M. Geller, "West Meets East: Early Greek and Babylonian Diagnosis," in H. F. J. Horstmanshoff and M. Stol (eds.), *Magic and Rationality in Ancient Near Eastern and Graeco-Roman Medicine* (Leiden: Brill, 2004), 11–61, and J. Scurlock, "From Esagil-kīn-apli to Hippocrates," *Le Journal des Médecines Cunéiformes* 3 (2004): 10–30, at 10–11 (see n. 10 for criticism of Geller's approach).

[6] The hypothesis of the two competing medical schools should be discarded: see V. Langholf, *Medical Theories in Hippocrates: Early Texts and the "Epidemics"* (Berlin: De Gruyter, 1990), 12–36, following W. D. Smith; H. von Staden, "Women and Dirt," *Helios* 19 (1992): 7–30, at 13 n. 38.

affectionibus, *De morbis mulierum*, and two texts on diseases quoted by Galen as "Cnidian."[7] They all focus on descriptions and classifications of diseases, adopt an impersonal, non-polemical style, and are mostly free from arguments or elaborate etiological speculation.[8] It is generally believed that these texts reflect the style or even preserve the content of earlier, pre-Hippocratic, non-speculative Greek medicine. Assuming that some exchange between the two cultures and their respective medical traditions must have existed long before the fifth century,[9] Assyriologists have looked for parallels between this group of writings and roughly contemporary Mesopotamian medical literature, especially in two series of tablets called the *Diagnostic* and the *Therapeutic Handbook*, respectively.[10] There are few such parallels,[11] and their significance is disputed. I give here only the more significant ones:

1. Egyptian, Babylonian, and Greek sources share a common structure to arrange lists of symptoms, diseases, wounds, treatments, etc.: the so-called schema *a capite ad calcem*, which is still used today.[12]
2. *morb.* II 47b.4, 7.70–72 L. (= 181.16–182.19 Jouanna) describes in a detailed manner how, when all pharmaceutical approaches have failed, to drain pus from a diseased lung over the course of ten days, including the day of prognosis. There exists a close parallel in Akkadian medicine.[13]

Nonetheless, it is still commonly accepted that these writings form the oldest stratum in the Hippocratic Corpus.

7 Langholf, *Medical Theories*, 20 quotes both passages; J. Jouanna, *Hippocrates*, trans. M. B. DeBevoise [orig. 1992] (Baltimore: The Johns Hopkins University Press, 1999), 145–46; Geller, "West Meets East," esp. 29–30. For historical differentiation within this group, see H. Grensemann, *Knidische Medizin, Teil II: Versuch einer weiteren Analyse der Schicht A in den pseudohippokratischen Schriften "De Natura Muliebri" und "De Muliebribus" I und II* (Stuttgart: Steiner, 1987), 66–73.

8 Langholf, *Medical Theories*, 12–36.

9 Even the assumption of continuities has met with considerable resistance, especially among historians of Greek science (for medicine, see P. J. van der Eijk, "Introduction," in Horstmanshoff and Stol, *Magic and Rationality*, 1–10, at 4 n. 10).

10 Key studies are D. Goltz, *Studien zur altorientalischen und griechischen Heilkunde. Therapie – Arzneibereitung – Rezeptstruktur* (Wiesbaden: Steiner, 1974); M. Stol, "An Assyriologist Reads Hippocrates," in Horstmanshoff and Stol, *Magic and Rationality*, 63–78; Geller, "West Meets East."

11 Goltz, *Studien*, 240–42 gives a list of fifteen matches of Greek and Mesopotamian medicine most of which she dismisses herself as insignificant. Stol, "Assyriologist," 65–67 discusses her remaining items and adds five more (71–76).

12 M. Asper, *Griechische Wissenschaftstexte. Formen, Funktionen, Differenzierungsgeschichten* (Stuttgart: Steiner, 2007), 58, 114 n. 149; similar cases: 267 n. 337, 371.

13 Translated and discussed by Stol, "Assyriologist," 71–72.

3. Late Babylonian texts advise the physician in certain cases of lung diseases to "seize the tongue" of the patient and then to give some medication to drink, apparently under the impression that the drug will reach the lungs that way.[14] In *De morbis* II one frequently finds the same technique mentioned, again applied to the treatment of lung diseases (for example, in II 47b.2, 7.66 L. [= 180.4–5 Jouanna]: ἐξειρύσας τὴν γλῶσσαν).[15] For Galen, to seize the tongue was a typical technique of "Cnidian" medicine, which may also point towards an older stratum of Greek medicine.
4. Both *De morbis* II and late Babylonian medicine advise the physician to shave the patient's head in the case of headaches or other pains that afflict the head (although they do *not* require shaving for technical reasons).[16]
5. Late Babylonian physicians count days, just as *De morbis* II does (e.g., 63, 7.96 L. [= 202.10 Jouanna]; 67, 7.102 L. [= 205.22 Jouanna]; 69, 7.106 L. [= 208.19 Jouanna]), and as, of course, does later "Hippocratic" medicine. More significantly, days are counted for the same purpose, namely in order to identify the disease and to give a valid prognosis. Tablet 16 of the *Diagnostic Handbook* gives a long row of prognoses, arranged according to the number of days that a certain symptom prevails.[17]
6. *morb.* II 72.1, 7.108–10 L. (= 211.15–20 Jouanna) gives two unusual symptoms for the disease *phrontis:* Φροντίς· δοκεῖ ἐν τοῖσι σπλάγχνοισι εἶναι οἷον ἄκανθα καὶ κεντεῖν ... καὶ δείματα ὁρᾷ καὶ ... τοὺς τεθνηκότας ἐνίοτε. ("Fright disease: There appears to be a thorn in his [i.e., the patient's] viscera and to sting ... and he sees terrifying visions ... and occasionally the dead.") The comparison of a certain pain with a thorn appears in the *Diagnostic Handbook*.[18] To "see the dead" is a frequent symptom in Akkadian prognosis.[19] Although in the Akkadian tradition the two symptoms do not appear in the diagnosis of the same disease, at least Greeks and Babylonians apparently used the same descriptions of symptoms at the same time, within roughly the same conceptual framework.

---

14 Goltz, *Studien*, 245; compare Stol, "Assyriologist," 74.
15 See the parallels collected by Goltz, *Studien*, 125 n. 149.
16 Goltz, *Studien*, 241–42 n. 16.
17 N. P. Heeßel, *Babylonisch-assyrische Diagnostik* (Münster: Ugarit-Verlag, 2000), 171–86.
18 Tablet 13, 42'–44' (R. Labat, *Traité akkadien de diagnostics et pronostics médicaux* [Leiden: Brill, 1951], 114) or, even closer, the text quoted by Geller, "West Meets East," 44. See also Langholf, *Medical Theories*, 54.
19 See Heeßel, *Babylonisch-assyrische Diagnostik*, tablet 28, 35' (trans. at 314). This tablet shows how close general prognosis is to omen texts in Akkadian literature.

There are more parallels of this sort, but it is more difficult to prove that they are not coincidental.[20] A typical case is trepanation.[21] In some cases, Hippocratic medicine appears to apply what is known in Near Eastern and Greek religious contexts as purification in a "secularized" manner as medical therapy.[22]

The most significant class of parallels, however, does not consist of definite borrowings of therapy or diagnostic know-how but of shared assumptions of disease and prognosis,[23] in a similar use of texts that results in strikingly similar

---

20 See Geller, "West Meets East," 22–23 on similarities between the *Diagnostic Handbook* and the Hippocratic *Prognosticon*; ibid. 32 on similar etiologies for a hip disease (*ischias*) in *De internis affectionibus* and the *Diagnostic Handbook*; ibid. 48 on similar words for pain that involve the notion of "biting" and "sharpness." Goltz, *Studien*, 242, 247 quotes a "magic" recipe in *Mul.* I 77 (8.172 L.), which shows a close resemblance with Babylonian amulets (see A. E. Hanson, "Uterine Amulets and Greek Uterine Medicine," *Medicina nei Secoli. Arte e Scienza* 7 [1995]: 281–99, 288 for amulets in gynecology). Geller, "West Meets East," 47 contends that these parallels are not arbitrary, but fails to give a methodologically sound rationale of how to tell arbitrary from non-arbitrary parallels. M. J. Geller, "Phlegm and Breath—Babylonian Contributions to Hippocratic Medicine," in I. L. Finkel and M. J. Geller (eds.), *Disease in Babylonia* (Leiden: Brill, 2007), 187–99 discusses the two etiological concepts of the four humors and of "breaths" and their parallels in Babylonian medicine (vague conclusions). Compare Scurlock's differing views ("From Esagil-kīn-apli to Hippocrates," 14–15). Klaus-Dietrich Fischer brings an interesting paper by Tanja Pommerening to my attention ("βούτυρος '*Flaschenkürbis*' und κουροτόκος im Corpus Hippocraticum, *De sterilibus* 214: Entlehnung und Lehnübersetzung aus dem Ägyptischen," *Glotta* 86 [2010]: 40–54). Pommerening demonstrates close parallels between an Egyptian twelfth-century B.C. text dealing with a prognosis of pregnancy and a fourth-century B.C. Hippocratic one. These parallels combine both linguistic and medical aspects on more than one level; thus, there is no doubt that the knowledge concerning the prognosis has somehow been handed down for more than eight hundred years and crossed the boundaries between two cultures.

21 See Stol, "Assyriologist," 75–76. Trepanation is briefly mentioned in *morb.* II 15, 7.26–28 L. (= 149.1, 150.6–7 Jouanna): ἢν ὕδωρ ἐπὶ τῷ ἐγκεφάλῳ γένηται ... τρυπῆσαι πρὸς τὸν ἐγκέφαλον ("if water occurs next to the brain ... drill up to the brain"). Stol quotes a Late Babylonian passage which begins "if a man's skull holds water" and then goes on to recommend "scrape the scull" (*gulgullašu teserrem*); to me, this does not seem to imply trepanation. See also G. Majno, *The Healing Hand: Man and Wound in the Ancient World* (Cambridge, Mass.: Harvard University Press, 1975), 59. Trepanation techniques, however, may well emerge without cultural contacts: compare the Inca cases discussed in V. A. Andrushko and J. W. Verano, "Prehistoric Trepanation in the Cuzco Region of Peru: A View into an Ancient Andean Practice," *American Journal of Physical Anthropology* 137 (2008): 4–13 and *passim*, and, in general, Dan Potts in this volume.

22 Von Staden, "Women and Dirt," esp. 16–18 on fumigation in gynecological therapy. Generally, see also a (rather disappointing) article by P. Demont, "L'ancienneté de la médecine Hippocratique: un essai de bilan," in A. Attia and G. Buisson (eds.), *Advances in Mesopotamian Medicine from Hammurabi to Hippocrates*, Cuneiform Monographs 37 (Leiden: Brill, 2009), 129–49, esp. 148–49.

23 See, e.g., on classifications, Scurlock, "From Esagil-kīn-apli to Hippocrates," 14.

textual structures. Such similarities point, ultimately, towards similar forms of how medicine is institutionalized in Near Eastern and archaic Greek society. It has often been observed that these Hippocratic treatises list and describe diseases in a format quite similar to Mesopotamian and Egyptian texts: name of the disease, symptoms (these two can sometimes fall into one category, when the disease is not defined by name but by its symptoms),[24] therapy, and prognosis following one another in the same order, within a similar textual frame, and employing a similar impersonal rhetoric (Jouanna's "schéma nosologique").[25] These four parts often even use similar phraseologies, e. g., conditionals or recipe-like structures in the same places. Consider the textual patterns of the following three examples, one Egyptian, one Assyrian, and one Greek:

> Another. If you see a man with bruises in his neck, suffering from the two members (i.e., the joint) of his neck, suffering from his head (*dp.t*), the vertebra of his neck being strong (i. e., stiff), his neck being heavy and it being impossible for him to regard his belly, it being difficult for him, you are to say: "One with bruises in his neck." You should let him anoint (*wrh*) himself, rub himself (i. e., with ointment) (*sdm*), so that he will be well immediately.[26] P. Ebers 295 (New Kingdom, ca. 1550 B.C.)

> If a man [suffers from] colic [..........] (and) food and drink are regurgitated, his bowel ... his abdomen is cramped ... he drinks *taramuš* in premium beer, crush juniper, *kukru*, ... and mix (them) in fat, [make] a pessary and insert it into his anus and he will improve.[27] Ass. 13955bu (Late Assyrian, ca. 9th–7th c. B.C.)

> Another disease: excessive pain grips his head and when he moves even a little, he vomits bile. Sometimes he has trouble urinating and is delirious. When the seventh day comes, he sometimes dies. If he survives this day, he dies on the ninth or the eleventh, if he does not bleed from the nose or the ears. When this is the case, during the headache ... let him drink. ... When he bleeds from his ears and the fever recedes and the pain, let him eat laxative foods. ... If the ears do not dry up this way but the hemorrhage continues, wash them carefully and pour into them "silver-blossom," and, realgar, white lead, of all the same amount, mix them until smooth. Fill the ear completely. ... They also die, if after the sharp pain has

---

24 See Geller, "West Meets East," 44.
25 J. Jouanna, *Hippocrate: pour une archéologie de l'école de Cnide* (Paris: Les Belles Lettres, 1974), 85–87; J. Jouanna, *Hippocrate, Maladies II*, t. X.2 (Paris: Les Belles Lettres, 1983), 15–23; H. Grensemann, *Knidische Medizin, Teil I: Die Testimonien zur ältesten knidischen Lehre und Analysen knidischer Schriften im Corpus Hippocraticum* (Berlin: De Gruyter, 1975), 177–94; Grensemann, *Knidische Medizin, Teil II*, 67; Langholf, *Medical Theories*, 55–70. For a wider context, see Asper, *Griechische Wissenschaftstexte*, 378–80.
26 I am very much indebted to Alexandra von Lieven for graciously providing this translation. Compare W. Westendorf, *Handbuch der altägyptischen Medizin*, 2 vols., Handbuch der Orientalistik 1.36 (Leiden: Brill, 1999), 2:602.
27 Ed. and trans. M. J. Geller, *Renal and Rectal Disease Texts*, Die babylonisch-assyrische Medizin in Texten und Untersuchungen 7 (Berlin: De Gruyter, 2005), 267, no. 54.

wandered into the ear, there is no bleeding within seven days.[28]
morb. II 14, 7.24–26 L. (= 147.8–148.14 Jouanna) (ca. 450–400 B.C.)

The structural parallels between these texts are evident: they all present identification, symptoms, therapy, and prognosis in a similar way, in the same order, and with a similar partly impersonal, partly imperatival, rhetoric. The Egyptian and the Akkadian texts at least belong to traditions that extend further into the past and future and could be matched with many more—in the Akkadian case, even with hundreds of—texts that show the exact same structure (which I cannot even list here). In Greek, the number of instances preserved in the "Cnidian" writings[29] warrants the assumption that, at some time, this method of textual organization, the "schéma nosologique," must have been a conventional way to structure medical texts.

Apparently, all three traditions share the same concept of "disease" as an entity, defined by a set of observable symptoms, caused by external factors, having the same trajectory in all patients, which allows for a general prognosis. In all three traditions, Egyptian, Assyrian, and Greek, individual case histories are apparently unknown. They must be a "Hippocratic" invention—one more reason to believe that with "Cnidian" writings we have access to an older tradition of medicine in Greece. The disease is an individual, identifiable entity that behaves in the same, predictable way with all humans it attacks. The similar importance of prognosis in all three traditions indicates that patients and physicians interacted in a comparable way, which allowed the physician to use prognosis both to enhance his reputation and to protect himself (thus, occasionally, we read similar warnings against treating the terminally ill in Greek, Mesopotamian, and Egyptian medical texts).[30]

Beyond these shared notions, the textual similarities point toward similar functions for their respective authors and addressees: their anonymity and their impersonal style, for example, hint at a closed group within which medical knowledge was thought to be collective, protected probably by guild-like institutions.[31] These texts were obviously not intended to act as vehicles for

---

[28] The text is longer. I have concentrated on the main structural elements.
[29] See the discussion in Jouanna, *Maladies II*, 15–32.
[30] Compare Heeßel, *Babylonisch-assyrische Diagnostik*, 61 nn. 89–91 with Jouanna, *Maladies II*, 251 n. 3 (on *morb.* II 48.3, 7.72 L.); Geller, "West Meets East," 39. Qualifications for the Greek case in Laskaris, *Art Is Long*, 8.
[31] I have argued that style and structure in scientific writing provide some information about the social structure of the communication involved: see Asper, *Griechische Wissenschaftstexte*, 43–45, 371–74.

physicians to advance personal ambitions, perhaps because they served merely as works of reference, as indicated by their additive structure, and of schooling, as indicated by their authoritative, succinct rhetoric that leaves no room for doubt.

These parallels seem on the one hand close enough to rule out any notion that early Greek medicine was entirely independent of Near Eastern and Egyptian medicine.[32] On the other hand, almost all the details of, for example, specific treatments, clusters of symptoms, drugs, and terminology are sufficiently different to preclude any assumption of very close, direct, and synchronic contacts. Apart from the fact that Greek medicine must have been at the receiving end of the tradition, which is evident from the dates of some of the texts, the age and the stability of the Near Eastern-Egyptian traditions, especially the time in which and the way by which knowledge was transmitted, are open to discussion. Neither problem, that is, the time and the way of transmission, is solved by assuming that, with respect to medicine, and as with many other discourses, there ever existed an Eastern Mediterranean *koinē*, an assumption which seems to be widely accepted now.[33] As for the time of transmission, the ninth to seventh centuries B.C. seem probable: in that case, medical knowledge would have crossed the Aegean in ways similar to alphabetic writing techniques and similar to so many crafts, practices, and narratives, ranging from techniques of decoration to time-reckoning and motifs in folk-narrative (whatever these ways were precisely), that is, in the context of the "Orientalizing revolution."[34]

Near Eastern and Egyptian luxury products and techniques, however, had found their way into mainland Greece already in the second millennium, especially from the fifteenth to the thirteenth centuries.[35] Mycenaean palace medicine *might* well have employed Egyptian or Babylonian physicians—a case which has recently been argued, with due stress on the social stratification of medicine.[36]

---

32 The extent to which Egyptian and Mesopotamian medicine relate to one another is difficult to pin down. In any case, they are much closer to one another than even to early Greek medicine: see Goltz, *Studien*, 251–57.
33 Geller, "West Meets East," 59.
34 The term was coined by Burkert in *Orientalizing Revolution*. He was probably inspired by Kuhn's famous concept of scientific revolutions. The evidence is summed up admirably by Burkert, *Orientalizing Revolution*, 14–33 and West, *East Face of Helicon*, 8–9, who does not, however, mention medicine.
35 See West, *East Face of Helicon*, 507; Demont, "L'ancienneté," 135.
36 R. Arnott, "Minoan and Mycenaean Medicine and its Near Eastern Contacts," in Horstmanshoff and Stol, *Magic and Rationality*, 153–73, esp. 155–63; 159 on medical instruments found in Nauplion (1450–1400 B.C.) that bear a strong resemblance to instruments described in a contemporary text from Ugarit. See also Laskaris, *Art Is Long*, 34–35.

Such an early acculturation would account for the scant similarities of content but similar concepts and textual structures between the two traditions.[37] (As the mathematical material may show, these structures are amazingly conservative and even cross language borders, provided there exists some institutional continuity.)[38] Nonetheless, although one should certainly assume that many remedies and therapies described in the Hippocratic Corpus are traditional and may, perhaps, even go back to Mycenaean times, it is impossible to know which ones do.[39] Admittedly, medical knowledge presents a different case than, say, categories of physical speculation or motifs in mythical narrative: those who can afford it will likely have the best medical care they can get. Therefore, Egyptian or Babylonian physicians might have conceivably practiced their art in Knossos or Mycenae in 1300 B.C. It is more difficult to see, however, how the knowledge of this elite palace medicine could have trickled down, as it were, to become an indigenous tradition as appears to be the case in the early Hippocratic writings mentioned above, roughly nine hundred years later. On the whole, it seems thus more natural to assume that an acculturation of medical concepts from the East took place at some time between the ninth and seventh centuries B.C. (the great reputation of Egyptian medicine among the Greeks seems to reflect the "Egyptophilia" of a slightly later period).[40]

As for the mode of transmission, we are hardly better off: knowledge does not travel by itself. It depends on carriers or some form of media. In this case, different from later times, texts or texts alone cannot have been the carriers, because it is difficult to see how they would have overcome the barriers of different languages, writing-systems, and terminologies. Essentially, the transmission

---

37 Compare the dream omens discussed by Geller, "West Meets East," 53: only the if-part is parallel, that is, the "system": the interpretation is completely different.
38 See Asper, "Two Cultures," 125–29.
39 Despite the fascinating approach taken by Arnott, "Minoan and Mycenaean Medicine," his article makes painfully clear that virtually nothing is known about the practices (let alone concepts) of Mycenaean court or folk medicine. On some long-standing medical traditions in Greece (trepanation, herbalists), see Laskaris, *Art Is Long*, 35–44. Compare A. E. Hanson, "Continuity and Change: Three Case Studies in Hippocratic Gynecological Therapy and Theory," in S. B. Pomeroy (ed.), *Women's History and Ancient History* (Chapel Hill: University of North Carolina Press, 1991), 73–110, who investigates the assimilation of already existing medical knowledge by "Hippocratic" medicine (esp. 78, 89, 95). The older traditions' social setting and provenance, however, remain unclear.
40 See, e.g., Hom. *Od.* 4.228–32 and Hdt. 2.84. Compare R. Thomas, "Greek Medicine and Babylonian Wisdom: Circulation of Knowledge and Channels of Transmission in the Archaic and Classical Periods," in Horstmanshoff and Stol, *Magic and Rationality*, 175–85, at 181–85; Nutton, *Ancient Medicine*, 40–41.

must have been oral and thus personal.[41] In the light of parallel acculturations,[42] it is most plausible to think of migrant physicians as "carriers." Greek traditions knew of individual physicians who traveled to the East, e.g., Democedes of Croton in the sixth century (Hdt. 3.129–30) or the Cnidian physician Ctesias, who, as of 405 B.C., had already lived for a long time at the Persian court.[43] These physicians may have come into contact with local or Babylonian medicine there and brought back some knowledge of it. I find it difficult to imagine, however, that such one-person contacts should be responsible for the *structural* similarity of pre-"Hippocratic" medicine (that is, Cnidian medicine) with the Near Eastern-Egyptian traditions. It seems therefore more probable that neither texts nor individuals but entire groups of medical practitioners migrated, groups who guarded their expert knowledge by trading on it in an institutionalized, controlled way: that is, by apprenticeship within an established group structure (it is tempting to use the medieval "guild" as an analogy).[44] Since successful medicine contains a dominant practical element *and* relies on writing for storage of information, reference, and, presumably, transmission, the "carrier" of the tradition—here a group—must have been itself institutionalized, involving a social structure that determined the forms of texts and their use, institutionalized recruitment, and "education."

As is typical for even much later practitioners in the Mediterranean (and elsewhere: compare Shapin's "invisible technicians"),[45] there is hardly any direct evidence for these groups. Two peculiarities of the Greek versus the Mesopotamian-Egyptian traditions can, however, perhaps lead to some inferences. First, there are no obvious Akkadian (or Egyptian) loanwords in early Greek medicine;[46] at the same time there are clear parallels as far as key concepts

---

41 See already Goltz, *Studien*, 239.
42 Compare Burkert, *Orientalizing Revolution*, 14–25, 41–45. On the problem of ethnicity in the transmission of knowledge, see the brief remarks in F. Rochberg-Halton, "The Cultures of Ancient Science: Some Historical Reflections," *Isis* 83 (1992): 547–53, 549–50.
43 Compare Stol, "Assyriologist," 66.
44 Thomas, "Circulation of Knowledge," 180–81 is greatly hampered by her concentration on "highly specialized or technical knowledge such as astronomical observational data or theories" and her disregard of actual practices. The same is true for Nutton, *Ancient Medicine*, 41–42, whose only model of transmission is of Greek individuals traveling east.
45 See S. Shapin, *A Social History of Truth: Civility and Science in Seventeenth-Century England* (Chicago: University of Chicago Press, 1994), 360 on the triple invisibility of technicians. S. Cuomo, *Technology and Culture in Greek and Roman Antiquity* (Cambridge: Cambridge University Press, 2007), 77 applies the concept to ancient practitioners.
46 *Pace* Geller, "West Meets East," 29 and Stol, "Assyriologist," 70, a "loanword" σίαλος/*suālu* ("coughing disease") does not exist, I am afraid. Σίαλος in the Hippocratic Corpus and elsewhere

and structures of medical knowledge and texts are concerned. By contrast, in other realms of early Greek culture, Semitic loanwords or "loan-concepts" do exist, mostly for acculturated techniques or luxury goods.[47]

If one can build anything on this evidence (I am not sure one can and, thus, the following is speculative), the acculturation of medicine must have been unlike the exchange of luxury goods in significant aspects: it must have been slow and taken place within small and closed groups.[48] The slow pace would have excluded a stage in which many foreign words were adopted, that is, a stage in which the practitioners crossed the language barrier quickly. Rather, they must have worked in a bilingual context for a considerable time. Occasionally, literal translation might have occurred: so far, the only known instance of this is the word for "suppository" for which both Greek and Akkadian use the word for "acorn" (βάλανος and *allānu*, respectively).[49] The groups' "closed" character would have ensured that it did not become part of the wider lexicon while the knowledge was in transition. Both criteria could be compared to the most blatant case of acculturation, namely, the adoption of a West Semitic alphabet. Here, the knowledge concerned remains firmly tied to its Semitic origins (even until today): the names, basic shapes, and the order of the letters; their phonetic equivalents; the whole concept of alphabetic (as opposed to syllabic or iconographic) writing; and probably the primary ways to handle the new technology (writing directions, inscriptions and their functions, tablets, leather-books, letters, etc.).[50] It spread rapidly and became at least passively part of the cultural competence of wider groups. Apparently, there was neither time nor need for a translation of any sort, unlike in medicine. Second, the almost complete

---

means "fat" (adj.), σίαλον "spittle." There is not a trace of "cough" (see J.-H. Kühn and U. Fleischer, *Index Hippocraticus* [Göttingen: Vandenhoeck & Ruprecht, 1986–1989], 732–33). In general, however, Greek tends to "Hellenize" words taken over from non-Greek languages. Thus, in theory words in the medical lexicon for which no plausible Indo-European etymology exists would qualify as candidates for being Semitic loanwords.

47 See the list in West, *East Face of Helicon*, 12–14 and the discussion in Burkert, *Orientalizing Revolution*, 33–40.

48 Compare Ulf's diagram ("Rethinking Cultural Contacts," 87) and his attempts to draw conclusions about the "contact zone" from observable features of the acts and products of acculturation. His concept of "heterarchy" (100), for example, probably applies to my case.

49 See J-H. Kühn and U. Fleischer, *Index Hippocraticus*, 119. In Akkadian, the word for "finger" is used, too (*ubānu*), for which no Greek parallel exists (compare, however, δάκτυλος IV LSJ = βάλανος). See Goltz, *Studien*, 75–76.

50 See M. Asper, "Medienwechsel und kultureller Kontext. Die Entstehung der griechischen Sachprosa," in J. Althoff (ed.), *Philosophie und Dichtung im antiken Griechenland* (Stuttgart: Steiner, 2007), 67–102, at 77 on early Greek "acculturated prose."

lack of congruence in medical detail and, at the same time, close resemblances in concepts and structures (which tend, apparently, to be far more conservative) point to the same picture: our model "carrier" is not a well-traveled, multilingual, highly paid expert of tongue-seizing therapy or *namburbî* rituals with an international group of customers, but a closely-knit group of practitioners sticking to their own traditions, traditions that take several generations to cross cultural and language boundaries, thereby avoiding the need for rapid translation. Ironically, the best—nonetheless, indirect—evidence for such groups comes from the polemics against them that some later Hippocratic writings have preserved.

II Polemics: Who Is Attacked in "Hippocratic" Writing?

The group of Hippocratic writings discussed above (*De internis affectionibus, De morbis II, De affectionibus, De morbis mulierum*) shares, among other features, an impersonal rhetoric, just as Egyptian or Babylonian medical texts do:[51] not only do the authors avoid revealing their names, they do not even refer to themselves or their audiences as individuals (except for the occasional, rather generic, imperative in recipe-like structures), for example, by making use of the authorial "I" and, more generally, by employing a rhetoric of expert authority, the "je scientifique."[52] All these strategies, however, are employed by another set of Hippocratic treatises that are particularly fond of "boundary-work":[53] here, the authors take pains to distinguish themselves polemically from a group of competitors. As the attack unfolds, an authorial construction of both author and

---

[51] For this category as applied to the reading of science writing in general, see Asper, *Griechische Wissenschaftstexte*, 43–45; in mathematics, Asper, "Two Cultures," 118.

[52] Compare the "authorial personae" of Celsus as analyzed by H. von Staden, "Author and Authority: Celsus and the Construction of the Scientific Self," in M. E. Vázquez Buján (ed.), *Tradición e innovación de la medicina latina de la antigüedad y de la alta edad media* (Santiago de Compostela: Universidade de Santiago de Compostela, 1994), 103–17, esp. 110–14. For the "je scientifique," see A. Debru, "La suffocation hystérique chez Galien et Aétius: réécriture et emprunt du 'je,'" in A. Garzya (ed.), *Tradizione e ecdotica dei testi medici tardoantichi e bizantini* (Naples: D'Auria, 1992), 79–89, at 85–87 and my remarks on Galen's "I" (Asper, *Griechische Wissenschaftstexte*, 333–34).

[53] See T. Gieryn, "Boundaries of Science," in S. Jasanoff (ed.), *Handbook of Science and Technology Studies* (Thousand Oaks: SAGE Publications, 1995), 393–443; Asper, *Griechische Wissenschaftstexte*, 166; Geller, "West Meets East," 15–16 on the Hippocratic Corpus as "a transition period" (that is, from anonymity and impersonality to personality and then, later, to very personal and name-dropping authors such as Galen).

competitors takes place. At the same time, these treatises show a new approach to medical problems, both regarding therapies[54] and, most conspicuously, explanations and logical argument. For competitive purposes, they explicitly stress the latter and have, therefore, often been claimed as "rational" or as the beginning of "rational" medicine.[55] It may suffice here to discuss only three instances of this boundary-work,[56] all three taken from treatises usually dated to the late fifth century B.C. In all three cases, the attacks offer us brief, and both indirect and heavily biased, glimpses of other forms of medical practice. What is more, in all three cases, these non-Hippocratic practitioners closely resemble their Mesopotamian colleagues.

De victu acutorum discusses how patients with "acute," that is, the most dangerous, diseases should be treated. It begins with an attack against "the Cnidians" (οἱ ξυγγράψαντες τὰς Κνιδίας καλεομένας γνώμας, *acut.* 1, 2.224 L.) who have allegedly neglected regimen, as was generally the case with "the old guard" (οἱ ἀρχαῖοι, *acut.* 1, 2.226 L.) who, according to this author, focused too much on (useless) nosological definitions and classifications. In addition, the author attempts to reorganize the traditional taxonomy of diseases, on the grounds that his predecessors did not know classes of diseases.[57] Both objections fit perfectly the group of writings discussed above (that are termed "Cnidian" according to polemics like these)[58] and, by implication, Babylonian-Egyptian medical texts. By means of polarization, the author implicitly presents himself as "modern" against the background of inadequate "old-timers." Similarly, Diocles of Carystus, a fourth-century B.C. physician, criticizes the "old-timers" for using the phases of the moon as a means of prognosis[59]—a method similar to Late Babylonian medicine, which sometimes used stars in prognosis.[60] Here, too, a direct line from Greek "old" medicine to first-millennium Mesopotamian medicine ex-

---

54 Geller, "West Meets East," 60 gives a list of fifth-century Greek therapeutic innovations (diet and regimen, purging/evacuation, blood-letting).
55 Discussion and refutation in van der Eijk, "Introduction," 3–7.
56 One can glimpse the same structure also in Hipp. *virg.* 1, 8.468.17–20 L.: the author lists some conditions typical for young women, among them seizures and fits of all kinds, including the "sacred disease" (8.466.4 L.). When the epileptic girls regain consciousness, they are prone to consider religious causes, namely Artemis, for their ailments and act accordingly, deceived by "seers."
57 Jouanna, *Hippocrates*, 153; compare Scurlock, "From Esagil-kīn-apli to Hippocrates," 22.
58 See Geller, "West Meets East," 19.
59 Text in P. J. van der Eijk, *Diocles of Carystus: A Collection of the Fragments with Translation and Commentary*, 2 vols. (Leiden: Brill, 2000–2001), 1:130, fr. 64.
60 See BM 56605 (from the late Seleucid period) as discussed in Heeßel, *Babylonisch-assyrische Diagnostik*, 112–27; Geller, "West Meets East," 38.

ists. In both cases, the author, employing a personal rhetorical stance, attacks competitors who remain anonymous, but are clearly part of a collective. The competition, however, is mainly diachronic, as it seems: the self-styled newcomer sets himself up against what he perceives as the "old guard."

A Hippocratic book on dreams, the so-called treatise *de victu* IV or *de insomniis*, presents a parallel case: against the current and widespread practice of dream-interpreters, the Hippocratic author tries to establish secure grounds for how one may use patients' dreams for prognosis. In this case, we learn more about the people the treatise attacks: first, it is again an anonymous collective against which the author competes directly. His main strategy of boundary-work is to divide dreams into separate categories. First, there are dreams sent by gods (*theia*) that portend favorable or unfavorable events for city-states or individuals (*vict.* IV 87.1, 6.640 L. [= 98.1–2 Joly]). This field is rightly claimed by professionals who command a pertinent (and, apparently, widely accepted) body of expert knowledge (*technē*).[61] But then there is a second class of dreams that originate in the soul, are caused purely by physical factors, and indicate, accordingly, physical states of the body (*vict.* IV 87.1, 6.642 L. [= 98.4–5 Joly]: ὁκόσα ... ἡ ψυχὴ τοῦ σώματος παθήματα προσημαίνει, "all the bodily ailments that the soul foretells"). These must be the object of the true physician, then. According to the Hippocratic author, the dream-professionals falsely claim to interpret both. Then, the final blow: not only are they sometimes right, sometimes wrong, but either way they do not understand the reason why (*vict.* IV 87.1, 6.642 L. [= 98.8 Joly]: οὐδέτερα τούτων γινώσκουσι δι' ὅ τι γίνεται, "of none of which they understand the reason of why it happens"). Instead, the dream-professionals suggest prayers.[62]

The Hippocratic author proceeds by describing a long series of "things that appear" (*ta phainomena*) in dreams and how to treat the dreamer. As Philip van der Eijk has recently shown,[63] both the practices of the group attacked and some features of the treatise itself show close similarities with Late Babylonian dream

---

[61] For a discussion of the term, see Cuomo, *Technology and Culture*, 7–40 (with many medical examples).

[62] The author recommends prayers himself (e.g., *vict.* IV 89.14, 6.652 L. [= 104.17–21 Joly]; IV 90.7, 6.656–58 L. [= 107.6–7 Joly]), but he complements them with therapy. On his dualistic approach see IV 87.2, 6.642 L. (= 98.12–13 Joly). As for his piety, compare his last sentence: εὕρηταί μοι δίαιτα ὡς δυνατὸν εὑρεῖν ἄνθρωπον ἐόντα σὺν τοῖσι θεοῖσιν, "I have found a healthy way of life as a human is able to find it, with the help of the gods" (IV 93.6, 6.662 L. [= 109.19–20 Joly]). See P. J. van der Eijk, "Divination, Prognosis, and Prophylaxis: The Hippocratic Work "On Dreams" (*De victu* 4) and its Near Eastern Background," in Horstmanshoff and Stol, *Magic and Rationality*, 187–218, at 213.

[63] Van der Eijk, "Divination, Prognosis, and Prophylaxis."

literature,⁶⁴ sometimes even in style and syntax.⁶⁵ At one point, "seeing the dead" in a dream (*vict*. IV 92.1, 6.658 L. [= 107.21–22 Joly] points to a certain, generally positive, prognosis, just as in Late Babylonian diagnostic literature. The order in which celestial bodies seen in dreams are treated (ch. 89) corresponds to Mesopotamian omen texts.⁶⁶ In Mesopotamia, dream interpreters (*šā'ilu*) could be part of diagnosis and prognosis.⁶⁷ The Hippocratic author operates upon the same simple principle of matching dream vision with prognosis (e. g., "crossing rivers and soldiers and wars and strange apparitions signify illness or madness," *vict*. IV 93.5, 6.662 L. [= 109.12–14 Joly]). In this case, the suggested therapy consists of a light diet, emetics, five days rest, then appropriate physical exercise, etc.). In the Babylonian tradition, there is no fundamental difference between divination and prognosis. The Hippocratic author newly constructs this difference as a fundamental one⁶⁸ that applies to two different kinds of dreams, the treatment of which falls under the expertise of two different groups of professionals. I do not suggest that the Hippocratic author directly targets Babylonian *šā'ilu*. The practices, however, and the knowledge of the professional group attacked show so many resemblances with older Mesopotamian traditions that they must be ultimately derived from them.

My third example, the famous Hippocratic treatise *De morbo sacro*, fits into the same pattern: by attacking a group of competitors, the author not only sketches out his own "rationalist" agenda as opposed to "magical" practices but also gives an impression of the practitioners, practices, and concepts he criticizes. Some of these practices and concepts show Near Eastern influences. He denounces his opponents as "sorcerers, purifiers, mendicants, and charlatans" (μάγοι τε καὶ καθάρται καὶ ἀγύρται καὶ ἀλαζόνες, 6.354 L. [= 3.20–4.1 Jouanna]; cf. 6.396 L. [= 33.3–4 Jouanna]). I will refer to them as "healers." Mainly, the Hippocratic author criticizes the healers on two grounds: first, for their religious etiologies of epilepsy that claim the disease is caused by the gods, an explanation

---

64 Van der Eijk, "Divination, Prognosis, and Prophylaxis," 214. The tradition reaches back into the second millennium, but most texts are from seventh-century Ninive. See N. P. Heeßel, *Divinatorische Texte I. Terrestrische, teratologische, physiognomische und oneiromantische Omina* (Wiesbaden: Harrassowitz, 2007), 10.
65 See the long conditional passages in, e. g., *vict*. IV 89, esp. 4–7, 6.646–48 L. (101.1–102.4 Joly).
66 Compare Asper, *Griechische Wissenschaftstexte*, 294 n. 514.
67 See Heeßel, *Babylonisch-assyrische Diagnostik*, 76–77, tablet 18, 12'–14' (trans. at 220) and compare texts quoted in 93 n. 94, 223–24.
68 The two tend to coalesce nonetheless: G. E. R. Lloyd, *The Revolutions of Wisdom: Studies in the Claims and Practice of Ancient Greek Science* (Berkeley: University of California Press, 1987), 41–42; Langholf, *Medical Theories*, 232–54; van der Eijk, "Divination, Prognosis, and Prophylaxis," 187.

which, he believes, contradicts certain theological assumptions that he takes for granted among his audience.[69] Admittedly, to explain certain diseases as punishment by a certain god for an individual's transgression is so widely attested all over the world that one cannot use it as an argument for the healers' Babylonian background.[70] Nonetheless, the healers' method of identifying definite "symptoms" in order to diagnose a certain god's wrath as the cause establishes a closer similarity.[71]

> If they [the patients] imitate a goat and if they roar and if they have convulsions in the right side, they [the healers] say the Mother of Gods is the cause. If he [the patient] shouts shriller and louder, they compare him to a horse and say Poseidon is the cause. (1.11, 6.360 L. [= 8.1-5 Jouanna])

In the following lines (6.360–62 L. [= 8.5–12 Jouanna]), Enodia, Apollon, Ares, and Hecate are all identified as causes for the disease, based on the symptoms of stool, a foaming mouth, and panic attacks. Not only do the concept and the conditional structure of the passage closely resemble Babylonian diagnostics, but the parallels are even more suggestive. In a Neo-Assyrian collection of diagnostics that is not part of the canonical Late Babylonian *Diagnostic Handbook*'s section (tablets 26–30) on "epilepsy," we find the following passages on *miqtu* ("fall" or perhaps "falling sickness"):[72]

> If, at the time it overwhelms him ..., he growls like a dog ...: Lord of the Roof has seized him. ... If he brays like a donkey ... an.ta.šub.ba ... his disease ...; he will not be saved. ... If a fall falls upon him and [...] like an ox, he roars ... [...]: an.ta.šub.ba has seized him.[73] ... If, at the time it overwhelms him, he moans like a dove ... his disease will [...]. (trans. Stol)[74]

---

69 P. J. van der Eijk, "The 'Theology' of the Hippocratic Treatise *On the Sacred Disease*," *Apeiron* 23 (1990): 87–119.
70 W. Burkert, *Creation of the Sacred: Tracks of Biology in Early Religions* (Cambridge, Mass.: Harvard University Press, 1996), 102–28.
71 On this passage, see R. Parker, *Miasma: Pollution and Purification in Early Greek Religion*, 2nd ed. (Oxford: Clarendon Press, 1996), 244–45 and Geller, "West Meets East," 20–21: "If one simply imagines the phrase 'hand of' the particular god here, one has a reasonable replica of a text resembling the Akkadian *Diagnostic Handbook*."
72 M. Stol, *Epilepsy in Babylonia* (Groningen: Styx Publications, 1993), 91: "It seems to be an older version and is known from only two fragments, the one from Middle Babylonian Nippur, the other from Neo-Assyrian Sultantepe." Stol believes that the *Diagnostic Handbook* reacts to this text.
73 "An.ta.šub.ba" and the enigmatic Lugal-gìr.ra ("Lord of the Roof") are two of several diseases (or disease-causing demons) in Babylonian diagnostics that are usually understood as epilepsy. On the problem of identifying "epilepsy" in Babylonian texts, see H. Avalos, "Epilepsy

The parallels are obvious and close. Furthermore, they include not only etiology and diagnostics, but also therapy which consists, mainly, in purifying rituals. The Hippocratic author ridicules the healers for the following practices:

> They perform purifications (καθαίρουσι) and of the purifying objects (καθαρμοί, perhaps καθαρμάτων, "offscouring," is preferable) some they hide in the earth, others they throw into the sea, others they carry away into the mountains, where nobody will touch or come across them. (1.12, 6.362 L. [= 9.4–8 Jouanna])[75]

Mark Geller has observed that these practices show parallels with parts of the so-called *namburbî*-ritual, practiced in the first millennium in Mesopotamia.[76] These rituals are purifications which partly consist of "sending off" the curse that has befallen the sick onto a magic substitute, a living scapegoat,[77] or an effigy of the sick person, or some other substance that is then hidden or buried or sent away.[78] After the house has been purified, the magician destroys the offscourings, by throwing them into the river, hiding them somewhere, or burning them.[79] Thus, for the aetiology, diagnosis, *and* therapy for epilepsy, there exists a first-millennium Mesopotamian background of the healers' concepts and practices, which strongly suggests a continuous tradition of some sort.[80] Against this tradition, the Hippocratic author unfolds his rationalist etiology and treatment, both of which rest on the assumptions of humoral pathology. Epilepsy is caused

---

in Mesopotamia Reconsidered," in Finkel and Geller, *Disease in Babylonia*, 131–36 *contra* Stol, *Epilepsy in Babylonia*.

74 I have excerpted a longer passage in order to bring out the parallels. The whole text is published, translated, and discussed in Stol, *Epilepsy in Babylonia*, 91–98, quote from lines 133–40, 141–47, 148–51, and 152–58 (pp. 93–95). Compare Heeßel, *Babylonisch-assyrische Diagnostik*, 222–23 and the objections of Scurlock, "From Esagil-kīn-apli to Hippocrates," 12–13.

75 Parker, *Miasma*, 210–11, 229–30.

76 Geller, "West Meets East," 24.

77 *De morbo sacro* does not mention living scapegoats. There is a proverbial curse, however, in Greek (κατ' αἶγας ἀγρίας "On to wild goats!") that refers to such practices, especially in the treatment of epilepsy. See Callimachus fr. 75.13 (Pfeiffer) and the material quoted by Pfeiffer *ad loc.*

78 See S. Maul, *Zukunftsbewältigung. Eine Untersuchung altorientalischen Denkens anhand der babylonisch-assyrischen Löserituale (Namburbi)* (Mainz: Verlag Philipp von Zabern, 1994), 91–92, 94–100. For methodological criticism of Maul's useful work, see N. Feldhuis, "On Interpreting Mesopotamian Namburbi Rituals," *Archiv für Orientforschung* 42/43 (1995/1996): 145–54, esp. 146–51; for background, see P. Attinger, "La médecine mésopotamienne," *Le Journal des Médecines Cunéiformes* 11/12 (2008): 1–96, at 2–6.

79 Maul, *Zukunftsbewältigung*, 99.

80 For therapeutic "secularizations" of Greek/Near Eastern religious purifications in the Hippocratic Corpus, see von Staden, "Women and Dirt," 13–18.

by an excess of phlegm in the brain and the able physician can diagnose the problem, treat, and heal it accordingly.

III Practitioners and Theoreticians

In the three cases discussed, the knowledge of the groups attacked by the authorial "I"—that is "old" or "Cnidian" physicians, dream-interpreters, and healers—shows affinities with Mesopotamian concepts, many of which are known to us from texts written not too much earlier. I believe that, considered together, the medical parallels are both too close and too wide-ranging to be explained by anything other than some kind of continuous tradition. As agents or "media" of this tradition—since the tradition cannot have consisted in trafficking decontextualized texts, oral or written,[81] nor in "international exchange," as knowledge might travel today—people are the most probable. Some of the parallels regard the structure of knowledge (e. g., the schema *a capite ad calcem*), which, preserved along with knowledge as a means of application, suggest not traveling individuals, but groups. Linguistic evidence or, rather, the lack of such evidence, might, as indicated above, point towards the slowness of such travel from East to West. For all these reasons I believe that guild-like groups of practitioners, partly itinerant and originating from Mesopotamia, spread this knowledge and its structures all over the (Eastern) Mediterranean. The famous Hippocratic Oath, as well as Near Eastern parallels, indicate how these groups tried to control their knowledge by institutionalizing family-structures.[82] As has often been pointed out, besides "rational" medicine, many other forms of medical care were still on offer in Greece in classical and Hellenistic times.[83] In the *Odyssey*, Eumaeus famously counts the "seer" (*mantis*) and the "healer of evils" (*iētēra kakōn*) as professional experts (itinerant *dēmoiergoi*).[84] The mythical case of Melampus and the historical ones of Epimenides, Thaletas, and Empedocles would fit the latter category.[85] In Plato's *Republic*, incantations (*epōdai*) appear as part

---

[81] *Pace* Scurlock who even assumes verbatim quotation from Akkadian in Greek ("From Esagil-kīn-apli to Hippocrates," 24 and elsewhere).
[82] For the Oath and Mesopotamian–Egyptian parallels (*Diagnostic Handbook*, Papyrus Ebers), see Geller, "West Meets East," 14. These artificial family structures are not only known from the Coan "Asclepiads" but even from such unlikely groups as the "Ouliadai" ("Parmenideans": see Nutton, *Ancient Medicine*, 46).
[83] See van der Eijk, "Introduction," 6.
[84] Hom. *Od.* 17.382–85; see Burkert, *Orientalizing Revolution*, 41 and my remarks in *Griechische Wissenschaftstexte*.
[85] See Geller, "West Meets East," 54; Burkert, *Orientalizing Revolution*, 42–43.

of medical practice just as drugs, cauterization, and surgery (4.426b1). These must have been only the proverbial tip of the iceberg, the hidden part which probably had more in common with the integrative approach of the Babylonian *āšipu* than with the strictly rationalist-empiricist method of the Hippocratic physician.[86] Magic *and* medicine are part of Babylonian and Greek socio-medical reality. Magic *versus* medicine, however, is a polemical-rhetorical invention of "Hippocratic" medicine, very much in the style of the similar polemical distinction of *historiē/logos* versus *muthos* made by contemporary Greek historians.[87]

As the consistent polemic shows, Hippocratic writers perceived these people as serious competitors. I suggest that instead of viewing the healers, dream-interpreters, and others like them as marginal and thereby buying into Hippocratic rhetoric, one should put things into a more probable perspective and, instead, assign marginality to the empiricist-rationalist approach and its promoters, at least in the fifth century.[88] The healers vilified by Hippocratic authors, who are usually invisible from our perspective (this invisibility is caused by the bias of our sources and, in the end, largely contingent on their social status),[89] provided the medicine that was most pervasive in the day. Their knowledge, concepts, structures, and treatments were part of an "undercurrent"—at least viewed from the theorists' tradition in which modern scholars usually include themselves—that formed a continuous tradition reaching from, at least, second-millennium Mesopotamia to Roman Imperial times and that was all-pervasive in the Eastern Mediterranean. This undercurrent surfaces occasionally: for example, in Hippocratic polemics, in occasional therapies in Pliny, and in the Talmudic tradition.[90]

---

[86] S. Maul, "Die 'Lösung vom Bann': Überlegungen zu altorientalischen Konzeptionen von Krankheit und Heilkunst," in Horstmanshoff and Stol, *Magic and Rationality*, 79–95, at 78 and Avalos, "Epilepsy," 135 have, from different angles, criticized the neat distinction between *asû* and *āšipu* as anachronistic. In fact, it probably reflects the modern dominance of "rationalist" medicine.

[87] For Hecataeus, Herodotus, and Thucydides, see Asper, *Griechische Wissenschaftstexte*, 39; G. E. R. Lloyd, *Demystifying Mentalities* (Cambridge: Cambridge University Press, 1990), 46. Compare T. E. Rihll and J. V. Tucker, "Practice Makes Perfect: Knowledge of Materials in Classical Athens," in C. J. Tuplin and T. E. Rihll (eds.), *Science and Mathematics in Ancient Greek Culture* (Oxford: Oxford University Press, 2002), 274–305, at 297–304.

[88] See Laskaris, *Art Is Long*, 32–33. On a similar project of reversing marginality, see Cuomo, *Technology and Culture*, 164–68.

[89] Shapin's comments (*Social History of Truth*, 359–61) on the invisibility of technicians in seventeenth-century England partly apply to the ancient world, too.

[90] For the latter two, see M. J. Geller, *Akkadian Healing Therapies in the Babylonian Talmud*, Preprint 259 (Berlin: Max Planck Institute for the History of Science, 2004), and Westendorf,

Thus, I assume that these practitioners are both the starting-point and the contrast against which early Greek theoretical medicine unfolded as an epiphenomenon, that is, the result of a development both tiny and local when compared with the broad and "global" traditions of practitioners. Here, as is often the case elsewhere in Greek writing, polemics tend to camouflage conceptual debt.[91] I have suggested a similar constellation as emerging in the realm of mathematical knowledge, that is, tiny groups of theoretical mathematicians positioning themselves against a background of mighty practitioners.[92]

If this is really what happened, the question of why it happened remains. The explanation that is, to me, still the most satisfying but nonetheless rather vague[93] refers to the "openness" of the Greek medical market that called for competitive strategies, among which writing, a "personal" style, and logical rather than empirical arguments formed a strategic union. In an open-market situation and a world where all medical practitioners enjoy more or less the same success rate, the emphasis on causal explanation, theoretical explanation, and refutation of opponents by way of logic might have been the most successful way to persuade patients to sign up with one's own group. From this angle, Hippocratic medicine, especially in its textual aspects, appears close to both sixth-century natural speculation and fifth-century sophists. Conceivably, one may understand these medical texts as an overlap of the two. To me, the crucial factor appears to be "openness." To our ears, "openness" and the lack of monopolization have a positive ring. In early Greece, however, the lack of medical authority that allowed medical practitioners of all sorts to thrive in free competition with one another is probably best understood as an effect of the earlier breakdown of the structure of palace societies, and especially its designated space for an elite and highly specialized palace medicine that was sufficiently protected against outside competitors by being part of the palace administration.[94] Far from being "elite," Hippocratic medicine is, thus, a typical offshoot of archaic Greece with its comparative

---

Handbuch, 2:571 n. 32 on a striking parallel in the Egyptian medical Papyrus Ebers (ca. 1550 B.C.) and Plin. *nat.* 30.70 (first c. A.D.).

91 See Laskaris, *Art Is Long*, 4–5; M. Asper, "*Un personaggio in cerca di lettore:* Galens *Großer Puls* und die 'Erfindung' des Lesers," in T. Fögen (ed.), *Antike Fachtexte—Ancient Technical Texts* (Berlin: De Gruyter, 2005), 21–39, esp. 32; *Griechische Wissenschaftstexte*, 361–63.

92 Asper, "Two Cultures," 129.

93 I have tried to give an account of the discussion in Asper, *Griechische Wissenschaftstexte*, 27–45, 377–83.

94 See Nutton, *Ancient Medicine*, 40 and 329 n. 23.

lack of social stratification (the same is true for Greek discourses on political power).[95]

Conclusion

This brief story of how a culturally distinctive form of medicine emerged in fifth-century B.C. Greece could and should be supplemented by parallel stories focusing on mathematics, astronomy, possibly "grammar," and so forth. I believe that they would all yield similar plots, part of which might be the following elements:
1. The oldest Greek forms of medical discourse that we can identify are long-term products of acculturation and in my view contingent upon what Burkert has called the "Orientalizing revolution."
2. The theoreticians put writing to new uses. Hippocratic medicine produces a whole new host of genres, from display speeches to collections of case studies.[96] Mathematics develops the stylistically odd but efficient Euclid-style treatise. If one can generalize from the colophons in the *Diagnostic Handbook*, Near Eastern medical traditions used texts mainly as storage devices, to be used by practitioners either for instruction or for reference.[97] Both functions remain popular among Greek medical and mathematical writers, too (for example, in lists of diseases or symptoms). Nonetheless, the group of "Hippocratic" writings presents arguments to audiences they apparently cannot reach by other means of communication. In mathematics, one possible explanation for the conceptual rigor of Euclid-style treatises is that they were developed exclusively for written communication (as is evident in the case of Archimedes and Apollonius).[98]

---

[95] For Bronze Age Greece, Egypt, and ancient Mesopotamia, see Arnott, "Minoan and Mycenaean Medicine," 155–63 ("largely confined to the elite," 156). The Asclepiad clan at Cos and Cnidos certainly was part of the local social elite (see Langholf, *Medical Theories*, 25–28), but there is no reason to assume that they restricted their therapies to peers.
[96] See, especially, P. J. van der Eijk, "Towards a Rhetoric of Ancient Scientific Discourse: Some Formal Characteristics of Greek Medical and Philosophical Texts (Hippocratic Corpus, Aristotle)," in E. J. Bakker (ed.), *Grammar as Interpretation: Greek Literature in its Linguistic Contexts* (Leiden: Brill, 1997), 77–129, and R. Wittern, "Gattungen im Corpus Hippocraticum," in W. Kullmann, J. Althoff, and M. Asper (eds.), *Gattungen wissenschaftlicher Literatur in der Antike* (Tübingen: Narr, 1998), 17–36.
[97] See Heeßel, *Babylonisch-assyrische Diagnostik*, 186, 314, 364.
[98] R. Netz, *The Shaping of Deduction in Greek Mathematics: A Study in Cognitive History* (Cambridge: Cambridge University Press, 1999), 271–312; Asper, *Griechische Wissenschaftstexte*, 147–56.

3. "Boundary work" accompanies the self-differentiation from practitioners' traditions. In medicine, the authors engage in extensive polemics that construct both the opponent and the writer himself (certainly in the eyes of the modern reader).
4. One might think of understanding both the emergence of rationalist medicine and of theoretical mathematics as a transition from "social" technologies of trust, that is, a rhetoric based on social authority (for example, the guild's pristine tradition, the specialist status of its practitioners, and the knowledge's commonly accepted usefulness) to "epistemic" technologies of trust, that is, logically compelling or, at least, persuasive arguments. Actually, it is the preference for epistemic rather than social authority that makes the two traditions, and any other "scientific" tradition, appear so similar.

Oddly, the seemingly clear divide of Eastern versus Western practices of argument that the modern historiography of science has found so compelling for so long was never really acknowledged by Greek theoreticians and their doxographers: quite the opposite, they constructed continuities where we moderns cannot perceive any. Greek theorists had their founding fathers spend time in Egypt or in the East (Pythagoras and Thales in Egypt, Democedes in Persia, Hippocrates in Egypt, and so on) whence they brought the main elements of their knowledge to Greece.[99]

---

99 For Hippocrates in Egypt, see J. R. Pinault, *Hippocratic Lives and Legends* (Leiden: Brill, 1992), 132. The so-called "Brussels Life of Hippocrates," a Latin translation of a Greek text with unclear provenance states: *eodem tempore accepit septem libros de Memfis ciuitate a Polibio, filio Apollonii, quos secum inde portauit et ex his libris suis canonem medicinae recte ordinauit* ("at the same time he received seven books in (?) the city of Memphis from Polybius, the son of Apollonius, which he took with him from there and with the help of which he prepared the canon of medicine in the right way," fol. 52v 38–43). In addition, Hippocrates spends some time with the Persians: *postquam reuersus est a Medis de Batchana ciuitate ab Arfaxath rege Medorum* ("after he returned from the Medes, from the city of Batchana and from Arfaxath, the king of the Medes," fol. 52v, 34–37). Admittedly, this is a late and somewhat sub-scholarly fiction. One finds nothing like that in Soranus' *Vita Hippocratis* (ed. Ilberg, CMG IV). The traditions about early Greek philosophers present a similar, more prominent case, conveniently summed up by Diogenes Laërtius (Thales [Diog. Laërt. 1.24], Solon [Diog. Laërt. 1.50], Plato [Diog. Laërt. 3.6–7], Pythagoras [Diog. Laërt. 8.2–3]). M. R. Lefkowitz, "Visits to Egypt in the Biographical Tradition," in M. Erler and S. Schorn (eds.), *Die griechische Biographie in hellenistischer Zeit* (Berlin: De Gruyter, 2007), 101–13 discusses the case with respect to Euripides', Plato's, and Eudoxus' fictional visits to Egypt, which she understands as "testimony to the desire on the part of *Hellenistic* Greeks to be associated with Egyptian learning" (111, emphasis added).

The big divide that *they* apparently were concerned about and tried to gloss over in silence or, in the case of Aristotle's narrative of discovery in *Metaphysics* A, explicitly tried to marginalize, was, instead, the one of practitioners and theorists *in Greece*. It seems that a great deal of attention and ingenuity went into ignoring the obvious question in what respects theory was influenced by and dependent on practice.

So what we see going on here is a reversal: instead of acknowledging their debt to a practitioner *koinē* that reaches to the far East, the theorists write these practitioners out of the picture and invent for themselves an "Eastern" pedigree, but a purely theoretical one that never existed. The substitution results in a fictitious, socially immaculate past, instead of a real past that leads down into lower social strata. An acculturated East is substituted for a fictitious East, or: what we see as an East-West divide not only is not a divide, it is even constructed into *one* fabricated narrative of linear transmission and perfection (e. g., Proclus on the emergence of Euclid-style mathematics). The divide that really counts, however, is the one between practitioners and theorists, and this one is solved by ignoring it.

Thus, allegedly typically Greek "rationalist" medicine, a core discourse of what is traditionally understood as "the emergence of science," turns out to be a case study of Greek acculturation. This acculturation resulted in a certain duplicity: solutions to medical and mathematical problems were on offer from two quite different perspectives. One wonders whether more case studies could enrich the picture in parallel ways: in the fields of, say, calculation and metrics, astronomy, music, architecture, and expert storytelling one might expect similar constellations. In all these cases, specific Greek forms did not directly differentiate themselves from Near Eastern or Egyptian traditions but rather from acculturated Greek adaptations of the former, based on practitioners' groups. Thus, instead of the two dominant narratives about the relations between Greece and her neighboring cultures that offer either seamless continuity or radical breaks, I suggest a third one that locates the break within Greek culture.

Bibliography

Andrushko, V. A., and J. W. Verano. "Prehistorical Trepanation in the Cuzco Region of Peru: A View into an Ancient Andean Practice," *American Journal of Physical Anthropology* 137 (2008): 4–13.

Arnott, R. "Minoan and Mycenaean Medicine and its Near Eastern Contacts," in Horstmanshoff and Stol, *Magic and Rationality*, 153–73.

Asper, M. "The Two Cultures of Mathematics in Ancient Greece," in E. Robson and J. Stedall (eds.), *The Oxford Handbook of the History of Mathematics* (Oxford: Oxford University Press, 2009), 107–32.

Asper, M. *Griechische Wissenschaftstexte. Formen, Funktionen, Differenzierungsgeschichten* (Stuttgart: Steiner, 2007).

Asper, M. "Medienwechsel und kultureller Kontext. Die Entstehung der griechischen Sachprosa," in J. Althoff (ed.), *Philosophie und Dichtung im antiken Griechenland* (Stuttgart: Steiner, 2007), 67–102.

Asper, M. "*Un personaggio in cerca di lettore*: Galens Großer Puls und die 'Erfindung' des Lesers," in T. Fögen (ed.), *Antike Fachtexte—Ancient Technical Texts* (Berlin: De Gruyter, 2005), 21–39.

Attinger, P. "La médecine mésopotamienne," *Le Journal des Médecines Cunéiformes* 11/12 (2008): 1–96.

Avalos, H. "Epilepsy in Mesopotamia Reconsidered," in Finkel and Geller, *Disease in Babylonia*, 131–36.

Burkert, W. *The Orientalizing Revolution: Near Eastern Influence on Greek Culture in the Early Archaic Age* (Cambridge, Mass.: Harvard University Press, 1992).

Burkert, W. *Creation of the Sacred: Tracks of Biology in Early Religions* (Cambridge, Mass.: Harvard University Press, 1996).

Cuomo, S. *Technology and Culture in Greek and Roman Antiquity* (Cambridge: Cambridge University Press, 2007).

Debru, A. "La suffocation hystérique chez Galien et Aétius: réécriture et emprunt du 'je,'" in A. Garzya (ed.), *Tradizione e ecdotica dei testi medici tardoantichi e bizantini* (Naples: D'Auria, 1992), 79–89.

Demont, P. "L'ancienneté de la médecine hippocratique: un essai de bilan," in A. Attia and G. Buisson (eds.), *Advances in Mesopotamian Medicine from Hammurabi to Hippocrates*, Cuneiform Monographs 37 (Leiden: Brill, 2009), 129–49.

Ervin, A. M. "A Review of the Acculturation Approach in Anthropology with Special Reference to Recent Change in Native Alaska," *Journal of Anthropological Research* 36 (1980): 49–70.

Feldhuis, N. "On Interpreting Mesopotamian Namburbi Rituals," *Archiv für Orientforschung* 42/43 (1995/1996): 145–54.

Finkel, I. L., and M. J. Geller, eds. *Disease in Babylonia* (Leiden: Brill, 2007).

Geller, M. J. "West Meets East: Early Greek and Babylonian Diagnosis," in Horstmanshoff and Stol, *Magic and Rationality*, 11–61.

Geller, M. J. *Akkadian Healing Therapies in the Babylonian Talmud*, Preprint 259 (Berlin: Max Planck Institute for the History of Science, 2004).

Geller, M. J. *Renal and Rectal Disease Texts*, Die babylonisch-assyrische Medizin in Texten und Untersuchungen 7 (Berlin: De Gruyter, 2005).

Geller, M. J. "Phlegm and Breath—Babylonian Contributions to Hippocratic Medicine," in Finkel and Geller, *Disease in Babylonia*, 187–99.

Gieryn, T. "Boundaries of Science," in S. Jasanoff (ed.), *Handbook of Science and Technology Studies* (Thousand Oaks: SAGE Publications, 1995), 393–443.

Goltz, D. *Studien zur altorientalischen und griechischen Heilkunde. Therapie – Arzneibereitung – Rezeptstruktur* (Wiesbaden: Steiner, 1974).

Gotter, U. "'Akkulturation' als Methodenproblem der historischen Wissenschaften," in S. Altekamp, M. R. Hofter, and M. Krumme (eds.), *Posthumanistische Klassische Archäologie. Historizität und Wissenschaftlichkeit von Interessen und Methoden* (Munich: Hirmer, 2001), 255–80.

Grensemann, H. *Knidische Medizin, Teil I: Die Testimonien zur ältesten knidischen Lehre und Analysen knidischer Schriften im Corpus Hippocraticum* (Berlin: De Gruyter, 1975).

Grensemann, H. *Knidische Medizin, Teil II: Versuch einer weiteren Analyse der Schicht A in den pseudohippokratischen Schriften "De Natura Muliebri" und "De Muliebribus" I und II* (Stuttgart: Steiner, 1987).

Hanson, A. E. "Continuity and Change: Three Case Studies in Hippocratic Gynecological Therapy and Theory," in S. B. Pomeroy (ed.), *Women's History and Ancient History* (Chapel Hill: University of North Carolina Press, 1991), 73–110.

Hanson, A. E. "Uterine Amulets and Greek Uterine Medicine," *Medicina nei Secoli. Arte e Scienza* 7 (1995): 281–99.

Heeßel, N. P. *Babylonisch-assyrische Diagnostik* (Münster: Ugarit-Verlag, 2000).

Heeßel, N. P. *Divinatorische Texte I. Terrestrische, teratologische, physiognomische und oneiromantische Omina* (Wiesbaden: Harrassowitz, 2007).

Horstmanshoff, H. F. J., and M. Stol, eds. *Magic and Rationality in Ancient Near Eastern and Graeco-Roman Medicine* (Leiden: Brill, 2004).

Ilberg, J. *Sorani Gynaeciorum libri IV, De signis fracturarum; De fasciis; Vita Hippocratis secundum Soranum*, CMG IV (Leipzig: Teubner, 1927).

Joly, R. *Hippocrate, Du régime*, t. VI.1 (Paris: Les Belles Lettres, 1967).

Jouanna, J. *Hippocrate : pour une archéologie de l'école de Cnide* (Paris: Les Belles Lettres, 1974).

Jouanna, J. *Hippocrate, Maladies II*, t. X.2 (Paris: Les Belles Lettres, 1983).

Jouanna, J. *Hippocrates*, trans. M. B. DeBevoise [orig. 1992] (Baltimore: The Johns Hopkins University Press, 1999).

Kühn, J.-H., and U. Fleischer. *Index Hippocraticus* (Göttingen: Vandenhoeck & Ruprecht, 1986–1989).

Labat, R. *Traité akkadien de diagnostics et pronostics médicaux* (Leiden: Brill, 1951).

Langholf, V. *Medical Theories in Hippocrates: Early Texts and the "Epidemics"* (Berlin: De Gruyter, 1990).

Laskaris, J. *The Art Is Long: "On the Sacred Disease" and the Scientific Tradition* (Leiden: Brill, 2002).

Lefkowitz, M. R. "Visits to Egypt in the Biographical Tradition," in M. Erler and S. Schorn (eds.), *Die griechische Biographie in hellenistischer Zeit* (Berlin: De Gruyter, 2007), 101–13.

Lefkowitz, M. R, and G. MacLean Rogers, eds. *"Black Athena" Revisited* (Chapel Hill: University of North Carolina Press, 1996).

Lloyd, G. E. R. *The Revolutions of Wisdom: Studies in the Claims and Practice of Ancient Greek Science* (Berkeley: University of California Press, 1987).

Lloyd, G. E. R. *Demystifying Mentalities* (Cambridge: Cambridge University Press, 1990).
Majno, G. *The Healing Hand: Man and Wound in the Ancient World* (Cambridge, Mass.: Harvard University Press, 1975).
Maul, S. *Zukunftsbewältigung. Eine Untersuchung altorientalischen Denkens anhand der babylonisch-assyrischen Löserituale (Namburbi)* (Mainz: Verlag Philipp von Zabern, 1994).
Maul, S. "Die 'Lösung vom Bann': Überlegungen zu altorientalischen Konzeptionen von Krankheit und Heilkunst," in Horstmanshoff and Stol, *Magic and Rationality*, 79–95.
Netz, R. *The Shaping of Deduction in Greek Mathematics: A Study in Cognitive History* (Cambridge: Cambridge University Press, 1999).
Nutton, V. *Ancient Medicine* (London: Routledge, 2004).
Palter, R. "Black Athena, Afrocentrism, and the History of Science," in Lefkowitz and MacLean Rogers, *"Black Athena" Revisited*, 209–66.
Parker, R. *Miasma: Pollution and Purification in Early Greek Religion*, 2nd ed. (Oxford: Clarendon Press, 1996).
Pfeiffer, R., ed. *Callimachus*, 2 vols. (Oxford: Clarendon Press, 1949–1953).
Pinault, J. R. *Hippocratic Lives and Legends* (Leiden: Brill, 1992).
Pommerening, T. "βούτυρος '*Flaschenkürbis*' und κουροτόκος im Corpus Hippocraticum, *De sterilibus* 214: Entlehnung und Lehnübersetzung aus dem Ägyptischen," *Glotta* 86 (2010): 40–54.
Rihll, T. E., and J. V. Tucker. "Practice Makes Perfect: Knowledge of Materials in Classical Athens," in C. J. Tuplin and T. E. Rihll (eds.), *Science and Mathematics in Ancient Greek Culture* (Oxford: Oxford University Press, 2002), 274–305.
Ritter, J. "Translating Rational-Practice Texts," in A. Imhausen and T. Pommerening (eds.), *Writings of Early Scholars in the Ancient Near East, Egypt, Rome, and Greece* (Berlin: De Gruyter, 2010), 349–83.
Rochberg-Halton, F. "The Cultures of Ancient Science: Some Historical Reflections," *Isis* 83 (1992): 547–53.
Scurlock, J. "From Esagil-kīn-apli to Hippocrates," *Le Journal des Médecines Cunéiformes* 3 (2004): 10–30.
Shapin, S. *A Social History of Truth: Civility and Science in Seventeenth-Century England* (Chicago: University of Chicago Press, 1994).
Stol, M. "An Assyriologist Reads Hippocrates," in Horstmanshoff and Stol, *Magic and Rationality*, 63–78.
Stol, M. *Epilepsy in Babylonia* (Groningen: Styx Publications, 1993).
Thomas, R. "Greek Medicine and Babylonian Wisdom: Circulation of Knowledge and Channels of Transmission in the Archaic and Classical Periods," in Horstmanshoff and Stol, *Magic and Rationality*, 175–85.
Ulf, C. "Rethinking Cultural Contacts," *Ancient West & East* 8 (2009): 81–132.
van der Eijk, P. J. "The 'Theology' of the Hippocratic Treatise *On the Sacred Disease*," *Apeiron* 23 (1990): 87–119.
van der Eijk, P. J. "Towards a Rhetoric of Ancient Scientific Discourse: Some Formal Characteristics of Greek Medical and Philosophical Texts (Hippocratic Corpus, Aristotle)," in E. J. Bakker (ed.), *Grammar as Interpretation: Greek Literature in its Linguistic Contexts* (Leiden: Brill, 1997), 77–129.
van der Eijk, P. J. *Diocles of Carystus: A Collection of the Fragments with Translation and Commentary*, 2 vols. (Leiden: Brill, 2000–2001).

van der Eijk, P. J. "Introduction," in Horstmanshoff and Stol, *Magic and Rationality*, 1–10.
van der Eijk, P. J. "Divination, Prognosis, and Prophylaxis: The Hippocratic Work 'On Dreams' (*De victu* 4) and its Near Eastern Background," in Horstmanshoff and Stol, *Magic and Rationality*, 187–218.
von Staden, H. "Women and Dirt," *Helios* 19 (1992): 7–30.
von Staden, H. "Author and Authority: Celsus and the Construction of the Scientific Self," in M. E. Vázquez Buján (ed.), *Tradición e innovación de la medicina latina de la antigüedad y de la alta edad media* (Santiago de Compostela: Universidade de Santiago de Compostela, 1994), 103–17.
West, M. L. *The East Face of Helicon: West Asiatic Elements in Greek Poetry and Myth* (Oxford: Oxford University Press, 1997).
Westendorf, W. *Handbuch der altägyptischen Medizin*, 2 vols., Handbuch der Orientalistik 1.36 (Leiden: Brill, 1999).
Wittern, R. "Gattungen im Corpus Hippocraticum," in W. Kullmann, J. Althoff, and M. Asper (eds.), *Gattungen wissenschaftlicher Literatur in der Antike* (Tübingen: Narr, 1998), 17–36.

Han Baltussen
"Hippocratic" Oaths?: A Cross-Cultural Exploration of Medical Ethics in the Ancient World

Abstract: This paper considers the cross-cultural similarities and differences between medical covenants from different ancient cultures (Greek, Indian, Chinese) which claimed to regulate doctor-patient interactions. While the differences are clearly determined by the cultural and social environments in which these covenants were formulated, detailed comparison reveals the presence of shared moral values which seem to be universal. It is suggested that the reason for this "universality" must lie in the doctor-patient relationship, in particular the intimacy of the specific interpersonal interactions, which transcends particular cultural contexts. This comparative approach also allows us to clarify the importance of such "mission statements": they are attempts to make the most of a delicate situation, in which physicians—as healer and confidante—tried to gain the trust of actual and potential patients, despite the limitations of their scientific knowledge.

I The Continuing Importance of the "Hippocratic" *Oath*

Recently declared the second most authoritative text today,[1] the so-called "Hippocratic" *Oath* continues to draw the attention of physicians and historians: on the one hand, it retains its intriguing and enduring value as a fundamental text of the Western European medical tradition, on the other, it keeps being referred

---

It is my honor and pleasure to dedicate this essay to a distinguished colleague in ancient world studies as a token of my respect, but also of gratitude for his hospitality and kind advice during my stay at IAS. Some of the ideas presented here are also based on my course "Ancient Medicine and its Legacy" and H. Baltussen, "Hippocratic Corpus," in M. Gagarin (ed.), *Oxford Encyclopedia of Ancient Greece and Rome*, 7 vols. (Oxford: Oxford University Press, 2010), 4:1–4, and idem, "Hippocratic Oath," in Gagarin, *Oxford Encyclopedia*, 4:4–6.
1 V. Nutton, *Ancient Medicine* (London: Routledge, 2004), 53: "Except for the Bible, no document and no author from Antiquity commands the authority in the twenty-first century of Hippocrates of Cos and the Hippocratic *Oath*."

back to when medical scandals appear in the news.² What is remarkable about the "Hippocratic" *Oath* is that it articulated its rules at the very start of a new field of knowledge. The oath consists of three broad sections (see Appendix A): first, the physician calls upon certain gods to witness the oath (1); next, he offers a number of clear injunctions and prohibitions to protect the life of the patient as much as possible (2–7); and finally he calls down blessings (8.i) or curses (8.ii) upon himself, as a consequence of observing or violating the oath. It can thus boast having separated killing and curing, something which was not always guaranteed in the earliest healing methods. For Greece the new medical approach was coming into view as a more "scientific" branch of knowledge with Hippocratic medicine in the fifth century B.C. The "Hippocratic" *Oath* is thought to have been written some time after, and constitutes a remarkable and remarkably economical declaration of self-regulation by a budding profession. In approximately four hundred words it defines principles of proper behavior and confidentiality, as well as the core responsibilities of the physician. Galen would later express his misgivings at how obvious one of its central tenets ("do no harm") was, but its novelty and the perceived need for a declaration of principles to ensure trust are clearly intimately linked and a major factor in its continuing success.

The excellent scholarly analyses that have clarified the "Hippocratic" *Oath*'s philological aspects as well as its immediate cultural and medical contexts have as a rule emphasized its special place and value in Greece and the European medical tradition. They have placed less emphasis on the fact that it is not unique.³ This is not to say that scholars are unaware of other ethical codes and "mission statements" for medicine, but the focus tends to be on the *Oath*'s foundational value and lasting influence across the ages in Western cul-

---

2 I refer to the Oath as "Hippocratic" because of its uncertain authorship: although it was claimed for Hippocrates by the first c. A.D. (Scribonius Largus), there is "no independent corroboration for this claim" (O. Temkin, "What Does the Hippocratic Oath Say? *Translation and Interpretation*," in *"On Second Thought" and Other Essays in the History of Medicine and Science* [Baltimore: The Johns Hopkins University Press, 2002], 21–28, at 21). The recent paper by H. von Staden, "'The *Oath*,' the Oaths, and the Hippocratic Corpus," in V. Boudon-Millot, A. Guardasole, and C. Magdelaine (eds.), *La science médicale antique: nouveaux regards* (Paris: Beauchesne, 2007), 425–66, offers a fascinating analysis of the language of the *Oath* in relation to the Hippocratic Corpus. The modern responses have also not abated: see, e.g., S. H. Miles, *Oath Betrayed: America's Torture Doctors*, 2nd ed. (Berkeley: University of California Press, 2009) in relation to the Abu Ghraib scandal.
3 See especially Thomas Rütten, "Receptions of the Hippocratic Oath in the Renaissance: The Prohibition of Abortion as a Case Study in Reception," *Journal of the History of Medicine and Allied Sciences* 51 (1996): 456–83 and previous note.

ture, even if its transmission up to the Renaissance is rather poorly documented.[4] In this paper I would like to compare the "Hippocratic" *Oath* to two further cases of medical oaths in other ancient cultures (India, China), which exhibit many interesting parallels with the Greek *Oath*. It is my claim that a comparison will illuminate how the core requirements for a set of regulatory rules in the medical profession are universal, and for very specific reasons. Once we take a closer look at the three texts, we are confronted with two questions: (1) What do the cross-cultural parallels teach us about the establishment of ethical codes for doctor-patient relationships?; and (2) Would we consider any of these oaths as being "ahead of their time"?

The first issue raises the additional question of whether the various oaths and pledges represent a universal of human thought (this is what a medical *oath* looks like) or whether the core notions underlying them are simply related to the central issues surrounding doctor-patient interaction (this is what a *medical* oath looks like): from the patient's point of view it is about the trust to be given to a stranger, and the courage to rely on new, "alternative" medicine when traditional ways still held sway. The second issue is intended as a mild criticism of retrospective accounts of medical ethics, but it also deals with the broader point of positivist interpretations of the history of medicine as it has arisen in the twentieth century.[5] To say that someone is "ahead of their time" more often than not involves a (misguided) value judgment about progress as a linear process, but usually also means that we measure earlier achievements by modern standards. In this case the ethical principles seem to transcend time, as both diachronic and synchronic evidence suggests.

II The "Hippocratic" *Oath* in Context

As a fundamental statement of prohibitions and injunctions for medical practice the "Hippocratic" *Oath* (HO) is an important marker of ancient medical ethics,

---

[4] In his pioneering history of the *Oath*'s transmission, Thomas Rütten, *Geschichten vom Hippokratischen Eid* (Wiesbaden: Harrassowitz, 2007), CD-ROM, illustrates how sparing the concrete evidence for the *Oath*'s use really is. It may well be a twentieth-century perception that it is of such fundamental importance, because of the increased role ethical questions have played in recent times, often in lockstep with technical advances, which force us to ask whether we ought to do certain things just because we can.

[5] See, for instance, P. J. van der Eijk, *Medicine and Philosophy in Antiquity: Doctors and Philosophers on Nature, Soul, Health and Disease* (Cambridge: Cambridge University Press, 2005), 3–5 for the rise of this type of research and the turning away in recent decades towards a history of medicine under the influence of cultural anthropology, social history, and comparativism.

but the document is not unique, nor can we be sure that it was universally accepted. It may well have come to represent the idea of medical ethics in the West, but there are several other, very similar pledges or "mission statements" of the medical profession in other ancient cultures, such as India and China, which deserve our attention.[6] The striking similarities among these oaths and pledges also raise the question of whether such declarations of principle may have come about through connections between them. The kind of similarities one finds are all closely linked to the intimate nature of the interactions between doctor and patient, and perhaps more importantly, to the very special nature of healing as the continuous effort to preserve and prolong life.

It is worthwhile to note that, so far as we know, Egypt has not produced an oath, despite the important role that Egyptian medical knowledge played in the development of ancient medicine.[7] The presence of a confidentiality clause imposing trustworthiness in the HO illustrates the need for proper behavior in the case of the doctor-patient relationship, which is an asymmetrical one. By "asymmetrical" I simply mean the imbalance of power which exists between the doctor and patient, as one between expert and lay customer. What is at stake is the vulnerability of the patient, in particular the integrity of women.[8] The HO contains other rules which ensure patient confidentiality (e.g., an injunction of non-disclosure). All these recommendations come in a tightly arranged solemn pledge of around four hundred words. Such remarkable brevity was no doubt dictated by the need for memorization, which would enhance its power as a constant guide for behavior.

---

[6] It is noteworthy that Greek medicine was introduced into Tibet in Late Antiquity and given priority over Indian or Chinese medicine (C. I. Beckwith, "The Introduction of Greek Medicine into Tibet in the Seventh and Eighth Centuries," *Journal of the American Oriental Society* 99.2 [1979]: 297–313, at 301). One text alludes to principles of ethics which can be considered "a version of the HO" (Beckwith, "Introduction of Greek Medicine," 304–305 with nn. 72 and 73). According to W. H. S. Jones, *The Doctor's Oath* (Cambridge: Cambridge University Press, 1924), 31 and 33, the "Arabic Oath" found in Ibn abi Usaybia's *Lives of Physicians* ('*Uyun al-anba*, 13th c.) is a descendent of the HO; for a translation see F. Rosenthal, *The Classical Heritage in Islam*, trans. Emile and Jenny Marmorstein (Berkeley: University of California Press, 1975), 183–84 [= *Das Fortleben der Antike im Islam* (Zürich: Artemis Verlag, 1965), 250–51].

[7] The peculiar "Oath of Imhotep" (quoted in S. G. Pérez, R. J. Gelpi, and A. M. Rancich, "Doctor-Patient Sexual Relationships in Medical Oaths," *Journal of Medical Ethics* 32 [2006]: 702–705, at 704) may seem close to the HO (e.g., "I shall refrain from sexual practices with my patients and others under my guard"), but is a modern construct motivated by (post)modern ideas derived from the HO.

[8] Pérez, Gelpi, and Rancich, "Doctor-Patient Sexual Relationships," 704.

The issue of trust is naturally one of intimacy, including potential embarrassment over exposing one's naked body; the gender issue of a female being examined by a male (physicians were as a rule male);[9] the risk of the abuse of power in the imbalance the relationship represents; and the delicate question of whether a reward should play a role. It is of course important to note that these texts see things very much from the point of view of the physician, but with the aim of establishing a code of behavior in working *with* the patient as well as maintaining trustworthy reputation with the public at large.

The right ethical behavior of doctors has clearly been a concern of patients and doctors themselves in all periods of history, even when the healing process was a complex matrix involving divine and human factors. Whereas a covenant may seem mostly focused on the encounter itself, both parties are also very much concerned about the *consequences* for their reputation beyond the walls of the surgery or the home: the patients will want their private matters kept private, and the physician wants to keep his reputation in order to maintain a clientele.[10] In aiming to provide a professional code, oaths of this kind take care of both concerns. At the core is a concern for trust, reliability, and continuing business. A second important aspect is the emphasis on reliable prognosis, as emphasized in *Prognostic* and *Art*.[11]

III Cross-Cultural Comparison: An Exploration

1 "Hippocratic" Oath

The general similarities among medical practices and theories across the ancient world is striking and calls for an explanation. It is tempting to look for one in deep-seated psychological mechanisms (one is tempted to speak of Jungian archetypes), a need for such ethical codes arising in places so far apart geographically. Although Jung may have wanted to argue that humans produce behavior

---

9 On female physicians see Holt Parker, "Women Doctors in Greece, Rome, and the Byzantine Empire," in Lilian R. Furst (ed.), *Women Healers and Physicians: Climbing a Long Hill* (Lexington: University of Kentucky Press, 1997), 131–50, and idem, "Galen and the Girls: Sources for Women Medical Writers Revisited," *Classical Quarterly* 62.1 (2012): 359–86. His brief account and "database" of known female doctors indicates that there were quite a few female physicians who were accepted as colleagues, but the numbers he presents (55 female physicians for the classical and Byzantine periods are listed in *Women Healers*, 140–47) also show that the majority of physicians were male—which is why I continue to speak of "he/his" in this context.
10 The same holds for many cultures. See below on China, §III.3.
11 A. R. Jonsen, *A Short History of Medical Ethics* (New York: Oxford University Press, 2000), 5.

of a universal type, it is not easy to uphold that claim for this specific context of medicine; surely his notion of archetype was related to human character and psyche at the most general level. Another possibility for health and sickness is perhaps to maintain an essentialist view about the human response. But the differences between cultures militate against such a line of argument. Yet it is not difficult to see the interaction between a physician and patient as a very special one, which, due to its confidential nature, may generate very similar responses to cope with, either before, during, or after the doctor's visit: the emotions such as anticipatory anxiety, potential embarrassment, pain, and possible lasting impairment or mutilation would make the decision to see a travelling healer (often the norm in ancient Greece), either at home (Hipp. *epid.* II 2, 5.84–98 L., case studies include domestic location; Hipp. *decent.* 12–13, 9.238–40 L., cf. Plut. *de prof. virt.* 11.81F) or in a surgery of some kind (e.g., Hipp. *medic.* 2, 9.206–208 L.). Besides, medical healing of the Hippocratic kind was a new and alternative treatment to traditional ways.

The crucial aspects of the HO can be summed up as follows (text in appendix A):

1. Its form (an *oath*) belongs firmly to a religious tradition of solemn statements as distinct from a mere promise or simple act of self-promotion.
2. The teacher-student relationship is one that transcends the formal educational set-up and resembles that of father and son (as was often the case).
3. Principles of etiquette and ethics go hand in hand and promote the importance of the patient, confidentiality, discretion, and trust.
4. Specific treatments and medications which involve treatments with high risk and ethical dilemmas should be avoided (abortion, surgery,[12] poison); one is tempted to suggest that the HO was intended for the "general practitioner," but it may also be a cautionary note for the novice physician (the difference between "generalist" and "specialist" can not be corroborated by good evidence for the archaic and classical periods).
5. The invocation of a higher authority to ensure compliance with the rules; an appeal to a higher being is considered important, which in a sense gives the gods a role equivalent to that of a modern "regulatory body."[13]

---

[12] An interesting parallel is found in China, where Confucianism declared itself against surgery (P. U. Unschuld, *Medicine in China: A History of Ideas* [Berkeley: University of California Press, 1985], 152).

[13] It is this religious context which made Edelstein argue that the HO has a Pythagorean origin and therefore a very narrow following originally (L. Edelstein, "The Hippocratic Oath," in O. Temkin and C. L. Temkin [eds.], *Ancient Medicine: Selected Papers of Ludwig Edelstein* [Balti-

The Oath establishes values for the profession, which in the fifth century B.C. was quite new. The Hippocratic approach to medicine was trying to replace the traditional healers, who were an unregulated group of practitioners, ranging from priest-healers to quacks. Clearly the new science was best served with a new ethic.

The other two "mission statements" to be compared are from India and China. The Indian "Oath of Initiation" (IO) is taken from the *Carakasamhita*, one of the oldest medical handbooks in Sanskrit, thought to originate around 1000 B.C. and still influential to this day.[14] Its core values come very close to the HO in several respects, as we shall see. The third oath, which is from the Chinese encyclopedic work *A Thousand Golden Remedies*, has a declaration of medical ethics of very similar nature. All three are deontological, although the Greek writings do not *present* the HO as such, but rather as a "morality of aspiration and virtue."[15]

2 Indian Oath

The IO, dated to ca. 100 A.D. but surviving in a version from around 300–500 A.D., is an intriguing text worth considering as a parallel (references are to the text in Appendix B). It is part of the sixth chapter of the *Carakasamhita*, entitled "Cikitsasthana," and the relevant section discusses therapies (§6.1) and practitioner-sages, "Asvins," who are miracle workers of sorts (§6.1.3 lists their extraordinary achievements), with particular comments on Indra's teachings (§6.1.4.6–7). The section ends with comments on the outcome of medical training:

> A physician who has completed a full course of training obtains a sattva (mental disposition) of the brāhma or ārsa type and is called thrice-born (trija) (4.52–54). The ethical principles he should adhere to are outlined 4.55–62.[16]

---

more: The Johns Hopkins Press, 1967], 3–63, at 17–20 and 53–54). This view is no longer generally accepted.

14 I am using the translation in I. A. Menon and H. F. Haberman, "The Medical Students' Oath of Ancient India," *Medical History* 14 (1970): 295–99, esp. 295–96 (the same authors discussed the date of the text in "Dermatological Writings of Ancient India," *Medical History* 13.4 [1969]: 387–92). Dating and section numbering come from G. J. Meulenbeld, *A History of Indian Medical Literature*, 3 vols. (Groningen: E. Forsten, 1999–2002), 1 A:52–53.
15 Jonsen, *A Short History*, 123 n. 22.
16 Meulenbeld, *A History*, 1 A:53.

The "Oath of Inititation" considers themes so close to the HO that comparison imposes itself even from a cursory reading. This is not to say there are no differences (on which see below), but it is best to sum up briefly the similarities in order to understand the rationale behind my exploration of these texts as well as to illustrate why one is tempted to think of universals. Major aspects and themes are its religious context (§§1–2 and 9), the demand of obedience (§3), and the authority of the teacher (§4). Then there are the stipulative rules about proper behavior, availability, exchange of presents (from women), and the notion of confidentiality (§8).

The IO starts by mentioning teaching and taking of the oath in the presence of a sacred fire (§1) and encouraging the student to follow certain restrictions in diet ("no meat, pure articles of food," §2—a form of moral and physical purity implied in the HO, §§3–7). Total obedience to the master is demanded by way of an absolute command ("thou shalt regard me as thy chief," IO §4). The relationship between teacher and student is described as that of father and son or master and slave/supplicant. This is a more authoritarian position than in the HO, where the relationship is described more in terms of "family" (HO §2–3, "regard him as equal to my parents"). Although to modern eyes the HO's formulation seems preferable and more benign, the emphasis is here on the considerable power the teacher has over his student—in the ancient family fathers often had absolute power over their dependents. A very strong injunction is stated regarding the importance of the patient. This includes a principle of around-the-clock availability ("day and night," §6), stipulations about appropriate behavior (not to be "a drunkard or a sinful man"), and a strong emphasis on truthfulness and purity in word. We may compare HO §5, in which the physician is encouraged to perform his duties in a "pure and holy way."[17]

There are also some marked differences which should not come as a surprise: the cultural contexts would almost certainly lead to other forms of managing the doctor's activities and professional self-definition. But as cultural context includes more than deliberate self-presentation, we should also be on the lookout for signs which result from the time, place, and societal circumstances in which an oath is written. Such a "cultural signature" does not, however, detract from the core ethical message of the covenant.

For instance, there is an unusual detail in §6 concerning the importance of the patient: it mentions the prohibition on treating people who "hate the king or

---

[17] For a detailed interpretation of the phrase see H. von Staden "'In a Pure and Holy Way': Personal and Professional Conduct in the Hippocratic Oath," *Journal of the History of Medicine and Allied Sciences* 51 (1996): 406–37.

who are hated by the public or who are haters of the public." This constitutes the "right and obligation to deny services."[18] It has no counterpart in the HO. But since it is motivated by political or social reasons, this aspect seems less relevant to setting the parameters of medical ethics, even if this is perhaps a point of doubtful morality on the side of the (potential) patient. But it also indicates how medical decisions could become politicized in certain contexts. In this case, then, striking differences turn out to be based on the immediate political and historical context.[19] The IO acknowledges the importance of the king as the highest authority, over and above the usual authority that a physician has over his students.

Furthermore, in the IO there is greater emphasis on the availability of the physician ("day and night," §6), and more detail regarding the avoidance of inappropriate and unlawful behavior. Special attention is paid to the "time and place" and "past experience" of a patient (§6). A whole section is dedicated to the house visit, prescribing certain attire (cf. Hippocratic etiquette in *Precepts* or *The Physician*), full focus on assistance of the ill, and a specific comment about the way in which one should respect the "peculiar customs of the patient's house" as confidential (§8). The oath ends with a religious section, placing great store in the proper relation to the gods (§9). The significant similarities are clear: a religious framework, asceticism, full dedication, moral rectitude, and a patient-centered outlook.

3 Chinese "Oath"

By bringing in a third example (not an oath in the strict sense) I am hoping not only to diversify the investigation but also to reinforce the argument that doctor-patient relationships have been regulated similarly in different cultures. If two examples begin to offer validity to the scholarly principle of generalization, on which we base our hypotheses and conclusions—one being hardly enough to base significant conclusions on, and two being the start of a pattern—a third should surely allow us to claim an even firmer basis for generalization. At least this approach goes beyond the *dictum* "Einmal ist keinmal, zweimal ist immer," attributed to Ulrich von Wilamowitz-Moellendorff (1848–1931). Together

---

18 As Menon and Haberman put it in their commentary ("The Medical Students' Oath," 297).
19 A point also argued by S. Aksoy, "Ancient Indian and Chinese Medical Oaths and the Comparison of their Medical Rules," *Yeni Tip Tarihi Arastirmalari* 7 (2001): 65–76, which was only available to me in abstract (article in Turkish). Menon and Haberman characterize it as "an indigenous product of Indian thought and culture" ("The Medical Students' Oath," 298).

the three examples suggest that a universal moral understanding emerges in such basic interactions between humans in which trust is fundamental to make this intimate relationship work, in particular when decisions of life and death may be required.[20]

In his compilation of all known Chinese medical knowledge to his day, the *Ch'ien Chin Yao Fang* ("A Thousand Golden Remedies"), Sun Ssu-miao (581–673 A.D.) writes the following (paragraph numbering is mine):

> [1] Medicine is an art which is difficult to master. If one does not receive a divine guidance from God, he will not be able to understand the mysterious points. A foolish fellow, after reading medical formularies for three years, will believe that all diseases can be cured. But after practicing for another three years, he will realize that most formulae are not effective. A physician should, therefore, be a scholar, mastering all the medical literature and working carefully and tirelessly.
>
> [2] A great doctor, when treating a patient, should make himself quiet and determined. He should not have covetous desire; he should have bowels [sic] of mercy on the sick and pledge himself to relieve suffering among all classes. Aristocrat or commoner, poor or rich, aged or young, beautiful or ugly, enemy or friend, native or foreigner, and educated or uneducated, all are to be treated equally. He should look upon the misery of the patient as if it were his own and be anxious to relieve the distress, disregarding his own inconveniences, such as night-call, bad weather, hunger, tiredness, etc. Even foul cases, such as ulcer, abscess, diarrhea, etc., should be treated without the slightest antipathy. One who follows this principle is a great doctor, otherwise, he is a great thief.
>
> [3] A physician should be respectable and not talkative. It is a great mistake to boast of himself and slander other physicians.
>
> [4] Lao Tze, the father of Taoism, said, "Open acts of kindness will be rewarded by man while secret acts of evil will be punished by God." Retribution is very definite. A physician should not utilize his profession as a means for lusting. What he does to relieve distress will be duly rewarded by Providence.
>
> [5] He should not prescribe dear and rare drugs just because the patient is rich or of high rank, nor is it honest and just to do so for boasting.[21]

Though of a relatively late date, it is likely that this list of injunctions (a moral code rather than an oath) goes back to much older versions, given that the work in which it appears summarizes a long medical tradition which tradition-

---

20 Note that the argument in favour of the transferability of the ancient *Chinese* ethical principles has recently been made by D. Fu-Chang Tsai, "Ancient Medical Ethics and the Four Principles of Biomedical Ethics," *Journal of Medical Ethics* 25 (1999): 315–21.
21 Translation by T. Lee, "Medical Ethics in Ancient China," *Bulletin of the History of Medicine* 13 (1943): 268–77, at 268–69.

ally starts in ca. 1800 B.C. in China.²² That this tradition goes a long way back but was not maintained is clear from, e.g., Hu-Szu-hui, the court-physician, who wrote in his medical manual in 1330 A.D.:

> Those of the very ancients who knew the Way had methods based in *yinyang*, and kept in harmony with magical calculations. They practiced moderation in their drinking and eating, and there was a regimen to their activities and repose. They were not disorderly in their actions. Therefore they could attain to a great age. People of today are not like that at all. There is no regimen in their activity and repose. They do not know how to avoid things which should be shunned in their eating and drinking, and are also not careful about moderation. They are much addicted to lust. They like strongly-flavored food; cannot keep the mean; and do not know how to be satiated. Therefore most, I think, will be decrepit at fifty.²³

Apart from the fact that we still recognize this sentiment, it clearly shows the difficulty that physicians had in influencing their patients in living a healthy life. The negation of all the things regarded as normal informs us about the views doctors had on the matter of healthy life style. Notably, it also gives us a hint about life expectancy in relation to health.

As we have to assume that mutual influence among the oaths is highly improbable, the similarities are quite striking. Like the HO the Chinese "oath" describes medicine as an art, and a difficult one at that (cf. Hipp. *aph.* 1.1, 4.458 L.); there is a need for empathy (§2) and a strong stomach when dealing with the more "foul cases" (cf. Hippocrates' *Epidemics*, which speaks *passim* of bilious and other discharges related to vomiting, diarrhea, and abscesses); the physician's guidance comes from god; books are important, but they cannot be a substitute for experience;²⁴ it emphasizes proper conduct and makes no distinction

---

22 See P. U. Unschuld, "Illness and Healing in Shang Culture," in *Medicine in China*, 17–28, on the Shang dynasty, "the first Chinese dynasty to leave traces of therapeutic activities ... approximately during the eighteenth through sixteenth century BC" (17).
23 Husihui Zhuan, *Yin Shan Zheng Yao*, Ch. 1, 14B in P. D. Buell and E. N. Anderson (eds.), *A Soup for the Qan: Chinese Dietary Medicine of the Mongol Era as Seen in Hu Szu-Hui's Yin-Shan Cheng-Yao* (Leiden: Brill, 2010), 258.
24 The misconception about learning medicine from books appears soon after the rise of written medical records (see esp. L. Dean-Jones, "Written Texts and the Rise of the Charlatan in Ancient Greek Medicine," in H. Yunis [ed.], *Written Texts and the Rise of Literate Culture in Ancient Greece* [Cambridge: Cambridge University Press, 2003], 97–121). The importance of hands-on experience is also part of the debate on mastering the full craft (it is, in modern terms, "embodied" knowledge). On the importance of the teacher-student relation in an oral culture as fundamental and the idea of *viva vox*, see J. Mansfeld, *Prolegomena: Questions to be Settled before the Reading of an Author, or a Text* (Leiden: Brill, 1994), 123–26 (with further literature).

by class, advocating respect for all patients; last but not least, here, too, poor conduct will be punished by god (appeal to higher authority).

The differences with the HO are also of interest. In the comment relating to class a role for money is implied, which becomes explicit in the last comment of §5; cutting, abortion, and dangerous drugs are not mentioned, either because the instruments and medicaments were not available or because no specific restrictions regarding specialized treatments seemed to hold; lastly, good conduct is rewarded by Providence, but punishment is left out.

What all three traditions share is an attempt to set up a covenant which can convince future patients to choose treatment from doctors who abide by an ethical code, but can also secure an income for the physician. Thus there seem to be moral and economical reasons for this "ethical turn" in medicine.

The following table maps out the results with particular emphasis on the similarities, with separate columns for the modern "descendants" of the HO.

	Ancient			Modern	
Origin	Greece	India	China	Declaration of Geneva	L. Lasagna, M.D.
Date (approx.)	400 B.C.	100 A.D.	600 A.D.	1948	1964
Oath	x	x		x	x
Respect for teachers	x	x	x	x	x
Share with brothers	x	x	x	x	?
Protect/pass on knowledge	x	x	x	x	x
No fee for teaching	x	-	x	n.a.	n.a.
Regimen	x	x	x	(x)	(x)
Keep patients from harm	x	x	x	x	x
Abstain from deadly drug	x	x	x	(implicit)	x
Abstain from abortive drug	x	(x)	-	(implicit)	x
Be pure in thought	x	x	x		
Abstain from surgery (defer)	x	-	-	(x)	(x)
Abstain from sexual relations	x	x	x	(implicit)	x
Maintain confidentiality	x	x	x	x	x
Consequences if not compliant	x	-	x	(x)	(x)
Duty to patient irrespective of race, nationality, politics, social standing	(x) partially		x	x	x

Table 1 Comparative table of oaths/aspirational pledges

The core principles to emerge here are to do with the central role of the teacher, the profession as a group, the knowledge of the group, and the patient. Aspects about which these cultures have concerns are all to do with causing harm (invasive and dangerous procedures; physical and mental safety; confidentiality) and ensuring that the profession looks professional. To what extent these rules were upheld or enforced, and by whom (in either case), is of course a very different matter, which cannot be dealt with here.

Conclusion

The current-day use of an oath or solemn statement regarding one's duties and behavior in medical practices has changed considerably: compared to the ancient world we no longer follow the spirit of the HO in all its aspects. A 1993 survey of 150 U.S. and Canadian medical schools, for example, showed that only 14 percent of modern oaths prohibit euthanasia, 11 percent hold covenant with a deity, 8 percent foreswear abortion, and a mere 3 percent forbid sexual contact with patients—all maxims held sacred in the classical version.[25] The original calls for free tuition for medical students and for doctors never to "use the knife" (that is, conduct surgical procedures) are of course no longer sustainable in modern times.

It is easy to find some of the reasons for these discrepancies. Designed to fit the very early stage of Greek medicine, the HO cannot be expected to make mention of such issues as the ethics of experimentation, team care, or a doctor's societal or legal responsibilities. Today's technology has created a completely new perspective on what is possible and *acceptable*. One consequence is that ethical considerations often seem to lag behind the technological developments, while social and legal issues have transformed modern ethical codes into "mission statements" full of legalese, concerned with the liabilities and societal pressures on the medical profession.[26] Doctors have begun to ask pointed questions regarding the feasibility of a single oath: Does increasing medical specialization not force us to diversify? Can the entanglement of government and health-care organizations still guarantee a patient's privacy? Do physicians have a moral obligation to treat patients with potentially contagious diseases as AIDS or the

---

[25] R. D. Orr et al., "Use of the Hippocratic Oath: A Review of Twentieth Century Practice and a Content Analysis of Oaths Administered in Medical Schools in the U.S. and Canada in 1993," *The Journal of Clinical Ethics* 8 (Winter 1997): 377–88, at 385–86. Cf. C. Schubert and R. Scholl, "Der Hippokratische Eid: Wie viele Verträge und wie viele Eide?" *Medizinhistorisches Journal* 40 (2005): 247–73.
[26] Both for the American and the Australian Medical Associations (short AMA) the mission statement illustrates this well.

Ebola virus? These are important questions, but they cannot be taken as evidence that the long standing principles of medical ethics should be abandoned; rather, they suggest that we should carefully update and adjust the central message of the ancient medical professions, as has been done for centuries.[27]

Evidently the concern for a patient's well-being and privacy has not disappeared. The three ancient "aspirational pledges" discussed here illustrate this well. Their timeless qualities seem to suggest that something very fundamental was enshrined at the core of these compact texts: the protection of life, the importance of trust, and the safeguarding of confidentiality—in short, *the sanctity of life and the viability of the profession*. But we should also acknowledge that we may be committing a form of "cultural chauvinism" if we privilege the ancient Greek oath and measure all others against it. I would maintain that, despite all the modern complications and difficulties of scale, the core business of interacting with patients in a responsible and professional way is still as much at the heart of the doctor-patient relationship as it was thousands of years ago. It would seem that this kind of insight is not time- or culture-specific, and could therefore arise independently in Greece, India, and China within the context of the new approaches in healing, ones that regulated behavior and set limits to treatments on the basis of the assessment of risk to the patient's health and the physician's reputation. If this is the case, there is no justification for calling this kind of moral code "ahead of its time"; rather it is the result of the special nature of the doctor-patient interaction. That we have ended up with a "four principles approach to biomedical ethics (autonomy, beneficence, non-maleficence, justice)"[28] is thus neither a complete novelty nor mere traditionalism. Any rational method towards healing would produce such basic principles, motivated by the fundamental need for mutual trust and a proper basis for managing conduct and risk.

---

27 This is also argued in S. H. Miles, *The Hippocratic Oath and the Ethics of Medicine* (Oxford: Oxford University Press, 2005), 112–13; cf. Rütten, "Receptions of the Hippocratic Oath," *passim*.
28 Fu-Chang Tsai, "Ancient Medical Ethics," 315.

Bibliography

Aksoy, S. "Ancient Indian and Chinese Medical Oaths and the Comparison of their Medical Rules," *Yeni Tip Tarihi Arastirmalari* 7 (2001): 65–76 (article in Turkish, abstract in English at PubMed US National Library of Medicine, accessed Dec. 8, 2011 at http://www.ncbi.nlm.nih.gov/pubmed/14570011).

Baltussen, H. "Hippocratic Corpus," in Gagarin, *Oxford Encyclopedia*, 4:1–4.

Baltussen, H. "Hippocratic Oath," in Gagarin, *Oxford Encyclopedia*, 4:4–6.

Beckwith, C. I. "The Introduction of Greek Medicine into Tibet in the Seventh and Eighth Centuries," *Journal of the American Oriental Society* 99 (1979): 297–313.

Buell, P. D., and E. N. Anderson, eds. *A Soup for the Qan: Chinese Dietary Medicine of the Mongol Era as Seen in Hu Szu-Hui's Yin-Shan Cheng-Yao*, 2nd ed. (Leiden: Brill, 2010).

Dean-Jones, L. "Literacy and the Charlatan in Ancient Greek Medicine," in H. Yunis (ed.), *Written Texts and the Rise of Literate Culture in Ancient Greece* (Cambridge: Cambridge University Press, 2003), 97–121.

Dunstan, G. R. "The Moral Status of the Human Embryo: A Tradition Recalled," *Journal of Medical Ethics* 10 (1984): 38–44.

Edelstein, L. "The Hippocratic Oath," in O. Temkin and C. L. Temkin (eds.), *Ancient Medicine: Selected Papers of Ludwig Edelstein* (Baltimore: The Johns Hopkins Press, 1967), 3–63.

Gagarin, M., ed. *Oxford Encyclopedia of Ancient Greece and Rome*, 7 vols. (Oxford: Oxford University Press, 2010).

Jones, W. H. S. *The Doctor's Oath* (Cambridge: Cambridge University Press, 1924).

Jonsen, A. R. *A Short History of Medical Ethics* (New York: Oxford University Press, 2000).

Lee, T. "Medical Ethics in Ancient China," *Bulletin of the History of Medicine* 13 (1943): 268–77.

Mansfeld, J. *Prolegomena: Questions to be Settled before the Reading of an Author, or a Text* (Leiden: Brill, 1994).

Menon, I. A., and H. F. Haberman. "Dermatological Writings of Ancient India," *Medical History* 13 (1969): 387–92.

Menon, I. A., and H. F. Haberman. "The Medical Students' Oath of Ancient India," *Medical History* 14 (1970): 295–99.

Meulenbeld, G. J. *A History of Indian Medical Literature*, 3 vols. (Groningen: E. Forsten, 1999–2002).

Miles, S. H. *The Hippocratic Oath and the Ethics of Medicine* (Oxford: Oxford University Press, 2005).

Miles, S. H. *Oath Betrayed: America's Torture Doctors*, 2nd ed. (Berkeley: University of California Press, 2009).

Nutton, V. *Ancient Medicine* (London: Routledge, 2004).

Orr, R. D., et al. "Use of the Hippocratic Oath: A Review of Twentieth Century Practice and a Content Analysis of Oaths Administered in Medical Schools in the U.S. and Canada in 1993," *The Journal of Clinical Ethics* 8 (Winter 1997): 377–88.

Parker, H. N. "Women Doctors in Greece, Rome, and the Byzantine Empire," in L. R. Furst (ed.), *Women Healers and Physicians: Climbing a Long Hill* (Lexington: University of Kentucky Press, 1997), 131–50 [bibliography missing due to printer's error, but available on author's homepage].

Parker, H. N. "Galen and the Girls: Sources for Women Medical Writers Revisited," *Classical Quarterly* 62 (2012): 359–86.
Pérez, S. G., R. J. Gelpi, and A. M. Rancich. "Doctor-Patient Sexual Relationships in Medical Oaths," *Journal of Medical Ethics* 32 (2006): 702–705.
Rosenthal, F. *The Classical Heritage in Islam*, trans. Emile and Jenny Marmorstein (Berkeley: University of California Press, 1975) [= *Das Fortleben der Antike im Islam* (Zürich: Artemis Verlag, 1965)].
Rütten, T. "Receptions of the Hippocratic Oath in the Renaissance: The Prohibition of Abortion as a Case Study in Reception," *Journal of the History of Medicine and Allied Sciences* 51 (1996): 456–83.
Rütten, T. *Geschichten vom Hippokratischen Eid* (Wiesbaden: Harrassowitz, 2007), CD-ROM.
Schubert, C., and R. Scholl. "Der Hippokratische Eid: Wie viele Verträge und wie viele Eide?" *Medizinhistorisches Journal* 40 (2005): 247–73.
Temkin, O. "What Does the Hippocratic Oath Say? Translation and Interpretation," in *"On Second Thought" and Other Essays in the History of Medicine and Science* (Baltimore: The Johns Hopkins University Press, 2002), 21–28.
Tsai, Daniel Fu-Chang. "Ancient Medical Ethics and the Four Principles of Biomedical Ethics," *Journal of Medical Ethics* 25 (1999): 315–21.
Unschuld, P. U. *Medicine in China: A History of Ideas* (Berkeley: University of California Press, 1985).
van der Eijk, P. J. *Medicine and Philosophy in Antiquity: Doctors and Philosophers on Nature, Soul, Health and Disease* (Cambridge: Cambridge University Press, 2005).
von Staden, H. "'In a Pure and Holy Way': Personal and Professional Conduct in the Hippocratic Oath," *Journal of the History of Medicine and Allied Sciences* 51 (1996): 406–37.
von Staden, H. "'The *Oath*,' the Oaths, and the Hippocratic Corpus," in V. Boudon-Millot, A. Guardasole, and C. Magdelaine (eds.), *La science médicale antique: nouveaux regards* (Paris: Beauchesne, 2007), 425–66.

Appendix A: "Hippocratic" *Oath*

[trans. von Staden, "'In a Pure and Holy Way,'" 406–408]

1. i. I swear
   ii. by Apollo the Physician and
       by Asclepius and
       by Health and Panacea and
       by all the gods as well as goddesses,
       making them judges [witnesses],
   iii. to bring the following oath and written covenant to fulfillment,
   iv. in accordance with my power and my judgment;
2. i. to regard him who has taught me this craft as equal to my parents, and
   ii. to share, in partnership, my livelihood with him and to give him a share when he is in need of necessities, and
   iii. to judge the offspring [coming] from him equal to [my] male siblings, and
   iv. to teach them this *technē*,
       should they desire to learn [it],
       without fee and written covenant, and
   v. to give a share
      both of rules
      and of lectures,
      and of all the rest of learning,
         to my sons and
         to the [sons] of him who has taught me and
         to the pupils who have
         both made a written contract
         and sworn by a medical convention but by no other.
3. i. And I will use regimens
      for the benefit of the ill
      in accordance with my ability and my judgment,
   ii. but from [what is] to their harm or injustice
       I will keep [them].
4. i. And I will not give a drug that is deadly to anyone if asked [for it],
   ii. nor will I suggest the way to such a counsel.
   iii. And likewise I will not give a woman a destructive pessary.
5. i. And in a pure and holy way
   ii. I will guard my life and my *technē*.
6. i. I will not cut,

	ii.	and certainly not those suffering from stone, but I will cede [this] to men [who are] practitioners of this activity.
7.	i.	Into as many houses as I may enter, I will go for the benefit of the ill,
	ii.	while being far from all voluntary and destructive injustice, especially from sexual acts both upon women's bodies and upon men's, both of the free and of the slaves.
8.	i.	And about whatever I may see or hear in treatment, or even without treatment, in the life of human beings —things that should not ever be blurted out outside—
	ii.	I will remain silent, holding such things to be unutterable [sacred, not to be divulged],
9.	i. a.	If I render this oath fulfilled, and if I do not blur and confound it [making it to no effect]
	b.	may it be [granted] to me to enjoy the benefits both of life and of *technē*.
	c.	being held in good repute among all human beings for time eternal.
	ii. a.	If, however, I transgress and perjure myself,
	b.	the opposite of these.

Appendix B: The (Indian) Oath of Initiation

[trans. Menon and Haberman, "The Medical Students' Oath"]

1. The teacher then should instruct the disciple in the presence of the sacred fire, Brahmanas [Brahmins] and physicians.

2. [saying] "Thou shalt lead the life of a celibate, grow thy hair and beard, speak only the truth, eat no meat, eat only pure articles of food, be free from envy and carry no arms.

3. There shall be nothing that thou should not do at my behest except hating the king, causing another's death, or committing an act of great unrighteousness or acts of leading to calamity.

4. Thou shalt dedicate thyself to me and regard me as thy chief. Thou shalt be subject to me and conduct thyself forever for my welfare and pleasure. Thou shalt serve and dwell with me like a son or slave or supplicant. Thou shalt behave and act without arrogance, with care and attention

and with undistracted mind, humility, constant reflection and ungrudging obedience. Acting either at my behest or otherwise, thou shalt conduct thyself for the achievement of thy teacher's purposes alone, to the best of thy abilities.

5. If thou desirest success, wealth and fame as a physician and heaven after death, thou shalt pray for the welfare of all creatures beginning with the cows and the Brahmanas.

6. Day and night, however thou mayest be engaged, thou shalt endeavour for the relief of patients with all thy heart and soul. Thou shalt not desert or injure thy patient for the sake of thy life or thy living. Thou shalt not commit adultery even in thought. Even so, thou shalt not covet others' possessions. Thou shalt be modest in thy attire and appearance. Thou shouldst not be a drunkard or a sinful man nor shouldst thou associate with the abettors of crimes. Thou shouldst speak words that are gentle, pure and righteous, pleasing, worthy, true, wholesome, and moderate. Thy behaviour must be in consideration of time and place and heedful of past experience. Thou shalt act always with a view to the acquisition of knowledge and fullness of equipment.

7. No persons, who are hated by the king or who are haters of the king or who are hated by the public or who are haters of the public, shall receive treatment. Similarly, those who are extremely abnormal, wicked, and of miserable character and conduct, those who have not vindicated their honour, those who are on the point of death, and similarly women who are unattended by their husbands or guardians shall not receive treatment.

8. No offering of presents by a woman without the behest of her husband or guardian shall be accepted by thee. While entering the patient's house, thou shalt be accompanied by a man who is known to the patient and who has his permission to enter; and thou shalt be well-clad, bent of head, self-possessed, and conduct thyself only after repeated consideration. Thou shalt thus properly make thy entry. Having entered, thy speech, mind, intellect, and senses shall be entirely devoted to no other thought than that of being helpful to the patient and of things concerning only him. The peculiar customs of the patient's household shall not be made public. Even knowing that the patient's span of life has come to its close, it shall not be mentioned by thee there, where if so done, it would cause shock to the patient or to others.

Though possessed of knowledge one should not boast very much of one's knowledge. Most people are offended by the boastfulness of even those who are otherwise good and authoritative.

9. There is no limit at all to the Science of Life, Medicine. So thou shouldst apply thyself to it with diligence. This is how thou shouldst act. Also thou shouldst learn the skill of practice from another without carping. The entire world is the teacher to the intelligent and the foe of the unintelligent. Hence knowing this well, thou shouldst listen and act according to the words of instruction of even an unfriendly person, when his words are worthy and of a kind as to bring to you fame, long life, strength and prosperity."

10. Thereafter, the teacher should say this—"Thou shouldst conduct thyself properly with the gods, sacred fire, Brahmanas, the guru, the aged, the scholars and the preceptors. If thou hast conducted thyself well with them, the precious stones, the grains and the gods become well disposed towards thee. If thou shouldst conduct thyself otherwise, they become unfavourable to thee." To the teacher that has spoken thus, the disciple should say, "Amen."

Alan C. Bowen
Simplicius in Thirteenth-Century Paris: A Question

Abstract: The debate in the sixth century between the Christian philosopher John Philoponus and the Platonist philosopher Simplicius about whether the cosmos was created or eternal was of momentous importance not only to their understanding of the world and of the means to salvation from its trials but also to their views of what astronomical science was and how it should proceed in making its arguments. This brief chapter outlines this debate and then explores the main lines of attack to be taken in determining how Thomas Aquinas, who was supplied by William of Moerbeke with a translation of the text in which Simplicius responds to Philoponus, dealt with Simplicius' reading of Aristotle in advancing a vigorous polemic against his Christian faith.

What follows is not an argument for some specific conclusion but an inquiry or, more precisely, an attempt to define a new inquiry which has arisen from my recent work and which is still in its earliest stages.[1] In responding in this way to the editors' very kind request to contribute to this volume honoring Heinrich von Staden, I reflect that it is very much in the spirit of the many pleasant conversations that I have enjoyed with Heinrich over the years since his arrival at the Institute for Advanced Study in 1998, conversations in which the focus has often been on how to contextualize and approach some issue raised in ancient documents. Moreover, it is my view as well that, if research is defined by its questions, then, to understand the answers and the arguments supporting them, we must first appreciate the questions themselves, their context(s), and their presuppositions.

At its most general, then, my subject is the reception of Simplicius' commentary on Aristotle's *De caelo* in Paris of the thirteenth century, especially, in the works of Thomas Aquinas (1224/5–1274). Yet, though this remark places my inquiry under the general rubric of an interest evident in Heinrich's own writings, it hardly gets to the substance of what is at issue. Allow me to explain.

In subsequent editions to his *Principia mathematica*, there is a concluding, general scholium in which Newton claims, "& hypotheses non fingo" ("And I

---

[1] Specifically, I shall draw on arguments made in Alan C. Bowen, "The Demarcation of Physical Theory and Astronomy by Geminus and Ptolemy," *Perspectives on Science* 15 (2007): 327–58 and Alan C. Bowen, *Simplicius on the Planets and Their Motions: In Defense of a Heresy* (Leiden: Brill, 2013). The translations in this chapter are mine.

make no hypotheses").[2] This claim, which is clearly meant to enhance the credibility and impact of Newton's analysis, is momentous. Yet, at first glance, it is most puzzling, given that there are in fact numerous propositions in the *Principia* which well deserve the epithet "hypothesis." So what is at issue? In a word or two, Newton's point is that the *Principia* not only allows one to compute the motion at any time of any body, in particular, the motion of any celestial body, it does so on the basis of a proper, causal physical theory. This is the claim by which the *Principia* purports to take its place as the culmination of a concern with the relationship of astronomy and physical theory that was in fact first enunciated by Posidonius of Apamea, a Stoic philosopher of the first century B.C.

This concern arises from the fact that, from the time of Posidonius to Newton himself, astronomers had presented divergent, inconsistent, and even physically implausible accounts of how the planets move. This was a serious problem for astronomers and even more so for philosophers, given that, as *they* saw it, these astronomers were effectively presenting their accounts (called "hypotheses") as real and offering as proof of this no more than the fact that their hypotheses saved (accounted for or explained away) the phenomena. To the philosopher *qua* physical theorist, then, the challenge was to determine which if any of these rival astronomical hypotheses was actually the case and, thus, to lay the groundwork for a new astronomy that was *non*-hypothetical in that it was ultimately rooted through demonstration in the highest principles of physical theory itself.

Now, from this standpoint, it is plain that Newton's "& hypotheses non fingo" succeeds only if the terms of this concern are radically changed by effectively folding what used to be called astronomy into a physical theory that has been reduced mainly to the study of mechanical causation. Moreover, though Newton's solution stands only with Ptolemy's in opening the way to an impressive technical advance in the science itself, we should not forget that there were earlier solutions and that these too enjoyed success, albeit on different terms and thus in different ways.

---

[2] See, e.g., Isaac Newton, *Philosophiae Naturalis Principia Mathematica* (Glasgow: University of Glasgow, 1871), 530.

1 Simplicius of Cilicia

To the Hellenist, the key source is Simplicius. But, though it has to date been customary to draw on his works in so far as they bear on Posidonius, one should not overlook that such references to this issue as one finds are there only as part of a prolonged and complex *apologia* mounted in response to an attack on critical tenets of Late Platonist religion by the Christian philosopher, John Philoponus.

Indeed, Simplicius, a Late Platonist of the sixth century A.D., was very nearly the last of a line of thinkers for whom the pursuit of philosophy had taken on the trappings of a religion, with its own associated rituals and regimen and the promise of salvation. This promise, so far as one can tell for Simplicius, was to be realized by the true understanding of sacred texts, texts organized in a sequence of study which, at a fairly advanced stage, went from the treatises of Aristotle to the dialogues of Plato. Indeed, for Simplicius, Aristotle was the disciple of Plato, who was in turn the prophet of the Craftsman God; and Aristotle's works were to be read as fully harmonious with those of Plato in all matters of philosophical substance.

Thus, when Philoponus (drawing on Alexander of Aphrodisias) pointed out that Aristotle's fifth simple body, aether, was inconsistent with the phenomena and current astronomical hypotheses, because aether is that substance which moves by nature in a circle about the center of the cosmos, the Earth, his aim was to deny that there was an eternal substance in the cosmos—it being granted that what moves by nature in a circle has no opposite and is thus ungenerated and imperishable—and, accordingly, to refute the claim that the cosmos is without beginning or end rather than created. In short, Philoponus was proposing to uphold the account of creation in *Genesis* by an attack on the philosophical underpinnings of Late Platonic religion using recent astronomical theorizing as his weapon.

To Simplicius, this attack was an acute embarrassment. For, though his manner of reading Plato and Aristotle allowed him to take the creation story in the *Timaeus* as an account of the work of the Craftsman God *before* time and to dispatch Philoponus' other criticisms of the aether quite easily, this particular criticism was telling because, like Philoponus, he and his fellow Platonists followed Ptolemy's eccentric and epicyclic hypotheses and not Aristotle's homocentric ones in accounting for the planetary motions. And so he was obliged to mount an *apologia* which, though aimed in part at Philoponus, was really for the benefit of the members of his school. This *apologia* is to be found in the digression that concludes Simplicius' commentary on Arist. *cael.* 2.12.[3]

This digression itself aims to establish the following:

(1) the homocentric hypotheses developed in *metaph.* Λ 8 are inadequate to the phenomena.
(2) Aristotle was aware of this. Indeed, there was at the time an effort to improve on these hypotheses, albeit one that proved unsuccessful.
(3) The fact is, however, that though there may be numerous hypotheses purporting to save the phenomena, *none* is demonstrably the case.
(4) Moreover, Aristotle was unique in his time for recognizing that the proper account of the planetary motions would have to come from physical theory and not from astronomy.
(5) Thus, when Aristotle assumes that the planets make more than two motions in *cael.* 2.12 and draws on current astronomical hypotheses in *metaph.* Λ 8 to establish an answer to the *philosophical* question of the number of unmoved movers, his treatment of planetary motion is merely provisional. That is, his acceptance of the *explananda* and the *explanantia* of homocentric theory was only apparent and not real because he was writing as a philosopher *qua* physical theorist addressing questions about the heavens by drawing on what was known or thought to be known of the heavens at the time.

By this reading of Aristotle, Simplicius hoped to avert Philoponus' polemic, to rescue Aristotle's authority in the school of Late Platonism, and to uphold the value of the *De caelo* in its program of education. At the same time, Simplicius indicates that the Late Platonists themselves have no real commitment to the "modern" planetary hypotheses, though following them is plainly the responsible course. Indeed, their situation proves to be the same as Aristotle's: since philosophical discussion of the heavens is to be constrained by astronomical observation and theorizing, then until astronomy is grounded in a proper physical theory of the planetary motions, reflections on the heavens in physical theory will be provisional or heuristic and not wholly demonstrative of unshakable truth. In this way, Simplicius clears the Late Platonist path from Aristotle to Plato and, ultimately, to a divine understanding of both how and why the heavenly bodies move as they do which is unshakably true.

The essential substantive proposition in this *apologia*, one which Simplicius elaborates in his commentary on *phys.* 2.2 (written later, but meant to be read before his *In de caelo*), is that no astronomical hypothesis is demonstrably the

---

[3] See Simp. *in de cael.* 492.25–510.35 (Heiberg).

case.[4] The fact that these hypotheses save the phenomena is not to be construed as proof of their truth. And this is a problem. That is, the sheer variety of the astronomers' hypotheses confronts the philosopher with a challenge that goes to the very heart of his faith and understanding of the world—to determine which, if any, of these hypotheses is the case.

It is striking that Simplicius makes no effort to determine the starting points of a true account of all the apparent planetary motions. Instead, at the close of the digression, he simply writes:

> Now, if this is more fitting to chapters about the heavens than to ones about first philosophy, none of us will criticize the rather lengthy digression from the [present] chapter, since it has come about at the right time. But we must return to what comes next in Aristotle's chapters. (Simp. *in de cael.* 510.31–35 [Heiberg])

I take this to mean that the answer to this question of how the planets really move is to come not at this stage in the process of Late Platonic education but at a later stage, one in which there is a much fuller assimilation to the Craftsman God. But to attain this assimilation and so to understand how and why the planets move as they really do, one must first read Aristotle.

2 Thomas Aquinas

Since, as it turns out, this question about planetary hypotheses was not just a problem in technical astronomy but one that reached to the foundations of philosophical accounts of what there is, to say nothing of religious faith, one really must inquire if subsequent readers of Simplicius' writings were aware of this *apologia* and addressed its concerns in their own work. And so I turn to the thirteenth century, when Simplicius' commentary on the *De caelo* first reappears, and to Thomas Aquinas in particular, given his role in the formation of Christian thought over the next centuries.

Now, there were two translations of Simplicius' *In de caelo* made in the thirteenth century. The earlier was by Robert Grosseteste (1168–1253) and the later, by William of Moerbeke (ca. 1215–ca. 1286). Moerbeke, a Dominican, made his own translation of the *In de caelo* in two stages. In the first (sometime in 1264–1266), he translated a section of the digression which Thomas Aquinas

---

4 See Simp. *in phys.* 290.3–293.15 (Diels). The lemma is Arist. *phys.* 2.2, 193b22–35.

used in his commentary on Aristotle's *metaph.* Λ 8.⁵ In the second (in 1271), Moerbeke completed a version of the whole text which included revising his earlier effort at the digression in the light of other Greek manuscripts.⁶ It was this translation that Thomas used in preparing his commentary on Aristotle's *De caelo*, which he began in 1271/2 and left incomplete at his death in 1274.

It would seem that Moerbeke, for his part, had no clear sense of what the digression was actually about. Though his rendering of the section on the planetary hypotheses is quite good (allowing for what are presumably the deficiencies of his Greek source),⁷ the summary of the rest⁸ is mistaken in some points and misleading in others:

> In this way, Simplicius pursues the fact that they cannot explain these [appearances], not even what they observe in eclipses. After that, he shows that not even Aristotle considered this position secure. Accordingly, [Simplicius] posits hypotheses in accordance with eccentric and epicyclic [circles]—he says that the first discoverers of this were the Pythagoreans. (I have not taken care to relate these [hypotheses] because they are handed down rather fully by Ptolemy.) Then, he raises certain objections against these hypotheses, concerning which he says that there will be time to consider others; he does not pursue anything bearing on them.⁹

Moreover, his colophon to the full translation of 1271 hints at a failure to appreciate what Simplicius is actually trying to accomplish in the commentary:

> But I, William of Moerbeke, of the Order of Brothers-Preachers of the Lord [sc. the Dominicans], Penitentiary of the Pope and Chaplain, with much bodily discomfort and much weariness of mind, offer this to Christendom in the belief that in this work of translation I have contributed more to studies of works in Latin.¹⁰

---

5 Whether he did this specifically for Thomas is a nice question, which need not detain us now. On the date, see Fernand Bossier, Christine Vande Vere, and Guy Guldentops, eds., *Simplicius. Commentaire sur le traité du ciel d'Aristote. Traduction de Guillaume de Moerbeke*, vol. 1 (Leuven: Leuven University Press, 2004), xxxviii–xli.
6 On the date, see Bossier, Vande Vere, and Guldentops, *Simplicius*, li.
7 That is, of Simp. *in de cael.* 492.25–504.4.32 (Heiberg).
8 That is, of Simp. *in de cael.* 504.4.33–510.35 (Heiberg).
9 For the Latin text, see Fernand Bossier, *Filologisch-historische navorsingen over die middeleeuwse en humanistische Latijnse vertalingen van de commentaren van Simplicius*, 3 vols., PhD diss., Katholieke Universiteit te Leuven, 1975, 3:20.1–8.
10 For the Latin text, see Fernand Bossier, "Traductions latines et influences du commentaire *In de caelo* en occident (XIIIe–XIVe s.)," in Ilsetraut Hadot (ed.), *Simplicius, sa vie, son œuvre, sa survie. Actes du Colloque International de Paris (26 Sept.–1er Oct. 1985)* (Berlin: De Gruyter, 1987) 288–325, at 305, or Bossier, Vande Vere, and Guldentops, *Simplicius*, li.

But, though Thomas certainly drew on Moerbeke's translations for his own work, there is no doubt that he saw Simplicius in a very different light, as the numerous references to him in the commentaries testify. And so, under the rubric of the reception of Simplicius, the question guiding my inquiry takes its final form:

> How does Thomas draw on Simplicius in addressing the problem of divergent planetary hypotheses as it bears on his own reading of Aristotle, on his considered views of astronomical science and what it tells us of the natural world, and on his Christian faith?

The answer to this question will, of course, be complex and colored by the nature of Thomas' goals in writing his own commentaries and as well as by the context in which he taught Aristotelian texts and promoted philosophy more generally at the University of Paris. It will also be forged in the light of the Averroist reading of Aristotle published in 1270 by his contemporary at the University, Siger of Brabant, and tempered significantly as well by the restrictions imposed in that same year when Bishop Étienne Tempier condemned as heretical thirteen Aristotelian propositions, several of which were essential to Simplicius' own faith and reading of Aristotle.

Bibliography

Bossier, Fernand. *Filologisch-historische navorsingen over die middeleeuwse en humanistische Latijnse vertalingen van de commentaren van Simplicius*, 3 vols., PhD diss., Katholieke Universiteit te Leuven, 1975.

Bossier, Fernand. "Traductions latines et influences du commentaire *In de caelo* en occident (XIIIe–XIVe s.)," in Ilsetraut Hadot (ed.), *Simplicius, sa vie, son œuvre, sa survie: actes du Colloque International de Paris (26 Sept.–1er Oct. 1985)* (Berlin: De Gruyter, 1987), 288–325.

Bossier, Fernand, Christine Vande Vere, and Guy Guldentops, eds. *Simplicius, Commentaire sur le traité du ciel d'Aristote. Traduction de Guillaume de Moerbeke*, vol. 1 (Leuven: Leuven University Press, 2004).

Bowen, Alan C. "The Demarcation of Physical Theory and Astronomy by Geminus and Ptolemy," *Perspectives on Science* 15 (2007): 327–58.

Bowen, Alan C. *Simplicius on the Planets and Their Motions: In Defense of a Heresy* (Leiden: Brill, 2013).

Diels, Hermann. *Simplicii in physicorum libros commentaria*, Commentaria in Aristotelem Graeca 9–10 (Berlin: Reimer, 1882–1895).

Heiberg, Johan L. *Simplicii in Aristotelis de caelo commentaria*, Commentaria in Aristotelem Graeca 7 (Berlin: Reimer, 1894).

Newton, Isaac. *Philosophiae Naturalis Principia Mathematica* (Glasgow: University of Glasgow, 1871).

Andrea Falcon
Aristotle and the Study of Animals and Plants

Abstract: While an impressive corpus of writings on animals is extant, no authentic work on plants has reached us. If Aristotle ever wrote on plants, his work was lost at an early date. There is nevertheless a great deal that Aristotle says on the topic of plants in the extant writings on animals. In this essay, I concentrate on these writings in order to learn more about how Aristotle conceived of his study of animals and plants.

I

My title is meant to make contact with the opening lines of the *Meteorology*. There, Aristotle makes it crystal clear that the explanatory project he has outlined is not complete until a study of plants is also in place:

> After we have dealt with all these subjects, let us then see if we can give some account, along the lines we have laid down, of *animals and plants, both in general and separately* (*peri zōōn kai phytōn, katholou te kai chōris*); for when we have done this we may perhaps claim that the whole investigation which we set before ourselves at the outset has been completed. (*meteorol.* 1.1, 339a5–8, trans. H. D. P. Lee, slightly modified)

While an impressive corpus of writings on animals is extant, no authentic work on plants has reached us. If Aristotle ever wrote on plants, his work was lost at an early date.[1] There is nevertheless a great deal that Aristotle says on this topic

---

I have fond memories of my time at the Institute for Advanced Study at Princeton (January–August, 2008). During that period Heinrich von Staden was a wonderful mentor, always generous in advice and suggestions. I am grateful to be invited to honor him with an essay, and I hope that this essay will interest him.

An earlier version of this essay was discussed with Allan Gotthelf. At the time when his health was deteriorating rapidly, Allan gave me his feedback on the question that is at the very heart of the essay, whether there is a unified study of life in addition to separate studies of animals and plants in the explanatory project that we call Aristotle's natural science. I remember our exchange of ideas with great pleasure and the sadness that comes from the realization that there will be no more. Allan died on August 30, 2013.

1 How early? Marwan Rashed has recently suggested that a work on plants may have circulated under the name of Aristotle as late as the second century A.D. He finds a reference to such a

in the surviving writings on natural philosophy.[2] I am interested in what can be learned from these writings about how Aristotle conceived of the study of "animals and plants, both in general and separately."

Before moving on, however, I would like to make a couple of remarks in connection with Aristotle's choice of words in the opening lines of the *Meteorology*. My first remark has to do with the order in which animals and plants are listed, namely, *first* animals and *then* plants. Given that the references to plants in the extant corpus are typically to a study to come, it is natural to ask whether the sequence outlined in the opening lines of the *Meteorology* is indicative of the order in which animals and plants are to be studied. I will answer this question in the affirmative. In the explanatory project pursued by Aristotle, the study of animals comes before the study of plants, and comes before it in the order of explanation.

While others speak of "order of learning," "order of exposition," or "order of teaching," I prefer to speak of "order of explanation," as the ultimate goal of the project outlined in the *Meteorology* is to deliver causal explanations. Since the opening lines of the *Meteorology* are intended to provide an outline of the structure of a complete science (*epistēmē*), we cannot make any inference regarding the actual order of discovery. It is quite possible that the relative order in which animals and plants are to be studied mirrors Aristotle's actual order of discovery. But this would be the exception rather than the rule. Consider, for instance, how the study of the heavens is listed in the outline offered at the beginning of the *Meteorology*. This study comes before the study of any part of the sublunary world, including the study of animals and plants. However, it is very unlikely that this is the order in which Aristotle proceeded in the study of

---

work in Galen's *Peri alupias* [*De indolentia*]. In explaining how he overcame the distress following the loss of his library in the fire of 192, Galen makes a reference to a number of precious and rare books he owned and lost in the fire, including a work on plants by Aristotle. See M. Rashed, "Aristote à Rome au IIe siècle: Galien, *De indolentia* §§ 15–18," *Elenchos* 32 (2011): 55–77. The lost work on plants mentioned by Galen should not be confused with the Ps-Aristotle, *On Plants*. The latter is (in the words of Lotte Labowsky) "a mediaeval Greek translation of a Latin rendering of an Arabic text translated, probably *via* a Syriac intermediary, from a lost Greek original" ("Aristotle's *De plantis* and Bessarion: Bessarion Studies II," *Medieval and Renaissance Studies* 5 [1961]: 132–53, at 132). The Greek original was a compendium on plants by Nicolaus of Damascus (second half of the first century B.C.). On the whole question, I refer the reader to H. J. Drossaart Lulofs and E. L. J. Poortman, *Nicolaus Damascenus, De plantis. Five Translations* (Amsterdam: North-Holland Publishing Press, 1989).

2 For a convenient collection of all the passages where Aristotle speaks of plants, I refer the reader to F. Wimmer, *Phytologiae Aristotelicae fragmenta* (Breslau: Grass, Barth, and Comp., 1838).

the physical world. Suffice it to recall that the empirical data that Aristotle had at his disposal in the study of the celestial world were, by his own admission, limited. In order to make progress, he often had to appeal to what was known about the sublunary world, and in particular about animals.[3]

My second remark has to do with the fact that Aristotle is not content to list animals and plants, and to list them in this order. He adds that animals and plants are to be studied in general *(katholou)* and separately *(chōris)*. This language is usually taken to be evidence that the complete *epistēmē* envisioned by Aristotle contains separate studies of animals and plants. It is also taken to be evidence that each of these two studies has a general as well as a special component. The rationale for such a complex explanatory structure is often found in the first book of the *Parts of Animals*. There, Aristotle says that the explanation of certain features of animal life should be given across different species of animals. He adds that such explanations are needed to avoid tedious repetitions. His examples are sleep, respiration, growth, decline, and death.[4] But avoiding needless repetitions cannot be the sole motivation for the explanatory strategy envisioned in the *Meteorology* and adopted in the biological works. The theory of scientific explanation outlined in the *Posterior Analytics* requires that explanations be given at the proper level of generality. This requirement is central to Aristotle's explanatory project. It is only by giving explanations at the right level of generality that we can capture salient features which might be otherwise missed. While this explanatory concern is clearly at work in the study of animals, it is not limited to that study; the study of the physical world as a whole is informed by it.[5]

What Aristotle has in mind is best illustrated by recalling a famous example introduced in the *Posterior Analytics*. We have proper knowledge of the fact that the sum of the triangle's internal angles is equal to two right angles if, and only if, we know that this property belongs to all triangles *qua* triangles. Since this geometrical property belongs to all triangles, it also belongs to equilateral, isosceles, and scalene triangles. But it does not belong to them in virtue of the fact

---

[3] For more on the place of the study of the heavens in Aristotle's explanatory project, I refer the reader to A. Falcon, *Aristotle and the Science of Nature: Unity without Uniformity* (Cambridge: Cambridge University Press, 2005), 1–30 and 85–112.
[4] *Part. anim.* 1.1, 639a19–20.
[5] For how this explanatory concern controls the study of animals, I refer the reader to J. G. Lennox, "Divide and Explain: The Theory of the *Analytics* in Practice," in A. Gotthelf and J. G. Lennox, *Philosophical Issues in Aristotle's Biology* (Cambridge: Cambridge University Press, 1987), 90–119 (reprinted in J. G. Lennox, *Aristotle's Philosophy of Biology* [Cambridge: Cambridge University Press, 2001], 7–38).

that they are equilateral, isosceles, or scalene. Rather, it belongs to them because they are triangles (in virtue of the fact that they are three-sided figures). Aristotle employs this example to show that there is a common explanatory level beyond that of equilateral, isosceles, and scalene triangles, and that this common explanatory level is reached by ignoring those facts that are specific to equilateral, isosceles, and scalene triangles.

We cannot rule out, it seems to me, that the explanatory principle that this example is meant to illustrate is at work in the opening lines of the *Meteorology*. More directly, we cannot rule out that these lines announce a general study of plants and animals taken together in addition to separate studies of animals and plants (a study analogous to the general study of triangles that must accompany narrower observations about particular kinds of triangles). As readers of the *Meteorology*, we do not know how much there is to be explained in common for animals and plants. I will return to this question in due course. For the time being, suffice it to say that the project of a general study of life, namely a study that treats certain aspects of animal and plant life in common, should be taken seriously. It is demanded by the theory of scientific explanation advanced in the *Posterior Analytics*.

II

The debate on the proper order of explanation within the study of life was quite intense in the Middle Ages and the Renaissance. In the Aristotelian tradition, plants have often been studied before animals. In the thirteenth century, for instance, this order was adopted (among others) by Albert the Great. In the opening lines of his *Physics*, Albert tells us that his intention is to follow not only the doctrine (*sententia*) but also the order (*ordo*) adopted by Aristotle.[6] When Albert comes to the part of the study of nature dealing with life, he is quite explicit in stating that the correct sequence is first plants and then animals, and that the study of the physical world ends with the study of animals.[7] Albert was influenced by Avicenna. In the Avicennian corpus of natural philosophy, the discussion of the soul (*liber sextus naturalium*) comes before the study of plants (*liber septimus naturalium*) and animals (*liber octauus naturalium*). Another notable supporter of the view that plants ought to be studied before animals was Averroes. In the introduction to his middle commentary on the *Meteorology*, Averroes

---

6 *Phys.* 1.1, 1.23–24.
7 *Phys.* 1.1, 4.58–61.

endorses the traditional sequence of writings (*Physics—On the Heavens—On Generation and Corruption—Meteorology*). Once the study of inanimate bodies is in place, it is possible to turn to the study of living bodies. Within the study of living bodies, plants are the first topic on the explanatory agenda, according to Averroes.⁸ A reason for adopting this relative order of study is not difficult to find: if the serial study of the soul adopted by Aristotle in his treatise *On the Soul* is taken as an indication of how living beings are to be studied, then plants are to be studied before animals, and animals before human beings.

Interestingly enough, some representatives of the Aristotelian tradition have looked elsewhere for instructions on how to proceed in the explanation of life. They have taken the outline of the complete science offered in the opening lines of the *Meteorology* to be evidence that Aristotle consciously postponed the study of plants until the conceptual apparatus for the study of animals was firmly in place. To my knowledge, the most elaborate defense of this approach to the study of life is found in a short but interesting essay written by Francesco Cavalli and printed in Venice by Matteo Capcasa between 1490 and 1495.⁹ One good reason to take this essay seriously is that it provides the reasoning informing the Aldine edition of Aristotle (Venice, 1495–1498). In the prefatory letter to the third volume, Aldo Manuzio tells us that, for the selection and arrangement of the writings concerned with natural philosophy, he sought the advice of Francesco Cavalli, whom he describes as a man of great study, very learned in philosophy, and an excellent doctor in Venice.¹⁰

---

8 This plan is adopted by the editors of the Juntine edition of Aristotle and Averroes. See *Aristotelis Stagiritae omnia quae extant opera ... Averrois Cordubensis in ea opera omnes qui ad nos pervenere commentarii* ..., 12 vols. (Venice: Giunta, 1562; facsimile reprint, Frankfurt a. Main: Minerva, 1962). The two books of the Ps.-Aristotle *On Plants* are printed right after Aristotle's *Meteorology* in volume 5. The study of animals can be found in volume 6. Note that Averroes does not agree with Avicenna in placing the study of the soul before the study of plants and animals. Averroes places the study of the soul within the study of animals. Following him, the editors of the Juntine edition print the treatise *On the Soul* in volume 6, right after the *Parts of Animals* and before the short essays on natural philosophy.
9 The exact title of the book by Francesco Cavalli is *De numero et ordine partium ac librorum physicae doctrinae Aristotelis* (Venice: Matteo Capcasa, 1490–1495) [GW 5832].
10 For more on the connection between Cavalli and the Aldine edition, I refer the reader to C. B. Schmitt, "Aristotelian Textual Studies at Padua. The Case of Francesco Cavalli," in A. Poppi (ed.), *Scienza e filosofia all'università di Padova nel Quattrocento* (Trieste: Edizioni Lint, 1983), 287–313 [reprinted in C. B. Schmitt, *The Aristotelian Tradition and Renaissance Universities* (London: Variorum Reprints, 1984), Chapter 13]. For the prefatory letters to the five volumes of the Aldine edition of Aristotle, see B. Botfield, *Prefaces to the First Editions of Greek and Roman Classics* (London: H. G. Bohn, 1861), 194–204. These letters can be found (with an Italian

In his essay on the order and parts of Aristotle's physical theory, Cavalli argues that, as a general methodological rule, what is simple should be studied before what is complex (for instance, the simple bodies should be studied before the mixed inanimate bodies, and the mixed inanimate bodies before the living bodies). According to Cavalli, however, Aristotle does not follow this methodological rule in the case of animals and plants. In this case, Cavalli argues, animals are to be studied first and are to be taken as a model (*exemplar, modulus*) in the subsequent study of plants. Cavalli has in mind certain passages from the biological writings where Aristotle seems to have adopted this explanatory strategy within the study of animals. Aristotle often begins this study with blooded animals, and specifically with human beings. It is only when a study of the paradigmatic case (blooded animals) is firmly in place that Aristotle ventures into the study of the remaining animals (bloodless animals).

An example of this explanatory strategy is found in the study of animal generation offered in the second and third book of *Generation of Animals*. Aristotle makes it clear that we must begin from the things that are first. Accordingly, his study begins with the live-bearing animals, and specifically with human beings.[11] The study of the live-bearing animals takes most of the second book (*gen. anim.* 2.4–8). It is only when this project is in place that Aristotle turns to the egg-laying animals. The study of the egg-laying animals is conducted in *gen. anim.* 3.1–8. Here too Aristotle adopts the methodological principle spelled out in *gen. anim.* 2.4 ("first the things that are first"). He begins with the animals that lay a complete, hard-shelled egg and continues with those that lay an incomplete egg. He ends with a discussion of what is specific about the bloodless animals that lay eggs. The soft-bodied animals (sepias and the like) and the soft-shelled animals (*karaboi* and creatures akin to them) are discussed in *gen. anim.* 3.8. This chapter marks the transition to the study of bloodless animals. This study continues with insects. Most insects produce a grub. What is distinctive about their mode of reproduction is discussed in *gen. anim.* 3.9. The generation of bees poses a difficulty (*aporia*). Aristotle deals with this *aporia* in *gen. anim.* 3.10. When this *aporia* is solved, Aristotle turns to the bloodless animals whose generation is not from a male and a female but is spontaneous. Along with a few of the insects, *gen. anim.* 3.11 is concerned with the generation of the hard-shelled animals and other stationary animals that are considered close to plants. With this chapter, the treatment of what is specific about the gen-

---

translation by Carlo Dionisotti) in G. Orlandi, *Aldo Manuzio editore. Dediche, prefazioni, note ai testi*, 2 vols. (Milan: Il Polifilo, 1976).

11 *Gen. anim.* 2.4, 737b25–27: "we must begin from the things that are first, and the complete animals are first, and the live-bearers are of this sort, and of these man is first."

eration of each kind of animals is complete.[12] Pragmatic reasons may have suggested adopting this explanatory strategy. Note, however, that Aristotle does not invoke them. His official justification for starting with blooded animals, and more specifically with human beings, is that this order of explanation mirrors the order of nature.[13]

It is not difficult to see that this explanatory strategy is fairly common in Aristotle's study of animals. Consider how the study of the non-uniform parts is introduced in *part. anim.* 2.10. Here, too, Aristotle begins with blooded animals, and specifically with human beings. He turns to the internal parts in bloodless animals only when the study of the corresponding parts in blooded animals is in place. In this case, Aristotle does mention a pragmatic reason for this explanatory strategy. We should start with human beings, Aristotle says, because we are more familiar with them. Note, however, that this order of explanation is not justified solely on pragmatic grounds. The primary reason for adopting it is that in the human being alone the non-uniform parts are arranged in the natural order as a consequence of their erect posture.[14]

In *Parts of Animals* as well as in *Generation of Animals*, Aristotle says that we must begin from the things that are first.[15] This means beginning from blooded animals, and specifically from human beings. The results achieved in the study of the paradigmatic case are adopted, and indeed adapted, as Aristotle moves on to the study of bloodless animals. We do not need to look at how Aristotle applies

---

[12] Note that the explanation of animal generation proceeds not only from blooded to bloodless animals but also from animals that reproduce sexually to animals that do not reproduce sexually. The study of the latter is conducted on the basis of the theoretical framework developed in the course of the study of sexual generation. Aristotle is quite explicit regarding this aspect of his strategy ("Concerning the generation of the other animals [sc. animals that do not reproduce sexually], we must speak about each of them according to the ongoing argument (*logos*), building it from what has been said" (*gen. anim.* 1.1, 716a2–4). This is exactly how Aristotle proceeds in his account of animal generation: he begins with a study of sexual generation and ends with a discussion of the generation of those animals whose coming into being is spontaneous.
[13] See *gen. anim.* 2.1, 732b32–33, where Aristotle states that it is necessary for us, in our study of animal generation, to realize "how well and orderly nature has rendered generation."
[14] *Part. anim.* 2.10, 656a11–13: "the parts are disposed in a natural order in this case alone, namely the upper part is oriented toward the upper part of the whole; for the human being alone among the animals is upright."
[15] *Gen. anim.* 2.4, 737b25: "we must begin from the things that are first" (*apo tōn prōtōn arkteon prōton*). Cf. *part. anim.* 2.10, 655b28–29: "Let's speak again, as it were from the beginning, beginning first from the things that are first" (*prōton apo tōn prōtōn*). Aristotle invokes this methodological principle also at the end of the first book of *Parts of Animals*, when he says that we should try to give the causes of what is common and what is specific, "beginning from the things that are first" (*part. anim.* 1.5, 646a4: *prōton apo tōn prōtōn*).

his knowledge of blooded animals to the study of bloodless animals. The crucial point, for our purposes, is that bloodless animals are studied in light of, and with reference to, blooded animals.

Cavalli thinks that the same explanatory strategy should be applied as we move from the study of animals to the study of plants. I will return to the implications of employing this strategy in the study of plants shortly. For the time being, I would like to add that Cavalli was not content to argue that Aristotle postponed the study of plants until the study of animals was firmly in place because his intention was to use the results achieved in the study of animals as a model for the study of plants; he also argued that Aristotle did not write on plants, and that this study was accomplished by Theophrastus. By Cavalli's lights, the *History of Plants* and the *Causes of Plants* contribute directly to the project outlined in the opening lines of the *Meteorology*. To get a feel for how radical this view is, I refer the reader to the comparative study of the first books of the *History of Plants* and the *History of Animals* offered by Allan Gotthelf.[16] The upshot of this study is that Theophrastus modeled his investigation of plants on Aristotle's study of animals. But this is not to say that he adopted it unchanged; rather, he adapted the conceptual resources developed by Aristotle for the study of animals to his subject matter. It is very telling that, in the opening book of the *History of Plants*, Theophrastus says that one should not expect to find in plants a complete correspondence with animals (1.1.3, repeated at the start of 1.1.4). But saying that Theophrastus worked within the broad theoretical framework developed by Aristotle, as Gotthelf does, is not quite the same as saying that Aristotle and Theophrastus contributed to a single explanatory project, and saying that this project is the one outlined in the opening lines of the *Meteorology*. And yet, this is exactly the view defended by Cavalli.[17]

---

16 A. Gotthelf, "*Historiai* I: *plantarum et animalium*," in W. W. Fortenbaugh and R. W. Sharples (eds.), *Theophrastean Studies*, vol. 3: *Natural Science, Physics, Metaphysics, Ethics, Religion, and Rhetoric* (New Brunswick: Transaction Publishers, 1988), 100–33 [reprinted as Chapter 14 in A. Gotthelf, *Teleology, First Principles, and Scientific Method in Aristotle's Biology* (Oxford: Oxford University Press, 2012), 307–42].

17 I note, in passing, that this is also the view implied by the selection and arrangement of writings adopted in the Aldine edition, which is (we should not forget) an edition of Aristotle, and not an edition of Aristotle and Theophrastus. We can safely assume that Aldo did not include the *On Plants* in his edition of Aristotle on the advice of Cavalli. Instead, he included the works on plants by Theophrastus. He printed them in volume 4, after the works on animals (volume 3).

III

On the interpretation recommended by Cavalli, Aristotle made animals the paradigmatic living beings. It follows from this methodological decision that plants are to be studied in light of the results achieved in the study of animals. It is nevertheless one thing to employ this explanatory strategy within the *genos* of animals and another to apply it as we move from the *genos* of animals to the *genos* of plants. By treating the former as a model for the latter, we do not simply suggest that both animals and plants fall under a larger *genos*, the *genos* of living beings; we also imply that a unified study of this larger *genos* is possible. To put it differently, if animals and plants are part of the same *genos*, then there must be facts that are true of all living beings. Those facts may become the object of a unified study of life. What are the prospects of such a study within the explanatory project pursued by Aristotle?

This is a large topic and will occupy us for the rest of this essay. It is best approached by returning to the beginning of *part. anim.* 2.10, where the transition from the study of the uniform parts to that of the non-uniform parts is made. There, Aristotle argues that there are two parts that are absolutely indispensable to animals. They are the part for taking in nourishment and the part for discharging useless residue, as it is not possible to grow without nourishment.[18] Since nutrition is common to all living beings, the part for taking in nourishment is found also in plants. According to Aristotle, however, plants do not have a part for eliminating useless residue, as they take in concocted nourishment from the soil.[19] But this also means that plants do not have the part between the two most indispensable ones. This part is dedicated to receiving and processing unconcocted nourishment and is found in all animals. This stretch of text is interesting for at least two reasons. Firstly, Aristotle seems to be engaged in the sort of exercise that I have called a unified study of life. Both animals and plants *qua* living beings need nourishment to survive. Both actively maintain their own being by taking in nourishment. The difference is that animals take in unconcocted nourishment while plants take in concocted nourishment.[20] Secondly, and more importantly, there is not much else Aristotle is able, or willing, to say on the topic of plants and animals *qua* living beings. It is very telling that he

---

18 *Part. anim.* 2.10, 655b30–32.
19 *Part. anim.* 2.10, 656a32–35.
20 Elsewhere in the *Parts of Animals*, Aristotle says that plants use the earth and the heat within it as if it were a stomach (*part. anim.* 2.3, 650a21–23).

goes on to say that plants have fewer organs as they are engaged in fewer activities, and for this reason they have to be studied separately.[21]

This passage can be taken as evidence that animals and plants did not constitute a single investigative domain for Aristotle, as he was not able to find explanatory starting points common to both of them, which is to say *archai* that could justify a unified treatment of animal and plant life.

IV

Aristotle conceived of his investigation of animals as a discrete and relatively self-contained explanatory project. He thought of this project as prior to the study of plants in the order of explanation, as he is often content to promise a separate study of plants, or so I have argued. And yet, it should not be overlooked that Aristotle offers a basic study of plants, and offers it in the context of his study of animals. In the process of developing concepts for the study of animals, Aristotle arrives at a basic understanding of plants as, so to speak, upside-down animals.

Admittedly, Aristotle never speaks of plants as upside-down animals, but this image can be extracted from how Aristotle talks about the organization of their bodies in the biological corpus. At first sight, it is tempting to dismiss this image as a catchy metaphor.[22] On reflection, however, it becomes clear that it conveys an important message. The message is that a living body is always a structured body. To my knowledge, Aristotle is the first to link life to structure and the first to develop the conceptual resources to describe the relevant structure in purely functional terms. The work *On the Progression of Animals* is especially important in this respect, as it contains the fullest discussion of the link Aristotle establishes between the structure of a living body and its biological functions. There, Aristotle argues that the dimensions (*diastaseis*) by which the bodies of animals are naturally bounded are six in number and are organized in three pairs, namely up and down, front and back, right and left.[23] Moreover, he argues that while both animals and plants display the division into an upper

---

21 *Part. anim.* 2.10, 656a2–4: "The nature of plants being stationary, does not have a variety of non-uniform parts, for the use of few organs (*organa*) is required for few activities (*praxeis*); this is why their visible aspect should be studied separately."
22 I note, in passing, that this image is used in the title of the most recent and comprehensive study of ancient Greek thought on the topic of plants. See L. Repici, *Uomini capovolti: le piante nel pensiero greco* (Bari: Laterza, 2000).
23 *Incess. anim.* 4, 705a26–28.

and a lower part, the other four dimensions are present in animals alone. The sensory apparatus is always implanted in the front of the animal, while the actual mechanism of locomotion involves the existence of a right and a left side of the body. Among other things, this means that the animals that can engage in locomotion are the maximally organized living bodies, whereas plants are the minimally organized living bodies. In *On the Progression of Animals*, Aristotle is very clear that the division of the living body into an upper and a lower part has nothing to do with the orientation of such a body with respect to the earth and the heaven. Rather, this division is connected with the biological function of taking in nourishment.[24] The entry point of nourishment is always the upper part of the living body. While in animals this part is the mouth, in plants it is the roots.[25] In functional terms, the roots are to the plant what the mouth is to the animal. Branches, leaves, flowers, and fruit appear to us to be the upper part of a plant, but they are in fact its lower part.

The claim that the body of all perishable living beings, including that of plants, is divided into an upper and a lower part is recalled several times in the biological corpus. There is no need to examine all the relevant passages. What matters for our discussion is that the link between life and structure, as well as the characterization of structure in terms of functional dimensions, is developed in the context of the study of animals, but it is emphatically not meant to be restricted to animals.[26]

V

I have argued that for Aristotle the study of animals comes before the study of plants in the order of explanation and that the results reached in the study of animals ought to be taken as our starting point for the subsequent study of plants. This strategy suggests a question about the prospects for a unified study of living beings within Aristotle's explanatory project. I have offered some evidence that the prospects for such a study are not very good. And yet,

---

24 *Incess. anim.* 4, 705a28–32.
25 *Incess. anim.* 4, 705b6–8.
26 Arguably, the most interesting discussion of this doctrine is found outside the biological corpus. In *cael.* 2.2, Aristotle makes an interesting attempt to apply this doctrine to the case of the outer sphere of the heavens. Aristotle considers this sphere a living body. For a recent discussion of the theoretical significance of this attempt, I refer the reader to J. G. Lennox, "De caelo II 2 and its Debt to the *De incessu animalium*," in A. C. Bowen and C. Wildberg (eds.), *New Perspectives on the De caelo* (Leiden: Brill, 2009), 187–214.

the opening lines of the *Meteorology* have been very important for my argument. They contain the promise of a common study of animals and plants, where Aristotle speaks of studying "animals and plants, both in general and separately." Does Aristotle ever fulfill that promise?

I would like to approach this question by translating the very beginning of the work *On Sense Perception,* which is meant to be a general introduction to the short essays collectively known as *Parva naturalia:*

> Since it was determined earlier about the soul considered in itself and about each of the powers of the soul considered from the point of view of the soul, the next thing is to investigate about *animals and all things that have life (peri tōn zōōn kai tōn zōēn echontōn hapantōn),* as to what are their specific and what are their common activities. So let us assume what has been said about the soul, and let us speak about the rest, and first about the things that are first (*sens.* 1, 436a1–6, my translation)

At least three remarks can be made in connection with this passage. First, Aristotle has already moved away from the study of the soul and is about to engage in a new investigation. His language suggests that this is not simply another investigation; it is another *kind* of investigation. Ensouled beings, and not the soul, are his object of study.[27] Second, Aristotle introduces a distinction between specific and common activities (*praxeis*). This distinction is best understood in light of the theory of scientific explanation outlined in the *Posterior Analytics*. As discussed earlier, that theory requires Aristotle not only to avoid needless repetitions but also to find the salient explanatory features. In some cases, this means giving explanations in common for most, or even all, animals; in other cases, it entails going beyond the case of animals in order to offer explanations that apply to all living beings, including plants. This is why the object of study is described by Aristotle as "animals and all things that have life." Third, Aristotle invokes the methodological principle that we ought to speak "first about the things that are first" (*prōton peri tōn prōtōn*). We have already seen that this principle plays a significant role both in *Parts of Animals* and in *Generation of Animals*. It is not immediately clear how this principle is to be understood in the context of the *Parva naturalia*. It is tempting to read it as dictating a progression in the order of explanation from what is common to most, or even all, animals to what is common to all living beings. But even if this suggestion is not accepted,

---

27 The claim that the investigations conducted in the treatise *On the Soul* and in the *Parva naturalia* are different in kind does not preclude that the *On the Soul* and the *Parva naturalia* are to be read together. Our passage tells us that what is achieved in the investigation of the soul is not simply taken for granted in the *Parva naturalia*; it is explicitly assumed as its explanatory starting point.

the existence of two explanatory levels within the collection of the *Parva naturalia* cannot be denied. More directly, the essays that constitute our collection can be further divided into two groups. While the first group deals with activities pertaining to some, most, or even all animals (*On Sense Perception, On Memory and Recollection, On Sleep, On Dreams, On Divination in Sleep*), the second is concerned with aspects of life that are not restricted to animals (*On Length and Shortness of Life*, and *On Youth and Old Age, Life and Death, and Respiration*). It is in this second group of essays that we find the most obvious examples of a common study of animals and plants.[28]

Aristotle's treatment of longevity is exemplary. At the beginning of *On Length and Shortness of Life*, Aristotle presents his task as twofold: to search for both the causes of why some animals are long-lived and others are short-lived and the causes of length and shortness of life in general.[29] In other words, his primary goal is the explanation of the phenomenon of longevity in animals, but his interest is not restricted to this case, as he promises an explanation of longevity in general. It is telling that the very first question that Aristotle raises, right at the beginning of the treatise, is whether the same cause is responsible for animals and plants being long-lived or short-lived.[30] Aristotle does not offer a direct answer to this question, but the rest of the essay suggests an implicit, positive answer. The cause of longevity, both in animals and plants, has to do with the capacity of the living body to maintain the correct amount and proportion of inner moisture and heat, as aging and death are causally linked by Aristotle to the progressive drying up and becoming cold of the living body. This explains why larger animals and plants generally live longer. The more inner moisture they have, the more resistant they are to drying up. But the quantity of the inner moisture alone is not sufficient to explain why some creatures live longer than others. Otherwise, Aristotle could not explain why some smaller animals, including human beings, live longer than significantly larger animals. Hence, other causal factors have to be taken into account. For Aristotle, the quality of

---

[28] At first sight, it is surprising to find an investigation of respiration along with these essays. Note, however, that this investigation is embedded in the larger project of looking for the causes of life and death. In the opening lines of *On Youth and Old Age*, Aristotle tells us that the plan is to speak about youth and old age, and life and death – and *at the same time* (*hama*) about the causes of respiration, since for some animals life depends on respiration (*iuv.* 1, 467b10–13). Evidently, the discussion of respiration falls within the discussion of youth and old age, life, and death.

[29] *Long. uit.* 1, 464b19–21.

[30] *Long. uit.* 1, 464b21–26.

the inner moisture is equally important. In particular, the moisture has to be hot in order to be resistant to the action of the cold.

Among the facts that get explained in *On Length and Shortness of Life* is why the longest lifespans are found amongst plants.[31] Giving an explanation of this fact entails giving an account of the distinctive manner in which plants preserve their life. The quality of the inner moisture in plants is the first causal factor that Aristotle mentions. This moisture is less watery, and hence it is less liable to the action of the cold. It has also an oiliness and viscosity that makes it more resistant to drying up.[32] Among plants, trees are typically the longest-lived. In connection with their longevity, Aristotle mentions a second causal factor: they have capacity to regenerate themselves: as one part goes out of existence, another one comes into existence.[33]

VI

If the interpretation developed so far is on the right track, there is not much that Aristotle is able, or willing, to say in general about animals and plants. This is a surprising result, especially if we bear in mind that the theory of scientific explanation advanced in the *Posterior Analytics* provides Aristotle with a strong theoretical motivation to develop a unified study of living beings. It confirms that Aristotle envisioned distinct and relatively self-contained studies of animals and plants.

In addition to the *On Length and Shortness of Life*, the essay known as *On Youth and Old Age, Life and Death, and Respiration* can certainly be regarded as contributing to a study that deals with certain aspects of animal and plant life in common. From the opening lines of this treatise, it is clear that animals are the primary but not the exclusive focus of the project. It is by focusing on the case of animals that Aristotle is able to establish that not just animal life but life in general coincides with the preservation of internal heat. In fully developed living beings, Aristotle says, death is due to a loss (*ekleipsis*) of internal heat.[34] Other definitions that Aristotle puts forward are those of generation, life, youth, and old age. While generation (*genesis*) is the first participation in the heat of the vegetative soul, life (*zōē*) is its maintenance. Youth (*neotēs*) is

---

31 *Long. uit.* 4, 466a9.
32 *Long. uit.* 6, 467a6–9.
33 *Long. uit.* 6, 467a10–12.
34 *Iuv.* 23, 478b31–32. The qualification "fully developed" is introduced to exclude the case of eggs and plant seeds (478b30–31).

Religion, and Rhetoric (New Brunswick: Transaction Publishers, 1988), 100–33 [Reprinted as Chapter 14 in A. Gotthelf, *Teleology, First Principles, and Scientific Method in Aristotle's Biology* (Oxford: Oxford University Press, 2012), 307–42].

Hossfeld, P. *Alberti Magni Opera omnia: Physica I–IV* (Münster i. W.: Aschendorff, 1987).

Kilwardby, R. *De ortu scientiarum*, ed. A. G. Judy (London: The British Academy, 1976).

Labowsky, L. "Aristotle's *De plantis* and Bessarion. Bessarion Studies II," *Medieval and Renaissance Studies* 5 (1961): 132–53.

Lee, H. D. P., trans. *Aristotle, Meteorologica* (Cambridge, Mass.: Harvard University Press, 1952).

Lennox, J. G. "*De caelo* II 2 and its Debt to the *De incessu animalium*," in A. C. Bowen and C. Wildberg (eds.), *New Perspectives on the De caelo* (Leiden: Brill, 2009), 187–214.

Lennox, J. G. "Divide and Explain: The *Posterior Analytics* in Practice," in A. Gotthelf and J. G. Lennox (eds.), *Philosophical Issues in Aristotle's Biology* (Cambridge: Cambridge University Press, 1987), 90–119 [Reprinted in J. G. Lennox, *Aristotle's Philosophy of Biology* (Cambridge: Cambridge University Press, 2001), 7–38].

Orlandi, G. *Aldo Manuzio editore. Dediche, prefazioni, note ai testi*, 2 vols. (Milan: Il Polifilo, 1976).

Rashed, M. "Aristote à Rome au IIe siècle: Galien, *De indolentia* §§ 15–18," *Elenchos* 32 (2011): 55–77.

Repici, L. *Uomini capovolti: le piante nel pensiero greco* (Bari: Laterza, 2000).

Scaliger, J. C. *In libros de plantis Aristotelis inscriptos commentarii* (Paris: Michel de Vascosan, 1556).

Schmitt, C. B. "Aristotelian Textual Studies at Padua: The Case of Francesco Cavalli," in A. Poppi (ed.), *Scienza e filosofia all'università di Padova nel quattrocento* (Trieste: Edizioni Lint, 1983), 287–313 [Reprinted in C. B. Schmitt, *The Aristotelian Tradition and Renaissance Universities* (London: Variorum Reprints, 1984), Chapter 13].

van Riet, S., ed. *Avicenna Latinus, Liber de anima seu sextus de naturalibus. Édition critique de la traduction latine médiévale. Introduction sur la doctrine psychologique d'Avicenne par G. Verbeke*, 2 vols. (Leuven: Peeters, 1968–1972).

Wimmer, F. *Phytologiae Aristotelicae Fragmenta* (Breslau: Grass, Barth, and Comp., 1838).

Zabarella, J. *De rebus naturalibus libri XXX quibus quaestiones, quae ab Aristotelis interpretibus hodie tractari solent, accurate discutiuntur* (Frankfurt a. Main: L. Zetzner, 1607) [Facsimile reprint, Frankfurt a. Main: Minerva, 1966].

Zimara, M. A., et al., eds. *Aristotelis Stagiritae omnia quae extant opera ... Averrois Cordubensis in ea opera omnes qui ad nos pervenere commentarii ...*, 12 vols. (Venice: Giunta, 1562) [Facsimile reprint, Frankfurt a. Main: Minerva, 1962].

the existence of two explanatory levels within the collection of the *Parva naturalia* cannot be denied. More directly, the essays that constitute our collection can be further divided into two groups. While the first group deals with activities pertaining to some, most, or even all animals (*On Sense Perception, On Memory and Recollection, On Sleep, On Dreams, On Divination in Sleep*), the second is concerned with aspects of life that are not restricted to animals (*On Length and Shortness of Life,* and *On Youth and Old Age, Life and Death, and Respiration*). It is in this second group of essays that we find the most obvious examples of a common study of animals and plants.[28]

Aristotle's treatment of longevity is exemplary. At the beginning of *On Length and Shortness of Life,* Aristotle presents his task as twofold: to search for both the causes of why some animals are long-lived and others are short-lived and the causes of length and shortness of life in general.[29] In other words, his primary goal is the explanation of the phenomenon of longevity in animals, but his interest is not restricted to this case, as he promises an explanation of longevity in general. It is telling that the very first question that Aristotle raises, right at the beginning of the treatise, is whether the same cause is responsible for animals and plants being long-lived or short-lived.[30] Aristotle does not offer a direct answer to this question, but the rest of the essay suggests an implicit, positive answer. The cause of longevity, both in animals and plants, has to do with the capacity of the living body to maintain the correct amount and proportion of inner moisture and heat, as aging and death are causally linked by Aristotle to the progressive drying up and becoming cold of the living body. This explains why larger animals and plants generally live longer. The more inner moisture they have, the more resistant they are to drying up. But the quantity of the inner moisture alone is not sufficient to explain why some creatures live longer than others. Otherwise, Aristotle could not explain why some smaller animals, including human beings, live longer than significantly larger animals. Hence, other causal factors have to be taken into account. For Aristotle, the quality of

---

[28] At first sight, it is surprising to find an investigation of respiration along with these essays. Note, however, that this investigation is embedded in the larger project of looking for the causes of life and death. In the opening lines of *On Youth and Old Age,* Aristotle tells us that the plan is to speak about youth and old age, and life and death – and *at the same time (hama)* about the causes of respiration, since for some animals life depends on respiration (*Iuv.* 1, 467b10–13). Evidently, the discussion of respiration falls within the discussion of youth and old age, life, and death.
[29] *Long. uit.* 1, 464b19–21.
[30] *Long. uit.* 1, 464b21–26.

the inner moisture is equally important. In particular, the moisture has to be hot in order to be resistant to the action of the cold.

Among the facts that get explained in *On Length and Shortness of Life* is why the longest lifespans are found amongst plants.[31] Giving an explanation of this fact entails giving an account of the distinctive manner in which plants preserve their life. The quality of the inner moisture in plants is the first causal factor that Aristotle mentions. This moisture is less watery, and hence it is less liable to the action of the cold. It has also an oiliness and viscosity that makes it more resistant to drying up.[32] Among plants, trees are typically the longest-lived. In connection with their longevity, Aristotle mentions a second causal factor: they have capacity to regenerate themselves: as one part goes out of existence, another one comes into existence.[33]

VI

If the interpretation developed so far is on the right track, there is not much that Aristotle is able, or willing, to say in general about animals and plants. This is a surprising result, especially if we bear in mind that the theory of scientific explanation advanced in the *Posterior Analytics* provides Aristotle with a strong theoretical motivation to develop a unified study of living beings. It confirms that Aristotle envisioned distinct and relatively self-contained studies of animals and plants.

In addition to the *On Length and Shortness of Life*, the essay known as *On Youth and Old Age, Life and Death, and Respiration* can certainly be regarded as contributing to a study that deals with certain aspects of animal and plant life in common. From the opening lines of this treatise, it is clear that animals are the primary but not the exclusive focus of the project. It is by focusing on the case of animals that Aristotle is able to establish that not just animal life but life in general coincides with the preservation of internal heat. In fully developed living beings, Aristotle says, death is due to a loss (*ekleipsis*) of internal heat.[34] Other definitions that Aristotle puts forward are those of generation, life, youth, and old age. While generation (*genesis*) is the first participation in the heat of the vegetative soul, life (*zōē*) is its maintenance. Youth (*neotēs*) is

---

[31] *Long. uit.* 4, 466a9.
[32] *Long. uit.* 6, 467a6–9.
[33] *Long. uit.* 6, 467a10–12.
[34] *Iuv.* 23, 478b31–32. The qualification "fully developed" is introduced to exclude the case of eggs and plant seeds (478b30–31).

the growth of the organ dedicated to the control of the bodily temperature, whereas age (*gēras*) is its decay.[35] Note that, as soon as these definitions are in place, Aristotle returns to animals, with a focus on the function of the heart and the lungs. The impression is that animals remain his primary concern, and that he ventures into a study of life only when he is able to find a common explanation for animals and plants. Put differently, Aristotle is prepared to say what he can about what is common to animals and plant *in the course of his study of animals*. This impression is reinforced by how our investigation is announced at the end of *On Length and Shortness of Life*. In looking forward to an investigation on the topic of youth and old age, life and death, Aristotle tells us that as soon as this investigation is in place, his *study of animals* (*methodos tōn zōōn*) will have reached its goal.[36] If the cross-reference at the end of *On Length and Shortness of Life* is taken seriously, then it is possible to consider not only *On Youth and Old Age, Life and Death, and Respiration* but also *On Length and Shortness of Life* as contributing, directly and immediately, to the study of animals.

VII

In these pages, I have reflected on an approach to the study of life via a study of animals. This approach is based on a reading of Aristotle defended by Cavalli, who opposes Avicenna, Averroes, and the majority of the Latin interpreters.[37] The subsequent debate on the relative order of the study of animals and plants cannot be discussed here. I am content to note that the order recommended by Cavalli was adopted by Jacopo Zabarella. His essay *On the Constitution and Parts of Natural Science* (*De constitutione scientiae naturalis liber*) is a lucid presentation of the contents and explanatory structure of the whole of Aristotle's natural science. It is also the culmination of a long and illustrious exegetical tradition. Zabarella accepts several key aspects of the reading promoted by Averroes, including the emphasis that the latter placed on the role that the theory of scientific explanation outlined in the *Analytics* plays in Aristotle's actual study of na-

---

[35] *Iuv.* 24, 479a29–32.
[36] *Long. uit.* 6, 467b6–9.
[37] In the thirteenth century, in addition to Albert the Great, a notable defender of the view that plants ought to be studied before animals was Robert Kilwardby. He too was influenced by Avicenna and his view that the study of living bodies was a three-part investigation of (1) the soul, (2) plants, and (3) animals. Cf. Robert Kilwardby, *De ortu scientiarum*, edited by A. G. Judy (London: The British Academy, 1976), §§ 48–52.

ture. In dealing with the study of life, however, Zabarella distances himself from what he calls the mistake of Averroes, who thought that the books on plants ought to be studied before the books on animals on the grounds that plants are comparatively simple living beings.[38] For Zabarella, in the correct order of explanation, animals come before plants. As Cavalli does, Zabarella takes the *Parva naturalia* to contribute to the study of animals even when they offer explanations in common for animals and plants.[39] The obvious question for him is then why Aristotle opted to offer the common explanation of animals and plants in the context of his study of animals. The reason that motivated Aristotle to follow this explanatory strategy, Zabarella says, is that those aspects that are common to animals and plants can be more easily studied in animals; it is only when the study of animals is firmly in place that Aristotle can move to the study of what is comparatively simpler but also more difficult to study.[40]

Bibliography

Botfield, B. *Prefaces to the First Editions of Greek and Roman Classics* (London: H. G. Bohn, 1861).
Cavalli, F. *De ordine et numero partium ac librorum physicae doctrinae Aristotelis* (Venice: Matteo Capcasa, ca. 1490–1495).
Drossaart Lulofs, H. J., and E. L. J. Poortman, eds. *Nicolaus Damascenus, De plantis. Five Translations* (Amsterdam: North-Holland Publishing, 1989).
Falcon, A. *Aristotle and the Science of Nature: Unity without Uniformity* (Cambridge: Cambridge University Press, 2005).
Gotthelf, A. "*Historiai* I: *plantarum et animalium*," in W. W. Fortenbaugh and W. R. Sharples (eds.), *Theophrastean Studies*, vol. 3: *Natural Science, Physics, Metaphysics, Ethics,*

---

38 J. Zabarella, *De naturalis scientiae constitutione liber*, in J. Zabarella, *De rebus naturalibus libri XXX* (Frankfurt a. Main: L. Zetzner, 1607; facsimile reprint, Frankfurt a. Main: Minerva, 1966), 90 C-D: "from here we can gather the mistake of Averroes, who thought that the books on plants are to be placed before the books on animals because plants have a simpler nature than animals."
39 Here is how Zabarella reads the opening lines of the *On Sense Perception* translated and discussed in section V: "Aristotle says that that book [= *On Sense Perception*], as well as the others that follow it [= the rest of the *Parva naturalia*], follow immediately after the books on the soul [= *On the Soul*]; Aristotle says, explicitly, that he will speak not only of animals but also of all the living beings and, moreover, that he will deal with nothing other than the activities that are proper to animals and also common to all living beings" (*De naturalis scientiae constitutione*, 96 B).
40 Zabarella, *De naturalis scientiae constitutione*, 90 B: "The reason which motivated Aristotle was that those things that belong in common to animals and plants are clearer (*distinctiora*), and therefore easier to know (*faciliora cognitu*) in animals, whereas they are more rudimentary (*rudiora*) and more obscure (*obscuriora*) in plants."

Religion, and Rhetoric (New Brunswick: Transaction Publishers, 1988), 100–33 [Reprinted as Chapter 14 in A. Gotthelf, *Teleology, First Principles, and Scientific Method in Aristotle's Biology* (Oxford: Oxford University Press, 2012), 307–42].

Hossfeld, P. *Alberti Magni Opera omnia: Physica I–IV* (Münster i. W.: Aschendorff, 1987).

Kilwardby, R. *De ortu scientiarum*, ed. A. G. Judy (London: The British Academy, 1976).

Labowsky, L. "Aristotle's *De plantis* and Bessarion. Bessarion Studies II," *Medieval and Renaissance Studies* 5 (1961): 132–53.

Lee, H. D. P., trans. *Aristotle, Meteorologica* (Cambridge, Mass.: Harvard University Press, 1952).

Lennox, J. G. "*De caelo* II 2 and its Debt to the *De incessu animalium*," in A. C. Bowen and C. Wildberg (eds.), *New Perspectives on the De caelo* (Leiden: Brill, 2009), 187–214.

Lennox, J. G. "Divide and Explain: The *Posterior Analytics* in Practice," in A. Gotthelf and J. G. Lennox (eds.), *Philosophical Issues in Aristotle's Biology* (Cambridge: Cambridge University Press, 1987), 90–119 [Reprinted in J. G. Lennox, *Aristotle's Philosophy of Biology* (Cambridge: Cambridge University Press, 2001), 7–38].

Orlandi, G. *Aldo Manuzio editore. Dediche, prefazioni, note ai testi*, 2 vols. (Milan: Il Polifilo, 1976).

Rashed, M. "Aristote à Rome au IIe siècle: Galien, *De indolentia* §§ 15–18," *Elenchos* 32 (2011): 55–77.

Repici, L. *Uomini capovolti: le piante nel pensiero greco* (Bari: Laterza, 2000).

Scaliger, J. C. *In libros de plantis Aristotelis inscriptos commentarii* (Paris: Michel de Vascosan, 1556).

Schmitt, C. B. "Aristotelian Textual Studies at Padua: The Case of Francesco Cavalli," in A. Poppi (ed.), *Scienza e filosofia all'università di Padova nel quattrocento* (Trieste: Edizioni Lint, 1983), 287–313 [Reprinted in C. B. Schmitt, *The Aristotelian Tradition and Renaissance Universities* (London: Variorum Reprints, 1984), Chapter 13].

van Riet, S., ed. *Avicenna Latinus, Liber de anima seu sextus de naturalibus. Édition critique de la traduction latine médiévale. Introduction sur la doctrine psychologique d'Avicenne par G. Verbeke*, 2 vols. (Leuven: Peeters, 1968–1972).

Wimmer, F. *Phytologiae Aristotelicae Fragmenta* (Breslau: Grass, Barth, and Comp., 1838).

Zabarella, J. *De rebus naturalibus libri XXX quibus quaestiones, quae ab Aristotelis interpretibus hodie tractari solent, accurate discutiuntur* (Frankfurt a. Main: L. Zetzner, 1607) [Facsimile reprint, Frankfurt a. Main: Minerva, 1966].

Zimara, M. A., et al., eds. *Aristotelis Stagiritae omnia quae extant opera ... Averrois Cordubensis in ea opera omnes qui ad nos pervenere commentarii ...*, 12 vols. (Venice: Giunta, 1562) [Facsimile reprint, Frankfurt a. Main: Minerva, 1962].

Christopher A. Faraone
A Case of Cultural (Mis)translation?: Egyptian Eyes on Two Greek Amulets for Ophthalmia

Abstract: Two puzzling Greek amulets seem to combine the image of a lizard – a common appearance on Greek ophthalmia amulets – and the *udjat*-eye, or "eye of Horus," one of the most popular Egyptian amulets from Pharaonic to Roman times. I argue that the presence of the *udjat*-eyes on these gems involves a curious kind of cultural translation of this protective Egyptian symbol, because on these gems it seems to act not as metonymy for a powerful god, but rather as an effigy of the eyes of the human patient. Or to put it another way: although the Egyptian *udjat*-eye normally signals the protective presence of the powerful god Horus, here it seems to have been co-opted as a convenient image of any human eye. The argument proceeds in two parts: in the first, I discuss a series of well-known rituals that use the regenerating eyes of a green lizard as models for the successful healing of human eye disease. In the second part, I turn to some unlikely parallels: recipes used to identify an unknown thief by striking or irritating the representation of an eye of Horus, with the expectation that the thief's eye will likewise be damaged or irritated. Both rituals, I shall argue, imagine the eyes of Horus not in their usual native Egyptian role as a symbol of protection but more simply as representations of human eyes in persuasively analogical rituals that aim in the case of the amulets to heal a damaged eye but, in the case of the thief-detection spells, to damage a healthy one.

Two puzzling Greek amulets seem to combine the image of a lizard – a common appearance on Greek ophthalmia amulets – and the *udjat*-eye, or "eye of Horus," one of the most popular Egyptian amulets from Pharaonic to Roman times. The first is a gem from Florence that has the usual lizard and a crescent moon on the

---

I would like to thank Véronique Dasen, Jacco Dieleman, Patricia Gaillard-Seux, François Gaudard and Sofia Torallas Tovar for their comments on and corrections to earlier drafts of this paper and express my gratitude to the honoree of this volume, who has known all along that in the ancient Mediterranean it is not always easy to separate the worlds of magic and medicine or those of Egypt and Greece.

obverse, but on the other side we find a pair of angel names, Ouriēl-Souriēl, encircling an odd triangular shape (fig. 1):[1]

Fig. 1 Attilio Mastrocinque, ed., *Sylloge Gemmarum Gnosticarum*, Bollettino di Numismatica, Monografia 8.2.1 and 2, 2 vols. (Rome: Istituto poligrafico e Zecca dello Stato, 2003–2008), Fig. 101

The engraving on the reverse is poorly done, but clear enough to make out a right triangle with some kind of short protuberance jutting out from the hypotenuse.[2] I suggest, in fact, that the central triangle here is a crudely executed eye of Horus, because we find a similar assemblage on a gem published many years ago by Goodenough (fig. 2),[3] where we see the lizard flanked by two of these Egyptian eyes and on the reverse, the same two angel names: Ouriēl-Souriēl.

---

1 Attilio Mastrocinque, ed., *Sylloge Gemmarum Gnosticarum*, Bollettino di Numismatica, Monografia 8.2.1 and 2, 2 vols. (Rome: Istituto poligrafico e Zecca dello Stato, 2003–2008), Fig. 101; this photograph is used with his permission.
2 Neither Mastrocinque, *Sylloge, ad loc.*, nor Simone Michel, *Die magischen Gemmen: Zu Bildern und Zauberformeln auf geschnittenen Steinen der Antike und Neuzeit*, Studien aus dem Warburg-Haus Band 7 (Berlin: Akademie-Verlag, 2004), 157–58 and 264, discuss the triangle.
3 Erwin Goodenough, *Jewish Symbols in the Greco-Roman Period*, 3 vols. (New York: Pantheon Books, 1953–1968), no. 1064 ("obsidian"). Goodenough purchased the gem in Athens (see 2:238), but its present whereabouts are unknown.

Fig. 2 Erwin Goodenough, *Jewish Symbols in the Greco-Roman Period*, 3 vols. (New York: Pantheon Books, 1953–1968), no. 1064

I suggest that the presence of the *udjat*-eyes here and on the reverse of the Florentine gem involves some kind of cultural translation of this protective Egyptian symbol, because on these gems it seems to act not as metonymy for a powerful god, but rather as an effigy of the eyes of the human patient. Or to put it another way: although the Egyptian *udjat*-eye normally signals by metonymy the protective presence of the powerful god Horus,[4] here it seems to have been co-opted as a convenient image of any human eye. My argument shall proceed in two parts: in the first, I discuss a series of well-known rituals that use the regenerating eyes of a green lizard as models for the successful healing of human eye disease. In the second part, I turn to some unlikely parallels: recipes used to identify an unknown thief by striking or irritating the representation of an eye of Horus, with the expectation that the thief's eye will likewise be damaged or irritated. Both rituals, I shall argue, imagine the eyes of Horus not in their usual native Egyptian role as a symbol of protection but more simply as representations of human eyes in sympathetically magical or persuasively analogical rituals that aim in the case of the amulets to heal a damaged eye but, in the case of the thief-detection spells, to damage a healthy one.[5] I shall nevertheless close

---

[4] Carol Andrews, *Amulets of Ancient Egypt* (London: British Museum Press, 1994), 43, calls it "the best known of all protective amulets" and explains how the eye belonged to Horus the Elder, the falcon-headed Horus, who was brother to Isis and Osiris (i.e., not their son). The elder Horus was a creator god, whose right eye was the sun and the left the moon. Over time, however, the eye also gets connected with Horus the Younger, the son of Isis and Osiris, and for the purpose of this essay, with its focus on Roman imperial and late-antique texts and objects, I will not distinguish between the two.

[5] Stanley J. Tambiah, "Form and Meaning of Magical Acts: A Point of View," in Robin Horton and Ruth Finnegan (eds.), *Modes of Thought* (London: Faber, 1973), 199–229, dismisses the

this essay with a suggestion that the Egyptian myth of the origins of the eye of Horus may provide a good rationale for why the figure was added to the Greek eye-healing amulets.

I The Eye of Horus on Two Greek Amulets for Eye-Disease

Regarding the lizard gems designed to cure diseases of the eye,[6] we are especially fortunate to have an unusually full set of literary testimonia, including recipes from handbooks and the reports of natural historians. Pliny the Elder is our earliest witness:

> alii terram substernunt lacertae viridi excaecatae et una in vitreo vase anulos includunt e ferro solido vel auro. cum recepisse visum lacertam apparuit per vitrum, emissa ea anulis contra lippitudinem utuntur. (*nat.* 29.130)
>
> Others put earth under a green lizard after blinding it, and shut it in a glass vessel with rings of solid iron or gold. When they can see through the glass that the lizard has recovered its sight, they let it out and use the rings for ophthalmia. (trans. Jones)

This ritual aims to transfer to the amulet the lizard's natural ability to regenerate its eyes, by placing the amulet in close contact with the animal as it heals. Here in the earliest extant recipe, there is no image or text, and the amulet consists solely of gold or iron rings. A little earlier in the *Natural History* (29.129), Pliny reports that if small black pebbles – probably obsidian and perhaps a pair,

---

common view that "sympathetic magic" is based on poor observation of empirical analogies. He distinguishes instead between the operation of "empirical analogies" (used in modern scientific discourse to *predict* future actions) and "persuasive analogies" (used in rituals in traditional societies to *encourage* future action). Such rituals do not betray inferior observation skills, but rather they reveal a profound belief in the extraordinary power of language. Cf. G. E. R. Lloyd, *Magic, Reason and Experience* (Cambridge: Cambridge University Press, 1979), 2–3 and 7.

6 The fundamental discussions are Wilhelm Drexler, "Alte Beschwörungsformeln," *Philologus* 58 (1899): 594–616, at 610–16; Armand L. Delatte, "Études sur la magie grecque IV: amulettes inédites des Musées d'Athènes," *Le Musée Belge* 18 (1914): 21–96, at 64–67; Campbell Bonner, *Studies in Magical Amulets Chiefly Graeco-Egyptian*, University of Michigan Studies, Humanistic Series 4 (Ann Arbor: University of Michigan Press, 1950), 69–71; and Michel, *Die magischen Gemmen*, 157–58. For a recent collection and renewed discussion of the ancient sources, see Patricia Gaillard-Seux, "Les maladies des yeux et le lézard vert," in Guy Sabbah (ed.), *Nommer la maladie: Recherches sur le lexique gréco-latine de la pathologie* (Saint-Étienne: Publications de l'Université de Saint-Étienne, 1997), 93–105.

one for each eye – are left with a lizard recovering its sight, they relieve pain in the eyes.⁷ According to the earliest evidence, then, three media – gold, iron, and black pebbles – can absorb the regenerative power of the lizard and then be used as amulets.

Aelian testifies, as an eyewitness a few centuries later, to a more complicated process that produces a more elaborate amulet: he watched someone put a green lizard in a new earthenware pot along with an unnamed herb and an iron ring with a lignite or jet gemstone carved with the image of a lizard. He tells us that as a result of this procedure the ring was "good for the eyes."⁸ In this second description, then, we find that an image of the lizard inscribed on a black stone has been added to the iron ring, but there is still no text.⁹

The most detailed account, however, is the latest: a recipe from a Byzantine handbook for amulets:¹⁰

> If you, after catching a solar lizard (ἡλιακὴ σαύρα) with your right hand, fashion two nails, either of gold or silver, and blind the lizard, throw them into a pot that has virgin earth in it. Let it be for nine days and after that upon opening the pot you will find that the lizard is seeing. Let this lizard go free alive and make the nails into rings and wear them, on the right hand the nail that struck out the lizard's right eye and on the left hand the nail that struck out the lizard's left eye, after enclosing in these rings a jasper stone with a carved lizard stretched out on its belly and beneath the following name χουθεσουλε¹¹

---

7 *Cyranides* 1.10.18–22 (Kaimakis) identifies the κιναίδιος λίθος as a black stone called ὀψιανός, which is probably obsidian; see LSJ s.v. Aelian (see below) says to put a black gemstone (lignite or jet) in with the lizard, and the Goodenough gem is said to be obsidian.

8 *Nat. anim.* 5.47: "In this matter I shall have no need of any witness from antiquity but shall narrate what I myself perceived (αὐτὸς ἔγνων). A man captured a lizard of the excessively green and unusually large species and with a point made of bronze pierced [sc. its eyes] and made it blind. And after boring some fine holes in a newly fashioned earthenware vessel so as to admit the air, but small enough to prevent the creature from escaping, he heaped very moist earth into it and put the lizard inside together with a certain herb, of which he did not divulge the name, and an iron ring with a bezel of lignite (λίθος Γαγάτης) engraved with a figure of the lizard. … And when he [sc. after nine days] … opened the vessel, I myself saw (ἔγωγε εἶδον) the lizard having its sight and its eyes seeing perfectly well, eyes which until then had been blinded. And we released the lizard on the spot where it had been captured and the man who had done these things asserted that that ring of his was good for the eyes" (trans. Scholfield).

9 Lignite (λίθος Γαγάτης) seems to be a form of petrified coal or jet; Goodenough, *Jewish Symbols*, 1: no. 1064, described his gem as "obsidian" (see n. 3 above).

10 The *Cyranides* is actually a Byzantine compilation, but much of it can probably be traced back to early Roman imperial times; see Gaillard-Seux, "Les maladies," 95 and D. M. Bain, "Koiraniden," in *Reallexikon für Antike und Christentum* 21 (2006): cols. 224–32.

11 The manuscripts are in disagreement here about the inscriptions and I translate the text of Kaimakis: ἔχοντα γλυφὴν σαύραν ἐπὶ κοιλὰν ἡπλωμένην καὶ ὑποκάτωθεν τὸ ὄνομα τοῦτο "χουθεσουλε." The variations are, however, instructive: four of the twelve MSS (AGHF) finish the

and wear them. For (i.e., in this way) you will be unharmed in your eyes for all your years and you will heal ophthalmia sufferers by fastening the ring to them. (*Cyranides* 2.14.22–31 [Kaimakis])

According to this recipe, then, we make a ring for each hand of the patient that is set with a jasper carved with the image of a lizard, a magical word, and one of the blinded lizard eyes: the right eye for the ring worn on the right hand and the left eye for the one worn on the left.

This passage gives us, in fact, a good description of a common type of amulet of Roman imperial date, variations of which you can see in the photo in fig. 1 and the drawing in fig. 3 of the obverse of a gem in the British Museum.[12] These gems usually depict a lizard from above with his head pointed into the concave side of a crescent moon; on about half of the extant gems it is surrounded by the Greek letters pi, eta, rho, and alpha, one near each foot, as we see in fig. 3.

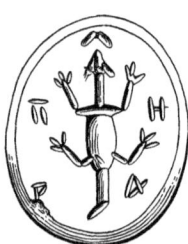

Fig. 3 Lizard gem

The *Cyranides* recipe does not mention the crescent moon, but in another late version of this recipe, the early fifth-century A.D. medical writer Marcellus Empiricus tells us that "the lizard must be caught and the remedy prepared in the old moon," a stipulation that suggests that the crescent moon on the stones

---

sentence with καὶ γραφόμενον πεῖραν ὑποκάτωθεν and a fifth (R) with καὶ γεγραμμένοις πεῖραν τοὔνομα τοῦτο ὑποκάτωθεν χουθουσελε. The word πεῖρα, "attempt," in the accusative case makes little sense here and Kaimakis ignores it. But more than a century ago Fernand de Mély and C. E. Ruelle, *Les lapidaires de l'antiquité et du moyen âge*, 3 vols. (Paris: E. Leroux, 1896– 1902), 2:61, printed a different text that has remained influential: καὶ γραφόμενον ΠΕΙΡΑΝ, ὑποκάτωθεν τὸ ὄνομα τοῦτο ΧΟΥΘΕΣΟΥΛΕ, "and being inscribed with PEIRAN [and] underneath CHOUTHESOULE." Drexler and others – see discussion in Bonner, *Studies*, 69 – came to think (probably rightly) that PEIRAN was a mistake for another word: πηρά, "maimed" or in this case "blinded," which is often (see below) inscribed around the image of the lizard, as we see on the gem in fig. 3.

12 Simone Michel, *Die magischen Gemmen im Britischen Museum* (London: British Museum Press, 2001), no. 424 (a "moss-agate").

refers to the time the lizard was caught.¹³ Scholars have suggested, moreover, that the standard inscriptions on these lizard gems may be interpreted as Greek words that refer to the different stages of the ritual: the four letters surrounding the lizard seem to add up to the word πηρά, an adjective meaning "maimed" or "blinded" in the feminine case, which probably describes the lizard (a feminine noun in Greek: σαύρα) and thus refers to the initial blinding of the lizard in the recipes quoted earlier.¹⁴

Many of these gems have two words inscribed on their reverse: κανθε σουλε, which have been interpreted as the command κανθέ, σ' οὖλε⟨ι⟩ "eye, heal yourself!" or "it (i.e., the lizard) heals you, eye!"¹⁵ This interpretation is not entirely satisfactory, because the verb οὖλειν usually does not (*pace* Festugière) mean "to heal" actively, but rather "to be healthy"; it is usually glossed by the verbs ὑγιαίνειν and χαίρειν and like them is used in the imperative to mean "fare well" or "good bye."¹⁶ Therefore, the preferred translation should be "Be healthy, eye!" The word κανθός, however, remains a problem, because in prose it usually means "corner of the eye" and only in poetry does it refer to the eye as a whole.¹⁷ But since 11 of the 22 examples in Michel's catalogue have κανθε σουλε on the reverse and (with two exceptions) πηρά on the obverse, this

---

13 *Med.* 8.49: *lacertam viridem excaecatam acu cuprea in vas vitreum mittes cum anulis aureis, argenteis, ferreis et electrinis ... deinde vas gypsabis diligenter, ... et post quintum vel septimum diem aperies, lacertam sanis luminibus invenies, quam vivam dimittes, anulis contra lippitudinem ita uteris, ut non solum digito gestentur, sed etiam oculis crebrius adplicenter ... observandum etiam, ut luna vetere ... capiatur lacerta atque ita remedium fiat....* ("A green lizard, blinded by a copper needle, you will place in a glass vase with rings of gold, silver, iron and electrum. ... Then carefully plaster up the vessel ... and after the fifth or seventh day you will open it. You will find the lizard with healthy eyes, which you will release living, so that you may use the rings against the inflammation of the eyes, not only when they are worn on the finger, but also when they are applied often to the eyes. ... It must also be observed, that the lizard be captured and the remedy be made during the old moon.") See Bonner, *Studies*, 71 for discussion. Marcellus allows for a greater variety of metals for the rings (gold, silver, iron, electrum, and copper) and that to be cured one could wear the rings on one's fingers or look through them.
14 See Bonner, *Studies*, 71 for earlier bibliography. The *Cyranides* tells us to inscribe πειρα as a single word on the gem, which is either an error or a variation that does not appear on the extant stones. For πηρά meaning "blind" see LSJ Suppl. s.v.
15 See André-Jean Festugière, "Amulettes magiques à propos d'un ouvrage récent," *Classical Philology* 46 (1951): 81–92, at 83; Michel, *Die magischen Gemmen*, 264; and Gaillard-Seux, "Les maladies," 94, for full discussion and earlier bibliography. The magical name is rendered differently as χουθεσουλε in the *Cyranides* recipe (see above).
16 See LSJ s.v.
17 See LSJ s.v.

would seem to be the standard version.¹⁸ The *Cyranides* recipe suggests, moreover, two details that we would have never suspected from the extant gemstones: that the ring itself was fashioned from the gold or silver nail used to blind the lizard, and that the eyes of the animal themselves had been placed in the setting beneath the gemstone.

How, then, are we to understand the magical principle behind these rituals? Our earliest witness, Pliny, suggested that only the media (gold and iron rings or black pebbles) were crucial for absorbing and transferring the regenerative power of the lizard, but gradually an image of the lizard becomes important, albeit on a black stone (Aelian), which is then replaced with the mottled jasper gems that are extant in our museums – dappled green presumably being a naturalistic color in which to depict the green lizard.¹⁹ Simpler contemporary recipes suggest, in fact, that over time the inscribed gems may have replaced the messy and transient organic material of the lizard eyes themselves.²⁰ A recipe from a late-antique papyrus handbook, for example, reads: "gouge out the righ[t eye] of a lizard [and th]row it in a go[at skin] and attach..."²¹ The precise goal for this amulet is lost or unstated, but Marcellus of Bordeaux (8.50) tells us that, if we wear the eyes of a green lizard sealed in a *bulla* or gold amulet-case (*lupinus aureus*), our eyes will not hurt, as long as we release the lizard alive in the same spot we found him. The *Cyranides* also preserves an abbreviated version of this recipe (2.14.8–9 [Kaimakis]): "The eyes of it (i.e., the lizard), if

---

18 Michel, *Die magischen Gemmen*, 264. But there are other patterns: four gems of similar color have versions of the magical word θυλωβρις – see Christopher A. Faraone, "Scribal Mistakes, Handbook Abbreviations, and Other Peculiarities on Some Ancient Greek Amulets," *MHNH: Revista internacional de investigación sobre magia y astrología antiguas* 13 (2013): 139–56 – but none of these gems have πηρά and two lack the crescent moon; seven have Jewish angel names, usually Souriēl or Ouriēl (as we saw in figs. 1 and 2), but again all but one lack πηρά; and three are on yellow jaspers with scorpions on their reverse side and lack the letters πηρά and the crescent moon. The yellow color of the last three stones suggests that they may have been used against scorpions; see Christopher A. Faraone, "Text, Image, and Medium: The Evolution of Greco-Roman Magical Gemstones," in Chris Entwistle and Noël Adams (eds.), *Gems of Heaven: Recent Research on Engraved Gemstones in Late Antiquity*, British Museum Research Papers no. 179 (London: British Museum, 2011), 50–61, at 55.
19 The *Cyranides* passage, which tells us how to produce an amulet that comes closest to the extant lizard gems, tells us to use jasper, but unfortunately does not stipulate the color.
20 Ann E. Hanson, "Uterine Amulets and Greek Uterine Medicine," *Medicina nei secoli* 7 (1995): 281–99, has suggested a similar move in the case of gynecological amulets from the organic material (wool smeared with red blood?) recommended by the Hippocratics to the red jasper stones recommended by Dioscorides.
21 Robert W. Daniel and Franco Maltomini, eds., *Supplementum Magicum*, Papyrologica Coloniensia 16.1 and 2, 2 vols. (Opladen: Westdeutscher Verlag, 1990–1992), no. 78 col. ii.3–6.

removed from it while it is still living, ... heal all ophthalmia. The lizard is released alive." These recipes, then, are simple versions of the more elaborate recipes known to Pliny, Aelian, and the *Cyranides:* there is no mention of nails or gems to which the healing power of the eyes is to be transferred: here the eyes themselves must be carried in order to cure human eye disease. Marcellus adds that, even if we just catch the blood from the lizard's eyes on clean wool, wrap it in purple cloth, and wear it on our neck, we will find this to be "a most efficacious remedy for eye pain."[22]

There are a number of other examples of the use of animal eyes or matter as an amulet for eye-disease: Julius Africanus, a Greek writer who died sometime after 240 A.D., apparently claimed that the eye of a partridge worn in dog-skin (presumably a pouch) cured ophthalmia and that the eyes of a frog worn in a linen rag on the left arm or the neck healed eye-disease more generally; in the same passage he also suggests that two gizzard stones of a swallow placed within a gold necklace will keep off ophthalmia.[23] Pliny (*nat.* 28.29) tells us that Licinius Mucianus, a Roman senator who was three times consul in the time of Vespasian, wore a living fly (*musca*) wrapped in white linen to cure his ophthalmia. Here one might speculate that, as in the case of the green lizard or the frog, the exaggerated size of the insect's eyes creates the focus for the healing of human eyes.[24] Animal eyes were also used to combat other eye problems: the right eye of a snake (Plin. *nat.* 29.131), for example, when "tied on" was useful for treating *epiphora*, a general medical term for flux, but one used most often to describe eyes that are constantly producing tears (LSJ s.v. ii.5),[25] and the eye

---

22 *Med.* 8.50; see Gaillard-Seux, "Les maladies," 94 for discussion.
23 Martin Wallraff et al., eds., *Iulius Africanus, Cesti: The Extant Fragments* (Berlin: De Gruyter, 2012), fr. D27.11–14 (gizzard stones of a swallow), 23–24 (eye of the partridge) and 27–30 (eyes of a frog). In his discussion of the partridge, Africanus says that the eyes of the partridge bound in pure linen and applied to the eyes each day will prevent dim-sightedness and cataracts. In the edition of Jean-René Vieillefond, *Les "Cestes" de Julius Africanus. Étude sur l'ensemble des fragments avec édition, traduction et commentaires* (Paris: Librairie M. Didier, 1970) these recipes appear on 224–27, where Viellefond identifies the *perdix* not as a partridge, but rather as a "vautour" ("vulture"). For the magical use of various parts of the swallow, see Patricia Gaillard-Seux, "De l'Orient à l'Occident: les recettes médico-magiques tirées de l'hirondelle," in Véronique Dasen (ed.), *Les savoirs magiques et leur transmission de l'Antiquité à la Renaissance* (Florence: Edizioni del Galluzzo, 2014) 169–94, and for the partridge as a proverbially sharp sighted bird, see Aristoph. *ran.* 523, with scholia.
24 But apparently text alone could do the job as well: in the same passage Pliny says that a man of senatorial rank with a similar affliction, M. Servilius Nonianus, once wore on his neck a papyrus inscribed with the Greek letters PA and fastened with a thread.
25 In this same passage Pliny says that weasels also have the power to regenerate their eyes and that people "use the animal as they use the lizard and the rings" (i.e., for eye cures).

of a crab wrapped in a purple cloth and suspended from the neck cured *lippitudo*, a disease of watery or bleary eyes (Marcellus 8.51).

Within this tradition of eye-healing amulets, then, we can see that the image of the lizard, its actual eyes, the metal nails used to gouge them out, or even the blood from the eyes could each be used to create a persuasive analogy[26] between the naturally regenerating eyes of the green lizard and the diseased eyes of the patient. The variations in these recipes are displayed in the table below:

Table 1 Variations in recipes for eye-healing amulets

Author	Step 1	Step 2	Amulet	Patient	Effect
RINGS OR STONES:					
Pliny (*nat.* 29.130)	blind a green lizard	release seeing	iron, gold rings	[wear ring]	cures *ophthalmia*
Pliny (*nat.* 29.129)	blind a green lizard	release seeing	(two?) black pebbles	NA	for pains in the eye
Pliny (*nat.* 29.130)	blind a weasel	release seeing	like the lizard in 29.130	NA	like the lizard in 29.130
Aelian (*nat. anim.* 5.47)	blind a green lizard	release seeing	iron ring w/ black stone engraved w/ lizard	wear ring	"good for the eyes"
J. Africanus (D27.11–14)	NA	NA	two gizzard stones of a swallow in a gold necklace	[worn on neck?]	prevents *ophthalmia*
Cyranides (2.14.22–31)	blind a lizard with two nails	release seeing	two silver or gold rings w/ jaspers engraved w/ lizard and letters; enclose w/ eyes	wear a ring on each hand	heals *ophthalmia*
Marcellus (8.49)	blind a green lizard at the old moon	release seeing	gold, silver, iron, electrum, or copper ring	wear rings or look through them	cures *lippitudo*
ANIMAL EYES ALONE:[27]					

---

26 See n. 5 above.

27 I leave out the many other passing mentions of animal eyes as *ophthalmia*-amulets in the *Cyranides*, where it seems to be a *topos* of a sort, e. g., the eyes of the viper (2.12.11–12 [Kaimakis]) or the quail (2.35.2–3 [Kaimakis]) "heal all *ophthalmia* if tied on (sc. to the body)"; the eyes of a swallow (2.50.18–19 [Kaimakis]) "stop *ophthalmia* if tied on to the forehead and heal all kinds of shivering fever"; the stones in an owl's head (3.9.4–5 [Kaimakis]) "heal *ophthalmia* if tied on (sc., to the body)"; the eyes of the sea perch (3.49.6–7 [Kaimakis]), the seal (3.67.14 [Kaimakis]), and the gilt head of a fish (3.74.6–7 [Kaimakis]) "heal all *ophthalmia* if carried." For the special use of parts of the swallow in Greek healing (for sore throat and epilepsy, as well as *ophthalmia*) see Gaillard-Seux, "De l'Orient à l'Occident."

Author	Step 1	Step 2	Amulet	Patient	Effect
Pliny (*nat.* 29.131)	blind a snake	release alive	right eye	"tied on"	for ἐπιφορά (eye flux)
Cyranides (2.14.8–9)	lizard	release alive	eyes (plural)	NA	heals all ophthalmia
Marcellus (8.50)	blind a green lizard at the old moon	release seeing	both eyes in *bulla* or gold *lupinus*	[wear on neck?]	"eyes will not hurt"
Marcellus (8.50)	blind a green lizard at the old moon	release seeing	eye blood on clean wool wrapped in purple cloth	wear on neck	"most efficacious remedy for eye-pain"
Marcellus (8.51)	NA	NA	single eye of a crab wrapped in purple cloth	wear on neck	cures *lippitudo*
J. Africanus (D27.27–30)	blind a frog	release alive	eyes of a frog in linen rag	worn on left shoulder or neck	cures eye-disease
J. Africanus (D27.23–24)	NA	NA	single eye of a partridge in dog-skin	[worn on neck?]	cures ophthalmia
Supplementum Magicum no. 78	gouge out right eye of a lizard	NA	throw it into goat-skin	wear [on neck?]	NA
OTHER ANIMAL MATTER:					
Pliny (*nat.* 28.29)	NA	NA	living fly tied on in white linen	NA	wards off ophthalmia

We can see, then, a general pattern. All of these recipes are used either to cure or prevent eye-disease, eye-pain, or three named diseases: ophthalmia (most frequently), *lippitudo*, or *epiphora* (all three, in fact, seem to be general terms for eye-disease or the rheum that they produce).[28] In all these spells the prominent eyes of an animal (e. g., of a lizard, frog, crab, and fly) or the animal's power according to ancient sources to regenerate its eyes (e. g., the lizard, weasel, swallow, and snake) apparently make it a valuable model for curing human

---

[28] Muriel Pardon, "La *lippitudo* dans la littérature classique: de l'oeil qui dégoutte à l'oeil qui dégoûte," in Maurizio Baldin, Marialuisa Cecere, and Daria Crismani (eds.), *Lingue tecniche del greco e del latino*, 4. *Testi medici latini antichi. Le parole della medicina. Lessico e storia; atti del settimo convegno internazionale, Trieste, 11–13 ottobre 2001* (Bologna: Pàtron, 2004), 651–62. I thank Patricia Gaillard for this reference.

eyes. The outliers here are the pair of gizzard stones, which, as suggested earlier, were perhaps meant to resemble a pair of human eyes.[29]

These closely related spells for creating various kinds of rings or other amulets seem to work almost exclusively on the idea of magical contagion: raw material (metal nails or pebbles), inscribed gems (lignite or jasper), or simple rings (from a variety of metals) are brought into contact with the lizard as it regenerates its eyes and then these materials are given to patients, who wear them or (in one case) look through them. All of the other amulets that use the actual eyes of animals seem, in contrast, to be worn not on the fingers, but rather elsewhere on the body in some kind of special case, either in the skin of an animal (goat or dog), in some special cloth (linen or "purple cloth"), or in a container of gold (*bulla*, *lupinus*, or "necklace"). Of the amulets that carry eyes alone, five (snake, partridge, crab, and lizard) require a single eye, while only Marcellus (in his description of the *bulla* or *lupinus* amulets), Africanus (frog), and the *Cyranides* recipe (for the rings worn on the left and right hands) stipulate that a pair of eyes must be used. In the two outlying cases (the whole fly and the two gizzard-stones), the model also uses pair of eyes or spherical objects. This variation is important, of course, because of the two gemstones that have generated this inquiry the Florentine has a single *udjat*-eye, while the Goodenough gem has two.

II Thief-Catching Spells that Employ *Udjat*-Eyes

The "eye of Horus" appears in somewhat more incongruous circumstances in two magical recipes that tell us how to catch someone who has stolen from us but who remains undetected. The fullest version is preserved in a recipe in a fourth-century A.D. magical handbook, where we are directed to draw a single eye on a wall and then strike the eye with a hammer, saying: "As long as I strike (κρούω) the eye (οὐάτιον) with this hammer, let the eye (ὀφθαλμός) of the thief be struck and swell up (κρουέσθω καὶ φλεγμαινέσθω) until it betrays him."[30] The

---

29 The body parts of a swallow are elsewhere used to cure ophthalmia. According to the *Cyranides* the eyes of a swallow (2.50.18–19 [Kaimakis]) "stop *ophthalmia* if tied on to the forehead," and Pliny says that if swallow chicks are blinded at the full moon and allowed to regain their sight, their heads burnt to ash and mixed with honey are an effective salve for a variety of eye-problems, including ophthalmia (*nat.* 29.128). For the special power of the swallow, see *supra*, n. 23.

30 Karl Preisendanz and Albert Henrichs, eds., *Papyri Graecae Magicae*, 2 vols. (Stuttgart: Teubner, 1973–1974), no. V 70–95.

recipe uses the word *ouation* here to indicate the image to be drawn on the wall and then hammered: it is an unusual Greek word based on the Egyptian term *udjat* and in Greek seems to refer solely to the *udjat*-eye.[31] In fact, this same recipe includes a fairly standard drawing of an *udjat*-eye for us to copy out and then strike with a hammer (see the facsimile in fig. 4, where the eye stands between two pyramids constructed from Greek vowels).

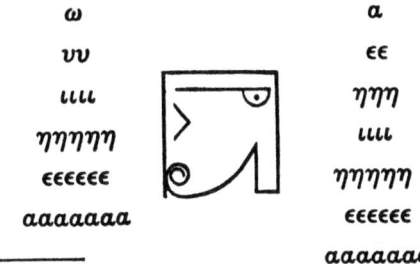

ω   α
υυ   εε
ιιι   ηηη
ηηηη   ιιιι
εεεεε   ηηηηη
ααααααα   εεεεεε
―――――   ααααααα

Fig. 4 *Udjat*-eye

It seems, therefore, that this schematized Egyptian design was used in a simple ritual of persuasive analogy to inflict harm on a human eye. There is, moreover, little native Egyptian content here,[32] and, in fact, it would presumably be sacreligious for an Egyptian, at least, to attack the eye of Horus in such a manner.[33]

---

31 See LSJ s.v. and Maltomini's comments *ad Supplementum Magicum*, no. 86 col. ii.17, where he quotes further bibliography.
32 The recipe says to write the letters ΧΟΩ on the wall as well, and Robert K. Ritner, in Hans-Dieter Betz, ed., *The Greek Magical Papyri in Translation*, 2nd ed. (Chicago: University of Chicago Press, 1996), 102 n. 9, suggests that this could be the Coptic word for "hand over" or "put." But since in the drawing the eye is accompanied by vowels, I wonder whether the scribe simply miscopied the first letter here, which should have been an alpha or some other vowel. Otherwise the spell uses a version of the Jewish exorcism formula, magic words of no obvious Egyptian character, and many vowels. A little farther down, the same papyrus has another thief-catching spell (Preisendanz and Heinrichs, *Papyri*, no. V 172–212) that invokes "Hermes, finder of thieves, Helios, the pupils (i. e., eyes) of Helios, Themis, Erinys, Ammon and Parammon to take control of the thief's throat and single him out in this hour!" Here the addition of two Egyptian gods at the end of a list of Greek ones probably indicates the Egyptian adaptation and extension of a Greek spell to a new context: for a parallel case in the invocation of gods, see Christopher A. Faraone, "The Ethnic Origins of a Roman-Era *Philtrokatadesmos* (*PGM* IV 296–434)," in Paul Mirecki and Marvin Meyer (eds.), *Magic and Ritual in the Ancient World* (Leiden: Brill, 2002), 319–43.
33 The demotic handbooks contain a number of Egyptian thief-detecting spells, but they are quite different: see below.

Two fragments from an Oxyrhynchus papyrus roll of a similar or slightly earlier date seem to preserve a number of similar spells for identifying an unknown thief, including this more threatening version of the recipe discussed above:

> ... and you will take the hammer and you will strike (κατακρούσεις) the eye (τὸν ὀφθαλμόν) and while striking say: "Thief, I will put out (ἐκβαλ[ῶ) your eye." And add: "Let the thief speak, before I put out his eye (πρὶν ἐκ[βαλεῖν με] τὸν ὀφθαλμόν)." (Daniel and Maltomini, *Supplementum Magicum*, no. 86 col. ii.1–8)

At line 9, there follows a different version of the operation, in which we are instructed to place an onion on the image of the eye (rather than strike it) and say: "Thief, let the onion bite (δακνέτ[ω) you," to which is added the comment: "If it (i.e., the irritation) does not subside, it is clearly him." In other words, if afterwards we see someone in our neighborhood with a bruised or irritated eye, we can conclude that this person is the hitherto unknown thief. The instructions for inscribing the eye appear at the very end of the fragment (line 21): "Take a nail and inscribe an *udjat*-eye (οὐάτι[ον)."

Both recipes, then, aim at rooting out a thief in a small face-to-face community. The first version in *Supplementum Magicum* no. 86 threatens, however, to blind the eye. Presumably in this case the eye is drawn and struck in a public place (in *Papyri Graecae Magicae* V it is drawn on a wall) and the goal is, I suggest, to make a public threat to damage or even blind the eye of the thief in the hope that he will panic and speak up to avoid punishment, a procedure which also seems to lie behind many of the public "prayers for justice," especially those aimed at unknown thieves.[34] There are, moreover, no prayers, magical words, or divine names – Egyptian or otherwise – anywhere on these two papyrus fragments: the magical operation is a purely mechanical one and thus here, too, the use of the *udjat*-eye seems to be devoid of any explicit Egyptian or theological significance: it is simply an effigy for a human eye.

In the case of these thief-detection spells, however, we may be able to trace how this mistranslation of the Egyption icon occurred. A Demotic Egyptian recipe in a bilingual Egyptian and Greek magical handbook in the British Museum

---

[34] For a full bibliography of curses against thieves, see Christopher A. Faraone, Brien Garnand, and Carolina López-Ruiz, "Micah's Mother (Judges 17:1–4) and a Curse from Carthage (*KAI* 89): Evidence for the Semitic Origin of Greek and Latin Curses against Thieves?" *Journal of Near Eastern Studies* 64 (2005): 161–86, and Christopher A. Faraone, "Curses, Crime Detection, and Conflict Resolution at the Festival of Demeter Thesmophoros," *Journal of Hellenic Studies* 131 (2011): 25–44. For Jewish magical rites for the same purpose, see Gideon Bohak, "Catching a Thief: The Jewish Trials of a Christian Ordeal," *Jewish Studies Quarterly* 13 (2006): 1–19.

has a rubric that begins "A way of finding a thief."[35] The recipe requires that we grow flax over the buried head of a drowned man, harvest the flax, and speak an incantation over it. Later we are to take the individual flax leaves and to utter (presumably *sotto voce*) one by one the names of various people in the village. If we hear a person speaking at the same moment we say his name, then he is the thief. The incantation includes a foundation myth for the ritual:

> I shall give them the words of Geb, which he gave to Isis, when Shu(?) concealed them in the papyrus (swamp) of Buto, she bringing the small amount of flax in her hand, she forming a knot, she tying these forelegs(?), until he was revealed to Horus in the papyrus (swamp). I will bring this small amount of flax in my own hand, I making it into a knot until So-and-So is revealed.... After answering, he will lift(?)...

At the very end of this spell we find a drawing of an *udjat*-eye, just as we did in the *Papyri Graecae Magicae* thief-catching recipe (fig. 4). Here, then, we learn that Isis invented or at least used this same flax ritual when she and Horus were hiding from Seth in the swamp and that she used it to reveal someone (Seth?) to Horus. The penultimate sentence translated above is, however, lacunose, and we cannot know whether the thief will be revealed *to* or *by* the *udjat*-eye, or even how this revelation will occur. But regardless of the restoration, we have here a recipe of purely Egyptian content that is composed entirely in Demotic with one short Coptic invocation. It is also completely unlike the two Greek spells that use the *udjat*-eyes to reveal a thief, but one wonders if a Greek using this bilingual handbook knew enough Demotic to understand the rubric and drawing and thereby conclude that Egyptian thief-catching spells employed a drawing of the eye of Horus, which he then mistranslated into his own native Greek rituals, which involved striking in effigy the eye of the thief. Some process such as this would explain, at least, how Greeks might have come up with the bizarre idea that you would strike the eye of one of the most important Egyptian deities in order to bruise the eye of a human thief.

Conclusion

The two lizard-amulets described at the start of the essay – the ones in the Florentine and Goodenough collections – provide us with evidence for a similar repurposing of the *udjat*-eye, but they seem to reveal a deeper understanding and

---

[35] The recipe is *PDM* xii 79–94; the incantation quoted below is in lines 90–94 in the translation of Janet H. Johnson in Betz, *Greek Magical Papyri*, 288–89.

appreciation of Egyptian myth. We saw how all of the recipes for eye disease – both those for creating rings with carved gemstones and those that simply use animal eyes tied in a sack on the neck – are, like the thief-catching spells, almost entirely automatic or mechanical in operation. There are, of course, some Jewish elements in both types: the *Papyri Graecae Magicae* ritual for revealing a thief, for example, uses a Jewish exorcism formula at one point, and a minority of the lizard gems (including the two with *udjat*-eyes) are inscribed with the angel names Ouriēl and Souriēl, the latter of which may have been attracted to these amulets on account of its similarity to the second half of the magic word κανθε σουλε. But aside from the *udjat*-eyes, there are no other signs of any Egyptian cultural influence. The focus of all of the lizard-gem recipes, moreover, is on using the actual eyes of the green lizard or objects placed in contact with them (e.g., the blood or the bloody nails) as effigies for the diseased eyes of the patient.

At first glance, then, it would seem that instead of using the actual eyes of a lizard, the sorcerer or gem-cutter who created the Florentine and Goodenough gems simply added the eyes of Horus as an alternative for the real eyes of the lizard. But Egyptian myths about Horus suggest otherwise, for at several points Horus loses his eye or suffers temporary blindness and is then healed. In some myths, the injured eye of this Horus is the moon, which waxes and wanes in a cycle that is explained by the mythical narrative of damage and then healing.[36] But it is, in fact, the creation story for the *udjat*-eye of Horus itself that is most pertinent:[37] in one of the most popular versions of these stories Seth plucks out the eye of Horus, and it is lost, but Thoth is able to recover and regenerate the eye, so that it can be replaced in Horus' head. This replacement eye is, in fact, from that time onwards called the *udjat*-eye, a term that comes from a related group of Egyptian words that mean "health" or "healthy" – the term *udjat* means, in short, "the healthy eye" or "the eye restored to health." Thus it seems possible that the creators of the two lizard gems under discussion may have known their Egyptian mythology fairly well and that they added the *udjat*-eye to their gems because, like the eyes of the green lizard in the rituals discussed earlier, Horus' eye was at one point destroyed and was regenerated. Indeed, the fact that κανθε σουλε, the most popular inscription on the backs of these gems – albeit not on the two under discussion – means "be healthy, eye!" might indeed have encouraged this kind of cultural translation. The

---

36 See n. 4 *supra*.
37 Edmund S. Meltzer, "Horus," in Donald B. Redford (ed.), *Oxford Encyclopedia of Ancient Egypt*, 3 vols. (New York: Oxford University Press, 2001), 2:119–22; and Andrews, *Amulets*, 43.

Horus myth might also help to explain the crescent moon over the head of the lizard on the gems, because some Egyptian myths equated the *udjat*-eye with the moon, which is constantly damaged (the waning moon) and then restored (the waxing moon): the crescent moon, the damaged eye of Horus, and the Greek adjective πηρά ("maimed") thus all coincide nicely with the blinding of the green lizard.[38]

Bibliography

Andrews, Carol. *Amulets of Ancient Egypt* (London: British Museum Press, 1994).
Bain, D. M. "Koiraniden," in *Reallexikon für Antike und Christentum* 21 (2006): cols. 224–32.
Betz, Hans-Dieter, ed. *The Greek Magical Papyri in Translation*, 2nd ed. (Chicago: University of Chicago Press, 1996).
Bohak, Gideon. "Catching a Thief: The Jewish Trials of a Christian Ordeal," *Jewish Studies Quarterly* 13 (2006): 1–19.
Bonner, Campbell. *Studies in Magical Amulets Chiefly Graeco-Egyptian*, University of Michigan Studies, Humanistic Series 4 (Ann Arbor: University of Michigan Press, 1950).
Daniel, Robert W., and Franco Maltomini, eds. *Supplementum Magicum*, Papyrologica Coloniensia 16.1 and 2, 2 vols. (Opladen: Westdeutscher Verlag, 1990–1992).
Delatte, Armand L. "Études sur la magie grecque IV: amulettes inédites des Musées d'Athènes," *Le Musée Belge* 18 (1914): 21–96.
de Mély, Fernand, and C. E. Ruelle. *Les lapidaires de l'antiquité et du moyen âge*, 3 vols. (Paris: E. Leroux, 1896–1902).
Drexler, Wilhelm. "Alte Beschwörungsformeln," *Philologus* 58 (1899): 594–616.
Faraone, Christopher A. "The Ethnic Origins of a Roman-Era *Philtrokatadesmos* (PGM IV 296–434)," in Paul Mirecki and Marvin Meyer (eds.), *Magic and Ritual in the Ancient World* (Leiden: Brill, 2002), 319–43.
Faraone, Christopher A. "Curses, Crime Detection, and Conflict Resolution at the Festival of Demeter Thesmophoros," *Journal of Hellenic Studies* 131 (2011): 25–44.
Faraone, Christopher A. "Text, Image, and Medium: The Evolution of Greco-Roman Magical Gemstones," in Chris Entwistle and Noël Adams (eds.), *Gems of Heaven: Recent Research on Engraved Gemstones in Late Antiquity*, British Museum Research Papers no. 179 (London: British Museum, 2011), 50–61.
Faraone, Christopher A. "Scribal Mistakes, Handbook Abbreviations, and Other Peculiarities on Some Ancient Greek Amulets," *MHNH: Revista internacional de investigación sobre magia y astrología antiguas* 13 (2013): 139–56.
Faraone, Christopher A., Brien Garnand, and Carolina López-Ruiz. "Micah's Mother (Judges 17:1–4) and a Curse from Carthage (*KAI* 89): Evidence for the Semitic Origin of Greek and Latin Curses against Thieves?," *Journal of Near Eastern Studies* 64 (2005): 161–86.

---

[38] I thank François Gaudard for this suggestion, although he points out that the two eyes on the Goodenough gem suggest that the creator of that amulet, at least, did not fully understand the Egyptian myths, because Horus never is blinded in both eyes.

Festugière, André-Jean. "Amulettes magiques: à propos d'un ouvrage recent," *Classical Philology* 46 (1951): 81–92.
Gaillard-Seux, Patricia. "Les maladies des yeux et le lézard vert," in Guy Sabbah (ed.), *Nommer la maladie: recherches sur le lexique gréco-latine de la pathologie* (Saint-Étienne: Publications de l'Université de Saint-Étienne, 1997), 93–105.
Gaillard-Seux, Patricia. "De l'Orient à l'Occident: les recettes médico-magiques tirées de l'hirondelle," in Véronique Dasen (ed.), *Les savoirs magiques et leur transmission de l'Antiquité à la Renaissance* (Florence: Edizioni del Galluzzo, 2014): 169–94.
Goodenough, Erwin. *Jewish Symbols in the Greco-Roman Period*, 3 vols. (New York: Pantheon Books, 1953–1968).
Hanson, Ann E. "Uterine Amulets and Greek Uterine Medicine," *Medicina nei secoli* 7 (1995): 281–99.
Jones, William H. S., trans. *Pliny, Natural History*, vol. 8 (Cambridge, Mass.: Harvard University Press, 1963).
Kaimakis, Dimitris. *Die Kyraniden* (Meisenheim am Glan: Hain, 1976).
Kenyon, Frederic G. *Greek Papyri in the British Museum* (London: British Museum, 1893).
Lloyd, Geoffrey E. R. *Magic, Reason and Experience* (Cambridge: Cambridge University Press, 1979).
Mastrocinque, Attilio, ed. *Sylloge Gemmarum Gnosticarum*, Bollettino di Numismatica Monografia 8.2.1 and 2, 2 vols. (Rome: Istituto poligrafico e Zecca dello Stato, 2003–2008).
Meeks, Dimitri, and Christine Favard-Meeks. *Daily Life of the Egyptian Gods*, trans. G. M. Goshgarian (Ithaca: Cornell University Press, 1996).
Meltzer, Edmund S. "Horus," in Donald B. Redford (ed.), *Oxford Encyclopedia of Ancient Egypt*, 3 vols. (New York: Oxford University Press, 2001), 2:119–22.
Michel, Simone. *Die magischen Gemmen im Britischen Museum* (London: British Museum Press, 2001).
Michel, Simone. *Die magischen Gemmen. Zu Bildern und Zauberformeln auf geschnittenen Steinen der Antike und Neuzeit*, Studien aus dem Warburg-Haus Band 7 (Berlin: Akademie-Verlag, 2004).
Pardon, Muriel. "La *lippitudo* dans la littérature classique: de l'oeil qui dégoutte à l'oeil qui dégoûte," in Maurizio Baldin, Marialuisa Cecere, and Daria Crismani (eds.), *Lingue tecniche del greco e del latino, 4. Testi medici latine antichi. Le parole della medicina. Lessico e storia; atti del settimo convegno internazionale, Trieste, 11–13 ottobre 2001* (Bologna: Pàtron, 2004), 651–62.
Preisendanz, Karl, and Albert Henrichs, eds. *Papyri Graecae Magicae*, 2 vols. (Stuttgart: Teubner, 1973–1974).
Sambin, Chantal. "La purification de l'oeil divin ou les deux vases de Kom Ombo," *Revue d'Egyptologie* 48 (1997): 185–99.
Scholfield, Alwyn F., trans. *Aelian, On the Characteristics of Animals* (Cambridge, Mass.: Harvard University Press, 1968).
Tambiah, Stanley J. "Form and Meaning of Magical Acts: A Point of View," in Robin Horton and Ruth Finnegan (eds.), *Modes of Thought* (London: Faber, 1973), 199–229.
Vieillefond, Jean-René. *Les "Cestes" de Julius Africanus. Étude sur l'ensemble des fragments avec édition, traduction et commentaires* (Paris: Librairie M. Didier, 1970).
Wallraff, Martin, et al., eds. *Iulius Africanus, Cesti: The Extant Fragments* (Berlin: De Gruyter, 2012).

Klaus-Dietrich Fischer
Gesund durchs Jahr mit Dr. Hippokrates – Monat für Monat!

Abstract: Not surprisingly, advice for staying healthy was a tremendously popular genre of medical literature for lay people, from the Hellenistic period to the seventeenth and eighteenth centuries. First organised according to the four seasons, these writings were later brought into line with the Roman calendar of twelve months, the older type still starting with March, the first month of the traditional Roman calendar. After Groenke, who was the first to attempt a classification of these texts and who, in his Berlin thesis, published numerous witnesses, scholars have paid little attention to these short texts. Here, I publish some additional witnesses for the first time, demonstrating that single types seem to have preserved their original shape rather well over several centuries. Finally, in the first appendix, I give a fuller account of the contents of perhaps the most important Western medical manuscript held at the National Library of Medicine in Bethesda (E 8, written around 1150). In the second appendix, I try to elucidate the vagaries of a term present in these regimens, namely *gersa*, *garsa*, or *geresis*.

Sicher hörte sich diese Überschrift, die in vielen lateinischen Handschriften mit dem Namen des Hippokrates verbunden wurde, ähnlich werbewirksam an wie die tatsächlich überlieferten, stark variierenden Titel, und genau das war ja der Grund gewesen, warum der berühmte Arzt aus Kos wieder einmal den Autor abgeben sollte[1]. Aber nicht nur dieses überaus illustren Namens bediente man sich, sondern neben Galen (the usual suspects) begegnen wir auch Herophilos als angeblichem Verfasser. Der Umstand, daß in der griechischen Überlieferung ein kleines Werk mit demselben Inhalt wie die Pseudohippocratica, nun aber dem Hierophilos [sic] zugeschrieben, existierte, war Heinrich von Staden natürlich

---

Diese Arbeit ist Teil des Forschungsprojekts FFI2013-42904-P (Ministerio Español de Economía y Competividad).

1 Vgl. Pearl Kibre, *Hippocrates Latinus. Repertorium of Hippocratic writings in the Latin Middle Ages*, revised ed. (New York: Fordham University Press, 1985), XI De cibis, diaeta (Calendarium dieteticum [sic]), 124–28. Kibre gliedert nach Jahrhunderten und ordnet nicht nach dem Inhalt, wie das Groenke (s. unten Anm. 2) tut, der seltsamerweise Kibre nicht zitiert. Das *Mittellateinische Wörterbuch* benutzt als Sigle für diese Quelle *Praecept. diaet.*

nicht verborgen geblieben[2], und somit ergibt sich eine recht entfernte Verbindung zwischen meiner hier vorgelegten kleinen Arbeit und seinem großen Werk über Herophilos.

Prophylaktische Ratschläge hatten sich zunächst (man denke an den diätetischen Brief unter dem Namen des Diokles von Karystos) an den vier Jahreszeiten orientiert, wofür in der Literatur z.T. der Begriff ‚Quatemberdiätetik' gebraucht wird; die Einteilung nach den zwölf Monaten des römischen Kalenders (*Regimen XII mensium*) ist zeitlich gewiß später anzusetzen, ohne daß sich über die geschichtliche Entwicklung gegenwärtig Genaueres sagen ließe[3]. Beide Formen, die Quatemberdiätetik und das *Regimen XII mensium*, finden wir vereint als *Praeceptum Galeni de humani corporis constitutione; de diæta quatuor anni tempestatum & duodecim mensium* (Chartier Band 6, 440–442). Daß bei diesem Genre oft

---

[2] Heinrich von Staden, *Herophilus. The art of medicine in early Alexandria. Edition, translation and essays* (Cambridge: Cambridge University Press, 1989), 582. Letzte mir bekannte Ausgabe ist die von Roberto Romano, „Il calendario dietetico di Ierofilo," *Atti della Accademia Pontaniana*, nuova serie 47 (Anno Accademico 1998): 197–222. Frank-Dieter Groenke, *Die frühmittelalterlichen lateinischen Monatskalendarien. Text–Übersetzung–Kommentar* (Berlin: Diss. med. dent. Freie Universität Berlin, 1986), 263–76, behandelt den „Inhalt der bei Ideler abgedruckten byzantinischen diätetischen Monatskalendarien" im Vergleich; speziell zum Pseudo-Herophilos Groenke, *Monatskalendarien*, 272–73.

[3] Die ausführliche Studie von Marilyn Nicoud, *Les régimes de santé au moyen âge. Naissance et diffusion d'une écriture médicale (XIIIe-XVe siècle)*, 2 Bände (Rome: École française de Rome, 2007) (Bibliothèque des Écoles françaises d'Athènes et de Rome. Fascicule 333) behandelt die Zeit vor dem 13. Jahrhundert nicht; der Untertitel suggeriert, daß es vorher keine diätetische Literatur gab und übergeht die Tatsache, daß das spätantike *Regimen XII mensium* natürlich auch im 13., 14. und 15. Jahrhundert weiter überliefert wurde. Auch in Vivian Nuttons Behandlung der Diätetik im *Neuen Pauly*, Band 3 bzw. 13, Stuttgart und Weimar, 1997 bzw. 1999, sind diese Regimina nicht erwähnt; Gleiches gilt für die Enzyklopädie *The Classical Tradition*, hrsg. von A. Grafton, G. W. Most und S. Settis (Cambridge, Mass.: Harvard University Press, 2010), wo man einen Artikel „Dietetics" vergebens sucht. Bei R. E. Bjork (Hg.), *Oxford Dictionary of the Middle Ages* (Oxford: Oxford University Press, 2010), gibt es zwar einen Artikel „regimen of health" (1390b–1391b) aus der Feder Linda E. Voigts, doch würde man auch hier vermuten, daß die reich überlieferten *Regimina XII mensium* nicht existierten; selbst Literaturhinweise darauf fehlen. Die medizinische Habilitationsschrift von Wolfram Schmitt, in Heidelberg unter der Leitung von Heinrich Schipperges entstanden (*Theorie der Gesundheit und ‚Regimen sanitatis' im Mittelalter*, Heidelberg, 1973), ist leider ungedruckt geblieben; das Frühmittelalter wird nur kurz gestreift (10–11), im Mittelpunkt stehen die Übersetzungen aus dem Arabischen vom 11. bis 13. Jahrhundert, weshalb es überrascht, daß Wolfram Schmitt in Nicouds Bibliographie nicht einmal genannt wird. Konzis, aber mit einer Reihe weiterführender Literaturangaben vor allem zum germanistischen Bereich behandelt die „Monatsregeln" Gundolf Keil, in Werner E. Gerabek, Bernhard D. Haage, Gundolf Keil und Wolfgang Wegner (Hgg.), *Enzyklopädie Medizingeschichte* (Berlin: De Gruyter, 2007), 1003b–1004b.

von diätetischen Kalendern gesprochen wird (*calendari dietetici* bei Beccaria[4]), ist naheliegend, daran Anstoß zu nehmen, reichlich pedantisch[5].

Diese Kleinform medizinischer Literatur für Laien ist im westlichen Mittelalter (erst lateinisch und später auch in die Landessprachen übersetzt) so stark verbreitet gewesen, daß es ständig weitere Funde von Textzeugen geben wird[6], genau wie die, über die ich an dieser Stelle berichten möchte. Während die eigentlichen medizinischen Handschriften, für das 9. bis 11. Jahrhundert bei Beccaria, für das 9. bis 12. bei Wickersheimer[7] (allerdings auf französische Bibliotheken beschränkt),

---

[4] Augusto Beccaria, *I codici di medicina del periodo presalernitano. Secoli IX, X e XI* (Roma: Edizioni di Storia e Letteratura, 1956) (Storia e Letteratura. 53), 443–44. Übersicht mit dem Versuch der Typisierung und lateinisch-deutschem Text bei Groenke, *Monatskalendarien*.
[5] Vgl. Gundolf Keil, „Monatsregeln," (wie Anm. 3) zur deutschsprachigen Rezeption.
[6] Z. B. Matrit. 19 fol. 90$^{rb\text{-}va}$, Groenke Typ I (ihm aber unbekannt; aufgeführt bei Kibre, *Hippocrates Latinus*, p. 128a). München, Bayerische Staatsbibliothek, clm 7999, fol. 143ra und 143$^{ra\text{-}b}$ stehen zwei weitere mit dem Monat Januar beginnende anonyme Regimina, die bei Groenke fehlen und die sich auch in seine Systematik nicht einordnen lassen, obgleich das zweite seinem Typ IIa ähnelt. Die Texte standen schon in der ausführlichen Beschreibung der Handschrift bei Steinmeyer-Sievers.
[7] Ernest Wickersheimer, *Les manuscrits latins de médecine du haut moyen âge dans les bibliothèques de France* (Paris: Éditions du Centre national de la recherche scientifique, 1966). Groenke hat gleich den ersten Eintrag bei Wickersheimer, 13, zur Handschrift Amiens, fonds Lescalopier 2, fol. I–XII, 11. Jahrhundert, übersehen. Der Text wurde publiziert von Heather Stuart, „A ninth-century account of diets and *dies Aegyptiaci*," *Scriptorium* 33 (1979): 237–44, hier 243. Stuart druckt auch den Text des gleichfalls übersehenen Fragments Leiden, Voss. Lat. F. 96 A, fol. 2^v, ab. Daß dort in der Überschrift des Regimens die *dies Aegyptiaci* erscheinen, konnte sich Stuart nicht erklären (Stuart, 240: „Only discovery of a more direct source will solve this problem"). Sie hat nicht erkannt, daß die letzten Worte bei jedem Monat die verworfenen Tage nennen, genau wie in dem Text aus Amiens, nur daß dort andere Tage als verworfen aufgeführt werden. (Der Text in Amiens ist in Hexametern, deshalb ist zu lesen *Martis prima necat cuius de cuspide quarta est*; *sic cuspide* Groenke, 210 = Typ IV, nur nach London, British Library, Sloane 475, fol. 4^v–5^v. Diese Verse stammen aus der *Anthologia Latina* 680a, vgl. Hennig, 82–83.) Da im Palatinus die *dies Aegyptiaci* vorangehen, lag es nahe, daß man beides kombinierte. Am Beginn des Regimen im Vossianus vermute ich *dieta* statt *dicta*, entweder von der Herausgeberin falsch gelesen, oder vom Schreiber der Handschrift selbst. Groenke scheint auch clm 17403 (13. Jahrhundert, aus dem Kloster Scheyern in Bayern), fol. 4^v übergangen zu haben. Gleiches gilt für Glasgow, University Library, Hunter. 414, fol. 61^r–63^r, s. 14/15. Auf den Groenke unbekannten Zeugen Bethesda, National Library of Medicine, E 8, fol. 88$^{r\text{-}v}$, geschrieben gegen 1150, der bis vor kurzem verschollen war, gehe ich später (und ausführlich in der Appendix Bethesdana) ein. Im Par. lat. 14025, s. 12/13, fol. 85vb–86ra, steht das Regimen Typ IIa Groenke, daran anschließend auf fol. 85$^{ra\text{-}b}$ Typ IIb; nach einem weiteren Rezept endet Buch 2 des Petro(n)cellus, von dem de Renzi nur einige Kapitel nach dieser Handschrift publiziert hat (diese Stücke fehlen in de Renzis Kapitelliste, *Collectio Salernitana*, Band 4, 314). Im Kodex 2169 des Salzburger Museum Carolino-Augusteum, geschrieben im 2. Viertel des 12. Jahrhunderts in Deutschland (so die Angabe der

sehr ordentlich erschlossen sind[8], gilt das für Handschriten außerhalb dieses zeitlichen, geographischen und sachlichen Bereiches weit weniger[9]. Derartige Kleintexte wie unser Regimen werden nämlich häufig irgendwo in einem kleinen, auf der Seite zufällig freigebliebenen Raum, manchmal am Rande, untergebracht, wie es auch beim Vaticanus Palatinus Latinus 485, von dem ich hier ausgehe, der Fall war. Deshalb werden sie dann leicht bei der Suche nach Handschriften mit klar medizinischem Inhalt übersehen.

Die Handschrift Vat. Palat. Lat. 485 wurde im Kloster Lorsch östlich von Worms (somit keine Autostunde entfernt von meinem Wohnort Wiesbaden) geschrieben. Sie wird in die Jahre 860 bis 875 datiert, also gegen Ende des ersten Jahrhunderts nach der Gründung dieser für die Handschriftenüberlieferung so wichtigen Pflanzstätte (Weihe der Klosterkirche am 1. September 774). Unserem Regimen gehen auf fol. 13^v zwei weitere medizinische Kleintexte voraus, ein Verzeichnis der verworfenen Tage (*dies Aegyptiaci*[10]) und ein Krankheitslunar[11]; auf

---

Hill Monastic Microfilm Library, andere datieren später), fol. 26^{r-v}, treffen wir wieder Typ IIa (mit den Tränken, doch hier mit dem Monat Januar beginnend), auf fol. 26^v–27^r Typ III. London, British Library, Harley 4977, fol. 103^{r-v}, bringt ebenfalls Typ IIa, endet jedoch bereits im Monat Dezember (Beginn gleichfalls im März). Mit den Worten *mense Iulio* bricht der Text von *Praecept. diaet.* IIa ab in der Handschrift München, Bayerische Staatsbibliothek, clm 13002 (geschrieben 1158), am Schluß von fol. 3rb; voraus geht *Praecept. diaet.* IIb und VII; für letzteres kennt Groenke nur den Laudunensis 426*bis* (der lateinische Text wurde bereits von Wickersheimer S. 40–41 gedruckt). Die Münchener Handschrift liest *bislingua* für *pislingua* im Laudunensis, bei Groenkes Übersetzung ‚Zungenmäusedort' hat sich ein Fehler eingeschlichen, lies ‚Zungen-Mäusedorn' (Ruscus hypoglossum). Zu verbessern ist ferner *cardo panę cum quinque grana piper*; es handelt sich um ein Wort (auch: *cardopan*; *cardopanus*, wofür man drei verschiedene Bedeutungen findet), irrig deshalb die Übersetzung „Distel, Brot mit fünf Kernen [sic] Pfeffer;" *cardone cum quinque granis piperis* clm 13002, wo auch *sarmina* (*sarminia* Wickersheimer beim Laudunensis) steht; Groenke las dort †*niniа*†; *sarminium* heißt ‚Kerbel', die Form *sarmina* ist CGL 3,595,18 belegt.

8 Speziell zu Hippokrates und den Pseudohippocratica s. das Verzeichnis von Kibre, *Hippocrates Latinus* (wie Anm. 1).

9 Die medizinischen Handschriften und Einzeltexte in den *Codices Latini Antiquiores* sind unter paläographischen Gesichtspunkten behandelt, der (in diesem Fall: medizinische) Inhalt spielte eine untergeordnete Rolle.

10 Wickersheimer, *Manuscrits*, druckt mehrere davon ab, vgl. dort 40 (in einer Sequenz von Regimina), Übersicht ebenda: 195. Umfassende Literaturhinweise zu den verworfenen Tagen bei John Hennig, „Versus de mensibus," *Traditio* 11 (1955): 65–90.

11 Handschrift Vp$_3$L bei Christoph Weißer, *Studien zum mittelalterlichen Krankheitslunar. Ein Beitrag zur Geschichte laienastrologischer Fachprosa* (Pattensen/Han.: Wellm, 1982) (Würzburger medizinhistorische Forschungen. 21); Edition dort 273. Weißer nicht bekannt war der Hunter. 96, fol. 18^{ra-b} (Inc. *Item ut essias quidquid in hac luna inciderit quid sper⟨a⟩uit* [= *sperabit*]), der dem

fol. 14ʳ folgt ein griechisches Alphabet samt Zahlzeichen[12]. Während dieses Alphabet in der Beschreibung der Handschrift, die samt dem farbigen Digitalisat im Internet steht, erwähnt wird, sind die medizinischen Kleintexte mit Ausnahme des Lunars nicht einzeln aufgeführt.

Man mag sich fragen, ob wir nach der kritischen Ausgabe verschiedener Versionen der Regimina durch Frank-Dieter Groenke[13], einen Schüler des um die Erforschung der Medizin im frühen Mittelalter so verdienten Gerhard Baader, den Abdruck eines weiteren Zeugen brauchen. Ich habe diese Frage bejahen zu können geglaubt, und das nicht nur, weil Groenkes kritischer Apparat ziemlich unübersichtlich geraten ist[14], sondern weil wir von diesem Text im Vat. Palat. Lat. 485 ausgehend Gelegenheit zu einigen darüber hinausführenden Beobachtungen haben werden. Beginnen wir also mit der Umschrift des Textes unter Beibehaltung der originalen Interpunktion und mit sparsamer Auflösung der wenigen Abkürzungen[15], woran sich meine deutsche Übersetzung anschließt:

Vaticanus bibliothecae apostolicae Palatinus Latinus 485
Incipiunt tempora pro sanitate corporis quae obseruare debeant
  1. Mens. mar.[16] bibat dulce. usitat agramen[17]. usitet radices confectas manducare. Asso balneo usitare. sanguinem minuare. solutionem non accipere quia frigores generat ipsa solutio.

Lunar entspricht, das er S. 174–175 nach Berol. Phillipp. 1790, fol. 41ᵛ, Aug. CLXXII, fol. 76ᵛ und Par. Lat. 11218, fol. 101ʳ ediert.
12 Für Parallelen in St. Gallen vgl. Bernice M. Kaczynski, *Greek in the Carolingian Age: The St. Gall Manuscripts*, (Cambridge, Mass.: The Mediaeval Academy of America, 1988), und allgemein Walter Berschin, *Griechisch-lateinisches Mittelalter. Von Hieronymus zu Nikolaus von Kues* (Bern: Francke, 1980). (Die aktualisierte englische Übersetzung dieses Werkes habe ich nicht eingesehen.)
13 Groenke, *Monatskalendarien* (wie Anm. 2).
14 Er hat sich für einen negativen Apparat entschieden, vermutlich unter dem Einfluß seines Betreuers Gerhard Baader, der nach der Vorgabe des *CMG* verfahren wollte. Dort kann das allerdings wesentlich leichter durchgeführt werden, da man sich anfangs oft auf zwei oder drei Handschriften beschränkt hatte. Jetzt ist man dort zum Glück in dieser Frage flexibler, was die Übersichtlichkeit für den Benutzer sehr erhöht.
15 Nicht immer habe ich Groß- und Kleinbuchstaben sicher unterscheiden können. Eine bequeme Kontrolle ist anhand des Digitalisats gegeben.
16 In der Beschreibung der Handschrift wird das als *Mensis martius* aufgelöst, es muß natürlich der Ablativ stehen.
17 *agramen* hängt – hierin folge ich Groenke – wohl mit *acer* ‚scharf' zusammen; er schließt sich Baaders Übersetzung im *Mittellateinischen Wörterbuch* s.v. ‚scharfes Würzgemüse' an, während Stuart, *Diets*, 243, das lateinische Wort ganz ohne Erklärungsversuch in ihre Übersetzung übernimmt.

2. Mens. apr. sanguinem minuare. potionem bibere. carnes recentes manducare. sanguinem intercodinium minuare. calidos usitare. dolorem stomachi purgare unguentos calasticos usitare. et si sic factum fuerit omnia membra sanare debent.

3. Mens. mai calidum bibat calidum usitat caput purget quia calidus in calore in praecordia ponit frigidum. licet in mense mad. uena epatica incidere. et potionem ad soluendum bibere. cataplasma in capite inponi oculos turbulentos sanare. pruri⟨gi⟩nem mundare. urinam curare. olera frigidas usitare. agramen manducare.

4. Mens. iun. Omni die mane ieiunus †in orodit† aquam bibere. ceruisam non. nisi spica *[fol. 14ʳ]* usitare. lactuces*(!)* manducare. acetum bibere.

5. Mens. Iul. non minuetur sanguis de uena in hoc tempore. nec potionem in ipso mense bibat. saluia et ruta usitet.

6. Mens. aug. nullo penitus caule manducet. agramen manducet ceruisam et me[m]ttum non bibet.

7. Mens. Sep. Omnia quae uis accipere debes. quia omnes escae cum tempore pro fructa*(!)* confecta sunt.

8. Mens. oct. racemis et musto usitare quia corpora sanat. et solutionem facit.

9. Mens. nou. et dec. bonum est studium habere. uena epatica incidere. gersis uentusarum inponere. quia in ipso tempore omnes humores sunt parati.

10. Mens. Ian. nullus penitus sanguinem minuare debet. nisi potionem contra officationem bibat. debeat et lectuariam accipere.

11. Mens. feb. de police[18] sanguinem minuare debes. et oxomelli*(!)* conficiendum facit. ad colera deducenda et fleoma per uentrem et melancolico quod plus inuenitur adbundare expellit per uentrem et causam in capite circum soluet. et qui ex humore nascuntur aut calico aut colera. et ipsum non permittit generare. uesicam curat et renes. cibos bene accipere facit et bene digerere. in omnibus aptissimum est.

Beginn der Zeitpunkte, die für die Gesundheit des Körpers Beachtung verdienen

1. Im Monat März nimm ein süßes Getränk zu dir, iß sauer eingelegtes Gemüse und eingelegten Rettich, nimm ein Schwitzbad, laß zur Ader, nimm kein Abführmittel, weil (starkes) Abführen ohne weiteres Schüttelfrost hervorruft.

2. Im Monat April lasse zur Ader, nimm einen Abführtrank, iß frisches Fleisch, entferne das Blut, das sich unter der Haut angesammelt hat, halte dich an warme Sachen, entferne die Magenschmerzen, benütze lockernde Salben. Wenn das so durchgeführt wird, müssen alle Teile des Körpers gesunden.

---

18 pollice *m²*.

3. Im Monat Mai trinke ein warmes Getränk und halte dich an Warmes, entleere den Kopf, denn wenn der heiß ist, erzeugt es Kälte in der Brust. Im Mai darf man aus der Lebervene zur Ader lassen und einen Abführtrank zu sich nehmen, ein Pflaster auf den Kopf auflegen, Sehstörungen heilen, juckende Hautstellen zum Abheilen bringen, sich um das Wasserlassen kümmern, kaltes Gemüse und sauer Eingelegtes zu sich nehmen.

4. Im Monat Juni trinke jeden Morgen nüchtern Wein mit Wasser, kein Bier, höchstens wenig [mit der Korrektur *pauca* statt *spica*], iß Lattich und trinke Essig.

5. Im Monat Juli, zu dieser Zeit soll nicht zur Ader gelassen werden, nimm auch keinen abführenden Trank zu dir; nimm Salbei und Raute.

6. Im Monat August iß auf gar keinen Fall Kohl, iß sauer Eingelegtes, trinke kein Bier und kein Met.

7. Im Monat September mußt du alles zu dir nehmen, worauf du Appetit hast, weil alle Speisen zu dieser Jahreszeit mit reifen Dingen zubereitet werden.

8. Im Monat Oktober iß Weintrauben und trinke Most, denn das macht den Körper gesund und führt ab.

9. Im Monat November und Dezember ist es gut darauf zu achten, daß man aus der Lebervene zur Ader läßt und blutig schröpft, weil zu dieser Zeit alle Säfte dafür bereit sind.

10. Im Monat Januar darfst du auf keinen Fall zur Ader lassen und einen Arzneitrank nur dann nehmen, wenn er gegen die Erstickung gerichtet ist. Man soll auch eine Latwerge (Elektuar) zu sich nehmen.

11. Im Monat Februar mußt du aus dem Daumen zur Ader lassen und Essighonig zubereiten, der dazu dient, die Galle hinauszubefördern und den Schleim durch den Darm, und er treibt den schwarzgalligen Saft, der, wie man feststellt, im Übermaß vorhanden ist, durch den Darm hinaus und löst alles Schlechte im Kopf ringsherum und alles, was aus einem Saft entsteht, entweder Sehstörungen oder gallige Krankheit, und verhindert die Erzeugung (dieses Saftes). Er heilt Blase und Nieren, macht Speisen bekömmlich und fördert die Verdauung. Er ist in jeder Hinsicht sehr angebracht.

Dieses Regimen wird man am ehesten Groenkes Typ IIa zuweisen, der ebenfalls mit dem Monat März beginnt; für ihn listet Groenke nicht weniger als sechsundzwanzig Zeugen auf; das sind so viele wie bei keinem anderen seiner Typen! Die erste Handschrift bei ihm ist zufällig Bamb. med. 1, das inzwischen in das Weltkulturerbe der UNESCO aufgenommene *Lorscher Arzneibuch* (Recept. Lauresh.). Hier ist das Regimen, beginnend oben auf fol. 41^v, von einer wohl kaum späteren Hand[19] geschrieben und bricht mitten im Monat Juni nach wenigen Zeilen ab; der Rest der betreffenden Seite ist leer geblieben[20]. Daraus ergibt sich klar, daß unser Regimen im Vat. Palat. Lat. 485 trotz der gleichen Schriftheimat[21] nicht aus Bamb. med. 1 kopiert worden sein kann; in dieselbe Richtung weist die Verderbnis †in orodi† aquam, wo der der Bamb. med. 1 gut lesbar mero de aqua hat.

Nun überspringen wir drei Jahrhunderte, zunächst den Ärmelkanal und anschließend den Atlantik, denn seit 1920 befindet sich die Handschrift, die den Text des Regimens ganz ähnlich überliefert wie unser Palatinus, in der National Library of Medicine in Bethesda, Maryland (Signatur E 8, früher nur 8 [Ricci]). Obwohl man sich dort ihrer Bedeutung bewußt ist, sind gegenwärtig nur wenige Blätter auf der Seite der Bibliothek einsehbar, und die detaillierte Beschreibung ist nicht allein unvollständig (das kann nach Lage der Dinge kaum anders sein), sondern bietet manches Mal ein fehlerhaftes (das heißt: falsch abgeschriebenes und nicht korrigiertes) Latein, das die Notwendigkeit, das Bildungswesen der USA zu reformieren, deutlich unterstreicht. Entgegen der Beschreibung („including many of the treatises constituting the compilation known as *Articella*") finden wir hier als zur *Articella* gehörend nur Philarets Pulsschrift[22], und unser Regimen ist, wen wollte es wundern, überhaupt nicht verzeichnet. Hier nun der Text von fol. 88$^{r\text{-}v}$:

Bethesdanus bibliothecae medicae nationis Americanae E 8

1. Mense mart. bibat dulce. usitet agriamen. ⟨asso⟩ balneo usitare. sanguinem minuere. solutionem non accipere. quia frigoras generat ipsa solutio.

---

19 Sie ist regelmäßiger und ausgeglichener als die Haupthand. Nichts Näheres zu den Händen bei Ulrich Stoll, *Das Lorscher Arzneibuch. Ein medizinisches Kompendium des 8. Jahrhunderts (Codex Bambergensis medicinalis 1). Text, Übersetzung, Fachglossar* (Stuttgart: Steiner, 1992) (Sudhoffs Archiv. Beihefte 28), und in der online-Beschreibung von M. Kautz. Das Regimen wäre als eigener Text aufzunehmen, da es sich um einen Nachtrag handelt, der in der Kapiteltafel, soweit ich sehe, fehlt.
20 Bamb. med. 1, fol. 38^v bringt ein Regimen mit einer abweichenden Zeiteinteilung. Groenke übergeht es bei seiner Beschreibung der Handschrift (Groenke, *Monatskalendarien*, 63). Es handelt sich um Groenke Typ VI, wofür er Vat. Reg. Lat. 1143, fol. 107$^{r\text{-}v}$ und 153$^{r\text{-}v}$ heranzieht. Auf fol. 153^v *trascalamer id est ippirico* ist, soweit ich sehe, der einzige Beleg für *triscalamus* (*trascalamer* ist die verballhornte Form, wie Groenke zutreffend festhält) außerhalb der Glossen.
21 Hier muß verwiesen werden auf die Behandlung durch Bernhard Bischoff, *Die Abtei Lorsch im Spiegel ihrer Handschriften*, 2. Auflage (Lorsch: Laurissa, 1989), 55.
22 Vgl. Ivan Garofalo, „Il *De pulsibus* di Philaretus e il Περὶ σφυγμῶν di Philaretos (con in appendice l'edizione del *De pulsibus*)," in María Teresa Santamaría Hernández (Hg.), *Textos médicos grecolatinos antiguos y medievales: estudios sobre composición y fuentes* (Cuenca: Ediciones de la Universidad de Castilla-La Mancha, 2012) (Humanidades. 123), 55–94, Edition = 81–94, mit neuer Einteilung in Abschnitte. In Garofalos Liste der lateinischen Handschriften auf S. 59 fehlt Bethesda.

2. Mense apl. sanguinem minuere. potionem bibere. carnes recentes manducare. *[lacuna ca. 20 litt.]* uenam epaticam incidere. calid. usitare. et si sic factum fuerit. omnia membra sanari debent.

3. Mense maio calidum bibat. uenam epaticam incidat. calidum usitet. caput purgabit. quia calidum in calorem precordia ponit. frigida licet. In mense mai uenam epaticam incidere et potionem ad soluendum bibere. cataplasma in capite poni. oculos turbulentos sanare. pruriginem mundare. urinam curare. olera frigida usitare. agriamen manducare.

4. Mense iunio omni die mane ieiunus mero de aqua bibat. nisi puscas usitare. lactucas manducare. acetum bibere.

5. Mense iulio non minuatur sanguis. nec de uenis in illo tempore nec potionem in ipso mense bibat. saluia et ruta usitet.

6. Mense augusto nullo penitus caule manducare. Agriamen manducare. ceruisiam non bibere.

7. Mense septemb. omnia que uis accipere debes. quia omnes escę cum tempore pro fructu confectę sunt.

8. Mense octob. racemis et musto usitare quia corpora sanat et solutionem faciunt.

9. Mense nouemb. et decemb. bonum est studium habere uenam epaticam incidere. gersis uentosarum imponere. quia in ipso tempore omnes humores sunt parati.

10. Mense ianuario *[fol. 88^v]* nullum penitus sanguinem minuere. nisi potionem contra officationem bibere debeant. et electuarium accipere.

11. Mense febr. de pollice sanguinem minuere debes. et oximel conficiendum facit ad colera deducenda. et flegma per uentrem et melancolicum quod plus inuenerit habundare expellit per uentrem et causam in capite circum soluet. et que ex humore nascuntur. aut caligo aut colera et ipsum non permittit generare. uesicam curat et renes. cibos bene accipere facit et bene digerere. in omnibus aptissimum est.

Der klassische Philologe in seiner Eigenschaft als Editor beklagt sich oft über die Entstellungen, die sich bei der Überlieferung eines Textes ergeben. Mich überrascht dagegen manchmal, wie groß die Übereinstimmung einzelner Textzeugen sein kann, wie wir das auch bei dem Vergleich von Vat. Palat. Lat. 485 und Bethesda, NLM E 8 sehen. Der für die Überlieferung gefährliche Zeitabschnitt war, wie mir scheinen will, der zwischen der Spätantike und der Karolingischen Reform, dunkle Jahrhunderte, in denen es keine Schulen mehr gab, die korrekte

Kenntnisse der lateinischen Schriftsprache vermittelt hätten und als viele Kopisten wohl die Texte, die sie kopieren sollten, selbst nicht mehr verstanden[23].

Offensichtlich wird das bei dem dritten Zeugen dieses Textes, nach neuerer Ansicht in der Gallia Narbonensis in einem Gebiet unter westgostischem Einfluß im späten 8. oder frühen 9. Jahrhundert (also ehe die Karolingische Reform sich auswirken konnte) niedergeschrieben, der berühmte, heute in der Universitätsbibliothek Glasgow aufbewahrte Hunterianus 96 (Codices Latini Antiquiores 156)[24]. Auf fol. 177$^{r\cdot v}$ treffen wir wieder auf unser Regimen:

Glasg. Hunter. 96
Incipit de anno ⟨c⟩erculo per omnes menses

1. mense marcium bibat dulce uset agriamen et radices confectas manducare et balneo usitare sanguine menuare solucionem non accipere quia ipsa solucio frigoras ienerat
   leuisticum et ruta frequencius bibere :,

2. mense [a]aprile sanguine minuare pocionem bibere carnem recentem manducare sanguine intercotaneo minuare calidum usare dolorem sthomaci purgare unguentu calasticum usare et si factum fuerit omnia membra sanare debent bethonica cum pipinella bibat +

3. mense maiu calidum bibat calidum uset capitem purget quia calidus in calore in precordia ponit et uena capitinale incidere et pocionem ad soluendum bibere cataplasmo inponi oculos turbulentus sanare purigine mundare hurina curare oleras frigidas non usare agramen manducare capite nullum penitus manducare
   absencium et fenuculi semen bibere ,

4. Mense iunio calicem pleno de aqua frigida bibere ceruisa non bibere lactucas manducare cum acet⟨o⟩ et pusca usare et depos caldum bibat
   saluia et flores de sauina bibat :,

5. Mense iulio sanguinem non minuare solucionem non accipere nec de uenis in illum tempus quia dies caniculares sunt
   saluia et ruta et flores appii ⟨bibere⟩

6. Mense augusto nullum penitus caulos non manducare agramen manducare ceruisa et metus non bibere
   pocione de pulegio bibe⟨re⟩ ,

---

[23] Ich habe den (nicht überprüften) Eindruck, daß in der Karolingerzeit meist junge, wenig erfahrene Kopisten die medizinischen Texte abschrieben, weil man die anderen, wertvolleren (nämlich alle kirchlich relevanten Texte und antike Literatur) ihnen nicht überlassen wollte.
[24] Auch er verdiente, in das Weltkulturerbe aufgenommen zu werden.

7. Mense septembre omnia quecumque ⟨h⟩abes signa et comed⟨e⟩[25] propter Infirmitatem carnis quia omnes fructi confecti sunt

costum cum mastice bibat:, *[fol. 177ᵛ]*

8. Mense octuber uua et mustum et mora comede quia corpus sanatur et [ab] solucionem[26] faciunt

et poti[c]onem cum gariofilis bibe :,

9. Mense nouember et december nouember non bagnare bonum est [e]studium ⟨h⟩abere uena epatica incidere ienesis[27] uentosa inponere quia omnes humores sunt parati Inde omnes morbos abstraere debent

pocionem de spico bibere debent :,

10. Mense ienuarium sanguinem non menuare solucionem non accipere nisi pocionem contra ofocaciones electuareia accipere debent

pocionem piper cum reuponticum bibat .,

11. Mense februariu de pollice sanguine minuare aximelli conficiendu facit ad colera deducenda et fleuma per uentrem melanconicum quod plus inuenitur expellit per uentrem et causa in capite circum soluet et que ex umore nascuntur aut calico aut colera et ipsa non permittit ienerare uisica curat et renes ciuum bene accipere facit et bene diierere in omnibus

hec probatum est ab ypograte :,

Nicht nur der womöglich nicht auf die oder eine Vorlage zurückgehende Verweis auf Hippokrates am Schluß, sondern mehr noch die in jedem Monat empfohlenen Tränke, die nur zum Teil mit den bei Groenke Typ IIa genannten übereinstimmen, sind der Grund, den Vat. Palat. Lat. 485 und Bethesda E 8 als untereinander enger verwandt anzusehen. Ich habe mich sehr gewundert, daß in Groenkes Studie ein Hinweis darauf fehlt, daß das Regimen Typ IIa später (als Kapitel 19 in der Druckfassung) Bestandteil der pseudosoranischen *Isagoge*[28] geworden ist, dort

---

25 *comed* natürlich *come* gesprochen, wie die 2. Person Imperativ Präsens Aktiv zu *comer* ‚essen' auch noch im heutigen Spanisch lautet.

26 Dem Schreiber war die christliche *absolutio* natürlich besser bekannt als *solutio*, geschweige denn der Sinn ‚abführen'. Ähnliches in meinem Aufsatz „Überlieferungs- und Verständnisprobleme im medizinischen Latein des frühen Mittelalters," *Berichte zur Wissenschaftsgeschichte* 17 (1994): 153–65.

27 Wohl *ieresis* herzustellen.

28 Ich gestatte mir den Hinweis auf meine ausführliche Studie „The *Isagoge* of Pseudo-Soranus. An Analysis of the Contents of a Medieval Introduction to the Art of Medicine," *Medizinhistorisches Journal* 35 (2000): 3–30. Die *Dieta Hippocratis per singulos menses anni obseruanda*, in *Physica S. Hildegardis ... Oribasii Medici de Simplicibus libri quinque, Theodori ... dieta ... Esculapii liber unus ...* (Argentorati: apud Ioannem Schottum, 1533), ist mit Typ IIa verwandt, beginnt aber mit dem Monat Januar statt März.

übrigens ebenfalls ohne die Tränke, die seinerseits der Hunter. 96 wiederum überliefert. Und wie wir im Laud. 426 bis gleich mehrere verschiedene Regimina hintereinander finden, steht auch in der Druckfassung der *Isagoge* von Albanus Torinus als Kapitel 20 ein weiteres Regimen, Groenke Typ III, was Groenke nicht vermerkt hat. Beide Regimina (IIa und III) bringt auch die Handschrift 1358 des Prager Metropolitankapitels aus dem 12. Jahrhundert hintereinander auf fol. 52ᵛ–53ᵛ. Es ist vielleicht instruktiv, ihren Text hier abzudrucken und anschließend den frühesten Zeugen, Hunter. 96, mit den beiden aus dem 12. Jahrhundert, Bethesda, NLM E 8 und Prag. Capitul. Metrop. 1358, zu vergleichen.

Prag. Capitul. Metrop. 1358

1. Mense martio dulci⟨b⟩us *[fol. 53ʳ]* oportet utere. dulcius bibere. acrum coctum comedere sanguinem non minuere solutionem uentris non accipere quia ipsa solutio frigora adtenuat.

2. Mense april. sanguinem minuere. potionem solutionis accipere carnes recentes comedere· balnea calida usitare. dolorem stomachi purgare. Unguentum chalasticum usitare.

3. Mense maio calida usitare· caput purgare. uenam epaticam incidere· potionem solutionis accipere· balneis calidis usitare· cataplasma in capite ponere· turbulentos oculos sanare· holeribus frigidis utere· acrum comedere.

4. Mense iun. omni mane ieiunus aquam binere· uinum quantum uolueris· ceruisam non bibere·

5. Mense iul. adiutoriis non usitare· sanguinem non minuere. saluia et ruta usitare.

6. Mense aug. caulos non comedere. acrum manducare. ceruisam et medonem recens bibere.

7. Mense sept. omnia quę uis comede. quia tunc omnis esca cum tempore suo confecta est.

8. Mense oct. racemis usitare mustum bibere quia corpori prodest et solutionem facit.

9. Mense nouemb. cinnamum uel cinamomum bibere. balneis non lauare quia tunc sanguis coagulatus est. et de calore humorem mouens· uenam epaticam incidere. generis uentosarum imponere.

10. Mense decemb. bonum studium impendere uenam capitalem incidere. generis uentosarum imponere. quia in ipso tempore omnes humores preparantur.

11. Mense ianuario sanguinem non minuere. nec adiutoria accipere. potionem propter suffocationem bibere. aliis uero potionibus quę uentrem laxant non utere.

12. Mense febr. de pollice *[fol. 53ᵛ]* manu⟨s⟩ sanguinem minuere. purgatoriis non utere.

Hunter. 96	Vat. Palat. Lat. 485	Bethesda NLM E 8
Incipit de anno ⟨c⟩erculo per omnes menses	Incipiunt tempora pro sanitate corporis quae obseruare debeant	
1. mense marcium bibat dulce uset agriamen et radices confectas manducare et ⟨asso⟩ balneo usitare sanguine menuare solucionem non accipere quia ipsa solucio frigoras ienerat	1. Mens. mar. bibat dulce. usitat agramen. usitet radices confectas manducare. Asso balneo usitare. sanguinem minuare. solutionem non accipere quia frigores generat ipsa solutio.	1. Mense mart. bibat dulce. usitet agriamen. ⟨asso⟩ balneo usitare. sanguinem minuere. solutionem non accipere. quia frigoras generat ipsa solutio.
leuisticum et ruta frequencius bibere		
2. Mense [a]aprile sanguine minuare pocionem bibere carnem recentem manducare sanguine intercotaneo minuare calidum usare dolorem sthomaci purgare unguentu calasticum usare et si factum fuerit omnia membra sanare debent	2. Mens. apr. sanguinem minuare. potionem bibere. carnes recentes manducare. sanguinem intercodinium minuare. calidos usitare. dolorem stomachi purgare unguentos calasticos usitare. et si sic factum fuerit omnia membra sanare debent.	2. Mense apl. sanguinem minuere. potionem bibere. carnes recentes manducare. *[lacuna]* uenam epaticam incidere. calid. usitare. et si sic factum fuerit. omnia membra sanari debent.
bethonica cum pipinella bibat		
3. mense maiu calidum bibat calidum uset capitem purget quia calidus in calore in precordia ponit et uena capitinale incidere et pocionem ad soluendum bibere cataplasmo inponi oculos turbulentus sanare purigine mundare hurina curare oleras frigidas non usare agramen manducare capite nullum penitus manducare	3. Mens. mai calidum bibat calidum usitat caput purget quia calidus in calore in praecordia ponit frigidum. licet in mense mad. uena epatica incidere. et potionem ad soluendum bibere. cataplasma in capite inponi oculos turbulentos sanare. pruri⟨gi⟩nem mundare. urinam curare. olera frigidas usitare. agramen manducare.	3. Mense maio calidum bibat. uenam epaticam incidat. calidum usitet. caput purgabit. quia calidum in calorem precordia ponit. frigida licet. In mense mai uenam epaticam incidere et potionem ad soluendum bibere. cataplasma in capite poni. oculos turbulentos sanare. pruriginem mundare. urinam curare. olera frigida usitare. agriamen manducare.
absencium et fenuculi semen bibere		
4. Mense iunio calicem pleno de aqua frigida bibere ceruisa non bibere lactucas manducare cum acet⟨o⟩ et pusca usare et depos caldum bibat	4. Mens. iun. Omni die mane ieiunus †in orodit† aquam bibere. ceruisam non. nisi spica usitare. lactuces manducare. acetum bibere.	4. Mense iunio omni die mane ieiunus mero de aqua bibat. nisi puscas usitare. lactucas manducare. acetum bibere.
saluia et flores de sauina bibat		

5. Mense iulio sanguinem non minuare solucionem non accipere nec de uenis in illum tempus quia dies caniculares sunt	**5.** Mens. Iul. non minuetur sanguis de uena in hoc tempore. nec potionem in ipso mense bibat.	**5.** Mense iulio non minuatur sanguis. nec de uenis in illo tempore nec potionem in ipso mense bibat.
saluia et ruta et flores appii ⟨bibere⟩	saluia et ruta usitet.	saluia et ruta usitet.
6. Mense augusto nullum penitus caulos non manducare agramen manducare ceruisa et metus non bibere	**6.** Mens. aug. nullo penitus caule manducet. agramen manducet ceruisam et me[m]tum non bibet.	**6.** Mense augusto nullo penitus caule manducare. Agriamen manducare. ceruisiam non bibere.
pocione de pulegio bibe⟨re⟩		
7. Mense septembre omnia quecumque ⟨h⟩abes signa et comed⟨e⟩ propter Infirmitatem carnis quia omnes fructi confecti sunt	**7.** Mens. Sep. Omnia quae uis accipere debes. quia omnes escae cum tempore pro fructa confecta sunt.	**7.** Mense septemb. omnia que uis accipere debes. quia omnes escę cum tempore pro fructu confectę sunt.
costum cum mastice bibat		
8. Mense octuber uua et mustum et mora comede quia corpus sanatur et [ab]solucionem faciunt	**8.** Mens. oct. racemis et musto usitare quia corpora sanat. et solutionem facit.	**8.** Mense octob. racemis et musto usitare quia corpora sanat et solutionem faciunt.
et poti[c]onem cum gariofilis bibe		
9. Mense nouember et december nouember non bagnare bonum est [e]studium ⟨h⟩abere uena epatica incidere ienesis uentosa inponere quia omnes humores sunt parati Inde omnes morbos abstraere debent	**9.** Mens. nou. et dec. bonum est studium habere. uena epatica incidere. gersis uentusarum inponere. quia in ipso tempore omnes humores sunt parati.	**9.** Mense nouemb. et decemb. bonum est studium habere uenam epaticam incidere. gersis uentosarum imponere. quia in ipso tempore omnes humores sunt parati.
pocionem de spico bibere debent		
10. Mense ienuarium sanguinem non menuare solucionem non accipere nisi pocionem contra ofocaciones electuareia accipere debent	**10.** Mens. Ian. nullus penitus sanguinem minuare debet. nisi potionem contra officationem bibat. debeat et ⟨e⟩lectuariam accipere.	**10.** Mense ianuario nullum penitus sanguinem minuere. nisi potionem contra officationem bibere debeant. et electuarium accipere.
pocionem piper cum reuponticum bibat		

11. Mense februariu de pollice sanguine minuare aximelli conficiendu facit ad colera deducenda et fleuma per uentrem melanconicum quod plus inuenitur expellit per uentrem et causa in capite circum soluet et que ex umore nascuntur aut calico aut colera et ipsa non permittit ienerare uisica curat et renes ciuum bene accipere facit et bene diierere in omnibus	11. Mens. feb. de police sanguinem minuare debes. et oxomelli conficiendum facit. ad colera deducenda et fleoma per uentrem et melancolico quod plus inuenitur adbundare expellit per uentrem et causam in capite circum soluet. et qui ex humore nascuntur aut calico aut colera. et ipsum non permittit generare. uesicam curat et renes. cibos bene accipere facit et bene digerere. in omnibus aptissimum est.	11. Mense febr. de pollice sanguinem minuere debes. et oximel conficiendum facit ad colera deducenda. et flegma per uentrem et melancolicum quod plus inuenerit habundare expellit per uentrem et causam in capite circum soluet. et que ex humore nascuntur. aut caligo aut colera et ipsum non permittit generare. uesicam curat et renes. cibos bene accipere facit et bene digerere. in omnibus aptissimum est.

hec probatum est ab ypograte

Die Gegenüberstellung läßt deutlich erkennen, daß es beim Typ IIa eine Untergruppe gibt, die den einzelnen Monaten nicht, wie es die vermutlich älteste Handschrift Glasgow, University Library, Hunter. 96, tut, als Trank einzunehmende Heilpflanzen zuordnet[29]. (Nur diese Tränke, allerdings mit jeweils zusätzlichen Pflanzen, bringt Groenkes Typen IIb und VI, in anderer Weise Typ III.) Auch weitere von Groenkes Typen bringen solche Heiltränke, doch verrät uns Groenke leider nirgends, welchen Kriterien er bei seiner Einteilung in Typen folgte, was gerade bei der Einordnung neuer Textzeugen hilfreich wäre. Es ist verwunderlich, daß nach Typ III der Rest (IVa, IVb, IVc, V, VI, VII, VIII und IX) nur durch jeweils eine[30] Handschrift vertreten ist. Es wäre auch interessant zu wissen, welche Typen in die Volkssprachen übersetzt wurden und so weitergewirkt haben.

Vorbeugung als der bessere Weg der Pflege der Gesundheit wird gerade wiederentdeckt, doch auch bei den Vorläufern von der Spätantike bis zur Renaissance (Druck der pseudosoranischen *Isagoge* 1528) bleibt noch viel zu tun. Auch wenn die hier behandelten Texte kaum zu heutiger Anwendung einladen, präsentiere ich sie Heinrich von Staden mit allen guten Wünschen für Gesundheit und Schaffenskraft.

---

29 So auch, mit teilweise anderen Pflanzen, Groenke Typ IX.
30 Vat. Reg. Lat. 1143 überliefert denselben Text von *Praecept. diaet.* VI zweimal, fol. 107$^{r\text{-}v}$ und 153$^{r\text{-}v}$.

Appendix Bethesdana

fol. 1ʳ *emplastrum* similar to Vindoc. 109 fol. 118ᵛᵃ.

fol. 1ᵛ-7ᵛ *emplastra* from Oribasius *syn.* book 3, beginning with 3.14 (in a version different from that printed in Molinier p. 853). Orib. *syn.* 3.13 follows, then *syn.* 3.12, Orib. *syn.* 3 add. p. 897 Mol. (*emplastrum usia*), Orib. *syn.* 3.17 (like Mol. p. 854) and a hitherto unidentified *emplastrum*. The ms. continues with Orib. *syn.* 3.2; 3.3; 3.4; 3.5; 3.6.3-4; Orib. *syn.* 3 add. p. 894 Mol. (*emplastrum pentatheum*; rather than Orib. *syn.* 3.7, which is much shorter); Orib. *syn.* 3.8; 3.9.1; 3.10; Orib. *syn.* 3 add. p. 894 Mol. (two recipes); Orib. *syn.* 3.9.2; Orib. *syn.* 3 add. p. 894 Mol.; Orib. *syn.* 3.9.3 (missing from the Greek text); Orib. *syn.* 3 add. p. 894 Mol. (three recipes); Orib. *syn.* 3.11. Next, two unidentifed *emplastra di airon*, followed by Orib. *syn.* 3.21; 3.23; 3.24; 3.25; 3.26; 3.30; 3.27; 3.28; 3.29; 3.31; 3.32; ends imperfectly at the end of the quire (fol. 7ᵛ) after the first words of Orib. *syn.* 3.33.

fol. 8ʳ shortened versions of Diosc. *mat. med.* 4.148 (4.143 lat.) and 4.162, followed by some lines *De pulmone* attributed to Constantinus (Africanus) and on the bite of a mad dog.

fol. 8ᵛ Short text on the humours, beginning *Sanguis naturaliter est calidus et humidus. cuius quedam pars in arteriis subtilis est* (not in Thorndike-Kibre[31]), continuing with *flegma* and *colera rubea*. A new hand copies a few lines taken from a phlebotomy treatise, inc.: *Sanguis si niger fuerit usque ad fus⟨c⟩um calorem emittatur*, then breaks off, leaving the last third of the page blank.

fol. 9ʳ-10ʳ recipes and short excerpts from medical works.

fol. 10ʳ *Incipiunt urine mulierum* = Thorndike-Kibre 1149.10. Remainder of page and fol. 10ᵛ blank.

fol. 11ʳ-12ʳ religious content, but verses on the four temperaments in a much later medieval hand added on fol. 12ʳ (*Largus amans hillaris* [sic] ...), fol. 12ᵛ is blank.

fol. 13ʳ⁻ᵛ preparation of medicinal substances, beginning with *cerussa*, and more recipes.

fol. 14ʳ-77ᵛ First Latin commentary on Hippocrates, *Aphorisms* (Lat-A), cf. Kibre, *Hippocrates Latinus*, 35a and 37a. Breaks off after the first words of Hipp. *aph.* 7.16.

---

31 Lynn Thorndike and Pearl Kibre, *A catalogue of incipits of mediaeval scientific writings in Latin*, rev. ed. (Cambridge, Mass.: The Mediaeval Academy of America, 1963) (now searchable online on the websites of the Mediaeval Academy of America and the National Library of Medicine).

fol. 78ʳ–87ᵛ *Incipit liber alexandri*, a collection of recipes similar to London, British Library, Arundel 166, fol. 15ʳ–44ᵛ (also found in Hunter. 96), not related to Alexander of Tralles. This collection contains as well interspersed excerpts from Theodorus Priscianus, books 1 and 2.

fol. 88ʳ–95ᵛ *Confectio qua utebatur aristopholus rex*, followed by a *Regimen XII mensium*, Groenke Type IIa (transcribed above), and more recipes.

fol. 96ʳ–97ʳ Various texts on Weights and Measures: Thorndike-Kibre 1506.10; 1506.8; 1506.9; 1555.2; followed by an item not in Thorndike-Kibre, inc.: *Chorus enim modii XXX*.

fol. 97ʳ *Uomitus iohannis scarpelli salernitani* and, in a new hand, more recipes.

fol. 100ᵛ Prayers to be recited when administering medicines, the second edited by Ernest Wickersheimer[32] from Vindoc. 109 fol. 98ʳᵃ.

fol. 101ʳ *Incipit prologus Galieni De pulsu et urinis*; Thorndike-Kibre identify this, wrongly, with Ps.-Alex. Trall. *puls. ur.* ed. Stoffregen. The beginning of the text that follows was edited by Leisinger[33], 36–42 according to München, Bayerische Staatsbibliothek, clm 11343, fol. 1ᵛ–3ʳ; the portion in the Bethesda ms. corresponds to Leisinger, p. 36–37 (middle).

It is followed *by De urinis gallienus. Urina subalba et uernitia in homine satis comedente et bibente quasi subaurea. incolumem esse significat.* (not in Thorndike-Kibre) and ends *non sanabilem significat* on fol. 102ʳ.

fol. 102ʳ *De pulsibus. Pulsus senis et iuuenis crassi in estatem idem sunt.* (not in Thorndike-Kibre)

fol. 102ʳ *Pronostica galieni. Species et uultus infirmi si bene compositi fuerint ...* (not in Thorndike-Kibre), ending *propriam linguam mente captus mandet. et difficile sanabitur* on fol. 103ᵛ.

fol. 103ᵛ–120ᵛ Oribasius, *euporista* (version La), book 2, Molinier p. 425.

fol. 120ᵛ *Inquisitiones uenarum sunt multe* (cf. Trier, Stadtbibliothek, ms. 40, fol. 21ᵛ–24ʳ, ed. Ferckel[34] p. 135–136, for a fuller version).

fol. 120ᵛ *Ad febres. Antid. pigra.*

fol. 120ᵛ–128ᵛ *Incipit dinamedia ypocratis*; fol. 121ʳ = Rose[35] p. 132.15.

---

[32] Ernest Wickersheimer, „Bénédiction des remèdes au Moyen Age," *Lychnos* (1952): 96–101, at 99.

[33] Hermann Leisinger, *Die lateinischen Harnschriften Pseudo-Galens* (Zürich: Orell Füssli, 1925) (Beiträge zur Geschichte der Medizin. 2).

[34] Christ. Ferckel, „Medizinische Marginalien aus dem Cod. Trevirens. Nr. 40," *Archiv für die Geschichte der Medizin* 7 (1913): 129–43.

[35] Valentin Rose, *Anecdota Graeca et Græcolatina. Mitteilungen aus Handschriften zur Geschichte der griechischen Wissenschaft*, Zweites Heft (Berlin: Dümmler, 1870).

fol. 128ᵛ–129ʳ *De passionibus unde eueniunt* = part 3 of the *Sapientia artis medicinae*³⁶.

fol. 129ʳ–130ʳ *De diuersitate potuum* = Isid. *orig.* 20.3.1–21.

fol. 130ʳ–130ᵛ *De signis ponderum secundum grecos*, inc.: *Ponderis signa plerisque ignota sunt.*

fol. 130ᵛ–131ʳ *De propriis nominibus arborum* = Isid. *orig.* 17.7.1–6.

fol. 131ʳ–146ᵛ Oribasius, *synopsis*, book 1 (version La).

fol. 146ᵛ–147ʳ Ps.-Hipp. *epist. ad Maecenatem*.

fol. 147ʳ⁻ᵛ Ps.-Hipp. *epist. ad Antig.*, shortened (= Beda, *temp. rat.* 30, p. 372.15–374.47).

fol. 148ʳ *Sphaera Pythagorae* (also called, more often, *Sphaera Apulei*).

fol. 148ᵛ list of popes, beginning with Pope Alexander.

fol. 149ʳ–158ᵛ Constantinus Africanus, *Pantegni*, Practica, book 9 (surgery, ed. Pagel³⁷).

fol. 158ᵛ–159ʳ A short treatise (probably an excerpt from a longer work) on urines, *De significatione ypostasis*, inc.: *Oportet ut intelligas quia ypostasis lenis et alba in fundo.*

fol. 159ʳ⁻ᵛ *Ut scias si homo sit sanus an infirmus*, inc.: *Oportet medicum corpus cognoscere sanum.*

fol. 160ʳ *Medicamen fortissimum*; remainder of page and fol. 160ᵛ blank.

fol. 161ʳᵃ–168ʳᵃ *Incipit alphabetum herbarum grece et latine*, inc.: *Agoriocannaui (m) id est Cantabrum* (ed. García González³⁸).

fol. 168ʳᵇ–169ᵛᵇ *Incipit epla. Galieni in antiballomen(on)* (Ps.-Gal. *de succedaneis*, here including preface).

fol. 170ʳ–175ʳ Isid. *orig.* 17.8–11.

fol. 175ʳ–176ʳ *Incipit de herbis Galieni Apolloni et Ciceronis* (ed. Ferraces Rodríguez³⁹, who did not know this ms.).

---

36 Momtschil Wlaschky, „*Sapientia artis medicinae*. Ein frühmittelalterliches Kompendium der Medizin," Kyklos (Leipzig) 1 (1928): 103–13.

37 Julius Pagel, „Eine bisher unveröffentlichte lateinische Version der Chirurgie der Pantegni nach einer Handschrift der Königl. Bibliothek zu Berlin [lat. fol. 74, s. XII]," *Archiv für klinische Chirurgie* 81 (Teil 1) (1906): 735–86.

38 Cf. Alejandro García González, „*Agriocanna*, a new medico-botanical glossary of pre-Salernitan origin," in David Langslow and Brigitte Maire (Hgg.), *Body, Disease and Treatment in a Changing World. Latin texts and contexts in ancient and medieval medicine* (Lausanne: Bibliothèque d'Histoire de la Médecine et de la Santé, 2010), 223–35, without the Bethesda ms.

39 Arsenio Ferraces Rodríguez, *Estudios sobre textos latinos de fitoterapia entre la antigüedad tardía y la alta edad media* (A Coruña: Universidade da Coruña, 1999) (Monografías. 73), Apéndice 2: Fragmentos de los *Dynamidia* pseudohipocráticos, 401–17, at 403–406, from Par. lat. 11219, fol. 207ʳᵇ–209ᵛᵇ.

fol. 176ᵛ *Urina est sanguinis maxime et ceterorum humorum colamentum. quę oritur de naturali operatiua uirtute.*

fol. 177ᵛ–179ᵛ Philaretus *de pulsibus* (ed. Garofalo).

fol. 179ᵛ *Curatio elefantiosorum. In primis fleuothomentur et clistere purgentur* (Esculapius 38.7 ff. Manzanero Cano[40] = 39 Schott).

fol. 180ʳ⁻ᵛ Ps.-Hipp. *capsula eburnea.*

fol. 180ᵛ–181ʳ *Prognostica Turicensia* (unedited, inc.: *Item. De tissicis et pleureticis. Sic probabis ęorum uitam. Quod expuerint mitte in carbonibus. Si fetuerit. non euadet).*

fol. 182ʳ⁻ᵛ Ps.-Alex. Trall. *puls. ur.* ed. Stoffregen[41], beginning with line 269 and ending with line 368, then continued on top of fol. 184ʳ and ending there with line 372.

fol. 183ʳ⁻ᵛ Thorndike-Kibre 552.3 Isaac *de febribus* trad. Constantinus Africanus (fragment).

fol. 183ᵛ preparation of various medicinal substances etc., starting with *calcu cecaumenu*, continued on fol. 185ʳ.

fol. 184ʳ⁻ᵛ *Ypocras dixit. Urina naturalis quę significat corporis sanitatem exit sine labore et dolore* (not in Thorndike-Kibre).

fol. 185ʳ Ps.-Hipp. *epist. ad Maecen.* 5.

fol. 185ᵛ *Electuarium Karoli Regis*, some more *electuaria* follow. This is the last page with medical content.

Appendix lexicographica: *gersa / garsa / geresis*

Im Calendarium (= Praecept. diaet. des *Mittellateinischen Wörterbuchs*) I Groenke, Zeile 59 (15 Handschriften), und in Praecept. diaet. IIa Groenke, Zeile 46, taucht ein Terminus auf, der trotz der Erläuterung bei Groenke (s. unten) und den Ausführungen von Johannes Staub im Lemma *garsa* des *Mittellateinischen Wörterbuchs* Anlaß zu weiteren Bemerkungen gibt, nachdem sich die Romanisten schon wiederholt dazu geäußert haben, offensichtlich auch, ohne eine abschließende Einigung zu erzielen. (Mit Herrn Staub habe ich auch über die Sache im September 2013 korrespondiert.)

---

40 Francisco Manzanero Cano, *Liber Esculapii* (Anonymus Liber Chroniorum). *Edición crítica y estudio* (Madrid: Diss. phil. Universidad Complutense de Madrid, 1996).
41 Malte Stoffregen, *Eine frühmittelalterliche Übersetzung des byzantinischen Puls- und Urintraktats des Alexandros. Text, Übersetzung und Kommentar* (Berlin: Diss. med. Freie Universität Berlin, 1977).

Die Romanisten – und ebenso J. Staub im *Mittellateinischen Wörterbuch*, Band 4, s. v. *garsa (gersa) – gehen von altfranzösisch jarse ‚Schröpfmesser, Messerchen zum Aderlaß' (so die Übersetzung bei Gamillscheg[42] s. v. gerce 1., vgl. REW 1863b) aus; genauer ist Groenke (250): „gersis] Bezeichnung für das Skarifikationsmesser, vgl. dazu ἐγχάραξις Skarifikation und ἐγχαράσσω eine Inzision mit dem Skarifikationsmesser vornehmen sowie [? K.-D. Fi.] Skarifikationsmesser[43]. Vgl. dazu Du Cange, Glossarium, Bd. 4 S. 38b." Allerdings setzte Groenke in Praecept. diaet. I *gersit ventosarum imponere* in den Text (im Apparat: gersit] genus Ha: generis[44] MpPh: genera omnia Va: genera BrHu: grassis Var3: et a venere *post* 56 *pro* non larvare [Tippfehler für lavare? K.-D. Fi.] *pos.* Cp: Var3: *om.* Pa9), während im Kommentar die Lesart *gersis* steht, vermutlich ein Flüchtigkeitsfehler, der den Gutachtern der zahnmedizinischen Dissertation, Gerhard Baader und Johanna Bleker, nicht aufgefallen war. In Praecept. diaet. IIa lautet Groenkes Text *gersis ventosarum inponere*[45] (im Apparat: garsas Vap2: garsis La[46]: *om.* Lh1). Halten wir an dieser Stelle immerhin fest, daß die Variante des Wortes mit einem *a* (*garsis*) gegenüber der mit *e* (*gersis*) nur schwach bezeugt ist, nochmals nämlich in *Praecept. diaet.* V Zeile 53 *garsas*[47]. Um welche grammatische Form des Wortes es sich handeln mag, lassen wir noch unentschieden.

Es ist unumgänglich, daß wir zunächst die Sache selbst, um die es geht und die mit *gersis uentosarum imponere* bezeichnet wird, klären. Unter Schröpfen verstehen wir das Aufsetzen von Schröpfköpfen auf die Haut an verschiedenen

---

42 Emil Gamillscheg, *Etymologisches Wörterbuch der französischen Sprache* (Heidelberg: C. Winter, 1969).
43 In *Praecept. diaet.* 5 Zeile 53 übersetzt Groenke (217) ‚Aderlaßmesser', ohne diese Abweichung weiter zu begründen.
44 Das ist auch die Lesung (zweimal) in der Groenke nicht bekannten Handschrift von Typ IIa, Praha, Capitul. Metrop. 1358, s. 12, fol. 52^v–53^r. Auf fol. 53^v folgt dort Groenke Typ III. Typ IIa und III sind Ps.-Sor. *isag.* 19–20; Torinus (1528, eine neuere Ausgabe gibt es nicht!) druckt *genibus uentosas imponere*, vermutlich die Korrektur eines Humanisten, wenn nicht von ihm selbst. Hätten wir nicht die frühen Handschriften dieses Textes, wäre man geneigt, sich der Lesung *genibus* anzuschließen, zumal sie gut zur Überschrift im lateinischen Oribasius *syn.* 1,14 *De charaxatione tibiarum* paßt und dem antiken Prinzip der Antispasis (ableitende Maßnahme von einem entfernten Ort) entspricht.
45 Groenke erklärt nicht, warum er in *Praecept. diaet.* 1 *imponere*, hier aber *inponere* bevorzugt, aber das ist für Philologen wie für Zahnärzte eher belanglos.
46 La ist der Laudunensis 426*bis*, das *bis* ist in Groenkes Siglenliste (116) versehentlich ausgelassen.
47 Der eng verwandte Text in London, British Library, Sloane 2839, fol. 109^v–108^r (sic! Blattversetzung) hat die betreffende Anweisung nicht. Groenke bringt ihn bei IIa, ohne das näher zu begründen.

Stellen des Körpers. In solchen aus verschiedenen Materialien[48] hergestellten Schröpfköpfen wird ein Unterdruck erzeugt (durch eine Flamme oder, in der primitivsten Form, indem ein ausgehöhltes Horn mit dem Mund ausgesaugt und dann z. B. mit Wachs verschlossen wird). Durch diesen Unterdruck bleibt der Schröpfkopf haften, und es wird Blut in das Gewebe unter der Haut angesaugt. Wenn die Haut geschlossen und nicht eingeschnitten worden ist (*scalpello* ‚mit dem Messer' bei Celsus), bleibt es bei einer lokalen Hyperämie (das ist das trockene Schröpfen). Werden in die Haut Einschnitte gemacht und danach der Schröpfkopf aufgesetzt, tritt infolge der Saugwirkung Blut aus (das ist das blutige Schröpfen). In der antiken Medizin handelt es sich um eine ableitende (oder entleerende) Maßnahme, übrigens genau wie beim Aderlaß aus der Lebervene, der in beiden Fassungen des Textes unmittelbar vorangeht.

Einschnitte oder Einritzungen vornehmen heißt in den griechischen Texten (am ausführlichsten wohl dargestellt bei Antyllos bei Orib. *coll. med.* 7,18 und bei Apollonios bei Orib. *coll. med.* 7,19) ἐγχαράσσειν (das davon abgeleitete Substantiv ist ἐγχάραξις; andere Synonyme s. unten); im Zusammenhang mit dem Schröpfen (σικύασις, der Schröpfkopf heißt σικύη) verwendet Antyllos (Orib. *coll. med.* 7,16,3) ἐγχαράσσειν. Von einem speziellen Instrument zum Einritzen der Haut hören wir, soweit ich weiß, in den antiken Quellen nichts, weshalb Groenkes Interpretation ‚Skarifikationsmesser' ebenso wie die bei Gamillscheg (‚Schröpfmesser') noch zu beweisen wäre; auch Meyer-Lübkes Übersetzung (REW 1863b) „frz. jarce „Schröpfkopf"" scheint mir zweifelhaft, denn das Französische verfügt über ein anderes, auch in unseren lateinischen Texten gut belegtes Wort, ventouse, m. E. zu spätlateinisch *uentosa*[49] (klassisch *cucurbita*[50], wohl als Lehnübersetzung von σικύη).

---

48 Cels. 2,11,1–2: *Cucurbitularum duo uero genera sunt, aeneum et corneum. Aenea altera parte patet, altera clausa est: altera cornea parte aeque patens altera foramen habet exiguum. In aeneam linamentum ardens coicitur, ac sic os eius corpori aptatur inprimiturque, donec inhaereat... Utraque non ex his tantum materiae generibus, sed etiam ex quolibet alio recte fit: ac si cetera defecerunt, caliculus quoque aut pultarius oris compressioris ei rei commode aptatur.* Vgl. die Erläuterungen in Aulus Cornelius Celsus, *Über die Arzneiwissenschaft in acht Büchern*, übersetzt und erklärt von Eduard Scheller, 2. Auflage von Walther Frieboes (Braunschweig: Fried. Vieweg und Sohn, 1908), 495.
49 Daß Gamillscheg mit seiner Herleitung von ventouse als „postverb. Subst. von ventouser „einen Schröpfkopf ansetzen", 12. Jh. dieses aus vlat. *ventosare ..." das Richtige trifft, wage ich zu bezweifeln (vgl. kurz *REW* 9207a), mir scheint es eher umgekehrt zu sein. Orib. *syn.* 1,13 Aa hat als Überschrift *De cocurbitis id est uentosis*. (Ich zitiere Aa nach der Ausgabe von Federico Messina, *La redazione Aa della traduzione latina della Synopsis ad Eustathium di Oribasio (libro primo). Introduzione, testo critico, traduzione e commento*, tesi di laurea, Università degli Studi di Catania, anno accademico 2003/2004.) Ps.-Democr. *med.* 77 tit. hat *De uentosis quas cufas*

Nun haben wir zwar keine lateinische Übersetzung der auch im Griechischen nur zum Teil überlieferten *Collectiones medicae* des Oribasius, aber gleich mehrere der in der *Synopsis* vorliegenden komprimierten Fassung (*syn.* 1,14) für seinen ebenfalls als Arzt praktizierenden Sohn Eustathios[51], unter dem Titel Περὶ ἐγχαράξεως. Das wird uns helfen, das Problem *gersis/garsa* weiter zu klären.

Sowohl die Version Aa wie La geben den Titel des Kapitels mit *De incaraxatione* wieder, Aa hat *syn.* 1,14,1 auch das Verb *incaraxare* (zu beiden Wörtern vgl. ThlL s. v. incha-). Es ist instruktiv, hier die Fassung im *Liber medicinalis* des Pseudo-Democritus zu vergleichen, dessen Ausgabe ich seit langem vorbereite:

> De charaxationibus. Apollonii.
>
> 78,1 Venae si saepius inciduntur, pariter uitalis spiritus multus digeritur et corpora debilitantur. Ideoque illis utendum est ut corpora non exanimentur. Hoc magnum est adiutorium, quibus opus est sanguinem euacuari.
>
> 78,2 Geresis iuuat in †curat† et oculorum reumatismo et aliis circa caput passionibus uel synanchis uel cynanchis, in ceruice si acceperint charaxationes siue in thorace, uel in aliis locis si fuerit constrictio, soluit uentosa cum charaxatione.

Im Titel *De charaxationibus* habe ich mich enger an die Überlieferung angeschlossen als seinerzeit Heeg, der wie in den Fassungen Aa und La des Oribasius hier und an den beiden folgenden Stellen *incaraxationibus* usw. vorzog, doch

---

uocant; *cufa* (zu griechisch κοῦφος ‚leicht') sehe ich hier nicht als Synonym von *uentosa*, sondern verstehe ‚leichtes (und somit auch trockenes, ohne Einritzen der Haut vorgenommenes) Schröpfen'; Antyllos (Orib. *coll. med.* 7,16,1) unterscheidet zwischen „leichten (ἐλαφραῖς) Schröpfköpfen, und zwar denen ohne Einritzen", und „starken (bzw. heftigen) Schröpfköpfen, und zwar mit Einritzen" (ich möchte p. 215,22 das δὲ streichen und erwäge auch, statt τοῖς ... ἐνοχλουμένοις eher τῶν ... ἐνοχλουμένων zu lesen). Cael. Aur. *acut. pass.* 3,21,201 *infigimus praeterea cucurbitas leues, quas Graeci* κούφας *uocant, scilicet sine scarificatione* kann den Sachverhalt weiter klären. (Die Übersetzung Cael. Aur. *tard. pass.* 5,4,74 *cucurbitae appositio leuis, quam Graeci cuphen appellant* „das leichte Aufsetzen eines Schröpfkopfes, das die Griechen κούφη nennen", würde ich mit dem Index der Ausgabe S. 1008 anders verstehen, nämlich als ‚das Aufsetzen eines leichten Schröpfkopfes, den ...' Hier folgt allerdings *cum adiecta scarificatione*, was ich mir nicht erklären kann.)

50 Cels. benutzt für den Schröpfkopf (bis auf eine Stelle, die vermutlich zu korrigieren ist) immer *cucurbitula*, und *cucurbita* für das Gemüse. Zu *cucurbita* das Verb *cucurbito, are* (griechisch σικυάζω) und das nur einmal, bei Caelius Aurelianus belegte Substantiv *cucurbitatio*. Theodorus Priscianus *eupor.* 2,27 *stomacho uentosas impono frequenter cucurbitas* möchte ich *cucurbitas* streichen; womöglich ist das Wort auch Theodorus Priscianus *eupor.* 2,29 *utimur uentosis cucurbitis* eingedrungen, doch findet sich die Verbindung auch noch Theodorus Priscianus *eupor.* 2,33; 2,53; 2,86 und 2,117.

51 Die Form *eustadium* in Bethesda, NLM E 8 ist eine im Spätlateinischen nicht selten anzutreffende intervokalische Sonorisierung.

schreibt auch Messina am Schluß dieses Kapitel *caraxationes. charaxo* und Ableitungen sind erstaunlich gut belegt (vgl. ThlL s.v.), sodaß man sich wundert, warum Meyer-Lübke, *REW* 1863b ein Sternchen bei *charassare für erforderlich hielt. Daß *-x-* zwischen Vokalen als *-ss-* ausgesprochen wurde, braucht nicht betont zu werden, ja Meyer-Lübke ist vielleicht sogar bei der Schreibung *charassare* zu folgen; *caraps-* ist bei Pseudo-Democritus handschriftlich gut bezeugt, vertritt natürlich ebenfalls *c(h)arass-*, und selbst der Schreiber der jüngsten, im Jahre 1516 kopierten Handschrift M, der an den beiden ersten Stellen noch zu *captionibus* und *cathaplasmationes* verschlimmbessert hatte, bringt am Ende, bei der dritten, *carapsatione* wie die übrigen vier[52].

Wie gelangen wir von *c(h)arass-* zu *gars-*? Da das erste *a* den Nebenton trug, leuchtet es ein, daß das zweite *a* vor dem dritten *a* mit dem Haupton ausfiel[53]; *c* zu *g* ist, gerade bei griechischen Wörtern, die in die lateinische Sprache aufgenommen worden sind, eher die Regel als eine Ausnahme.

Für Ps.-Democr. med. 78,2 *geresis* haben Aa und La übereinstimmend *incaraxatio* und der griechische Text ἐγχάραξις, doch möchte ich nicht daran denken, daß hierin eine Form von ἐγχάραξις zu sehen ist (*ieresis* der Handschrift O steht m. E. für *geresis*, genau wie *goresis* in M; *aferesis* wollte Heeg lesen). Es spricht viel dafür, daß wir in *geresis* eine vulgäre Schreibung für διαίρεσις sehen sollten[54], vgl. die lateinische Übersetzung von Ps.-Gal. introd. s. medic. 19,8 (14.784.14 Kühn), wo τριῶν διαιρέσεων *tres geresis* entspricht[55].

Weniger einfach liegt der Fall bei Orib. syn. 7,14,3 p. 147 Molinier, wo Aa *mox autem in primis diereses dandas sunt* schreibt, La *dieresis autem datus in initio*, denn im griechischen Text heißt es ἀποσχάζοντες δ' ἐν ἀρχῇ τὰ ἐκχυμώματα. Doch sowohl ἀπο- wie κατασχάζω samt ihren Ableitungen bedeuten ‚skarifizieren, Einschnitte in die Haut machen'. Daraus schließen wir, daß in den lateinischen Oribasiusübersetzungen Aa und La (und wir dürfen jetzt hinzufügen: ebenso beim Pseudo-Democritus) *diaeresis* speziell das Skarifizieren bedeuten kann. Damit eröffnet sich nun die Möglichkeit, *gersis* in *Praecept. diaet.* ebenfalls zu *diaeresis* zu stellen und eine Entwicklung *geresis* → *gersis* anzunehmen; an διαίρεσις hatten auch Romanisten schon gedacht. Und es könnte so sein, daß die scheinbaren

---

52 In der *Bamberger Chirurgie* (*Tract. de chirurg.*) Zeile 1575 Sudhoff (S. 379) lesen wir ebenfalls *caraxacio*.
53 Anders Gamillscheg s.v. gerce 1. (S. 476): „Unmittelbar zu lat. *charaxare* ... ist lautlich wegen des Schwundes des zweiten *-a-* schwierig."
54 *diaeresis* → *dier-* (zweisilbig) → *dier-* einsilbig mit Palatalisierung, wie *diurnus* → italienisch *giorno*.
55 Gleich danach die Schreibvariante *zeresim*.

Varianten *garsis* und *gersis* gar keine Varianten im eigentlichen Sinne darstellen, sondern auf zwei verschiedene, aber hier gleichbedeutende Wörter zurückgehen.

Betrachten wir noch eine zweite Oribasiusstelle, die in Aa nach *syn.* 9,45 eingeschoben ist (*syn.* 9 add. S. 360–362 Molinier), betitelt *Item ex alio auctore de suffocatione matricis, quod Graeci pnigmos*[56] *[matricis] dicunt*. Dort lesen wir (S. 361,9 Molinier): *Quibus autem accessio post cibum fit, uomere cogis. Post septem autem dies catharticum dia colocynthidos hiera⟨n⟩ dabis, et post tres dies cucurbitulas imponis in lumbos et subuentrale uel ilia et diereses damus, et post haec damus castorium bibere. Saepius enim per haec liberatae sunt.* Dieser lateinische Text ist eng verwandt mit dem, der bei Aetios im 16. Buch steht, als Kapitel 67 (S. 420 Romano = S. 100–101 Zervos): Τὰς δὲ μετὰ τροφὴν ἐμπεσούσας τῷ παροξυσμῷ, ἐμεῖν ἀναγκαστέον· πάσας μὲν γὰρ κουφίζει ὁ ἔμετος, μάλιστα δὲ τὰς μετὰ τροφὴν παροξυνομένας· τῇ δὲ τρίτῃ σικύαν[57] τῷ ἤτρῳ λέγω δεῖν[58] προσφέρειν καὶ τῇ ὀσφύι, καὶ ἀμυχαῖς κατακνίζοντα τοὺς τόπους καὶ ἁλσὶν ἐπιψύχειν τὰς ἀμυχάς, καὶ ἀναλαβόντα ἀπὸ τῶν βοηθημάτων τούτων ἀπολύειν ἐπὶ τὴν τοῦ καστορίου πόσιν· συνεχῶς γὰρ παραλαβανομένη πολλάκις τῆς διαθέσεως ἀπήλλαξε. Hier treffen wir ein weiteres griechisches Verb für ,skarifizieren', κατακνίζω; daß man die Einschnitte mit Salz kühlen soll (ἐπιψύχειν), wie wir es bei Zervos und ebenso bei Romano lesen, wird keiner glauben, der sich vorstellen kann, welche Wirkung in eine Wunde geriebenes Salz hat – Kühlen dürfen wir ausschließen! Viel eher denke ich an ἐπισμήχειν, doch ist dieser Teil leider im Lateinischen ausgelassen.

Welche Form sollen wir in *gersis/garsis* in *Praecept. diaet.* sehen? Die Konstruktion mit *imponere* verlangt nach einem Akkusativobjekt, wozu auch *garsas et uentosas ad⟨h⟩ibe Praecept. diaet.* V Zeile 53 (genauso die Handschrift Vap2 bei *Praecept. diaet.* IIa Zeile 46) paßt. Altfranzösisch *jarse* erlaubt wohl keine Entscheidung, ob wir von einem Substantiv der 3. oder der 1. lateinischen Deklination ausgehen sollten. Ich würde mich allerdings im Lichte der beiden Oribasiusstellen für die 3. lateinische Deklination aussprechen; Wechsel von -*is* und -*es* kommt in spät- und vulgärlateinischen Texten häufig vor. Ganz unstrittig dürfte allerdings die Interpretation der Passage sein, nämlich daß es um das blutige Schröpfen (mit vorherigen Einschnitten also) geht, was Groenkes Übersetzung (z. B. S. 115, 141 „setz(e) Schröpfköpfe an", S. 217 „verwende ... Schröpfköpfe", S. 237 „lege

---

56 πνιγμός, gewöhnlich aber ὑστερικὴ πνίξ *suffocatio* oder *offocatio matricis*.
57 σηκυίαν (eine *vox nihili?*) Romano, ohne Angabe einer Variante! Er übersetzt (S. 421) „applicare zucca al basso ventre (!)"; Antonio Garzya, der als Herausgeber des Bandes zeichnet, kann das nicht gelesen haben, und man kann sich des Gefühls nicht erwehren, daß Text und Übersetzung bei Romano nicht in den besten Händen waren.
58 δὴ Romano σοι Zervos; *correxi*.

Schröpfköpfe an") nicht erkennen läßt. Aber wenn Zahnärzte (wie Groenke) heutzutage ans Schröpfen denken, dann nur im übertragenen Sinn!

Appendix 3: Bei Groenke nicht genannte Zeugen von Praecept. diaet.

Praecept. diaet. I
Matrit. 19, fol. 90^{rb-va}

Praecept. diaet. IIa (= Ps.Sor. isag. 19)
Ambianus, fonds Lescalopier 2
Bethesda, National Library of Medicine, E 8, fol. 88^{r-v}
Glasgu. Hunter. 96, fol. 177^{r-v}
Glasgu. Hunter. 414, fol. 61^{r}–63^{r}
Lond. Harleianus 4977, fol. 103^{r-v}
Lugd. Voss. Lat. F. 96 A, fol. 2^{v}
clm 17403 , fol. 4^{v}
Par. lat. 14025, fol. 85vb–86ra
Prag. Capitul. Metrop. 1358, fol. 52^{v}–53^{r}
Salisburg. Carolino-Augusteus 2169, fol. 26^{r-v}
Vat. Palat. Lat. 485, fol. 13^{v}–14^{r}
Vat. Palat. Lat. 1449, fol. 120^{v}

Praecept. diaet. IIb
clm 13002, fol. 3rb
Par. lat. 14025, fol. 85^{ra-b}

Praecept. diaet. III (= Ps.Sor. isag. 20)
Prag. Capitul. Metrop. 1358, fol. 53^{v}
Salisburg. Carolino-Augusteus 2169, fol. 26^{v}–27^{r}

Praecept. diaet. VII
clm 13002, fol. 3rb

Praecept. diaet. (abweichend von Groenkes Typen)
clm 7999, fol. 143ra
clm 7999, fol. 143^{ra-rb} (ähnlich IIa)

Literaturverzeichnis

Beccaria, Augusto. *I codici di medicina del periodo presalernitano. Secoli IX, X e XI* (Roma: Edizioni di Storia e Letteratura, 1956) (Storia e Letteratura. 53).

Berschin, Walter. *Griechisch-lateinisches Mittelalter. Von Hieronymus zu Nikolaus von Kues* (Bern: Francke, 1980).

Bischoff, Bernhard. *Die Abtei Lorsch im Spiegel ihrer Handschriften*, 2. Auflage (Lorsch: Laurissa, 1989).

de Renzi, Salvatore. *Collectio Salernitana* ..., Band 4 (Napoli: Filiatre-Sebezio, 1856).

Ferckel, Christ., „Medizinische Marginalien aus dem Cod. Trevirens. Nr. 40," *Archiv für die Geschichte der Medizin* (= Sudhoffs Archiv) 7 (1913): 129–43.

Ferraces Rodríguez, Arsenio. *Estudios sobre textos latinos de fitoterapia entre la antigüedad tardía y la alta edad media* (A Coruña: Universidade da Coruña, 1999) (Universidade da Coruña. Monografías. 73).

Fischer, Klaus-Dietrich. „Überlieferungs- und Verständnisprobleme im medizinischen Latein des frühen Mittelalters," *Berichte zur Wissenschaftsgeschichte* 17 (1994): 153–65.

Fischer, Klaus-Dietrich. „The *Isagoge* of Pseudo-Soranus. An Analysis of the Contents of a Medieval Introduction to the Art of Medicine," *Medizinhistorisches Journal* 35 (2000): 3–30.

Gamillscheg, Emil. *Etymologisches Wörterbuch der französischen Sprache* (Heidelberg: C. Winter, 1969).

García González, Alejandro. „*Agriocanna*, a new medico-botanical glossary of pre-Salernitan origin," in David Langslow and Brigitte Maire (Hgg.), *Body, Disease and Treatment in a Changing World. Latin texts and contexts in ancient and medieval medicine* (Lausanne: Bibliothèque d'Histoire de la Médecine et de la Santé, 2010), 223–35.

Garofalo, Ivan. „Il *De pulsibus* di Philaretus e il Peri sphygmon di Philaretos (con in appendice l'edizione del *De pulsibus*)," in María Teresa Santamaría Hernández (Hg.), *Textos médicos grecolatinos antiguos y medievales: estudios sobre composición y fuentes* (Cuenca: Ediciones de la Universidad de Castilla-La Mancha, 2012) (Humanidades. 123), 55–94.

Groenke, Frank-Dieter. *Die frühmittelalterlichen lateinischen Monatskalendarien. Text—Übersetzung—Kommentar* (Berlin: Diss. med. dent. Freie Universität Berlin, 1986).

Hennig, John. „Versus de mensibus," *Traditio* 11 (1955): 65–90.

Kaczynski, Bernice M. *Greek in the Carolingian Age: The St. Gall Manuscripts* (Cambridge, Mass: The Mediaeval Academy of America, 1988).

Keil, Gundolf. „Monatsregeln," in W. E. Gerabek, Bernhard D. Haage, G. Keil und W. Wegner (Hgg.), *Enzyklopädie Medizingeschichte* (Berlin: De Gruyter, 2007), 1003b–1004b.

Kibre, Pearl. *Hippocrates Latinus. Repertorium of Hippocratic writings in the Latin Middle Ages*, revised ed. (New York: Fordham University Press, 1985).

Leisinger, Hermann. *Die lateinischen Harnschriften Pseudo-Galens* (Zürich: Orell Füssli, 1925) (Beiträge zur Geschichte der Medizin. 2).

Manzanero Cano, Francisco. *Liber Esculapii (Anonymus Liber Chroniorum). Edición crítica y estudio* (Madrid: Diss. phil. Universidad Complutense de Madrid, 1996) (ungedruckt).

Messina, Federico. *La redazione Aa della traduzione latina della Synopsis ad Eustathium di Oribasio (libro primo). Introduzione, testo critico, traduzione e commento* (Catania: tesi di laurea, Università degli Studi di Catania, anno accademico 2003/2004) (ungedruckt).

Meyer-Lübke, Wilhelm. *Romanisches etymologisches Wörterbuch*, 3. Aufl. (Heidelberg: C. Winter, 1935).

Nicoud, Marilyn. *Les régimes de santé au moyen âge. Naissance et diffusion d'une écriture médicale (XIIIe-XVe siècle)*, 2 Bände (Rome: École française de Rome, 2007) (Bibliothèque des Écoles françaises d'Athènes et de Rome. Fascicule 333).

Pagel, Julius. „Eine bisher unveröffentlichte lateinische Version der Chirurgie der Pantegni nach einer Handschrift der Königl. Bibliothek zu Berlin [lat. fol. 74, s. XII]," *Archiv für klinische Chirurgie* 81 (Teil 1) (1906): 735–86.

REW s. Meyer-Lübke.

Romano, Roberto. „Il calendario dietetico di Ierofilo," *Atti della Accademia Pontaniana*, nuova serie 47 (Anno Accademico 1998): 197–222.

Romano, Roberto. „Aezio Amideno, a cura di R. R.," in Antonio Garzya (Hg.), *Medici bizantini. Oribasio di Pergamo, Aezio d'Amida, Alessandro di Tralle, Paolo d'Egina, Leone medico* (Torino: Unione Tipografico-Editrice Torinese, 2006) (Classici greci), 253–553 (griechisch und italienisch).

Rose, Valentin. *Anecdota Graeca et Græcolatina. Mitteilungen aus Handschriften zur Geschichte der griechischen Wissenschaft*, Zweites Heft (Berlin: Dümmler, 1870).

Scheller, Eduard. *Aulus Cornelius Celsus, Über die Arzneiwissenschaft in acht Büchern, übersetzt und erklärt von E. S.*, 2. Auflage von Walther Frieboes (Braunschweig: Fried. Vieweg und Sohn, 1908).

Schmitt, Wolfram. *Theorie der Gesundheit und 'Regimen sanitatis' im Mittelalter* (Heidelberg: medizinische Habilitationsschrift Heidelberg, 1973) (ungedruckt).

Steinmeyer-Sievers = *Die althochdeutschen glossen, gesammelt und bearbeitet von Elias Steinmeyer und Eduard Sievers*, Band 4 (Berlin: Weidmann, 1882).

Stoffregen, Malte, *Eine frühmittelalterliche Übersetzung des byzantinischen Puls- und Urintraktats des Alexandros. Text, Übersetzung und Kommentar* (Berlin: Diss. med. Freie Universität Berlin, 1977).

Stoll, Ulrich. *Das Lorscher Arzneibuch. Ein medizinisches Kompendium des 8. Jahrhunderts (Codex Bambergensis medicinalis 1. Text, Übersetzung, Fachglossar* (Stuttgart: Steiner, 1992) (Sudhoffs Archiv. Beiheft 28).

Stuart, Heather. „A ninth-century account of diets and *dies Aegyptiaci*," *Scriptorium* 33 (1979): 237–44.

Th-K = Thorndike, Lynn, und Pearl Kibre. *A catalogue of incipits of mediaeval scientific writings in Latin*, rev. ed. (Cambridge, Mass.: The Mediaeval Academy of America, 1963).

von Staden, Heinrich. *Herophilus. The art of medicine in early Alexandria. Edition, translation and essays* (Cambridge: Cambridge University Press, 1989).

Weißer, Christoph. *Studien zum mittelalterlichen Krankheitslunar. Ein Beitrag zur Geschichte laienastrologischer Fachprosa* (Pattensen/Han.: Wellm, 1982) (Würzburger medizinhistorische Forschungen. 21).

Wickersheimer, Ernest. „Bénédiction des remèdes au Moyen Age," *Lychnos* (1952): 96–101.

Wickersheimer, Ernest. *Les manuscrits latins de médecine du haut moyen âge dans les bibliothèques de France* (Paris: Éditions du Centre national de la recherche scientifique, 1966).

Wlaschky, Momtschil. „Sapientia artis medicinae. Ein frühmittelalterliches Kompendium der Medizin," *Kyklos* (Leipzig) 1 (1928): 103–13.

Allan Gotthelf
Teleology and Embryogenesis in Aristotle's *Generation of Animals* 2.6

I Two Formulations of the Aim of *Generation of Animals*

Because Aristotle's *Generation of Animals* (*gen. anim.*) is to a large extent "one long argument" (to borrow Charles Darwin's characterization of his *Origin of Species*),[1] let us place *gen. anim.* 2.6 in the context of the whole of *gen. anim.*,

---

I am glad to have this opportunity to thank Heinrich von Staden publicly for his support of, and intellectual company during, my visit to the Institute for Advanced Study in the first half of 2001 —and for his many valuable writings, from which I have learned much. I thank also the editors of this volume for permitting me to contribute to this volume a paper which, while written very recently (the great bulk of it in 2010), will already have appeared in my collected papers volume (Allan Gotthelf, *Teleology, First Principles, and Scientific Method in Aristotle's Biology*, Oxford Aristotle Studies [Oxford: Oxford University Press, 2012]); health problems have not permitted the preparation of a newer paper. (It appears here with the kind permission of Oxford University Press.) Still, this paper bears on a matter of special interest to Heinrich: teleology and mechanism in Aristotle and the early Aristotelian tradition. While I take issue with a thesis of Heinrich's in the Additional Note below, I offer this paper in the spirit of the intellectual exchange he so enjoys—and with congratulations on his retirement. First thoughts on this topic were presented to seminars in 1994 at Oxford University and Tokyo Metropolitan University. While preparing an early version of this paper for presentation at the University of Texas at Austin, I received helpful written comments from Devin Henry. Although the thesis and much of the argument of the paper as it now stands are significantly different from that of those earlier versions (from which it will be differentiated as the paper progresses), they are in the same broad spirit, and I remain thankful both to Devin and to those who participated in discussion on those various occasions. More recently, comments from Mariska Leunissen on specific passages in 2.6, as noted, were very helpful. Extensive discussion of later drafts of the paper with David Charles, and then of the penultimate version with Andrea Falcon, significantly improved the final version.
Editors' note: Sadly, Allan Gotthelf died before the publication of this volume. We are very pleased to have been given his permission to include his paper.
1 Charles Darwin, *On the Origin of Species by Means of Natural Selection* (London: John Murray, 1859), 459. I developed this theme in a presentation under that name to the Fifth Pittsburgh/London Workshop on Aristotle's *Generation of Animals:* Understanding the Methodology at Work in *GA*, at the University of Western Ontario, 27 May 2010. Andrea Falcon and I are currently carrying that project further, as we work together on a co-authored paper, provisionally called "'One Long Argument?—The Unity of Aristotle's *Generation of Animals*."

identifying what this treatise's overall aim is and how teleology is embedded in that aim, and thus in this treatise.

We are told of *gen. anim.*'s aim in at least two places. The first is in *Parts of Animals* (*part. anim.*) 1.1, in a familiar passage; the second is in the opening paragraph of *gen. anim.*

In *part. anim.* 1.1, having argued that coming-to-be is for the sake of being, and not the reverse, Aristotle writes of what follows from that proposition for the study both of animal parts and of their generation. With good reason, scholars standardly take these statements as governing the forms treatises like our *part. anim.* 2–4 and *gen. anim.* are to take. First *part. anim.* 2–4:

> **(T1)** Hence we should if possible say that because this is what it is to be a man, therefore he has these things; for he cannot be without these parts. Failing that, we should get as near as is possible to it: we should either say altogether that it cannot be otherwise, or that it is at least good thus. And these follow. (640a33–b1, trans. Balme)[2]

Then *gen. anim.*:

> **(T2)** And because he is such a thing, his coming-to-be necessarily happens so and is such. And that is why this part comes to be first, and then this. (640b1–3, trans. Balme)[3]

From the nature of what is to be produced, certain things follow (of conditional necessity) regarding how that product must come into being. *And that is why* the parts come into being in the order in which they do. I will connect the first of these two lines with the opening stretch of *gen. anim.* in a moment, but it is striking to notice, in view of the second line, how much of *gen. anim.* 2.6 (and indeed 2.1–6) is indeed, and explicitly, concerned with the order in which the parts come to be.[4] Perhaps this shouldn't be a surprise, since one can't say much about *how* or *why* an animal's *genesis* occurs as it does without knowing in what order the

---

2 For the ways in which this passage models the explanations in *part. anim.* 2–4, see Gotthelf, *Teleology, First Principles and Scientific Method*, 12 n. 18, 35, 38, 124 n. 16, 154, 175, 191, 219–20 nn. 4 and 6, and *passim*.

3 The Greek reads: ἐπεὶ δ' ἔστι τοιοῦτον, τὴν γένεσιν ὡδὶ καὶ τοιαύτην συμβαίνειν ἀναγκαῖον. διὸ γίγνεται πρῶτον τῶν μορίων τόδε, εἶτα τόδε. The next sentence is presumably retrospective, covering the instructions for *part. anim.* and *gen. anim.*: "And this is the way we should speak of everything that is composed (*sunistamenōn*) naturally."

4 2.1, 733b23–734b4, 735a12–26; 2.5–6, 741b15–37, 742a16–b17, 742b32–743a21, 743b18–36.

parts come to be. Still, I am struck by the extent of Aristotle's preoccupation with this question, and I will return to it later.[5]

The opening stretch of *gen. anim.* (715a1–18) announces that we have already studied the final, formal, and material causes of each of the parts of animals except the generative ones. That leaves a *part. anim.*-like, three-cause study of the generative parts—and an overall study of the efficient cause. This latter study and the study of the generation of each animal, Aristotle says, somewhat enigmatically, "are, in a way (*pōs*), *the same thing.*"[6] A primary aim of our treatise, *On the Generation of Animals*, then, is a study of "the source of the moving cause" of animals and their parts, that is: a study of *what agent(s), and of what actions of theirs, bring animals and their parts into existence?*

If we now bring to bear on this conclusion the two *gen. anim.*-aim-stating sentences we isolated in *part. anim.* 1.1 as **T2**, what can we add? Starting from the second sentence, we may add the words: *And in what order?*—a sensible addition, which keeps us within the ambit of an efficient-causal account. The first sentence, however, reads: "And because he is such a thing, his coming-to-be necessarily happens so and is such." This is, as already mentioned, an account in terms of conditional necessity, and this brings our attention to the (immediate) *final cause* of generation: the production, or existence, of an animal of a certain nature. A full account of the generation of animals will, I propose, *not only* include *but also* intimately relate the final and the efficient causes of generation. The need for the final-cause dimension of the account will explain why Aristotle says that an account of the efficient cause by itself is an account of generation only *pōs*, "in a way." With the teleological dimension in place, the account of generation is, in Aristotle's eyes, complete.

---

[5] This preoccupation of Aristotle in *gen. anim.* 2 (and especially in 2.6) is especially brought home in Mariska Leunissen's "Order and Method in Aristotle's *Generation of Animals* II," her as-yet unpublished presentation to the Fifth Pittsburgh/London *GA* Workshop (May 2010). She gives a full account of the different sorts of priority at work in Aristotle's account in 2.6, and of how those different senses shape the construction of the chapter. For my own briefer discussion, see below, 152–65.

[6] *Gen. anim.* 1.1, 715a14–15. I have purposely not supplied an object for "the efficient cause"— whether of the animals or of their parts (or both)—since the Greek reads only: περὶ αἰτίας δὲ τῆς κινούσης τίς ἀρχή. τὸ δὲ περὶ ταύτης σκοπεῖν καὶ τὸ περὶ τῆς γενέσεως τῆς ἑκάστου τρόπον τινὰ ταὐτόν ἐστιν. But I am going to assume it is both, since 715a9–11 says as much for the other three causes.

II Toward the Theory in *gen. anim.* 2.6

That there is an intimate connection, at least in the case of generation, between the final and the efficient cause will not be a surprise to readers familiar with my own interpretation of Aristotelian final causality.[7] For, if we connect the operation of a final cause with the operation, as efficient cause, of a *dunamis* directed upon such an end, then that connection of causes is just what we should expect. And if we are to get illumination on the nature of a final cause from a study of *gen. anim.*, the route to this illumination will be to follow closely *gen. anim.*'s account of the efficient cause of generation, looking in particular for any move that is made, via conditional necessity, from the nature (or essence) of the animal produced to the nature of the process required to produce it.

Yet it is striking that explicit mentions of final causation occur rarely in *gen. anim.*, once the review of the generative parts is completed by the end of 1.16. A final cause of generation itself is offered at the beginning of 2.1, followed by an account of the final cause of the separation of the male and female principles. Final causes are not mentioned again until 2.4, at 738b1, when a final cause is given for the diversion of female nutritive residue to the uterus to serve as generative material, and then at 739b29 for the formation of membranes around the outer surface of the *katamēnia* in the conception process. However, when Aristotle turns to his account of embryological development in 2.6, he draws at 742a16–b17 on a set of general theses about various types of priority involved in final causation to guide him in the empirical identification of the sequence in which parts are formed in the embryo and to help explain why the sequence occurs in the order it does. The aim here is to account for the efficient cause, but the efficient *archē* within the embryo (viz., the heart or its analogue), Aristotle claims, must stand in various relationships to the end, the *telos*, of the generative process—although quite what those relationships are is difficult to determine from this very difficult passage. Still this passage, so outlined, appears to give us *precisely* what the *part. anim.* 1.1 passage said should be our guide in determining the character and the order of the process by which the animals come into being. So this 2.6 passage will have to be our focus. But let us build up to it step by step, starting from *gen. anim.* 1.1, with our eye on what Aristotle's account of the efficient cause of generation has to tell us about the nature of final causation, as operative in animal generation.

---

[7] See Allan Gotthelf, *Teleology, First Principles, and Scientific Method*, Part 1, and the taxonomy of recent interpretations that I offer there in Chapter 3.

First, some context. In my earlier work on Aristotle's teleology,[8] I made two general claims about the *gen. anim.* theory: (i) that Aristotle does not attempt to offer (the outlines of) a full material-level account of the generative process, but rather appeals at key places to a *dunamis* that is *primitively* directed upon the form the offspring is to embody; and (ii) that the material bearer or basis of that *dunamis*, sometimes identified as a distinctive "vital" heat (*psuchikē thermotēs, zōtikē thermotēs*)[9] and sometimes as "movements" (*kinēseis*),[10] has no material-level description of its vital potentials, being different in this respect from "ordinary" heat found in ordinary air and fire. In his valuable 1988 paper, "Aristotle on Hypothetical Necessity and Irreducibility,"[11] David Charles questioned both of these claims, and, in doing so, he invites us to take a close look at the passages in question to see if there is some way to decide the matter. Though a reasonably plausible case can be made for a reading like Charles', in the end the weight of the evidence in *gen. anim.* 2.6, I shall argue, is firmly in favor of my two claims.

---

8 Gotthelf, *Teleology, First Principles, and Scientific Method*, Ch. 1.
9 732a18–19, 739a11, 762a20; 739b23, cf. 737a2–5. See, too, A. L. Peck, trans., *Aristotle, Generation of Animals* (London: Heinemann, 1942), liv n. c, and Gad Freudenthal, *Aristotle's Theory of Material Substance: Heat and Pneuma, Form and Soul* (Oxford: Oxford University Press, 1995), 19–34 and *passim*.
10 729b5–6 and ff., and **T3, T4, T7**, and **T10** below; cf. Peck, *Aristotle: Generation of Animals*, liv, and Freudenthal, *Aristotle's Theory of Material Substance*, 27–29. In a very few passages, and always in passing, Aristotle indicates what the relationship between these two *archai* might be. Here are two such passages:
  (i) "[T]he nutritive soul-capacity, just as, in the animals and plants themselves later on, it produces growth from the food, using as tools heat and cold (for the soul-capacity's movement is *in these* ...)" (2.4, 740b29–32, part of **T7** below).
  (ii) "The heat is present in the spermatic residue *having its* movement and actuality in an amount and character" (2.6, 743a26–28, part of **T10** below).
These passages give rise, however, to several interrelated questions which I don't know how to answer: What is it for heat either to have movements *simpliciter*, or to have them *in* it? Is every case of (vital) heat*ing* the action of "movements" that are either of or in the agent's heat? Is every generative moving of another thing a case of heating it? The best discussion of the relationships among the Aristotelian notions involved in these questions is in Freudenthal, *Aristotle's Theory of Material Substance*, Ch. 1, § 1 and Ch. 3, § 2 (though he says relatively little about the generative "movements"); however, there is more that still needs to be said, if we are to fully understand Aristotle's conceptions of heat in general, of "vital" heat in particular, and of generative "movements"—and thereby his precise view of the efficient causation involved in generation.
11 David Charles, "Aristotle on Hypothetical Necessity and Irreducibility," *Pacific Philosophical Quarterly* 69 (1988): 1–53.

A focus on these claims makes clear the need, right off the bat, to make a distinction we have not yet made in regard to efficient causes. Scholars, myself included, often tend to view efficient causes as *material-efficient causes*, cases where the elemental bodies or powers act or interact in a certain way of (what we often call) material necessity. But Aristotle at least as often speaks of *form* and *soul* as source of change, and there are what we might call *formal-efficient causes*, cases where the agent is a formal nature, a *dunamis* for form, and so forth. It is precisely an open question whether in every case in which the agent is said to be a formal nature there is a material-efficient causal process —a process operating wholly of material necessity—that underlies it. So, in following Aristotle's account of the efficient cause of generation through *gen. anim.* 1 and 2 in hopes of gaining light on his account of final causation, we will need always to be clear whether we think Aristotle is speaking of material-efficient or of formal-efficient causation (or somehow of both).

Let us return, then, to *gen. anim.* 1.1. At 715a19 Aristotle begins his argument from the three-part *phainomenon* ("appearance") that some animals are produced from the (i) *coupling* of (ii) *male* and (iii) *female*, i.e., that generation sometimes occurs inside one animal ("female") as a result of some sort of action upon it by another animal ("male") during a process in which relevant parts of each come into contact ("copulation"). This often, though not always, involves a distinct active agent, a "seed" (*sperma*), supplied at least in part by the male. We will thus understand such generations (and, it turns out, the spontaneous ones as well) when we understand more precisely the respective contributions to this process of the female and the male, including the extent to which each contributes "seed." The female evidently contributes some sort of material, and argument from observable data confirms this; this material counts as "seed" to some extent (which will be important in the discussion of family resemblance in Book 4), but it is evidently incomplete, since females do not generate by themselves. By contrast, although in most kinds of animals the male contributes "seed," sometimes it does not; this is a crucial observation, since it supports the general claim, extending to all animal kinds that reproduce (a claim which Aristotle argues for on other grounds as well), that the male's power (*dunamis*) to act on the female material is not exercised by its contributing any additional *material*. Aristotle consolidates his results and states his hypothesis in a key statement late in Book 1 at 730b8–23:

> **(T3)** One may also grasp from these cases how the male contributes to generation. For not every male emits seed, and in those that do emit it, the seed is no part of the fetus that is being produced, just as nothing comes away from the carpenter to the matter of the timber, nor is there any part of carpentry in the thing produced, but the shape and the form is pro-

duced *in* (*enginetai*) the matter *from* the carpenter *through* the movement (*kinēsis*): his soul (in which is the form) and his knowledge move his hands or some other part in a certain sort of movement—different when that which is produced from them is different, the same when it is the same—the hands move the tools, and the tools move the matter. Similarly the male's nature, in those that emit seed, uses the seed as a tool containing movement in actuality (*energeiāi*), just as in the things produced by art the tools are in movement; for the movement of the art is in a way in them.[12]

Further argument in Book 1 Chapter 1 supports and substantially elaborates this hypothesis, moving from the complexity and specificity of the generative product to the nature of its immediate efficient cause. The key passage is 734b17-735a4:

(T4) It is clear, then, that there is something which makes ⟨the parts⟩, but not by being a certain "this," present in a completed state at the beginning [sc. but rather by being a *dunamis* and *kinēsis*, 734b8, 11; cf. 729b5-6].

Now, we must grasp how in the world each part is produced from that beginning, by taking first of all as our starting point that everything that is produced by nature or by art is produced *by* a thing that is ⟨whatever it is⟩ in actuality, *out of* what is potentially of that sort. Now the seed and the movement and source it has are such that when the movement stops each part is produced having soul. For it is not face nor flesh unless it has soul: after their death it will be equivocal to say that the one is a face and the other flesh, as it would be if they were made of stone or wood. The uniform parts and the instrumental parts are produced simultaneously. And just as we would not say that fire alone makes an axe or any other tool, neither should we ⟨say this⟩ of foot or hand. Nor, likewise, of flesh, for it too has a function. Now, hard and soft and supple and brittle, and whatever other such conditions (*pathē*) that belong to the parts containing soul—heat and cold would make those, but would not go so far as to make the definition (*logos*) in virtue of which the one is now flesh and the other bone: that is due to the movement (*kinēsis*) from the generator who is in *full actuality* (*entelecheiāi*) what the thing out of which the product comes is *potentially*, just as is the case with the things produced according to an art; for the hot and the cold make the iron hard or soft, but what makes it a *sword* is the movement of the tools which contains a definition (*logos*) belonging to the art. For the art is source

---

12 (730b8) λάβοι δ' ἄν τις ἐκ τούτων καὶ τὸ ἄρρεν πῶς συμβάλλεται πρὸς τὴν γένεσιν· οὐδὲ (b10) γὰρ τὸ ἄρρεν ἅπαν προΐεται σπέρμα, ὅσα τε προΐεται τῶν ἀρρένων, οὐθὲν μόριον τοῦτ' ἔστι τοῦ γιγνομένου κυήματος, ὥσπερ οὐδ' ἀπὸ τοῦ τέκτονος πρὸς τὴν τῶν ξύλων ὕλην οὔτ' ἀπέρχεται οὐθέν, οὔτε μόριον οὐθέν ἐστιν ἐν τῷ γιγνομένῳ τῆς τεκτονικῆς, ἀλλ' ἡ μορφὴ καὶ τὸ εἶδος ἀπ' ἐκείνου ἐγγίγνεται (b15) διὰ τῆς κινήσεως ἐν τῇ ὕλῃ, καὶ ἡ μὲν ψυχὴ ἐν ᾗ τὸ εἶδος καὶ ἡ ἐπιστήμη κινοῦσι τὰς χεῖρας ἤ τι μόριον ἕτερον ποιάν τινα κίνησιν, ἑτέραν μὲν ἀφ' ὧν τὸ γιγνόμενον ἕτερον, τὴν αὐτὴν δὲ ἀφ' ὧν τὸ αὐτό, αἱ δὲ χεῖρες τὰ ὄργανα, τὰ δ' ὄργανα τὴν ὕλην. ὁμοίως δὲ καὶ ἡ φύσις ἡ ἐν τῷ ἄρρενι (b20) τῶν σπέρμα προϊεμένων χρῆται τῷ σπέρματι ὡς ὀργάνῳ καὶ ἔχοντι κίνησιν ἐνεργείᾳ, ὥσπερ ἐν τοῖς κατὰ τέχνην γιγνομένοις τὰ ὄργανα κινεῖται· ἐν ἐκείνοις γάρ πως ἡ κίνησις τῆς τέχνης. The Greek text of *gen. anim.* provided is from H. J. Drossaart Lulofs, ed., *Aristotelis De generatione animalium: Recognovit brevique adnotatione critica instruxit*, Oxford Classical Texts (Oxford: Clarendon Press, 1965).

and form of the thing produced, but in another; the movement of the nature is in the thing itself, being from another nature which has the form in actuality (*energeiāi*).[13]

This distinction between the power of heat and cold to produce the *pathē* of the parts and the power of the "movement" supplied by the father to produce the parts themselves will be crucial, and we will return to it shortly. At this point in the text it raises the question for Aristotle of what the material nature of the semen (the male seed, *gonē*) must be for it to be able to carry the generative *dunamis*. In a long and careful discussion in 2.2, which makes use of the chemical theory presented in *Meteorology* 4, Aristotle infers from the unusual reactions of semen to heat and cold that it must contain "*pneuma*," which in its ordinary usage refers to breath or warm wind, but which had apparently begun to take on a role in the medical tradition as an important constituent of living bodies, given the evident centrality to life of breath and its warmth.[14] At 735b37–736a16 Aristotle concludes:

> (T5) The seed, then, is a combination of *pneuma* and water, and the *pneuma* is hot air; that is why it is moist in nature, because it is composed of water. ... [W]hile it is inevitable that one seed is earthier than another and it is so most of all in those which have much earthy

---

13 (734b17) ὅτι μὲν οὖν ἔστι τι ὃ ποιεῖ, οὐχ οὕτως δὲ ὡς τόδε τι οὐδ᾽ ἐνυπάρχον ὡς τελεσμένον τὸ πρῶτον, δῆλον. Πῶς δέ ποτε ἕκαστον γίγνεται (b20) ἐντεῦθεν δεῖ λαβεῖν, ἀρχὴν ποιησαμένους πρῶτον μὲν ὅτι ὅσα φύσει γίγνεται ἢ τέχνῃ ὑπ᾽ ἐνεργείᾳ ὄντος γίγνεται ἐκ τοῦ δυνάμει τοιούτου. τὸ μὲν οὖν σπέρμα τοιοῦτον, καὶ ἔχει κίνησιν καὶ ἀρχὴν τοιαύτην ὥστε παυομένης τῆς κινήσεως γίγνεσθαι ἕκαστον τῶν μορίων καὶ ἔμψυχον. οὐ γάρ ἐστι πρόσωπον (b25) μὴ ἔχον ψυχὴν οὐδὲ σάρξ, ἀλλὰ φθαρέντα ὁμωνύμως λεχθήσεται τὸ μὲν εἶναι πρόσωπον τὸ δὲ σάρξ, ὥσπερ κἂν εἰ ἐγίγνετο λίθινα ἢ ξύλινα. ἅμα δὲ τὰ ὁμοιομερῆ γίγνεται καὶ τὰ ὀργανικά· καὶ ὥσπερ οὐδ᾽ ἂν πέλεκυν οὐδ᾽ ἄλλο ὄργανον φήσαιμεν ἂν ποιῆσαι τὸ πῦρ μόνον οὕτως οὐδὲ πόδα (b30) οὐδὲ χεῖρα. τὸν αὐτὸν δὲ τρόπον οὐδὲ σάρκα· καὶ γὰρ ταύτης ἔργον τί ἐστιν. σκληρὰ μὲν οὖν καὶ μαλακὰ καὶ γλίσχρα καὶ κραῦρα καὶ ὅσα ἄλλα τοιαῦτα πάθη ὑπάρχει τοῖς ἐμψύχοις μορίοις θερμότης καὶ ψυχρότης ποιήσειεν ἄν, τὸν δὲ λόγον ᾧ ἤδη τὸ μὲν σὰρξ τὸ δ᾽ ὀστοῦν οὐκέτι, ἀλλ᾽ ἡ κίνησις ἡ (b35) ἀπὸ τοῦ γεννήσαντος τοῦ ἐντελεχείᾳ ὄντος ὅ ἐστι δυνάμει ἐξ οὗ γίγνεται, ὥσπερ καὶ ἐπὶ τῶν γιγνομένων κατὰ τέχνην· σκληρὸν μὲν γὰρ καὶ μαλακὸν τὸν σίδηρον ποιεῖ τὸ θερμὸν (735a1) καὶ τὸ ψυχρόν, ἀλλὰ ξίφος ἡ κίνησις ἡ τῶν ὀργάνων ἔχουσα λόγον [τὸν] τῆς τέχνης. ἡ γὰρ τέχνη ἀρχὴ καὶ εἶδος τοῦ γιγνομένου, ἀλλ᾽ ἐν ἑτέρῳ· ἡ δὲ τῆς φύσεως κίνησις ἐν αὐτῷ ἀφ᾽ ἑτέρας οὖσα φύσεως τῆς ἐχούσης τὸ εἶδος ἐνεργείᾳ.

14 D. M. Balme, *Aristotle, De partibus animalium I and De generatione animalium I (with Excerpts from II.1–3), with a Report on Recent Work and Additional Bibliography by Allan Gotthelf*, Clarendon Aristotle Series (Oxford: Oxford University Press, 1992), 161. Cf. Philip J. van der Eijk, *Diocles of Carystus: A Collection of the Fragments with Translation and Commentary*, 2 vols. (Leiden: Brill, 2000–2001), Index s.v. (in each volume) breath (*pneuma*); and Philip J. van der Eijk, *Medicine and Philosophy in Classical Antiquity: Doctors and Philosophers on Nature, Soul, Health and Disease* (Cambridge: Cambridge University Press, 2005), Index s.v. *pneuma*.

stuff in virtue of the bulk of their bodies, it is thick and white because *pneuma* is mixed in. ... The cause of the whiteness of the seed is that the semen is foam, and foam is something white, especially that which has the smallest constituents, each bubble too small to be seen.[15]

Since air is already warm and moist, *pneuma* must carry an additional heat. The nature of this heat is the subject of a crucial passage in the next chapter, at 736b29–737a7 and 18–24:

(T6) Now, every soul-potential seems to be associated with a body different from and more divine than the so-called elements; and as souls differ from each other in value and lack of value, so too this sort of nature differs. For within the seed of all things there is present that which makes the seed fertile, what is called "the hot." But this is not fire or that sort of potentiality but rather the *pneuma* enclosed within the seed, that is, within the foamy part, and more precisely the nature in the *pneuma*, being analogous to the element of the stars. That is why fire generates no animal, and none is seen to be constituted in things subjected to fire, whether wet things or dry; but the heat of the sun and that of animals do— not only that conveyed through the seed but also if there should be any other residue of their nature, even this too will contain a life-source (*zōtikē archē*). Such things make it evident, then, that the heat in the animals is neither fire nor has its origin from fire.

...

Since the seed is a residue and is moving with the same movement as that with which the body grows when the final nutriment is being apportioned out, when it goes into the uterus it constitutes and moves the female's residue with the same movement as that with which it itself is actually moving. For that too is residue and contains all the parts potentially, though none actually.[16]

---

15 (735b37) ἔστι μὲν οὖν τὸ σπέρμα κοινὸν πνεύματος καὶ ὕδατος, τὸ δὲ πνεῦμά ἐστι θερμὸς ἀήρ· διὸ ὑγρὸν τὴν φύσιν ὅτι ἐξ ὕδατος. ... (736a5) μᾶλλον μὲν γὰρ ἕτερον ἑτέρου σπέρμα γεωδέστερον ἀναγκαῖον εἶναι, καὶ μάλιστα τοιοῦτον ὅσοις πολὺ γεῶδες ὑπάρχει κατὰ τὸν ὄγκον τοῦ σώματος, παχὺ δὲ καὶ λευκὸν διὰ τὸ μεμῖχθαι πνεῦμα. ... (a13) ... αἴτιον δὲ τῆς λευκότητος τοῦ σπέρματος ὅτι ἐστὶν ἡ γονὴ ἀφρός, ὁ δ' ἀφρὸς λευκόν, (a15) καὶ μάλιστα τὸ ἐξ ὀλιγίστων συγκείμενον μορίων καὶ οὕτω μικρῶν ὥσπερ ἑκάστης ἀοράτου τῆς πομφόλυγος οὔσης ...

16 (736b29) πάσης μὲν (b30) οὖν ψυχῆς δύναμις ἑτέρου σώματος ἔοικε κεκοινωνηκέναι καὶ θειοτέρου τῶν καλουμένων στοιχείων· ὡς δὲ διαφέρουσι τιμιότητι αἱ ψυχαὶ καὶ ἀτιμίᾳ ἀλλήλων οὕτω καὶ ἡ τοιαύτη διαφέρει φύσις. πάντων μὲν γὰρ ἐν τῷ σπέρματι ἐνυπάρχει ὅπερ ποιεῖ γόνιμα εἶναι τὰ σπέρματα, τὸ καλούμενον θερμόν. τοῦτο δ' οὐ πῦρ οὐδὲ τοιαύτη δύναμίς ἐστιν ἀλλὰ τὸ (b35) ἐμπεριλαμβανόμενον ἐν τῷ σπέρματι καὶ ἐν τῷ ἀφρώδει πνεῦμα καὶ ἡ ἐν τῷ πνεύματι φύσις, ἀνάλογον οὖσα τῷ (737a1) τῶν ἄστρων στοιχείῳ. διὸ πῦρ μὲν οὐθὲν γεννᾷ ζῷον, οὐδὲ φαίνεται συνιστάμενον ἐν πυρουμένοις οὔτ' ἐν ὑγροῖς οὔτ' ἐν ξηροῖς οὐθέν· ἡ δὲ τοῦ ἡλίου θερμότης καὶ ἡ τῶν ζῴων οὐ μόνον ἡ διὰ τοῦ σπέρματος, ἀλλὰ κἄν τι περίττωμα τύχῃ τῆς φύσεως (a5) ὂν ἕτερον, ὅμως ἔχει καὶ τοῦτο ζωτικὴν ἀρχήν. ὅτι μὲν οὖν ἡ ἐν τοῖς ζῴοις θερμότης οὔτε πῦρ οὔτε ἀπὸ πυρὸς ἔχει τὴν ἀρχὴν ἐκ τῶν τοιούτων ἐστὶ φανερόν. ... (a18) τοῦ δὲ σπέρματος ὄντος περιττώματος καὶ κινουμένου κίνησιν τὴν αὐτὴν καθ' ἥνπερ τὸ σῶμα αὐξάνεται (a20) μεριζομένης τῆς ἐσχάτης τροφῆς, ὅταν ἔλθῃ εἰς τὴν ὑστέραν συνίστησι καὶ κινεῖ

It is *prima facie* odd that, having distinguished at 2.1, 734b19ff. (**T4** above) the causal power of the "movement" supplied by the father from the causal power of heat and cold, Aristotle here presents that movement as somehow resident in the semen's distinctive *heat*.¹⁷ Let us for the moment, however, continue with our narrative and note that Aristotle has now completed the presentation of the broad lines of his theory of the efficient cause of generation: the father, as ultimate efficient cause, acts on a material supplied by the mother, the (useful part of the) *katamēnia*, but does so through his semen which carries, embedded somehow in a distinctive heat, a set of movements which initiate a process leading in normal circumstances to the production of an offspring one in form with the father. What makes that production possible is that the semen and *katamēnia* are both residues of nutritive material; as a result, the set of movements supplied by the father are the very same ones that within himself carry the potentiality to maintain his own form by converting his food into more of himself in the nutritive and growth process. That last point is a stroke of genius on Aristotle's part, explaining how the offspring can—and will in the paradigm case—be like the father; and it provides, incidentally, one part of the basis for the claim which he makes, several times in *gen. anim.* as well as in the *De anima*, that the nutritive soul and the generative soul are one and the same potentiality.¹⁸

In 2.4 Aristotle analyzes in more detail the formation of the male and female contributions, as residues of the nutritive process, discusses the initial constitution or "setting" (*kuēsis, sunistēsis*) of the embryo and its early formation, and turns to the question of its development and differentiation (*diakrisis*, 740b13). Another crucial passage follows, with marked similarities to **T4** above (734b19ff.), which is not surprising given their similar aims; the present passage, however, builds on the results of the theory developed since that earlier passage. The reference is 740b12–741a2, a passage that gets rather convoluted towards the end, but whose sense remains clear. I translate literally:

τὸ περίττωμα τὸ τοῦ θήλεος τὴν αὐτὴν κίνησιν ἥνπερ αὐτὸ τυγχάνει κινούμενον κἀκεῖνο. καὶ γὰρ ἐκεῖνο περίττωμα, καὶ πάντα τὰ μόρια ἔχει δυνάμει, ἐνεργείᾳ δ' οὐθέν.

17 I have suggested in Gotthelf, *Teleology, First Principles, and Scientific Method*, 19 n. 32 that the distinction between these two kinds of heat (or between these two aspects of the vital heat) reflects the fact that vital heat is indeed *heat* and so contains both material and formal powers or potentials. Compare the illuminating discussion in Freudenthal, *Aristotle's Theory*, 31–34 and cf. n. 7 above. I return to this *prima facie* puzzle below.

18 E.g., *gen. anim.* 2.1, 735a17–18; 2.4, 740b35–37 [most of which is quoted in **T8** below]; and *de an.* 2.4, 416a19. The other part of the basis for identifying the nutritive and generative faculties is their having the same goal, which is the maintenance in being (in whatever way possible) of the organism. Cf. Gotthelf, *Teleology, First Principles, and Scientific Method*, 32–33 and 58.

(T7) The differentiation (*diakrisis*) of the parts is produced, not as some think, because it is natural for like to be carried to like, ... but because the residue of the female is potentially such as the animal is by nature, and the parts are in it potentially and not at all actually; it is due to this cause that each of them is produced, and because when the active and the to-be-acted-upon come into contact, in the manner in which the one is active and the other to-be-acted-upon (I mean by "manner" the how and where and when), straightaway the one acts and the other is acted upon. [Cf. *metaph.* Θ 5, 1047b35–1048a7.] The female, then, provides the matter, the male the source of the movement. And just as things which are produced by the agency of an art are produced by means of tools—it's truer in fact to say "by the *movement* of the tools" (this ⟨movement⟩ being the actuality of the art, and the art being the shape of that which comes to be in another)—in this way the nutritive soul-capacity, just as, in the animals and plants themselves later on, it produces growth from the food using as tools heat and cold (for the soul-capacity's movement is *in* these, and each thing is produced according to a certain definition [*logōi tini*]), in this way also at the very beginning the thing produced by a nature is constituted. For it is the same matter by which it grows and from which it is first constituted, so that the productive capacity ⟨for growth⟩ is also the same as the initial one (but greater than it). If then this is the nutritive soul, it is also that which generates, and this is the nature of each ⟨organism⟩ and present in all plants and animals ... [19]

---

[19] I take the passage's long and convoluted antepenultimate sentence (b25–34) to be drawing, within an overarching parallel between artistic and natural (i.e., biological) production, a parallel between the production of growth in an already formed animal or plant and the initial production—i.e., constituting—of that animal or plant. Aristotle's main claim here is the overarching parallel—that just as an art produces its products via tools, that is, via the movement of its tools, so too the nature of the organism works via *its* tools, viz. heat and cold, or more precisely, now, via the movement it imparts to heat and cold. Aristotle wants to make this latter claim in regard to the formation of the offspring from its very conception, but seems to think the claim is clearer in the case of the development of an organism already given birth to; so he draws a second parallel, on the natural side of the overarching art-nature parallel, between, in effect, the forming embryo and the growing child. Just as the relevant aspect of the child's nature, its nutritive soul-capacity, imparts movements to its body's heat and cold, which movements maintain and develop its body and faculties, thus using its heat and cold as "tools," so too this same nutritive capacity, in its initial form, uses heat and cold as tools to form the embryo. The sentence reads as if Aristotle has switched focus away from the art-nature parallel with which he began, but because the nature side of that parallel is a nature's use of tools in the formation of the embryo, and that use, he thinks, is for the listener clearer in the case of a child's growth, he reaches the former by paralleling it to the latter. The last sentence of the passage then offers some justification for the growth-embryogenesis parallel. Here is the Greek of the entire passage: (740b12) ἡ δὲ διάκρισις γίγνεται τῶν μορίων οὐχ ὥς τινες ὑπολαμβάνουσι διὰ τὸ πεφυκέναι φέρεσθαι τὸ ὅμοιον πρὸς τὸ ὅμοιον ...· ἀλλ' ὅτι τὸ περίττωμα τὸ τοῦ θήλεος δυνάμει τοιοῦτόν ἐστιν οἷον φύσει τὸ (b20) ζῷον καὶ ἔνεστι δυνάμει τὰ μόρια ἐνεργείᾳ δ' οὐθέν, διὰ ταύτην τὴν αἰτίαν γίγνεται ἕκαστον αὐτῶν, καὶ ὅτι τὸ ποιητικὸν καὶ τὸ παθητικὸν ὅταν θίγωσιν, ὃν τρόπον ἐστὶ τὸ μὲν ποιητικὸν τὸ δὲ παθητικὸν (τὸν δὲ τρόπον λέγω τὸ ὣς καὶ οὗ καὶ ὅτε), εὐθὺς τὸ μὲν ποιεῖ τὸ δὲ πάσχει. ὕλην μὲν οὖν (b25) παρέχει τὸ θῆλυ, τὴν δ' ἀρχὴν τῆς κινήσεως τὸ ἄρρεν. ὥσπερ δὲ τὰ ὑπὸ τῆς τέχνης γιγνόμενα γίγνεται διὰ τῶν ὀργάνων—ἔστι δ' ἀληθέστερον εἰπεῖν

How shall we translate *logōi tini*—*by* or *according to* a certain definition? In line with 735a1–2 (in **T4** above), *hē kinēsis hē tōn organōn echousa logon ton tēs technēs*, perhaps the meaning is "each thing is produced by a movement whose definition (and thus nature) specifies the shape and form of the product."[20]

The latter part of this passage, where the parallels with **T4** are found, is important for both of our questions: it bears on the question of the nature of the heat which carries the "movement" responsible for generation and development, and it bears on the question of whether Aristotle intends to offer a full material account of the generation of animal parts to complement his teleological account, or whether the teleological account, though it makes use of material necessities, is predicated on the unavailability of a full material-level causal account. We will return to both questions shortly.

Since female birds can produce unfertilized eggs that have some life in them ("wind-eggs"), and since a few animal kinds have females and no males, Aristotle rounds out his presentation with a closer examination of the female contribution to the formal side of generation. He turns finally in ch. 6 to a detailed account of the formation of individual animal parts. This is the *diakrisis* ("articulating out," "differentiation") that was described generally in the first part of **T7**. It is here, at 741b7, indicated with the verb *diorizetai* (perhaps "delineates" or even, in its physical meaning, "defines"). And it is here, in the detailed accounts of the formation of individual parts, that I am thinking we might look for the reach that material-efficient causation has (and doesn't have) in Aristotle's explanatory account of the generation of animals and their parts.

One view, then, would be roughly this. It would hold that 2.6 is meant to point towards the basic pattern of a full material-level account, in which sufficient conditions are given for the production of a sufficient number and range of animal parts to make clear that such an account is always in principle possible. Material-efficient causes would *necessitate* what are clearly treated as teleo-

---

διὰ τῆς κινήσεως αὐτῶν· αὕτη δ' ἐστὶν ἡ ἐνέργεια τῆς τέχνης, ἡ δὲ τέχνη μορφὴ τῶν γιγνομένων ἐν ἄλλῳ—οὕτως ἡ τῆς θρεπτικῆς ψυχῆς δύναμις, ὥσπερ (b30) καὶ ἐν αὐτοῖς τοῖς ζῴοις καὶ τοῖς φυτοῖς ὕστερον ἐκ τῆς τροφῆς ποιεῖ τὴν αὔξησιν, χρωμένη οἷον ὀργάνοις θερμότητι καὶ ψυχρότητι (ἐν γὰρ τούτοις ἡ κίνησις ἐκείνης, καὶ λόγῳ τινὶ ἕκαστον γίγνεται), οὕτω καὶ ἐξ ἀρχῆς συνίστησι τὸ φύσει γιγνόμενον. ἡ γὰρ αὐτή ἐστιν ὕλη ᾗ αὐξάνεται καὶ ἐξ ἧς (b35) συνίσταται τὸ πρῶτον, ὥστε καὶ ἡ ποιοῦσα δύναμις ταὐτὸ τῷ ἐξ ἀρχῆς· μείζων δὲ αὕτη ἐστίν. εἰ οὖν αὕτη ἐστιν ἡ θρεπτικὴ ψυχή, αὕτη ἐστὶ καὶ ἡ γεννῶσα· καὶ τοῦτ' ἔστιν ἡ (741a1) φύσις ἡ ἑκάστου ἐνυπάρχουσα καὶ ἐν φυτοῖς καὶ ἐν ζῴοις πᾶσιν …

20 Cf. Gotthelf, *Teleology, First Principles, and Scientific Method*, 16.

logical outcomes, but would *explain* those outcomes only insofar as they are identified as formal and there because these outcomes are good, etc.[21]

To anticipate, the alternative view we will look at holds that 2.6 aims to provide not the "bottom-up" material-level side of embryogenesis (as I will call the development of the embryo and its parts), but rather a single, unified account of the entire embryogenesis—an account in which the embryo's formal nature (its potential for form) is playing the central efficient-causal role, by "making use of" material-efficient agents such as heat and cold and the necessary natures of the materials from which the embryo is being constructed. According to this view, these material-level contributions, though necessary, are by no means sufficient by themselves for embryogenesis. On this view the respective contributions of the formal nature and of the material agency of heat and cold correspond more or less to a natural reading of what we have already been given in general terms in **T4** above, and their interplay corresponds to what we have already been given in T7 (though we get useful amplification of them both in a passage from 2.6 itself that we will call below "**T10**"). I will defend this alternate view.

Advocates of the first view would maintain that **T4** and **T7** are consistent with their thesis that Aristotle sees the relevant material agency as *sufficient* for embryogenesis. We shall see. I provide in a footnote the relevant parts of the two passages, for ease of reference.[22]

---

21 I mean such a general formulation to capture the views both of David Charles, "Aristotle on Hypothetical Necessity and Irreducibility," and Susan Sauvé Meyer, "Aristotle, Teleology, and Reduction," *Philosophical Review* 101 (1992): 791–825 (on which more shortly). See Gotthelf, *Teleology, First Principles, and Scientific Method*, Ch. 3, for a fuller and more precise characterization of their views, and of some differences between them.

22 (T4–excerpt) Now, hard and soft and supple and brittle, and whatever other such conditions (*pathē*) that belong to the parts containing soul—heat and cold would make those, but would not go so far as to make the definition (*logos*) in virtue of which the one is now flesh and the other bone: that is due to the movement (*kinēsis*) from the generator who is in *full actuality* (*entelecheiāi*) what the thing out of which the product comes is *potentially*, just as is the case with the things produced according to an art; for the hot and the cold make the iron hard or soft, but what makes it a *sword* is the movement of the tools which contains a definition (*logos*) belonging to the art (734b31–735a4).

(T7–excerpt) And just as things which are produced by the agency of an art are produced by means of tools—it's truer in fact to say "by the *movement* of the tools" (this ⟨movement⟩ being the actuality of the art, and the art being the shape of that which comes to be in another)—in this way the nutritive soul-capacity, just as, in the animals and plants themselves later on, it produces growth from the food using as tools heat and cold (for the soul-capacity's movement is *in* these, and each thing is produced according to a certain definition [*logōi tini*]), in this way also at the very beginning the thing produced by a nature is constituted. ... If then this is the nutritive

III The Embryogenetic Account in 2.6

Gen. anim. 2.6 is a lengthy chapter. My tendency for a long time had been to focus in initially at the point in the chapter at which Aristotle has established that the first part to be produced is the heart, which then serves as the "source" for the rest, because it is in this stretch of text that he appears to be identifying the parts he discusses as being formed of material necessity. Indeed, the discussion opens at 743a1–5 with an indication that the focus will be on the formation, by material agency, of the uniform parts:

> **(T8)** From the heart the blood-vessels extend ⟨throughout the body⟩, as when artists sketch out preliminary figures on the walls; for the parts are situated around the blood-vessels, because they are generated out of them. The generation [out] of the uniform parts occurs by the agency of cooling and heat, for some things are constituted and solidified by the cold and some by the hot.[23]

It ends at 745b21–22, however, with the remark:

> **(T9)** We have stated, then, how *each* of the parts is constituted, and what the cause of their generation is.[24]

I underscore "each": not each of the uniform parts, but each of the *parts*. And yet, in the discussion that takes place between **T8** and **T9**, there is no mention of heterogeneously structured nonuniform parts, such as hand or foot. That fact also helps firm up the OCT editor's excision of *ek* ("out") at a4 in **T8**. Though found in all the main manuscripts, except for the earliest, Z, *ek* is not read in the early ninth-century Arabic translation nor in William of Moerbeke's thirteenth-century Latin translation (itself thought to rest on MSS earlier than any we have). And the *ek* is just not plausible. Aristotle's standard examples when he is distinguishing between uniform and nonuniform parts, as **T4** above exemplifies, are, for the uniform parts, flesh and bone, and for the nonuniform parts, heterogeneously structured ones such as hand and foot. But, in the lines imme-

---

soul, it is also that which generates, and this is the nature of each ⟨organism⟩ and present in all plants and animals (740b25–741a2).

23 (743a1) ἐκ δὲ τῆς καρδίας αἱ φλέβες διατεταμέναι, καθάπερ οἱ τοὺς κανάβους γράφοντες ἐν τοῖς τοίχοις· τὰ γὰρ μέρη περὶ ταύτας ἐστίν, ἅτε γιγνόμενα ἐκ τούτων. ἡ δὲ γένεσίς ἐστιν [ἐκ] τῶν ὁμοιομερῶν ὑπὸ ψύξεως καὶ θερμότητος· (a5) συνίσταται γὰρ καὶ πήγνυται τὰ μὲν ψυχρῷ τὰ δὲ θερμῷ.

24 πῶς μὲν οὖν ἕκαστον συνίσταται τῶν μορίων εἴρηται, καὶ τί τῆς γενέσεως αἴτιον.

diately following **T8,** his attention is directed to the formation of flesh (and other uniform parts that harden out of it, such as nail and hoof) and of bone (and sinew), the former resulting from a cooling process, the latter from a kind of internal "baking"—precisely what **T8**, without the *ek*, is announcing he will attend to. Whether or not, in the remainder of 2.6, Aristotle ever turns to heterogeneously structured nonuniform parts like hand or foot, there is certainly no project of systematically explaining their formation *out of* such uniform parts as flesh and bone. The nonuniform parts may be said to be formed *along with* the uniform ones, i.e., *as* the latter are forming, but not *out of* them.[25] So the *ek* must go.[26]

One reason why the formation of nonuniform parts is not discussed between **T8** and **T9** *may* be that Aristotle does not think they are formed of necessity by the agency of heat and cold. Indeed, at one time I hypothesized that Aristotle's view was that uniform parts were formed wholly by material necessity but nonuniform parts were not; they, *à la* **T4**, were due to the formal nature (via specific movements, directed to form, that the formal nature initiates in the material). This view was appealing, since it appeared to deny that the embryo as a whole is formed of material necessity; instead, only its simpler (i.e., uniform) parts are.

But there are two problems with that picture, as I have come to realize. First, note that **T4** says, of the divide between what is caused by the material agency of heat and cold alone and what is caused by the formal nature, that flesh falls on the latter side, along with hand and foot ("for even [flesh] has a function," 734b30–31). *And,* in the lines that immediately follow in **T4**, lines which I have singled out above, what is ascribed to the material agency of heat and

---

[25] Cf., e.g., 2.1, 734b27–28 (in **T4** above): they come to be *hama*, simultaneously. At best, one could say that the stretch between **T8** and **T9** addresses nonuniform parts that are composed of one uniform part (cf. *part. anim.* 2.1 on sense-organs and viscera, and Montgomery Furth, "Aristotle's Biological Universe: An Overview," in Allan Gotthelf and James G. Lennox [eds.], *Philosophical Issues in Aristotle's Biology* [Cambridge: Cambridge University Press, 1987], 21–52, at 34–36). But these are not discussed under that heading (in which case one could imagine Aristotle explicitly beginning a turn to the treatment of the nonuniform parts), and only the eye is discussed at any length, at 743b32–744b12; even there the main focus is on why the eye, an important part formed out of the best of the nourishment and thus, one would expect, first, is in fact formed last. The other sense organs are mentioned once in that discussion (743b34–744a6) and then briefly just after the eye, in a passage asserting generally that more and less valuable parts are formed, respectively, from higher and lower qualities of nutriment ("the bodies of the other sense organs," 744b23).

[26] It is most likely a case of dittography—see the *ek* in the previous line—encouraged by an interpretative confusion. We mustn't, however, forget that earlier *ek:* we will need to explain which parts are generated *out of the blood-vessels*. On the face of it, these are the uniform parts mentioned in the next line, the one we have been discussing; but see 157–58 below.

cold is the production, not of the whole of each uniform part, but only of their *pathē*—their *conditions*, such as their being "hard and soft and supple and brittle"; and, to be precise, that passage does not reflect the uniform/nonuniform distinction, since what is produced by heat and cold are the *pathē* of each of "the parts ... having soul" (b31–33), i.e., the *pathē* of *both* the uniform *and* the nonuniform parts.

That is the first problem for thinking that our passage 2.6 might be attributing the uniform parts (though not the nonuniform) to material agency alone: it is the *pathē* (of both the uniform and the nonuniform parts), and not the uniform parts *per se*, that are caused by the agency of heat and cold alone. The second problem has to do with what actually happens in the stretch of text from **T8** to **T9**. For not only is there no attempt by Aristotle in that stretch to show that nonuniform parts are formed of material necessity from uniform parts (or in any other way), but even the crucial part of the causation in the explanation of uniform parts is not said to be due to material necessity, but to "the nature" and "the source of movement." Some aspects of the formation of the "eye body" are ascribed to (material) necessity (e.g., at 744a13) but the eye itself is formed when "the source of movement takes control (*kratei*)" (744a33–34), which is late for an upper part (744a33–35); the eyelid, needed only when there is an eye, is also formed late, since "nature makes nothing superfluous or without a point" (744a36–38). Likewise, the use of higher-quality nutriment for more valuable parts, just mentioned, is said to be the work of a nature internal to the forming animals.[27] And notice the agency of the nature (*hē phusis*) at 745a32, in the discussion of teeth. There is really very little that is ascribed to material necessity alone. For, even in Aristotle's own context, quite apart from any consideration of the views of his predecessors, accounts of formations wholly in terms of material necessity are in danger of treating those formations as accidental (743a21–23, to be discussed momentarily as part of **T10**). This danger is avoided, we will see, not by an appeal, such as Susan Sauvé Meyer makes in her 1992 paper, "Aristotle, Teleology, and Reduction," to a "reliable material mechanism"—there does not seem to be any such thing to appeal to, and in any case we hear of none. (We'll return to this shortly.[28]) The danger is avoided by appeal—in the very same causal account that incorporates such material necessity as is at work—*to the agency of the developing animal's formal nature*. In addition to

---

[27] οὕτως ἐν τοῖς γιγνομένοις αὐτοῖς ἡ φύσις ἐκ μὲν τῆς καθαρωτάτης ὕλης σάρκας καὶ τῶν ἄλλων αἰσθητηρίων τὰ σώματα συνίστησιν (744b22).
[28] See Gotthelf, *Teleology, First Principles, and Scientific Method*, 74–75, and *infra*, n. 47.

the instances of such agency I have already just cited, there is the important explicit general remark, at 743a17–b5, that I will label "**T10**."

To set up this passage, before I quote it, we need to return to a puzzle I noted earlier in comparing **T4** and **T6** but did not try to resolve.[29] **T4** distinguishes the agency of the "movement" supplied by the formal nature from the material agency of heat and cold (734b31–36). **T6** speaks of a soul- (or form-) specific generative heat which is somehow associated with those formative movements (736b29–737a1, 18–22). The question of whether heat and cold are separate material agents or, rather, express the agency of the formal nature, arises again in subsequent passages. In **T7** the formal nature is said to "use heat and cold" as tools, because its formative movement and definition is resident in the animals that have that formal nature 740b25–35). In **T8**, as we noted when we introduced that passage, heat and cold appear to be spoken of as material agents, having certain causal powers as such, e.g., to solidify (743a1–5); however, their being said in the same breath also to "constitute" (a5) leaves open the possibility that here, too, they may be "tools" of the formal nature; the appearance of "constituted" again in **T9** (745b21) perhaps suggests as much. The puzzle, then, is twofold: (a) to what extent, and when, are references to the agency of hot and cold references to a material-efficient agency that is to be distinguished from any agency (or explanatory role) of the formal nature; and (b) in those cases where heat and cold are said to be "tools" of the formal nature, being "used" by it for its own more global ends, how is the agency of heat and cold here to be understood, and how does its material dimension stand in relation to its role as tool of the formal nature?

Our next passage, 743a6–b5, which I am calling "**T10**," is a very long one, but it should help us with these questions. It follows immediately upon **T8** (which, again is 743a1–5), and is the first part of the stretch we have been discussing, from **T8** to **T9**. To set the context, I repeat the immediately preceding lines, which we have called "**T8**," and then identify the beginning of this new material:

> (**T8**) From the heart the blood-vessels extend ⟨throughout the body⟩, as when artists sketch out preliminary figures on the walls; for the parts are situated around the blood-vessels, because they are generated out of them. The generation [out] of the uniform parts occurs by the agency of cooling and heat, for some things are constituted and solidified by the cold and some by the hot.
>
> (**T10**) We have spoken previously elsewhere concerning the differences between these, and have stated what sorts of things are dissolvable by fluid and by fire and what sorts of things are not dissolvable by fluid and cannot be melted by fire. Now, as the nourishment oozes

---

29 Other than to make some back references (in n. 17), which were not meant to be decisive.

through the blood-vessels and the passages in each of them—just as water does when it stands in unbaked earthenware—flesh, or its counterpart, is formed (*gignontai*), being constituted by the cold (which is why it is dissolvable by fire). As that nourishment wells up, the excessively earthy stuff in it, having little fluidity and heat, becomes cooled while the fluid is evaporating together with the hot, and is formed into parts that are hard and earthy, e. g., nails, horns, hoofs, and bills; that is why these can be softened, but none of them can be melted by fire; though some, e. g., eggshell, can be melted by fluids.

It is by the internal heat that both the sinews and the bones are formed as the moisture dries up. That is in fact why bones, like pottery, are indissolvable; for they have been baked by the heat present in their generation, as though in an oven. But this ⟨heat⟩ does not make any chance thing flesh or bone, nor in any chance place nor at any chance time; rather ⟨it makes into flesh or bone⟩ that which is naturally so ⟨made⟩ and where it is naturally so ⟨made⟩ and when it is naturally so ⟨made⟩. For neither will that which is potentially *be* by the agency of a motive source that doesn't have the actuality, nor will that which has the actuality produce ⟨a thing⟩ out of any chance thing, just as the carpenter would not make a chest out of something other than wood and just as there will not be a chest ⟨made⟩ out of the wood without him.

The heat is present in the spermatic residue having its movement and actuality in an amount and character suitably proportioned (*summetros*) to each of the parts. To the extent to which it is deficient or excessive, the thing being formed will come out inferior or deformed, similarly to what happens with things that are constituted outside the body, via boiling, for culinary enjoyment or any other use. But there we supply a suitable proportion (*summetria*) of heat to the movement, while here it is provided by the nature of the generator. In spontaneous generations the cause is a seasonal movement and heat.

Cooling is a deprivation of heat. The nature uses both [sc. cold and heat], as having a potentiality to make, of necessity, one thing this and another that; in the generations ⟨under discussion⟩, however, their cooling and heating occur for the sake of something, that is, each of the parts is formed such ⟨as it is⟩—the flesh soft, the sinew dry and elastic, the bone dry and crackable—in one respect of necessity, in another respect for the sake of something.[30]

---

[30] My thanks to Devin Henry for reminding me some time ago of the importance of this passage for the issues of this chapter, and to Alan Bowen and Jim Lennox for valuable advice on the passage's translation. The Greek of this important passage reads (beginning with **T8**):

(**T8**) (743a1) ἐκ δὲ τῆς καρδίας αἱ φλέβες διατεταμέναι, καθάπερ οἱ τοὺς κανάβους γράφοντες ἐν τοῖς τοίχοις· τὰ γὰρ μέρη περὶ ταύτας ἐστίν, ἅτε γιγνόμενα ἐκ τούτων. ἡ δὲ γένεσίς ἐστιν [ἐκ] τῶν ὁμοιομερῶν ὑπὸ ψύξεως καὶ θερμότητος· (a5) συνίσταται γὰρ καὶ πήγνυται τὰ μὲν ψυχρῷ τὰ δὲ θερμῷ. (**T10**) (743a6) περὶ δὲ τῆς τούτων διαφορᾶς εἴρηται πρότερον ἐν ἑτέροις, ποῖα λυτὰ ὑγρῷ καὶ πυρὶ καὶ ποῖα ἄλυτα ὑγρῷ καὶ ἄτηκτα πυρί. διὰ μὲν οὖν τῶν φλεβῶν καὶ τῶν ἐν ἑκάστοις πόρων διαπιδύουσα ἡ τροφή, καθάπερ ἐν τοῖς ὠμοῖς κεραμίοις (a10) τὸ ὕδωρ, γίγνονται σάρκες ἢ τὸ ταύταις ἀνάλογον ὑπὸ τοῦ ψυχροῦ συνισταμέναι, διὸ καὶ λύονται ὑπὸ πυρός. ὅσα δὲ γεηρὰ λίαν τῶν ἀνατελλόντων, ὀλίγην ἔχοντα ὑγρότητα καὶ θερμότητα, ταῦτα δὲ ψυχόμενα ἐξατμίζοντος τοῦ ὑγροῦ μετὰ τοῦ θερμοῦ γίγνεται σκληρὰ καὶ γεώδη τὴν μορφήν, οἷον (a15) ὄνυχες καὶ κέρατα καὶ ὁπλαὶ καὶ ῥύγχη· διὸ μαλάττεται μὲν πυρί, τήκεται δ' οὔθέν, ἀλλ' ἔνια τοῖς ὑγροῖς, οἷον τὰ κελύφη τῶν ᾠῶν.

ὑπὸ δὲ τῆς ἐντὸς θερμότητος τά τε νεῦρα καὶ τὰ ὀστᾶ γίγνεται ξηραινομένης τῆς ὑγρότητος.

The formation of parts subsequent to the heart is described in this passage in the following order (which may or may not be, according to Aristotle, the order of their formation):
(i) blood-vessels;[31]
(ii) flesh;
(iii) parts that are hard and earthy, e.g., nails, horns, hoofs, and bills;
(iv) sinews and bones.

This remark about the agency by which sinews and bones are formed is brought home with an apparently more general example that speaks of
(v) flesh or bone,
and then elaborates on the agency involved, and the conditions of that agency; the final lines of the passage then take as their subject flesh, sinew, *and* bone. This suggests that when Aristotle switches to flesh or bone he is switching to commenting on the agency involved in the formation of *all* uniform parts. I will provide confirmation of this in a moment. But let us return to the start of this list to note the agency to which each of these formations is ascribed:

(i) The blood-vessels are not clearly assigned to any agency, but the process is analogized to the sketching out by artists of preliminary figures on walls, and it is said that on that framework certain (unspecified) parts form "around" the blood vessels; the text goes on to speak about "constitution" and "solidification" of the uniform parts, and this is ascribed to the agency of heat or cold. There is no mention in this discussion of the action of a formal nature, though this *might*

διὸ καὶ ἄλυτά ἐστι τὰ ὀστᾶ ὑπὸ τοῦ πυρὸς καθάπερ κέραμος· οἷον (a20) γὰρ ἐν καμίνῳ ὠπτημένα ἐστὶν ὑπὸ τῆς ἐν τῇ γενέσει θερμότητος. αὕτη δὲ οὔτε ὅ τι ἔτυχε ποιεῖ σάρκα ἢ ὀστοῦν οὔθ' ὅπου ἔτυχεν οὔθ' ὁπότ' ἔτυχεν, ἀλλὰ τὸ πεφυκὸς καὶ οὗ πέφυκε καὶ ὅτε πέφυκεν. οὔτε γὰρ τὸ δυνάμει ὂν ὑπὸ τοῦ μὴ τὴν ἐνέργειαν ἔχοντος κινητικοῦ ἔσται, οὔτε τὸ τὴν ἐνέργειαν ἔχον ποιήσει ἐκ τοῦ τυχόντος, (a25) ὥσπερ οὔτε κιβωτὸν μὴ ἐκ ξύλου ὁ τέκτων ποιήσειεν ἂν οὔτ' ἄνευ τούτου κιβωτὸς ἔσται ἐκ τῶν ξύλων.

ἡ δὲ θερμότης ἐνυπάρχει ἐν τῷ σπερματικῷ περιττώματι τοσαύτην καὶ τοιαύτην ἔχουσα τὴν κίνησιν καὶ τὴν ἐνέργειαν ὅση σύμμετρος εἰς ἕκαστον τῶν μορίων. καθ' ὅσον δ' ἂν ἐλλείπῃ ἢ (a30) ὑπερβάλλῃ ἢ χεῖρον ἀποτελεῖ ἢ ἀνάπηρον τὸ γιγνόμενον, παραπλησίως τοῖς ἔξω συνισταμένοις διὰ τῆς ἑψήσεως πρὸς τροφῆς ἀπόλαυσιν ἤ τινα ἄλλην ἐργασίαν. ἀλλ' ἐνταῦθα μὲν ἡμεῖς τὴν τῆς θερμότητος συμμετρίαν εἰς τὴν κίνησιν παρασκευάζομεν, ἐκεῖ δὲ δίδωσιν ἡ φύσις ἡ τοῦ γεννῶντος. (a35) τοῖς δὲ αὐτομάτως γιγνομένοις ἡ τῆς ὥρας αἰτία κίνησις καὶ θερμότης.

ἡ δὲ ψύξις στέρησις θερμότητός ἐστιν. χρῆται δ' ἀμφοτέροις ἡ φύσις ἔχουσι μὲν δύναμιν ὥστε (743b1) τὸ μὲν τοδὶ τὸ δὲ τοδὶ ποιεῖν, ἐν μέντοι τοῖς γιγνομένοις ἕνεκά τινος συμβαίνει τὸ μὲν ψύχειν αὐτῶν τὸ δὲ θερμαίνειν καὶ γίγνεσθαι τῶν μορίων ἕκαστον, τὴν μὲν σάρκα μαλακὴν τῇ μὲν ἐξ ἀνάγκης ποιούντων τοιαύτην τῇ δ' ἕνεκά τινος, τὸ δὲ (b5) νεῦρον ξηρὸν καὶ ἑλκτὸν τὸ δ' ὀστοῦν ξηρὸν καὶ θραυστόν.

[31] I put aside for the moment those parts, whatever they are, that are said there to be generated "out of the blood-vessels" (see n. 26 above).

be implied by the craft analogy, and by the fact that the blood-vessels are formed first after the heart, which is the *seat* (*archē*) of the nutritive soul and formal nature.[32]

(ii) & (iii) Flesh is "constituted" by the agency of the cold as the blood "wells up" through the walls of the blood vessels; the cold also produces hard and earthy parts, such as nails and horns; in both cases there are frequent references to "chemical" facts that support the assignment of material agency, with no (immediate) reference to a formal nature.[33]

(iv) Sinews and bones likewise have "chemical" properties that show that they are formed by some sort of heating; the heating is assigned to the "internal heat."[34]

(v) Since "flesh or bone" is introduced as subject in the midst of a statement about the formation of "sinews and bones," and since "flesh or bone" becomes the subject for the rest of what is said, I think it is clear that "flesh or bone" is a stand-in for "all the uniform parts." The immediate agent for the formation of flesh or bone is "the heat present in the spermatic residue";[35] this heat does not make into flesh or bone "any chance thing," "at any chance place" and "at any chance time," but only a "natural" material at a "natural" place, and a "natural" time (*to pephukos kai hou pephuke kai hote pephuken*). The remainder of the passage then makes the following four points about that heat:

(a) It must be such as to "have the (full) actuality" of that which is produced, at least in the way that the semen and *katamēnia* have it in full actuality;[36]

---

[32] Indeed, as Mariska Leunissen has suggested to me, the agent of this process has been mentioned earlier in 742a37–b3 and 742b33–35: it is the heart—the part that "has the source and end of the whole nature" (742b1).

[33] By "chemical facts" I mean the sort of material causal connections that may be found in *meteorol*. 4.1–11. These have to do with the connection between a material body's perceptible properties and the elementary powers from which the body is constituted. Compare, e. g., 743a11–17 with *meteorol*. 4.6, esp. 383a26–32 (n.b. *keras* [in sing. or pl.] at 743a15 and 383a32).

[34] It is not clear whether the mention here of *internal* heat (*hupo de tēs entos thermotētos*) brings in a reference to formal powers, as "vital-heat" or "soul-heat" does. Although I suspect it may (cf. 742a14–16, and Peck *ad loc.*, on the internal formation of fresh *pneuma*), I will not appeal in my argument to this mention.

[35] As Mariska Leunissen has also pointed out to me, this "spermatic residue" must be what remains of the mother's contribution, since the father's residue is no part of the *kuēma*, while the mother's residue has already been identified (at 744b28–38) as the first, and best, nutriment, used for the most important parts. On the other hand, see the end of the next note.

[36] 743a28. Matters here are a bit tricky: as my previous note suggests, the "actuality" mentioned in (a) is ascribed to the heat in the mother's residue, i.e., the *katamēnia*. By contrast, **T3** above, the first of our passages to speak of actuality, does not mention heat, or the mother, but says that the semen, *qua* tool of "the nature in the male," contains its movement "in actuality" (unlike a

(b) This heat, which was said to be present in the spermatic residue, is now said to be due to "the nature of the generator";[37]

(c) The heat must be there "in an amount and character suitably proportioned (*summetros*) to" (the movement needed to produce) each particular part;[38]

(d) The heat (and cold) which supplies the already mentioned movement and (thereby) material agency is *made use of* by the (formal) nature that is at work,[39] in such a way that, while the immediate effects of the material agency—"the flesh soft, the sinew dry and elastic, the bone dry and crackable"—are *of necessity*, the formal nature making use of it is producing that result (under some description) *for the sake of something*.

Our question is what the respective roles are of the formal and material natures in the formation of the uniform parts, as this is described in this passage. A close look makes it clear that those roles are at least very close to what they are said to be in passages **T4** and **T7** above. Here, again, are the operative parts of **T4** and **T7**;[40] to show the parallels in their accounts of the roles of heat and cold both in their own right and as tools of the formal nature, I identify via the same letters in square brackets the portions of **T4** corresponding to the portions of **T10** that are given letters in round brackets in the summary just above.

> **(T4)** Now, hard and soft and supple and brittle, and whatever other such conditions (*pathē*) that belong to the parts containing soul—**[d]** heat and cold would make those, but would not go so far as to make the definition (*logos*) in virtue of which the one is now flesh and the

---

craftsman's tool which, when not in touch with the craftsman, has the movement of the art only potentially). **T4**, to be cited again momentarily, likewise speaks only of movement when it speaks of actuality, this time of the movement of "generator," which is said to have that movement "in full actuality" (*entelecheiāi*). **T7** also speaks of the movements imparted to the craftsman's tools as containing the actuality of the art, but, in speaking of animal generation, analogizes to those tools the heat and cold which serve as agents of the nutritive soul. What links our claim in **T10** with those of **T3**, **T4**, and **T7** even more closely than it might otherwise seem to be linked is (i) the fact that both **T10** and **T7** speak of the movement as carrying, and using as a tool, an appropriate heat; and (ii) the fact that the "spermatic residue" in **T10**, though originally the mother's *katamēnia*, is *katamēnia* that has been "set" or "constituted" by the heat from the *sperma*; the *kuēma*'s own heat is thus closer in character or quantity to the father's heat than that heat was before the *katamēnia* was "set." This, I take it, helps explain why Aristotle can go on immediately to speak of the "spermatic heat" as the heat of "the generator."

37 743a34. See the previous note.
38 The fact that Aristotle now speaks of a *summetria* for "each of the parts" (743a29) further supports my suggestion that he is now speaking of the formation of uniform parts in general. For more on this *summetria*, see below, 160–61.
39 743a36–37.
40 For the full passages, see above, 145–46 and 148–49.

other bone: that is due to **[b]** the movement (*kinēsis*) from the generator who is **[a]** in *full actuality* (*entelecheiāi*) what the thing out of which the product comes is *potentially*, just as is the case with the things produced according to an art; for **[d]** the hot and the cold make the iron hard or soft, but what makes it a *sword* is the movement of the tools which contains a definition (*logos*) belonging to the art.

(T7) And just as things which are produced by the agency of an art are produced by means of tools—it's truer in fact to say **[d]** "by the *movement* of the tools" (this ⟨movement⟩ being **[a]** the actuality of the art, and the art being the shape of that which comes to be in another)—in this way **[d]** the nutritive soul-capacity, just as, in the animals and plants themselves later on, it produces growth from the food using as tools heat and cold (for **[a]** the soul-capacity's movement is *in* these, and each thing is produced according to a certain definition [*logōi tini*]), in this way also at the very beginning the thing produced by a nature is constituted. ... If then this is the nutritive soul, **[b]** it is also that which generates, and this is the nature of each ⟨organism⟩ and present in all plants and animals.

(a), **(b)**, and **(d)** show clearly that the same causal theory that is outlined in **T4** and **T7** lies behind **T10**—and that the primary generative agent of the uniform parts in **T10** is the formal nature of the embryo, aimed at the production of an offspring one in form with the parents. **T4** attributed to heat and cold as agents in their own right only the *pathē* of the ensouled parts, giving as examples their hard or soft and their supple or brittle character. The last lines of **T10** provide as examples of what is produced of necessity *pathē* characteristic of the uniform parts under discussion: "the flesh soft, the sinew dry and elastic, the bone dry and crackable." **(c)**, which we find in **T10** but not in **T4** and **T7**, introduces a concept we have not seen before in *gen. anim.*: the *summetria* (which I have translated "suitable proportion") between the "amount and character" of the "movement and actuality" of the internal generative heat that embodies the formal nature and the parts being produced. This is a difficult notion, made more difficult to pin down by its long and varied history from ancient Greece to the present day.[41]

I take it, however, that this requirement is a specification, or narrowing, of the requirement conveyed in **T10** at 743a21–26: The immediate agent for the formation of flesh or bone is "the heat present in the spermatic residue."[42] This heat does not make into flesh or bone "any chance thing," "at any chance place" and

---

[41] See Giora Hon and Bernard R. Goldstein, *From Summetria to Symmetry: The Making of a Revolutionary Scientific Concept* (New York: Springer, 2008). Their discussion of *summetria* in Aristotle is not of great help for our purposes, although I do make use of their English translation for the relevant meaning of the word. Oddly, they do not discuss these two *gen. anim.* occurrences.

[42] On this "spermatic residue," see nn. 35–37 above. This is the female residue *after* it has been acted upon by the heat and movements of the male *sperma*.

"at any chance time," but only a "natural" material at a "natural" place, and a "natural" time (*to pephukos kai hou pephuke kai hote pephuken*). **(c)** presumably stresses that the heat is "targeted" to the specific material, place, and time needed. Such targeting is a function of the formal nature, so once again we have **T10** paralleling **T4** and **T7**: the formings of the uniform parts are ascribed by Aristotle to the formal nature. Only the *pathē* of the nonuniform parts, as such, are due to heat and cold, as such.[43]

This does not mean, as I already said, that 2.6 is concerned only with the formation of uniform parts. In fact, this will be confirmed if we, now, widen our focus to the entire chapter, and note that the discussion of uniform parts in **T10** is embedded in a larger, single, partly chronological narrative of the order of the formation of the entire embryo and its parts. This narrative begins very early in 2.6. It sets as its primary focus "what parts come into being after what" (742a16–17), and though it names very few nonuniform parts, it speaks of, for example, the upper and lower parts of the body, the head, the eyes, and the thighs. And it is in this context that attention is given, in the passage we have already discussed (**T10**), to the formation of the uniform parts, such as flesh, bone, and sinew, and, in particular, to the formation of their generation.

This narrative is structured overall by an extended passage, very early in the chapter, that aims to connect various priority relationships of the parts, in different senses of priority, with the order of their coming-to-be. The passage is surprisingly cumbersome, repetitive, and over-long (so far as I can tell). I will quote its first two-thirds or so in full, in part to show that the narrative includes the wide range of parts I have mentioned, and in part to allow us to see how 2.6 carries out a good part of what we identified at the beginning of this chapter as the overall aim of *gen. anim.*—the integration of the efficient and final causes of generation. The portion I will quote is 742a16–b17.

> **(T11)** Some of the early students of nature made an attempt to state what comes to be after what among parts, but had little experience of what actually happens. For among parts as

---

43 This reading is confirmed by the concluding lines of **T10**, 743a37–b5:
"Cooling is a deprivation of heat. The nature uses both [sc. cold and heat], as having a potentiality to make, of necessity, one thing this and another that; in the generations ⟨under discussion⟩, however, their cooling and heating occur for the sake of something, that is, each of the parts is formed such ⟨as it is⟩—the flesh soft, the sinew dry and elastic, the bone dry and crackable—in one respect of necessity, in another respect for the sake of something."
Heat and cold, *acting as such, from material necessity*, produce the *pathē* of the parts—"the flesh soft, the sinew dry and elastic, the bone crackable." They have no larger direction, as it were. But, in another respect—i.e., (I supply) *qua used by the formal nature*—they are directed *heneka tou*—for the sake of the functional parts they are tools in creating.

in other things one is naturally prior to another, and there are many sorts of priority, for the thing for the sake of which and the thing for the sake of this differ, and the latter is prior in generation but the former is so in being. The thing for the sake of this also has two differentiae—the thing from which motion originates *and* the thing which that for the sake of which uses. I mean, for example, the thing which can generate and the thing which is instrumental to the thing generated; for one of these must be present earlier—the one that makes (e.g., the *aulos* teacher *before* the learner and the *auloi after* the one learning to play the *aulos*, since it is superfluous that there be *auloi* present for those who do not know how to play them).

Given that there are three [sc. sorts of relata in the various priority relations we have distinguished][44]—one being the end which we say is the thing for the sake of which, a second (belonging to things for the sake of this) being that which initiates movement and is generative (for that which makes and is generative, in so far as they are such are relative to the thing made and generated), and a third being useful, that is, being the thing the end uses—it is necessary that there be present first a part in which is the source of movement (for from the start this is one part of the end and indeed the most governing part), then, after this, is the whole and the end; and third, and last, the parts instrumental to these in relation to some uses.

As a result, if there is any such thing which must be present in animals, in that it contains a source and end of its whole nature, then this must come to be in so far as it is first a mover and in so far as it is part of the end, along with the whole. As a result, all of the instrumental parts that can generate the nature must themselves always be present earlier (for they are for the sake of another *as a source*), but all the instrumental parts that are not of this sort must be present later than the ones for the sake of another. That is why it is not easy to distinguish which of the parts are prior, all those which are for the sake of another or that for the sake of which these are. For the parts that initiate movement complicate things, in that they are prior in the generation to the end, and it is not easy to distinguish the movement-initiating parts from the instrumental.

And yet in accordance with this method one must seek what comes to be after what, for the end is later than some parts but earlier than others. And on account of this the part which has the source comes to be first, then following it the upper body. That is why the parts about the head and the eyes in embryos appear largest, while those below the umbil-

---

44 I add this phrase to bring out the progression of Aristotle's argument. He starts from a general account of different priority relationships in a wide class of comings-to-be—cf. "among parts *as in other things*" (742a18–19). These "other things" include craft comings-to-be, such as in the case of the coming-to-be of a trained music student presented at a26–28. Having indicated the range of comings-to-be he will consider, he then (a20–23) identifies the different sorts of priority relationships present in these comings-to-be and, incidentally, the different relata encompassed by these relationships. At a28–32, in a genitive absolute construction, he focuses in on those three distinct relata, still at the general level of comings-to-be. Only then, at a32–33 does he return to animals parts in particular, to cash out this analysis of priority relationships that determine *ti meta ti ginetai*. This manner of progression helps to some extent to explain the slow, lumbering, and to my mind repetitive progression of the text here and especially in what follows—a progression which is not even completed at the point where I leave off in my quotation.

icus, e.g., the thighs, appear small. The reason is that the lower parts are for the sake of the upper, and are neither parts of the end nor generative of it.[45]

As I mentioned, this passage aims at providing a framework for creating an account of the order in which the parts of an embryo come to be, an account which will include both uniform and nonuniform parts. In identifying the generative parts as coming-to-be first, and the source of these—the heart or its analogue—as that which comes-to-be first of all,[46] this passage identifies an efficient causal sequence, though only in the broadest strokes. It is only late in the passage, and then again later in the chapter, that there is any specificity in the efficient

---

45 I thank Alan Bowen for his help with the translation of this passage. Here is the Greek: (742a16) τῶν δ' ἀρχαίων τινὲς φυσιολόγων τί μετὰ τί γίγνεται τῶν μορίων ἐπειράθησαν λέγειν, οὐ λίαν ἐμπειρικῶς ἔχοντες τῶν συμβαινόντων. τῶν γὰρ μορίων ὥσπερ καὶ ἐπὶ τῶν ἄλλων πέφυκεν ἕτερον ἑτέρου πρότερον. τὸ δὲ πρότερον (a20) ἤδη πολλαχῶς ἐστιν· τό τε γὰρ οὗ ἕνεκα καὶ τὸ τούτου ἕνεκα διαφέρει, καὶ τὸ μὲν τῇ γενέσει πρότερον αὐτῶν ἐστι τὸ δὲ τῇ οὐσίᾳ. δύο δὲ διαφορὰς ἔχει καὶ τὸ τούτου ἕνεκα· τὸ μὲν γάρ ἐστιν ὅθεν ἡ κίνησις, τὸ δὲ ᾧ χρῆται τὸ οὗ ἕνεκα. λέγω δ' οἷον τό τε γεννητικὸν καὶ τὸ ὀργανικὸν τῷ γεννωμένῳ· (a20) τούτων γὰρ τὸ μὲν ὑπάρχειν δεῖ πρότερον, τὸ ποιητικόν, οἷον τὸ διδάξαν τοῦ μανθάνοντος, τοὺς δ' αὐλοὺς ὕστερον τοῦ μανθάνοντος αὐλεῖν· περίεργον γὰρ μὴ ἐπισταμένοις αὐλεῖν ὑπάρχειν αὐλούς· τριῶν δ' ὄντων—ἑνὸς μὲν τοῦ τέλους ὃ λέγομεν εἶναι οὗ ἕνεκα, δευτέρου δὲ τῶν τούτου ἕνεκα τῆς ἀρχῆς (a30) τῆς κινητικῆς καὶ γεννητικῆς (τὸ γὰρ ποιητικὸν καὶ γεννητικόν, ᾗ τοιαῦτα, πρὸς τὸ ποιούμενόν ἐστι καὶ γεννώμενον), τρίτου δὲ τοῦ χρησίμου καὶ ᾧ χρῆται τὸ τέλος—πρῶτον μὲν ὑπάρχειν ἀναγκαῖόν τι μόριον ἐν ᾧ ἡ ἀρχὴ τῆς κινήσεως (καὶ γὰρ εὐθὺς τοῦτο τὸ μόριόν ἐστι τοῦ τέλους ἓν καὶ κυριώτατον), (a35) ἔπειτα μετὰ τοῦτο τὸ ὅλον καὶ τὸ τέλος, τρίτον δὲ καὶ τελευταῖον τὰ ὀργανικὰ τούτοις μέρη πρὸς ἐνίας χρήσεις. ὥστ' εἴ τι τοιοῦτόν ἐστιν ὅπερ ἀναγκαῖον ὑπάρχειν ἐν τοῖς (742b1) ζῴοις, τὸ πάσης ἔχον τῆς φύσεως ἀρχὴν καὶ τέλος, τοῦτο γίγνεσθαι πρῶτον ἀναγκαῖον, ᾗ μὲν κινητικὸν πρῶτον, ᾗ δὲ μόριον τοῦ τέλους μετὰ τοῦ ὅλου. ὥστε τῶν μορίων τῶν ὀργανικῶν ὅσα μέν ἐστι γεννητικὰ τὴν φύσιν, ἀεὶ πρότερον (b5) δεῖ ὑπάρχειν αὐτά (ἄλλου γὰρ ἕνεκά ἐστιν ὡς ἀρχή), ὅσα δὲ μὴ τοιαῦτα τῶν ἄλλου ἕνεκα ὕστερον. διὸ οὐ ῥᾴδιον διελεῖν πότερα πρότερα τῶν μορίων, ὅσα ἄλλου ἕνεκα ἢ οὗ ἕνεκα ταῦτα. παρεμπίπτει γὰρ τὰ κινητικὰ τῶν μορίων πρότερον ὄντα τῇ γενέσει τοῦ τέλους, τὰ δὲ κινητικὰ πρὸς (b10) τὰ ὀργανικὰ διελεῖν οὐ ῥᾴδιον. καίτοι κατὰ ταύτην τὴν μέθοδον δεῖ ζητεῖν τί γίγνεται μετὰ τί· τὸ γὰρ τέλος ἐνίων μὲν ὕστερον ἐνίων δὲ πρότερον. καὶ διὰ τοῦτο πρῶτον μὲν τὸ ἔχον τὴν ἀρχὴν γίγνεται μόριον, εἶτ' ἐχόμενον τὸ ἄνω κύτος. διὸ τὰ περὶ τὴν κεφαλὴν καὶ τὰ ὄμματα μέγιστα (b15) κατ' ἀρχὰς φαίνεται τοῖς ἐμβρύοις, τὰ δὲ κάτω τοῦ ὀμφαλοῦ, οἷον τὰ κῶλα, μικρά. τοῦ γὰρ ἄνω τὰ κάτω ἕνεκεν καὶ οὔτε μόρια τοῦ τέλους οὔτε γεννητικὰ αὐτοῦ.

46 See the concluding lines of the passage which follows this quotation: Now with those things that do not change the origin or source (*archē*) is the essence (*ti estin*); but in things that come into being, immediately there are several ⟨sources⟩, and another manner ⟨of source⟩ and not all the same. Among them, one in number, is the source from which the movement comes, and that is why the heart is the first part which all blooded animals have, as we have said at the beginning; in others the analogue to the heart comes to be first (742b33–743a1).

causal account of the formation of parts, and even there, as we have seen, primarily in the account of the uniform parts coming to have the *pathē* they do. The focus of this passage is rather on the logical movement from the *end* of the development—that for the sake of which it takes place—to that which comes to be *for that*: first the source or origin of the development—the parts that form first and facilitate the formation of subsequent parts. This is a movement via *conditional necessity*—and it mirrors surprisingly well what is called for in the *part. anim.* 1.1 statement of the aims of a proper treatment of generation that we looked at, at the beginning of this chapter:

> **(T2)** And because he is such a thing, his coming-to-be necessarily happens so and is such. And that is why this part comes to be first, and then this.

In **T11** the conditionally necessary inference starts from "the thing for the sake of which"—the *end* (*telos*) of generation, which is, roughly, the integrated set of parts of the organism which are essential to its functioning, all of which are found in the upper part of the body. The inference then moves to two sorts of parts which are for the sake of that end: the generative parts, including the heart (or its analogue), and the instrumental parts, which serve the more fundamental parts and the organism as a whole. Those include the parts found in the lower part of the body. The situation is complicated, as the passage says (742b6–10), for several reasons; the one we will stop over is that the generative *source* (*archē*) of the embryonic development, the heart (or its analogue), is also the *source* for the organism as such at each of its stages—the seat of its soul—through the rest of its life. As such it is part of the end—the thing for the sake of which—as well as, in its generative capacity, a key part that comes to be and is for the sake of that end. But, that understood—and the relationship between the lower, instrumental parts and the upper, more defining parts understood as well—our passage is easily interpreted as satisfying the first sentence of **T2**: "And because he is such a thing, his coming-to-be happens so and is such." The passage likewise satisfies the second sentence of **T2**: "And that is why this part comes to be first, and then this"—as the opening words of **T11** make clear: our aim is "to state what comes to be after what among parts." **T11**, then, not only gives us a method (*methodos*, 742b10–11) for identifying the order in which the parts come to be, it also gives us the teleological tools for understanding *why* they come to be in the order they do.

It is within this framework that Aristotle goes on, in the rest of the chapter, to assign agency for the formation of the parts in the embryo, and to assign it primarily to a formal-efficient cause and only secondarily to material-efficient causes, as per such earlier passages as **T4** and **T7**. What **T11** does is to place that complex efficient-causal account in the framework of a broader teleological account, thereby completing, in principle anyway, the full account of generation promised in the opening passage of *gen. anim.* itself.[47]

IV Size and Shape: Indirect Confirmation for Our Reading of *gen. anim.* 2.6

I have said I could not find any place in 2.6 where Aristotle attempts to give an account, at least in outline, in terms of material agency, of the formation of a heterogeneous nonuniform part. I would like to close with a look at one possible exception, since its ultimate failure to work is instructive as well. We might consider that the first move from uniform to nonuniform involves adding shape, and there is a case to be made that Aristotle thinks that the generation of shape can be shown to be necessitated at the material level. For, in animals which possess bones (and the same can be said for other supporting materials), the shape is

---

[47] In saying this, I don't mean to imply, of course, that the presentation of the theory of generation is completed at *gen. anim.* 2.6. As Andrea Falcon and I will show in the paper mentioned in n. 1 above (80–86), *gen. anim.* 3–5 are integral to the overall project of a theory of generation described in **T2** and in the opening lines of *gen. anim.* 1.1.

This is perhaps a good place to take up again an objection I raised, in *Teleology, First Principles, and Scientific Method*, ch. 3 (80 ff.) to one key aspect of the influential account of Aristotelian natural teleology given by Susan Sauvé Meyer in her 1992 paper, "Aristotle, Teleology, and Reduction." As I explained there, Meyer claims that for Aristotle the intrinsic (or *per se*) efficient cause of the generation of animals and their parts is a "reliable mechanism" operating in the generative process so as to "overdetermine" that result. Meyer, I noted, simply assumes this to be the case and doesn't cite any texts from *gen. anim.* to support it. I complained that she effectively begged the question against my own view that the relevant intrinsic efficient cause is precisely the formal nature, the *dunamis* for form, that underwrites my own interpretation of Aristotle's teleology (for *it* is responsible for the regularity of the outcome), and that she was being anachronistic in taking for granted the presence, in Aristotle's view, of any sort of material mechanism. We see the soundness of both of these charges more clearly, now that we have looked more closely at these 2.6 passages. There simply is no such "mechanism" and indeed no material necessitation of the generative outcome. Aristotle's account in this chapter of the generation of animals and their parts is unequivocally based primarily on the action of the formal nature, what I have called the animal's *potential for form*, a *dunamis* which, once set up in the newly formed embryo, is the embryo's *soul*, some of it still potential but its initial part—the nutritive capacity—actual; and this is the formal-efficient cause of the embryo's subsequent development, as we have seen. There is no material-efficient cause of that development *per se*. For more on the question of whether for Aristotle "material mechanisms" could underlie, or exist alongside, teleological causation, see the Additional Note below (168–73).

significantly (if not fully) determined by the differing lengths of the bones, and Aristotle *does* in this chapter talk about the length of bones. Here is what he says, at 745a4–9:

> **(T12)** ... and because of this the bones continue growing until a certain point; for all animals have a certain limit to their size, for which reason there is one as well for the growth of their bones. For if they continued growing forever, then whatever animals contain bone or its analogue would go on growing as long as they lived, for these are the determinant of the size in these animals.[48]

In this passage's first line, it is not clear what the referent of the "this" (*touto*) in "because of this" (*dia touto*) is. Perhaps Aristotle is thinking, while discussing in the preceding lines the special ("seminal" or "nutritive") nutriment from which new bone growth must come, that there is only a *fixed amount* of such nutriment available; and if limiting material conditions fixed the length of each bone, then indeed they would fix the size—and thus, presumably, the shape—of the animal.

I would stress in response, however, that this account is at best speculative: we see no actual hint of it in the text of 2.6, nor any general mathematical perspective on Aristotle's part from which it would be natural to project, D'Arcy-Thompson-like,[49] that radial lines mapping skeletons would fix sufficiently specific shapes for Aristotle's purposes here. Nor must we forget that an insistence on quantitative limit is not necessarily understood by Aristotle fully quantitatively: he may be thinking of the limit in question not as a number or measurable amount, but irreducibly in terms of *form*, the bones growing to *just the limit called for by* the form.[50] And, indeed, although **T12** doesn't mention the shape

---

[48] καὶ διὰ τοῦτο τὰ μὲν ὀστᾶ μέχρι τινὸς λαμβάνει τὴν αὔξησιν· ἔστι γάρ τι πᾶσι τοῖς ζῴοις πέρας τοῦ μεγέθους, διὸ καὶ τῆς τῶν ὀστῶν αὐξήσεως. εἰ γὰρ ταῦτ' εἶχεν αὔξησιν ἀεὶ καὶ τῶν ζῴων ὅσα ἔχει ὀστοῦν ἢ τὸ ἀνάλογον ηὔξάνετ' ἂν ἕως ἔζη· τοῦ γὰρ μεγέθους ὅρος ἐστὶ ταῦτα τοῖς ζῴοις.
[49] Cf. Thompson's magisterial *On Growth and Form* (D'Arcy W. Thompson, *On Growth and Form*, rev. ed. [Cambridge: Cambridge University Press, 1942]).
[50] Even if one takes **T12** to be saying that the amount of material flowing in a definite direction and solidifying is fixed by material factors alone, is it clear that this would for Aristotle fix the precise situating of bones and flesh, and so forth needed for the highly specific shapes, densities, and so forth that are required by the form and function of the part in question? That seems entirely unlikely, as does the idea that material factors could set the bones at the angles they need to be relative to each other, and the curvature of the ribs, and so forth. As I and others have noted a long time ago, in most of the many passages in which Aristotle speaks of a nature's diverting (*katachrētai, parakatachrētai*) materials to some teleological use, he specifically does *not* say that that diversion itself occurs of material necessity: cf., e.g., *part. anim.* 3.2, 663b20–35: "So then, what the nature of the horns is for the sake of has been stated, and owing to what cause some have such things while others do not; but, since there is a necessary nature, we must

of the whole animal, it does mention size and says that animals have a limit to the growth of their bones *because* they have a limit to their size.[51] Given that, we should expect him to say the same for shape: shape as well as size is inherent in the nature of an animal,[52] and it is shape along with size that is responsible for the length the bones grow to (as well as their directions, something crucial for shape). Finally, advocates of any such account will have to remember the general fact about Aristotelian matter that, as one moves gradually downward from organisms to their nonuniform and then uniform constituents and then on down to the elements and then elemental powers out of which the uniform parts are built, one finds less and less structure and fewer and fewer resources to explain precise, complex structures and other mathematical properties at the level of organisms.[53]

Returning to Aristotle's text, it is important also to remember that there is no sign in 2.6 of any general concern with explaining the sorts of features of nonuniform parts that distinguish them from uniform parts as such.[54] This makes my projection as to how one might try to read the passage about bone size all the less likely. One would expect to be told that one was being given a paradigm for the necessary formation of a vast swath of animal features, if that were the case.

Finally, the proposed way of taking **T12** would appear to run against the grain of *de an.* 2.4, 415b8–416a19, where in the context of insisting that *soul* is the formal, final, and efficient cause of functioning living bodies (with the matter

---

say how the nature according to the account makes use of things present of necessity for the sake of something. ... For the residual surplus of this sort of body, being present in the larger of the animals, is used by ⟨the⟩ nature for protection and advantage, and the surplus, which flows of necessity to the upper region, in some cases it distributes to teeth and tusks, in other cases to horns" (663b20–24, 31–35; trans. Lennox, with one modification, in pointed brackets). See also James G. Lennox, *Aristotle on the Parts of Animals I–IV*, translated with an Introduction and Commentary, Clarendon Aristotle Series (Oxford: Oxford University Press, 2001), *ad loc.*, and my own discussions in *Teleology, First Principles, and Scientific Method*, 133 and 390.

51 Mariska Leunissen reminds me that Aristotle identifies size as something that is part of the definition of the substantial being of animals in the following passages: *part. anim.* 4.9, 685b12–15 (cf. *incess. anim.* 8, 708a9–20); *part. anim.* 4.6, 683a18–19; *incess. anim.* 8, 708a9–20; and *pol.* 5.9, 1309b18–35.

52 See Gotthelf, *Teleology, First Principles, and Scientific Method*, 232–33, 235–38.

53 E.g., Sarah Waterlow [Broadie], *Nature, Change, and Agency in Aristotle's* Physics (Oxford: Oxford University Press, 1982), 82–88; John M. Cooper, "Hypothetical Necessity and Natural Teleology," in Gotthelf and Lennox, *Philosophical Issues in Aristotle's Biology*, 243–74, at 270–71; Sheldon M. Cohen, "Aristotle on Hot, Cold, and Teleological Explanation," *Ancient Philosophy* 9 (1989): 255–70, at 258; cf. Montgomery Furth, "Aristotle's Biological Universe," esp. 24–25, 42–46.

54 For a list of such features, see Furth, "Aristotle's Biological Universe," 46–49.

qua efficient cause only a sort of *contributory* cause [*sunaition pōs*, 416a14]), Aristotle insists that Empedocles' view that fire is the cause *simpliciter* of growth cannot be right,

> for the growth of fire is unlimited while there is something to be burnt, but in all things which are naturally constituted there is a limit and a *logos* both for size and for growth; and these belong to soul, but not to fire, and to *logos* rather than to matter.[55]

Here, once again, a feature of an animal that is more than just a *pathos* of one of its parts is attributed not to material but to formal agency.

On the basis of our review of **T1–T12** and of *gen. anim.* 1–2 at large, then, I conclude that Aristotle's discussion of embryogenesis in *gen. anim.* 2.6 is entirely compatible with the interpretation of his natural teleological theory that I have developed in my earlier work—and indeed lends strong support to that interpretation.

Additional Note

In n. 47 above, I reflected on the implications of our study of *gen. anim.* 2.6 in this paper for the claim of Susan Sauvé Meyer that for Aristotle the intrinsic (or *per se*) efficient cause of the generation of animals and their parts is a "reliable mechanism"—a material mechanism operating in the generative process so as to "overdetermine" that result.[56] I observed that there is no evidence in *gen. anim.* 2.6 of such a mechanism, and that the intrinsic efficient cause of animal generation is in fact a formal nature or *dunamis* which is primitively directed

---

55 416a15–18, trans. D. W. Hamlyn, *Aristotle, De Anima Books II and III (with Passages from Book I), with a Report on Recent Work and a Revised Bibliography by C. Shields*, Clarendon Aristotle Series (Oxford: Oxford University Press, 1993). Hamlyn gives "proportion" for the first "*logos*" and "principle" for the second. I think rather that the first "*logos*" points to the fact that an animal kind's size (range) is inherent in its definition, so that "a limit and a *logos*" could loosely be translated as "a definable limit." Hamlyn does better with the second "*logos*," since this is clearly that familiar usage in which "*logos*," though it carries the sense of "definition," tends to signify what is identified in the definition, something here like "definable form." Since the line of thought in this *de an.* passage is fully consonant with the different causal roles the *gen. anim.* texts we are studying assign to material factors, on the one hand, and form, *logos*, (formal) nature and so forth, on the other, it is not unreasonable to look for consonance also between the *de an.* passage and our **T11** and thus not unreasonable to reject an interpretation of **T12** which sees it as assigning to material factors the same causal responsibility for embryonic shape as it does for embryonic *pathē*.
56 In Meyer, "Aristotle, Teleology, and Reduction."

upon the animal form—an "irreducible potential for form" as I have called it. I referred in that note to my fuller critical discussion of Meyer's theses in her 1992 paper in Chapter 3 of my *Teleology, First Principles, and Scientific Method*.

In that chapter, I took issue with the view, held by some, that Aristotle's use twice in *gen. anim.* 2 of an analogy of certain aspects of animal generation to the "marvelous automata" settles affirmatively the question of whether there were available to Aristotle the outlines of a "reliable mechanism" underlying animal generation. I cited work by Sylvia Berryman, Jean De Groot, and Devin Henry in support of my conclusion.[57] I discussed as well Berryman's very interesting paper, "Teleology Without Tears: Aristotle and the Role of Mechanistic Conceptions of Organisms,"[58] in which, as I wrote, she

> explains convincingly that the term "mechanism" should not be used in situations where a materially necessary sequence or set of sequences results in accidental outcomes, but only in cases where regular outcomes are (for the most part) guaranteed by an internal structure analogous in its operation to a machine (even if there is no account of how such a structure first came to be). Such an internal mechanism, she observes, would count as an intrinsic efficient cause of the sort that might license talk of, and explanation in terms of, goals. Citing Henry 2005 and her own 2003 paper,[59] Berryman seems to agree that Aristotle's comparison of aspects of the early generative process in animals to the "automatic marvels" is not meant as an analogy of that sort. Such analogies only come post-Aristotle, with the tradition of writings on the art of mechanics leading up to the time of Hero of Alexandria in the first century A.D. (a tradition that begins with Ps.-Arist. *Mechanica*), and the concurrent development of complicated, somewhat "self-moving," machines, such as are referred to by Hero.

I went on to suggest that even such mechanical analogies would not have dislodged Aristotle from the view that essential to teleological processes is a primitive directiveness upon the end, for which there is no fully material-efficient causal counterpart, on the grounds that even with such ancient machine analogies to hand, Aristotle's models of qualitative change would not have lent themselves to "the building up of precise, delimited structures." For these analogies do not involve the production or development into something different and more complicated than the machines themselves are, as embryos do develop

---

57 Gotthelf, *Teleology, First Principles, and Scientific Method*, 81; references to the mentioned scholarship on the "marvelous automata" may be found in n. 54.
58 Sylvia Berryman, "Teleology without Tears: Aristotle and the Role of Mechanistic Conceptions of Organisms," *Canadian Journal of Philosophy* 37 (2007): 351–70.
59 Devin Henry, "Embryological Models in Ancient Philosophy," *Phronesis* 51 (2005): 1–42; Sylvia Berryman, "Ancient Automata and Mechanical Explanations," *Phronesis* 48 (2003): 345–69.

into something much more complicated than they initially are.[60] In a footnote, I observed however that "Some remarks in von Staden 1997 might suggest that my claim here is overstated,"[61] and I promised to discuss those remarks in my contribution to the present volume.

The paper I was referring to is "Teleology and Mechanism: Aristotelian Biology and Early Hellenistic Medicine."[62] Its primary aim is to show that Erasistratus, one of the primary figures in early Hellenistic medicine, integrated "teleological and mechanistic perspectives within a single coherent philosophical-scientific system," building significantly on a similar integration within Aristotle's biology.[63] The primary evidence is a series of attributions by Galen, many of which represent Erasistratus as viewing nature as "capable of expert craftsmanship" and even "of forethought," as it "makes all things for the sake of something and nothing without a point," attributions which, Professor von Staden argues, place Erasistratus clearly in Aristotle's court.[64]

The Galenic (and other) evidence supports also the attribution to Erasistratus of certain theses regarding the material basis of living (as well as non-living) things, and support the view that

> [on] this material basis, Erasistratus consistently applies principles that also appear in Alexandrian mechanism, notably in pneumatics, hydraulics, and hydrostatics. Physiological processes are explicable, at one level, in terms of the material properties and structures of the parts of a mechanistically operating body.

Yet, von Staden observes, "Erasistratus also insists on a teleological framework by depicting these natural parts and processes ... as purposive."[65]

He illustrates the mechanistic side of this picture with Erasistratus's depiction of the heart "as an automatic double-action suction-and-force 'bellows.'" The evidence as to how precisely Erasistratus views the relationship of such a

---

60 Gotthelf, *Teleology, First Principles, and Scientific Method*, 81–83.
61 Gotthelf, *Teleology, First Principles, and Scientific Method*, 83 n. 61.
62 Heinrich von Staden, "Teleology and Mechanism: Aristotelian Biology and Early Hellenistic Medicine," in Wolfgang Kullmann and Sabine Föllinger (eds.), *Aristotelische Biologie. Intentionen, Methoden, Ergebnisse* (Stuttgart: Steiner, 1997), 183–208.
63 Von Staden, "Teleology and Mechanism," 184–85.
64 Von Staden, "Teleology and Mechanism," 186–92. Von Staden makes clear that Aristotle's use of the language of forethought is more guarded and "as if" (although he perhaps presses the evidence more than is warranted by Aristotle's clearly metaphorical or analogical terminology). His subsequent cautions regarding the use of Galenic evidence (197–99) are instructive, but I certainly agree with him that the evidence unequivocally supports his attribution to Erasistratus of the above-summarized view of nature.
65 Von Staden, "Teleology and Mechanism," 200–201.

mechanistic account to the teleological character of the heart—its character as existing, and having come to be, for the sake of the heart's functional role in the life of the organism—he observes "is quite simply, too fragmentary, although, as indicated above, in several respects it strongly suggests that Erasistratus' teleologism was closer to the Aristotelians' teleologism than to that of other schools.'" And, indeed, he goes on to ascribe to the Hellenistic medical scientist a conception of that relationship between teleology and mechanism which he finds in Aristotle.[66] It is this conception that, in my view, the present study suggests ought to be reconsidered.

Von Staden refers to Aristotle's frequent assertion, in regard to various biological phenomena, that they are *both for the sake of something and from necessity*. He takes these assertions to be evidence that Aristotle happily offers the two sorts of explanation side by side (even if giving some sort of priority to the teleological one), rather than—as for instance I have argued in the present paper—as parts of a *single* unified explanation. Indeed, he speaks of this approach of Aristotle's as a sort of "double accounting":

> This kind of "double accounting" ("*both* because of x *and* for the sake of y") is, of course, a familiar pattern in Aristotle's biological treatises.
>
> Erasistratus thus could readily have discerned in Aristotle two types of explanation of the same natural phenomenon co-existing more or less peacefully, one a mechanistic account in terms of the natural properties and interactions of matter ("natural necessity" or "hypothetical necessity," the αἰτία ἐξ ἀνάγκης associated with the "mechanical" behavior of matter), and the other a for-the-sake-of-something (ἕνεκά τινος) or "final cause" account, often also described in terms of how "nature habitually uses" (φύσις καταχρῆται) a natural part or process, even though the explanatory primacy of final cause is reiterated often enough.[67]

Now, if Erasistratus did in fact draw this conclusion from his reading of Aristotle's biological writings, he would, of course, retrospectively have been in good company. But if the alternative reading of Aristotle defended in the present study (and in other work in my recent book) is sound, or at least equally plausible, it invites a reconsideration of the Erasistratean evidence—evidence of which no one has greater command than von Staden.[68]

---

66 Von Staden, "Teleology and Mechanism," 204–205 and ff.
67 Von Staden, "Teleology and Mechanism," 205–206.
68 One can hope that "Teleology and Mechanism" is a foreshadowing of the full work on Erasistratus that many of us have been eagerly awaiting since reading some years ago von Staden's magisterial study of Herophilus (*Herophilus: The Art of Medicine in Early Alexandria* [Cambridge: Cambridge University Press, 1989]).

I cannot repeat even the heart of the argument above, in which I make the case that the generation of animals and their parts is provided in *gen. anim.* 2.6 with one unified explanation, in which only the *pathē* of the forming parts are attributed to material-efficient causation alone, and that even that causation is embedded in the formal-efficient (and goal-directed) production by a formal nature of the organism or part as a whole. But allow me to say at least this.

First, as I have stressed, passages in which Aristotle speaks of a nature as "making use of" (or "diverting") materials present of necessity—such as in the *part. anim.* discussion of horns—do not ever say that the particular use by nature itself occurs from necessity.[69]

Secondly, at least some (and possibly all) of the passages that speak of a phenomenon as occurring both of necessity and for the sake of something, speak of the interactions of the materials as occurring of necessity *once the formal nature is at work*. There are not two explanations "side by side" but, as in the horns case, one unified explanation in which a material-efficient necessary sequence occurs within an overall process initiated by a formal nature.[70]

Relevant too, to both of these points, thirdly, is the rich, but non-sufficient role played by the matter in what Mariska Leunissen, and I following her, have called "secondary teleology." Here, too, the very significant role that the material-efficient cause has is understood to be embedded in a larger teleological account, not in a separate material-efficient account that exists "side by side" with a teleological one.[71]

Finally, let me address the question one might ask regarding the "mechanical" action of parts already formed and now working as fully developed parts in mature organisms, parts such as the heart. I don't think Aristotle would object to such "mechanistic" accounts, had he seen reason to accept them, so long as two things were recognized: first, that these parts must be understood as having come to be by a teleological process to which no fully sufficient underlying material-efficiently caused process corresponds, *and* that these parts are *maintained* in existence, not by "mechanistic" processes alone but by a formal-efficient ongoing process which is due to *soul* (nutritive soul in particular), as is implied by the passage in *de an.* 2.4 I discussed above, and surrounding texts thereof. The material-efficient cause is at best a *sunaition*.[72]

---

69 See *supra* n. 50.
70 See the discussion of the last portion of **T10** above, especially item **(d)** on 159.
71 See *Teleology, First Principles, and Scientific Method*, 85 n. 66, 130–32 (in an essay co-written with Leunissen; see also the references to Leunissen's prior writings on the subject therein).
72 Above, 168.

For all these reasons, it is no surprise at all that Aristotle would frequently use language pointing to the analogy between nature and craft.[73] It is the *formal nature* he is speaking of, and his point is that the formation and subsequent activity of the organism and its parts is under the efficient (as well as final) causal control *of that formal nature*. That Erasistratus picks up, and perhaps stresses, such language may well suggest a sensitivity to this (in my view more accurate) picture of the relation between "mechanistic" sequences (if that is in fact the right adjective to use) and teleological causation. But that determination I leave to our worthy honoree.

Bibliography

Balme, D. M. *Aristotle, De partibus animalium I and De generatione animalium I (with Excerpts from II.1–3)*, with a Report on Recent Work and Additional Bibliography by Allan Gotthelf, Clarendon Aristotle Series (Oxford: Oxford University Press, 1992).

Berryman, Sylvia. "Ancient Automata and Mechanical Explanations," *Phronesis* 48 (2003): 345–69.

Berryman, Sylvia. "Teleology without Tears: Aristotle and the Role of Mechanistic Conceptions of Organisms," *Canadian Journal of Philosophy* 37 (2007): 351–70.

Charles, David. "Aristotle on Hypothetical Necessity and Irreducibility," *Pacific Philosophical Quarterly* 69 (1988): 1–53.

Cohen, Sheldon M. "Aristotle on Hot, Cold, and Teleological Explanation," *Ancient Philosophy* 9 (1989): 255–70.

Cooper, John M. "Hypothetical Necessity and Natural Teleology," in Gotthelf and Lennox, *Philosophical Issues in Aristotle's Biology*, 243–74.

Darwin, Charles. *On the Origin of Species By Means of Natural Selection* (London: John Murray, 1859).

Drossaart Lulofs, H. J., ed. *Aristotelis De generatione animalium: Recognovit brevique adnotatione critica instruxit*, Oxford Classical Texts (Oxford: Oxford University Press, 1965).

Freudenthal, Gad. *Aristotle's Theory of Material Substance: Heat and Pneuma, Form and Soul* (Oxford: Oxford University Press, 1995).

Furth, Montgomery. "Aristotle's Biological Universe: An Overview," in Gotthelf and Lennox, *Philosophical Issues in Aristotle's Biology*, 21–52.

Gotthelf, Allan. *Teleology, First Principles, and Scientific Method in Aristotle's Biology*, Oxford Aristotle Studies (Oxford: Oxford University Press, 2012).

---

[73] Citing Aristotle's use in *part. anim.* and *gen. anim.* of "the *mēchanē* group of words, especially the verb μηχανάομαι," with nature as the subject, von Staden writes: "Erasistratus may well also have been impressed by the fact that Aristotle, in an extension of his 'technical' figuration of nature, often uses the language of mechanical, technical construction to characterize nature and natural entities" (von Staden, "Teleology and Mechanism," 207).

Gotthelf, Allan, and James G. Lennox, eds. *Philosophical Issues in Aristotle's Biology* (Cambridge: Cambridge University Press, 1987).

Hamlyn, D. W. *Aristotle, De anima Books II and III (with Passages from Book I), with a Report on Recent Work and a Revised Bibliography by C. Shields*, Clarendon Aristotle Series (Oxford: Clarendon Press, 1993).

Henry, Devin. "Embryological Models in Ancient Philosophy," *Phronesis* 51 (2005): 1–42.

Hon, Giora, and Bernard R. Goldstein. *From Summetria to Symmetry: The Making of a Revolutionary Scientific Concept* (New York: Springer, 2008).

Lennox, James G. *Aristotle on the Parts of Animals I–IV*, Clarendon Aristotle Series (Oxford: Clarendon Press, 2001).

Meyer, Susan Sauvé. "Aristotle, Teleology, and Reduction," *Philosophical Review* 101 (1992): 791–825.

Peck, A. L. *Aristotle, Generation of Animals* (Cambridge, Mass.: Harvard University Press, 1942).

Thompson, D'Arcy W. *On Growth and Form*, rev. ed. (Cambridge: Cambridge University Press, 1942).

van der Eijk, Philip J. *Diocles of Carystus: A Collection of the Fragments with Translation and Commentary*, 2 vols. (Leiden: Brill, 2000–2001).

van der Eijk, Philip J. *Medicine and Philosophy in Classical Antiquity: Doctors and Philosophers on Nature, Soul, Health and Disease* (Cambridge: Cambridge University Press, 2005).

von Staden, Heinrich. *Herophilus: The Art of Medicine in Early Alexandria* (Cambridge: Cambridge University Press, 1989).

von Staden, Heinrich. "Teleology and Mechanism: Aristotelian Biology and Early Hellenistic Medicine," in Wolfgang Kullmann and Sabine Föllinger (eds.), *Aristotelische Biologie: Intentionen, Methoden, Ergebnisse* (Stuttgart: Steiner, 1997), 183–208.

Waterlow [Broadie], Sarah. *Nature, Change, and Agency in Aristotle's Physics* (Oxford: Oxford University Press, 1982).

Danielle Gourevitch, Philippe Charlier
Un ex-voto ophtalmologique inscrit à Rougiers (Var) : pour une étude des maladies des yeux dans le monde gallo-romain

Abstract: The authors describe an unpublished Gallo-Roman ex-voto from Rougiers (Var in Southern France). It represents ailing eyes (or perhaps eyes that had been ailing) and bears a puzzling inscription. The authors aim at providing a comprehensive description of the object and shedding light on the meaning of the inscription (ARITO) by considering it in relation to other ophthalmological ex-votos from the same period and the same region, with and without inscriptions, drawing as well on the famous collyrium stamps that are a mine of information about ancient eye diseases, their management, and the fears they inspired.

I Un nouvel ex-voto pour les yeux découvert dans le Var

Des fouilles confiées à l'INRAP sous la direction de Philippe Chapon, à Rougiers dans le Var, ont permis la découverte d'un ex-voto ophtalmologique gallo-romain d'un intérêt exceptionnel, car il est inscrit et permet quelques avancées dans la connaissance des maladies des yeux, réelles ou craintes, dans le monde gallo-romain (fig. 1).[1] La découverte a été faite au lieu-dit le Clos Sainte-Anne, à l'occasion de brèves fouilles préventives (1er mars–3 mai 2005) pour la création d'un lotissement, menées par Philippe Chapon et son équipe de l'INRAP-Méditerranée. Il s'agit d'une villa active du Ier siècle après J.-C. à la fin du IIème, avec *pars urbana* et *pars rustica*, essentiellement viticole, sans dédaigner les oliviers et la production d'huile d'olive : de celle-là 400 m2 environ ont été fouillés ; de celle-ci, 2000 m2, auxquels s'ajoute un très grand cellier qui contenait une soixantaine de *dolia* permettant un stockage de 250 à 300

---

[1] Philippe Chapon, J. M. Michel (dir.), P. Aycard, Yvon Lemoine, M. Pasqualini, et J.-P. Sargiano, *Le Clos Saint-Anne – Un établissement rural antique à Rougiers (Var)*, Rapport final d'opération – Fouille archéologique (INRAP : 2006) [Archives du SRA DRAC PACA, Aix-en-Provence].

hectolitres de vin. Les fouilleurs ne semblent malheureusement pas avoir noté où exactement a été découvert l'ex-voto ophtalmologique qui va nous occuper, ni dans quelle pièce, ni à quelle profondeur : il n'y a donc aucun moyen de comprendre comment il s'est retrouvé là. On ne sait donc s'il a été jeté ou s'il a été perdu, ni s'il a réellement été offert à une divinité – ce qui supposerait qu'il ait été récupéré par la suite, ce qui ne semble pas conforme aux usages connus –, ou si on prévoyait seulement de l'offrir. De plus, le contexte n'étant pas très net, la datation ne peut non plus être plus précise que celle de l'établissement en général (Ier-IIème s.).

Fig. 1 L'ex-voto de Rougiers (© Inrap)

Toujours est-il qu'il s'agit d'une mince tôle rectangulaire en bronze recouvert d'une fine couche argentée ; mutilée, elle mesure actuellement 39 sur 25 mm ; ses contours sont irréguliers, et il est probable qu'elle a été abimée après avoir été offerte. Elle est très ouvragée et même un peu surchargée : le bas porte une sorte de guirlande de 4 à 6 mm de haut, composée de six arceaux jointifs, sauf les deux de droite, ce qui permet d'en imaginer aussi à gauche après un petit espace vide. La partie manquante du haut peut avoir porté un motif analogue. Un encadrement de traits incisés la surmonte et se prolonge sur le côté droit de la photo, le gauche étant mutilé, nous l'avons dit (ce qui prend une réelle importance pour l'interprétation de l'inscription). La frise est ciselée, et l'on distingue bien les marques d'impact du ciselet, d'environ 1 mm de diamètre, travail d'un artisan spécialisé. Incisées de la même façon, une haste (au moins) à gauche de l'image, deux au milieu, mais aucune à droite, semblent purement décoratives, alors que, s'il n'y avait que les deux du milieu, on aurait pu penser à une représentation symbolique du nez, qui n'est pas constante mais fréquente. Les yeux, de forme allongée, semblent avoir été d'abord dessinés (coins de l'œil gauche, coin extérieur de l'œil droit), puis grossièrement découpés, presque déchiquetés, ce qui contraste avec la finesse du reste de l'objet ; ils ne sont ni symétriques ni au même niveau horizontal. Dans le coin intérieur de l'œil gauche deux « virgules », et une au-dessus sont énigmatiques : plutôt que des précisions sur la maladie du donateur, on y verra plutôt des maladresses ou accidents postérieurs à la réalisation. Il s'agit incontestablement d'un ex-voto ophtalmologique, mais il a bien des traits qui intriguent, et en particulier sous les yeux son inscription, parfaitement lisible mais non moins mystérieuse, ARITO.

II Ex-voto oculaires anépigraphes, ex-voto oculaires inscrits

On répertorie habituellement depuis l'exposition-colloque de Lons-le-Saunier[2] trois grandes séries de tels ex-voto gallo-romains, confectionnés à partir d'une tôle de bronze découpée. Dans la première, on a un œil unique (droit ou gauche, ou peu différencié anatomiquement), sur une plaque ovale ou en losange. Dans la seconde, peut-être la plus fréquente, une plaque rectangulaire porte les deux yeux, en relief selon la technique du repoussé : ainsi au sanctuaire d'Apollon Moritasgus à Alésia (Côte d'Or),[3] au *fanum* de Montot (Haute-Saône) ou encore au centre thermal et cultuel de Chassey-lès-Montbozon (Haute-Saône).[4] Dans la troisième, toujours sur une plaque allongée, des yeux évidés : au sanctuaire des Sources de la Seine (Côte d'Or),[5] à Mirebeau-sur-Bèze (Côte d'Or),[6] au temple de Genainville,[7] au temple de la forêt d'Halatte (Oise).[8] L'ex-voto de Rougiers est à

---

2 Anne-Sophie de Cohën et Laurence Bailly, éds., *L'Œil dans l'Antiquité romaine*, catalogue (Lons-le-Saunier : Centre Jurassien du Patrimoine, 1994). Suivi de J. Royer, M.-J. Roulière-Lambert, et A.-S. de Cohën, *L'Œil dans l'Antiquité : approche pluridisciplinaire ; table ronde de Lons-le-Saunier – Jura – 11–12 février 1994* (Lons-le-Saunier : Centre Jurassien du Patrimoine, 2002), en particulier 61–74. Nous ne saurions trop remercier Monique Dondin-Payre pour son attention éclairée et patiente à ce nouvel objet.
3 De Cohën et Bailly, *L'Œil dans l'Antiquité romaine*, 73.
4 De Cohën et Bailly, *L'Œil dans l'Antiquité romaine*, 72.
5 De Cohën et Bailly, *L'Œil dans l'Antiquité romaine*, 116. Voir aussi Simone Deyts, *Ex-voto de bois de pierre et de bronze du sanctuaire des sources de la Seine : art celte et gallo-romain* (Dijon : Musée archéologique, 1983), et Simone Deyts, *Un peuple de pélerins : offrandes de pierre et de bronze des sources de la Seine*, Revue archéologique de l'Est et du Centre-Est, Suppl. 13 (Dijon: Revue archéologique de l'Est et du Centre-Est, 1994).
6 Martine Joly et Pierre-Yves Lambert, « Un ex-voto dédié à Minerve trouvé sur le sanctuaire de Mirebeau-sur-Bèze (Côte d'Or) », *Revue archéologique de l'Est* 53 (2004): 233–37.
7 Musée archéologique de Guiry-en-Vexin, Val d'Oise, n. inv. G. 493.

rattacher à cette troisième catégorie, mais si cette classification est commode, elle est assez grossière, comme le montrent des découvertes récentes. Ainsi les deux ex-voto d'Allones (Sarthe) sont l'un un groupe de deux yeux, l'autre un œil aujourd'hui unique, mais sur lequel des trous pourraient marquer une ancienne jointure ; ce sont deux plaques qui ne sont pas rectangulaires mais suivent la forme prêtée aux yeux, pourvus d'une pupille évidée bien ronde (fig. 2).

Fig. 2 Ex-voto d'Allones

Ces plaques oculaires sont en général anépigraphes, comme ces « oggetti muti » auxquels s'est intéressé Olivier de Cazanove.[9] En fait, la mutité des ex-voto pour la santé n'est pas totale, et cet auteur remarque précisément parmi les ex-voto d'yeux[10] une plaquette de bronze, provenant du sanctuaire des sources de la Seine,[11] aux deux yeux symétriques, porteurs de longs cils relevés avec au-

---

8 Marc Durand, *Un sanctuaire gallo-romain dans la forêt d'Halatte* (Amiens : Drac, 2000).
9 Olivier de Cazanove, « Oggetti muti? : le iscrizioni degli ex voto anatomici nel mondo romano », in J. Bodel et M. Kajava (éds.), *Dediche sacre nel mondo greco-romano : diffusione, funzioni, tipologie* (Roma : Institutum Romanum Finlandiae, 2009), 355–71.
10 En dehors du monde gallo-romain que nous prenons seul en considération, on peut renvoyer à l'ex-voto en bronze d'*Aurelia Artemisia* provenant d'Éphèse et conservé à Hambourg, Museum für Kunst und Gewerbe, pour lequel J. Benedum, « Zu einem Augenvotiv aus römischer Zeit », in O. Baum et O. Glandien (éds.), *Zusammenhang. Festschrift für Marielene Putscher*, Bd. 1 (Köln : Wienand Verlag, 1984), 23–32 ; R. Merkelbach, « Aurelia Artemisia aus Ephesos, eine geheilte Augenkranke », *Epigraphica Anatolica* 20 (1992): 55–56 ; et C. Habicht, *Die Inschriften des Asklepieions*, Altertümer von Pergamon VIII 3 (Berlin : De Gruyter, 1969), 127–28 (« Augenvotiv aus Pergamon »). Ainsi qu'à des ex-voto de pierre, en Grèce et en Turquie, avec B. Forsén, *Griechische Gliederweihungen. Eine Untersuchung zu ihrer Typologie und ihrer religions- und sozialgeschichtlichen Bedeutung*, Papers and Monographs of the Finnish Institute at Athens 4 (Helsinki : Suomen Ateenan-instituutin säätiö, 1996), fig. 54 et 55a, ces deux exemplaires étant d'époque romaine impériale. Enfin un ex-voto inscrit, des deux yeux, à Zeus Petarenos, en Phrygie, est visible à Pergame : cf. Doris Pinkwart et al., *Pergamon. Ausstellung in Erinnerung an Erich Boehringer* (Ingelheim a. Rh. : G. H. Boehringer, 1972), 37 b.
11 De Cohën et Bailly, *L'Œil dans l'Antiquité romaine*, 116 ; Cazanove, « Oggetti muti », 360–61.

dessus d'eux une inscription énigmatique, *MATTA* : ce serait un anthroponyme d'origine celte, le nom du dévot, ou plutôt de la dévote (fig. 3).[12]

Fig. 3 Ex-voto des sources de la Seine (d'après Cazanove)

Deux autres[13] sont porteurs aussi de la formule abrégée, typiquement romaine, *v.s.l.m.* pour *votum solvit libens merito* : l'un provient du sanctuaire de Mirebeau-sur-Bèze, sans datation précise mais en contexte flavien, avec une inscription très soignée, comportant le nom du dédicant *Olitius* et le nom de la divinité, Minerve[14] (après la révision de Dondin-Payre pour la lecture du nom). Il s'agit d'une découverte de 2001,[15] à l'occasion de travaux préalables à un projet de construction, dans le sanctuaire celtique puis gallo-romain de Mirebeau-sur-Bèze, en territoire lingon. C'est une *tabella ansata* en alliage cuivreux, pourvue de deux queues d'aronde et d'un trou de fixation en son milieu. Les deux yeux sont en relief selon la technique du repoussé : au-dessus de ceux-ci le nom du dédicant et le nom de la divinité (sans sa marque casuelle *-ae* et ainsi réduit à *MINERV*) ; en-dessous la formule *VS* d'un côté, *LM* de l'autre (fig. 4).

L'autre est en argent et provient de Pannes (fig. 5),[16] dans le Loiret, sans nom de divinité cette fois et dans un sanctuaire auquel aucune divinité tutélaire ne peut pour l'instant être attribuée, mais avec le nom du dédicant, une femme encore, *Prisceia Auiola*, à l'onomastique très particulière. Nous ne voyons aucune raison de douter avec certains archéologues et historiens des religions qu'il s'agisse là d'un ex-voto pour les yeux, malgré sa beauté insolite avec ses

---

12 Yann Le Bohec, *Inscriptions de la cité des Lingons* (Paris : Comité des travaux historiques et scientifiques, 2003), M 12 (= *CIL* XIII 2867). Puis Iiro Kajanto, *The Latin Cognomina* (Helsinki : Keskuskirjapaino, 1965 [repr. 1985]), 348.
13 Cazanove, « Oggetti muti », 359–61.
14 Cf. n. 6.
15 Le site a également livré une cinquantaine d'ex-voto oculistiques anépigraphes (représentant habituellement deux yeux, mais exceptionnellement quatre), pour une centaine à Alésia et environ 120 aux sources de la Seine.
16 Monique Dondin-Payre et Christian Cribellier, « Un ex-voto oculaire inscrit trouvé au ‹ Clos du Détour › à Pannes (Loiret), sanctuaire du territoire sénon », *Revue archéologique de centre de la France* 50 (2011) : 555–68.

Fig. 4 L'ex-voto de Mirebeau-sur-Bèze (© Martine Joly)

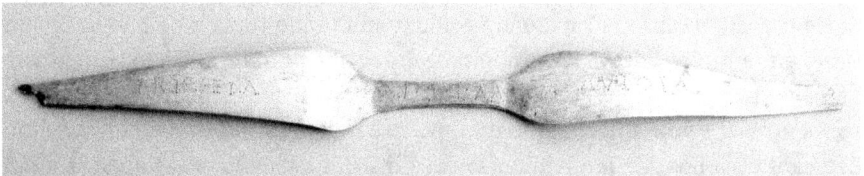

Fig. 5 L'ex-voto de Pannes (© SRA Centre, 2010)

lignes parfaitement épurées et son inscription exceptionnelle ; sa forme générale, très schématique, est à rapprocher d'un type reconnu par Isabelle Fauduet sur le site de Bû.[17] Il a été découvert au lieu-dit le Clos du Détour, à Pannes dans le Loiret, sur l'ancien territoire des Sénons, dans une fosse creusée dans un sanctuaire. Comme pour l'objet principal de notre étude, la circonstance favorable a été celle de fouilles préventives, cette fois en 1996/1997, préalablement à la construction d'un tronçon de l'autoroute A 77. On ignore le dieu concerné, mais cela n'a pas vraiment d'importance pour notre propos puisqu'on sait bien aujourd'hui que le dévot, faute de divinité spécialisée dans son entourage, peut adresser sa supplique puis ses remerciements à peu près à

---

[17] Isabelle Fauduet, « Ex-voto métalliques en forme d'yeux du sanctuaire gallo-romain des Bois du Four à Chaux à Bû », *Bulletin de la Société archéologique d'Eure-et-Loir* 15 (1988) : 31–36, et Isabelle Fauduet, « Les ex-voto anatomiques du sanctuaire gallo-romain du Bû », *Revue archéologique de l'Ouest* 7 (1990) : 93–100 (94, fig. 3).

n'importe quelle divinité. L'objet est superbe tant pour son matériau, l'argent, que pour sa forme portant à l'extrême les principes des autres ex-voto oculaires connus, et exceptionnel non seulement parce qu'il est inscrit, mais aussi parce que l'inscription indique le nom du dédicant, que ce dédicant est une femme, et que cette femme est *civis romana* : *Prisceia Aviola*. Elle ne dit malheureusement pas de quoi elle se plaint. Comme ne le disait pas non plus *Matta*, mais comme semble le dire au premier abord notre dévot anonyme du Var, auquel nous allons revenir.

III Un nom de maladie ?

Le don qu'a fait le dévot varois serait donc le quatrième ex-voto ophtalmologique inscrit. Mais le mot *ARITO* est-il un mot complet ou un mot abrégé, en cinq lettres parfaitement lisibles ? La publication de fouille a interprété le terme *arito* comme une forme en rapport avec l'adjectif *aridus*, sec, desséché, une sorte d'abréviation du dérivé *aritudo*, sécheresse, et plus précisément sécheresse oculaire ou moderne xérophtalmie, une atteinte des glandes lacrymales qui peut être aiguë (inflammation cornéenne) ou chronique.[18] En règle générale et principalement dans les sociétés pré-industrielles, l'empoussiérage en effet, notamment au moment des récoltes chez les agriculteurs, sinon lors de la pratique de travaux à risque (boulanger, maçon, menuisier, etc.) peut être cause d'une telle xérophtalmie, maladie professionnelle, ici bilatérale selon le dessin. Or certes la présence exagérée ou l'absence des larmes peuvent être signalées comme pathologiques mais ni l'adjectif *aridus* ni l'abstrait *aritudo*[19] ne sont signalés sur les cachets ou dans la littérature médicale antique. Alors faut-il envisager un autre nom commun qui fournirait une autre explication diagnostique ? Celui d'ASPRITUDO sous une forme abrégée est très tentant, si l'on veut bien s'aider d'autres sources, épigraphiques également, que constituent les cachets à collyres friands de cette maladie, qu'ils présentent sous diverses abréviations, pas toujours claires si on ne connaît pas l'ensemble des occurrences.

---

18 Philippe Chapon, J.-M. Michel (dir.), P. Aycard, Yvon Lemoine, M. Pasqualini, et J.-P. Sargiano, « Le Clos Saint-Anne – Un établissement rural antique à Rougiers (Var) », *Revue archéologique de Narbonnaise* 42 (2009) : (à paraître).
19 Semble remonter à Plaute.

IV Ex-voto oculaires et cachets à collyres

En effet cachets et ex-voto intéressent une même clientèle, qui, surtout loin des grandes villes, fait appel tantôt aux dieux tantôt au médecin rationnel, selon les caractères ou selon les circonstances,[20] ce qui justifie un rapprochement entre ces deux types de sources. Nous proposons donc de répertorier complètement un autre nom en *-tudo*, très bien attesté celui-ci dans la médecine antique, *aspritudo*, et particulièrement dans la même zone géographique et historique : la lecture (très probable ou tout à fait sûre) des cachets répertoriés par Voinot[21] permet de dresser le tableau qui suit:

1	Mandeure	*ad aspri*
5	Ingwiller	*ad aspritudinem tolle*
6	Inconnu	*ad asprit*
10	Inconnu	*ad aspri*
		at at[22]
18	Apt	*ad aspritudi*
20	Inconnu	*ad aspr*
22	Inconnu	*ad ... udinem oculo*[23]
25	Iéna	*ad asprit*
24	inconnu	*ad aspritud*
29	Maastricht	*ad aspritudine*
		at aspritudines
31	Inconnu	*ad aspr*
33	Bourg-en-Bresse	*ad aspr*

---

20 Cf. la présentation « Accidents de voiture dans l'Empire romain : recourir au médecin ou au dieu ? » de Danielle Gourevitch au colloque de médecine légale organisé en novembre 2011 par Philippe Charlier, parue depuis dans *Histoire des sciences médicales* 46 (2012) : 125–31.

21 Jacques Voinot, *Les cachets à collyres dans le monde romain* (avec une préface de Ralph Jackson), 2ᵉ édition (Montagnac : Monique Mergoil, 1999).

22 *At at* est mystérieux. Étant donné la fréquence d'*aspritudo*, et habituellement le peu de rigueur orthographique, on peut considérer ce mot comme possible ici.

23 Le génitif *oculorum* est inattendu et rare (cf. Voinot, *Les cachets à collyres*, 62, 106, 125, 140, 268). Quant à *-udinem*, il peut correspondre à *aspritudinem* ou à *lippitudinem*, très fréquent également : 5, 17, 23, 25, 29, 31, 32, 38 (4), 39, 40, 50, 59, 61, 62, 72, 84, 103 (2), 104 (2), 111, 115 (2), 123, 126 (*sicca l.*), 132 (2), 137, 157, 159 (2), 184, 187 (4), 188, 198, 199, 203 (2), 208 (*sicca l.*), 211, 220, 227, 231, 234, 242, 249, 253, 276. La *lippitudo* est une cause de mise à l'écart des soldats au camp de *Vindolenda*, en Bretagne (*Tab. Vindol.* II.154, cf. A. K. Bowman et J. D. Thomas, *The Vindolanda Writing-Tablets [Tabulae Vindolandenses II]*, with contributions by J. N. Adams [London : British Museum, 1994], 90–98).

35	Naix-en-Barrois	*ad aspritudin*
36	Naix-en-Barrois	*ad aspr*
37	Naix-en-Barrois	*ad asprit*
46	Lillebonne	*ad aspr*
62	Cessey-sur-Tille	*ad asprit*
66	Bavai	*ad asprit*
67	Riegel	*ad spritud*
72	Mayence	*ad asprit*
76	inconnu	*ad aspr*
79	Orange	*ad aspritud*
84	Metz	*ad as*
91	Bavai	*ad asp*
92	Néris	*ad aspr*
93	Tranent	*ad cicatrices et aspritudin*
97	Vienne	*ad aspritudinem*
100	Sens	*ad a*[24]
101	Lambèse	*ad asprit*
102	Fontaine-Valmont	*ad asprit et cicatrices*
109	Dalheim	*ad aspr*
115	Mandeure	*ad aspr et claritates*
116	Lyon	*ad aspritudin*
120	Mandeure	*ad aspritudin*
123	Saint-Privat d'Allier	*ad asp*
140	Reims	*ad aspritudine*
141	Le Plessis-Brion	*ad aspritu*
142	Mandeure	
	(sur une grande face)	*lipp asp*
145	Ratisbonne	*ad asprit et caligin*
		ad aspr et dia
		ad asprit et cicatric vet
148	Trèves	*ad asprit*
150	Lavigny	*ad aspritud*
		ad aspritu
154	Collanges	*ad asp*

---

[24] *Ad a* est mystérieux ; dans le corpus de Voinot, le seul nom de maladie en dehors d'*aspritudo* commençant par *a* est *albae* dans *at albas*, à Saint-Alban en Angleterre, taches blanches de la cornée peut-être.

159	Reims	*ad aspritudi*
		ad asp
160	Lyon	*ad aspr*
		ad aspri
		ad aspr et cal
165	Merdirgnac	*ad as*
168	Dourdan	*ad a*
		ad aspr
175	Reims	*ad aspritudines*
182	Houtain l'Evêque	*ad aspritudinem et sycosis*
		ad aspret syco
183	Amiens	*ad aspr*
188	Rome	*at aspritudine*
192	Neuville-sur-Sarthe	*ad asprit*
196	Bitburg	*ad asp*
197	Cologne	*ad aspr et cicat*
		ad aspr et cicat
198	Bonn	*ad asp*
199	Reims	*ad asprit*
203	Montcy Saint-Pierre	*ad asprit et cica*
		ad aspr ve
206	Châtelans	*ad aspri*
209	Harrold	*ad aspr et c*
213	Gand	*ad aspritud*
220	Cirencester	*ad aspritudin*
224	Saalburg	*ad asp*
227	Este	*ad aspritudines toll*
		ad aspritudines
228	Beaumont	*ad asp*
230	Reims	*ad as*
231	Langres	*ad aspritud*
233	Ciachi-Garbou	*ad asprite*
234	Rottweil	*ad aspritudine*
239	Berne	*ad aspri et cl*
		ad aspritudi
240	Berne	*ad aspri*
244	Bavay	*ad as*
247	York	*ad asp*

249	Londres	*ad aspritudines*
		ad d es[25] *et cicat*
250	Trèves	*ad a*
251	Courbehaye	*ad aspr*
252	Cáceres	*ad aspr*
259	Syracuse	*conr asprit*
261	Rhus	
	(latin translittéré)	*ad aspritudinem*
267	Windisch	*ad aspr*
271	Chester	*ad aspri*
277	Amay	*ad aspritudi*
281	Vendeuil-Caply	*ad asprit*
288	Maastricht	*ad aspetu*[26]
		ad aspritudines

On peut ajouter deux cachets que Voinot ne connaissait pas :[27]

| Strasbourg[28] | *ad asprit* |
| Orléans[29] | *ad aspi* |

On comprend donc qu'on puisse être tenté par une telle hypothèse, qui permet de faire coïncider deux types de documents de médecine populaire, ex-voto et cachets. Mais nous convenons qu'il est toujours risqué de proposer un *unicum* au jugement de ses pairs, ici un ex-voto anatomique porteur d'une forme abrégée, selon les besoins de l'espace, du nom de la maladie incriminée!

---

25 *Ad d es : ad aspritudines* ? *Ad lippitudines* ? Les deux sont possibles, mais comme souvent deux faces des cachets portent le même nom de maladie, nous préférons *aspritudines*, mais les deux mots sont parfois associés (cf. n° 142).

26 *Aspetu* est mystérieux : erreur pour *aspritu*, nom de maladie? Pour *impetu*, moment de la maladie, son début ?

27 Mais il est évident qu'on ne peut être certain de l'exhaustivité, tant les découvertes sont, curieusement, devenues fréquentes ces dernières années ! Même situation pour les ex-voto, et c'est ainsi que la carte publiée par Fauduet (« Les ex-voto anatomiques », 96, fig. 3) est déjà périmée.

28 Sylvie Dardaine, « Nouvelles inscriptions découvertes rue du Donon à Strasbourg-Koenigshoffen », *Cahiers alsaciens d'archéologie, d'art et d'histoire* 43 (2000) : 45–54 (pour le cachet, 49–51).

29 Sébastien Jesset, « Orléans (Loiret) : un cachet d'oculiste du Ier siècle découvert en centre-ville », *Revue archéologique du Loiret* 30-31 (2005-2006) : 106.

V Un nom propre, divinité ou dédicant ?

Alors il faut bien tenter aussi l'hypothèse d'un nom propre, personne ou dieu. Le nom de la divinité avec laquelle affaire aurait été faite serait un masculin au datif, *Aritus, Ariti* : dieu inconnu actuellement, mais la révélation d'un nouveau théonyme, nom ou surnom, n'est pas exclue : on découvre parfois des divinités à rayonnement purement local. De plus *ario/areo* est identifié comme une racine celtique, dont la présence est tout à fait possible dans cette région.

Alors s'agirait-il de l'anthroponyme du dédicant, forcément au nominatif, donc *Arito, Aritonis,* nom à ce jour inconnu dans cette déclinaison, alors qu'*Aritus* est attesté en Espagne et retenu dans les *onomastica*.[30] Si on opte en fin de compte pour l'hypothèse d'un anthroponyme, et si l'on tient compte du manque à gauche de l'image, on peut envisager un nom auquel manque son début, le nom CH]ARITO, peut-être C]ARITO avec perte de l'aspiration, ce qui entrerait mieux dans l'espace, nom grec bien attesté en milieu romain.[31]

VI Conclusion

On peut dire avec certitude en croisant l'apport des ex-voto oculaires avec celui des cachets d'oculiste ou cachets à collyre[32] que les Gallo-Romains, hommes et femmes, éprouvaient de grandes craintes quant à leur vision. On peut dire aussi que la crainte, ou la plainte, ou l'attente médicale, la plus explicitement fréquente en matière de maladie des yeux est celle qui touche à l'*aspritudo*. Si notre proposition d'interprétation de l'abréviation portée sur cet ex-voto était correcte, cet *unicum* (mais un *unicum* fait toujours trembler !) serait d'une extrême importance pour l'épidémiologie de la maladie, réelle ou fantasmée, et s'intègrerait parfaitement dans ce que nous savons de l'épidémiologie romaine des maladies des yeux. Nous ne croyons pas en effet qu'il faille, comme on le fait pourtant régulièrement, traduire *aspritudo* (qui désigne seulement une rugosité,

---

30 *Onomasticon provinciarum Europae latinarum* ou *OPEL*, notamment I.171, recueil relativement fiable, quoique fondé sur les indices.
31 Heikki Solin, *Die griechischen Personennamen in Rom. Ein Namenbuch*, 2ᵉ édition (Berlin : De Gruyter, 2003), 489–93.
32 Celui de Trilport ne nous intéresse pas directement ici puisque l'*aspritudo* n'y apparaît pas, mais mérite d'être cité pour le nom du médecin, pour le médicament à base de poisson (*diox* ?), et pour la présence de l'inscription sur l'une des grandes faces (IIème-IIIème siècle). Cf. Jean-Pierre Laporte, « Trilport : un cachet d'oculiste gallo-romain », *Bulletin du Groupement Archéologique de Seine-et-Marne* 24–25 (1983–1984) : 7–8.

ou une sensation de rugosité) par « trachome », mot qui désigne une infection bactérienne très contagieuse causée par *Chlamydia trachomatis* ; touchant au départ la paupière, la maladie évolue en l'absence de traitement vers des lésions cornéennes irréversibles pouvant mener à la cécité : il serait grotesque d'imaginer « nos ancêtres les Gaulois » comme autant de malheureux aveugles dévorant de la viande de sanglier, faute de voir ce qu'ils ont dans l'assiette. En outre, les conditions météorologiques et sanitaires ne sont pas du tout les mêmes en Gaule et en Égypte, où l'*aspritudo* est également fréquente, et il fort probable que ce n'est pas la même réalité diagnostique qui se cache sous le même nom selon les régions, un symptôme (la rugosité) ne suffisant pas à faire un diagnostic : « les yeux grattent » aussi en cas d'allergie, de poussière dans l'œil ou de sable du désert (qui d'ailleurs peut arriver jusqu'à Rome quand les vents lui sont favorables), ou d'autres corps étrangers – parfois d'origine professionnelle, comme dans les fameuses carrières d'Égypte –, de diverses maladies infectieuses ou d'états carentiels (en particulier par carence en vitamine A).[33] En tout cas l'iconographie des ex-voto,[34] bien qu'elle soit souvent inventive et belle,[35] n'apporte rien qui puisse permettre d'affiner le diagnostic. Et si le nom qui intrigue n'est pas celui de la maladie, mais celui du dédicant ou de la divinité, on doit se borner à affirmer la présence très répandue de diverses maladies des yeux, sources de crainte et objets de soin, parmi lesquels les soins médicaux proprement dits ne paraissent pas suffisants aux malades.

---

33 Cf. Philippe Charlier, chapitre et encadré sur « les maladies de l'œil et de l'orbite », dans Philippe Charlier (dir.), *Ostéo-archéologie et techniques médico-légales, tendances et perspectives : pour un « Manuel pratique de paléopathologie »* (Paris : De Boccard, 2008).
34 Le plus souvent, on l'aura remarqué, le matériau est bon marché, et l'objet probablement fabriqué d'avance pour satisfaire tous les désirs. Ces témoignages de piété ont encore leurs héritiers dans les églises de Grèce, mais l'œil est concurrencé par l'automobile !
35 L'ex-voto en or (ce qui est un cas unique) de Wroxeter, découvert en décembre 1967, probablement au bord d'une route antique, représente deux yeux (avec six petits trous de fixation) dont les détails sont joliment figurés : les iris sont comme deux gros boutons, et les cils, inférieurs et supérieurs, marqués de traits en creux. Un cachet a également été découvert dans cette ville antique de *Viroconium* (*CIL* VII 1308). Cf. Kenneth S. Painter, « A Roman Gold Ex-Voto from Wroxeter », *Antiquaries Journal* 51 (1971) : 329–31.

Bibliographie

Benedum, Jost. « Zu einem Augenvotiv aus römischer Zeit », in O. Baum et O. Glandien (éds.), *Zusammenhang. Festschrift für Marielene Putscher*, Bd. 1 (Köln : Wienand Verlag, 1984), 23–32.

Bourgeois, Claude. *Divona I. Divinités et ex-voto du culte gallo-romain de l'eau* (Paris : De Boccard, 1991).

Bowman, Alan K., et J. David Thomas. *The Vindolanda Writing-Tablets (Tabulae Vindolandenses II)*, with contributions by J. N. Adams (London : British Museum, 1994).

Cazanove, Olivier de. « Oggetti muti? : le iscrizioni degli ex voto anatomici nel mondo romano », in J. Bodel et M. Kajava (éds.), *Dediche sacre nel mondo Greco-Romano : diffusione, funzioni, tipologie* (Roma : Institutum Romanum Finlandiae, 2009), 355–71.

Chapon, Philippe, J. M. Michel (dir.), P. Aycard, Yvon Lemoine, M. Pasqualini, et J.-P. Sargiano. *Le Clos Saint-Anne – Un établissement rural antique à Rougiers (Var)*, Rapport final d'opération – Fouille archéologique (INRAP : 2006) [Archives du SRA DRAC PACA, Aix-en-Provence].

Chapon, Philippe, J. M. Michel (dir.), P. Aycard, Yvon Lemoine, M. Pasqualini, et J.-P. Sargiano. « Le Clos Saint-Anne – Un établissement rural antique à Rougiers (Var) », *Revue archéologique de Narbonnaise* 42 (2009) : (à paraître, communication des auteurs).

Charlier, Philippe, dir. *Ostéo-archéologie et techniques médico-légales, tendances et perspectives : pour un « Manuel pratique de paléopathologie »* (Paris : De Boccard, 2008).

de Cohën, Anne-Sophie, et Laurence Bailly, éds. *L'Œil dans l'Antiquité romaine*, catalogue (Lons-le-Saunier : Centre Jurassien du Patrimoine, 1994).

Dardaine, Sylvie. « Nouvelles inscriptions découvertes rue du Donon à Strasbourg-Koenigshoffen », *Cahiers alsaciens d'archéologie, d'art et d'histoire* 43 (2000) : 45–54.

Deyts, Simone. *Ex-voto de bois de pierre et de bronze du sanctuaire des sources de la Seine : art celte et gallo-romain* (Dijon : Musée archéologique, 1983).

Deyts, Simone. *Un peuple de pèlerins : offrandes de pierre et de bronze des sources de la Seine*, Revue archéologique de l'Est et du Centre-Est, Suppl. 13 (Dijon : Revue archéologique de l'Est et du Centre-Est, 1994).

Dondin-Payre, Monique, et Christian Cribellier. « Un ex-voto oculaire inscrit trouvé au ‹ Clos du Détour › à Pannes (Loiret), sanctuaire du territoire sénon », *Revue archéologique de centre de la France* 50 (2011) : 555–68.

Dumontet, Monique, et Anne-Marie Romeuf. *Ex-voto gallo-romains de la Source des roches à Chamalières Musée Bargoin (Clermont-Ferrand)* (Clermont-Ferrand : Musée Bargoin, 1980).

Dumontet, Monique, et Anne-Marie Romeuf. *Les ex-voto gallo-romains de Chamalières (Puy de Dôme) : bois sculptés de la source des Roches* (Paris : Ed. de la maison des sciences de l'homme, 2000).

Durand, Marc, éd., *Le temple gallo-romain de la forêt d'Halatte (Oise)*, Revue archéologique de Picardie numéro spécial 18 (Amiens : Revue archéologique de Picardie, 2000).

Fauduet, Isabelle. « Ex-voto métalliques en forme d'yeux du sanctuaire gallo-romain des Bois du Four à Chaux à Bû », *Bulletin de la Société archéologique d'Eure-et-Loir* 15 (1988) : 31–36.

Fauduet, Isabelle. « Les ex-voto anatomiques du sanctuaire gallo-romain du Bû », *Revue archéologique de l'Ouest* 7 (1990) : 93–100.
Fauduet, Isabelle, et Elisabeth Rabeisen. « Ex-voto de bronze d'Argentomagus et d'Alesia : à propos des offrandes métalliques des sanctuaires gallo-romains », in J. Arce et F. Burkhalter (éds.), *Bronces y religión romana : actas del XI Congreso Internacional de Bronces Antiguos* (Madrid : Consejo superior de investigaciones científicas, 1993), 141–52.
Forsén, Börn. *Griechische Gliederweihungen. Eine Untersuchung zu ihrer Typologie und ihrer religions- und sozialgeschichtlichen Bedeutung*, Papers and Monographs of the Finnish Institute at Athens 4 (Helsinki : Suomen Ateenan-instituutin säätiö, 1996).
Gourevitch, Danielle. « Accidents de voiture dans l'Empire romain : recourir au médecin ou au dieu ? » *Histoire des sciences médicales* 46.1 (2012) : 125–31.
Gourevitch, Danielle. *Pour une archéologie de la médecine romaine* (Paris : De Boccard, 2011).
Habicht, Christian. *Die Inschriften des Asklepieions*, Altertümer von Pergamon VIII 3 (Berlin : De Gruyter, 1969).
Jannet-Vallat, Monique, Simone Deyts, et Elisabeth Rabeisen. *Il était une fois la Côte d'Or ... 20 ans de recherches archéologiques* (Dijon : Musée archéologique, 1990) [pour les Bolards, 146–47, 262–69].
Jesset, Sébastien. « Orléans (Loiret). Un cachet d'oculiste du Ier siècle découvert en centre-ville », *Revue archéologique du Loiret* 30-31 (2005–2006) : 106.
Joly, Martine, et Pierre-Yves Lambert. « Un ex-voto dédié à Minerve trouvé sur le sanctuaire de Mirebeau-sur-Bèze (Côte d'Or) », *Revue archéologique de l'Est* 53 (2004) : 233–37.
Kajanto, Iiro. *The Latin Cognomina* (Helsinki : Keskurskirjapaino, 1965 [repr. 1985]).
Labonnelie, Muriel. « Annexe IV : Séquence pédagogique consacrée aux cachets à collyres », dans Gourevitch, *Pour une archéologie*, 221–23.
Labonnelie, Muriel. « Les trois cachets à collyres retrouvés à Berne (Bernisches Historisches Museum, inv. n° 27848, 27849 et 27850) », dans Gourevitch, *Pour une archéologie*, 132–35 (dans Chap. VII *Instruments et spécialités*).
Laporte, Jean-Pierre. « Trilport : un cachet d'oculiste gallo-romain », *Bulletin du Groupement Archéologique de Seine-et-Marne* 24-25 (1983–1984) : 7–8.
Le Bohec, Yann. *Inscriptions de la cité des Lingons* (Paris : Comité des travaux historiques et scientifiques, 2003).
Merkelbach, Reinhold. « Aurelia Artemisia aus Ephesos, eine geheilte Augenkranke », *Epigraphica Anatolica* 20 (1992) : 55–56.
Painter, Kenneth S. « A Roman Gold Ex-Voto from Wroxeter », *Antiquaries Journal* 51 (1971) : 329–31.
Pinkwart, Doris, et al. *Pergamon. Ausstellung in Erinnerung an Erich Boehringer* (Ingelheim a. Rh. : C. H. Boehringer, 1972).
Royer, Jean, Marie-Jeanne Roulière-Lambert, et Anne-Sophie de Cohën. *L'Œil dans l'Antiquité : approche pluridisciplinaire ; table ronde de Lons-le-Saunier – Jura – 11–12 février 1994* (Lons-le-Saunier : Centre Jurassien du Patrimoine, 2000) [Avec notamment L. Bailly, « Les ex-votos oculistiques : témoins du culte de l'eau guérisseuse en Gaule romaine », 61–74; Danielle Gourevitch et Mirko Grmek, « Les yeux malades dans l'art antique », 95–108; et Isabelle Fauduet, « Les ex-voto en forme d'yeux en Gaule », 143–52].

Sikora, Éva, et Claude Bourgeois. « Chapitre III. Médecine des yeux dans le sanctuaire de l'eau de Pouillé (Loir-et-Cher) », *Revue archéologique du centre de la France*, n° spécial 21 : *Médecine, villes d'eaux, sanctuaires des eaux, en Gaule* (1982) : 241–48.

Solin, Heikki. *Die griechischen Personennamen in Rom. Ein Namenbuch*, 2ᵉ édition (Berlin : De Gruyter, 2003).

Voinot, Jacques. *Les cachets à collyres dans le monde romain* (avec une préface de Ralph Jackson), 2ᵉ édition (Montagnac : Monique Mergoil, 1999).

Brooke Holmes
Medicine and Misfortune: *Symptōma* in Greek Medical Writing

Abstract: The Hippocratic writers have a rich vocabulary to talk about signs and proofs, but they lack a word for symptom. The term *symptōma* does begin to appear in post-Hippocratic writers from the third century B.C. (and some late Hippocratic texts) but it does not easily map onto the semantic field of "symptom" in English and other modern languages. In this paper, I consider the use of *symptōma* in non-medical texts; examine the evidence for the appearance of *symptōma* in post-Hippocratic medical writing; and work through Galen's definition of the term in relationship to three related terms, *epigennēma*, *pathos*, and *pathēma*. In closing, I suggest that although *symptōmata* remain important primarily for what they communicate to the physician, the term also helps create a space in medical writing for the misfortunes suffered by the patient, that is, illness as opposed to disease.

The word *symptōma* first appears in extant Greek literature in the fourth book of Thucydides' *Histories* under the sign of misadventure. A series of unfortunate events has stranded a large group of Spartan hoplites on the island of Sphacteria. But the Athenian general, Demosthenes, hesitates to attack, fearing that the thick forests of the island will put his troops unnecessarily at risk of being assailed from an "invisible position" (ἐξ ἀφανοῦς χωρίου, 4.29.3). The situation abruptly changes, however, when another chance event, an accidental fire, clears the island of most of its trees, emboldening the Athenians to attack. Lightly armed troops storm the island under cover of darkness and quickly seize the higher ground so that the Spartans, wherever they turn, have enemies behind them. Unable to see for all the dust and arrows and stones and too weighted down with heavy armor to pursue their attackers, the Spartans are like sitting ducks until finally they manage to retreat to their last refuge, a small fort at the edge of a cliff. Here, however, the nightmare is replayed. The Athenian archers somehow manage to climb around behind them, escaping observation until they suddenly appear at the point where the Spartans least expect them. "The

---

I am grateful to Emilio Capettini, Klaus-Dietrich Fischer, and Vivian Nutton for their very careful reading of the paper and helpful comments. It is a great pleasure to be able to dedicate this paper, which grew out of my doctoral dissertation, to Heinrich von Staden, for many years of intellectual generosity, incisive feedback, moral support, professional advice, and friendship.

Lacedaemonians," Thucydides concludes, "were now assailed on both sides, and to compare a smaller thing to a greater, were in the same predicament as at Thermopylae" (καὶ οἱ Λακεδαιμόνιοι βαλλόμενοί τε ἀμφοτέρωθεν ἤδη καὶ γιγνόμενοι ἐν τῷ αὐτῷ ξυμπτώματι, ὡς μικρὸν μεγάλῳ εἰκάσαι, τῷ ἐν Θερμοπύλαις, 4.36.3). The "predicament" here is the *symptōma*.

Were we to base ourselves on this one case, we might tentatively see a *symptōma* as a quandary or a catastrophe arising from a string of ill-starred events.[1] It may or may not be important that the emphasis in the Sphacteria episode is on being caught off-guard by an attack from left field (the Thermopylae episode also turns on the enemy catching the Spartans unawares from an unexpected path). In any event, the more general sense of "misfortune" or "catastrophe" is what we would expect from a noun formed from the verb *sympiptō*, "to fall upon," "to happen to or concurrently with," or "to collapse."

It is surprising, however, given the meaning of "symptom" in English and other modern European languages, that the first extant occurrence of the word *symptōma* is not found in a medical context. More surprising still is that the word is not found in any of the fifth- and fourth-century B.C. texts gathered in the Hippocratic Corpus.[2] It is true that if you pick up an English translation of one of these texts, you may find references to "symptoms." Yet a quick glance at the Greek will show that the translator has just fleshed out simple demonstrative pronouns in the original (*tade*, *tauta*).[3] Nor is there an obvious candidate for what will later be called *symptōma*.[4] While we are obviously hampered by the

---

[1] I have argued elsewhere that the attack from left field is crucial to the sense of the symptom, understood in our sense, in archaic and classical Greek literature: see B. Holmes, *The Symptom and the Subject: The Emergence of the Physical Body in Ancient Greece* (Princeton: Princeton University Press, 2010), esp. 48–58.

[2] Although it does appear as a likely *falsa lectio* at *flat.* 3.2, 6.94 L. (= 106.2–3 Jouanna, printing πάντων): συμπτωμάτων M : πάντων A : συμπάντων Ermerins.

[3] Nor is there any attempt to define what a symptom is in the Hippocratic Corpus: see I. Johnston, *Galen, On Diseases and Symptoms* (Cambridge: Cambridge University Press, 2006), 66.

[4] Some scholars see *sēmeion* or *tekmērion* fulfilling this role, e.g., B. Gundert, "Symptom," in Karl-Heinz Leven (ed.), *Antike Medizin. Ein Lexikon* (Munich: Beck, 2005), 840. Yet semiotic language in the Corpus is used relatively infrequently and in marked ways to cue that a phenomenon is being enlisted in an argument or a prognosis: see V. Langholf, "Zeichenkonzeptionen in der Medizin der griechischen und römischen Antike," in R. Posner, K. Robering, and T. A. Sebeok (eds.), *Semiotik. Ein Handbuch zu den zeichentheoretischen Grundlagen von Natur und Kultur*, 4 vols. (Berlin: De Gruyter, 1997), 912–21, at 914; R. Thomas, *Herodotus in Context: Ethnography, Science, and the Art of Persuasion* (Cambridge: Cambridge University Press, 2000), 193, 195–98. The Hippocratic words for "sign" cannot therefore be seen as synonyms of the later *symptōma*. I would suggest rather that the word *pathēma* comes closest to the later *symptōma*, especially in the plural. It appears roughly

loss of much fifth- and fourth-century B.C. medical writing, our evidence suggests that the word *symptōma*, rather than originating as a medical term, enters Greek medical writing at some point after the fourth century B.C.

In this paper, I sketch the appearance of the term *symptōma* in our corpus of postclassical Greek medical texts, taking into account, too, the word's meaning in non-medical contexts from the fourth century B.C., particularly the sense of "misfortune." While such an inquiry is hampered by the fragmentary state of many of the sources, it is nevertheless possible to glimpse the emergence of *symptōma* as a technical term as early as Erasistratus. Its technical status will eventually be confirmed by Galen, who expends some effort in trying to pin down its precise meaning within a larger semantic network, thereby rescuing it from the muddier waters of his predecessors. Yet despite the fact that the word appears to have been eventually integrated into the medical vocabulary, as we will see below, the ancient *symptōma* cannot be conflated with our symptom. In the last section, I examine whether the medical distinction between signs and symptoms in modern British and American medicine is relevant to the semantic field of the *symptōma*, suggesting that the term does carve out a space for the sufferings generated by the disease independent of the semiotic value attached to these events by the physician.

I

While the word *symptōma* does not appear in our fifth- and fourth-century B.C. Hippocratic texts, the same is not true of the verb *sympiptō*.[5] It shows up in these texts a number of times, often with the quite literal sense of "to collapse" or "to fall in," with possible subjects ranging from the uterus and the belly to the patient himself.[6] The noun *symptōsis* ("collapse" or "shrinkage") belongs to

---

sixty times in fifth- and fourth-century B.C. medical writing: see, e.g., *epid.* I 2, 2.606 L. (= 1:182.1 Kühlewein); *mul.* I 1, 8.10 L. (= 88.12 Grensemann); *prog.* 1, 2.110 L. (= 193.7 Alexanderson). The plural *algēmata* can also be used of the patient's sufferings: see, e.g., *aff.* 27, 6.240 L. (= 48 Potter); *flat.* 9, 6.104 L. (= 115.10 Jouanna).

5 The verb συμπίπτω/ξυμπίπτω is found approximately sixty times in the Hippocratic Corpus as a whole.

6 For συμπίπτω/ξυμπίπτω as "collapse" or "shrink," see, e.g., *epid.* IV 23, 5.164 L.; *epid.* V 11, 5.210 L. (= 7.19 Jouanna); *epid.* VI 3.1, 5.292 L. (= 52.6 Manetti and Roselli); *fist.* 4, 6.452 L. (= 140.28 Joly); *foet. exsect.* 1, 8.512 L. (= 368.16 Potter); *int.* 34, 7.252 L. (= 186.16 Potter); *morb.* IV 55, 7.602 L. (= 118.11 Joly); *mul.* I 27, 8.70 L.; *mul.* II 133, 8.282 L.; *superf.* 7, 8.480 L. (= 76.5 Lienau). The verb still retains this sense in medical writing of the Hellenistic and imperial periods: see, e.g., Erasistr. fr. 147 (Garofalo); Sor. *gyn.* 1.33 (23.9 Ilberg), 1.44 (31.12 Ilberg), 1.58 (43.24 Ilberg).

the same semantic field.⁷ In some cases, however, and especially in *Epidemics* I and III, the verb is used in contexts that do not seem so far removed from those appropriate to modern symptoms: events such as coughing, fever, and headaches, sometimes labeled "signs" (*sēmeia*), *happen* to the patient, as at *epid.* I 19: "all the aforementioned signs happened to mature women and unmarried women" (γυναιξὶ δὲ καὶ παρθένοισι συνέπιπτε μὲν καὶ τὰ ὑπογεγραμμένα σημεῖα πάντα).⁸ Already in Alcmaeon we see the expression "disease happens" (νόσον συμπίπτειν, DK 24 B4). Yet these usages of the verb do not seem to have generated a noun, *symptōma*, with a medical meaning in the fifth and fourth centuries.

If we turn to the handful of postclassical texts in the Hippocratic Corpus, however, we do come across the word *symptōma* on two occasions, although in only one case does the meaning seem specifically medical. To take the first case: in the treatise *Decorum*, usually dated to the later Hellenistic period, we find the following statement:⁹

> καὶ γὰρ μάλιστα ἡ περὶ θεῶν εἴδησις ἐν νόῳ αὐτῇ ἐμπλέκεται· ἐν γὰρ τοῖσιν ἄλλοισι πάθεσι καὶ ἐν συμπτώμασιν εὑρίσκεται τὰ πολλὰ πρὸς θεῶν ἐντίμως κειμένη ἡ ἰητρική, οἱ δὲ ἰητροὶ θεοῖσι παρακεχωρήκασιν. οὐ γὰρ ἔνι περιττὸν ἐν αὐτέῃ τὸ δυναστεῦον. καὶ γὰρ οὗτοι πολλὰ μὲν μεταχειρέονται, πολλὰ δὲ καὶ κεκράτηται αὐτέοισι δι' ἑωυτέων. (Hipp. *decent.* 6, 9.234 L. (= 27.13–18 Heiberg])

> αὐτῇ M : αὐτὴ Littré, with one ms. : αὐτῷ Foës, Ermerins

In fact, it is especially the knowledge of the gods that is embedded in the mind (by medicine?).¹⁰ For in affections generally, and especially in accidents (*symptōmata*), medi-

---

See also Pl. *Phd.* 80c7–8; Arist. *gen. anim.* 2.6, 744a14; 3.2, 754a10. For the more general sense of collapse, see, e.g., Eur. *Herc. fur.* 905 (συμπίπτει στέγη).

7 See, e.g., *aph.* 1.3, 4.460 L.; *epid.* II 1.6, 5.76 L.; *epid.* IV 35, 5.178 L.; *epid.* VI 3.1, 5.292 L. (= 52.6 Manetti and Roselli).

8 *Epid.* I 19, 2.658 L. (= 1:196.6–8 Kühlewein). See also, e.g., *epid.* I 18, 2.654 L. (= 1:195.8–9 Kühlewein); *epid.* III 3, 3.70 L. (= 1:225.9 Kühlewein), 6, 3.80 L. (= 1:227.1–2 Kühlewein). The verb is also found in this construction outside these two treatises: see, e.g., *acut. (sp.)* 6, 2.404, Ch. 4 L. (= 70.21 Joly); *Coac.* 130, 5.610 L. (= 132.14 Potter); *epid.* VII 97, 5.452 L. (= 107.8 Jouanna); *mul.* I 2, 8.16 L. (= 92.4 Grensemann).

9 On the treatise's date, see U. Fleischer, *Untersuchungen zu den pseudohippokratischen Schriften* ΠΑΡΑΓΓΕΛΙΑΙ, ΠΕΡΙ ΙΗΤΡΟΥ *und* ΠΕΡΙ ΕΥΣΧΗΜΟΣΥΝΗΣ. Neue deutsche Forschungen, Abt. Klassische Philologie, 10 (Berlin: Junker und Dünnhaupt Verlag, 1939), 59, 67, 108, and *passim*; Fleischer dates the treatise to the first- or second-century A.D. He is followed by J. Jouanna, *Hippocrates*, trans. M. B. DeBevoise (Baltimore: The Johns Hopkins University Press, 1999), 380, 405–406.

10 Heiberg (followed by Jones) retains M's αὐτῇ, presumably standing in for ἡ ἰατρική. Perhaps more attractive is the αὐτῷ conjectured by Foës (who is followed by Ermerins on the grounds that it refers to ὁ ἰητρός at the end of the previous chapter).

cine is held in honor by the gods for the most part; physicians have yielded to the gods. For in medicine, that which is powerful is not in excess. While physicians treat many things, many diseases are also overcome for them spontaneously. (trans. adapted from Jones)

There has not been a consensus on the meaning of *symptōmata* here, in part due to the fact that the treatise as a whole is abstruse in both style and content, in part due to the uncertain status of the word itself in this period. Littré translates "symptômes"; in his Loeb translation, Jones opts for "accidents," adding the note "surely not 'symptoms,' as Littré translates it."[11] Jones is no doubt right that the Greek word is what the French would call a "faux ami." But the context, with its reference to other *pathē*, supports an interpretation more specifically medical than Jones' own "accidents" implies. Fleischer and Pohlenz, for example, both recognize that in the Hellenistic period, *symptōma* can have the sense of disease or affection, much like *pathos*, as we will see further below.[12]

Fleischer, in fact, goes further, pointing out that the construction used (ἐν γὰρ τοῖσιν ἄλλοισι πάθεσι καὶ ἐν συμπτώμασιν) suggests that *symptōma* is a particular kind of *pathos*. The context lends further support to his claim. The author has been discussing the notion of wisdom in medicine and, in particular, the wisdom associated with medicine. Here, such wisdom expands to deal, in some way, with the gods. The gist of the chapter seems to be that the gods play a role in the success of medicine. The *symptōmata*, then, may be cases where the gods particularly favor medicine, presumably the most difficult and challenging cases. If these are the cases where the physician is most at a loss —and so most in need of divine aid—it may be precisely because of the suddenness or mysteriousness of the affection; indeed, Fleischer hypothesizes that *symptōma* is an affection that is accidental and unexpected ("etwa das Zufällige und Unerwartete").[13] These connotations recall the Sphacteria episode, where the idea of being caught off-guard is dominant.

The association of *symptōma* with the accidental and unexpected turns out to be consistent with what we see in a number of non-medical authors. Polybius, for example, frequently uses *symptōma* with the sense of "mischance" or "disas-

---

11 É. Littré, ed. and trans., *Œuvres complètes d'Hippocrate*, 10 vols. (Paris: J. B. Baillière, 1839–1861), 9:235; W. H. S. Jones, *Hippocrates*, vol. 2 (Cambridge, Mass.: Harvard University Press, 1923), 288.
12 Fleischer, *Untersuchungen*, 91; M. Pohlenz, *Hippokrates und die Begründung der wissenschaftlichen Medizin* (Berlin: De Gruyter, 1938), 86. For the meaning of "disease" in a nontechnical context, see ps.-Pl. *Ax.* 364c8, which has been dated to the last two centuries B.C.: see J. P. Hershbell, *Pseudo-Plato, Axiochus* (Chico, Ca.: Scholars Press, 1981), 20–21. For its meaning in technical contexts, see further below.
13 Fleischer, *Untersuchungen*, 91.

ter," as do other postclassical historians such as Diodorus Siculus and Josephus.[14] The relationship between *symptōma* and contingency is found, too, in Aristotle. In *On Prophesy in Sleep*, for example, he sets it in a triad of possible relationships between events, here what happens in dreams: x is either a cause of y, a sign of y, or else their relationship is only coincidental, in which case we have a *symptōma* (462b31–32). Elsewhere, the adverbial expression *apo symptōmatos* occurs together with expressions such as *apo tychēs* and *apo tautomatou*, reinforcing the idea of coincidence.[15] In still other Aristotelian passages, *symptōma* has the sense of an "accidental" rather than an essential property.[16] It is to designate such "accidents" that it is taken up by Epicurus, for whom nothing besides body and void can be thought that is neither a property nor an accident of a body.[17]

It is precisely the sense of something like "mischance" or more simply "chance event" that is dominant when we look to the other (late) case of *symptōma* in the Hippocratic Corpus, in the so-called "Letters to Democritus" that form part of the pseudo-Hippocratic letters and are usually dated to the last centuries

---

[14] The word appears 43 times in Polybius' *Histories* with this meaning: see, e.g., 1.22.1, 1.35.2, 2.7.11, 3.81.7, 5.88.2, 6.53.3, 9.10.9, 21.22.6). See also Diodor. Sic. 15.48.4, 16.46.5, 19.11.7, 20.21.3; Ios. ant. Iud. 15.144.3. The sense is usually negative but see, e.g., Polyb. 9.6.5, where τι καὶ τυχικὸν σύμπτωμα is a "stroke of luck."

[15] E.g., Arist. *phys.* 2.8, 198b36–199a1 (οὐ γὰρ ἀπὸ τύχης οὐδ᾽ ἀπὸ συμπτώματος), 199a4–5 (μήτε ἀπὸ συμπτώματος μήτ᾽ ἀπὸ ταὐτομάτου). In the biological writings, however, *symptōma* seems to be a necessary accident, rather than a random occurrence: see, e.g., *hist. anim.* 8, 620b35; *gen. anim.* 4.4, 770b6.

[16] See Arist. *cat.* 8, 9b20, 10a3–4; *top.* 4.5, 126b36, 126b39. The claim in K. Kapparis, "Review of Brooke Holmes, *The Symptom and the Subject: The Emergence of the Physical Body in Ancient Greece*," *Journal of the History of Medicine and Allied Sciences* 66 (2010): 249–51, that these instances exhibit a usage of the word exactly like that found in postclassical medical writers is not tenable. It is thus wrong to conclude, as Kapparis hastily does, that "Aristotle seems to be the author who deserves credit for introducing the usage of 'symptom' as we know it in medicine" (250), not least of all because even postclassical medical usage does not coincide with "our" understanding of the symptom.

[17] Epic. *ep. Hdt.* 40; see also Lucr. 1.430–32, 445–48. It seems that the "property" is an essential part of a body's existence, although it lacks any reality at the atomic level, whereas the "accident" is the contingent capacity of a body, something which befalls it regularly but does not inherently belong to it. Munro argued that the terms were indistinguishable (H. A. J. Munro, *T. Lucreti Cari De rerum natura, libri sex*, 3 vols., 4th ed. [Cambridge: Deighton Bell, 1886], 2:69–70 *ad* 1.449) but cf. C. Bailey, *Epicurus, The Extant Remains* (Oxford: Clarendon Press, 1926), 235–36, and A. A. Long and D. N. Sedley, *The Hellenistic Philosophers*, 2 vols. (Cambridge: Cambridge University Press, 1987), 1:36–37. Bodily sensation, for example is a *symptōma* because it is essential neither for the body nor for the soul but is produced by their interaction (*ep. Hdt.* 68).

B.C.[18] The passage concerns the conditions under which medicine falls short of its goals, a *tychē* vs. *technē* moment (he is talking about instances where pharmaceutical plants have happened to be contaminated by venomous snakes). Smith translates *symptōma* accordingly as "hazards of fortune." Here, then, the word does not exhibit a specific medical meaning.

If we cast our net wider over postclassical medical writing, however, we begin to see further evidence of the status of *symptōma* as something like a technical term.[19] We find the word three times, for example, in the fragments of Erasistratus. One of these, fr. 284 (Garofalo), in which we find Erasistratus' opinion on what the Greeks called "ox-hunger" (*boulimos*) in a discussion drawn from Aulus Gellius (16.3.10), refers to ox-hunger as *to symptōma*; in Latin, it is a *morbus*. The word here is undeniably medical, but its precise meaning is hard to pin down. Much as in *Decorum*, we seem to be in the general region of diseases and affections, but it is difficult to know how the "symptom" is different from a disease or affection. The second fragment suggests there is a difference without specifying what it is:

> ἐννοεῖν δὲ χρὴ καὶ τὰ τοιαῦτα διότι οὐ πάντες ἄνθρωποι ἐπὶ ταὐτὰ φέρονται πάθη, ἀλλὰ γενομένου περὶ πλείους τοῦ αὐτοῦ συμπτώματος, λέγω δὲ πληθώρας, οὐ πᾶσιν ἐπὶ τοὺς αὐτοὺς τόπους εἴθισται ἡ ὁρμὴ φέρεσθαι, ἀλλὰ τοῖς μὲν ἐπὶ τὸ ἧπαρ, ἐνίοις δὲ ἐπὶ τὴν κοιλίαν, ἄλλοις [δὲ εἰς ἐπιληπτικὰ πάθη, τοῖς] δὲ ἐπὶ τὰ ἄρθρα. (fr. 162, 39–44 [Garofalo] = Gal. *de venae sect. adv. Erasistrateos* 11.239 K.)
>
> δὲ εἰς ἐπιληπτικὰ πάθη, τοῖς del. Garofalo
>
> One must understand these matters because not all people end up with the same affections, but, despite the fact that for most the same "symptom" is present—that is, plethora—the attack does not usually affect the same parts in all of them, but for some it is the liver, for others, the bowel, for others [an epileptic condition comes about, for still others] it is the joints.

The *symptōma* here may be distinguished from the *pathos*, although given that the "symptom" is specified as plethora, they may simply be synonyms; at the same time, it is possible for the same symptom (i.e., plethora) to affect different

---

[18] *Epist.* 16, 9.346 L. (= 72.19 Smith). On the dating of the "Letters," see W. D. Smith, *Hippocrates, Pseudepigraphic Writings* (Leiden: Brill, 1990), 20–29.
[19] The word also acquires a technical meaning in Hellenistic mathematical writing: see, e.g., Apollon. Perg. *con.* 1 (4.3 Heiberg); Archim. *sph. cyl.* 1 (8.18 Mugler), *con. sph.* 13 (187.4 Mugler), with M. N. Fried and S. Unguru, *Apollonius of Perga's "Conica": Text, Context, Subtext* (Leiden: Brill, 2001), 13 and 80–97.

parts of the body, suggesting that we are dealing with a condition that can affect various locations.[20]

The last Erasistratean fragment in which the word appears, also from Galen, complicates matters further. Anyone who wants to cure correctly, it reads, should be well-trained in the art of medicine and leave none of the symptoms that occur (μηδὲν τῶν γινομένων συμπτωμάτων) unexamined.[21] Erasistratus goes on to give an example of just such a "symptom": the secretion of dark urine from a woman who feels faint and feverish but gives no other sign of distress. Here, in contrast to the first two examples, the *symptōma* would seem to be an evident sign of what is happening inside the body and is thus closer to our own sense of "symptom." The use of the plural also leaves open the possibility of a single affection or disease being accompanied by a number of "symptoms."

These few examples from Erasistratus are tantalizing, confirming that the word *symptōma* could be used in a specifically medical sense by the third century B.C. (without that sense coinciding with our own understanding of the symptom).[22] Interestingly, the word appears to have a relatively neutral status: the sense of an unforeseen accident is not discernible. It also seems polyvalent, sometimes suggesting the affection or disease, sometimes the phenomena that accompany it.

Later sources offer further confirmation of the word's status as a technical term in medicine while also implying an ongoing range of uses. Soranus, probably writing in the second century A.D., uses the singular of pathological conditions such as pica (a pregnant woman's appetite for non-nutritive stuffs like clay

---

[20] Cf. I. Garofalo, *Erasistrati fragmenta* (Pisa: Giardini, 1988), 121: "Se qui *symptoma* ha il valore ordinario la pletora è sintomo di incapacità di utilizzare o disperdere il nutrimento." See also Sor. *gyn.* 3.26 (109.11 Ilberg): uterine suffocation is named after both the affected part and one particular "symptom"—namely, suffocation. See also Gal. *de meth. med.* 2.2, 10.81–85 K. (= 126.20–132.11 Johnston and Horsley).

[21] Erasistr. fr. 222 (Garofalo) (= Gal. *de atra bile* 5.138 K. [= 88.20–23 de Boer]): καλῶς οὖν ἔχει τὸν βουλόμενον ὀρθῶς ἰατρεύειν ἐν τοῖς κατ' ἰατρικὴν γυμνάζεσθαι καὶ μηδὲν τῶν γινομένων συμπτωμάτων περὶ τὸ πάθος ἀζήτητον ἀφεῖναι, ἀλλ' ἐπισκοπεῖσθαί τε καὶ πραγματεύεσθαι ("Someone wishing to cure correctly would do well to be trained in the art of medicine and leave none of the 'symptoms' that occur in connection with the affection unexamined, but must observe them systematically").

[22] Cf. D. Fausti, "Malattia e normalità: il medico ippocratico e l'inferenza dei segni non verbali," in A. Thivel and A. Zucker (eds.), *Le normal et le pathologique dans la Collection hippocratique : actes du Xème colloque international hippocratique* (Nice: Faculté des lettres, arts et sciences humaines de Nice-Sophia Antipolis, 2002), 229–44, at 236 n. 22. Fausti claims that in Erasistratus the word acquires "il senso odierno," which she glosses with a contemporary medical definition ("il fenomeno organico che si manifesta come indizio di una malattia e può essere individuato dall'osservatore" [235]).

or chalk) or uterine hemorrhage in his *Gynecology*, implying that the word is again analogous to, if not synonymous with, terms like *pathos* or *nosos*.[23] When he uses the plural, however, he is referring to the sufferings or events that accompany a pathological condition (and seem to be caused by it).[24] The plural is used in a similar fashion in a fragment from Archigenes, also usually dated to the second century A.D.:

> Εἰ δὲ πρὸς μεταβολὴν ἐπιτηδείως ἔχοι ἡ φλεγμονή, τουτέστιν ἀρχὴν μεταβολῆς ἤδη δέξοιτο πάντα τὰ προειρημένα, ἐπὶ τῆς φλεγμονῆς συμπτώματα ἐπιταθήσεται. λέγω δὴ τὰ ἀλγήματα ἢ πυρετοὶ ἢ παραφροσύνη ἢ ἄση ἢ ἀπορία κτλ.[25]
>
> If the inflammation should take its expected turn, that is to say if all of the things aforementioned have already undergone the beginning of the change, the symptoms become more intense on top of the inflammation. I am referring of course to pain, fever, delirium, nausea, malaise, etc.

Archigenes makes it clear here that the "symptoms" are the events that follow upon the inflammation (pains, fevers, delirium, and so on). Fragments from other imperial-age medical writers yield similar evidence but no explicit account of the word's meaning.[26] For that we have to turn to Galen, whose attempts to pin down the sense of *symptōma* confirm both its polyvalence in earlier writers and the difficulty of stabilizing its meaning within the physician's vocabulary.

---

23 On pica: Sor. *gyn.* 1.46 (32.17 Ilberg); 1.48 (35.13 Ilberg); 1.54 (39.9 Ilberg); on uterine hemorrhage: 3.40 (119.7 Ilberg). See also *gyn.* 3.1 (97.27 Ilberg).
24 Sor. *gyn.* 1.4 (5.17 Ilberg), 1.28 (18.22 Ilberg), 1.46 (32.17 Ilberg), 3.5 (97.2 Ilberg), 3.24 (108.21 Ilberg), 3.47 (126.4 Ilberg), 3.50 (128.3 Ilberg). At 2.9 (57.15 Ilberg), the word refers to things that befall a newborn: Temkin translates "mishaps," but the meaning may also be closer to "affections."
25 G. Larizza Calabrò, "Frammenti inediti di Archigene," *Bollettino del Comitato per la preparazione dell'Edizione nazionale dei Classici greci e latini* 9 (1961): 67–72, at 68.2–4; see also 68.13–14, with M. Wellmann, *Die pneumatische Schule bis auf Archigenes* (Berlin: Weidmann, 1895), 161.
26 See also Anon. Paris. *de morb. acut. et chron.* 7.2 (50.26 Garofalo); Antyll. 44.10 Dietz; Aret. 3.3.1 (38.2 Hude), 3.16.1 (60.8 Hude); ps.-Diosc. *alex. pr., bis* (10.9–10; 10.15 Sprengel), 18 (29.8 Sprengel); Erot. fr. 33 (109.15 Nachmanson); Marcellin. *puls.* 14 (464.1–2 Schöne), Philumen. *ven.* 4.11 (8.25 Wellmann), 22.6 (29.11 Wellmann), 25.3 (32.3 Wellmann), 34.1 (38.5 Wellmann); Ruf. *satyr. gon.* 32 (76.6 Daremberg); Sever. *clyst.* 16.2–3 Dietz, 32.25 Dietz. Note that *symptōma* does not occur among the definitions given in the first section of the Anonymous Londiniensis papyrus. The concept that is sometimes named by *symptōma* in later texts is there either *to hepomenon* or *to parakolouthon pathos* (IV 9–12, 7–8 Manetti).

II

Despite his contempt for hair-splitters, Galen seems to have been seduced throughout his career by the promise of technical precision.[27] That promise proves somewhat elusive in the opening chapter of *On the Differentiae of Symptoms*, where Galen attempts to delineate the proper territory of the *symptōma* by differentiating it from a constellation of related words.[28]

Galen begins with a triad of physical conditions that depart from what is according to nature: the disease, the cause of disease, and the "symptom" of disease.[29] But before examining the relationship between these terms, it is worth seeing how Galen disentangles the *symptōma* from a series of words that he says are habitually used as its synonyms: *epigennēma*, *pathos*, and *pathēma*.[30] For despite the common usage, Galen believes that these words "do not entirely signify the same thing" (σημαίνεται μὲν οὖν οὐ πάντῃ ταὐτὸν ἐκ τῶν ὀνομάτων).[31]

The first of these, *epigennēma*, from the verb *epigennaō*, "to grow upon [something]" or "to generate after," is the least familiar of these terms to us; and, indeed, Galen says it is not especially common even among physicians.[32]

---

[27] For Galen's views on language, see R. J. Hankinson, "Usage and Abusage: Galen on Language," in S. Everson (ed.), *Language* (Cambridge: Cambridge University Press, 1994), 166–87; H. von Staden, "Science as Text, Science as History: Galen on Metaphor," in P. J. van der Eijk, H. F. J. Horstmanshoff, and P. H. Schrijvers (eds.), *Ancient Medicine in its Socio-cultural Context*, 2 vols. (Amsterdam: Rodopi Press, 1995), 499–517.

[28] Galen undertakes a similar task in *de meth. med.* 1.3–1.9; I make reference to this discussion in the notes, observing where it departs from the discussion in *de sympt. diff.*

[29] See also Gal. *de meth. med.* 1.9, 10.75 K. (= 116.22–25 Johnston and Horsley), 2.3, 10.86 K. (= 134.17–19 Johnston and Horsley). For Galen's understanding of the *symptōma*, see also L. García Ballester, *Galeno en la sociedad y en la ciencia de su tiempo (c. 130–c. 200 d. de C.)* (Madrid: Ediciones Guadarrama, 1972), 179–84.

[30] Cf. ps.-Gal. *def. med.* 170, 19.395 K.: σύμπτωμά ἐστι τοῦ πάθους ἐπιγέννημα (the symptom is the "aftereffect" of the affection); see also *def. med.* 415, 19.445 K.

[31] Gal. *de sympt. diff.* 1.1, 7.43 K. (= 198.12 Gundert); see also *de meth. med.* 2.3, 10.86 K. (= 134.24–25 Johnston and Horsley).

[32] It is, however, a technical term in Stoicism meaning "byproduct": e.g., pleasure is an *epigennēma* rather than an end in itself (Diog. Laërt. 7.86, 94–95; Cic. *fin.* 3.32). See also ps.-Gal. *de hist. philos.* 131 (647.19 Diels; cf. Ch. 39, 19.343 K.), where, in a discussion of fever, Diocles is said to have made fever an *epigennēma* (Διοκλῆς δέ φησιν ἐπιγέννημα εἶναι τὸν πυρετόν) for ἐπιγίνεται δὲ τραύματι καὶ βουβῶνι; in his edition, van der Eijk doubts that Diocles used the term *epigennēma* (*Diocles of Carystus: A Collection of the Fragments with Translation and Commentary*, 2 vols. [Leiden: Brill, 2000], 2:124–25). The term appears once in the Hippocratic Corpus, at *Coac.* 225, 5.634 L., where it refers to a coating on the tongue.

Galen defines the *epigennēma* here, aiming to distinguish it from the *symptōma*, as that which *necessarily* (ἐξ ἀνάγκης) follows diseases (and only diseases). It thus has a narrower meaning than symptom, which Galen defines as everything that happens to an organism that is contrary to nature (πᾶν, ὅ τι περ ἂν συμβεβήκῃ τῷ ζῴῳ παρὰ φύσιν).[33]

The aspect of the symptom that is contrary to nature turns out to be the clue to the difference that Galen draws here between *symptōma* and *pathos*. Galen here defines *pathos* as a change (*alloiōsis*) from any state (*diathesis*) to another as a result of the body being acted upon; less precisely, he says, some people use it to designate the resulting state.[34] The *pathos* thus acquires its meaning in this context through the opposition of active and passive. The opposition of "contrary to nature" and "according to nature" is, by contrast, irrelevant to the *pathos* in Galen's eyes: for example sensation, which is according to nature, is also a *pathos* (here Galen is avowedly following Plato).[35] The *symptōma*, on the other hand, is defined, at least in *On the Differentiae of Symptoms*, as being contrary to nature.[36] The same event, then—say, a cut or, to take Galen's own example, a tremor—can be both a *pathos* and a *symptōma*, depending on how you look at it: insofar as it is a change of state resulting from the body being acted upon, it is a *pathos*; insofar as it is contrary to nature, it is a *symptōma*.[37]

Here the symptom begins to look like the disease (*nosos, nosēma*), which is also, of course, contrary to nature. Galen at times seems willing to admit this, allowing at one point that a disease is just one type of *symptōma* (in fact, by this reasoning, even the cause of a disease is a kind of *symptōma*).[38] On other

---

[33] σύμπτωμα μὲν γὰρ εἶναι πᾶν, ὅ τι περ ἂν συμβεβήκῃ τῷ ζῴῳ παρὰ φύσιν, ἐπιγέννημα δὲ οὐ πᾶν, ἀλλὰ τὸ μόνοις τοῖς νοσήμασιν ἐξ ἀνάγκης ἑπόμενον ("For a symptom is everything that happens to an organism contrary to nature, while an 'epiphenomenon' isn't everything [sc. that happens contrary to nature] but what necessarily supervenes on diseases alone," Gal. *de sympt. diff.* 1.21, 7.51 K. [= 210.16–18 Gundert]).
[34] Gal. *de sympt. diff.* 1.3–5, 7.44–45 K. (= 200.9–202.12 Gundert).
[35] See also Gal. *de meth. med.* 2.3, 10.89 K. (= 138.21–140.9 Johnston and Horsley).
[36] Cf. Gal. *de meth. med.* 1.8, 10.64 K. (= 100.14–26 Johnston and Horsley), where the *symptōma* is initially defined as a condition that is either in accord with nature or contrary to nature but does not help or harm the functions of the body (Galen goes on to give a definition more consistent with that in *de sympt. diff.*: see *infra*, n. 38).
[37] On the tremor, see Gal. *de sympt. diff.* 1.23, 7.51 K. (= 210.23–26 Gundert).
[38] See esp. Gal. *de sympt. diff.* 1.26–27, 7.53 K. (= 212.20–214.1 Gundert), where Galen, having defined the symptom as "anything that should happen to the animal that is contrary to nature" (σύμπτωμα δέ, πᾶν, ὅπερ ἂν συμπίπτῃ τῷ ζῴῳ παρὰ φύσιν) continues: ὥστε καὶ ἡ νόσος ὑπὸ τὴν τοῦ γενικοῦ συμπτώματος ἀναχθήσεται προσηγορίαν· ἔστι γάρ πως καὶ αὕτη σύμπτωμα ("therefore, a disease can be referred to under the designation of the class 'symptom,' for it is in

occasions, however, he strives to establish a difference between the disease and the *symptōma*, bringing us back to the triad of the cause, the disease, and the *symptōma*. Galen regularly defines the disease as a condition that impedes a function (*energeia*) of the body.[39] The *symptōma*, on the other hand, while contrary to nature, is not held responsible for harming the body's capacity to function (although it may *be* that harm).[40]

Taken together with the cause (*aitia*) of the disease, the disease and the *symptōma* thus form a series: we begin with the cause, proceed to the disease, and end with the symptom.[41] From this perspective, the *symptōma* appears to have the narrower sense of something that supervenes on a disease (although again, unlike the *epigennēma*, it is not limited to supervening on a disease). Galen evocatively says at one point that symptoms are a kind of shadow (*skiai*), an analogy he elsewhere attributes to Archigenes.[42] The more general idea of symptoms as events that supervene on disease is one that Galen himself suggests was found among other imperial-age writers as well.[43]

Galen's definitional work at the beginning of *On the Differentiae of Symptoms* thus locates the *symptōma* within two different fields of meaning, one where it broadly designates anything that befalls an animal that is contrary to nature, one where it more narrowly captures those misfortunes, including but not limited to damage to function, that supervene on a disease. The dual emphasis corresponds nicely to the two ways we have seen the word *symptōma* function in some of our more fragmentary postclassical medical sources (usually in the singular or the plural, respectively). Moreover, while the second sense begins to approach the modern understanding of "symptom," it does not coincide with it.

---

a way itself also a symptom"); cf. above, n. 31. He goes on to classify antecedent causes as a type of symptom as well.

39 E.g., Gal. *de sympt. diff.* 1.18–19, 7.50 K. (= 210.3–4 Gundert).

40 Gal. *de sympt. diff.* 1.20, 7.50 K. (= 210.11–12 Gundert). See also *de meth. med.* 1.8, 10.65–66 K. (= 102.3–6 Johnston and Horsley), where the *symptōma* is first a condition that does not harm function, then the actual damage to a function; 1.9, 10.71 K. (= 112.2–3 Johnston and Horsley).

41 See also *de const. art. med.* 14, 1.272 K. (= 98.3–5 Fortuna); *de meth. med.* 1.9, 10.70 K. (= 110.1–2 Johnston and Horsley), 2.3, 10.90 K. (= 140.26–142.2 Johnston and Horsley).

42 *De sympt. diff.* 1.19–20, 7.50 K. (= 210.5–7 Gundert). At *de loc. aff.* 1.2, 8.20 K., he attributes the analogy with shadows to Archigenes.

43 See esp. the definitions that Galen attributes to the Methodist Olympicus at *de meth. med.* 1.9, 10.68 K. (= 104.25–106.3 Johnston and Horsley), where a *pathos* is "a condition of the body that is contrary to nature and persists" (διάθεσις παρὰ φύσιν τοῦ σώματος ἐπίμονος) while a *symptōma* is "that which happens contingent on the affection" (ὃ τῷ πάθει συμβαίνει); it has a form that is "more specific and particular" among things that are contrary to nature. Despite Galen's objections, the definition of symptom here is close to his own (narrower) definition.

But one aspect of the word's ancient semantic field—namely, the sense of the *symptōma* as unexpected or calamitous—is less pronounced in Galen than it was in other medical texts. I want to close by seeing whether that sense may be restored, at least implicitly, to the semantic field of the *symptōma* in postclassical Greek medical writing by introducing a distinction that is important to the modern sense of "symptom"—namely, that between subjective and objective perspectives on suffering.

III

If the affection and the symptom had offered us different angles on the same event (the cut, the tremor), the idea of the shadow opens up the possibility of two perspectives on the same "symptom," as it were. From the physician's perspective, the shadow is that which follows, the trace of an event happening elsewhere. But from the perspective of the patient, the shadow could be understood quite differently, perhaps along the lines of the "symptom" that we first encountered in Thucydides: as a misfortune, daemonic and disruptive, like an eclipse suddenly blocking the light.

The contrast between a subjective and an objective perspective is important to how the symptom is seen in modern Anglo-American medicine. Whereas the symptom primarily designates what the patient experiences without the phenomenon necessarily being observable to others (e.g., anxiety, backache), the sign is taken to be public (a bloody nose, or a lesion) and therefore deemed objective.[44] More broadly, the contrast can be between the two perspectives (so that a rash might be both subjectively experienced and objectively observed). Such a distinction, however, does not initially seem relevant to the ancient symptom. The definitional work that we have just followed in Galen, for example, treats the symptom as something objective that happens to the body. In a passage from Rufus of Ephesus' treatise on the interrogation of patients, he ac-

---

44 For example, s.v. "human disease," *Encyclopædia Britannica. Encyclopædia Britannica Online Academic Edition.* Encyclopædia Britannica Inc., 2013. Web. 27 May. 2013. <http://www.britannica.com/EBchecked/topic/275628/human-disease>: "Diseases usually are indicated by signs and symptoms. A sign is defined as an objective manifestation of disease that can be determined by a physician; a symptom is subjective evidence of disease reported by the patient"; s.v. "symptom," *Taber's Medical Dictionary Online.* Unbound Medicine, Inc., 2000–2013. Web. 9 June 2013: "A symptom represents the subjective experience of disease. Symptoms are described by patients in their complaint or history of the present illness. By contrast, signs are the objective findings observed by health care providers during the examination of patients."

tually criticizes a reliance on *symptōmata* at the expense of listening to the patient.[45] Here we would seem to have further confirmation of the distance between the ancient and the modern symptom.

Nevertheless, if we consider the relationship of the symptom to a term we have not yet looked at, the sign (*sēmeion*), we can perhaps uncover a less obvious aspect of the ancient "symptom." In Galen, it is clear that symptoms come before signs, to the extent that they cover everything that the body undergoes that is contrary to nature, usually as a result of a disease. These symptoms are the raw material to which the physician applies the knowledge of his experience and his training. Galen's two major texts on symptoms, *On the Differentiae of Symptoms* and *On the Causes of Symptoms*, are written to supply just this kind of knowledge, allowing the physician to move backwards from the observed symptom to the damaged function: like shadows, symptoms point to the hidden disease. In light of the intimate relationship between symptoms and signs, it is not surprising that Galen frequently uses the two words together, especially in his commentaries on Hippocratic texts.[46] To take just one example, in his commentary on *Prognostic*, he writes that "it is possible to predict that there will be a crisis from 'critical symptoms,' which one can also call 'critical signs'" (ὥσθ' ὅτι μὲν ἔσται κρίσις ἐκ τῶν κρισίμων συμπτωμάτων, ἃ δὴ καὶ σημεῖα κρίσιμα καλεῖν ἐγχωρεῖ, προγνῶναι δυνατόν ἐστιν). He then goes on to add "remember that critical signs are symptoms that indicate by their class impending secretions or apostases" (μέμνησο δὲ ὅτι καὶ τὰ κρίσιμα σημεῖα συμπτώματά ἐστι τῷ γένει δηλοῦντα τὰς ἐσομένας ἐκκρίσεις ἢ ἀποστάσεις).[47]

But notice in this last example that critical signs are drawn from a larger body of symptoms. It is possible that all these symptoms are meaningful in their own way. Yet even so, the *symptōma* always has the potential to exceed whatever signifying work is attached to it, to the extent that it also designates just what happens to the body or to a part of the body as a result of the disease (or indeed, just what happens contrary to nature). Consider a passage from *On the Affected Parts*, where Galen differentiates between a *symptōma* that befalls

---

[45] *Quaest. med.* 21–22 (30.28–32.19 Gärtner).
[46] See, e.g., Gal. *in Hp. aph. comm.* 17B.390 K.; *in Hp. epid. I comm.* 3.18, 17A.256 K. (= 128.28 Wenkebach), 3.19, 17A.261 K. (= 131.15–16 Wenkebach); *in Hp. epid. III comm.* 1.5, 17A.535 K. (= 33.20–21 Wenkebach), 1.6, 17A.539 K. (= 35.17 Wenkebach), 2.8, 17A.638 K. (= 102.22 Wenkebach), 3.74, 17A.754 K. (= 169.17–18 Wenkebach); *in Hp. prorrhet. I comm.* 1.4, 16.514 K. (= 15.20 Diels), 2.36, 16.590 K. (= 52.25–26 Diels). See also *de cris.* 1.14, 9.614 K. (= 109.17–21 Alexanderson), 3.10, 9.748 K.; *de diebus decr.* 1.13, 9.837 K.
[47] Gal. *in Hp. prog. comm.* 3.39, 18B.312–13 K. (= 376.12–13, 18–20 Heeg).

the stomach and the kinds of *symptōmata* that participate in the reconstruction of a causal narrative:

> οὐχ ἁπλῶς οὖν προσήκει σκοπεῖσθαι τοῦτο μόνον, εἰ ἡ γαστὴρ πέπονθεν, ἤ τι τῶν ἐντέρων, ἀλλὰ καὶ τί τὸ πάθος ἐστί, καὶ διορίσασθαί γε, τίνα μὲν ἴδια τῶν παθῶν ἐστι σημεῖα, τίνα δὲ τῶν πασχόντων μορίων· οἷον ὅτι τὸ μὲν ἀπεπτεῖν γαστρός ἐστι σύμπτωμα, τὸ δ' ἐπὶ τὸ κνισῶδες ἢ ὀξῶδες ἐκτρέπεσθαι τὰ ἐδηδεσμένα τῶν κατ' αὐτὴν αἰτίων τε καὶ παθημάτων. (*de loc. aff.* 1.4, 8.42–43 K.)
>
> It is not appropriate, then, to simply consider this alone, whether the stomach has been affected, or something of the viscera; but also what kind of affection is present, which signs are specific for the affections, and which are specific to the affected parts. For example, not digesting is a "symptom" of the stomach, but that food taken in turns fatty or acidic is a "symptom" of the causes and conditions associated with the stomach.

If all symptoms have causes, then, not all symptoms are equally revelatory for a physician trying to give an account of what has happened. Yet regardless of their status in the etiological account, *all* symptoms are events that happen to the body contrary to nature. If we view the *symptōma* in these terms, we grant some independence to what befalls the body and the patient—unhappily, perhaps unexpectedly—from the perspective not of diagnostic meaning but of the event itself.

Perhaps, then, for all the diagnostic work that symptoms can support, they carve out space for the misfortunes that the patient experiences apart from the physician's interpretive story. It is interesting that if we look at Soranus, where the word *symptōma* occurs sixteen times, signs are almost always signs of something: they are revelatory for the physician. The *symptōmata*, on the other hand, are regularly objects for the physician to assuage (*parēgorein*).[48] They are from this perspective what the body—and, more importantly, an embodied person—undergoes, rather than what makes sense to the physician, misfortunes as much as medical data. They belong to the patient's distress.[49]

Such misfortunes already exist in the Hippocratic Corpus, despite the fact that there is no word for them. There are certainly cases, for example, where

---

[48] See Sor. *gyn.* 1.46 (32.17 Ilberg), 3.24 (108.21 Ilberg). See also, e.g, Gal. *de meth. med.* 11.11, 10.764–65 K. (= 156.18–158.2 Johnston and Horsley), 12.1, 10.811–14 K. (= 226.8–228.24 Johnston and Horsley), 12.7, 10.849–50 K. (= 282.5–14 Johnston and Horsley) on the treatment of symptoms versus the treatment of the underlying disease.

[49] See, e.g., Gal. *de meth. med.* 7.8, 10.506 K. (= 308.25–28 Johnston and Horsley).

the physician pursues a treatment solely to alleviate pain.[50] Yet at the same time, if the various phenomena produced by a disease make it into the text, it is because they are seen as meaningful for the physician: everything in the text, in other words, is already a sign. Is it possible that the designation of events and states as *symptōmata* in postclassical medicine complicates this picture, by creating a space within medicine, however limited, for these events as misfortunes of a kind? One must admit, of course, that these unhappy events continue to be incorporated into medical interpretation, at least in Galen.[51] If a space for the experience of the patient does emerge, I freely admit that this is a byproduct of a terminological distinction that serves the physician. But the same might be said of the distinction between symptom and sign in modern medicine. As the physician Richard Baron has written:

> We seem to have a great deal of difficulty taking seriously any human suffering that cannot be directly related to an anatomic or pathophysiologic derangement. It is as if this suffering had a value inferior to that associated with real disease.[52]

Perhaps medical writing, by its very nature, codes whatever happens to the patient in ways that are meaningful or potentially meaningful to the physician. Still, between the sudden calamity that strikes the Spartans on Sphacteria and, far in the distance, the modern distinction between the subjective symptom and the objective sign, it may be possible to stake out a bit of terrain for what medical anthropologists call illness, to differentiate it from the medical phenomenon of disease, and to place that terrain as it is fleetingly glimpsed in ancient medical writing under the figure of the *symptōma*.

---

50 E.g., *aff.* 31, 6.244 L. (= 54.15–18 Potter). See also H. King, "The Early Anodynes: Pain in the Ancient World," in R. D. Mann (ed.), *The History of the Management of Pain: From Early Principles to Present Practice* (Carnforth, N.J.: Parthenon Publishing Group, 1988), 51–62, at 54–57.
51 The emphasis on symptoms as part of a causal account is probably less important in Empiricism or Methodism, where internal causality does not have the prominence it does in Rationalist or Dogmatist writers.
52 R. Baron, "An Introduction to Medical Phenomenology: I Can't Hear You While I'm Listening," *Annals of Internal Medicine* 103 (1985): 606–11.

Bibliography

Alexanderson, B. *Die hippokratische Schrift Prognostikon* (Göteborg: Elander, 1963).
Alexanderson, B. *ΠΕΡΙ ΚΡΙΣΕΩΝ, von Galenos* (Uppsala: Almquist and Wiksell, 1967).
Bailey, C. *Epicurus, The Extant Remains* (Oxford: Clarendon Press, 1926).
Baron, R. "An Introduction to Medical Phenomenology: I Can't Hear You While I'm Listening," *Annals of Internal Medicine* 103 (1985): 606–11.
Daremberg, C., and C. E. Ruelle. *Œuvres de Rufus d'Éphèse* (Paris: Imprimerie nationale, 1879).
De Boer, W. *Galeni De propriorum animi cuiuslibet affectuum dignotione et curatione, De animi cuiuslibet peccatorum dignotione et curatione, De atra bile*, CMG V 4,1,1 (Leipzig: Teubner, 1937).
Diels, H. *Doxographi Graeci*, 4th ed. (Berlin: De Gruyter, 1965).
Diels, H. *Galeni In Hippocratis Prorrheticum I commentaria III*, CMG V 9,2 (Leipzig: Teubner, 1915).
Dietz, F. R., ed. *Severi iatrosophistae De clysteribus liber*, Diss. Med. (Königsberg: Fratres Borntraeger, 1836).
Fausti, D. "Malattia e normalità: il medico ippocratico e l'inferenza dei segni non verbali," in A. Thivel and A. Zucker (eds.), *Le normal et le pathologique dans la Collection hippocratique : actes du Xème colloque internationale hippocratique* (Nice: Publication de la Faculté des lettres, arts et sciences humaines de Nice-Sophia Antipolis, 2002), 229–44.
Fleischer, U. *Untersuchungen zu den pseudohippokratischen Schriften ΠΑΡΑΓΓΕΛΙΑΙ, ΠΕΡΙ ΙΗΤΡΟΥ und ΠΕΡΙ ΕΥΣΧΗΜΟΣΥΝΗΣ* (Berlin: Junker und Dünnhaupt Verlag, 1939).
Fortuna, S. *Galeni De constitutione artis medicae ad Patrophilum*, CMG V 1,3 (Berlin: Akademie-Verlag, 1997).
Fried, M. N., and S. Unguru. *Apollonius of Perga's "Conica": Text, Context, Subtext* (Leiden: Brill, 2001).
García Ballester, L. *Galeno en la sociedad y en la ciencia de su tiempo (c. 130–c. 200 d. de C.)* (Madrid: Ediciones Guadarrama, 1972).
Garofalo, I. *Erasistrati fragmenta* (Pisa: Giardini, 1988).
Garofalo, I., and B. Fuchs. *Anonymi medici De morbis acutis et chroniis* (Leiden: Brill, 1997).
Gärtner, H. *Rufus von Ephesos, Die Fragen des Arztes an den Kranken*, CMG Suppl. 4 (Berlin: Akademie-Verlag, 1962).
Grensemann, H. *Hippokratische Gynäkologie. Die gynäkologischen Texte des Autors C nach den pseudohippokratischen Schriften "De Muliebribus I, II" und "De Sterilibus"* (Wiesbaden: Steiner, 1982).
Gundert, B. "Symptom," in Karl-Heinz Leven (ed.), *Antike Medizin. Ein Lexikon* (Munich: Beck, 2005), 840.
Gundert, B. *Galeni De symptomatum differentiis*, CMG V 5,1 (Berlin: Akademie-Verlag, 2009).
Hankinson, R. J. "Usage and Abusage: Galen on Language," in S. Everson (ed.), *Language* (Cambridge: Cambridge University Press, 1994), 166–87.
Heeg, J. *In Hippocratis Prognosticum commentaria III*, CMG V 9,2 (Leipzig: Teubner, 1915).
Heiberg, J. L. *Apollonii Pergaei quae Graece extant cum commentariis antiquis*, ed. stereotypa ed. anni 1893 (Stuttgart: Teubner, 1974).
Heiberg, J. L. *Hippocratis Opera*, CMG I 1 (Leipzig: Teubner, 1927).

Hershbell, J. P. *Pseudo-Plato, Axiochus* (Chico, Ca.: Scholars Press, 1981).
Holmes, B. *The Symptom and the Subject: The Emergence of the Physical Body in Ancient Greece* (Princeton: Princeton University Press, 2010).
Hude, C. *Aretaeus*, CMG II (Leipzig: Akademie-Verlag, 1958).
Ilberg, J. *Sorani Gynaeciorum libri IV, De signis fracturarum; De fasciis; Vita Hippocratis secundum Soranum*, CMG IV (Leipzig: Teubner, 1927).
Johnston, I. *Galen, On Diseases and Symptoms* (Cambridge: Cambridge University Press, 2006).
Johnston, I., and G. H. R. Horsley. *Galen, Method of Medicine*, 3 vols. (Cambridge, Mass.: Harvard University Press, 2011).
Joly, R. *Hippocrate, De la génération; De la nature de l'enfant; Des maladies IV; Du fœtus de huit mois*, t. XI (Paris: Les Belles Lettres, 1970).
Joly, R. *Hippocrate, Des lieux dans l'homme; Du système des glandes; Des fistules; Des hémorroïdes; De la vision; Des chairs; De la dentition*, t. XIII (Paris: Les Belles Lettres, 1978).
Joly, R. *Hippocrate, Du régime des maladies aiguës; Appendice; De l'aliment; De l'usage des liquides*, t. VI.2 (Paris: Les Belles Lettres, 1972).
Jones, W. H. S. *Hippocrates*, vol. 2 (Cambridge, Mass.: Harvard University Press, 1923).
Jouanna, J. *Hippocrate, Des vents; De l'art*, t. V.1 (Paris: Les Belles Lettres, 1988).
Jouanna, J. *Hippocrates*, trans. M. DeBevoise (Baltimore: The Johns Hopkins University Press, 1999).
Jouanna, J. *Hippocrate, Epidémies V et VII*, t. IV.3 (Paris: Les Belles Lettres, 2000).
Kapparis, K. "Review of Brooke Holmes, *The Symptom and the Subject: The Emergence of the Physical Body in Ancient Greece*," *Journal of the History of Medicine and Allied Sciences* 66 (2010): 249–51.
King, H. "The Early Anodynes: Pain in the Ancient World," in R. D. Mann (ed.), *The History of the Management of Pain: From Early Principles to Present Practice* (Carnforth: Parthenon Publishing Group, 1988), 51–62.
Kühlewein, H. *Hippocratis opera quae feruntur omnia*, 2 vols. (Leipzig: Teubner, 1894–1902).
Langholf, V. "Zeichenkonzeptionen in der Medizin der griechischen und römischen Antike," in R. Posner, K. Robering, and T. A. Sebeok (eds.), *Semiotik. Ein Handbuch zu den zeichentheoretischen Grundlagen von Natur und Kultur*, 4 vols. (Berlin: De Gruyter, 1997), 912–21.
Larizza Calabrò, G. "Frammenti inediti di Archigene," *Bollettino del Comitato per la preparazione dell'Edizione nazionale dei Classici greci e latini* 9 (1961): 67–72.
Lienau, C. *Hippocratis De superfetatione*, CMG I 2,2 (Berlin: Akademie-Verlag 1973).
Littré, É., ed. and trans. *Œuvres complètes d'Hippocrate*, 10 vols. (Paris: J. B. Baillière, 1839–1861).
Long, A. A., and D. N. Sedley. *The Hellenistic Philosophers*, 2 vols. (Cambridge: Cambridge University Press, 1987).
Manetti, D. *Anonymus Londiniensis De medicina* (Berlin: De Gruyter, 2011).
Manetti, D., and A. Roselli. *Ippocrate: Epidemie, libro sesto* (Florence: La Nuova Italia, 1982).
Mugler, C. *Archimède, De la sphère et du cylindre; La mesure de cercle; Sur les conoïdes et les sphéroïdes*, t. 1 (Paris: Les Belles Lettres, 1970).
Munro, H. A. J. *T. Lucreti Cari De rerum natura, libri sex*, 3 vols., 4th ed. (Cambridge: Deighton Bell, 1886).

Nachmanson, E. *Erotiani vocum Hippocraticarum collectio, cum fragmentis* (Uppsala: Appelbergs boktryckerei-aktiebolag, 1918).
Pohlenz, M. *Hippokrates und die Begründung der wissenschaftlichen Medizin* (Berlin: De Gruyter, 1938).
Potter, P. *Hippocrates*, vol. 5 (Cambridge, Mass.: Harvard University Press, 1988).
Potter, P. *Hippocrates*, vol. 6 (Cambridge, Mass.: Harvard University Press, 1988).
Potter, P. *Hippocrates*, vol. 9 (Cambridge, Mass.: Harvard University Press, 2010).
Schöne, H. "Markellinos' Pulslehre. Ein griechisches Anekdoton," *Festschrift zur 49. Versammlung deutscher Philologen und Schulmänner* (Basel: E. Birkhäuser, 1907), 448–72.
Smith, W. D. *Hippocrates, Pseudepigraphic Writings* (Leiden: Brill, 1990).
Sprengel, K. *Pedanii Dioscoridis Anazarbei De materia medica libri quinque*, 2 vols. (Leipzig: prostat in officina libraria Car. Cnoblochii, 1829–1830).
Thomas, R. *Herodotus in Context: Ethnography, Science, and the Art of Persuasion* (Cambridge: Cambridge University Press, 2000).
van der Eijk, P. J. *Diocles of Carystus: A Collection of the Fragments with Translation and Commentary*, 2 vols. (Leiden: Brill, 2000).
von Staden, H. "Science as Text, Science as History: Galen on Metaphor," in P. J. van der Eijk, H. F. J. Horstmanshoff, and P. H. Schrijvers (eds.), *Ancient Medicine in its Socio-cultural Context*, 2 vols. (Amsterdam: Rodopi Press, 1995), 499–517.
Wellmann, M. *Die pneumatische Schule bis auf Archigenes* (Berlin: Weidmann, 1895).
Wellmann, M. *Philumeni De venenatis animalibus eorumque remediis*, CMG X 1,1 (Leipzig: Teubner, 1908).
Wenkebach, E. *Galeni In Hippocratis Epidemiarum Libr. III Comm. III*, CMG V 10,2,1 (Leipzig: Teubner, 1936).
Wenkebach, E., and F. Pfaff. *Galeni In Hippocratis Epidemiarum Libr. I Comm. III; In Hippocratis Epidemiarum Libr. II Comm. V*, CMG V 10,1 (Leipzig: Teubner, 1934).

Carl A. Huffman
Mathematics in Plato's *Republic*

Abstract: Scholars have been divided as to whether mathematics plays only an instrumental role in the guardians' search for knowledge in Plato's *Republic* (e.g., Annas) or whether it is constitutive of that knowledge (e.g., Burnyeat). Does mathematics serve simply to turn the eye of the soul to the intelligible realm by accustoming it to *a priori* reasoning or is the knowledge that the guardians find in the intelligible realm, which guides their governance of the state, also mathematical in nature? The discussion of the guardians' ten year training in mathematics in Book 7 has been the central text in this controversy, but evidence from Books 8 and 9, which has been largely ignored, is crucial to deciding the question. That evidence, and in particular the famous passage on the nuptial number, shows that mathematics does play a prominent role both in governance and moral decision making so that it is constitutive of knowledge of the good. Read in light of this later evidence Book 7 can be seen to treat mathematics as playing both an instrumental and a constitutive role. Mathematics and dialectic both study the intelligible realm, which is mathematical in nature. The difference between mathematics and dialectic resides in method: dialectic can give an account of its starting points, whereas mathematics does not.

Introduction

The role of mathematics in Plato's *Republic* is a many-faceted issue, and I cannot treat all of those facets in this short paper. My goal is to address two related central questions. First, what is the primary purpose of the ten years of training in advanced mathematics mandated for the philosopher kings in Book 7? Second, is that mathematical training directly employed in their governance of the state and their practical moral decision-making?

My paper will fall into three parts. First, I will set out some of the most prominent answers to these questions in recent scholarship and show that scholars are divided as to whether mathematics plays only an instrumental role in the guardians' search for the knowledge of the good or whether it is constitutive of that knowledge. Second, I will examine evidence from Books 8 and 9, which has been largely ignored in scholarship on these questions. This evidence shows that advanced mathematics does play a prominent role both in governance and also in moral decision-making and hence that mathematics is constitutive of knowledge of the good. Third, I will return to Book 7 in order to explain the features of the text that have led scholars mistakenly to suppose that mathematics plays only an instrumental role in the education of the philosopher kings. Plato, in fact, portrays it as both instrumental and constitutive. My goal is to show what the text of the *Republic* indicates about Plato's views on these issues, and I will not have time to examine how philosophically plausible those views are.

I Overview of the Scholarship

In Book 7 of the *Republic*, Plato describes in considerable detail the advanced education of the philosopher kings. He makes clear that the highest study is dialectic (534e–535a) and that the highest object of study is the Good (540a). Nonetheless, he devotes a considerable portion of Book 7 to discussion of the five studies that he calls the prelude to dialectic (531d): 1) the study of number; 2) geometry; 3) solid geometry; 4) astronomy; and 5) harmonics.[1] The would-be rulers are introduced to mathematics in the form of play early in their upbringing (536d–537a). At the age of twenty, however, they will draw together these subjects, which were studied unsystematically in their youth, and study them synoptically, paying attention to, in Socrates' words, "their connection with one another and the nature of reality" (537c2–3). The philosopher kings devote ten years to this systematic study of mathematics, from age twenty to age thirty, twice as many years as are devoted to dialectic, which they study from age thirty to age thirty-five. Plato clearly indicates that they will be studying mathematics at the highest level. In the case of geometry, Socrates explicitly says that they will study not just a modicum of it but "the greater and more advanced part of it" (526d6–e1, trans. Shorey).[2] Glaucon's responses to Socrates' visionary versions of astronomy and harmonics are designed to emphasize their complexity and difficulty. He exclaims that the proposed harmonics is "a superhuman task" (531c5, trans. Shorey). The central question about mathematics in the *Republic* is

---

[1] He, in fact, spends twice as many pages discussing these mathematical disciplines (522c–531d) as dialectic (531d–535a). This disproportion is partly to be explained by the relative ease of describing the nature of the mathematical disciplines, whereas Socrates tells Glaucon that he would not be able to follow a similarly detailed account of dialectic (533b). Yet, the emphasis on mathematics cannot be explained on these grounds alone, since, in the final curriculum, the philosopher kings devote more years of study to mathematics than to dialectic.

[2] All translations are my own unless otherwise indicated. The solid geometry that Plato envisages is so advanced that it is said not even to have been established as a discipline yet (528b).

thus, as Myles Burnyeat has put it, "why are [Plato's philosopher kings] required to study so much mathematics, for so long?"[3]

The dominant scholarly view has been that mathematics plays an instrumental role in the education of the guardians; the guardians do not study mathematics for its own sake but because it provides good training for dialectic and the study of the form of the Good. It is important to get clear about exactly what sort of instrumental role mathematics plays.[4] Socrates points out that those who have studied number and geometry are immeasurably more receptive to all other studies than those who have not (526b and 527c). Thus, Plato clearly thinks that mathematics does play an instrumental role in all higher learning by sharpening the intellect in a general way. This cannot be the true value of mathematics, however, since Socrates explicitly labels this role as "an incidental benefit" (*parergon*, 527c3). If mathematics' true benefit is instrumental in nature, it must be found in a more specific instrumental role. A clear statement of such a view is found in Annas' *An Introduction to Plato's "Republic"*:

> It is made very clear that the chief point of these [mathematical] studies is to encourage the mind in non-empirical and highly abstract reasoning. Plato is not so concerned to produce experts in these subjects as he is to produce people who are accustomed to *a priori* reasoning about subjects where most people comprehend only an empirical approach.[5]

Thus mathematics is studied not for its own sake, not to produce experts in the field, but to accustom the guardians to "non-empirical and highly abstract reasoning," which is the sort of reasoning employed in dialectic. So mathematics does not just sharpen the intellect in a general sort of way but specifically accustoms the mind to *a priori* reasoning. Annas adds yet another way in which mathematics is instrumental to the highest sort of knowledge. Plato thinks that philosophical knowledge requires understanding, and for something to be understood, it must be part of an explanatory system. On Annas' view, mathematics provides the model of an explanatory system, although the explanatory system that grounds dialectic is not mathematical.[6]

Burnyeat's question—"Why are they required to study so much mathematics, for so long?"—implicitly challenges all such instrumentalist approaches, since it

---

[3] Myles F. Burnyeat, "Plato on Why Mathematics is Good for the Soul," in Timothy Smiley (ed.), *Mathematics and Necessity* (Oxford: Oxford University Press, 2000), 1–81, at 1.
[4] For a subtle analysis of the various ways in which mathematics might play an instrumental role, see Burnyeat, "Mathematics," 1–6.
[5] Julia Annas, *An Introduction to Plato's "Republic"* (Oxford: Oxford University Press, 1981), 273.
[6] Annas, *Introduction*, 242–43.

is *prima facie* difficult to see why ten years of such advanced mathematics are needed to get the guardians used to the idea of *a priori* reasoning or the idea of an explanatory system. Burnyeat himself argues that mathematics is not merely instrumental in coming to an understanding of the highest object of knowledge, the form of the Good, but that it is "a constitutive part of ethical understanding."[7] According to Burnyeat, the Good is in fact a mathematical concept—namely, unity—and what is distinctive about Plato is his systematic exploitation of the fact that Greek value-concepts like concord, proportion, and order are also central to contemporary mathematics. The fundamental concepts of mathematics are the fundamental concepts of ethics and aesthetics as well.[8]

If Burnyeat is right, it is, of course, no wonder that Plato should think ten years study of advanced mathematics appropriate. Several other scholars have recently adopted similar views of mathematics as having a constitutive role in ethical understanding. Sedley, for example, argues that "mathematical thinking is not just a propaedeutic training for philosophical dialectic about values, but stands at the very heart of the discipline's methodology," and, although he rejects Burnyeat's suggestion that the Good is unity, he suggests that the Good may have been viewed as "something like an ideal proportionality, intelligible only through the conceptual framework of high level mathematics."[9]

The split between an instrumental and a constitutive view of the role of mathematics is closely tied to the issue of whether mathematics plays any practical role in the reasoning used by the philosopher kings in governing the state and making moral judgments. If one supposes that mathematics has a constitutive role in knowledge of the Good, then it would seem to follow that decisions made in accordance with the Good would have a mathematical basis. Thus, Burnyeat thinks that through study of advanced mathematics the guardians will learn mathematical structures that they will then embody in the state.[10] On the other hand, Annas argues that, while Plato grants great instrumental value to mathematical studies in developing the intellect, he assigns them only a crude practical value to the rulers; they are "useful for counting and measuring

---

[7] Burnyeat, "Mathematics," 6.
[8] Burnyeat, "Mathematics," 76.
[9] David Sedley, "Philosophy, the Forms, and the Art of Ruling," in G. R. F. Ferrari (ed.), *The Cambridge Companion to Plato's "Republic"* (Cambridge: Cambridge University Press, 2007), 256–83, at 270–71. Nicholas Denyer seems to suggest that the Good is a certain sort of orderly beauty, which is made possible only if mathematical Forms are as they are: see Nicholas Denyer, "Sun and Line: The Role of the Good," in Ferrari, *Companion to Plato's "Republic,"* 284–309, at 307–308. Here, too, mathematical knowledge would seem to be constitutive of the form of the Good.
[10] Burnyeat, "Mathematics," 55–56.

armies. In modern jargon they are useful for technology."[11] Annas is disappointed that "Plato does not claim that they are of practical value because they lead to rational and informed decisions on practical and ethical matters."[12] She describes Plato's odd coupling of mathematics' instrumental role in turning the soul to truth and reality with this degraded practical use in the technology of war as "grotesque" and compares it to "the philosophy behind a lot of NATO research funding."[13] At the core of Annas' unease with Plato's use of mathematics is her concern that it marks "a sad fall from the idea that knowledge of the highest kind has the practical force of making one just in action and decision."[14]

Thus, a stark contrast emerges between interpretations of the role that mathematics plays in Plato's *Republic*. For Burnyeat, the philosopher kings are doing applied mathematics when they set up the basic structures that govern the moral and political life of the ideal city, whereas for Annas, such practical decisions are not mathematical. The philosopher kings use mathematics for two important purposes during their education: 1) to get them accustomed to *a priori* reasoning and 2) as a model for an explanatory system; but as mature rulers they then make no further use of it other than for the technology of war.[15]

---

11 Annas, *Introduction*, 275.
12 Annas, *Introduction*, 275.
13 Annas, *Introduction*, 275.
14 Annas, *Introduction*, 275.
15 Plato does allude to the practical utility of mathematics in conducting war but he distances himself more and more from the importance of this role as the discussion of mathematics develops, so I doubt that he, in the end, thought this a very important justification for its study. When the dialogue's interlocutors begin to look for the kind of study that would lead the soul from the world of becoming to the world of being, Socrates immediately adds that this study must not be useless for soldiers (521d). Even here the negative formulation "not useless" does not highlight its value. When they identify the study of number as a study that might turn the soul, it is also immediately identified as also necessary for the art of war (522c). Once Socrates turns to geometry, however, he makes clear that its use in war is not at all on a level with its use in turning the soul towards being and he labels the military use as just a by-product (527c). Finally, when Glaucon's first response to the proposal that they add astronomy as a third mathematical discipline is to praise its value to the military, Socrates expresses amusement that he should be so concerned to appease the multitude by appealing to such crude utility; the relative insignificance of such utility is emphasized by stressing that the eye of the soul which mathematics helps to purify is worth ten thousand physical eyes that a general might use in war (527d-e). One might wonder why Socrates introduced the issue of military utility in the first place if it is so insignificant, and Glaucon might well feel that Socrates has set him up to be the butt of his joke. I would suggest that the initial reference to a possible military use for mathematics is largely a rhetorical strategy. Plato wants to introduce mathematics into the discussion of education and to form a connection between the new higher education and earlier descriptions of the guardians' education, which had emphasized their military leadership. Mathematics is thus

Even some of the scholars who adopt a constitutive view of the value of mathematics have been very hesitant to see moral and political decision-making as mathematical in nature. Thus, while Sedley argues that "the first principles invoked and applied in the course of decision-making would regularly exhibit mathematical features," he also stresses that "this is by no means to suggest that all the detailed decision-making in the ideal city will be mathematical in form."[16] It is not surprising, then, that other instrumentalists are as skeptical as Annas. Thus Ferrari rejects precisely the role for mathematics that Burnyeat adopts when he says, "We need not regard the education of the philosopher king ... as an internalization of mathematical structures that function as blueprints for applying his knowledge of the good to the social world."[17] He similarly seems to dismiss a view like Sedley's when he says that the philosopher kings are not "embodying in society a mathematical proportion whose structure they first discovered in abstraction."[18] It is clear, then, that no consensus has

---

first introduced as a discipline used by generals as it is by all other crafts (522c), but after introducing mathematics to the discussion Plato then endeavors to show that the mathematics he has in mind is something far different from that conceived of by the multitude and that its true value is thus also something far different from its use in war. Thus, Plato is not the NATO administrator who demands that all mathematical research be useful for the technology of war but rather the shrewd researcher who is willing to posit some limited use for his research in war in order to attract the attention of the granting agency, in this case the city-state, before revealing that its ultimate value is something quite different and higher.

16 Sedley, "Art of Ruling," 271.
17 G. R. F. Ferrari (ed.), *Plato, The Republic*, trans. Tom Griffith (Cambridge: Cambridge University Press, 2000), xxx.
18 Ferrari, *Republic*, xxx. Ferrari suggests, instead, that the philosopher king uses his mathematical knowledge "in resolving particular problems that arise while he is taking his turn running the city" and compares Aristotle's use of the concept of proportionate equality in the discussion of justice in Book 5 of the *Nicomachean Ethics*. The problem with this approach is that the sort of mathematics used by Aristotle is quite simple and would hardly require the advanced mathematics prescribed for the guardians by Plato. Ferrari falls back on an instrumental explanation and regards the ten years of mathematics as necessary to transmit to the soul of the learner the rational order and proportion of mathematics, comparing the influence on the soul of the study of cosmic harmony that is described in the *Timaeus*. Ferrari is certainly right that Plato thought that studying the ordered realm of mathematics and the forms would help produce order in the soul. Yet, in Book 7, while Plato emphasizes that *mousikē* plays an important role in the habituation of the soul in the earlier education of the guardians, this is contrasted with education in mathematics (522a), which is not described in the language of habituation; Plato simply does not describe the value of mathematics in *Republic* 7 in terms of the order it produces in the soul, and it must therefore be for some additional reason that the guardians study it for so long and at such an advanced level.

emerged as to what role Plato envisaged for mathematics in the governance of the state or in making moral decisions.

II The Neglected Evidence of Books 8 and 9

Part of the problem is that scholars focused too exclusively on Books 6 and 7 overlooked evidence from Books 8 and 9, where Plato describes the functioning, or rather malfunctioning, of the five basic types of constitution and shows quite conclusively that mathematics does "lead to rational and informed decisions on practical and ethical matters," as called for by Annas.[19] Plato's discussion of these constitutions begins and ends with an important reference to advanced mathematics. He begins by describing the crucial degeneration of the ideal state in terms of a failure in mathematics; this is the passage describing the famous nuptial number. He ends by giving a mathematical answer to the central question of the *Republic*, i.e., whether the life of the just or unjust person is happier; we, along with Glaucon, are somewhat bemused to learn that the just person lives a life that is exactly 729 times more pleasant than that of the most unjust person, the tyrant (587e). Now both of these passages are clearly partly humorous in intent, but Plato's humor, so I will argue, does not do away with his vision of government and moral decision-making by the numbers. I will discuss each of the passages in turn.

I hesitate to raise the issue of the nuptial number for fear that my readers will very quickly find that they have something more pressing to do. Rest assured, however, that my intention is not to drag you through arcane mathematical procedures in order to propose yet another candidate for the mysterious number. While some mathematical sense can be made of what Plato says (see, e.g., the appendices in Adam's commentary),[20] I do not think that he expected us to derive a specific number from what he says. Indeed, it would be amazing if he thought we could, since the whole point of the passage is that even the guardians with all of their advanced training in mathematics would fail to grasp the elusive number.[21] The real function of the passage on the nuptial number becomes clear from its context.

---

19 Annas, *Introduction*, 275.
20 James Adam, ed., *The Republic of Plato: Edited with Critical Notes, Commentary and Appendices*, 2 vols. (Cambridge: Cambridge University Press, 1929), 2:264–312.
21 Plato clearly presents the guardians' training as going beyond anything actually available in his day in the cases of stereometry (528b-d), astronomy (530c), and harmonics (531c).

After finishing his picture of the ideal state and the ideally just man in Book 7, Socrates turns to four inferior types of state and four corresponding types of individuals. The first step is to examine how the ideal state becomes corrupted and devolves into the next best state, the timocracy. He makes the basic point that all revolutions start from dissension in the ruling class itself (545c). But how is it that dissension arises among the guardians in the ideal state? It arises because of a mathematical error. Plato's state is based on a breeding program designed to ensure that the best male and female guardians mate to produce the best possible rulers. Plato now reveals that this breeding program is based on a mathematical formula; it is the failure to grasp and implement this formula properly that leads to dissension among the rulers (546d–547a). There are four points that I would like to make about this nuptial number.

First, Plato clearly indicates that his account of the number that governs good births is to be taken with a grain of salt; it is only partly serious. He does this by presenting his account as a statement of the Muses, whom Socrates invokes to explain the origin of the discord in his state, just as Homer did at the beginning of the *Iliad*. The Muses are described (545e1–3) as acting playfully (παιζούσας), joking (ἐρεσχηλούσας) with us as with children, and addressing us in a tragic (τραγικῶς) and highfalutin style (ὑψηλολογουμένας) that is mock serious (ὡς δὴ σπουδῇ λεγούσας).[22]

Second, what is playful in the Muses' response is the lofty tragic tone of the diction. It has not been emphasized enough that Plato's description of the number is not in the typical language of Greek mathematics but in language that would have sounded arcane and pompous to the contemporary Greek audience.[23] The number that governs human begettings is described as

> ... the first in which augmentations dominating and dominated when they have attained to three distances and four limits of the assimilating and the dissimilating, the waxing and the waning, render all things conversable and commensurable with one another, whereof a basal four-thirds wedded to the pempad yields two harmonies at the third augmentation, the one the product of equal factors taken one hundred times, the other of equal length one way but oblong, – one dimension of a hundred numbers determined by the rational diameters of the pempad lacking one in each case, or of the irrational lacking two; the other dimension of a hundred cubes of the triad. (546b5–c7, trans. Shorey)

---

[22] That the whole account of the number is part of the Muses' mock serious discourse is confirmed by Glaucon's reference to it as the statement of the Muses at the end (547a).

[23] See Adam, *Republic*, 2:202: "Moreover, the style of the whole passage, though extraordinarily rhythmical and highly-wrought, acquires a touch of fantastic humour from the bewildering parade of mathematical terms, at some of which Plato's own contemporaries would probably have smiled."

Modern translators often miss Plato's purposefully obscure language and try to render the passage in clear English, thus undercutting his intent. For example, Grube's translation of one clause runs, "Of these the lowest numbers in the ratio of four to three, married to five, give two harmonies when multiplied three times."[24] Shorey's rendering is much better, "Whereof a basal four thirds wedded to the pempad yields two harmonies at the third augmentation."[25] Shorey's "basal" translates *puthmēn*, which does refer to the lowest numbers in a given ratio, as Grube's translation indicates. It is, however, an unusual word in this mathematical sense, appearing with this meaning nowhere else in Plato and in Aristotle only in his reference to this passage in the *Republic* (*pol.* 1316a6).[26] Its literal reference to the base of something such as a cup or the bottom of something such as the sea is applied metaphorically to numbers. Shorey nicely catches the odd flavor of this metaphor with his translation "basal." The word for the number five, *pempas*, is equally odd. It is found twice in this description of the nuptial number but nowhere else in Plato and only once in Aristotle, again when he quotes this passage of the *Republic* (*pol.* 1316a6).[27] Shorey seems to be right on target in reflecting Plato's unusual diction by transliterating the Greek to produce the obscure sounding "pempad" rather than using the ordinary "five."[28]

Third, despite the playful language the topic is serious: the failure of Plato's ideal state. Glaucon affirms that along with playfulness there is truth in the Muses' speech, and Socrates agrees that this is necessary, since they are

---

[24] G. M. A. Grube, trans., *Plato's Republic* (Indianapolis: Hackett, 1974). Reeve's revision is an improvement ("Of these elements, four and three, married with five, give two harmonies when thrice increased"), since the archaic sounding "when thrice increased" reflects Plato's unusual use of αὐξάνω to indicate multiplication, whereas Grube's "multiplied three times" wrongly suggests that Plato was using the normal language for multiplication. See G. M. A. Grube, trans., *Plato's Republic*, rev. ed. by C. D. C. Reeve (Indianapolis: Hackett, 1992).
[25] Paul Shorey, trans., *Plato, The Republic* (Cambridge, Mass.: Harvard University Press, 1935), 247.
[26] It does appear in its mathematical sense in a fragment from Speusippus' work *On Pythagorean Numbers* (fr. 28 Tarán), but its appearance in that obscure work just underlines its arcane nature. It is more common in the arithmetical tradition of later antiquity, e.g., Theon Smyrnaeus 80.15.
[27] It is not found elsewhere in the fourth century or earlier except in Xenophon, who uses it to refer to a squad of five men (e.g., *hist. Graec.* 7.2.6).
[28] As I have argued elsewhere, the best parallel for the diction of this passage as a whole is found in some otherwise unknown harmonic theorists alluded to in one of the testimonia for Archytas, so that Plato seems to be drawing on a very obscure source to make the Muses sound appropriately arcane. See Carl A. Huffman, *Archytas of Tarentum: Pythagorean, Philosopher and Mathematician King* (Cambridge: Cambridge University Press, 2005), 428–43.

Muses (547a). Plato makes clear that the failure is a result of an error on the part of the guardians. Adam argued that "the fault lies not with the rulers, but with the inevitable law of Change."[29] It is true that Plato says that the state declines because of a universal law that there is destruction for everything that has come to be (546a). This assertion, however, does not tell us how the destruction comes about, and Plato's account of the mechanism of decline emphasizes the failure of the guardians. It is the guardians who are the subject of the verbs expressing failure, it is they who "cause children to be begotten out of season" (γεννήσουσι παῖδάς ποτε οὐ δέον, 546b3–4), it is they who are ignorant (ἀγνοήσαντες, 546d1) of which births are better and which worse and thus "bring together brides and grooms at inopportune times" (συνοικίζωσι νύμφας νυμφίοις παρὰ καιρόν, 546d1–2).[30]

Fourth, the guardians' error is presented as a failure in applied mathematics. At the end of the description of the nuptial number quoted above, Socrates says that "this entire geometrical number is determinative of better and inferior births" (546c7–8, trans. Shorey). He has earlier maintained that the state will degenerate because "those whom you have educated as leaders of the city will not by calculation with sensation ascertain good births and births that should not be allowed" (546b1–3). Plato thus indicates that there are two factors that lead to the failure of the guardians. First, there is *logismos*, which can simply mean "reasoning," but in the context of the nuptial number clearly has its more specialized meaning of "mathematical reasoning" or "calculation." The grand language used in the description of the nuptial number underlines the difficulty of the mathematics. The second factor is sensation (*aisthēsis*). Bringing about the best human offspring requires an understanding of abstract mathematical laws but also proper perception of how those laws apply to specific individuals in the state, i.e., it requires applied mathematics.[31]

---

29 Adam, *Republic*, 2:204.
30 Adam argued that, since Plato has created a perfect state, the seeds of its dissolution cannot lie in the state itself, since this would mean that it is not, in fact, perfect (Adam, *Republic*, 2:202). The problem with this argument is that Plato, so far as I can see, never claims that his state is perfect or that his guardians are infallible, just that they are the best that can be achieved.
31 One might argue that the guardians grasp the number that governs births adequately and that their failure is only in applying it properly. Thus, Annas says that "Plato is symbolically expressing the idea that no ideal can ever be fully realized; the world we experience can never exemplify what is perfectly, unqualifiedly, and stably just" (Annas, *Introduction*, 296). Plato's language literally says not that they do not grasp the nuptial number but rather that they do not ascertain good births. Plato lists both factors, calculation and perception, side by side, however, which *prima facie* indicates that they both play a role. Moreover, one of the points of the grand and complex language that Plato uses to describe the nuptial number is surely to make plausible

The twofold failure of the guardians corresponds to the structure of their education as set out by Socrates in Book 7. The guardians devote ten years to abstract mathematics (537b-d), but another fifteen years to gaining practical experience governing in the sensible world (540a). It is important to emphasize again that Plato ascribes the failure to the guardians and not to nature and that the failure is one of wisdom. This is shown by the concessive clause that Plato inserts in his description of their failure. They do not ascertain the proper births, "despite all their wisdom" (καίπερ ὄντες σοφοί, 546b1). Their wisdom is clearly relevant; they simply do not have enough of it. Ten years of abstract mathematics cannot guarantee that they will grasp the complexities of the nuptial number, nor can fifteen years of experience in the sensible world guarantee that they will apply it properly. Plato clearly regarded the successful application of mathematical structures to the sensible world as crucial to good governance of the state. Mathematics must thus be constitutive of at least part of the knowledge that the guardians need to rule and not just play the instrumental role of practice in *a priori* reasoning.

A passage at the end of Book 9 indicates that mathematics was also used in making practical moral decisions and can, in fact, answer the central moral question of the *Republic:* Is the life of the just person or the completely unjust tyrant happiest? Socrates demonstrates that the philosopher king lives exactly 729 times more happily than the tyrant (587e). Is this calculation simply a joke? Plato is, to be sure, having fun at several levels. Glaucon responds to this calculation by exclaiming that Socrates has introduced an *amēchanon logismon*. As in the passage on the nuptial number, *logismos* clearly refers to a mathematical calculation, and its use here may be intended to remind us of that earlier *amēchanos logismos*. The word *amēchanos* indicates that the calculation is something that reduces one to helplessness, that is "overwhelming" (Shorey) or "extraordinary" (Grube). One can imagine Glaucon delivering the line with eyes wide open in mock amazement, an amazement that has two grounds. First, the word for 729 times in Greek is a real mouthful, containing no less than 35 letters and taking up half a line of Greek text, ἐννεακαιεικοσικαιεπτακοσιοπλασιάκις. Second, it is far from clear how Socrates arrives at this number. The tyrant is said to be three times removed from the oligarch, who is in turn three times removed from the philosopher king, so one might conclude that he was 3 x 3 and hence 9 times removed (587c-d). Socrates then says, however, that the number should be squared and then cubed so that the final number is 9

---

the guardians' failure. It is no wonder that they cannot fully grasp mathematics of this complexity.

cubed or 729, but it is not at all obvious why the cube of 9 is used rather than 9 itself.[32] So Glaucon may be smiling at the hand waving that Socrates has just done.

Plato's humor shows that he has not given us a convincing calculation; at the same time, as in the case of the nuptial number, it stresses the complexity of the mathematics involved and hence points to the need for advanced education in mathematics. For Plato again indicates that he does have a serious point to make with his extraordinary calculation. Socrates is adamant in the face of Glaucon's amazement that "it is a true number and pertinent to the lives of men" (588a3–4, trans. Shorey). As scholars have observed, Plato wants to arrive at the number 729 because it is the double of 364.5, which was regarded as the length of the year by some fifth-century thinkers.[33] The tyrant can be thought of as having less pleasure than the philosopher king on every day and every night of the year and thus to be 364.5 days plus 364.5 nights or 729 times less happy. That Plato has something like this point in mind is clear, because he says that the number 729 is "true and pertinent to the lives of men, if days and nights and months and years are" (588a3–5, trans. Shorey). He is thus connecting the number to the days and nights in a year. I suggest that Plato thought the number was true in that, although he would not insist on the exact number, it shows 1) the appropriate level of magnitude of difference between the happiness of the just and unjust life and 2) that the happiness of the just life results from the mathematical structure of the world. He does think that, if we knew enough about mathematics and enough about the world, we could calculate the relative happiness of people of different moral profiles. Plato thought that important

---

[32] For some suggestions see Adam, *Republic*, 2:360.

[33] See Adam, *Republic*, 2:361. Philolaus, whose influence on Plato is clear in other dialogues such as the *Timaeus* and *Philebus*, accepted such a year. For the evidence concerning the length of the year according to Philolaus and for his influence on Plato, see Carl A. Huffman, *Philolaus of Croton: Pythagorean and Presocratic* (Cambridge: Cambridge University Press, 1993), 276–78 and 149, and Carl A. Huffman, "The Philolaic Method: The Pythagoreanism Behind the *Philebus*," in Anthony Preus (ed.), *Before Plato: Essays in Ancient Greek Philosophy* (Albany: SUNY Press, 2001), 67–85. Philolaus also argued for a great year, a period in which the lunar and solar periods can be reconciled, of 59 years. There are 708 months in 59 years, and we are told that Philolaus added 21 intercalary months so that there are also a total of 729 months in Philolaus' great year. Thus 729 refers not just to the number of days and nights in a year but also to the number of months in Philolaus' great year. Since Socrates says that the number applies to men "if days and nights and months and years pertain to them" (trans. Shorey), we can so far see how days, nights, and months apply. In order to account for Plato's mention of years, Adam suggests that Philolaus might have had a "greatest year" consisting of 729 regular years but there is no direct evidence for this.

moral decisions as well as crucial decisions in governing depend on our grasp of advanced mathematics, a mathematics so advanced, in fact, that he laughs at his own inability to grasp and express it.

III Mathematics in Book 7

If the passages in Books 8 and 9 discussed above show that mathematics does have a constitutive role to play in the guardians' education, why have so many scholars concluded, on the basis of Book 7, that its role is only instrumental? Clearly it is because much of the language used in Book 7 to describe the role of mathematics emphasizes its instrumentality. Plato's focus in Book 7 is on method, on how one becomes educated, rather than on the ultimate content of the education. The allegory of the cave is explicitly introduced to describe "our nature regarding education and lack of education" (514a1–2). Immediately after the allegory, Plato draws the conclusion that education does not, as some people claim, put knowledge in the soul of the ignorant like putting sight in the eyes of the blind (518c). Rather, there is an organ in the soul that is analogous to the eye in the body, and education involves turning this eye of the soul from the world of becoming to the world of being (518c). Socrates and Glaucon then look for an art that turns the soul in the right direction, and this art turns out to be mathematics. The language used here and in many passages in Book 7 thus does not talk of mathematics as constructing a specific sort of knowledge in the soul. Instead the language is instrumental. The two dominant metaphors are turning (e.g., στρέφειν, 518c7; περιακτέον, 518c8; μεταστραφήσεται, 518d5; περιστροφή, περιαγωγή, 521c5–6) and drawing/dragging (e.g., ὁλκόν, 521d4; ἑλκτικός, 523a2), not of constituting.[34]

There is, however, a second less prominent strand in Plato's presentation of mathematics in Book 7, which suggests that it does have a constitutive role in knowledge of the form of the Good and indeed of all forms. At 533a8 Socrates

---

[34] In some passages Socrates does seem to go further and suggest that mathematics might not just serve the instrumental role of turning the eye of the soul in the right direction but might also help us master important knowledge about reality. Thus, at 526e2, he asks if advanced geometry facilitates "the apprehension of the idea of the good" (trans. Shorey). We might think that, if advanced geometry facilitates understanding the form of the good, the form of the good is mathematical in nature, and that this passage is a point for the constitutive view of mathematics. In his answer to the question, however, Socrates gives no indication that the idea of the good is mathematical in nature; rather all studies that play the instrumental role of compelling the soul to turn to the region where the most blessed part of reality resides are declared suitable. So this passage supports Annas and the instrumentalists.

asserts that only the power of dialectic reveals the truth, but adds the important condition that it only does so "to the one who is experienced in all the things we have gone through," where the reference is clearly to the mathematical studies. Similarly, at 536d, the study of calculation, geometry, and the rest of the propaedeutic studies are declared to be necessary preliminary training for dialectic. It is hard to see how this necessity can refer just to mathematics' instrumental role, since, as Annas points out, Plato's argument shows that non-mathematical concepts can turn us to the intelligible realm just as well as the mathematical.[35] Plato's call for the study of things that provoke or summon thought and that thus lead us to the intelligible realm includes all things that "fall upon the senses along with their opposites" (524d2–3), so it includes mathematical concepts such as the one and the many but also non-mathematical opposites such as the hard and the soft and the light and the heavy (524a). If the mathematical studies are really necessary to grasp the truth, that necessity must arise from something other than their ability to turn the soul; it must arise from their role in revealing the constitution of the truth. Plato has, in fact, all along presented mathematics not just as instrumental in turning us towards the truth but also as studying the constitution of it. Thus, at 526b, not only is the study of number assigned the instrumental role of forcing the soul to use pure thought, but that thought is also said to be directed at truth itself. Mathematics is not just practice in *a priori* reasoning: it is *a priori* reasoning about the truth. Similarly at 527b4 geometry is said to be "knowledge of what always is." Finally, at 531c6–7, the abstract study of concordant numbers that Socrates proposes as the harmonics of the inaudible intelligible realm is praised as " useful ... for inquiry into the beautiful and the good." This formulation suggests that advanced mathematics does not just turn us towards the realm of the beautiful and good but actually helps us inquire into that realm and hence that the beautiful and the good have a structure that is revealed by mathematics.

This interpretation is supported by the imagery of the cave. Mathematics does turn the soul from the sensible to the intelligible and draw it up out of the cave into the realm of eternal being (532b), but it does more than that: it is also described as studying part of the intelligible realm. It studies the reflections and shadows in the world outside the cave (532c). It does not study the highest reality directly but, as Socrates says, it dreams about reality (532c). Mathematics effects the transition from the cave to the external world. It starts in the sensible realm by relying on visible diagrams (511a), but it takes us out of the cave to the intelligible realm. This is its instrumental role. However, it also begins

---

[35] Annas, *Introduction*, 273–74.

the study of the intelligible world that we find there.[36] This is its constitutive role. It is crucial to recognize that, for Plato, there is just one intelligible realm, which is a coherent whole. He stresses that true knowledge will see the connections between all things and achieve a synoptic view (537c, see also 531d).

Thus, mathematics does not deal with an isolated set of *a priori* ideas that provide us with practice in a certain sort of reasoning but that do not reveal the intelligible realm with which the guardians must be familiar to rule well. For Plato, mathematics deals with the one and only intelligible realm so that it is studying the good, the beautiful, and the just, albeit not directly but through images. Plato makes clear that the difference between mathematics and dialectic resides in method: dialectic can give an account of its starting points whereas mathematics does not (533c).[37] This is again borne out by the imagery of the allegory of the cave. The shadows and reflections that mathematics studies in the intelligible world do reflect the basic form of their originals, which strongly suggests that those originals are mathematical in nature as well. The originals differ from their reflections and shadows not by being non-mathematical but rather by being known through an unhypothetical principle and not through the unexamined hypotheses of mathematics. The philosopher kings are studying the intelligible world when they study mathematics, although their grasp of that world will be made more secure when they add the study of dialectic. They are trying to grasp the ideal mathematical structures that the demiurge used in constructing the physical world and that they need to incorporate into the ideal state, if it is going to reflect the intelligible idea of justice. Advanced mathematics is serious ethical business, since failure to grasp it properly will lead to the failure of the state. Given human fallibility, that failure may be inevitable, but Plato seems convinced that the better the rulers understand advanced mathematics, the better the state will be governed and the longer it will last.

---

[36] I take it that this is what Burnyeat means when he says that "mathematics provides the lowest-level articulation of objective value" (Burnyeat, "Mathematics," 45).

[37] There is a rich scholarship on the difference in method between mathematics and dialectic. See Ian Mueller, "Mathematical Method and Philosophical Truth," in Richard Kraut (ed.), *The Cambridge Companion to Plato* (Cambridge: Cambridge University Press, 1992), 170–99; Ian Mueller, "On the Notion of a Mathematical Starting Point in Plato, Aristotle, and Euclid," in Alan C. Bowen (ed.), *Science and Philosophy in Classical Greece* (New York and London: Garland, 1991), 59–97; Denyer, "Sun and Line"; and Dominic T. J. Bailey, "Plato and Aristotle on the Unhypothetical," *Oxford Studies in Ancient Philosophy* 30 (2006): 101–26.

Bibliography

Adam, James, ed. *The Republic of Plato: Edited with Critical Notes, Commentary and Appendices*, 2 vols. (Cambridge: Cambridge University Press, 1929).
Annas, Julia. *An Introduction to Plato's "Republic"* (Oxford: Oxford University Press, 1981).
Bailey, Dominic T. J. "Plato and Aristotle on the Unhypothetical," *Oxford Studies in Ancient Philosophy* 30 (2006): 101–26.
Burnyeat, Myles F. "Plato on Why Mathematics Is Good for the Soul," in Timothy Smiley (ed.), *Mathematics and Necessity* (Oxford: Oxford University Press, 2000), 1–81.
Denyer, Nicholas. "Sun and Line: The Role of the Good," in Ferrari, *Companion to Plato's "Republic,"* 284–309.
Ferrari, G. R. F., ed. *The Cambridge Companion to Plato's "Republic"* (Cambridge: Cambridge University Press, 2007).
Ferrari, G. R. F., ed. *Plato, The Republic*, trans. Tom Griffith (Cambridge: Cambridge University Press, 2000).
Grube, G. M. A., trans. *Plato's Republic* (Indianapolis: Hackett, 1974; rev. ed. by C. D. C. Reeve published in 1992).
Huffman, Carl A. *Archytas of Tarentum: Pythagorean, Philosopher and Mathematician King* (Cambridge: Cambridge University Press, 2005).
Huffman, Carl A. *Philolaus of Croton: Pythagorean and Presocratic* (Cambridge: Cambridge University Press, 1993).
Huffman, Carl A. "The Philolaic Method: The Pythagoreanism Behind the *Philebus*," in A. Preus (ed.), *Before Plato: Essays in Ancient Greek Philosophy* (Albany: SUNY Press, 2001), 67–85.
Mueller, Ian. "Mathematical Method and Philosophical Truth," in Richard Kraut (ed.), *The Cambridge Companion to Plato* (Cambridge: Cambridge University Press, 1992), 170–99.
Mueller, Ian. "On the Notion of a Mathematical Starting Point in Plato, Aristotle, and Euclid," in Alan C. Bowen (ed.), *Science and Philosophy in Classical Greece* (New York: Garland, 1991), 59–97.
Sedley, David. "Philosophy, the Forms, and the Art of Ruling," in Ferrari, *Companion to Plato's "Republic,"* 256–83.
Shorey, Paul, trans. *Plato, The Republic* (Cambridge, Mass.: Harvard University Press, 1935).
Tarán, Leonardo. *Speusippus of Athens: A Critical Study with a Collection of the Related Texts and Commentary* (Leiden: Brill, 1981).

Katerina Ierodiakonou
Hellenistic Philosophers on the Phenomenon of Changing Colors

Abstract: Observe the color of the pigeon's neck, or of the peacock's tail, as it turns around. At one moment one has the impression of a certain color, and a moment later the impression of a quite different one. Is it true, then, that in these and similar cases the object which we perceive actually is multi-colored, or is it because of the way we perceive it that it looks as if it were multi-colored, though in reality it just has a single color? This question puzzled the philosophers of the Hellenistic dogmatic schools, namely, the Epicureans and the Stoics. The ways in which they tried to answer it were very different, for they had different views about the nature and perception of color. The Epicureans claimed that the atoms which constitute an object are not colored; rather, colors are due to the atomic surface structure of an object and its orientation relative to the source of light. By contrast the Stoics suggested that colors are properties of the four elements, fire, air, water, and earth, and hence of the objects themselves which are composed from these elements. My aim in this paper is (i) to get a better understanding of how the Epicureans and the Stoics settled the issue about the color of the dove's neck or of the peacock's tail; (ii) to consider their theories as reactions to, or advances on, other ancient theories; and (iii) to assess their relative plausibility.

Observe the color of the pigeon's neck, or of the peacock's tail, as it turns around. At one moment one has the impression of a certain color, and a moment later the impression of a quite different one. And it is not that one has the impression of just a different shade of the same color, that is the same color just lighter and paler, or darker and more vivid; one rather has the impression of an altogether different color, i.e., of a different hue. Is it true, then, that in these and similar cases the object which we perceive actually is multi-colored, or is it because of the way we perceive it that it looks as if it were multi-colored, though in reality it just has a single color? This question puzzled the philosophers of the Hellenistic dogmatic schools, namely, the Epicureans and the Stoics.

This paper is just a small token of my gratitude to Heinrich von Staden for his pioneering work on the Hellenistic theories of perception, and also for his warm encouragement of my own work on ancient theories of colors during my stay at the Institute of Advanced Study in Princeton in the autumn of 2003.

And there were more examples in antiquity which were supposed to raise similar problems. For instance, the changing color of the sea, when it is rough and then when it is calm; the changing color of the air and of the clouds, especially during sunrise and sunset; the changing color of the skin of human beings, when they are feeling hot or cold, angry or excited; finally, the changing color of chameleons and of other animals, depending on their close environment. However, the examples of the changing color of the pigeon's neck and of the peacock's tail seem to have really gripped the ancients' imagination. For there are at least ten ancient philosophers, all from the period after Aristotle, who, in various contexts, refer to these examples and comment on them, some more extensively than others.[1]

Interestingly enough, the example of the changing color of the peacock's tail, in particular, can also be found in modern popular introductions to color theories, in chapters in which the authors attempt to illustrate what nowadays is known as the phenomenon of shimmering or unsteady colors.[2] Perhaps our modern intuitions immediately grasp what shimmering or unsteady colors are by a simple reference to the iridescent colors of some CDs; but it seems that the peacock's tail is still considered an illuminating example of this special phenomenon of changing colors. The phenomenon in modern terms, briefly stated, is the following. The color of the peacock's tail changes, because the light rays reflected by the tail's surface, which is not perfectly smooth but has tiny ridges, travel different distances; they thus interfere with each other, and sometimes they cancel each other out. Hence, the iridescent color of the peacock's tail that we perceive is not produced simply by the light rays which have not been absorbed by its surface; they rather constitute a certain kind of mixture of light which changes both with the slightest movement of the peacock's tail, and thus the variation of the angle in which the light falls on it, and with the position of the observer relative to the tail.

Needless to say, the ways in which the Epicurean and the Stoic philosophers tried to account for this phenomenon differ greatly from modern explanations. They also differ considerably from each other, since the Epicurean and the Stoic theories of the nature and the perception of color clearly depend on very different views about the material constitution of things and about sense perception. In what follows, I do not of course want to show that the accounts of the

---

1 Apart from the texts discussed here, see also: ps.-Arist. *col.* 792a24, 792b28, 793a15, 793b9; Alex. Aphr. *in de sens.* 51.3; Philop. *in de an.* 315.28; *in de gen. et corr.* 23.9; Olymp. *in cat.* 98.25; cf. Sophonias, *In libros Aristotelis de anima paraphrasis* 121.5.
2 E.g., Hazel Rossotti, *Colour: Why the World Isn't Grey* (Princeton: Princeton University Press, 1985), 44–47, 91–101.

Hellenistic philosophers are more convincing than that given by modern scientists. My aim is to reconstruct the Epicureans' and the Stoics' explanations of the phenomenon of the changing color of the pigeon's neck, or of the peacock's tail, and to try to understand better the grounds on which they defended them. Besides, if we consider the Hellenistic doctrines concerning this special phenomenon as reactions to, or advances on, earlier theories of color and color perception, we may be able to throw some light on the Epicurean and the Stoic views on color in order to assess their relative plausibility.

First, though, let us examine why and how Hellenistic philosophers got engaged in a discussion about the phenomenon of changing colors. Cicero, in his *Academica*, twice refers to the example of the pigeon's neck, and in both cases he presents it as an example introduced into philosophy by the Sceptics with the intention of questioning the reliability of our sense perceptions.[3] In one of these passages Cicero, who consistently throughout this work adopts a Sceptical stance, asks:

> Quid ergo est quod percipi possit, si ne sensus quidem uera nuntiant? quos tu Luculle communi loco defendis. quod ne id facere posses, idcirco heri non necessario loco contra sensus tam multa dixeram; tu autem te negas infracto remo neque columbae collo commoueri. primum cur? nam et in remo sentio non esse id quod uideatur, et in columba pluris uideri colores nec esse plus uno. (Cic. *acad.* 2.79)

> So what *is* apprehensible, if not even the senses give true reports? You defend them, Lucullus, with a stock argument – though it was precisely to stop you doing that that I exceeded my brief yesterday and said so much against the senses. But you say that you weren't moved by the bent oar or the pigeon's neck. First, why not? I recognize, after all, that my impressions misrepresent the oar and show several colors of the pigeon's neck, though there isn't more than one. (trans. Brittain)

That is to say, the Sceptic claims that, if at one time the pigeon's neck appears to have one particular color and a moment later a quite different one, our sense perceptions cannot be relied upon to tell us which the color is that it actually has; it can be one or the other, but not both. Hence, on the basis of sense perception we cannot be certain about the color of the pigeon's neck, and we should rather suspend judgement.

---

3 In Elias' commentary of Aristotle's *Categories* (*in cat.* 204.2–5) the example of the pigeon's neck is not used to illustrate a Sceptical strategy but Protagoras' relativism: according to Protagoras, Elias claims, each thing's character is relative to the other things surrounding it; for instance, the color of the pigeon's neck is black in the shadow and red in the sunlight. There is no reason, however, to suppose that the example of the pigeon's neck, which Elias chooses to illustrate Protagoras' doctrine, was already used by Protagoras long before the Sceptics.

The Sceptics after Carneades used this particular example, together with that of the oar which looks bent in water and that of the large square tower which looks small and round from the distance, as paradigmatic cases which show that we cannot rely on our sense perceptions. For they indicate, according to the Sceptics, that we perceive an object differently depending, for instance, on its location, the distance, or the position from which we perceive it, without our being able to judge which sense perception, if any, is veridical. The example of the changing color of the pigeon's neck was in fact used as an illustration in one of Aenesidemus' modes, perhaps even by Aenesidemus himself.[4] In general, Aenesidemus' modes offered the Sceptics strategies to counter any given claim; this mode, in particular, could be applied in cases in which the claim arguably depends on the relative position of the person who makes it. So, in the case of the pigeon's neck, the same object appears different depending on the position from which we perceive it, i.e., depending on how the pigeon turns its neck relative to us.

To meet the Sceptics' challenge, the Epicureans and the Stoics tried to explain the phenomenon of the changing color of the pigeon's neck in such a way as to restore our trust in sense perception, and thus to defend the possibility of certain knowledge. Accordingly, in the other passage from the *Academica* in which Cicero refers to the changing color of the pigeon's neck, he sketches both the Epicurean and something like the Stoic account of the phenomenon. This time, though, the account is put into the mouth of Cicero's interlocutor, Lucullus, who objects to Cicero's Scepticism and endorses a view which is supposed to be that of Antiochus, and hence a view we may assume to be at least heavily influenced by the Stoic position:

> Ordiamur igitur a sensibus. quorum ita clara iudicia et certa sunt, ut, si optio naturae nostrae detur et ab ea deus aliqui requirat contentane sit suis integris incorruptisque sensibus an postulet melius aliquid, non uideam quid quaerat amplius. nec uero hoc loco expectandum est dum de remo inflexo aut de collo columbae respondeam; non enim is sum qui quidquid uidetur tale dicam esse quale videatur; Epicurus hoc uiderit et alia multa. meo autem iudicio ita est maxima in sensibus ueritas, si et sani sunt ac ualentes et omnia remouentur quae obstant et inpediunt. itaque et lumen mutari saepe uolumus et situs earum rerum quas intuemur, et interualla aut contrahimus aut diducimus, multa que facimus usque eo dum adspectus ipse fidem faciat sui iudicii. (Cic. *acad.* 2.19)

> Let's start with the senses. Their judgements are so clear and certain that if human nature were given the choice – if a god demanded of it whether it is satisfied with its senses when they are sound and undamaged or whether it requires something better – I can't see what more it could ask for. But don't expect me to give counterarguments here dealing with the

---

4 Sext. Emp. *Pyrrh. subfig.* 1.120; Diog. Laërt. 9.86; Philo Alex. *ebr.* 173.

bent oar or the pigeon's neck: I'm not someone who claims that everything is exactly as our impressions represent it. Epicurus can see to that (and a lot more as well)! Still, in my judgment, there is a great deal of truth in the senses, providing they are healthy and properly functioning and all obstacles and impediments are removed. That's why we often want the light changed or the positions of the things we're looking at, and we reduce or increase their distance from us and alter many conditions until our vision itself provides the warrant for its own judgment. (trans. Brittain)

It seems, then, that both the Epicureans and the Stoics advocated that we can indeed have knowledge of the color of the pigeon's neck. According to Epicurus, both impressions, i.e., both the initial impression that we have of the color of the pigeon's neck and the impression that we have when it turns around, are true. According to the Stoics, on the other hand, one of these impressions is true and the other is false; the true impression is the one which we have under normal conditions, i.e., when the pigeon is at the right distance from the observer, in the right position, and under the right kind of light. With this brief description of the Epicureans' and the Stoics' views in mind, let us now try to get a firmer understanding of how they supported their respective doctrines against the Sceptics, as well as how the discussion of this phenomenon fits in with the Epicurean and the Stoic theories of the nature and perception of color.

As Lucullus says in the passage just quoted, the Epicureans were committed to the view that all impressions are true.[5] They had to claim, therefore, that both our initial impression of the pigeon's neck as having a certain color as well as our impression of it a moment later, when it turns around, as having a different color are true. But how did the Epicureans manage to argue in favor of such a counterintuitive view?

There is a passage in the second book of Lucretius' *De rerum natura* in which Lucretius mentions the example of the changing color of the pigeon's neck and of the peacock's tail in the course of defending the Epicurean doctrine that atoms have no color. After having established the standard atomist thesis that only atoms and the void exist, Lucretius here talks about the nature of atoms and claims that, although atoms have size, shape, and weight, they have no

---

[5] On the Epicurean doctrine that all impressions are true, see for instance: Gisela Striker, "Epicurus on the Truth of Sense Impressions," *Archiv für Geschichte der Philosophie* 59 (1977): 125–42; Christopher C. W. Taylor, "All Perceptions are True," in Malcolm Schofield et al. (eds.), *Doubt and Dogmatism: Studies in Hellenistic Epistemology* (Oxford: Oxford University Press, 1980), 105–24; Stephen Everson, "Epicurus on the Truth of the Senses," in Stephen Everson (ed.), *Epistemology* (Cambridge: Cambridge University Press, 1990), 161–83.

color. To support this, he gives seven arguments. Our passage introduces the third argument; in his highly poetic language Lucretius says:

> Praeterea quoniam nequeunt sine luce colores
> esse neque in lucem existunt primordia rerum,
> scire licet quam sint nullo uelata colore;
> qualis enim caecis poterit color esse tenebris?
> lumine quin ipso mutatur propterea quod
> recta aut obliqua percussus luce refulget;
> pluma columbarum quo pacto in sole uidetur,
> quae sita ceruices circum collumque coronat;
> namque alias fit uti claro sit rubra pyropo,
> inter dum quodam sensu fit uti uideatur
> inter caeruleum uiridis miscere zmaragdos.
> caudaque pauonis, larga cum luce repleta est,
> consimili mutat ratione obuersa colores;
> qui quoniam quodam gignuntur luminis ictu,
> scire licet, sine eo fieri non posse putandum est.
> (Lucr. 2.795–809)
>
> Besides, since colors depend upon light for their existence, and the primary elements of things do not emerge into the light, it is evident that they are not robed in any color. For what color can there be in blinding darkness? Why, even in the light color varies according as the incident ray that it reflects is perpendicular or oblique. Consider the iridescence imparted by sunlight to the plumage that rings and garlands the neck of the dove: sometimes it is glossed with red garnet, sometimes, as seen from a different angle, it appears to blend green emeralds with blue lazuli. Likewise, when the tail of a peacock is drenched in a flood of light, it exhibits changing colors as it is turned about. Since these colors are produced by a certain incidence of light, obviously we must not suppose that they can be produced without it. (trans. Smith)

Lucretius here argues that atoms are colorless on the basis of two premises: first, colors depend upon the light for their existence; and second, atoms do not emerge into the light. Though there are obvious problems with the interpretation of this argument, for instance with the interpretation of the puzzling claim that atoms do not emerge into the light, this is not my concern here.[6] I will rather take it for granted that on Epicurus' view atoms do not have colors. For I am interested in what this passage positively tells us about how the Epicureans explained

---

[6] Lucretius' claim that atoms cannot emerge into the light has been interpreted by scholars in different ways. For instance, it has been suggested that, because atoms are very small, they are invisible, or again, because it is impossible to alter the arrangement of the parts in atoms, atoms are not subject to the action of the light; cf. Cyril Bailey, *Lucretius, De rerum natura*, 3 vols. (Oxford: Clarendon Press, 1947), 2:927–28.

the fact that we perceive objects as colored, although atoms themselves have no color. In particular, I want to focus on what Lucretius says about the phenomenon that some objects appear to change color in a special way, and on how this phenomenon is explained by the Epicurean theory of color and color perception.

As I said before, the standard atomist thesis is that only atoms and the void exist. But if Epicurus believed that only atoms and the void exist, he would have regarded colors as unreal and subjective in the sense that the perceiver projects them onto reality. It seems, however, that on Epicurus' view the transient attributes of objects, their "accidents" (*sumptōmata*), also exist, though they do not have the independent existence of atoms and of the void, and are not like objects or like the "permanent attributes" (*sumbebēkota*) of objects:

> καὶ οὐκ ἐξελατέον ἐκ τοῦ ὄντος ταύτην τὴν ἐνάργειαν, ὅτι οὐκ ἔχει τὴν τοῦ ὅλου φύσιν ᾧ συμβαίνει ὃ δὴ καὶ σῶμα προσαγορεύομεν, οὐδὲ τὴν τῶν ἀΐδιον παρακολουθούντων, οὐδ' αὖ καθ' αὑτὰ νομιστέον – οὐδὲ γὰρ τοῦτο διανοητὸν οὔτ' ἐπὶ τούτων οὔτ' ἐπὶ τῶν ἀΐδιον συμβεβηκότων –, ἀλλ', ὅπερ καὶ φαίνεται, συμπτώματα πάντα ⟨κατὰ⟩ τὰ σώματα νομιστέον, καὶ οὐκ ἀΐδιον παρακολουθοῦντα οὐδ' αὖ φύσεως καθ' ἑαυτὰ τάγμα ἔχοντα, ἀλλ' ὃν τρόπον αὐτὴ ἡ αἴσθησις τὴν ἰδιότητα ποιεῖ, θεωρεῖται. (Epicur. *ep. Hdt.* 71 [= Diog. Laërt. 10.71])

> And we should not banish this self-evident thing from the existent, just because it does not have the nature of the whole of which it becomes an attribute – "body," as we also call it – nor that of the permanent concomitants. Nor should we think of them as *per se* entities: that is inconceivable too, for either these or the permanent attributes. For we should think of all the accidents of bodies as just what they seem to be, and not as permanent concomitants or as having the status of a *per se* nature either. They are viewed in just the way that sensation itself individualizes them. (trans. Long and Sedley)

Sextus Empiricus, too, confirms that Epicurus did not think that only atoms and the void exist. There really also are objects; moreover, there really also are the attributes which belong to these objects, attributes which are either permanent or transient:

> καθόλου γάρ, ἵνα μικρὸν ἄνωθεν προλάβωμεν εἰς τὴν τοῦ λεγομένου παρακολούθησιν, τῶν ὄντων τὰ μέν τινα καθ' ἑαυτὰ ὑφέστηκεν, τὰ δὲ περὶ τοῖς καθ' ἑαυτὰ ὑφεστῶσι θεωρεῖται. καὶ καθ' ἑαυτὰ μὲν ὑφέστηκε πράγματα οἷον αἱ οὐσίαι (ὡς τὸ σῶμα καὶ κενόν), περὶ δὲ τοῖς καθ' ἑαυτὰ ὑφεστῶσι θεωρεῖται τὰ καλούμενα παρ' αὐτοῖς συμβεβηκότα. τούτων δὲ τῶν συμβεβηκότων τὰ μέν ἐστιν ἀχώριστα τῶν οἷς συμβέβηκεν, τὰ δὲ χωρίζεσθαι τούτων πέφυκεν. ἀχώριστα μὲν οὖν ἐστι τῶν οἷς συμβέβηκεν ὥσπερ ἡ ἀντιτυπία μὲν τοῦ σώματος, εἶξις δὲ τοῦ κενοῦ· οὔτε γὰρ σῶμα δυνατόν ἐστί ποτε νοῆσαι χωρὶς τῆς ἀντιτυπίας οὔτε τὸ κενὸν χωρὶς εἴξεως, ἀλλ' ἀΐδιον ἑκατέρου συμβεβηκός, τοῦ μὲν τὸ ἀντιτυπεῖν, τοῦ δὲ τὸ εἴκειν. οὐκ ἀχώριστα δέ ἐστι τῶν οἷς συμβέβηκε καθάπερ ἡ κίνησις καὶ ἡ μονή. τὰ γὰρ συγκριτικὰ τῶν σωμάτων οὔτε κινεῖται διὰ παντὸς ἀνηρεμήτως οὔτ' ἀκινητίζει διὰ παντός, ἀλλὰ ποτὲ μὲν συμβεβηκυῖαν ἔχει τὴν κίνησιν, ποτὲ δὲ τὴν μονήν. (Sext. Emp. *adv. math.* 10.220–23)

> For – to start at a slightly earlier point so as to make the account intelligible – it is a universal principle that of things that exist some are *per se* while others are viewed as belonging to *per se* things. What exists *per se* are things like the substances, namely body and void, while what are viewed as belonging to *per se* things are what they call "attributes." Of these attributes some are inseparable from the things of which they are attributes, others are of a kind to be separated from them. Inseparable from the things of which they are attributes are, for example, resistance from body and non-resistance from void. For body is inconceivable without resistance, and so is void without non-resistance: these are permanent attributes of each – resisting of the one, yielding of the other. Not inseparable from the things of which they are attributes are, for example, motion and rest. For compound bodies are neither always in ceaseless motion nor always at rest, but sometimes have the attribute of motion, sometimes that of rest. (trans. Long and Sedley)

Yet neither Sextus Empiricus nor Lucretius (1.445–58), who also makes a similar distinction between permanent attributes (*coniuncta*) and accidents (*eventa*), mention color in their classifications of attributes. What did Epicurus have in mind, then, when he talked about "the nature of color" (τὴν φύσιν τοῦ χρώματος: Diog. Laërt. 10.49; cf. Philod. *sign.* 18.1)?

In his letter to Herodotus, Epicurus listed colors among the permanent attributes of objects together with sizes, shapes, and weights:

> Ἀλλὰ μὴν καὶ τὰ σχήματα καὶ τὰ χρώματα καὶ τὰ μεγέθη καὶ τὰ βάρη καὶ ὅσα ἄλλα κατηγορεῖται σώματος ὡσανεὶ συμβεβηκότα ἢ πᾶσιν ἢ τοῖς ὁρατοῖς καὶ κατὰ τὴν αἴσθησιν αὐτὴν γνωστά, οὔθ' ὡς καθ' ἑαυτάς εἰσι φύσεις δοξαστέον – οὐ γὰρ δυνατὸν ἐπινοῆσαι τοῦτο – οὔτε ὅλως ὡς οὐκ εἰσίν, οὔθ' ὡς ἕτερ' ἄττα προσυπάρχοντα τούτῳ ἀσώματα, οὔθ' ὡς μόρια τούτου ἀλλ' ὡς τὸ ὅλον σῶμα καθόλου μὲν ⟨ἐκ⟩ τούτων πάντων τὴν ἑαυτοῦ φύσιν ἔχον ἀΐδιον, οὐχ οἷον δὲ εἶναι συμπεφορημένον – ὥσπερ ὅταν ἐξ αὐτῶν τῶν ὄγκων μεῖζον ἄθροισμα συστῇ ἤτοι τῶν πρώτων ἢ τῶν τοῦ ὅλου μεγεθῶν τοῦδέ τινος ἐλαττόνων, – ἀλλὰ μόνον, ὡς λέγω, ἐκ τούτων ἁπάντων τὴν ἑαυτοῦ φύσιν ἔχον ἀΐδιον. (Epicur. *ep. Hdt.* 68–69 [= Diog. Laërt. 10.68–69])

> Now as for the shapes, colors, sizes, weights, and other things predicated of a body as permanent attributes – belonging either to all bodies or to those which are visible, and knowable in themselves through sensation – we must not hold that they are *per se* substances: that is inconceivable. Nor, at all, that they are non-existent. Nor that they are some distinct incorporeal things accruing to the body. Nor that they are parts of it; but that the whole body cannot have its own permanent nature consisting *entirely* of the sum total of them, in an amalgamation like that when a larger aggregate is composed directly of particles, either primary ones or magnitudes smaller than such-and-such a whole, but that it is only in the way I am describing that it has its own permanent nature consisting of the sum total of them. (trans. Long and Sedley)

But although it makes perfect sense that size, shape, and weight are permanent attributes of an object, for they are the attributes of its underlying atoms, the classification of color as a permanent attribute is somewhat surprising, since atoms are colorless. It is reasonable, though, to suggest that Epicurus thought of color as among the permanent attributes of objects, because it is inconceivable for a visible object not to be colored. That is to say, perhaps it is color, i.e., being colored in some way or other, which is a permanent attribute of objects. But it is obvious that an object may not always have the same particular color, as hair turns grey or leaves turn yellow. For this reason, colors should be counted among the transient attributes of objects, since the particular color which an object has is clearly separable from the object of which it is an attribute.

In any case, whether they are permanent or transient attributes of objects, colors are not, according to Epicurus, unreal and subjective constructions of our mind. They do exist, and when we say that an object has a particular color, we say something about the object itself. Interestingly enough, our ancient sources suggest that Epicurus disagreed on this point with Democritus, who is famously reported to have stated that colors exist "by convention" (*nomōi*: e.g., DK 68 B9; B125). In fact, it is Plutarch, in his work *Adversus Colotem*, who underlines the difference of opinion between Democritus and Epicurus, and then accuses Epicurus of inconsistency:

> ἐγκλητέος οὖν ὁ Δημόκριτος οὐχὶ τὰ συμβαίνοντα ταῖς ἀρχαῖς ὁμολογῶν ἀλλὰ λαμβάνων ἀρχὰς αἷς ταῦτα συμβέβηκεν. ἔδει γὰρ ἀμετάβλητα μὴ θέσθαι τὰ πρῶτα, θέμενον δὲ δὴ συνορᾶν ὅτι ποιότητος οἴχεται πάσης γένεσις. ἀρνεῖσθαι δὲ συνορῶντα τὴν ἀτοπίαν ἀναισχυντότατον· ⟨ὥστ' ἀναισχυντότατον⟩, ὃ Ἐπίκουρός φησιν, ἀρχὰς μὲν ὑποτίθεσθαι τὰς αὐτάς, οὐ λέγειν δέ 'νόμῳ χροιὴν καὶ γλυκὺ καὶ λευκόν' καὶ τὰς ἄλλας ποιότητας. (Plut. *adv. Colot.* 1111A-B)

> So Democritus should be charged, not with drawing conclusions which agree with his principles, but with choosing principles from which those conclusions follow. For he ought not to have posited changeless primary substances, but having posited them he should have seen that all qualities disappear. But it is altogether shameless to see the absurdity and then deny it. So Epicurus is being altogether shameless when he says that he posits the same principles, but does not say that color is by convention, and sweet and bitter and the other qualities. (trans. Taylor)

In other words, Plutarch believes that Democritus was wrong to claim that only atoms and the void exist, but Epicurus in addition was also inconsistent; on the one hand, he followed Democritus' principle that only atoms and the void exist and, on the other, he claimed that colors exist, too. Is Plutarch's criticism of Epicurus' inconsistency fair? As I explained before, it seems that on Epicurus' view there are not just atoms and the void; on the basis of them there are also ordinary objects with their own nature, and it is part of their nature that they are colored. For it was crucially important for Epicurus to defend the veridicality of our sense perceptions. He did not want to say that something has a particular color only by convention; rather, he insisted that the objects we perceive actually have the color we perceive them to have.[7]

So, if Epicurus believed that colors are not subjective constructions of our mind and our impressions of colors are always true, what is it exactly that we actually perceive when we perceive an object as having a particular color? In his second argument in favor of the thesis that atoms have no color, Lucretius gives us another example of the phenomenon of the changing color of an object, namely the changing color of the sea. He claims that great winds change the color of the sea from black to white, because they change the "arrangement" (*taxis*) and the "position" (*thesis*) of the atoms at the surface of the sea:

> Praeterea si nulla coloris principiis est
> reddita natura et uariis sunt praedita formis,
> e quibus omne genus gignunt uariantque colores
> propterea, magni quod refert semina quaeque
> cum quibus et quali positura contineantur
> et quos inter se dent motus accipiantque,
> perfacile extemplo rationem reddere possis
> cur ea quae nigro fuerint paulo ante colore,
> marmoreo fieri possint candore repente;
> ut mare, cum magni commorunt aequora uenti,
> uertitur in canos candenti marmore fluctus.
> (Lucr. 2.757–67)

> Moreover, if the elements are destitute of any color, but are endowed with various shapes by means of which they produce every kind and variation of color, it being of great consequence in what groupings and positions all the seeds are combined, and what motions they reciprocally impart and receive, you can easily explain at once why something that was dark a little while ago can suddenly become as white as marble: as the sea, when mighty winds have disturbed its smooth surface, is turned into white-flecked billows of gleaming marble. (trans. Smith)

And there is more evidence indicating that Epicurus believed that the color of an object is generated not only by the attributes of its underlying atoms, i.e., by their character, but also by their arrangement and their position at the surface

---

[7] On this difference between Democritus and Epicurus, see David N. Sedley, "Epicurean Anti-Reductionism," in Jonathan Barnes and Mario Mignucci (eds.), *Matter and Metaphysics* (Naples: Bibliopolis, 1988), 295–327; David Furley, "Democritus and Epicurus on Sensible Qualities," in Jacques Brunschwig and Martha Nussbaum (eds.), *Passions and Perceptions: Studies in Hellenistic Philosophy of Mind* (Cambridge: Cambridge University Press, 1993), 72–94; Timothy O'Keefe, "The Ontological Status of Sensible Qualities for Democritus and Epicurus," *Ancient Philosophy* 17 (1997): 119–34; Luca Castagnoli, "Democritus and Epicurus on Sensible Qualities in Plutarch's *Against Colotes* 3–9," *Aitia* 3 (2013), http://aitia.revues.org/622; DOI: 10.4000/aitia.622.

of the object, i.e., by their configuration (e.g., Diog. Laërt. 10.44). Moreover, it seems that he believed, at least on Plutarch's account, that the colors of objects vary in accordance with the arrangement and the position of their atoms "in relation to the sight" (πρὸς τὴν ὄψιν), i.e., relative to a perceiver:

> αὐτὸς γὰρ οὖν ὁ Ἐπίκουρος ἐν τῷ δευτέρῳ τῶν πρὸς Θεόφραστον οὐκ εἶναι λέγων τὰ χρώματα συμφυῆ τοῖς σώμασιν, ἀλλὰ γεννᾶσθαι κατὰ ποιάς τινας τάξεις καὶ θέσεις πρὸς τὴν ὄψιν, οὐ μᾶλλόν φησι κατὰ τοῦτον τὸν λόγον ἀχρωμάτιστον σῶμα εἶναι ἢ χρῶμα ἔχον. ἀνωτέρω δὲ κατὰ λέξιν ταῦτα γέγραφεν· 'ἀλλὰ καὶ χωρὶς τούτου τοῦ μέρους οὐκ οἶδ' ὅπως δεῖ τὰ ἐν τῷ σκότει ταῦτ' ὄντα φῆσαι χρώματα ἔχειν.' καίτοι πολλάκις ἀέρος ὁμοίως σκοτώδους περικεχυμένου οἱ μὲν αἰσθάνονται χρωμάτων διαφορᾶς οἱ δ' οὐκ αἰσθάνονται δι' ἀμβλύτητα τῆς ὄψεως· ἔτι δ' εἰσελθόντες εἰς σκοτεινὸν οἶκον οὐδεμίαν ὄψιν χρώματος ὁρῶμεν, ἀναμείναντες δὲ μικρὸν ὁρῶμεν. (Plut. adv. Colot. 1110C-D)

> Now Epicurus himself, in the second book of *Reply to Theophrastus*, says that colors are not inherent in the nature of bodies but are generated according to certain arrangements and positions in relation to the sight; and with this statement he concedes that a body is no more uncolored than colored. And he has already written earlier on (and I quote): "But also, quite apart from this section, I don't know how one should say that these things have color when they are in the dark." Yet often, when the surrounding air is of the same degree of darkness, some perceive a distinction of color and others don't, because of weakness of sight; moreover, on entering a dark house we see no sight of color, but after a short interval we do. (trans. Furley)

But why would the perceiver have anything to do with the color of an object, if this depends on the atomic surface structure of the object? To explain perception, Epicurus stated, following in this Democritus, that very fine effluences of atoms, in Greek the so-called *eidōla*, and in Latin *simulacra*, are constantly emitted from the surface of objects, reach our sensory apparatus with enormous velocity, and affect our sense organs.[8] Note here that in normal conditions the *eidōla* emitted from the object preserve the arrangement and the position of the atoms as found at its surface. Thus, when we say that we see the color of an object, Epicurus took this to mean that we must be referring both to the character and to the configuration of the atoms transmitted by the *eidōla*. Indeed, Epicurus seems to have claimed, in contrast to Democritus, that the *eidōla* are colored as much as the objects themselves:

---

[8] On the Atomists' theory of perception, see for instance: Walter Burkert, "Air-Imprints or *Eidola*: Democritus' Aetiology of Vision," *Illinois Classical Studies* 2 (1977): 97–109; Richard W. Baldes, "Democritus on the Nature and Perception of 'Black' and 'White,'" *Phronesis* 23 (1978): 87–100; Kelli C. Rudolph, *Reading Theophrastus: A Reconstruction of Democritus' Physics of Perception*, Ph.D. thesis, University of Cambridge, 2008; Kelli C. Rudolph, "Democritus' Perspectival Theory of Vision," *Journal of Hellenic Studies* 131 (2011): 67–83.

Δεῖ δὲ καὶ νομίζειν ἐπεισιόντος τινὸς ἀπὸ τῶν ἔξωθεν τὰς μορφὰς ὁρᾶν ἡμᾶς καὶ διανοεῖσθαι· οὐ γὰρ ἂν ἐναποσφραγίσαιτο τὰ ἔξω τὴν ἑαυτῶν φύσιν τοῦ τε χρώματος καὶ τῆς μορφῆς διὰ τοῦ ἀέρος τοῦ μεταξὺ ἡμῶν τε κἀκείνων, οὐδὲ διὰ τῶν ἀκτίνων ἢ ὡνδήποτε ῥευμάτων ἀφ' ἡμῶν πρὸς ἐκεῖνα παραγινομένων, οὕτως ὡς τύπων τινῶν ἐπεισιόντων ἡμῖν ἀπὸ τῶν πραγμάτων ὁμοχρόων τε καὶ ὁμοιομόρφων κατὰ τὸ ἐναρμόττον μέγεθος εἰς τὴν ὄψιν ἢ τὴν διάνοιαν ὠκέως ταῖς φοραῖς χρωμένων, εἶτα διὰ ταύτην τὴν αἰτίαν τοῦ ἑνὸς καὶ συνεχοῦς τὴν φαντασίαν ἀποδιδόντων καὶ τὴν συμπάθειαν ἀπὸ τοῦ ὑποκειμένου σῳζόντων κατὰ τὸν ἐκεῖθεν σύμμετρον ἐπερεισμὸν ἐκ τῆς κατὰ βάθος ἐν τῷ στερεμνίῳ τῶν ἀτόμων πάλσεως. (Epicur. *ep. Hdt.* 49–50 = Diog. Laërt. 10.49–50)

And we must indeed suppose that it is on the impingement of something from outside that we see and think of shapes. For external objects would not imprint their own nature, of both color and shape, by means of the air between us and them, or by means of rays or of effluences passing from us to them, as effectively as they can through certain delineations penetrating us from objects, sharing their color and shape, of a size to fit into our vision or thought, and traveling at high speed, with the result that their unity and continuity then results in the impression, and preserves their co-affection all the way from the object because of their uniform bombardment from it, resulting from the vibration of the atoms deep in the solid body. (trans. Long and Sedley)

So, under normal conditions, the *eidōla* provide us with the desired information about the color of objects. Even the fact that different people sometimes think that they perceive the same object as having a different color should not surprise us. For this can be explained, according to Epicurus, in the following way. The *eidōla* which are emitted from the surface of an object initially preserve the configuration of its atoms. However, the *eidōla* which actually affect the perceiver do not always preserve this configuration. Sometimes it happens that the configuration of the atoms in the *eidōla* changes due to external factors; for instance, when the perceiver sees an object from a distance, the configuration of the atoms may have been affected by the journey over this distance, and hence the object's color looks paler (Lucr. 4.353–63). Sometimes it happens that the configuration of the atoms in the *eidōla* changes due to the physical constitution of the perceiver's eyes; for instance, the perceiver may suffer from jaundice (Lucr. 4.332–36), or the perceiver's eyes may be of a kind that does not allow certain sizes of atoms to enter them (Plut. *adv. Colot.* 1109C-E). Hence, the perceiver who does not perceive the object under normal conditions, having an impression rather like the impression obtained from an object with a different color, wrongly takes this impression to be the impression of an object with a different color. But, in fact, it is the true impression of the color of the same object as perceived by a perceiver from a distance or with eyes of a certain kind.

To summarize: Epicurus claimed that colors are not attributes of atoms, nor permanent attributes of objects which are aggregates of atoms. Rather, colors are transient attributes of objects in the sense that they are transient dispositions of objects in virtue of their atomic surface structures to affect perceivers in ways that depend on the conditions under which the *eidōla* of objects reach the perceivers.

But let us return to the special case of the pigeon's neck. Epicurus insisted that both the impression of the pigeon's neck as being red and the impression of it as being green-blue are true. We might think, following his general account of color, that the reason why they are both true is that, as the pigeon turns its neck, the arrangement and the position of the atoms at the surface of the neck also change. However, why would the mere fact that the pigeon turns its neck result in the change of the configuration of its atoms relative to each other? Moreover, if this were the case, why would Lucretius stress that the change in color has something to do with the angle of incidence of the light, i.e., whether the light falls at a right or at an oblique angle?

It is not at all clear from Lucretius' text or, for that matter, from our other ancient sources what role light plays in the Epicurean theory of visual perception. We are only told that light, too, consists of atoms (Lucr. 2.150–56; 4.185–90), and that without it we would not of course be able to perceive objects and their colors (Philod. *sign.* 17.37–18.10; Lucr. 2.795–98; Plut. *adv. Colot.* 1110C-D). It is unclear, though, whether the atoms of light change the atomic structure of the surface of the pigeon's neck, whether they affect the *eidōla* that are emitted from the pigeon's neck and reach the perceiver, or whether they change the atomic structure of the eyes of the perceiver (Lucr. 4.337–52). But even if the exact function of light in visual perception remains a puzzle, there should be no doubt that Epicurus explained the changing color of the pigeon's neck, or of the peacock's tail, as a result of how the light falls on it. That is to say, Epicurus believed that, when the pigeon turns its neck, its color changes, because the light now falls on the pigeon's neck at a different angle. Hence, when we first see the pigeon's neck as red and immediately afterwards we see it as green-blue, both impressions are true. For the color of the pigeon's neck has changed, given its different orientation relative to the source of light.

The same example of the phenomenon of changing colors, namely, the changing color of the pigeon's neck, or of the peacock's tail, puzzled Stoic philosophers too. And they also tried to explain it in such a way as to be able to defend the possibility of certain knowledge against the Sceptics' attacks. However, if we follow what Lucullus implies in the second passage from Cicero's *Academica* (2.19), which I quoted at the beginning, the Stoics seem to have claimed, in contrast to the Epicureans, that there is only one true impression of the color of the pigeon's neck. For the pigeon's neck has just one color; and, according to the Stoics, we can reliably perceive this one color. But in order to better understand why the

Stoics claimed that the pigeon's neck, or the peacock's tail, has just one color, although it appears to have different colors when it turns around, we need to closely study the few things our ancient sources report about the Stoic theory of the nature and the perception of color. Unfortunately, though, the relevant material is even more meagre and scattered than that concerning the Epicurean views on color.

Among the texts on Stoic physics we find two definitions of color, which are both attributed to the founder of the Stoic school: Zeno is said to have defined colors as "πρώτους σχηματισμούς τῆς ὕλης" (Aët. 1.15.6) and as "ἐπίχρωσιν τῆς ὕλης" (Ps.-Gal. *de hist. philos.* 616.2–3 Diels); these definitions have been understood to mean that colors are, respectively, "the primary shapes of matter" and "the surface coloration of matter."[9] Nevertheless, the term "*schēmatismoi*" does not have to mean "shapes" in the sense of geometrical figures, as the more common term "*schēmata*" often does. It can also have a more general meaning, namely it can mean the "characteristics" or "features" which something exhibits. In fact, the other Stoic text in which we find the term "*schēmatismoi*," a Chrysippean fragment on rhetoric preserved in Plutarch's *De Stoicorum repugnantiis* (1047A), uses it to refer to the features or expressions of the face and the hands of the rhetorician.

Most interestingly, again in Plutarch's *De Stoicorum repugnantiis*, but this time in the context of Stoic physics, we find the verb "*schēmatizein*" used alongside the verb "*eidopoiein*"; and in this particular text, both terms are used together meaning "to give form and shape" to matter, in the general sense of giving matter its characteristic features or qualities:

> καίτοι πανταχοῦ τὴν ὕλην ἀργὸν ἐξ ἑαυτῆς καὶ ἀκίνητον ὑποκεῖσθαι ταῖς ποιότησιν ἀποφαίνουσι, τὰς δὲ ποιότητας πνεύματ' οὔσας καὶ τόνους ἀερώδεις, οἷς ἂν ἐγγένωνται μέρεσι τῆς ὕλης, εἰδοποιεῖν ἕκαστα καὶ σχηματίζειν. (Plut. *Stoic. repug.* 1054B)
>
> Yet they [the Stoics] maintain that matter, which is of itself inert and motionless, is everywhere the substrate for qualities, and that qualities are breaths and aeriform tensions which give form and shape to the parts of matter in which they come to be. (trans. Long and Sedley)

In addition, there are a lot of texts in the Aristotelian commentaries, in which matter as such is characterized as "*aschēmatistos*," but also as "*apoios*," i.e., unqualified, or as "*aneideos*" and "*amorphos*," i.e., without form (e.g., Alex. Aphr.

---

9 David Hahm, "Early Hellenistic Theories of Vision and the Perception of Colour," in Peter K. Machamer and Robert G. Turnbull (eds.), *Studies in Perception: Interrelations in the History of Philosophy and Science* (Columbus: Ohio State University Press, 1978), 60–95, at 85.

de an. 3.28–4.4; *quaest.* 49.30–33; 52.21–23). Hence, when colors are defined by Zeno as "πρῶτοι σχηματισμοὶ τῆς ὕλης" (the first characteristics of matter), what he must refer to is the fact that matter as such is without any shape, in the sense that it does not have any form or quality; and colors are said to be the first characteristics matter receives, or at least among its first characteristics.

But what does it actually mean that colors are the first characteristics of matter? In a much discussed passage from Diogenes Laertius, in which he presents basic Stoic doctrines in physics, matter is again described as "unqualified" (*apoios*) and "without form" (*amorphos*) (cf. also Posidon. Phil. fr. 92 Edelstein-Kidd); on the other hand, the four elements, i.e., fire, air, earth, and water, are said to be "endowed with form" (*memorphōsthai*):

> Δοκεῖ δ' αὐτοῖς ἀρχὰς εἶναι τῶν ὅλων δύο, τὸ ποιοῦν καὶ τὸ πάσχον. τὸ μὲν οὖν πάσχον εἶναι τὴν ἄποιον οὐσίαν τὴν ὕλην, τὸ δὲ ποιοῦν τὸν ἐν αὐτῇ λόγον τὸν θεόν· τοῦτον γὰρ ἀίδιον ὄντα διὰ πάσης αὐτῆς δημιουργεῖν ἕκαστα. ... διαφέρειν δέ φασιν ἀρχὰς καὶ στοιχεῖα· τὰς μὲν γὰρ εἶναι ἀγενήτους ⟨καὶ⟩ ἀφθάρτους, τὰ δὲ στοιχεῖα κατὰ τὴν ἐκπύρωσιν φθείρεσθαι. ἀλλὰ καὶ σώματα [MSS: ἀσωμάτους Suda] εἶναι τὰς ἀρχὰς καὶ ἀμόρφους, τὰ δὲ μεμορφῶσθαι. (Diog. Laërt. 7.134)

> They [the Stoics] think that there are two principles of the universe, that which acts and that which is acted upon. That which is acted upon is unqualified substance, i.e., matter; that which acts is the reason in it, i.e., god. For this, since it is everlasting, constructs every single thing throughout all matter. ... They [the Stoics] say there is a difference between principles and elements: the former are ungenerated and indestructible, whereas the elements pass away at the conflagration. The principles are also bodies ["incorporeal," in the version preserved in the Suda] and without form, but the elements are endowed with form. (trans. Long and Sedley)

Moreover, Plutarch seems to suggest that on the Stoic view the four elements are "primarily" (*prōtōs*) characterized not only by whether they are hot or cold, but also by their color; fire, for instance, is hot and bright, while air is cold and dark (*Stoic. repug.* 1053E; *prim. frig.* 952C).

The doctrine that each of the four elements should be associated with a specific color can be found in the writings of ancient philosophers already before the Stoics. In particular, there is plenty of textual evidence indicating that, according to Empedocles, the color of an object depends on the mixture of the colors of its constitutive elements, though it is debatable whether in his case all four elements are indeed colored or only some of them (e.g., DK 31 B23; B71; Theophr. *sens.* 59).[10] In Aristotle's works, too, we find passages in which the four elements

---

10 Cf. Katerina Ierodiakonou, "Empedocles on Colour and Colour Vision," *Oxford Studies in Ancient Philosophy* 29 (2005): 1–37.

are characterized by specific colors, so that the color of objects is generated by the mixture of the colors of their constitutive elements; for instance, in the *De generatione animalium* (e.g., 735b33–37; 779b28–33; 786a2–7; 786a12–21) and in *Meteorologica* (e.g., 374a1–3), the elements of fire and air are said to be hot and white, whereas the elements of earth and water are cold and black.[11]

Similarly, the Stoics, having posited that the four basic elements are colored, came to argue that the color of an object depends on the mixture of the elements which constitute this object. There seems to be, however, an important difference between the color of the elements and the color of ordinary objects; in the case of the elements color should be regarded as a permanent, essential attribute, just like the opposites hot and cold are, whereas in the case of ordinary objects color is rather a transient, accidental attribute. For, although it is true that every object cannot but be colored, since it is the mixture of colored elements, it is not part of the essence of a human being, for instance, to have a specific color. But if the color of ordinary objects characterizes them only accidentally, is there a category, like the Aristotelian category of quality, under which the Stoics classified colors?

The so-called Stoic doctrine of the four categories, or the four "primary genera" (*prōta genē*), is quite difficult to understand, even at a basic level, and hence an issue of considerable controversy.[12] To get at least a rough idea of what the four Stoic categories represent, let me list them and then give examples which the Stoics themselves most probably used in order to illustrate them. The four categories are the following: "substances" or "substrates" (*hupokeimena*), "qualified" (*poia*), subdivided into "commonly qualified" (*koinōs poia*) and "peculiarly qualified" (*idiōs poia*), "somehow disposed" (*pōs echonta*), and "relatively somehow disposed" (*pros ti pōs echonta*):

> Οἱ δέ γε Στωικοὶ εἰς ἐλάττονα συστέλλειν ἀξιοῦσιν τὸν τῶν πρώτων γενῶν ἀριθμὸν καί τινα ἐν τοῖς ἐλάττοσιν ὑπηλλαγμένα παραλαμβάνουσιν. ποιοῦνται γὰρ τὴν τομὴν εἰς τέσσαρα, εἰς ὑποκείμενα καὶ ποιὰ καὶ πῶς ἔχοντα καὶ πρός τί πως ἔχοντα. (Simp. *in cat.* 66.32–67.2)

---

[11] There are, of course, problems as to how such a doctrine fits in with the Aristotelian account of color in the *De anima* (418a31–b1) and in the *De sensu* (439b11–12), but this is not what I am concerned with here.

[12] Cf. Andreas Graeser, "The Stoic Categories," in Jacques Brunschwig (ed.), *Les Stoiciens et leur logique* (Paris: Vrin, 1978), 199–222; Stephen Menn, "The Stoic Theory of Categories," *Oxford Studies in Ancient Philosophy* 17 (1999): 215–47; Jacques Brunschwig, "Stoic Metaphysics," in Brad Inwood (ed.), *The Cambridge Companion to the Stoics* (Cambridge: Cambridge University Press, 2003), 206–32.

> The Stoics see fit to reduce the number of the primary genera, and others they take over with minor changes. For they make their division a fourfold one, into substrates, the qualified, the disposed, and the relatively disposed. (trans. Long and Sedley)

It seems, then, that the Stoic categories, leaving aside substance which is generally understood as matter, are a classification of characterizations of objects: either we characterize an object as a certain matter being qualified in a certain way, for instance as a human being (Simp. *in cat.* 212.26: "the grammarian"), and this is the commonly qualified, or as Socrates, and this is the peculiarly qualified; or as a certain matter being somehow disposed, for instance as running (Sen. *ep.* 113.2: "virtue" here being the soul disposed in a certain way); or finally, as a certain matter being relatively somehow disposed, for instance as being the man on the right or the son of somebody (Simp. *in cat.* 166.23–26).

Hence, though quality in itself is not a Stoic category, it is clear that form or quality when characterizing matter makes it formed or qualified, and then matter can be thought of in terms of the second category, i.e., the qualified. For instance, when the white and the hot characterize matter they form the element of fire. In this case, the quality whiteness makes the element fire what it is, and thus whiteness is regarded as an essential quality. On the other hand, in the case of whiteness characterizing Socrates, the Stoics would not have regarded it as a quality which makes Socrates what he is. For the Stoics seem to have claimed, at least according to Simplicius (*in cat.* 214.34–35), that matter is regarded as qualified, when its constituents are organized in such a way that their cooperation contributes towards the fulfillment of a single function; but the quality whiteness, being transient and accidental in the case of Socrates, does not play any role in the fulfillment of his function as a human being.

Should the quality whiteness, then, and in general color, be understood on the Stoic view both as what makes something qualified, but also as what makes an already qualified individual something disposed in a certain way? It is not easy to settle this issue, especially because of the lack of relevant evidence. Still, there is something more we can say with regard to the Stoic account of the nature of color. Qualities of bodies are, according to the Stoics, themselves corporeal, and hence colors are corporeal, too (e.g., Gal. *qualit. incorp.* 19.467, 473, 483 K.; Plut. *comm. not.* 1085E; Simp. *in cat.* 217.32–33). They are corporeal, because incorporeals cannot affect anything, since it is only a body which can affect another body, and colors do affect our sense organs. Furthermore, they are corporeal, because incorporeals cannot be in anything, since being in something requires contact, and colors are in things. Therefore, the Stoics claimed, colors are qualities or "tenors" (*hexeis*) that are corporeal; in particular, they are "breaths" (*pneumata*) and "aeriform tensions" (*tonoi aerōdeis*), which sustain the bodies and give them their particular characteristics (cf. also Plut. *Stoic. repug.* 1054B):

> πάλιν ἐν τοῖς περὶ Ἕξεων οὐδὲν ἄλλο τὰς ἕξεις πλὴν ἀέρας εἶναί φησιν· 'ὑπὸ τούτων γὰρ συνέχεται τὰ σώματα· καὶ τοῦ ποιὸν ἕκαστον εἶναι τῶν ἕξει συνεχομένων αἴτιος ὁ συνέχων ἀήρ ἐστιν, ὃν σκληρότητα μὲν ἐν σιδήρῳ πυκνότητα δ' ἐν λίθῳ λευκότητα δ' ἐν ἀργύρῳ καλοῦσι.' (Plut. *Stoic. repug.* 1053F)
>
> In his books *On tenors* he [Chrysippus] again says that tenors are nothing but currents of air: "It is by these that bodies are sustained. The sustaining air is responsible for the quality of each of the bodies which are sustained by tenor; in iron this quality is called hardness, in stone density, and in silver whiteness." (trans. Long and Sedley)

To summarize: colors, according to the Stoics, are intrinsic qualities or attributes of objects which may be either essential, as in the case of the four elements, or accidental, as in the case of the ordinary objects we perceive. An ordinary object has the color it has because of the mixture of elements which are its constituents; and as to the elements themselves, they have the colors they have in virtue of the breaths, or aeriform tensions, permeating them.

But what did the Stoics have to say about the way we perceive objects and their colors? Most of our sources suggest that, according to the Stoics, we see when the air between our eye and an object is pricked by the *pneuma* emitted from the eye, and is thus stretched into the shape of a cone with its base contiguous with the object. As soon as the cone of stretched air undergoes a change of state induced by the state of the object, for example the particular color it has, this state is communicated to the eye, and the state of the eye is registered by the rational part of the soul:[13]

> Χρύσιππος κατὰ συνέντασιν [τὰ ὄντα] τοῦ μεταξὺ ἀέρος ὁρᾶν ἡμᾶς, νυγέντος μὲν ὑπὸ τοῦ ὀπτικοῦ πνεύματος, ὅπερ ἀπὸ τοῦ ἡγεμονικοῦ μέχρι τῆς κόρης διήκει, κατὰ δὲ τὴν πρὸς τὸν περικείμενον ἀέρα ἐπιβολὴν ἐντείνοντος αὐτὸν κωνοειδῶς, ὅταν ᾖ ὁμογενὴς ὁ ἀήρ· προχέονται δ' ἐκ τῆς ὄψεως ἀκτῖνες πύριναι, οὐχὶ μέλαιναι καὶ ὁμιχλώδεις· διόπερ ὁρατὸν εἶναι τὸ σκότος. (Aët. 4.15.3)
>
> Chrysippus says we see by virtue of the stretching of the intervening air. This air is pricked by the visual *pneuma*, which advances from the principal part [of the soul] to the pupil. Upon its impact against the surrounding air the visual *pneuma* stretches the air conically,

---

[13] On the Stoic theory of perception, see David Hahm, "Early Hellenistic Theories of Vision"; Heinrich von Staden, "The Stoic Theory of Perception and Its 'Platonic' Critics," in Machamer and Turnbull, *Studies in Perception*, 96–136; Håvard Løkke, "The Stoics on Sense-Perception," in Simo Knuuttila and Pekka Kärkkäinen (eds.), *Theories of Perception in Medieval and Early Modern Philosophy*, Studies in the History of Philosophy of Mind 6 (Dordrecht: Springer, 2008), 35–46.

whenever the air is homogeneous. Fiery rays, not black misty ones, are poured forth from the sight. Hence darkness is visible. (trans. Hahm)

Hence, the process of visual perception is explained by the Stoics in parallel with touch. For it involves a continuous substance between our eye and the perceived object, a visual body which often is compared with a walking stick, or with the net and the rod through which a shock passes from the electric eel to the hands of the fisherman (e.g., Alex. Aphr. *mantissa* 130.14–17; Gal. *de plac. Hipp. et Plat.* 5.642 K. [= 7.7.20 De Lacy]; Chalc. *in Tim.* 237):

> ὁρᾶν δὲ τοῦ μεταξὺ τῆς ὁράσεως καὶ τοῦ ὑποκειμένου φωτὸς ἐντεινομένου κωνοειδῶς, καθά φησι Χρύσιππος ἐν δευτέρῳ τῶν Φυσικῶν καὶ Ἀπολλόδωρος. γίνεσθαι μέντοι τὸ κωνοειδὲς τοῦ ἀέρος πρὸς τῇ ὄψει, τὴν δὲ βάσιν πρὸς τῷ ὁρωμένῳ· ὡς διὰ βακτηρίας οὖν τοῦ ταθέντος ἀέρος τὸ βλεπόμενον ἀναγγέλλεσθαι. (Diog. Laërt. 7.157)
>
> Seeing takes place when the light between the visual faculty and the object is stretched into the shape of a cone, as Chrysippus says in the second book of his *Physics* and Apollodorus, too. The air adjacent to the pupil forms the tip of the cone with its base next to the visual object. What is seen is reported by means of the stretched air, as by a walking-stick. (trans. Long and Sedley)

There are, of course, problems in grasping this Stoic doctrine in all its details. For instance, it is not clearly specified what the *pneuma* emitted from the eye exactly does, when it is said that it stretches or intensifies the air between the eye and the object; presumably, it turns it into a cone of pneumatic air that is somehow receptive to the color of the object. It thus seems that the Stoic theory constitutes a further development of the part of Plato's theory of visual perception, which involves a body formed in the air by the sunlight and the particles of fire emitted from the eyes of the perceiver.[14] In the case of color perception, in particular, the Stoics described the change of state of the stretched air in terms which suggest that it should be thought of as a change in density; namely, they used the terms "confuse" (*sunchutikon esti*) and "diffuse" (*diachei*), which remind us again of Plato's corresponding use of the terms "compression" (*sunkrisis*) and "division" (*diakrisis*):

> ἐπεὶ τὸ πῦρ θερμὸν ἅμα καὶ λαμπρόν ἐστι, δεῖ τὴν ἀντικειμένην τῷ πυρὶ φύσιν ψυχράν τ' εἶναι καὶ σκοτεινήν· ἀντίκειται γὰρ ὡς τῷ λαμπρῷ τὸ ζοφερὸν οὕτω τῷ θερμῷ τὸ ψυχρόν· ἔστι γὰρ ὡς ὄψεως τὸ σκοτεινὸν οὕτω τὸ ψυχρὸν ἁφῆς συγχυτικόν, ἡ δὲ θερμότης διαχεῖ

---

[14] On Plato's theory of color perception, see for instance: Katerina Ierodiakonou, "Plato's Theory of Colours in the *Timaeus*," *Rhizai* II.2 (2005): 219–33.

τὴν αἴσθησιν τοῦ ἁπτομένου καθάπερ ἡ λαμπρότης τοῦ ὁρῶντος. τὸ ἄρα πρώτως σκοτεινὸν ἐν τῇ φύσει πρώτως καὶ ψυχρόν ἐστιν. (Plut. *prim. frig.* 948D-E)

> Since fire is simultaneously hot and bright, the nature opposite to fire must be both cold and dark; for as dark is the opposite of bright, so is cold of hot, and as dark confuses sight, so cold has the same effect on touch. But heat diffuses the sense of the person touching, just as brightness does that of the person seeing. Therefore what is primarily dark in its nature is also primarily cold. (trans. Long and Sedley)

Assuming, therefore, that the Stoics thought of colors as intrinsic attributes of objects, and of color perception as the result of the stretched air between the eyes and the object getting more or less dense, how would they have explained the phenomenon of the changing color of the pigeon's neck, or of the peacock's tail? According to the Stoics, the pigeon's neck has just one color, namely, the color of the mixture of elements which constitute it. In normal conditions this color affects the stretched air between the object and the eyes in such a way that it reaches our eyes, and the information about the kind of color it is proceeds from there to the rational part of our soul. As to the supposed change of color, the Stoics seem to have claimed that it is only apparent. Seneca, for instance, explains how the color of the pigeon's neck appears changing as a result of the smoothness and moisture of its surface which functions as a mirror; it thus can come to reflect other colors from the close environment:

> quod apparet [sc. arcus] a sole fieri; quod apparet leue quiddam esse debere et simile speculo quod solem repercutiat; deinde quod apparet non fieri ullum colorem sed speciem falsi coloris, qualem, ut dixi, columbarum ceruix et sumit et ponit, utcumque deflectitur. hoc autem et in speculo est, cui nullus inditur color, sed simulatio quaedam coloris alieni. (Sen. *nat. quaest.* 1.7.2)
>
> Obviously, a rainbow is caused by the sun. Obviously there must be something smooth and like a mirror which may reflect the sun. Finally, it is obvious that no actual color is formed but only the appearance of a counterfeit color, the sort that the neck of a pigeon alternately takes on or puts aside whenever it changes position, as I have already said. This also is the case in a mirror, which assumes no color but only a kind of copy of the color of something else. (trans. Corcoran)

In fact, the earlier passage to which Seneca alludes here suggests that the light, and in particular its angle of incidence, plays an important role in the way the pigeon's, or the peacock's, neck reflects as a mirror the colors from its close environment:

> et uariis coloribus pauorum ceruix, quotiens aliquo deflectitur, nitet. numquid ergo specula dicemus eiusmodi plumas, quarum omnis inclinatio in colores nouos transit? Non minus nubes diuersam naturam speculis habent quam aues quas rettuli et chamaeleontes et reli-

qua animalia quorum color aut ex ipsis mutatur, cum ira uel cupidine incensa cutem suam uariant umore suffuso, aut positione lucis, quam prout rectam vel obliquam receperunt, ita colorantur. (Sen. *nat. quaest.* 1.5.6–7)

And the neck of the peacock gleams with many colors whenever it moves one way or another. Are we going to say that feathers of this kind are mirrors, whose every tilting movement changes into new colors? Clouds have a nature no less different from a mirror than the birds I mentioned, and chameleons, and other animals whose color changes; either those which alter from within themselves their color by an infusion of moisture when they are aroused by anger or rut, or those which take on a hue from the position of the light, in so far as they receive it directly or obliquely. (trans. Corcoran)

Although it is unclear how, according to the Stoics, too, the angle of incidence of light influences our color perception, at least this passage clarifies the context in which Seneca raises the issue of the changing color of the pigeon's, or of the peacock's, neck. The discussion is about the way rainbows are produced, and in particular about Posidonius' theory that rainbows are the result of the clouds functioning as concave mirrors which give us rough representations of the sun (Sen. *nat. quaest.* 1.5.10; 1.5.13). Seneca stresses that, though he generally agrees with Posidonius' explanation, he still thinks that clouds, unlike mirrors, do have their own color, even if we cannot perceive it. Similarly, Seneca argues, the pigeon's neck functions as a mirror reflecting different colors from its environment; and it is in this sense that its changing color is apparent. The fact, however, that the changing color of the pigeon's neck in this sense is apparent should not deceive us into believing that the pigeon's neck does not have a single color due to its constitutive elements. On the contrary, the Stoics insisted, the pigeon's neck does have a single color of which we can have reliable knowledge. Therefore, on the Stoic view the Sceptics were wrong to use the phenomenon of the changing colors of the pigeon's neck and of the peacock's tail in order to argue that in these and similar cases sense perception is unreliable.

But can we actually, according to the Stoics, acquire knowledge of the single color of the pigeon's neck? If not, does this undermine the Stoic position about the reliability of sense perception? We may also wonder about this in the case of the Epicurean view: We may grant the Epicureans that all impressions are true, and hence that the impressions we have of an object's color are absolutely reliable. For they are exactly the impressions which we have of an object's color given the conditions under which we receive the *eidōla* from the object.

Does this mean, though, that we, on the basis of the impressions we have, can tell what the color of the object is? This would be the case, if we somehow could tell that the impression we have is precisely the sort of impression we would have of such and such a color under the given conditions; but it is questionable whether we always can tell this. These and further issues, like for instance the role of light in the Hellenistic theories of color perception, remain unsettled. However, I think that both the Epicureans and the Stoics undertook the task to explain, against the Sceptics' arguments, the phenomenon of the changing color of the pigeon's neck, or of the peacock's tail, with a certain ingenuity.

Bibliography

Arrighetti, Graziano. *Epicuro, Opere* (Turin: Einaudi, 1973).
Bailey, Cyril. *Lucretius, De rerum natura*, 3 vols. (Oxford: Clarendon Press, 1947).
Baldes, Richard W. "Democritus on the Nature and Perception of 'Black' and 'White,'" *Phronesis* 23 (1978): 87–100.
Brittain, Charles, trans. *Cicero, On Academic Scepticism* (Indianapolis: Hackett, 2006).
Bruns, Ivo, ed. *Alexandri Aphrodisiensis praeter commentaria scripta minora*, CAG suppl. 2.1 (Berlin: Reimer, 1887).
Bruns, Ivo, ed. *Alexandri Aphrodisiensis praeter commentaria scripta minora*, CAG suppl. 2.2 (Berlin: Reimer, 1892).
Brunschwig, Jacques. "Stoic Metaphysics," in Brad Inwood (ed.), *The Cambridge Companion to the Stoics* (Cambridge: Cambridge University Press, 2003), 206–32.
Burkert, Walter. "Air-Imprints or *Eidola*: Democritus' Aetiology of Vision," *Illinois Classical Studies* 2 (1977): 97–109.
Busse, A. *Eliae in Porphyrii isagogen et Aristotelis categorias commentaria*, CAG 18.1 (Berlin: Reimer, 1900).
Busse, A. *Olympiodori prolegomena et in categorias commentarium*, CAG 12.1 (Berlin: Reimer, 1902).
Castagnoli, Luca. "Democritus and Epicurus on Sensible Qualities in Plutarch's *Against Colotes* 3–9," *Aitia* 3 (2013), http://aitia.revues.org/622; DOI: 10.4000/aitia.622.
Corcoran, Thomas H., trans. *Seneca, Naturales quaestiones*, vol. 1 (Cambridge, Mass.: Harvard University Press, 1971).
De Lacy, Phillip. *Galeni De placitis Hippocratis et Platonis*, CMG V 4,1,2 (Berlin: Akademie-Verlag, 1978).
De Lacy, Phillip, and Estelle De Lacy. *Philodemus, On Methods of Inference: A Study in Ancient Empiricism*, 2nd ed. (Naples: Bibliopolis, 1978–1984).
Diels, Hermann. *Doxographi Graeci* (Berlin: Reimer, 1879).
Edelstein, Ludwig, and I. G. Kidd. *Posidonius, The Fragments*, vol. 1 (Cambridge: Cambridge University Press, 1972).
Everson, Stephen. "Epicurus on the Truth of the Senses," in Stephen Everson (ed.), *Epistemology* (Cambridge: Cambridge University Press, 1990), 161–83.
Furley, David. "Democritus and Epicurus on Sensible Qualities," in Jacques Brunschwig and Martha Nussbaum (eds.), *Passions and Perceptions: Studies in Hellenistic Philosophy of Mind* (Cambridge: Cambridge University Press, 1993), 72–94.
Graeser, Andreas. "The Stoic Categories," in Jacques Brunschwig (ed.), *Les Stoiciens et leur logique* (Paris: Vrin, 1978), 199–222.

Hahm, David. "Early Hellenistic Theories of Vision and the Perception of Color," in Machamer and Turnbull, *Studies in Perception*, 60–95.
Hayduck, Michael. *Sophonias, In libros Aristotelis de anima paraphrasis*, CAG 23.1 (Berlin: Reimer, 1883).
Hayduck, Michael. *Philoponus, In Aristotelis de anima libros commentaria*, CAG 15 (Berlin: Reimer, 1897).
Hense, Otto. *Seneca, Epistulae morales* (Leipzig: Teubner, 1938).
Hine, Harry M. *Seneca, Naturales quaestiones* (Leipzig: Teubner, 1996).
Hubert, Curt E. H. *Plutarchi De primo frigido*, in *Plutarchi Moralia*, vol. 5.3, 2nd ed. (Leipzig: Teubner, 1960).
Ierodiakonou, Katerina. "Empedocles on Color and Color Vision," *Oxford Studies in Ancient Philosophy* 29 (2005): 1–37.
Ierodiakonou, Katerina. "Plato's Theory of Colors in the *Timaeus*," *Rhizai* II.2 (2005): 219–33.
Kalbfleisch, Karl. *Simplicius, In Aristotelis categorias commentarium*, CAG 8 (Berlin: Reimer, 1907).
Løkke, Håvard. "The Stoics on Sense-Perception," in Simo Knuuttila and Pekka Kärkkäinen (eds.), *Theories of Perception in Medieval and Early Modern Philosophy*, Studies in the History of Philosophy of Mind 6 (Dordrecht: Springer, 2008), 35–46.
Long, Anthony A., and David N. Sedley. *The Hellenistic Philosophers*, vol. 1: *Translations of the Principal Sources with Philosophical Commentary* (Cambridge: Cambridge University Press, 1987).
Long, Herbert S. *Diogenes Laertius, Vitae philosophorum* (Oxford: Clarendon Press, 1964).
Machamer, Peter K., and Rober G. Turnbull, eds. *Studies in Perception: Interrelations in the History of Philosophy and Science* (Columbus: Ohio State University Press, 1978).
Menn, Stephen. "The Stoic Theory of Categories," *Oxford Studies in Ancient Philosophy* 17 (1999): 215–47.
Mutschmann, Hermann. *Sexti Empirici opera*, 3 vols. (Leipzig: Teubner, 1912–1961).
O'Keefe, Timothy. "The Ontological Status of Sensible Qualities for Democritus and Epicurus," *Ancient Philosophy* 17 (1997): 119–34.
Plasberg, Otto. *Cicero, Academica priora siue Lucullus* (Leipzig: Teubner, 1922).
Rossotti, Hazel. *Colour: Why the World isn't Grey* (Princeton: Princeton University Press, 1985).
Rudolph, Kelli C. *Reading Theophrastus: A Reconstruction of Democritus' Physics of Perception*, Ph.D. thesis, University of Cambridge, 2008.
Rudolph, Kelli C. "Democritus' Perspectival Theory of Vision," *Journal of Hellenic Studies* 131 (2011): 67–83.
Sedley, David N. "Epicurean Anti-Reductionism," in Jonathan Barnes and Mario Mignucci (eds.), *Matter and Metaphysics* (Naples: Bibliopolis, 1988), 295–327.
Smith, Martin F., trans. *Lucretius, On the Nature of Things* (Indianapolis: Hackett, 2001).
Striker, Gisela. "Epicurus on the Truth of Sense Impressions," *Archiv für Geschichte der Philosophie* 59 (1977): 125–42.
Taylor, Christopher C. W. "All Perceptions are True," in Malcolm Schofield et al. (eds.), *Doubt and Dogmatism: Studies in Hellenistic Epistemology* (Oxford: Oxford University Press, 1980), 105–24.
Taylor, Christopher C. W. *The Atomists: Leucippus and Democritus* (Toronto: University of Toronto Press, 1999).

Vitelli, Girolamo. *Philoponus, In Aristotelis libros de generatione et corruptione commentaria*, CAG 14.2 (Berlin: Reimer, 1897).
von Staden, Heinrich. "The Stoic Theory of Perception and its 'Platonic' Critics," in Machamer and Turnbull, *Studies in Perception*, 96–136.
Waszink, Jan H. *Chalcidius, In Platonis Timaeum commentarium* (Leiden: Brill, 1962).
Wendland, Paul. *Philonis Alexandrini opera quae supersunt*, vol. 2 (Berlin: Reimer, 1897, repr. 1962).
Wendland, Paul. *Alexander, In librum de sensu commentarium*, CAG 3.1 (Berlin: Reimer, 1901).
Westman, Rolf. *Plutarchi Moralia*, vol. 6.2, 2nd ed. (Leipzig: Teubner, 1959).

Jacques Jouanna
Erotian, Reader of Hippocrates' *Prognostic:* A New Discovery

Abstract: The paper has the aim of presenting a new solution for Erotian's gloss Φ 2, which he read not in *Prorrhetic* I, as Nachmanson proposed, but in a section of *Prognostic* that is now lost but preserved indirectly through *Coan Prenotions*. The account of the discovery proceeds in four parts: 1) A comparison of the parallel passages between *Prognostic* and *Coan Prenotions*, leading to the hypothesis of a lacuna in *Prognostic* preserved by the epitomizer of *Coan Prenotions*. 2) It is in this lacuna that the gloss Φ 2 may be observed. It follows that in the original order of the glosses found since Heringa, the gloss can be inserted perfectly into the sequence of glosses on *Prognostic*. 3) A confirmation of the existence of the gloss in *Prognostic* can be found in Rhazes' *Continens*, which reads this gloss in the lemmata of Galen's *Commentary* on *Prognostic*. 4) The situation of the new discovery within the history of discoveries relating to Erotian's glosses after the two great discoveries made by Heringa (1749) and Daremberg (1849) a century later. With Heringa, the main glosses on *Prognostic* were identified. The beginning of the twenty-first century has added two more: a. gloss I 1 (Anastassiou/ Irmer 2006) b. gloss Φ 2 (Jouanna, in this article).

In honor of Heinrich von Staden, I would like to present a development in the identification of the source of a gloss from Erotian's Hippocratic *Glossary*, dedicated to Nero's physician, the *archiatros* Andromachus. Erotian's is a glossary that Professor Heinrich von Staden knows well through having worked closely with one of its sources—namely, Bacchius of Tanagra, a follower of Herophilus —in his magisterial work on Herophilus.[1] The advance does not result from a study of Erotian's *Glossary* itself. Rather, it arises from a problem that has been posed independently by the process of establishing the text of the *Prognostic* attributed to Hippocrates in view of a new critical edition of the treatise. The development thus stems from what is fashionable in the history of science to call "serendipity." As opposed to what we might call "natural" discoveries, insofar as they are made on the basis of material not yet edited or underutilized—in the case of *Prognostic*, we see this, for example, with the Arabic and Syriac traditions

---

[1] H. von Staden, *Herophilus: The Art of Medicine in Early Alexandria* (Cambridge: Cambridge University Press, 1989).

and even sometimes Greek papyri—those developments made in the absence of new material are perhaps even more valuable because they were not so readily available for discovery and the vagaries of research never happen in the same way twice.

The gloss in question is presented in the edition of reference for Erotian, the 1918 edition of Ernst Nachmanson, as follows:

> Φ 2 φακῶν ἐρέγματα· ἀντὶ τοῦ διαιρήματα· καὶ γὰρ ἐρεγμός κυρίως λέγεται ὁ δίχα διῃρημένος κύαμος.
>
> ἐρέγματα of lentils, that is, pieces; in fact, ἐρεγμός is properly said of the bean divided into two. (90.6–7 Nachmanson)

In his edition, Nachmanson proposes to trace the gloss back to the treatise *Prorrhetic* I: "e Prorrh. I 21 (V 516.2 L.) aut 53 (V 524.1 L.) fluxisse putaverim" (*Test.* 2). Here are the two passages in question:

> ἐπὶ τοῖσι χολώδεσι διαχωρήμασι τὸ ἀφρῶδες ἐπάνθισμα, κακόν, ἄλλως τε καὶ ὀσφὺν προηλγηκότι καὶ παρενεχθέντι. (*prorrh.* I 21, 5.516.2–4 L.) [*Coac.* 595]
>
> In bilious stools, a frothy scum is a bad sign, especially for someone who previously suffered from pain in the lumbar region and delirium.

> ἐν ὀξέσι χολώδεσιν ἔκλευκα, ἀφρώδεα, περίχολα διαχωρήματα, κακόν.
> (*prorrh.* I 53, 5.524.1–3 L.) [*Coac.* 590]
>
> In acute bilious diseases, stools that are very white, frothy, and mixed with plenty of bile are a bad sign.

It is true that the expression φακῶν ἐρέγματα is not found elsewhere. But Nachmanson's hypothesis is that the expression glossed by Erotian refers to one of these two passages where it is a question of bilious stools; the comparison of the stools with pounded lentils then disappeared from the text of *Prorrhetic* at some point after Erotian was writing.

I will present a new proposal regarding the source of the gloss by laying out the major stages of research involved, here organized into three, before adding a fourth stage concerning the place of this development in the history of the rediscovery of Erotian's *Glossary* from the Renaissance up to the twenty-first century.

I

The first stage of research, which has been occasioned by the new edition of *Prognostic*, is the establishment of the text of *prog.* 11, where we find a discussion of the observation of stools.[2] The author of *Prognostic*, who we should understand as an established master writing for advanced students or colleagues, first gives a tripartite definition of prognosis as a practice that bears at once on the past, the present, and the future before surveying the different domains in which the physician should undertake an examination of the bedridden patient with the aim of giving a correct diagnosis and prognosis, whether favorable or unfavorable, during the course of an acute illness. The Hippocratic physicians, who before the advent of systematic human dissection were unable to observe the interior of the human body, compensated for this shortcoming by making extremely precise observations of everything that was evacuated by the body both above and below, believing that these emissions would reveal information about health or disease. After a review of the favorable signs, the discussion of stools in chapter 11 turns to the unfavorable signs.

The text as it has been transmitted by the direct tradition does not appear to pose any major problems insofar as it follows the text of the Galenic lemmata, that is, the words of Hippocrates cited by Galen in his commentary on *Prognostic*. With the benefit of the Galenic commentary, we gain access, at least in principle, to the state of the text in the second century A.D., when Galen was writing. But to the extent that this passage from *Prognostic* was also used by the epitomizer responsible for *Coan Prenotions*, which are earlier than Galen—for Galen is aware of the treatise[3]—we can go even further back in the history of the text by comparing the parallel passages in *Prognostic* and *Coan Prenotions*.

---

[2] Jacques Jouanna, with Anargyros Anastassiou and Caroline Magdelaine, *Hippocrate, Pronostic*, t. III.1 (Paris: Les Belles Lettres, 2013). References are to the new edition and the classic edition of Littré. The text is drawn from the new edition. See also Hugo Kühlewein, *Hippocratis opera*, 2 vols. (Leipzig: Teubner, 1894–1902), 1:72–108; W. H. S. Jones, *Hippocrates*, vol. 2 (Cambridge, Mass.: Harvard University Press, 1923), 1–55; Bengt Alexanderson, *Die hippokratische Schrift Prognosticon: Überlieferung und Text* (Göteborg: Elanders Boktryckeri Aktiebolag, 1963).
[3] Galen knows the text both by the title Κῳακαὶ προγνώσεις (Gal. *gloss.* 19.69.10 K.) and by the abbreviated title Κῳακαί (Gal. *gloss.* 19.81.13 K.). He mentions it not only in his *Glossary* but also in his *Commentary on Epidemics III* 17A.500 K. (= 13.5 Wenkebach); 17A.574 K. (= 59.5 and 8 Wenkebach); 17A.575 K. (= 59.11 Wenkebach); 17A.579 (= 62.8 and 12 Wenkebach); 17A.580 (= 63.8 Wenkebach). Galen considers the text to be composed of extracts from the treatises of Hippocrates, including *Prognostic*. See A. Anastassiou and D. Irmer, *Testimonien zum Corpus Hippocraticum*, 4 pts. in 3 vols. (Göttingen: Vandenhoeck and Ruprecht, 1997–2012), 2.1:167. Outside of the commentaries, the title is not cited in any other Galenic treatise extant in Greek. Nevertheless, it is discussed in a text conserved in Arabic entitled *On Medical Names* (23.13 ff. Meyerhof and Schacht); see Anastassiou and Irmer, *Testimonien*, 2.2:118.

prog. 11, 2.136.10–138.6 L.
(= 11a § 5–6, 29.6–31.2 Jouanna)

Coac. 600, 5.724.3–5 L.

1. Ὑδαρὲς δὲ κάρτα ἢ **λευκὸν** ἢ **χλωρὸν** ἰσχυρῶς ἢ **ἀφρῶδες** διαχωρέειν, πονηρά ταῦτα πάντα.

1. φλαῦρον δὲ καὶ τὸ σφόδρα **χλωρὸν** ἢ **λευκὸν** ἢ **ἀφρῶδες** ἢ ὑδαρές.

2. Πονηρὸν δὲ καὶ **σμικρόν** τε ἐὸν καὶ **γλίσχρον** καὶ **λεῖον** καὶ **λευκὸν** καὶ **ὑπόχλωρον**.

2. Καὶ τὸ **σμικρόν** τε καὶ **γλίσχρον** καὶ **λεῖον** καὶ **ὑπόχλωρον** κακόν.

Coac. 621, 5.728.7–13 L.

3. Τουτέων δ' ἔτι **θανατωδέστερα** ἂν εἴη **τὰ μέλανα** ἢ **λιπαρὰ** ἢ **πελιὰ** ἢ ἰώδεα καὶ κάκοσμα.

3. **Θανατώδεά** ἐστι τῶν διαχωρημάτων **τὸ λιπαρόν**, καὶ **τὸ μέλαν** καὶ **τὸ πελιὸν** μετὰ **δυσωδίης** καὶ τὸ χολῶδες ἔχον ἐν ἑωυτῷ **φακῶν** ἢ ἐρεβίνθων **ἐρίγμασι** παραπλήσια, ἢ οἷον θρόμβους αἵματος εὐανθεῖς, κατὰ τὴν ὀδμὴν ὅμοιον τῷ τῶν νηπίων,

4. Τὰ δὲ **ποικίλα**, **χρονιώτερα** μὲν τουτέων, ὀλέθρια δὲ οὐδὲν ἦσσον· ἔστι δὲ ταῦτα **ξυσματώδεα** καὶ **χολώδεα** καὶ **αἱματώδεα** καὶ **πρασοειδέα** καὶ **μέλανα**, ποτὲ μὲν **ὁμοῦ** διεξερχόμενα ἀλλήλοισι, ποτὲ δὲ κατὰ μέρος.
Then follows a discussion of the winds (c. 11, 2.138.6 L. [= 11b § 1, 31.3 Jouanna]) beginning with **Φῦσαν**

4. καὶ **τὸ ποικίλον**, τὸ δ' αὐτὸ καὶ **χρόνιον**· γίνοιτο δ' ἂν τοιοῦτον **αἱματώδες**, **ξυσματώδες**, **χολώδες**, **μέλαν**, **πρασοειδές**, καὶ **ὁμοῦ** καὶ ἐναλλάξ.

Table 1 Comparison of parallel passages in *Prognostic* and *Coan Prenotions*

The text on the left, *Prognostic*, is the source text from which the text on the right, *Coan Prenotions*, is drawn.[4] The numbers (1–4) in each column correspond to the four types of stool that pose a threat, in some cases portending death. I have used bold for those words that are identical; words that are synonymous are underlined.

Everything in the two columns is by and large comparable. Nevertheless, in (3), we find an important difference in the length of the discussion. *Coan Prenotions* present a rather long passage which is absent from the text of *Prognostic* as

---

4 This is the relationship between the two treatises that is most widely accepted. The relationship between the two texts is too precise to be due to chance. Three hypotheses are theoretically possible, although they do not share the same degree of likelihood: either *Prognostic* is the source for *Coan Prenotions*, or the reverse is true, or both works are based on a common source. All three hypotheses have been defended by scholars: see O. Poeppel, *Die hippokratische Schrift Κῳακαὶ προγνώσεις und ihre Überlieferung*, Diss. maschinenschriftlich (Kiel, 1959), 54–67, and more recently V. Langholf, "Symptombeschreibungen in *Epidemien* I und III und die Struktur des *Prognostikon*," in F. Lasserre and P. Mudry (eds.), *Formes de pensée dans la collection hippocratique: actes du IV*ᵉ *Colloque international hippocratique; Lausanne, 21–26 septembre 1981* (Geneva: Droz, 1983), 109–20, at 115 ff., who advocates the third hypothesis.

it reads in the extant editions. The passage has been italicized in the right-hand column: καὶ τὸ χολῶδες ἔχον ἐν ἑωυτῷ **φακῶν** ἢ ἐρεβίνθων **ἐρίγμασι** παραπλήσια, ἢ οἷον θρόμβους αἵματος εὐανθεῖς, κατὰ τὴν ὀδμὴν ὅμοιον τῷ τῶν νηπίων. One can translate it thus: "a fatal sign is a bilious stool that contains material like pounded lentils or chickpeas or like fresh clots of blood, and whose smell is like that of the stool of infants." Insofar as the text appears to be the natural continuation of the beginning of the sentence, we might ask whether the epitomizer behind *Coan Prenotions* took his text from a copy of *Prognostic* that is earlier than the text in our possession, a text that was more complete. This phrase would then have dropped out at some point between the text used by the epitomizer of *Coan Prenotions* and the text we have today. In terms of the first stage of research, we are still at the stage of a hypothesis that is exactly like the hypothesis put forth by Nachmanson in 1918.

II

The second stage of research takes its departure from the observation of the unusual expression included in the phrase from *Coan Prenotions* that is believed to originate in *Prognostic* itself—namely, φακῶν ... ἐρίγμασι, which leads us back to the gloss from Erotian that I presented at the outset, φακῶν ἐρέγματα. Does this mean the gloss is a gloss on the passage from *Coan Prenotions*? Beginning in the sixteenth century, such was the solution adopted by Foës in his Hippocratic dictionary, *Œconomia Hippocratis*, published in 1588, which admirably sums up the whole of Renaissance erudition concerning Hippocrates.[5]

This hypothesis appeared highly reasonable. But the progress made by philologists in the late nineteenth and early twentieth centuries has cast doubt on this attribution: on the one hand, the title *Coan Prenotions* does not figure in the list of Hippocratic treatises that Erotian gives at the start of his *Glossary*; on the other hand, Erotian does not appear to have glossed this treatise. Such a situation explains why the most recent editor of Erotian, Nachmanson, did not even raise the possibility of a relationship with *Coan Prenotions* when editing this gloss.[6] For this reason, the prevailing opinion is that this gloss of Erotian still has not been securely traced back to its source.[7]

---

[5] A. Foës, *Œconomia Hippocratis* (Francofurdi: Apud Andreae Wecheli heredes, Claudium Marnium, & Jo. Aubrium, 1588), s.v. ἐρείκειν (238a, ll. 23ff.) and s.v. φακῶν ἐρέγματα (650).
[6] In his *Erotianstudien* (Uppsala: Akademiska bokhandeln, 1917), 277, Nachmanson had surmised that the words might have been in an earlier text of *Prorrhetic* I different from ours.
[7] See Anastassiou and Irmer, *Testimonien*, 1:495 and n. 4: "Ausdruck im CH nicht belegt."

But once we accept that the expression could have come from *Prognostic*, we may ask whether Erotian, who did not know *Coan Prenotions*, could have found the expression that he glossed in his copy of *Prognostic*, just as the epitomizer of *Coan Prenotions* could have found it in his own copy of that work.

In order to pursue this route, it is appropriate to offer some clarification about the history of the transmission of Erotian's *Glossary*. I summarize here everything that specialists in Hippocrates consider accepted knowledge. First, what we possess is not Erotian's original work, which had glossed the Hippocratic treatises in order according to genre. Erotian says very clearly at the end of his *Preface*, having given a list of Hippocratic treatises organized by genre, that he will begin by glossing the semiotic writings, and he ends his introduction with the phrase "Let us now start with *Prognostic*."[8] Yet what follows does not at all correspond to this. It is, rather, the work of an epitomizer, who reorganizes glosses that had been initially organized treatise by treatise in alphabetical order, based on the first letter of the word. What we find, then, is that the glossary begins with alpha and goes to the end of the alphabet, all the way to omega. Nevertheless, within the entries for each letter of the alphabet, the epitomizer, rather than reclassifying the glosses according to a strict alphabetical order, preserves within each of the columns he creates the glosses just as he encountered them, that is, in the original order in which Erotian had listed them. As a result, by comparing the order of the glosses as they appear for each letter of the alphabet, we can reestablish the order in which the treatises had been arranged by Erotian when he glossed them. It follows, then, that if Erotian began with *Prognostic*, the glosses taken from this treatise will appear first in each letter of the alphabet.

Where, then, do we find our gloss, which is listed under the letter Φ (because of the presence of φακῶν, although the word that really needed explaining was ἐρέγματα)? It is the second entry under the heading of Φ, what we would call Φ 2. It is therefore a gloss that appears at the beginning of the Φ series in a section that corresponds precisely to *Prognostic*. As far as the gloss Φ 1 is concerned, which comes before our entry Φ 2 in Erotian's *Glossary*, everyone would agree that it pertains to the φῦσα that appears in the same chapter of *Prognostic*, chapter 11. Therefore, Erotian would have found these two words in need of an explanation practically in the same place. If we fill in the lacuna in *Prognostic* with the help of the phrase from *Coan Prenotions*, we can reconstruct the state of the text as Erotian would have read it. We must, however, use the plural instead of the singular to reflect the modifications made by the epitomizer to his

---

8 Erot. 9.25–26 Nachmanson.

model, which are evident at the beginning of the series in (3) above. Here, then, is how this rediscovered text of *Prognostic* should appear:

> *prog.* 11, 2.138.1–6 L. (= 11a § 6, 30.3–31.2 Jouanna)
>
> **3.** Τουτέων δ' ἔτι θανατωδέστερα ἂν εἴη τὰ μέλανα ἢ λιπαρὰ ἢ πελιὰ ἢ ἰώδεα καὶ κάκοσμα. ⟨καὶ τὰ χολώδεα ἔχοντα ἐν ἑωυτοῖσι   Erotian's
> **φακῶν** ἢ ἐρεβίνθων **ἐρίγμασι** παραπλήσια, ἢ οἷον θρόμβους αἵμα-   gloss Φ 2
> τος εὐανθεῖς, κατὰ τὴν ὀδμὴν ὅμοια τοῖσι τῶν νηπίων⟩.
>
> **4.** Τὰ δὲ ποικίλα, χρονιώτερα μὲν τουτέων, ὀλέθρια δὲ οὐδὲν
> ἧσσον· ἔστι δὲ ταῦτα ξυσματώδεα καὶ χολώδεα καὶ αἱματώδεα
> Erotian's καὶ πρασοειδέα, καὶ μέλανα, ποτὲ μὲν ὁμοῦ διεξερχόμενα ἀλλή-
> gloss Φ 1   λοισι, ποτὲ δὲ κατὰ μέρος.
>
> *prog.* 11, 2.138.6 L. (= 11b § 1, 31.3 Jouanna) **Φῦσαν** δὲ ἄνευ ψόφου κτλ.

Table 2 Reconstruction of Erotian's text of *prog.* 11

It is certainly the case that if the order of the text were more strictly observed, we would expect that the gloss Φ 2 would come before Φ 1. But the passages are so close to one another that we can say the order of the text has been generally respected. Therefore, in terms of a demonstration that makes no pretensions to mathematical precision, the mystery of the source of Erotian's gloss Φ 2 appears to be solved. With the same move, the editor fills the lacuna and gets as close as possible to *Prognostic* as it would have been read by Erotian in the first century B.C. and, before him, by the epitomizer of *Coan Prenotions*. Of course, the restoration of the lacunose passage remains approximate, since the epitomizer of *Coan Prenotions* does not give a literal citation from *Prognostic*.

III

The third stage of research is offered as a verification of sorts. After having read my commentary on this passage, Anargyros Anastassiou, who had collaborated with me on the new edition of *Prognostic* and who had completed the third volume of the *Testimonien zum Corpus Hippocraticum*, discovered, with the help of the digital collections of the BIUSanté, the following citation attributed to Hippocrates in the vast medical encyclopedia of Rhazes (865–ca. 925), which was written in Arabic and translated into Latin under the title *Continens*:

Ypo⟨crates⟩: si in egestione fuerint cortices ad modum corticum lupinorum in omni morbo mortales aderunt et si egestio fuerit fetens (setens trad., corr. Anastassiou) ad modum egestionis puerorum, mala erit.[9]

Hippocrates: if in the excrement there are husks similar to the husks of lupine seeds, these are fatal signs in all diseases and if the excrement smells like the excrement of infants, this is bad.

The similarity between these two concrete comparisons used to describe stools cannot be by chance: it is a citation of our passage. But even more important is identifying the context in which Rhazes makes the citation. It is found in his Book 16 on prognostic signs, where Rhazes cites the Hippocratic *Prognostic* on a number of occasions. Rhazes, as one might have suspected, did not read *Prognostic* directly but through the lemmata of Galen's *Commentary on Prognostic*, whose traces one can follow. It is at the moment when Rhazes reaches the discussion of stools that corresponds to *prog.* 11 that our citation appears.[10] It is therefore certain that he found the phrase in *Prognostic* and not in *Coan Prenotions*.

IV

A fourth stage of research should now be added—namely, the placement of these philological advances in the history of other discoveries made regarding the identification of Erotian's glosses in *Prognostic* following the publication of the *editio princeps* of Erotian's *Glossary* by Henri Estienne in 1564.[11] Who was responsible for discovering the original order of the glosses modified by the epitomizer and who on that basis first established the list of glosses in Erotian drawn from *Prognostic*? The answer that one would have given until very recently would be that the discovery goes back to Johannes Ilberg, at the end of the nineteenth century, in his long study entitled "Das Hippokrates-Glossar des Erotianos und

---

[9] *Continens*, fol. 326 verso a 26–33 (= Anastassiou and Irmer, *Testimonien*, 3:170).

[10] There remains one point that I have not resolved. In Rhazes, the citation corresponding to Erotian Φ 2 comes after the one that corresponds to Φ 1. Therefore, Rhazes' order corresponds to Erotian's order, as opposed to that which was reconstructed through *Coan Prenotions*. But the continuation of the text in Rhazes is not obvious.

[11] H. Estienne, *Dictionarium medicum … Lexica duo in Hippocratem huic Dictionario praefixa sunt, unum Erotiani, nunquam antea editum, alterum, Galeni, …* (Paris: Fuggeri, 1564), 5 ff.

seine ursprüngliche Gestalt," published in Leipzig in 1893.¹² But a reconsideration of the history of the tradition of Erotian's text has brought the true discoverer, who is much earlier, out of the shadows. He is the Dutchman Adriaan Heringa, whose massive study was published in Leeuwarden in 1749 under the overly general title *Observationum criticarum liber singularis in quo passim veteres Auctores, Graeci maxime, emendantur.*¹³ In the first, ten-page chapter, Heringa offers a magisterial demonstration of what I have presented above as common knowledge. Then, taking *Prognostic* as an example, he gives the original order of thirty rare words that are glossed in the treatise and shows how these thirty glosses were redistributed in alphabetical order in such a way that they are located at the head of each letter of the alphabet. He even establishes the order in which Erotian originally read the treatises in noting that this order does not correspond exactly to the classification given by Erotian in his *Preface*. Heringa certainly admits that the majority of the glosses from *Prognostic* had already been attributed to that treatise by Foës in his *Œconomia Hippocratis*, cited above. What Heringa adds is method. What had belonged to the empirical domain crosses over into the order of philological science. The list of the glosses from *Prognostic* that was established by Heringa in the middle of the eighteenth century has remained unchanged for two-and-a-half centuries.¹⁴

One could ask why Heringa's discovery remained unnoticed or undervalued for such a long time. Why were Heringa's conclusions not exploited thoroughly by the great Littré in his edition of *Prognostic*, which remains one of the great successes of his monumental edition of Hippocrates? He certainly knew Heringa, whom he cites in his general introduction.¹⁵ The reason is due in part to the fact

---

12 J. Ilberg, "Das Hippokrates-Glossar des Erotianos und seine ursprüngliche Gestalt," *Abhandlungen der Königlich Sächsischen Gesellschaft der Wissenschaften, Philologisch-historische Classe* 34 (1893): 101–48.
13 A. Heringa, *Observationum criticarum liber singularis in quo passim veteres Auctores, Graeci maxime, emendantur* (Leovardia: Coulon, 1749). I am pleased to note that before my own inquiry, Heringa's discovery had already been rehabilitated, as I noted already, by Anastassiou and Irmer in *Testimonien*, 1:xxvii. They give a clear exposition of Heringa's contribution before those of Ilberg and Nachmanson. Nevertheless, insofar as they are not primarily concerned with *Prognostic*, they do not note the details regarding the glosses that I have given. Moreover, Heringa's contribution to Erotian's *Glossary* cannot be reduced to the first ten pages, although these are the most important. Heringa returned to some of the glosses at *Observationum criticarum*, 104–10 (Ch. 13) and 111–20 (Ch. 14).
14 It is enough to compare the list of glosses from *Prognostic* found in Heringa, *Observationum criticarum*, 6–7 with those in Nachmanson, *Erotianstudien*, 273 to see that they are identical.
15 É. Littré, ed. and trans., *Œuvres complètes d'Hippocrate*, 10 vols. (Paris: J. B. Baillière, 1839–1861), 1:100 ("Heringa … a fait clairement voir que l'ordre d'Érotien avait été celui-ci …"), with n. 1: "Obs. p. 3, seq."

that he knew Heringa's work only indirectly, by means of citations in the edition of Erotian in use in that period, the 1780 edition of Johann Franz.[16] Nevertheless, the passage was long enough that Littré could have become aware of all the glosses that Heringa attributes to *Prognostic*. It is clear that when Littré edited *Prognostic* in 1840, he had recourse to Heringa directly, as one can determine from what is on that occasion a precise reference.[17] Still, he did not exploit the most essential elements of that discovery, neither as regards the glosses on *Prognostic* nor as regards the original order of the treatises glossed by Erotian.

Nor was Heringa's discovery exploited to the full by the second scholar to make an important discovery regarding Erotian's *Glossary*. Exactly a century after the discovery of Heringa was first made, Charles Daremberg, a younger contemporary of Littré, while on an official mission to Italy in 1849, discovered in two manuscripts in the Vatican Library two scholia that he declared were for the most part fragments from Erotian's *Glossary*. He published these at the end of the first volume of his *Notices et extraits des manuscrits médicaux* in an appendix entitled "Scholies inédites sur Hippocrate."[18] One of these scholia, a gloss on a word from the *Oath*, preserved the final portion of the *Glossary*. In effect, the end of the *Glossary* found by Daremberg included a final address to the

---

[16] J. Franz, *Erotiani, Galeni et Herodoti glossaria in Hippocratem ex recensione Henrici Stephani Graece et Latine* (Lipsiae: Johannis Friderici Junii, 1780). Indeed, the reference to Heringa in Littré is mysterious. He does not cite Heringa's book but only an abbreviation of the title with a reference to p. 3 ff. He must have found this reference (which is inaccurate) in Franz's edition. On p. 28 of his edition, in line 5 of the lefthand column, Franz, discussing Erotian's order and having first cited Stephanus (= Henri Estienne, the author of the *editio princeps*), continues: "Iuvabit de eodem argumento apponere verba Heringae in Obss. p. 3. seq. Fundamenti autem loco, inquit, pauca praemonenda sunt de Erotiani Lexico etc." Yet the beginning of his long citation is not found on p. 3 of Heringa but at the bottom of p. 2. The citation continues for two columns in the edition of Franz to pp. 28–30 and ends at line 20 of the lefthand column of p. 31 (*restat* etc.). Thus the long citation corresponds to pages 2.32–7.26 of Ch. 1 of Heringa's *Observationum criticarum*. The reference given by Franz is therefore inaccurate. It is repeated by Littré, who did not consult the original on this occasion.

[17] Littré, *Œuvres complètes*, 2:132–33, n. 26 : "Heringa (p. 9) a remarqué qu'il fallait substituer περιεστικόν aux leçons plus ou moins altérées que présentent les mss. et les imprimés. J'ai adopté sa correction etc." This is essentially the exact reference that Heringa gives on p. 9: "Ubi pro περιεστηκὸν lege περιεστικόν." This is, however, the only reference that Littré makes to Heringa in his edition of *Prognostic*. He therefore did not exploit Heringa's discovery fully, despite the fact that in this text he makes direct reference to his study.

[18] C. Daremberg, *Notices et extraits des manuscrits médicaux grecs, latins, et français des principales bibliothèques de l'Europe* (Paris: Imprimerie impériale, 1853), 198 ff. He had already published these in his *Archives des missions scientifiques* from 1851 ("Notices et extraits des manuscrits médicaux grecs et latins des principales bibliothèques d'Angleterre," *Archives des missions scientifiques et littéraires* 2 [1851]: 113–68, 470–71, 484–545).

doctor Andromachus, which picks up the initial address that has been known to us since the *editio princeps* from the Renaissance. The parallels between the beginning and the end of the *Glossary* make the discovery especially illuminating:

The beginning of the *Glossary* preserved in the revised version known since the Renaissance	The end of the original *Glossary* discovered by Daremberg in 1849 in a scholion to *Vat. gr.* 277 (R) and *Vat. Urb.* 68 (U)
τὴν Ἱπποκράτους πραγματείαν, ἀρχιατρὲ **Ἀνδρόμαχε**, οὐκ ὀλίγα συμβαλλομένην πᾶσιν ἀνθρώποις ὁρῶν κτλ. (Erot. 3.3–4 Nachmanson)	τοιαῦται μὲν λέξεις εἰσίν, ἃς συναγαγεῖν ἠδυνήθημεν ... θαυμαστότατε **Ἀνδρόμαχε** κτλ. (Erot. 116.11–16 Nachmanson)

It is true that this gloss from the *Oath* followed by the end of the *Glossary* had already been published by Littré in his edition of the *Oath*.[19] He had found it in a late Paris manuscript, the *Parisinus gr.* 2255 (which we now know to be a copy of one of the two manuscripts that Daremberg consulted in the Vatican). He had simply added the following remark: "Cette glose paraît être empruntée au *Glossaire* d'Érotien ; cependant il ne s'en retrouve rien dans ce qui nous reste de cet auteur." Littré did not give any justification for his opinion. He therefore did not advance the argument about the parallel addresses to Andromachus at the beginning and the end of Erotian's work.

It was Daremberg, therefore, who was the first to move things along:

> M. Littré dit seulement que cette scolie paraît empruntée au *Glossaire* d'Érotien, mais cet emprunt est indubitable. Il y a plus, nous avons dans ces lignes la fin même, et comme l'Épilogue du *Glossaire* d'Érotien qui était, on le sait dédié à Andromaque le Jeune, médecin de Néron.[20]

But Daremberg's next remark demonstrates that he did not fully grasp the conclusions of Heringa's pioneering work.

> Nous savons, de plus, que le *Serment* figurait un des derniers parmi les ouvrages compris dans le Canon hippocratique dressé par Érotien ; après lui venaient l'opuscule *De l'art* et le traité *De l'ancienne médecine*. Si donc nous pouvons nous en rapporter à cette clausule, si, d'un autre côté, nous rappelons l'ordre dans lequel a été rédigé le *Glossaire*, Érotien n'aurait expliqué aucun des mots propres aux deux ouvrages dont je viens de rappeler les titres.[21]

---

[19] Littré, *Œuvres complètes*, 4:628 n. 5.
[20] Daremberg, *Notices et extraits*, 220.
[21] Daremberg, *Notices et extraits*, 221.

This conclusion does not do justice to the facts that Heringa had uncovered through his method. Heringa had already shown that the order in which Erotian had arranged his material did not correspond exactly to the classification of the treatises grouped in the Preface. Heringa had indicated that *On Ancient Medicine* and *On the Art* had been glossed previously by Erotian and he concluded by saying he could not locate the *Oath:*

> Si quis vero diligenter *Erotianum* cum *Hippocrate* contenderit, is videbit eum in Glossario suo tenuisse ordinem Librorum non praecise eum, quem in praefatione; sed Libros ita fere explicuisse, ut his subjeci.[22]

Heringa then gives a list of the treatises in the order they were read by Erotian: *Prognostic; Prorrhetic* I and II; *Humors; Epidemics; Nature of a Human Being; Aphorisms; Nature of Child; Airs, Waters, Places; On Breaths; On the Sacred Disease; On the Use of Liquids; Regimen; Places in a Human Being; On Ancient Medicine; On the Art; In the Surgery; Instruments of Reduction; Ulcers; On Head Wounds; On Fractures; On Joints; Regimen in Acute Diseases; Sterile Women; On Diseases* I–II; *Diseases of Women* I–II; *On the Embassy; On the Altar.* Heringa concludes by saying that he was unable to determine the order of five of the treatises glossed by Erotian: *On Wounds and Weapons; Hemorrhoids and Fistulas; Nutrition; Oath;* and *Law,* of which the first is now lost.[23]

If Daremberg had referred to this part of Heringa's discovery, he would not have advanced a deduction that, while logical on the surface, was in fact false. He could have instead completed the discovery of Heringa, since he would have been able to establish henceforth the certain place of the *Oath* at the end of the collection of texts glossed by Erotian, which Heringa had been unable to do himself. Daremberg certainly knew what "le savant Héringa" had said about *Prognostic*.[24] But no more than Littré had, he failed to pay attention to what Heringa had concluded, thanks to his method, about the difference between the order of Hippocratic texts in Erotian's Preface and the order of the texts that were actually glossed by Erotian.

When one arrives at the significant contribution made by Ilberg at the end of the nineteenth century in his "Das Hippokrates-Glossar des Erotianos und seine ursprüngliche Gestalt," which takes up the very subject that Heringa had treated,

---

[22] Heringa, *Observationum criticarum liber singularis*, 9.
[23] See now Mathias Witt, *Weichteil- und Viszeralchirurgie bei Hippokrates. Ein Rekonstruktionsversuch der verlorenen Schrift* Περὶ τρωμάτων καὶ βελῶν, Beiträge zur Altertumskunde 270 (Berlin: De Gruyter, 2009).
[24] Daremberg, *Notices et extraits*, 221.

one is astonished to see the difference in the treatment he gives the two scholars. While he rightly gives credit to Charles Daremberg, he is disconcertingly quick in passing over the discovery of Heringa. Although he recognizes that Heringa was the first to speak of the matter, he disposes of his predecessor's work in two lines by citing his slightly ridiculous expression "*interpolator nefarius, maleferiatus,*" following it with "*und so weiter.*"[25] But it is in this "*und so weiter*" that one finds what is most essential. By declaring, after having cited Heringa (1749) and the other two editors, Franz (1780) and Klein (1865), that "keiner hat den Versuch gemacht, die echte Form des ganzen Buches, soweit möglich, wiederzufinden," Ilberg plays on the expression "*soweit möglich*" ("as far as is possible"). Although there was room for improvement, Heringa had already made a proposal about the original order of the texts glossed by Erotian that was basically correct, having analyzed all the glosses by means of his method. Ilberg's list, compared to that of Heringa, is on the whole identical.[26] Ilberg's merit was mainly to have stated explicitly what had already been implicitly established by Heringa.

Ernst Nachmanson is less evasive than Ilberg when it comes to the work of Heringa. Discussing the original order of the glosses in Ch. 4 in his *Erotianstudien*, he recognizes Heringa's contribution more openly:

> Der erste, der die entscheidende Umänderung in der Überlieferung durchschaute und einen energischen Vorstoss unternahm, die ursprüngliche Anlage aufzuhellen, war Adrian Heringa. Er hat S. 6 ... diejenigen Glossen zusammengestellt, welche zum Prognostikon, der ersten Hippokrateschrift der von Erotian im Vorwort gegebenen Liste, gehören.[27]

Moreover, he explains that Heringa had no successors before the end of the nineteenth century. According to Nachmanson, it was Ilberg who had found the exact solution to the problem. But what Nachmanson seems to neglect or has not made explicit is that Heringa had already established the original order of the glosses in a manner essentially analogous to that of Ilberg. The solution to the problem, then, had already been discovered by Heringa. The epistemological

---

25 Ilberg, "Das Hippokrates-Glossar," 106.
26 Ilberg's list ("Das Hippokrates-Glossar," 141–42) is as follows: *Prognostic, Prorrhetic* I, *Humors, Epidemics, Aphorisms, On the Nature of the Child, Nature of a Human Being, Airs, Waters, Places, On Breaths, On the Sacred Disease, Law* [?], *On the Use of Liquids, On Regimen in Health, Places in a Human Being, On Ancient Medicine, On the Art, In the Surgery, Instruments of Reduction, Ulcers, On Head Wounds, On Wounds and Weapons, On Fractures, On Joints, Regimen in Acute Diseases, On Diseases* I, *Sevens, On Diseases* II–III, *Internal Affections, Diseases of Women* I–II, *On the Nature of Woman, Hemorrhoids and Fistulas, Sterile Women, (On the Altar), On the Embassy, Oath.*
27 Nachmanson, *Erotianstudien*, 262.

leap had been made not at the end of the nineteenth century but in the middle of the eighteenth—hardly the same thing.

The situation only began to advance further at the start of the twenty-first century. Anastassiou and Irmer, in the first volume of their *Testimonien zum Corpus Hippocraticum*, were the first, as I have already said, to give pride of place to Heringa's discovery, and they do so clearly. They were also the first to reproduce the list of treatises in the original order proposed by Heringa, which no one had done previously, since the long citation of Heringa by Franz in his 1780 edition stops after the list of glosses from *Prognostic*.[28] We may now compare very conveniently the lists of Heringa, Ilberg, and Nachmanson.[29]

Moreover, as far as our primary object of interest at present, *Prognostic*, is concerned, the two scholars from Hamburg have proposed, following a reexamination of Erotian's *Glossary*, to add a new gloss to *Prognostic*:[30] the gloss I 1 ἰητρείην· θεραπείαν (Erot. 45.9 Nachmanson), attributed by Nachmanson to the treatise *Ulcers* (6.408.12 L.). Although ἰητρείην does not appear in our text of *Prognostic*, after the first chapter (2.110.6 ff. L. = 2.2 Jouanna) we find the word θεραπείαν, which could have replaced the rare word ἰητρείην by a well-attested process that the same Heringa had already drawn attention to when discussing other glosses of Erotian.[31]

Seven years later, I propose to now add gloss Φ 2. If Nachmanson had sought to connect this gloss with *Prorrhetic* I, it was because Erotian, after dealing with the glosses in *Prognostic*, had moved on to the rare terms in *Prorrhetic* I, as Heringa had already observed.[32] The intuition to look for the source in a lost passage on the observation of stools was right. But his hypothesis did not take into account the fact that the gloss Φ 2 corresponds to a passage that we also find in *Coan Prenotions*. Erotian had, in fact, found this gloss in the passage from *Prognostic* that in his time was as complete as the copy of *Prognostic* used by the epitomizer of *Coan Prenotions*.

Every new discovery poses new problems. If Rhazes had read, in a context that is clearly Galenic, a phrase from *Prognostic* that is omitted from the direct tradition and in the Greek lemmata of Galen, is it not necessary then to conclude that Rhazes read this phrase in the copy of the *Commentary on Prognostic* at his

---

28 See *supra* n. 16.
29 Anastassiou and Irmer, *Testimonien*, 1:xxvii–xxviii.
30 Anastassiou and Irmer, *Testimonien*, 1:403 and n. 1.
31 See Heringa, *Observationum criticarum*, 104: Hippocrates et Erotianus passim emendantur et illustrantur. Multa Glossemata in textum irrepsisse, lectione vera expulsa. (Summary of Ch. 13).
32 It is also at the beginning of *Prorrhetic* I that Ilberg had tentatively tried to locate the gloss: see "Das Hippokrates-Glossar," 128.

disposal? We would then have proof that the Arabic translation of Rhazes preserves parts of Galen's *Commentary on Prognostic* that are lost in the Greek version. This idea remains to be explored. But what is certain is that the Arabic translation of Galen's lemmata made by Ḥunayn ibn Isḥāq in the ninth century, which Oliver Overwien and Uwe Vagelpohl have been kind enough to acquaint me with, does not include the passage discussing the pieces of lentils or the smell of an infant's stools. Did Rhazes have access to a Greek source that was different from Ḥunayn's? This is the most likely answer. But this new problem should be taken up by specialists in the Arabic tradition, since philologists who are familiar with the Hippocratic tradition but without a knowledge of Arabic can know the work of Rhazes only indirectly, by means of the medieval Latin translation.

To conclude, I wish to stress that philology today, taken in the full sense of the word, in taking advantage of the resources of the information revolution which allow for the instantaneous consultation of primary sources and the rapid exchange of information among specialists, will not be advanced by a neglect of past scholarship, however distant. Rather philology should, if it is to remain scientific, be even more firmly grounded in history than before. It should locate new discoveries in the context of previous ones, thereby minimizing the idea of an epistemological leap—without denying it altogether—while rehabilitating the notions of both complexity and progress, judging the quality of earlier discoveries in the scientific and historical context within which they were made.[33]

Bibliography

Alexanderson, B. *Die hippokratische Schrift Prognosticon: Überlieferung und Text* (Göteborg: Elanders Boktryckeri Aktiebolag, 1963).

Anastassiou, A., and D. Irmer. *Testimonien zum Corpus Hippocraticum*, 4 pts. in 3 vols. (Göttingen: Vandenhoeck and Ruprecht, 1997–2012).

Daremberg, C. *Notices et extraits des manuscrits médicaux grecs, latins, et français des principales bibliothèques de l'Europe* (Paris: Imprimerie impériale, 1853).

Daremberg, C. "Notices et extraits des manuscrits médicaux grecs et latins des principales bibliothèques d'Angleterre," *Archives des missions scientifiques et littéraires* 2 (1851): 113–68, 470–71, 484–545.

Estienne, H. *Dictionarium medicum ... Lexica duo in Hippocratem huic Dictionario praefixa sunt, unum Erotiani, nunquam antea editum, alterum, Galeni, ...* (Paris: Fuggeri, 1564).

---

[33] Translated from the French by Brooke Holmes.

Foës, A. *Œconomia Hippocratis* (Francofurdi: Apud Andreae Wecheli heredes, Claudium Marnium, & Jo. Aubrium, 1588).
Franz, J. *Erotiani, Galeni et Herodoti glossaria in Hippocratem ex recensione Henrici Stephani Graece et Latine* (Lipsiae: Johannos Friderici Junii, 1780).
Heringa, A. *Observationum criticarum liber singularis in quo passim veteres Auctores Graeci maxime emendantur* (Leovardia: Coulon, 1749).
Ilberg, J. "Das Hippokrates-Glossar des Erotianos und seine ursprüngliche Gestalt," *Abhandlungen der Königlich Sächsischen Gesellschaft der Wissenschaften, Philologisch-historische Classe* 34 (1893): 101–48.
Jones, W. H. S. *Hippocrates*, vol. 2 (Cambridge, Mass.: Harvard University Press, 1923).
Jouanna, J., with A. Anastassiou and C. Magdelaine. *Hippocrate, Pronostic*, t. III.1 (Paris: Les Belles Lettres, 2013).
Kühlewein, H. *Hippocratis opera*, 2 vols. (Leipzig: Teubner, 1894–1902).
Langholf, V. "Symptombeschreibungen in *Epidemien* I und III und die Struktur des *Prognostikon*," in F. Lasserre and P. Mudry (eds.), *Formes de pensée dans la collection hippocratique: actes du IVᵉ Colloque international hippocratique; Lausanne, 21–26 septembre 1981* (Geneva: Droz, 1983), 109–20.
Littré, É., ed. and trans. *Œuvres complètes d'Hippocrate*, 10 vols. (Paris: J. B. Baillière, 1839–1861).
Meyerhof, M., and J. Schacht. *Galen, Über die medizinischen Namen: arabisch und deutsch* (Berlin: Verlag der Akademie der Wissenschaften, 1931).
Nachmanson, E. *Erotianstudien* (Uppsala: Akademiska bokhandeln, 1917).
Nachmanson, E. *Erotiani vocum Hippocraticarum collectio, cum fragmentis* (Uppsala: Appelbergs boktryckerei-aktiebolag, 1918).
Poeppel, O. *Die hippokratische Schrift Κῳακαὶ προγνώσεις und ihre Überlieferung*, Diss. maschinenschriftlich (Kiel, 1959).
von Staden, H. *Herophilus: The Art of Medicine in Early Alexandria* (Cambridge: Cambridge University Press, 1989).
Wenkebach, E. *Galeni, In Hippocratis Epidemiarum librum III commentaria III*, CMG V 10,2,1 (Leipzig: Teubner, 1936).
Witt, M. *Weichteil- und Viszeralchirurgie bei Hippokrates. Ein Rekonstruktionsversuch der verlorenen Schrift Περὶ τρωμάτων καὶ βελῶν*, Beiträge zur Altertumskunde 270 (Berlin: De Gruyter, 2009).

Joshua T. Katz
Aristotle's Badger

Abstract: It is sometimes said that the badger was unknown to the ancient Greeks. This is incorrect, though it is the case that they spoke comparatively little about the shy, nocturnal mustelid. In this paper, which is intended as a contribution to both classical philology and the history of biology, I examine the lexical evidence for the animal in Greece, in particular affirming that the word τρόχος in *De generatione animalium* 3.6 means "badger" (*Meles meles*) and explaining why Aristotle mentions it in connection with the hyena—the striped kind (*Hyaena hyaena*) rather than the spotted (*Crocuta crocuta*). The hyena's hermaphroditism was and remains a widespread topos, and it turns out that the badger is a pseudohermaphrodite as well.

In a passage from *De generatione animalium* 3.6 that, were it not for its content, any teacher of Greek could profitably use as a paradigmatic example of the difference between the active voice and the passive, Aristotle speaks impatiently of the supposed hermaphroditism of two animals, the hyena and the τρόχος:

> εὐηθικῶς δὲ καὶ λίαν διεψευσμένοι καὶ οἱ περὶ τρόχου καὶ ὑαίνης λέγοντες. φασὶ γὰρ τὴν μὲν ὕαιναν πολλοί, τὸν δὲ τρόχον Ἡρόδωρος ὁ Ἡρακλεώτης δύο αἰδοῖα ἔχειν, ἄρρενος καὶ θήλεος, καὶ τὸν μὲν τρόχον αὐτὸν αὑτὸν ὀχεύειν, τὴν δ' ὕαιναν ὀχεύειν καὶ ὀχεύεσθαι παρ' ἔτος. (757a2-7)

> There is another silly and extremely wrong-headed story which is told about the *trochos* and the hyena, to the effect that they have two pudenda, male and female (there are many who assert this of the hyena; Herodorus of Heraclea asserts it of the *trochos*), and that whereas the *trochos* impregnates itself, the hyena mounts and is mounted in alternate years. (trans. Peck)

The word τρόχος, which A. L. Peck in his Loeb edition declines to translate and which the standard Greek dictionary, citing only Herodorus, defines simply and unhelpfully as "an animal" (LSJ s.v. τροχός, ὁ B [τρόχος, ὁ] II), is rendered by

---

Heinrich von Staden knows far more than I do about both Aristotle and animals, though perhaps not badgers. I offer this paper to him as a token of thanks for his engaged friendship over nearly two decades, including during the year (2002–2003) I had the privilege of thinking about ancient animals as a Member of the School of Historical Studies at the Institute for Advanced Study. Belated gratitude goes to the American Council of Learned Societies and the National Endowment for the Humanities.

those who are a bit more daring as "badger."¹ My purpose in this brief paper is to re-examine and affirm the evidence for τρόχος as badger and to show that the author of the entry s.v. *Dachs* in the recent and much-consulted *Neuer Pauly* is quite wrong to write that the badger is "[e]in den Griechen wahrscheinlich unbekanntes nachtaktives Raubtier."²

Aristotle's comment appears in a section of his culminating work of zoology in which he deals with the sexual organs and (sometimes unusual) reproductive methods of various animals. After discussing the eggs of birds and then fish (749a10–755a37) and directly before comparing the generation of these two kinds of animals (757a14–b30), the great philosopher and natural scientist gives scathing accounts of a number of beliefs that he views as old wives' tales: that certain fish lack sex-distinctions (755b1–756a5) and conceive by swallowing milt (756a5–b12), the latter a foolish story (τὸν εὐήθη ... λόγον, 756b5–6) found also in Herodotus (2.93); that ravens and ibises copulate by mouth while weasels give birth orally (756b13–757a2), as Anaxagoras and some other philosophers hold, very simplistically and without adequate reflection (λίαν ἁπλῶς καὶ ἀσκέπτως, 756b17);³ and—to round out these erroneous tales—that the hyena and the τρόχος are hermaphroditic (757a2–13).

---

1 Notably, Pierre Louis gives "blaireau" in his Budé edition (*Aristote, De la génération des animaux* [Paris: Les Belles Lettres, 1961], 116, adding in n. 2 that "[l]'identification est incertaine"); he is followed by Luc Brisson (*Le sexe incertain: androgynie et hermaphrodisme dans l'Antiquité gréco-romaine*, 2nd ed. [Paris: Les Belles Lettres, 2008], 120), while Diego Lanza (*La riproduzione degli animali / De generatione animalium*, in Diego Lanza and Mario Vegetti [eds.], *Opere biologiche di Aristotele* [Turin: Unione Tipografico-Editrice Torinese, 1971], 775–1042, at 951) does not offer a translation and wonders whether "si tratti di un animale immaginario" (n. 27). André Wartelle ("Brèves remarques de vocabulaire grec," *Revue des études grecques* 113 [2000]: 211–19, at 214) is skeptical but offers no opinion of his own; see also Elwira Kaczyńska and Krzysztof Tomasz Witczak, "*Mustelidae* in the Cretan Dialect of Modern Greek," *Eos* 94 (2007): 289–305, at 302.
2 Christian Hünemörder, "Dachs," in Hubert Cancik and Helmuth Schneider (eds.), *Der neue Pauly: Enzyklopädie der Antike. Altertum*, vol. 3 (Stuttgart: J. B. Metzler, 1997), col. 257, citing in support the section on badgers in Otto Keller's standard but outdated work on ancient animals: *Die antike Tierwelt*, vol. 1: *Säugetiere* (Leipzig: Engelmann, 1909), 173–75, with notes at 427. As for τρόχος specifically, Keller's statement that "[d]ie gewöhnliche Ansicht, daß das Wort τρόχος bei Aristoteles den Dachs bezeichne, ist grundlos" (174) is itself insufficiently backed up, based principally on the claim, "also ἀρκόμυς d. h. Bärenmaus, nannte das griechische Volk den Dachs"; even if ἀρκόμυς, a medieval form, does mean "badger" (see below in the text), it does not follow that there could not be other names.
3 There is an especially large literature on Greco-Roman weasels, thanks above all to Maurizio Bettini, who has written one wonderful book and many articles on them: see in the first place *Nascere: storie di donne, donnole, madri ed eroi* (Turin: Einaudi, 1998); on oral birth, see 162–72 and index s.v. "concezione, dalla bocca."

How does Aristotle know that these last two animals are in fact normally sexed? He goes on immediately to explain:

> ὦπται γὰρ ἡ ὕαινα ἓν ἔχουσα αἰδοῖον· ἐν ἐνίοις γὰρ τόποις οὐ σπάνις τῆς θεωρίας· ἀλλ' ἔχουσιν αἱ ὕαιναι ὑπὸ τὴν κέρκον ὁμοίαν γραμμὴν τῷ τοῦ θήλεος αἰδοίῳ. ἔχουσι μὲν οὖν καὶ οἱ ἄρρενες καὶ αἱ θήλειαι τὸ τοιοῦτον σημεῖον, ἀλλ' ἁλίσκονται οἱ ἄρρενες μᾶλλον· διὸ τοῖς ἐκ παρόδου θεωροῦσι ταύτην ἐποίησε τὴν δόξαν. ἀλλὰ περὶ μὲν τούτων ἅλις τὰ εἰρημένα. (757a7–13)
>
> In some localities, however, there is ample opportunity for inspection, and the hyena has been observed to possess one pudendum only; but hyenas have under the tail a line similar to the female pudendum. Both male and female ones have this mark, but as males are captured more frequently, casual inspection has given rise to this erroneous idea. I have now said enough on these subjects. (trans. Peck)

In fact, there is plenty more to say, and as far as the hyena is concerned—an animal that is widely viewed as a repulsive witch's familiar—scholars and laymen alike have been saying it, from thousands of years ago right up to the present.[4] To quote Aristotle, the hyena's supposed hermaphroditic nature was a topos already in his day (φασὶ γὰρ τὴν μὲν ὕαιναν πολλοί ...), and the Stagirite brings up the matter in another work as well: *hist. anim.* 6.32 (579b15–29), discussed below. Among the ancient accounts of the hyena's sexual ambiguity, Ov. *met.* 15.408–10 is likely to be the most familiar;[5] for contemporary African folk-beliefs, the article

---

4 The classic book on hyenas (just the spotted kind; see below in the text) is Hans Kruuk, *The Spotted Hyena: A Study of Predation and Social Behavior* (Chicago: University of Chicago Press, 1972); see also Kruuk's *Hyaena* (London: Oxford University Press, 1975). Experts on the sexual morphology and behavior of the spotted hyena include Stephen E. Glickman and Kay E. Holekamp, both of whom have written many fascinating publications on the social as well as biological differences between males and females. Glickman's paper on the history of the animal's bad reputation from Aristotle to the 1994 Disney movie *The Lion King* ("The Spotted Hyena from Aristotle to the Lion King: Reputation is Everything," *Social Research* 62 [1995]: 501–37)—among other things, it is said to be sexually ambiguous, to be a scavenger, to be ugly, to stink, and to laugh like, well, a hyena—is both delightful and authoritative; Holekamp's homepage (http://hyenas.zoology.msu.edu) and the website of the IUCN Hyaena Specialist Group (http://www.hyaenidae.org), which Holekamp helps maintain, give outstanding accounts of the latest research. See also Gus Mills and Heribert Hofer (eds.), *Hyaenas: Status Survey and Conservation Action Plan* (Gland, Switzerland: IUCN, 1998) (scientific and policy-oriented), Holger Funk, *Hyaena: On the Naming and Localisation of an Enigmatic Animal* ([Munich]: GRIN, 2010) (*rezeptionsgeschichtlich*), Gus and Margie Mills, *Hyena Nights & Kalahari Days* (Auckland Park, South Africa: Jacana, 2010) (popular), and Mikita Brottman, *Hyena* (London: Reaktion, 2012) (semipopular).

5 *Si tamen est aliquid mirae nouitatis in istis,* | *alternare uices et, quae modo femina tergo* | *passa marem est, nunc esse marem miremur hyaenam.* See also Ael. *nat. anim.* 1.25 and Opp. *cyn.* 3.288–

of first resort is T. O. Beidelman's "Ambiguous Animals," in which it is reported that the Kaguru of east-central Tanzania, who have no love for the creatures, "believe that hyenas are hermaphroditic, a common belief in Africa and elsewhere."[6]

The reason people both ancient and modern view the hyena as they do is not at all mysterious: high on the list of wonders of the animal kingdom is the fact that the female spotted hyena has an extraordinary hypertrophied (ca. 8-inch) and erectile clitoris, through which she urinates, copulates, and gives birth.[7] This pseudo-penis is very easy to mistake for a phallus, even by experts, and the female has no vagina as such but only an accumulation of fused labial tissue near the anus that Hans Kruuk calls a "sham scrotum."[8] Under the circumstances, what is perhaps surprising is that Aristotle *rejects* the idea of hermaphroditism. Be that as it may, there is a potentially serious problem with Aristotle's

---

92, as well as Diodor. Sic. 32.12.2 (joining Aristotle in denying the widespread belief) and Plin. *nat.* 8.105 (*uulgus credit, Aristoteles negat*). Note, too, Aesop's two fables that involve hyenas: 242 and 243 Perry (= respectively 341 and 340 Chambry).

6 T. O. Beidelman, "Ambiguous Animals: Two Theriomorphic Metaphors in Kaguru Folklore," *Africa* 45 (1975): 183–200, at 190.

7 Besides the material cited in n. 4 above, see Glickman et al., "Mammalian Sexual Differentiation: Lessons from the Spotted Hyena," *Trends in Endocrinology and Metabolism* 17.9 (2006): 349–56, with comparative photographs at 350 (Figure 1) of a semi-erect penis and an erect clitoris. Credit for the determination that the spotted hyena is sexually dimorphic after all—i.e., that it is not a hermaphrodite but rather a pseudohermaphrodite—is generally given to the British anatomist Morrison Watson, who beginning in 1877 wrote a series of papers on the animal's anatomy (on the female, see in the first place "On the Female Generative Organs of *Hyæna crocuta*," *Proceedings of the Zoological Society of London* [1877]: 369–79 + Plates XL and XLI and also "On the Male Generative Organs of *Hyæna crocuta*," *Proceedings of the Zoological Society of London* [1878]: 416–28 + Plates XXIV and XXV," at 423–25 and "Additional Observations on the Anatomy of the Spotted Hyæna," *Proceedings of the Zoological Society of London* [1881]: 516–21 + Plate XLIX, at 516–19—all still well worth reading and with exquisite plates); see, however, now Holger Funk, "R. J. Gordon's Discovery of the Spotted Hyena's Extraordinary Genitalia in 1777," *Journal of the History of Biology* 45 (2012): 301–28 on the observations of the Dutch explorer Robert Jacob Gordon from a century earlier (and also the work of Thomas Bewick; see below in the text, with n. 37). (It may also be noted that Watson opens his 1877 paper with the following charming statement: "Before entering on the subject of this communication, I desire to state that to Mr. A. H. Garrod is undoubtedly due the merit of having first recognized the peculiar characteristics of the female generative organs of *Hyæna* [now *Crocuta*] *crocuta*, he having some time since examined a specimen which died in the [Zoological] Society [of London]'s garden. The structure of these organs, however, struck him as so peculiar that he hesitated to publish the account of his dissection until such time as he should be able to verify his observations by the examination of a second specimen" ["On the Female Generative Organs," 369].)

8 Kruuk, *The Spotted Hyena*, 211 and *Hyaena*, 75.

hyena, and researchers since Morrison Watson have not hesitated to point it out:[9] Aristotle would seem in his "inspection" to be confusing two of the four extant species in the family Hyaenidae, namely the striped hyena (*Hyaena hyaena*), which has (and since Neolithic times has had) a reasonably wide geographical distribution, and the spotted hyena (*Crocuta crocuta*[10]), which is known only from sub-Saharan Africa.[11] Stephen E. Glickman sums up the situation thus:

> As has been noted by a number of biologists (Watson, 1878 …), Aristotle was surely dealing with striped hyenas, when the rumor must have originated with spotted hyenas.[12] In his initial general description of the hyena, Aristotle writes, "The hyena in colour resembles the wolf, but is more shaggy, and is furnished with a mane running all along the spine" (*HA*, VI, 3[2], 579b, 15[–16]). Such a mane is characteristic of the striped hyena, not the spotted

---

9 See Watson, "On the Male Generative Organs," 425–28.
10 The evidently non-native word κροκόττας (also κοροκότ(τ)ας and κροκούττας) is first attested in Ctesias (fr. 76 Lenfant), who uses it to describe an "Ethopian" animal commonly known as κυνόλυκος (literally "dog-wolf"). The κροκόττας and the ὕαινα are brought together in Ael. *nat. anim.* 7.22, and while it might seem a good rule of thumb, when one encounters either word in a Greek source, to start from the assumption that the latter refers to the striped hyena while the former (when it refers to a hyena at all) designates the spotted variety, this may actually not be a good idea (*pace* Andrew Nichols, *Ctesias on India and Fragments of his Minor Works* [London: Bristol Classical Press, 2011], 170): for one thing, Oppian refers to both animals as ὕαινα in *cyn.* 3.262–92; and for another, it is possible that Ctesias' Ethopian animal is actually Indian, in which case (see immediately below in the text), if it is describing a hyena, it almost certainly has to be a striped one. For discussion, see now Nichols, *Ctesias on India*, 170–72.
11 Heribert Hofer and Gus Mills ("Worldwide Distribution of Hyaenas," in Mills and Hofer, *Hyaenas: Status Survey*, 39–63) provide maps and discussion of the "worldwide distribution of hyaenas"—now and in the recent past as well (for the distant past, see Nadin Rohland et al., "The Population History of Extant and Extinct Hyenas," *Molecular Biology and Evolution* 22 [2005]: 2435–43 and n. 17 below). Whereas "[t]he historical distribution of the striped hyena encompasses Africa north of and including the Sahel zone, eastern Africa south into Tanzania, the Arabian Peninsula and the Middle East up to the Mediterranean shores, Turkey, Iraq, the Caucasus (Azerbaidjan, Armenia, Georgia), Iran, Turkmenistan, Uzbekistan, Tadzhikistan, Afghanistan (excluding the higher areas of the Hindukush), and the Indian subcontinent" (Hofer and Mills, "Worldwide Distribution of Hyaenas," 44), the spotted hyena is "[h]istorically widespread throughout Africa south of the Sahara" (54). As for the other two species of hyena, the brown hyena (*Hyaena brunnea*) and the aardwolf (*Proteles cristata*), these are found almost exclusively in southern Africa, though there are records of the latter all along the eastern part of the continent.
12 I note that Beidelman's account of the Kaguru, cited above in the text, definitely concerns spotted hyenas, for the author specifically states that "[t]he hyena's spotted and rough coat may also be a source of repulsion, for mottled surfaces are repugnant to Kaguru" ("Ambiguous Animals," 190). On the difficulties sometimes with knowing from a report whether an African hyena is spotted or striped, see Jürgen W. Frembgen, "The Magicality of the Hyena: Beliefs and Practices in West and South Asia," *Asian Folklore Studies* 57 (1998): 331–44, at 332–35.

hyena. From this single anatomical characteristic, it is clear that Aristotle was not considering the species on which the original rumor was based. The description of a "line beneath the tail" might also fit the striped hyena better than the spotted. Although all hyenids have anal scent glands in this position, both brown and striped hyenas "... possess a large glandular pouch below the tail which largely obscures the external genitalia ..." ([E. P.] Walker [et al., *Mammals of the World*]). Finally, in Aristotle's time, striped hyenas would have been common in North Africa and the Middle East, while the range of the spotted hyena had largely, if not completely, retreated to sub-Saharan Africa. D'Arcy Thompson concluded that Aristotle did much of his writing on biological subjects while residing on the Isle of Lesbos, off the coast of Turkey, and did some travelling on the mainland as well .... Turkey could have been prime striped hyena habitat in those times .... It is likely that the hunters that Aristotle consulted would have been familiar with striped hyenas, and, perhaps, he even had the opportunity for direct examination of this animal. His anatomical descriptions are certainly quite precise.[13]

Glickman and others could be right that the parazoological rumor about the hyena's sexuality began with the spotted variety; certainly this animal's anatomy is overtly remarkable. But is Jürgen W. Frembgen, in a paper about the (striped) hyena in western and southern Asian folklore that is intended as something of a counterweight to the frequent reports from Africa, right to say that "the striped hyena is an ordinary mammal without any peculiar anatomy" and that "[w]hereas the spotted hyena is loaded with such symbolic meanings against the background of biological facts, the same does not hold true for the striped hyena, whose anatomy has no androgynous characteristics at all"?[14] And if he is right, then what can explain the widespread association that he documents between the sexual organs and rectum of the striped hyena and love magic—heterosexual, homosexual, and bisexual—especially in Afghanistan and Pakistan?[15]

---

13 Glickman, "The Spotted Hyena," 510–11 (footnote omitted).
14 Frembgen, "The Magicality of the Hyena," 340.
15 Here are two examples from Frembgen, "The Magicality of the Hyena": "In Afghanistan ... if a man looks through the vulva [of a striped hyena that has been treated in a certain ritual fashion] ..., he will definitely get hold of the one he is longing for. A particular aspect of this belief and custom is the alleged inevitability of its successful use. This has led to the proverbial expression in Dari (Afghan Persian) of *kus-e kaftar bay*, as well as in Pashto of *kus-e kaftar ware*, both invoking the 'vulva of the hyena' and roughly meaning 'it's going like clockwork' (i.e., it happens as smoothly as if you would look through the vulva of a female striped hyena)" (339–40) and "Among the Pakhtun of the North-West Frontier Province ... and Baluchistan, the vulva is kept in red *sandur* powder (vermilion), itself having the connotations of marriage, sexuality, and fertility. To attract young men, homosexuals and bisexuals likewise use the rectum of a striped hyena, if possible cut out of a freshly dead animal while the sphincter of the anus is still moving. The expression 'to possess the anus of a [striped] hyena' therefore denotes somebody who is attractive and has many friends (lovers). For the same reason a striped hyena's penis can

Frembgen ends his discussion with two ideas, neither of which, he admits, has terribly much support: perhaps "hearsay on the distinct sexuality of the African spotted hyena traveled northwards and eastwards ... and reached western and southern Asia ... [where it] could have been orally transmitted by traders, dervishes, migrants, etc., and subsequently transferred to the local striped hyena"; or perhaps we should assume the "earlier existence of the spotted hyena in parts of western and southern Asia."[16]

It is possible, to be sure, that one of these ideas is correct;[17] also possible is that the putative sexual connection between the two kinds of hyenas is actually not all that interesting since love magic can involve the body parts, including genitalia, of all sorts of animals.[18] But I think there is a more convincing explanation, and to get there, I suggest that we return to the passage of Aristotle with which we began and move away from the hyena and back to the mysterious τρόχος. It is clear that Aristotle's principal wish in *gen. anim.* 3.6 is to disabuse his readers of any belief that they might have in an annually sexually reorienting hyena; he is just not that interested in the self-copulating τρόχος, an animal about which his reticence and the invocation of a comparatively obscure mytho-

---

be kept in a small box filled with *sandur*" (340); compare Brottman, *Hyena*, 65–66, part of her lengthy discussion of "hyena magic" (61–93, with notes at 151–52). On the use of parts of the striped hyena as an aphrodisiac, see also Heribert Hofer and Gus Mills, "Population Size, Threats and Conservation Status of Hyaenas," in Mills and Hofer, *Hyaenas: Status Survey*, 64–79, at 68 (Table 5.3) and Marion L. East and Heribert Hofer, "Cultural and Public Attitudes: Improving the Relationship between Humans and Hyaenas," in Mills and Hofer, *Hyaenas: Status Survey*, 96–102, at 96.

16 Frembgen, "The Magicality of the Hyena," 341.
17 Funk ("R. J. Gordon's Discovery," 324–25) is appropriately skeptical of the first idea, which goes back to Watson, "On the Male Generative Organs," 427–28. In support of the second, I note the claim of Ishwar Prakash ("Hyena," in R. E. Hawkins [ed.], *Encyclopedia of Indian Natural History* [Delhi: Oxford University Press, 1986], 300–301, at 300) that "[r]emains of the African spotted *Crocuta* have been found in caves near Karnool, Tamil Nadu." But when do these remains date from? The paleospecies *Crocuta crocuta spelaea*, otherwise known as the cave hyena, is well attested in the Eurasian fossil record, but the evidence does not seem to extend south of Pakistan and is in any case from the Pleistocene era; see Rohland et al., "The Population History."
18 "Pliny the Elder, in particular, provides abundant evidence ... that men believed they could increase their sexual potency or stamina by eating or wearing parts of ... strange animals such as the crane and the skink" (Christopher A. Faraone, *Ancient Greek Love Magic* [Cambridge, Mass.: Harvard University Press, 1999], 20; n. 90 provides references to classical sources—above all Plin. *nat.* 28.261–62—and to secondary literature on ancient Mesopotamia and medieval Europe).

grapher from ca. 400 B.C., Herodorus of Heraclea, suggest that he knows little.[19] Can we do better?

The first thing to be sure of is that there is actually a word τρόχος to do better with. The apparatus to 757a4 and 6 in the OCT reports, "τροχίλος Ξ,"[20] but τροχίλος is the word for quite another animal. At least two other animals in fact: both "wren" and "Egyptian plover."[21] But this is no normal *varia lectio*: Ξ stands for "translatio Arabica a Yaḥyā ibn al-Biṭrīq confecta, s. ix[in.],"[22] and what we find in the Arabic is *uṭrūshīlūs*,[23] that is to say, an attempt to render τροχίλος.[24] Clearly τρόχος is the *lectio difficilior*, and since neither Aristotle nor anyone else points to peculiar sexual features of or behavior in either kind of τροχίλος, further comment is largely unnecessary. Still, it is worth adding that another word for τροχίλος (sometimes τρόχιλος) in the meaning "wren" is ὀρχίλος (sometimes ὄρχιλος).[25] Since the latter word is very likely to

---

19 On Herodorus, see F. Jacoby, "Herodoros 4)," in *Paulys Real-Encyclopädie der classischen Altertumswissenschaft*, vol. 8.1 (Stuttgart: J. B. Metzler, 1912), cols. 980–87, with 982 on Aristotle's use of his work; for a summary account, see Fritz Graf, "Herodoros," in Hubert Cancik and Helmuth Schneider (eds.), *Der neue Pauly: Enzyklopädie der Antike. Altertum*, vol. 5 (Stuttgart: J. B. Metzler, 1998), col. 469.
20 H. J. Drossaart Lulofs, *Aristotelis De generatione animalium* (Oxford: Clarendon Press, 1965), 116.
21 For details, see D'Arcy Wentworth Thompson, *A Glossary of Greek Birds*, new ed. (London: Oxford University Press, 1936), 287–89 and W. Geoffrey Arnott, *Birds in the Ancient World from A to Z* (London: Routledge, 2007), 247–49, the latter with much recent secondary literature; according to Arnott, there may be three or even four kinds of τροχίλος. The name probably means "runner" (thus Arnott, *Birds in the Ancient World*, 247; see also José Luis García Ramón, "Anthroponymica Mycenaea: I. Mykenisch *o-ki-ro*, alph.-gr. ὀρχίλος; II. Mykenisch *da-te-wa /Daitēwās/* und *e-u-da-i-ta*, alph.-gr. Δαίτας, Πανδαίτης," *Minos* 35–36 [2000–2001 (publ. 2002)]: 431–42, at 434–35 and "Der Name *o-ki-ro /ork^hilos/*: Eine überzeugende Deutung von Prof. Ilievski," *Minos* 37–38 [2002–2003 (publ. 2006)]: 371–72, at 371); in view of the roadrunner, Nan Dunbar (*Aristophanes, Birds* [Oxford: Clarendon Press, 1995], 159) is obviously mistaken in her claim that "we [in English] have no bird-name derived from 'run'."
22 Drossaart Lulofs, *Aristotelis De generatione animalium*, xxxii.
23 See J. Brugman and H. J. Drossaart Lulofs, *Aristotle, Generation of Animals: The Arabic Translation Commonly Ascribed to Yaḥyā ibn al-Biṭrîq* (Leiden: Brill, 1971), 117 and 205. I am grateful to my colleague Michael Cook for his gracious assistance with the Arabic.
24 Around 1220, Michael Scot translated Aristoteles Arabus into Latin, which led for our passage to the following (amusing) outcome: *Et hoc quod dicunt de ave qui dicitur graece atrosilus et de alzabot falsum est. Et Aradoroz qui erat ex Hirkala dicit quod alzabo habet duo membra, scilicet membrum maris et feminae, et quod acriziloz coit secum et quod zabo coit et coitur in quolibet anno econtra* ... (Aafke M. I. van Oppenraaij, *Aristotle, De animalibus: Michael Scot's Arabic-Latin Translation*, vol. 3: *Books XV–XIX: Generation of Animals* [Leiden: Brill, 1992], 140–41, with apparatus).
25 See Thompson, *A Glossary of Greek Birds*, 219–20 and Arnott, *Birds in the Ancient World*, 158.

be a derivative of ὄρχις "testicle" (and come from Proto-Indo-European *$h_1erĝ^h$- "mount sexually"),[26] one may wonder whether the context in Aristotle has contributed to the substitution of τροχίλος for τρόχος—all the more so since, as we are about to see, badgers are renowned for their nether parts.

As already noted, those who discuss the passage sometimes identify our word as a badger, an animal that one prominent scholar has suggested was unknown to the Greeks. The best evidence for the equation between Aristotle's τρόχος and "badger" seems to be that it is the *terminus technicus* for the Eurasian badger (*Meles meles*) in katharevousa[27]—not the usual Modern Greek word, ασβός, to which I shall return. But even if the identification is correct, as I believe it is, it does not help us understand why anyone would believe that the badger impregnates itself. I suggest that my prior work on badgers helps solve the puzzle,[28] for the idea that an animal known to us only because of an unusual genital practice would be a badger fits neatly with the main contention of my original paper on these creatures, namely that the Hittite noun *tašku-*, though attested with the meaning of a genito-anal body part (something like "anus" or "scrotum"), is etymologically the *Wanderwort* *tasku- (*vel sim.*), meaning "badger," a designation maintained, for the most part, by its manifestations elsewhere in Indo-European, notably Germanic (e.g., German *Dachs*),

---

[26] See García Ramón, "Anthroponymica Mycenaea," 431–36 and also "Der Name *o-ki-ro*," with special reference to P. Hr. Ilievski. The literature on Indo-European testicles and the root *$h_1erĝ^h$*- has grown hugely in recent years—indeed, I have contributed to it myself—but this is not the place for further comment. There is no reason to follow Thompson, *A Glossary of Greek Birds* in believing that τροχίλος may "not [be] connected with τρέχω" (287)—see n. 21 above; that said, the claim in Robert Beekes, *Etymological Dictionary of Greek*, 2 vols. (Leiden: Brill, 2010), 1507 that "[t]he basic meaning of the verb ... is confirmed by ... τροχίλος 'birds [sic] that run'" is bizarre—or that it and ὀρχίλος are "probably identical words and of foreign origin" (220).

[27] See, e.g., Kaczyńska and Witczak, "*Mustelidae* in the Cretan Dialect," 302: "Mod. Gk. τρόχος 'badger' ... is a literary word, introduced to the scientific nomenclature from the Aristotelian work *De generatione animalium* III 6 (757a). However, the semantics of Anc. Gk. τρόχος remains far from being established with certainty."

[28] See Joshua T. Katz, "Hittite *tašku-* and the Indo-European Word for 'Badger'," *Historische Sprachforschung* 111 (1998): 61–82 and also "How the Mole and Mongoose Got their Names: Sanskrit *ākhú-* and *nakulá-*," *Journal of the American Oriental Society* 122 (2002) [= Joel P. Brereton and Stephanie W. Jamison (eds.), *Indic and Iranian Studies in Honor of Stanley Insler on his Sixty-fifth Birthday*]: 296–310. Both articles cite ample secondary literature on the scientific matters discussed below. Two important books have appeared since I submitted the first one: Ernest Neal and Chris Cheeseman, *Badgers* (London: Poyser, 1996) and now Timothy J. Roper, *Badger* (London: Collins, 2010), the latter written mainly because "during the last 15 years an enormous amount of new research on badgers has been published" (ix).

Celtic (e.g., the Gaulish onomastic element *Tasc/go[-]*),²⁹ Iranian (Ossetic *tæxsal* [æ]),³⁰ and possibly Indic³¹ (Sanskrit *nakulá-* "mongoose").³² While a semantic shift from "badger" to "private parts" may seem peculiar, I argue that the badger is so closely associated with its habitual activity of "musking"—i.e., marking territory by squatting—that the word for the animal comes to be used for its musky bits, in an uncharacteristic, but explicable, whole-for-part relationship.³³ Most

---

29 Late Latin *taxo* (gen. *taxonis*), whence such Romance words for "badger" as Spanish *tejón* and *taisson* in dialectal French, is a borrowing from Celtic or, possibly, Germanic.
30 According to V. I. Abaev, *Istoriko-ètimologičeskij slovar' osetinskogo jazyka*, vol. 3 (Leningrad: Nauka, 1979), 284–85, it is not clear exactly what kind of animal a *tæxsal(æ)* is, but it is "similar to a marten" (284: "poxož na kunicu"). (I missed the Ossetic form in my previous papers.)
31 See Katz, "How the Mole and Mongoose Got their Names."
32 Like most etymological proposals, this one has been evaluated both positively and negatively: see Joshua T. Katz, "To Turn a Blind Eel," in Karlene Jones-Bley et al. (eds.), *Proceedings of the Sixteenth Annual UCLA Indo-European Conference, Los Angeles, November 5–6, 2004* (Washington, D.C.: Institute for the Study of Man, 2005), 259–96, at 282 n. 53 for some references, to which add, e.g., Xavier Delamarre, *Dictionnaire de la langue gauloise: une approche linguistique du vieux-celtique continental*, 2ⁿᵈ ed. (Paris: Errance, 2003), 293 (pro) and Alwin Kloekhorst, "Hittite *ḫāpūša(šš)-* (Formerly Known as *ḫapuš-* 'Penis')," *Journal of Indo-European Studies* 33 (2005): 27–39 (contra, proposing that the Hittite word means rather "thighbone"). The existence of *tasku-* does not rule out the possibility that other words for "badger" were known in and around Indo-European territory in ancient times, especially since the badger is one of a few species "found so widely over Eurasia that they should have been familiar to the Proto-Indo-Europeans irrespective of where their homeland lay" (J. P. Mallory and D. Q. Adams, *The Oxford Introduction to Proto-Indo-European and the Proto-Indo-European World* [Oxford: Oxford University Press, 2006], 153). Mallory and Adams (*The Oxford Introduction*, 141) themselves reconstruct only *meli-*, on the basis of Latin *meles* and "Slovenian *melc*," by which they presumably mean Slovincian *məlc*; see Katz, "How the Mole and Mongoose Got their Names," 297 n. 7, with references, above all to V. Blažek. According to Elwira Kaczyńska and Krzysztof Tomasz Witczak ("Remarks on Some Doric Elements in the Modern Greek Dialects of Crete," *Eos* 92 [2005]: 112–20, at 113–14 and "*Mustelidae* in the Cretan Dialect," 298–301), the Modern Cretan dialect word ἄρκαλος "badger"—connected to Aelian's ἄρκηλος (*nat. anim.* 7.47), which means "young leopard" or some kind of leopard other than the usual πάρδαλις, and to Hesychius' gloss ἀρκήλα· ⟨τὸ⟩ ζῷον. Κρῆτες τὴν ὕστριχα—goes back to a Proto-Indo-European form *u̯r̥ko-* "badger," is related to such words as Nepali *bharsia* (and, via Turkish, Polish *borsuk* and Russian *barsuk*), and instantiates the *Benennungsmotiv* of the "fat" badger; I do not have a firm opinion on these ideas but am not persuaded by the evidence they adduce for the naming of badgers after fat.
33 Compare now the proposal of Irene Balles ("Some New Celtic and Other Etymologies," in Giovanni Belluscio and Antonio Mendicino [eds.], *Scritti in onore di Eric Pratt Hamp per il suo 90. compleanno* [Rende, Italy: Università della Calabria, Centro Editoriale e Librario, 2010], 15–19, at 15–16) that Proto-Celtic *brokko-* "badger" (whence, e.g., Old Irish *brocc* and Welsh *broch*, then borrowed into English as *brock*) reflects Proto-Indo-European *bʰrog-ko-*, which she defines as "stinking, the stinking one, skunk" and with which she compares Old Irish *broimm* and Welsh

striking of all is that badgers have some extra equipment down there that they use in musking: they employ not just their anal scent gland, like most animals that scent-mark territory (e. g., skunks), but also a unique "subcaudal gland," between the tail and the anus, which was properly identified only in recent years and which a leading expert on badgers—not incidentally, also an authority on hyenas—describes as "a very large pocket ..., enormous and very unusual."[34] It would, I believe, be quite natural for someone—Herodorus, say—to believe that this extra opening in an identifiably male badger was a vagina and, then, to come up with a tale about how this animal, which displays both male and female genitalia, mounts itself.

Now, it happens that, despite Frembgen's words about the normality of the striped hyena's anatomy (see above), there is in fact something curious about its nether parts—and I am by no means the first person to point this out. The peculiarity in question is very different from, and certainly less spectacular than, the spotted hyena's long clitoris; on the other hand, it is similar to the peculiarity just noted for the badger. Here is what none other than Aristotle himself writes about it in *hist. anim.* 6.32, his other treatment of the genitalia of the hyena besides *gen. anim.* 3.6:

> ἡ δὲ ὕαινα ... περὶ δὲ τῶν αἰδοίων ὃ λέγεται, ὡς ἔχει ἄρρενος καὶ θηλείας, ψεῦδός ἐστιν. ἀλλ' ἔχει τὸ μὲν τοῦ ἄρρενος ὅμοιον τῷ τῶν λύκων καὶ τῶν κυνῶν, τὸ δὲ δοκοῦν θηλείας εἶναι ὑποκάτω μὲν ἔχει τῆς κέρκου, παραπλήσιον δ' ἐστὶ τῷ σχήματι τοῦ θήλεος, οὐκ ἔχει μέντοι οὐδένα πόρον· ὑποκάτω δ' ἐστὶν αὐτοῦ ὁ τῆς περιττώσεως πόρος. ἡ δὲ θήλεια ὕαινα ἔχει μὲν καὶ τὸ ὅμοιον τῷ τῆς θηλείας λεγομένῳ αἰδοίῳ, ἔχει δ' ὥσπερ ὁ ἄρρην αὐτὸ ὑποκάτω τῆς κέρκου, πόρον δὲ οὐδένα ἔχει· μετὰ δὲ τοῦτο ὁ τῆς περιττώσεώς ἐστι πόρος, ὑποκάτω δὲ τούτου τὸ ἀληθινὸν αἰδοῖον. (579b16–26)

> The statement is made that the hyena has both male and female sexual organs; but this is untrue. In fact the sex-organ of the male resembles that of the wolf and the dog; the organ which is thought to be like that of the female is under the tail, and in shape does resemble that of the female, but there is no passage connected with it, and the passage for the residue is situated underneath it. The female hyena indeed has the organ which is said to be similar to a female organ, but, as in the male, it is underneath the tail and has no passage connected with it; after that is placed the passage for the residue, and underneath this the true genital organ. (trans. Peck)

Let us jump twenty-three centuries. In her standard work *The Carnivores*, R. F. Ewer discusses the perianal region of the badger and the hyena (both striped

---

bram "fart": "This is semantically well-motivated, because the badger, like all *Mustelidae*, emits a foul-smelling secretion" (16).

34 Hans Kruuk, *The Social Badger: Ecology and Behaviour of a Group-living Carnivore* (Meles meles) (Oxford: Oxford University Press, 1989), 130. For Kruuk's work on hyenas, see n. 4 above.

and spotted) together, even providing side-by-side drawings of the area in the (male) badger and the (female) striped hyena.³⁵ She concludes:

> It seems to me ... that the circum-anal pouches of Herpestinae and *Cryptoprocta* [i.e., mongooses and the fossa], the largely super-anal but entirely comparable pocket of hyaenids, the badger supra-anal or sub-caudal pocket and the anal pouches of *Mellivora* and *Ailurus* [i.e., the ratel and the red panda], although they clearly have more than a single evolutionary origin, are all basically similar. ... There is little justification for distinguishing further between them.³⁶

Already the great naturalist and engraver Thomas Bewick (1753–1828) writes of the (specifically) striped hyena that

> immediately underneath the tail, and above the anus, there is an orifice like that of the Badger, which opens into a kind of pouch, and contains a substance of the consistence of civet, but of a rank, disagreeable odour. This opening may probably have given rise to the error of the ancients, who asserted, that the Hyena was every alternate year male and female.³⁷

In his 1910 Oxford translation of *De generatione animalium*, Arthur Platt, who says of τρόχος that "[i]t is impossible to guess what this animal may have been,"³⁸ states that Aristotle's story about the hyena "probably arose" from consideration of the striped creature's "post-anal median glandular pouch."³⁹ And as we have seen, Glickman, who relies on Platt, follows him in mentioning the pouch—which the hyena employs in what is called "pasting," analogous to the badger's use of the subcaudal gland in musking—though he does not make explicit the interest I believe it holds.⁴⁰ In short, I propose that whatever information about the spotted hyena may lurk in the background of Aristotle's report, it is likely that in *gen. anim.* 3.6 he is in fact bringing the badger and

---

35 See R. F. Ewer, *The Carnivores* (Ithaca: Cornell University Press, 1973), 98–100, with Figure 3.11. See now the photograph of the badger's anal region in Roper, *Badger*, 188 (Figure 130).
36 Ewer, *The Carnivores*, 99; the glands of various other musky animals, including the ones she mentions here, are considered at 90–100.
37 Thomas Bewick, *A General History of Quadrupeds* (Chicago: University of Chicago Press, 2009), 299—a reprint of the 1885 edition of a work that first appeared in 1790.
38 Arthur Platt, *De generatione animalium* (Oxford: Clarendon Press, 1910), *ad* 757a3, n. 1.
39 Platt, *De generatione animalium*, *ad* 757a3, n. 2, with reference to W. H. Flower and R. Lydekker's classic study of mammals. Compare Watson, "On the Male Generative Organs," 426 and esp. 427–28, though the discussion is somewhat tangled; see also Funk, *Hyaena*, 22, as well as the "additional note" to our passage in *gen. anim.* 3.6 in A. L. Peck, *Aristotle, Generation of Animals*, rev. ed. (Cambridge, Mass.: Harvard University Press, 1953), 565–66.
40 See Glickman, "The Spotted Hyena," 510.

the striped hyena together in an anatomically coherent fashion.⁴¹ To put it simply, while the female spotted hyena sports a pseudo-penis, both the male striped hyena and the male badger have a pseudo-vagina.⁴²

At this point, it makes sense to return to the question of the very existence of badgers in ancient Greece. Since *Meles meles* is indigenous to the region, there is no *a priori* reason to doubt that there was an old word—or more than one old word—in Greek, and yet it is the case that, to quote Otto Keller, "[v]om Dachs ... ist weder bei den Griechen noch bei den Römern viel die Rede, obgleich er gewiß nicht selten war."⁴³ Still, there is an important difference between saying that badgers were not prominent and claiming that they were "unknown" and thus never spoken of. So far we have seen some evidence—good evidence, I believe, but circumstantial—that Herodorus of Heraclea and, secondarily, Aristotle, called the badger τρόχος, a word that leads a marginal existence even today in scientific parlance. But what other support is there for ancient Greek talk about badgers?

The short answer is that there is not much, and all the forms that we have are in one way or another as problematic as τρόχος.⁴⁴ However, the accumulation of possible words for this perfectly common animal, whose nocturnal habits appear to have made it nonetheless largely invisible to humans, suggests that some of them should be probative.⁴⁵ Perhaps the least interesting word is

---

41 For a capital rather than genital connection between some kind of hyena(-like creature) and a badger, see Plin. *nat.* 8.72, who writes that the *leucrocotas* (a hapax, presumably to be connected to *corocottas* [*nat.* 8.107] and κροκόττας [see n. 10 above]) is endowed *capite melium*.
42 In her splendid book *The Ravenous Hyenas and the Wounded Sun: Myth and Ritual in Ancient India* (Ithaca: Cornell University Press, 1991), Stephanie W. Jamison considers the role of (striped) hyenas in Vedic myth (68–81 and *passim*), concluding that "[t]he gender agreement between Greek [ὕαινα] and Sanskrit [*sā́lāvr̥kī-*, generally translated as "hyena"] might at first seem merely a curious grammatical accident, but in fact it seems to reflect important features of hyena physiology and social organization[, namely that] ... among striped hyenas the sexes look very much alike ... [and t]he prominence of the female physiologically is matched in the organization of hyena societies, which seem to show female dominance" (73–74).
43 Keller, *Die antike Tierwelt*, 173.
44 Much of the material in this paragraph and the next is to be found already in Katz, "How the Mole and Mongoose Got their Names," 298–301.
45 The idea of Angela Steinmeyer-Schareika (*Das Nilmosaik von Palestrina und eine ptolemäische Expedition nach Äthiopien* [Bonn: Habelt, 1978], 64–65, with 142 [Figures 19 and 20]) that the rare word χοιροπίθηκος—literally "hog-monkey" and known from Aristotle (*hist. anim.* 2.11 [503a19]) and the Nile mosaic of Palestrina—means "badger" (or "ein den Griechen bisher unbekanntes Tier, das sie stark an einen Dachs erinnerte" [65]) has nothing to recommend it; see P. G. P. Meyboom, *The Nile Mosaic of Palestrina: Early Evidence of Egyptian Religion in Italy* (Leiden: Brill, 1995), 26 and esp. 125–26, with Figure 16.

ἀρκόμυς,⁴⁶ glossed into Latin in a tenth-century codex as *meles*, which in Classical sources appears at least most of the time to mean "badger." Keller implies that ἀρκόμυς, literally "bear-mouse,"⁴⁷ is our earliest Greek word for "badger,"⁴⁸ but there is no reason to accept that this is so. More striking is the accusative plural πυκτίδας (*v. l.* πικτίδας), attested in Aristophanes nearly a millennium and a half earlier, where it shows up in *Acharn.* 879 in a list of animal-wares that the Boeotian-speaking merchant from Thebes is hawking: "geese, hares, foxes, moles, hedgehogs, cats, πυκτίδας, martens, otters, Copaic eels." Even if the word does refer to animals with a pelt, as seems likely, "the traditional translation 'badgers' is only a guess";⁴⁹ furthermore, it is possible that its sense is entirely different.⁵⁰

This leaves the Cretan dialect form ἄρκαλος, which a pair of Polish scholars have recently suggested is an archaic name for the badger,⁵¹ and the standard Modern Greek word ασβός (sometimes ἀσβος).⁵² Now, ασβός has been taken to be a borrowing from Slavic (cf., e.g., Czech *jezevec* and Slovak *jazvec*), and the resemblance certainly is hard to deny.⁵³ In an earlier paper, however, I suggested—without, I confess, having noticed the Slavic forms—that ασβός is a very old word indeed, reflecting directly *$azg\underset{\sim}{u}$-o-*, the thematization of *$azg\underset{\sim}{u}$-*, a form

---

46 See n. 1 above.
47 Kaczyńska and Witczak ("Remarks on Some Doric Elements," 113–14 and "*Mustelidae* in the Cretan Dialect," 298–302) do not cite ἀρκόμυς, but perhaps the first element is better connected to Modern Cretan ἄρκαλος (see n. 32 above)?
48 Keller, *Die antike Tierwelt*, 174.
49 Thus S. Douglas Olson, *Aristophanes, Acharnians* (Oxford: Oxford University Press, 2002), 293.
50 Mutually incompatible alternative solutions can be found in Carmen Morenilla-Talens, "Aprosdoketon oder Hapax? Aristophanes, Ach. 879," *Glotta* 64 (1986): 216–21 (πυκ-: "writing tablets") and Ove Hansen, "A Note on πικτίδας/πυκτίδας of Aristophanes' Ach. 879," *Philologus* 134 (1990): 158–59 (πικ-: "timber from pine trees"); see the discussion in Olson, *Aristophanes, Acharnians*, 293 (πικ-: "presumably another animal with an attractive or exotic coat"), as well as Wartelle, "Brèves remarques," 214.
51 See nn. 32 and 47 above. Kaczyńska and Witczak ("Remarks on Some Doric Elements," 113–14 and "*Mustelidae* in the Cretan Dialect," 299 and 302) suggest, in part on the basis of Modern Pontic ξυστρίγκι, that the Classical form ὕστριξ (cf. the Hesychian gloss cited in n. 32 above), which is generally translated as "porcupine," may have the primary sense "badger."
52 The contemporary word κουνάβι, though occasionally translated as "badger" (see, e.g., Henriette Walter and Pierre Avenas, *L'étonnante histoire des noms des mammifères: de la musaraigne étrusque à la baleine bleue* [Paris: Laffont, 2003], 378), really means "ferret."
53 See Wanda Budziszewska, *Zapożyczenia słowiańskie w dialektach nowogreckich* (Warsaw: Omnitech, 1991), 9 (supplying many dialect forms), cited with seeming approval by Kaczyńska and Witczak ("Remarks on Some Doric Elements," 113 and "*Mustelidae* in the Cretan Dialect," 301).

of the *Wanderwort* for "mole" (cf. Greek σκάλοψ and (ἀ)σπάλαξ, Sanskrit *ākhú-* "mole-like rat," and Hittite *āšku-* "mole (?)") that travels in concert with **tasku-* "badger" and gets confused with it also in Galatian (there is a Christian sect known as both Τασκοδρουγῖται and *Ascodrogitae/Ascodrobi*, literally probably "Badger-" and "Mole-noses") and Basque (*azkoin* "badger").[54] My etymology of ασβός is phonologically impeccable and, I believe, semantically perfectly justifiable; it is, however, not easy to square with the Slavic data and of course requires us to imagine that the word in question (first attested in Medieval Greek, as ἄσβος[55]) was part of Hellenic vocabulary for thousands of years while remaining unrecorded, or virtually so. This is not impossible since the Greeks simply do not seem to be interested enough in badgers to do more than leave the occasional—fascinating—trace. In the end: *non liquet*.

With that we can return to τρόχος and move toward a summing-up with a few sadly inconclusive thoughts about where this fascinating word may have come from, a matter that the etymological dictionaries ignore. The usual meaning of τρόχος, with barytone accentuation, is of course "(race)course"; like its oxytone counterpart τροχός "wheel," it is a derivative of τρέχω "run" and the Proto-Indo-European verbal root **dʰreĝʰ-* "turn (around an axis)."[56] And so the obvious first thought about our animal is that it is somehow the same as these: after all, τὰ ζῷα τρέχει.[57] But πολλὰ ζῷα τρέχει, and badgers are not among the better trekkers. A second idea—possible but essentially impossible to evaluate—is that τρόχος is somehow based on the *Wanderwort* **tasku-*,[58] with later folk-etymological crossing with τρέχω.[59] And third and finally, I put

---

54 Katz, "How the Mole and Mongoose Got their Names," 300–301.
55 A further connection to Hesychius' ἄψοος (or ἀψός Latte), defined as "some animal that eats vines," is not particularly compelling; see now Kaczyńska and Witczak, "*Mustelidae* in the Cretan Dialect," 300–301.
56 The shape of the root and especially its semantics are the subject of debate, but I follow Françoise Létoublon and Charles de Lamberterie, "La roue tourne," *Revue de philologie* 54 (1980): 305–26 (also Charles de Lamberterie and Françoise Létoublon, "Deux points de vocabulaire homérique: éclairage comparatif," *LALIES* 2 [1980 (publ. 1983)]: 13–16, at 15–16).
57 Compare Walter and Avenas, *L'Étonnante Histoire*, 377, where the repeated claim at 377–78 that the word for "badger" is *"trekhos"* or *"treko"* does not inspire confidence. Incidentally, Michael Weiss (*Outline of the Historical and Comparative Grammar of Latin* [Ann Arbor: Beech Stave, 2009], 211 n. 23) notes that "the illustrative example [τὰ ζῷα τρέχει] is not a true Ancient Greek sentence," seemingly first appearing in the grammatical literature in the Renaissance.
58 As ασβός may perhaps go back to **azgu-*, the *Wanderwort* with which **tasku-* seems to be paired (see above in the text and also immediately below).
59 Similar to this is the idea that τρόχος might be based on something like the Biblical Hebrew word *taḥaš*, which is often said to mean "badger," again presumably with influence from τρέχω. But this is impossible, and not merely because there are so many other animals that *taḥaš* has

forth the possibility, which I would like to be right but have little confidence in, that the -χ- of τρόχος goes back to an aspirated labiovelar *g^{uh}, inherited from Proto-Indo-European, and that the word is related to Trophonius (Τροφώνιος), the cult figure (and, according to some, son of Apollo) who built the first temple of Apollo at Delphi. This idea, at which Luc Brisson may be subtly hinting,[60] might (?) work phonologically and presents no morphological difficulties.[61] But what about the semantics? Why would Trophonius, whose name is more typically connected to the family of τρέφω "nurse, nourish; bring up," be called "The One who Has (or is Associated with) a τρόχος"? I can do no more here than point to the firm bond between Trophonius and Asclepius (Ἀσκληπιός), the god of healing, who is definitely Apollo's son and who may be the "Mole-God" since a number of scholars have connected his name to the *Wanderwort* for "mole" mentioned above.[62] Once again, *non liquet*.

Colleagues sometimes express surprise at the attention I pay in my publications not just to lower body parts, and not just to animals, but to the combination of the two: besides the 1998 paper on badgers and their subcaudal gland and the 2002 follow-up on moles and mongooses, the two are intertwined in a paper from 2005 on the mythopoetic nexus of snaky eels, visual impairment,

---

been proposed to be, from "dugong" to "giraffe" and now to "crocodile" (thus, cautiously, Oded Borowski, *Every Living Thing: Daily Use of Animals in Ancient Israel* [Walnut Creek, Calif.: AltaMira, 1998], 206, with a brief *historia quaestionis*): as Stephanie Dalley ("Hebrew *taḥaš*, Akkadian *duhšu*, Faience and Beadwork," *Journal of Semitic Studies* 45 [2000]: 1–19) demonstrates with great verve, *taḥaš*, whose translation as "badger; Dachs" goes back to Martin Luther, is not an animal at all but rather "denotes beading and attaching pendants, and inlaying in stone, metal, faience and glass, and is usually made on leather but sometimes also wool or linen, or as cloisonné in precious metals, timber, etc." (16).

60 See Brisson, *Le sexe incertain*, 119–21.

61 A parallel for the variation between -χ- and -φ- in a word that has undergone paradigmatic split is ἔχις ~ ὄφις, both from Proto-Indo-European *h_1e/og^{uh}-i- "snake"; in this case, however, there is a straightforward phonological explanation (see Jay H. Jasanoff and Alan Nussbaum, "Word Games: The Linguistic Evidence in *Black Athena*," in Mary R. Lefkowitz and Guy MacLean Rogers [eds.], *"Black Athena" Revisited* [Chapel Hill: University of North Carolina Press, 1996], 177–205, at 198). As for the morphology, Τροφώνιος could easily be derived from a base noun τρόχ^{h}ος "badger" with the addition of the so-called Hoffmann suffix plus the familiar -ιος.

62 However, this view of Asclepius' name is by no means the *communis opinio*: for some literature pro and contra, see, e.g., Katz, "How the Mole and Mongoose Got their Names," 297 n. 5 and also 306. Recent discussions, with other suggestions, include Jan N. Bremmer, "Anaphe, Aeschrology and Apollo Aigletes: Apollonius Rhodius 4.1711–1730," in Annette Harder and Martijn Cuypers (eds.), *Beginning from Apollo: Studies in Apollonius Rhodius and the Argonautic Tradition*, Caeculus 6 (Louvain: Peeters, 2005), 18–34, at 21–27 and *passim* and Andreas Willi, "Νόσος and ὀσίη: Etymological and Sociocultural Observations on the Concepts of Disease and Divine (Dis)Favour in Ancient Greece," *Journal of Hellenic Studies* 128 (2008): 153–71, at 160–61.

and illicit sex (with attention to penises, sodomizing weapons, hermaphroditism, and the evil eye) and in one from 2006 on the Sphinx's anal background.[63] But I would argue that my vision is not idiosyncratic and warped and that the research in which I have been engaged reflects a coherent view of the world. In his book on hermaphroditism, Brisson has a chapter titled "Le Médiateur" in which he discusses the myth of Tiresias[64] and then presents a "Petit bestiaire associé au mythe."[65] This bestiary contains six liminal creatures: the mouse, the mole, the badger, the hyena, the snake, and the shrew.[66] Independently of Brisson (at least at the start), I have over the years moved from badger to mole to snake and now to the link between hyena and badger—and almost to shrew, for in the words of Brisson "la musaraigne évoque irrésistiblement la mangouste."[67] As for what comes next, I intend to build on the conclusions about Aristotle's τρόχος in the present paper and publish an account of what really is our earliest attested Greek badger: an unrecognized one in Hesiod's *Works and Days*. But for now, περὶ μὲν τούτων ἅλις τὰ εἰρημένα.

Among the most interesting questions that people have been raising in recent decades—both in academic Festschriften and in public life—are those that concern the nature of sex and gender and of the human/animal interface:

---

[63] Respectively Katz, "Hittite *tašku-*," "How to Mole and Mongoose Got their Names," "To Turn a Blind Eel," and "The Riddle of the *sp(h)ij-*: The Greek Sphinx and her Indic and Indo-European Background," in Georges-Jean Pinault and Daniel Petit (eds.), *La langue poétique indo-européenne: actes du Colloque de travail de la Société des Études Indo-Européennes (Indogermanische Gesellschaft / Society for Indo-European Studies), Paris, 22–24 octobre 2003* (Louvain: Peeters, 2006), 157–94.

[64] On which see Katz, "To Turn a Blind Eel," 276–81, as well as "The Riddle of the *sp(h)ij-*" on the Riddle of the Sphinx.

[65] Brisson, *Le sexe incertain*, 103–27, with notes at 157–61; for discussion of Aristotle's τρόχος in connection with Tiresias—and with the hyena and the hare (see the following footnote)—see already Luc Brisson, *Le mythe de Tirésias: essai d'analyse structural* (Leiden: Brill, 1976), 73–76.

[66] Surprisingly omitted is the hare, aside from brief mentions at 121, 126, and 127: this may be the animal most associated cross-culturally with hermaphroditism. For connections between the trickster figures Hare (Lapin) and Hyena (Bouki) in African and Louisiana African-American tales, see Beidelman, "Ambiguous Animals" (and "The Moral Imagination of the Kaguru: Some Thoughts on Tricksters, Translation and Comparative Analysis," in William J. Hynes and William G. Doty [eds.], *Mythical Trickster Figures: Contours, Contexts, and Criticisms* [Tuscaloosa: University of Alabama Press, 1993], 174–92, at 177–78, with many additional references) and Marcia Gaudet, "Bouki, the Hyena, in Louisiana and African Tales," *Journal of American Folklore* 105 (1992): 66–72.

[67] Brisson, *Le sexe incertain*, 125. By further implication it evokes the mouse as well since Brisson (116–17) discusses Apollo the "Mouse-God" (Σμινθεύς) right before moving on to the mole, which he refers to as "la souris aveugle" and describes as "l'animal sacré d'Asklépios" (117).

for example, what is biological and what is socially constructed? There has been an explosion of interest in what we can learn from so-called anatomical peculiarities (the more one looks around, the less the world appears to conform to binary oppositions) and non-standard sexual proclivities (ditto), and it would not be very sensible in a paper of this kind to try to select for highlighting even a few works on how ethical and practical issues concerning such matters as hermaphroditism and bisexuality are dealt with in antiquity and the modern world.[68] But the scholarship of Heinrich von Staden continues to lead the way —for gender, for animals, and for their combination[69]—and I hope that this small contribution to classical philology and the history of biology, which answers some but not all of the questions it raises, is worthy of its dedicatee.

Bibliography

Abaev, V. I. *Istoriko-ètimologičeskij slovar' osetinskogo jazyka*, vol. 3 (Leningrad: Nauka, 1979).
Arnott, W. Geoffrey. *Birds in the Ancient World from A to Z* (London: Routledge, 2007).
Balles, Irene. "Some New Celtic and Other Etymologies," in Giovanni Belluscio and Antonio Mendicino (eds.), *Scritti in onore di Eric Pratt Hamp per il suo 90. compleanno* (Rende, Italy: Università della Calabria, Centro Editoriale e Librario, 2010), 15–19.
Beekes, Robert. *Etymological Dictionary of Greek*, 2 vols. (Leiden: Brill, 2010).

---

[68] Still, it is appropriate to cite the outstanding recent book on gender by one of the editors of this volume, my colleague Brooke Holmes. On "Aristotle's thoughts on the matter of sexual difference," which she argues are "marked by ... a tension between fixity and fluidity" (45), see Holmes, *Gender: Antiquity and its Legacy* (London: I. B. Tauris, 2012), 40–46 and index s.v. "Aristotle, views on males and females in."

[69] See, e. g., Heinrich von Staden, "*Apud nos foediora verba*: Celsus' Reluctant Construction of the Female Body," in Guy Sabbah (ed.), *Le Latin médical: la constitution d'un langage scientifique; réalités et langage de la médecine dans le monde romain. Actes du IIIe Colloque international "Textes médicaux latins antiques" (Saint-Étienne, 11–13 septembre 1989)*, Centre Jean-Palerne, Mémoires 10 (Saint-Étienne: Publications de l'Université de Saint-Étienne, 1991), 271–96; idem, "Women and Dirt," *Helios* 19 (1992) [= David Konstan (ed.), *Documenting Gender: Women and Men in Non-literary Classical Texts*]: 7–30; idem, "Animals, Women, and *pharmaka* in the Hippocratic Corpus," in V. Boudon-Millot, V. Dasen, and B. Maire (eds.), *Femmes en médecine: actes de la Journée internationale d'étude organisée à l'Université René-Descartes-Paris V, le 17 mars 2006 en l'honneur de Danielle Gourevitch* (Paris: De Boccard, 2008), 171–204 (with a substantial discussion at 189–99 of the use of beaver testicles in the treatment of "hysteria"); and idem, "The Living Environment: Animals and Humans in Celsus' *Medicina*," in Nicoletta Palmieri (ed.), *Conserver la santé ou la rétablir: le rôle de l'environnement dans la médecine antique et médiévale; actes du Colloque international, Saint-Étienne, 23–24 octobre 2008*, Centre Jean-Palerne, Mémoires 35 (Saint-Étienne: Publications de l'Université de Saint-Étienne, 2012), 161–92.

Beidelman, T. O. "Ambiguous Animals: Two Theriomorphic Metaphors in Kaguru Folklore," *Africa* 45 (1975): 183–200.
Beidelman, T. O. "The Moral Imagination of the Kaguru: Some Thoughts on Tricksters, Translation and Comparative Analysis," in William J. Hynes and William G. Doty (eds.), *Mythical Trickster Figures: Contours, Contexts, and Criticisms* (Tuscaloosa: University of Alabama Press, 1993), 174–92.
Bettini, Maurizio. *Nascere: storie di donne, donnole, madri ed eroi* (Turin: Einaudi, 1998). [English translation: *Women & Weasels: Mythologies of Birth in Ancient Greece and Rome*, trans. Emlyn Eisenach (Chicago: University of Chicago Press, 2013).]
Bewick, Thomas. *A General History of Quadrupeds* (Chicago: University of Chicago Press, 2009). [For further details, see n. 37 above.]
Borowski, Oded. *Every Living Thing: Daily Use of Animals in Ancient Israel* (Walnut Creek, Calif.: AltaMira, 1998).
Bremmer, Jan N. "Anaphe, Aeschrology and Apollo Aigletes: Apollonius Rhodius 4.1711–1730," in Annette Harder and Martijn Cuypers (eds.), *Beginning from Apollo: Studies in Apollonius Rhodius and the Argonautic Tradition*, Caeculus 6 (Louvain: Peeters, 2005), 18–34. [Reprinted in Jan N. Bremmer, *Greek Religion and Culture, the Bible and the Ancient Near East* (Leiden: Brill, 2008), 249–65.]
Brisson, Luc. *Le mythe de Tirésias: essai d'analyse structural* (Leiden: Brill, 1976).
Brisson, Luc. *Le sexe incertain: androgynie et hermaphrodisme dans l'Antiquité gréco-romaine*, 2nd ed. (Paris: Les Belles Lettres, 2008). [English translation (of 1st ed. [1997]): *Sexual Ambivalence: Androgyny and Hermphroditism in Graeco-Roman Antiquity*, trans. Janet Lloyd (Berkeley: University of California Press, 2002).]
Brottman, Mikita. *Hyena* (London: Reaktion, 2012).
Brugman, J., and H. J. Drossaart Lulofs. *Aristotle, Generation of Animals: The Arabic Translation Commonly Ascribed to Yaḥyâ ibn al-Biṭrîq* (Leiden: Brill, 1971).
Budziszewska, Wanda. *Zapożyczenia słowiańskie w dialektach nowogreckich* (Warsaw: Omnitech, 1991).
Dalley, Stephanie. "Hebrew *taḥaš*, Akkadian *duhšu*, Faience and Beadwork," *Journal of Semitic Studies* 45 (2000): 1–19.
Delamarre, Xavier. *Dictionnaire de la langue gauloise: une approche linguistique du vieux-celtique continental*, 2nd ed. (Paris: Errance, 2003).
Drossaart Lulofs, H. J. *Aristotelis De generatione animalium* (Oxford: Clarendon Press, 1965).
Dunbar, Nan. *Aristophanes, Birds* (Oxford: Clarendon Press, 1995).
East, Marion L., and Heribert Hofer. "Cultural and Public Attitudes: Improving the Relationship between Humans and Hyaenas," in Mills and Hofer, *Hyaenas: Status Survey*, 96–102.
Ewer, R. F. *The Carnivores* (Ithaca: Cornell University Press, 1973).
Faraone, Christopher A. *Ancient Greek Love Magic* (Cambridge, Mass.: Harvard University Press, 1999).
Frembgen, Jürgen W. "The Magicality of the Hyena: Beliefs and Practices in West and South Asia," *Asian Folklore Studies* 57 (1998): 331–44.
Funk, Holger. *Hyaena: On the Naming and Localisation of an Enigmatic Animal* ([Munich]: GRIN, 2010).
Funk, Holger. "R. J. Gordon's Discovery of the Spotted Hyena's Extraordinary Genitalia in 1777," *Journal of the History of Biology* 45 (2012): 301–28.

García Ramón, José Luis. "Anthroponymica Mycenaea: I. Mykenisch *o-ki-ro*, alph.-gr. ὀρχίλος; II. Mykenisch *da-te-wa* /*Daitēwās*/ und *e-u-da-i-ta*, alph.-gr. Δαίτας, Πανδαίτης," *Minos* 35–36 (2000–2001 [publ. 2002]): 431–42.

García Ramón, José Luis."Der Name *o-ki-ro* /*orkʰilos*/: Eine überzeugende Deutung von Prof. Ilievski," *Minos* 37–38 (2002–2003 [publ. 2006]): 371–72.

Gaudet, Marcia. "Bouki, the Hyena, in Louisiana and African Tales," *Journal of American Folklore* 105 (1992): 66–72.

Glickman, Stephen E. "The Spotted Hyena from Aristotle to the Lion King: Reputation is Everything," *Social Research* 62 (1995): 501–37.

Glickman, Stephen E., et al. "Mammalian Sexual Differentiation: Lessons from the Spotted Hyena," *Trends in Endocrinology and Metabolism* 17.9 (2006): 349–56.

Graf, Fritz. "Herodoros," in Hubert Cancik and Helmuth Schneider (eds.), *Der neue Pauly: Enzyklopädie der Antike. Altertum*, vol. 5 (Stuttgart: J. B. Metzler, 1998), col. 469. [English translation: "Herodorus," in Hubert Cancik and Helmuth Schneider (eds.), *Brill's New Pauly: Encyclopaedia of the Ancient World. Antiquity*, vol. 6 (Leiden: Brill, 2005), cols. 264–65.]

Hansen, Ove. "A Note on πικτίδας/πυκτίδας of Aristophanes' Ach. 879," *Philologus* 134 (1990): 158–59.

Hofer, Heribert, and Gus Mills. "Worldwide Distribution of Hyaenas," in Mills and Hofer, *Hyaenas: Status Survey*, 39–63.

Hofer, Heribert, and Gus Mills. "Population Size, Threats and Conservation Status of Hyaenas," in Mills and Hofer, *Hyaenas: Status Survey*, 64–79.

Holmes, Brooke. *Gender: Antiquity and its Legacy* (London: I. B. Tauris, 2012).

Hünemörder, Christian. "Dachs," in Hubert Cancik and Helmuth Schneider (eds.), *Der neue Pauly: Enzyklopädie der Antike. Altertum*, vol. 3 (Stuttgart: J. B. Metzler, 1997), col. 257. [English translation: "Badger," in Hubert Cancik and Helmuth Schneider (eds.), *Brill's New Pauly: Encyclopaedia of the Ancient World. Antiquity*, vol. 2 (Leiden: Brill, 2003), col. 458.]

Jacoby, F. "Herodoros 4)," in *Paulys Real-Encyclopädie der classischen Altertumswissenschaft*, vol. 8.1 (Stuttgart: J. B. Metzler, 1912), cols. 980–87.

Jamison, Stephanie W. *The Ravenous Hyenas and the Wounded Sun: Myth and Ritual in Ancient India* (Ithaca: Cornell University Press, 1991).

Jasanoff, Jay H., and Alan Nussbaum. "Word Games: The Linguistic Evidence in Black Athena," in Mary R. Lefkowitz and Guy MacLean Rogers (eds.), *"Black Athena" Revisited* (Chapel Hill: University of North Carolina Press, 1996), 177–205.

Kaczyńska, Elwira, and Krzysztof Tomasz Witczak. "Remarks on Some Doric Elements in the Modern Greek Dialects of Crete," *Eos* 92 (2005): 112–20.

Katz, Joshua T. "*Mustelidae* in the Cretan Dialect of Modern Greek," *Eos* 94 (2007): 289–305.

Katz, Joshua T. "Hittite *tašku-* and the Indo-European Word for 'Badger'," *Historische Sprachforschung* 111 (1998): 61–82.

Katz, Joshua T. "How the Mole and Mongoose Got their Names: Sanskrit *ākhú-* and *nakulá-*," *Journal of the American Oriental Society* 122 (2002) [= Joel P. Brereton and Stephanie W. Jamison (eds.), *Indic and Iranian Studies in Honor of Stanley Insler on his Sixty-fifth Birthday*]: 296–310.

Katz, Joshua T. "To Turn a Blind Eel," in Karlene Jones-Bley et al. (eds.), *Proceedings of the Sixteenth Annual UCLA Indo-European Conference, Los Angeles, November 5–6, 2004* (Washington, D.C.: Institute for the Study of Man, 2005), 259–96.

Katz, Joshua T. "The Riddle of the *sp(h)ij-*: The Greek Sphinx and her Indic and Indo-European Background," in Georges-Jean Pinault and Daniel Petit (eds.), *La langue poétique indo-européenne: actes du Colloque de travail de la Société des Études Indo-Européennes (Indogermanische Gesellschaft/Society for Indo-European Studies), Paris, 22–24 octobre 2003* (Louvain: Peeters, 2006), 157–94.
Keller, Otto. *Die antike Tierwelt*, vol. 1: *Säugetiere* (Leipzig: Engelmann, 1909).
Kloekhorst, Alwin. "Hittite ḫāpūša(šš)- (Formerly Known as ḫapuš- 'Penis')," *Journal of Indo-European Studies* 33 (2005): 27–39.
Kruuk, Hans. *The Spotted Hyena: A Study of Predation and Social Behavior* (Chicago: University of Chicago Press, 1972).
Kruuk, Hans. *Hyaena* (London: Oxford University Press, 1975).
Kruuk, Hans. *The Social Badger: Ecology and Behaviour of a Group-living Carnivore (Meles meles)* (Oxford: Oxford University Press, 1989).
Lamberterie, Charles de, and Françoise Létoublon. "Deux points de vocabulaire homérique: éclairage comparatif," *LALIES* 2 (1980 [publ. 1983]): 13–16.
Lanza, Diego. *La riproduzione degli animali / De generatione animalium*, in Diego Lanza and Mario Vegetti (eds.), *Opere biologiche di Aristotele* (Turin: Unione Tipografico-Editrice Torinese, 1971), 775–1042.
Létoublon, Françoise, and Charles de Lamberterie. "La roue tourne," *Revue de philologie* 54 (1980): 305–26.
Louis, Pierre. *Aristote, De la génération des animaux* (Paris: Les Belles Lettres, 1961).
Mallory, J. P., and D. Q. Adams. *The Oxford Introduction to Proto-Indo-European and the Proto-Indo-European World* (Oxford: Oxford University Press, 2006).
Meyboom, P. G. P. *The Nile Mosaic of Palestrina: Early Evidence of Egyptian Religion in Italy* (Leiden: Brill, 1995).
Mills, Gus, and Heribert Hofer, eds. *Hyaenas: Status Survey and Conservation Action Plan* (Gland, Switzerland: IUCN, 1998).
Mills, Gus, and Margie Mills. *Hyena Nights & Kalahari Days* (Auckland Park, South Africa: Jacana, 2010).
Morenilla-Talens, Carmen. "Aprosdoketon oder Hapax? Aristophanes, *Ach*. 879," *Glotta* 64 (1986): 216–21.
Neal, Ernest, and Chris Cheeseman. *Badgers* (London: Poyser, 1996).
Nichols, Andrew. *Ctesias on India and Fragments of his Minor Works* (London: Bristol Classical Press, 2011).
Olson, S. Douglas. *Aristophanes, Acharnians* (Oxford: Oxford University Press, 2002).
Peck, A. L. *Aristotle, Generation of Animals*, rev. ed. (Cambridge, Mass.: Harvard University Press, 1953).
Peck, A. L. *Aristotle, Historia animalium*, vol. 2 (Cambridge, Mass.: Harvard University Press, 1970).
Platt, Arthur. *De generatione animalium* (Oxford: Clarendon Press, 1910). [Collected in J. A. Smith and W. D. Ross (eds.), *The Works of Aristotle*, vol. 5 (Oxford: Clarendon Press, 1912).]
Prakash, Ishwar. "Hyena," in R. E. Hawkins (ed.), *Encyclopedia of Indian Natural History* (Delhi: Oxford University Press, 1986), 300–301.
Rohland, Nadin, et al. "The Population History of Extant and Extinct Hyenas," *Molecular Biology and Evolution* 22 (2005): 2435–43.
Roper, Timothy J. *Badger* (London: Collins, 2010).

Steinmeyer-Schareika, Angela. *Das Nilmosaik von Palestrina und eine ptolemäische Expedition nach Äthiopien* (Bonn: Habelt, 1978).
Thompson, D'Arcy Wentworth. *Historia animalium*, in J. A. Smith and W. D. Ross (eds.), *The Works of Aristotle*, vol. 4 (Oxford: Clarendon Press, 1910).
Thompson, D'Arcy Wentworth. *A Glossary of Greek Birds*, new ed. (London: Oxford University Press, 1936).
van Oppenraaij, Aafke M. I. *Aristotle, De animalibus: Michael Scot's Arabic-Latin Translation*, vol. 3: *Books XV–XIX: Generation of Animals* (Leiden: Brill, 1992).
von Staden, Heinrich. "*Apud nos foediora verba*: Celsus' Reluctant Construction of the Female Body," in Guy Sabbah (ed.), *Le Latin médical: la constitution d'un langage scientifique; réalités et langage de la médecine dans le monde romain. Actes du IIIe Colloque international "Textes médicaux latins antiques" (Saint-Étienne, 11–13 septembre 1989)*, Centre Jean-Palerne, Mémoires 10 (Saint-Étienne: Publications de l'Université de Saint-Étienne, 1991), 271–96.
von Staden, Heinrich. "Women and Dirt," *Helios* 19 (1992) [= David Konstan (ed.), *Documenting Gender: Women and Men in Non-literary Classical Texts*]: 7–30.
von Staden, Heinrich. "Animals, Women, and *pharmaka* in the *Hippocratic Corpus*," in V. Boudon-Millot, V. Dasen, and B. Maire (eds.), *Femmes en médecine: actes de la Journée internationale d'étude organisée à l'Université René-Descartes-Paris V, le 17 mars 2006 en l'honneur de Danielle Gourevitch* (Paris: De Boccard, 2008), 171–204.
von Staden, Heinrich. "The Living Environment: Animals and Humans in Celsus' Medicina," in Nicoletta Palmieri (ed.), *Conserver la santé ou la rétablir: le rôle de l'environnement dans la médecine antique et médiévale; actes du Colloque international, Saint-Étienne, 23–24 octobre 2008*, Centre Jean-Palerne, Mémoires 35 (Saint-Étienne: Publications de l'Université de Saint-Étienne, 2012), 161–92.
Walter, Henriette, and Pierre Avenas. *L'étonnante histoire des noms des mammifères: de la musaraigne étrusque à la baleine bleue* (Paris: Laffont, 2003).
Wartelle, André. "Brèves remarques de vocabulaire grec," *Revue des études grecques* 113 (2000): 211–19.
Watson, M. "On the Female Generative Organs of *Hyæna crocuta*," *Proceedings of the Zoological Society of London* (1877): 369–79 + Plates XL and XLI.
Watson, M. "On the Male Generative Organs of *Hyæna crocuta*," *Proceedings of the Zoological Society of London* (1878): 416–28 + Plates XXIV and XXV.
Watson, M. "Additional Observations on the Anatomy of the Spotted Hyæna," *Proceedings of the Zoological Society of London* (1881): 516–21 + Plate XLIX.
Weiss, Michael. *Outline of the Historical and Comparative Grammar of Latin* (Ann Arbor: Beech Stave, 2009).
Willi, Andreas. "Νόσος and ὁσίη: Etymological and Sociocultural Observations on the Concepts of Disease and Divine (Dis)Favour in Ancient Greece," *Journal of Hellenic Studies* 128 (2008): 153–71.

W. R. Laird
Heron of Alexandria and the Principles of Mechanics

Abstract: Heron of Alexandria's express purpose in the *Mechanica* was to meet the mechanical challenge – how to move any weight with any power by means of a machine. In the process, Heron reduced all complex machines to five simple powers or machines – the lever, wheel and axel, pulley, wedge, and screw – and sought their common principles. From the pseudo-Aristotelian *Mechanical Problems* he adopted the principle of circular movement, which, combined with Archimedes' law of the balance, explained the lever and the wheel and axel. But the pulley, wedge, and screw were not susceptible to analysis into circular movements. In all five machines, however, the multiplication of power is accompanied by what Heron called retardation, such that the greater their power the slower their movement. Some historians of science have seen in this a germinal form of the principle of virtual work, and so have credited Heron with anticipating the theoretical mechanics of the seventeenth century. But to the other machines, Heron applied a new principle, which I shall call the principle of dividing and sharing the load. He showed that the power of the *polyspaston* or compound pulley is as the number of supporting ropes that share the load divided among them. And he analysed the wedge (and by extension, the screw) by effectively dividing the load among thinner wedges and smaller blows, each of which shares in the overall movement. So Heron in fact appealed to several general principles to explain the marvellous effects of mechanical devices – the pseudo-Aristotelian principle of circular movement, the Archimedean law of the balance, and his new principle of dividing and sharing the load. Retardation, then, although it is common to all five simple machines, was seen by Heron not as the principle of their effectiveness, but rather as a result or effect of their operation.

The *Mechanica* of Heron of Alexandria (fl. ca. 60 A.D.) was the most comprehensive account of mechanical theory and mechanical practice in the ancient

---

The research for this paper was done at the Institute for Advanced Study, Princeton, N.J., with the help of the Fund for Historical Studies, and at the Huntington Library, San Marino, Ca., with the help of a Dibner Research Fellowship; their support is gratefully acknowledged.

world.[1] Drawing on the Aristotelian *Mechanica* and the works of Archimedes, as well as relying on his own experience and practical intuition, Heron attempted to explain how the various machines familiar to mechanical practitioners accomplished their remarkable effects. His overarching purpose was to meet the ancient mechanical challenge—to move a given weight with a given power—and much of the *Mechanica* was given over simply to describing the arrangement and use of machines to accomplish this. But since it was necessary to this purpose to calculate the effectiveness of such machines, Heron sought to reduce them to their principles, to explain exactly how they enabled large resistances to be moved by small powers. In the process, Heron was apparently the first to distinguish five "powers" (δυνάμεις)—the windlass (or wheel and axle), the lever, the pulley (πολύσπαστον, the compound pulley or block and tackle), the wedge, and the screw—as idealized machine types, as the elements that compose more complex practical machines. The effectiveness of each of these five powers, Heron claimed, can be explained through a single common principle, the principle of concentric circles, which is a simplified version of the principle of circular movement found in the Aristotelian *Mechanica*. Further, in one passage he also claimed that the principle of concentric circles itself is explained by the balance.

Despite these claims, Heron was in fact able to reduce only the windlass and the lever to the principle of concentric circles, by showing how small powers moving through large circles can move large powers through smaller. Although he also attempted to explain the inclined plane and the pulley by reducing them to circles and balances, ultimately the pulley, wedge (which he did not identify with the inclined plane), and screw could not be explained this way. But Heron did note that all of the powers exhibit what he called retardation or "slowing up," such that when a small power moves a large resistance with the help of a machine, it does so more slowly than a large power could. Historians of early mechanics, including Giovanni Vailati, René Dugas, Marshall Clagett,

---

[1] On Heron's life and works, see Aage G. Drachmann, "Hero of Alexandria," in Charles C. Gillispie, *Dictionary of Scientific Biography*, 18 vols. (New York: Scribner, 1970–1990), 6:310–14. There was once considerable controversy over when Heron of Alexandria actually lived, estimates spanning some five hundred years, from the time of Archimedes (d. 212 B.C.), whom Heron mentions, to the time of Pappus of Alexandria (fl. ca. 310 A.D.), who mentions him. The controversy was finally put to rest by Otto Neugebauer, who dated an eclipse mentioned by Heron in his *Dioptra* to the year A.D. 62 ("Über eine Methode zur Distanzbestimmung Alexandria-Rom bei Heron," *Kongelige Danske Videnskabernes Selskabs Skrifter* 26 [1938]: 21–24), so that we can now say with some confidence that Heron lived in the first century of the modern era; but for a reexamination of Neugebauer's use of the eclipse, see Nathan Sidoli, "Heron of Alexandria's Date," *Centaurus* 53 (2011): 55–61.

and François De Gandt, have generally seen this to be Heron's fundamental principle and recognized in it an early form of the principle of virtual work.[2] In a recent article, however, Mark Schiefsky has argued that virtual work was not in fact the fundamental principle explaining the phenomenon of slowing up, and that slowing up itself was for Heron not the cause of a machine's effectiveness, but rather an incidental result, which he had observed by experience.[3]

But if the principle of concentric circles cannot explain the pulley, the wedge, and the screw, and if virtual work or slowing up is not the principle of their action, how did Heron explain them? In his treatment of the pulley, the wedge, and (implicitly) the screw, he in fact invoked a completely different principle, a principle that he made explicit only towards the end of Book 2 of the *Mechanica*, in his treatment of seventeen miscellaneous mechanical questions. These questions were introduced by a remarkable methodological and theoretical statement, which laid out the requirements for research into mechanics, stated explicitly the physical postulates to be used, and revealed the hidden mechanical principle that underlies all marvelous mechanical effects. This principle was what I call the principle of dividing and sharing the load, and its role in Heron's mechanics has received relatively little attention. In what follows, then, I should like to show how this principle gradually emerged in the

---

[2] Giovanni Vailati, "Il principio dei Lavori Virtuali da Aristotele a Erone d'Alessandria," *Atti della R. Accademia delle Scienze di Torino* 32 (1897): 940–62, at 958–62 [repr. in Giovanni Vailati, *Scritti di G. Vailati (1863–1909)* (Leipzig: J. A. Barth 1911), 91–106 (at 103–106)]; and G. Vailati, "Heronis Alexandrini Opera quae supersunt omnia," *Bollettino di Bibliografia e Storia delle Scienze Matematiche* 4 (1901): 79–81 [repr. in Vailati, *Scritti*, 374–76]; René Dugas, *A History of Mechanics*, trans. J. R. Maddox (New York: Routledge & Kegan Paul, 1957), 14–18 [originally published as *Histoire de la mécanique* (Neuchâtel: Éditions du Griffon, 1950)]; Marshall Clagett, *The Science of Mechanics in the Middle Ages* (Madison: University of Wisconsin Press, 1959), 17–18, 51; François De Gandt, "Force et science des machines," in J. Barnes et al. (eds.), *Science and Speculation: Studies in Hellenistic Theory and Practice* (Cambridge: Cambridge University Press, 1982), 96–127, at 113–16. Pierre Duhem, in his *Origines de la statique*, 2 vols. (Paris: A. Hermann, 1905–1906), attributed the principle of virtual velocities to the Aristotelian *Mechanica*, but was more modest in his evaluation of Heron, allowing him only "une sorte d'indication de la modification que Galilée apportera au principe des vitesses virtuelles formulé par Aristote et par Chariston [i.e., Duhem's suppositious author of the *Liber karastonis*]; mais cette indication est extrêmement vague et indécise" (2:317) [= *The Origins of Statics*, trans. G. F. Leneaux, V. N. Vagliente, and G. H. Wagener, Boston Studies in the Philosophy of Science 123 (Dordrecht: Kluwer, 1991), 471]; De Gandt, in turn, argued against Duhem's identification of the principle in the Aristotelian *Mechanica* ("Force et science," 116–27).

[3] Mark J. Schiefsky, "Theory and Practice in Heron's *Mechanics*," in Walter Roy Laird and Sophie Roux (eds.), *Mechanics and Natural Philosophy before the Scientific Revolution*, Boston Studies in the Philosophy of Science 254 (Dordrecht: Springer, 2008), 15–49, at 39–41.

course of Heron's *Mechanica*, how it was implicitly applied to explain the pulley, the wedge, and the screw, and how it was finally stated and used explicitly to explain some of the seventeen questions.[4]

The principle of dividing and sharing the load seems to have developed in Heron's mind gradually, perhaps as a result of practical experience and the intuitive analysis of mechanical effects. It makes its first tentative appearance in the course of his argument in Book 1 that any load can be moved along a smooth horizontal surface by any power however small. Now since this is clearly contrary to our general experience of moving loads, Heron must explain how it can in fact be true. As evidence he notes that water will flow down the slightest incline, which suggests that on a smooth horizontal surface it can be moved by the smallest power. The reason he gives for this is that the parts of water do not adhere together, but are easily divided. In the case of solid bodies, in contrast, because their parts resist separation and because they are not perfectly smooth, the parts grip the surface like teeth and, being joined together, they hold back

---

[4] Heron's *Mechanica* survives today only in the Arabic translation made in the late eleventh century by Qusta Ibn Luqa, the Greek text probably having been lost shortly after this translation was made. Pappus of Alexandria (early fourth century A.D.) quoted several excerpts in the eighth book of his *Mathematical Collection*, though even then he noted that many of the Greek copies he found were incomplete. The Arabic text has been edited twice, first from a single manuscript with a translation into French by B. Carra de Vaux, "Les mécaniques ou L'élévateur de Héron d'Alexandrie, publiées pour la première fois sur la version arabe de Qostâ ibn Lûqâ, et traduites en français," *Journal Asiatique* ser. IX t. 1 (1893): 386–472, and ser. IX t. 2 (1893): 152–269, 420–514; this edition has been recently reprinted in facsimile with a preface by Donald R. Hill and commentary in the form of extracts from A. G. Drachmann, *The Mechanical Technology of Greek and Roman Antiquity: A Study of the Literary Sources* (Copenhagen: Munksgaard, 1963), as B. Carra de Vaux, *Héron d'Alexandrie, Les mécaniques, ou, L'élévateur des corps lourds*, Collection Sciences et Philosophie Arabes (Études et Reprises) (Paris: Les Belles Lettres, 1988); the other edition of the Arabic text, from all four extant manuscripts, with a translation into German, was made by Ludwig Nix, to which Wilhelm Schmidt added the fragments in Greek from Pappus, in L. Nix and W. Schmidt, *Heronis Alexandrini Opera quae supersunt omnia*, vol. 2, fasc. 1: *Mechanica et catoptrica* (Leipzig: Teubner, 1900). There is still no complete English translation. Drachmann, in his *Mechanical Technology of Greek and Roman Antiquity* (19–140), has translated into English excerpts from Nix's edition of the Arabic, interspersed with textual and technical commentary, though he often passes over those parts of the text not directly concerned with technics. Excerpts are also translated from Nix's German translation in Morris R. Cohen and I.E. Drabkin, eds., *A Source Book in Greek Science* (Cambridge, Mass.: Harvard University Press, 1948), 197–200, 223–28; and from Nix's Arabic edition in Marshall Clagett, *The Science of Mechanics in the Middle Ages*, 38–48. I have relied on Nix's and Carra de Vaux's translations, citing the text by book and chapter number, which are consistent between the two editions, and citing English translations where they are available.

one another, so that a greater power is necessary to move solid bodies.[5] Thus even the smallest power can move the various parts of easily divisible loads such as water, whereas a greater power is necessary to move cohesive solids. The capacity to be easily divided, then, allows water to be moved more easily than solid bodies.

The inchoate principle of dividing and sharing the load also appeared in the course of Heron's discussion of how a single static weight is divided and distributed among several supports. In a chapter that seems out of place in this discussion of static supports, Heron observes that it is evident that a small force cannot move a large weight except by means of a machine. Now two men can easily move a weight that one man alone cannot move at all; but although the weight is not moved until the force of the second man is added to that of the first, the second man alone does not move it. This is obvious because, if the first man were to leave off, the second man alone would also not be able to move it. But if the weight were divided in half, one man alone could move one of the halves, leaving the other at rest. For the same reason, when a number of forces combine to move a weight, and one of them leaves off, the rest cannot move it; but after the addition of the last force, they can move it easily. The case is similar with impact: when after a number of blows the last blow finally breaks an object into pieces, this is the effect not only of all the previous blows but also of the last one. And this is shown by experience, for when we finally with great effort move a weight, our power in the end equals the weight.[6] This reasoning, which is similar to Aristotle's in the shiphaulers' argument in the *Physics*,[7] was prompted here by the discussion of how the weight of a beam is shared by several supports. Heron's general point was that great effects are sometimes achieved by a kind of division and sharing of the load among many small powers, or by a kind of addition and accumulation of small powers until together they become equal to and then overcome the resistance.

The same principle can be found more explicitly at work in Heron's explanation of three of the five powers in Book 2, for although he applied the principle of concentric circles to the windlass and lever, he applied the principle of dividing and sharing the load to the pulley, the wedge, and the screw. His general purpose in Book 2 was to explain the construction and use of the five powers so that they could then be used to meet the mechanical challenge –to move a

---

5 *Mech.* 1.20–21; trans. Drachmann, *Mechanical Technology*, 46–47.
6 *Mech.* I.29; Drachmann dismisses this chapter, without any discussion, as simply out of place (A. G. Drachmann, "Fragments from Archimedes in Heron's Mechanics," *Centaurus* 8 [1963]: 91–146, at 120).
7 Arist. *phys.* 7.5, 249b27–250b7 and 8.3, 253b14–26.

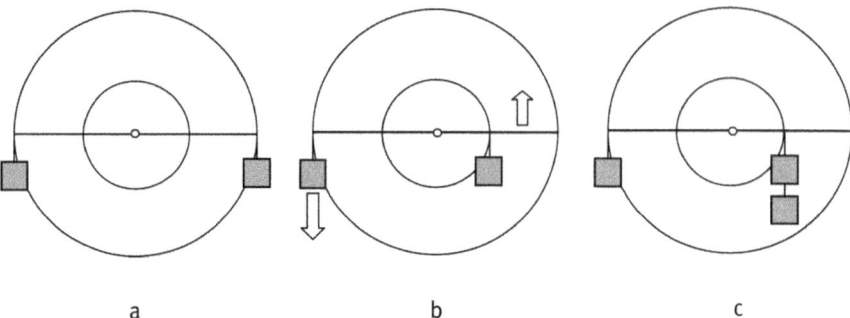

a   b   c

Fig. 1

given weight with a given power. But he also promised to show how, despite being very different in form or design, they all owe their effectiveness to a single common nature or principle.[8] He then listed them by name as the axle in a wheel (or windlass), the lever, the *polyspaston* or pulley, the wedge, and the screw, and described in turn the construction and use of each.[9] In the chapters that follow these descriptions, Heron went on to explain the effectiveness of these five powers by giving the general principle that explains them and then showing in each case how that principle applies. The general principle he gave here at the beginning was the principle of concentric circles derived from the Aristotelian *Mechanica* (although he does not cite it as his source). Imagine two vertical circles able to rotate together about a common center; if two equal weights are hung from the ends of the horizontal diameter of the larger circle, the circle will be in equilibrium, since the diameter is effectively a balance beam (see fig. 1a). Now, if one of the weights is moved to the circumference of the smaller circle, the balance will incline towards the other weight (fig. 1b). Equilibrium can be restored, however, by adding an additional weight to the weight on the smaller circle such that the resulting weights will be inversely as their distances from the center (fig. 1c), which Archimedes had proved in his "book on the equilibrium of weights," presumably *On the Equilibrium of Planes*. From this it is clear that any heavy body can be balanced by a weak power when they are placed on con-

---

[8] *Mech.* 2.1; trans. Drachmann, *Mechanical Technology*, 50, as modified by Schiefsky, "Theory and Practice," 22; see also the translation in Cohen and Drabkin, *Source Book*, 224, from Pappus's Greek version (Friedrich Hultsch, *Pappi Alexandrini Collectionis quae supersunt*, 3 vols. [Berlin: Weidmann, 1875–1878], 1114).

[9] *Mech.* 2.1–6; trans. with commentary in Drachmann, *Mechanical Technology*, 50–61; Cohen and Drabkin, *Source Book*, 224–28, give a slightly abridged translation of Pappus's Greek version of these chapters (Hultsch, *Pappi Alexandrini*, 1114–30).

centric circles such that the ratio of their horizontal distances from the center is inversely as the ratio of the great weight to the weak power, and it can be moved when the power is greater.[10]

Notice that Heron does not present the equality of the ratios of weights and distances as the cause of equilibrium, but only as its condition, and that he merely cites but does not give Archimedes' proof. Notice too that to move the weight requires a slightly greater ratio of power to weight than the inverse ratio of the distances, though how much greater is not specified. Finally, one might have expected Heron to give some version of the Aristotelian principle of circular movement, with its analysis into natural and violent motion, as the cause of motion and equilibrium, since his analysis of the balance into concentric circular motions was almost certainly inspired by the beginning of the Aristotelian *Mechanica*, but that, too, is lacking here.[11]

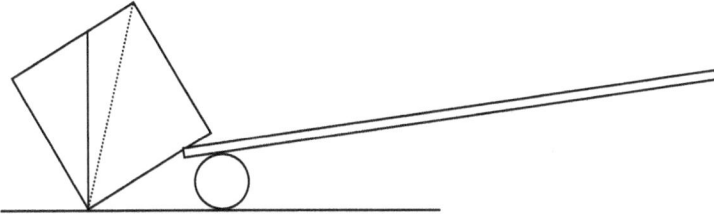

Fig. 2

If the common nature that explains the mechanical powers is concentric circles, then Heron succeeded nicely in reducing the lever and the windlass to it. The principle of concentric circles is immediately obvious in the windlass, where the axle or drum is the smaller circle and the wheel with handles is the larger.[12] In the case of the lever, the concentric circles are less obvious but easily discovered: the fulcrum of the lever is the center of rotation, the shorter arm of the lever, which lifts the weight, is a radius of the smaller circle, and the longer arm where the power is applied is a radius of the larger. Heron also suggested that concentric circles can explain the curious effect when a lever lifts only one end of a weight while the other end remains on the ground as its point of

---

10 *Mech.* 2.7; trans. in Clagett, *Science of Mechanics*, 45; and trans. in Drachmann, *Mechanical Technology*, 61–62.
11 Ps.-Arist. *mech.* 1, 848b1–850a1, ed. and trans. in Walter S. Hett, *Aristotle, Minor Works* (Cambridge, Mass.: Harvard University Press, 1936).
12 *Mech.* 2.10; trans. in Drachmann, *Mechanical Technology*, 67.

rotation (see fig. 2). In this case, as the weight rotates through part of a circle about this point, the load on the lever decreases, since a vertical plane drawn through the point of contact between the weight and the ground divides the weight into two parts, which counterbalance. When these two parts become equal, the weight is held in equilibrium without the addition of any power from the lever. There are thus two sets of concentric circles at work here: one set centered at the fulcrum, the other centered at the point of rotation of the weight.[13]

The other powers—the pulley, the wedge, and the screw—cannot be readily reduced to concentric circles or to the balance. But since the pulley is compounded of sheaves or wheels, each of which is circular, Heron began his explanation of it by equating the horizontal diameter of a pulley wheel with an equal-armed balance in equilibrium, and then divided the weight and added pulley wheels in such a way as to preserve the equilibrium. He begins with a weight hung over a fixed pulley from a rope attached at both ends to the weight, which will of course be in equilibrium (fig. 3a). If the weight is divided into two equal halves, each supported by one end of the rope, they are suspended effectively at the ends of an equal-armed lever and thus will also be in equilibrium (fig. 3b). Now with such an arrangement, Heron notes, one cannot counterbalance a great weight with a small power. But if, instead of both ends of the rope being attached to the weight, one end of it is passed through a second pulley, which is itself attached to the weight, and then that end of the rope is attached to the support above, the weight will be supported by three ropes, each of which supports only one third of the weight (fig. 3c). If the weight is then divided into two unequal parts, one twice the other, then the one rope attached to the smaller part will support the smaller part in equilibrium with the larger part supported by the two parts of the rope passing through the second pulley (fig. 3d). If the smaller weight is replaced with an equal power, then it will also hold the larger weight in equilibrium (fig. 3e). And again, if the fixed end of the rope is passed over a second fixed pulley and attached to the weight, there will now be four ropes, each supporting one quarter of the weight (fig. 3f). And finally, if the weight is divided into four, one of the quarter weights (or a power equal to it) will balance the other three together (fig. 3 g, 3 h). In general, then, in equilibrium, the ratio of weight to power is as the ratio of supporting ropes; to raise the weight, one must increase either the number of supporting

---

13 *Mech.* 2.8–9; abridged trans. and paraphrase in Clagett, *Science of Mechanics*, 46–47; trans. in Drachmann, *Mechanical Technology*, 63–65; see Schiefsky, "Theory and Practice," 43–45.

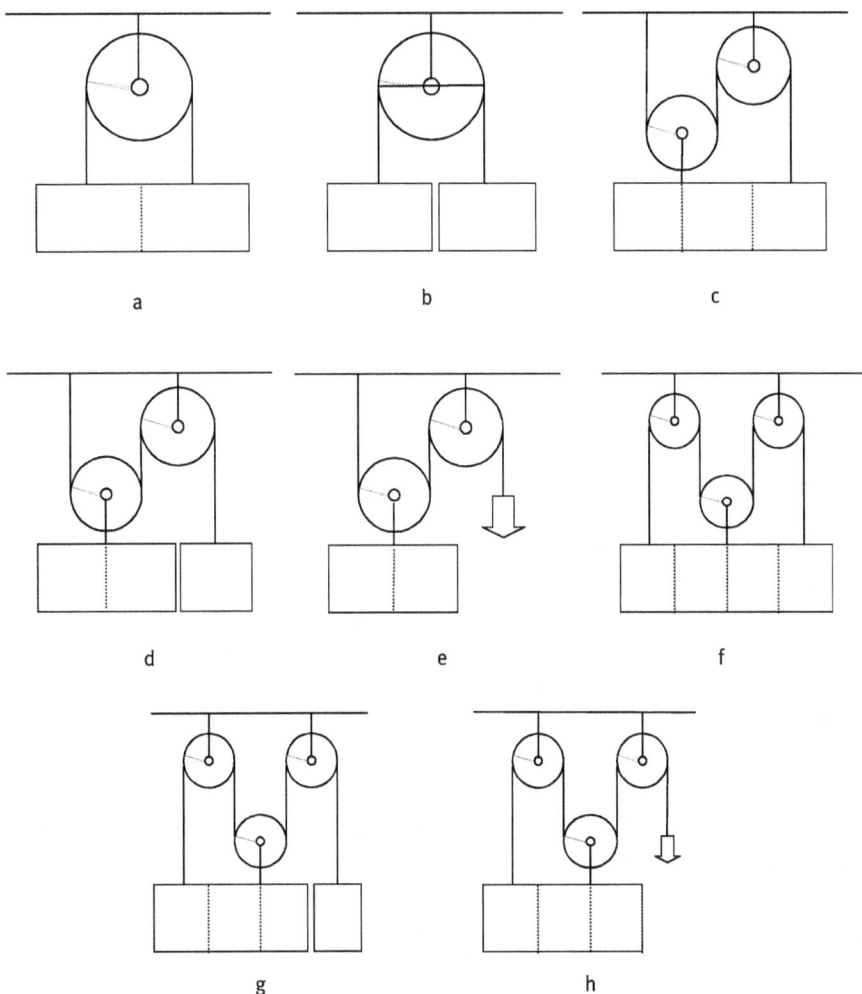

Fig. 3

ropes (by adding more pulley wheels) or the power.[14] Although he began by analysing the pulley as though it were a balance beam in equilibrium, Heron ended up revealing the true underlying reason for the lifting power of the pulley, not by

---

14 *Mech.* 2.11–13; trans. in Drachmann, *Mechanical Technology*, 67–71; I have redrawn the figures; for an analysis of the pulley in terms of retardation or "slowing up," see Schiefsky, "Theory and Practice," 37–39.

reducing it to concentric circles, but by dividing the weight and distributing it over or sharing it among a number of ropes.

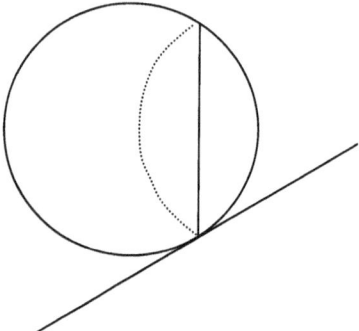

Fig. 4

As for the inclined plane, Heron had already attempted in Book 1 to find the power needed to draw a weight up an inclined plane by reducing the whole problem to the balance, in a way similar to his treatment of a weight lifted at one end by a lever. Stipulating that the surfaces of contact between the inclined plane and the weight be smooth to minimize friction, he reduced the problem to that of finding a power sufficient to hold the weight in equilibrium on the inclined plane, to which a minimal power could then be added to draw it up. He takes the ideal case of a cylindrical weight with a horizontal axis touching the plane along only one line, so that instead of sliding on the plane, the weight will roll and thus be almost entirely free of friction (see fig. 4). Heron then treats the cylinder in cross-section as though it were a balance in disequilibrium: a vertical plane drawn from the line of contact between the cylinder and the inclined plane will divide the cylinder into two unequal parts. With the line of contact being the center of the balance and the two parts as the weights, the larger part will overpower the smaller, causing the cylinder to roll down the plane. Now the smaller part can sustain in equilibrium that portion of the larger equal to itself, so that to sustain the whole cylinder, one must subtract from the larger a power equal to the difference between the larger and the smaller. Thus a power equal to this difference is needed to sustain the weight in equilibrium on the plane; to raise it one need add only the smallest additional power.[15] Heron has thus reduced the problem of the inclined plane to a problem of a bal-

---

[15] *Mech.* 1.23; trans. in Drachmann, *Mechanical Technology*, 47–48; trans. in Cohen and Drabkin, *Source Book*, 199–200; abridged trans. in Clagett, *Science of Mechanics*, 41–42; see also Schiefsky, "Theory and Practice," 46–47.

ance in equilibrium; once in equilibrium, a weight on an inclined plane is effectively on a horizontal plane, so that it can then be moved, as he had previously shown, by the addition of any power, however small.[16] He has thus reduced (however unsatisfactorily) the inclined plane to the balance.

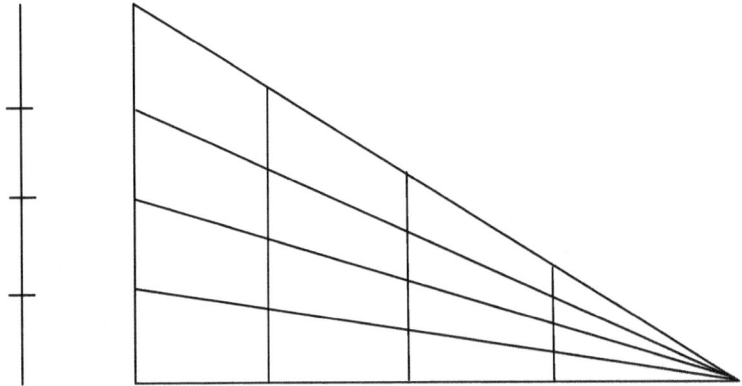

Fig. 5

Curiously, Heron never identified the inclined plane with the wedge, perhaps because their applications were so different: the inclined plane was used for raising heavy bodies with traction, whereas the wedge, actuated by a blow, was used for splitting. And despite the fact that in the Aristotelian *Mechanica* the wedge was analysed as two opposing levers, Heron's explanation of the wedge in Book 2 has nothing whatever to do with the lever, the balance, or concentric circles, but everything to do with dividing the load. Heron first assumes that because a wedge is driven in some finite time (since all motion is in time), but the blow that drives it is instantaneous, a wedge must continue to be moved for a short time after the instantaneous blow has ceased, in the same way that a projectile continues to be moved after leaving its thrower. Although it is not true that the blow is instantaneous or that the wedge is moved as a projectile, nevertheless Heron has realized that the motion of a wedge takes place in a short but finite time, and its effectiveness to move loads is measured by the relation of this time to the strength of the blow and to the distance it is moved. Again echoing Aristotle's shiphaulers' argument from the *Physics*, Heron argues first that if a

---

16 *Mech.* 1.20; trans. in Drachmann, *Mechanical Technology*, 46; the diagram follows Drachmann's suggestion, since there is no diagram in the manuscript and Cohen and Drabkin in their diagram seem to overinterpret the text.

given blow moves a wedge a certain distance, then some fraction of that blow will move the wedge the same fraction of the distance, or will move the same fraction of the wedge the whole distance, and that the separation of the load will be in the same ratio (see fig. 5). On the assumption that the movement caused by each blow takes the same amount of time, in a given time one can drive the whole wedge with one great blow, or one can drive it the same distance with multiple smaller blows in the same multiple of time, so that "the ratio of time to time is like the [inverse] ratio of the blow to the blow, and the head of the whole wedge to the head of one of the smaller wedges."[17] By dividing the power of a single blow into many smaller blows, and dividing a single large wedge into many smaller wedges, Heron has shown how even the smallest blow can drive any wedge, albeit the smallest distance; that many such blows will move it as far as one large blow, but more slowly; and that the smallest blow will drive a similar fraction of the wedge the same distance, but with proportionally less separation. Although retardation is again the result of the mechanical action of the wedge, the origin and cause of the wedge's effectiveness lie in dividing and sharing the load.

The fifth and final power is the screw. Rather than giving a full explanation of the screw, Heron merely shows how a screw is simply a wedge wrapped around a cylinder and then describes how to make a template for laying one out. But he does note that, unlike the wedge, the screw is actuated not by a blow but by turning; and, like the wedge, where the smaller its angle the less power is needed to drive it, in a screw the smaller the angle of the threads, the less power is needed to turn it and so to lift the weight, though he gives no precise ratio. He then describes a worm gear driving a cogwheel, observing that "as many teeth as there are on the wheel, so many turns the screw has to make for the wheel to make one turn." The final chapter on the five powers describes how a screw can move the *tylos* (τύλος), a sliding block that engages with half the screw to lift a weight or move a load.[18] But in neither application of the screw does Heron give any precise ratio of weights and loads, or even state explicitly that there is a multiplication of power.

---

17 *Mech.* 2.15–16; trans. in Drachmann, *Mechanical Technology*, 72–74 (my insertion); the vertical line in the diagram represents the force of the blows, divided in the same ratios as the thicknesses of the wedges and the distances they are driven; Schiefsky sees the ratio of times and forces in Heron's explanation of the wedge as analogous to the ratio of force and weight in concentric circles, and thus as another instance of the general phenomenon of "slowing up" ("Theory and Practice," 28–31).

18 *Mech.* 2.17–19; trans. in Drachmann, *Mechanical Technology*, 76–77, 79–81 (the quotation is on 77).

Thus only the windlass, the lever, and the inclined plane have actually been reduced to concentric circles; the pulley and the wedge have been explained rather by dividing their loads and distributing the parts among smaller movers, resulting in slower movements; the screw, being merely a twisted wedge, is given no separate explanation at all. So it is surprising when Heron asserts that he has reduced all five powers to concentric circles, or more fundamentally to the balance, which he says provides the principles of concentric circles and shows that the ratio of the weights is inversely as the ratio of the lengths of their arms.[19]

The next dozen chapters of Book 2 are devoted to meeting the mechanical challenge of moving a given weight with a given power by means of the five powers. Because each of the powers has practical limitations, Heron describes how they can be combined in entrainments, each stage of which is within the practical limits of the simple machine.[20] Now all such entrainments, Heron notes, bring with them a delay or retardation, in that more time is required in proportion to the ratio of weight to power. If one is required to move a 1000-talent weight with a five-talent power with a series of windlasses, for example, the power would have to move through 200 times the distance the weight is moved, and thus would take 200 times as long as a 1000-talent power would take to move the same weight the same distance. Similarly, a system of pulleys can be arranged to lift a weight of 1000 talents with a power of five, but it will also experience the same proportional retardation, since the ratio of the height that the weight is lifted to the length of rope drawn through it will be the same as the ratio of the weight to the power. And in an entrainment of levers, a retardation occurs in the same ratio, since levers are equivalent to wheels and axle. The wedge and the screw, as we saw above, entail a similar retardation, for the lighter the blow on the wedge, the more blows it takes to move a weight, and the finer the threads of a screw, the more turns it requires, both of which take more time. In all cases, then, the ratio of the times is as the inverse ratio of the power to the load. But, as Schiefsky has argued, Heron saw this retardation as an incidental and disadvantageous result of using such machines, and not as embodying the cause and principle of their effectiveness.[21]

At this point in Book 2, after having described the five powers, explained their effectiveness, and shown how they can be combined to move any given weight with any given power, Heron started on a new tack altogether. In obvious imitation of the Aristotelian *Mechanica*, he listed a series of seventeen miscella-

---

19 *Mech.* 2.20; trans. in Drachmann, *Mechanical Technology*, 81.
20 *Mech.* 2.20; trans. in Drachmann, *Mechanical Technology*, 81–82.
21 *Mech.* 2.21–29; trans. in Drachmann, *Mechanical Technology*, 82–90; see Schiefsky, "Theory and Practice," 32–42.

neous mechanical questions or problems.[22] Some of these he had taken over from the Aristotelian *Mechanica*; others were either original with him or perhaps drawn from other sources, though in every case where he had taken over a question from the *Mechanica*, his treatment was innovative. His purpose here was not to meet the mechanical challenge—to move a given weight with a given power—but rather to explain some mechanical marvels that are commonly met with. More importantly, these questions were introduced by a remarkable methodological and theoretical statement, which laid out the requirements for research into mechanics, stated explicitly the physical postulates to be used, and revealed the hidden mechanical principle that he claimed underlies all marvelous mechanical effects.

One who concerns himself with mechanics, Heron begins, must know the causes at work in mechanical movements, just as they have been explained earlier for the lifting of heavy bodies. Alluding to certain ancients who had presented some things bearing on mechanics, among whom he must particularly mean the Aristotelian *Mechanica*, Heron observes that we are amazed by things that appear contrary to what we already know. In an allusion to *phys.* 1.1, he states that we begin our search for knowledge from what is already clear to us; and echoing the beginning of the Aristotelian *Mechanica*, he adds that our amazement increases when things seem to contradict this. To discover the true causes of such marvels, one must begin from one or more physical principles or postulates and use them to reveal to us the causes, which are things already known in themselves. The postulates that one needs here, Heron says, are 1) that the light is easily moved and the heavy is moved with difficulty; and 2) that the same load can be more easily moved with a greater power than with a lesser. These two postulates arise one from the other and are clear and obvious to us, so that they constitute the clear and certain knowledge that we begin with when we seek the cause of mechanical movements. Now mechanical movements at first seem marvelous and obscure because, by allowing us to move a great weight with a small power, they seem to contradict these postulates. The cause of this obscurity is our failure to recognize the fundamental mechanical principle at work: "One should notice that all questions that are treated in mechanics and of which there is a obscurity with respect to the cause arise because *we cannot see how heavy bodies are shared among the bodies that move them*" (emphasis added).

---

[22] Drachmann omits almost all of this section from his translation and commentaries, dismissing it as merely the *pro forma* treatment of puzzles by a teacher of mechanics (*Mechanical Technology*, 93).

This principle in itself is known to us, both from experience and because it follows directly from the two postulates: what is shared is easier to move because either the load is less or the mover is greater. The obscurity arises through our failure to recognize this principle in mechanical movements. It is nonetheless clear to us in other cases, and here Heron gives the example of several men moving a load. A load that one man alone cannot move, or move only with difficulty, many men can move easily. The reason is that a part of the load falls to each man so that the load is shared among those who carry it.[23] This is the principle I have called dividing and sharing the load.

To solve the seventeen questions that follow, Heron generally looked for one of the two postulates—that lighter weights are more easily moved, and that a given weight is more easily moved with a greater power. In two cases where these postulates seem be contradicted—when a heavy weight is easily moved by a small power—he applied the principle of dividing and sharing the load to show that, in fact, there is no contradiction. Question 5 asks why a flat body falls more slowly than a spherical body of the same weight. Heron dismisses the obvious answer, believed by many, that a flat body offers more surface to the resistance of the air than a spherical. Rather, he argues, it is because a flat body is effectively composed of numerous parts, each of which receives in proportion to its extent only a part of the moving power and not the whole. He seems to mean that the moving power is divided and shared among the parts of a flat body and thus is less powerful. Here it is the power that is divided and thus rendered less effective in moving a coherent and unified load. And in question 17, Heron asks why liquids, which are naturally heavy, are so easily moved. The reason he gives is the same as we saw earlier in Book I, that water is easily moved because it is easily divided, so that when one moves a small quantity, the rest flows after it. The capacity to be easily divided makes water easy to move.[24]

This set of miscellaneous questions did not give Heron much scope to apply the principle of dividing and sharing the load, but as we saw above he had already successfully employed it in his explanations of the pulley and of the wedge. The pulley divides the load between a number of ropes, so that each rope, including the one drawn by the moving power, sustains only a fraction

---

[23] *Mech.* 2.33.
[24] *Mech.* 2.34 (Ch. 34 begins with the first question and contains all 17; the numeral "34" designating the beginning of the Chapter is missing in Carra De Vaux's translation, 144 [164]); on question 17, see also Book 1.21, trans. in Cohen and Drabkin, *Source Book*, 198–99, where Heron attributes the ability of water to flow down the slightest declivity to the fact that its parts are easily separated (which I have discussed above).

of the weight. In the case of the wedge, both the power, in the form of blows, and the load, in the form of the resistance to separation, are divided, so that many small blows on the large wedge or on many smaller wedges can have the same effect over time as one large blow has all at once on the large wedge. Heron's insight into the principle of dividing and sharing the load seems to have arisen not from abstract and theoretical considerations of the Aristotelian principle of circular movement or of Archimedes' rule of equilibrium, but from practical experience with the effectiveness and the limitations of various machines.

In the *Mechanica*, then, Heron ended up appealing to two different fundamental principles to explain the effects of machines: the principle of concentric circles and the balance to explain the windlass, lever, and inclined plane; and the principle of dividing and sharing the load to explain the pulley, the wedge, and several miscellaneous problems. Both principles result in retardation, in that a small power must be continually or repeatedly applied over time to produce the same result that a large power could produce over a shorter time or all at once. But for Heron, this retardation was not the cause of the effectiveness of machines, but rather the result of it, for with the help of a machine one can move a great weight with a small power, though it will take a longer time. Consequently, it cannot be the case that his fundamental principle is the principle of work, virtual or not, however inchoate. The principle of work is of course implicit in any mechanical analysis that gets the ratios of weights, powers, distances, and times right; in Heron's mechanics, those ratios were merely the result of the more fundamental principles of concentric circles or the balance, and dividing and sharing the load, and not the principles themselves of the mechanical effects.

Bibliography

Carra de Vaux, Bernard. "Les mécaniques ou L'élévateur de Héron d'Alexandrie, publiées pour la première fois sur la version arabe de Qostâ ibn Lûqâ, et traduites en français," *Journal Asiatique* ser. IX t. 1 (1893): 386–472, and ser. IX t. 2 (1893): 152–269, 420–514.

Carra de Vaux, Bernard. *Héron d'Alexandrie, Les mécaniques, ou, L'élévateur des corps lourds*, Collection Sciences et Philosophie Arabes (Études et Reprises) (Paris: Les Belles Lettres, 1988).

Clagett, Marshall. *The Science of Mechanics in the Middle Ages* (Madison: University of Wisconsin Press, 1959).

Cohen, Morris R., and I. E. Drabkin, eds. *A Source Book in Greek Science* (Cambridge, Mass.: Harvard University Press, 1948).

De Gandt, François. "Force et science des machines," in J. Barnes et al. (eds.), *Science and Speculation: Studies in Hellenistic Theory and Practice* (Cambridge: Cambridge University Press, 1982), 96–127.

Drachmann, Aage G. *The Mechanical Technology of Greek and Roman Antiquity: A Study of the Literary Sources* (Copenhagen: Munksgaard, 1963).

Drachmann, Aage G. "Hero of Alexandria," in Charles C. Gillispie, *Dictionary of Scientific Biography*, 18 vols. (New York: Scribner, 1970–1990), 6:310–14.

Dugas, René. *A History of Mechanics*, trans. J. R. Maddox (New York: Routledge & Kegan Paul, 1957). [Originally published as *Histoire de la mécanique* (Neuchâtel: Éditions du Griffon, 1950).]

Duhem, Pierre. *Origines de la statique*, 2 vols. (Paris: A. Hermann, 1905–1906). [English translation: *The Origins of Statics*, trans. G. F. Leneaux, V. N. Vagliente, and G. H. Wagener, Boston Studies in the Philosophy of Science 123 (Dordrecht: Kluwer, 1991).]

Hett, Walter S. *Aristotle, Minor Works* (Cambridge, Mass.: Harvard University Press, 1936).

Hultsch, Friedrich. *Pappi Alexandrini Collectionis quae supersunt*, 3 vols. (Berlin: Weidmann, 1875–78).

Neugebauer, Otto. "Über eine Methode zur Distanzbestimmung Alexandria–Rom bei Heron," *Kongelige Danske Videnskabernes Selskabs Skrifter* 26 [1938]: 21–24.

Nix, Ludwig, and Wilhelm Schmidt. *Heronis Alexandrini, Opera quae supersunt omnia*, vol. 2, fasc. I: *Mechanica et catoptrica* (Leipzig: Teubner, 1900).

Schiefsky, Mark J. "Theory and Practice in Heron's *Mechanics*," in Walter Roy Laird and Sophie Roux (eds.), *Mechanics and Natural Philosophy before the Scientific Revolution*, Boston Studies in the Philosophy of Science 254 (Dordrecht: Springer, 2008), 15–49.

Sidoli, Nathan. "Heron of Alexandria's Date," *Centaurus* 53 (2011): 55–61.

Vailati, Giovanni. "Il principio dei Lavori Virtuali da Aristotele a Erone d'Alessandria," *Atti della R. Accademia delle Scienze di Torino* 32 (1897): 940–62 [repr. in Vailati, *Scritti*, 91–106].

Vailati, Giovanni. "*Heronis Alexandrini Opera quae supersunt omnia*," *Bollettino di Bibliografia e Storia delle Scienze Matematiche* 4 (1901): 79–81 [repr. in Vailati, *Scritti*, 374–76].

Vailati, Giovanni. *Scritti di G. Vailati (1863–1909)* (Leipzig: J. A. Barth 1911).

Helen Lang
Plato on Divine Art and the Production of Body

Abstract: A number of Platonic texts concerning body are regularly referred to as, or even translated by, "dimension"; in some cases, the expression "all three dimensions" is taken as the equivalent of "length, breadth, and depth." But "dimension" is a Latin word first found in Cicero, who is strongly influenced by Stoic philosophy. Thus *prima facie* it is unclear that "dimension" translates Plato's texts or expresses his conclusions. This paper examines a number of texts from the *Sophist, Statesman, Laws* 10, and the *Timaeus,* in which Plato defines body as the outcome of divine art and governed by soul. In each case, length, breadth, and depth, which define body, are produced by a divine artist using proper measure. Proper measure in its turn explains how an original through divine art comes to be present in a copy, itself a work of art. I argue that body is a copy made through divine art, and so a divine maker, that depends on proper measure and so order; indeed, as a perfect work of art, body is first and foremost orderly. This conclusion returns us to "dimension." While "dimension," as a translation and a concept, may appear to be unambiguous, it is not. Associated with late medieval or early modern philosophy, it can refer to body or to the space occupied by a body. The expression "all three dimensions," taken to signify "length, breadth, and depth," suppresses any sense of order among these three. Since according to Plato body is first and foremost orderly, the expression "all three dimensions," which suppresses any reference to order, cannot be linguistically equivalent to "length, breadth, and depth" as they express Plato's view of body and the production of body.

"Dimension," from a Latin word (*dimensio*) first found in Cicero, has been used to render several etymologically unrelated Greek words and phrases. For example, Cornford translates Plato's phrase in the *Sophist* "length, breadth, and depth" as "all three dimensions."[1] Aristotle locates place in the genus of things

---
This paper has vastly benefited from the criticisms and suggestions of a number of readers. I would like to take this opportunity to thank them.

1 Pl. *soph.* 235d6–e2; the translation is that of F. M. Cornford in E. Hamilton and H. Cairns, eds., *The Collected Dialogues of Plato, Including the Letters,* Bollingen Series LXXI (New York: Bollingen, 1961). There are other examples, e.g., his account of education as moving from plane figures to solid figures is translated as moving from two-dimensional to three-dimensional

having three "intervals" (*ta diastēmata*), which is also regularly rendered as "dimensions."[2] "Dimension" means "(spatial) measure" and as a translation might be defended as capturing the sense of the Greek. Here we reach an interpretive, or philosophical, problem: What is the conceptual force of "dimension" and does it capture Plato's (or Aristotle's) meaning? Interpreters of Plato (and his predecessors) or Aristotle regularly turn to a notion of dimensionality. Zeyl, introducing his translation of the *Timaeus*, comments "In the physical theories of Plato's predecessors, the ultimate *archai*, or primary principles of the physical world, were three-dimensional bodies of some sort."[3] But can we be sure that "Plato's predecessors," even the atomists, thought in terms of "three-dimensional bodies"?

Translators and interpreters alike may think "dimension" is unambiguous. But it is not. Associated with late medieval or early modern philosophy, dimension can refer to body or to the space occupied by a body; "all three dimensions" can refer directly to body, i.e., as synonymous with "length, breadth, and depth," or "all three dimensions" can stand as a general designation that refers to "length, breadth, and depth," which in turn refer to body. However clear "dimension" may initially seem, it introduces these issues into the already difficult arguments of ancient philosophy. In this paper, I shall set aside "dimension" and consider Plato's arguments concerning body, taking body as it later becomes part of physics. Body, defined as whatever has length, breadth, and depth, appears in a number of dialogues, including the *Sophist*, *Statesman*, *Laws* 10, and *Timaeus*. I turn to these dialogues to analyze the arguments where body appears. While I shall consider their literary setting and language, my primary analysis concerns their logical structure and its implications. Only such an

---

figures in *resp.* 7, 528a9–b3; here the translation is that of Paul Shorey, also in Hamilton and Cairns: "After plane surfaces, said I, we went on to solids in revolution before studying them in themselves. The right way is next in order after the second dimension to take the third. This, I suppose, is the dimension of cubes and of everything that has depth." This paper, for reasons of length, will deal only with the problems raised by the *Sophist* and related accounts. All references to Plato are to the O.C.T. and all translations are my own unless otherwise indicated.
2 Cf. Arist. *phys.* 4.1, 209a4–6. For the translation "dimension," cf. J. Barnes, ed., *The Complete Works of Aristotle: The Revised Oxford Translation*, vol. 1, Bollingen Series LXXI (Princeton: Princeton University Press, 1984). The original translation (this line is unchanged by Barnes) is that of W. D. Ross. For the same line, cf. also E. Hussey, trans. *Aristotle's Physics, Books III and IV* (Oxford: Clarendon Press, 1983); also note, at 102: "It is natural to think of the place of a three-dimensional body as being itself three-dimensional." All references to Aristotle are to the O.C.T. and all translations are my own unless otherwise indicated.
3 D. J. Zeyl, trans. *Plato, Timaeus* (Indianapolis: Hackett, 2000), liv.

analysis, I shall argue, can uncover the force, both philosophical and linguistic, of Plato's account of body and the issues raised by it.

In Plato's dialogues, length, breadth, and depth, which ultimately define body, always appear within some other context, e.g., defining the sophist or the statesman. These contexts regularly introduce the notion of "proper measure"; as I shall argue, proper measure is the criterion that defines body; all body, both the cosmos as a whole and the various bodies contained within it, come to be through divine art. Divine art relies on soul both in producing body and in governing it. Indeed, I shall conclude that politics, not science, provides the model and the metaphors of Plato's account of the cosmos and all body (or bodies) contained within it.

After considering proper measure in the *Sophist* and *Statesman*, I shall turn to the arguments concerning soul and body in *Laws* 10 and the *Timaeus*. Proper measure, I shall conclude, always signifies the relation (produced by a maker) of a likeness to the good or true and never signifies "dimension" as a "(spatial) measure." Insofar as "dimension" signifies a consideration of body (or space) apart from its relation to the good or soul, it is not only absent from Plato but is misleading when introduced into his arguments. Finally, I shall offer an analysis of the turn to "dimension" by Plato's translators and interpreters.

I Imitation and The Problem of Proper Measure

Sophists, e.g., Protagoras or Gorgias, are Plato's living targets. The sophist, unlike the philosopher, takes on many guises; he lives in the realm of shadows and appearances, and slips through the net of argument, "hiding in the art of imitation."[4] Consequently, the hunt for him must separate out several distinct kinds that are mixed up and so confused: first the different kinds of imitation, then the different kinds of makings required by different kinds of imitation, and lastly the outcome of each kind of imitation. The Stranger's process of separating out establishes kinds that cannot be further subdivided. As a consequence, it unambiguously differentiates members of the realm of becoming. Length, breadth, and depth appear here.

The *Sophist* opens with an elaboration of the host/guest relationship (*soph.* 216c5–6; cf. 231b7, 235c5). A stranger visiting from Elea, where he associates with Parmenides and Zeno, is honored and compared to a god (216a2–5). The difference between him and the sophists suggests a distinction between

---

4 Cf. *soph.* 231b7; 235c5.

the true philosopher and an apparent philosopher, the sophist. Acceding to his hosts' wishes, the Stranger agrees to take up a difficult question: How do we define "sophist," "statesman," and "philosopher" (217a3)?[5] First he will hunt down the sophist and give an account of him.. Because the task is so difficult, the Stranger wishes to practice his method of tracking on an easier prey, recognizable by all: an angler (218d1–e3). Then he will turn to the sophist.

The search for a definition of an angler is sometimes dismissed as methodological or philosophically inferior.[6] But the sophist resembles the angler: both are hunters with expertise in acquisition. Furthermore, since the Stranger's method is itself a form of hunting, even the philosopher may be a hunter of sorts.[7] In fact, the hunt for the angler reaches conclusions central to defining the sophist.

The method of division used to define the angler leads the Stranger to ask: When something being moved – something itself possessing an impulse toward what is proper for it – continually misses the mark and goes astray, do we say the going astray is produced by proper measure of one thing in relation to the other, or by the opposite, that is, by the absence of measure (228c4–5)?[8] It is produced by the absence of measure that signals an unintelligent soul, ignorant of the truth (228c6–d1).[9] But all soul possesses an impulse toward knowledge, i.e., hit-

---

[5] For a fuller discussion of how the "prologue" sets the problem of the dialogue, along with the distinction among "kinds" here, cf. N. Notomi, *The Unity of Plato's "Sophist": Between the Sophist and the Philosopher* (Cambridge: Cambridge University Press, 1999), 21–27.

[6] Cornford for example omits this "inferior passage" (218d1–230e) altogether from his translation. P. Pellegrin, "Le *Sophiste* ou *De la division*, la diérétique et le son-être. Aristote – Platon – Aristote," in P. Aubenque (ed.), *Études sur le "Sophiste" de Platon* (Naples: Bibliopolis, 1991), 389–416, at 403, notes that the Stranger speaks of "the method" without further explanation and speculates that it could represent a kind of "scholarly exercise" pursued within the Academy. Notomi, *The Unity of Plato's "Sophist,"* 77, comments: "Practice is important to dialectic, and through practice we can acquire the method of inquiry."

[7] Cf. S. Rosen, *Plato's "Sophist": The Drama of Original and Image* (New Haven: Yale University Press, 1983), 91. Cf. soph. 222a2. R. S. Bluck, *Plato's "Sophist": A Commentary*, ed. by G. C. Neal (Manchester: University Press, 1975), 37, concludes: "It looks as though there was a special reason other than its easiness to define for the choice of Angling as an 'example'"; he goes on the compare the divisions here in the *Sophist* to those in the *Statesman*, *Theaetetus*, and other dialogues.

[8] The contrast here is between ὑπὸ συμμετρίας and ὑπὸ ἀμετρίας. For the claim that everything that comes to be recognizes the good, desires it, and is aimed at it, cf. Pl. *Phlb.* 20d1–10. The notion of an "impulse" (ὁρμή) is also found in Aristotle (*phys.* 2.1, 192b18–19) where the absence of an innate impulse for change (οὐδεμίαν ὁρμὴν ἔχει μεταβολῆς ἔμφυτον) distinguishes things that are by art from things that are by nature.

[9] Cf. also Pl. *resp.* 6, 486b10–d11 where the philosophical soul must possess due measure because truth is akin to due measure and not the absence of measure.

ting the good and the true; therefore no soul is voluntarily ignorant; ignorance can only be an aberration away from soul's true impulse, an aberration that leaves the soul deformed and devoid of proper measure (228d4). Ignorance is a problem that can be cured only by knowledge, and knowledge is acquired through education; the sophist, who claims to educate the young, claims to have knowledge. Yet he possesses not true knowledge, but only an imitation of real things (τῶν ὄντων, 233c10–11; 234e5–235a9). "Imitation" is an ambiguous term and so requires a distinction. There are two kinds of imitation (δύο ... εἴδη τῆς μιμητικῆς, 235d1), one true, the other a fraud, i.e., an imitation that pretends to be true even though it is not. Because of the pretense to truth, the two kinds of imitation are hard to sort out. Proper measure defines the difference between them.

One kind of imitation involves "likeness-making" (ἡ εἰκαστική, 235d6; 236b1), which produces a "likeness" (ἡ εἰκών, 236a8) that conforms to "the proper measure of the original in length, breadth, and depth and in addition to these in giving the proper color to every part."[10] The other kind of imitation is a "semblance" (τὸ φάντασμα, 236b7)";[11] it requires semblance-making (ἡ φανταστική, 236c4) by a maker interested neither in what is beautiful or good, nor in proper measure, nor in an imitation of them; rather, dismissing the truth, he seeks only what appears to be beautiful.[12] In short, while a likeness enters into a relation to the original through proper measure, a semblance is an imitation that fails to be a likeness because it bears no relation to the good. Since proper measure relates a likeness to the original that serves as its target and since the maker of likenesses gives proper measure to his works, a likeness is the kind of imitation

---

10 Soph. 235d6–e2: ἔστι δ' αὕτη μάλιστα ὁπόταν κατὰ τὰς τοῦ παραδείγματος συμμετρίας τις ἐν μήκει καὶ πλάτει καὶ βάθει, καὶ πρὸς τούτοις ἔτι χρώματα ἀποδιδοὺς τὰ προσήκοντα ἑκάστοις, τὴν τοῦ μιμήματος γένεσιν ἀπεργάζηται. Here we see Cornford's translation: "The perfect example of this consists in creating a copy that conforms to the proportions of the original in all three dimensions and giving moreover the proper color to every part." M. Villela-Petit, "La question de l'image artistique dans le *Sophiste*," in Aubenque, *Études sur le "Sophiste*," 53–90, speaks of length, breadth, and depth, but in the next sentence takes them as three dimensions: "c'est-à-dire les rapports de commensurabilité entre longueur, largeur, et profondeur. Déjà l'allusion aux trois dimensions suffit à indiquer que l'Etranger ... " (74). For a more recent translation, using "length, width, and depth," cf. Notomi, *The Unity of Plato's "Sophist*," 148.
11 We may note that the word φάντασμα is quite late, apparently coined by Aeschylus at *sept*. 710.
12 Soph. 235e5–236c6. The key expression here about the true, at 236a4–6, reads: Ἆρ' οὖν οὐ χαίρειν τὸ ἀληθὲς ἐάσαντες οἱ δημιουργοὶ νῦν οὐ τὰς οὔσας συμμετρίας ἀλλὰ τὰς δοξούσας εἶναι καλὰς τοῖς εἰδώλοις ἐναπεργάζονται; Cornford makes this question a statement, but gets the meaning precisely: So artists, leaving the truth to take care of itself, do in fact put into the images they make, not the real proportions, but those that will appear beautiful.

that hits the mark, i.e., what is good and true. A semblance is the opposite: because its maker cares nothing for the good, it only appears to be good while in fact it suffers absence of measure.

The distinction between these kinds of imitation, one true, the other fraudulent, may seem surprising, given its absence elsewhere in the corpus. Consideration of this point will allow us to see the full force of the Stranger's argument. In the *Phaedo*, for example, sensation and its objects are inferior and hinder knowledge; only when the soul acts independently of body does it know "true reality" (again τῶν ὄντων, *Phd.* 65c3). True reality alone is permanent and knowable by soul; therefore, all imitations alike fail to meet the standards of being and fail to be knowable; no distinction among imitations is possible and so all are dismissed (65a9–d2). Justice itself, beauty, and goodness do meet these standards and so are knowable (65d4–7). When we turn to the *Sophist*, there is a distinction among imitations. The Stranger presupposes what is true (and real) and in this sense his argument agrees with that of the *Phaedo*. But his hunt for the sophist leads him to distinguish between imitations that are aimed at them (and so made true and real) and those that are not, i.e., semblances. Thus, the distinction in the *Sophist* between a likeness and a semblance presupposes the arguments of the *Phaedo* concerning real things;[13] the Stranger establishes that proper measure is required by a relation to the true; when proper measure is present, the making results in a likeness of real things, but when proper measure is absent, the making results in a phantasm, a fraudulent appearance of the good and true.[14]

---

[13] The point lies beyond the bounds of this paper, but S. Menn, "On Socrates' First Objections to the Physicists (*Phaedo* 95e8–97b7)," *Oxford Studies in Ancient Philosophy* 38 (2010): 37–68, at 47, makes an important point on the rejection of physics in the *Phaedo*; his argument anticipates Plato's arguments about the cosmos in the later dialogues, especially the *Timaeus* and *Laws* 10: "Plato's point against the physicists is not that such statements of identity, through time or across possible situations, are never true, but rather that the physicists cannot explain when and why they are true, and that, lacking such explanations, they cannot explain becoming or being larger." Cf. also 58–68, where Menn makes a comparison with *Tim.* 30b3. A somewhat different take on this question is offered by S. K. Strange, "The Double Explanation in the *Timaeus*," in G. Fine (ed.), *Plato*, vol. 1: *Metaphysics and Epistemology* (Oxford: Oxford University Press, 1999), 397–415, who argues that in Socrates' critique of the physicists in the *Phaedo*, mind alone is the cause of nature while in the *Timaeus* an additional cause, necessity, is admitted (cf. esp. 400).

[14] T. Johansen, "Body, Soul, and Tripartition in Plato's *Timaeus*," *Oxford Studies in Ancient Philosophy* 19 (2000): 87–111. Johansen argues throughout that the *Timaeus* answers questions about the soul/body relation raised by the *Phaedo* (and the *Phaedrus*). He concludes: "The dialogue forces us to rethink the image of Plato as enemy of the body" (111).

Although the distinction between a likeness and a semblance is clearly established by this argument, the two kinds of making remain less clear. Later in the *Sophist*, the Stranger addresses this problem by identifying two kinds of making (δύο ποιητικῆς γένη), one divine, the other human (265e5–266d3). Divine art makes divine products, including us, all other living things, and the elements; these in turn produce their own offspring, i.e., the images we experience in perception and dreams. Human art makes human products, including a thing itself, e.g., a house, and its offspring, e.g., a painting of a house. The Stranger does not pursue these distinctions further, but divine art is central to the *Timaeus*, where god makes a perfect likeness, namely, the cosmos.[15]

Hunting the sophist, the Stranger returns to semblance-making, which involves mimicry and which also divides into kinds, although, he says, there is a severe shortage of names (*soph.* 267a1; 267d4–8). The sophist, along with the ignorant demagogue, lurks here (268b7–9). The sophist cannot be defined apart from the philosopher and the statesman because he is nothing other than a semblance, a fraudulent appearance, (268b11–d5). The demagogue is the fraudulent appearance of a statesman. He, and the further semblances of which he is the maker, the Stranger concludes, are but the fleeting shadows of the kind of imitation that bears no relation to the good or true, despite appearing to do so. Being a semblance, the sophist has no "true reality." Names signify the real thing; what has no "true reality," like the sophist, has no name.

In the *Sophist*, the Stranger's account of proper measure is complete. The sophist contrasts with the maker of likenesses, who gives each icon the "proper measure" of the original in length, breadth, and depth (and color). "Dimension" means "measure" or "measurement," originally in a spatial sense (cf. Cic. *Tusc.* 1.57). At least *prima facie*, the relation seems clear: length, breadth, and depth are "the measured," thereby yielding a thing's dimensions. Does "proper measure" in the *Sophist* take us to "dimension" as a measure of things? If so, then the translation of "length, breadth, and depth" as "three dimensions" seems justified. If not, then what does "proper measure" – and indeed "length, breadth, and depth" – mean?

Measure appears throughout Plato's dialogues. Several closely related words concerning measure are first found in Plato, i.e., "the art of measurement" (ἡ μετρητική), "proper measure (or symmetry)" (ἡ συμμετρία), and its opposite "absence of measure" (ἡ ἀμετρία) (all three appear in the *Sophist*).[16] These words,

---

[15] Here in the *Sophist*, cf. 266b2–4. God is also mentioned, although the details of his work are not pursued, at *resp.* 10, 597b5–d8.
[16] Cf. C. D. Buck and W. Petersen, *A Reverse Index of Greek Nouns and Adjectives: Arranged by Terminations with Brief Historical Introductions* (Chicago: University of Chicago Press, 1945), 155.

together with "measure" (τὸ μέτρον), "due measure" (τὸ ἔμμετρον), and the act of measuring (μετρεῖν), appear hundreds of times in Plato's dialogues. But what exactly does Plato mean by "measure"?

"Measure," along with its role in the arts of making, is central to the sister dialogue of the *Sophist*, the *Statesman*, where the Stranger from Elea reappears to define the statesman. After an extended argument, the Stranger asks if the argument has not been excessively roundabout (*pol.* 283b1–3). Young Socrates insists that it contains nothing pointless but the Stranger insists that they examine excess and deficiency to find the length required for an argument to be praiseworthy or blameworthy (283b4–c1).

The discussion of length and brevity, excess and deficit, returns us to the art of measuring (ἡ μετρητική) (283c11–d2). One part of this art measures greatness and smallness according to what relation things have to one another, while another part concerns the "fixed being" (ἀναγκαίαν οὐσίαν) of measure for becoming (283d7–9). In short, there is relative measure and fixed measure. Furthermore, relative measure presupposes fixed measure and so "fixed measure" is primary whenever a thing must be measured (284a1–c3). The Stranger concludes that the art of measurement applies to all things that become while all becoming involves art and so requires not only relative measure but also fixed measure (285a1–4). Lastly, the Stranger divides the art of measurement into two parts.[17] One part measures "number and length and depth and breadth and quickness." The other concerns due occasion, due time, due performance (284e4–285a4). Such measurement is involved in all things that come to be and the task involved in knowing such things (insofar as they are knowable) is to understand their measurements.

This account of measures complements that found in the *Sophist*. There, through proper measure, human making and divine making both produce a likeness of the original in length, breadth, and depth (and color). In the *Statesman*, fixed measure is required for all art because through it things that come to be reflect the good and true, those targets at which, through their making, they are aimed.[18] In effect, fixed measure is the standard used to produce the proper

---

[17] Because my focus here lies with "measure," I do not consider issues raised by division as a procedure in the *Statesman*. For an analysis of this problem and various options that have been pursued in the literature, cf. D. De Chiara-Quenzer, "The Purpose of the Philosophical Method in Plato's *Statesman*," *Apeiron* 31 (1998): 91–126.

[18] The point lies beyond the bounds of this paper, but "all art" also includes that of ruling. Hence its role in the *Statesman*, where the true statesman will be defined, and the *Sophist*, who is connected with the demagogue, the "false" statesman. The art of ruling is also part of the *Republic*, which is traditionally thought to be the "entertainment" referred to at the opening of

measure of the original in the likeness. This account returns us to "length, breadth, and depth." Plato mentions neither body nor the measurement of body in these arguments; rather length, breadth, and depth are the outcome of due measure, the principle that allows things made through art to aim at the good and true. Recalling the account of likenesses and semblances in the *Sophist*, we can see that length, breadth, and depth occur only in contexts involving fixed measure and a relation to the good through the art of making. Length, breadth, and depth have no reality apart from this relation and the role of measure. As a result, they have no reality apart from art or from the maker, whether a god, a philosopher, or a statesman, who produces likenesses through art.

An important implication appears here: the Stranger's account of measures applies to *all* things that come to be because they require art. Therefore, all things that come to be presuppose fixed measure. In the *Sophist* and *Statesman*, there is no reference to measurement as spatial, no mention of body as such, and no measure of things that come to be outside the relation to fixed measure. Indeed, Plato's view here could hardly differ more from the spatial sense associated with "dimension."

II Leadership and the Generation of Body

The *Statesman* concludes with Socrates proclaiming that next to the Stranger's definition of a sophist has been placed a picture of the true king and statesman. The *Sophist* and the *Statesman* share a common account of the true political life, an account that acknowledges the causal role played by the good and the true. As we have seen, the semblance-making sophist and the demagogue alike fail to recognize the good – letting it take care of itself – and so must be rejected; fixed measure, defined in the *Statesman*, makes whatever comes to be both good and true. In the *Sophist*, human art contrasts with divine art, which produces likenesses, including animals, plants, and the elements.[19]

But Plato is not finished with sophists, or any relativists, who deny fixed measure. They reappear in *Laws* 10 as atheists, who deny not only the gods but also art, attributing all apparent order to nature and chance (*leg.* 10, 889c5–6). *Laws* 10 opens with the Athenian identifying a comprehensive princi-

---

the *Timaeus*, where the Demiurge practices the art of making the cosmos (see below). For an interesting account of ruling as art, cf. R. D. Parry, "The Craft of Ruling in Plato's *Euthydemus* and *Republic*," *Phronesis* 48 (2003): 1–28.

19 In the *Timaeus* the cosmos turns out to be an animal (Aristotle, too, calls the heavens "ensouled": *cael.* 2.2, 285a29: ὁ δ' οὐρανὸς ἔμψυχος).

ple of law for all acts of violence (884a,1–2). No one, he claims, who believes in the gods, as the law prescribes, engages in such acts (885b4–6). Therefore, the good legislator must make a case for the gods; in so doing, as we shall now see, he implicitly makes a case for divine art, as we have seen it in the *Sophist* and the *Statesman* (885e1–6; cf. 890b3–d8). Divine art takes him to a "true account" of sensible things.[20]

In *Laws* 10, the Athenian combats crimes of violence by proving god. He explains that those wishing to impress the young as wise claim that nature is due to chance and law to convention (889b1–890a1).[21] Chance and convention, trotted out as true causes, exclude first the good or god as causes and then divine art or fixed measure. The legislator, after calling on god's help, intends to defeat atheism by proving that soul is prior to body – a point we shall also see in the *Timaeus* (893a6–b4).

The Athenian's proof of soul's priority rests on a technical analysis of motion. This analysis is often compared to Aristotle's treatment of motion in the *Physics*. But we may note that while Aristotle's argument forms part of the science of nature, Plato's argument is ultimately political, i.e., part of a comprehensive principle regarding crimes of violence.

The argument itself is compact. Some things are moving and some are at rest; they move and rest "in some sort of space" (ἐν χώρᾳ τινί, 893c2). Some things are moving in circles about their center, some are moving continuously to different places, some are moving through a point of support (893c3–e1). Moving objects experience not only change of place, but also (through collisions) augmentation and diminution (893e1–894a1). All such motions, the Athenian argues, require a prior cause, or condition, i.e., soul. Since the things exhibiting these motions are bodies, body requires soul.

What "condition" (πάθος) does the coming to be of all things (and so these different motions in them) require (894a2)? Before a thing can be perceived, it must be acted upon in three increments or increases (αὔξη) by soul: there must be a beginning that, receiving an increase, comes to a second stage and then to a third when a moved thing becomes perceptible to whatever can per-

---

[20] Cleary makes the point that this argument "overturns" the sophistic distinction between *nomos* and *physis*: J. J. Cleary, "The Role of Theology in Plato's *Laws*," in F. L. Lisi (ed.), *Plato's "Laws" and its Historical Significance: Selected Papers of the I International Congress on Ancient Thought, Salamanca 1998* (Sankt Augustin: Academia, 2001), 125–40, at 125.

[21] Cf. both *soph.* 265c7–9 and the so-called autobiography of Socrates at *Phd.* 96a6–100a7, in which he describes his disillusionment with materialism, his excitement at hearing that Anaxagoras calls on mind as the cause of all things, and his subsequent disappointment on discovering that Anaxagoras, too, is in fact a materialist (98b7–99d2).

ceive (894a1–5). Since the third increase yields perceptible objects, the first and second increases are "prior" to perception; being prior to perception, they are completely obscure. Here there is no mention of length, breadth, and depth, but we might speculate about these increases: a point moved by soul produces a line, a line moved by soul produces a plane and a plane, rotated or folded, produces body with length, breadth, and depth, which can be perceived. We might wish that Plato would tell us more about these three increases produced by soul. But because this argument concerns soul as mover and its productivity, rather than what is produced by it, the three increases are left unspecified and so unexplained.

All motion, the Athenian continues, may be divided into two kinds: (1) the motion (belonging to body) that is able to move other things, but unable to move itself; and (2) the motion (belonging to soul) that is able to move itself and other things with respect to integration and disintegration, augmentation and diminution, genesis and corruption (894b8–c1). The motion of soul, i.e., self-motion, must be first (894c10–895a3). Why is self-motion, and so soul, prior to what is moved by another, body?

The Athenian answers this question next. If all things came together and stood still, that which is self-moved would arise first, before that which is moved by another. Soul, by setting itself in motion, can also move all things – sky, earth, and sea. Because body cannot move itself, it must depend on soul for its motion. The independent is prior to the dependent; therefore, "we say that soul came to be prior to body and body is both secondary and posterior, soul governing, body governed according to nature."[22] Indeed, we must say that soul controls the heavens (896e1–2).

In fact, "self-movement" is the very definition of soul; consequently, soul is the source of motion and so the first-born of all things. And that motion which, wherever it arises, is produced by another because it is never able to be moved itself in itself, is second, i.e., after soul. If soul is prior to body, then the characteristics of soul, "moods and habits and wishes and calculations and true opinions and purposes and memories would be prior to the length and breadth and depth and strength of body" (896c9–d3).

This claim completes Plato's account of body in the context of the *Laws*. The generation of what is moved by another, i.e., body, including length, breadth,

---

22 Cf. 896c1–3: εἶμεν ψυχὴν μὲν προτέραν γεγονέναι σώματος ἡμῖν, σῶμα δὲ δεύτερόν τε καὶ ὕστερον, ψυχῆς ἀρχούσης, ἀρχόμενον κατὰ φύσιν. Exactly the same argument appears at *Phdr.* 245c5–246a2. See R. Demos, "Plato's Doctrine of the Psyche as a Self-Moving Motion," *Journal of the History of Philosophy* 6 (1968): 133–45, who argues that the *Laws* 10 "version" is prior and that of the *Phaedrus* a late addition.

and depth, is a consequence of the activity of soul that moves itself by itself. Body depends upon soul in two ways. First body receives three increases (perhaps as I suggest above length, breadth, and depth) that soul alone can originate and, after receiving the third increase, body becomes perceptible. Secondly, body depends on soul for its motion. Self-moving soul originates motion while body, which depends on another, requires the self-mover for its motion. Body, in length, breadth, and depth, and strength, depends upon soul, with its wishes, habits, etc. Consequently, all bodily motion, even as it must be in some space, originates with soul as its cause and its ruler.

Laws 10 opens with the problem of assault and more generally crimes of violence. In this context, the Athenian proves the priority of soul by analyzing motion. His argument rests consistently and exclusively on political metaphors, e.g., soul "governs" body. Like Socrates, he presupposes a distinction between soul and body to affirm the superiority and the causal role of soul (and the just life) over body (and materialism/atheism, with its appeals to chance and convention). If we accept the Athenian's argument, atheism or materialism is not defeated by showing that there are no bodily collisions, augmentations, and diminutions. It is defeated by showing that these "physical" events are not primary; rather, they presuppose the presence of soul, the self-mover as a ruling principle. We may see bodies collide; but now we understand that such collisions presuppose a prior invisible cause, self-moving soul. It hardly seems necessary to add that a notion of spatial measure is not relevant to the Athenian's argument for the dependence of body on soul.

Indeed, in *Laws* 10 soul and body lead the ultimate political life: soul governs and controls all things while body depends upon the governance of soul for its coming into being, i.e., generation and motion. As we shall now see, this argument works with that of the *Timaeus*, where the cosmos originates in the divine art of making that turns to the best of all paradigms. If the *Sophist* and the *Statesman* give us proper measure and due measure and *Laws* 10 gives us the agency of soul governing body, the *Timaeus* gives us divine art and soul as the messenger of god in the art of divine image-making.

III Divine Art and Image Making

All likenesses become, as we have seen, and so require either human art or divine art. Divine art, as the *Sophist* defines it, produces living things and the elements, along with their offspring, dreams and perception. Divine art as image-making is central to the *Timaeus*. A long introduction, concerning hosts and guests and recalling yesterday's argument, opens the dialogue. Critias, who summarizes that argument, explains the order of the upcoming speeches. First, Timaeus will explain the generation of the cosmos from its beginning through the generation of man; then Critias will speak again, returning to political themes of perfect rule and citizenship (*Tim.* 27a2–b6).

Timaeus is "the best astronomer among us," and so the most qualified to speak about the becoming of the cosmos (27a3–4). No atheist, he first calls upon the gods; next he distinguishes being from becoming (27c1–3; 27d5–28a1). All things that become, we know from the *Statesman*, do so by art and so require a maker. Indeed, it has been argued that the *Timaeus* "explicitly harks back" to the *Republic* and *Phaedo*, as well as utilizing the conclusions of the *Statesman*.[23] Timaeus reflects the Stranger's view: it is necessary that everything that becomes come to be by some cause, "the Demiurge" (also called "god") (28a4–5).

In effect, the *Timaeus* gives us the divine maker and the divine art of making. In the *Sophist*, the Stranger explains that divine art produces us, all other living things, and the elements. In the *Timaeus*, the Demiurge makes the best of all likenesses.[24] Desiring that all things be as good as possible and so looking to the best possible model, the Demiurge is the cause of the genesis for everything that is generated (28a6–29b1). In order that the cosmos be the most beautiful of things that become, the Demiurge looks to what is eternal, unchanging, and apprehended by reason (29a6–b2). If we think in the terms of the *Statesman*, the Demiurge engages in divine art, using proper measure: he looks to the good and the true in order to bring the proper measure of the original into the imitation and thereby produce the best possible likeness. Throughout the *Timaeus*, proper measure relates what is made by the Demiurge to its original and so maintains the order implied by the division of making.[25]

Timaeus begins his account with a portrait of the Demiurge: free of jealousy and the best of causes, he desires that the image made by him be the fairest and best possible; therefore, he brings order out of disorder by putting "mind in soul and soul in body."[26] Speaking in the language proper to images, we must say that

---

23 O. Goldin, "Plato and the Arrow of Time," *Ancient Philosophy* 18 (1998): 125–43, at 137.
24 Throughout the *Timaeus*, the cosmos is called a "likeness"; for one example, cf. 29b2: τὸν κόσμον εἰκόνα τινὸς εἶναι.
25 Cf. 29b1–d3. Timaeus concludes this opening speech by reflecting on his own status as a mortal, who can only produce words. In the division of the *Sophist*, his art would be a human likeness-making. Words follow what they describe: when they bear on the eternal they should be eternal; when they bear on what comes into being, they are probable rather than exact.
26 *Tim.* 30a2–c1: ... κάλλιστον εἴη κατὰ φύσιν ἄριστόν τε ἔργον ... (b5–6) νοῦν μὲν ἐν ψυχῇ, ψυχὴν δ' ἐν σώματι ... (b4–5).

the cosmos was generated as a living thing that is truly alive and intelligent by "the forethought of god" (30b6–c1: τὴν τοῦ θεοῦ πρόνοιαν). Next Timaeus speaks of the paradigm and the relation of the copy to the paradigm (31a4).[27] Wishing the cosmos to be as like as possible to its paradigm, god made it like the paradigm as whole to whole and he made it unique: there is and will be one only-begotten, generated heaven (31b3).[28] He makes soul prior to body and older than body, the ruler and mistress of body.[29] In *Laws* 10, the Athenian proves the priority of soul and characterizes it as ruling body; here we see the divine art that makes it such. The accounts agree on the political metaphors that express the relation between soul and body.

In Timaeus' account of divine art, the model is the best, the maker is the best, and the made is the best possible likeness. The good is the sole model used by the Demiurge, who takes nothing away and adds nothing of his own, except the wish that all things be as good as possible.[30] The Demiurge is the exact opposite of the maker of semblances who "dismisses the truth" in the *Sophist*. He attends to the good and the true in every way: in the act of making, nothing intervenes between the good and the Demiurge as maker or between the good as paradigm and the cosmos as its unique copy. The maker of images in the *Sophist* produces a likeness with the proper measure of the original in length, breadth, and depth (and color); here in the *Timaeus*, divine art is nothing other than taking the good as the sole model and reproducing it in the best possible way: not bit by bit, which might introduce disorder, but whole to whole and unique. Now we turn to body.

---

27 The word here, *paradeigma*, is first found in Hdt. 5.62, and then comes into its own in Plato's extensive usage. Buck and Petersen, *A Reverse Index*, 221, note that this form was "one of the most productive," with almost half these words found in the philosophers and the Hippocratic writers and the rest largely coined by tragedians. They note that this form was satirized by the comic poets through mock formations because of its "flavor of pompous tragic grandeur and pretentious intellectualism."
28 The Demiurge is referred to as "father" a number of times in the dialogue (e. g., 28c3; 41a7). Here the heaven is characterized as a unique biological offspring. "Only-begotten" (μονογενής) sounds odd (and Christian) in English, but there is no other direct translation of this term.
29 *Tim.* 34c4–35a1: ὁ δὲ καὶ γενέσει καὶ ἀρετῇ προτέραν καὶ πρεσβυτέραν ψυχὴν σώματος ὡς δεσπότιν καὶ ἄρξουσαν ἀρξομένου συνεστήσατο ἐκ τῶνδέ τε καὶ τοιῷδε τρόπῳ.
30 In *resp.* 1, 342e6–11, Socrates argues against Thrasymachus that the just ruler considers only the ruled, for whom he works.

1 Timaeus' First Account of The Cosmos

In *Laws* 10, soul generates body; here the Demiurge constructs the body of "the all."[31] Timaeus gives two additional accounts of the generation of the cosmos. The first opens with the word "bodily form" (*sōmatoeidēs*)[32] and concerns "the body of the all," i.e., of the heavens.[33] Because it is a bodily form, the heaven must be "visible and tangible";[34] being visible requires fire, while being tangible requires solidity and therefore earth (31b5–6). Any two things, e.g., fire and earth, require a bond, or mean, to unite them; the most beautiful bond is that which, most of all, makes a thing one; proportion (*analogia*) is by nature best suited to complete this task (31c1–4). If the body of the all had been flat (*epipedon*) and without depth (*bathos*), a single bond would have sufficed to unite it with the other terms; but being solid (*stereon*), it requires two bonds. Therefore the Demiurge placed air and water between fire and earth, bonding each to the other with the same proportion insofar as was possible.[35] Through harmonized proportion, i.e., making a multitude one by fitting the parts together, he formed the body of the all with the elements earth, air, fire, and water; he bound them together with a love so powerful that no one except its maker could dissolve it.[36] Thus he made the body of the all as perfectly one as possible.

This account of the likeness produced by the Demiurge is cast in the language of form, number, and proportion – the language of being. Timaeus contrasts what is flat with what is solid, and as such requires depth. The body of the all is a likeness that becomes one by virtue of being made through divine art; insofar as it is possible, divine image-making produces unity through

---

31 Plato and Aristotle both regularly use *to pan* to refer to the cosmos, conceived as a whole. On the relation between soul in *Laws* 10 and the Demiurge of the *Timaeus*, cf. R. Mayhew, "The Theology of the *Laws*," in C. Bobonich (ed.), *Plato's "Laws": A Critical Guide* (Cambridge: Cambridge University Press, 2010), 197–216, at 207.
32 *Tim.* 31b4. See below n. 34 for the Greek text. The striking word *sōmatoeidēs* occurs only nine times in the entire Platonic corpus: *Tim.* 31b4; 36d9; *Phd.* 81b5, c4, e1; 83d5; 86a2; *pol.* 273b4; *resp.* 532c7.
33 *Tim.* 32a8: τὸ τοῦ παντὸς σῶμα.
34 *Tim.* 31b4: σωματοειδὲς δὲ δὴ καὶ ὁρατὸν ἁπτόν τε δεῖ τὸ γενόμενον εἶναις. Cf. 32b8: οὐρανὸν ὁρατὸν καὶ ἁπτόν.
35 *Tim.* 32a7–b8. Zeyl, *Plato: Timaeus*, xxxix, cites *stereoeidē* at 32b1 as meaning a "three-dimensional object."
36 *Tim.* 32c2–4; Buck and Petersen, *A Reverse Index*, 120, give an interesting note on nouns ending in -ια, using φιλία as an example: "Abstracts in -ια do not always designate a quality or condition. Not infrequently they suggest action, being secondarily associated with a verb, e.g., φιλία 'friendliness': φίλος 'friend' became 'love' when connected with φιλέω."

harmonized proportion. The body of the all, even as visible and tangible, is not thought of in terms of measurable extent, i.e., length, breadth, and depth; even as visible and tangible, body is explained by Timaeus in geometrical terms of construction and proportion. Proportion in the construction of the cosmos resembles proper measure in the *Sophist*; that is, proportion makes a likeness hit its proper mark, i.e., becoming as completely one as possible. Thus, the requirements of body are met in the best way possible and insofar as is possible, including having depth and being bound together to become one. In short, divine art makes the body of the all the most perfect possible likeness of being and ultimately the good.

Timaeus' account furthers Plato's systematic rejection of materialism and atheism, which exclude the causal roles of the good, god, and soul in the generation of the cosmos. The assertion that the body of the all must be visible and tangible does not imply that body possesses extent or is measurable. Visibility and tangibility imply that body must be a solid; since body must be a solid, Timaeus introduces proportion into his account because proportion produces unity within plurality. In the *Sophist*, the likeness maker – and hence the likeness too (as opposed to the semblance maker and hence the semblance too) – reproduces in the copy "the proper measure of the original in length, breadth, and depth." Measure is at the heart of moved things when the relations among them are properly realized because only through proper measure can likenesses hit their mark, the good and the true. The cosmos is the best likeness possible, produced without jealousy by a divine maker using the best model and proper proportion as the instrument of his divine image-making. The proper measure of the original in length, breadth, and depth is expressed not by a measurable extent but by a proper proportion in relation to the model.

Here we can see how the account of the cosmos in the *Timaeus* and the account of likeness-making in the *Sophist* complement one another. Proportion is the measure that makes body hit its proper mark: body expresses the good by being a solid that is perfectly unified (insofar as it is possible) by god's use of proportion to impart the good to it. Timaeus' account of body not only agrees with the account of likeness-making found in the *Sophist*, but also explains it. Both accounts rest on unity and the relation of the cosmos to the good. The body of the all is not flat but solid and so has depth; Timaeus' account focuses on the unity produced by proportion because it is focused on the good at the heart of divine making, i.e., the good at which divine making unerringly and without exception aims.

Timaeus details the perfections of the body of the heavens: insofar as it is possible, it suffers no defect, lack, or want; it comprises all the elements; it is the perfect shape, i.e., spherical, and exhibits perfect motion, i.e., circular locomotion (32c5–34a7). This body is itself both an animal – it has soul – and a "truly happy god" (εὐδαίμονα θεόν, 34b8). The entire account of the heavens requires the priority of soul; in excellence and nobility, soul is prior to and older than body (καὶ γενέσει καὶ ἀρετῇ προτέραν καὶ πρεσβυτέραν ψυχὴν σώματος); so god combined soul and body so that the older would rule the younger; therefore, Timaeus' story would be incomplete without an account of soul, body's mistress and ruler (34b10–c5).

Here we can see that the proof of the soul's priority in *Laws* 10 reappears in the *Timaeus*. Timaeus takes up the construction of the cosmos by god, given soul's priority. Like the Athenian, Timaeus calls on political metaphor: soul rules body.[37] Together they constitute the heavens and Timaeus adds a further point: time came into being at the same moment as did the heavens (38b6–7).[38] The heavens exhibit intelligent circular periodicity precisely because the proper relation obtains between soul and body, mover and moved, ruler and ruled.[39] God then made the various kinds of animals that would be included "within the best likeness" (εἰς ὁμοιότητα); he then hands off the making of mortal things, including humans, to the lesser gods, who are secondary cooperative causes – his ministers.[40]

The making of the cosmos, which Critias said that Timaeus would explain (27a3–6), now seems complete. Timaeus's account includes the maker, the ruling soul, the form, the body, and the making itself of the cosmos. But Timaeus's story (itself an entertainment provided to guests) is an account of divine art and the perfect likeness that it produces. And in this story, body is the perfect likeness, ruled by soul; it is secondary, not primary. The questions of spatial extent and its measurement simply disappear from an account that is cast entirely in terms of proportion, perfection, and a relation to the best model. The making of the cosmos requires a second account.

---

[37] For an extended discussion of the compatibility of the "psychology" of the *Timaeus* and *Laws*, cf. R. Kamtekar, "Psychology and the Inculcation of Virtue in Plato's *Laws*," in C. Bobonich (ed.), *Plato's "Laws": A Critical Guide* (Cambridge: Cambridge University Press, 2010), 127–48, at 130–43.

[38] As we saw in the *Statesman*, the art of measure divides into two kinds, one that concerns "number and length and depth and breadth and quickness" and the other that concerns due occasion, due time, due performance. The account of time lies beyond the bounds of this paper but we may note that it may follow this distinction.

[39] *Tim.* 38c2–39e2; cf. esp. 39c2. This intelligent circularity is the first motion after that of soul in *Laws* 10.

[40] *Tim.* 39e2–47e2; cf. esp. 39e3 and 46c7–d3.

2 Timaeus' Second Account of the Cosmos

With small exceptions, Timaeus explains, the things spoken about in the first account of the making of the cosmos are those crafted "through mind" (τὰ διὰ νοῦ); but one must also speak of those things that come to be "through necessity" (τὰ δι' ἀνάγκης) because the genesis of the cosmos results from a mixing of these two (47e2–48a2). Mind "persuaded" necessity to enter the mix; so anyone wishing to explain the whole must account for both its more perfect aspect, i.e., mind, which produces unity, and its less perfect aspect, "the form of its variable cause" (τὸ τῆς πλανωμένης εἶδος αἰτίας), i.e., necessity, which causes variation within unity (48a2–7). So Timaeus begins again in order to consider the variable cause, or necessity, taking up first the nature of fire, water, air, and earth, as they were prior to the generation of the heavens, and then their generation (48a7–c2).

Because Timaeus casts his first account of "the body of the all" in the language of form, number, and proportion, body appears as determined by the good and ruled by soul; to this extent it becomes as intelligible and unified as possible. Here is the cosmos, and body, as it comes to be through mind. But body may also be thought to appear independently of the good and soul, i.e., it appears to be "necessary" and variable rather than unified. Herein lies the considerable challenge of Timaeus' second account: to explain the cosmos, and body, as it comes to be through necessity. He must explain (insofar as it is possible) the phenomena dear to the heart of every materialist or atheist: simple body as such, the elements – earth, air, fire, and water – characteristics such as heavy and light, and the generation of body, including presumably its length, breadth, and depth.[41] Given the priority of the perfect paradigm, divine causality, and soul in his first account (not to mention the argument of *Laws* 10), Timaeus cannot appeal to any form of materialism. Therefore, he begins again, calling on the gods a second time, to take up body as it comes to be through necessity, i.e., as it is inherently variable.

Timaeus' second invocation of the gods reiterates his rejection of atheism and so marks a true beginning: he returns to the original division between being and becoming.[42] But the account of things that are by necessity requires

---

[41] Indeed, this part of the *Timaeus* has been subjected to analyses concluding both that it affirms Plato's earlier critiques of materialism and that it suggests that Plato has turned "into a sort of proto-Aristotelian." For a review of these issues and the literature that they give rise to, cf. D. P. Hunt, "The 'Problem of Fire': Referring to Phenomena in Plato's *Timaeus*," *Ancient Philosophy* 18 (1998): 69–80.

[42] *Tim.* 48d4–e2. Cf. R. Mayhew, "On Prayer in Plato's *Laws*," *Apeiron* 41 (2008): 45–62. Mayhew links the argument of *Laws* 10 "that gods exist and that they are good" to the actual use of prayer

a third category: a receptacle, or nurse, of all becoming (49a1–51b7: πάσης εἶναι γενέσεως ὑποδοχὴν αὐτὴν οἷον τιθήνην). The receptacle, he explains, has no character of its own, but temporarily receives the appearance of whatever comes to be in it. Thus there are three natures: (1) what comes to be (the likeness), (2) where it comes to be (in the receptacle), and (3) what the likeness comes to be like (the paradigm).[43] The relations among these kinds are straightforward: the receptacle has no form and makes no contribution to a likeness, which is entirely caused by its paradigm (50d2–51a3); rather, the receptacle receives temporary characteristics from whatever likeness comes to be in it (51b4–6). Consequently, the paradigm causes the likeness and the likeness appears or comes to be *in* the receptacle; the receptacle itself appears to be a third kind, being a kind of space (γένος ὂν τὸ τῆς χώρας) that is eternal and indestructible in and of itself but unknowable except for a kind of illegitimate reason (52a8–b2).[44] In *Laws* 10, as we saw above, some things move and rest "in some sort of space" (893c2). Here Timaeus explains that all things that come to be must be somewhere (ἀναγκαῖον εἶναί που), in some place and having some space.[45]

The likenesses that appear in the receptacle, Timaeus explains, are continuously changing and so should not be called a "this" or a "that" but rather "of such a nature" or a "such" (49d4–e4). That is, the likeness is always changing, change is "in" the receptacle, and the likeness is not a thing in itself but rather of such a nature, i.e., it is in relation to the paradigm. And so the likeness is like a form, but possesses a temporary and changing set of characteristics that get their names from the forms. For example, the form "dog" can be defined and is unchanging; the likeness "Fido" sometimes grows or sheds or barks and sometimes does not. These characteristics are like adjectives, coming and going, and as such they require a space "where" such changes occur. These changing attributes are all that Timaeus can point to because there is no form or paradigm in

---

in the dialogues; although he mentions Timaeus' first invocation of the gods (50 n. 15), he fails to mention the second.
43 *Tim.* 50c7–d2: ἐν δ' οὖν τῷ παρόντι χρὴ γένη διανοηθῆναι τριττά, τὸ μὲν γιγνόμενον, τὸ δ' ἐν ᾧ γίγνεται, τὸ δ' ὅθεν ἀφομοιούμενον φύεται τὸ γιγνόμενον.
44 This last point is beyond the scope of this paper. But it is an intrinsic part of the argument. Space itself cannot be conceived, except by thinking of some likeness that is in space; we then remove the likeness in our mind's eye (so to speak) and think of the space that remains, "where" the likeness was. But without the likeness there is nothing there. So thinking about space is like thinking about nothing. It is illegitimate.
45 52a8–d1. The phase εἶναί που also occurs in Arist. *phys.* 4.1, 208a29, where he says that anyone who thinks about it sees that all things that are "must be somewhere."

the receptacle. Things that become are here described as collections of attributes, forever changing their relations to the forms.

As we saw in the *Sophist*, the opposite of proper measure is absence of measure. Timaeus now calls on this notion. Having distinguished being, becoming, and the receptacle, Timaeus approaches body by describing "what was there" before god came to be present, i.e., before divine art ordered things with a view to the best model by means of form and number (53b4–5). Speaking of before god was present, he uses the past tense. There were elements that were "irrational and without measure" (ἀλόγως καὶ ἀμέτρως, 53a7–8). In the *Sophist*, a thing misses its mark "by absence of measure" (ὑπὸ ἀμετρίας, *soph*. 228c5); here in the *Timaeus*, things without measure are irrational and have no way to hit the mark. The receptacle, since it is nothing but "where" things appear, takes on whatever likeness momentarily happens along; hence the receptacle took on whatever completely diverse appearances result from being full of powers that are neither like nor balanced – things were scattered and shaken (52e1–5).[46] "Completely diverse appearances" signals the absence of measure, order, or unity. Not soul, with reason, number, or measure, but being heavy or light moved the elements every which way (52d2–53a8); being heavy and light, the elements had only faint traces of themselves and to this extent had distinct places but no actual order (53b1–2). In short, the world was what one expects when god —and hence divine making and its model—are absent (53b3–4). Here is the world of things that are by necessity: the accidental, strange, out of balance, shifting chance world of the materialist/atheist.[47]

The materialist/atheist presupposes that body is prior while order, including soul, results from various accidental conjunctions. But for Plato the case is exactly the opposite: soul is the senior ruling partner and the order it provides allows body to become. Without soul (and hence without order or unity), there is no body; there are only the completely diverse, disorganized, appearances of necessity. Being heavy and light, the elements appear as "faint traces"; that is, they are prior to body (which is produced by soul). As traces, they act to make a thing be here rather than there; but in the absence of the order produced by

---

[46] Cf. 52e1: παντοδαπὴν μὲν ἰδεῖν φαίνεσθαι ("to appear altogether diverse to see").

[47] There is an extensive argument in the literature, which lies beyond the bounds of this paper, as to whether soul is the cause of this kind of motion or whether it is "pre-cosmic." On this problem (and a review of the relevant literature), cf. R. D. Parry, "The Cause of Motion in *Laws* X and of Disorderly Motion in *Timaeus*," in S. Scolnicov and L. Brisson (eds.), *Plato's "Laws": From Theory into Practice; Proceedings of the VI Symposium Platonicum* (Sankt Augustin: Academia, 2003), 268–75. Parry concludes (rightly I believe) that this account accords with the view attributed to the atheist at *leg.* 10, 889b-c.

soul, everything is mixed together. Since order is prior to body and required by body, without sufficient order there can be neither elements nor body. "Motion" here resembles that described in *Laws* 10 as the opposite of intelligent motion: never regular or uniform, never in the same compass or around the same center, or in one place, having neither order, plan, or law but only kinship with folly of every kind (898b5–8). The folly of the materialist view is clear: maintaining the primacy of body, materialism fails to explain either body or its supposed powers because it fails to recognize that body, being ruled by soul, expresses order, partial unity, indeed the presence of the perfect model brought to it by divine image-making.

Timaeus' first account starts with form and number: god brings order out of disorder by putting mind in soul and soul in body; he thereby makes the cosmos as good as possible. Being flat or having depth may appear to the uneducated as immediate characteristics of body but in fact, as Timaeus, whom Critias called "the best astronomer among us" (ἀστρονομικώτατον ἡμῶν, 27a4) explains, these qualities are produced by proper proportion. Timaeus' second account starts from the atheists' world, i.e., a world marked by the absence of god and order (53b5–7). This world does not even have body and *a fortiori* cannot have length, breadth, and depth. It lacks what the Athenian explains in *Laws* 10: the three increases produced by soul that result in perceptible body. Things are "somewhere," i.e., in a receptacle that cannot be known except by a kind of "illegitimate reasoning"; being heavy or light may be somehow associated with distinct places, but everything is unclear and faint. In the absence of god, there is neither knowledge nor opinion nor even sensation because there is nothing to be known or even sensed. No body, just traces.

Body, in this second account, must be generated from the completely diverse appearances that are prior to body. This generation is Timaeus' next step. It is neither the production of something absolutely new nor the production of something absolutely real; rather, it is the shift from "completely diverse appearances" to what is perceptible and so to some extent ordered. Because the generation of body is nothing other than order, measure, and divine governance being introduced into diverse appearances, insofar as they can be received there, it does not require (or imply) materialism, i.e., the assumption of material parts, such as atoms. Hence the meaning of "through necessity" is clear: body originates in these diverse appearances. The order and measure that mark the divine art represent that very moment when, by introducing soul into the world, god makes "proper measure" supersede "absence of measure."[48] The cosmos is gen-

---

[48] Cf. 53b1: ὅτε δ' ἐπεχειρεῖτο κοσμεῖσθαι τὸ πᾶν ("when the all began to be ordered").

erated by mixing things that are through mind (as we saw in Timaeus' first account of the order and unity of body) with things that are through necessity (as we see in his second account of the diverse appearances of body) because such mixing – the mixing of unity with diversity – produces the best possible outcome for the completely diverse appearances that mark necessity and the absence of god (53e6–54a7). It produces body.

Prima facie, Timaeus' second account may seem opposed to his first account. The first starts with the distinction between being and becoming, takes up god, and yields things that are through mind, i.e., through what produces order and unity. The second starts with the absence of god, completely diverse appearances, and things that are through necessity, i.e., through what is completely diverse and changing. But with respect to the cosmos and body, the two accounts work together to explain why body is as beautiful and good as possible. Consequently, the account of the cosmos as formed by god, in the best possible way through divine art, is completed by an account of the generation of body that signals the very moment of god's arrival when things that are through necessity are mixed with things that are by mind to the supreme advantage of the former: completely diverse traces become body that is able to take on appearances.

Timaeus explains the generation of body by means of what he calls "an unaccustomed argument" (ἀήθει λόγῳ, 53c1) in which length, breadth, and depth reappear as the outcome of soul's activity. Fire, earth, water, and air are bodies, Timaeus explains, and every kind of body has depth (53c4–6).[49] Depth must always be bounded by surface and surface is composed of triangles, which in turn are composed of angles (53c6–d1). The principles prior to these are known only to god and whoever among men is the friend of god (53d6–7). In *Laws* 10, three increases lead to body; here the order is reversed, setting out from body: solids, flat things, and a first (known-only-to-god) principle.[50] An extended account of the construction of bodies, starting from the "most beautiful bodies" (*kallista sōmata*), follows (53d7–e2); after this account Timaeus explains how they mix together to form visible bodies (53c7–69a5).

The construction of solids, i.e., bodies, results from combinations of triangles. Three flat triangles can be combined to make one solid triangle and from the combination of these four triangles comes the first kind of solid form; the second kind of solid is formed from a different combination of triangles (54e3–55a8). Altogether five solid figures can be generated with the proper arrangement of triangles (55a8–c7). The first is used for the simplest construction

---

[49] Zeyl, *Plato: Timaeus*, lxvi, remarks: "All bodies have 'depth,' that is, are three-dimensional."
[50] For a specific example of this language, cf. *Tim.* 55b3–4.

while the last brings us to complex construction, i.e., figures of animals. In effect, body is conceived not as perceptible, but as intelligible, a solid constructed from triangles on a geometrical model.

The elements, too, are generated in exactly this way, e.g., earth has a cubic form, compounded by triangles (55d8–57d6). The cube is the clearest example of a solid composed of "flat things," i.e., triangles. Each of its six sides is composed of two identical isosceles triangles and so a cube may be thought of as composed of twelve isosceles triangles. The other elements are also generated from "flat things" and can combine with (or dissolve into) one another so as to compose (or dissolve) larger masses.

Motion and rest, Timaeus proceeds, must also be understood.[51] Motion in bodies requires that there be a mover and a moved (because, we recall from *Laws* 10, body is not a self-mover but is moved by another); the division between mover and moved implies that motion is impossible whenever things are "uniform" (ὁμαλά), i.e., not differentiated as mover and moved (57e3–6).[52] Non-uniformity produces recognizable outcomes for the cosmos.[53] The cosmos exhibits revolution; such motion compresses everything as much as possible, leaving no room for a void, and producing the transformation of each element into the other elements (58a7). Change of size produces changes in position always maintaining non-uniformity; consequently, there is always a differentiation between what moves and what is moved with the result that motion among the elements is also eternal (58c2–4).

An account of the mixing of the elements follows, including their diversification by form, and combination and changes into one another. An explanation of sensation appears next, and it ends with an explanation of color (67e4–68d2). With this end, we have an argument that explains the cosmos as a likeness, in

---

51 57d7–58c6. At 57e1, Timaeus comments that something has been said on this subject already, which may be a reference to *Laws* 10.

52 This view is repeated by Aristotle, according to whom all motion implies a division between the mover, i.e., the agent that produces the motion, and the patient, i.e., the moved. As we have seen in *Laws* 10, for Plato the motion of self-moving soul is unlike physical motion because it moves itself immediately and directly as an identity of mover and moved.

53 This argument is quite technical. It helps to keep in mind that Plato and Aristotle always use the verb "move," *kineō*, transitively (unlike English, where "move" can be either transitive, i.e., move something, or intransitive, i.e., move about). Therefore, in Plato and Aristotle, it always means "produce motion in another." For a study of the different expressions using this verb, including *Laws* 10, *Phaedrus*, and a number of texts in Aristotle, cf. H. Lang, *The Order of Nature in Aristotle's "Physics": Place and the Elements* (Cambridge: Cambridge University Press, 1998), 40–46.

regard to length, breadth, depth, and color, for things that come to be and so ultimately come to be good, insofar as is possible.

This account is indeed an "unaccustomed argument": it explains the generation of body and the features of the cosmos associated with body, e.g., the elements, motion, and absence of a void, but it does so without presupposing materialism or any kind of material parts, although it does set out from the completely diverse appearances in the receptacle. The argument is "unaccustomed" to formalists because an explanation of body is its goal and it is "unaccustomed" to materialists because it is an account of body that, although it starts out from "completely diverse appearances," itself rests on the arrival of order and structure through which what is completely diverse becomes measured.[54] What is completely diverse becomes body.

Although the three increases found in *Laws* 10 do not appear in the "unaccustomed argument" for the generation of body, Timaeus, too, identifies three stages. The first stage is known only to god and the one who is a friend of god. That is, it is not knowable in the ordinary (or mortal) sense of the word because it lies in the division between the absence of god and the presence of god. The presence of god, as yet invisible, divides the world of completely diverse appearances that are prior to body from the beginning, or origin, of body. This division represents the very moment when proper measure begins to replace absence of measure with the result that the generation of body, i.e., the process of measure, begins. In and of itself this stage, or division, is also prior to body and so cannot be known or even sensed; nonetheless, it is posterior to the complete absence of god and proper measure precisely because it signals the arrival of god. The first stage between absence of measure and presence of proper measure can only, as Timaeus indicates, be characterized as a divine and mysterious beginning.

The second stage of the generation of body comprises flat things, such as surfaces, without which depth is impossible. Timaeus hardly pauses over flat things, perhaps because there are no such things in the cosmos as we see it – indeed such objects are neither generated nor perceptible and his argument is intended to account for the generation of body, which has depth, or is solid, and so is perceptible to touch. Flat things, e.g., triangles, are presupposed by body because depth presupposes length and breadth; consequently, they appear in the account to this extent: body is made of triangles.

---

[54] Here the *Timaeus* is in complete agreement with the "battle of the gods and the giants" in the *Sophist* (246a4–248e6).

At the third stage, things having depth come from flat things. This stage gives us perceptible body, first the elements and then all body formed from them. Motion appears through inequality and ultimately yields the endless variety of all body, including bodies of different length, breadth, and depth. Here a clear difference emerges between Timaeus' first and second arguments. In the first, proportion produces a solid having depth because soul brings order and unity to body; here in the second, Timaeus proposes a sequence of three stages resulting in body that exhibits motion; these stages produce the cosmos as we see it – it exhibits revolution, it is as compressed as possible, the elements transform into one another, and this transformation produces bodies that exhibit diverse movements and change of place.

This argument, Timaeus explained at the outset, explicitly concerns things that come about through necessity as they are mixed with those of soul. Necessity represents the completely diverse appearances and their random scattering that resemble folly and are produced by absence of measure prior to divine presence. The generation of body results from a mixing of these things with those that come to be through soul and that we saw in the first account. It yields the details, insofar as one can speculate about them, of how order comes from disorder when one considers change and body not insofar as they presuppose unity but insofar as they originate in disorder and inequality: completely diverse appearances are transformed into body in the best way possible. The fact that body is the primary object of perception, i.e., touch, in no way means that it can claim the status of primary reality. In fact, just the opposite: the fact that body is the primary object of perception means that it is the outcome of a dynamic process in which completely diverse, irrational, unmeasured appearances are mixed with measure insofar as is possible. If we conjoin Timaeus' earlier claim – soul, being senior to body and its ruler, is sent by the Demiurge to bring order out of chaos – to the account of motion and soul as the ruler and mistress of body in *Laws* 10, then the process leading to the generation of body can be understood and the questions raised earlier may be answered.

IV Conclusion

The question that started with the *Sophist* and the translation of the expression "in length, breadth, and depth" as "in all three dimensions" has now led us through parts of the *Sophist*, *Statesman*, *Laws* 10, and *Timaeus* and raised a number of related issues. Before turning to them, two points should be made. First, these dialogues share an important community of interests that connect them and suggest a strong sense in which they work together. In the most general terms, each dialogue occurs within a host/guest setting – a setting to which Plato pays careful attention – and each concerns the related themes of a likeness in contrast to the thing itself, of the good, of justice and wisdom, and of the role of god and soul in nature and human affairs.[55] Kahn notes:

> The issue of the interaction between literary and philosophical readings of Plato is perhaps the most burning question in Platonic scholarship today. Neglect of literary form tends to be accompanied by a certain insensitivity to the historical and cultural context of Plato's work, and hence by the risk of conceptual anachronism.[56]

The *Timaeus* in particular has been read as "Plato's physics" since at least the twelfth century.[57] But here, as with the *Phaedo*, we can see that Plato is not

---

[55] *Laws* 10 may be a partial exception to the claim of a host/guest setting; it appears to be a private conversation among friends who are not distinguished as host and guest (cf. *leg.* 1, 635a2–5).

[56] C. H. Kahn "From *Republic* to *Laws:* A Discussion of Christopher Bobonich, *Plato's Utopia Recast,*" *Oxford Studies in Ancient Philosophy* 26 (2004): 337–62, at 343–44. Also, on the growing recognition of the importance of the setting of the dialogues, cf. G. Boys-Stones, "Phaedo of Elis and Plato on the Soul," *Phronesis* 49 (2004): 1–23, esp. 1–3 and 17. We might note that the role of the dialogue form is a longstanding concern in Platonic studies. Four articles, dating from 1962–1992, form a representative group of studies and comprise "Part II" in N. D. Smith (ed.), *Plato: Critical Assessments*, vol. 1: *General Issues of Interpretation* (London: Routledge, 1998). For an engaging consideration of the dialogue form, or dramatic setting, as it explicitly bears on the *Timaeus*, cf. D. J. Schoos, "Timaeus' Banquet," *Ancient Philosophy* 19 (1999): 97–107 and C. Osborne, "Space, Time, Shape, and Direction: Creative Discourse in the *Timaeus*," in C. Gill and M. M. McCabe (eds.), *Form and Argument in Late Plato* (Oxford: Clarendon Press, 1996), 179–211; for the *Sophist*, cf. M. Frede, "The Literary Form of the *Sophist*," in Gill and McCabe, *Form and Argument*, 135–51; for the *Laws*, cf. Bobonich, "Reading the *Laws*," in Gill and McCabe, *Form and Argument*, 249–82; for a general account of the dialogue form in late Plato, cf. C. Gill "Afterword: Dialectic and the Dialogue Form in Late Plato," in Gill and McCabe, *Form and Argument*, 283–311.

[57] C. Steel, "The Moral Purpose of the Human Body: A Reading of *Timaeus* 69–72," *Phronesis* 46 (2001): 105–28, at 105–106, gives a short but clear history of this problem. Cf. also Zhmud, "Plato as 'Architect of Science,'" 211–44, who traces the tendency to portray Plato as the architect of science back to the early Academy and its interpretation, rather than to Plato himself. We may note that without comment J. M. Cooper repeats this view in his edition of Plato (often used by students): "Timaeus sets out the foundations of the sciences of astronomy, physics, chemistry, and physiology, including diseases of body and soul and provisions for their treatment" (*Plato: Complete Works*, edited with introduction and notes; associate editor D. S. Hutchinson [Indianapolis: Hackett, 1997], 1224). For an example in the literature, cf. D. L. Guetter, "Celestial Circles in the *Timaeus*," *Apeiron* 36 (2003): 189–203. At 189 Guetter says, without defense or reference: "There is a good deal of poetry in the *Timaeus*, but it is also an early work in natural

"doing physics" so much as offering an alternative to physics. The first clue is the dialogue setting, that is, the host/guest relation, and the second is Plato's consistent emphasis of certain topics throughout the entire corpus.[58]

The absence of the good – or letting the truth take care of itself in the *Sophist* – leads to semblance, to demagoguery, to sophistry, and to materialism and its correlative, atheism. The presence of the good informs the order of intelligible things and is the sole model in the *Timaeus* for the formation of the perfect likeness by the Demiurge through soul. Recognition of the good is the mark of a citizen and a leader, who is wise and just. As *Laws* 10 makes the point, no atheists here.

Second, the account of a likeness (as opposed to a semblance) in the *Sophist*, along with the accounts of body in both *Laws* 10 and the *Timaeus*, rests on the proper relation between soul and body. Likeness-making is the activity of a maker, divine or human, who looks to the good and is without jealousy; body, which is "younger," is ruled, as is proper, by soul, which is "senior." Consequently, insofar as it is possible, body becomes what the intelligible always is. Against the social setting of these dialogues and their shared commitment to the good, I shall offer some conclusions concerning the issues that have been raised and the larger problem of the relation between intelligible objects and body.

(1) The notion of dimension arises in conjunction with imitation because a copy may imitate an original in length, breadth, and depth (and Plato adds, as we saw above, in color). All things that hit their mark do so by proper measure and likewise all that miss their mark do so by the absence of measure. A likeness (as opposed to a semblance) is produced by a maker who takes the good and true as the ultimate paradigm with the result that likenesses hit their mark insofar as it is possible. The cosmos is a perfect likeness, constructed by a divine maker, who uses the best model. So, too, the body of the cosmos is as perfect as possible because it possesses proper proportion, by which it is measured and unified and, therefore, in the language of the *Sophist*, hits the mark, i.e., imitates the good.

Length, breadth, and depth appear as a consequence of the presence of proportion. Hence they do not characterize body as such; they characterize body in-

---

philosophy." Osborne, "Space, Time, Shape," concludes that the *Timaeus* "is essential to understanding what Plato's view of science is. ... It seems clear that the kind of studies required to instill in the philosopher an understanding of reliable scientific models has not changed fundamentally from that envisaged in the *Republic*" (210).

58 We might note a recent interpretation of ὑποδοχή, "the receptacle" or "third kind," in Timaeus' second account as "a place of welcoming and refuge" [that functions to sustain whatever comes to be in it]. Cf. J. J. Reynolds, "How Is the Third Kind in Plato's *Timaeus* a Receptacle," *Ancient Philosophy* 28 (2008): 87–104.

sofar as it is made perfect by proportion, which itself is a relation to the good. Unlike either flat things or number, there is nothing that can be added to body: by virtue of the proportion present to it, the body of the cosmos is complete, as, too, is any particular body within the cosmos. Even looked at as originating through necessity, body, coming out of completely diverse and strange appearances, emerges with the arrival of order. Prior to god's arrival there is no body, only faint traces. Body, completed by depth, emerges from the order produced by god (and soul) starting from the moment of their arrival. Again, body becomes as perfect as possible, i.e., as unified and complete as it can be.

(2) The generation of body, whether seen through the unity provided by soul (as Timaeus' first account explains) or seen through the diversity provided by necessity (as Timaeus' second account explains), appears to raise the problem of dimension because, just as in *Laws* 10, body increases from a mysterious starting point to what is flat and from there to what has depth. In each case, the argument returns us to length, breadth, and depth. In Timaeus' first account, body is not flat but has depth because it is unified and made complete by the presence of proper proportion. Hence, having depth (and so presumably length and breadth) is nothing other than the way body comes to be by being measured. In his second account, Timaeus starts from the absence of god, i.e., the completely diverse appearances in the receptacle. This account of becoming chronicles the moment of god's (and soul's) arrival and the consequent production of body. This process is nothing other than the unfolding of proper proportion and it imitates the true order of intelligible objects insofar as is possible for body in the receptacle: progressing in order from an indefinite starting point, through flat things to the addition of depth. If we mix Timaeus's two accounts, "through soul" with "through necessity," body becomes, insofar as it is capable of becoming, what the intelligible always is: the best possible for its kind. Body is the temporary and ever changing presence of attributes reflecting temporarily what the intelligible is always. And Timaeus' account ends with color. The cosmos, and all body within it, is a likeness of the proper measure of the original in length, breadth, and depth as well as in color.[59]

(3) A final question remains. The *Sophist*, the *Statesman*, and the *Timaeus* all take place in a host/guest setting that Plato emphasizes. Does this consistent setting contribute to, or in some way reflect, Plato's larger position? The answer lies in the political character of Plato's arguments and metaphors. In short, educated soul comes to know the order of intelligible objects and is able to lead body to

---

[59] Color, of course, appeared in the *Sophist* along with length, breadth, and depth; cf. 235d6–e2.

the good. Soul brings order to body by generating proper proportion and through the presence of this proportion body is able to hit its proper mark.

The true legislator in *Laws* 10 refutes atheism by explaining soul as the ruler and mistress of body. The Demiurge, or god, of the *Timaeus*, acts without jealousy, constructing the cosmos as the best likeness possible. He does so by sending soul to persuade sociable nature and to force unsociable nature. An unsociable nature is whatever nature lies completely outside social structure and so completely outside the possibility of persuasion. And the construction of the cosmos does occasionally require force, as when god brings together the unsociable nature of the different into the same (*Tim.* 35a8). But the more important cosmic elements are social in nature and ruled through persuasion. After mind persuades necessity, soul rules body; the lesser gods, after listening to god's oration, act as his ministers.[60] In fact, the appeal to force in the *Timaeus* is strictly limited and minimal. Finally, by explaining how body is constructed to be the best, Plato can claim to give a better – more persuasive and true – account of body, than do the materialists or atheists.[61]

Here we may return to dimension as "(spatial) measure." It is often remarked that as a concept dimension has its origins in Aristotle's category of quantity.[62] Aristotle defines body as "complete magnitude" and this definition rests squarely on his account of substance, which in its turn requires the rejection of Plato's "separate forms."[63] Not only is there no causal role for dimension in Plato's account of body and no language that reflects its meaning, dimension originates in a view that explicitly rejects Plato's account. How then does it come to translate "length, depth, and breadth" in Plato? "Dimension" is native to the translator. A review of *Plato: Complete Works* begins:

> "'Stick to Jowett – you'll easily spot where he's going wrong.'" Such used to be the advice given to Greekless graduate students by the late G. E. L. Owen. Owen was afraid, of course,

---

[60] For god's oration to the heavens he has just made, cf. *Tim.* 41a7–d3. It begins famously with the words: Θεοὶ θεῶν, ὧν ἐγὼ δημιουργὸς πατήρ τε ἔργων. For mind persuading necessity, cf. 48a2–3.
[61] Cf., for example, *Republic* 1. Thrasymachus is a sophist who presents an unsociable nature and who is beaten by force and led from the room so that the true search for justice may be conducted by a society of friends working together by agreement within a mutual relation.
[62] For an excellent study of this problem, cf. D. Des Chene, *Physiologia: Natural Philosophy in Late Aristotelian and Cartesian Thought* (Ithaca: Cornell University Press, 1996), *passim*, e.g., 86, 361–62.
[63] For Aristotle's account of body, cf. *cael.* 1.1, 268a1–268b5; for the characterization of body as "complete magnitude," cf. 268b2–3. On quantity as divided into plurality and magnitude, i.e., discrete and continuous, cf. *Categories* 6.

that if Greekless readers used supposedly more sophisticated translations, in particular those of Cornford, they would be more subtly and unconsciously misled by modern-sounding renderings than they would be by the occasional Hegelianisms (or obvious non-sequiturs) of Jowett.[64]

Dimension is so ubiquitously present in modern views of body that it has achieved the status of a fact – a fact that is given in experience that can be referred to, that is value-neutral, non-interpretive, and available to be called on in any context. On these grounds alone it is brought to the dialogues of Plato (and, although we do not see it here, to Aristotle's physics as well). The analysis of Plato's account of body yields two important conclusions about "dimension" as a concept: (1) it is not present in Plato; (2) "dimension" is neither fundamental as a conceptual category nor even, necessarily, given as an immediate measure of body.

"Dimension" tells us not about Plato but about modern conceptions of body. The assumption that as a concept "dimension" bears directly on body presupposes the primacy of body. It also presupposes a direct relation to body with nothing intervening, e. g., god, soul, or the good. In short, dimension represents the very materialism rejected by Plato. If it seems "given," as a fundamental fact of experience, then from Plato's point of view, this status bears witness not to its primacy but to the continuing prevalence of the materialism or atheism rejected in *Laws* 10 – as well as the continuing need for a legislator who can by persuasion educate those who are sociable. And the same point may be made for length, breadth, and depth: however immediate in experience they seem to us, according to Plato they signal that an act of interpretation, indeed an act of divine art, has occurred. And like all things open to interpretation, they may be construed in different ways. Materialists construe them as the primary characteristics of body. But Plato construes them as body's way, when ruled by soul, of being measured and through proper measure of hitting the mark: far from being dimensions, length, breadth, and depth are body's way of expressing the good.

---

64 D. B. Robinson, "After Jowett, Now Hackett: A New Plato," *Apeiron* 31 (1998): 285–92, at 285; after a largely favorable review of Cooper, *Plato: Complete Works*, Robinson concludes, at 291, that it is best to learn Greek.

Bibliography

Aubenque, P., ed. *Études sur le "Sophiste" de Platon* (Naples: Bibliopolis, 1991).
Barnes, J., ed. *The Complete Works of Aristotle: The Revised Oxford Translation,* vol. 1, Bollingen Series LXXI (Princeton: Princeton University Press, 1984).
Bluck, R. S. *Plato's "Sophist": A Commentary,* ed. by G. C. Neal (Manchester: University Press, 1975).
Bobonich, C. "Reading the *Laws*," in Gill and McCabe, *Form and Argument,* 249–82.
Bobonich, C., ed. *Plato's "Laws": A Critical Guide* (Cambridge: Cambridge University Press, 2010).
Boys-Stones, G. "Phaedo of Elis and Plato on the Soul," *Phronesis* 49 (2004): 1–23.
Buck, C. D., and W. Petersen. *A Reverse Index of Greek Nouns and Adjectives: Arranged by Terminations with Brief Historical Introductions* (Chicago: University of Chicago Press, 1945).
Burnet, J. *Platonis Opera,* 5 vols. (Oxford: Oxford University Press, 1900–1907).
Cleary, J. J. "The Role of Theology in Plato's *Laws*," in F. L. Lisi (ed.), *Plato's "Laws" and its Historical Significance: Selected Papers of the I International Congress on Ancient Thought, Salamanca 1998* (Sankt Augustin: Academia, 2001), 125–40.
Cooper, J. M. *Plato: Complete Works,* edited with introduction and notes; associate editor D. S. Hutchinson (Indianapolis: Hackett, 1997).
De Chiara-Quenzer, D. "The Purpose of the Philosophical Method in Plato's *Statesman*," *Apeiron* 31 (1998): 91–126.
Demos, R. "Plato's Doctrine of the Psyche as a Self-Moving Motion," *Journal of the History of Philosophy* 6 (1968): 133–45.
Des Chene, D. *Physiologia: Natural Philosophy in Late Aristotelian and Cartesian Thought* (Ithaca: Cornell University Press, 1996).
Frede, M. "The Literary Form of the *Sophist*," in Gill and McCabe, *Form and Argument,* 135–51.
Gill, C. "Afterword: Dialectic and the Dialogue Form in Late Plato," in Gill and McCabe, *Form and Argument,* 283–311.
Gill, C., and M. M. McCabe, eds. *Form and Argument in Late Plato* (Oxford: Clarendon Press, 1996).
Goldin, O. "Plato and the Arrow of Time," *Ancient Philosophy* 18 (1998): 125–43.
Guetter, D. L. "Celestial Circles in the *Timaeus*," *Apeiron* 36 (2003): 189–203.
Hamilton, E., and H. Cairns, eds. *The Collected Dialogues of Plato, Including the Letters,* Bollingen Series LXXI (New York: Bollingen, 1961).
Hunt, D. P. "The 'Problem of Fire': Referring to Phenomena in Plato's *Timaeus*," *Ancient Philosophy* 18 (1998): 69–80.
Hussey, E., trans. *Aristotle's Physics, Books III and IV* (Oxford: Clarendon Press, 1983).
Johansen, T. "Body, Soul, and Tripartition in Plato's *Timaeus*," *Oxford Studies in Ancient Philosophy* 19 (2000): 87–111.
Kahn, C. H. "From *Republic* to *Laws*: A Discussion of Christopher Bobonich, *Plato's Utopia Recast*," *Oxford Studies in Ancient Philosophy* 26 (2004): 337–62.
Kamtekar, R. "Psychology and the Inculcation of Virtue in Plato's *Laws*," in Bobonich, *Plato's "Laws,"* 127–48.

Lang, H. *The Order of Nature in Aristotle's "Physics": Place and the Elements* (Cambridge: Cambridge University Press, 1998).

Mayhew, R. "On Prayer in Plato's *Laws*," *Apeiron* 41 (2008): 45–62.

Mayhew, R. "The Theology of the *Laws*," in Bobonich, *Plato's "Laws,"* 197–216.

Menn, S. "On Socrates' First Objections to the Physicists (*Phaedo* 95e8–97b7)," *Oxford Studies in Ancient Philosophy* 38 (2010): 37–68.

Notomi, N. *The Unity of Plato's "Sophist": Between the Sophist and the Philosopher* (Cambridge: Cambridge University Press, 1999).

Osborne, C. "Space, Time, Shape, and Direction: Creative Discourse in the *Timaeus*," in Gill and McCabe, *Form and Argument*, 179–211.

Parry, R. D. "The Cause of Motion in *Laws* X and of Disorderly Motion in *Timaeus*," in S. Scolnicov and L. Brisson (eds.), *Plato's "Laws": From Theory into Practice; Proceedings of the VI Symposium Platonicum* (Sankt Augustin: Academia, 2003), 268–75.

Parry, R. D. "The Craft of Ruling in Plato's *Euthydemus* and *Republic*," *Phronesis* 48 (2003): 1–28.

Pellegrin, P. "Le *Sophiste* ou *De la division*, la diérétique et le non-être. Aristote – Platon – Aristote", in Aubenque, *Études sur le "Sophiste,"* 389–416.

Reynolds, J. J. "How Is the Third Kind in Plato's *Timaeus* a Receptacle," *Ancient Philosophy* 28 (2008): 87–104.

Robinson, D. B. "After Jowett, Now Hackett: A New Plato," *Apeiron* 31 (1998): 285–92.

Rosen, S. *Plato's "Sophist": The Drama of Original and Image* (New Haven: Yale University Press, 1983).

Schoos, D. J. "Timaeus' Banquet," *Ancient Philosophy* 19 (1999): 97–107.

Smith, N. D., ed. *Plato: Critical Assessments*, vol. 1: *General Issues of Interpretation* (London: Routledge, 1998).

Steel, C. "The Moral Purpose of the Human Body: A Reading of *Timaeus* 69–72," *Phronesis* 46 (2001): 105–28.

Strange, S. K. "The Double Explanation in the *Timaeus*," in G. Fine (ed.), *Plato*, vol. 1: *Metaphysics and Epistemology* (Oxford: Oxford University Press, 1999), 397–415.

Villela-Petit, M. "La question de l'image artistique dans le *Sophiste*," in Aubenque, *Études sur le "Sophiste,"* 53–90.

Zeyl, D. J., trans. *Plato, Timaeus* (Indianapolis: Hackett, 2000).

Zhmud, L. "Plato as 'Architect of Science,'" *Phronesis* 43 (1998): 211–44.

Roberto Lo Presti
"For sleep, in some way, is an epileptic seizure" (*somn. vig.* 3, 457a9–10)

Empirical Background, Theoretical Function, and Transformations of the Sleep/Epilepsy Analogy in Aristotle and Medieval Aristotelianism

Abstract: In his treatise *De somno* Aristotle establishes a somewhat puzzling analogy between sleep and the epileptic attack, going so far as to end up by proposing an assimilation of sleep to epilepsy as far as their material cause is concerned. The plausibility of this assimilation is confirmed, in Aristotle's eyes, by the fact that children are especially prone to epilepsy and that they sleep longer and deeper than adults. Surprising though this definition may appear to us, it finds interesting parallels in some medical treatises of the Hippocratic Corpus. This seems to suggest that a link between sleep and the epileptic attack was recognized and described already by the medical authors. The picture becomes more and more interesting if we think of the fact that in modern and contemporary times, too, studies which investigate the connections between certain forms of epileptic (or epileptiform) fit and sleep represent an important strand of the neurological research on the causes and the "historia naturalis" of epilepsy. It seems clear, therefore, that in Aristotle's definition of sleep as "a kind of epilepsy" we may trace the nucleus of a medical problem, which is far from being solved or fully explained. In this paper I investigate how medieval (Byzantine, Arab, and Latin) commentators of *De somno et vigilia* read and explained Aristotle's definition of sleep: I focus in particular on Michael of Ephesus' commentary, on Averroes' *Compendium De somno et vigilia*, on a passage of Avicenna's *Liber de anima*, and, as for Latin Aristotelianism, on the first Latin commentaries on Aristotle's *De somno* (the only extant commentaries based on the *translatio vetus* of the *Parva naturalia*), namely the commentaries by Adam of Buckfield and Albert the Great. By discussing an intriguing case study, this paper aims at providing a contribution to the history of the reception of the Aristotelian biology put in the wider context of a history of medical ideas.

This paper is part of a research project on Aristotelianism and medicine, which I have been working on since September 2010 within the framework of the AvH-Stiftung funded research program "Medicine of the Mind, Philosophy of the Body," directed by Philip van der Eijk. Heartfelt thanks to the colleagues and friends whose suggestions and remarks have helped me to better elucidate some key points of my argument.

"For sleep, in some way, is an epileptic seizure."

Introduction

The Aristotelian treatise *De somno et vigilia*, one of the short writings included in the *Parva naturalia*, contains the first systematic account of the physiology of sleep of classical antiquity. In the first chapter of the *De sensu*, which also serves as an introduction to what we may call the "*Parva naturalia* project,"[1] Aristotle had already included sleep and waking among the activities that are "common to soul and body" and are therefore to be investigated by exploring the material conditions and the psycho-physiological processes through which the connection and interaction between body and soul materialize.[2]

In order to carry out such an exploration, Aristotle seems to make a large, sometimes sophisticated, sometimes more superficial, use of empirical claims that denote his interest in, and acquaintance with, a variety of issues debated within the medical circles of the fifth and fourth century B.C.[3] To my eyes, an example of Aristotle's complex approach to, use of, and eventually influence on aspects of the medical tradition is to be found in the association that Aristotle establishes between sleep and epilepsy while discussing the physiological proc-

---

1 Of course, the title "*Parva naturalia*" is not Aristotle's own creation, but the product of the medieval reception of these works (see David Ross, *Aristotle, Parva naturalia* [Oxford: Clarendon Press, 1955], 1). Nonetheless, the existence and the internal coherence of a well-defined "research project" behind the writings included in the *Parva naturalia* is not in doubt, even if we still do not have a satisfactory answer as to whether Aristotle originally conceived these writings as parts of a unitary corpus (Ingemar Düring, *Aristoteles* [Heidelberg: Winter, 1966], 560–62, and Alberto Jori, *Aristotele* [Milan: Bruno Mondadori, 2003], 228, are quite sceptical about that; see contra Andrea Carbone, *Aristotele: l'anima e il corpo* [Milan: Bompiani, 2002], 7–8) and, above all, "as to what is excluded and on what grounds" (Philip J. van der Eijk and Maithe Hulskamp, "Stages in the Reception of Aristotle's Works on Sleep and Dreams in Hellenistic and Imperial Philosophical and Medical Thought," in Christophe Grelland and Pierre-Marie Morel [eds.], *Les 'Parva naturalia' d'Aristote: fortune antique et médiévale* [Paris: Publications de la Sorbonne, 2010], 47–75, at 48). For further discussion of these questions see Philip J. van der Eijk, *Aristoteles. De insomniis. De divinatione per somnum* (Berlin: Akademie Verlag, 1994), 69–73, and Pierre-Marie Morel, *Aristote, Petits traités d'histoire naturelle* (Paris: Flammarion, 2000), 10–24.
2 *Sens.* 436a8–17. See also *part. anim.* 645b33–35, where sleep is included among "the most important activities and affections" (πράξεις καὶ πάθη μέγιστα) that all living beings have in common.
3 See Geoffrey E. R. Lloyd, "The Empirical Basis of Physiology in the *Parva naturalia*," in Geoffrey E. R. Lloyd and Gwilym E. L. Owen (eds.), *Aristotle on Mind and the Senses: Proceedings of the Seventh Symposium Aristotelicum* (Cambridge: Cambridge University Press, 1978), 215–40.

esses that cause a subject to fall asleep. To put it briefly, Aristotle argues that sleep "is not any and every incapacity of the sensitive faculty," as it is the result of a temporary, reversible incapacity of the *prōton aisthētikon*. Such incapacitation, says Aristotle, is directly connected with digestion, insofar as it is caused by a flow of hot vapors that are produced during digestion. After reaching the brain and being subject to a process of cooling and densification there, these vapors flow back, causing the innate heat to be dispelled from the outer and upper regions of the body and trapped in its inner part, and so impeding the *prōton aisthētikon* from carrying out its own function.[4]

Aristotle also observes that sleep can be a consequence of fatigue, insofar as fatigue acts as a solvent; the dissolved matter, unless it is cold, has the same effect as undigested food. Finally, sleep can be induced by certain diseases such as those that arise from moist and hot residues, as is the case with fever and lethargy, for they produce the same result as undigested food.[5]

It is for the same reason, as Aristotle sees it, that "children sleep a great deal, because all the food is borne upwards. The greater size of the upper parts in comparison with the lower in early youth proves this."[6] It is at this point in Aristotle's argument that we read what follows:

> Διὰ ταύτην δὲ τὴν αἰτίαν καὶ ἐπιληπτικὰ γίνεται· ὅμοιον γὰρ ὁ ὕπνος ἐπιλήψει, καὶ ἔστι τρόπον τινὰ ὁ ὕπνος ἐπίληψις. Διὸ καὶ συμβάνει τισὶν ἡ ἀρχὴ τούτου τοῦ πάθους καθεύδουσιν, καὶ καθεύδοντες μὲν ἁλίσκονται, ἐγρηγορότες δ' οὔ· ὅταν γὰρ πολὺ φέρηται τὸ πνεῦμα ἄνω, καταβαῖνον πάλιν τὰς φλέβας ὀγκοῖ, καὶ συνθλίβει τὸν πόρον δι' οὗ ἡ ἀναπνοὴ γίνεται. (*somn. vig.* 457a8–14)
>
> For the same reason they also become epileptic, for sleep is like an epileptic fit; indeed in a certain way sleep is an epileptic fit. Consequently in some people this affection begins in sleep, and they are seized with it while asleep, but not while awake. For when a large volume of *pneuma* is carried up, as it descends again it swells the veins and compresses the passage through which respiration passes. (trans. Hett, modified)

In establishing an analogy between sleep and epilepsy, Aristotle makes a move that cannot be fully appreciated if one does not contextualize it in a framework

---

[4] *Somn. vig.* 456b18–29. For a discussion of the main interpretative problems posed by Aristotle's physiological explanation of sleep, see Jürgen Wiesner, "The Unity of the Treatise *De somno* and the Physiological Explanation of Sleep in Aristotle," in Lloyd and Owen, *Aristotle on Mind and the Senses*, 241–80; Henriette Wijsenbeek-Wijler, *Aristotle's Concept of Soul, Sleep and Dreams* (Amsterdam: Hakkert, 1978), 176–201; Luciana Repici, *Aristotele: il sonno e i sogni (Il sonno e la veglia, I sogni, La divinazione durante il sonno)* (Venice: Marsilio, 2003), 12–22.
[5] *Somn. vig.* 456b34–457a4.
[6] *Somn. vig.* 457a4–8.

of medical observations and theories regarding these two affections. For the medical literature, both anterior to and coeval with Aristotle, had already hinted at the existence of some kind of similarity or even connection between them. In the Hippocratic treatise *De flatibus* (14, 6.110–14 L. [= 121.6–124.10 Jouanna]), for example, the author gives an account of the so-called "sacred disease," explaining that this affection, like many others, is caused by an alteration of the blood (which can be more or less severe according to different affections), which this medical author considers the center and main source of thinking in a human being. In particular, sleep and drunkenness are mentioned as conditions which can be also explained by referring to certain alterations of blood: its cooling and slowing down, in the case of sleep;[7] its excess in the case of drunkenness. In *Epidemics* V, we are told the story of a patient who suffers from epileptic seizures more at night, during sleep, than during the day (ἐλαμβάνετο δὲ τὰς νύκτας μᾶλλον τῶν ἡμερέων ἐν τῷ ὕπνῳ, 22, 5.222 L. [= 14.1–18 Jouanna]). This observation clearly fits Aristotle's claim that some people "are seized with epilepsy when asleep, but not when awake." On the one hand, we might even imagine that Aristotle had in mind, if not this specific case history from the *Epidemics*,[8] other observations of that kind when structuring his own argument.[9]

---

[7] Hipp. *flat.* 14, 6.110–12 L. (= 121,13–122,6 Jouanna): "There are many testimonies that this is the case. In the first place, sleep, which is common to all the animals, bears witness to the truth of my words. When sleep comes upon the body, the blood is chilled, as it is of the nature of sleep to cause chill. When the blood is chilled, its passages become more sluggish. This is evident; the body grows heavy and sinks (all heavy things naturally fall downwards); the eyes close; the intelligence alters, and certain other fancies linger, which are called dreams" (trans. Jones). The principle according to which sleep consists in, or is connected with, a bodily process of cooling is also to be found in Empedocles (DK 31 A 85), Parmenides (DK 28 A 46b), and in other places of the Hippocratic Corpus (*epid.* VI 4.12, 5.310 L.). Also in Aristotle, as we have already seen, processes of cooling play an important role in the causal account of sleep. For a discussion of the place of *De flatibus* among the theories of sleep of the classical period see Cesare Marelli, "Place de la *Collection hippocratique* dans les théories biologiques sur le sommeil," in François Lasserre and Philippe Mudry (eds.), *Formes de pensée dans la Collection hippocratique: actes du IVᵉ Colloque international hippocratique* (Geneva: Droz, 1983), 331–39, esp. 334–35.

[8] J. Jouanna, *Hippocrate, Epidémies V et VII*, t. IV.3 (Paris: Les Belles Lettres, 2003), xxxiv–xlv has suggested as the date of composition of *Epidemics V–VII* a period spanning from 358 to 347 B.C., while the *terminus post quem* for the redaction of the works of the *Parva naturalia* has been traditionally set in 348–347 B.C. (see Ross, *Aristotle, Parva naturalia*, 17).

[9] Another passage suggesting the existence of a link between sleep and epilepsy is to be found in Diocles of Carystus (fr. 182 van der Eijk), who warns that "difficulty of breathing" (*duspnoia*) ... "suffocation" (*pnigmoi*), "epileptic seizures" (*epilēptika*) and "effusions during sleep" (*exoneiriasmoi*) threaten those who sleep on their backs.

On the other hand, the fact that Aristotle looked at these two phenomena as related may have influenced, to a greater or lesser extent, the medical writers who came after him and who had, indeed, a clear perception of the relevance of Aristotle's ideas to medicine, as well as to biology and natural philosophy.[10] It is, in fact, from the Imperial Age onward that a number of medical or medically informed accounts of epilepsy start to explicitly characterize the epileptic's state of unconsciousness as a deep sleep or a sleep-like condition.[11] Some writers even draw a definite clinical distinction between convulsive seizures and *non*-convulsive seizures, the latter being likened to sleep.[12]

A significant exception is represented by Galen, who, while having a clear knowledge of sleep-like diseases or, to put it in other words, of pathological forms of sleep,[13] neither includes epilepsy among such diseases nor adopts

---

[10] See van der Eijk and Hulskamp, "Stages in the Reception of Aristotle's Works," 65–66.
[11] See Cael. Aur. *tard. pass.* 1.4.63 (480 Drabkin) where "continued wakefulness or else excessive and unprofitable sleep [*plurimus et inanis somnus*]" are counted among the manifestations characteristic of the epileptic seizure; Apul. *apol.* 43.8-10: "Thallus, whom you have named, needs a doctor more than a magician. For epilepsy has made him so pitiful that three or four times a day he falls without any incantations and convulsions leave all of his limbs powerless. His face is ulcerous, his forehead and the back of his head bruised, his eyes dull, his nostrils gaping and his feet unsteady. The best magician ever is the man in whose presence Thallus remains standing for a long time, since he's inclined to fall because of this illness just like someone on the verge of sleep [*ita plerumque morbo ceu somno uergens inclinatur*]" (trans. Butler, modified); Theodorus Priscianus *eupor.* 2.15 (148.3–4 Rose).
[12] See Anon. Paris. *de morb. acut. et chron.* 3.2.1 (20 Garofalo): "Some of the patients suffering epilepsy seem to be held in a deep sleep [οἱ μὲν ὕπνῳ βαθεῖ κατέχεσθαι δοκοῦσιν], while others have convulsions, tremble, and undergo contractions; the patients are dumb and insensible when asleep, have breathing which is heavy and noisy, pulse large, slow, or like that in lethargy, do not remember anything of what happens during the affection" (trans. Fuchs); Cael. Aur. *tard. pass.* 1.4.60–61 (478 Drabkin): "Epilepsy takes its name from the fact that it is a seizure both of the senses and of the mind. … There are two forms of epilepsy. One appears to be like a deep sleep [*somno similis altissimo*]; the other racks the body with various convulsions. The former type is considered more serious, since it is like apoplexy" (trans. Drabkin). See also Cass. Fel. 71 (168.16–19 Rose); Esculapius, *De morborum, infirmitatum, passionumque corporis humani caussis, descriptionibus et cura* (Strasbourg: Schott, 1533), 3.
[13] See, for example, *de sympt. caus.* 1.8, 7.142 K.: "in cases of torpor, deep sleepiness, lethargic states and in all the diseases inducing an unnatural sleep" (ἐν κάροις καὶ ληθάργοις καὶ κώμασι καὶ πᾶσι τοῖς παρὰ φύσιν ὑπνώδεσι νοσήμασιν). In Galen's view, these affections clearly represent pathological manifestations that have features in common with sleep, both from the clinical and from the etiological point of view, but that always remain distinct from "natural," "physiological" sleep (τὸν κατὰ φύσιν ὕπνον, as Galen defines it few lines below in the same chapter of *De symptomatum causis*).

the sleep/epilepsy analogy for argumentative or demonstrative purposes.¹⁴ Rather, in a passage of the *De symptomatum causis* (1.8, 7.144 K.), in which it is argued that apopleptic and epileptic attacks are both caused by the brain (αἵ τε γὰρ ἀποπληξίαι καὶ ἐπιληψίαι διὰ τὸν ἐγκέφαλον γίγνονται), Galen reverses the analogy and claims that apoplexy and epilepsy differ to the extent that apoplexy "is an affection regarding the activities of voluntary movement, *like a deep sleep* in the perceptual activities [οἷον ὁ βαθὺς ὕπνος ἐν ταῖς αἰσθητικαῖς ἐνεργείαις]," while "the convulsion occurring in the epileptic fits is *like sleeplessness* in the perceptual activities [οἷον ἐν ταῖς αἰσθητικαῖς ἡ ἀγρυπνία]."¹⁵

Pointing out the medical origin of the association between sleep and epilepsy, as Aristotle puts it, can help us form a clearer picture of the epistemological framework in which this association is to be contextualized (in this respect, one should not forget that the section of the *De somno* in which Aristotle refers to epilepsy is less theoretical in tone and very much focused on collecting and somehow connecting empirical data and medical evidence). Nevertheless, there still remains in Aristotle's account a number of unanswered questions as well as areas of opacity and ambiguity, which the comparison with the ancient medical versions of the sleep/epilepsy connection cannot solve, but only make more prominent.

There are two points—actually two substantial differences—between the rationale underlying Aristotle's characterization of sleep "as a kind of epileptic fit" and that of the previously mentioned medical accounts, which are worth looking at: 1) While Aristotle makes use of a pathological state as a term of comparison to strengthen and further clarify a specific point his causal account of a physiological process, the texts of the medical tradition always adopt the opposite strategy, for they refer to sleep in order to better characterize certain clinical manifestations of a pathological entity such as epilepsy. This is anything but a trivial

---

14 The only place in Galen's writings in which sleep and epileptic seizure are somehow connected is in *Hp. aph. comm.* 3.29, 17B.643 K., in which Galen reports the belief according to which sleeping on the earth might induce the disease in adolescents.

15 This use of sleep and sleeplessness as an analogue, respectively, of apoplexy and of epilepsy is particularly interesting as it is part of a wider discussion on the physiological causes of sleep and other sleep-like pathological affections (see *infra*, n. 17) in which Galen had already explicitly mentioned Aristotle's *De somno et vigilia*, acknowledging Aristotle's explanation of sleep as caused by a concentration and condensation of vaporous matter in the brain but also overtly criticizing the explanatory inconsistencies that, in Galen's eyes, result from locating the *sensorium commune* in the heart (see 7.141.17–142.12 K.). For this reason, one might even read Galen's rethinking and, in fact, reversal of the analogy between sleep and epilepsy as an intentional polemical move, as an indirect reference to and thus a rejection, or at least a correction, of the analogy as it had been established by Aristotle.

point, as it suggests different ways of conceiving the relationship between "physiological" and "pathological" phenomena,[16] as well as different ways of conceptualizing the causal structure of a biological process *qua* change undergone by a material substance.[17] 2) While Aristotle's claim has an *explanatory* function, as it is part of an exercise in reasoning about the causes of sleep, the medical accounts that establish a link between sleep and the epileptic fit always characterize clinical evidence and therefore have a strictly *descriptive* function. This makes the Aristotelian version of the sleep/epilepsy association particularly strong, although, on closer inspection, the explanatory framework within which the association belongs is not entirely perspicuous. For Aristotle assimilates sleep to the

---

[16] On the dialectic between the categories of "the physiological" and "the pathological" in Aristotle's characterization of sleep "as an epileptic fit" see Armelle Debru, "L'épilepsie dans le De somno d'Aristote," in Guy Sabbah (ed.), *Médecins et médecine dans l'Antiquité* (Saint-Étienne: Publications de l'Université de Saint-Étienne, 1982), 25–41, at 29–31.

[17] In this respect it is important to observe that Aristotle's reference to the epileptic fit as an *analogon* of sleep is framed within his overall explanation of sleep as a byproduct of digestion. More generally, one might say that "Aristotle's theory of sleep belongs within a strand in Greek thought in which sleep was defined negatively as the absence of a number of activities and abilities" (Philip J. van der Eijk, "Theoretical and Empirical Elements in Aristotle's Treatment of Sleep, Dreams and Divination in Sleep," in *Medicine and Philosophy in Classical Antiquity* [Cambridge: Cambridge University Press, 2005], 169–205, at 171). Although Aristotle's overall treatment of sleep is not at all exhausted by this negative definition (since to this negative definition corresponds a positive evaluation of the affections "sleep": see Philip J. van der Eijk, "Aristotle on Cognition in Sleep," *Nottingham Classical Literature Studies* 8 [2003]: 25–40, at 28; Roberto Lo Presti, "Le sommeil dans les *Epidémies hippocratiques*," in V. Leroux, N. Palmieri, and C. Pigné [eds.], *Le sommeil: approches philosophiques et médicales de l'Antiquité à la Renaissance*, forthcoming), insofar as sleep is said to be *necessary* for the animal's preservation (454a20–b4; 455b16–29), there is a clear difference between Aristotle's explanatory strategy and, for example, Galen's account of sleep, according to which the brain *intentionally* activates sleep for its own preservation. The two main sources for Galen's physiological theory of sleep are *de caus. puls.* 3.9–10, 9.131–42 K. and *de sympt. caus.* 1.8, 7.139–44 K. For a complete list of Galenic sources on sleep, see van der Eijk and Hulskamp, "Stages in the Reception of Aristotle's Works," 68 n. 72; for an overall interpretation of Galen's account of sleep see Maithe Hulskamp, "Sleep and Dreams in Ancient Medical Diagnosis and Prognosis," PhD diss., University of Newcastle, 2008, 94–131. See also Malcolm Lowe, "Aristotle's *De somno* and His Theory of Causes," *Phronesis* 23 (1978): 279–91, at 286–87, who further emphasizes some peculiar traits of Aristotle's accounts of the causes of sleep, by pointing out that, strictly speaking, "the final cause of sleep may not exist." *Contra* Lowe see Stephen Everson, "The *De somno* and Aristotle's Explanation of Sleep," *Classical Quarterly* 57 (2007): 502–20, at 519, who argues that, in Aristotle's account, "sleep is susceptible of explanation in terms of a final cause, but this does not require that it should have an efficient cause as well." For a comparative approach to Aristotle's and Galen's accounts of sleep, see van der Eijk and Hulskamp, "Stages in the Reception of Aristotle's Works," 66–74.

epileptic fit (*ho hupnos epilēpsis*), even if—he specifies—such an assimilation is to be taken only "in a certain way" or "in a qualified way." It is likely that beyond the claim that, in order to be valid, the assimilation of sleep to the epileptic fit must be qualified in some way, lies the idea that sleep and the epileptic fit are analogous simply with respect to a certain kind of cause—namely, the material one[18]—and also, in this case, perhaps in relation to some aspects of the causal mechanism. For, while the processes of exhalation, the cooling and condensation of vapors originated from digestion, and the descent of the cooled matter seem to be quite the same both in the case of an epileptic seizure and in the case of sleep,[19] it is not entirely clear from the text of the *De somno* whether the physical changes underlying epilepsy actually culminate in an incapacitation of the *prōton aisthētikon*, as is the case with sleep; nor is the question further raised in other Aristotelian writings. For, as Aristotle puts it, the epileptic seizure seems to be primarily connected with respiratory trouble, which results from the condensed vapors swelling the veins of the neck and obstructing the windpipe (τὸν πόρον δι' οὗ ἡ ἀναπνοὴ γίνεται).[20]

Therefore, if the material processes that produce sleep and the epileptic fit are merely similar, rather than identical, on what grounds can Aristotle suggest that sleep *is*, in a way, an epileptic fit? Why does he not instead limit himself to likening the former to the latter? For if sleep *is* an epileptic fit only "in some respect," there must be other respects in which it is not. How can we explain, then,

---

18 In referring to the material cause as the object of Aristotle's discussion in *De somno*, I accept Everson's reading of the Aristotelian argument (Everson, "The *De somno*," 516–20) against Lowe (Lowe, "Aristotle's *De somno*") and Gallop (David Gallop, *Aristotle on Sleep and Dreams* [Peterborough, Ont.: Broadview Press, 1990], reviewed by Michael J. Woods, "Aristotle on Sleep and Dreams," *Apeiron* 25 [1992]: 179–88). On this point, see also Ross, *Aristotle, Parva naturalia*, 260.
19 As Debru, "L'épilepsie dans le *De somno*," 28–29 and 32–33 remarks, this physiological explanation, according to which the epileptic seizure is connected to the digestive processes and results from a vaporous substance (and not from a humor), has no parallel in the ancient medical literature, nor is it further developed in other Aristotelian writings. In fact, the only other text of the *Corpus Aristotelicum* that puts forward an etiological principle for epilepsy is the Pseudo-Aristotelian *pr.* 30.1, in which epilepsy is counted among the affections to which the melancholic subjects are subject (see 953a16, 953b6, 954b30). But, interestingly, melancholics are also mentioned in *De somno* (457a27–29) as subjects who are *not* prone to sleep, due to the fact that their inner parts are chilled and therefore limit evaporation of food (see Philip J. van der Eijk, "Aristotle on Melancholy," in *Medicine and Philosophy*, 139–68, at 142–43). In *De somno*, the defining feature of melancholics is coldness, but in *part. an.* 672b29, it is heat; in the Pseudo-Aristotelian *pr.* 30.1, 954a14–15, it is the possibility of great changes in temperature, from the very hot to the very cold.
20 On the link between epilepsy and respiration in Aristotle's *De somno* and in the ancient medical literature, see Debru "L'épilepsie dans le *De somno*," 34–36.

why, in connecting these two affections, Aristotle is not at all concerned with further explaining the nature, conditions, and limits of their similarity? For this is exactly what he does elsewhere in the *De somno* (455b3–13; 456b9–19), where he carefully distinguishes sleep from other, apparently sleep-like, incapacities of the sensitive faculty—there he mentions unconsciousness, throttling, and faintness, but, interestingly, *not* epilepsy—with respect to the sense organ affected (sleep is an incapacitation of the primary sense organ, while other affections are not) and to the material cause of the affection (sleep is the product of the exhalation involved in digestion, while the other affections are not).[21]

* * *

It is important to turn our attention to these unanswered questions regarding the physiological link that Aristotle establishes between sleep and epilepsy, first in order to get a fuller understanding of Aristotle's own ideas on these two phenomena. They can also help us grasp the further developments of these ideas in the centuries-long, multifaceted history of Aristotelianism, as well as in the reception of these ideas among those parts of the medical tradition which influenced, and were in turn most significantly influenced by, the understanding, the transformation(s), and the re-codification(s) of the Aristotelian theoretical paradigm. For, on the one hand, it is precisely at those places where Aristotle's account of "the activities common to soul and body" becomes opaque that it was easiest for medical authors and Aristotelian commentators (in many cases these two figures coincide) *to go beyond*, *rework*, and *transform* Aristotle's account by combining it with theoretical and empirical materials of various provenance and eventually *to absorb* it into their own research or teaching agenda. On the other hand, it is just the need to explain portions of Aristotle's physiological theories, in which empirical, viz. medical, elements seem to have played an important, yet somewhat unclear, role, that prompts medical authors and commentators to ground their discussion and interpretation of these theories on what they assumed to be a firmer empirical basis.

In what follows, I will investigate how some key figures within Byzantine, Arabic, and Latin medieval Aristotelianism understood and explained Aristotle's analogy between sleep and the epileptic fit. I ask what logical or theoretical arguments they put forward and what empirical evidence, if any, they adduced in order to justify, further develop, redefine, or reject the analogy as they found it in the *De somno et vigilia*. By taking this fragment of the Aristotelian tradition as a case study, I aim to cast fresh light on whether, to what degree, and in which

---

21 *Somn. vig.* 455b3–13; 456b9–11, 17–19.

terms the understanding of Aristotle's physiological theories incorporated or was mediated by issues, modes of reasoning, concepts, and explanatory patterns arising from medicine and whether the reception of Aristotle may have influenced, or have been influenced by, a specific commitment to medical research and/or teaching. I will start by examining Michael of Ephesus' commentary on the *De somno et vigilia* (fl. early or mid-twelfth c.), making reference, as well, to the work of other Byzantine commentators, namely Sophonias (fl. ca. 1300) and Theodorus Metochites (1270–1332), who is the author of the most extensive commentary on Aristotelian philosophy in the late Byzantine period.[22] As far as medieval Aristotelianism is concerned, I shall focus both on the Arabic and on the Latin reception of the sleep/epilepsy analogy. I will therefore discuss the relevant passages of the following works: the Latin translations of Averroes' *Compendium libri de sompno et vigilia*,[23] Avicenna's *Liber de anima*,[24] and the only two extant Latin commentaries that are based on the anonymous *translatio*

---

[22] Michael of Ephesus' and Sophonias' commentaries have been edited within the "*Commentaria in Aristotelem Graeca*" series, respectively, vol. 22.1 (Paul Wendland, *Michaelis Ephesii in Parva naturalia commentaria*, CAG 22.1 [Berlin: Reimer, 1903]) and 5.6 (Paul Wendland, *Themistii [Sophoniae] in Parva naturalia commentarium*, CAG 5.6 [Berlin: Reimer, 1903]); Theodorus Metochites' commentary has been edited by Hendrik J. Drossaart Lulofs (*Aristotelis "De somno et vigilia liber" adiectis veteribus translationibus et Theodori Metochitae commentario* [Leiden: Burgersdijk & Niermans, 1943]), together with Aristotle's original text, the anonymous *translatio vetus*, and its revised version (*translatio nova*), which William of Moerbeke brought to completion before 1284. On the figure of Theodorus Metochites as a Byzantine intellectual and Aristotelian commentator, see Hans-Georg Beck, *Theodoros Metochites. Die Krise des byzantinischen Weltbildes im 14. Jahrhundert* (Munich: C. H. Beck, 1952) and Eva de Vries-van der Velden, *Théodore Métochite: une réévaluation* (Amsterdam: Rodopi, 1987).

[23] Averroes' *Compendium* was translated into Latin by Michael Scotus between 1220 and 1235 (see Peter De Leemans, "*Secundum viam naturae et doctrinae:* lire le *De motu animalium* et les *Parva naturalia* d'Aristote au Moyen Âge," in Grelland and Morel, *Les Parva naturalia d'Aristote*, 197–220). This translation has been edited by Emily Ledyard Shields, *Averrois Cordubensis Compendia Librorum Aristotelis qui Parva naturalia vocantur*, recensuit Aemilia Ledyard Shields adiuvante Henrico Blumberg (Cambridge, Mass.: The Medieval Academy of America, 1949). On the early reception of Averroes within Western Latin Aristotelianism, see Roland de Vaux, "La première entrée d'Averroès chez les Latins," *Revue des sciences philosophiques et théologiques* 22 (1933): 193–243.

[24] The Latin translation of Avicenna's *Liber de Anima* belongs to a corpus of translations from Arab into Latin made in Toledo in the twelfth century A.D. This translation has been edited by van Riet–Verbeke (S. van Riet and G. Verbeke, *Avicenna Latinus, Liber de Anima seu Sextus de Naturalibus, IV–V* [Louvain: Ed. Orientalistes, 1968] and *Avicenna Latinus, Liber de Anima seu Sextus de Naturalibus, I–III* [Louvain: Ed. Orientalistes, 1972]).

vetus of the *De somno et vigilia* (second half of the twelfth c.),[25] namely Albert the Great's paraphrase;[26] and a commentary—perhaps the oldest extant Latin commentary on the *De somno*—which has been transmitted as part of Thomas Aquinas' exegetical work on Aristotle but which was in fact very probably written by the *Oxoniensis Magister* Adam of Buckfield (fl. 1238–1278) between the late 1230s and the 1240s.[27]

---

[25] Other (lost) commentaries of the *De somno* based on the *translatio vetus* were written by Galfridus de Aspale, Roger Bacon, and Robert Kilwardby (see Charles Lohr, "The New Aristotle and 'Science' in the Paris Arts Faculty (1255)," in O. Weijers and L. Holtz [eds.], *L'enseignement des disciplines à la Faculté des arts [Paris et Oxford, XIIIe-XVe siècles]* [Turnhout: Brepols, 1997], 257). On the anonymous *translatio vetus* of *De somno* see Drossaart Lulofs, *Aristotelis De somno et vigilia*, xi–xii and Jozef Brams, "Der Einfluß der Aristoteles-Übersetzungen auf den Rezeptionsprozeß," in L. Honnefelder et al. (eds.), *Albertus Magnus und die Anfänge der Aristoteles-Rezeption im lateinischen Mittelalter* (Münster i. W.: Aschendorff, 2005), 27–43, at 41.

[26] In A. Borgnet, ed., *Alberti Magni Opera Omnia*, 38 vols. (Paris: apud Ludovicum Vivès, 1890–1895), 9:213–55.

[27] For example, the 1566 Venice edition of Thomas Aquinas' commentaries on the *Parva naturalia* (with Moerbeke's *translatio nova* and a new translation by Niccolò Leonico Tomeo of Aristotle's text) included the commentaries on *De somno, De insomniis,* and *De divinatione* among Thomas' authentic works. In modern times Adam of Buckfield's commentary has been edited as part of the *corpus thomisticum* (Thomas Aquinas, *Opera omnia*, 25 vols. [Parma: Fiaccadori, 1852–1873], 20:215–44; Thomas Aquinas, *Opera omnia*, 34 vols. [Paris: Vivès, 1871–1882], 24:293–331). On the authorship of this commentary on the *De somno et vigilia* see Franz Pelster, "Adam von Bocfeld (Bockingfold), ein Oxforder Erklärer des Aristoteles um die Mitte des 13. Jahrhunderts: sein Leben und seine Schriften," *Scholastik* 11 (1936): 1992–2224, at 2220–21; Martin Grabmann, *Die Werke des heiligen Thomas von Aquin* (Münster i. W.: Aschendorff, 1949), 296–97; James A. Weisheipl, "Science in the Thirteenth Century," in J. I. Catto (ed.), *The History of the University of Oxford* (Oxford: Clarendon Press, 1984), 435–69, at 462–64; Julie Brumberg-Chaumont, "La première reception du *De memoria et reminiscentia* au Moyen Âge latin: le commentaire d'Adam de Buckfield," in Grelland and Morel, *Les Parva naturalia d'Aristote*, 121–41, at 122. According to a medieval catalogue of the manuscripts of Beauvais Cathedral, compiled in 1664, Alfred of Sareshel, who was a translator and commentator of the Aristotelian *libri naturales* active a few decades before Adam of Buckfield, might have written a commentary on the *De somno et vigilia* already between the late twelfth and early thirteenth century. On this point see Charles Burnett, "The Introduction of Aristotle's Natural Philosophy into Great Britain: A Preliminary Survey of the Manuscript Evidence," in John Marenbon (ed.), *Aristotle in Britain during the Middle Ages* (Turnhout: Brepols, 1996), 21–50, at 31–32; Daniel A. Callus, "Introduction of Aristotelian Learning to Oxford," *Proceedings of the British Academy* 29 (1943): 229–81, at 236–37; Henri Omont, "Recherches sur la bibliothèque de l'église cathédrale de Beauvais," *Mémoires de l'Académie des Inscriptions et Belles-Lettres* 40 (1914): 1–91, at 48 n. 143.

I Interpreting and Rethinking the Sleep/Epilepsy Analogy in Byzantine Aristotelianism: Michael of Ephesus' Commentary on the *De somno et vigilia*

As an increasing number of scholars have recognized in recent years, Michael of Ephesus must be considered a key figure within Aristotelianism, especially with respect to the reception of Aristotle as a natural philosopher. Working in Constantinople within the intellectual circle and under the patronage of Anna Comnena,[28] Michael wrote commentaries on many of Aristotle's works, aiming to provide coverage of those texts of the Aristotelian corpus that had been neglected by earlier commentators.[29] He was the first to offer extensive commentary on Aristotle's zoological and biological works and, consequently, the first to produce an in-depth discussion of the *De somno et vigilia* as part of a comprehensive commentary on those treatises which Latin Aristotelianism would collectively refer to as the *Parva naturalia*.[30] His importance within the Aristotelian tradition

---

[28] On the figure of Anna Comnena and on the role of the circle of intellectuals that she created in Constantinople, see Georgina Buckler, *Anna Comnena: A Study* (Oxford: Clarendon Press, 1929); Robert Browning, "An Unpublished Funeral Oration of Anna Comnena," *Proceedings of the Cambridge Philological Society* 188 (1962): 1–12; Anthony Preus, *Aristotle and Michael of Ephesus, On the Movement and Progression of Animals* (Hildesheim: Olms, 1981), 8–14; Thalia Gouma-Peterson (ed.), *Anna Komnene and her Times* (New York: Garland, 2000).

[29] This is what Michael himself suggests at the end of his commentary on the *Parva naturalia* (149.8–16 Wendland), when he makes a list of Aristotle's works on which he intends to comment (*De partibus animalium, De incessu animalium, De memoria, De motu animalium, De generatione animalium, De longitudine et brevitate vitae, De iuventute*) or on which he had already commented (*Metaphysica*). On the other hand, Michael's own commitment to Aristotle's works without commentaries entirely belongs to Anna Comnena's project, as it is described by Georgios Tornikes in his funeral oration for Anna (see Jean Darrouzes, *Georges et Demetrios Tornikes, Lettres et Discours* [Paris: CNRS, 1970], 283). I quote the English translation of Tornikes' text as one finds it in Preus, *Aristotle and Michael of Ephesus*, 10: "Evidence on her love of learning are the works which our philosophers dedicated to her; these concern the works of Aristotle which had not been commented upon before her day, but had been handed down by word of mouth in all sort of forms, without certainty and unworthily. ... I myself have heard the philosopher from Ephesos blame her as the cause of his blindness, because he had worked night after night, without sleep, commanded by her to write commentaries on the works of Aristotle."

[30] See above, n. 1. On the textual tradition of the *Parva naturalia* see Drossaart Lulofs, *Aristotelis de somno et vigilia*, ix-xxxvii; H. J. Drossaart Lulofs, *Aristotelis De insomniis et De divinatione per somnum* (Leiden: Burgersdijk & Niermans, 1947), 1:xlv–lxxvii; Hendrik J. Drossaart Lulofs, "Aristoteles Arabus," *Forum der letteren* 1 (1960): 169–82; Aurél Förster, *Konstruktion und*

lies not only in the fact that he was, so to speak, a pathbreaking commentator, but also in the fact that in more than one case, when discussing, clarifying, or rethinking Aristotle's psycho-physiological theories, Michael seems to have somehow taken into account, without making explicit reference to, theoretical and empirical materials elaborated within the medical tradition, and in particular by Galen.[31] In this respect Michael seems to have developed an exegetical method that presents some points of contact with the method adopted by the Arab commentators on Aristotle's biological works. It is in fact also through Michael, and not only through the Arab commentators, that thirteenth-century Latin Scholasticism "discovered" Aristotle's biological works and took them as one of its major focus of interest.[32]

Michael's treatment of the analogy between sleep and epilepsy is exactly one of those places in the exegetical work of the Ephesian commentator in which it is possible to explore whether, in what terms, and with what "theoretical" consequences the acquaintance with the contents of knowledge transmitted within the medical tradition may have affected the reception of Aristotle's biological ideas. On the one hand, when discussing the passage in which Aristotle

---

Entstehung der aristotelischen sogenannten Parva naturalia (Budapest: Akadémia, 1932); Georges Lacombe, "The Medieval Latin Versions of the *Parva naturalia*," *The New Scholasticism* 5 (1931): 289–314; René Mugnier, "Les manuscrits des *Parva naturalia* d'Aristote," in *Mélanges offerts à A. M. Desrousseaux* (Paris: Hachette, 1937), 327–33 and idem, "La filiation des manuscrits des *Parva naturalia* d'Aristote," *Revue de philologie* 26 (1952): 36–46; Philip J. van der Eijk, *Aristoteles, De insomniis; De divinatione per somnum* (Berlin: Akademie-Verlag, 1994), 94–101.

31 In his commentary on the *Parva naturalia*, Michael mentions Galen four times, once (52.19–20 Wendland) while commenting upon *De somno*, once in the commentary on *De insomniis* (67.21 Wendland). On these two references to Galen see van der Eijk and Hulskamp, "Stages in the Reception of Aristotle's Works," 74–75. Galen is mentioned also in the commentary on *De iuventute* (109.20 and 135.28 Wendland). On the role of Galenic medical ideas in the Greek Aristotelian commentators (with references also to Michael of Ephesus), see Robert B. Todd, "Galenic Medical Ideas in the Greek Aristotelian Commentators," *Symbolae Osloenses* 52 (1977): 117–34 (especially 126–28).

32 To get an idea of the channels through which this influence may have been exerted it is sufficient to recall that the first corpus of Latin translations of Aristotle's works was produced between the first half of the twelfth and the beginning of the thirteenth centuries and that it is possible, although not always probable, that some of the most important translators—such as James of Venice, Robert Grosseteste, and William of Moerbeke—were in contact with the Byzantine milieu, when not directly with Michael and his commentaries: see Lorenzo Minio-Paluello, "Jacobus Venetus Graecus," *Traditio* 8 (1952): 265–304; Martin Grabmann, *Guglielmo di Moerbeke O.P., il traduttore delle opere di Aristotele* (Rome: Pontificia Università Gregoriana, 1946); Preus, *Aristotle and Michael of Ephesus*, 13–21. On the medieval Latin reception of Aristotle's scientific corpus, see Sybil D. Wingate, *The Medieval Latin Versions of the Aristotelian Scientific Corpus, with Special Reference to the Biological Works* (London: Courier, 1931).

characterizes sleep as epilepsy, Michael seems to agree with, and further develop, the overall structure of Aristotle's argument:

> ἐπεὶ δὲ ἡ ἐπίληψις ἀργία ἐστὶ πασῶν τῶν αἰσθήσεων (ἀργοῦσι δ' αὗται καὶ ἐν τοῖς ὕπνοις), εἶπεν ὅτι ⟨ὅμοιον ὁ ὕπνος ἐπιλήψει.⟩ ὥστε οὐχ ἁπλῶς γίνονται ἐπιληπτικοί, εἰ καί ποτε συμβαίνει τοῦτο διὰ τὸ μάλιστα ἐν τοῖς ὕπνοις γίνεσθαι πολλὴν ἀναθυμίασιν, ἀλλὰ τρόπον τινά. ὡς γὰρ ὁ ὕπνος ὅμοιον ἐπιλήψει, οὕτω καὶ τὸ ἀκινητίζειν τὰ παιδία ὅμοιον ἐπιλήψει. καὶ ἐπεὶ πολλὴ γίνεται ἡ ἀπὸ τῆς τροφῆς ἀναθυμίασις ἐν τοῖς ὕπνοις διὰ τὸ καὶ τότε μάλιστα ἐνεργεῖν τὸ θρεπτικόν, ὀλίγη δὲ ἐν τῇ ἐγρηγόρσει, εὐλόγως ἁλίσκονται ἐπιληψίαις καθεύδοντες οἱ ἄνθρωποι μᾶλλον ἤπερ ἐγρηγορότες. πολὺ γάρ, ὡς εἴρηται, φερόμενον τὸ πνεῦμα ἄνω ἐν τοῖς ὕπνοις, καταβαῖνον πάλιν ὥσπερ εὔριπος ὀγκοῖ τὰς φλέβας· ... φερόμενον οὖν ἄνω καὶ κατιὸν κάτω τὸ πνεῦμα ὀγκοῖ τὰς φλέβας, αἷς συνθλίβει καὶ συστέλλει καὶ στενοῖ τὸ ἐν αὐταῖς πνεῦμα τὴν τραχεῖαν ἀρτηρίαν· ταύτῃ γὰρ εἶπε πόρον, ᾗ ἡ ἀναπνοὴ γίνεται· διὰ ταύτης γὰρ ἀναπνέομεν. ταύτης δὲ στενωθείσης ἀνάγκη μὴ γίνεσθαι εἰσπνοήν, καὶ ταύτης μὴ γινομένης ἀναγκαῖον ἐπιλήψεις συμβαίνειν. (*In parva naturalia commentaria* 54.25–55.3 Wendland)
>
> But since epilepsy is an idleness of all the senses (these are inactive also in sleep), he said that "sleep is like epilepsy." And so people do not become epileptics "in an unqualified way" but, even if it happens above all because of much evaporation being produced during sleep, in a certain way. For just as sleep is similar to epilepsy, in the same way the immobility of children is similar to epilepsy. And since much evaporation from food is produced during sleep due to the fact that the nutritive faculty is most active during sleep, while it is less active during wakefulness, people are seized by epileptic fits when they sleep more than when they are awake. For, as it has been said, much *pneuma* is carried upwards during sleep, and when it flows back, it swells the vessels like a strait. ... Therefore in its upward and downward movement the *pneuma* swells the vessels, through which the *pneuma*, which is contained in them, compresses, restricts, and narrows the windpipe. By this he means the passage through which respiration takes place. For we breathe through this. When this is constrained, it is necessarily impossible for breath to be taken in, and when this cannot take place, epileptic seizures necessarily follow.[33]

Since epilepsy is an incapacitation of all the senses (ἀργία πασῶν τῶν αἰσθήσεων), sleep is like epilepsy. Like sleep, epilepsy, too, is prompted by vapors arising from food, which means that epilepsy is to be explained essentially as a pathological byproduct of digestion; respiration is the physiological activity which comes to be hindered when streams of vapors and cooled matter ascend and descend and whose hindrance results in the epileptic seizure.

The picture gets more complicated and somewhat contradictory when we consider an earlier passage in Michael's commentary. For, while discussing Aristotle's treatment of visions (*phantasmata*) that occur in people who have fainted and the distinction that he draws between these visions and proper dreams (*en*-

---

33 Translations unless otherwise noted are my own.

hupnia) that one has while being asleep,[34] Michael quite unexpectedly interpolates a reference to epilepsy in order to reject its assimilation to sleep:

> Ὅτι μὲν οὖν φαντάσματά τινα ὁρῶσι, καὶ οὐ μόνον ὁρῶσιν ἀλλὰ καὶ λέγουσιν οἱ λειποψυχήσαντες ἰσχυρῶς (λέγουσι δὲ ταῦτα οὐκ εἰδότες· οὐ γὰρ οἴδασιν ἢ ὅλως αἰσθάνονται τί ὁρῶσιν ἢ λέγουσιν), δῆλον ἐκ τῆς ἐπαγωγῆς· ὁρῶμεν γὰρ τοῦτο πολλοὺς πάσχοντας. ὅτι δὲ οὔκ ἐστιν ὁ ὕπνος ἐπιληψία, εἴρηται πολλάκις. εἰσὶ δὲ τὰ λεγόμενα λύσις ἀπορίας τινός· εἰπὼν γὰρ ὅτι ἤδη γεγένηταί τισι καὶ φαντασία λειποψυχήσασιν ἰσχυρῶς, ὥσπερ ἀναμνησθεὶς ἐπήγαγε· ⟨τοῦτο μὲν οὖν ἔχει ἀπορίαν.⟩ ἠπόρησε γὰρ ἄν τις, ὅτι, εἴπερ γίνονται φαντάσματα καὶ ἐν ταῖς λειποψυχίαις καὶ ἐν τοῖς ὕπνοις, ἀκόλουθόν ἐστι λέγειν τὴν λειποψυχίαν ὕπνον. εἰ δὲ τοῦτο, ἔστιν ἄρα ὁ ὕπνος ἡ τυχοῦσα τῶν αἰσθήσεων ἀδυναμία. Λύει οὖν αὐτὴν λέγων· ⟨εἰ γὰρ ἐνδέχεται καταδαρθεῖν τὸν λειποψυχήσαντα⟩ καὶ τὰ ἑξῆς. ἔστι δὲ τὸ λεγόμενον· εἰ γὰρ ὑπνώττει ὁ λειποψυχῶν καὶ ὅλως ὕπνος ἐστὶν ἡ λειποψυχία, ἔσται καὶ τὸ φάντασμα τοῦ λειποψυχήσαντος ἐνύπνιον. ἀλλὰ τῶν μὲν ἐνυπνίων, εἰ καὶ μὴ πάντων, ἀλλὰ τῶν πλείστων ἐγερθέντες μνημονεύομεν, οὐδενὸς δὲ τῶν ἐν ταῖς λειποψυχίαις φαντασμάτων μετὰ τὴν ἀπαλλαγὴν μνημονεύουσι· πολλὰ γὰρ ἐν τῷ πάθει λέγοντες οὐδενὸς ἐγερθέντες μνημονεύουσιν. οὐκ ἄρα τὰ φαντάσματα ἐνύπνια. ὥστε οὐδ' ἡ ἐπίληψις ὕπνος. τὸ δὲ ⟨περὶ ὧν τὸν αὐτὸν λόγον ὑποληπτέον εἶναι πάντων⟩ ἴσον ἐστὶ τῷ 'τὰ αὐτὰ δεῖ λέγειν περὶ πνιγμοῦ καὶ ἐκνοίας καὶ τῶν ὁμοίων.' (*In parva naturalia commentaria* 53.3–22 Wendland)

The fact that those who suffer from a violent fainting have visions of some sort and that they not only see something but also speak (they speak, but they do not know that; for they neither know nor are they aware of what they see and say) is clear by induction. For we see that many people experience that. As to the fact that sleep is not epilepsy, this has been often said. What is being said provides the solution for a certain difficulty. For having said that visions appear to certain people who suffer from violent fainting, he concludes by recalling something to his mind: ⟨this gives rise to a difficulty⟩. For one could raise the question whether, if visions occur both in cases of fainting and in sleep, the consequence is to define fainting as sleep. If so, then sleep is any incapacity of the senses whatsoever. Therefore he solves the difficulty by saying: ⟨if it is possible for him, who is fainted, to fall asleep⟩ and so on. What he means is this: For if the person who has fainted sleeps and in general if fainting is sleep, then the vision of the person who has fainted will be a dream. But we do remember dreams, if not all, then most of them, after awaking, while they do not remember the visions they had while they were unconscious after the recovery. For, although they speak a lot during the time in which they are affected, they do not remember anything once they have awakened. Therefore the visions are not dreams. And therefore epilepsy is not sleep. To say ⟨the same explanation must be supposed for all these things⟩ is the same as to say "one must say the same things concerning choking, the loss of senses, and the things similar to these."

---

[34] Arist. *somn. vig.* 3, 456b9–16: "For sleep, as has been said, is not any and every incapacity of the sensitive faculty; for such incapacity is produced by unconsciousness, throttling, and faintness. Also imagination has been known to occur even in a severe faint. This fact involves a difficulty; if it is possible for one who has fainted to fall asleep, his imaginary vision [*to phantasma*] might be a dream [*enhypnion*]. Again, words are often spoken by people who are in so deep a trance that they seem to be dead" (trans. Hett).

In this passage of Michael's commentary, we find a distinction between sleep and epilepsy that is not found in the corresponding passage of Aristotle's *De somno*. Indeed, the distinction clashes with Aristotle's own argument of 457a8–14, which, as we have seen, Michael seems far from rejecting or even questioning.[35] When reading Michael's text, one immediately has the impression that there has been some (intentional or involuntary) shift from *leipopsuchia* to *epilēpsia* and that the distinction between sleep and fainting, which Aristotle actually draws, has been reinterpreted as a distinction between sleep and epilepsy. For Michael's remark that "it has been often said, that sleep is not an epileptic fit" (ὅτι δὲ οὐκ ἔστιν ὁ ὕπνος ἐπιληψία, εἴρηται πολλάκις) finds no confirmation in Aristotle's *De somno*, while it is true that the distinction between *hupnos* and other forms of perceptual inactivity (among which we do find *leipopsuchia* but not epilepsy) had been already drawn by Aristotle a few lines above (455a31–b13). Moreover, the logical structure of Michael's text appears to be quite inconsistent. He first puts forward that "it is clear [δῆλον] that *those who have fainted* [οἱ λειποψυχήσαντες] are strongly affected by visions ... *but* [δὲ] it has been often said that sleep is not epilepsy," and then remarks that "people do not remember visions occurring in a *faint* [τῶν ἐν ταῖς λειποψυχίαις φαντασμάτων] ... and indeed visions are not dreams [οὐκ ἄρα τὰ φαντάσματα ἐνύπνια], *so that* [ὥστε] *epilepsy is not sleep*."[36]

This passage has to be read together with another passage of Michael's commentary where the Ephesian discusses the section of the *De somno* in which Aristotle for the first time distinguishes sleep from other forms of perceptual inactivity:

> ἀλλὰ καὶ τὰς ἐν τῷ αὐχένι δύο φλέβας εἴ τις δήσει, ἀναισθητεῖ τὸ ζῷον, ὕπνος δὲ οὐκ ἔστιν. ἀλλ' ὅταν ἡ ἀδυναμία τοῦ αἰσθάνεσθαι μήτε ἔν τινι τῶν αἰσθητηρίων γένηται (ἐνδέχεται γὰρ πηρωθῆναι τὰ αἰσθητήρια, ἀλλ' ὅμως οὐδὲ ἡ ἀκινησία τῶν αἰσθήσεων ἡ διὰ πήρωσιν τῶν αἰσθητηρίων γινομένη ὕπνος ἐστίν), ὅταν οὖν μήτε ἔν τινι τῶν αἰσθητηρίων μήτε δι' ἣν ἔτυχεν αἰτίαν γένηται, οἷον δι' ἐπίληψιν ἢ κατάληψιν τῶν ἐν τῷ αὐχένι φλεβῶν, ἀλλ' ἐν τῷ πρώτῳ αἰσθητηρίῳ καὶ τῇ πρώτῃ αἰσθήσει, τότ' ἐστὶν ὁ ὕπνος. (*In parva naturalia commentaria* 49.8–15 Wendland)

---

35 The very same distinction between sleep and epilepsy is to be found in Sophonias' commentary on the *De somno et vigilia*, which to a large extent consists in a kind of paraphrase of Aristotle's text, enriched with insertions of Michael's comments. See Sophonias, *In parva naturalia commentarium* 24.6–18 Wendland.

36 The use of the syntagma εἴρηται πολλάκις as well as of other syntagms such as πολλάκις εἴρηται, ὡς πολλάκις εἴρηται, καθάπερ πολλάκις εἴρηται is quite a typical feature of Michael's argumentative technique (the TLG database lists 47 occurrences, 22 in the commentary on the *PN*), which he may have derived directly from Aristotle (21 occurrences listed in the TLG).

But also if one binds the two vessels in the neck, the animal becomes insensible; nonetheless it is not sleep. When the incapacity to perceive does not occur in any sense organ whatsoever (for it is possible that the sense organs are incapacitated, but nevertheless the immobility of the senses due to the incapacitation of the sense organs is not sleep), when therefore it neither occurs in any sense organ whatsoever nor is engendered by any chance cause, just like (when it occurs) with the stoppage and the compression of the vessels in the neck [οἷον δι' ἐπίληψιν ἢ κατάληψιν τῶν ἐν τῷ αὐχένι φλεβῶν], but when it takes place in the primary sense organ and the primary sense faculty, then it is actually sleep.

This is the first of three references to *epilēpsis* occurring in less than six pages of Michael's commentary.[37] In this case, however, there is no comparison between sleep and an epileptic fit at stake, as the term *epilēpsis* is used here with a completely different, and almost unheard of, meaning;[38] the term indicates, together with *katalēpsin*, the "stoppage" of the vessels in the neck,[39] which is indeed one of the possible causes to which Aristotle ascribes those forms of *anaisthēsia* such as *leipopsuchia*, *pnigmos*, and *eknoia*, which he considers to be *different* from sleep. It is also interesting to note that in the corresponding passage of Aristotle's text, one finds the verb *katalambanesthai* to indicate the stoppage of the vessels of the neck.[40] Aristotle's use of the verb may justify the choice of the noun *katalēpsis* by Michael, but it does not license the use of a somewhat redun-

---

37 See 49.14 Wendland; 53.6 and 20 Wendland; 54–55 Wendland and passim.
38 The only occurrence of *epilēpsis* I have found in the medical literature with a similar technical meaning is in Aëtius Amid. lib. med. 9.6, where we read μετ' ἐπιψύξεως τῶν ἄκρων καὶ τῆς τοῦ πνεύματος ἐπιλήψεως.
39 One could argue that, although in this passage the terms *epilēpsis* and *katalēpsis* do not refer to specific affections, they most likely *allude* to them. For, when using the syntagma δι' ἐπίληψιν ἢ κατάληψιν, Michael could not be unaware of the fact that from late antiquity onward medical terminology had long speculated on the terms "epilepsy" and "catalepsy" as connoting distinct affections (see Cael. Aur. diaet. pass. 57 = Valentin Rose, Anecdota Graeca et Graecolatina, 2 vols. [Berlin: Dümmler, 1864–1870], 2:231) or distinct forms of the same affection (Esculapius, De morborum, 4). Some doctors maintained the same distinction between epilepsy and catalepsy up to the Renaissance: see Arnold of Villanova, De epilepsia 1, col. 1602 (Basel, 1585) and Antonius Guainerius, Tractatus de egritudinibus capitis (ca. 1470–1480), fol. 11ᵛ. On the distinction between epilepsy and catalepsy see Owsei Temkin, The Falling Sickness: A History of Epilepsy from the Greeks to the Beginnings of Modern Neurology, 2nd ed. (Baltimore: The Johns Hopkins University Press, 1971), 120–23; on the notion of "catalepsy" in antiquity, see Evert D. Baumann, "Die Katalepsie der Antiken," Janus 1942 (1938): 7–24.
40 455b7-8: ἔτι δ'οἱ τὰς ἐν τῷ αὐχένι φλέβας καταλαμβανόμενοι ἀναίσθητοι γίνονται ("those who have the vessels in the neck compressed lose consciousness"). Interestingly, the noun *katalēpsis* occurs also at the end of Aristotle's De somno (458a29), where Aristotle claims to have demonstrated that "sleep is a *katalēpsis* of the first sense organ to prevent it from functioning" (καὶ τί ἐστιν ὁ ὕπνος, ὅτι τοῦ πρώτου αἰσθητηρίου κατάληψις πρὸς τὸ μὴ δύνασθαι ἐνεργεῖν).

dant syntagm such as δι' ἐπίληψιν ἢ κατάληψιν, especially in consideration of the fact that the same term *epilēpsis* had been used by Aristotle and will be used by Michael, too, after a few lines with an entirely different meaning.

In a way, one is tempted to read the passage just quoted in *direct* connection to, and as the *explicit* premise of, the strong claim that Michael puts forward shortly afterward (53.6 Wendland) when he draws a clear distinction between sleep and epilepsy. If this were the case, however, we should conclude that in commenting on *somn. vig.* 456b11, Michael did not draw a distinction between sleep and an epileptic fit but simply reaffirmed the fact the sleep is *not* a "stoppage." His decision not to draw the distinction appears, on the one hand, to be methodologically problematic, as it implies a rather rough shift from a form of *causal* reasoning (something is or is not a *cause of* sleep) to a *predicative* one (something *is* or *is not* sleep). On the other hand, that decision does not explain why Michael came back to the "technical" meaning of *epilēpsis* as "epileptic attack" in his commentary on *somn. vig.* 457a8, nor why he accepted the analogy between *hupnos* and *epilēpsis* as it is put forward in Aristotle's text, without even recognizing the contradiction between his own claim and Aristotle's.

In reality, when claiming that "it has been often said that sleep is not epilepsy," Michael probably refers to his entire commentary on *somn. vig.* 455a31 (48.29–49.15 Wendland). At the same time, he is most likely referring not specifically to his mention of the word *epilēpsis*, but rather to the negative clause οὐκ ἔστιν ὁ ὕπνος, which occurs four times with reference to the fact that sleep *is not* simply an inactivity of all the senses (48.32), *is not* a fainting (49.2), *is not* whatever state of unconsciousness (49.5), and *is not* the state of insensibility which follows the compression of the vessels of the neck (49.8–10). The use of the syntagma εἴρηται πολλάκις in relationship to the claim that sleep and epilepsy are not the same thing seems to be more prosaically the result of some overlap between phenomena erroneously associated with sleep, which Michael (following Aristotle) has already dealt with, and epilepsy, which Michael (in this case departing from Aristotle's text) is now introducing in his argument.

The question of course is whether such overlapping was intentional or unintentional or, to put it another way, whether Michael just confused *leipopsuchia* with *epilēpsia* or consciously shifted from the former to the latter, so implying that these two phenomena should be considered similar in some respects. Actually, it is my impression that Michael's overall treatment of the Aristotelian analogy between sleep and epilepsy maintains a certain degree of *intentional* ambiguity as a consequence of an exegetical strategy, insofar as Michael, instead of overtly arguing with Aristotle, tries to present his own rejection of the sleep/epilepsy analogy as an entirely Aristotelian move. Put briefly, he seems to adopt a twofold strategy: on the one hand, in explaining on what physiological grounds Aristotle may have compared sleep to epilepsy, Michael entirely accepts Aristotle's theory of the movements of vapors ascending to and descending from the brain. At the same time, he lays greater stress on the idea that epilepsy results from respiratory problems and is thus a kind of suffocation caused by the clash between these vapors.[41] In so doing, Michael, despite hewing closely to Aristotle's account, plays on the ambiguities of that account and ends up developing a physiological account of the epileptic fit in which *no reference at all* is made to the incapacitation of the *sensorium commune*.

On the other hand, Michael's *explicit* distinction between sleep and epileptic attack results from his elaboration of Aristotle's own distinction between sleep and fainting. Michael explains that sleep and fainting differ to the extent that someone who is asleep has "dreams" (*enhupnia*), which he remembers after waking up, while people who have fainted have "visions" (*phantasmata*), which they do not remember after recovering consciousness.[42] Michael's distinction also arises from the contextual association and overlap between *epilēpsia* and *leipopsuchia*, which is not found in Aristotle's text. This conclusion is supported by a passage of Theodorus Metochites' commentary on Aristotle's *De somno*, in which Theodorus discusses the very distinction between sleep and fainting and the difficulties to which such a distinction may give rise. At one point in the commentary we read:

---

41 To get a clear idea of how far Michael was committed to establishing the physiological grounds of a link between epilepsy and blockage of respiration, suffice it to compare the length of his whole argument on the epileptic fit as a consequence of a hindrance to breathing (ten lines: 55.3–12 Wendland, see above) with the length of Aristotle's corresponding passage (just three lines, 457a12–14: ὅταν γὰρ πολὺ φέρηται τὸ πνεῦμα ἄνω, καταβαῖνον πάλιν τὰς φλέβας ὀγκοῖ, καὶ συνθλίβει τὸν πόρον δι'οὗ ἡ ἀναπνοὴ γίνεται.)

42 In commenting on *insomn.* 458b15 (62.10–15 Wendland), Michael grounds the distinction between dreams (*enhupnia*) and visions (*phantasmata*) on a completely different basis: a vision that appears to us while asleep is a dream if we cannot exert any kind of "rational" control over it, that is to say if we are not aware of the "oneiric" nature of the vision we are having; if, while sleeping, we have visions which we rationally recognize as dreams, these visions—Michael explains, respecting Aristotle's own distinction between actual dreams and other forms of visions and sensations which may occur in a sleeping subject—are not *enhupnia* but *phantasmata*. Michael accounts for further differences between dreams and visions at the end of his commentary on *De insomniis*. In discussing 462a8 (75.11–17 Wendland), he refers to two kinds of visions in particular, both of which are mentioned by Aristotle himself, that cannot be ascribed to the category of dreams: visions that appear when someone is ill and even while the senses are still active and visions that actually appear while the subject is asleep but in a phase in which the discerning faculty is not entirely restrained.

> οὕτω γὰρ δὴ καὶ μανίαι καὶ παραφροσύναι καὶ λειποψυχίαι ὕπνοι ἂν εἶεν, ἐπειδὴ κἄν τούτοις ἀργία καὶ κατάκλεισίς ἐστι τοῦ αἰσθητικοῦ. Μᾶλλον δ' ἐν τούτοις ἀπορεῖ μήποτε ὕπνος ἐστὶν ἡ λειποψυχία, ὅτι καὶ φαντάσματα ὁρῶσι καὶ λόγους λέγουσιν οἱ λειποψυχοῦντες ἐνίοτε καὶ κατεσχημένοι πάθεσί τισιν ἐπιλήπτοις, ὃ δὴ καὶ ἐν τοῖς καθεύδουσι γίνεται. (Commentarium in parva naturalia 18.28–32 Drossaart Lulofs)
>
> *Maniai* as well as *paraphrosunai* and *leipopsuchiai* could be forms of sleep, as one can also observe in these phenomena inactivity and blockage of the perceptual activity. Above all, one may ask whether *leipopsuchia* is not sleep. For sometimes those affected by fainting [οἱ λειποψυχοῦντες] *and those who are affected by certain epileptic affections* [κατεσχημένοι πάθεσί τισιν ἐπιλήπτοις] *have visions and pronounce discourses, something that also happens with those who are asleep.*

Here, Theodorus Metochites seems to have explicitly taken into account—and, in fact, to have corrected—Michael's argument (which he certainly bore well in mind and took as a model while writing his own commentary), while implicitly revealing its rationale. For the fact that fainting and certain epileptic affections were thought to bring about similar phenomena (visions and discourses) explains what brought Michael to affirm that "sleep is not an epileptic fit" in a context focused on the distinction between sleep and fainting and the different nature of the visions that may follow these affections.

If this was the line of reasoning that Michael was actually following in introducing a reference to epilepsy, he must have grounded it in a cluster of medical doctrines and shared opinions concerning the appearance of visions in epileptic subjects. In truth, both the Byzantine and the Latin medical traditions had long speculated about the appearance of dream-like visions, mostly dreadful and thus classifiable as nightmares, as a premonitory symptom of an epileptic fit. We first find such an observation in Oribasius' account of nightmares (*ephialtēs*).[43] Oribasius goes even further with the association of this affection with epilepsy, explaining that those who suffer from nightmares experience, while being asleep, the very same things (suffocation, speechlessness, heaviness) that epileptics experience during the day.[44] Along the same lines as Oribasius' account, the chapters on *ephialtēs* in Aëtius of Amida and Paul of Aegina contain observations on nightmares as signs of an impending epileptic fit.[45] Aëtius most clearly defines nightmares as "a threatening symptom and a forerunner [μελέτη καὶ προοίμια] of epileptic fit, mania, or apoplectic attack" (Aëtius Amid. *lib. med.* 6.11 [152.14

---

[43] Orib. *eupor.* 4.117.1 (491.10–16 Raeder).
[44] *Syn.* 8.2 (245.2–8 Raeder).
[45] Both Aëtius and Paul use the same descriptive pattern as Oribasius. But while Paul's text (Paul. Aeg. 3.15 [159.7–11 Heiberg]) is exactly the same as Oribasius', Aëtius (Aëtius Amid. *lib. med.* 6.11 [152.13–26 Olivieri]) adopts Oribasius with some variations.

Olivieri]), while Paul of Aegina, in his chapter on epilepsy, refers to "dreadful visions of dreams" (ἐνυπνίων ὄψεις ταραχώδεις) that precede the epileptic seizures.⁴⁶ We find the same situation on the Latin side in Caelius Aurelianus' treatment of the incubus in his *Tardae passiones*.⁴⁷

Moreover, as the passage of Oribasius that I have just referred to indicates, there is textual evidence showing that medical authors had already described suffocation (*pnix*) as a recurrent manifestation both of the nightmare and of the epileptic fit. This hints that Michael probably took into account materials of medical provenance either in addressing the analogy between sleep and the epileptic fit or in trying to make sense out of an inconsistent reading of the manuscript(s) of the *De somno*, which he may have had at his disposal.

It is also important to note that, in addressing the relationships between affections such as sleep, fainting, and epilepsy and in distinguishing them on the basis of their material cause as well as the different nature of the visions they engender, Michael never raised the question of the "supernatural" or "divine" origin of such visions. In fact, he openly rejects the idea of the divine nature of "prophetic" dreams at the beginning of his commentary on *De divinatione per somnum*.⁴⁸ In so doing, he keeps to Aristotle's strictly physiological approach to this matter.⁴⁹ At the same time, he remains within the "naturalistic" tradition

---

46 See Paul. Aeg. 3.13 (153.2 Heiberg).
47 Cael. Aur. *tard. pass.* 1.3.55 (474 Drabkin): *est autem supradicta passio [incubus] epilepsiae tentatio*. See also *tard. pass.* 1.4.64 (480 Drabkin): *aut plurimus et inanis somnus, insomnia turbata*, listed among the manifestations of the epileptic seizure.
48 *In parva naturalia commentaria* 77.20–78.2 Wendland: ἀλλὰ τὸ μὲν ἔχειν τινὰ σημειῶδες δῆλον, τὸ δὲ μὴ λέγειν αἰτίας εὐλόγους πῶς τοῦτο συμβαίνει, ἀλλ' ἀποφαίνεσθαι, ὅτι θεός ἐστιν ὁ ταῦτα πέμπων, ἀπιστεῖν ἡμᾶς ποιεῖ. ἄλογον γὰρ τὸ τὸν θεόν, ὃς ἑαυτὸν μόνον ὁρᾷ (αὐτὸς γάρ ἐστι τὸ ἀεὶ νοοῦν καὶ ἀεὶ νοούμενον), εἰς τοιαῦτα κατάγειν καὶ κατασπᾶν ("But, on the one hand, it is clear that it [sc. the dream] has some 'sign-like' nature; on the other hand, the fact that it does not express reasonable causes, how this happens, but just shows it, makes us not believe, that God is he who sends these dreams. For it is not reasonable to bring down and reduce the divinity, who looks only at himself [for he is the same who always thinks and is always thought] to such things").
49 Interestingly, Aristotle and Michael agree in denying any "divine" origin for prophetic dreams, even if their respective arguments rest on quite different grounds: Aristotle (463b12–18) claims that, if one admits that prophetic dreams are sent by a god, it becomes impossible to explain why other animals also have dreams and why people of low intellectual capacities are especially subject to prophetic dreams. The underlying rationale of Aristotle's claim is that gods, if they actually played the role of dream-senders, would send dreams only to the best and most intelligent of people. Michael (77.20–78.2 Wendland) turns to a "theological" argument, that of self-knowledge as the only possible knowledge of God, which will be debated within Latin Scholasticism (for example, by Thomas Aquinas in the *Summa Theologica*, Q. 14, article 1):

of Greek medical thought, which, from the Hippocratic *De morbo sacro*, had approached epilepsy as a "natural" disease, that is to say, entirely explained by physical and, one might say, immanent causes.

On the other hand, while Michael most likely draws from the medical tradition when distinguishing between sleep and epilepsy, his overall views on the physiology of sleep seem to be very much in line with Aristotle's own views and therefore not to be affected by the contrast between Aristotelian cardiocentrism and the encephalocentrism characterizing all the medical (Galenic and post-Galenic) theories, which explain both sleep and epilepsy as affections having their origin or their main cause in the brain. This point is worth emphasizing, because it marks a substantial difference between Michael's way of engaging with and rethinking the sleep/epilepsy analogy, on the one hand, and the Arabic and Latin reception of that analogy, on the other.

II The Reception of the Sleep/Epilepsy Analogy in Arabic and Latin Medieval Aristotelianism

What characterizes many medieval and early modern accounts of epilepsy in general and, more specifically, accounts of the Aristotelian analogy between sleep and epilepsy is that these accounts show different degrees of acquaintance with Galenic physiological theories, by which I mean Galen's own theories as well as the theories developed within the frame of Arab Galenism and then transmitted to the Latin world.[50] In some cases, the accounts also show quite

---

insofar as God only looks at and knows himself in his perfection, he cannot be involved in such "human" forms of knowledge.

50 See Mark D. Jordan, "The Construction of a Philosophical Medicine: Exegesis and Arguments in Salernitan Teaching on the Soul," *Osiris* 2 (1990): 42–61; Danielle Jacquart, "The Role of Arabic Medicine in the Medieval West," in Roshdi Rashed and Régis Morelon (eds.), *Encyclopedia of the History of Arabic Science*, 3 vols. (London: Routledge, 1996), 3:963–84; Danielle Jacquart and Françoise Micheau, *La médecine arabe et l'Occident médiéval* (Paris: Maisonneuve et Larose, 1990), 32–45; Charles Burnett and Danielle Jacquart, eds., *Constantine the African and ʿAlī ibn al-ʿAbbās al-Maǧūsī: The Pantegni and Related Texts* (Leiden: Brill 1994); Ivan Garofalo, "La traduzione araba dei compendi alessandrini delle opere del canone di Galeno," *Medicina nei secoli* n.s. 6 (1994): 329–48; Gotthard Strohmaier, "Galen in Arabic: Problems and Prospects," in Vivian Nutton (ed.), *Galen: Problems and Prospects* (London: Wellcome Institute, 1981), 187–96, and idem, "The Uses of Galen in Arabic Literature," in Vivian Nutton (ed.), *The Unknown Galen* (London: Institute of Classical Studies, 2002), 113–20; Luis García Ballester, ed., *Galen and Galenism: Theory and Medical Practice from Antiquity to the European Renaissance* (Aldershot: Ashgate, 2002).

a strong and explicit commitment to fitting certain elements of these theories within an otherwise Aristotelian theoretical, conceptual, and categorical framework.

In other respects, and insofar as they start describing epilepsy as a kind of ecstatic state and the epileptic as a subject liable to prophetic dreams or dreamlike visions,[51] many of these accounts provide an example of how "new" theories, medical conceptions, and even rationalized forms of popular beliefs may have been drawn from specific, culturally determined ways of interpreting, vulgarizing, and contaminating Aristotelian and Peripatetic doctrines with each other and with other non-Aristotelian materials. In this context, one text in particular from the Aristotelian corpus is worth mentioning: *De divinatione per somnum*.[52] In this treatise, Aristotle claims that dreams, while not godsent, indeed have a "daemonic" nature (they are *daimonia*). He goes on to say that melancholics, insofar as their physical constitutions cause them to experience a particularly wide range of visions and dreams and to pass very quickly from one dream to another, are among those people who are likely to experience some kind of "divinatory" dream, in the sense that their dreams are more likely to correctly predict actual events or events that will happen, even in places and times

---

[51] For a general survey of the medieval and Renaissance representations of the epileptic as a prophet, see Temkin, *Falling Sickness*, 148–61.

[52] This treatise, together with Cicero's *De divinatione*, provided medieval and Renaissance doctors and commentators with "scientific" and authoritative models for investigating the nature of prophetic dreams on physiological and psychological grounds and, through substantial processes of reworking and rethinking, it also provided the basis for further exploring the question concerning the "freeing" of the soul from waking bodily activities during sleep. On Aristotle's *De divinatione per somnum*, see van der Eijk, *Aristoteles, De insomniis* (introduction, 52–67 and 87–93; commentary, 251–338) and idem, "Theoretical and Empirical Elements," especially 186–205; see also Ross, *Aristotle, Parva naturalia*, 279–85. The most comprehensive study of Cicero's *De divinatione* and its relationship with Aristotle's *De divinatione per somnum* and the other ancient theories of divination has been provided by François Guillaumont, *Le "De divinatione" de Cicéron et les théories antiques de la divination* (Brussels: Latomus, 2006). Other important contributions are Olaf Gigon, "Cicero und Aristoteles," *Hermes* 87 (1959): 143–62; Mary Beard, "Cicero and Divination," *Journal of Roman Studies* 76 (1986): 33–46; Sebastiano Timpanaro, *Marco Tullio Cicerone, Della divinazione* (Milan: Garzanti, 1988), xxvii–xciv; José Kany-Turpin and Pierre Pellegrin, "Cicero and the Aristotelian Theory of Divination by Dreams," in William W. Fortenbaugh and Peter Steinmetz (eds.), *Cicero's Knowledge of the Peripatos* (New Brunswick: Transaction Publishers, 1989), 220–45; Philip J. van der Eijk, "Aristotelian Elements in Cicero's *De divinatione*," *Philologus* 137 (1993): 223–31; David Wardle, *Cicero on Divination: De divinatione, Book I* (Oxford: Clarendon Press, 2006), 1–43.

remote from the dreamer.[53] It is clear, also, that the Arabic-Latin accounts of epilepsy as a demonic divinatory state stem from a substantial rethinking of this text, as well as from its contamination by a number of medical and other Aristotelian sources. For these accounts ascribe to the epileptics what Aristotle had explicitly ascribed to the melancholics, that is, the capacity to have some kind of self-fulfilling dreams.

This move was not inconceivable within an Aristotelian frame of thought, at least not for those who were acquainted with the characterization of the melancholic type as we find it in ps.-Arist. *pr.* 30.1 (953a10–955a40). In this pseudo-Aristotelian text,[54] epilepsy is included among the diseases that characteristically affect the melancholic subject[55] (earlier claims about the link between melancholy and epilepsy can also be found in the Hippocratic Corpus).[56] The melancholic type is in turn described in *pr.* 30.1 as particularly susceptible to ecstatic states.[57] In other respects, the idea that some "superior," meta-corporeal factor is entirely or partially responsible for epilepsy, as well as for all the bodily and mental states considered to predispose the subject to ecstasy and prophecy,

---

53 See especially 463b12–19. For a thorough discussion of this passage, see van der Eijk, *Aristoteles, De insomniis*, 290–96 (with further bibliography); see also van der Eijk, "Theoretical and Empirical Elements," 190, and idem, "Aristotle on Melancholy," 163.

54 Helmut Flashar, *Aristoteles, Problemata physica* (Berlin: Akademie-Verlag, 1962); Raymond Klibansky, Erwin Panofsky, Fritz Saxl, *Saturn and Melancholy: Studies in the History of Natural Philosophy, Medicine, Religion and Art* (Edinburgh: Nelson, 1964), 15–40; Jackie Pigeaud, *Aristote: l'homme de genie et la mélancolie* (Paris: Rivage, 1988) and van der Eijk, "Aristotle on Melancholy" form the basis for the interpretation of this text and, more generally, for the discussion of Aristotle's views on melancholy. See also Helmut Flashar, *Melancholie und Melancholiker in den medizinischen Theorien der Antike* (Berlin: Akademie-Verlag, 1966); Jackie Pigeaud, *La maladie de l'âme: étude sur la relation de l'âme et du corps dans la tradition medico-philosophique antique* (Paris: Les Belles Lettres, 1981), 122–38; idem, "Prolégomènes à une histoire de la mélancolie," *Histoire Economie Société* 3 (1984): 501–10; idem, *Folie et cures de la folie chez les médecins de l'antiquité Gréco-romaine: la manie* (Paris: Les Belles Lettres, 1987); Walter Müri, "Melancholie und schwarze Galle," *Museum Helveticum* 10 (1953): 21–38; Bennett Simon, *Mind and Madness in Ancient Greece* (Ithaca: Cornell University Press, 1978), 228–37. On the various steps of the reception of the *Problemata*, see Pieter De Leemans and Michèle Goyens, eds., *Aristotle's "Problemata" in Different Times and Tongues* (Leuven: Leuven University Press, 2006) and John Monfasani, "The Pseudo-Aristotelian *Problemata* and Aristotle's *De animalibus* in the Renaissance," in Anthony Grafton and Nancy Siraisi (eds.), *Natural Particulars: Nature and the Disciplines in Renaissance Europe* (Cambridge, Mass.: MIT Press, 1999), 205–47.

55 Ps.-Arist. *pr.* 30.1, 954b28–32.

56 Hipp. *epid.* VI 8.31, 5.356 L. (= 192–94 Manetti-Roselli).

57 See 954a21–26 and 954a34–38. On the connection of melancholy with prophetic states in the Pseudo-Aristotelian *Problema* 30.1 see Klibansky, Panofsky, and Saxl, *Saturn and Melancholy*, 35–36.

largely relied on the strongly "spiritualizing" reading of Aristotle's theories of dreams and divination.[58] Such a reading spread among entire generations of physicians and Aristotelian commentators in the Muslim world, thanks to the Arabic version of *De divinatione per somnum*,[59] a text that was paired in the Western Latin world with theories and beliefs designed to explain a number of bodily and mental states by means of the weighted category of the "daemonic." In this context, one should note that the Latin *daemonicus* not only translates but also radically transforms the meaning of the Greek *daimonios* as we find it in *De divinatione per somnum*. While the Latin adjective *daemonicus* alludes, already in the Imperial Age (in authors like Apuleius, for example), to a system of liminal spaces through which human life and the region of the gods are put in communication[60] and may even qualify as something "sent or caused by a

---

[58] The reception of Aristotle's theory of dreams and divination should in turn be contextualized by the wider debate that arose within Arab Aristotelianism and was further developed by Scholastic Aristotelianism on the nature, powers, and articulations of the intellectual and imaginative faculties. On the articulation of this debate in the Muslim world (and especially on the three key figures of Islamic Philosophy: Alfarabi, Avicenna, and Averroes), see Herbert A. Davidson, *Alfarabi, Avicenna and Averroes on Intellect: Their Cosmologies, Theories of the Active Intellect, and Theories of Human Intellect* (New York: Oxford University Press, 1992), especially 116–23, 340–51 on the Arabic accounts of prophecy. For the reception and further development of this debate within Western Latin Aristotelianism, see Edward P. Mahoney, "Sense, Intellect, and Imagination in Albert, Thomas, and Siger," in Norman Kretzmann, Anthony Kenny, and Jan Pinborg (eds.), *The Cambridge History of Later Medieval Philosophy* (Cambridge: Cambridge University Press, 1982), 602–22; Peter King, "Scholasticism and Philosophy of the Mind: The Failure of Aristotelian Psychology," in Allen I. Janis and Tamara Horowitz (eds.), *Scientific Failure* (Lanham, Md.: Rowman & Littlefield Publishers, 1994), 109–38.
[59] This version is part of the Arabic version of the *Parva naturalia* known as *Kitāb al-Ḥiss wa-l-maḥsūs*, the title of the first treatise (*De sensu et sensibilibus*) being used for the collection as a whole. A copy of this text was discovered in 1985 by Hans Daiber in a seventeenth-century manuscript in the Raza Library, Rampur (India). For an analysis of the contents and the reception of the Arabic version of *De divinatione per somnum*, see Rotraud E. Hansberger, "How Aristotle Came to Believe in God-Given Dreams: The Arabic Version of *De divinatione per somnum*," in Louise Marlow (ed.), *Dreaming Across Boundaries: The Interpretation of Dreams in Islamic Lands* (Boston: Center for Hellenic Studies, 2008), 50–77. Avicenna is one of the authors whose theory of dreams, as part of a broader spiritualized conception of man, reflects an acquaintance with the Arabic *De divinatione* (Hansberger, "How Aristotle," 65–66). On Avicenna's "anthropology," see Verbeke in van Riet and Verbeke, *Avicenna Latinus*, IV–V, 1–73.
[60] Patricia C. Miller, *Dreams in Late Antiquity: Studies in the Imagination of a Culture* (Princeton: Princeton University Press, 1997), 56, defines a daemonic language as "a medium of exchange between an ideal world (of the gods) and a flawed world (of human beings)." See also Jonathan Z. Smith, "Towards Interpreting Demonic Powers in Hellenistic and Roman Antiquity," in Wolfgang Haase (ed.), *Aufstieg und Niedergang der römischen Welt*, II 16.3 (Berlin: De Gruyter, 1986), 425–39, at 429.

demon," the Greek *daimonios* seems rather to have been used by Aristotle in the sense of "beyond the boundaries of human control" for the purpose of qualifying a *natural* agent that escapes human understanding.[61]

1 Avicenna and Averroes

Paradigmatic examples of the reception of Aristotle's claim about the connection between sleep and epilepsy can already be found in the tradition of Arab Aristotelianism and its initial reception in the Latin world. The first is a passage from the Latin translation of Avicenna's *Liber de anima* (4.2 [18 Van Riet]) on the activation of the imaginative faculty and the consequent appearance of prophetic visions during sleep and under some other specific circumstances.[62] The second is in a passage of Averroes' *Compendium Libri Aristotelis de Sompno et Vigilia* (202va66–202vb11 [114–15 Shields]) on the *etiological* link between prophecy and epilepsy.[63] In fact, this is the only passage of his *Compendium*, which also encompasses the themes discussed by Aristotle in *De insomniis* and *De divinatione per somnum*, where Averroes mentions epilepsy. For in Averroes' text, there is no trace of Aristotle's analogy between sleep and epilepsy.

The reference to epilepsy comes at the end of a quite detailed physiological account within which Averroes explains why the inactivation of the external

---

[61] As suggested by van der Eijk, "Theoretical and Empirical Elements," 191–92, the adjective *daimonios* as it is used by Aristotle in *div. somn.* 463b14 must be intended as the opposite of *anthrōpinos*, "human," insofar as it indicates what is beyond human control, "just as it is beyond our control what kind of natural, physiological constitution we have." However, it seems to me that the semantic evolution of the Greek adjective *daimonios* in the Imperial Age (above all in the intellectual milieu of Middle Platonism) tended to mirror that of the Latin *daemonicus*. Proof of this may be the use that Plutarch makes of the adjective *daimonios* to indicate the capacity of every person to construct a framework within which she or he can interpret experience and through which she or he can fill the gap between the human experience and the otherwise inaccessible regions of transcendence (see Miller, *Dreams in Late Antiquity*, 55–56, with textual references).

[62] The Latin translation of Avicenna's *Liber de anima* was carried out in the twelfth century by an *équipe* of translators within the so-called Toledan School of Translators. On this translation and the intellectual milieu in which it was made, see Verbeke in van Riet and Verbeke, *Avicenna Latinus*, I–III, 91–105. On Avicenna's doctrine of prophecy and its reception in the Latin West, see Nikolaus Hasse, *Avicenna's "De anima" in the Latin West: The Formation of a Peripatetic Philosophy of the Soul* (London: Warburg Institute, 2000), 154–73.

[63] On the Arabic translations and commentaries of the *Parva naturalia*, see Francis E. Peters, *Aristoteles Arabus: The Oriental Translations and Commentaries on the Aristotelian Corpus* (Leiden: Brill, 1968), 45–47.

senses, together with the resulting "freeing" of the "internal faculties" (*virtutes interiores*), processes that take place naturally during sleep and under some other circumstances (like deep concentration), are the essential conditions for the appearance of prophetic visions.[64] It is at this point that we find Averroes' claim that:

> Ideo prophetia venit in dispositione simili epilepsie: iste enim virtutes interiores, quando movebuntur forti motu, contrahentur virtutes exteriores, adeo quod forte accidet ex hoc syncopis. (202vb 8–11 [115 Shields])
>
> Prophesy accompanies a state such as epilepsy. For, when the internal senses are put in motion by a strong movement, the external senses contract themselves, so that a syncope results.

This claim relies entirely on the post-Aristotelian distinction between external "corporeal" senses (the five senses, to put it roughly) and internal, "spiritual," and "cerebral" ones (according to the most basic classification, these comprise the imaginative and the cogitative faculties, as well as the faculties of memory and recollection),[65] which work with the materials coming from external sensation (at least this is the task of the lower forms of imagination) but are activated

---

[64] Davidson, *Alfarabi, Avicenna, and Averroes*, 344, has summarized Averroes' argument in this way: "in the waking state, perceptions enter from the outside and ascend through the hierarchy of internal faculties, or internal senses, of the soul. At the top, they are processed by the imaginative faculty and transmitted to the memory. In the dream state, a man seems to perceive with his five senses, although no sense objects are present outside the soul. Since the percepts do not originate from without, they must originate from within and travel in the opposite direction. Memory is not the initiator, since memory is quiescent in sleep. The inner sense that remains awake and in constant motion when the other faculties are asleep is the imaginative faculty, and it must be the faculty responsible for dreams. The imaginative faculty recovers impressions from the memory, recombines them, and projects them out through the sense faculties, so that the dreamer seems to perceive sense objects, although none are in actuality present outside the soul." The concept of *virtutes interiores* is post-Aristotelian (see Harry A. Wolfson, "The Internal Senses in Latin, Arabic, and Hebrew Philosophic Texts," *The Harvard Theological Review* 28 [1935]: 69–133, at 69).

[65] There is no doubt that these "internal senses" correspond to the *phantastikon*, the *dianoētikon*, and the *mnēmoneutikon* as discussed by Aristotle, respectively in *de an.* 3.3 and 3.4–6, and in *De memoria*. However, nowhere in the Aristotelian Corpus are these faculties defined as a *unitary* system of faculties distinguished from the external senses. For a thorough analysis of the various classifications of the internal senses provided by Latin, Arabic, and Hebrew philosophical texts, see Wolfson, "The Internal Senses."

by an "external" meta-corporeal transcendent agent, namely the active intellect.[66]

As concerns the terms "spiritual" and "cerebral," one can explain their use by the fact that the faculties to which these terms are applied reside *within* the *brain*—insofar as it is in the brain that their activity becomes evident—and operate *without* corporeal organs,[67] but rather through various kinds and degrees of "spiritual forms," as Averroes makes clear in a passage of his commentary on Aristotle's *De sensu*:[68]

> And the common sense, when it receives the form, gives it back to the informing agent, that is the imaginative power (*virtus imaginabilis*); and then this informing agent will receive it [sc. the form] through a more spiritual kind of reception (*receptione magis spirituali*); at that point the form will be in the third level. For forms have three levels, only one of which is corporeal (*corporalis*); the second is in the common sense and is spiritual (*spiritualis*); the third is in the imaginative faculty and is *more* spiritual (*magis spiritualis*). (*Compendium Libri Aristotelis De sensu et sensato* 193rb52–59 [= 38 Shields])

In order to understand what these movements of the internal senses consist in, one must refer to Averroes' psychological account of dreams, which provides the general framework for the discussion of prophecy. For Averroes explains dreams as a kind of movement, and more precisely as a *reverse* movement of the intelligible forms (*intentiones*). When someone is awake, the sense objects, after being

---

[66] On the Arabic theories of the active intellect, see Davidson, *Alfarabi, Avicenna, and Averroes*. For an in-depth discussion of Averroes' account of the human intellect (which was subject to substantial reassessments in the course of Averroes' life, so that scholars currently speak of an initial, a transitional, and a final account), see Richard C. Taylor, *Averroes (Ibn Rushd) of Cordoba: Long Commentary on the "De anima" of Aristotle* (New Haven: Yale University Press, 2009), xix-lxxvi. See also Michael Blaustein, "Averroes on Imagination and the Intellect," PhD diss., Harvard University, 1984.

[67] Wolfson, "The Internal Senses," 70. See Averroes, *Colliget* 1.20: *Quamquam imaginatio, cogitatio, recordatio, conservatioque neque partes, neque membra sint, habent nihilo minus in cerebro sedes, in quibus earum functiones sunt evidentiores* ("Although imagination, thinking, recollection and memory are neither parts nor limbs, nevertheless they have their seat in the brain, where their functions are more evident").

[68] For a thorough discussion of Averroes' theory of the internal senses, see Helmut Gätje, "Die 'inneren Sinne' bei Averroes," *Zeitschrift der deutschen morgenländischen Gesellschaft* 115 (1965): 255–93. A point raised by Gätje, which is worth stressing, concerns the distinction between the categories of "corporeal" and "spiritual" in Averroes. The German scholar argues that, within an apparently rigid dichotomous framework, Averroes actually describes the whole system of the internal senses as a progressive process of spiritualization, decorporalization, and universalization of the forms, a process that starts already in the sensory organs and culminates in the highest forms of intellection.

caught by the external senses, first affect the common sense and before arriving in the memory and the faculty of recollecting, pass through the imaginative and the cogitative faculty. In sleep, explains Averroes, this movement must take place in the opposite direction, as the external senses are inactive.[69] But since both memory and the cogitative faculty are also inactive during sleep, it must be the imaginative faculty, whose activity is continuous, that serves as the source and center of this movement.[70] For imagination passes from one *intentio* to another—running sometimes into a form held within the common sense, sometimes into a form "stored" in the *rememorativa virtus*—and, after "processing" this form and shaping it as an "imagined form," sends it back to the common sense, which in turn transmits this form to the individual senses. This, concludes Averroes, happens not only during sleep but may also happen in the case of fear or illness.[71]

Now, Averroes' whole theory of the internal senses incorporates materials of various provenance, above all doctrinal elements of the Galenic tradition and psychological theories that had been developed within Neoplatonic Aristotelianism.[72] For, on the one hand, notions like "*spiritus*" and "*forma spiritualis*" should be understood in light of a century-long process through which both Aristotle's concept of soul and Galen's notion of "psychic *pneuma*" were gradually spiritualized. On the other hand, Averroes localized imagination, cogitation, and memory in the anterior, middle, and posterior ventricles of the brain, respectively.[73] This localization, which in a way provides a concrete qualification of the "back to front" movement of the internal senses during sleep and sleep-like states, follows an anatomophysiological tripartition that at the time of

---

69 *Compendium Libri Aristotelis De Sompno et Vigilia* 201vb48–54 (96–97 Shields).
70 *Compendium Libri Aristotelis De Sompno et Vigilia* 201vb55–58 (97 Shields).
71 *Compendium Libri Aristotelis De Sompno et Vigilia* 201vb59–63 (97–98 Shields), and 202ra2–15 (98–99 Shields). On Averroes' explanation of the "psychophysiology" of dreams, see Gätje, "Die inneren Sinne," 286–87.
72 On the sources of Averroes' commentaries (with a special focus on Averroes' *Long Commentary on De anima*), see Taylor, *Averroes, Long Commentary*, lxxvi–xcvi. On the influence of Neo-platonism on the Arab reception of Aristotle, see Cristina D'Ancona, "From Late Antiquity to the Arab Middle Ages: The Commentaries and the Harmony between the Philosophies of Plato and Aristotle," in Honnefelder et al., *Albertus Magnus*, 45–69.
73 *Colliget* 1.20. The same theory of the internal senses is to be found in the *Kitab Kāmil aṣ-ṣinā'a aṭ-ṭibbiyya* ("*The Complete Art of Medicine*") of the Persian physician 'Alī ibn al-'Abbās al-Maǧūsī (died 982–994): see Jacquart and Micheau, *La médecine arabe*, 127 and E. Ruth Harvey, *The Inward Wits: Psychological Theory in the Middle Ages and Renaissance* (London: Warburg Institute, 1975).

Averroes was already well-established in the Arabic medical and philosophical tradition.

Although the tripartition was usually connected—by ar-Rāzī (ninth to tenth century), for example—to Galen's theory of the ventricular system[74] (Averroes wants to trace it back to Aristotle, too[75]), it was in fact first formulated within a Neoplatonic milieu and, more specifically, by Nemesius of Emesa, whose treatise *On the Nature of Man* indicates its author's deep acquaintance with, and reworking of, Galenic and other post-Galenic medical materials.[76]

In truth, a proper theory of the ventricular localization is not to be found in Galen, whose doctrine of the ventricular system represents the brain's functioning in much more plastic terms.[77] One could say, however, that ventricular localization is a "doctrine that Galen hints at rather than specifies"[78] or, more cautiously, that Galen's "delineation of the ventricular system provided a suitable template from which to elaborate an increasingly formalized and specialized role for each ventricle."[79] One Galenic passage that served as a template for later theories about the ventricular localization of the faculties—the very passage to which ar-Rāzī refers—is Book 3, Ch. 9 of *De locis affectis*, where Galen makes the ventricles of the brain the seat of the *pneuma* and orders them in a kind of descending hierarchical order from the posterior to the two anterior ventricles:

---

74 See Wolfson, "The Internal Senses," 73.

75 Cf. his *Commentarium magnum* on *de an.* 3.6 (Taylor, *Averroes, Long Commentary*, 331): "Some people were uncertain about what was said (namely, that the intellect does not have an instrument) because it was said that the imaginative power is in the anterior of the brain, the cogitative power in the middle, and the power of memory in the posterior. This was not only said by physicians but is said in *Sense and Sensibilia*."

76 For Nemesius' theory of localization, see *nat. hom.* 56.2 (imagination), 64.2 (sensation), 68.12 (thought), 69.18 (memory), 69.21–70.11 (empirical evidence for the overall theory of localization). For a discussion of the sources and/or the parallels of Nemesius' theory and of its connection with Galen's ideas on the ventricular system, see Robert W. Sharples and Philip J. van der Eijk, *Nemesius, On the Nature of Man* (Liverpool: Liverpool University Press, 2008), 121 n. 607. For a general study of the ancient theories of the cerebral ventricles, see Peter Grunert, "Die Bedeutung der Hirnkammer in der antiken Naturphilosophie und Medizin," *Antike Naturwissenschaft und ihre Rezeption* 12 (2002): 151–82.

77 On Galen's theory of the ventricular system, see Julius Rocca, *Galen on the Brain: Anatomical Knowledge and Physiological Speculation in the Second Century AD* (Leiden: Brill, 2003), 113–67 (on the anatomy of the ventricular system) and 245–47 (on the reception of Galen's doctrine of the ventricular system and its further development in the theories of ventricular localization).

78 Robert B. Todd, "Philosophy and Medicine in John Philoponus' Commentary on Aristotle's *De anima*," *Dumbarton Oaks Papers* 38 (1984): 103–10, at 107.

79 Rocca, *Galen on the Brain*, 246.

τοῖς γὰρ ἐκ τῆς ἀνατομῆς φαινομένοις ἀκολουθοῦσιν ἡμῖν εὔλογον ἐφαίνετο, τὴν μὲν ψυχὴν αὐτὴν ἐν τῷ σώματι τοῦ ἐγκεφάλου κατῳκῆσθαι, καθ' ὃ καὶ τὸ λογίζεσθαι γίγνεται, καὶ ἡ τῶν αἰσθητικῶν φαντασιῶν ἀπόκειται μνήμη· τὸ πρῶτον δ' αὐτῆς ὄργανον εἰς ἁπάσας τὰς αἰσθητικάς τε καὶ προαιρετικὰς ἐνεργείας εἶναι, τὸ κατὰ τὰς κοιλίας αὐτοῦ πνεῦμα, καὶ μᾶλλόν γε κατὰ τὴν ὄπισθεν· οὐ μὴν οὐδὲ περὶ τῆς μέσης κοιλίας ἀπογινώσκειν προσῆκεν ὡς οὐ κυριωτάτης· πολλὰ γὰρ εὔλογα καὶ πρὸς ταύτην ἡμᾶς ἄγει, καθάπερ γε τῶν ἐμπροσθίων δυοῖν ἀπάγει. (de loc. aff. 3.9, 8.174 K.)

For it seemed reasonable to us, who follow what appears from dissection, that the soul itself is settled in the corporeal substance of the brain, in which also reasoning takes place and the memory of the perceptual images is stored. The first instrument of this [the soul] for all the perceptual and motor activities is the *pneuma* contained in its cavities, and above all in the back cavity. Indeed, it would not be convenient to depart from this judgment concerning the middle ventricle, as if it were not the most important one. For many reasonable arguments make us incline toward this ventricle, just as they lead us away from the two anterior ventricles.[80]

This is the same chapter of the *De locis affectis* in which Galen defines epilepsy as a convulsion of the whole body, accompanied by damage to thought and perception. It is also where he explains the epileptic fit as the result of the shaking of the origin of the nerves in reaction to the accumulation of a thick humor and the blockage of the psychic *pneuma* in the ventricles of the brain:

Ἀλλὰ καὶ ἡ ἐπιληψία σπασμός ἐστιν ἁπάντων τῶν τοῦ σώματος μορίων, οὐ συνεχὴς ὡς ἐμπροσθότονός τε καὶ ὀπισθότονος καὶ τέτανος, ἀλλ' ἐκ διαστημάτων χρόνου γιγνόμενος· οὐ μόνον δὲ τούτῳ διενήνοχεν τῶν εἰρημένων σπασμῶν, ἀλλὰ καὶ τῇ βλάβῃ τῆς διανοίας καὶ τῶν αἰσθήσεων· ᾧ καὶ δῆλον ὡς ἄνω πού κατ' αὐτὸν ἐγκέφαλον, ἡ τούτου τοῦ πάθους ἐστὶ γένεσις. ἐπειδὴ δὲ καὶ ταχὺ παύεται, κατὰ τὰς κοιλίας αὐτοῦ μᾶλλον εὔλογόν ἐστι παχὺν χυμὸν ἐμφράττοντα τὰς διεξόδους τοῦ πνεύματος ἐργάζεσθαι τὸ πάθος, ἑαυτὴν κλονούσης τῆς ἀρχῆς τῶν νεύρων ὑπὲρ τοῦ διώσασθαι τὰ λυποῦντα. (de loc. aff. 3.9, 8.173 K.)

But also epilepsy is a spasm of all the parts of the body, not a continuous one like emprosthotonus or opisthotonus and tetanus, but one which occurs at intervals of time: and it differs not only in this respect from the spasms that I just mentioned, but also because of the damage to the faculty of thought and to the senses. With this, it is also clear that the origin of this affection is in the upper part, somewhere in the brain. And since it also comes to an end quickly, it is very reasonable that a thick humor obstructing the passages of *pneuma* in the brain ventricles produces this affection, when the origin of the nerves shakes itself to push away those harming materials.

---

80 An Arabic translation of Galen's *De locis affectis* was provided by Ḥubayš ibn al-Ḥasan al-Aʿsam by the mid-ninth century A.D. (before 857/858). An anonymus Latin translation of this Arabic translation of *De locis affectis* has been transmitted by several manuscripts. See Manfred Ullmann, *Die Medizin im Islam* (Leiden: Brill, 1970), 41–42 and Ivan Garofalo, "La traduzione araba del *De locis affectis* di Galeno," *Studi classici e orientali* 45 (1995): 13–63.

It therefore seems plausible, at least in my view, that this chapter of Galen's *De locis affectis* or other sources which incorporated these Galenic materials may have provided part of the basis on which Averroes built the connection between prophecy and epileptic fits in the passage cited above from his *Compendium*. Support for this argument can be found not only in the association between epilepsy and prophetic vision as products of the imaginative faculty (the background of this association is the ventricular localization of both phenomena), but also in Averroes' physiological explanation of the loss of consciousness in the epileptic seizure, insofar as this is explained as the final outcome of the violent movement suffered by the internal senses during the attack. I say "in some regard" as the parallel between Averroes' and Galen's explanations can be drawn only up to a certain point. For in Averroes the loss of consciousness seems to be the direct consequence of the strong movement of the internal senses, while Galen distinguishes the cause of the convulsions (the shaking of the nerves) from that of the loss of consciousness (the blockage of the psychic *pneuma* in the brain's ventricles). Moreover, while the movement to which Galen refers is clearly physical, things are definitely more complicated in the case of Averroes. For it is not clear whether Averroes is referring to a corporeal movement of the animal spirits, in which case the similarity with Galen's text would be stronger,[81] or whether he is referring to the movements of the internal senses as incorporeal "seelische Bewegungen," to use an apt expression of H. Gätje, "which in some way constitute the form of determined bodily reactions of a physical kind."[82]

The acquaintance with doctrines elaborated within the Galenic tradition and the attempt to build an internally consistent theoretical framework from Aristotelian and Galenic materials are features that we can identify elsewhere in Averroes' epitome of the *Parva naturalia*, as well as, of course, in his collection of medical writings, the *Kitāb al-Kullīyāt* (lat. *Colliget*).[83] For, as Mark Jordan has

---

[81] Averroes' concept of *spiritus animalis* as a substance flowing from the brain into the nerves and providing the "physiological medium" of the soul processes is clearly modeled on Galen's notion of *pneuma psychikon* (see Gätje, "Die inneren Sinne," 290–91).

[82] Gätje, "Die inneren Sinne," 275: "Obwohl das Vorstellungsvermögen in gewisser Weise auf ein im Gehirn liegendes körperliches Substrat (vorderer Hirnventrikel) angewiesen ist, darf man diese Vorgänge nicht als rein materielle Prozess ansehen, handelt es sich doch um seelische Bewegungen, die gewissermaßen die Form von bestimmten körperlichen Reaktionen stofflicher Art bilden."

[83] The title of the Latin translation of this collection of medical essays—published as the tenth volume in the Venetian Latin edition of Aristotle's works of 1527; it first appeared in Padua in the second half of the thirteenth century (see Nancy Siraisi, *Arts and Sciences at Padua: The "Studium" of Padua before 1350* [Toronto: Pontifical Institute of Medieval Studies, 1973], 115)—elo-

pointed out, Averroes incorporates Galen (or, better to say, doctrines that he took to be Galen's) fully as an interlocutor in his reading of Aristotle. For this reason, "whoever reads Averroes on Aristotle reads about Galen."[84]

This approach sometimes takes the form of an attempt to explain aspects of Aristotelian doctrines through Galenic categories and vice versa; at other times, it materializes as an effort to reconcile different or even contrasting aspects of Aristotle's and Galen's doctrines; many other times, it results in a clearer stand *pro* Aristotle and *contra* Galen.[85] A perfect example of the approach is provided by Averroes' attempt to give the brain a clearer and stronger role in the physiology of sleep and wakefulness by accounting for these activities as the result of two concurrent and complementary causal principles, namely the heart and the brain.[86] It seems, in fact, that, in clarifying the role of the brain in sleep, Averroes was trying to adjust the Aristotelian account, in which the brain's role is limited to one circumscribed phase of the sleep induction and is not part of the very definition of sleep,[87] to the criticisms that Galen had made against it in *de sympt. caus.* 1.8 (see above, n. 17), criticisms with which Averroes must have been very well acquainted, as he himself wrote a commentary on this Galenic treatise.[88]

---

quently expresses Averroes' aims: *Collectaneorum de re medica Averrhoi philosophi post Aristotelem atque Galenum facile doctissimi sectiones tres*. On Averroes' medical thought, see Paul Mazliak, *Avicenne et Averroès: médecine et biologie dans la civilisation de l'Islam* (Paris: Vuivert, 2004); Miguel Cruz Hernández, "Esencia y estructura de la medicina de Averroes," in C. Álvarez de Morales and E. Molina López (eds.), *La medicina en al-Andalus* (Granada: El Legado Andalusí, 1999), 237–54.

84 Mark D. Jordan, "The Disappearance of Galen in Thirteenth-Century Philosophy and Theology," in Albert Zimmermann and Andreas Speer (eds.), *Mensch und Natur im Mittelalter*, 2 vols. (Berlin: De Gruyter, 1991–1992), 2:703–17, at 708.

85 This is the case, for example, with Averroes' criticisms of Galen's physiology of respiration as we find it in *Colliget* 1.19 (see J. Christoph Bürgel, *Averroes "contra Galenum." Das Kapitel von der Atmung im Colliget des Averroes als ein Zeugnis mittelalterlich-islamischer Kritik an Galen* (Göttingen: Vandenhoeck und Ruprecht, 1967). On Averroes' criticisms of Galen's doctrines, see also Taylor, *Averroes, Long Commentary*, xciii–xciv.

86 See *Compendium De sompno et Vigilia* 201rb54–65 (84–85 Shields), and 201va43–51 (90–91 Shields). Cf. the chapter on sleep of Averroes' *Colliget* (1.21) and the treatment of Aristotle's and Galen's views on the main center of perception in *Colliget* 1.11.

87 On the "functional marginality" of the brain in Aristotle's psychophysiology, see Tullio Manzoni, *Aristotele e il cervello: Le teorie del più grande biologo dell'antichità nella storia del pensiero scientifico* (Rome: Carocci, 2007), 98–105.

88 Gerhard Endress, "Averrois Opera," in Gerhard Endress and Jan A. Aertsen (eds.), *Averroes and the Aristotelian Tradition: Sources, Constitution, and Reception of the Philosophy of Ibn Rushd (1126–1198)* (Leiden: Brill, 1999), 339–81, at 375. As has been suggested to me by Oliver Overwien, whom I thank, it is also possible that some of the contaminations between Aristotelian and

Let us move now to the discussion of the imaginative and the prophetic faculties in Avicenna's *Liber de anima*.[89] In a passage from the second chapter of the fourth part—a discussion of the activities of the internal senses, sleep and wake, dreams, and the faculty of divination—Avicenna claims that there are four categories of physically or mentally impaired subjects whose imaginative faculty is released in such a way that they may see things and hear sounds as vivid and real as those that they actually see and hear under normal conditions (*in tempore salutis*): the epileptic (*epilepticus*), someone who is frightened (*perterritus*), someone who is weak (*dissolutus*), and someone in a deep sleep (*soporatus*).

Now this statement can be seen as an Aristotelian claim in two respects. First, it is entirely consistent with Aristotle's conception of sleep as a somewhat "defective" condition in comparison with the waking state, which in a teleologically oriented explanatory framework represents the normative physiological state. Second and even more interestingly, it is a kind of rewriting of Aristotle's own claim in *de an.* 3.3, 429a5–9 that visions produced by the imaginative faculty persist in us and resemble sensations and that men may therefore be guided by what they imagine, "because their mind is temporarily clouded over by *emotion*, or *disease*, or *sleep*" (τὰ δὲ διὰ τὸ ἐπικαλύπτεσθαι τὸν νοῦν ἐνίοτε πάθει ἢ νόσῳ ἢ ὕπνῳ).[90]

It is interesting to note, however, that the parallelism between Aristotle's and Avicenna's claims is only partial, in that Avicenna does not refer to the generic category of the "ill person," but introduces a further qualification insofar as he mentions two separate kinds of people: those who are "weak" or "physically impaired" (in this case *dissolutus* should be understood as a synonym of *infirmus* without any moral connotation, as it translates the Arabic ḍaʿīf, "weak,"

Galenic materials that we find in Averroes' *Compendium* of the *Parva naturalia* should not be ascribed directly to Averroes but should be assigned to the Arabic adaptation of the *Parva naturalia*, which Averroes used as a source text for his commentary and which displays a certain receptivity to the Galenic doctrines. A critical edition with English translation and commentary by R. Hansberger, Akademische Rätin am Lehrstuhl VI für spätantike und arabische Philosophie der Universität München, is in preparation. Hopefully, this edition will also make it possible to cast fresh light on whether and to what extent the Arab translator of the *Parva naturalia* was acquainted with Galenic doctrines and to what degree his translation was actually permeated by such doctrines.

89 On Avicenna's account of prophecy, see Davidson, *Alfarabi, Avicenna*, 116–23. For a general discussion of Avicenna's treatment of imagination in the wider context of his "science of the soul," see Verbeke in van Riet and Verbeke, *Avicenna Latinus*, I–III, 51–90.

90 Aristotle's *De anima* had been translated into Arabic by Ḥunayn ibn Isḥāq in the ninth century. For a comprehensive study of Avicenna's reception and integration of the Aristotelian tradition, see Dimitri Gutas, *Avicenna and the Aristotelian Tradition* (Leiden: Brill, 1988).

"infirm")[91] and "the epileptics." The second category hints here, as will be clear in a while, at the more general category of the mentally ill subjects.

In fact, the reference to a specific affection testifies to two distinct steps in the reception and eventually the transformation of this specific point of Aristotle's theory of imagination. For Avicenna does mention the "insane person" (the original text reads *al-insānu l-maǧnūnu*),[92] but by chance, intent, or misunderstanding his Latin translator rendered this as "the epileptic."[93] On the one hand, and insofar as a centuries-long medical tradition had seen both epilepsy and the wide-ranging ensemble of mental diseases as either affecting or originating in the brain, the choice to render the Arabic *al-insānu l-maǧnūnu* with the Latin *epilepticus* was on the whole a move consistent with Avicenna's encephalocentric account of the imaginative faculty.[94] More generally, the translation is consistent with his psychology, which stemmed directly from a deep acquaintance with Galen's account of the nervous system and with the psychological theories that were transmitted as part of the Galenic tradition, even if their origin was often post-Galenic and Neoplatonic.[95] It must be noted, however, that *epilep-*

---

91 See Hans Wehr, *A Dictionary of Modern Written Arabic (Arabic-English)* (Wiesbaden: Harrassowitz, 1979), 633. Thanks to Matteo Martelli and Oliver Overwien for the help with Avicenna's Arabic text.
92 See Wehr, *Dictionary*, 164.
93 As noted by Jacquart and Micheau, *La médecine arabe*, 42, there are two Arabic translations of the Greek *epilēpsia*: the first – *ibīlimsiyā* – is a simple transliteration from the Syriac *'īfī-limsiyā*, which is in turn a transliteration of the Greek word. Over the course of time, the transliterated form tended to be replaced by the Arabic term *aṣ-ṣar'*, which is derived from the verb "to drop down."
94 Avicenna, *Canon* 1.8, §559 (Laleh Bakhtiar, *Avicenna, Canon of Medicine* [Chicago: KAZI Publications, 1999], 164): "Imagination. This preserves the percepts of the composite sense after they have been so conjoined and holds them after the sense impressions have subsided. Common sense is the recipient and imagination is the preserver. The proof of this belongs to the philosopher. The main place of the activities of these two faculties is the anterior part of the brain." On the localization of the internal senses in Avicenna's theory, see Verbeke in van Riet and Verbeke, *Avicenna Latinus*, IV–V, 49–52.
95 Avicenna, *Canon* 1.5, §389–90 (Bakhtiar, *Avicenna*, 108–109): "There are direct as well as indirect advantages of the nerves. The direct advantage is that the brain receives sensation and controls movement through them. The indirect advantage is that the nerves give strength to the muscles and thus to the body as a whole. ... Nerves arise from the brain and spread out as a network of fine branches in the tissues and the superficial skin. Some arise directly from the brain and are known as nerves of the brain, while others arise indirectly, emerging from the spinal cord, which runs downward in the vertebral column. The nerves which arise directly from the brain are responsible for movements of the head, face and internal organs, while spinal nerves supply the remaining organs. Galen has pointed out that it is one of God's great favors that He has especially protected the nerve which supplies the internal organs." On the Galenic

ticus is neither the most obvious nor the most literal rendering of *al-insānu l-maǧnūnu*, even in an Avicennian context. For *al-insānu l-maǧnūnu* is, in Avicenna, a very generic term indicating mental derangement, which can be applied to the definition, or the description of the effects, of a wide spectrum of affections in the brain. Epilepsy is just one of these affections, and perhaps not the most characteristically related to a state of madness. For in Avicenna's *Canon*, the treatment of epilepsy is included in a section devoted to the injuries of the brain that have an effect on voluntary movement. Psychic disturbances are also mentioned, but they are not presented as the most prominent and distinctive feature of epilepsy, as we see, on the contrary, in the case of melancholy.[96]

How can we try to make sense out of this translation, then? An answer to this question may be found by looking at the cultural background of the translator of Avicenna's *Liber de anima*, who was most likely the archdeacon of Cuéllar and the first director of the Toledan School of Translators, Dominicus Gundissalinus.[97] For his choice to render the Arabic *al-insānu l-maǧnūnu* with *epilepticus* is consistent with—and might in fact have been influenced by—the distinction between three forms of diseases of the brain—namely epilepsy, melancholy, and mania—drawn in the chapter on chronic affections in Isidore of Sevilla's *Origines* (ca. 560–636 A.D.), that is to say in "the" medieval encyclopedic text *par excellence:* a text, which in its medical and anthropological sections —respectively Books 4 and 11—combines and rearranges materials coming from a number of ancient sources (especially manuals, compendia, and other technical literature) and which also served as a fundamental source for most of the late medieval investigations into various aspects of the *philosophia naturalis* (includ-

---

and Neoplatonic influences on Avicenna's theory of perception, see Simo Knuuttila, "Aristotle's Theory of Perception and Medieval Aristotelianism," in Simo Knuuttila and Pekka Kärkkäinen (eds.), *Theories of Perception in Medieval and Early Modern Philosophy* (Berlin: Springer, 2008), 1–22, at 9–11.

96 See Michael W. Dols, *Majnūn: The Madman in Medieval Islamic Society* (Oxford: Clarendon Press, 1992), 85–86.

97 On the composition and working method of the *équipe* of translators of Avicenna's *Liber de Anima* and on the identity of the Latinist who contributed to this translation, see van Riet and Verbeke, *Avicenna Latinus*, I–III, 91–103 and Charles Burnett, "Translating from Arabic into Latin in the Middle Ages: Theory, Practice, and Criticism," in Steve G. Lofts and P. W. Rosemann (eds.), *Editer, traduire, interpréter: essais de méthodologie philosophique* (Louvain-la-Neuve: Institut Supérieur de Philosophie, 1997), 55–78, at 65–67. See, more generally, on the so-called "School of Toledo" and the institutional context of Arabic-Latin translations Charles Burnett, "The Institutional Context of Arabic-Latin Translations of the Middle Ages: A Reassessment of the School of Toledo," in Olga Weijers (ed.), *Vocabulary of Teaching and Research Between Middle Ages and Renaissance* (Turnhout: Brepols, 1995), 214–35.

ing the inquiry into human nature), from the twelfth century up to the fifteenth century;[98] a text, moreover, with which Gundissalinus was very well acquainted (four codices of Isidorus' *Origines* were available at Gundissalinus' times in Toledo's *Biblioteca Capitular*) and which he made use of as a reference tool in his work as a translator and in his own thinking.[99]

In drawing a distinction between various chronic diseases of the brain, Isidore defines epilepsy as an affection *specific to the imaginative faculty*, arguing that "epilepsy occurs in the imagination (*phantasia*), melancholia in reason, and folly (*mania*) in memory" (*epilemsia autem in phantasia fit; melancholia in ratione; mania in memoria*).[100] Now, this distinction is functional and spatial at the same time, as it clearly implies the doctrine of the cerebral ventricles, according to which imagination is located in the anterior ventricle, reason in the middle ventricle, and memory in the posterior one.[101] But, above all, it establishes the principle according to which a "mental" disease having its origin in the "imaginative" part of the brain takes shape in the form of epilepsy. This principle could therefore provide a good explanation of why the translator of Avicenna's *Liber de anima* thought that, in this specific context where imagination is at stake, the Latin *epilepticus* was the best possible translation of an Arab word meaning "insane."

---

[98] On the main features and the sources of Isidore's anthropology, see Fabio Gasti, *L'antropologia di Isidoro: le fonti del libro XI delle "Etimologie"* (Como: New Press, 1988) and *Isidoro di Siviglia, Etimologie, Libro XI: de homine et portentis* (Paris: Les Belles Lettres, 2010), x-xxxii (see especially xii–xiii on Isidore's medical sources); more specifically on the medical sources of Isidore see Klaus-Dietrich Fischer, "Neue oder verlachlässigte Quellen der *Etymologien* Isidors von Sevilla (Buch 4 und 11)," in Arsenio Ferraces Rodríguez (ed.), *Isidorus medicus: Isidoro de Sevilla y los textos de medicina* (La Coruña: Universidade de Coruña, 2005), 131–74. On the reception of Isidore of Sevilla's *Origines* in the late Middle Ages, see Jacques Elfassi and Bernard Ribémont, eds., *La réception d'Isidore de Séville durant le Moyen Âge Tardif (XIIe-XVe s.)* (Paris: special issue of the *Journal of Medieval and Humanistic Studies* 16, 2008).

[99] Gundissalinus was not only a translator but also an original thinker. For he wrote a treatise *On the Soul* taking Avicenna's *De anima*, which he himself had translated, as a model (see Hasse, *Avicenna's "De anima,"* 13–18). On Gundissalinus' acquaintance with and use of Isidorus (especially in his treatise *De divisione philosophiae*), see Alexander Fidora, "La Recepción de San Isidoro de Sevilla por Domingo Gundisalvo (ca. 1110–1181): Astronomía, Astrología y Medicina en la Edad Media," *Estudios Eclesiásticos* 75 (2001): 663–77 and José Martínez Gázquez, "Isidoro de Sevilla y la medicina en los enciclopedistas hispanos: D. Gundisalvo y Juan Gil de Zamora," in Ferraces Rodríguez, *Isidorus medicus*, 215–25, at 218–22.

[100] Isid. *orig*. 4.7.9.

[101] See above, n. 76. Isidorus must have drawn this theory directly from Nemesius of Emesa and Augustine of Hippo, as suggested by Angelo Valastro Canale, *Isidoro, Etimologie o origini*, 2 vols. (Turin: UTET, 2004), 1:364 n. 22.

In light of these remarks, this passage, both in its original version and in its Latin translation, can be taken as evidence of the fact that Avicenna's theory of the internal senses resulted from a deep contamination of elements belonging to the Aristotelian tradition by elements belonging to the medical-Galenic tradition.[102] For the shift from "insane" to "epileptic" is not entirely arbitrary or inexplicable even from an Aristotelian point of view. In fact, as Oswei Temkin has already suggested, "since Aristotle himself had likened the epileptic seizure to sleep, one of the conditions in which the mind failed to restrain the imaginative faculty, the Latin translation, consciously or not, remained within the sphere of Aristotelian thought."[103]

Furthermore, the Arab *maǧnūn* can mean the "insane" but also the "lunatic" and the "possessed";[104] epilepsy was acceptable as a disease that could lead to possession-like ecstatic states marked by visions. This is what we have already seen in Averroes and what one may conclude from many other textual sources —mainly twelfth-century Latin translations of Arab texts dating from the ninth to the eleventh century—containing evidence of a belief in a link between epileptic seizures, "ecstatic" or "demonic" states, "melancholic nature," and "divination." For example, the Arab physician Abulqasis (Abū l-Qāsim Ḫalaf ibn al-ʿAbbās az-Zahrāwī, 936–1013 A.D.) enumerates five efficient causes of epilepsy: four of them are entirely "natural," while the fifth is that epilepsy "is caused by some outside agent whose mode of action is not known, and it is said that it is caused by demons" (*aut causatur ex re extranea cuius modus ignoratur et dicitur quod est a causa demonum*).[105] In Constantinus Africanus' (ca. 1020–1087) Latin translation of the *De melancholia*, we find the claim that:

> epilepsia ... a vulgo divinatio appellatur, quia morbus est absconsus, dicentes demoniacos esse hunc morbum patientes. Sed medicorum alii epilepsiam, alii dicunt esse maniam.

---

102 Cf. Gotthard Strohmaier, "Avicennas Lehre von den 'inneren Sinnen' und ihre Voraussetzungen bei Galen," in Paola Manuli and Mario Vegetti (eds.), *Le opere psicologiche di Galeno: atti del terzo colloquio galenico internazionale (Pavia: 10–12 settembre 1986)* (Naples: Bibliopolis, 1988), 231–42. More generally, on the influence of Galenism on Avicenna's theories, see Danielle Jacquart, "Avicenne et le galénisme," in Ivan Garofalo and Amneris Roselli (eds.), *Galenismo e medicina tardoantica: fonti greche, latine e arabe; atti del seminario internazionale di Siena, Certosa di Pontignano (9–10, IX, 2002)* (Naples: Istituto Universitario Orientale, 2003), 265–82.
103 Temkin, *Falling Sickness*, 156–57.
104 Wehr, *Dictionary*, 164.
105 Abulqasim, *Liber theoricae nec non practicae*, fol. 34ʳ (wrongly numbered XXXIII, like the preceding folio).

The crowd calls epilepsy "divination" [*divinatio*] because it is a hidden disease, and they say that the sufferers from this disease are demoniacs [*demoniacos*]. But among the physicians, some say that it is epilepsy and some (call it) mania.[106]

The term *divinatio* also occurs in Gerard of Cremona's Latin translation of the *Practica* of the Syrian physician Serapion (Yuḥannā ibn Sarābiyūn, ninth century) as one of the popular names used to refer to epilepsy.[107] And it is in commenting on Serapion's text that a fifteenth-century physician, Antonius Guainerius, tries to explain on the basis of "rational" arguments why the name *divinatio* should be understood in relation to the prophetic faculties developed by epileptics during seizures rather than arising from the "demonic" nature of epilepsy.[108]

2 Adam of Buckfield and Albert the Great

With Adam of Buckfield's commentary on the *De somno et vigilia* we move into thirteenth-century Latin Aristotelianism and more precisely into the Aristotelian school that developed in Oxford, where it made a decisive contribution to the early reception of Aristotle's natural philosophy in the Western world.[109] As a matter of fact, Adam (fl. 1238–1278) was probably the first Latin commentator on Aristotle's natural philosophy in its entirety. He was therefore the first to comment on all the treatises of the *Parva naturalia* that were available at that time (*De sensu, De memoria, De somno, De longitudine et brevitate vitae*), treatises which had been available in Latin translation (*translatio vetus*) since the end of the twelfth century. Adam could read and understand these treatises through the mediation of Averroes' *Compendium*, whose Latin translation by Michael Scot

---

106 Constantinus Africanus, *Opera omnia*, 2 vols. (Basel, 1536–1539), 1:289–90. Also in another of his works—*Pantegni*—Constantinus discusses the points of contact between epilepsy and demonic possession and puts forward empirical strategies to distinguish them. On this point, see Temkin, *Falling Sickness*, 107. A German translation, with an edition of the Arabic text and Constantine's Latin version was published by Karl Garbers, *Isḥāq Ibn ʿImrān, Maqāla fī-'l-mālīḫūliyā (Abhandlung über die Melancholie) und Constantini Africani libri duo de melancholia* (Hamburg: Buske, 1977).
107 Serapion, *Practica Io. Serapionis dicta breuiarium: Liber Serapionis de simplici medicina. Liber de simplici medicina, dictus circa instans. Practica Platearii* (Venice, 1503), fol. 8ᵛ.
108 Antonius Guainerius, *opus preclarum*, fol. 17ʳ.
109 See Callus, "Introduction of Aristotelian Learning"; Weisheipl, "Science in the Thirteenth Century"; and Burnett, "The Introduction of Aristotle's Natural Philosophy."

had already started to circulate within the Aristotelian circles of Oxford University at the time when Adam was active as a lecturer.[110]

In Adam's commentary on the *De somno* we find three references to epilepsy, each addressed to a different point raised by Aristotle in his account of sleep. In the fifth *lectio* of his commentary, Adam deals with Aristotle's discussion of the bodily processes that give rise to sleep, and it is here that he explicitly addresses Aristotle's analogy between sleep and the epileptic fit. In so doing, however, Adam somehow inverts the terms of the analogy:

> Et est, quod quidam morbus, qui vocatur epilepsia, qui est morbus caducus, fit propter multas evaporationes ascendentes ad cerebrum. Talis morbus est similis somno, et est quodammodo somnus: quia cum deberet somnus naturalis fieri propter superabundantiam vaporum densorum descendentium et obturantium venas, intantum tumefiunt venae quod arteria, per quam fit respiratio vitae, intantum constringitur, quod pene deficit spiramen vitae, et ita fit per somnum morbus. Unde et frequentius dormientibus accidit talis passio, et raro vigilantibus.[111]

> And it happens that a certain disease, called epilepsy, which is the falling sickness, originates because of the copious evaporations that ascend to the brain. *This disease is like sleep, and in a way it is sleep.* For since natural sleep has to occur as a result of an excess of dense vapors descending and obstructing the vessels, the vessels swell to such a point that the artery, through which the respiration of life takes place, is obstructed to such a point that the breath of life withdraws and in this way the disease occurs through sleep.

This inversion testifies to a remarkable attempt to clarify while at the same time developing Aristotle's account, whose characterization of sleep "as a kind of epileptic fit" was based on three claims, the first of a theoretical kind, the other two most likely based on empirical observations: 1) sleep and the epileptic attack can

---

110 See Brumberg-Chaumont, "La première reception," 122–24. For a biography of Adam of Buckfield, see Alfred B. Emden, *A Biographical Register of the University of Oxford to A.D. 1500*, vol. 1: *A.–E.* (Oxford: Clarendon Press, 1957), 297 and Palémon Glorieux, *La faculté des arts et ses maîtres à Paris au XIIIe siècle* (Paris: Vrin, 1971), 64–66. For a list of Adam's works, see Charles Lohr, "Medieval Latin Aristotle Commentaries: Adam of Buckfield," *Traditio* 23 (1967): 317–23 and Olga Weijers, *Le travail intellectuel à la faculté des arts de Paris: textes et maîtres (ca. 1200–1500; répertoire des noms commençant par G* (Turnhout: Brepols, 1998), 24–30. Two dissertations have been consecrated to the works and thought of Adam of Buckfield, to which I could not gain access: H. Powell, "The Life and Writing of Adam of Buckfield." BLitt diss., University of Oxford, 1964; Edmund John French, "Adam of Buckfield and the Natural Philosophy of Early Universities," PhD diss., University of London, 1998.
111 The text of Adam's commentary is quoted from the digitized version of *Sancti Thomae Aquinatis In Aristotelis Stagiritae nonnullos libros commentaria: adjectis brevibus adnotationibus* (Parma: Fiaccadori, 1865–1867), which is available at http://www.corpusthomisticum.org/xsv.html.

be likened to one another with respect to the bodily processes by which they are engendered; 2) small children sleep more and are, at the same time, more prone to epilepsy; 3) in some epileptics, seizures occur only or mainly during sleep. Adam's explanatory strategy consists in bringing these three premises, and especially the last one, to their extreme consequences and in making explicit points that in Aristotle's account remain somewhat vague or undeveloped.

More specifically, Adam seems to force the boundaries of the analogical reasoning, in that he describes sleep and epilepsy not just as analogous phenomena: rather, he describes the former (sleep) as a precondition of the latter (epilepsy). For he argues that, since the material process that brings about "natural sleep" and epilepsy is the same, the very same conditions that activate sleep can cause a seizure as an incidental result, so that *fit per somnum morbum*. In this way Adam's account clarifies Aristotle's analogy between sleep and epilepsy by giving it a firmer explanatory basis. At the same time, it suggests the *biological* priority of sleep as a "natural" condition over the epileptic attack: it is now epilepsy that should be likened to sleep, and it is epilepsy that occurs *through sleep*.

A kind of "essential tension" between assimilation and differentiation also characterizes Adam's treatment of the analogy between sleep and epilepsy in two additional passages of his commentary. In commenting on Aristotle's distinction between sleep and other cases where the perceptual faculty is inactive (*impotentia sentiendi*), Adam accepts the argument that, since people who suffer from various kinds of lack or alteration of consciousness (Adam mentions *amentia*, *suffocatio*, and *animae defectus*) also have visions, there should be no difference between these visions (*phantasmata*) and proper dreams (*somnia*)—which is actually not the case according to both Aristotle and Adam—if all the affections previously mentioned were sleep.[112] In connection with this, Adam claims:

> Eadem ratio est de laborantibus morbo caduco, qui sunt quasi mortui, et aliis hujusmodi passionibus defectum sensus facientibus aliqua passione gravi. Nullus enim talis dormit.
>
> The same reasoning holds good concerning those laboring under the falling sickness, who are, as it were, dead, and concerning other affections of this kind which by some grave affliction cause a defect of sensation. *For no such person is asleep.*

Here Adam seems to follow the same line of reasoning which we have already seen in Michael of Ephesus and, more explicitly, in Theodorus Metochites, according to which the epileptic fit, together with other forms of loss of consciousness, can engender the appearance of visions, which are nonetheless not to be

---

112 *Commentarium de somno et vigilia*, lectio V.

confused with dreams. In contrast to Michael and Theodorus, however, Adam of Buckfield substantiates such a statement by explicitly referring to epilepsy as an affection that implies (it is not clear whether as cause or as effect) an alteration of the intellect and the imaginative faculty. For in his sixth *lectio*, in commenting on Aristotle's discussion of whether sleep is caused by cold or heat and, more specifically, in paraphrasing *somn. vig.* 3, 457b20–27, Adam claims:

> Et sic minuitur calor naturalis, et fugatur a partibus exterioribus per vaporis frigiditatem descendentem, qui calor sursum ferebatur a corde, et cadunt homines. Et hoc potest intelligi de casu somni sive epilepsiae. Et dicuntur homines cadere, quia solus homo est animal rectae staturae. Et ideo de solis illis proprie dicitur cadere. Et cum cadit, primo habet virtutem intellectivam alteratam, et deinde phantasiam. Et illud proprie de epilepsia est verum.
>
> And thus the natural heat that was carried upwards from the heart is diminished and driven away from the peripheral parts by the descending coldness of the vapor, and people fall down. *This can be understood to also be the case with the falling of those overtaken by sleep as well as of epilepsy.* And man is said to fall, because he alone is an animal of erect stature, and therefore he alone can properly be said to fall. And when he falls, his intellectual power is the first to be changed, and his imagination next. *And this is especially true of epilepsy.*

It is interesting to note that Adam not only makes a reference to epilepsy here which cannot be found in Aristotle's text, but he also lays special stress on the connection between the changes occurring in the cogitative and imaginative faculties and the epileptic fit. This passage might perhaps be understood once again in light of the above mentioned chapter on the diseases of the brain from Isidore of Sevilla's *Origines*, as well as Adam of Buckfield's intellectual indebtedness to Averroes.[113] For, as we have seen, Averroes in his *Compendium libri De somno et vigilia* had referred to the epileptic seizure and the syncope resulting from it while discussing the relation between sleep and the activity of the imaginative faculty.

★ ★ ★

An example of how readings and rethinkings of the Aristotelian analogy between sleep and epilepsy may have been contaminated both by a variety of different sources and by the avowed desire to go beyond and complete Aristotle's own theories in circumstances when they had been deemed defective is provided by Albert the Great's paraphrase of the *De somno*. There is no need here to recall

---

[113] See Callus, "Introduction of Aristotelian Learning," 256 and Weisheipl, "Science in the Thirteenth Century," 462.

the tremendous impact that Albert's methods of intellectual inquiry, together with his exegetical and theoretical achievements, had on the constitution and the development of Scholastic Aristotelianism and, more generally, on the philosophical discourse of the Western medieval and early modern world.[114] In this context, it has been often remarked that his treatises *De homine* and *De anima* are still two of the most valuable pieces of secondary literature on the Peripatetic psychological tradition.[115]

But Albert's paraphrase of the *De somno et vigilia* is also worthy—I would say *especially* worthy—of consideration. For at the end of the first chapter, Albert puts forward a kind of methodological statement, in which he informs the reader that he will make digressions from Aristotle's text when it appears to be faulty or obscure (*facientes digressiones ab ipso ubicumque videbitur aliquid imperfectum vel obscurum dictum*) and that, in so doing, he will follow only the opinions of the Peripatetics, and especially of Avicenna, Averroes, Alfarabi, and Algazali (*sequemur tantum Peripateticorum sententias et praecipue Avicennae et Averrois et Alfarabii et Algazelis*), and sometimes will also take into account Galen's opinion (*tangemus etiam quandoque opinionem Galeni*).[116]

In fact, this is not the only place in Albert's *Opera* where he explicitly and somewhat programmatically acknowledges Galen's authority in matters that also pertain to medicine.[117] Nor is the paraphrase of the *De somno* the only exegetical work in which Albert either cites or implicitly draws on Galenic views with which he had a direct or indirect acquaintance.[118] By saying "indirect" acquaintance, I refer to the fact that in a number of cases, Albert actually deals with Galenic materials without recognizing them as Galenic. For, as Mark Jordan

---

114 See Ludger Honnefelder, "Die philosophische Bedeutung Alberts des Großen," in Honnefelder et al., *Albertus Magnus*, 249–79.
115 See, for example, Hasse, *Avicenna's "De anima,"* 60.
116 Albertus Magnus, *De somno et vigilia* 1.1 (9.123 Borgnet).
117 Albertus Magnus, *Super II Sententiarum*, Lib. 2, d. 13, art. 2 (27.247a Borgnet): *sed si de medicina loqueretur, plus ego crederem Galeno, vel Hipocrati: et si de naturis rerum loquatur, credo Aristoteli plus vel alii experto in rerum naturis* ("But if we spoke of medicine, I would trust more Galen or Hippocrates; and if we speak of the nature of things, I trust Aristotle more than any other expert in the nature of things").
118 Albert cites Galen in the *Physica* for his views on time (4.3.3–4 [264.18–20; 265.28–36; 266.79–81; 267.8–18 Hossfeld]); in *De generatione* for the distinction between qualities and their intensities or actions (1.6.8 [172.41–48 Hossfeld]); in *De anima* for the role of heat in growth and taste (2.2.7 and 2.3.29 [91.15–29; 140.35–40/63–67 Stroick]); in the *Quaestiones super de animalibus* on a large number of anatomical and physiological points (91.60–62; 106.65; 171.55; 206.53–56; 233.48; 237.42–44 Filthaut).

has pointed out,[119] Albert seems to take most of his Galenic doctrine from Avicenna's *Canon* or *Liber de anima*. But while the *Canon* is full of direct references to Galen, the *De anima*—which Albert most heavily depends on—contains none, even when it offers Galenic doctrine. In this regard, Albert's paraphrases of and commentaries on Aristotle's works—and the *De somno* is no exception—provide clear evidence of how complex the interactions between the Aristotelian and the medical traditions could be, as well as their bidirectional and stratified nature. These features of Albert's exegetical strategy also materialize in his account of the physiology of sleep and in the way in which the reference to epilepsy is placed in the context of the account.

On the one hand, Albert explains the alternation between sleep and waking as the result of two opposite processes by which the *spiritus* and the heat expand from the center to the periphery of the body (waking) and retract from the external to the internal parts (sleep); on the other hand, he substantiates Aristotle's general claim that no animal having the perceptual faculty can uninterruptedly exercise it and that periods of activity must be necessarily followed by periods of inactivation and restoration of the perceptual faculty (*somn. vig.* 454a25–30) in terms that one might define as Galenic, as they have very much in common with Galen's account of sleep as put forward in *de sympt. caus.* 1.8 (see above, §1). For Albert explains that the use of the senses produces a debilitation and ultimately an exhaustion of the vital spirit (*spiritum qui est vitae vehiculum*), which would be entirely consumed (so provoking death) if there were no physiological mechanism capable of restoring and replenishing it. Sleep is precisely such a mechanism, insofar as during sleep the heat retracts toward the internal parts of the body and "the source and principle of the spirit" (*ad fontem et ad principium spiritus*) and, in so doing, replenishes the exhausted spirit.[120]

Now, not only does Albert's very concept of *spiritus* as the medium of the perceptual activity ultimately stem from the Galenic notion of *pneuma psychikon*, but his way of localizing the various physiological processes that engender or are engendered by sleep reflects the same "Aristotelian-Galenic syncretism" that is characteristic of his two most important sources, namely, Avicenna and Averroes.[121] For as it had been the case with Averroes and Avicenna, Albert

---

119 Jordan, "The Disappearance of Galen," 713.
120 Albertus Magnus, *De somno et vigilia* 1.8 (9.134 Borgnet).
121 The passage in which Albert most clearly departs from Aristotle (who gives a "negative" definition of sleep as "necessitated" by waking, which instead is for the sake of good, 455b17–28) to follow Averroes is found in *De somno et vigilia* 1.9 (9.135 Borgnet). Here Albert expressly cites the great Arab commentator (cf. Averroes, *Compendium Libri Aristotelis De Sompno et Vigilia* 201ra17–24 [76–77 Shields]) in putting forward a distinction between *sensus naturalis* and *sensus*

too aims to make a vast body of acquired anatomical knowledge on the sense organs and the brain fit Aristotle's cardiocentric theoretical framework. In so doing, he is inclined to give the brain a more prominent role in the physiology of sleep and waking than Aristotle had done, while trying to maintain the ontological and functional priority of the heart. It is in light of this general theoretical aim that one must understand Albert's argument that there is only one actual source and principle that provides the "substance" of these three spirits (natural, vital, and animal), which is the heart, and that there are three different bodily regions in which these spirits are located and where they accomplish their respective functions: the liver, the region of the heart, and the brain (here it is clear that Albert is adopting the Platonic-Galenic tripartite scheme).[122]

With specific regard to the cause of sleep, that is to say, the process by which vapors ascend to the brain, are cooled, and descend back down to render the *sensorium commune* inactive, Albert applies this dual scheme heart–brain by claiming that "the first cause of the cooling comes from the heart, although the proximate efficient cause comes from the brain" (*illius enim refrigerationis causa prima est a corde, licet causa proxima efficiens a cerebro sit*).[123]

In fact, it is precisely in accounting for the inactivation of the *sensorium commune* that Albert's views diverge most significantly from Aristotle's and display doctrinal elements of medical, and above all Galenic, provenance. For, if Aristotle believes that the brain is just the bodily region in which the cooling and condensation of the vapors arising from digestion take place, while the inactivation of the *sensorium commune* results from these condensed and cooled vapors flowing back and making the heat gather in the region of the heart, according to Albert it is *in the brain itself as coordinating center and point of departure of the sensorial nerves* that the process of inactivation of the perceptual faculty takes place. To Albert's eyes, this happens because the cooled humor (*infrigidatus humor*) touches at first the organ of the *sensus communis*, which is located in the anterior part of the head and the brain, and then it touches the nerves of the particular sense organs which are directed to the *sensus communis*, so hin-

---

spiritualis (which corresponds to the distinction between external and internal senses) and in explaining that the *sensus spiritualis* is worthier than the *sensus naturalis* insofar as the objects of the former are spiritual and universal while the objects of the latter are corporeal and particular and consequently that sleep gives access to a worthier kind of knowledge than wakefulness because sleep is *vigor et confortatio* of the spiritual sense and *debilitas et vinculum* of the corporeal sense.
122 Albertus Magnus, *De somno et vigilia* 2.3 (9.142 Borgnet).
123 Albertus Magnus, *De somno et vigilia* 2.3 (9.142 Borgnet).

dering the perceptual activity as a whole and immobilizing each of the individual sense organs.[124]

This process is further described in the fourth chapter of the second part of Albert's paraphrase, which contains a digression on the four causes of sleep according to Averroes. Here, when discussing the material cause of sleep, Albert argues that this consists in two kinds of coldness (*frigiditas duplex*). The first arises from food, in that the vapors generated from the digestion of food spread through the whole body and cause the heat to leave the external parts of the body;[125] the second kind of coldness is to be connected with the bodily region to which the evaporation of food is primarily directed, that is, the brain. For it is in the brain that these vapors get cold; once they have become cold and started flowing back, they touch and immobilize the sensory nerves.[126]

It is in such a theoretical and explanatory context, characterized by the co-centrality of the brain in the whole etiological account of sleep, that we must consider Albert's only reference to the connection between sleep and epilepsy.[127] For we have seen that previous commentators had rethought the analogy sleep/epilepsy in the context of a wider physiopathology of the imaginative faculty, mainly as a consequence of their attempt to make Aristotle's cardiocentric theoretical framework compatible and consistent with a well-established corpus of medical doctrines on the primary role of the brain as the center of the cognitive and perceptual faculties. In a way, Albert's treatment of such a connection sticks to Aristotle's text much more than had been the case with the previous commentaries. For neither in the paraphrase of *De somno et vigilia* nor in that of the *De divinatione per somnum* does Albert mention epilepsy with reference to the affections of the imaginative faculty and the appearance of dreams and dream-like visions. Rather, he confines himself to paraphrasing the very passage in which Aristotle had put forward that "sleep is in a way an epileptic fit":

> Et ipse somnus quaedam epileptia videtur esse. Propter quod etiam pluribus hominibus principium epileptiae accidit in somno, licet postea forte evigilent: et dormientes quandoque in epileptia consumuntur et moriuntur, suffocato nimis humido: vigilantes eniam quoniam non habent humidum nisi ex causa aegritudinis, et ideo minus habent quam illi qui habent ex somno et ex morbo: cum enim multus spiritus evaporationis tam in somno quam

---

124 Albertus Magnus, *De somno et vigilia* 2.1 (9.138 Borgnet). In this account Albert entirely follows Avicenna, who had located the *sensus communis* in the anterior ventricle of the brain, and departs from Averroes, who on this specific point had maintained an orthodox Aristotelian view, locating the *sensus communis* in the heart.
125 Albertus Magnus, *De somno et vigilia* 2.4 (9.144 Borgnet).
126 Albertus Magnus, *De somno et vigilia* 2.4 (9.144 Borgnet).
127 Albertus Magnus, *De somno et vigilia* 2.8 (9.150–151 Borgnet).

> in epileptia feratur sursum ad caput, descendens per venas colli et gutturis, facit intumescere venas, et illis intumescentibus coarctatur et praefocatur porus cannae per quem fit respiratio, et sic suffocatur. (*De somno et vigilia* 2.8 [9.150–51 Borgnet])
>
> And sleep itself seems to be a kind of epilepsy. For this reason also in most people the epileptic fit starts when they are asleep, although they wake afterwards. And sometimes those who are caught by epilepsy while sleeping are killed and die, as they suffocate from the excess of the humor. For when they are awake, they do not have humors if not because of the disease and on that account have less humor than those who have it as a consequence both of sleep and of the disease. For since much vapor both in sleep and in epilepsy moves upwards to the head, then moving backwards through the veins of the neck and the throat, it makes the veins swell. When these veins swell, the passage of the windpipe, through which respiration takes place, is constricted and strangled, and in this way one suffocates.

It is interesting to observe, however, that Aristotle's argument and Albert's paraphrase can be assimilated only at a superficial level of analysis, to the extent that they both connect this flow of cool and condensed vapors, which is regarded as part of the material process engendering both sleep and the epileptic fit, with symptoms of suffocation. But Aristotle's own account seems to suggest that sleep and the epileptic fit are in a way the same thing, since the entire causal chain by which they are brought about—that is, a flow of cold and condensed vapors that descend from the brain and *inactivate the sensorium commune located in the region of the heart*—is the same. Aristotle does not make this assumption explicit in his account, but we can identify it by virtue of the fact that Aristotle is otherwise very careful in distinguishing between sleep and other forms of loss of consciousness that result from different causes or that affect sense organs other than the common sense.

Yet that assumption cannot be taken for granted—at least not from this passage—in Albert's account, insofar as he locates the common sense in the brain. For, in order to consistently affirm that sleep and epilepsy are to some extent analogous with respect to the material cause, Albert should have reconsidered the terms of this analogy in connection with those physio- or physiopathological processes *of the brain* which are held to be responsible for sleep and for the epileptic fit.

That Albert conceived of the epileptic fit as a disease of the brain caused by the upward movement of vapors from the abdominal region to the head seems, in fact, to be confirmed by a passage of the *Ethica*, in which he connects epilepsy to an "obstruction of the brain" (*ex oppilatione cerebri*).[128] Further support for

---

[128] *Ethica* 7.1.12 (7.492a Borgnet). This reference to epilepsy occurs in a context in which Albert establishes a double analogy between *intemperantia* and dropsy or phtisis, on the one hand, and

this inference is provided by a passage of *De vegetabilibus*, in which, discussing the properties of the *coriandrum sativum*, Albert points out that "one of its properties is to prevent vapor from ascending to the head; for this reason it is added to the food of those who suffer from epilepsy caused by vapors coming from the stomach" (*Eius autem proprietas est prohibere vaporem, ne ad caput ascendat: et ideo ponitur in cibum eius, qui epilepsiam patitur ex vapore stomachi*).[129]

What is not clear, however, from the passages of Albert's *Opera* in which epilepsy is discussed is whether these vapors were thought to cause epilepsy by directly affecting the nerves and eventually the *sensorium commune*. For we have just one passage, once again from *De vegetabilibus*, in the light of which one might suppose that Albert connected epilepsy with some kind of impairment of the nerves. Here Albert mentions all the possible therapeutic uses of an herb called *stycados*, which is said to have the property "to strengthen the nerves" when they are debilitated (*confortat nervorum debilitatem*) and also to be of help to those who are affected by melancholy or epilepsy (*confert etiam herba haec melancholiae et epilepsiae*).[130] Generally speaking, these passages and the few others in which Albert makes explicit reference to epilepsy do not provide, at least in my view, sufficient evidence to try to reconstruct Albert's overall views on the causes and the whole phenomenology of the epileptic seizure.

Nevertheless, it is interesting to observe that, if these few references are to be taken as a reliable indication that Albert conceived of epilepsy as an incapacitation of the *sensorium commune* resulting from the flow of condensed cooled vapors and humors toward the nerves, then we would have to accept an epistemological scenario in which the applicability of the sleep/epilepsy analogy would be intact, although it should be entirely referred to processes taking place in the brain and not in the region of the heart.

As a consequence of this radical shift from the heart to the brain, the link between suffocation and epileptic seizure would also be substantially reconfigured. For, in Aristotle's etiological schema, suffocation seems to be part of the very definition of the material cause of epilepsy (the vapors obstruct the wind-

---

concupiscentia and epilepsy, on the other. Actually this passage re-elaborates materials that are to be found in Aristotle's *eth. Nic.* 7.5, 1149b10–22. Here Aristotle makes reference to epilepsy to distinguish between those who are mad by nature and those whose madness is the result of a disease (such as epilepsy or mania). After this distinction Aristotle actually goes on to explain the difference between a bestial and morbid form of human incontinence (*akrasia*) and a "simple" form of incontinence corresponding to human intemperance (*akolasia*).

129 *De vegetabilibus* 6.2.4 (494 Meyer-Jessen).
130 *De vegetabilibus* 6.2.17 (562 Meyer-Jessen).

pipe and hinder respiration while flowing downwards toward the region of the heart, where they presumably cause the inactivation of the *sensorium commune*). On the contrary, in the hypothetical etiological scheme, which one may try to reconstruct from Albert's references to epilepsy, suffocation should be intended just as a byproduct, a secondary concomitant result, of a pathological process, whose effects at a cognitive/perceptual level were entirely determined by the anatomophysiology of the brain.

Conclusions

To conclude, I would like to emphasize once more what I consider to be a characteristic feature of the Aristotelian treatment of the sleep/epilepsy analogy (the adjective "Aristotelian" here refers both to Aristotle's own account as we find it in *De somno* and to the reception, understanding, and rethinking of that account within the rich history of Aristotelianism). This feature consists of a constant and pervasive entwinement between theoretical claims, on the one hand, and, on the other hand, a set of empirical observations, together with the doctrines that stem from those observations. The provenance of these doctrines is in some cases distinctively medical, while in other cases it is *originally* medical, in the sense that they first took shape in a medical context but were soon incorporated into the philosophical debates of late antiquity (for example within Neoplatonism) and theoretically (re)shaped before converging in the wide-ranging framework of medieval (Byzantine, Arab, Latin) Aristotelianism.

We have seen that a cluster of observations and accounts pointing to various kinds of empirically appreciable connections or similarities between sleep and epileptic manifestations most probably represented the primary source from which Aristotle himself drew the "rough materials," so to say, with which he built his analogy between sleep and epilepsy. Yet, far from making his own account passively depend on such claims, Aristotle conceived of the connection between sleep and epilepsy in quite a peculiar way. For he represents them as analogous phenomena with respect to the material cause, a feature that makes his account stand as quite an exception within the classical and late-antique scientific and medical traditions.

As to the ways in which the analogy between sleep and epilepsy was understood and rethought in the traditions of medieval Aristotelianism that I have reviewed, I have tried to reconstruct some exemplary cases. These cases can be also taken as important steps within a very complex historical and epistemological process, through which the three major late-antique traditions of natural philosophy—namely, the Aristotelian, the medical (Galenic and post-Galenic), and the Neoplatonic—"reacted" to one another and became reciprocally entwined in various forms and to different degrees. We see these interactions at times in the attempt to make sense of Aristotle's own argument and bring it to the highest possible degree of coherence and intelligibility according to the interpreter's theoretical paradigm and empirical background; at other times, we see them in the more or less deliberate attempts to apply the analogy between sleep and epilepsy quite *independently* of Aristotle's own conceptualization of it and therefore to *de*construct and *re*construct the terms of the analogy in light of, or as a key to, theoretical or empirical issues entirely intrinsic to the intellectual horizon and the "research and/or teaching agenda" reflected by each text, be it written in the form of a commentary or of an "original" medical or philosophical treatise.

Of course this analysis cannot be considered in any way exhaustive, as it is confined to texts belonging to a limited, although extremely important, phase of medieval Aristotelianism (from the tenth century up to mid-thirteenth century). Much work remains to be done to shed light on how the sleep/epilepsy analogy was thematized and in light of which new physiological theories and anatomical discoveries it was eventually accounted for within early modern Aristotelianism until at least the first half of the seventeenth century. At any rate, the texts discussed in this paper do represent a coherent and self-consistent ensemble and can in many ways be considered as foundational also of the later reception of and debate about this specific point of Aristotle's account of sleep. For they provided later Aristotelian commentators with a general framework—consisting of questions, different explanatory strategies, and interpretative approaches—which early modern commentators never really depart from, but which they rather rearrange in various ways according to their research or teaching aims and their theoretical/empirical background.

Bibliography

Adam of Buckfield. *Sancti Thomae Aquinatis In Aristotelis Stagiritae nonnullos libros commentaria: adjectis brevibus adnotationibus* (Parma: Fiaccadori,1865–1867) [digitized version available at http://www.corpusthomisticum.org/xsv.html].

Abulqasim, *Liber theoricae necnon practicae Alsharavii* ... (Augustae Vindelicorum: Grimm & Wirsung, 1519).

(Pseudo-)Arnaldus de Villanova. *De epilepsia.* in *Arnaldi Villanovani ... opera omnia* ... (Basileae: Waldkirch, 1585).

Bakhtiar, Laleh. *Avicenna, The Canon of Medicine* (Chicago: KAZI Publications, 1999).

Baumann, Evert D. "Die Katalepsie der Antiken," *Janus* 42 (1938): 7–24.

Beard, Mary. "Cicero and Divination," *Journal of Roman Studies* 76 (1986): 33–46.

Beck, Hans-Georg. *Theodoros Metochites. Die Krise des byzantinischen Weltbildes im 14. Jahrhundert* (Munich: C. H. Beck, 1952).
Blaustein, Michael. "Averroes on Imagination and the Intellect," PhD diss., Harvard University, 1984.
Borgnet, Auguste. *Alberti Magni Opera Omnia*, 38 vols. (Paris: apud Ludovicum Vivès, 1890–1899).
Brams, Jozef. "Der Einfluß der Aristoteles-Übersetzungen auf den Rezeptionsprozeß," in Honnefelder et al., *Albertus Magnus*, 27–43.
Browning, Robert. "An Unpublished Funeral Oration of Anna Comnena," *Proceedings of the Cambridge Philological Society* 188 (1962): 1–12.
Brumberg-Chaumont, Julie. "La première reception du *De memoria et reminiscentia* au Moyen Âge latin: le commentaire d'Adam de Buckfield," in Grelland and Morel, *Les Parva Naturalia d'Aristote*, 121–41.
Buckler, Georgina. *Anna Comnena: A Study* (Oxford: Clarendon Press, 1929).
Bürgel, J. Christoph. *Averroes "contra Galenum." Das Kapitel von der Atmung im Colliget des Averroes als ein Zeugnis mittelalterlich-islamischer Kritik an Galen* (Göttingen: Vandenhoeck und Ruprecht, 1967).
Burnett, Charles. "The Institutional Context of Arabic-Latin Translations of the Middle Ages: A Reassessment of the School of Toledo," in Olga Weijers (ed.), *Vocabulary of Teaching and Research Between Middle Ages and Renaissance* (Turnhout: Brepols, 1995), 214–35.
Burnett, Charles. "The Introduction of Aristotle's Natural Philosophy into Great Britain: A Preminilary Survey of the Manuscript Evidence," in John Marenbon (ed.), *Aristotle in Britain during the Middle Ages* (Turnhout: Brepols, 1996), 21–50.
Burnett, Charles. "Translating from Arabic into Latin in the Middle Ages: Theory, Practice, and Criticism," in Steve G. Lofts and P. W. Rosemann (eds.), *Editer, traduire, interpréter: essais de méthodologie philosophique* (Louvain-La-Neuve: Institut Supérieur de Philosophie, 1997), 55–78.
Burnett, Charles, and Danielle Jacquart, eds. *Constantine the African and ʿAlī Ibn-al-ʿAbbās al-Maǧūsī: The Pantegni and Related Texts* (Leiden: Brill, 1994).
Butler, Harold E. *The Apologia and Florida of Apuleius of Madaura* (Oxford: Clarendon Press, 1909).
Callus, Daniel A. "Introduction of Aristotelian Learning to Oxford," *Proceedings of the British Academy* 29 (1943): 229–81.
Carbone, Andrea. *Aristotele: l'anima e il corpo* (Milan: Bompiani, 2002).
Constantinus Africanus. *Opera omnia*, 2 vols. (Basel, 1536–1539).
Cruz Hernández, Miguel. "Esencia y estructura de la medicina de Averroes," in C. Álvarez de Morales and E. Molina López (eds.), *La medicina en al-Andalus* (Granada: El Legado Andalusí, 1999), 237–54.
D'Ancona, Cristina. "From Late Antiquity to the Arab Middle Ages: The Commentaries and the Harmony between the Philosophies of Plato and Aristotle," in Honnefelder, Wood, Dreyer, and Aris, *Albertus Magnus*, 45–69.
Darrouzes, Jean. *Georges et Demetrios Tornikes, Lettres et Discours* (Paris: CNRS, 1970).
Davidson, Herbert A. *Alfarabi, Avicenna, and Averroes, on Intellect: Their Cosmologies, Theories of the Active Intellect, and Theories of Human Intellect* (New York: Oxford University Press, 1992).

Debru, Armelle. "L'épilepsie dans le *De somno* d'Aristote," in Guy Sabbah (ed.), *Médecins et médecine dans l'antiquité* (Saint-Étienne: Publications de l'Université de Saint-Étienne, 1982), 25–41.

De Leemans, Pieter. "*Secundum viam naturae et doctrinae:* lire le *De motu animalium* et les *Parva naturalia* d'Aristote au Moyen Âge," in Grelland and Morel, *Les Parva Naturalia d'Aristote*, 197–220.

De Leemans, Pieter, and Michèle Goyens, eds. *Aristotle's "Problemata" in Different Times and Tongues* (Leuven: Leuven University Press, 2006).

De Vries-van der Velden, Eva. *Théodore Métochite: une réévaluation* (Amsterdam: Rodopi, 1987).

Dols, Michael W. *Majnūn: The Madman in Medieval Islamic Society* (Oxford: Clarendon Press, 1992).

Drabkin, Israel E., ed. and trans. *Caelius Aurelianus, On Acute Diseases and On Chronic Diseases* (Chicago: University of Chicago Press, 1950).

Drossaart Lulofs, Hendrik J. *Aristotelis De somno et vigilia liber adiectis veteribus translationibus et Theodori Metochitae commentario* (Leiden: Burgersdijk & Niermans, 1943).

Drossaart Lulofs, Hendrik J. *Aristotelis De insomniis et De divinatione per somnum*, 2 vols. (Leiden: Burgersdijk & Niermans, 1947).

Drossaart Lulofs, Hendrik J. "Aristoteles Arabus," *Forum der letteren* 1 (1960): 169–82.

Düring, Ingemar. *Aristoteles* (Heidelberg: Winter, 1966).

Elfassi, Jacques, and Bernard Ribémont, eds. *La réception d'Isidore de Séville durant le Moyen âge tardif (XIIe-XVe s.)* (Paris: Special Issue of the *Journal of Medieval and Humanistic Studies* 16, 2008).

Emden, Alfred B. *A Biographical Register of the University of Oxford to* A. D. *1500*, vol. 1: *A.-E.* (Oxford: Clarendon Press, 1957).

Endress, Gerhard. "Averrois Opera," in Gerhard Endress and Jan A. Aertsen (eds.), *Averroes and the Aristotelian Tradition: Sources, Constitution and Reception of the Philosophy of Ibn Rushd (1126–1198)* (Leiden: Brill, 1999), 339–81.

Esculapius. *De morborum, infirmitatum, passionumque corporis humani caussis, descriptionibus et cura*, in: *Physica S. Hildegardis ... Oribasii ... Theodori ... Esculapii* (Strasbourg: Schott, 1533).

Everson, Stephen. "The *De somno* and Aristotle's Explanation of Sleep," *Classical Quarterly* 57 (2007): 502–20.

Ferraces Rodríguez, Arsenio, ed. *Isidorus medicus: Isidoro de Sevilla y los textos de medicina* (La Coruña: Universidade da Coruña, 2005)

Fidora, Alexander. "La Recepción de San Isidoro de Sevilla por Domingo Gundisalvo (ca. 1110–1181): Astronomía, Astrología y Medicina en la Edad Media," *Estudios Eclesiásticos* 75 (2001): 663–77.

Filthaut, Ephrem. *Alberti Magni Opera omnia*, vol. 12: *Quaestiones super de animalibus* (Münster i. W.: Aschendorff, 1955).

Fischer, Klaus-Dietrich. "Neue oder verlachlässigte Quellen der *Etymologien* Isidors von Sevilla (Buch 4 und 11)," in Ferraces Rodríguez, *Isidorus medicus*, 131–74.

Flashar, Hellmut. *Aristoteles, Problemata physica* (Berlin: Akademie-Verlag, 1962).

Flashar, Hellmut. *Melancholie und Melancholiker in den medizinischen Theorien der Antike* (Berlin: de Gruyter, 1966).

Förster, Aurél. *Konstruktion und Entstehung der aristotelischen sogenannten Parva Naturalia* (Budapest: Akadémia, 1932).
French, Edmund J. "Adam of Buckfield and the Natural Philosophy of Early Universities," PhD diss., University of London, 1998.
Gallop, David. *Aristotle on Sleep and Dreams* (Peterborough, Ont.: Broadview Press, 1990).
García Ballester, Luis. *Galen and Galenism: Theory and Medical Practice from Antiquity to the European Renaissance* (Aldershot: Ashgate, 2002).
Garofalo, Ivan. "La traduzione araba dei compendi alessandrini delle opere del canone di Galeno," *Medicina nei Secoli* n.s. 6 (1994): 329–48.
Garofalo, Ivan. "La traduzione araba del *De locis affectis* di Galeno," *Studi classici e orientali* 45 (1995): 13–63.
Garofalo, Ivan, and Brian Fuchs. *Anonymi Medici De morbis acutis et chroniis*, ed. with comm. by I. Garofalo, trans. into English by B. Fuchs (Leiden: Brill, 1997).
Gasti, Fabio. *L'antropologia di Isidoro: Le fonti del libro XI delle "Etimologie"* (Como: New Press, 1998).
Gasti, Fabio. *Isidoro di Siviglia, Etimologie, Libro XI. De homine et portentis* (Paris: Les Belles Lettres, 2010).
Gätje, Helmut. "Die 'inneren Sinne' bei Averroes," *Zeitschrift der deutschen morgenländischen Gesellschaft* 115 (1965): 255–93.
Gigon, Olaf. "Cicero und Aristoteles," *Hermes* 87 (1959): 143–62.
Glorieux, Palémon. *La faculté des arts et ses maîtres à Paris au XIIIe siècle* (Paris: Vrin, 1971).
Gouma-Peterson, Thalia, ed. *Anna Komnene and her Times* (New York: Garland, 2000).
Grabmann, Martin. *Guglielmo di Moerbeke O.P., il traduttore delle opere di Aristotele* (Rome: Pontificia Università Gregoriana, 1946).
Grabmann, Martin. *Die Werke des heiligen Thomas von Aquin* (Münster i. W.: Aschendorff, 1949).
Grelland, Christophe, and Pierre-Marie Morel, eds. *Les Parva Naturalia d'Aristote: fortune antique et médiévale* (Paris: Publications de la Sorbonne, 2010)
Grunert, Peter. "Die Bedeutung der Hirnkammer in der antiken Naturphilosophie und Medizin," in *Antike Naturwissenschaft und ihre Rezeption* 12 (2002): 151–82.
Guaynerius, Antonius. *Antonii Guaynerii ... opus preclarum ...* (Lugduni: 1525).
Guillaumont, François. *Le "De divinatione" de Cicéron et les théories antiques de la divination* (Brussels: Latomus, 2006).
Gutas, Dimitri. *Avicenna and the Aristotelian Tradition* (Leiden: Brill, 1988).
Hansberger, Rotraud E. "How Aristotle Came to Believe in God-Given Dreams: The Arabic Version of *De divinatione per somnum*," in Louise Marlow (ed.), *Dreaming Across Boundaries: The Interpretation of Dreams in Islamic Lands* (Boston: Center for Hellenic Studies, 2008), 50–77.
Harvey, E. Ruth. *The Inward Wits: Psychological Theory in the Middle Ages and Renaissance*, Warburg Institute Surveys VI (London: Warburg Institute, 1975).
Hasse, Nikolaus. *Avicenna's "De anima" in the Latin West: The Formation of a Peripatetic Philosophy of the Soul (1160–1300)* (London: Warburg Institute, 2000).
Heiberg, Johan L. *Paulus Aegineta, Pars prior: libri I–IV*, CMG IX 1 (Leipzig: Teubner, 1921).
Hett, Walter S. *Aristotle, Minor Works* (Cambridge, Mass.: Harvard University Press, 1936).
Honnefelder, Ludger. "Die philosophische Bedeutung Alberts des Großen," in Honnefelder et al., *Albertus Magnus*, 249–79.

Honnefelder, Ludger, et al., eds. *Albertus Magnus und die Anfänge der Aristoteles-Rezeption im lateinischen Mittelalter* (Münster i. W.: Aschendorff, 2005).
Hossfeld, Paul. *Alberti Magni Opera omnia*, vol. 5.2: *De natura loci; De causis elementorum; De generatione et corruptione* (Münster i. W.: Aschendorff, 1980).
Hossfeld, Paul. *Alberti Magni Opera omnia*, vol. 4.1: *Physica I–IV* (Münster i. W.: Aschendorff, 1987).
Hulskamp, Maithe. "Sleep and Dreams in Ancient Medical Diagnosis and Prognosis," PhD diss., University of Newcastle, 2008.
Jacquart, Danielle. "The Role of Arabic Medicine in the Medieval West," in Roshdi Rashed and Régis Morelon (eds.), *Encyclopedia of the History of Arabic Science*, 3 vols. (London: Routledge, 1996), 3:963–84.
Jacquart, Danielle. "Avicenne et le galénisme," in Ivan Garofalo and Amneris Roselli (eds.), *Galenismo e medicina tardoantica: fonti greche, latine e arabe; atti del seminario internazionale di Siena, Certosa di Pontignano (9–10, IX, 2002)* (Naples: Istituto Universitario Orientale, 2003), 265–82.
Jacquart, Danielle, and Françoise Micheau. *La médecine arabe et l'Occident médiéval* (Paris: Maisonneuve et Larose, 1990).
Jones, William H. S. *Hippocrates*, vol. 2 (Cambridge, Mass.: Harvard University Press, 1923).
Jordan, Mark D. "The Construction of a Philosophical Medicine: Exegesis and Arguments in Salernitan Teaching on the Soul," *Osiris* (1990): 42–61.
Jordan, Mark D. "The Disappearance of Galen in Thirteenth-Century Philosophy and Theology," in Albert Zimmermann and Andreas Speer (eds.), *Mensch und Natur im Mittelalter*, 2 vols. (Berlin: De Gruyter, 1991–1992), 2:703–17.
Jori, Alberto. *Aristotele* (Milan: Bruno Mondadori, 2003).
Jouanna, Jacques. *Hippocrate, Des vents; De l'art*, t. V.1 (Paris: Les Belles Lettres, 1988).
Jouanna, Jacques. *Hippocrate, Epidémies V et VII*, t. IV.3 (Paris: Les Belles Lettres, 2003).
Kany-Turpin, José, and Pierre Pellegrin. "Cicero and the Aristotelian Theory of Divination by Dreams," in William W. Fortenbaugh and Peter Steinmetz (eds.), *Cicero's Knowledge of the Peripatos* (New Brunswick: Transaction Publishers, 1989), 220–45.
King, Peter. "Scholasticism and the Philosophy of the Mind: The Failure of Aristotelian Psychology," in Allen I. Janis and Tamara Horowitz (eds.), *Scientific Failure* (Lanham, Md.: Rowman & Littlefield Publishers, 1994), 109–38.
Klibansky, Raymond, Erwin Panofsky, and Fritz Saxl. *Saturn and Melancholy: Studies in the History of Natural Philosophy, Medicine, Religion and Art* (Edinburgh: Nelson, 1964).
Knuuttila, Simo. "Aristotle's Theory of Perception and Medieval Aristotelianism," in Simo Knuuttila and Pekka Kärkkäinen (eds.), *Theories of Perception in Medieval and Early Modern Philosophy* (Berlin: Springer, 2008), 1–22.
Lacombe, Georges. "The Medieval Latin Versions of the *Parva Naturalia*," *The New Scholasticism* 5 (1931): 289–314.
Lloyd, Geoffrey E. R. "The Empirical Basis of Physiology in the *Parva Naturalia*," in Lloyd and Owen, *Aristotle on Mind and the Senses*, 214–40.
Lloyd, Geoffrey E. R., and Gwilym E. L. Owen, eds. *Aristotle on Mind and the Senses: Proceedings of the Seventh Symposium Aristotelicum* (Cambridge: Cambridge University Press, 1978).
Lohr, Charles. "Medieval Latin Aristotle Commentaries: Adam of Buckfield," *Traditio* 23 (1967): 317–23.

Lohr, Charles. "The New Aristotle and 'Science' in the Paris Arts Faculty (1255)," in O. Weijers and L. Holtz (eds.), *L'enseignement des disciplines à la Faculté des arts (Paris et Oxford, XIIIe-XVe siècles): actes du colloque international* (Turnhout: Brepols, 1997), 251-69.

Lo Presti, Roberto. "Le sommeil dans les *Epidémies* hippocratiques," in V. Leroux, N. Palmieri, and C. Pigné (eds.), *Le sommeil: approches philosophiques et médicales de l'Antiquité à la Renaissance*, forthcoming.

Lowe, Malcolm. "Aristotle's *De somno* and his Theory of Causes," *Phronesis* 23 (1978): 279-91.

Mahoney, Edward P. "Sense, Intellect and Imagination in Albert, Thomas, and Siger," in Norman Kretzmann, Anthony Kenny, and Jan Pinborg (eds.), *The Cambridge History of Later Medieval Philosophy* (Cambridge: Cambridge University Press, 1982), 602-22.

Manetti, Daniela, and Amneris Roselli. *Ippocrate, Epidemie, libro sesto* (Florence: La Nuova Italia, 1982).

Manzoni, Tullio. *Aristotele e il cervello: le teorie del più grande biologo dell'antichità nella storia del pensiero scientifico* (Rome: Carocci, 2007).

Marelli, Cesare. "Place de la *Collection hippocratique* dans les théories biologiques sur le sommeil," in François Lasserre and Philippe Mudry (eds.), *Formes de pensée dans la Collection hippocratique: actes du IVe Colloque international hippocratique* (Geneva: Droz, 1983), 331-39.

Martínez Gázquez, José. "Isidoro de Sevilla y la medicina en los enciclopedistas hispanos: D. Gundisalvo y Juan Gil de Zamora," in Ferraces Rodríguez, *Isidorus medicus*, 215-25.

Mazliak, Paul. *Avicenne et Averroès: médecine et biologie dans la civilisation de l'Islam* (Paris: Vuivert, 2004).

Meyer, Ernst H. F., and Karl F. W. Jessen. *Alberti Magni De vegetabilibus libri VII* (Berlin: Reimer, 1867).

Miller, Patricia C. *Dreams in Late Antiquity: Studies in the Imagination of a Culture* (Princeton: Princeton University Press, 1997).

Minio-Paluello, Lorenzo. "Jacobus Veneticus Graecus," *Traditio* 8 (1952): 265-304.

Monfasani, John. "The Pseudo-Aristotelian *Problemata* and Aristotle's *De animalibus* in the Renaissance," in Anthony Grafton and Nancy Siraisi (eds.), *Natural Particulars: Nature and the Disciplines in Renaissance Europe* (Cambridge, Mass.: MIT Press, 1999), 205-47.

Morel, Pierre-Marie. *Aristote, Petits traités d'histoire naturelle* (Paris: Flammarion, 2000).

Mugnier, René. "Les manuscrits des *Parva Naturalia* d'Aristote," in *Mélanges offerts à A. M. Desrousseaux* (Paris: Hachette, 1937), 327-33.

Mugnier, René. "La filiation des manuscrits des *Parva Naturalia* d'Aristote," *Revue de philologie* 26 (1952): 36-46.

Müri, Walter. "Melancholie und schwarze Galle," *Museum Helveticum* 10 (1953): 21-38.

Olivieri, Alessandro. *Aetii Amideni Libri medicinales V-VIII*, CMG VIII 2 (Berlin: Akademie-Verlag, 1950).

Omont, Henri. "Recherches sur la bibliothèque de l'église cathédrale de Beauvais," *Mémoires de l'Académie des Inscriptions et Belles-Lettres* 40 (1914): 1-91.

Pelster, Franz. "Adam von Bocfeld (Bockingfold), ein Oxforder Erklärer des Aristoteles um die Mitte des 13. Jahrhunderts: sein Leben und seine Schriften," *Scholastik* 11 (1936): 196-224.

Peters, Francis E. *Aristoteles Arabus: The Oriental Translations and Commentaries on the Aristotelian Corpus* (Leiden: Brill, 1968).

Pigeaud, Jackie. *La maladie de l'âme: étude sur la relation de l'âme et du corps dans la tradition médico-philosophique antique* (Paris: Les Belles Lettres, 1981).
Pigeaud, Jackie. "Prolégomènes à une histoire de la mélancolie," *Histoire Economie Société* 3 (1984): 501–10.
Pigeaud, Jackie. *Folie et cures de la folie chez les médecins de l'antiquité gréco-romaine: la manie* (Paris: Les Belles Lettres, 1987).
Pigeaud, Jackie. *Aristote: l'homme de génie et la mélancolie* (Paris: Rivage, 1988).
Powell, H. "The Life and Writing of Adam of Buckfield." BLitt diss., University of Oxford, 1964.
Preus, Anthony. *Aristotle and Michael of Ephesus, On the Movement and Progression of Animals* (Hildesheim: Olms, 1981).
Raeder, Hans Henning. *Oribasii Synopsis ad Eustathium; Libri ad Eunapium*, CMG VI 3 (Leipzig: Teubner, 1926).
Repici, Luciana. *Aristotele: il sonno e i sogni (Il sonno e la veglia, I sogni, La divinazione durante il sonno)* (Venice: Marsilio, 2003).
Rocca, Julius. *Galen on the Brain: Anatomical Knowledge and Physiological Speculation in the Second Century AD* (Leiden: Brill, 2003).
Rose, Valentin. *Anecdota Graeca et Graecolatina*, 2 vols. (Berlin: Dümmler, 1864–1870).
Rose, Valentin. *Cassii Felicis De Medicina* (Leipzig: Teubner, 1879).
Rose, Valentin. *Theodori Prisciani Euporiston libri III, cum physicorum fragmento et additamentis pseudo-Theodoreis* (Leipzig: Teubner, 1894).
Ross, David. *Aristotle, Parva Naturalia* (Oxford: Clarendon Press, 1955).
Serapion, Ioannes, et al. *Practica Io. Serapionis dicta breuiarium: Liber Serapionis de simplici medicina; Liber de simplici medicina, dictus circa instans; Practica Platearii* (Venice: Andreas Toresanus de Asula, 1503).
Sharples, Robert W., and Philip J. van der Eijk. *Nemesius, On the Nature of Man* (Liverpool: Liverpool University Press, 2008).
Shields, Emily Ledyard. *Averrois Cordubensis Compendia Librorum Aristotelis qui Parva Naturalia vocantur* (Cambridge, Mass.: The Medieval Academy of America, 1949).
Simon, Bennett. *Mind and Madness in Ancient Greece* (Ithaca: Cornell University Press, 1978).
Siraisi, Nancy. *Arts and Sciences at Padua: The "Studium" of Padua before 1350* (Toronto: Pontifical Institute of Mediaeval Studies, 1973).
Smith, Jonathan Z. "Towards Interpreting Demonic Powers in Hellenistic and Roman Antiquity," in Wolfgang Haase (ed.), *Aufstieg und Niedergang der römischen Welt*, II 16.3 (Berlin: De Gruyter, 1986), 425–39.
Strohmaier, Gotthard. "Galen in Arabic: Problems and Prospects," in Vivian Nutton (ed.), *Galen: Problems and Prospects* (London: Wellcome Institute, 1981), 187–96.
Strohmaier, Gotthard. "Avicennas Lehre von den 'inneren Sinnen' und ihre Voraussetzungen bei Galen," in Paola Manuli and Mario Vegetti (eds.), *Le opere psicologiche di Galeno: Atti del terzo colloquio galenico internazionale (Pavia: 10–12 settembre 1986)* (Naples: Bibliopolis, 1988), 231–42.
Strohmaier, Gotthard. "The Uses of Galen in Arabic Literature," in Vivian Nutton (ed.), *The Unknown Galen* (London: Institute of Classical Studies, 2002), 113–20.
Stroick, Clemens. *Alberti Magni Opera omnia*, vol. 7.1: *De anima* (Münster i. W.: Aschendorff, 1968).

Taylor, Richard C. *Averroes (Ibn Rushd) of Cordoba: Long Commentary on the "De Anima" of Aristotle* (with Thérèse-Anne Druart, subeditor) (New Haven: Yale University Press, 2009).
Temkin, Owsei. *The Falling Sickness: A History of Epilepsy from the Greeks to the Beginnings of Modern Neurology*, 2nd ed. (Baltimore: The Johns Hopkins University Press, 1971).
Thomas Aquinas. *Opera omnia*, 25 vols. (Parma: Fiaccadori, 1852–1873).
Thomas Aquinas. *Opera omnia*, 34 vols. (Paris: Vivès, 1871–1882).
Timpanaro, Sebastiano. *Marco Tullio Cicerone, Della divinazione* (Milan: Garzanti, 1988).
Todd, Robert B. "Galenic Medical Ideas in the Greek Aristotelian Commentators," *Symbolae Osloenses* 52 (1977): 117–34.
Todd, Robert B. "Philosophy and Medicine in John Philoponus' Commentary on Aristotle's *De anima*," *Dumbarton Oaks Papers* 38 (1984): 103–10.
Ullmann, Manfred. *Die Medizin im Islam* (Leiden: Brill, 1970).
Valastro Canale, Angelo. *Isidoro, Etimologie o origini*, 2 vols. (Turin: UTET, 2006).
van der Eijk, Philip J. "Aristotelian Elements in Cicero's *De divinatione*," *Philologus* 137 (1993): 223–31.
van der Eijk, Philip J. *Aristoteles, De insomniis; De divinatione per somnum* (Berlin: Akademie-Verlag, 1994).
van der Eijk, Philip J. *Diocles of Carystus: A Collection of the Fragments with Translation and Commentary*, 2 vols. (Leiden: Brill, 2000–2001).
van der Eijk, Philip J. "Aristotle on Cognition in Sleep," *Nottingham Classical Literature Studies* 8 (2003): 25–40.
van der Eijk, Philip J. *Medicine and Philosophy in Classical Antiquity* (Cambridge: Cambridge University Press, 2005).
van der Eijk, Philip J. "Aristotle on Melancholy," in *Medicine and Philosophy*, 139–68.
van der Eijk, Philip J. "Theoretical and Empirical Elements in Aristotle's Treatment of Sleep, Dreams, and Divination in Sleep," in *Medicine and Philosophy*, 169–205.
van der Eijk, Philip J., and Maithe Hulskamp. "Stages in the Reception of Aristotle's Works on Sleep and Dreams in Hellenistic and Imperial Philosophical and Medical Thought," in Grelland and Morel, *Les Parva Naturalia d'Aristote*, 47–75.
van Riet, Simone, and Gerard Verbeke. *Avicenna Latinus, Liber de Anima seu Sextus de Naturalibus*, IV–V (Louvain: Ed. Orientalistes, 1968).
van Riet, Simone, and Gerard Verbeke. *Avicenna Latinus, Liber de Anima seu Sextus de Naturalibus*, I–III (Louvain: Ed. Orientalistes, 1972).
Vaux, Roland de. "La première entrée d'Averroès chez les Latins," *Revue des sciences philosophiques et théologiques* 22 (1933): 193–243.
Wardle, David. *Cicero on Divination: De Divinatione, Book I* (Oxford: Clarendon Press, 2006).
Wehr, Hans. *A Dictionary of Modern Written Arabic (Arabic-English)* (Wiesbaden: Harrassowitz, 1979).
Weijers, Olga. *Le travail intellectuel à la faculté des arts de Paris: textes et maîtres (ca. 1200–1500) ; répertoire des noms commençant par G* (Turnhout: Brepols, 1998).
Weisheipl, James A. "Science in the Thirteenth Century," in J. I. Catto (ed.), *The History of the University of Oxford* (Oxford: Clarendon Press, 1984), 435–69.
Wendland, Paul. *Michaelis Ephesii in Parva naturalia commentaria*, CAG 22.1 (Berlin: Reimer, 1903).
Wendland, Paul. *Themistii (Sophoniae) in Parva naturalia commentarium*, CAG 5.6 (Berlin: Reimer, 1903).

Wiesner, Jürgen. "The Unity of the Treatise *De somno* and the Physiological Explanation of Sleep in Aristotle," in Lloyd and Owen, *Aristotle on Mind and the Senses*, 241–80.

Wingate, Sybil D. *The Mediaeval Latin Versions of the Aristotelian Scientific Corpus, with Special Reference to the Biological Works* (London: Courier, 1931).

Wijsenbeek-Wijler, Henriette. *Aristotle's Concept of Soul, Sleep, and Dreams* (Amsterdam: Hakkert, 1978).

Wolfson, Harry A. "The Internal Senses in Latin, Arabic, and Hebrew Philosophic Texts," *The Harvard Theological Review* 28 (1935): 69–133.

Woods, Michael J. "Aristotle on Sleep and Dreams," *Apeiron* 25 (1992): 179–88.

Arnaldo Marcone
Il *Numen Augusti* nel *senatus consultum de Cn. Pisone patre*

Abstract: This paper addresses the question of the meaning of *numen Augusti* in the *s. c. de Cn. Pisone patre,* where the senate says that the "numen" of Divus Augustus was violated by Piso, and of the worship of the imperial *numen* in general. It considers how *numen,* always an elevated word, seems to have been applied to the living Augustus first by poets, later also on inscriptions, some of them official. It also discusses I. Gradel's re-reading of the *Fasti Praenestini* that eliminates the altar to the *numen* of Augustus putatively set up in 6 A.D. and the nature of the altar in Narbo.

La locuzione *numen Augusti,* che compare due volte nel testo del *senatus consultum de Cn. Pisone patre* (l. 46 e l. 69), ha già meritato speciale considerazione.[1] I passi in questione sono questi:

> (ll. 46–48): bellum etiam civile ex/citare conatus sit iam pridem <u>numine divi Aug(usti)</u> virtutibusq(ue) Ti(beri) Caesaris Aug(usti) / omnibus civilis belli sepultis malis repetendo provinciam Syriam post / mortem Germanici Caesaris...

> (ll. 68–70): <u>numen quoq(ue) divi Aug(usti)</u> violatum esse ab eo arbitrari senatum / omni honore, qui aut memoriae eius aut imaginibus, quae, antequam in / deorum numerum referre{n}tur ei r[....]tae erant, habeba{n}tur, detracto.

Gli editori del testo hanno dato prova di prudenza nel non tradurre il termine *numen* evidentemente in considerazione della sua indubbia problematicità:

> Auch einen Bürgerkrieg habe er zu entfachen versucht, nachdem schon seit langem durch das *numen* des vergöttlichten Augustus und die vielfältige Tüchtigkeit des Ti. Caesar Aug. alle Übel des Bürgerkrieges zu Grabe getragen seien, indem er nach dem Tode des Germanicus Caesar die Provinz Syrien zurückzugewinnen versuchte.

> Ferner sei der Senat der Ansicht, daß auch das *numen* des vergöttlichten Augustus von ihm verletzt worden sei, indem ihm jegliche Ehrung entzogen wurde, die *seinem* Andenken oder

---

[1] D. Fishwick, "Cn. Piso Pater and the Numen Divi Augusti," *Zeitschrift für Papyrologie und Epigraphik* 159 (2007): 297–300.

auch seinen Bildern, die ihm (errichtet) worden waren, noch bevor er unter die Götter versetzt wurde, erwiesen wurde.[2]

In particolare gli editori hanno messo in evidenza come la seconda ricorrenza del termine abbia un valore più pregnante della prima. E hanno sottolineato come "an beiden Stellen, an denen ein *numen* des Augustus im s. c. erwähnt wird, ist der noch lebende Augustus gemeint," vale a dire che nel *s. c.* ogni riferimento ad Augusto come *divus* riflette la sua qualifica ufficiale dopo la *consecratio* del 14.[3] Si tratta di una considerazione importante che non sembra essere stata pienamente recepita da alcuni studiosi che successivamente si sono occupati della questione.

Con riferimento alla prima occorrenza Fishwick scrive: "since the text states that the evils of the civil wars were buried by the *numen* of Divus Augustus, not just the mention of Divus Augustus himself, is clearly an historical anachronism."[4]

Secondo Fishwick il *numen* dell'imperatore, in vita o divinizzato, non avrebbe potuto essere coinvolto nella vittoria di Azio (cui il testo del *s. c.* fa riferimento) non foss'altro perché Augusto ricevette *ufficialmente* il *numen* solo attorno al 6 d.C.

Che un testo ufficiale, dalle implicazioni tanto importanti, possa contenere un anacronismo storico è già di per sé poco plausibile. Ma ancor meno accettabile è immaginare un conferimento (ufficiale?) del *numen* sulla base della sua prima attestazione nei *Fasti Praenestini*, ove, almeno secondo la ricostruzione generalmente accolta del testo, troviamo una dedica di un altare al *numen Augusti*.[5]

In realtà, accertata, come gli editori hanno fatto, la rilevanza per quanto viene deciso nel *s. c.* in merito alle offese recate da Pisone al *numen* imperiale, merita cercare di verificare l'evoluzione semantica del termine e la sua valenza

---

2 W. Eck, A. Caballos e F. Fernández, *Das senatus consultum de Cn. Pisone patre* (München: Beck 1996), 43. Gli editori dell'*Année épigraphique* traducono *numen* alternativamente "puissance divine" o "divinité" (*AE* 1996, no. 885, pp. 293 e 294).
3 Eck, Caballos e Fernández, *Das senatus consultum*, 188.
4 Fishwick, "Cn. Piso," 297–98.
5 *Inscr. It.* XIII 2, 115, no. 17. La lettura del testo non è peraltro sicura anche se la ricostruzione proposta da Mommsen e dalla Taylor appare tuttora quella preferibile. Proposte alternative di ricostruzione in R. T. Scott, "Providentia Aug.," *Historia* 31 (1982): 436–59 e in I. Gradel, *Emperor Worship and Roman Religion* (Oxford: Clarendon Press, 2002), 236–39 (con una chiara riproduzione della ricostruzione del testo generalmente accolta) per il quale l'*ara numinis Augusti* a Roma sarebbe "a ghost of modern scholarship" (238). *Contra*: D. Fishwick, *The Imperial Cult in the Latin West*, vol. 3, pt. 4 (Leiden: Brill, 2005), 237–45.

specifica in un momento tanto delicato di definizione del potere imperiale nelle sue valenze religiose. Tra le proposte di spiegazione la più radicale appare quella di Ittai Gradel secondo il quale, dal momento che non si poteva tributare un culto a un *numen*, l'adorazione della "divinità" (*numen* = qualità divina) dell'imperatore vivente era equivalente ad adorarlo direttamente come dio: il culto del *numen* era cioè semplicemente sinonimico del culto diretto.[6] Una questione apparentemente rilevante riguarda il momento in cui le offese di Pisone, in particolare verso le *imagines* di Augusto associate a quelle recate alla sua *memoria*, avrebbero avuto luogo. Il testo non fornisce elementi precisi ma il fatto che tali *imagines* siano state realizzate prima della divinizzazione di Augusto (*quae antequam in deorum numerum referrentur*) induce gli editori a ritenere che anche l'azione (forse solo verbale) di Pisone abbia avuto luogo prima della divinizzazione.[7] Il testo, in effetti, sembra deliberatamente evitare di fissare un preciso riferimento cronologico, quasi a voler suggerire che le azioni scellerate di Pisone avessero preceduto e seguito la morte di Augusto.

Gli argomenti di Fishwick appaiono sofisticati ma, non per questo, condivisibili. A suo parere le offese recate alle *imagines* avrebbero semplicemente amplificato quelle recate alla *memoria* di Augusto (*aut memoriae eius aut imaginibus*), e quindi devono avuto luogo necessariamente tra il 19 agosto del 14 (giorno della morte) e il successivo 17 settembre (giorno della deificazione). Il decreto rifletterebbe la prospettiva dell'anno 20 e, per questo motivo, parlare di offese recate al *numen* del *divus* Augusto è prolettico se non anacronistico *tout court*.[8]

La trasformazione del culto pubblico del detentore del potere a Roma aveva conosciuto una trasformazione decisiva negli ultimi decenni. È appena il caso di ricordare come nei mesi critici successivi all'assassinio di Cesare e, dunque, in una situazione suscettibile di iniziative destabilizzanti per l'assetto della Repubblica, Cicerone si preoccupasse ancora di ricordare con forza la distinzione tra gli onori che si debbono ai "morti," pure eccellenti, e quelli che si rendono agli dei.[9] Egli manifestava in questi termini il proprio dissenso per le decisioni prese nella seduta del senato del 1 settembre cui non aveva partecipato:

---

6 *Contra:* Fishwick, *The Imperial Cult*, 238. Sui vari significati attribuiti e attribuibili a *numen* cf. in generale W. Pötscher, "'Numen' und 'Numen Augusti,'" in W. Haase (ed.), *Aufstieg und Niedergang der römischen Welt*, II 16.1 (Berlin: De Gruyter, 1978), 355–92.
7 Eck, Caballos e Fernández, *Das senatus consultum*, 187.
8 Fishwick, "Cn. Piso," 299.
9 Cf. soprattutto A. Fraschetti, "L'eroizzazione di Germanico," in A. Fraschetti (ed.), *La commemorazione di Germanico nella documentazione epigrafica. Tabula Hebana e Tabula Siarensis* (Roma: L'Erma di Bretschneider, 2000), 139–60 (spec. 146–47).

> Pensate dunque, padri coscritti, che io avrei votato il decreto da voi adottato a malincuore, che mescolava i *parentalia* con le *supplicationes*, che introduceva nella Repubblica pratiche assolutamente sacrileghe, che decretava *supplicationes* a un morto? ... mai potei essere indotto ad associare un qualsiasi morto con le pratiche religiose che competono agli dei immortali in modo tale che si facciano pubbliche suppliche a uno di cui esiste da qualche parte una tomba dove gli sono resi onori funebri.[10]

Una corretta considerazione del significato di *numen* e delle sue implicazioni appare indispensabile anche alla luce di varie proposte che trovano riscontro in un'ampia bibliografia.[11] Secondo Varrone *numen* significherebbe semplicemente e direttamente *imperium*: *numen dicunt esse imperium* (*ling. Lat.* 7.85).[12] Dunque *numen* sarebbe connesso all'idea di potere, di forza intrinseca (misteriosa?) che appartiene, in linea di principio, alla divinità.[13] Proprio per questo non è necessario interpretare *numen* come sinonimo di *genius* o, come ora vuole Gradel, di divinità *tout-court*. Virgilio, Orazio e soprattutto Ovidio sono ricchi di riferimenti alla "divinità" dell'imperatore e al suo *numen* con varie sfumature di significato. Si tratta di voci poetiche ma variamente indicative di un clima e di un'aspettativa, oltre che di un'evoluzione del valore semantico del termine. Particolarmente importante in questo senso appare Orazio (*ep.* 2.1.16), che usa l'espressione *iurare per numen* che sembra ricalcare quella, ben attestata, *iurare per Iovem*.[14] In buona sostanza Orazio sembra suggerire che Augusto ha conseguito ormai uno *status* semidivino che rende ammissibile che si facciano giuramenti al suo altare per il suo *numen*. Meno pregnante è invece l'altra

---

10 Cic. *Phil.* 1.13.
11 Questa la definizione di Festo: *quasi nutus dei ac potestas* (178.9–10 Lindsay). Cf. Gradel, *Emperor Worship*, 235.
12 Sappiamo di un detto popolare, tramandatoci da Artemidoro (2.36), secondo cui "governare vuol dire avere il potere di un dio" (τὸ κρατοῦν δύναμιν ἔχει θεοῦ). Ma già in un frammento di una commedia di Menandro troviamo lo stesso concetto: τὸ κρατοῦν γὰρ νῦν νομίζεται θεός (Καρίνη, fr. 201.3 Kassel-Austin), cf. A. Marcone, "Un 'dio presente': osservazioni sulle premesse ellenistiche del culto imperiale romano," in S. Bussi e D. Foraboschi (edd.), *Roma e l'eredità ellenistica*, Studi Ellenistici 23 (Pisa: Fabrizio Serra Editore, 2010), 205–15.
13 È interessante la conferma che si ricava da un papiro del II secolo, che ci ha conservato, nella forma del quesito a domanda e risposta, la versione banalizzata, ma proprio per questo realistica, di come fosse percepito a livello popolare il rapporto tra la divinità e l'imperatore: "Che cos'è un dio?" "È regnare". "Che cos'è un re?" "Uguale a un dio": (P. Heidelberg inv. 1716 verso). Cf. P. Bilabel, "Fragmente aus der Heidelberger Papyrussammlung," *Philologus* 80 (1924): 331–41 (339).
14 Sembra da escludersi la lezione *nomen* presente in alcuni codici: cf. C. O. Brink, *Horace on Poetry: Epistles Book II; The Letters to Augustus and Florus* (Cambridge: Cambridge University Press, 1982), 54–55.

ricorrenza (*carm.* 4.5.34–36) ove si allude genericamente all'equiparazione del *numen* imperiale a figure greche semidivine.

Anche in questo caso di tratta di verificare se, e in quale misura, formule come *iurare per genium* possano essere considerate identiche come supposto da Mommsen e da altri dopo di lui.[15] Sembra invece potersi escludere che il verso di Orazio faccia riferimento al primo tentativo di introdurre il culto del *numen Augusti*.[16] Converrà accettare le attestazioni poetiche per quello che sono senza peraltro sottovalutarne le implicazioni. Ma la persistenza dei riferimenti al culto del *numen* (a cominciare da Verg. *georg.* 1.29–30: *ac tua nautae / numina sola colant*) non può essere pura coincidenza. E Ovidio, pur apparentemente banalizzante, deve essere tenuto in conto.[17]

E, soprattutto, non si può trascurare la *praefatio* del *De Architectura* di Vitruvio, che non è databile con sicurezza ma che risalirà a un anno vicino alla battaglia di Azio,[18] in cui, tra l'altro, si riecheggia il famoso enunciato del cap. 34 delle *Res Gestae* secondo la corretta restituzione del testo originale (*potens rerum omnium*): *cum divina tua mens et numen, imperator Ceasar, imperio potiretur orbis terrarum* (Vitr. *arch.* lib. 1 praef. 1).

In proposito si deve notare come la locuzione *divina mens*, che non poco disturbava i primi commentatori, ricorra in Cicerone a proposito del giovane Ottaviano, un indizio non secondario della flessibilità e della creatività terminologica in un'epoca di grande innovazioni sul piano politico e ideologico: *C. Caesar Octavianus adulescens, paene potius puer, incredibili ac divina quadam mente atque virtute* (Phil. 3.3).

Il riconoscimento ufficiale – che è altra cosa, ovviamente, rispetto a un presunto "conferimento" – del *numen Augusti* e la sua menzione dovettero tardare ma non di molto. I riscontri epigrafici appaiono sufficientemente eloquenti. Sappiamo di un altare, probabilmente una copia di età antonina, dedicato al *numen Augusti* dalla *plebs Narbonensium* nel 12/13 d.C. per adempiare a un voto fatto nell'11:

---

15 T. Mommsen, "Die Literaturbriefe des Horaz," in *Gesammelte Schriften*, 8 voll. (Berlin: Weidmann, 1905–1913), 7:175–86. Altra letteratura in Brink, *Horace*, 56.
16 Come vuole S. Weinstock, *Divus Julius* (Oxford: Clarendon Press, 1971), 213 n. 7.
17 Si veda, ad esempio, *trist.* 3.8.13: *Augusti numen adora*. E nei *Fasti* Ovidio "neatly encapsulates the invisible presence of Augustus": S. F. R. Price, "The Place of Religion: Rome in the Early Empire," in A. K. Bowman, E. Champlin e A. Lintott (edd.), *The Cambridge Ancient History*, new ed., vol. 10 (Cambridge: Cambridge University Press: 1996), 812–47 (838).
18 P. Fleury, l'editore di Vitruvio per le Belles Lettres (*Vitruve, De l'Architecture*, vol. 1 [Paris: Les Belles Lettres, 1990]), propende per un periodo compreso tra il 35 e il 25 a.C.

> T(ito) Statilio Taur[o] / L(ucio) Cassio Longino / co(n)s(ulibus) X K(alendas) Octobr(es) / numini Augusti votum / susceptum a plebe Narbo/nensium in perpetuom
>
> ... Pléps Narbonen/sium áram Narbone in foro posuit ad / quam quot annis VIIII K(alendas) Octobr(es) qua die / eum saeculi felicitas orbi terrarum / réctorem édidit tres equites Romaní / á plebe et tres líbertini hostias singu/las inmolent et colónis et incolís ad / supplicandum númini eius thús et vínum / dé suo eá die praestent et VIII K(alendas) Octobr(es) / thus vínum colonís et incolís item prae/stent ...[19]

Dunque l'ara è dedicata ad Augusto ancora in vita. È possibile, ma non strettamente necessario, presupporre un modello di riferimento a Roma per tale iniziativa di cui si avrebbe un riscontro nella dedica di un altare monumentale registrata nei *Fasti Praenestini* per il 17 gennaio forse del 6 d.C. (in ogni caso tra il 5 e il 9 d.C.).[20]

A Leptis Magna tra l'11 e il 12 d.C. il locale collegio dei *quindecemviri* dedica un *chalcidicum* e altro al *numen* di Augusto:

> Numini Imp(eratoris) Caesaris divi f(ilii) Aug(usti) pont(ificis) m[ax(imi)] imp(eratoris) XX co(n)s(ulis) XII]I tr(ibunicia) pot(estate) XXXIIII calchidicum et porticus et porta et via ab XVvir(is) sac(rorum) [... dedica]ta est.[21]

In questa prospettiva il problema dell'evoluzione della formula nella locuzione aggettivale *numen Augustum* non sembra porre interrogativi particolari.[22] Nel caso dell'altare eretto a Forum Clodii, ove si parla per tre volte di un'*ara* (dedicata) *numini Augusti*[23] a colpire è l'organizzazione del rituale, che ormai sembra precisamente codificato in rapporto al *dies natalis* di Augusto (23–24 settembre) e a quello di Tiberio (il 16 novembre).[24] Ed è notevole che si precisi che i decurioni devono provvedere *thure ac vino*, prima di andare a desinare a loro volta, che i Geni dei due imperatori pranzino all'altare.

---

19 *CIL* XII 4333 = *ILS* 112 (cf. V. Ehrenberg e A. H. M. Jones, *Documents Illustrating the Reigns of Augustus and Tiberius*, 2nd ed. [Oxford: Clarendon Press, 1955], 85–86 [no. 100]).
20 *Inscr. It.* XIII 2, 115, no. 17 (Ehrenberg e Jones, *Documents*, 46). Cf. A. Alföldi, *Die zwei Lorbeerbäume des Augustus* (Bonn: R. Habelt, 1973), 42–43.
21 *IRT* 342 (a) = *AE* 1948, no. 8 (Ehrenberg e Jones, *Documents*, 91 [no. 105a]). Cf. M. Pentiricci, "L'attività edilizia a Leptis Magna tra l'età tetrarchica e il V secolo: una messa a punto," in I. Tantillo e F. Bigi (edd.), *Leptis Magna: una città e le sue iscrizioni in epoca tardoromana* (Cassino: Università di Cassino, 2010), 112–15.
22 Come sembra a D. Fishwick, "Numen Augustum," *Zeitschrift für Papyrologie und Epigraphik* 160 (2007): 247–55 (spec. 249–55).
23 *CIL* XI 3303 = *ILS* 154 (Ehrenberg e Jones, *Documents*, 87 [no. 101]).
24 Cf. S. De Maria, *Gli archi onorari di Roma e dell'Italia romana* (Roma: L'Erma di Bretschneider, 1988).

Pare verosimile che la formula *numen Augustum* debba considerarsi una semplice variante di *numen Augusti*.[25] È proposta suggestiva che esso "no se considerara como el poder divino 'personal' de Augusto (o de Tiberio), sino como el poder divino del soberano en general, un poder que desde la formación de la monarquía recaía en los sucesores."[26]

Un ultimo elemento da tener presente riguarda la dedica di archi onorari, recanti statue dell'imperatore, nei quali poteva esserci esplicito riferimento al culto per il suo *numen*. Le statue degli imperatori potevano essere oggetto di culto allo stesso modo delle divinità. Per l'inizio del Principato un esempio è rappresentato dall'arco, fatto costruire da un *sevir et Augustalis* a Parma mentre per l'età domiziano-traianea merita considerazione l'arco di Medinaceli (Soria, *Hispania citerior*). Secondo il testo dell'iscrizione, peraltro ora perduta, il *sevir* di Parma Munatius Apsyrtus dedicò il monumento al *numen Augusti* e provvedette alla pavimentazione della strada di accesso in città.[27] Si tratta di una serie di indicazioni che concorrono a precisare il fondamento religioso del regime imperiale che ha – conviene ribadirlo – il suo punto di partenza nell'appellativo di Augusto attribuito a Ottaviano dal senato nella seduta del 16 gennaio del 27. Non è semplice cogliere con precisione l'evoluzione del culto imperiale nei suoi riscontri puntuali. Ma sembra possibile individuare una sicura evoluzione in un'indiscutibile, coerente linea di continuità.[28] Rifiutata la consacrazione che Agrippa avrebbe voluto nel pantheon da lui fatto costruire nel 27 abbiamo riscontri obiettivi che Augusto in prima persona promosse un culto indiretto della sua persona e della sua famiglia. La stessa articolazione di Roma in *regiones* fu da lui utilizzata per organizzare il culto dei Lari Augustali e unire così più strettamente a sé i ceti inferiori della plebe urbana e dei liberti.[29] Non è implausibile pensare che oggetto di culto fosse inizialmente il *genius* e quindi, forse per evoluzione interna spontanea, in qualche modo anticipata dai poeti, questo fosse esteso al *numen*

---

25 Cf. G. Alföldy e J. M. Abascal, "La inscripción de l'arco," in J. M. Abascal e G. Alföldy (edd.), *El arco de Medinaceli (Soria, Hispania citerior)* (Madrid: Real Academia de la Historia; Alicante: Universidad de Alicante, 2002), 71–115 (101). Cf. Ov. *trist.* 5.11.20: *Caesareum numen.*
26 Alföldy e Abascal, "La inscripción," 106.
27 De Maria, *Gli archi onorari*, 166–67 e 248–49.
28 Sulle scelte graduali che portarono alla definizione e all'organizzazione del culto imperiale a Roma cf. J. Scheid, "Honorer le prince et vénérer les dieux: culte public, cultes des quartiers et culte impérial dans la Rome augustéene," in N. Belayche (ed.), *Rome, les Césars et la Ville aux deux premiers siècles de notre ère* (Rennes: Presses universitaires de Rennes, 2001), 85–105.
29 Cf. D. Kienast, *Augustus Prinzeps und Monarch* (Darmstadt: Wissenschaftliche Buchgesellschaft, 1992), 255.

dell'imperatore.[30] L'ara *numinis Augusti* dedicata da Tiberio il 17 gennaio del 6 d. C., il giorno del compleanno di Livia, assumerebbe così il significato di completamento di un processo: anche i ceti superiori risultano così vincolati al culto del sovrano. Certo è che le formulazioni contenute in un testo così rilevante dal punto di vista politico, come quello del s. c. de Cn. Pisone, sembrano segnalarsi come un punto di arrivo, che appare definitivo, di un'evoluzione di un elemento dell'ideologia imperiale già presupposto nei testi epigrafici ricordati. L'offesa recata al *numen*, che intenderei quindi nel senso di "autorità (= potere) sacra," significa l'offesa alla sacralità della funzione e della figura dell'imperatore oltre che della sua persona.[31] Così è senz'altro nella prima delle due attestazioni nel testo del *s. c.* in cui l'offesa recata ad Augusto deve essere genericamente intesa come un'azione politica che contravveniva alle sue decisioni e alla sua volontà di pacificazione. La seconda, invece, che presuppone una valutazione puntuale da parte del senato (*numen... quoque arbitrari senatum,* ove il *quoque* risulta avere un valore del tutto particolare), sembra presupporre azioni (o prese di posizione) specifiche gravemente lesive dell'autorità dell'imperatore in un arco di tempo prolungato, tanto contro le sue *imagines* quanto contro la sua *memoria*. È un passo importante nella via del riconoscimento della peculiare funzione del monarca romano, della sua eccezionalità e della sua inviolabilità. La formula *devotus numini maiestatique*, così diffusa come manifestazione di lealtà verso l'imperatore nel corso del terzo e del quarto secolo, può considerarsi un ulteriore sviluppo in questa direzione.[32]

C'è un'iscrizione, su cui ha richiamato di recente l'attenzione Ignazio Tantillo,[33] che merita di essere ricordata per il valore esemplare che ha in merito all'evoluzione semantica del concetto di *numen* e del suo significato religioso nella Tarda Antichità. Essa fu fatta apporre tra il 324 e il 326 dal governatore della Tripolitania, Laenatius Romulus, sulla base di una statua per Costantino

---

30 Con la riforma del 7 a.C. i *magistri vici* si occupano del culto dei *Lares* e del *Genius Augusti*: si veda Suet. *Aug.* 31.4 con il commento di N. Louis, *Commentaire historique et traduction du "Divus Augustus" de Suètone* (Bruxelles: Éditions Latomus, 2010), *ad loc.* (272). Cf. G. Niebling, "Laribus Augustis Magistri Primi. Der Beginn des Compitalkultes der Lares und des Genius Augusti," *Historia* 5 (1956): 303–31 e R. Duthoy, "Les *Augustales," in W. Haase (ed.), *Aufstieg und Niedergang der römischen Welt*, II 16.2 (Berlin: De Gruyter, 1978), 1254–309 (spec. 1291).

31 Tra le gravissime accuse rivolte contro Pisone c'era "the shocking and killing fact that the *numen* of Augustus had been violated by Piso's removing 'all signs of honour from his memory and his portraits'" (B. Levick, *Augustus: Image and Substance* [Harlow: Longman, 2010], 292).

32 Cf. A. Sánchez-Ostiz, "Notas sobre *numen* y *maiestas* en Apuleyo," *Latomus* 62 (2003): 844–63.

33 Cf. R. Tantillo, "L'impero della luce: riflessioni su Costantino e il sole," *MEFRA* 115 (2003): 985–1048.

eretta nel foro di Leptis Magna per commemorare il restauro del foro medesimo.[34] Si deve considerare la formula usata per descrivere la statua del principe: *statua marmorea suo numine radians*.[35] La *iunctura* è particolarmente ardita non solo linguisticamente: nei casi paralleli è il materiale (il marmo o il metallo) che dà luminosità al monumento. In questo caso è il *numen* di Costantino, dunque dell'imperatore che ha la forza di un dio, che conferisce alla statua un peculiare effetto di irraggiamento.

Numen è, d'altra parte, un concetto centrale della teologia politica dioclezianea: là dove erano raffigurati gli imperatori opera la loro *vis divinitatis* (*demens qui nesciebat...ubique vim vestrae divinitatis esse, ubi vultus vestri, ubi signa colerentur*).[36] Al *numen* di Costantino è dedicato il panegirico pronunciato a Treviri nel 310: il retore dedica intenzionalmente il suo discorso al solo *numen* di Costantino (*Paneg.* 6.1.4: *tuo modo, Constantine, numini dicabo*).[37]

Bibliografia

Alföldi, A. *Die zwei Lorbeerbäume des Augustus* (Bonn: R. Habelt, 1973).

Alföldy, G., e J. M. Abascal. "La inscripción de l'arco," in J. M. Abascal e G. Alföldy (edd.), *El arco de Medinaceli (Soria, Hispania Citerior)* (Madrid: Real Academia de la Historia; Alicante: Universidad de Alicante, 2002), 71–105.

Bilabel, P. "Fragmente aus der Heidelberger Papyrussammlung," *Philologus* 80 (1924): 331–41.

Brink, C. O. *Horace on Poetry: Epistles Book II; The Letters to Augustus and Florus* (Cambridge: Cambridge University Press, 1982).

De Maria, S. *Gli archi onorari di Roma e dell'Italia romana* (Roma: L'Erma di Bretschneider, 1988).

Duthoy, R., "Les *Augustales," in W. Haase (ed.), *Aufstieg und Niedergang der römischen Welt*, II 16.2 (Berlin: De Gruyter, 1978), 1254–309.

Eck, W., A. Caballos, e F. Fernández. *Das senatus consultum de Cn. Pisone patre* (München: Beck, 1996).

Ehrenberg, V., e A. H. M. Jones. *Documents Illustrating the Reigns of Augustus and Tiberius*, 2nd ed. (Oxford: Clarendon Press, 1955).

---

34 *IRT* 467.

35 Rr. 8–9. Cf. Tantillo, "L'impero della luce," 993.

36 *Paneg.* 11.13.5 del 297/8. È notevole come il concetto di *numen* si presti a essere utilizzato nella fase di transizione verso la cristianizzazione dell'Impero (*Paneg.* 12.1.1, 4.1, 13.2, 19.1): cf. K. Rosen, *Constantin der Große, die Christen und der Donatistenstreit 312-314. Eine Untersuchung zu Optatus von Mileve, Appendix V, und zum Verhältnis von Staat und Kirche im 4. Jahrhundert*, Nordrhein-Westfälische Akad. d. Wiss. und der Künste, Vorträge G 432 (Paderborn: Ferdinand Schöningh, 2011), 20–21.

37 Tantillo, "L'impero della luce," 1000–1001.

Fishwick, D. *The Imperial Cult in the Latin West*, vol. 3, pt. 4 (Leiden: Brill, 2005).
Fishwick, D. "Cn. Piso Pater and the Numen Divi Augusti," *Zeitschrift für Papyrologie und Epigraphik* 159 (2007): 297–300.
Fishwick, D. "Numen Augustum," *Zeitschrift für Papyrologie und Epigraphik* 160 (2007): 247–55.
Fleury, P. *Vitruve, De l'Architecture*, vol. 1 (Paris: Les Belles Lettres, 1990).
Fraschetti, A. "L'eroizzazione di Germanico," in A. Fraschetti (ed.), *La commemorazione di Germanico nella documentazione epigrafica. Tabula Hebana e Tabula Siarensis* (Roma: L'Erma di Bretschneider, 2000), 139–60.
Gradel, I. *Emperor Worship and Roman Religion* (Oxford: Clarendon Press, 2002).
Kassel, R., e C. Austin. *Poetae comici Graeci*, vol. VI 2: *Menander: Testimonia et fragmenta apud scriptores servata* (Berlin: De Gruyter, 1998).
Kienast, D. *Augustus Prinzeps und Monarch* (Darmstadt: Wissenschaftliche Buchgesellschaft, 1992).
Levick, B. *Augustus: Image and Substance* (Harlow: Longman, 2010).
Lindsay, W. M. *Sexti Pompei Festi De verborum significatu quae supersunt cum Pauli epitome* (Leipzig. Teubner, 1913; rist., Stuttgart: Teubner, 1997).
Louis, N. *Commentaire historique et traduction du "Divus Augustus" de Suétone* (Bruxelles: Éditions Latomus, 2010).
Marcone, A. "Un 'dio presente': osservazioni sulle premesse ellenistiche del culto imperiale romano," in S. Bussi e D. Foraboschi (edd.), *Roma e l'eredità ellenistica*, Studi Ellenistici 23 (Pisa: Fabrizio Serra Editore, 2010), 205–15.
Mommsen, T. "Die Literaturbriefe des Horaz," in *Gesammelte Schriften*, 8 voll. (Berlin: Weidmann, 1909), 7:175–86.
Niebling, G. "Laribus Augustis Magistri Primi. Der Beginn des Compitalkultes der Lares und des Genius Augusti," *Historia* 5 (1956): 303–31.
Pentiricci, M. "L'attività edilizia a Leptis Magna tra l'età tetrarchica e il V secolo: una messa a punto," in I. Tantillo e F. Bigi (edd.), *Leptis Magna: una città e le sue iscrizioni in epoca tardoromana* (Cassino: Università di Cassino, 2010), 112–15.
Pötscher, W. "'Numen' und 'Numen Augusti,'" in W. Haase (ed.), *Aufstieg und Niedergang der römischen Welt*, II 16.1 (Berlin: De Gruyter, 1978), 355–92.
Price, S. F. R. "The Place of Religion: Rome in the Early Empire," in A. K. Bowman, E. Champlin, e A. Lintott (edd.), *The Cambridge Ancient History*, new ed., vol. 10 (Cambridge: Cambridge University Press: 1996), 812–47.
Rosen, K. *Constantin der Große, die Christen und der Donatistenstreit 312–314. Eine Untersuchung zu Optatus von Mileve, Appendix V, und zum Verhältnis von Staat und Kirche im 4. Jahrhundert*, Nordrhein-Westfälische Akad. d. Wiss. und der Künste, Vorträge G 432 (Paderborn: Ferdinand Schöningh, 2011).
Sánchez-Ostiz, A. "Notas sobre *numen* y *maiestas* en Apuleyo," *Latomus* 62 (2003): 844–63.
Scheid, J. "Honorer le prince et vénérer les dieux: culte public, cultes des quartiers et culte impérial dans la Rome augustéene," in N. Belayche (ed.), *Rome, les Césars et la Ville aux deux premiers siècles de notre ère* (Rennes: Presses universitaires de Rennes, 2001), 85–105.
Scott, R. T. "Providentia Aug.," *Historia* 31 (1982): 436–59.
Tantillo, R. "L'impero della luce: riflessioni su Costantino e il sole," *MEFRA* 115 (2003): 985–1048.
Weinstock, S. *Divus Julius* (Oxford: Clarendon Press, 1971).

Stephen Menn
How Archytas Doubled the Cube

Abstract: Archytas' doubling of the cube, or more generally his construction of two mean proportionals between two given lines, is perhaps the only achievement of Greek science which respectable historians of science continue to attribute to something like divine inspiration. Building on the work of B. L. van der Waerden and Wilbur Knorr, I reconstruct an analysis which could have led to Archytas' synthetic construction. I show how Archytas could be led to this analysis by trying to extend techniques of analysis in plane geometry to solid geometry using some elementary propositions of spherics, i.e., of the mathematical foundations of astronomy. Although the sphere is not explicitly mentioned in Archytas' construction, it is crucial to understanding the underlying analysis, and to making the construction seem non-miraculous. Archytas' work allows us a glimpse into the history of Greek geometry before the discovery of the conic sections (credited to Menaechmus two generations later). For later Greek geometry, a "solid problem," i.e., one solved by "solid methods," is one solved using conic sections. But this is a narrowing of an earlier practice when "solid problems" such as the doubling of the cube were solved using techniques of solid geometry, always using surfaces of revolution and their intersections with each other and with planes to find the locus of the point being sought, but with no restriction to cones or their plane sections.

Hippocrates of Chios reduced the problem of doubling the cube to the problem of finding two mean proportionals: that is, given two straight lines A and D, of finding straight lines B and C such that A:B::B:C::C:D. Hippocrates did not himself solve the problem of finding two mean proportionals, but subsequent mathematicians devoted themselves to finding ingenious solutions to the problem. Eratosthenes in giving his own solution described earlier solutions as well, and Eutocius in his commentary on Archimedes, *On the Sphere and the Cylinder* 2.1 (where Archimedes simply tells his reader to find two mean proportionals, without saying how) reports twelve solutions in detail. Of these, the earliest and the most brilliant is the solution attributed to Archytas of Tarentum in the early fourth century B.C., which finds the crucial

---

This paper is dedicated to Heinrich von Staden, an intellectual hero of mine since our first meeting at Yale in 1995. I would like to thank G. E. R. Lloyd, Henry Mendell, Reviel Netz, and Heike Sefrin-Weis for comments on earlier versions, and Tim Wagner for producing the diagrams. This paper was written in the framework of the Topoi Excellence Cluster, as part of a collaborative project within Topoi research group D–3 on the origin and history of the conic sections.

point as the intersection of a cylinder, a cone, and a torus, or rather a quasi-torus with a hole of radius 0. There seems to be nothing else like this, where a point is found as the intersection of three curved surfaces, anywhere else in extant Greek mathematics, and the construction is astonishingly sophisticated for so early in the development of mathematics. At the same time, there is no reason to doubt the authenticity of Archytas' construction. Eutocius reports it on the excellent authority of Eudemus' *History of Geometry*, and there is no reason why Eutocius should not (like Proclus in his commentary on Euclid's *Elements* 1, and Simplicius in his commentary on Aristotle's *Physics* 1) have had direct access to Eudemus' work. Furthermore, while the construction is sophisticated, it is so in a distinctive way, alien to the later development of Greek geometry, so that it is implausible that a later discovery is being projected back onto Archytas' time: the anomaly is not just the heavy reliance on motion, but the use of curved surfaces as auxiliaries to solve a problem set out in terms of straight lines, and in particular the use of the torus. We seem to have here a witness of a promising direction that fourth-century geometry could have gone in, but ultimately did not.[1]

I want to try to reconstruct some context that will make Archytas' work seem less miraculous (and this will also support the case for authenticity). This is in part a matter, as so often in Greek geometry, of rationally reconstructing the analysis behind a transmitted synthesis.[2] But it is also a matter of reconstructing what Ken Saito calls the "toolbox," the set of techniques ready to be applied to the analysis of any given problem, that was in the hands of early Greek geometers. Earlier attempts at reconstruction, especially those of B. L. van der Waerden and Wilbur Knorr, have contributed valuable elements to the solution, but every reconstruction I have seen leaves an arbitrary step in Archytas' construction, not justified by any previous analysis, so that it would apparently be pure luck that this arbitrary step would lead to a successful construction. We can do better.

★ ★ ★

---

[1] Some later geometers developed the study of the torus (Dionysodorus, in the Archimedean spirit, determined the volume of the torus, and, as we will see, Perseus studied plane sections of a torus), but all their work is lost except for a few references. There is no evidence that anyone after Archytas used solid geometry to find a desired point as the intersection of three curved surfaces.

[2] On Greek geometrical analysis, some starting-points are my "Plato and the Method of Analysis," *Phronesis* 47 (2002): 193–223; Jaakko Hintikka and Unto Remes, *The Method of Analysis: Its Geometrical Origin and its General Significance* (Dordrecht: D. Reidel, 1974); and Alexander Jones, *Pappus of Alexandria: Book 7 of the Collection*, edited with translation and commentary, 2 vols. (New York: Springer, 1986).

Archytas' construction, as Eutocius reports it from Eudemus, is as follows:[3]

---

[3] I am translating the text in Carl A. Huffman, *Archytas of Tarentum: Pythagorean, Philosopher, and Mathematician King* (Cambridge: Cambridge University Press, 2005), 342–43 (= 84.12–88.2 Heiberg-Stamatis). Huffman translates the text at 343–44; Reviel Netz also translates the text as part of his complete translation of Eutocius' commentary on *On the Sphere and the Cylinder*: Reviel Netz, *The Works of Archimedes: Translated into English, Together with Eutocius' Commentaries, with Commentary, and Critical Edition of the Diagrams*, vol. 1: *The Two Books on the Sphere and the Cylinder* (Cambridge: Cambridge University Press, 2004), 290–93.

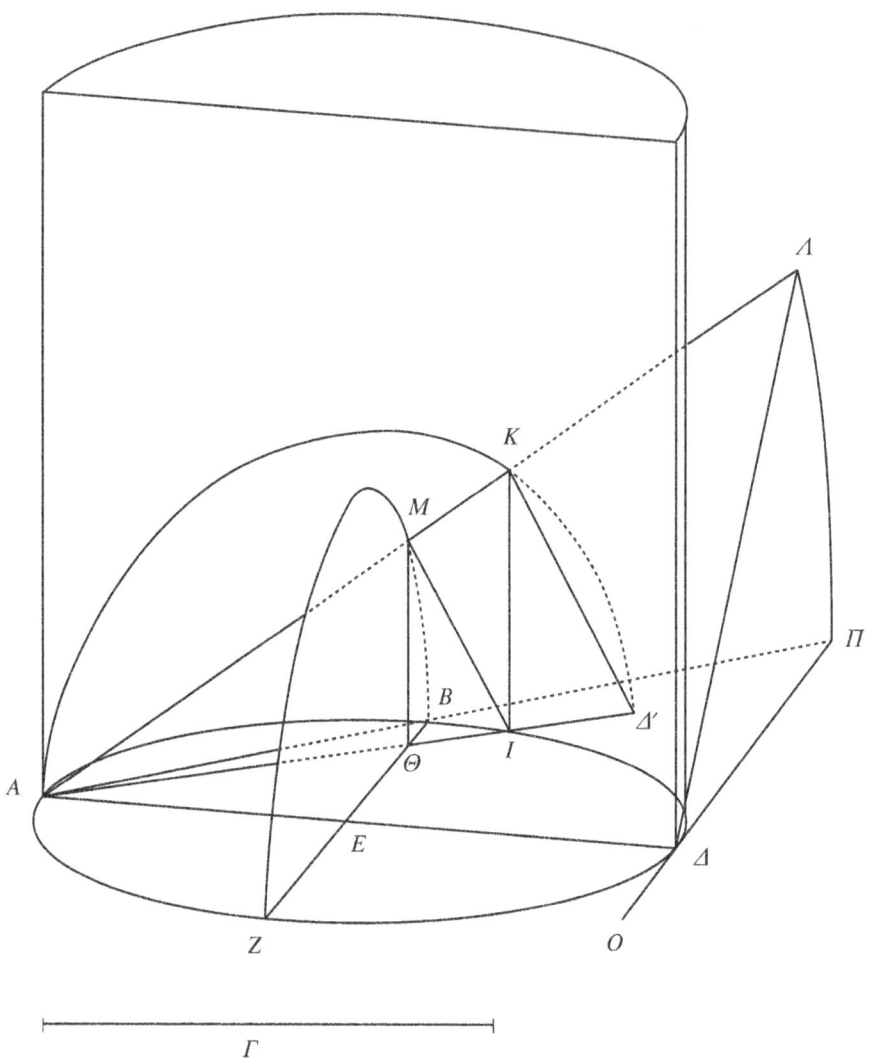

Diagram 1 (after van der Waerden)

Let the two given straight lines be AΔ and Γ. One must then find two mean proportionals between AΔ and Γ. Let a circle ABΔZ be circumscribed around the greater straight line AΔ, and let AB be fitted in [to the circle, as a chord] equal to Γ, and, having been produced, let it meet at Π the tangent to the circle from Δ. Let BEZ be drawn parallel to ΠΔΟ, and let a semicylinder be conceived at right angles on the semicircle ABΔ, and on AΔ a semicircle lying at right angles in the parallelogram of the semicylinder [i.e., in the vertical rectangle, at right angles to the base plane ABΔZ, which divides the cylinder over circle ABΔZ into two semicylinders]. When this semicircle is rotated from Δ to B, with the end of the diameter, A, remaining fixed, it will cut the cylindrical surface in its rotation and draw a certain [curved] line on it. Again, if while AΔ remains fixed the triangle AΠΔ is rotated with a motion contrary to that of the semicircle,[a] with its straight line AΠ it will produce a conic surface; and this line, while it is being rotated [around axis AΔ], meets the line on the cylinder at some point. At the same time, the point B [which lies on AΠ and is moved with AΠ] will circumscribe a semicircle in the surface of the cone. At the place where the two lines meet,[b] let the moving semicircle have position Δ'KA, and let the triangle rotating in the contrary direction have position AΛA,[c] and let the point of this meeting be K. Let the semicircle drawn by B [as it rotates around axis AΔ] be BMZ, and let the common section of it and of circle BΔZA [which are in two different planes] be BZ, and let a perpendicular be dropped from K to the plane of semicircle BΔA: it will fall on the circumference of the circle, since the cylinder is at right angles [and since point K is on this cylinder]. Let it fall, and let it be KI, and let the line joined from I to A meet BZ at Θ, and let line AΛ meet semicircle BMZ at M, and let KΔ', MI and MΘ be joined. Then since each of the semicircles Δ'KA and BMZ is at right angles to the base plane [i.e., to the plane of circle ABΔZ],[d] therefore their common section MΘ is also at right angles to the plane of the circle [i.e., the same base plane]; so MΘ is also at right angles to BZ. Therefore rectangle BΘ.ΘZ, which is [equal to] rectangle AΘ.ΘI, is equal to the square on MΘ. Therefore triangle AMI is similar to each of triangles MIΘ and MAΘ, and angle IMA is right.[e] And angle Δ'KA is also right [since it is in a semicircle]. Therefore lines KΔ' and MI are parallel, and it will be proportional, as Δ'A to AK (that is, as KA to AI) so IA to AM, because of the similarity of the triangles [sc. the right triangles AMI, AIK, and AKΔ', all in the same angle at A, since AMK lie on one straight line and AIΔ' on another]. Therefore the four lines Δ'A [which is equal to ΔA], AK, AI and AM are in continuous proportion. And AM is equal to Γ, since it is also equal to AB. Therefore, the two lines AΔ and Γ having been given, two mean proportionals AK and AI have been found.

Notes to Archytas' Construction
  [a] The semicircle on AΔ is being rotated around an axis passing through A and perpendicular to the base-plane ABΔZ, while the triangle AΠΔ is being rotated around the axis AΔ. Since the two axes of rotation are at right angles to each other, neither direction that the triangle could rotate could be properly "contrary" to the direction of rotation of the semicircle. If we think of lines AΔ and ABΠ as lying on a horizontal plane, and the semicircle on AΔ as lying in a vertical plane above its base AΔ, then as the semicircle pivots horizontally around point A, its base moving from AΔ to AB produced, the triangle AΠΔ is turning "upwards" on its hinge AΔ, and so mov-

ing into the region of space across which the semicircle is moving. If triangle AΠΔ stayed in its initial position, it (and in particular its side AΠ) would never encounter the semicircle at any stage in its motion (except for the base of the semicircle, which stays in the plane AΠΔ); again, if triangle AΠΔ moved "down," away from the semicircle, it would not encounter it. Since it moves "up," into the path of the semicircle, it can be described as moving contrary to the semicircle, i.e., moving so as to encounter the semicircle. But it makes no difference to Archytas' construction how fast the triangle and the semicircle move: his construction turns not on where the semicircle is at time t and where the triangle (or its side AΠ) is at the same time t, but rather on the encounter between the moving line AΠ and the curve on the cylinder conceived as the static result of the moving semicircle's cutting the cylinder. We would describe the curve on the cylinder as the intersection between the surfaces of the semitorus (the solid of revolution generated by the semicircle) and the cylinder; and we would describe the point K which Archytas constructs as the intersection between the surfaces of the cone (the solid of revolution generated by the triangle), the semitorus and the cylinder.

(b) The "two lines" are the curved line on the cylinder, i.e., the intersection of the surface of the (semi-)cylinder with the surface of revolution produced by the rotation of the semicircle (the semitorus, or, strictly speaking, only a segment of the semitorus, since Archytas considers only part of the rotation of the semicircle, as its base pivots around point A from diameter AΔ to chord AB produced, not all the way in a full circle around point A), and the straight line AΠ (i.e., AB produced) as it rotates around axis AΔ. Archytas considers the line on the cylinder as the fixed result of the motion of the semicircle, while he considers the line AΠ as in motion.

(c) Archytas uses the letter Δ, without differentiation, both for the original position of Δ, diametrically opposite to A in the circle ABΔZ, and for the position that point Δ comes to occupy when the semicircle on AΔ is rotated around point A within the base-plane ABΔZ. This has the confusing result that when he says that the moving semicircle has position ΔKA, and the moving triangle has position ΔΛA, the Δ of the triangle is the original Δ while the Δ of the semicircle has moved to another position. To avoid this, I say Δ′ for the new position of point Δ, and keep Δ as the label for the original position.

(d) Semicircle Δ′KA is at right angles to the base plane because the initial semicircle on AΔ is at right angles to the base plane, and it remains at right angles to the base plane as it is rotated around A. Semicircle BMZ is at right angles to the base plane because it is produced by rotating point B around the axis AΔ, which lies in the base plane.

(e) Lines BΘZ and AΘI are chords crossing in circle ABΔZ, so by Euc. elem. 3.35 the rectangles BΘ.ΘZ and AΘ.ΘI are equal. MΘ is perpendicular to the plane of circle ABΔZ, and hence to both of these chords. Since circle-segment BMZ is a semicircle, the square on MΘ is equal to rectangle BΘ.ΘZ (we have here the diagram of Euc. elem. 6.8 and 6.13). So the square on MΘ is also equal to rectangle AΘ.ΘI, and so circle-segment AMI will also be a semicircle, with angle AMI right, and with lesser triangles AΘM and MΘI each similar to greater triangle AMI, as in Euc. elem. 6.8 and 6.13.

* * *

Before we offer rational reconstructions of Archytas' analysis and of his toolbox, some historical reflections may help. Although Archytas' procedure in finding point K as the intersection of three surfaces has no precise parallel in extant Greek mathematics, it is not entirely isolated. Archytas' synthesis would have been preceded by an analysis showing, from one property of the desired point K, that it lies on a given cone, from another that it lies on a given cylinder, from another that it lies on a given torus. So the cone, the cylinder, and the torus would be loci, described in locus-theorems "if a point stands in a given relation to a given figure, it lies on a given locus," namely the cone or the cylinder or the torus; also, if the point has *two* of the three required properties, it will lie on a further locus, the intersection of the cone with the cylinder or with the torus or of the cylinder with the torus. Locus-theorems are always important in Greek analyses, and while the locus in question is typically a curve in the plane, we know that some geometers were interested in loci which are surfaces of solids or curves on such surfaces. There were whole treatises on such loci. These texts are lost, and there is some confusion about what they meant by the different kinds of loci, but Pappus says that an "Aristaeus the Elder" sometime before Euclid wrote a treatise on solid loci, "immediately following on [συνεχῆ] the conics," in five books, and that Apollonius analyzed the problem of finding two mean proportionals "by conic sections, but others by the solid loci of Aristaeus."[4] Whatever solid loci are here, they were not solids (there won't be a locus-theorem saying that a point with a certain property is *inside* a given sphere or cube). A priori, they might be the surfaces of solids, such as cones, but the whole later ancient mathematical/interpretive tradition takes them to be, rather, a kind of curved line. But if they were curved lines, they were called "solid" because they were curved lines "whose origin is from the cutting of a solid figure, like the cylindrical helix or the conic lines" (Procl. *in Euc.* 394–95 Friedlein): thus in some cases the intersection of a curved surface with a plane, like the conics, but in other cases non-planar curves on a curved surface, like the cylindrical helix, or Archytas' intersection of a cylinder with a torus. So those who analyzed two mean proportionals "by the solid loci of Aristaeus," whether surfaces or curves on surfaces, will have been doing something like Archytas' construction (less like Archytas if Aristaeus' curves on surfaces were plane sections of surfaces than if they were non-planar curves on surfaces). Meanwhile Euclid is said to have written a treatise in two books on "surface loci" (τόποι πρὸς ἐπιφανείᾳ, listed right after Aristaeus on solid loci). Here again it is not entirely clear whether, as most of the later ancient tradition thinks, these loci are themselves surfaces (so Proclus in the passage already cited, and Papp. 7, 660.18–662.10 Hultsch)

---

4 For Aristaeus writing on solid loci, immediately following on the conics, cf. Papp. 7, 672.20–21 Hultsch (for other references to his treatise on solid loci cf. 7, 636.23 Hultsch, and on conics cf. 672.11–13 Hultsch). For the chronological sequence Aristaeus-Euclid-Apollonius cf. 676.19–678.12 Hultsch. For Apollonius solving the problem of two mean proportionals by conic sections, others by the solid loci of Aristaeus, cf. 3, 56.4–6 Hultsch.

or whether they are curves on surfaces—but if they were curves on surfaces, they were *not* plane sections of surfaces (the conics, which are the paradigmatic solid loci, are never described as surface loci).[5] In any case, these kinds of loci were used in the analysis of problems, and solid and/or surface loci were not restricted to plane curves, much less to conic sections. Someone named Perseus wrote on "spiric sections," i. e., plane sections of the torus (σπεῖρα, Procl. *in Euc.* 111.19–112.8 Friedlein), some of which are said to resemble hippopedes. The Eudoxian hippopede, too, can be represented as the intersection of a sphere with a cylinder or a cone, and the theorem that a point borne on the equator of a uniformly rotating sphere, itself carried inside a concentric sphere rotating uniformly with the same speed around a different axis, remains on a given hippopede, can easily be formulated as a solid- or surface-locus theorem, although we have no direct evidence that any ancient writer in fact pursued this line of thought.

These texts on solid and surface loci, however sparse our witnesses to them, give some context for what Archytas did. They also help to bring out an underlying tension in what becomes the standard ancient classification of problems as planar, solid, or linear. A planar problem is one that can be solved by planar methods; a solid problem is a problem that cannot be solved by planar methods but can be solved by solid methods; a linear problem is a problem that cannot be solved by planar or solid methods but can be solved by linear methods; and it is supposed to be a fault to solve a planar problem by solid or linear methods, or a solid problem by linear methods, when they could and should have been solved by simpler methods. According to Pappus, planar methods are those that use only straight lines and circles; solid methods are those that in addition use con-

---

[5] Pappus 7 has a brief and unsatisfactory series of lemmas to Euclid on surface loci (1004.16–1014.24 Hultsch), after the much longer series of lemmas to the *Conics*. Jones, in his discussion of what little can be reconstructed of Euclid's *Surface Loci* from this text (*Pappus*, 2:591–95), thinks that Euclid's surface loci were surfaces, rather than curves on surfaces; but, if so, a locus theorem "if a point stands in a given relation to a given figure, it lies on a given surface" will also have immediate consequences "if a point on a given surface stands in a given relation to a given figure, it lies on a given curve on the surface." But at Papp. 4, 258.20–262.2 Hultsch (in Heike Sefrin-Weis, *Pappus of Alexandria: Book 4 of the Collection*, edited with translation and commentary [London: Springer, 2010], 58–59 and 137–39), it looks as if the "surface locus" mentioned is a cylindrical helix. At Papp. 4, 270.17–24 Hultsch (Sefrin-Weis, *Pappus*, 63–65 and 164–66) the "lines found on so-called surface loci" might be either themselves surface loci (in which case surface loci are curves) or intersections of two surface loci (in which case surface loci are surfaces). Eutoc. *in con.* 184.26–8 Heiberg is ambiguous. On balance, surface loci seem to be surfaces. Wilbur R. Knorr, however, was convinced that they were curves on surfaces (*The Ancient Tradition of Geometric Problems* [Boston: Birkhäuser, 1986], 129, 166, 230, and 342–43).

ics; and linear methods are those that use other kinds of curves.[6] On reflection, this does not really make sense: why are constructions involving conics, and no others, "solid constructions"? Conics are curves in the plane, not curves in space. It is certainly true that conics are curves in the plane that can be generated by a plane section of a solid, namely a cone. But many other kinds of curves can also by generated by plane sections of solids, such as spiric sections and also the quadratrix—why should conics be "solid" while these are "linear"? The text we have cited from Proclus includes the cylindrical helix alongside the conics as a solid locus, and so presumably problems solved using it are solid problems; and Eutocius says in his commentary on Apollonius' *Conics* that "the so-called solid loci have received their name from the fact that the lines through which the problems about them are constructed have their origin from the cutting of solids, like the sections of the cone *and many other* [curved lines]" (184.21–6 Heiberg, my emphasis).[7] And we have seen that even Pappus, the witness most insistent that solid problems must be solved exclusively through conics, distinguishes "the solid loci of Aristaeus" from the conics, as different methods for finding two mean proportionals. I suggest that these witnesses are preserving an earlier and broader sense of "solid problem," and that Archytas' methods for finding two mean proportionals (a preeminently solid problem) were "solid methods" in this older sense.

The conics, according to Proclus (*in Euc.* 111.21–2 Friedlein), were invented by Menaechmus, a student of Eudoxus, so roughly two generations after Archytas:[8]

---

[6] Papp. 3, 54.7–56.17 and 4, 270.1–272.14 Hultsch (extended and very closely parallel discussions of the three kinds of problems; on the latter, cf. Sefrin-Weis, *Pappus*, 63–65 and 144–46); more briefly in the same sense 7, 662.10–15 Hultsch. It is possible that Pappus is taking this classification from Apollonius. From the fragments of Apollonius' *Plane Loci* it seems that just those loci are planar which turn out to be circles (or, presumably, straight lines), and the dedicatory letter of the *Conics* says that *Conics* 3 contains many theorems "useful for the synthesis of solid loci" (4.11–12 Heiberg): the loci in question (probably mainly the "three-line locus" and "four-line locus") will be conics, and *perhaps* all solid loci will be conics. Nothing hangs on the notion of "linear" loci or constructions or problems: this is just a grab-bag name for whatever cannot be constructed by planar or solid methods, whatever exactly these include.
[7] Strictly speaking the quadratrix is generated by projecting onto a plane the intersection between another plane and a plectoid surface, see Sefrin-Weis, *Pappus*, 137–39 and 258–60. When Proclus (cited above) says that the cylindrical helix arises "from the cutting of a solid figure," either he means that it is the intersection of a cylinder with a plectoid, or he is using "cutting" for something broader than the intersection of two surfaces—the "inscribing" of a curve on a surface. On plectoid surfaces, see discussion below.
[8] For Menaechmus as a student of Eudoxus: Procl. *in Euc.* 67.9–10 Friedlein. Stob. 2.31.115 has a story of him talking with Alexander the Great. The references are collected in François Lasserre,

the only thing we know that Menaechmus did with them was to solve the problem of finding two mean proportionals (for Eutocius reports Menaechmus' solution to the problem along with solutions advanced by Archytas and others), and very likely this is the context in which he invented them. (Menaechmus gave two alternative constructions, one by intersecting two parabolas, one by intersecting a parabola with a right hyperbola. The idea can be seen using a Cartesian coordinate system. We find X and Y such that A:X::X:Y::Y:B by solving $X^2 = AY$ and $Y^2 = BX$, the graphs of the equations being two parabolas, or $X^2 = AY$ and $XY = AB$, the graphs being a parabola and a right hyperbola.) Although far less wonder-inducing than Archytas' construction, this is a major technical improvement. After the introduction of the conics and a collection of locus-theorems about them (saying that a given conic is the locus of a point satisfying a given σύμ-πτωμα, e.g., a relation between a line drawn from the point to the axis of the conic and the segment of the axis it cuts off, or a relation between lines drawn from the point to each asymptote parallel to the other asymptote), it became possible to systematically analyze and then construct, within the plane, problems that had previously been solved *ad hoc* by complicated three-dimensional constructions: the difference between Archytas' and Menaechmus' analyses and constructions of the two mean proportionals is paradigmatic. Three-dimensional considerations are now needed only once, for the initial construction of the conics and the proof of their most basic συμπτώματα, from which everything else can be derived by considerations taking place entirely within the plane, and far easier to diagram. Once this happens, the older Archytas-style constructions drop out of the "toolbox," and since the only solid loci actively in use are the conics, "solid problem" will now name only things constructed with conics; so much so that Archytas' construction could be criticized as using inappropriate methods, as a linear solution to a solid problem.[9]

De Léodamas de Thasos à Philippe d'Oponte: témoignages et fragments, édition, traduction et commentaire (Naples: Bibliopolis, 1987), 117–24.

[9] W. R. Knorr, "The Ancient Classification of Problems," in *Ancient Tradition*, 341–48, points out the unsatisfactory character of Pappus' definition of solid problems, and the tensions between the different ancient accounts, and rightly concludes that Pappus is not reflecting the original meaning of the classification of problems or of loci. But Knorr's alternative proposal (at 348) seems to be that ancient geometers were somehow instinctively groping toward the idea of the degree of an equation that could be solved using a certain type of locus; this is going in completely the wrong direction. Knorr in his "Observations on the Early History of the Conics" (*Centaurus* 26 [1982]: 1–24) suggests that Menaechmus introduced the parabola and right hyperbola in the context of finding two mean proportionals, not as sections of a cone, but as defined purely in the plane by συμπτώματα whose graphs could be plotted as a series of dots. This involves dissociating Eratosthenes' talk of "cutting the cone in the triads of Menaechmus"

If Archytas was working in the context of a freer use of solid methods than what prevails in most extant Greek mathematics, this does not mean that anything went. Even when the torus is added to the more usual Greek repertoire of curved surfaces, the sphere and cylinder and cone (Archimedes will add "spheroids" [ellipsoids] and "conoids" [paraboloids and hyperboloids]), it is still striking how limited these are. The solids treated by Greek solid geometry are almost always bounded either by planes or by surfaces of revolution—that is, the solids are either polyhedra or solids of revolution, or result from cutting solids of revolution by planes (like Archytas' semicylinder). This tendency is even stronger than might be suggested by the names I have been using for these surfaces and solids. Euc. *elem.* 11 defines the sphere, cone, and cylinder all as solids of revolution, generated by revolving respectively a semicircle, a right triangle, and a rectangle around a straight-line edge. This means that the cones and cylinders he considers are exclusively *right* circular cones and cylinders, i.e., in the case of the cone, that the vertex of the cone is "directly above" the center of the base-circle (that it lies on the line passing through the center of the base-circle and perpendicular to the plane of the base-circle); and so, for instance, when Euclid proves in *elem.* 12 that the volume of the cone is one-third the volume of a cylinder with the same base and height, he is proving it in far less than its full generality. Archimedes' "spheroids" and "conoids" are always ellipsoids and paraboloids and hyperboloids *of revolution*. As far as I know no Greek geometer ever discusses ellipsoids with three unequal axes; Archimedes is the first person who clearly mentions oblique cones and cylinders, and Apollonius the first who uses them systematically.[10] Apart from oblique cones and cylinders, the only

---

from what Menaechmus himself did, and also assuming that the "older names" for the conics (section of a right-angled cone etc., in use through the time of Euclid and Archimedes but abandoned by Apollonius in favor of parabola, hyperbola, ellipse) are later than Menaechmus, and that Menaechmus himself spoke of parabola etc., using terms from the "application of areas." Even if Knorr is right, he must admit that at least by the time of Aristaeus it had been discovered by means of locus-theorems that Menaechmus' curves are sections of cones, and thus that sections of cones can be used to find two mean proportionals. Knorr finds it implausible that Menaechmus would have worked out a theory of conics just for solving the problem of finding two mean proportionals, but Archytas had already worked out even more complicated techniques for analyzing the problem using various surfaces of revolution, and Menaechmus would be showing how to streamline these techniques of analysis so as to use only right circular cones and their plane sections. (It will be true on anyone's story that whoever introduced the conics did so for purposes of analysis, and that a systematic synthetic theory of conics was worked out only later.) Jones, *Pappus*, 2:573–77 defends the attributions of the conics to Menaechmus.
10 The study of pyramids, and Eudoxus' procedure for determining the volume of a cone by inscribing pyramids in it (preserved in Euc. *elem.* 12), would have led to the realization that the

curved surface I know of in Greek geometry *not* generated by revolution is the plectoid surface of Papp. 4, 262.18–19 and 270.21–22 Hultsch, and this is the exception that proves the rule, being generated by uniformly rotating a line segment around one endpoint while simultaneously uniformly translating it in a direction perpendicular to the plane of rotation.[11]

The study of these surfaces of revolution would have been heavily concentrated on the study of the sphere, which is most closely analogous to the study of a circle in the plane. The incentive for studying the sphere, and circles inscribed in the sphere, came naturally from astronomy: treatises like Autolycus' *On a Moving Sphere* and Theodosius' *Spherics* discuss thinly disguised abstract versions of the celestial equator, parallels, meridians, ecliptic, horizon, and arctic circle. Unlike the much later *Spherics* of Menelaus, these treatises do not especially privilege "greatest circles" on a sphere: they do not treat greatest circles, and triangles composed of greatest circles, as the analogues of straight lines and triangles in the plane. Rather, they treat circles as sections of the sphere, not "immanently" as curved lines on the surface of the sphere but as "ambient" results of cutting the solid by a plane: strikingly, they understand the angle between two circles in the sphere, not as the angle between two curved lines on the surface of the sphere, but as the angle between the two planes in space, which is *not* the same except in the special case of the angle between two greatest circles. Archytas said, according to Nicomachus, that those concerned with the μαθήματα made determinations "about geometry and arithmetic and spherics ... and not least also about music."[12] We should not rely too much on the wording, since parallel reports do not use the word "spherics," and Huffman has suggested that this word is Nicomachus' paraphrase: the parallel in Porphyry, which might stay closer to Archytas' words, has them making determinations "about the speed and risings and settings of the stars ... and about geometry and numbers and not least about music."[13] But in either case Archytas will be thinking of work on the mathematical foundations of the theory of the visible heav-

---

restriction to right cones was arbitrary. Jones, *Pappus*, 2:591–95 thinks that Euclid's surface loci included oblique cones and perhaps oblique cylinders.

11 Sefrin-Weis, *Pappus*, 60 and 64, 140 and 144–45. Papp. 4, 262.13 Hultsch (cf. Sefrin-Weis, *Pappus*, 60 and 140) also describes a cylindroid whose base is an Archimedean spiral; but we can describe this surface as generated, not by uniform rotation, but by uniform translation, of a curve in the plane. The real work of generating this surface happens with the generation of the spiral in the plane—and this too happens through the combination of a uniform rotation of a line around one endpoint with the uniform motion of a point along that line.

12 This is Archyt. fr. 1C in Huffman, *Archytas*, 108–109.

13 The Porphyry text is Huffman's fr. 1 A (*Archytas*, 103–107). See Huffman, *Archytas*, 114–24 and esp. 120–21 on Nicomachus' and Porphyry's manners of citation.

ens, of the same general nature as Autolycus' *On a Moving Sphere* and *On Risings and Settings* and Euclid's *Phaenomena*, based on a discussion of the sphere and the various circles on it. Autolycus in talking about a *moving* sphere assumes a series of propositions about resting spheres, many of them proved two hundred years later in the elementary part of Theodosius' *Spherics*, and Heath (following Tannery and Loria) argued that Autolycus and Euclid were both drawing on an earlier basic treatise on the sphere at rest, whose contents were later absorbed by Theodosius.[14] Tannery suggested Eudoxus as the author; it might equally have been Archytas. Whether Archytas wrote out and systematized the elements of spherics, or whether that waited until the next generation, he was certainly capable of using these elements as tools for solving problems.

* * *

With this historical context to suggest the kinds of techniques of analysis that might have been in Archytas' toolbox, we can try to reconstruct the analysis that might lie behind his construction of the two mean proportionals. As van der Waerden and Knorr have noted, it is the planar configuration AMIKΔ' which he is trying to analyze. Archytas' final conclusion is that the two lines AI and AK have been found as mean proportionals between the given line AΔ (equal to AΔ') and the given line Γ (equal to AB and thus to AM). It follows from examination of the plane figure AMIKΔ', since the angles AMI, AIK, and AKΔ' are right and the angle at A is common, that the triangles AMI, AIK, and AKΔ' are similar, and therefore that the ratios AM:AI, AI:AK, and AK:AΔ' are equal. All this can be understood without any reference to solid geometry, and it is the plane figure AMIKΔ' that we are trying to construct in order to find two mean proportionals between lengths AM and AΔ' given in magnitude. Solid geometry enters when we assume the plane figure AMIKΔ' *as if* it had already been constructed, and try to analyze it using locus-theorems, showing that some of the points of the figure lie on lines or surfaces which are given if other aspects of the figure are given. (Specifically, we will show that K lies on three surfaces which are given on the assumptions that the point A is given, the lines AM and AΔ' are given in magnitude, the line AI is a chord of a given circle, and the plane of that circle cuts the plane AMIKΔ' at right angles.) Once the analysis has been carried through, we can also use solid geometry to construct the loci, and thus the whole figure AMIKΔ', from the givens of the problem—given *outside* the plane AMIKΔ'.

---

14 See T. L. Heath, *History of Greek Mathematics*, 2 vols. (Oxford: Clarendon Press, 1921), 1:348–53, and references therein.

Archytas, and Hippocrates when he reduced the problem of doubling the cube to finding two mean proportionals, were trying to extend easier constructions in the plane: to construct a square equal in area to a given rectangle, we have to find just *one* mean proportional. Euc. *elem.* 6.8 says that "if in a right-angled triangle a perpendicular is drawn from the right angle to the base, then the triangles adjoining the perpendicular are similar both to the whole and to one another," with the corollary that "if in a right-angled triangle a perpendicular is drawn from the right angle to the base, then the straight line so drawn is a mean proportional between the segments of the base." We can use Archytas' diagram AMIKΔ', neglecting the point Δ'. So taking the right triangle AIK, if a perpendicular is drawn from the right angle at I to the opposite side AK, meeting the side AK at M, the triangles AMI and IMK are similar to each other and to the whole triangle AIK. Since sides of similar triangles are proportional, it follows that AM:MI::IM:MK; it equally follows that AM:AI::AI:AK. And the same reasoning, applied now to the diagram AIKΔ', yields the conclusion AI:AK::AK:AΔ'. So, in Archytas' diagram, AM:AI::AI:AK::AK:AΔ', and AI and AK are two mean proportionals between AM and AΔ'.

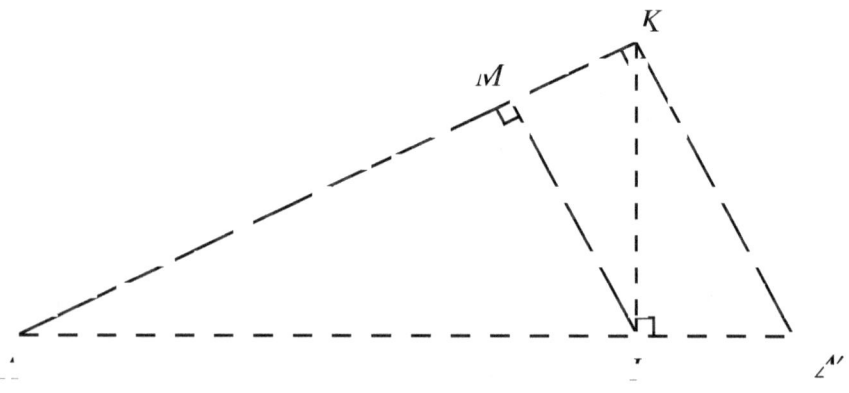

Diagram 2

In Archytas' diagram AMIKΔ', as more trivially in the diagram of *elem.* 6.8, we can assume some elements as given and try to show that other points in the diagram lie on given loci. For instance, in the simpler diagram AMIK, corresponding to *elem.* 6.8, if points A, M, and K are given in position, then the line MI is given in position, since it is the perpendicular at M to line AMK. (Line MI isn't given in *magnitude*, since we don't know how far from M it extends, but we know where M is and we know what direction line MI extends in.) Also, since the right angle AIK is an angle in a semicircle, the circle AIK will also be given, since it is the unique circle of which AK is a diameter; we could also put this by saying that the semicircle AIK is given. It follows that, if points A, M, and K are given in position, the point I is also given in position, since it is the intersection of line MI with the semicircular circumference AIK; or, at worst, it is one of the two intersections of line MI with the circumference of circle AIK. So we can reverse the analysis to find point I if points A, M, and K are given, and to prove that line MI is a mean proportional between AM and MK (which is what Euclid does in *elem*. 6.13), or that line AI is a mean proportional between AM and AK.

We can try imitating this kind of argument with Archytas' diagram AMIKΔ'. Indeed, Archytas must have done so, because there is clearly some kind of analogy between the way he finds point K and the way we found point I just above. In Archytas' three-dimensional diagram, point K will be located on cylinder ABIΔK, which is a right cylinder whose base is circle ABIΔ, and is therefore given in position (although not magnitude) if circle ABIΔ is given; this cylinder contributes to locating point K analogously to the way that line MI contributed to locating point I in the two-dimensional case. Similarly, point K is located on semicircle AKΔ', bisected at right angles by the plane ABIΔ; and this semicircle AKΔ' is in turn located on the semitorus which results from rotating semicircle AKΔ' around vertex A in the plane of circle ABIΔ (i.e., which results from rotating semicircle AKΔ' around an axis passing through A and perpendicular to the plane ABIΔ). Since this semitorus also includes the semicircle on diameter AΔ perpendicular to plane ABIΔ, the same semitorus can also be generated by taking this semicircle on diameter AΔ and rotating it around this same axis passing through A and perpendicular to plane ABIΔ; so the semitorus is given if points A and Δ and the circle ABIΔ (or just the plane it lies in) are given. Thus the semitorus contributes to locating point K analogously to the way that the semicircle AIK contributed to locating point I in the two-dimensional case. Just as point I was found at the intersection of given line MI with given semicircle AIK, so now point K will be found on the intersection of given cylinder ABIΔK with the given semitorus generated by rotating the semicircle on diameter AΔ. Of course, since the intersection of these two surfaces will be a curve rather than a single point, if we want to find point K we will have to locate it on a third surface as well.

The analogy I have just described has certainly been seen before, notably by van der Waerden and Knorr, even if they don't make it fully explicit, and they try to use it to motivate Archytas' construction.[15] But they have not really succeeded.

---

[15] Bartel L. van der Waerden, *Erwachende Wissenschaft* (Basel: Birkhäuser, 1956), 249–52; Knorr, *Ancient Tradition*, 50–52.

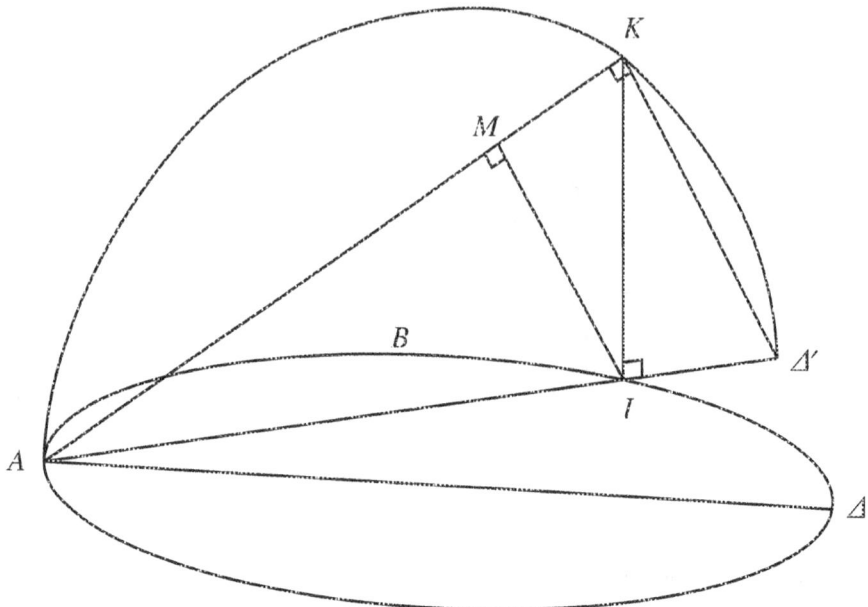

Diagram 3

There is a serious disanalogy between the two-dimensional case and the three-dimensional case as I have described them in the last two paragraphs. In the two-dimensional case, everything emerged from the analysis of the figure AMIK, and we saw that if some elements of that figure (say, the points A and M and K) were given, the point I lay on loci which were given in position. In the three-dimensional case, unfortunately, we needed to draw on things lying outside the plane AMIKΔ' which did not emerge from the analysis of the figure AMIKΔ', and especially the circle ABIΔ. Thus the point K lies on the cylinder ABIΔK, but the cylinder ABIΔK cannot be defined in terms of the figure AMIKΔ' alone: it also depends on the circle ABIΔ. The torus too cannot be defined without reference to the plane ABIΔ. It is true that everything else in Archytas' three-dimensional diagram can be derived from analyzing the diagram containing the plane figure AMIKΔ' together with the circle ABIΔ lying in a plane at right angles to plane AMIKΔ' and having line AI as a chord (and having diameter AΔ equal to AΔ'). But why would it ever have occurred to Archytas to construct such a diagram?

Van der Waerden proposes that Archytas wanted to let point I traverse some path on which line AI would pass through all lengths from a minimum equal to the shorter given line AM to a maximum equal to the longer given line AΔ', so that somewhere along the path the line AI would have to be equal to the first of the desired two mean proportionals between AM and AΔ'; and so he decided to constrain the point I to a circle including the point A and having a line equal to AΔ' as diameter, so that the line AI would traverse all the possible values (actually, it will traverse even more values, starting from zero), and for some reason he decided to constrain the figure AMIKΔ' to lie in a plane perpendicular to the circle AIΔ and passing through line AI as it sweeps around the circle. But these would be apparently arbitrary constraints, not emerging from an analysis of the figure AMIKΔ'; it would apparently be pure luck that M winds up being restricted to a given circle and thus K would be restricted to a given cone, and thus also pure luck that we would be able to find K as the intersection of this cone with the cylinder above the circle AIΔ and the semitorus generated by rotating the semicircle on diameter AΔ. Knorr rightly insists on the importance of starting from an analysis of figure AMIKΔ', but what he proposes is just as arbitrary as van der Waerden's suggestion. Like van der Waerden, he pivots figure AMIKΔ' around point A, in such a way that line AIΔ remains in the same base plane and figure AMIKΔ' stays at right angles to that plane; instead of restricting I to a given circle in the base plane he restricts M, with no justification at all, to a fixed circle in a plane orthogonal to the base plane, and it is apparently pure luck that I winds up being restricted to a circle in the base plane, and thus K to the cylinder over that circle.

Ultimately both van der Waerden and Knorr fall back on Archytas' prodigious spatial imagination,[16] which allows him to move elements of diagrams freely in space until they combine to form a figure AMIKΔ' of the required type. To give a genuine explanation, we need to start from the other end, by analyzing the figure AMIKΔ' and discovering what it entails, what else is given if some elements of this figure are given.

Van der Waerden has noticed, in passing and without further comment, a crucial point about Archytas' construction, namely that, if the point A is given and the point I lies on the given circle ABIΔ, the point M lies on a given sphere, the unique sphere on which circle ABIΔ is a greatest circle (the sphere whose center is the center of circle ABIΔ and whose radius is the radius of circle ABIΔ). This sphere is not mentioned in Eutocius' report and does not enter into Archytas' construction (although we will be able to see a "trace" of it there), but it is nonetheless crucial to the analysis that lies behind it. To see this, and to try to overcome the arbitrariness in van der Waerden's and Knorr's

---

[16] "Wahrhaft göttliche Eingebung" (van der Waerden, *Erwachende Wissenschaft*, 251); "stunning tour de force of stereometric insight" (Knorr, *Ancient Tradition*, 50).

constructions, let us go back to the analysis of the figure AMIKΔ', and see what else is given if some elements are given.

* * *

If the point I is given and the line AMK is given in position, then the point M is certainly given, since it is the base of the perpendicular from point I to line AMK. However, if points A and I are given, even if we have no further information about the line AMK, we do still have some information about point M, namely that (because angle AMI is right) it lies on a given sphere, the unique sphere of which line AI is a diameter. This remains true even if points A and I are not given, but remain diametrically opposite points on a given circle: since the sphere with diameter AI remains the same as the diameter AI moves around the circle, point M continues to lie on a given sphere, the unique sphere on which this circle is a greatest circle. We can put this in another way. First, note that, if B is any other point on the circle ABI, there is a unique sphere passing through the points ABIM (and thus containing circles ABI and AMI). Then, the theorem says that if AI is a diameter of circle ABI, and if angle AMI is right (equivalently: if AI is a diameter of circle AMI), then circle ABI is a greatest circle on sphere ABIM. This theorem is, in fact, Theodosius, *Spherics* 1.12, "circles in a sphere which bisect each other are greatest." As a consequence, since there is only one sphere on which circle ABI is a greatest circle, if circle ABI is given, then point M will lie on a given sphere.

However, these are not the only conditions under which point M will lie on the sphere on which circle ABI is a greatest circle—equivalently, under which circle ABI will be a greatest circle on sphere ABIM. Recall that, since angle AMI is right, circle AMI is bisected by circle ABI. When does it happen that a circle on a sphere is bisected by another circle which is a greatest circle on the sphere? It happens under precisely two conditions. First, the two circles might both be greatest circles on the sphere: "greatest circles in a sphere bisect each other" (*sph*. 1.11). However, "if a greatest circle in a sphere bisects another circle in the sphere which is *not* greatest, it cuts it at right angles [i.e., the planes of the two circles cut each other at right angles] and [passes] through its poles" (*sph*. 1.14). And, conversely, "if a greatest circle in a sphere cuts another circle in the sphere at right angles [i.e., their planes cut each other at right angles], it bisects it and [passes] through its poles" (*sph*. 1.13). Thus a greatest circle in a sphere bisects another circle in the sphere if and only if *either* the second circle is also a greatest circle, *or* the two circles cut each other at right angles. So, since circle ABI bisects circle AMI, if circle ABI is a greatest circle on the common sphere ABIM, then *either* circle AMI is also a greatest circle on the common sphere ABIM (so that circle AMI also bisects circle ABI, and chord AI is a diam-

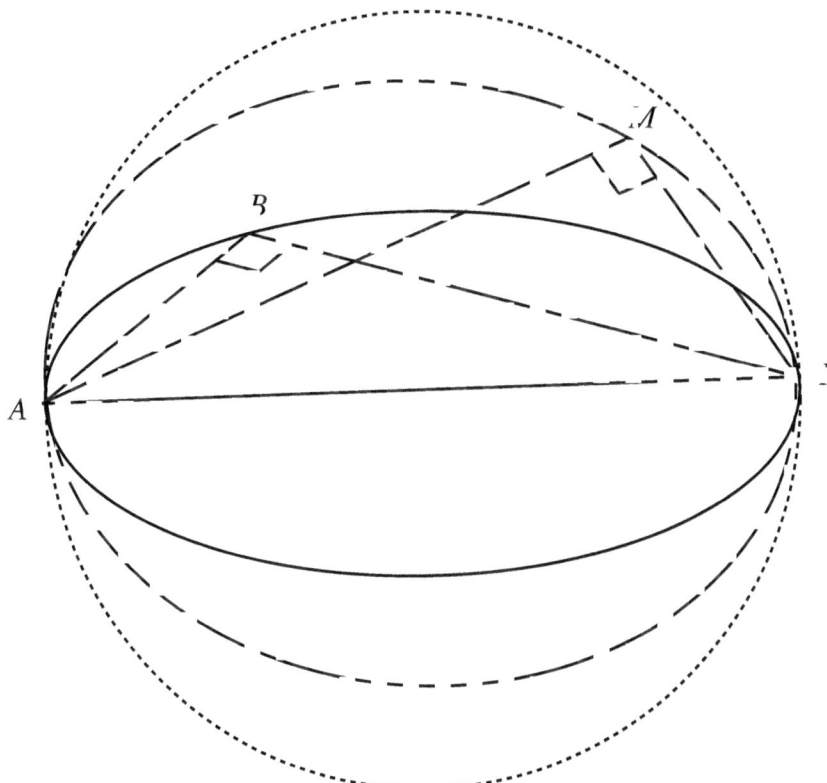

Diagram 4

eter of circle ABI as well as of circle AMI), *or* the plane of circle ABI and the plane of circle AMI cut each other at right angles.

The converse of this disjunctive theorem is also true: since circle ABI bisects circle AMI, if *either* circle AMI also bisects circle ABI, *or* the planes of the two circles cut each other at right angles, then circle ABI is a greatest circle on the common sphere ABIM. The first half of this is obvious: if two circles in a sphere, ABI and AMI, bisect each other, then the common chord AI is a diameter of both, and therefore both circles are greatest circles on the unique sphere of which AI is a diameter. The second half is slightly less obvious: if a circle in a sphere bisects another circle in the sphere at right angles, then the first circle is a greatest circle in the sphere. This proposition is not in Heiberg's edition of Theodosius (or in any of his manuscripts), but it is the third of the additional theorems which Christopher Clavius adds in his 1586 Rome translation of the *Spherics*, "ex alia

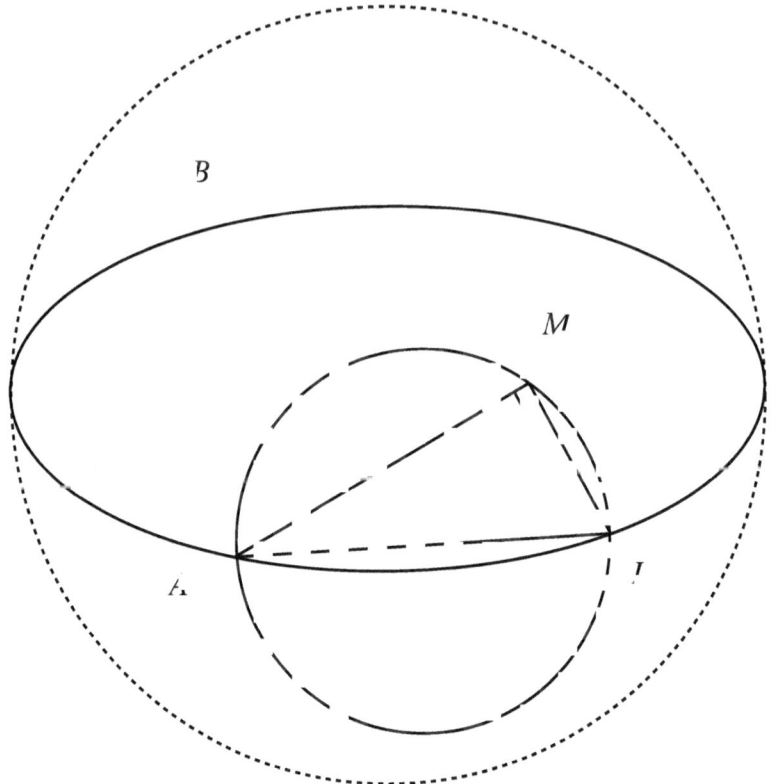

Diagram 5

versione," after 1.15.[17] Here is one way to prove it (we will see below another way, which is more likely to have been Archytas'). Suppose circle ABI bisects circle AMI, and suppose the planes ABI and AMI cut each other at right angles. If the common chord AI is a diameter of the common sphere ABIM, then we know that circle ABI is a greatest circle on the sphere. So suppose that chord AI is not a diameter of the sphere. It follows that circle AMI is not a greatest circle on the sphere (since, if it were, the chord AI, being a diameter of a greatest circle, would be a diameter of the sphere). It also follows that O, the center of the sphere, is not on line AI; thus the plane AIO cuts the sphere in a greatest circle

---

[17] I am not suggesting that Clavius took these theorems from an authentic, otherwise lost, part of Theodosius' treatise; but it would be very surprising if these elementary results, closely related to the ones Theodosius proves, were not known to Theodosius, and to earlier Greek writers on spherics.

AXI. Since greatest circle AXI bisects circle AMI, and since AMI is not a greatest circle on the sphere, circle AXI cuts circle AMI at right angles (by *sph.* 1.14). So planes ABI and AXI both cut the plane AMI at right angles, and they both cut it in the same line AI; so they are the same plane. So plane ABI passes through the center of the sphere; so it cuts the sphere in a greatest circle ABI, as was to be shown.

* * *

This completes an analysis of the assumption that M lies on the sphere on which ABI is a greatest circle, given that circle ABI bisects circle AMI (i.e., given that angle AMI is right): this assumption holds if, and only if, *either* circles ABI and AMI bisect each other (and common chord AI is a diameter of both circles and of the sphere) *or* planes ABI and AMI cut each other at right angles. From now on, instead of considering further the case where AI is a common diameter of circles ABI and AMI, let us consider the case where the plane of circle ABI and the plane of figure AMIKΔ' cut each other at right angles. Everything is now straightforward. If circle ABI is given, we have a great deal of information about the figure AMIKΔ'. This will already be enough to locate point K on one surface, and, if point A is given, and the lines AM and AΔ'—the lines between which we are trying to find two mean proportionals—are given in length, then we can locate point K on two more surfaces as well. And once point K is found, the remaining points MIΔ' are very easily found as well.

For, as we have seen, if the circle ABI is given, the point M lies on a given sphere, namely the unique sphere of which circle ABI is a greatest circle. If, in addition, the point A is given, and the line AM is given in length, the point M lies on the circumference of a given circle in this sphere (we can call it BMZ as in Eutocius' diagram, taking B and Z to be the two intersections of this circle with circle ABI on the surface of the sphere), and A is a pole of this circle BMZ. ("A pole of a circle is a point on the surface of the sphere, from which point all the straight lines falling on the circumference of the circle are equal to each other ...." [Theodosius, *sph.* 1, def. 5]; *sph.* 1.8–10 show that any circle on a sphere has exactly two poles, the endpoints of the diameter of the sphere drawn through the center of the circle and perpendicular to the plane of the circle.) Since A is on the line through the center of circle BMZ perpendicular to the plane of circle BMZ, the cone with vertex A and base circle BMZ is a right circular cone (the only kind of cone that would have been recognized in Archytas' time), and since both the point A and the circle BMZ are given, this cone ABMZ is given in position and magnitude. Since points AMK lie on a straight line, point K lies on the surface of a prolongation of cone ABMZ beyond its base circle BMZ: this cone is given in position although not in magnitude.

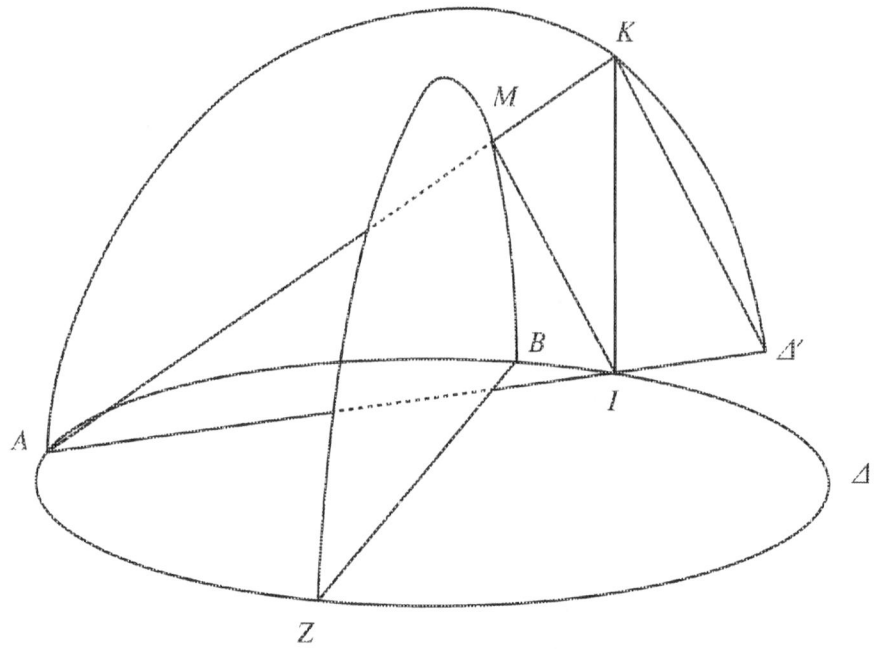

Diagram 6

Again, if circle ABI is given, then since angle AIK is right and plane AMIKΔ' cuts plane ABI at right angles, and line IK in the plane AMIKΔ' cuts the common line AI at right angles, line IK will cut plane ABI at right angles, and so point K will lie on the surface of a right cylinder on given base circle ABI: this cylinder is given in position although not in magnitude (except that we could set the cylinder up, arbitrarily, on either side of plane ABI).

Furthermore, if circle ABI and point A are given, and if, in addition, the line AΔ' is given in length, then, since the points AIΔ' lie on a straight line, the point Δ' will lie on the circumference of a given circle (the circle with given center A and given radius AΔ' in given plane ABI). Furthermore, since angle AKΔ' is right, and plane AMIKΔ' cuts plane ABI at right angles, point K will lie on the circumference of a semicircle AKΔ' set up on the radius AΔ' at right angles to plane AMIKΔ'; and so point K will lie on the surface of the semitorus obtained by rotating the semicircle AKΔ' around the given point A as the radius AΔ' sweeps around the given circle, and this semitorus is given (except that we could set it up, arbitrarily, on either side of plane ABI).

Once point K is found at the intersection of the surfaces of the cone, the cylinder, and the semitorus, the other points of figure AMIKΔ' are readily found as well. Point A is already given. Plane ABI is given, and point I is the base of a perpendicular dropped from point K to plane ABI. Line AK is given, and point M is the base of a perpendicular dropped from point I to line AK. And since angle AKΔ′ is right, and point Δ′ lies on line AI produced, point Δ′ is the intersection of a perpendicular at K to line AK, within plane AIK, with line AI produced. (As we will see, Archytas proceeds in a slightly different way.)

* * *

It is also straightforward to convert all this into a synthetic construction. The synthetic construction will look somewhat different from the analysis starting from the figure AMIKΔ′, since the synthesis must start only from the givens of the problem. Strictly speaking, what is given is just a length equal to AM and a length equal to AΔ′, between which we are seeking two mean proportionals. But *de facto* the circle ABI and a point A on that circle are also assumed as given. These are all that we need to begin the synthetic construction, and they may be taken entirely arbitrarily, except that circle ABI must be taken to have diameter at least equal to length AΔ′, to ensure that the circle does have a chord AI which is the first of two mean proportionals between AM and AΔ′. Archytas in fact chooses circle ABI to have a diameter exactly equal to length AΔ′; indeed he draws the circle around the given greater line AΔ (not merely around some other line equal to it): there is no need to do this, but it is the most obvious choice to make, and it gives us a determinate construction.

Archytas' construction, as reported in our passage from Eutocius, is for the most part the obvious one. First construct circle ABI as a circle whose diameter is the greater given length, the line AΔ, then draw the chord AB equal to the lesser given length, the line Γ. In terms of the figure from the analysis, line AΔ will be the diameter of the sphere ABIM, which is also the axis of the cone ABMZ, prolonged as necessary. The cone will thus be constructed by rotating chord AB around the axis AΔ, and prolonging as necessary (strictly speaking, a cone is not the surface but the solid produced by rotating a right triangle around one of its sides, so it is the solid of revolution of triangle ABE, or of its prolongation AΠΔ). The sphere ABIM could be constructed by rotating circle ABI around axis AΔ, but since Archytas will construct the desired point K from the intersection of the cone, the cylinder, and the semitorus, he does not need the sphere and does not explicitly construct it: he does construct the circle BMZ by rotating point B around axis AΔ, as a byproduct of constructing the cone, and if length Γ and thus point B were allowed to vary, the locus of the circle BMZ and of the point M would be the sphere ABIM. Archytas constructs the cylinder as a right cylinder, of whatever height is necessary, whose base is circle ABI (to avoid generating duplicate values of K, he considers only the semicylinder over the semicircle ABΔ, and not its reflection over the opposite semicircle AZΔ, as well as considering only the cylinder "above" plane ABI, and not its reflection "below" that plane).[18] He constructs the semitorus by taking a semicircle on base AΔ, perpendicular to the plane of circle ABI, and rotating it around point A as line AΔ sweeps out a circle, remaining in the plane of circle ABI. Point K will be found as the intersection of these three surfaces, and figure AMIKΔ' will be constructed around it. Archytas refers to the plane of circle ABI, not the plane of figure AMIKΔ', as "the base plane" (τὸ ὑποκείμενον ἐπίπεδον), and indeed he cannot take plane AMIKΔ' as the base plane, because it is not given, but found only in the course of the construction.

Once he has found point K, Archytas constructs point I in the obvious way, as the base of a perpendicular KI dropped to the plane of circle ABΔ: since K is on the surface of the right cylinder above circle ABΔ, I will be on the circumference of the circle. Rather than deriving point Δ' from point K by completing the right triangle AKΔ' in plane AIK (where Δ' lies on AI produced), Archytas simply lets Δ' be the position that point Δ occupies when the semicircle on AΔ, in being rotated around point A, intersects the cylinder and the cone so as to yield point K; and so, rather than giving this point a new name Δ', he just calls it Δ. Again, rather than constructing point M by a construction within plane AMIKΔ' (as the base of a perpendicular dropped from point I to line AK), Archytas constructs M as the intersection between line AK and the circle BMZ generated by rotating point B around axis AΔ. Since K was on the prolonged cone on circle BMZ from vertex A, this means that Archytas is taking M to be the point on circle BMZ where line AM prolonged, sweeping around circle BMZ to produce the cone, hits point K. Likewise he has taken Δ' to be the point where semicircle AΔ', being rotated around point A to produce the semitorus, hits point K; and he has taken I to be the point where line IK at right angles to plane ABΔ, sweeping around the circle ABΔ to produce the cylinder, hits point K. In other words, he derives points I, Δ', and M, not by constructions from point K in the plane AMIKΔ', but from the constructions by revolution of (respectively) the cylinder, the semitorus, and the cone, which intersect to yield point K. Of course the re-

---

18 This restriction is in fact sufficient to specify a unique K. From the point of view of modern algebraic geometry, the torus is a surface of degree 4, the cone and the cylinder each of degree 2, so they have intersection number 16. However, the torus and the cone each have double points at A, and the cylinder passes through A, so the three surfaces have a quadruple intersection at A, and intersect at only 12 points away from A. These 12 points come in three symmetric sets of four, generated by reflection in the plane ABΔ and in the plane through line AΔ cutting plane ABΔ at right angles. These three solutions correspond to the three possible values of length AK, i.e., the three cube-roots of $\Gamma \cdot (A\Delta)^2$. Since two of these cube-roots are imaginary, there is only one real value of K (together with its symmetric reflections).

sults are equivalent. But because he has derived points I, Δ', and M this way, he has to do a bit more work to show that figure AMIKΔ' does indeed have the desired properties, i.e., that angles AMI, AIK, and AKΔ' are right, so that AM:AI::AI:AK::AK:AΔ', so that lines AI and AK are mean proportionals between lines AM and AΔ' (it is obvious from the construction that AM = AB = Γ and that AΔ' = AΔ, so this will yield two mean proportionals between the two given lines Γ and AΔ). The only claim that is not immediately obvious is that angle AMI is right.

Since M is on the circumference of circle MBZ, and is thus on the (unmentioned) sphere with diameter AΔ and greatest circle AIΔ, and since AMI lies in the plane AKΔ' constructed at right angles to the plane of circle AIΔ, it is sufficient to prove that "if a greatest circle in a sphere cuts another circle in the sphere at right angles, it bisects it" (Theodosius, *sph.* 1.13), so that circle AIΔ will bisect circle AMI, segment AMI will be a semicircle, and angle AMI will be right. So the last step of Archytas' argument can be seen as his archaic version of *sph.* 1.13. Archytas proves it (in Eutocius' report) by intersecting the lines AI and BZ, both chords of circle ABIΔ, in a point Θ, an auxiliary point which was not mentioned in our analysis and also not in his construction of figure AMIKΔ'. Both segment BMZ and segment AMI, intersecting in straight line MΘ, are at right angles to the plane of circle ABIΔ: we know that segment BMZ is a semicircle, and we are trying to prove that segment AMI is also a semicircle. Because segment BMZ is a semicircle and line MΘ in this segment is at right angles to the base BΘZ (because it is at right angles to the whole plane of circle ABIΔ, in which line BΘZ is a chord), we have $MΘ^2 = BΘ·ΘZ = AΘ·ΘI$ (by Euc. *elem.* 3.35, since BΘZ and AΘI are chords crossing in circle ABIΔ). So since line MΘ in segment AMI is at right angles to base AΘI, and $MΘ^2 = AΘ·ΘI$, segment AMI is a semicircle, as was to be shown; and all the other properties of figure AMIKΔ' follow immediately.

We can restate this as a proof of our theorem from *sph.* 1.13: if in a sphere the greatest circle AΔI cuts circle AMI at right angles, it bisects it, i.e., segment AMI is a semicircle. For drop a perpendicular MΘ from point M on the surface of the sphere to the plane of circle AΔI; point Θ will lie on the chord AΘI which is the intersection of the circular areas AMI and AΔI. We can then find some chord XY of circle AΔI, passing through point Θ, such that circle AΔI bisects circle XMY, i.e., so that segment XMY is a semicircle: for instance, if we take XY to be the diameter of circle AΔI passing through point Θ, then plane XMY cuts the sphere in a circle passing through the diametrically opposite points X and Y; circle XMY is therefore a greatest circle, and greatest circle AΔI bisects it in points X and Y, so segment XMY is a semicircle. Since line MΘ is at right angles to the base of the semicircular segment, XΘY (because it is at right angles to the whole plane of

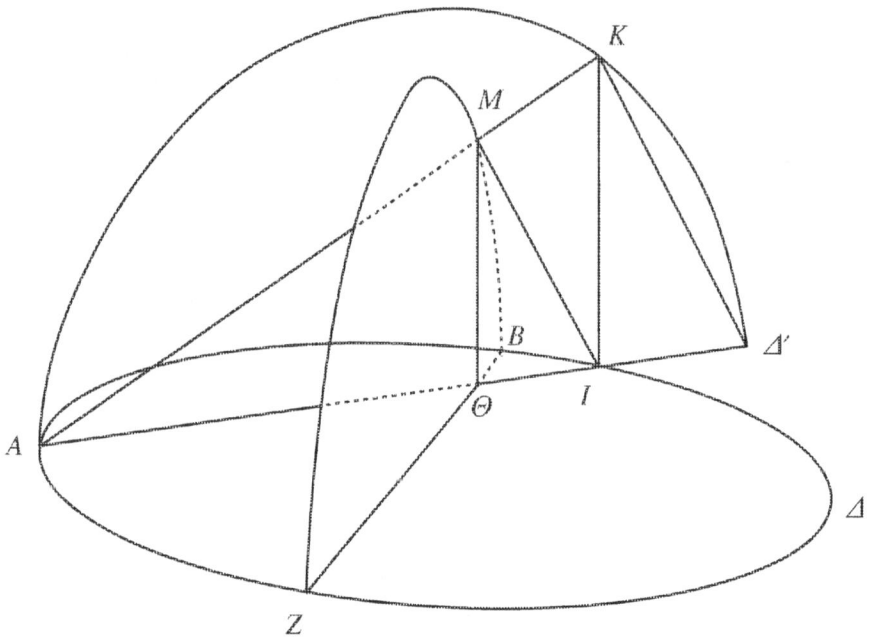

Diagram 7

circle AΔI, in which line XΘY is a chord), we have $MΘ^2 = XΘ·ΘY = AΘ·ΘI$ (since XΘY and AΘI are chords crossing in circle AΔI). So since line MΘ in segment AMI is at right angles to base AΘI, and $MΘ^2 = AΘ·ΘI$, segment AMI is a semicircle, as was to be shown.

We could also restate this as a proof of our (and Clavius') converse theorem that if circle AΔI (not assumed to be greatest) *bisects* circle AMI at right angles, then circle AΔI is a greatest circle in sphere AΔIM, or (equivalently) point M lies on the surface of the unique sphere on which circle AΔI is a greatest circle. For if we drop perpendicular MΘ to plane AΔI as before, then since segment AMI is a semicircle orthogonal to the plane of circle AΔI, $MΘ^2 = AΘ·ΘI$. Then, letting O be the center of circle AΔI and producing the line OΘ in both directions to yield the diameter XOΘY, we have $MΘ^2 = AΘ·ΘI = XΘ·ΘY = (XO + OΘ) · (XO − OΘ) = XO^2 − OΘ^2$. So $MO^2 = MΘ^2 + OΘ^2 = XO^2$, and so MO = XO, the radius of the circle AΔI, and so M is on the surface of the sphere. We can think of this as a σύμπτωμα of the sphere: M is on the surface of the sphere whose greatest circle is AΔI if and only if the square of the perpendicular MΘ dropped from M to plane AΔI is equal to what modern geometers will call the "power" of the point Θ, that is, the rectangle on the segments of any chord of the circle AΔI broken at point Θ. And Archytas is more likely to have thought of spherics in this way than in the more abstract and

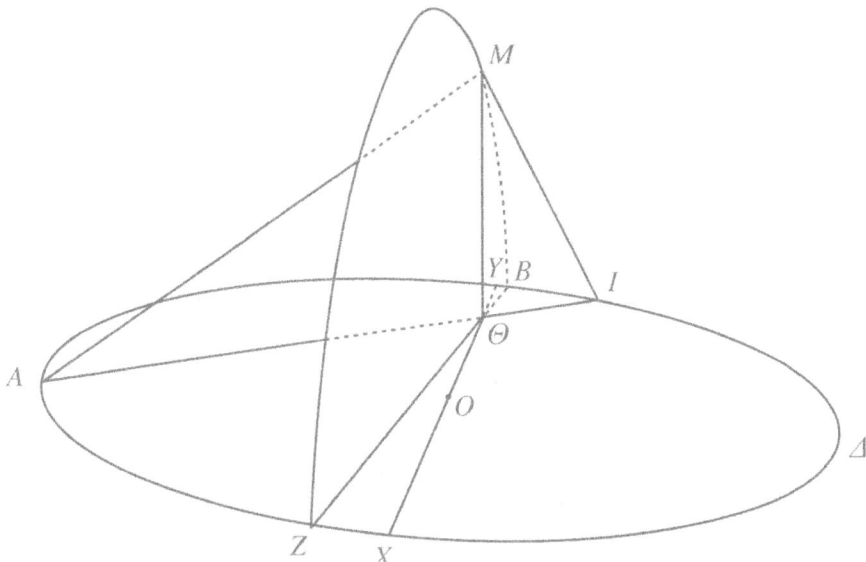

Diagram 8

less "computational" terms used by Theodosius. Indeed, Archytas' "computational" proof that angle AMI is right can be seen as an illustration of the claim in Archyt. fr. 4 that logistic "completes" (ἐπιτελεῖ) geometrical demonstrations, if we do not restrict logistic to *numbers*, but allow it to be concerned with ratios and with the "geometrical algebra" of Euc. *elem.* 2; conversely, as Archytas' whole construction illustrates, geometrical constructions serve to find proportions that purely "logistical" methods cannot.

* * *

We have a reconstruction of the analysis behind Archytas' argument that would make it non-miraculous: the analysis of figure AMIKΔ', generalizing the analysis in the plane of the simpler figures AMIK and AIKΔ', is a natural development from spherics. The sphere is never explicitly mentioned in Archytas' construction as Eutocius reports it, but it is implicitly there both in the construction of point M and in the proof that angle AMI is right, and it must have played a role in Archytas' analysis, in perhaps roughly the way I have reconstructed it. An examination of Archytas' proof that angle AMI is right has given us a glimpse of what spherics looked like for him. His "solid" method of analysis shows an archaic precursor of the methods of analysis through cones and conic sections which will end by eclipsing Archytas' methods for handling solid problems; more specifically, his apparent use of the σύμπτωμα of a sphere, or of the section of the sphere over chord AI, is a precursor of analysis through the συμπτώματα of cones and conic sections. Finally, we may be able to make some more sense of the passage where Plato, in *Republic* 7, introduces the discipline of stereometry. Having passed too hastily, in the series of μαθήματα, from (plane) geometry to astronomy, Plato says that we were wrong "in taking after the plane [surface], the solid already in rotation, before taking [the solid] just in itself" (528a9–b1). The study of the solid in rotation is certainly the study of the "moving sphere" as in Autolycus,[19] and so the study it presupposes, beyond plane geometry, would seem to be especially the study of the sphere at rest. But Plato describes this in-between discipline as being "about the growth [αὔξη] of cubes and what shares in depth" (528b2–3, cf. 528d8, "the discipline of the growth of depth"). Plato is conventionally translated as speaking of the second and third "dimension" (so especially 528b1–2, δευτέρα and τρίτη αὔξη, and also λαβοῦσα αὔξην at *leg.* 10, 894a3), but such an abstract concept of "dimension" is hard to justify, and is certainly not what αὔξη normally means. It has long been suspected that Plato intended some sort of allusion to Archytas' doubling of the cube. We can now see how spherics as the study of the sphere at rest, studied as the foundation of mathematical astronomy, could lead to the "growth" and doubling of the cube.

Bibliography

Friedlein, Gottfried. *Procli Diadochi in primum Euclidis elementorum librum commentarii* (Leipzig: Teubner, 1873).
Heath, Thomas L. *History of Greek Mathematics*, 2 vols. (Oxford: Clarendon Press, 1921).
Heiberg, Johann L. *Apollonii Pergaei quae Graece extant cum commentariis antiquis*, ed. stereotypa ed. anni 1893 (Stuttgart: Teubner, 1974).
Heiberg, Johann L., and Evangelos S. Stamatis. *Archimedis opera omnia cum commentariis Eutocii*, iterum ed. I. L. Heiberg, corrigenda adiecit E. S. Stamatis, vol. 3 (Stuttgart: Teubner, 1972).
Hintikka, Jaakko, and Unto Remes. *The Method of Analysis: Its Geometrical Origin and its General Significance* (Dordrecht: D. Reidel, 1974).
Huffman, Carl A. *Archytas of Tarentum: Pythagorean, Philosopher, and Mathematician King* (Cambridge: Cambridge University Press, 2005).
Hultsch, F. *Pappi Alexandrini Collectionis quae supersunt*, 3 vols. (Berlin: Weidmann, 1876–1878; repr. Amsterdam: Hakkert, 1965).

---

19 The study of the whole celestial sphere (or its abstract analogue) in rotation, not the study of a rotating *star*, since in Autolycus-style mathematical astronomy the stars are in fact treated as points rather than as solids.

Jones, Alexander. *Pappus of Alexandria: Book 7 of the Collection*, edited with translation and commentary, 2 vols. (New York: Springer, 1986).

Knorr, Wilbur R. "Observations on the Early History of the Conics," *Centaurus* 26 (1982): 1–24.

Knorr, Wilbur R. *The Ancient Tradition of Geometric Problems* (Boston: Birkhäuser, 1986).

Lasserre, François. *De Léodamas de Thasos à Philippe d'Oponte: témoignages et fragments*, édition, traduction et commentaire (Naples: Bibliopolis, 1987).

Menn, Stephen. "Plato and the Method of Analysis," *Phronesis* 47 (2002): 193–223.

Netz, Reviel. *The Works of Archimedes: Translated into English, Together with Eutocius' Commentaries, with Commentary, and Critical Edition of the Diagrams*, vol. 1: *The Two Books on the Sphere and the Cylinder* (Cambridge: Cambridge University Press, 2004).

Sefrin-Weis, Heike. *Pappus of Alexandria: Book 4 of the Collection*, edited with translation and commentary (London: Springer, 2010).

van der Waerden, Bartel L. *Erwachende Wissenschaft* (Basel: Birkhäuser, 1956).

Ian S. Moyer
A Revised Astronomical Dating of Thessalus' *De virtutibus herbarum*

Abstract: This paper uses astronomical data in Thessalus' treatise *De virtutibus herbarum* to arrive at a rough estimate of its date of composition. Dates of the sun's ingress into the signs of the zodiac that are given in the treatise are compared with modern calculations. The resulting differences are compared with the differences between modern calculations of the dates of the sun's ingress and those derived from other ancient sources and methods (Ptolemy's *Handy Tables*, Vettius Valens' *Anthology*, and *P. Oxy.* 4148) to determine the likely date of composition of Thessalus' treatise. The revised estimates suggest that the text was composed between the middle of the first century A.D. and the early third century A.D., with dates in the second century most probable.

While investigating the ancient cultural and intellectual dialogue between Greece and Egypt, I have on a few occasions found myself trespassing in that daunting terrain known as the History of Science. On one of these occasions, I was fortunate to be at the Institute for Advanced Study, where Professor Heinrich von Staden facilitated the work of members in various fields of ancient history. During my time there, I was delving deeper into the history of a work composed by a certain Thessalus, a work that the author claimed was a divine revelation of astrological influences on the potency of medicinal plants.[1] The pleasant and productive atmosphere in which I worked was a result not only of Professor von Staden's intellect and broad learning, but also of the kindly and amiable way in which he welcomed scholars gathered there, encouraged our work, and gently engaged us in conversations. So it is a real pleasure to offer Professor von Staden some revised estimates of the date of Thessalus that build on my work at the Institute.

The astro-botanical work of Thessalus is best known for its wonderful autobiographical prologue, in which the author, a Greek from Asia Minor, tells of his medical studies at Alexandria and his discovery of a book of remedies written by the Egyptian pharaoh Nechepso. According to this story, which is presented as a dedicatory epistle to a Roman emperor, Thessalus' lack of success with the rem-

---

[1] Hans-Veit Friedrich, ed. *Thessalos von Tralles, griechisch und lateinisch* (Meisenheim am Glan: Anton Hain, 1968).

edies precipitated a quest for magical knowledge in Egypt, a quest that ended in Thebes with a divine revelation procured for him by an Egyptian priest. The revealed wisdom of Asclepius supplemented and completed the original book of Nechepso, and is presented as an authoritative account of Egyptian magical knowledge. Elsewhere, I have examined the way in which Thessalus engages with the Nechepso-Petosiris tradition of astrological literature and how he represents and appropriates this tradition in reworking it for a wider audience in the Roman Empire.[2] A major obstacle to using the Thessalus text as a historical document is the date of its original composition. Dates ranging from Nero to Late Antiquity have been proposed, primarily on the basis of connections and associations with medical and magical figures ranging from Thessalus of Tralles to Hermes Trismegistos.[3] Franz Cumont, however, had already pointed out a different way of dating Thessalus – an astronomical approach that has fallen by the wayside in subsequent discussions.[4] The most critical part of Asclepius' revelation to Thessalus – the part that allowed him to complete Nechepso's imperfect wisdom – was the correct times at which to pick the plants used in the remedies of the treatise. Each plant was connected with a sign of the zodiac and was to be picked when the sun was in that sign. In essence, Asclepius provided Thessalus with the dates of the sun's ingress into each sign of the zodiac. To put it another way, the god gave a series of dates and longitudes of the sun, since the first point of Aries is 0°, the first point of Taurus is 30°, and so forth.

Owing to the ancient practice of using sidereal rather than tropical longitudes to measure the positions of heavenly bodies, there is a noticeable difference between modern and ancient longitudes calculated during the first three centuries of the common era. This difference, moreover, changes over time at a rate roughly commensurate with the precession of the equinoxes, so the astronomical data of Asclepius' revelation can be used to provide an estimate of the period in which the text was composed. This principle, originally suggested by Cumont, was the basis of my earlier work in which I proposed that Thessalus composed his text in the first half of the second century A.D.[5] In the present study, I expand on that discussion, and also offer a revised version of the argument based on a new method that has been made feasible by the efficiencies of astronomical software.

---

2 Ian Moyer, *Egypt and the Limits of Hellenism* (Cambridge: Cambridge University Press, 2011), 208–73.
3 For a fuller discussion, see Moyer, *Egypt*, 211–17.
4 Franz Cumont, "Écrits hermétiques II: le médecin Thessalos et les plantes astrales d'Hermès Trismégiste," *Revue de philologie* 42 (1918): 85–108, at 96–98.
5 Moyer, *Egypt*, 211–19, 293–97.

Table 1 Dates of the sun's ingress according to the Madrid manuscript of Thessalus

Zodiac sign		Alexandrian calendar	Asia Minor calendar	Roman calendar	Julian date
Aries	♈	Phamenoth 22	Dystros 18	XV. Kalend. Apr.	March 18
Taurus	♉	Pharmouthi 23	Xanthikos 18	XIV. Kalend. Mai.	April 18
Gemini	♊	Pachons 25	Artemisios 20	XIII. Kalend. Iun.	May 20
Cancer	♋	Pauni 25	Daisios 19	XIII. Kalend. Iul.	June 19

As in my previous study, I drew the dates of ingress from two manuscripts in the rather complex and varied textual tradition of Thessalus: Codex Matritensis Bibl. Nat. 4631 (T in Friedrich's edition), and a version of the Latin translation of Thessalus in Codex Vindobonensis 3124 (V in Friedrich). The Madrid manuscript is the only surviving version of the Thessalus text with the original prologue intact, but it breaks off in the midst of discussing comfrey, the plant to be harvested in the sign of Cancer. The manuscript therefore provides only four dates of ingress (table 1).

There is no complete series of dates in any of the other Greek manuscripts, and those dates that are preserved appear to be corrupt in places. The Vienna manuscript of the Latin translation of Thessalus includes a full series of dates that agree in the first four cases with those in the Madrid manuscript and appear more or less coherent as a set of dates of the sun's ingress into the signs of the zodiac (table 2).

Table 2 Dates of the sun's ingress according to the Vienna manuscript of Thessalus

Zodiac sign		Roman calendar	Julian date
Aries	♈	XV. Kalend. Apr.	March 18
Taurus	♉	XIV. Kalend. Mai.	April 18
Gemini	♊	XIII. Kalend. Iun.	May 20
Cancer	♋	XIII. Kalend. Iul.	June 19
Leo	♌	XI. Kalend. Aug.	July 22
Virgo	♍	XI. Kalend. Sept.	August 22
Libra	♎	X. Kalend. Oct.	September 22
Scorpio	♏	XII. Kalend. Nov.	October 21
Sagittarius	♐	XV. Kalend. Dec.	November 17
Capricorn	♑	XVI. Kalend. Ian.	December 17
Aquarius	♒	XVII. Kalend. Feb.	January 16
Pisces	♓	XII. Kalend. Mar.	February 18

In order to get a rough measure of the extent to which the dates in each series are inaccurately computed or corrupt, I have compared them to a hypothetical series of dates of solar ingress calculated according to the solar theory of Hipparchus and Ptolemy, which accounted for the anomalous motion of the sun over the course of the year. To reconstruct the hypothetical series, I took March 18 as the date of the sun's ingress into Aries, assumed (following Hipparchus) that the apogee of the sun's eccentric orbit was 65.5° from the vernal equinox, and calculated a series of longitudes to find the dates of ingress into each sign.[6] The dates from the Madrid manuscript differ by no more than a day from the reconstructed series – a variation that is not surprising, since the moment of ingress can fall on one day or another depending on the vagaries of the calendar relative to the sun's motion. The number of days that the sun spends in each sign suggests a proportionally longer spring and summer, as one would expect with a Hipparchan model, but this is only a rough impression based on the movement of the sun over the course of three signs in the spring and summer. The Vienna manuscript includes the complete series of dates of ingress, but they show somewhat more variation from the reconstructed dates. Though most are within a day, ingress into Libra (September 22) and Pisces (February 18) comes two days later than the reconstructed series. The summer and winter, moreover, are somewhat longer and the spring somewhat shorter than the standard Hipparchan figures for the lengths of the seasons (tables 3 and 4).

In brief, the dates preserved in the Madrid manuscript and the Vienna manuscript of the Latin translation provide a plausible series of dates of solar ingress, but there are some departures from the series reconstructed according to the Hipparchan and Ptolemaic method of calculating solar longitudes. It is not clear whether these departures are the result of error, differences in the method of calculation, or textual corruption.

As I mentioned above, the dates of the sun's ingress provide figures for the sun's longitude at each of these dates. These longitudes can be compared to modern calculations, and the differences between them can provide a rough estimate of the date at which they were calculated. The problem, of course, is that these longitudes are not very exact. The sun's longitude changes on average about 0°59' from the beginning to the end of the day. In my previous work, I arrived at an estimate by choosing an arbitrary time of the day for the modern calculations. I used Tuckerman's ephemeris tables and a simple linear interpolation

---

6 For the method, see James Evans, *The History and Practice of Ancient Astronomy* (New York: Oxford University Press, 1998), 210–35, and Otto Neugebauer, *The Exact Sciences in Antiquity* (New York: Dover, 1969), 191–93.

Table 3 Dates of the sun's ingress in Thessalus compared with dates reconstructed according to ancient solar theory

	Zodiac sign		Thessalus' dates of ingress	Reconstructed dates – using solar theory of Hipparchus	Thessalus relative to reconstructed dates
Madrid and Vienna mss	Aries	♈	March 18	March 18	–
	Taurus	♉	April 18	April 19	– 1 day
	Gemini	♊	May 20	May 19	+ 1 day
	Cancer	♋	June 19	June 20	– 1 day
Vienna ms	Leo	♌	July 22	July 21	+ 1 day
	Virgo	♍	August 22	August 21	+ 1 day
	Libra	♎	September 22	September 20	+ 2 days
	Scorpio	♏	October 21	October 20	+ 1 day
	Sagittarius	♐	November 17	November 17	0
	Capricorn	♑	December 17	December 18	– 1 day
	Aquarius	♒	January 16	January 17	– 1 day
	Pisces	♓	February 18	February 16	+ 2 days

Table 4 Lengths of the seasons according to Thessalus compared with those of Hipparchus

Season	Length according to Hipparchan Model	Length according to Thessalus (Madrid and Vienna mss)
Spring (Aries, Taurus, Gemini)	94.5 days	92 days
Summer (Cancer, Leo, Virgo)	92.5 days	94 days
Autumn (Libra, Scorpio, Sagittarius)	88.125 days	87 days
Winter (Capricorn, Aquarius, Pisces)	90.125 days	92 days

to calculate average modern longitudes of the sun at 7:00 am on Thessalus' dates of ingress for twenty years between 50 and 150 A.D.[7] The choice of 7:00 am was a rough approximation of the beginning of the day according to the Egyptian epoch. To find the difference, I subtracted the modern longitude from the initial

---

7 Bryant Tuckerman, *Planetary, Lunar, and Solar Positions 601 B.C. to A.D. 1 at Five-Day and Ten-Day Intervals* (Philadelphia: American Philosophical Society, 1962) and idem, *Planetary, Lunar, and Solar Positions A.D. 2 to A.D. 1649 at Five-Day and Ten-Day Intervals* (Philadelphia: American Philosophical Society, 1964).

degree of the zodiac sign for the date of ingress (i.e., 0°, 30°, 60°, etc.). The main limitations of the approach are (a) that it does not take into account the full possible range of longitude hidden behind Thessalus' dates, and (b) that it uses a limited number of samples for the modern calculations.

Astronomical software allowed me to dispense with the labor involved in using ephemeris tables and therefore to deal more adequately with the range of longitudes on a given date.[8] I gathered modern figures for the dates of ingress into each of the signs for every year from 1 to 200 A.D., assuming Alexandria as the point of observation. Rather than using modern epochs for the days (midnight to midnight), I reckoned them according to the Egyptian epoch (dawn to dawn). Owing to the variable relationship between the calendrical cycle and the sun's progress through the zodiac, modern dates of ingress fell within a range of about three days (i.e., a given date plus or minus one day). I compared the modern dates of ingress to the dates in the Madrid and Vienna manuscripts of Thessalus and calculated the mean difference between the two and also the standard deviation. The mean difference between modern dates and those in the Madrid manuscript is 3.83 days with a standard deviation of 0.82. The mean difference between modern dates and those in the Vienna manuscript is 3.54 days with a standard deviation of 1.28. As I noted above, the dates in the Vienna manuscript showed more deviation from the reconstructed series of dates than the Madrid manuscript dates, and that variability (whatever its cause) contributes to the greater standard deviation from the mean difference. The difference between Thessalus' dates and modern dates can be converted into a difference in solar longitude by multiplying the number of days by the sun's average daily progress along the ecliptic. The results appear in table 5.

Table 5 Average difference between modern calculations of solar longitude and those in Thessalus

	Average difference from modern calculations	Standard deviation
Madrid manuscript	3.78°	0.81°
Vienna manuscript	3.49°	1.26°

---

[8] All modern calculations in the present study were performed using *TheSkyX: Serious Astronomer Edition* (Software Bisque, 2010).

These differences are somewhat less than those calculated in my previous study, and also describe a wider range of variation, as was to be expected given the change in method.[9]

In order to arrive at the date when Thessalus made his calculations, their average difference from modern calculations must be compared to a temporal series of differences between ancient calculations and modern ones. The difficulty is that the method used to calculate the dates in Thessalus' treatise is unknown, so I have used three sets of comparanda: solar longitudes from the horoscopes in the *Anthology* of Vettius Valens (2nd c. A.D.), Ptolemy's *Handy Tables* (2nd c. A.D.), and a table of summer solstices from 161–237 A.D. preserved on papyrus (P. Oxy. 4148, 3rd c. A.D.).

In their study of Greek horoscopes, O. Neugebauer and H. van Hoesen examined the solar longitudes of 28 horoscopes in the *Anthology* of Vettius Valens where specific dates were mentioned and found that they were greater than modern longitudes by 7° to 2° over a period from 54 to 157 A.D. Though the precise methods of calculation used in these horoscopes are unknown, Neugebauer and van Hoesen suggested that they were based on tables that differed from modern calculations by ca. +5° in 50 A.D. and ca. +3.5° in 160 A.D.[10] Valens' figures themselves, however, do not follow such a predictable pattern. Owing to inaccuracies in calculation, there is considerable variation in the differences between the longitudes in Vettius Valens and modern figures. These differences are shown in chart 1, along with a trendline derived from the data. The rate of change in this trendline is more than twice what would be expected from the precession of the equinoxes. This may reflect inaccurate calculations or differences in method.[11]

The trendline can be extrapolated to cover the period from 1 to 250 A.D. (chart 2), thus providing a useful comparison for the differences between Thessalus' figures and modern calculations. Note that the average deviation of Vettius Valens' figures from the trendline is 0.84° (noted in chart 2).

---

9 In my previous study, the average difference based on the dates in the Madrid manuscript alone was: 4° ± 0.5°. The average difference based on the dates in the Vienna manuscript 3.8° ± 0.7° (see Moyer, *Egypt*, 294).
10 Otto Neugebauer and Henry B. van Hoesen, *Greek Horoscopes* (Philadelphia: American Philosophical Society, 1959), 171–73 and 179–82.
11 Compare chart 1 to Neugebauer and van Hoesen *Horoscopes*, 182, fig. 33 in which the differences are shown together with a skew line representing how much the decrease in difference over time would be owing to precession: 1° every 72 years. The actual rate of change in the trendline fitting the data from Vettius Valens is much steeper: 1° every 32.4 years.

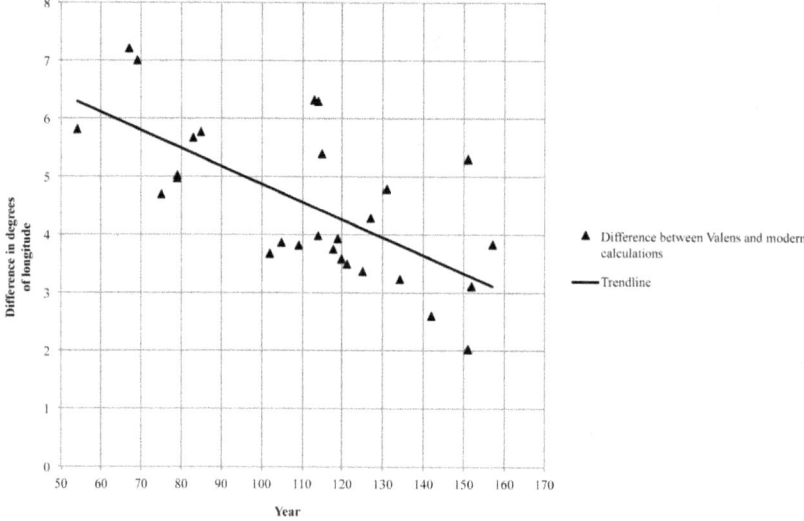

Chart 1 Differences between modern calculations of solar longitude and those of Vettius Valens

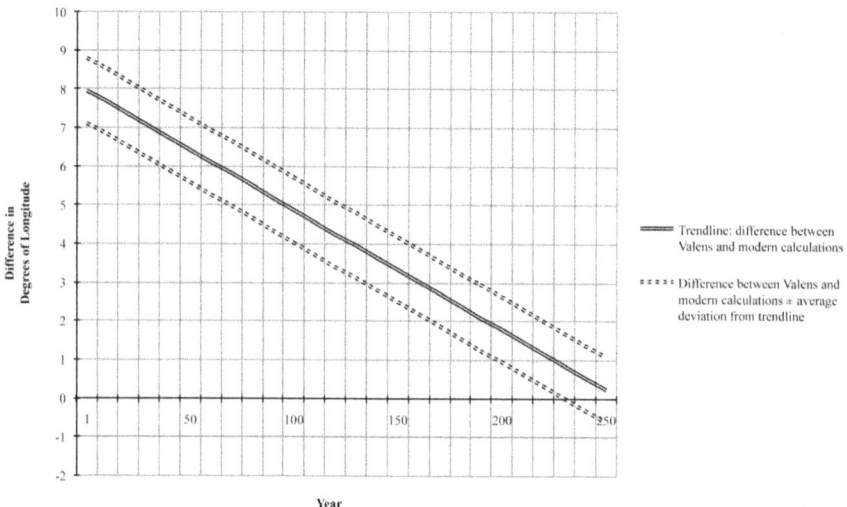

Chart 2 Vettius Valens compared with modern calculations: extrapolated trendline of changing differences in solar longitudes

This is comparable to the standard deviation in the differences between modern calculations and those in the Madrid manuscript of Thessalus. The standard deviation of the Vienna differences is, as noted above, even greater. The variability in the figures from Vettius Valens, an astrologer of the second century

A.D., puts Thessalus' variations in a context of roughly contemporary astrological practice. Both undoubtedly employed the exact sciences of antiquity, but perhaps not too exactly. The two sets of mean differences, along with the average range of deviation, are shown in charts 3 and 4. The intersection of the two sets of lines indicates a range of years in which the dates of ingress in Thessalus' treatise were most likely calculated.

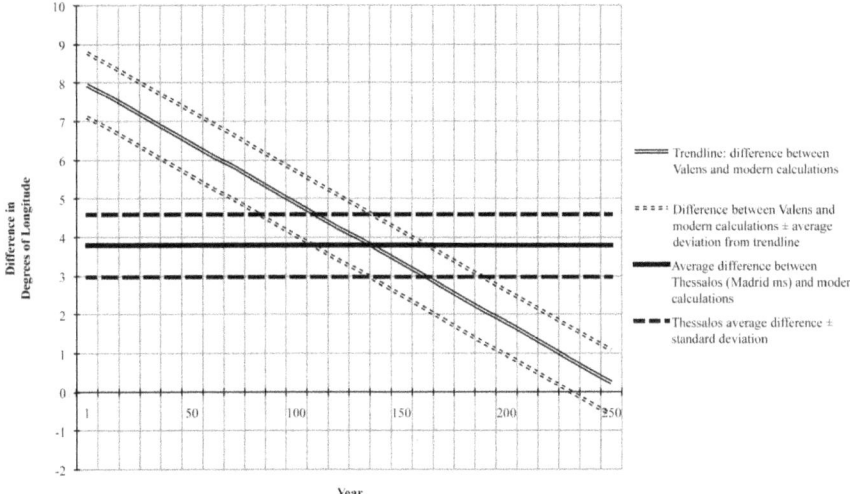

Chart 3 Vettius Valens vs. modern calculations compared with Thessalus (Madrid manuscript) vs. modern calculations

The other two comparanda for the differences between Thessalus and modern calculations are based not on ancient calculations (with all the possibilities of hidden errors) but on ancient methods of calculation. In retrospect, the most influential of these were the kinematic models embodied in Ptolemy's *Handy Tables*, a revised and enlarged version of the tables in the *Almagest* that Ptolemy created for the practical use of astronomers and astrologers. Aside from kinematic models and tables, the other major approaches to calculating the positions of heavenly bodies were arithmetic schemes. In ancient practice, this involved the use of an epoch table listing occurrences of a certain kind of event (such as the sun's solstice or perigee) over a certain period of years along with a template for reckoning longitudes in relation to the epoch event. One of the few surviving examples of this method for calculating solar longitudes is preserved in P. Oxy. 4148. This papyrus contains a table of exact times of the summer solstice for the years 161–237 A.D. The epoch table follows an ancient convention in assuming that the sun was at 8° of Cancer at the summer solstice, and therefore the

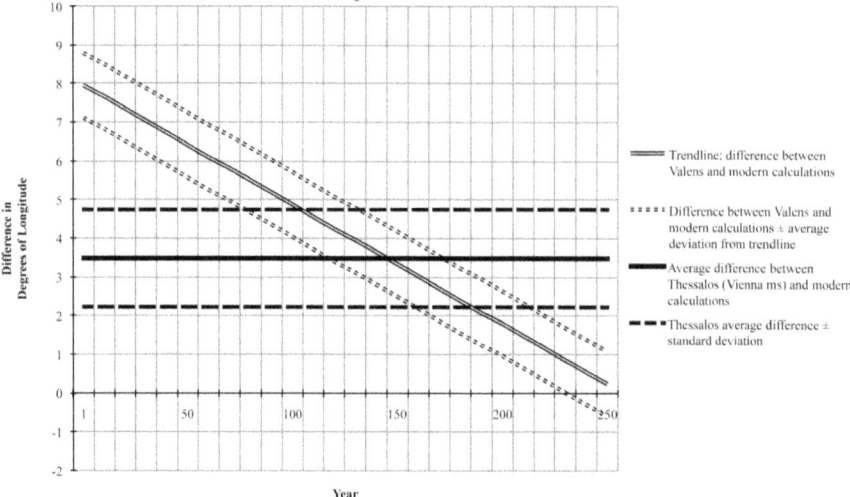

Chart 4 Vettius Valens vs. modern calculations compared with Thessalus (Vienna manuscript) vs. modern calculations

table provides a series of longitudes at precise dates.[12] Papyri attest that both kinematic and arithmetic methods of computation were used in Roman Egypt, and that arithmetic methods continued to compete with kinematic tables well into the fourth century A.D., long after Ptolemy wrote his definitive works.[13] Conversely, kinematic models for dealing with the anomalous motion of the sun (and other bodies) date back at least to Hipparchus (second half of second century B.C.). Either type of method could have been used to calculate the dates of

---

12 Alexander Jones, *Astronomical Papyri from Oxyrhynchus (P. Oxy. 4133–4300a)*, 2 vols., Memoirs of the American Philosophical Society 233 (Philadelphia: American Philosophical Society, 1999), 1:120 suggests that the longitude at the epoch dates could be either Cancer 0° or Cancer 8°, but after comparing the ancient longitudes with modern ones, only Cancer 8° provides a plausible series of differences. The table is also not explicit about the epoch from which hours of the day are counted. Jones suggested sunset (6:00 pm) on analogy with the Standard Lunar Scheme, but I assumed morning epoch, which makes the ancient-modern differences closer to those of the *Handy Tables*. If evening epoch is assumed, the differences between modern longitudes and those in P. Oxy. 4148 would be 0.48° less, which would increase the date ranges by ca. 29 years. Jones also noted that there is a significant discrepancy between the dates of solstices recorded in this text and actual solstices, and suggested that these discrepancies may be the result of extrapolating from observations earlier than the Augustan era. Nevertheless, it is one method that Thessalus may have used and so provides another comparandum.
13 For an overview of the varieties of astronomical papyri, see Jones, *Astronomical Papyri*, 1:35–47.

ingress in Thessalus' treatise. It is entirely possible that neither Ptolemy's *Handy Tables* nor the scheme of P. Oxy. 4148 was used by Thessalus, but they are at least two known possibilities. Charts 5 and 6 compare the differences between these ancient methods of calculation and modern calculations to the differences between Thessalus and modern calculations.

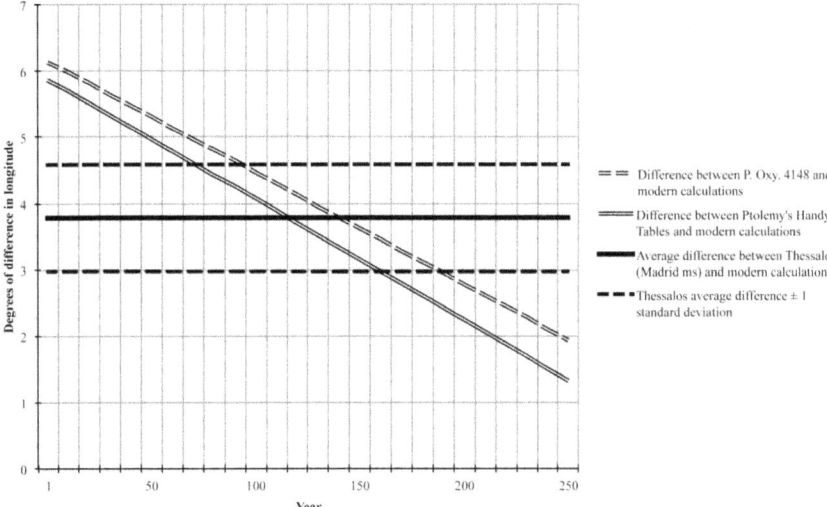

Chart 5 Ancient vs. modern methods of calculating solar longitude compared to Thessalus (Madrid manuscript) vs. modern calculations

The results of all three comparisons are summarized in table 11 below. Using the figures from the Madrid manuscript alone, the possible dates range from ca. 71 A.D. to ca. 189 A.D. Estimated dates based on the mean differences all fall in the first half of the second century A.D. When the figures from the Vienna manuscript are included, the possible dates range from ca. 63 A.D. to ca. 233 A.D., with estimates based on the mean differences falling around the middle of the second century. As noted above, ancient calculations show considerable variability due to errors, computational short-cuts, and the like – even when the object is to describe something as steady and predictable as the sun's motion along the ecliptic. Divinely revealed or not, the dates of ingress in Thessalus' treatise betray such variability. His dates, moreover, do not provide a precise correlation of time and longitude, but rather an implicit range defined by the sun's motion over the course of the day. In this revised approach to the problem of dating Thessalus by astronomical means, I have taken these variables into account more completely. The variability in Thessalus' figures is especially apparent

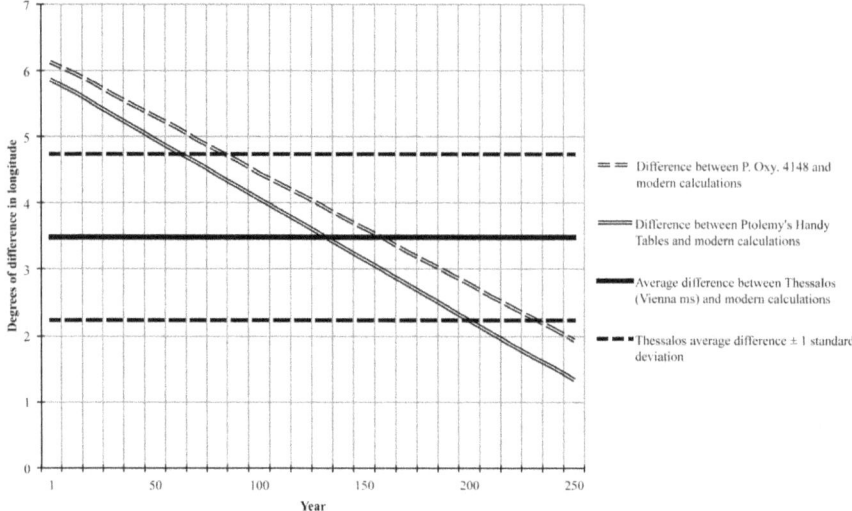

Chart 6 Ancient vs. modern methods of calculating solar longitude compared to Thessalus (Vienna manuscript) vs. modern calculations

when the dates of ingress from the Vienna manuscript are taken into account. These may have been subject to modification in the process of translation from Greek into Latin and textual transmission. In taking a more cautious approach, I have reached a more cautious conclusion. The range of dates of ingress as preserved in both the Madrid and Vienna manuscripts of Thessalus' *De virtutibus herbarum* show that it could well have been composed anywhere from the middle of the first century A.D. to the early third century A.D., but the somewhat more reliable dates preserved in the Madrid manuscript favor a date in the second century A.D.

Table 6 Summary of results

Comparison text	Date of Thessalus text, using figures from the Madrid ms only	Date of Thessalus text, using figures from the Vienna ms
Vettius Valens	136 CE ± 54 years (82–189 CE)	145 CE ± 68 years (77–213 CE)
Ptolemy's *Handy Tables*	116 CE ± 45 years (71–161 CE)	132 CE ± 69 years (63–201 CE)
P. Oxy. 4148	141 CE ± 48 years (93–189 CE)	158 ± 75 years (83–233 CE)

Bibliography

Cumont, Franz. "Écrits hermétiques II: le médecin Thessalos et les plantes astrales d'Hermès Trismégiste," *Revue de philologie* 42 (1918): 85–108.

Evans, James. *The History and Practice of Ancient Astronomy* (New York: Oxford University Press, 1998).

Friedrich, Hans-Veit, ed. *Thessalos von Tralles, griechisch und lateinisch* (Meisenheim am Glan: Anton Hain, 1968).

Jones, Alexander. *Astronomical Papyri from Oxyrhynchus (P. Oxy. 4133–4300a)*, 2 vols., Memoirs of the American Philosophical Society 233 (Philadelphia: American Philosophical Society, 1999).

Moyer, Ian S. *Egypt and the Limits of Hellenism* (Cambridge: Cambridge University Press, 2011).

Neugebauer, Otto. *The Exact Sciences in Antiquity*, repr. 2nd ed. (New York: Dover, 1969).

Neugebauer, Otto, and Henry B. van Hoesen. *Greek Horoscopes* (Philadelphia: American Philosophical Society, 1959).

Tuckerman, Bryant. *Planetary, Lunar, and Solar Positions 601 B.C. to A.D. 1 at Five-Day and Ten-Day Intervals* (Philadelphia: American Philosophical Society, 1962).

Tuckerman, Bryant. *Planetary, Lunar, and Solar Positions A.D. 2 to A.D. 1649 at Five-Day and Ten-Day Intervals* (Philadelphia: American Philosophical Society, 1964).

Vivian Nutton
What's in a *Nomen*?: Vlatadon 14 and an Old Theory Resurrected

Abstract: Until 1891 it was assumed that Galen's *gentilicium* was Claudius, but more recent scholars, following the discovery of new inscriptions from Pergamum, have preferred Aelius or Julius. The appearance of Claudius in two late Greek MSS and an early catalogue has provided fresh support for the previous theory. This paper re-examines the new evidence, which is unusual in many ways, as well as the epigraphic dossier on the Nicons of Pergamum. It concludes that only the discovery of a new inscription with Galen's full name will settle the issue finally, and that the epigraphic material, even if not related to Galen's own family, also offers an insight into the social conditions in which he was brought up.

The Vlatadon manuscript of Galen has produced many surprises – missing portions of Greek texts, traces of others surviving otherwise only or mainly in translation, and in *Avoiding Distress* a previously lost treatise that throws much light on Galen's career in Rome and on his library. Recently, Stefan Alexandru has drawn attention to another remarkable feature that is to be found towards the end of the codex, the presence of a Roman *gentilicium,* Claudius, in the heading of Galen's commentary on the Hippocratic *Regimen in Acute Diseases.*[1] In a remarkably erudite paper he argues strongly that this new information, contained in a manuscript written in Constantinople around 1450, supports the traditional view that our author was called Claudius Galenus. Although, as we shall see, other possibilities have been canvassed for more than a century, he rejects them on the grounds that none of the inscriptions on which these theories have been based relates unequivocally to Galen's father. The Vlatadon codex, in his view, bears witness to a much older Greek tradition and confirms the credibility of those Renaissance scholars who believed in a Claudius Galenus. Alexandru's extremely valuable observations on the modern history of this name throw much light on the methods of scholars in both Renaissance Italy and imperial Germany, while his comments on the text of Vlatadon 14 help to establish still further its importance as a witness to the textual tradition of Galen. But de-

---

[1] Stefan Alexandru, "Newly Discovered Witnesses Asserting Galen's Affiliation to the *Gens Claudia,*" *Annali della Scuola Normale Superiore di Pisa* ser. 5, 3/2 (2011): 385–433.

spite the abundant learning on display, there are still weaknesses in his argument, as well as neglected pieces of information, that impose hesitation before this resurrected theory can be fully accepted.

From the Renaissance onwards until the late nineteenth century, there was no reason to doubt that Galen's *gentilicium* was Claudius. It was known in Northern Italian circles by the end of the fifteenth century and appears in the printed versions of early translations by Leoniceno, Lorenzi, and Thomas Linacre. It is recorded as such as in the history of medicine begun by Gian Giacomo Bartolotti in Ferrara between 1498 and 1507, the first account of ancient medicine to use the Greek sources that were becoming available.[2] Earlier biographers, in Greek, Arabic, and Latin, did not report it, and, as far as is known, it does not appear in the heading or colophon of any medieval translation of Galen into Latin or, earlier, into Arabic. Hence Alexandru concludes, rightly, that the *gentilicium* became known in the Renaissance West through a Greek source that would have appeared trustworthy to scholars and translators of the caliber of Politian, Crinito, and Lorenzi in Florence or of Leoniceno in Ferrara.[3] But if this was a manuscript, as opposed to information passed on from one scholar to another, it quickly disappeared from view, and does not appear to have been the source of any of the Greek manuscripts of Galen copied in Renaissance Italy.[4]

That, with one exception, the name Claudius is not found in any extant medieval manuscript or biography, whether in Greek or in Latin, is not surprising. We know of the full Roman names of authors such as Plutarch, the historian Aulus Claudius Charax of Pergamum, and Alexander of Aphrodisias only thanks to the chance discoveries of inscriptions that can be linked to them. Even authors such as Aelius Aristides, Cassius Dio, and Claudius Aelianus, whose full names are preserved, are cited in the majority of their manuscripts and by Byzantine excerptors and biographers only by their *cognomina*. That the Greeks should have continued to refer in a similar way to Galen *tout court* is understandable and should not be used to argue that this bourgeois citizen of Pergamum was a *peregrinus*, a non-citizen of Rome, until he entered imperial service in the

---

2 Gian Giacomo Bartolotti, *De antiquitate medicinae, On the Antiquity of Medicine*, ed. Dorothy M. Schullian and Luigi Belloni (Milan: Industrie grafiche italiane Stucchi, 1954), 58; for the date of composition, see xx, xxv. The name is not in Giovanni Tortelli's *De medicina et medicis* of 1449, the first history of medicine to use directly sources from classical antiquity, albeit those in Latin. Lorenzi's preface to his 1494 version of the Hippocratic *Aphorisms* is the earliest reference so far known.

3 Alexandru, "Newly Discovered Witnesses," 405.

4 Nor can it have been a manuscript of the same type as Vlatadon 14, for Renaissance scholars show no knowledge of any other piece of information it alone transmits, despite its obvious interest.

late 160s. At a time when large numbers of Pergamum's upper classes had obtained Roman citizenship, it is extremely unlikely that a family as wealthy and as well-connected as Galen's had not already done so.

The weak evidence for Claudius would not have troubled scholars, had it not been for the discovery at Pergamum of a series of inscriptions, datable to the Antonine period, and referring to two architects, Aelius Nicon and Iulius Nicodemus "also known as young Nicon."[5] It was Hermann Schöne in 1891 who first drew attention to these inscriptions and suggested that the evidence of the inscriptions should take precedence over the weakly attested "Claudius" as the *gentilicium* of Galen's father, an architect by the name of Nicon.[6] His argument was further developed by Ernst Klebs in the long preamble to his discussion of "Claudius Galenus" in the first volume of the authoritative *Prosopographia Imperii Romani*.[7] Klebs had consulted at least one other Galenic scholar, Iwan von Müller, who had confirmed that he too had not come across the name in any surviving manuscript.[8] Klebs appealed for help in locating any such testimony, but did not go further in explaining the origin of the name. It was left to two other German classicists, Wilhelm Crönert and Karl Kalbfleisch, to propose alternative theories. Crönert suggested that it arose from confusion with the name of Claudius Ptolemy, Karl Kalbfleisch from a misunderstanding of the (Latin) abbrevia-

---

5 The inscriptions are easily accessible together in René Cagnat, *Inscriptiones Graecae ad res Romanas pertinentes*, 4 vols. (Paris: E. Leroux, 1901–1927), 4:nos. 502–506. Nos. 503 and 504 had been long known as *Corpus Inscriptionum Graecarum* 3546 and 3544. The others were found by the German excavators of Pergamum.
6 Hermann Schöne, "Galeniana," in *Schedae philologae Hermanno Usener a sodalibus Seminarii Regii Bonnensis oblatae* (Bonn: F. Cohen, 1891), 88–93, at 91–92, at that stage not knowing of the newly discovered inscriptions relating to Aelius Nicon.
7 Ernst Elimar Klebs, *Prosopographia Imperii Romani saec. I, II, III*, pars 1 (Berlin: Reimer, 1897), 379–80. It is worth noting, however, that the name Nicon does not occur in any surviving work by Galen, even though he refers to his father several times, but is found only in the entry on Galen in the Suda. Scholars have, *faute de mieux*, accepted that the name appeared somewhere in the vast Galenic Corpus, and was thence transmitted to the Suda, but the fact that the same compiler also can reveal the names of both the father and mother of Soranus of Ephesus, neither of whom is mentioned elsewhere, imposes caution. If the Suda entry is regarded as invention, the evidence of the Pergamene inscriptions is irrelevant. Indeed, if a Pergamene Nicon was a famous architect in his day, one might suggest that the naming of this man as Galen's father is owed to the source of the Suda, wishing to tie the two together.
8 At this time, the Greek manuscript tradition of Galen was poorly understood, the Latin tradition even more so, nor had much research been carried out into the correspondence and publications of the medical humanists of the Renaissance. Subsequent research has strengthened the *argumentum ex silentio*.

tion "Cl." for "clarissimus."[9] Kalbfleisch's theory was publicized widely by Karl Sudhoff, the dominant figure in the history of medicine for half a century, and, although neither explanation is entirely convincing, there were few writers who continued to refer to Claudius Galen.[10] Even those who disputed whether Galen was an Aelius, a Iulius, or a *peregrinus* agreed that he was unlikely to have been a Claudius.

This scholarly consensus has now been challenged by the recent discovery of the *nomen* Claudius in the remarkable Greek MS, Vlatadon 14.[11] This MS was copied around 1450 in Constantinople by a variety of scribes, many of whom were connected in some way with the circle of intellectual doctors at the Kral hospital. It contains several unique texts and fragments of texts in their original Greek, most notably *Avoiding Distress*, and frequently preserves a textual tradition that differs from that of the vulgate as represented in the Laurentian and Marcian copies of Galen. As Alexandru shows, following Antoine Pietrobelli, in the commentary on *Regimen in Acute Diseases*, it has many good readings of its own and is close to an Athos manuscript, Iviron 184. In other works, such as *On Prognosis*, its Greek text finds its closest relative in a fragmentary Milan manuscript from the same circle, Ambrosiana Q 3 sup., and more significantly in the excellent (lost) manuscript that formed the basis for many of the unusual Latin translations of Niccolò da Reggio (fl. 1308–1348). Vlatadon 14 was written by several scribes, and, had the name of Galen

---

[9] Wilhelm Crönert, "Klaudios Galenos," *Mitteilungen zur Geschichte der Medizin und der Naturwissenschaften* 1 (1902): 3–4, at 4; Karl Kalbfleisch, "'Claudius' Galenus," *Berliner philologische Wochenschrift* 22 (1902): 413. The controversy, and the views of subsequent scholars, are examined in detail by Alexandru, "Newly Discovered Witnesses," 383–92.
[10] Karl Sudhoff, *Kurzes Handbuch der Geschichte der Medizin* (Berlin: Karger, 1922), 108. Alexandru carefully lists the varying view of modern scholars.
[11] Thessalonica, Vlatadon 14, fol. 160 (fol. 149 in the old numbering). Detailed descriptions of the MS and of its history are given by Véronique Boudon-Millot and Antoine Pietrobelli, "De l'arabe au grec: un nouveau témoin du texte de Galien (le manuscrit *Vlatadon* 14)," *Comptes rendus des séances de l'Académie des Inscriptions et Belles-Lettres* 149 (2005): 497–534; eidem, "Galien ressuscité: édition *princeps* du *De propriis placitis*," *Revue des études grecques* 118 (2005): 168–233; Antoine Pietrobelli, "Variation autour du Thessalonicensis *Vlatadon* 14: un manuscrit copié au *xenon* du Kral, peu avant la chute de Constantinople," *Revue des études byzantines* 68 (2010): 95–126; Véronique Boudon-Millot, Jacques Jouanna, and Antoine Pietrobelli, eds., *Galien, Ne pas se chagriner*, t. IV (Paris: Les Belles Lettres, 2010), LXIII–LXVIII. I have not seen Pietrobelli's thesis, "Histoire du texte, édition critique et traduction annotée du livre I du commentaire de Galien au Régime des maladies aigües d'Hippocrate," Thèse de doctorat, Université de Paris IV – Sorbonne, 2008, vol. 1, clx–cxlvi, but I am grateful to him for his help in discussing problems with this manuscript. The observations on the relationship with the Milan manuscript and with Niccolò are made by me on the basis of work still unpublished.

appeared in the early part of it, which contains the most novelties, it would have carried even greater authority than Alexandru implies. As it is, it appears on fol. 160 (fol. 149 in the old numbering) only as part of a curious heading to Galen's commentary on *Regimen in Acute Diseases*. This heading, although not the text that immediately follows, is in a hand that was identified independently by Dieter Harlfinger and Antoine Pietrobelli as that of the scholar-scribe Constantine Lascaris (1434–1501), who copied other portions of the manuscript.

I give the title as transcribed by Alexandru:

Εἰς τὸ περὶ διαίτης ὀξέων Ἱπποκράτους, οἱ δὲ περὶ πτισάνης, οἱ δὲ πρὸς τὰς κνιδίας γνώμας ἐξήγησις γαλήνου κλαυδίου τοῦ περγαμηνοῦ.[12]

This heading displays several unusual features, leaving aside the name and the reversed order of *nomen* and *cognomen*, which is a literary rather than an epigraphic form.[13] No other manuscript of this commentary presents the variant titles that are given to the Hippocratic text, but they had long been familiar to scholars. They are noted in Athenaeus' *Deipnosophists*, in several passages by Galen himself, as well as in the important Hippocratic MS M, Venice, Marcianus gr. 269.[14] But their presence as the title of the book that is being commented upon is unusual, to say the least, especially as it is found nowhere else in the Galenic tradition. Alexandru suggests that traces might be found in the partly illegible heading in its closest relative, Iviron 184, but Pietrobelli, who has been able to examine the manuscript itself, confirms that such a hypothesis can be excluded. Most significant of all, Galen never refers to this commentary under this title, not

---

12 Alexandru, "Newly Discovered Witnesses," pl. 11–13, gives an excellent photograph of the heading.

13 The example of Cassius Dio Cocceianus offers a good parallel. Printed editions give the first two names in the reverse order, whereas the overwhelming majority of MSS and citations refer to him either as Dio or by adding some variant of Cocceianus.

14 Alexandru, "Newly Discovered Witnesses," 414–16. But although the textual tradition of M has been influenced by the Galenic commentary, and possibly vice-versa, there is little or no reason to assume that this interaction extends to the title of the commentary. For the history of M, see Jacques Jouanna, "L'Hippocrate de Venise (*Marcianus gr.* 269: coll. 533): nouvelles observations codicologiques et histoire du texte," *Revue des études grecques* 113 (2000): 193–210; Véronique Boudon-Millot, "À propos de l'Hippocrate de Venise (Marcianus gr. 269, coll. 533): nouvelles observations sur les signatures," *Revue des études grecques* 117 (2004): 759–66; Franco Giorgianni, *Hippokrates, Über die Natur des Kindes (De genitura und De natura pueri)* (Wiesbaden: Ludwig Reichert Verlag, 2006), 84. Lascaris may well have known of this manuscript in Constantinople before it came into Bessarion's possession.

least in his discussion of his commentaries in *On My Own Books*.[15] Whatever the virtues of the readings of Vlatadon 14 for the text of the commentary, this heading bears all the signs of a later scholarly confection.[16] When it was produced is far from clear, but the most economical hypothesis is that it was invented by the young Lascaris, eager to show off his own erudition.

This does not exclude the possibility that Lascaris had access to a genuine tradition going back to antiquity, but if he did, it seems not to have been known widely, if at all, to other Byzantine scholars, including those at the Kral hospital. Crönert's suggestion that it arose from some confusion with Claudius Ptolemy remains a possible option, as does that of Kalbfleisch, that it represents a *cl(arissimus)*, albeit in a modified form. Given the circumstances in which Vlatadon 14 was written, Lascaris cannot have obtained his information directly from a Latin source. But there is a possible intermediary in Johannes Argyropoulos, who had returned from Italy to teach at the Kral and who was the head of the group of scholars who copied Vlatadon 14 and other Galenic texts there. He could well have come across a reference in a Latin text to *cl(arissimus)* Galen during his time in Padua and passed this information on to his young follower.[17] But it is safest to confess ignorance as to the reason why Lascaris should have chosen to compose this unique heading.

Lascaris, however, could well have been the conduit whereby this information came to Northern Italy in the second half of the fifteenth century. He knew many of its leading scholars, and his knowledge of matters Greek, and especially matters codicological, would have lent authority to his pronouncements, even without the need to supply manuscript proof. Once in circulation, the *gentilicium* would have been easily accepted by the new humanists, and thus made its way into print.

There is, however, a second possible reference to a Greek manuscript containing the disputed name, albeit one played down by Alexandru. The learned Giacomo Filippo Tomasini in his catalogue of Venetian public and private libra-

---

15 Gal. *de libr. propr.* 9.6, 19.35 K. (= 160 Boudon-Millot).
16 Alexandru's discussion, "Newly Discovered Witnesses," 417–23, shows that Vlatadon 14 had many better readings than its twin, Iviron 184, and occasionally some uniquely good readings, but none is as striking as Lascaris' heading. But the extremely rare material, including passages not otherwise accessible in Greek, comes only in the first part of the MS and, almost certainly, from a different source to the commentary.
17 While Kalbfleisch's theory presupposes at some point the misunderstanding of a written Latin text, that need not have been the heading of a Galenic treatise, but could have been derived from the reading of a discussion of Galen, where such laudatory adjectives are common. The fact that "*clarissimus Galenus*" has not yet been reported is more a reflection on a lack of concern among medievalists, ignorant of the appeals for help by Klebs and Kalbfleisch.

ries reports a Greek manuscript of Claudius Galenus, *De moribus* in the collection of Giuseppe degli Aromatarii (1587–1660), a collection long dispersed and in part lost.[18] This is intriguing, not least because no Greek manuscript of this treatise survives today, although there are fragments and a summary in Arabic.[19] Tomasini lists both the previous manuscript in the collection, a Byzantine anthology of short anatomical and therapeutic works by or ascribed to Galen, and the succeeding one, the short tract on purgatives, simply as "Galenus," which might give added weight to his use of the fuller form for this treatise. But since no other Renaissance scholar appears to have known of this, this codex is unlikely to have been the source for their use of the name "Claudius." Given that we are dealing with a lost manuscript about which we know nothing, it would be unwise to make too much of what could be no more than a mistake by Tomasini himself, who would have been familiar with the traditional name. But, equally, this evidence should not be lightly disregarded unless a better alternative can be found. *Prima facie,* it offers stronger support to the theory than does Lascaris' unusual title, and may now be supported by the very recent discovery of the *nomen* in Phillipps 1524, another fifteenth-century manuscript.[20]

The modern case against Claudius was based on the discovery of five inscriptions from Pergamum, *IGRR* 502–506. Although none can be securely dated, they have usually been placed around the middle of the second century, and their content and style would also suit what is known of Galen's father and family. They form an obvious group, and name three individuals: Aelius Nicon, an architect, Aelius Isidotus, a land-surveyor, and another architect, Julius Nicodemus, also called Nicon or the new Nicon. All are written in a high-flown Greek, including an allusion to Euripides, and several take the form of isopsephic verses, poetry in which the numerical equivalents of the letters in each line add up to the same total. Most, if not all, are the work of Aelius Nicon, whose poems on the statue of a satyr, *IGRR* 502, and in praise of geometry, 503, and a hymn to the sun, 506, would appear to have been carved on public

---

[18] Giacomo Filippo Tomasini, *Bibliothecae Venetae manuscriptae publicae et privatae* (Udine: N. Schiratti, 1650), 94 = Alexandru, "Newly Discovered Witnesses," pl. 9–10.
[19] See the introduction to this treatise in Peter N. Singer, ed., *Galen, Psychological Writings* (Cambridge: Cambridge University Press, 2014), 45–76.
[20] Véronique Boudon-Millot, *Galien de Pergame, un médecin grec à Rome* (Paris: Les Belles Lettres, 2012), 27, which appeared after this article had been written, reports the existence of the *nomen* Claudius in Phillipps 1524, another Greek codex of the fifteenth century, but remains undecided as to its value; "Le prénom Claudius...n'a semble-t'il rien d'authentique et paraît d'origine beaucoup plus tardive." It is not yet clear to me whether this MS has any link with Lascaris or Northern Italy.

monuments for which he was responsible.²¹ A further poem, *IGRR* 504, praises Julius Nicodemus who, together with Aelius Isidotus, gave a portico that was the headquarters of the *astynomoi*, civic officials to whose number one or other of the benefactors may have belonged. The final inscription, 505, commemorates Nicodemus after his death and was either inscribed on his tomb or on some public monument he had had erected. The three men are clearly linked in some way. Aelius Isidotus may have been a male relative of Aelius Nicon, a father, brother, or uncle, but that explanation does not apply to Nicodemus, unless he is related on the female side of the family.²² Nor, since he has a different *gentilicium*, is he likely to have been a freedman or a client of Nicon's, even though the phrase that is used, ὁ καὶ (Νείκων) "also known as....," can denote a name used by a non-citizen before being granted citizenship.²³

Modern scholars, from Schöne to Susan Mattern, have rightly drawn parallels between the inscriptions and what Galen tells us of his own father, even if the more scrupulous have refused to be drawn into a preference for one or other man.²⁴ But they have failed to appreciate the significance of the addition in *IGRR* 504 of νέος, "young" or "new," to the name of Nicodemus. Galen's father died around 149 A.D., when Galen was in his twentieth year, i.e., presumably in his late forties and hence unlikely to be called "young."²⁵ On grounds of age, then, Nicodemus would appear to be excluded as Galen's father. But there is also another possible interpretation of the adjective that, even if not directly mentioning age, adds further weight to the candidacy of Aelius Nicon. On this interpretation, the author of the poem suggests that Julius regarded himself, or was regarded, as a successor to an older man, "a young Nicon," even if, in

---

21 There are valuable remarks on these inscriptions in Erwin Ohlemutz, *Die Kulte und Heiligtümer der Götter in Pergamon* (repr., Darmstadt: Wissenschaftliche Buchgesellschaft, 1968), 85, 169–70, 275; and Heinrich Schlange-Schöningen, *Die römische Gesellschaft bei Galen – Biographie und Sozialgeschichte* (Berlin: De Gruyter, 2003), 49–54.

22 For the numerous Aelii at Pergamum, see the list given by Christian Habicht, *Die Inschriften des Asklepieions*, Altertümer von Pergamon VIII.3 (Berlin: De Gruyter, 1969), 124.

23 Schlange-Schöningen, *Die römische Gesellschaft*, 49, suggests that Nicodemus was a new citizen. It often denotes a name used by family and friends.

24 Schlange-Schöningen, *Die römische Gesellschaft*, 52, who gives the most detailed account of Galen's family, notes the lack of a perfect fit between the inscriptions and Galen's own accounts, but the disparity is not as great as Alexandru implies. Richard Evans, *A History of Pergamum: Beyond Hellenistic Kingship* (London: Continuum, 2012), 150–51 (but very derivative), and Susan P. Mattern, *Prince of Medicine: Galen in the Roman World* (Oxford: Oxford University Press, 2013), both accept Aelius Galen, without further argument. Boudon-Millot, *Galien*, 24–25, remains agnostic.

25 Gal. *de bon. mal. suc.* 1, 6.756 K.

the event, he died first. The obvious explanation is that that man was Aelius Nicon, who, in his own eyes or in those of the wider community, had established such a reputation that he could be seen as a model for the younger man.

Even if there was no family link between the two men, the relationship must have been close. One Aelius, Isidotus, collaborated with Nicodemus in the gift of a portico, and Nicon composed a commemorative poem on this occasion and another after Nicodemus' death. Both Nicon and Nicodemus were architects, and the most likely explanation for their relationship is that the older man was either a colleague or the master of the younger. It is not difficult to imagine a talented young man from a wealthy Pergamene family joining with another, more established family to learn how to be an architect. The prestige to be gained, not only in the city but also in the province and, potentially, even in Rome, would have outweighed any considerations of the taint of trade. His joint gift of a portico would not only signal his sense of civic duty but also, if he was its designer, his ability as an architect. He was a young man on the make.[26]

Aelius Nicon is a much more plausible candidate for Galen's father. He already enjoyed a certain reputation at the time of Nicodemus' death, and was a man of wealth, with intellectual ambitions, and strong moral and religious beliefs. He praises justice, and his poem in honor of the sun is eloquent testimony to his piety. Christian Habicht's reconstruction of a very battered inscription from the Asclepieion reveals the dedication to the god made by a P. Aelius Nicon, which, if correct, would further establish links between the god Asclepius and Galen's family, but, as the editor himself admits, there are too many other possibilities for this restoration to be certain.[27] Of potentially greater significance is the profession of Aelius Isidotus, clearly a relative of some sort of the elder Nicon. He is called a land surveyor, a profession found going back several generations in Galen's family, as is revealed further in *Avoiding Distress*.[28] All this adds weight to the notion that Galen's Roman *gentilicium* was Aelius, presumably from a family that had received citizenship in the early years of Hadrian.

---

[26] If his relationship with Aelius Nicon was solely that of a junior colleague, without specific reference to his age, Nicodemus could still have been Galen's father, although the chronological fit is very difficult.

[27] Habicht, *Die Inschriften*, no. 104.

[28] Gal. *indol*. 59 (19 Jouanna). For the new material on his family, see Mattern, *Prince of Medicine*, Ch. 1, and Vivian Nutton, *Ancient Medicine*, rev. ed. (London: Routledge, 2012), 222–23. Assuming that these inscriptions were carved in the 140s, it may well have been his father who gained Roman citizenship from Hadrian. A date in the 150s or later would, of course, exclude Aelius Nicon's paternity.

Against this can be set the evidence provided by Alexandru, and, one must acknowledge, the vagaries of survival. We do not know how many architects worked in Pergamum, still less how many were called Nicon. The chance discovery of one (or two) with that name does not preclude the existence of still others. If we could be certain that Lascaris was drawing on a good source when he wrote his heading in Vlatadon 14, or if Tomasini's lost manuscript and Phillipps 1524 could be proved to go back to medieval Byzantium before the fifteenth century, the case for retaining Claudius Galen would be overwhelming. None of these possibilities, however, is more than that. Lascaris' heading bears all the hallmarks of an individual reconstruction by a budding intellectual, and Tomasini's report is inconclusive. In the absence of better evidence, Aelius Nicon, architect and poetaster, must still remain a strong candidate for Galen's father.

This debate about a name should not hide one important fact. Whichever *nomen* is preferred, Iulius, Claudius, or Aelius, Galen will have belonged to a family of Roman citizens, hardly surprising given Galen's family wealth and the opportunities for patronage in Pergamum. The inscriptions, if they belong to members of his family, also suggest participation by them in the civic euergetism typical of Asia Minor at the time. Galen himself refused to put his own wealth on a par with that of a correspondent from Pergamum who was among the elite in a city of some forty thousand citizens, but there can be little doubt that, even before he became an imperial physician in 168/9, he was far from poor.[29] His *peregrinatio academica* in Smyrna and Alexandria was in part made possible by family wealth, and differed only in length from that of other young members of the bourgeoisie of Asia Minor, including at least one from Pergamum. The benefactions of both Nicons, leaving aside Isidotus the surveyor, also place them firmly in the center of the city's intellectual and social life. It may be only chance that has so far robbed us of evidence for gifts by Galen to his native city comparable to those of Stertinius Xenophon to Cos or of the Statilii to Heraclea in Caria and for some inscription that would finally settle the question of his name.[30]

---

[29] Gal. *de an. aff. dign. et cur.* 9, 5.49 K., with the comments of Boudon-Millot, *Galien*, 23–25.
[30] Nutton, *Ancient Medicine*, 261–63. I am grateful for help received from Antoine Pietrobelli and Heikki Solin, and for Stefan Alexandru's email exchange with me at an early stage in his project. My disagreement should not obscure my respect for his scholarship, which has revealed much that was previously neglected.

Bibliography

Alexandru, Stefan. "Newly Discovered Witnesses Asserting Galen's Affiliation to the *Gens Claudia*," *Annali della Scuola Normale Superiore di Pisa*, ser. 5, 3/2 (2011): 385–433.

Bartolotti, Gian Giacomo. *De antiquitate medicinae, On the Antiquity of Medicine*, ed. Dorothy M. Schullian and Luigi Belloni (Milan: Industrie grafiche italiane Stucchi, 1954).

Boudon-Millot, Véronique. "À propos de l'Hippocrate de Venise (Marcianus gr. 269, coll. 533): nouvelles observations sur les signatures," *Revue des études grecques* 117 (2004): 759–66.

Boudon-Millot, Véronique. *Galien de Pergame, un médecin grec à Rome* (Paris: Les Belles Lettres, 2012).

Boudon-Millot, Véronique, Jacques Jouanna, and Antoine Pietrobelli, eds. *Galien, Ne pas se chagriner*, t. IV (Paris: Les Belles Lettres, 2010).

Boudon-Millot, Véronique, and Antoine Pietrobelli. "De l'arabe au grec: un nouveau témoin du *texte* de Galien (le manuscrit *Vlatadon* 14)," *Comptes rendus des séances de l'Académie des Inscriptions* 149 (2005): 497–534.

Boudon-Millot, Véronique, and Antoine Pietrobelli. "Galien ressuscité: édition *princeps* du *De propriis placitis*," *Revue des études grecques* 118 (2005): 168–233.

Cagnat, René. *Inscriptiones Graecae ad res Romanas pertinentes*, 4 vols. (Paris: E. Leroux, 1901–1927).

Crönert, Wilhelm. "Klaudios Galenos," *Mitteilungen zur Geschichte der Medizin und der Naturwissenschaften* 1 (1902): 3–4.

Evans, Richard. *A History of Pergamum: Beyond Hellenistic Kingship* (London: Continuum, 2012).

Giorgianni, Franco. *Hippokrates, Über die Natur des Kindes (De genitura und De natura pueri)* (Wiesbaden: Ludwig Reichert Verlag, 2006).

Habicht, Christian. *Die Inschriften des Asklepieions*, Altertümer von Pergamon VIII.3 (Berlin: De Gruyter, 1969).

Jouanna, Jacques. "L'Hippocrate de Venise (*Marcianus gr.* 269: coll. 533): nouvelles observations codicologiques et histoire du texte," *Revue des études grecques* 113 (2000): 193–210.

Kalbfleisch, Karl. "'Claudius' Galenus," *Berliner philologische Wochenschrift* 22 (1902): 413.

Klebs, Ernst Elimar. *Prosopographia Imperii Romani saec. I, II, III*, pars 1 (Berlin: Reimer, 1897).

Mattern, Susan P. *Prince of Medicine: Galen in the Roman World* (Oxford: Oxford University Press, 2013).

Nutton, Vivian. *Ancient Medicine*, rev. ed. (London: Routledge, 2012).

Ohlemutz, Erwin. *Die Kulte und Heiligtümer der Götter in Pergamon* (repr., Darmstadt: Wissenschaftliche Buchgesellschaft, 1968).

Pietrobelli, Antoine. "Histoire du texte, édition critique et traduction annotée du livre I du commentaire de Galien au Régime des maladies aigües d'Hippocrate," Thèse de doctorat, Université de Paris IV–Sorbonne, 2008.

Pietrobelli, Antoine. "Variation autour du Thessalonicensis *Vlatadon* 14: un manuscrit copié au *xenon* du Kral, peu avant la chute de Constantinople," *Revue des études byzantines* 68 (2010): 95–126.

Schlange-Schöningen, Heinrich. *Die römische Gesellschaft bei Galen – Biographie und Sozialgeschichte* (Berlin: De Gruyter, 2003).

Schöne, Hermann. "Galeniana," in *Schedae philologae Hermanno Usener a sodalibus Seminarii Regii Bonnensis oblatae* (Bonn: F. Cohen, 1891), 88–93.

Singer, Peter N., ed. *Galen, Psychological Writings* (Cambridge: Cambridge University Press, 2014).

Sudhoff, Karl. *Kurzes Handbuch der Geschichte der Medizin* (Berlin: Karger, 1922).

Tomasini, Giacomo Filippo. *Bibliothecae Venetae manuscriptae publicae et privatae* (Udine: N. Schiratti, 1650).

D. T. Potts
An Archaeological Meditation on Trepanation

Abstract: Trepanation is a surgical practice of great antiquity. Geographically widespread across many parts of the world, it was described by a number of ancient authors in some detail. The present study examines the varieties of trepanation attested in the archaeological record, concentrating principally on the Mediterranean and Near East, and examines the evidence of origins and medico-technological diffusion debated in the literature with particular reference to ancient Greek trepanation and its Near Eastern antecedents.

Introduction

As Heinrich von Staden once wrote:

> It often is said, quite vaguely, that religious, moral, and esthetic taboos, as well as their psychological concomitants, inhibited practically all ancient and medieval physicians from opening the human body for anatomical purposes. Indeed, before Herophilus and Erasistratus, relatively superficial surgical incisions and excisions prompted by pathological conditions constituted the usual limit of "cutting" human bodies, although there are a few notable exceptions.[1]

One of those exceptions was surely trepanation.

The voluminous archaeological literature on trepanation in antiquity[2] demonstrates that at least six techniques were used (table 1).[3]

For help with references unavailable to me while writing this in Sydney I would like to sincerely thank Prof. Elisabetta Borgna (Udine), Prof. Anagnostis Agelarakis (Adelphi), Prof. Mark Geller (Berlin), Dr. Rachael Sparks (UCL), and Dr. Judith Littleton (Auckland). Prof. Deborah Martin (Nevada) kindly read an earlier draft of the chapter and provided helpful feedback.
1 H. von Staden, "The Discovery of the Body: Human Dissection and its Cultural Contexts in Ancient Greece," *The Yale Journal of Biology and Medicine* 65 (1992): 223–41, at 225.
2 As Künzl remarked, "Die Literatur zur prähistorischen und antiken Trepanation ist unendlich." See E. Künzl, "Archäologische Beiträge zur Medizingeschichte: Methoden – Ergebnisse – Ziele," in G. Sabbah (ed.), *Études de médecine romaine*, Centre Jean-Palerne de l'Université de Saint-Étienne Mémoire VIII (Saint-Étienne: Université de Saint-Étienne, 1988), 61–80, at 71, n. 1.
3 Trepanation as understood here was a surgical procedure undertaken with a variety of tools, not just the classic, saw-toothed or serrated tool known as a *trepan*. Cf. J. Kirkup, "The Evolution

Table 1

Technique	Chief Characteristics	Tools Employed
drilling/cutting	drilling circle of small holes, cutting or chiselling interstices to make larger hole to enable removal of bone	bow drill
boring/cutting	boring circle of small holes, cutting or chiselling interstices to make larger hole to enable removal of bone	borer of stone or metal
grooving	circular groove cut into cranium, loosening roundel of bone; some edge bevelling	sharp stone flake, metal chisel/gouge
scraping	bone removed by scraping action; care taken to avoid disrupting dura mater, blood vessels, meninges and brain; circular with external, bevelled edges	sharp stone flake, metal blade
sawing	polygonal or rectangular cuts made so as to remove bone, often leaves crossing cut-marks	stone or metal blade
crown trepanning	disk of bone removed	"crown trepan" hollow metal cylinder with toothed edge, central pin and transverse handle

The recommendations of the ancient authors[4] regarding trepanation are well-known,[5] and trepanation continued to be discussed by a host of medieval,

---

of Cranial Saws and Related Instruments," in R. Arnott, S. Finger, and C. U. M. Smith (eds.), *Trepanation: History – Discovery – Theory* (Lisse: Swets & Zeitlinger, 2003), 289–304. This is important since a few scholars, when surveying the evidence of trepanation, seem to consider only the process to be attested where a cranial saw or trepan was involved. The majority of the literature, however, clearly attests to the application of a much broader definition of the procedure, admitting the use of a variety of different tools and techniques.

[4] E.g., Hipp. *vuln. cap.* 14.2–8, 3.236–42 L.; *loc. hom.* 32, 6.324 L; *morb.* II 23, 7.38 L. and 25, 7.38–40 L.; Aret. 7.4.3 (153.3 Hude); Gal. *de meth. med.* 10.445 K.; Cels. 8.3.1–2, 7–9.

[5] Cf. M. Hanson, *Hippocrates, On Head Wounds,* CMG I 4,1 (Berlin: Akademie-Verlag, 1999); J. Rocca, "Galen and the Uses of Trepanation," in Arnott, Finger, and Smith, *Trepanation,* 253–71; V. G. Dimopoulos, J. S. Robinson III, and K. N. Fountas, "The Pearls and Pitfalls of Skull Trephination as Described in the Hippocratic Treatise 'On Head Wounds,'" *Journal of the History of the Neurosciences* 17 (2008): 131–40; E. Tullo, "Trepanation and Roman Medicine: A Comparison of Osteoarchaeological Remains, Material Culture and Written Texts," *Journal of the Royal College of Physicians of Edinburgh* 40 (2010): 165–71.

Renaissance, early modern, and nineteenth-century authors as well.⁶ In addition, the terminology of the procedure and instruments involved,⁷ as well as a selection of actual instruments that may have been used to perform trepanations,⁸ have been well-studied. Some of the inferences made by modern historians of surgery on the basis of this evidence, however, require modification.

I Trepanation in Ancient Greece and its Origins

Thus, for example, in 2007 S. Missios suggested that trepanation "was a well-established procedure during the ancient classical period."⁹ Notwithstanding the Hippocratic evidence, a survey of the archaeological literature on trepanation in Greece (table 2) and Asia Minor (table 3) suggests precisely the opposite.

---

6 For a useful survey of the literature in all periods, see L. Gallez, *La trépanation du crâne: histoire, technique opératoire, indications et contre-indications, résultats* (Paris: Georges Carré, 1893), 23–141.
7 F. Adams, *The Seven Books of Paulus Ægineta, Translated from the Greek, with a Commentary Embracing a Complete View of the Knowledge Possessed by the Greeks, Romans, and Arabians on All Subjects Connected with Medicine and Surgery*, vol. 2 (London: The Sydenham Society, 1846), 434–36; J. S. Milne, *Surgical Instruments in Greek and Roman Times* (Oxford: Clarendon Press, 1907), 131–33; L. J. Bliquez, "Two Lists of Greek Surgical Instruments and the State of Surgery in Byzantine Times," *Dumbarton Oaks Papers* 38 (1984): 187–204, at 203; M. A. Liston and L. P. Day, "It Does Take a Brain Surgeon: A Successful Trepanation from Kavousi, Crete," in L. A. Schepartz, S. C. Fox, and C. Bourbou (eds.), *New Directions in the Skeletal Biology of Greece. Hesperia* Supplement 43 (Princeton: American School of Classical Studies at Athens, 2009), 57–73, at 67–70; P. F. Fabbri, N. Lonoce, M. Masieri, D. Caramella, M. Valentino and S. Vassallo, "Partial Cranial Trephination by means of Hippocrates' *Trypanon* from 5th Century BC Himera (Sicily, Italy)," *International Journal of Osteology* 20 (3 Sept 2010): DOI: 10.1002/oa.1197; Tullo, "Trepanation and Roman Medicine," 167–68. For evidence from the Talmud, see B. Arensburg and I. Hershkovitz, "Cranial Deformation and Trephination in the Middle East," *Bulletins et mémoires de la Société d'Anthropologie de Paris* XIVᵉ Série 5/3 (1988): 139–50, at 141.
8 Milne, *Surgical Instruments*, Pl. 43.3–5; L. J. Bliquez, "The Tools of Asclepius: The Surgical Gear of the Greeks and the Romans," *Veterinary Surgery* 11 (Oct-Dec, 1982): 150–56, at 155–56; L. J. Bliquez, *Roman Surgical Instruments and Other Minor Objects in the National Archaeological Museum of Naples* (Mainz: von Zabern, 1994), 28, 78, 90; Liston and Day, "It Does Take a Brain Surgeon," Fig. 4.4–8.
9 S. Missios, "Hippocrates, Galen, and the Uses of Trepanation in the Ancient Classical World," *Neurosurgical Focus* 23/1 (2007): E11, at 8.

Table 2

Date B.C.	Period	Site	Notes
2000–1650	Middle Helladic	Lerna	Grave BA-1, 33 Ler, male 20–25 years old; scraped with blunt tool, 60 × 40 mm[10] Delphi male; scraped, tear-drop shaped[11]
1900–1800	Middle Cypriot I	Vounous (Cyprus)	skull II, male, 18–20 years old; 4 × 6 mm[12]
1800–1700	Middle Minoan IIB	Hagios Charalambos	no. 8083, young adult male; +10 mm dia.[13] no. 8123, child; max. dia. 8.5 mm[14] no. 8124, male[15]
1650	Middle/Late Helladic transition	Asine, Barbouna Cemetery	49 As., 30 year old male; 107 As. 30–35 male, axe wound (?)[16]
1650–1550	Late Helladic IA	Mycenae	Grave Γ–51, male, 27 × 30 mm, scraped[17]

---

[10] J. L. Angel, *The People of Lerna: Analysis of a Prehistoric Aegean Population* (Princeton: American School of Classical Studies at Athens, 1971), 43–44, 93, Pl. 22.

[11] S. K. Manolis, M. J. Pagrigorakis, and C. Zafeiratos, "Trepanations in Greece: Observations on a Middle Bronze Age Skull," in M. Schultz, K. Kreutz, and W.-R. Teegen (eds.), *Xth European Meeting of the Paleopathology Association, Göttingen, 29th August–3rd September, 1994: Abstracts, Homo* 45 Supplement (Stuttgart: Gustav Fischer Verlag, 1994), S80; R. Arnott, "Surgical Practice in the Prehistoric Aegean," *Medizinhistorisches Journal* 32/3-4 (1997): 249–78, at 259–60.

[12] P. M. Fischer, *Prehistoric Cypriot Skulls: A Medico-Anthropological, Archaeological and Micro-Analytical Investigation*, Studies in Mediterranean Archaeology 75 (Göteborg: Paul Aströms, 1986), 7, 8–9, Fig. 51.

[13] S. Chlouveraki, P. B. Betancourt, C. Davaras, H. M. C. Dierckx, S. C. Ferrence, J. Hickman, P. Karkanas, P. J. P. McGeorge, J. D. Muhly, D. S. Reese, E. Stravopodi, and L. Langford-Verstegen, "Excavations in the Hagios Charalambos Cave: A Preliminary Report," *Hesperia* 77/4 (2008): 539–605, at 583–85.

[14] Chlouveraki et al., "Excavations in the Hagios Charalambos Cave," 586.

[15] Chlouveraki et al., "Excavations in the Hagios Charalambos Cave," 586.

[16] J. L. Angel, "Appendix 1. Ancient Skeletons from Asine," in S. Dietz (ed.), *Asine II: Results of the Excavations East of the Acropolis, 1970–1974*, fasc. 1. *General Stratigraphical Analysis and Architectural Remains*, Skrifter utgivna av Svenska Institutet i Athen, 4°, 24.1 (Stockholm: Svenska Institutet i Athen, 1982), 105–38, at 109; Arnott, "Surgical Practice," 259.

[17] J. L. Angel, "Human Skeletons from Grave Circles at Mycenae," in G. E. Mylonas, *Ο Ταφικός Κύκλος Β των Μυκηνών*, vol. A (Athens: Archaeological Society of Athens, 1973), 379–97, at 380–381, Pls. 244, 249; J. H. Musgrave, R. A. H. Neave, and A. J. N. W. Prag, "Seven Faces from Grave Circle B at Mycenae," *Annual of the British School at Athens* 91 (1996): 107–36, at 115–17, 131–32; Arnott, "Surgical Practice," 258–59.

Date B.C.	Period	Site	Notes
1480–1425	Late Minoan IB	Knossos (post-mortem)	adult, 6.3 mm long; grooved, post-mortem (?)[18]
1390–1050	Late Helladic IIIB/C	Kefalonia	23 Ce, 15 × 21 mm. scraped[19]
1190–1050	Late Helladic IIIC	Argos-Deiras	36-1, 36bis-1, age unknown, drilled[20]
		Achaea Klauss	Tomb ΣT, burial Z, scraped[21]
		Phourni Archanes	information lacking[22]
		Argos	Tombs 173-1, 176-1, 183, using trepan[23]
		Agia Triada	Grave 2, adult, male (?), 32.53 × 25.37 mm, scraped[24]
900–850	Geometric	Argos	Tomb 16, male 30–35 years old, cut[25]

---

18 S. M. Wall, J. H. Musgrave, and P. M. Warren, "Human Bones from a Late Minoan IB House at Knossos," *Annual of the British School at Athens* 81 (1986): 333–88, at 341, 349, 373; Arnott, "Surgical Practice," 257.
19 J. L. Angel, "Ancient Cephallenians: The Population of a Mediterranean Island," *American Journal of Physical Anthropology* 1 (1943): 229–56, at 247; Arnott, "Surgical Practice," 260.
20 R. P. Charles, *Étude anthropologique des nécropoles d'Argos: contribution à l'étude des populations de la Grèce antique* (Paris: Vrin, 1963), 67–69; Arnott, "Surgical Practice," 256–57; Fabbri et al., "Partial Cranial Trephination."
21 C. Paschalides and P. J. P. McGeorge, "Life and Death in the Periphery of the Mycenaean World at the End of the Late Bronze Age: The Case of the Achaea Klauss Cemetery," in E. Borgna and P. Cassola Guida (eds.), *Dall'Egeo all'Adriatico: organizzazioni sociali, modi di scambio e interazione in età postpalaziale (XII–XI sec. a.C.) / From the Aegean to the Adriatic: Social Organisations, Modes of Exchange and Interaction in Postpalatial Times (12th–11th B.C.)*, Studi e ricerche di protostoria mediterranea 8 (Rome: Quasar, 2009), 79–113, see Fig. 19.
22 Chlouveraki et al., "Excavations in the Hagios Charalambos Cave," 593 n. 127, noting that in the original publication by A. Poulianos (Η καταγωγή των Κρητών [Athens: Ανθρωπολογική Εταιρεία Ελλάδος, 1971]), "neither the nature nor the date of this trephination is clear since no illustration or find context is given."
23 Charles, *Étude anthropologique*, passim; Fabbri et al., "Partial Cranial Trephination"; C. Mountrakis, S. Georgaki, and S. K. Manolis, "A Trephined Late Bronze Age Skull from Peloponnesus, Greece," *Mediterranean Archaeology and Archaeometry* 11 (2011): 1–8, at 2.
24 Mountrakis et al., "A Trephined Late Bronze Age Skull," 3–4, Figs. 2–5.
25 R. P. Charles, "Étude anthropologique des nécropoles d'Argos: contribution à l'étude des populations de la Grèce antique," *Bulletin de Correspondance Hellénique* 82 (1958): 268–313, at 279, 310–311; Arnott, "Surgical Practice," 257 n. 19.

Date B.C.	Period	Site	Notes
800–700		Kavousi Vronda	Grave 5, burial 1, male 25–40 years old; cremated; 65 × 20 mm, scraped or grooved with curved blade[26]
700–600		Abdera	scraped, 14.78 × 9.19 mm[27]

Table 3

Approx. Date	Period	Site	Technique[28]
9600–7000 B.C.	Aceramic Neol.	Aşikli Höyük	drilling
		Çayönü	drilling
7000–6000 B.C.	Neolithic	Çatal Höyük	drilling (probably post-mortem)[29]
		Kurban Höyük	grooving, sawing?
6000–5000 B.C.	Early Chalcolithic	Kuruçay	drilling
3100–2000 B.C.	Early Bronze Age	Bakla Tepe	scraping
		Elmali-Karataş	sawing, grooving[30]
		Küçükhöyük	sawing, grooving
		İkiztepe	sawing (triangle, circle, rectangle); scraping
2000–1650 B.C.	Middle Bronze Age	Çavlum	sawing (rectangle)
		Kültepe	sawing, grooving
		Açemhöyük	drilling
		Lidarhöyük	scraping; sawing[31]

---

26 Liston and Day, "It Does Take a Brain Surgeon," 63–66.
27 A. P. Agelarakis, "Early Evidence of Cranial Surgical Intervention in Abdera, Greece: A Nexus to *On Head Wounds* of the Hippocratic Corpus," *Mediterranean Archaeology and Archaeometry* 6/1 (2006): 5–18, at 6–7.
28 Unless otherwise indicated below, all data, including full bibliography, can be found in Y. S. Erdal and Ö. D. Erdal, "A Review of Trepanations in Anatolia with New Cases," *International Journal of Osteoarchaeology* 21 (2011): 505–34, at 528–32 and Appendix A1.
29 J. L. Angel, "Early Neolithic Skeletons from Çatal Hüyük: Demography and Pathology," *Anatolian Studies* 21 (1971): 77–98, at 94.
30 J. L. Angel, "Appendix. Human Skeletal Remains at Karataş," *American Journal of Archaeology* 70 (1966): 255–57.
31 T. Koca and M. Schulz, "Trephination as a Medical Indication Following Trauma Observed in the Middle Bronze Age Population from Lidar Höyük (Turkey)," in M. Schultz, K. Kreutz, and W.-

Approx. Date	Period	Site	Technique[28]
1650–1200 B.C.	Late Bronze Age	Troy	sawing (post-mortem?)[32]
1200–600 B.C.	Iron Age	Dilkaya	boring, cutting
		Karagündüz	boring, cutting
		Hakkari	boring, cutting
		Van Kalesi	sawing, grooving
25 B.C.–300 A.D.	Roman	Gordion	sawing
100–300 A.D.		Perge	drilling; boring, cutting
100–500 A.D.		Cevizcioğlu Çiftliği	drilling; scraping
200–700 A.D.	Early Byzantine	Nicaea	scraping
800–1000 A.D.		Kovuklukaya	drilling
900–1000 A.D.		Allianoi	drilling
1222–1254 A.D.	Late Byzantine	İznik	scraping; sawing

It is always important to remember, of course, that archaeological samples are subject to a wide range of biases and, as a result, are rarely as representative as we would like them to be. On the other hand, Greece and Turkey have been investigated intensively and a clear trend emerges in tables 2 and 3, viz. that trepanation was most popular in the Bronze Age, became less so in the Iron Age, and is completely unrepresented in skeletal samples from the classical period.

The disjunction between the archaeological and the literary evidence in this case is extreme, particularly given Hippocratic writings on head injuries and other Greek sources on the treatment of war wounds to the head.[33] It is possible, however, that this incongruity is a reflection of the Greek mortuary evidence available from this era, when cremation was preferred to burial. This is a factor that has undoubtedly skewed the surviving skeletal sample so that, even if trepanation was practiced in Greece during the classical period, evidence of it may be very hard to find.[34]

R. Teegen (eds.), *X*th *European Meeting of the Paleopathology Association, Göttingen, 29*th *August–3*rd *September, 1994: Abstracts*, *Homo* 45 Supplement (Stuttgart: Gustav Fischer Verlag, 1994), S68.

32 H. Kiesewetter, "Ein trepanierter Schädel aus Troia VI," *Studia Troica* 12 (2002): 73–80.

33 C. F. Salazar, *The Treatment of War Wounds in Graeco-Roman Antiquity* (Leiden: Brill, 2000), 45.

34 O. T. P. K. Dickinson, "Ancient Greece," in D. J. Davies and L. H. Mates (eds.), *Encyclopedia of Cremation* (Aldershot: Ashgate, 2005), 6–8, at 7.

In Sicily and south Italy, on the other hand, trepanation is attested at this time and in the Hellenistic period. One case dating to the sixth to fifth century B.C. is known from Himera, the eastern necropolis of Pestavecchia, in Sicily, while in Campania two cases dating to the fifth to fourth centuries B.C. are reported from Pontecagnano, and in Apulia one is recorded from a fourth- to third-century context at Poggiardo.[35] Lest it be suggested that trepanation might have been introduced into south Italy and Sicily by Greek colonists, however, it is worth underscoring the fact that trepanation had been practiced in south Italy, e.g., at Trasano near Matera in Basilicata, since the Neolithic (ca. 5000 B.C.).[36]

Missios also speculated on the origins of ancient Greek trepanation, suggesting that Hippocrates "had some exposure to Celtic tradition and trepanation practices" and that in areas where "cultural exchange between the Celtic and Greek worlds" took place, e.g., near the Black Sea and "along the amber trade route from Western Europe to the Mediterranean," specialized knowledge of trepanation may have been transmitted from the Celts to the Greeks.[37] Further, he suggested that

> The contributions of Hippocrates and his contemporaries and the intellectual environment of Ancient Greece allowed trepanation, and with it ancient neurosurgery, to evolve from an ancient, oral, disorganized, and mystical tradition that was likely Celtic in origin, to a written, scientific, and systematic topic of study, devoid of superstitions and religious influences.[38]

Leaving aside the dubious characterization of Celtic mentality by Missios, it is clear that he was ignorant of the published evidence of trepanation in Greece during the Bronze and Iron Age. Perhaps this is no bad thing. Had he been aware of the practice of trepanation by the Minoans and Mycenaeans (table 2), Missios' chauvinism might have encouraged him to suggest that knowledge of the procedure diffused in the opposite direction, from Greece to the Celtic

---

35 Fabbri et al., "Partial Cranial Trephination."
36 F. Germanà and G. Fornaciari, *Trapanazioni, craniotomie e traumi cranici in Italia*, Collana di Studi Paletnologici 5 (Pisa: Giardini Editori e Stampatori, 1992); F. Mallegni and A. Valassina, "Secondary Bone Changes to a Cranium Trepanation in a Neolithic Man Discovered at Trasano, South Italy," *International Journal of Osteoarchaeology* 6 (1996): 506–11; L. Capasso, E. Michetti, L. Pierfelice, and R. D'Anastasio, "Neurosurgery 7000 Years Ago in Central Italy," *The Lancet* 359 (2002): 2206; É. Crubézy, J. Bruzek, J. Guilaine, E. Cunha, D. Rougé, and J. Jelinek, "The Antiquity of Cranial Surgery in Europe and in the Mediterranean Basin," *Comptes Rendus de l'Académie des sciences de Paris, sciences de la terre et des planètes* 332 (2001): 417–23, at 419–20.
37 Missios, "Hippocrates," 4.
38 Missios, "Hippocrates," 5.

world. One wonders, however, what he would have made of the Neolithic evidence from Italy dating to the fifth millennium B.C., let alone that of Taforalt in Morocco, the Epipalaeolithic inhabitants of which performed the world's earliest trepanations, so far as we know, eleven to twelve thousand years ago.[39] So much for the notion that trepanation was part of a "mystical tradition that was likely Celtic in origin."[40]

II Technological Transfer or Multiple Instances of Invention

Since trepanation was first identified in Peru[41] and France[42] during the mid-nineteenth century, the procedure has been documented archaeologically at hundreds of sites in Europe,[43] the Near East,[44] Egypt,[45] the Indian sub-continent,[46]

---

39 J. Dastugue, "Un orifice crânien préhistorique," *Bulletins et mémoires de la Société d'Anthropologie de Paris*, 10th sér. 10/4 (1959): 357–63; D. Ferembach, J. Dastugue, and M. J. Poitrat-Targowla, *La nécropole épipaléolithique de Taforalt (Maroc oriental): étude de squelettes humains* (Rabat: CNRS, 1962). For the date see V. Mariotti, B. Bonfiglioli, F. Facchini, S. Condemi, and M. G. Belcastro, "Funerary Practices of the Iberomaurusian Population of Taforalt (Tafoughalt; Morocco, 11–12,000 BP): New Hypotheses Based on a Grave by Grave Skeletal Inventory and Evidence of Deliberate Human Modification of the Remains," *Journal of Human Evolution* 56 (2009): 340–54; and M. G. Belcastro, S. Condemi, and V. Mariotti, "Funerary Practices of the Iberomaurusian Population of Taforalt (Tafoughalt, Morocco, 11–12,000 BP): The Case of Grave XII," *Journal of Human Evolution* 58 (2010): 522–32.
40 Missios, "Hippocrates," 5.
41 H. R. Fernando and S. Finger, "Ephraim George Squier's Peruvian Skull and the Discovery of Cranial Trepanation," in Arnott, Finger, and Smith, *Trepanation*, 3–18.
42 P. Broca, "Sur les trépanations préhistoriques," *Bulletin de la Sociétie d'Anthropologie de Paris* 11 (1876): 236–56; P. Broca, "Sur l'âge des sujets soumis à la trépanation chirurgicale néolithique," *Bulletin de la Sociétie d'Anthropologie de Paris* 11 (1876): 572–76; S. Finger and W. T. Clower, "On the Birth of Trepanation: The Thoughts of Paul Broca and Victor Horsley," in Arnott, Finger, and Smith, *Trepanation*, 19–42, at 19–30 with bibliography.
43 E.g., Künzl, "Archäologische Beiträge," 61–80; K. W. Alt, C. Jeunesse, C. H. Buitrago-Téllez, R. Wächter, E. Boës, and S. L. Pichler, "Evidence for Stone Age Cranial Surgery," *Nature* 387 (1997): 360; M. C. Lillie, "Cranial Surgery Dates Back to Mesolithic," *Nature* 391 (1998): 354; J. Weber and J. Wahl, "Neurosurgical Aspects of Trepanations from Neolithic Times," *International Journal of Osteoarchaeology* 16 (2006): 536–45; Z. Bereczki, E. Molnár, A. Marcsik, and G. Pálfi, "Evidence of Surgical Trephinations in Infants from the 7th–9th Centuries AD Burial Site of Kiskundorozsma-Kettőshatár," *Acta Biologica Szegediensis* 54/2 (2010): 93–98. This is merely a small selection of the literature on ancient trepanation in Europe. For another ten studies by various authors of trepanation in Britain, Ireland, Denmark, Portugal, Italy, Austria, the Czech Republic, Russia, and the Ukraine, see Arnott, Finger, and Smith, *Trepanation*, 55–198.

Mongolia,[47] China,[48] Siberia,[49] Australia (?),[50] Melanesia,[51] and the New World.[52] Scholars have often wondered whether this distribution pattern reflects multiple discoveries of a similar procedure across space and time (albeit using a variety of tools and techniques; cf. table 1) or diffusion. For de Nadaillac, the practice of trepanation by a multitude of different cultures from an early date was a sign, not of diffusion from a racially/intellectually superior civilization to the rest of the "uncivilized," "under-civilized," or "pre-logical" world, but rather of the

---

44 For surveys see, e. g., S. Mogliazza, "An Example of Cranial Trepanation Dating to the Middle Bronze Age from Ebla, Syria," *Journal of Anthropological Sciences* 87 (2009): 187–92, Table 1; Y. S. Erdal, "A Retrospective Study on Trepanation in Anatolia," in P. Matthiae, F. Pinnock, L. Nigro, and N. Marchetti (eds.), *Proceedings of the 6th International Congress of the Archaeology of the Ancient Near East*, vol. 1 (Wiesbaden: Harrassowitz, 2010), 261–76; Erdal and Erdal, "A Review," 533–34, Appendix A1; J. Littleton and K. Frifelt, "Trepanations from Oman: A Case of Diffusion?," *Arabian Archaeology and Epigraphy* 17 (2006): 139–51, Table 3; and H. Kiesewetter, "Analyses of the Human Remains from the Neolithic Cemetery at al-Buhais 18 (Excavations 1996–2000)," in H.-P. Uerpmann, M. Uerpmann, and S. A. Jasim (eds.), *Funeral Monuments and Human Remains from Jebel al-Buhais* (Tübingen: Kerns Verlag, 2006), 103–380, at 193–202 and Table 6.10.
45 A. G. Nerlich, A. Zink, U. Szeimies, H. G. Hagedorn, and F. W. Rösing, "Perforating Skull Trauma in Ancient Egypt and Evidence for Early Neurosurgical Therapy," in Arnott, Finger, and Smith, *Trepanation*, 191–201.
46 A. R. Sankhyan and G. H. J. Weber, "Evidence of Surgery in Ancient India: Trepanation at Burzahom (Kashmir) over 4000 Years Ago," *International Journal of Osteoarchaeology* 11 (2001): 375–80; A. R. Sankhyan and G. R. Schug, "First Evidence of Brain Surgery in Bronze Age Harappa," *Current Science* 100/11 (2011): 1621–22.
47 N. Bazarsad, "Four Cases of Trepanation from Mongolia, Showing Surgical Variation," in Arnott, Finger, and Smith, *Trepanation*, 203–208.
48 K. Han and X. Chen, "The Archaeological Evidence of Trepanation in Early China," *Bulletin of the Indo-Pacific Prehistory Association* 27 (2007): 22–27.
49 E. Murphy, "Trepanations and Perforated Crania from Iron Age South Siberia: An Exercise in Differential Diagnosis," in Arnott, Finger, and Smith, *Trepanation*, 209–21.
50 S. G. Webb, "Two Possible Cases of Trephination from Australia," *American Journal of Physical Anthropology* 75 (1988): 541–48.
51 G. Martin, "Trepanation in the South Pacific," *Journal of Clinical Neuroscience* 2/3 (1995): 257–64; D. A. K. Watters, "Skull Trepanation in the Bismarck Archipelago," *Papua New Guinea Medical Journal* 50/1–2 (2007): 20–24.
52 The literature on New World trepanation is voluminous but to cite just a few publications with extensive bibliography, see J. L. Stone and M. L. Miles, "Skull Trepanation among the Early Indians of Canada and the United States," *Neurosurgery* 26/6 (1990): 1015–19; J. W. Verano, "Trepanation in Prehistoric South America: Geographic and Temporal Trends over 2,000 Years," in Arnott, Finger, and Smith, *Trepanation*, 223–36; J. L. Stone and J. Urcid, "Pre-Columbian Skull Trepanation in North America," in Arnott, Finger, and Smith, *Trepanation*, 237–49; and V. A. Andruschko and J. W. Verano, "Prehistoric Trepanation in the Cuzco Region of Peru: A View into an Ancient Andean Practice," *American Journal of Physical Anthropology* 137 (2008): 4–13.

shared intelligence and similar intellect of all peoples throughout history.[53] In 1946 J. L. Loughborough evaluated the trepanations then known from Europe, Peru, and the Loyalty Islands (French New Caledonia). Even though many parts of the world were at that time devoid of cases (e.g., Africa, Asia, Australia, North America, England, all of which may now be added to the list of regions in which trepanation was practiced in antiquity), he observed that "the locations of these finds seem to follow no definite pattern."[54] In the end, he concluded, "There is no evidence of the diffusion of trepanning between these geographical areas; therefore we must accept it, if tentatively, as independent invention."[55]

In more recent times, however, diffusionist voices have made themselves heard. In 1994 the respected British physical anthropologist D. R. Brothwell stated that he found it difficult to believe the technique of trepanation had been discovered more than once. Suggesting that the practice of trepanation in Africa and Eurasia derived ultimately from a single "discovery," he attributed the appearance of the procedure in the New World to diffusion, suggesting three potential mechanisms: Eurasians bringing the knowledge of trepanation across the Bering landbridge in the original peopling of North America; trans-Pacific diffusion at a later date by seafaring groups from Polynesia; or trans-Atlantic diffusion by, e.g., Phoenician mariners.[56] Some years later, Brothwell suggested, apropos the practice of trepanation in Neolithic France, that

> this tradition...is likely to have arrived with the farming traditions from the Near East, if it hadn't arrived earlier with the expansion of Upper Palaeolithic society through Europe. And if we see this ancient surgical knowledge dispersed in Europe, North Africa, and Asia, then why not into the Americas? I am getting dangerously close to supporting world diffusionism, but the fact is that over the past 50,000 years or more, people spread around the world, and they could have taken a very basic surgical knowledge with them.[57]

In the same vein, Sankhyan and Schug have recently suggested that "striking similarities in trepanation techniques across the continents" permit one to "con-

---

[53] J. F. A. du P. de Nadaillac, "Mémoire sur les trépanations préhistoriques," *Comptes-rendus des scéances de l'Académie des Inscriptions et Belles-Lettres* 30/2 (1886): 280–94, at 294.
[54] J. L. Loughborough, "Notes on the Trepanation of Prehistoric Crania," *American Anthropologist* 48/3 (1946): 416–22, at 417.
[55] Loughborough, "Notes on the Trepanation of Prehistoric Crania," 421.
[56] D. R. Brothwell, "Ancient Trephining: Multi-Focal Evolution or Trans-world Diffusion?," *Journal of Paleopathology* 6 (1994): 129–38, at 136–37.
[57] D. R. Brothwell, "The Future Direction of Research," in Arnott, Finger, and Smith, *Trepanation*, 365–72, at 370–71.

sider it as…important evidence for prehistoric movements of people and for transfer of surgical skills from one society to another."[58]

Notwithstanding Brothwell's unfortunate allusion to putative Phoenician mariners crossing the Atlantic Ocean, his essential argument should not be dismissed out of hand. In fact, the question of whether or not trepanation was only ever invented once is reminiscent of the debate over the discovery of certain complex metallurgical procedures in antiquity and the domestication of certain cultivars and animals. In the early and mid-twentieth century, for example, scholars like Max Loehr and Cyril Stanley Smith argued that bronze metallurgy in China, which developed very differently from metalworking in the West, owed its ultimate origin to the Near East. As diffusionism went out of fashion and nationalism became a force in archaeological explanation, however, sentiment shifted. The 1960s and 1970s witnessed some vehement protestations of the independent invention of Chinese metallurgy. More recently, however, the pendulum has again swung back in favor of a Near Eastern origin and diffusion of metallurgical technology to China via Central Asia and Xinjiang.[59]

Faced with the currently available evidence, it would be possible to argue for a single origin of the trepanation procedure in north Africa (Morocco) during the terminal Pleistocene, followed by its diffusion into and elaboration in Europe, the Near East, and Asia, and its further diffusion from there, by land to North and eventually South America, and by land bridges and/or sea into Australia and Melanesia. Many more examples spanning the chronological range of attested trepanations beginning twelve to eleven thousand years ago, such as the example from Vasiliyevka in the Ukraine which has been dated to ca. 8020–7620 cal (2σ) B.C.,[60] would be required to flesh out the picture. On the other hand, as Crubézy et al. have recently stressed,

> The oldest examples of trephination…appear in populations of hunter-gatherers distant from one another in time and in space (North Africa, Ukraine, Portugal) and their small area together with the very simple techniques used (drilling and/or abrasion)…could argue for autonomous, unrelated centres.[61]

---

58 Sankhyan and Schug, "First Evidence of Brain Surgery," 1621.
59 D. T. Potts, "Technological Transfer and Innovation in Ancient Eurasia," in J. Renn (ed.), *The Globalization of Knowledge in History* (Berlin: Max Planck Research Library for the History and Development of Knowledge Studies 1, 2012), 105–23.
60 Lillie, "Cranial Surgery," 354.
61 Crubézy et al., "Antiquity of Cranial Surgery," 421–22.

Nevertheless, even if the procedure was discovered independently more than once, it is highly unlikely that trepanation was in all cases a spontaneous invention. It is certainly possible that, in many cases, trepanation appeared in one area because of the diffusion of the technical knowledge of it, as well as practitioners, from an adjacent or nearby region. Just as nobody seriously disputes the diffusion of numerous cultivars and domesticated animals, or many different technical procedures (particularly in metallurgy), across great distances, so too is there no *a priori* reason why trepanation, as a social and medical techno-complex, should have been any different. With that in mind, then, we return to the evidence from ancient Greece.

III Ancient Greek Trepanation in its Wider Near Eastern Context

From an archaeological perspective, it is striking that the earliest evidence of trepanation in Bronze Age Greece dates to the Middle Minoan period, i.e., the early second millennium B.C. Compared to the Moroccan evidence from the twelfth millennium B.C., the first Greek cases can only be described as "late." While we can safely dismiss Missios' hypothesis of a Celtic origin for ancient Greek trepanation, the evidence from Greece's nearest eastern and western neighbors is surely relevant. As noted above, trepanation had been practiced in Italy since about 5000 B.C.,[62] thousands of years before the first Greek colonists reached Sicily or south Italy, but it has an even longer history in Anatolia (table 3) and the Levant (table 4).

Table 4

Date B.C.	Period	Site	Notes
8350–6000	Pre-Pottery Neolithic	Jericho	healed wound, plastered[63]

---

62 Mallegni and Valassina, "Secondary Bone Changes," 506–11.
63 G. Kurth and O. Roher-Ertl, "On the Anthropology of the Mesolithic to Chalcolithic Human Remains from the Tell es-Sultan in Jericho, Jordan," in K. M. Kenyon (ed.), *Excavations at Jericho*, vol. 3 (Jerusalem: British School of Archaeology, 1981), 407–99; Arensburg and Hershkovitz, "Cranial Deformation," 140; J. Zias and S. Pomeranz, "Serial Craniectomies for Intracranial Infection 5.5 Millennia Ago," *International Journal of Osteoarchaeology* 2 (1992): 183–86, at 183.

Date B.C.	Period	Site	Notes
4500–3500	Chalcolithic	Azor	drilled[64]
4000–3000		Wadi Hebran (south Sinai), tomb 105	male, 35–40 years old; scraped; 23 mm dia.[65]
		Wadi Makuqh, 6 km NW of Jericho	young adult male, 28 × 24, 23 × 15, 25 × 50 mm, scraped[66]
3000–2700	Early Bronze II	Arad	male, 16–18 years old, 40 mm dia., scraped[67]
2000–1750	Middle Bronze I	Qatna	unpublished; cut[68]
2200	Middle Bronze Age	Dimona	child, 8–9 years old, 42 × 29 mm[69]
1800–1600	Middle Bronze	Ras Shamra	R2, female, 40 years old, 23 mm dia., scraped R4, female, 18 years old, 35 × 45 mm, scraped[70]
2000	Middle Bronze II	Jericho	male 20–30 years old, 30, 20.2, 23.3, 15.4 mm dia.; cut and drilled[71]
	Middle Bronze II	Tell Beit Mersim (Tomb 7)	adult male, circular, 35 mm dia.; healed[72]
1500–1300	Late Bronze	Ras Shamra	R6, female (?), three circular holes, c. 11–12 mm

---

64 D. Ferembach, "À propos d'une trépanation sur un fragment d'occipital humain Chalcolithique trouvé à Azor Israël," *Garcia de Orta, Série Antropobiologica* 3/1–2 (1984): 139–42.
65 I. Hershkovitz, "Trephination: The Earliest Case in the Middle East," *Journal of the Israel Prehistoric Society* 20 (1987): 128–35, at 131*.
66 Zias and Pomeranz, "Serial Craniectomies," 183–86, Figs. 1–4.
67 P. Smith, "The Trephined Skull from the Early Bronze Age Period at Arad," *Eretz-Israel* 21 (1990): 83*–93*, at 90*.
68 Mogliazza, "An Example of Cranial Trepanation," Table 1.
69 P. Mogle and J. Zias, "Trephination as a Possible Treatment for Scurvy in a Middle Bronze Age (ca. 2200 BC) Skeleton," *International Journal of Osteoarchaeology* 5/1 (1995): 77–81, at 77.
70 V. Vallois and D. Ferembach, "Les restes humains de Ras Shamra et Minet el Beida: étude anthropologique," *Ugaritica* 4 (1962): 565–622, at 567–68; J. Dastugue, "Les pièces pathologiques de Ras Shamra et Minet el Beida," *Ugaritica* 4 (1962): 623–30, at 623–24.
71 K. P. Oakley, W. M. A. Brooke, A. R. Akester, and D. R. Brothwell, "Contributions on Trepanning or Trephination in Ancient and Modern Times," *Man* 59 (1959): 93–96, at 95.
72 P. Smith and L. Dujovny, "Middle Bronze Age II Human Remains from Caves 7, 24, and 33," in S. Ben-Arieh (ed.), *Bronze and Iron Age Tombs at Tell Beit Mirsim*, Israel Antiquities Authority Reports 23 (Jerusalem: Israel Antiquities Authority, 2004), 203–205, at 204. My thanks to Dr. Rachael Sparks (UCL) for helping me find this source.

Date B.C.	Period	Site	Notes
			dia.[73]
		Minet el-Beidha	M1, 12 × 20 mm[74]
	Iron Age	Tell Duweir (Lachish)	three individuals, 17 × 21, 22 × 21, 29 × 17, sawn quadrilaterals[75]
		Tell el-Mazar	three adults, drilled?[76]
500 B.C.– 300 A.D.	Iron to Roman	Timna'	male, 26 × 19, sawn[77]
	Hellenistic- Roman	Akko	drilled[78]
		Jericho/Qaranthal	Homo I: female 25–35 years old, 55 × 42 mm Homo II: male 40–50 years old, 55 × 42 mm Homo III: male 40–50 years old, 83 mm long[79]

The history of interaction between Greece and its eastern neighbors during the Bronze and Iron Ages may well have afforded opportunities for a diffusion of the knowledge of trepanation, as well as the movement of healers trained in the procedure.

It might be tempting to try to infer influence from one or another area by looking at the particular trepanation techniques used. While this seems promising at first glance, it is difficult because, particularly in cases where the patient

---

73 Vallois and Ferembach, "Les restes humains," 571–77, Fig. 5.
74 Dastugue, "Les pièces pathologiques," 626–27.
75 J. L. Starkey, "Discovery of Skulls with Surgical Holing at Tell Duweir, Palestine," *Man* 36 (1936): 169–70, see Figs. 1–3; T. W. Parry, "Three Skulls from Palestine Showing Two Types of Primitive Surgical Holing: Being the First Skulls Exhibiting This Phenomenon That Have Been Discovered on the Mainland of Asia," *Man* 36 (1936): 170–71.
76 H. Kiesewetter, "Analyses," see Table 6.10, citing A. M. Disi, W. Henke, and J. Wahl, *Tell el-Mazar: Study of the Human Skeletal Remains* (Amman: University of Jordan, 1982), which I have not seen.
77 D. Ferembach, "À propos d'un crâne trépané trouvé à Timna: origine de certaines tribus berbères," *Bulletins et Mémoires de la Société d'Anthropologie de Paris*, 10th sér. 8/5–6 (1957): 244–75, at 253–75.
78 Z. Goldman, "Surgical Trepanation in Ancient Times," *Bulletin of the Israel Exploration Society* 25 (1961): 258–60 (in Hebrew); Fabbri et al., "Partial Cranial Trephination."
79 J. Zias, "Three Trephinated Skulls from Jericho," *Bulletin of the American Schools of Oriental Research* 246 (1982): 55–58.

survived and there has been a remodelling of bone, or where the osteological evidence is fragmentary, it is not always possible to ascertain the technique used. Nevertheless, the recent, exhaustive publication of data on forty ancient trepanations found in Anatolia (table 3)[80] provides some interesting food for thought. To begin with, it is noteworthy that the earliest trepanations in Anatolia, dating to the Aceramic (Aşikli Höyük, Çayönü) and Ceramic Neolithic (Çatal Höyük), used the drilling technique, presumably employing a simple bow-drill fitted with a flint or, more probably, given its prevalence in Anatolian prehistory, an obsidian drill-bit. Sawing may have been practiced at Kurban Höyük in the late Neolithic (ca. 5000 B.C.), but it only became the main technique used during the Bronze Age, when evidence of drilling is absent. During the Iron Age the technique of boring and cutting was dominant, to the exclusion of all other methods, apart from one instance of sawing. Then, during the Roman era, sawing, drilling, scraping, and possibly boring and cutting are all attested. Scraping, sawing, and drilling are also found on crania from the Byzantine period.

These observations raise a number of issues. Whether or not trepanation is labelled "esoteric knowledge," its practitioners are very likely to have passed it on from generation to generation, as well as spreading it within the geographical limits of their existence. Having said that, it nevertheless seems perfectly plausible to imagine the co-existence of different techniques in one period, depending on the training of any individual practitioner. Like most technical accomplishments, trepanation was probably learned by young healers working together on patients with older practitioners. Indeed, Galen suggested that it would take many operations before one became adept at the procedure.[81] As for the specific techniques used to perform a trepanation, a host of unrelated yet mechanically similar technologies, such as bead-making and stone vessel manufacture (both of which, in certain cases, employed the bow-drill), may have played a role in the appearance of the drilling technique. Similarly, the prevalence of sawing in the Bronze Age is probably a reflection of the increasing elaboration of the repertoire of copper and bronze tools at this time. The particular predilection for the boring and cutting technique at Dilkaya and Karagündüz, in Urartu, may reflect an isolated, local development in that region. Galen suggested that different trepanation procedures went in and out of fashion, noting,

---

[80] Erdal and Erdal, "A Review," Appendix A1.
[81] J. Rocca, *Galen on the Brain: Anatomical Knowledge and Physiological Speculation in the Second Century AD* (Leiden: Brill, 2003), 75–76, 183; Tullo, "Trepanation and Roman Medicine," 169.

> The ancients excised them [bone fragments resulting from cranial trauma] sawing with a [crown] trepan by twisting it. Those after them with trepans, adding a base to the chisels. Those [who live] now are satisfied with chisels alone. (trans. Salazar)[82]

On the other hand, scraping, which appeared in Anatolia during the Bronze Age and continued in use into the medieval (Late Byzantine) era, required little in the way of technology and may have been a technique favored by non-professionals, i.e., home practitioners who were not themselves trained in medicine. It was, after all, the technique used on all three of the trepanations recorded in the mid-fifth millennium B.C. at Neolithic Jabal Buhais 18 in the UAE,[83] showing that it was practiced in the Near East from an early date. Yet it would be wrong to consider the scraping technique primitive (given that an unworked stone flake can be used as a scraper), and characteristic only of "folk" surgery. Rather, there is ample evidence to show that scraping was also employed by trained medical practitioners. For example, in BAM V 480 III 57–64,[84] the only possible reference to trepanation in the extant corpus of Mesopotamian medical texts,[85] we read:

> If a man's skull holds water, you touch with your big finger the spot that holds water. If his...is...[and water (?)] descends from his skull, you open and you scrape (?) his skull; the water of his skull [you siphon off...] you wash, you sprinkle oil, you put (this) on the wound. (...) If his... is not..., you put a...of (amulet) stones around his head.[86]

---

82 *Introd. s. medic.* 19.5, 14.783 K. Translation at Salazar, *Treatment of War Wounds*, 46.
83 Kiesewetter, "Analyses," 199 and Table 6.10.
84 F. Köcher, *Die babylonisch-assyrische Medizin in Texten und Untersuchungen*, Band V. Keilschrifttexte aus Ninive 1 (Berlin: De Gruyter, 1980), xxiii–xxv.
85 Mark Geller (pers. comm.) suggests that this may simply refer to the treatment of a surface wound.
86 M. Stol, "An Assyriologist Reads Hippocrates," in H. F. J. Horstmanshoff and M. Stol (eds.), *Magic and Rationality in Ancient Near Eastern and Graeco-Roman Medicine* (Leiden: Brill, 2004), 63–78, at 75. This edition and Stol's translation supersede earlier treatments, e.g., R. Campbell Thompson, "Assyrian Prescriptions for Diseases of the Head (Continued)," *American Journal of Semitic Languages and Literatures* 24/4 (1908): 323–53, at 345; R. Campbell Thompson, "Assyrian Prescriptions for the Head," *American Journal of Semitic Languages and Literatures* 53/4 (1937): 217–38, at 234; or R. Labat, "À propos de la chirurgie babylonienne," *Journal Asiatique* 242 (1954): 207–18, at 212–14. As far as the date of this text goes, the cuneiform tablets from which it is known are Neo-Assyrian, all from Nineveh. For German translations see D. Goltz, *Studien zur altorientalischen und griechischen Heilkunde: Therapie – Arzneibereitung – Rezeptstruktur*, Sudhoffs Archiv Beiheft 16 (Wiesbaden: Steiner, 1974), 89; J. Fincke, "Spezialisierung und Differenzierung im Bereich der altorientalischen Medizin: Die Dermatologie am Beispiel der Symptome *simmū matqūtu, kalmātu (matuqtu), kibšu, kiṣṣatu* und *gurištu*," in G. Selz (ed.), *The Empirical Dimension of Ancient Near Eastern Studies / Die empirische Dimension altorientalischer Forschungen* (Vienna: Wiener Offene Orientalistik 6, 2011), 159–208, at 161, n. 11.

Moreover, in Greece, where scraping was the most prevalent trepanation technique used,[87] iron scrapers or raspatories (ξυστήρ/ξυστήριον, as described by Hippocrates, Oribasius, and Paul of Aegina)[88] were employed. Examples from an Iron Age grave at Vronda have been identified as surgical instruments used for trepanation,[89] and it is likely that their presence in a tomb reflects the expertise in life of the deceased with whom they were buried.

IV Experimentation, Demography, Causality

Experimental evidence gives us some idea of how long such a procedure may have taken. In 1912 Lucas-Championnière published the result of an experimental trepanation conducted using a flint tool that took forty minutes to complete.[90] No doubt stimulated by this, in 1913 T. W. Parry undertook "a series of some fifty experiments on both recent and dry human skulls with all the primitive implements that have been used for the purpose – namely, flint, obsidian, shark's teeth, shell, glass, and even slate."[91] He found "the average time taken...to do this operation on a fresh adult skull was half an hour" and, using a beach-worn shell, about 25 minutes for the cranium of a 9-month-old infant. Using a bow-drill on the cranium of a 68–year-old male, it took Parry 25 minutes just to drill a series of holes, and 65 minutes "to remove a rondel from the skull of a female 40 years of age."[92]

Finally, it is interesting to consider the demographics of trepanation. No one who has delved into the literature on trepanation can fail to have been impressed by the high survival rate (70–80 % worldwide)[93] amongst those who underwent the procedure, a point made in virtually every paper on the topic. In view of the issues of sterility and infection surrounding trepanation in the pre-modern, let alone in the prehistoric past, this statistic is truly staggering. The accomplish-

---

[87] R. Arnott, "Surgical Practice," 252.
[88] Bliquez, "Two Lists," 201; Missios, "Hippocrates," 3.
[89] Liston and Day, "It Does Take a Brain Surgeon," 61.
[90] J. Lucas-Championnière, *Trépanation néolithique pré-colombienne, trépanation des Kabyles, trépanation traditionelle* (Paris: G. Steinheil, 1912), 192; cf. F. Sanagustin, "La chirurgie dans le Canon de la medecine (Al-Qânûn fî-ṭ-ṭibb) d'Avicenne (Ibn Sinâ)," *Arabica* 33 (1986): 84–122, at 96, n. 54.
[91] T. W. Parry, "Trephination of the Living Human Skull in Prehistoric Times," *British Medical Journal* 1/3246 (1923): 457–60, at 458.
[92] Parry, "Trephination of the Living Human Skull," 459.
[93] J. Zias, "Death and Disease in Ancient Israel," *Biblical Archaeologist* 54/3 (1991): 146–59, at 156.

ments of ancient surgeons, moreover, are all the more impressive when we read recommendations on how the procedure was best performed from the nineteenth century.[94] As Martin observed in discussing the trepanations of Neolithic Jabal Buhais 18, these "reveal...an extraordinary knowledge of anatomy and physiology, precise and careful use of flint blades, medical knowledge regarding cause-effect relationships, an appreciation of life-saving interventions and careful post-surgery treatment" (table 5).[95]

Table 5

Date B.C.	Period	Site	Notes
8600	Proto-Neolithic	Shanidar Cave?	partial trepanation (?) on three crania with signs of healed depression fractures[96]
4500	Neolithic	Jabal Buhais 18 (UAE)	BC, female, 30–40 years old, 34 × 47 mm, scraped DW, female, 40+ years old, healed; scraped HD, male, 50–60 years old, 20 mm dia.; scraped[97]
4000	Middle Chalcolithic	Dum Gar Parchinah (Iran)	three individuals, drilled (?), scraped (?)[98]

---

94 F. Tyrrell, *The Lectures of Sir Astley Cooper, Bart. F.R.S., Surgeon to the King, &c. &c. on the Principles and Practice of Surgery; with Additional Notes and Cases*, vol. 1 (Boston: Wells and Lilly, 1825), 256–60; J. Lucas-Championnière, *Antiseptic Surgery: The Principles, Modes of Application, and Results of the Lister Dressing*, trans. F. H. Gerrish (Portland: Loring, Short, and Harmon, 1881), 190; Gallez, *Trépanation*, 161–91 for an exhaustive review of the procedures followed by dozens of leading European surgeons in the nineteenth century.
95 D. L. Martin, "Bioarchaeology in the United Arab Emirates," *Arabian Archaeology and Epigraphy* 18 (2007): 124–31, at 129.
96 D. Ferembach, "Étude anthropologique des ossements humains Proto-Néolithiques de Zawi Chemi Shanidar (Iraq)," *Sumer* 26 (1970): 21–65. No reference to trepanation is made, however, in either A. P. Agelarakis, "The Shanidar Cave Proto-Neolithic Human Population: Aspects of Demography and Paleopathology," *Human Evolution* 8/4 (1993): 235–53, or R. S. Solecki, R. L. Solecki, and A. P. Agelarakis, *The Proto-Neolithic Cemetery in Shanidar Cave* (College Station: Texas A & M University Press, 2004), 159–97. This, however, may be explained by the strict definition of trepanation employed by Agelarakis who notes (pers. comm., 6 November 2011), that a tool such as a true trepanon, i.e., a rotary sawing device or something similar, was not used at Shanidar, but careful scraping of the cranial bone surfaces was detected. According to the less rigid definition of trepanation adopted here, involving a variety of techniques, including scraping, the Shanidar individuals should thus be included in the inventory of ancient trepanations. Cf. A. P. Agelarakis, "Early Evidence," 5–18.
97 H. Kiesewetter, "Analyses of the Human Remains," 197–99.
98 E. Haerinck and B. Overlaet, *The Chalcolithic Period at Parchinah and Hakalan*, Luristan Excavation Documents 1 (Ghent: Royal Museums of Art and History, 1996), 11. Cf. E. Haerinck and B. Overlaet, "Pošt-e Kuh," in E. Yarshater (ed.), *Encyclopaedia Iranica Online* (15 November

3100–2900	Early Bronze Age	Jabal Hafit (UAE)	1309.1, female; nine lesions 12–15 × 21–32 mm 1315-C, four lesions, c. 14–16.5 × 32–34 mm, scraped[99]
2000–1700	Old Babylonian	Tell Biʿa (Syria)	male, advanced age, 25 × 20 mm quadrilateral cut male, 30–40 years old. 15 × 10 mm, bored[100]
1650–1600	Middle Bronze II	Ebla (Syria)	male, 45–55 years old, 45 × 50 mm, cut quadrilateral[101]
1600	Old Babylonian	Tell Mozan (Syria)	A15.52, adult, possibly female[102]
1100–800	Iron Age	Dinkha Tepe (Iran)	child, 11–12 years old, 14 × 25 mm, scraped[103]
800–600	Iron Age	Wadi Bawshar (Oman)	adult, sex undetermined, drilled (?)[104]
0–100 A.D.	Parthian	Isin (Iraq)	male, c. 40 years old; two 4–5 mm holes, drilled post-mortem[105]

As Martin has emphasized, there is no doubt that "most literature on trepanations shows a preponderance of males undergoing the surgery"[106] and, where trepanations are found on crania belonging to articulated skeletons, other signs of trauma (fractures, traumas caused by blunt instruments, cut, and gouge marks) are often present as well, suggesting that trepanations were performed commonly on males who had sustained blows to the head, resulting in depression fractures, in battle. Angel, for example, suggested that the trepanation on

---

2006), <http://www.iranicaonline.org/articles/post-e-kuh>: "No anthropological research on the human remains was made, but trepanation was noted."
99 Littleton and Frifelt, "Trepanations from Oman."
100 W. Wolska, "Zwei Fälle von Trepanation aus der altbabylonischen Zeit Syriens," *Mitteilungen der Deutschen Orient-Gesellschaft* 126 (1994): 37–50.
101 S. Mogliazza, "An Example of Cranial Trepanation," 187–88, Figs. 1–2.
102 L. Ramos, "Ancient Disease and Trauma: A Case from Tell Mozan, Syria, Dating to 1600 B.C.," *American Journal of Physical Anthropology Supplement* 44 (2007): 195.
103 R. Mallin and T. A. Rathbun, "A Trephined Skull from Iran," *Bulletin of the New York Academy of Medicine* 52/7 (1976): 782–87.
104 M. Kunter, "Individual Skeletal Diagnoses," in P. Yule (ed.), *Studies in the Archaeology of the Sultanate of Oman* (Rahden: Verlag Marie Leidorf, 1999), 75–78.
105 G. Ziegelmeyer, "Anthropologische Auswertung des Skelettmaterials," in B. Hrouda (ed.), *Isin-Išān Bahrīyāt II. Die Ergebnisse der Ausgrabungen 1975–1978*, Abhandlungen der Bayerischen Akademie der Wissenschaften, phil.-hist. Klasse NF 87 (Munich: Verlag der Bayerischen Akademie der Wissenschaften, 1981), 103–29, at 107–108; G. Ziegelmeyer and F. Parsche, "Ausgrabungen in der antiken Stadt Isin und am Oberlauf des Euphrat, Südosttürkei," *Homo* 35/3-4 (1984): 229–43, at 234–35, Abb. 4b.
106 Martin, "Bioarchaeology," 129.

the male 107 from Asine, like that of 51 from Mycenae, "may represent surgical removal of fragments after a battle axe wound...but with survival."[107] Similarly, a survey of trepanation in prehistoric Europe concluded that trepanations were most often performed on individuals who had sustained cranial fractures, and that most of these were male,[108] while in Melanesia during the early twentieth century, trepanation was regarded as "a piece of normal war-surgery, to remedy skull-wounds from sling-stones or clubs."[109] Certainly it is the case that cranial fractures and their treatment were discussed at some length by Hippocrates, Celsus, Rufus, Galen, Oribasius, and Paul.[110] The Persian physician Ibn Sīnā (Avicenna, 980–1037 A.D.) recommended trepanation exclusively in cases of cranial trauma, never for mental disorders,[111] but the Safavid (1502–1736) sources are silent on trepanation.[112] It was practiced in the Ottoman realm,[113] however, and in early modern Europe, as Joannes Scultetus' (1595–1645) treatise *Armamentarium chirurgicum*, based on his experiences during the Thirty Years' War, and many other sources attest.[114]

While there can be no doubt that this assessment seems, generally speaking, to be true, it is nevertheless the case that the trepanation of females and children is also attested. Two of the three Neolithic trepanations at Jabal Buhais 18 were on females, and female trepanations are also known from Burzahom (ca. 2300–2000 B.C.) in Kashmir (eleven attempts on one individual!); Elmali-Karataş (ca. 2300 B.C.) in Asia Minor; Timargarha (ca. 1500 B.C.) in Pakistan; and Roman Jericho.[115] It would be too simplistic to suggest that cranial fractures in antiquity were restricted to fighting males (even fighting male children), howev-

---

107 Angel, "Appendix 1," 109.
108 Kiesewetter, "Analyses," 195–96; cf. C. Kenesi, "Les trépanations dans les civilisations disparues," *Histoire des sciences médicales* 35/1 (2001): 51–56, at 54.
109 J. L. Myres, "Review of W.J. Wölfel, *Die Trepanation: Studien über Zusammenhänge und kulturelle Zugehörigkeit der Trepanation*," *Man* 25 (1925): 179–81, at 180.
110 Salazar, *Treatment of War Wounds*, 13–15, 45–46 with references.
111 Sanagustin, "La chirurgie," 97, n. 55; A. Aciduman, B. Arda, F. G. Özaktürk, and Ü. F. Telatar, "What Does *Al-Qanun fi al-Tibb* (The Canon of Medicine) Say on Head Injuries?," *Neurosurgical Review* 32 (2009): 255–63, at 260–61.
112 C. Elgood, *Safavid Surgery*, Analecta Medico-Historica 2 (Oxford: Pergamon Press, 1966), 62.
113 Erdal and Erdal, "A Review," 534, noting three cases.
114 L. Bakay, *The Treatment of Head Injuries in the Thirty Years' War (1618–1648): Joannis Scultetus and his Age* (Springfield, Ill.: Charles C. Thomas, 1971), 69–75; T. F. Dagi, "The Management of Head Trauma," in S. H. Greenblatt, T. F. Dagi, and M. H. Epstein (eds.), *A History of Neurosurgery in its Scientific and Professional Contexts* (Park Ridge, Ill.: American Association of Neurological Surgeons, 1997), 289–343, at 299–300. Cf. Gallez, *Trépanation*, 59–107.
115 See Kiesewetter, "Analyses," Table 6.10 with references.; J. Zias, "Three Trephinated Skulls from Jericho," *Bulletin of the American Schools of Oriental Research* 246 (1982): 55–58, at 55–56.

er, and that females underwent trepanation for other reasons. While a whole host of other non-traumatic causes of trepanation, from toothache, epilepsy, and fever through scurvy and sinusitis have been canvassed in the literature,[116] cranial trauma is present on females as well. Both of the trepanned females at Neolithic Jabal Buhais 18, for example, showed signs of depression fractures inflicted by blunt instruments.[117] Erdal estimates that 50 % of the forty ancient trepanations recorded in Turkey were prompted by cranial trauma.[118]

Children who underwent trepanation are attested, for example, at Lothal in Pakistan (ca. 2300–1900 B.C.),[119] Dimona (ca. 2200 B.C.) in Palestine,[120] Fidenae in suburban Rome (late first/early second century A.D.),[121] and Frankish Corinth (ca. 1312 A.D.),[122] but they are statistically rare, at least in the Old World literature surveyed here. Particularly striking is the recently published evidence of trepanation on the crania of three children, aged 2–3, 8–11, and 11–13, from the Avar (seventh- to ninth-century A.D.) cemetery of Szeged-Kiskundorozsma-Kettöshatár in southern Hungary.[123] In the case of the two older children, depression fractures suggest that trepanation was undertaken following instances of cranial trauma. In the case of the youngest child, no signs of trauma were visible but the cranium showed signs of hydrocephalus and trepanation may have been used to treat any number of illnesses, "from inflammations of infectious origin to metabolic deficiencies and developmental anomalies."[124] One should remember, too, that, in his discussion of bloodletting by means of a small trepan following a skull fracture, Hippocrates made particular mention of the fact that "the bone of young persons is thinner and more superficial than that of elder persons" (*vuln. cap.* 18.1, 3.250 L.).

---

116 Gallez, Trépanation, 363–441.
117 Kiesewetter, "Analyses," 197.
118 Erdal, "A Retrospective Study," 269.
119 Sankhyan and Weber, "Evidence of Surgery," 375, on a 9–10 year old.
120 Mogle and Zias, "Trephination as a Possible Treatment," 77, on an 8–9 year old.
121 R. Mariani-Costantini, P. Catalano, F. di Gennaro, G. di Tota, and L. R. Angeletti, "New Light on Cranial Surgery in Ancient Rome," *The Lancet* 355 (2000): 305–307, on a 5–6 year old using a chisel (?).
122 E. Barnes and D. J. Ortner, "Multifocal Eosinophilic Granuloma with a Possible Trepanation in a Fourteenth Century Greek Young Skeleton," *International Journal of Osteoarchaeology* 7 (1997): 542–47, at 543–44, Figs. 2–3; C. K. Williams II, L. M. Snyder, E. Barnes, and O. H. Zervos, "Frankish Corinth: 1997," *Hesperia* 67/3 (1998): 223–81, at 245 and Pl. 40a.
123 Bereczki et al. "Evidence of Surgical Trephinations," 93–98.
124 Bereczki et al. "Evidence of Surgical Trephinations," 97. For trepanning in case of hydrocephalus, cf. Paul. Aeg. 6.3, "On Hydrocephalus," trans. Adams, *The Seven Books*, 251.

There is no reason to agree with scholars who assert that the origins of trepanation "are not to be found in rational medical practice" but reside in "a ritual act,"[125] or that "the goals and purposes of prehistoric cranial operations.... included magical, ritual and religious motivations, along with practical treatment of head injury, with a considerable spiritual element also present."[126] Nor does it seem likely that Iron Age and earlier trepanations were conducted in "religious/ ritualistic contexts" and that "its purely medical nature" only became apparent from the time of Hippocrates onwards.[127] On the other hand, it is the case that some trepanations in both the past and the present have been performed "to obtain bone discs from people alive or dead for protection from demons."[128]

Conclusion

This meditation on trepanation has traversed a great deal of ground, much of it surveyed by other scholars already, though not necessarily with the same amount of attention to the archaeological, as well as the medical historical and surgical literature. Certainly a study of trepanation highlights the importance of considering neither the bioarchaeological nor the ancient literary evidence in isolation. In addition to honoring Heinrich von Staden, the aim of this modest study is simply to make historians of medicine more aware of the richness of the archaeological record and to urge bioarchaeologists and physical anthropologists to more actively exploit the available Greek, Latin, Arabic, and early modern European sources on trepanation. In the final analysis, all students of trepanation and ancient medicine more generally, regardless of their disciplinary background, can only benefit from increased cross-disciplinary dialogue.

---

125 R. Arnott, "Holes in the Head and More: Surgery in the Aegean Bronze Age," *Medical Historian (The Bulletin of the Liverpool Medical History Society)* 9 (1997): 21–34, at 24.
126 Weber and Wahl, "Neurosurgical Aspects," 537.
127 Mountrakis, Georgaki, and Manolis, "A Trephined Late Bronze Age Skull," 6.
128 Han and Chen, "The Archaeological Evidence," 26. Cf. P. Broca, *Sur la trépanation du crâne et les amulettes crâniennes à l'Époque néolithique* (Paris: Ernest Leroux, 1877).

Bibliography

Aciduman, A., B. Arda, F. G. Özaktürk, and Ü. F. Telatar. "What Does *Al-Qanun fi al-Tibb* (*The Canon of Medicine*) Say on Head Injuries?," *Neurosurgical Review* 32 (2009): 255–63.

Adams, F. *The Seven Books of Paulus Ægineta, Translated from the Greek, with a Commentary Embracing a Complete View of the Knowledge Possessed by the Greeks, Romans, and Arabians on All Subjects Connected with Medicine and Surgery*, vol. 2 (London: The Sydenham Society, 1846).

Agelarakis, A. P. "The Shanidar Cave Proto-Neolithic Human Population: Aspects of Demography and Paleopathology," *Human Evolution* 8/4 (1993): 235–53.

Agelarakis, A. P. "Early Evidence of Cranial Surgical Intervention in Abdera, Greece: A Nexus to *On Head Wounds* of the Hippocratic Corpus," *Mediterranean Archaeology and Archaeometry* 6/1 (2006): 5–18.

Alt, K. W., C. Jeunesse, C. H. Buitrago-Téllez, R. Wächter, E. Boës, and S. L. Pichler. "Evidence for Stone Age Cranial Surgery," *Nature* 387 (1997): 360.

Andruschko, V. A., and J. W. Verano. "Prehistoric Trepanation in the Cuzco Region of Peru: A View into an Ancient Andean Practice," *American Journal of Physical Anthropology* 137 (2008): 4–13.

Angel, J. L. "Ancient Cephallenians: The Population of a Mediterranean Island," *American Journal of Physical Anthropology* 1 (1943): 229–56.

Angel, J. L. "Appendix. Human Skeletal Remains at Karataş," *American Journal of Archaeology* 70 (1966): 255–57.

Angel, J. L. "Early Neolithic Skeletons from Çatal Hüyük: Demography and Pathology," *Anatolian Studies* 21 (1971): 77–98.

Angel, J. L. *The People of Lerna: Analysis of a Prehistoric Aegean Population* (Princeton: American School of Classical Studies at Athens, 1971).

Angel, J. L. "Appendix 1. Ancient Skeletons from Asine," in S. Dietz (ed.), *Asine II: Results of the Excavations East of the Acropolis, 1970–1974*, fasc. 1. *General Stratigraphical Analysis and Architectural Remains*, Skrifter utgivna av Svenska Institutet i Athen, 4°, 24.1 (Stockholm: Svenska Institutet i Athen, 1982), 105–38.

Angel, J. L. "Human Skeletons from Grave Circles at Mycenae," in G. E. Mylonas, *Ο Ταφικός Κύκλος Β των Μυκηνών*, vol. A. (Athens: Archaeological Society of Athens, 1973), 379–97.

Arensburg, B., and I. Hershkovitz. "Cranial Deformation and Trephination in the Middle East," *Bulletins et mémoires de la Société d'Anthropologie de Paris* XIVe Série 5/3 (1988): 139–50.

Arnott, R. "Holes in the Head and More: Surgery in the Aegean Bronze Age," *Medical Historian (The Bulletin of the Liverpool Medical History Society)* 9 (1997): 21–34.

Arnott, R. "Surgical Practice in the Prehistoric Aegean," *Medizinhistorisches Journal* 32/3–4 (1997): 249–78.

Arnott, R., S. Finger, and C. U. M. Smith, eds. *Trepanation: History – Discovery – Theory* (Lisse: Swets & Zeitlinger, 2003).

Bakay, L. *The Treatment of Head Injuries in the Thirty Years' War (1618–1648): Joannis Scultetus and his Age* (Springfield, Ill.: Charles C. Thomas, 1971).

Barnes, E., and D. J. Ortner. "Multifocal Eosinophilic Granuloma with a Possible Trepanation in a Fourteenth Century Greek Young Skeleton," *International Journal of Osteoarchaeology* 7 (1997): 542–47.

Bazarsad, N. "Four Cases of Trepanation from Mongolia, Showing Surgical Variation," in Arnott, Finger, and Smith, *Trepanation*, 203–208.

Belcastro, M. G., S. Condemi, and V. Mariotti. "Funerary Practices of the Iberomaurusian Population of Taforalt (Tafoughalt, Morocco, 11–12,000 BP): The Case of Grave XII," *Journal of Human Evolution* 58 (2010): 522–32.

Bereczki, Z., E. Molnár, A. Marcsik, and G. Pálfi. "Evidence of Surgical Trephinations in Infants from the 7th–9th Centuries AD Burial Site of Kiskundorozsma-Kettöshatár," *Acta Biologica Szegediensis* 54/2 (2010): 93–98.

Bliquez, L. J. "The Tools of Asclepius: The Surgical Gear of the Greeks and the Romans," *Veterinary Surgery* 11 (Oct-Dec, 1982): 150–56.

Bliquez, L. J. "Two Lists of Greek Surgical Instruments and the State of Surgery in Byzantine Times," *Dumbarton Oaks Papers* 38 (1984): 187–204.

Bliquez, L. J. *Roman Surgical Instruments and Other Minor Objects in the National Archaeological Museum of Naples* (Mainz: von Zabern, 1994).

Broca, P. "Sur les trépanations préhistoriques," *Bulletin de la Sociétie d'Anthropologie de Paris* 11 (1876): 236–56.

Broca, P. "Sur l'âge des sujets soumis à la trépanation chirurgicale néolithique," *Bulletin de la Sociétie d'Anthropologie de Paris* 11 (1876): 572–76.

Broca, P. *Sur la trépanation du crâne et les amulettes crâniennes à l'Époque néolithique* (Paris: Ernest Leroux, 1877).

Brothwell, D. R. "Ancient Trephining: Multi-focal Evolution or Trans-world Diffusion?," *Journal of Paleopathology* 6 (1994): 129–38.

Brothwell, D. R. "The Future Direction of Research," in Arnott, Finger, and Smith, *Trepanation*, 365–72.

Campbell Thompson, R. "Assyrian Prescriptions for Diseases of the Head (Continued)," *American Journal of Semitic Languages and Literatures* 24/4 (1908): 323–53.

Campbell Thompson, R. "Assyrian Prescriptions for the Head," *American Journal of Semitic Languages and Literatures* 53/4 (1937): 217–38.

Capasso, L., E. Michetti, L. Pierfelice, and R. D'Anastasio. "Neurosurgery 7000 Years Ago in Central Italy," *The Lancet* 359 (2002): 2206.

Charles, R. P. "Étude anthropologique des nécropoles d'Argos: contribution à l'étude des populations de la Grèce antique," *Bulletin de Correspondance Hellénique* 82 (1958): 268–313.

Charles, R. P. *Étude anthropologique des nécropoles d'Argos: contribution à l'étude des populations de la Grèce antique* (Paris: J. Vrin, 1963).

Chlouveraki, S., P. B. Betancourt, C. Davaras, H. M. C. Dierckx, S. C. Ferrence, J. Hickman, P. Karkanas, P. J. P. McGeorge, J. D. Muhly, D. S. Reese, E. Stravopodi, and L. Langford-Verstegen. "Excavations in the Hagios Charalambos Cave: A Preliminary Report," *Hesperia* 77/4 (2008): 539–605.

Crubézy, É., J. Bruzek, J. Guilaine, E. Cunha, D. Rougé, and J. Jelinek. "The Antiquity of Cranial Surgery in Europe and in the Mediterranean Basin," *Comptes Rendus de l'Académie des sciences de Paris, Sciences de la Terre et des planètes* 332 (2001): 417–23.

Dagi, T. F. "The Management of Head Trauma," in S. H. Greenblatt, T. F. Dagi, and M. H. Epstein (eds.), *A History of Neurosurgery in its Scientific and Professional Contexts* (Park Ridge, Ill.: American Association of Neurological Surgeons, 1997), 289–343.

Dastugue, J. "Un orifice crânien préhistorique," *Bulletins et Mémoires de la Société d'Anthropologie de Paris*, 10th sér. 10/4 (1959): 357–63.

Dastugue, J. "Les pièces pathologiques de Ras Shamra et Minet el Beida," *Ugaritica* 4 (1962): 623–30.

Dickinson, O. T. P. K. "Ancient Greece," in D. J. Davies and L. H. Mates (eds.), *Encyclopedia of Cremation* (Aldershot: Ashgate, 2005), 6–8.

Dimopoulos, V. G., J. S. Robinson III, and K. N. Fountas. "The Pearls and Pitfalls of Skull Trephination as Described in the Hippocratic Treatise 'On Head Wounds'," *Journal of the History of the Neurosciences* 17 (2008): 131–40.

Elgood, C. *Safavid Surgery*, Analecta Medico-Historica 2 (Oxford: Pergamon Press, 1966).

Erdal, Y. S. "A Retrospective Study on Trepanation in Anatolia," in P. Matthiae, F. Pinnock, L. Nigro, and N. Marchetti (eds.), *Proceedings of the 6th International Congress of the Archaeology of the Ancient Near East*, vol. 1 (Wiesbaden: Harrassowitz, 2010), 261–76.

Erdal, Y. S., and Ö. D. Erdal. "A Review of Trepanations in Anatolia with New Cases," *International Journal of Osteoarchaeology* 21 (2011): 505–34.

Fabbri, P. F., N. Lonoce, M. Masieri, D. Caramella, M. Valentino, and S. Vassallo. "Partial Cranial Trephination by means of Hippocrates' *Trypanon* from 5th century BC Himera (Sicily, Italy)," *International Journal of Osteology* 20 (3 Sept 2010): DOI: 10.1002/oa.1197.

Ferembach, D. "À propos d'un crâne trépané trouvé à Timna: origine de certaines tribus berbères," *Bulletins et Mémoires de la Société d'Anthropologie de Paris*, 10th sér. 8/5–6 (1957): 244–75.

Ferembach, D. "Étude anthropologique des ossements humains Proto-Néolithiques de Zawi Chemi Shanidar (Iraq)," *Sumer* 26 (1970): 21–65.

Ferembach, D. "À propos d'une trépanation sur un fragment d'occipital humain chalcolithique trouvé à Azor Israël," *Garcia de Orta, Série Antropobiologica* 3/1–2 (1984): 139–42.

Ferembach, D., J. Dastugue, and M. J. Poitrat-Targowla. *La nécropole épipaléolithique de Taforalt (Maroc oriental): étude de squelettes humains* (Rabat: CNRS, 1962).

Fernando, H. R., and S. Finger. "Ephraim George Squier's Peruvian Skull and the Discovery of Cranial Trepanation," in Arnott, Finger, and Smith, *Trepanation*, 3–18.

Fincke, J. "Spezialisierung und Differenzierung im Bereich der altorientalischen Medizin: Die Dermatologie am Beispiel der Symptome *simmū matqūtu, kalmātu (matuqtu), kibšu, kiṣṣatu* und *gurištu*," in G. Selz (ed.), *The Empirical Dimension of Ancient Near Eastern Studies / Die empirische Dimension altorientalischer Forschungen* (Vienna: Wiener Offene Orientalistik 6, 2011), 159–208.

Finger, S., and W. T. Clower. "On the Birth of Trepanation: The Thoughts of Paul Broca and Victor Horsley," in Arnott, Finger, and Smith, *Trepanation*, 19–42.

Fischer, P. M. *Prehistoric Cypriot Skulls: A Medico-anthropological, Archaeological and Micro-analytical Investigation*, Studies in Mediterranean Archaeology 75 (Göteborg: Paul Aströms, 1986).

Gallez, L. *La trépanation du crâne: histoire, technique opératoire, indications et contre-indications, résultats* (Paris: Georges Carré, 1893).

Germanà, F., and G. Fornaciari. *Trapanazioni, craniotomie e traumi cranici in Italia*, Collana di Studi Paletnologici 5 (Pisa: Giardini Editori e Stampatori, 1992).

Goldman, Z. "Surgical Trepanation in Ancient Times," *Bulletin of the Israel Exploration Society* 25 (1961): 258–60 (in Hebrew).
Goltz, D. *Studien zur altorientalischen und griechischen Heilkunde: Therapie – Arzneibereitung – Rezeptstruktur*, Sudhoffs Archiv Beiheft 16 (Wiesbaden: Steiner, 1974).
Gross, C. G. "A Hole in the Head," *The Neuroscientist* 5/4 (1999): 263–69.
Güleç, E. "Trepanation in Ancient Anatolia: Six New Case Studies (Abstract)," *Papers on Paleopathology Presented to the 22nd Meeting of the Paleopathology Association* (1995): 6.
Haerinck, E., and B. Overlaet. *The Chalcolithic Period at Parchinah and Hakalan*, Luristan Excavation Documents 1 (Ghent: Royal Museums of Art and History, 1996).
Haerinck, E., and B. Overlaet. "Pošt-e Kuh," in E. Yarshater (ed.), *Encyclopaedia Iranica Online*, 15 November 2006, <http://www.iranicaonline.org/articles/post-e-kuh>.
Han, K., and X. Chen. "The Archaeological Evidence of Trepanation in Early China," *Bulletin of the Indo-Pacific Prehistory Association* 27 (2007): 22–27.
Hanson, M. *Hippocrates, On Head Wounds*, CMG I 4,1 (Berlin: Akademie-Verlag, 1999).
Hershkovitz, I. "Trephination: The Earliest Case in the Middle East," *Journal of the Israel Prehistoric Society* 20 (1987): 128–35.
Kenesi, C. "Les trépanations dans les civilisations disparues," *Histoire des sciences médicales* 35/1 (2001): 51–56.
Kiesewetter, H. "Ein trepanierter Schädel aus Troia VI," *Studia Troica* 12 (2002): 73–80.
Kiesewetter, H. "Analyses of the Human Remains from the Neolithic Cemetery at al-Buhais 18 (Excavations 1996–2000)," in H.-P. Uerpmann, M. Uerpmann, and S. A. Jasim (eds.), *Funeral Monuments and Human Remains from Jebel al-Buhais* (Tübingen: Kerns Verlag, 2006), 103–380.
Kirkup, J., "The Evolution of Cranial Saws and Related Instruments," in Arnott, Finger, and Smith, *Trepanation*, 289–304.
Koca, T., and M. Schultz. "Trephination as a Medical Indication Following Trauma Observed in the Middle Bronze Age Population from Lidar Höyük (Turkey)," in Schultz, Kreutz, and Teegen, *Xth European Meeting*, S68.
Köcher, F. *Die babylonisch-assyrische Medizin in Texten und Untersuchungen*, Band V. *Keilschrifttexte aus Ninive 1* (Berlin: De Gruyter, 1980).
Kunter, M. "Individual Skeletal Diagnoses," in P. Yule (ed.), *Studies in the Archaeology of the Sultanate of Oman* (Rahden: Verlag Marie Leidorf, 1999), 75–78.
Künzl, E. "Archäologische Beiträge zur Medizingeschichte: Methoden – Ergebnisse – Ziele," in G. Sabbah (ed.), *Études de médecine romaine*, Centre Jean-Palerne de l'Université de Saint-Étienne Mémoire VIII (Saint-Étienne: Université de Saint-Étienne, 1988), 61–80.
Kurth, G., and O. Roher-Ertl. "On the Anthropology of the Mesolithic to Chalcolithic Human Remains from the Tell es-Sultan in Jericho, Jordan," in K. M. Kenyon (ed.), *Excavations at Jericho*, vol. 3 (Jerusalem: The British School of Archaeology, 1981), 407–99.
Labat, R. "À propos de la chirurgie babylonienne," *Journal Asiatique* 242 (1954): 207–18.
Lillie, M. C. "Cranial Surgery Dates Back to Mesolithic," *Nature* 391 (1998): 354.
Liston, M. A., and L. P. Day. "It Does Take a Brain Surgeon: A Successful Trepanation from Kavousi, Crete," in L. A. Schepartz, S. C. Fox, and C. Bourbou (eds.), *New Directions in the Skeletal Biology of Greece*, Hesperia Supplement 43 (Princeton: American School of Classical Studies at Athens, 2009): 57–73.
Littleton, J., and K. Frifelt. "Trepanations from Oman: A Case of Diffusion?," *Arabian Archaeology and Epigraphy* 17 (2006): 139–51.

Loughborough, J. L. "Notes on the Trepanation of Prehistoric Crania," *American Anthropologist* 48/3 (1946): 416–22.

Lucas-Championnière, J. *Antiseptic Surgery: The Principles, Modes of Application, and Results of the Lister Dressing*, trans. F. H. Gerrish (Portland: Loring, Short, and Harmon, 1881).

Lucas-Championnière, J. *Trépanation néolithique pré-colombienne, trépanation des Kabyles, trépanation traditionelle* (Paris: G. Steinheil, 1912).

Mallegni, F., and A. Valassina. "Secondary Bone Changes to a Cranium Trepanation in a Neolithic Man Discovered at Trasano, South Italy," *International Journal of Osteoarchaeology* 6 (1996): 506–11.

Mallin, R., and T. A. Rathbun. "A Trephined Skull from Iran," *Bulletin of the New York Academy of Medicine* 52/7 (1976): 782–87.

Manolis, S. K., M. J. Pagrigorakis, and C. Zafeiratos. "Trepanations in Greece: Observations on a Middle Bronze Age Skull," in Schultz, Kreutz, and Teegen, *Xth European Meeting*, S80.

Mariani-Costantini, R., P. Catalano, F. di Gennaro, G. di Tota, and L. R. Angeletti. "New Light on Cranial Surgery in Ancient Rome," *The Lancet* 355 (2000): 305–307.

Mariotti, V., B. Bonfiglioli, F. Facchini, S. Condemi, and M. G. Belcastro. "Funerary Practices of the Iberomaurusian Population of Taforalt (Tafoughalt; Morocco, 11–12,000 BP): New Hypotheses Based on a Grave by Grave Skeletal Inventory and Evidence of Deliberate Human Modification of the Remains," *Journal of Human Evolution* 56 (2009): 340–54.

Martin, D. L. "Bioarchaeology in the United Arab Emirates," *Arabian Archaeology and Epigraphy* 18 (2007): 124–31.

Martin, G. "Trepanation in the South Pacific," *Journal of Clinical Neuroscience* 2/3 (1995): 257–64.

Milne, J. S. *Surgical Instruments in Greek and Roman Times* (Oxford: Clarendon Press, 1907).

Missios, S. "Hippocrates, Galen, and the Uses of Trepanation in the Ancient Classical World," *Neurosurgical Focus* 23/1 (2007): E11.

Mogle, P., and J. Zias. "Trephination as a Possible Treatment for Scurvy in a Middle Bronze Age (ca. 2200 BC) Skeleton," *International Journal of Osteoarchaeology* 5/1 (1995): 77–81.

Mogliazza, S. "An Example of Cranial Trepanation Dating to the Middle Bronze Age from Ebla, Syria," *Journal of Anthropological Sciences* 87 (2009): 187–92.

Mountrakis, C., S. Georgaki, and S. K. Manolis. "A Trephined Late Bronze Age Skull from Peloponnesus, Greece," *Mediterranean Archaeology and Archaeometry* 11 (2011): 1–8.

Murphy, E. "Trepanations and Perforated Crania from Iron Age South Siberia: An Exercise in Differential Diagnosis," in Arnott, Finger, and Smith, *Trepanation*, 209–21.

Musgrave, J. H., R. A. H. Neave, and A. J. N. W. Prag. "Seven Faces from Grave Circle B at Mycenae," *Annual of the British School at Athens* 91 (1996): 107–36.

Myres, J. L. "Review of W.J. Wölfel, *Die Trepanation: Studien über Zusammenhänge und kulturelle Zugehörigkeit der Trepanation*," *Man* 25 (1925): 179–81.

de Nadaillac, J. F. A. du P. "Mémoire sur les trépanations préhistoriques," *Comptes-rendus des séances de l'Académie des Inscriptions et Belles-Lettres* 30/2 (1886): 280–94.

Nerlich, A. G., A. Zink, U. Szeimies, H. G. Hagedorn, and F. W. Rösing. "Perforating Skull Trauma in Ancient Egypt and Evidence for Early Neurosurgical Therapy," in Arnott, Finger, and Smith, *Trepanation*, 191–201.

Oakley, K. P., W. M. A. Brooke, A. R. Akester, and D. R. Brothwell. "Contributions on Trepanning or Trephination in Ancient and Modern Times," *Man* 59 (1959): 93–96.

Parry, T. W. "Trephination of the Living Human Skull in Prehistoric Times," *British Medical Journal* 1/3246 (1923): 457–60.

Parry, T. W. "Three Skulls from Palestine Showing Two Types of Primitive Surgical Holing: Being the First Skulls Exhibiting this Phenomenon that Have Been Discovered on the Mainland of Asia," *Man* 36 (1936): 170–71.

Paschalides, C., and P. J. P. McGeorge. "Life and Death in the Periphery of the Mycenaean World at the End of the Late Bronze Age: The Case of the Achaea Klauss Cemetery," in E. Borgna and P. Cassola Guida (eds.), *Dall'Egeo all'Adriatico: organizzazioni sociali, modi di scambio e interazione in età postpalaziale (XII–XI sec. a.C.) / From the Aegean to the Adriatic: Social Organisations, Modes of Exchange and Interaction in Postpalatial Times (12th–11th B.C.)*, Studi e ricerche di protostoria mediterranea 8 (Rome: Quasar, 2009), 79–113.

Potts, D. T. "Technological Transfer and Innovation in Ancient Eurasia," in J. Renn (ed.), *The Globalization of Knowledge in History* (Berlin: Max Planck Research Library for the History and Development of Knowledge Studies 1, 2012), 105–23.

Poulianos, A. *Η καταγωγή των Κρητών* (Athens: Ανθρωπολογική Εταιρεία Ελλάδος, 1971).

Ramos, L. "Ancient Disease and Trauma: A Case from Tell Mozan, Syria, Dating to 1600 B.C.," *American Journal of Physical Anthropology Supplement* 44 (2007): 195.

Rocca, J. *Galen on the Brain: Anatomical Knowledge and Physiological Speculation in the Second Century AD* (Leiden: Brill, 2003).

Rocca, J. "Galen and the Uses of Trepanation," in Arnott, Finger, and Smith, *Trepanation*, 253–71.

Salazar, C. F. *The Treatment of War Wounds in Graeco-Roman Antiquity*, Studies in Ancient Medicine 21 (Leiden: Brill, 2000).

Sanagustin, F. "La chirurgie dans le Canon de la médecine (*Al-Qânûn fî-ṭ-ṭibb*) d'Avicenne (*Ibn Sinâ*)," *Arabica* 33 (1986): 84–122.

Sankhyan, A. R., and G. R. Schug. "First Evidence of Brain Surgery in Bronze Age Harappa," *Current Science* 100/11 (2011): 1621–22.

Sankhyan, A. R., and G. H. J. Weber. "Evidence of Surgery in Ancient India: Trepanation at Burzahom (Kashmir) over 4000 Years Ago," *International Journal of Osteoarchaeology* 11 (2001): 375–80.

Schultz M., K. Kreutz, and W.-R. Teegen, eds. *Xth European Meeting of the Paleopathology Association, Göttingen, 29th August–3rd September, 1994: Abstracts, Homo* 45 Supplement (Stuttgart: Gustav Fischer Verlag, 1994).

Smith, P. "The Trephined Skull from the Early Bronze Age Period at Arad," *Eretz-Israel* 21 (1990): 89*–93*.

Smith, P., and L. Dujovny. "Middle Bronze Age II Human Remains from Caves 7, 24, and 33," in S. Ben-Arieh (ed.), *Bronze and Iron Age Tombs at Tell Beit Mirsim*, Israel Antiquities Authority Reports 23 (Jerusalem: Israel Antiquities Authority, 2004), 203–205.

Solecki, R. S., R. L. Solecki, and A. P. Agelarakis. *The Proto-Neolithic Cemetery in Shanidar Cave*, Texas A&M University Anthropology Series 7 (College Station: Texas A&M University Press, 2004).

Starkey, J. L. "Discovery of Skulls with Surgical Holing at Tell Duweir, Palestine," *Man* 36 (1936): 169–70.

Stol, M. "An Assyriologist Reads Hippocrates," in H. F. J. Horstmanshoff and M. Stol (eds.), *Magic and Rationality in Ancient Near Eastern and Graeco-Roman Medicine* (Leiden: Brill, 2004), 63–78.

Stone, J. L., and M. L. Miles. "Skull Trepanation among the Early Indians of Canada and the United States," *Neurosurgery* 26/6 (1990): 1015–19.

Stone, J. L., and J. Urcid. "Trepanation in Prehistoric South America: Geographic and Temporal Trends over 2,000 Years," in Arnott, Finger, and Smith, *Trepanation*, 237–49.

Tullo, E. "Trepanation and Roman Medicine: A Comparison of Osteoarchaeological Remains, Material Culture and Written Texts," *Journal of the Royal College of Physicians of Edinburgh* 40 (2010): 165–71.

Tyrrell, F. *The Lectures of Sir Astley Cooper, Bart. F.R.S., Surgeon to the King, &c. &c. on the Principles and Practice of Surgery; with Additional Notes and Cases*, vol. 1 (Boston: Wells and Lilly, 1825).

Vallois, H. V., and D. Ferembach. "Les restes humains de Ras Shamra et Minet el Beida: étude anthropologique," *Ugaritica* 4 (1962): 565–622.

Verano, J. W. "Trepanation in Prehistoric South America: Geographic and Temporal Trends over 2,000 Years," in Arnott, Finger, and Smith, *Trepanation*, 223–36.

von Staden, H. "The Discovery of the Body: Human Dissection and its Cultural Contexts in Ancient Greece," *The Yale Journal of Biology and Medicine* 65 (1992): 223–41.

Wall, S. M., J. H. Musgrave, and P. M. Warren. "Human Bones from a Late Minoan IB House at Knossos," *Annual of the British School at Athens* 81 (1986): 333–88.

Watters, D. A. K. "Skull Trepanation in the Bismarck Archipelago," *Papua New Guinea Medical Journal* 50/1–2 (2007): 20–24.

Webb, S. G. "Two Possible Cases of Trephination from Australia," *American Journal of Physical Anthropology* 75 (1988): 541–48.

Weber, J., and J. Wahl. "Neurosurgical Aspects of Trepanations from Neolithic Times," *International Journal of Osteoarchaeology* 16 (2006): 536–45.

Williams II, C. K., L. M. Snyder, E. Barnes, and O. H. Zervos. "Frankish Corinth: 1997," *Hesperia* 67/3 (1998): 223–81.

Wolska, W. "Zwei Fälle von Trepanation aus der altbabylonischen Zeit Syriens," *Mitteilungen der Deutschen Orient-Gesellschaft* 126 (1994): 37–50.

Zias, J. "Three Trephinated Skulls from Jericho," *Bulletin of the American Schools of Oriental Research* 246 (1982): 55–58.

Zias, J. "Death and Disease in Ancient Israel," *Biblical Archaeologist* 54/3 (1991): 146–59.

Zias, J., and S. Pomeranz. "Serial Craniectomies for Intracranial Infection 5.5 Millennia Ago," *International Journal of Osteoarchaeology* 2 (1992): 183–86.

Ziegelmeyer, G. "Anthropologische Auswertung des Skelettmaterials," in B. Hrouda (ed.), *Isin-Išān Bahrīyāt II: Die Ergebnisse der Ausgrabungen 1975–1978*, Abhandlungen der Bayerischen Akademie der Wissenschaften, phil.-hist. Klasse NF 87 (Munich: Verlag der Bayerischen Akademie der Wissenschaften, 1981), 103–29.

Ziegelmeyer, G., and F. Parsche. "Ausgrabungen in der antiken Stadt Isin und am Oberlauf des Euphrat, Südosttürkei," *Homo* 35/3–4 (1984): 229–43.

Christine Proust
Des listes pour apprendre, résoudre, classer, archiver, explorer ou inventer

Abstract: Mesopotamian mathematical texts tend to be written in the form of lists, as is the case with the majority of learned texts from the ancient Near East. The very nature of their structure conveys particular meanings that deserve to be studied in their own right. In this contribution, I am interested primarily in texts composed of long lists of problem statements. Although they exhibit formal resemblances, these lists are very diverse and reflect different intellectual projects. Certain lists of statements are catalogs, which appear to have as their goal the inventory, classification, and perhaps the archiving of existing mathematical material, largely utilized for teaching purposes. Others are very long lists of several thousand statements covering a series of many numbered tablets. By means of these highly developed structures, the scribes have produced entirely new mathematical material. The approach of these scribes seems to be more heuristic than classificatory.

Cette contribution résulte de recherches sur « l'art des listes » dans les mathématiques cunéiformes que j'ai pu développer à l'*Institute for Advanced Study* en 2009 dans le cadre d'un *membership* financé par la fondation Neugebauer. Ces travaux sont ou seront publiés dans différents articles et ouvrages. La présente contribution se limite à un aspect des listes mathématiques, leur statut pragmatique. Pour plus de détails sur d'autres aspects, je renvoie à ces publications. Ma réflexion s'est considérablement enrichie du contexte intellectuel de l'*IAS*, notamment au travers des discussions qui ont accompagné les deux exposés que j'ai pu donner pendant mon séjour. Le premier exposé, intitulé *Why Do Colophons Matter? : Old Babylonian Mathematical Tablets* était programmé dans le cadre des *Ancient Studies seminars* animé par Heinrich von Staden, que je remercie particulièrement. Le deuxième, intitulé *The Art of Lists*, est intervenu dans le cadre du séminaire intitulé *Historical Studies Lunchtime Colloquia* du département d'histoire de l'*IAS*. Le contenu de la présente contribution s'inspire largement de ces deux exposés et des échanges d'idées qu'ils ont suscités. Les outils plus strictement linguistiques que j'ai utilisés dans ce travail ont été forgés grâce à cinq ans de séminaires sur « *histoire des sciences, histoire du texte* » organisés à Paris par Jacques Virbel et Karine Chemla, envers qui ma dette intellectuelle est considérable. Les matériaux textuels, ainsi que le recul nécessaire à une approche globale des listes, ont été acquis grâce à l'accueil de l'*Institute for the Study of the Ancient Word* et l'Unité Mixte de Recherche (CNRS) de *New York University* en 2010, et à ma résidence à l'Institut Méditerranéen d'Études Avancées de Marseille en 2010–2011. Enfin, je n'ai pu étudier directement les sources qui sont à la base de mon travail sur les listes que grâce à l'accès aux collections de tablettes cunéiformes qui m'a été généreusement offert par Béatrice André-Salvini pour le Louvre, et par Benjamin Foster et Ulla Kasten pour la *Yale Babylonian Collection*.

L'art des listes

L'organisation du savoir sous forme de listes est si générale en Mésopotamie qu'elle a été perçue, dans les débuts de l'assyriologie, comme une forme de pensée caractéristique du Proche Orient Ancien, comme un trait de civilisation. Cette approche unificatrice est encore très présente dans les études actuelles. Pourtant, elle me paraît insatisfaisante pour au moins deux raisons. D'une part la façon dont les listes ont été créées et utilisées a connu des variations considérables au cours de la longue histoire de l'écriture cunéiforme, selon les contextes, les époques, les genres de textes. Des listes d'apparences très similaires peuvent résulter de démarches tout à fait différentes, et c'est ce point précis que je voudrais mettre en évidence ici dans le cas des listes mathématiques. D'autre part les listes ne sont pas spécifiques au Proche Orient Ancien. Elles sont omniprésentes dans toutes les traditions écrites, et, dans une certaine mesure dans les traditions orales. Aujourd'hui, les listes ont envahi notre quotidien, par exemple avec les bases de données et les documents stockés dans nos ordinateurs.

C'est sans doute Jack Goody qui, le premier, a souligné l'importance des listes dans les écrits anciens. Pour lui, la liste est même une sorte d'archétype de l'écrit à ses débuts.[1] Ce n'est pourtant que récemment que cette forme de texte banale et apparemment anodine est devenue un objet d'étude en tant que tel. Les études récentes des listes, et plus généralement des énumérations, portent sur des corpus de tous genres et de toutes époques.[2] De telles études me paraissent essentielles pour progresser dans la compréhension des textes anciens.

Pour étudier les structures de liste dans la documentation cunéiforme, il est intéressant d'accorder une attention particulière aux textes mathématiques. En

---

[1] Jack Goody, *The Domestication of the Savage Mind* (Cambridge : Cambridge University Press, 1977), 80–89.

[2] Voir les publications suivantes ainsi que les bibliographies qu'elles fournissent : Christophe Luc, Mustapha Mojahid, Marie-Paule Pery-Woodley, and Jacques Virbel, « Les énumérations : structures visuelles, syntaxiques et rhétoriques », dans M. Gaio and E. Trupin (eds.), *Document électronique dynamique : actes du troisième Colloque International sur le Document Electronique CIDE 2000* (Lyon : Europia Productions, 2000), 21–40; Bernard Sève, *De haut en bas : philosophie des listes*, L'ordre philosophique (Paris : Editions du Seuil, 2010); Jacques Virbel, « Aspects du contrôle des structures textuelles », dans J. Lambert and J.-L. Nespoulous (eds.), *Perception auditive et compréhension du langage : état initial, état stable, et pathologie* (Marseille : Solal, 1997), 251–72; Jacques Virbel, « Les énumérations textuelles (preprint en français) », dans Karine Chemla and Jacques Virbel, eds., *Texts, Textual Acts, and the History of Science* (Springer, à paraître).

effet, la logique interne aux mathématiques permet de saisir des mécanismes de création des listes qui ne sont pas toujours clairs dans d'autres genres de textes, notamment dans les textes de divination. De plus, les mathématiques cunéiformes nous offrent ce que les savants de Mésopotamie ont produit de plus élaboré en matière de liste. Dans certains textes mathématiques, on voit cette « science » textuelle se déployer avec une virtuosité spectaculaire, comme on le verra dans ce qui suit. Mais, curieusement, cet aspect n'a pas attiré l'attention des historiens des mathématiques. Ces derniers se sont concentrés sur les résultats mathématiques, mais se sont peu intéressés aux structures textuelles. L'unité de texte examinée par les chercheurs modernes est principalement le problème, considéré de façon isolée, indépendamment de l'ensemble des problèmes qu'un scribe a choisi de réunir et de noter sur une même tablette, dans un ordre précis. A de rares exceptions près, il n'a guère été tenu compte du fait que les problèmes constituaient des listes. Or certaines informations capitales résident précisément dans la structure de ces listes. Une analyse des structures de liste est particulièrement utile pour les textes mathématiques qui ne contiennent pas de procédure de résolution (on verra des exemples plus loin). Dans ces cas, le contenu mathématique ne réside pas dans la résolution des problèmes, puisque celle-ci est absente, mais dans la façon dont les énoncés sont construits et agencés. Or cette construction ne se décèle pas à l'échelle d'un seul énoncé, mais à l'échelle de tout un ensemble d'énoncés, ensemble qui résulte parfois d'un mécanisme générateur unique.

Mais qu'est-ce qu'une liste ? Les listes ou, plus généralement, les structures énumératives, sont si multiformes qu'elles défient les tentatives de définition. La diversité des formes et des usages des listes est extrême. Au-delà des problèmes de définition, qui ne sont peut-être pas essentiels, ce sont des questions plus pragmatiques qui me semblent pouvoir apporter une aide dans la compréhension des textes anciens : quel type de message un scribe voulait-il exprimer ou transmettre lorsqu'il notait une liste sur une tablette d'argile ? Il est bien sûr difficile de répondre à une telle question puisque les scribes n'ont guère laissé de traces explicites de leurs intentions. Cependant, je voudrais montrer, par les quelques analyses qui suivent, que le but n'était pas toujours, comme on le suppose souvent, de transmettre un savoir, c'est-à-dire d'exposer des connaissances ou de donner des instructions pour la résolution d'un problème. Les scribes pouvaient poursuivre des buts autres, comme par exemple l'archivage, ou certaines formes de réflexion théorique, ou encore des tentatives d'exploration de domaines qui leurs étaient inconnus.

I Les mathématiques cunéiformes : quelques repères et quelques questions

Les textes mathématiques écrits sur argile qui ont été découverts en Mésopotamie et dans les régions voisines représentent plus de 2000 tablettes ou fragments. L'immense majorité d'entre eux date de la période paléo-babylonienne (= pB, début du deuxième millénaire avant notre ère), dont il sera essentiellement question dans ce qui suit. Une partie considérable de ce corpus est constitué de « tablettes scolaires », c'est-à-dire de tablettes écrites par des écoliers au cours de leur formation élémentaire. Les autres documents mathématiques peuvent être qualifiés de « savants » par opposition aux écrits d'écoliers. Ce sont des textes plus ou moins érudits, sans doute écrits par des étudiants avancés ou des maîtres d'écoles de scribes. La fonction de certains de ces textes est reconnaissable par les caractéristiques matérielles des tablettes d'argile sur lesquels ils sont écrits. Cette typologie est bien connue pour les textes scolaires élémentaires, mais moins claire pour les textes d'enseignement avancés ou d'érudition. Dans le domaine des mathématiques et de la littérature, on peut néanmoins noter une certaine régularité. Il semblerait que, dans les niveaux avancés de l'enseignement, maîtres et étudiants utilisaient des tablettes écrites sur une seule colonne, dites de « type S ». Des tablettes multi-colonnes, dites de « types M », avaient des usages qui semblent plus variés, comme par exemple des examens ou des travaux de pure érudition. Cette typologie n'est en fait véritablement établie que pour les textes littéraires de Nippur,[3] mais elle peut apporter des indices, à recouper avec d'autres, pour identifier les auteurs de textes de provenance autre ou inconnue.

Du fait que les mathématiques d'époque pB sont essentiellement issues des activités des écoles de scribes, ces textes sont généralement interprétés par les historiens comme relevant d'une transmission de maître à élève. Pourtant, un tel modèle de communication, qui se limite à la transmission de haut en bas, ne permet pas de comprendre des phénomènes tels que la circulation des idées, la standardisation du cursus de formation, ou les voies originales de l'invention mathématique. Dans l'étude des listes mathématiques cunéiformes qui suit, je voudrais montrer au contraire la diversité des modèles de communication.

---

3 Steve Tinney, « On the Curricular Setting of Sumerian Literature », *Iraq* 61 (1999) : 159–72; Paul Delnero, « Sumerian Literary Catalogues and the Scribal Curriculum », *Zeitschrift für Assyriologie* 100 (2010) : 32–55.

Les textes mathématiques savants se partagent en plusieurs catégories selon leur contenu, leur structure, le type de tablettes sur lesquelles ils sont écrits, leur langue, leur provenance, leur datation, etc.[4] Je ne vais pas entrer ici dans le détail de ces classifications.[5] Je m'intéresserai seulement à trois catégories qui, d'une certaine façon, ont été reconnues comme telles par leurs auteurs dans la mesure où ces catégories portent des noms sumériens ou akkadiens. La première de ces catégories comprend les textes de procédure (*kibsum* en akkadien), qui sont des listes d'énoncés de problèmes suivis d'une procédure de résolution détaillée. La deuxième catégorie est formée des catalogues, qui sont des listes d'énoncés de problèmes mais ne contiennent aucune indication pour la résolution; ils se terminent généralement par une brève notice (colophon) indiquant le nombre d'énoncés (N im-šu = N sections) et, souvent, le sujet de ces énoncés (champs, ou briques, ou canaux, etc.). La dernière catégorie est, comme la précédente, composée de listes d'énoncés sans procédure de résolution. Cependant, contrairement à celles des catalogues, ces listes se déploient sur plusieurs tablettes (une dizaine ou plus), chacune portant un numéro, comme les pages d'un livre, et sont appelées « séries » en assyriologie. En examinant chacune de ces trois catégories de textes, j'essaierai de montrer la diversité des buts poursuivis par les scribes et des publics auxquels ils s'adressaient.

---

[4] Note sur les langues : les deux langues utilisées dans les textes mathématiques (et d'une façon générale dans les textes cunéiformes d'époque pB provenant de Mésopotamie), sont le sumérien et l'akkadien. Le sumérien est une langue de type agglutinant, de groupe linguistique inconnu, qui était parlée en Mésopotamie du sud au cours du 3e millénaire, et qui a probablement disparu des usages courants avant l'époque pB pour devenir une langue d'érudition enseignée dans les écoles de scribes. L'akkadien est une langue sémitique qui, à l'époque pB, était la langue maternelle d'une grande partie de la population de Mésopotamie.

[5] Pour un classement des sources mathématiques selon l'origine géographique, voir Albrecht Goetze, « The Akkadian Dialects of the Old-Babylonian Mathematical Texts », dans O. Neugebauer and A. J. Sachs (eds.), *Mathematical Cuneiform Texts*, American Oriental Society 29 (New Haven: American Oriental Series & American Schools of Oriental Research, 1945), 146–51 (Ch. 4), et Jens Høyrup, *Lengths, Widths, Surfaces : A Portrait of Old Babylonian Algebra and its Kin* (Berlin: Springer, 2002), Ch. 8 et 9. Pour une discussion sur les problèmes posés par la classification du matériel mathématique cunéiforme et une bibliographie associée, voir Christine Proust, « Reading Colophons from Mesopotamian Clay-Tablets Dealing with Mathematics », NTM Zeitschrift für Geschichte der Wissenschaften, Technik und Medizin (2012) : 123–56.

II Des listes pour résoudre des problèmes : les textes de procédure

Les textes mathématiques les plus nombreux, ou en tous cas les plus étudiés, sont des listes de problèmes résolus. Le plus souvent, les énoncés sont rédigés à la 1ère personne, et les procédures de résolution sont à la 2e personne. La procédure se présente comme une liste de consignes, ouverte par une formule figée : « toi, pour le savoir », et fermée par une autre formule figée : « telle est la façon d'opérer ». Les problèmes d'un texte de procédure portent généralement sur une même thématique : surfaces de champs, creusement de canaux, volume de tas de briques, opérations commerciales, ou autre. Une liste de problèmes peut être obtenue par différents procédés. Parfois, les problèmes sont emboîtés, chacun reprenant le précédent en ajoutant un élément de complexité. Parfois, les données du problème varient, ce qui permet d'explorer tous les aspects d'une situation donnée, ou bien tous les cas possibles. Parfois, la situation imaginée (par exemple le creusement d'un canal) est la même dans tous les problèmes de la liste (par exemple, mêmes dimensions du canal, même nombre d'ouvriers, mêmes salaires ...), et les différents énoncés sont obtenus par permutation circulaire des paramètres.

Un exemple de ce dernier procédé est fourni par la tablette YBC 4663, d'époque pB, d'origine inconnue et conservée à Yale. Elle contient 8 problèmes concernant le creusement d'une tranchée. Les énoncés sont écrits en sumérien (caractères droits dans la traduction ci-dessous), et la procédure est écrite en akkadien (mots en italiques). La traduction des deux premiers problèmes est la suivante :[6]

#1

1. Une tranchée. Sa longueur est 5 ninda, sa largeur est 1 ½ ninda, sa profondeur est ½ ninda. Le volume assigné (à chaque ouvrier) est 10 gin$_2$. Le salaire (d'un ouvrier) est 6 še.

2. Que sont la base, le volume, le nombre d'ouvriers et l'argent (salaire total) ? Toi, pour le savoir :

3. *La longueur et la largeur croise, cela te donnera 7.30.*

4. *7.30 à la profondeur élève, cela te donnera 45.*

---

[6] Traduction personnelle, basée sur Neugebauer et Sachs, *Mathematical Cuneiform Texts*, 70.

5. *L'inverse du volume assigné dénoue, cela te donnera 6. A 45 élève, cela te donnera 4.30.*

6. *4.30 au salaire élève, cela te donnera 9. Telle est la façon d'opérer.*

#2

7. *9 gin$_2$ l'argent (total) pour la tranchée. 1 ½ ninda sa largeur, ½ ninda sa profondeur. 10 gin$_2$ le volume assigné. 6 še (d'argent) le salaire (d'un ouvrier).*

8. *Sa longueur combien ? Toi, pour le savoir : la largeur et la profondeur croise.*

9. *9 te donnera. L'inverse du volume assigné dénoue,*

10. *à 9 élève. 54 te donnera.*

11. *54 au salaire élève. 1.48 te donnera.*

12. *L'inverse de 1.48, 33.20 te donnera. 33.20 à 9, l'argent (total), élève.*

13. *5 te donnera. 5 ninda sa longueur. Telle est la procédure.*

Le problème 2 dérive du problème 1 par une simple permutation circulaire des paramètres : certaines données deviennent des inconnues et vice-versa. Les six autres problèmes de la tablette résultent du même procédé. La procédure de résolution proprement dite est une liste de consignes encadrée par les deux formules qui ouvrent et ferment l'algorithme. Ces deux formules soulignent la nature prescriptive de la liste d'instructions. On le voit, les textes de procédure sont des listes (de problèmes) renfermant d'autres listes (de consignes).

Les textes de procédure ont généralement pour but de montrer comment on résout un type de problème donné, représenté par une collection de cas particuliers. Dans le cas de la tablette YBC 4663, une telle démarche reflète probablement une activité d'enseignement. Le fait que la tablette ne comporte qu'une seule colonne (type S) donne à penser qu'elle a été écrite par ou pour un étudiant avancé, et donc confirme le caractère didactique du texte. Il convient toutefois de noter que les textes de procédure ne reflètent pas toujours aussi clairement des activités d'enseignement et, dans certains cas, peuvent s'inscrire dans des réseaux de relations entre érudits non liés strictement à l'enseignement.[7]

---

[7] Voir à ce sujet Alain Bernard et Christine Proust, eds., *Scientific Sources and Teaching Contexts*

III Des listes pour classer et archiver : les catalogues

Les catalogues sont des listes d'énoncés de problèmes, ne donnant aucune indication sur la façon de les résoudre. Les problèmes listés dans les catalogues se retrouvent généralement dans les textes de procédure, sous des formes similaires ou même absolument identiques comme on va le voir dans ce qui suit. Les auteurs des catalogues rassemblaient dans une tablette des énoncés de problèmes relatifs à un même sujet.

Voici par exemple la traduction du début du catalogue YBC 4657 contenant 31 énoncés de problèmes concernant le creusement d'une tranchée :

#1 Une tranchée. 5 ninda sa longueur, 1 ½ ninda sa largeur, ½ ninda sa profondeur. 10 gin_2 le volume assigné (à chaque ouvrier). 6 še (d'argent) [le salaire d'un ouvrier].
La base, le volume, le nombre d'ouvriers et l'argent (des salaires) combien ? 7 ½ la base, 45 le volume.
4×60 + 30 [ouvriers], 9 gin_2 l'argent (des salaires).

#2 L'argent (total) pour la tranchée est [9 gin_2]. 1 ½ ninda sa largeur, ½ ninda sa profondeur. 10 gin_2 la tâche assignée. 6 še (d'argent) le salaire d'un ouvrier.
Sa longueur [combien] ? 5 ninda sa longueur.

#3 L'argent (total) [pour la tranchée est 9 gin_2]. 5 ninda sa longueur, ½ ninda sa profondeur. 10 gin_2 la tâche assignée. 6 še (d'argent) le salaire d'un ouvrier.
Sa largeur combien ? 1 ½ ninda la largeur.

#4 L'argent (total) pour la tranchée [est 9 gin_2]. 5 ninda sa longueur, 1 ½ ninda sa largeur. 10 gin_2 la tâche assignée.
6 še (d'argent) le salaire (d'un ouvrier). Sa profondeur combien ? [½ ninda] sa profondeur.

---

throughout History : Problems and Perspectives, Boston Studies in the Philosophy and History of Science 301 (Dordrecht : Springer, 2014).

La tablette se termine par un colophon ainsi libellé : 31 sections (im-šu) à propos de tranchées (ki-la$_2$). Le colophon précise donc le nombre de problèmes contenu dans la tablette et le thème des problèmes.[8]

Si le texte est homogène du point de vue du thème des énoncés, qui tous concernent le creusement d'une tranchée, ils ne le sont pas d'un point de vue mathématique. On distingue assez nettement quatre groupes différents.[9] Le premier de ces groupes contient exactement les énoncés du texte de procédure YBC 4663 vu précédemment, les énoncés étant cités dans les mêmes termes et dans le même ordre.[10] Le troisième de ces groupes correspond lui aussi à un texte de procédure connu, YBC 4662, qui se trouve également à l'Université Yale (voir fig. 1).

Les autres groupes d'énoncés de notre catalogue devaient probablement être associés de la même manière à un texte de procédure. Le lien entre les catalogues et certains textes de procédure est donc très étroit. Mais dans quel sens ce lien s'est-il établi ? Les énoncés des textes de procédures sont-ils extraits des catalogues, ou bien les catalogues sont-ils des compilations réalisées à partir de plusieurs textes de procédure ? La présence d'un colophon dans les catalogues, et l'organisation des énoncés en plusieurs groupes distincts conduit à privilégier la deuxième hypothèse. Mais la question qui nous intéresse ici est autre, c'est celle de la fonction des catalogues. Les catalogues ont-ils été écrits dans le même but que les textes de procédure ? Les catalogues sont généralement considérés par les historiens modernes comme des répertoires d'exercices pédagogiques, tout comme les textes de procédure. Pourtant, le fait qu'ils soient structurés différemment incite à considérer la question de plus près. Rappelons tout d'abord que les catalogues ne donnent pas de consignes précises pour résoudre les problèmes. Par contre, leurs auteurs prenaient soin de noter le nombre exact de problèmes, et leur thématique. Par ailleurs, les catalogues sont des ensembles composites de plusieurs collections de problèmes provenant de textes de procédure différents, quoique portant sur le même thème. Plusieurs hypothèses sur la fonction des catalogues sont dès lors possibles. Une première hypothèse est suggérée par le lien étroit entre les catalogues et les textes de procédure, et par le fait que tous les catalogues connus sont de type S. Ces deux particularités évoquent les activités d'enseignement. Les catalogues pourraient

---

[8] Ici, 31 « sections » signifie « 31 énoncés » car chacun d'eux est écrit dans une section. Cependant la relation entre sections et énoncés n'est pas toujours aussi simple.
[9] Pour la description précise de ces groupes et l'analyse de leur contenu mathématique, voir Proust, « Reading Colophons ».
[10] Ce parallèle remarquable a été mis en évidence par O. Neugebauer et A. Sachs, qui ont publié les trois textes ensembles (Neugebauer et Sachs, *Mathematical Cuneiform Texts*, 66).

Fig. 1 Catalogue et textes de procédure associés. Les problèmes 1–8 du catalogue YBC 4657 (au centre) correspondent aux problèmes résolus de YBC 4663 (à gauche), et les problèmes 19–28 du catalogue correspondent aux problèmes résolus de YBC 4662 (à droite).

résulter d'une volonté d'organiser le cursus mathématique et, peut-être, de le standardiser. Mais la présence des colophons, avec un compte des problèmes et une « étiquette » thématique, peut faire aussi penser à des pratiques d'archivage. En effet, au premier millénaire, soit dans des époques très postérieures à la période pB, la pratique du colophon s'est considérablement développée en liaison avec la gestion des bibliothèques. On peut donc penser que les catalogues reflètent des pratiques de classification et d'archivage liées à l'apparition des premières bibliothèques.[11]

En toutes hypothèses, les catalogues résultent d'une démarche différente de celle des textes de procédure. Il s'agit non pas de demander au lecteur de résoudre des problèmes, mais en quelque sorte de décrire un état de l'art. Alors que les textes de procédure ont un caractère prescriptif, les catalogues apparaissent plutôt comme descriptifs. Du point de vue de leur statut pragmatique, ils se rapprochent des index, des inventaires, des tables des matières, ou encore des programmes d'enseignement officiels.

IV Des listes pour explorer : les textes de séries

Souvent confondues avec les catalogues, les séries témoignent d'une toute autre démarche. Les séries mathématiques sont des listes d'énoncés de problèmes, tout comme les catalogues, mais ces listes sont beaucoup plus longues et couvrent plusieurs tablettes numérotées. La majorité des exemplaires se trouvent actuellement conservées dans la *Yale Babylonian Collection*, deux autres se trouvent à l'*Oriental Institute* de Chicago, deux autres au *Vorderasiatisches Museum* de Berlin ; tous ces textes ont été publiés par O. Neugebauer dans un chapitre de ses *Mathematische Keilschrifttexte* qui leur est consacré.[12] Le corpus s'est enrichi récemment de deux nouveaux exemplaires qui étaient restés ignorés dans les réserves du musée du Louvre.[13] Pour montrer les particularités des textes de séries, je vais me concentrer sur un des exemplaires du Louvre.

La tablette AO 9071 est une tablette multi-colonnes (type M). Elle se termine par un colophon qui donne le nombre de sections, comme dans les catalogues. Mais ici, ce nombre est plus important : 95. De plus, le colophon contient un numéro : « tablette numéro 7 » (voir fig. 2).

Le début du texte montre comment se met en place le processus qui a permis de générer les centaines d'énoncés de la série de tablettes. Il est donc intéressant de l'examiner de près.

---

[11] Cette idée est inspirée du travail de J.-J. Glassner sur les textes de divination (« Écrire des livres à l'époque paléo-babylonienne : le traité d'extispicine », *Zeitschrift für Assyriologie* 99 [2009] : 1–81.).
[12] Otto Neugebauer, *Mathematische Keilschrifttexte I–III* (Berlin : Springer, 1935–1937), Ch. 7.
[13] Christine Proust, « Deux nouvelles tablettes mathématiques du Louvre : AO 9070 et AO 9072 », *Zeitschrift für Assyriologie* 99 (2009) : 167–232.

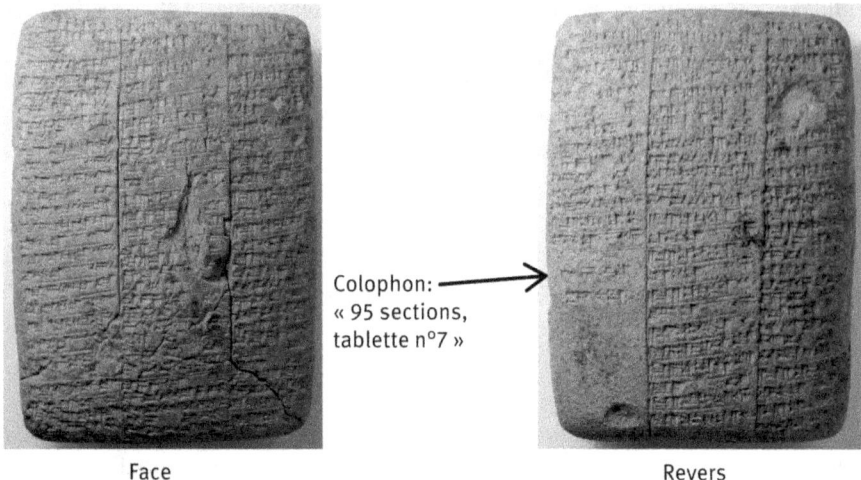

Fig. 2 Tablette AO 9071 (photo C. Proust, courtoisie Béatrice André-Salvini, Musée du Louvre)

Section	Traduction de AO 9071, face, col. 1
	(Proust, « Deux nouvelles tablettes »)
1a	La longueur et la largeur j'ai accumulé : 50 ninda.
b	La longueur excède la largeur de 10 ninda.
2	⅔ de la longueur : la largeur.
3	La moitié de la longueur et 5 ninda : la largeur.
4	Le tiers de la longueur et 10 ninda : la largeur.
5*	Le 5ème de la longueur et largeur à 10 ninda et la longueur j'ai ajouté, 15 ninda à la largeur j'ai ajouté ⟨ : 1.25⟩.
6	Le tiers de ce dont la longueur excède la largeur à la longueur j'ai ajouté : 33.20.
7*	2 fois répété, j'ai ajouté : 36.40.
8	J'ai soustrait : 26.40.
9	2 fois répété, j'ai soustrait : 23.20.
10	9 fois répété : la longueur.
11	12 fois répété : en excès de 10 ninda.
12	6 fois répété : en défaut de 10 ninda.
13	A la largeur j'ai ajouté : 23.20.
14	2 fois répété, j'ai ajouté : 26.40.
15	J'ai soustrait : 16.40.
16	6 fois répété : la largeur
17	A la longueur et la largeur j'ai ajouté : 53.20.
18	2 fois répété, j'ai ajouté : 56.40.

19	J'ai soustrait : 46.40.
20*	15 fois répété :
	j'ai égalisé.
21	12 fois répété : en défaut de 10 ninda.
22a	La longueur 3 fois répété,
	la largeur 2 fois répété, j'ai accumulé : 2.10.
b	[La longueur] et la largeur j'ai accumulé : 50.

Le texte commence par un énoncé simple, d'un genre bien attesté dans les mathématiques pB. Cet énoncé donne deux relations linéaires entre la longueur et la largeur inconnues d'un rectangle, plus précisément leur somme et leur différence. En désignant par u la longueur (uš) et s la largeur (sag), on peut exprimer comme suit cet énoncé en langage moderne :

u + s = 50 ninda et u − s = 10 ninda.

La solution est la suivante : la longueur est 30 ninda, la largeur est 20 ninda. On retrouve cette même solution pour tous les problèmes de la tablette ainsi que de la série (ainsi, du reste, que d'une bonne partie des textes mathématiques cunéiformes).

Dès la deuxième section, on constate qu'une seule relation entre la longueur et la largeur est donnée : ⅔ u = s. Une partie des informations manque donc. Il en est de même dans les sections 2 à 6. En fait, l'information manquante est la première relation entre la longueur et la largeur, donnée ligne 1 (u + s = 50 ninda que je désigne dans la suite, par commodité, équation E_1).

L'énoncé 7 est encore plus lacunaire :

#7 2 fois répété, j'ai ajouté : 36.40.

On ne sait pas ce qui est répété, ni à quoi on ajoute. Pour le savoir, comme précédemment, il faut rechercher des informations déjà données dans la section précédente :

#6 Le tiers de ce dont la longueur excède la largeur à la longueur j'ai ajouté : 33.20.

Dans le texte cunéiforme, aucun indice matériel ou typographique ne permet de déceler quels sont les éléments de la section 6 qui sont réutilisés dans la section 7. Cependant, l'habitude de ce genre de texte donne la clé assez facilement.[14] Ce qui

---

14 Neugebauer a décrit les étapes de l'enquête quasi policière qui l'a amené à percer le secret des textes de série, c'est-à-dire à comprendre les principes de génération des énoncés, dans un

est répété est « Le tiers de ce dont la longueur excède la largeur », soit une expression que je désigne par P (expression principale) dans ce qui suit. Ce à quoi on ajoute est « la longueur », une expression que je désigne par S (expression secondaire) dans ce qui suit. La première relation entre la longueur et la largeur est toujours la même, à savoir la relation E_1 donnée dans la section 1.

Les énoncés 8 à 12 sont construits de la même façon. Ils donnent des relations entre les expressions P et S, mais sans expliciter ni P, ni S, ni l'équation E_1. Ainsi, les sections 7 à 12 dépendent de la section 6 pour la définition de P et de S, et de la section 1 pour la définition de E_1.

Dans la section 13, la valeur de ce à quoi on ajoute (expression secondaire S) est changée : c'est maintenant « la largeur ». Les sections 14-15-16 donnent d'autres relations entre P et cette nouvelle valeur de S. Ainsi, les sections 13 à 16 dépendent des sections 13, 6 et 1. Dans la section 17, S change encore, et les sections 18 à 21 utilisent cette nouvelle valeur de S.

Dans la section 22, la valeur de P (et de E_1) est changée, et le même processus est itéré : variations de S, puis, pour chaque valeur de S, variations de la relation entre P et S.

Dans la section 35, la valeur de E_1 est encore changée, et le même processus s'enclenche à nouveau : variations de P, puis, pour chaque valeur de P, variations de S, puis, pour chaque valeur de S, variations de la relation entre P et S.

Les valeurs de E_1 et de P sont de plus en plus compliquées, les énoncés de plus en plus lacunaires. Ainsi, plus on avance dans le texte, plus les énoncé écrits sont brefs et les énoncés représentés compliqués. Par exemple, la section 59 est réduite à quatre signes (ba-zi-ma 45), qui signifient :

J'ai soustrait : 45

Une fois que toute l'information à laquelle renvoie cette formule des plus laconiques a été récupérée dans les différents points du texte où elle se trouve, on obtient un énoncé dont la représentation selon le formalisme moderne est la suivante :

$$\begin{cases} (3x+2y)\frac{1}{13}+x=40 \\ -\left\{\left\langle\left\{[(x+25)+(y+1.30-x)+(x+y+35)]\frac{1}{11}+4x\right\}\frac{1}{7}+y\right\rangle\times 2\times\frac{1}{16}-(x+2y)+x+y+(3x-2y)\right\}\frac{1}{7}+(x+y)=45 \end{cases}$$

---

article passionnant écrit dès 1934 (Otto Neugebauer, « Serientexte in der babylonischen Mathematik », *Quellen und Studien zur Geschichte der Mathematik, Astronomie und Physik* B 3 [1934–36] : 106–14). Voir aussi Otto Neugebauer, *The Exact Sciences in Antiquity* (London : Oxford University Press, 1951), § 33, 63–66.

On obtient ainsi une structure arborescente à 4 niveaux. Dans chacun de ces niveaux sont définis des segments de l'énoncé. Ces niveaux sont, en allant du tronc vers les extrémités de l'arborescence :

>Niveau 4 : définition de l'équation E_1
>Niveau 3 : définition de l'expression principale P
>Niveau 2 : définition de l'expression secondaire S
>Niveau 1 : définition de la relation entre P et S

Chaque segment subit des variations tour à tour.

L'équation E_1 subit plusieurs variations. A l'échelle de la tablette AO 9071, on trouve 5 variantes (d'autres variantes interviennent à l'échelle de la série) :

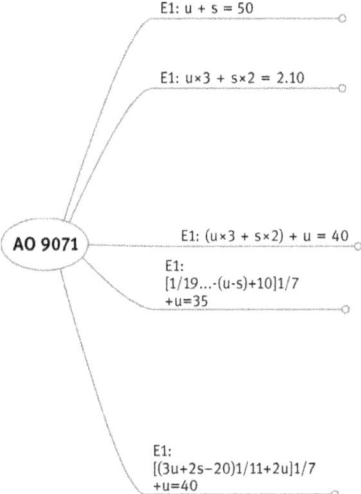

Fig. 3

Pour chaque variante l'équation E_1, l'expression principale P subit plusieurs variations (entre 1 et 4) :

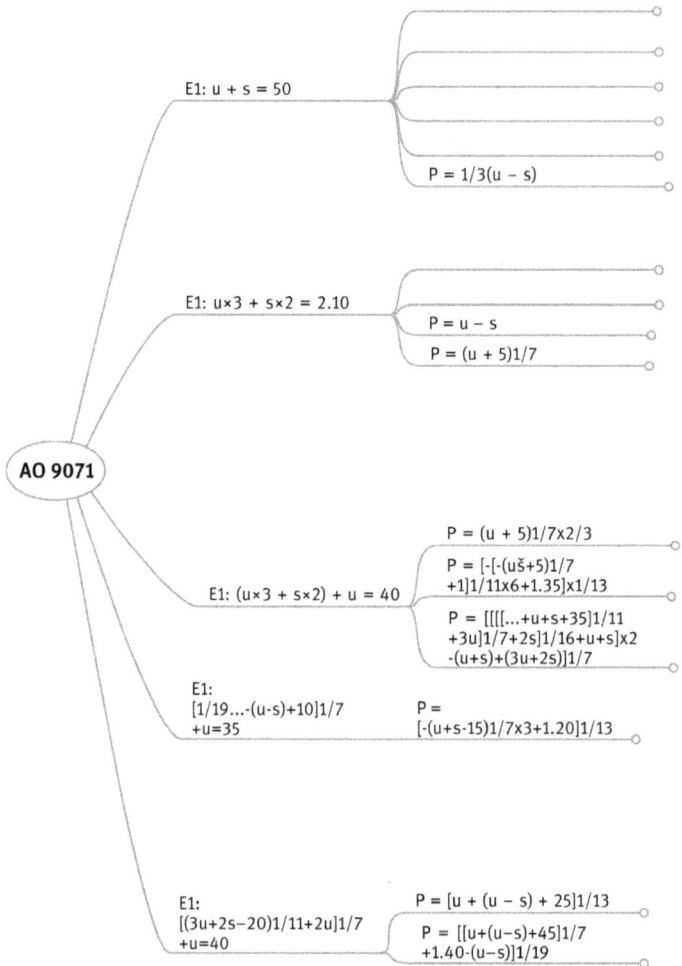

Fig. 4 Variations de niveaux 3

Pour chaque variante de l'expression principale P, l'expression secondaire S subit plusieurs variations (entre 1 et 4 variantes assez régulières) :

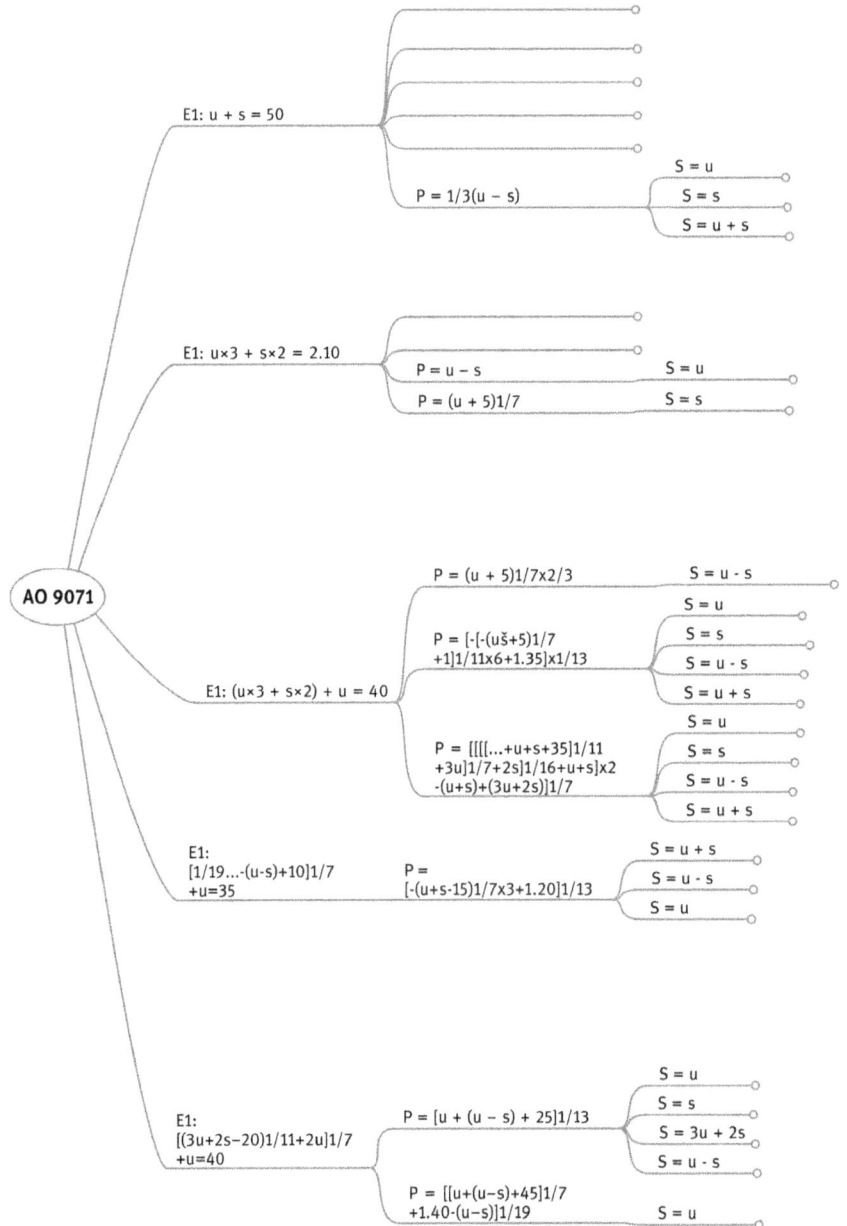

Fig. 5 Variations de niveaux 2

Pour chaque variante l'expression secondaire S, la relation entre S et P subit plusieurs variations, elles aussi assez régulières (entre 1 et 6) :

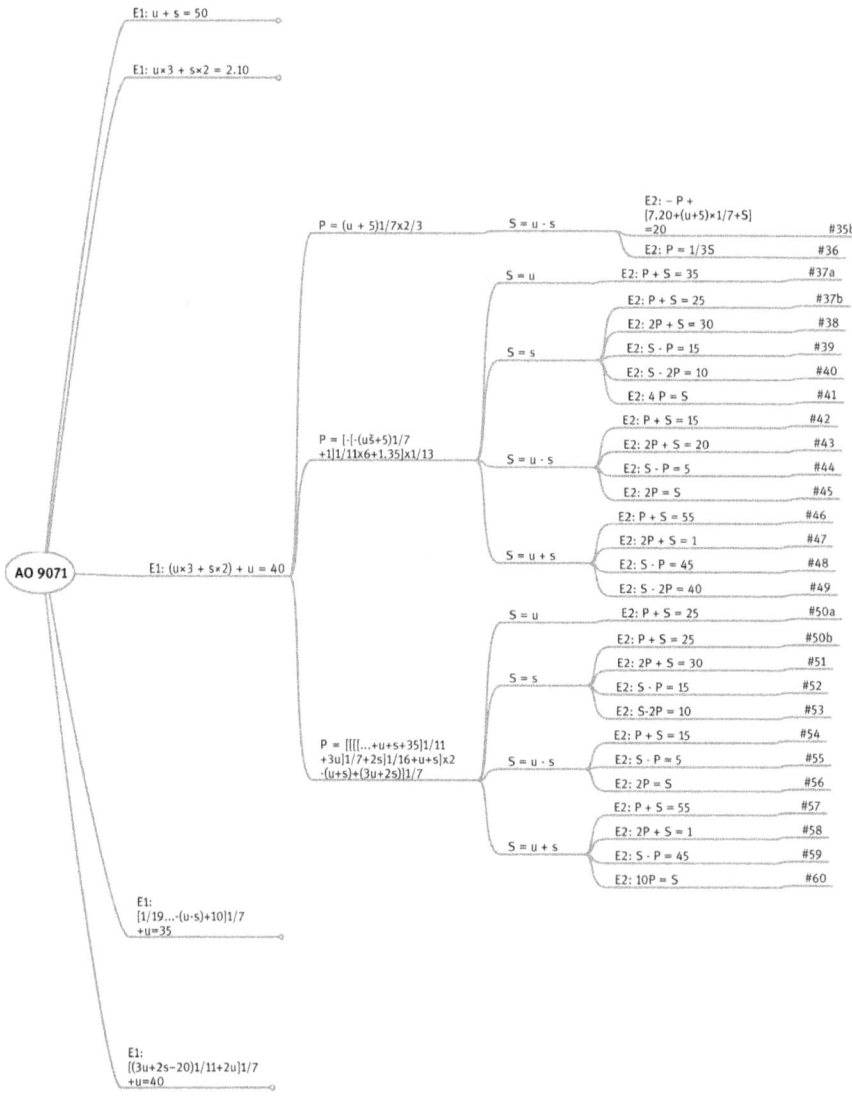

Fig. 6 Variations de niveaux 1 (pour des raisons de place, seule une branche est développée dans ce diagramme)

Sur cette structure arborescente, se greffe une information dont la répartition répond au principe d'économie qu'on a vu fonctionner ci-dessus dans le début du texte : toute information donnée à un niveau de l'arborescence est omise dans les niveaux qui en dépendent. En conséquence, lorsqu'on lit l'énoncé écrit dans une section, il faut remonter dans le texte pour chercher l'information donnée dans les niveaux parents. Dans le cas de la section 59, l'information se trouve en quatre points différents. Circuler dans le texte pour récupérer l'information utile n'était sans doute pas un exercice facile à la portée d'un scribe ordinaire. Les lecteurs ou usager des séries devaient baigner une culture du texte très particulière.

La combinaison d'une structure de liste arborescente avec une répartition rationalisée de l'information donne des potentialités à l'écriture mathématique très supérieure à celle que l'on trouve dans les autres textes mathématiques. Elle permet une compression considérable de l'information. Elle permet aussi l'expression d'opérations hiérarchisées sur plusieurs niveaux (jusqu'à 5 dans le # 59 de la tablette du Louvre) que nous représentons aujourd'hui dans les formules mathématiques par des niveaux de parenthèse.

D'un point de vue mathématique, le procédé de génération des énoncés revient à remplacer certaines combinaisons des inconnues par d'autres combinaisons de même nature (ici, ces combinaisons sont linéaires, mais dans d'autres textes, elles peuvent être quadratiques). Ces substitutions systématiques permettent de produire à l'infini de nouveaux énoncés mathématiquement équivalents (mêmes inconnues, même degré, mêmes solutions). Soulignons que si le procédé permet de contrôler la fabrication des énoncés, il ne permet pas de contrôler les procédures de résolution. En effet, la liste n'est pas organisée en fonction des méthodes de résolution, comme dans les textes de procédure et les catalogues, mais en fonction des caractéristiques formelles des énoncés. Et de fait, si les énoncés initiaux dans chaque tablette semblent relativement communs, on obtient au fil des listes des énoncés de plus en plus complexes, voire extravagants. Pour certains de ces problèmes, on peut douter sérieusement que les scribes aient été en mesure des les résoudre.[15]

Nous avons donc devant les yeux un exemple de liste qui semble poursuivre d'autres buts que ceux qui ont été repérés dans les cas précédents : ni prescriptifs, ni descriptifs, ils relèvent d'un autre statut pragmatique. Qu'était-il donc attendu du lecteur de ces gigantesques listes ? Ce n'est pas la résolution des problèmes, puisque celle-ci est manifestement impossible dans de nombreux cas. Ce n'est pas l'archivage du matériel existant, puisque la plupart des problèmes ne sont pas attestés ailleurs. Les séries se présentent plutôt comme un

---

[15] Le problème 59 évoqué ci-dessus en est un exemple. On trouve même des problèmes de degrés 4 ou 5 non réductibles à des degrés plus bas dans d'autres les textes de séries.

outil heuristique dont le but est d'explorer les propriétés d'un objet mathématique, l'énoncé de problème.

A qui s'adressaient les séries ? Il me paraît clair que les séries ne sont pas des collections de problèmes destinés à l'éducation mathématique des étudiants d'écoles de scribes. Les séries ont clairement été écrites par des érudits d'un très haut niveau, connaissant parfaitement la tradition mathématique de leur époque, et maîtrisant avec virtuosité l'écriture cunéiforme. Les destinataires n'étaient sans doute pas leurs étudiants, mais leurs pairs.

V Diverses raisons d'écrire des listes

Voir les listes de problèmes comme des manuels d'exercices pour l'enseignement est non seulement réducteur, comme indiqué en introduction, mais surtout conduit à privilégier certaines interprétations et à en occulter d'autres. Or les exemples qui ont été présentés ici montrent que le réseau de communication dans lequel s'inscrit un texte mathématique est beaucoup plus riche et complexe que la simple transmission de maître à élève. Quand on regarde les textes mathématiques en tant que listes, on constate qu'ils ne répondent pas tous aux mêmes « projets d'écriture », c'est-à-dire qu'ils ne s'adressent pas aux mêmes types de lecteurs, et qu'ils ne poursuivent pas les mêmes objectifs. Certains s'apparentent à des manuels, d'autres à des inventaires, d'autres encore ressemblent plutôt à des entreprises exploratoires.

Ces questions pourraient être importantes du point de vue de l'histoire intellectuelle de la Mésopotamie à la fin de la période paléo-babylonienne. En effet, les séries semblent ne pas avoir été élaborées dans le même contexte que les catalogues et les textes de procédure. Les indices paléographiques plaident pour une datation relativement tardive des séries, c'est-à-dire de la fin de l'époque paléo-babylonienne. D'autres indices, tels que l'organisation en séries numérotées, l'usage d'une langue complètement artificielle imitant le sumérien, suggère que ces textes pourraient venir de la région centrale de la Mésopotamie, par exemple Kiš ou Sippar. En même temps, leurs auteurs connaissaient parfaitement la tradition mathématique du Sud, dont ils reproduisaient la terminologie et les thématiques. Les séries ont sans doute été écrites alors que la tradition mathématique s'était déjà pleinement développée. Elles témoignent en tout cas d'une connaissance profonde de la tradition mathématique pB. En même temps elles sont, par certains aspects, totalement innovantes.

Bibliographie

Bernard, Alain, et Christine Proust, eds. *Scientific Sources and Teaching Contexts throughout History : Problems and Perspectives*, Boston Studies in the Philosophy and History of Science 301 (Dordrecht : Springer, 2014).

Chemla, Karine, et Jacques Virbel, eds. *Texts, Textual Acts, and the History of Science* (Springer, à paraître).

Delnero, Paul. « Sumerian Literary Catalogues and the Scribal Curriculum », *Zeitschrift für Assyriologie und Vorderasiatische Archäologie* 100 (2010) : 32–55.

Glassner, Jean-Jacques. « Écrire des livres à l'époque paléo-babylonienne : le traité d'extispicine », *Zeitschrift für Assyriologie und Vorderasiatische Archäologie* 99 (2009) : 1–81.

Goetze, Albrecht. « The Akkadian Dialects of the Old-Babylonian Mathematical Texts », dans Neugebauer et Sachs, *Mathematical Cuneiform Texts*, 146–51 (Ch. 4).

Goody, Jack. *The Domestication of the Savage Mind* (Cambridge : Cambridge University Press, 1977).

Høyrup, Jens. *Lengths, Widths, Surfaces : A Portrait of Old Babylonian Algebra and its Kin* (Berlin : Springer, 2002).

Luc, Christophe, Mustapha Mojahid, Marie-Paule Pery-Woodley, et Jacques Virbel. « Les énumérations : structures visuelles, syntaxiques et rhétoriques », dans M. Gaio et E. Trupin (eds.), *Document électronique dynamique : actes du troisième Colloque International sur le Document Electronique CIDE 2000* (Lyon : Europia Productions, 2000), 21–40.

Neugebauer, Otto. « Serientexte in der babylonischen Mathematik », *Quellen und Studien zur Geschichte der Mathematik, Astronomie und Physik* B 3 (1934–1936) : 106–14.

Neugebauer, Otto. *Mathematische Keilschrifttexte I–III* (Berlin : Springer, 1935–1937).

Neugebauer, Otto. *The Exact Sciences in Antiquity* (London : Oxford University Press, 1951).

Neugebauer, Otto, et Abraham J. Sachs. *Mathematical Cuneiform Texts*, American Oriental Society 29 (New Haven : American Oriental Series & American Schools of Oriental Research, 1945).

Proust, Christine. « Deux nouvelles tablettes mathématiques du Louvre : AO 9071 et AO 9072 », *Zeitschrift für Assyriologie und Vorderasiatische Archäologie* 99 (2009) : 167–232.

Proust, Christine. « Reading Colophons from Mesopotamian Clay-tablets Dealing with Mathematics », *NTM Zeitschrift für Geschichte der Wissenschaften, Technik und Medizin* 20 (2012) : 123–56.

Sève, Bernard. *De haut en bas : philosophie des listes*, L'ordre philosophique (Paris : Editions du Seuil, 2010).

Tinney, Steve. « On the Curricular Setting of Sumerian Literature », *Iraq* 61 (1999) : 159–72.

Virbel, Jacques. « Aspects du contrôle des structures textuelles », dans J. Lambert et J.-L. Nespoulous (eds.), *Perception auditive et compréhension du langage : état initial, état stable, et pathologie* (Marseille : Solal, 1997), 251–72.

Virbel, Jacques. « Les énumérations textuelles », preprint en français de « The Description of Enumerations », dans Chemla et Virbel, *Texts, Textual Acts, and the History of Science*.

Francesca Rochberg
Conceiving the History of Science Forward

Abstract: Telling the story of science is no longer simply about accounting for the creation of modern science in the early modern period. This is to conceive of the history of science backward. Cuneiform astronomical texts, in their role as direct antecedent to Western astronomy, have long been incorporated into telling the story of science backward. The singling out of astronomy as the Mesopotamian contribution to the history of science, however, does more to reinforce the conception of the history of science backward from modernity to antiquity than it does to elucidate what a Mesopotamian science might be from the perspective of cuneiform evidence. Primary among the criteria for knowledge from the standpoint of the cuneiform record is that of the phenomenon as sign. In this paper I argue that the semiotic nature of physical phenomena determined what proper objects of knowledge were for cuneiform scholar-scribes. From this standpoint, conceiving the history of science forward, I explore traces of the preoccupation with phenomena as signs in later pre-modern science.

It is a singular privilege to contribute to a Festschrift in honor of Heinrich von Staden. For years I have looked to his research as a model of historiographic method and reasoned argument on the basis of ancient texts. This contribution is but a small token of gratitude both for his great contributions to the ancient history of science and medicine and for the rare opportunity I had while a member at the Institute for Advanced Study at Princeton, occupying the office next to his, not only to enjoy leisurely discussions of ancient Greek and Near Eastern sciences, but also of horses, fires, and life in the American outback.

Nearly three thousand years ago a celestial omen was recorded on a Babylonian tablet that said "If in the heavens there are no stars, the land will be in mourning (lit.: will wear unkempt hair)."[1] This terse statement unmistakably evokes such desire and expectation for the heavens to contain, as it were, "word" of the future, that a heaven without stars was in itself a dire sign of national proportions. The view that the stars form a decipherable language written by the gods upon the sky is a trope that begins in ancient Mesopotamia, where the stars and constellations were called "heavenly writing," and the very same

---

[1] Erm 15639:15', in Wayne Horowitz, "Astral Tablets in the Hermitage Saint Petersburg," *Zeitschrift für Assyriologie* 90 (2000): 194–206, at 200.

idea continues in the history of astrology even into the sixteenth century. Anthony Grafton has pointed out that,

> Giovanni Gioviano Pontano, who published his treatise *On Celestial Things* in 1512, argued explicitly that the language of the stars conformed in all essential ways to the language of humans ... Stars and planets, Pontano argued, formed the letters of a cosmic alphabet ... Every planet ... played the role of a letter with defined qualities. Every astrologically significant configuration of two or more planets ... resembled a word or a phrase, the sense of which the astrologer could determine.[2]

Grafton has said that "the continuity of the astrological tradition is, perhaps, unmatched in the intellectual history of the West,"[3] and in another context noted "how astrological procedures outlasted more fragile creations of the human spirit, like the Roman Empire."[4] This is perfectly true. Astrological prognostication outlived many other forms of ancient Near Eastern prognosticative techniques, e. g., from the exta of a sacrificed sheep. What is at issue, however, is not merely the continuity of astrology back to celestial divination in ancient Mesopotamia,[5] but the continuity seen manifested more broadly in preoccupations with natural phenomena as signs from the deity (or deities). Ancient works that were produced for such purposes, such as the many cuneiform text series of omens (celestial and terrestrial, including dreams, abnormal births, physiognomy, and everyday occurrences in a person's house) dominated what we can classify generally as knowledge in the Babylonian and Assyrian scribal traditions of ancient Mesopotamia, though they may also be classified in a variety of other ways, as science, or as divination, and associated with other forms of practice, such as magic, or even religion.

Introducing his book *The Bible: Protestantism and the Rise of Natural Science*, Peter Harrison said,

> Many histories of science have tended to work backwards, beginning with the modern conception of the discipline of science, and projecting it back in time. Historians of science, as Paolo Rossi has observed, frequently concern themselves with 'an imaginary object', con-

---

2 See Anthony Grafton, *Cardano's Cosmos* (Cambridge, Mass.: Harvard University Press, 1999), 6. For the Babylonian evidence of the sky as a surface upon which gods could write, and the ominous phenomena as a written language, see Francesca Rochberg, *In the Path of the Moon*, Studies in Ancient Magic and Divination 6 (Leiden: Brill, 2010), 304–305, and 417–18.
3 Grafton, *Cardano's Cosmos*, 6.
4 Anthony Grafton, "Starry Messengers: Recent Work in the History of Western Astrology," *Perspectives on Science* 8 (2000): 70–83, at 82.
5 Of course this tradition is not characterized solely by continuities, and much development resulted in discontinuities when the texts are analyzed with an attention to the fine detail.

structing it from a variety of texts and heterogeneous disciplines. It is of course inevitable that in order to understand the past we must understand it through modern conceptions. At the same time, it needs to be recognised that 'science', as we understand it, does not have a history which can be traced back beyond the seventeenth century. Accordingly, what I have sought to do in this book is to work forwards, in as much as that is possible. For virtually the first fifteen hundred years of the common era the study of natural objects took place within the humanities, as part of an all-encompassing science of interpretation which sought to expound the meanings of words and things.[6]

Ancient Mesopotamian knowledge of what Harrison refers to here as "natural objects," roughly for the fifteen hundred years before the Common Era, constitutes very much the same thing that Harrison identifies for the first fifteen hundred years of the Common Era, i.e., it is, in his words, "part of an all-encompassing science of interpretation which sought to expound the meanings of words and things." This is represented in the cuneiform tradition of knowledge in the overwhelming focus by specialized scribe-scholars on the systematic and interpretive science of divination from signs. The principal qualification must be in designating the object of this knowledge, not as "natural objects," but as observed, imagined, and conceived objects in relation to physical things, and for the focus not on observation of the signs alone, but on their interpretation according to systematic textual compendia. Both sign and text together constituted the science of cuneiform scribal scholarship.

In our modern zeal to bring cuneiform texts within the purview of the history of science, this fact has been routinely minimized or pushed to the background in order to examine the more "scientific" aspects of this "science of interpretation," such as the empirical, the logical, or the predictive, all in order to tell the story of science in the anticipated way. Because "telling the story of science" was once primarily about accounting for the creation of modern science in the early modern period,[7] this perspective was set to conceive the history of science backward on the essentially nineteenth-century assumption that science is singular; it was always as we now understand and define it, one science having universal characteristics. The model of the singular history of science has undergone a series of challenges in tandem with revisions to the ahistorical notion of the singularity of science itself.[8]

---

[6] See Peter Harrison, *The Bible: Protestantism and the Rise of Natural Science* (Cambridge: Cambridge University Press, 1998), 8–9.
[7] Paula Findlen, "The Two Cultures of Scholarship?," *Isis* 96 (2005): 230–37, at 231.
[8] Hans-Jörg Rheinberger, *On Historicizing Epistemology: An Essay* (Stanford: Stanford University Press, 2010), 1.

Cuneiform astronomical texts, in their role as direct antecedents to Western astronomy, were readily incorporated into telling the story of science backward. The singling out of astronomy, particularly mathematical astronomy, as *the* Mesopotamian contribution to the history of science did its share in reinforcing the conception of the history of science backward from modernity to antiquity. Because the ancients described or predicted phenomena such as the synodic appearances of planets or the first visibility of the moon, the question of what the ancients observed and predicted about natural phenomena was raised, but not why or how the ancient astronomers viewed the phenomena as objects of knowledge, or how the phenomena fit within a system of local epistemic values.

Today scholars are concerned with just such questions, i.e., the questions of the nature of scientific objects of study, description, and prediction in various periods or cultures, and their accompanying values. Accordingly, the view that ideas have a discrete existence enabling them to be transmitted across time and space, along the lines of the history of ideas of Lovejoy's generation, has been rejected in favor of historicized epistemologies, where the interest is, in Hans-Jörg Rheinberger's words, "no longer how knowing subjects might attain an undisguised view of their objects, rather ... what conditions had to be created for objects to be made into objects of empirical knowledge under historically variable conditions."[9]

In the evolution of the field of the history of science, Heinrich von Staden was the first to take issue with the problem of backwards-facing historiography with respect to science in antiquity. Already in 1992, in his "Affinities and Elisions: Helen and Hellenocentrism," von Staden charted a nuanced path between presentism and historicism, arguing for laying aside modern categories and expectations when it comes to the natural or exact sciences so that the criteria for knowledge and the values of knowledge in the past may be revealed.[10] The problem, as he put it, was

> not, of course, "presentism," nor whether or not there are significant conceptual, methodological, and other continuities between ancient and modern science ... the issue is the distorting consequences entailed by the historiographic habit of brightly foregrounding elements of legitimate continuity, with which we tend to be culturally more at home, while relegating discontinuities to a shadowy obscure background at best.[11]

---

9 Rheinberger, *Historicizing Epistemology*, 3.
10 Heinrich von Staden, "Affinities and Elisions: Helen and Hellenocentrism," *Isis* 83 (1992): 578–95.
11 Von Staden, "Affinities," 584.

In the case of ancient Babylonian knowledge, it is not that the mathematical astronomical texts do not somehow reflect a continuity, or kinship, with later, even modern, science, especially in the goal to produce predictive algorithms for cyclical astronomical phenomena.[12] Meaningful continuities to Hellenistic Greek astronomy are to be identified in the cuneiform mathematical astronomical texts, which account for predictive methods and their parameters that were in fact transmitted to and implanted in the Western tradition. The history of science rightly establishes some causal connection between the Babylonian astronomy and its later Greek counterparts. A deeper understanding of the significance of those texts within their own culture, however, cannot come from analysis of their impact on later astronomy alone. When a kind of historical equivalent to the philosophical principle of entailment was applied to the growth of scientific knowledge whereby only those ideas that were logically continuous with the body of science considered valid "now" were viewed as relevant, the ground was established for reconstructing the history of science backward, and only those aspects of ancient knowledge that were progressively entailed in the forward march of science were counted.

Fortunately this is no longer the sole rationale for the study of historical sciences. Instead, many areas of the history of early science have turned to examinations of local knowledge and its various contexts, recognizing that what *epistēmē* meant to Plato or Aristotle, or *scientia* to Augustine, or the Scholastics, was different in each case. Of course the same should apply to Mesopotamian knowledge, though the Akkadian language lacks a true counterpart to our word for and conception of "knowledge." With this goal in mind, I offer to reflect briefly here on the nature of knowledge in cuneiform sources and the implications for the place of Mesopotamia in the broader scope of the history of science.

Conceiving the history of science forward one cannot help but notice a persistent preoccupation with phenomena as signs in later antique and mediaeval science, with descendants in both Western European and Eastern Byzantine and Islamic traditions, not to mention Indian science. Peter Harrison, in the statement quoted above, remarked on the "study of natural objects" being "part of an all-encompassing science of interpretation."[13] The transmission of this "science of interpretation" is well documented and well known,[14] but the reasons for widespread transmission of traditions of omens, horoscopy, or iatromathe-

---

12 Described in Otto Neugebauer, *A History of Ancient Mathematical Astronomy*, vol. 2 (Berlin: Springer, 1975).
13 See n. 6 above.
14 See David Pingree, *From Astral Omens to Astrology: From Babylon to Bikaner*, Serie Orientale Rome 78 (Rome: Istituto Italiano per l'Africa e l'Oriente, 1997).

matics have to do with how (natural and other) phenomena were regarded and understood.[15] Divination and astrology were transmitted among various cultures of knowledge because knowledge of celestial and terrestrial objects as signs assumed a central place, both from the point of view of prognostication of the future as well as of the philosophy of inference-making from signs.

In the West this is clear in the attention paid to the theory of signs in Stoic, late antique, and medieval philosophy. From Philodemus and Sextus Empiricus to St. Augustine, from Boethius to Roger Bacon and William of Ockham, all of whom wrote on the theory of signs, it is clear that signs and prognostication from signs was no marginal concern, though not all of it had to do with divinatory signs. Plotinus in the third century spoke of the world as "teeming with signs" and, "of the wise as those who are able to draw inferences from them," as John Magee points out in his study of the background to Boethius on signification.[16] Eventually knowledge of signs as a criterion for science was replaced by other epistemic criteria, such as accuracy, generality, simplicity, or consilience. But *scientia* did not carry fixed meaning throughout European natural philosophy, as it changed from having to do with reasoning to particular knowledge of the properties of things for Aristotle, to a thoroughgoing empiricism for Epicurus, to an instrument for learning of God for Augustine. Thus the history of science must accommodate attested variations in conceptions of knowledge.

What can be said, then, about an Assyro-Babylonian conception of knowledge, and where does it stand on the historical continuum to Western science? Primary among the criteria for knowledge, as we can infer criteria from the standpoint of the cuneiform record, was the fact that physical phenomena were signs (Akkadian *ittu*/pl. *ittātu/idātu*),[17] and knowledge of signs was for prognostication of the future. It was a primary technique for gaining access to the divine will, i.e., what the gods held in store for humankind. Omen texts constitute the overwhelming bulk of the holdings of the seventh century B.C. Nineveh library, if not originating with (and some would emphasize the Babylonian origin and importation of many of these texts), but developed extensively by and for the scholars and the scribal masters, the so-called *ummânū*, best known from correspondence from the Neo-Assyrian courts of Esarhaddon and Assurbanipal.

---

15 The question of what was or is "natural" is a separate inquiry outside the bounds of the present paper, but necessary for a newly historicized cuneiform epistemology. In a forthcoming study, "Before Nature: Cuneiform Knowledge and the History of Science," I raise the issue of the nature of a science that has no conception of such a domain as the "natural world."
16 John Magee, *Boethius on Signification and Mind* (Leiden: Brill, 1989), 59 and n. 30.
17 See CAD s.v. *ittu* A

With respect to the Akkadian cuneiform record overall, i.e., texts spanning the Old Babylonian period (ca. 1800 B.C.) to the Parthian period (2nd century B.C.–1st century A.D.), divinatory texts represent the largest category of sources, many thousands of tablets. In the field of Assyriology, omen lists have long been accepted as a principal part of "Mesopotamian science,"[18] yet this designation must be acknowledged as a convenient anachronism. As von Staden said, it is not simply a matter of "presentism," because the epistemic values inherent in observational, predictive, systematic, and rational knowledge are evident in cuneiform divinatory texts. These "scientific values," while identifiable in omen texts, account only partially for the particular character of this divinatory knowledge.

Evidence is best for the Neo-Assyrian period but arguably it was also the case in Old and Late Babylonian times that Babylonian divination played the key role in what constituted knowledge for the *ummânū*. In the Neo-Assyrian period several hundred letters between the *ummânū* and the kings Esarhaddon and his son, Assurbanipal, testify to the cultural value of divination and its political rationale. This we cannot simply assume for other periods, although the prognostications attached to signs, particularly in the celestial omen compendia, are always in the main directed toward "the king" and his "land," "people," "army," or other social objects of concern to a monarch, such as "the market," or "plague." Knowledge of divinatory things, the signs, and their stated consequences was intimately connected to texts. Skill and command of the textual repertoire of scholarship, expressed by the term *ṭupšarrūtu*, an abstract formation from the noun "scribe," is one way we can approach a Babylonian, or Assyro-Babylonian, concept of "knowledge." No Akkadian term existed for the sum of things known. However, mastery of words and of the textual repertoire of divination had a term, *le'ûtu* "knowledge," meaning ability or skill in scholarly literature and derived from the verb "to be able" (*lē'û*). A passage in one of the royal letters from the Neo-Assyrian corpus says in reference to the king's own rhetoric, "is not this (skillful rhetoric) the epitome of the scholarship/knowledge" (*la annû šû le'ûtu ša ṭupšarrūti*).[19] Simo Parpola dates this letter to the early years of Assurbanipal, a king whose claims to literacy seem to be borne out.[20]

---

18 See Francesca Rochberg, "Mesopotamian Science," in A. Jones and L. Taub (eds.), *Cambridge History of Ancient Science* (Cambridge: Cambridge University Press, forthcoming).
19 ABL 1277 rev. 9, see Parpola, LAS I, Text 318, pp. 274–75; Parpola SAA 10, no. 30, for a slightly different translation, and CAD s.v. *ṭupšarrūtu*.
20 Cf. Alisdair Livingstone, "Ashurbanipal: Literate or Not?," *Zeitschrift für Assyriology* 97 (2007): 98–118.

An abstract Akkadian noun "knowledge" from the verb "to know" *idû* occurs only in lexical lists (**idūtu/edūtu* = giš.á.zu.zu[21]) and nowhere in actual use in Akkadian context, while the abstract *mudûtu* from the verb *idû* is used to refer to knowledge as information in a more restricted sense than "the sum of what is known." Illustrative of this restricted sense in the Akkadian usage is a passage from the Code of Hammurabi that says, "the witnesses who know the lost property will tell what they know (*mudûšunu*, literally, "their knowledge") before the deity."[22] An adjective *mudû*, derived from the same verb *idû*, means "knowing," as in "expert in a specific craft, etc., wise, competent, learned, knowledgeable, expert,"[23] and is said of gods, kings, and artisans. This term is used in colophons to late Babylonian astronomical texts where interdictions against showing such texts to one "who does not know" (*la mūdû*) occur, expressed as "(only) the one who knows may show it (i.e., the tablet containing such knowledge) to another one who knows. The one who does not know may not see (it)."[24]

In the realm of textual learning, to be knowledgeable meant being skilled in the repertoire. Being knowledgeable was said in reference to the contents of the texts containing "wisdom" (*nēmequ*). The Akkadian term *nēmequ* "wisdom," or "knowledge," is found in colophons to scholarly texts and is associated principally with two deities, Ea, god of wisdom and esoteric knowledge par excellence (called *bēl nēmeqi* "lord of wisdom"), and Nabû, god of the scribal craft (*nēmeq Nabû tikip santakki* "the wisdom of Nabû, the cuneiform signs"[25]). As mentioned before, King Assurbanipal is said to have acquired literacy and knowledge, to have "comprehended the wisdom of Ea ... the entire craft of the scribes."[26] Specific mention of the nature of these esoteric texts includes the divinatory corpus of extispicy (*bārûtu* "inspection [of the liver]," also designated as *nēmeq Šamaš u Adad* "the wisdom of Shamash and Adad [lords of divination]"), celestial divination (UD.SAR ᵈ*Anum* ᵈEN.LÍL.LÁ "the series *Enūma Anu Enlil*"), and the craft of the exorcist (*kakugallūtu*[27]). This was a knowledge associated with training and expertise, hence characterized by another adjective *emqu*, meaning "experienced, skilled, educated." Accordingly, the object of knowledge was not strictly

---

21 CAD s.v. **idūtu*.
22 CH §9:35, see CAD s.v. *mudûtu*, usage a.
23 CAD s.v. *mudû*.
24 See, for example, *piristi ilāni rabûti mūdû mūdâ likallim la mūdû aj īmur*, in Hermann Hunger, *Babylonische und assyrische Kolophone*, Alter Orient und Altes Testament 2 (Kevelaer: Butzon & Bercker; Neukirchen-Vluyn: Neukirchener Verlag des Erziehungsvereins, 1968), no. 206.1.
25 CAD s.v. *nēmequ*, usage b, CT 27_38.43, and Hunger, *Kolophone*, no. 319.6.
26 CAD s.v. *nēmequ*, usage b, Hunger, *Kolophone*, no. 330.4 and 331.5.
27 See CAD s.v.

speaking the world of observed phenomena. Rather, it was the contents of the scholarly textual corpora.

Where classificatory statements are occasionally preserved, for example on Late Babylonian astronomical tablets, the scribes classified divinatory texts as "secret knowledge" (*pirištu*) accessible only to an exclusive group of literati, the *ummânū mūdû*, "learned masters." A Late Babylonian text giving, among other things, rules for calculating month lengths and intervals of lunar visibility around the full moon, begins with the statement: [*ṭu*]*ppi niṣirti šamê pirištu ilāni rabûti* ("tablet of the secret of heaven, secret knowledge (*pirištu*) of the great gods").[28] Other texts designated with the term *pirištu* are principally divinatory texts, including provoked divination by inspection of the exta and unprovoked divination by observation of physical and other phenomena.

Instead of focusing on a direct relation between knowledge and observation of the physical world, the Babylonians focused on the exclusivity of knowledge resulting from its origins with the divine and the transmission of the divine knowledge to humankind in a remote mythological time. A particularly vivid text describes the god's gift of knowledge directly to a king, here the god Shamash to King Enmeduranki of Sippar, and also stipulates that future savants will teach their sons "the secrets of the great gods" but only after an oath to the divinatory patron deities Shamash and Adad is sworn by the young scribe:

> Shamash in Ebabbarra [appointed] Enmeduranki [king of Sippar], the beloved of Anu, Enlil, and Ea. Shamash and Adad [brought him in] to their assembly ... and showed him how to observe oil on water, a mystery of Anu, Enlil and Ea, they gave him the tablet of the gods, the liver, a secret of heaven and the underworld. ... Then he in accordance with their word brought the men of Nippur, Sippar, and Babylon into his presence, ... showed them how to observe oil on water, a mystery of Anu Enlil and Ea, he gave them the tablet of the gods, the liver, a secret of heaven and underworld, ... The learned savant (*ummânu*, LÚ.UM.ME.A) who guards the secret of the great gods, will bind his son whom he loves with an oath before Shamash and Adad by tablet and stylus and will instruct him.[29]

As mentioned previously, injunctions against scribes who were not among the privileged with access to texts classified as *pirištu*[30] (or *niṣirtu* "secret" or "guard-

---

[28] BM 42282+42294 obv. 1, see Lis Brack-Bernsen and Hermann Hunger, "BM 42282+42294 and the Goal-Year Method," *SCIAMVS* 9 (2008): 3–23, at 6.

[29] K. 2486+ 1–22, with ellipses, see Wilfred G. Lambert, "Enmeduranki and Related Matters," *Journal of Cuneiform Studies* 21 (1967): 126–38, at 132.

[30] The word *pirištu* has been assumed to derive from the root *prs* "to cut" or "divide," and as well, "to decide," a verb associated with the divine decisions at the basis of divination. The derivation of *pirištu* from *parāsu* furthermore underscores the exclusivity of this form of knowledge, separate from all other forms. Recently, however, Alan Lenzi has discussed the

ed") are known from the Middle Babylonian and Middle Assyrian to the Late Babylonian periods, roughly 1500 to the first century B.C.[31]

The principal sources for a study of cuneiform knowledge derive from the corpora of five scholarly disciplines. The members of these disciplines were identified as *ummânu*, the honorific meaning "savant," "expert," or "master," who functioned in various official capacities within the court, libraries, or archives of the state and/or the temple. The *ummânū* were further specified as the *ṭupšar Enūma Anu Enlil*, expert in astral phenomena, the *bārû*, expert in divination by extispicy, the *āšipu*, who used incantations and rituals to treat human beings afflicted by divine disfavor, the *asû* who, like the *āšipu* treated the body in the grip of demonic or divine influence (both were thought to bring disease) by means of herbal remedies, and the *kalû* who was responsible for religious ritual performance (often songs of lamentation) and probably the calendrical regulation of ritual. Rigid distinctions did not apply between these scribal specializations and the texts they wrote, copied, and utilized. In the Late Babylonian period, omens (including astral, abnormal birth, and human physiognomic omens) and astronomical texts are, for example, within the professional domain of both *āšipu*s and *kalû*s.

Within the framework of scholarly corpora, knowledge of the phenomena may thus be divided into three areas of thought and activity reflected in the various texts of the five disciplines: 1) divinatory, including but not limited to astral divination, 2) empirical, and 3) predictive astral sciences. Many smaller divisions could be justified. These broad classifications conceal a greater diversity of interrelated forms of knowledge and practice. For example, the divinatory sciences were supplemented by an equally systematized body of incantational and ritual texts whose aim was the establishment, or re-establishment if disrupted, of the proper order and relationship between human and divine. If omen texts have a place in the landscape of the history of science, so too does this, what we call, magical literature.

Late Babylonian scholarly texts demonstrate much innovation in the divinatory literature. Whereas older forms of omen, magical, ritual, and divinatory texts are still preserved, new kinds of astral omens, such as zodiacal or nativity omens, as well as astrological medicine and horoscopy appear after 500 B.C.

---

possibility that the underlying root is in fact *pzr* "to hide" or "be hidden." This would argue for the meaning "secret (knowledge)" over "exclusive (knowledge)." See Alan Lenzi, *Secrecy and the Gods: Secret Knowledge in Ancient Mesopotamia and Biblical Israel*, State Archives of Assyria Studies 19 (Helsinki: Helsinki University Press, 2008), 64–66.

31 See Paul-Alain Beaulieu, "New Light on Secret Knowledge in Late Babylonian Culture," *Zeitschrift für Assyriologie* 82 (1992): 98–111, and Lenzi, *Secrecy and the Gods*.

Methods for predicting astral phenomena also take new directions. An older, so-called goal year method used observations from available records to project the phenomena of the moon and planets for a target or goal year. Though the extant "goal-year texts" all date from the Seleucid period, the method used in these texts is already attested in the sixth century, as Lis Brack-Bernsen and Hermann Hunger have shown with respect to the so-called lunar sixes, i.e., the various intervals between the risings and or settings of the moon and sun around conjunction and opposition.[32] Indication that this method may have roots in the Neo-Assyrian Period, at least that some of the later attested goal-year periods existed at that time, comes from the texts where omens are coded in numbers, the cryptic DT 72 and 78, originally published by C. J. Gadd.[33] Some of the later known goal-year periods, such as 71 years (associated with Jupiter), 59 years (associated with Saturn), or 8 years (associated with Venus) appear in these texts as numbers of years for various human activities or experiences: he made firm (for) 59 yrs; he lived long (for) 15 years; he rejoiced (for) 12 years; he set in order (for) 8 years.[34]

A different methodology from that of goal-year texts applied arithmetical functions to generate positions and dates of planetary and lunar phenomena, the so-called Systems A and B. What the predictions of phenomena in these tables were for is strictly a matter for us to infer. If it was for divinatory purposes – and we should include horoscopy in that category – this goal is not made explicit in the tables. One suspects that divination with its particular cultural and epistemic values cannot possibly account for all innovation in Late Babylonian astronomy, but prediction of celestial phenomena did not begin only in the Late Babylonian period in a discrete scientific world unrelated to previous tradition. Already within the context of *Enūma Anu Enlil* astronomical prediction was a part of divination. *Enūma Anu Enlil* Tablet 14,[35] giving durations of lunar visibility day by day for the equinoctial months, as well as the coefficients for calculating the visibility in other months, attests to the early construction of predictive schemes to be used in the context of omen divination. Late innovations in astronomical prediction in the form of goal-year methods and the computational schemes of Systems A and B can be seen as continuous with this early effort to understand and control not only what could be prognosticated from the signs, but the signs themselves.

---

32 See Brack-Bernsen and Hunger, "BM 42282+42294."
33 C. J. Gadd, "Omens Expressed in Numbers," *Journal of Cuneiform Studies* 21 (1967): 52–63.
34 71 years = 6 tropical yrs + 65 synodic periods; 59 years = 2 tropical years + 57 synodic periods; 8 years = 8 tropical years + 5 synodic periods.
35 Farouk Al-Rawi and Andrew George, "Enūma Anu Enlil XIV and Other Early Astronomical Tables," *Archiv für Orientforschung* 38–39 (1991–1992): 52–73.

While divinatory texts are generally separated by content (celestial, physiognomic, birth, etc.), their interconnections begin to be made more explicit in the last half of the first millennium. During this period a variety of texts establish an intimate connection between the celestial and the terrestrial realms. This is not an entirely new conception either as the complementarity between the two realms of signs is made plain in the so-called Diviner's Manual from the Nineveh Library.[36] The concern in this instructional text is to have the diviner predict astral signs and to establish properly the dates and timing of the signs in accordance with the calendar so that apotropaic rituals might be effectively performed.

The integration of astral with terrestrial divinatory knowledge seems to intensify after the development of birth astrology, that is, by the meaning of celestial signs for the human being (and the human body) in general, no longer focusing, as did *Enūma Anu Enlil*, predominantly on the king. In a commented text, human appearance, health, and births were brought into explicit relation with the stars. Its opening lines: (The omen series) "*If an izbu*," (the omen series) "*Symptoms*," (the omen series) "*Physical Characteristics*." Aries, Taurus, Orion are for predicting the appearance (*qabi lānu*). When they (the planets?) reach (the various zodiacal signs) it refers to physical characteristics. Preserve the secret of heaven and earth!"[37] The laconic nature of the commentary leaves open the question of exactly what the connection between birth, medical, physiognomic, and astral phenomena were, yet a decidedly genethlialogical turn has been taken. The reference to the planets reaching, presumably, the zodiacal signs, expressed with the phrase *kî ikšudu* "when they reach," relates no doubt to an important feature of later astronomical texts, the Diaries and Almanacs, which was to track when the planets entered ("reached") each zodiacal sign. The colophons of Almanacs identify these texts as "measurements of the 'reachings' of the (divine) planets" (*mešhi ša kašādī ša ^{d}bibbī*). Though the statements of the "reachings" in Diaries and Almanacs do not hint at an astrological meaning, the arrival of a planet into a sign must have been astrologically significant. Horoscopes seem to confirm this by taking the zodiacal positions of the planets from Diaries and/or Almanacs, most likely from the sections in which the planetary "reachings" are enumerated.[38]

---

[36] A. Leo Oppenheim, "A Babylonian Diviner's Manual," *Journal of Near Eastern Studies* 33 (1974): 197–220.

[37] See Barbara Böck, "'An Esoteric Babylonian Commentary' Revisited," *Journal of the American Oriental Society* 120 (2000): 615–20.

[38] Francesca Rochberg, "Babylonian Horoscopy: The Texts and their Sources," in *In the Path of the Moon*, 206–207.

Astrology finds a connection to extispicy in the Late Babylonian Period as well. A late Uruk text associates traditionally ominous parts of the liver with a god, one of the twelve months, and a heliacally rising star, thus: "the path (of the liver) is Šamaš, Ajāru, Taurus; the gall bladder is Anu, Tašrītu, Libra," and so on.[39] The term "sign" or "ominous part" – literally "flesh" UZU = šīru, used in liver omens – is found again in a Seleucid astrological context where the term refers not to ominous parts of the liver, but to ominous parts of zodiacal signs, thus 12 UZU.MEŠ HA.LA šá "the division of" sign such-and-such. Magical practice also establishes new connections to the zodiac, such as in a list of spells with their correlated "regions" in the zodiacal signs, for example "changing one's mind (is the) region of Leo; overturning a judgment (is the) region of Aquarius; loosening the grasp (is the) region of Virgo."[40] These are the first three lines of a long list of correspondences between signs of the zodiac and incantations to influence various people, or demons, or gods, though the purpose or implementation of these correspondences is unstated.

To varying degrees ancient Mesopotamian knowledge was imbued with meaning relating to the gods, to prognostication from signs and to magic. Mathematical astronomy was also produced by scribes who were identified with the scribal professions that specialized in divinatory and magical knowledge. In the case of the divinatory sciences, the ultimate aim and value of the signs was to obtain knowledge of the gods' "decisions" that made divination by signs possible. Similarly, the aim of magic was to maintain or rectify divine pleasure in human activity by incantation and ritual. Late Babylonian astronomy as the study of the periodic behavior of astral phenomena, even if its furthest elaborations no longer appear on the surface to share in the cultural or epistemic values of divinatory sciences, seems nonetheless to have had its origins in the systematic attention to the phenomena as signs. The dependence of the horoscopes on other astronomical texts and their methods points to a unity, or at least an interconnectedness, of astronomical and astrological knowledge in late Babylonia.

Astronomy and the preservation by the *ummânū* of esoteric learning in magical, medical, and liturgical texts dominated the activities of the Late Babylonian

---

39 SpTU 14, 159, see Ulla Koch-Westenholz, *Babylonian Liver Omens: The Chapters Manzazu, Padanu and Pan Takalti of the Babylonian Extispicy Series Mainly from Assurbanipals's Library*, Carsten Niebuhr Institute of Ancient Near Eastern Studies 25 (Copenhagen: Museum Tusculanum Press, 2000), 24–25.
40 BRM 4 20, with BRM 4 19 (and parallels with STT 300 although without the zodiacal references), see the discussion in M. J. Geller, *Look to the Stars: Babylonian Medicine, Magic, Astrology, and Melothesia*, Max-Planck-Institut für Wissenschaftsgeschichte Preprint 401 (Berlin: Max-Planck-Institut für Wissenschaftsgeschichte, 2010), 25–54.

literati in the last centuries of the cuneiform tradition. The astronomical content of the mathematical astronomical texts of the Late Babylonian Period may not appear to relate to celestial divination, yet the few surviving colophons show their copyists to be members or descendants of the expert classes of scribes, the *ṭupšar Enūma Anu Enlil*s, *kalû*s, and *āšipu*s. As already mentioned, a few preserved rubrics indicate that astronomical ephemerides were classed by the scribes along with other texts as *pirištu*, "secret" or "exclusive," knowledge of the great gods. Indeed, the importance of the divine in all the activities associated with cuneiform science seems inescapable, just as the insistence upon God's role and relationship to the physical world was integral to medieval and early modern science. Andrew Cunningham, in an effort to differentiate premodern from modern science observed that natural philosophy was fundamentally about "God's achievements, God's intentions, God's purposes, God's messages to man."[41]

The cuneiform knowledge corpus, essentially consisting of the interpretation of texts wherein the interpretation of divine messages in the form of signs in natural phenomena were recorded, belongs to the history of hermeneutics, in Harrison's terms, "broadly conceived as the general science of interpretation."[42] From a philological perspective, focusing on scribal commentaries that seek to explain or interpret words in the transmitted textual corpora, Eckart Frahm has offered a detailed study of cuneiform hermeneutics and has observed that "the enormous amount of thought that Mesopotamian scholars put into their divinatory tradition had a large impact on the rise of the Babylonian and Assyrian text commentary."[43] The interpretation of phenomena was therefore eventually accompanied by further interpretation in the form of commentaries to the texts in which the phenomena were systematically recorded.[44]

The characteristically Assyro-Babylonian scholarly emphasis on interpretation of observed, or imagined, phenomena makes for a significant continuity with later Western tradition, all the more clear when viewed from antiquity forward. A particularly cogent example of such continuity with antiquity in later European science can be seen in the persistent centrality of astrology and divination into the Renaissance and early modern period. The image of the world in medieval and Renaissance Europe was, of course, the bequest from diverse an-

---

[41] Andrew Cunningham, "Getting the Game Right: Some Plain Words on the Identity and Invention of Science," *Studies in History and Philosophy of Science* 19 (1988): 365–89, at 384.
[42] Harrison, *Bible*, 9.
[43] Eckart Frahm, *Babylonian and Assyrian Text Commentaries: Origins of Interpretation*, Guides to the Mesopotamian Textual Record 5 (Münster: Ugarit-Verlag, 2011), 23.
[44] For a typology of the cuneiform commentaries, see Frahm, *Commentaries*, 28–58.

cient sources, from Plato, Aristotle, the Stoics, and the Bible, and with these sources some traces as well of what were originally ancient Near Eastern ideas, perhaps most notably, the very idea of celestial prognostication. The ancient image, or ancient images, of the world had long supported the claim to influence from the stars, be it by physical rationale via planetary rays, or the more divinatory interpretive enterprise based on correspondences between macrocosm and microcosm. The persistent relevance of astrological forecasting points to continuities from antiquity to early modern science based on the search for knowledge of the meaning of celestial signs, even as a struggle over the ancient world picture, or one of them, viz., the earth-centered cosmos, was taking place.

The project of tracing back from modernity to antiquity a universal scientific attitude, in which only one particular kind of knowledge is deemed appropriate for understanding the world is not sustainable in the face of the contents, nature, and goals of many pre-modern sciences as they begin to be restored to standards of their own local knowledge. What we call science today is not what was called *scientia* in the early modern period or earlier. The English word "science" has taken on a succession of various meanings, especially since the seventeenth century, commensurate with the cultural and epistemic values held by successive groups of inquirers into the phenomena of nature. And, as Raymond Williams pointed out, "changes in ideas of Nature encouraged the further specialization of ideas of method and demonstration towards the 'external world', and the conditions for the emergence of **science** as the theoretical and methodological study of *nature* were then complete."[45] Even if many of the ominous phenomena attested in cuneiform were what we would categorize as "natural phenomena," for the world of ancient Mesopotamia, *nature* per se was not the object of knowledge, lending science in the cuneiform texts a different epistemic slant as compared with that of later Western traditions. Even the highly refined Babylonian mathematical astronomy, with its complex arithmetical predictive models, was still imbued with the sense of the divine astral gods and their messages to humankind via their appearances. As such, Babylonian astronomy was not engaged with nature in the same way as we understand astronomy to be today.

Though we can try to define a conception or conceptions of knowledge in the ancient Near East, no equivalent classification in cuneiform languages serves to unify particular forms of knowledge, reasoning-style, or social practice in quite the same way as does our classifier "scientific." What it was to have knowledge

---

[45] Raymond Williams, *Keywords: A Vocabulary of Culture and Society*, rev. ed. (Oxford: Oxford University Press, 1983), 278, boldface and italics in original.

in ancient Mesopotamia may only be partly translatable to later Western scientific values and practice, but it *is* in part translatable. Threads of continuity picked up from the perspective of history looking backward, such as one finds in theoretical astronomy, are different from those that show themselves more clearly by looking forward, as seen in the pervasive interest in prognostication and signs, and in the textual representation of such signs and their corresponding prognoses. But these too are legitimate elements with connections to later sciences, and the manifestation in Mesopotamia of the knowledge of signs and all their related preoccupations is necessary evidence for a more complete account of the history of science as a whole.

Bibliography

Al-Rawi, Farouk, and Andrew George. "Enūma Anu Enlil XIV and Other Early Astronomical Tables," *Archiv für Orientforschung* 38–39 (1991–1992): 52–73.

Beaulieu, Paul-Alain. "New Light on Secret Knowledge in Late Babylonian Culture," *Zeitschrift für Assyriologie* 82 (1992): 98–111.

Böck, Barbara. "'An Esoteric Babylonian Commentary' Revisited," *Journal of the American Oriental Society* 120 (2000): 615–20.

Brack-Bernsen, Lis, and Hermann Hunger. "BM 42282+42294 and the Goal-Year Method," *SCIAMVS* 9 (2008): 3–23.

Cunningham, Andrew. "Getting the Game Right: Some Plain Words on the Identity and Invention of Science," *Studies in History and Philosophy of Science* 19 (1988): 365–89.

Findlen, Paula. "The Two Cultures of Scholarship?," *Isis* 96 (2005): 230–37.

Frahm, Eckart. *Babylonian and Assyrian Text Commentaries: Origins of Interpretation*, Guides to the Mesopotamian Textual Record 5 (Münster: Ugarit-Verlag, 2011).

Gadd, C. J. "Omens Expressed in Numbers," *Journal of Cuneiform Studies* 21 (1967): 52–63.

Geller, M. J. *Look to the Stars: Babylonian Medicine, Magic, Astrology, and Melothesia*, Max-Planck-Institut für Wissenschaftsgeschichte Preprint 401 (Berlin: Max-Planck-Institut für Wissenschaftsgeschichte, 2010).

Grafton, Anthony. *Cardano's Cosmos* (Cambridge, Mass.: Harvard University Press, 1999).

Grafton, Anthony. "Starry Messengers: Recent Work in the History of Western Astrology," *Perspectives on Science* 8 (2000): 70–83.

Gurney, O., and J. J. Finkelstein. *The Sultantepe Tablets*, 2 vols. (London: British Institute of Archaeology at Ankara, 1957–1964).

Harrison, Peter. *The Bible: Protestantism and the Rise of Natural Science* (Cambridge: Cambridge University Press, 1998).

Horowitz, Wayne. "Astral Tablets in the Hermitage Saint Petersburg," *Zeitschrift für Assyriologie* 90 (2000): 194–206.

Hunger, Hermann. *Babylonische und assyrische Kolophone*, Alter Orient und Altes Testament 2 (Kevelaer: Butzon & Bercker; Neukirchen-Vluyn: Neukirchener Verlag des Erziehungsvereins, 1968).

Lambert, Wilfred G. "Enmeduranki and Related Matters," *Journal of Cuneiform Studies* 21 (1967): 126–38.

Lenzi, Alan. *Secrecy and the Gods: Secret Knowledge in Ancient Mesopotamia and Biblical Israel*, State Archives of Assyria Studies 19 (Helsinki: Helsinki University Press, 2008).

Livingstone, Alisdair. "Ashurbanipal: Literate or not?," *Zeitschrift für Assyriology* 97 (2007): 98–118.

Magee, John. *Boethius on Signification and Mind* (Leiden: Brill, 1989).

Neugebauer, Otto. *A History of Ancient Mathematical Astronomy*, 3 vols. (Berlin: Springer, 1975)

Oppenheim, A. Leo. "A Babylonian Diviner's Manual," *Journal of Near Eastern Studies* 33 (1974): 197–220.

Pingree, David. *From Astral Omens to Astrology: From Babylon to Bikaner,* Serie Orientale Roma 78 (Rome: Istituto Italiano per l'Africa e l'Oriente, 1997).

Rheinberger, Hans-Jörg. *On Historicizing Epistemology: An Essay* (Stanford: Stanford University Press, 2010).

Rochberg, Francesca. *In the Path of the Moon,* Studies in Ancient Magic and Divination 6 (Leiden: Brill, 2010).

Rochberg, Francesca. "Mesopotamian Science," in A. Jones and L. Taub (eds.), *Cambridge History of Ancient Science* (Cambridge: Cambridge University Press, forthcoming).

von Staden, Heinrich. "Affinities and Elisions: Helen and Hellenocentrism," *Isis* 83 (1992): 578–95.

Williams, Raymond. *Keywords: A Vocabulary of Culture and Society*, rev. ed. (Oxford: Oxford University Press, 1983).

Amneris Roselli
Galeno sull'autenticità del *Prorretico I*

Abstract: The *Commentary on Prorrhetic I* happens to be the only commentary Galen wrote on a work much appreciated in his time but which he himself considered as spurious and much inferior, at the level of doctrine, to the authentic writings of Hippocrates. My article analyzes the philological criteria employed by Galen to demonstrate the spurious character of *Prorrhetic I* and to show that its compiler has used and interpreted from an Empiricist perspective Hippocrates' *Epidemics*.

I Premessa

Il Commento di Galeno al *Prorretico I*[1] non ha avuto fortuna nella recente letteratura critica sui commenti antichi che si è preferibilmente rivolta allo studio di commenti ad altri trattati ippocratici.[2] Questa mancanza d'attenzione si

---

Dedico con molto piacere questo lavoro ad Heinrich von Staden, che nel corso degli anni, ad Arcavacata, a Pisa e specialmente durante il semestre che ho trascorso all'Institute for Advanced Study di Princeton, è stato per me guida generosa, sollecita e amichevole ad una più profonda comprensione della letteratura scientifica antica; ai suoi lavori su Galeno interprete di Ippocrate questa ricerca deve molto di più di quanto appaia dalle singole note. Ringrazio per le loro preziose osservazioni Philip van der Eijk e i membri del gruppo berlinese che lavora sul progetto *Medicine of the Mind and Philosophy of the Body*, tra i quali ho trascorso un periodo di ricerca generosamente sostenuto dall'Alexander von Humbolt-Stiftung e completato questo lavoro.

1 Il *Prorretico I* è costituito da 170 brevi unità testuali di contenuto prognostico che definisco aforismi. Nel corso di questo lavoro si utilizzeranno le edizioni di Potter (P. Potter, *Hippocrates*, vol. 8 [Cambridge, Mass.: Harvard University Press, 1995], 165–209) e di Littré (É. Littré, *Œuvres complètes d'Hippocrate*, vol. 5 [Paris: J. B. Baillière, 1846], 510–72). I lemmi ippocratici sono citati così come appaiono in Galeno nell'edizione Diels (H. Diels, *Galeni In Hippocratis Prorrheticum I*, CMG V 9,2 [Leipzig: Teubner, 1915], 3–178); quando necessario, si darà notizia delle divergenze tra tradizione ippocratica e tradizione galenica; le traduzioni italiane dei testi citati in questo lavoro sono sempre mie. Il Commento di Galeno è diviso in tre libri; esso è contemporaneo alla stesura del commento ad *Epidemie III* (infra, n. 21 e 26), e dunque fu redatto nel secondo periodo del soggiorno di Galeno a Roma.

2 Cf. per il commento a *Natura dell'uomo*, J. Jouanna, "La lecture du traité hippocratique de la Nature de l'homme par Galien," in M. O. Goulet-Cazé (ed.), *Le commentaire entre tradition et innovation; actes du Colloque international de l'Institut des traditions textuelles, Paris-Villejuif 1999* (Paris: Vrin, 2000), 273–92 (= "Galen's Reading of the Hippocratic Treatise *The Nature of Man*," in J. Jouanna, *Greek Medicine from Hippocrates to Galen: Selected Papers*, trans. by N. Allies, ed. with a preface by P. J. van der Eijk [Leiden: Brill, 2012], 313–33); per il commento ad

spiega facilmente: non solo il testo commentato non fa parte della rosa dei testi ippocratici oggi più letti, ma Galeno stesso ha fatto del suo meglio per isolare il *Prorretico I* dagli altri scritti ippocratici, in particolare mostrando la sua inferiorità dottrinale rispetto al *Prognostico* e alle *Epidemie I* e *III* (che, a suo giudizio, rappresentano il livello più alto raggiunto nel campo della raccolta dei segni clinici e della loro valutazione prognostica), e dichiarando la sua non autenticità. Ma, condannando il *Prorretico*, Galeno ha fornito riflessioni esplicite sul suo metodo di analisi filologica e linguistica dei testi e un interessante ed ampio documento della critica di autenticità.

Alla disattenzione dei moderni nei confronti del *Commento al Prorretico* questo lavoro vuole in parte rimediare, studiandolo come documento sui criteri di valutazione dell'autenticità degli scritti ippocratici applicati da Galeno e insieme come testimone di una lettura del *Prorretico* in chiave empirica.

II Galeno sull'autenticità degli scritti di Ippocrate

L'autenticità degli scritti ippocratici era per Galeno un problema non solo ineludibile, essendo già stato sollevato dalla filologia ippocratica precedente, ma anche cruciale per la definizione corretta dell'autentica dottrina di Ippocra-

---

Aforismi, e più in generale per il genere commento, H. von Staden, "'A Woman Does Not Become Ambidextrous': Galen and the Culture of Scientific Commentary," in R. Gibson e C. Kraus (edd.), *The Classical Commentary: Histories, Practices, Theory* (Leiden: Brill, 2002), 109–39, e idem, "Interpreting Hippocrates," in C. W. Müller, C. Brockmann, e C. W. Brunschön (edd.), *Ärzte und ihre Interpreten. Medizinische Fachtexte der Antike als Forschungsgegenstand der Klassischen Philologie*, Fachkonferenz zu Ehren von Diethard Nickel (München: Saur, 2006), 15–47; per il commento a *Arie, acque, luoghi*, ma in un'ottica più generale, G. Strohmaier, "Galen als Hippokrateskommentator," in W. Geerlings e C. Schulze (edd.), *Der Kommentar in Antike und Mittelalter. Beiträge zu seiner Erforschung* (Leiden: Brill, 2002), 253–74; per il commento a *Epidemie I* e *II*, P. J. van der Eijk, "Exegesis, Explanation and Epistemology in Galen's Commentaries in Epidemics, Books One and Two," in P. E. Pormann (ed.), *Epidemics in Context: Greek Commentaries on Hippocrates in the Arabic Tradition* (Berlin: De Gruyter, 2012), 25–48. Il commento al *Prorretico I* non aveva però mancato di suscitare l'attenzione di L. O. Bröcker, "Die Methoden Galens in der literarischen Kritik," *Rheinisches Museum* 40 (1885): 426–28. Si vedano anche V. Nutton, *Galen, On Prognosis*, CMG V 8,1 (Berlin: Akademie-Verlag, 1979), 175–76 e W. D. Smith, *The Hippocratic Tradition* (Ithaca: Cornell University Press, 1979), 155–60.

te.³ All'autenticità degli scritti ippocratici Galeno ha anche dedicato, non sappiamo in quale momento della sua lunga produzione, un trattato, ora perduto: *Sugli scritti autentici e spuri di Ippocrate*.⁴ Sarà utile allora far precedere l'analisi del *Commento al Prorretico* da una rapida rassegna di passi in cui Galeno, sulla scorta dei suoi predecessori, tocca il problema dell'autenticità di altri scritti ippocratici, nel loro complesso o in alcune loro parti:⁵ basterà ricordare l'interpolazione di una breve sezione sul sudore alla fine del cap. 6 del *Prognostico*, le interpolazioni massicce in *Epidemie II* e in *Natura dell'uomo*, quelle nella parte finale degli *Aforismi*, l'ultima parte di *Regime nelle malattie acute* (il cosiddetto *acut. [sp.]*), l'intero trattatello *De glandulis*. In tutti questi casi Galeno non si limita a segnalare interpolazioni e veri e propri falsi ma motiva il suo giudizio e individua le cause dell'interpolazione e della composizione di scritti interi: p. es., il malanimo nei confronti di Ippocrate avrebbe determinato l'intrusione in *Natura dell'uomo* di passi davvero risibili per il loro contenuto dottrinale (è il caso del cap. 11 sulle vene, troppo distante dal livello di conoscenza anatomica che si trova invece nel capitolo sulle vene di *Epidemie II*). Un caso diverso è costituito dalle inserzioni che i discendenti di Ippocrate hanno fatto rielaborando i suoi materiali;⁶ le loro interpolazioni sono di segno opposto, ispirate da pietà e rispetto per il grande Ippocrate, ma non all'altezza di Ippocrate.

I criteri fondamentali per scoprire le interpolazioni sono quelli della coerenza dottrinale o delle scelte di vocabolario, che tradiscono la loro recenziorità. Basterà fornire un paio di esempi. Nel suo commento ad *Articolazioni* 11 (dove Ippocrate annuncia un trattato sulle ghiandole: περὶ ἀδένων

---

3 Cf. I. Sluiter, "The Embarrassment of Imperfection: Galen's Assessment of Hippocrates' Linguistic Merits," in P. J. van der Eijk, H. F. J. Horstmanshoff, e P. H. Schrijvers (edd.), *Ancient Medicine in its Socio-cultural Context*, 2 voll. (Amsterdam: Rodopi, 1995), 2:519–35.
4 Cf. Gal. *in Hp. nat. hom.* 15.9.15–16 K. (= 7.19–20 Mewaldt): περὶ τῶν γνηcίων τε καὶ νόθων Ἱπποκράτουc cυγγραμμάτων. Quel che sappiamo sulle opinioni degli antichi a proposito dell'autenticità degli scritti ippocratici dipende in larghissima misura da Galeno e dai suoi commenti. Sul tema cf. D. Manetti, "Philology and Medicine: A Strategic Relationship," in S. Matthaios, F. Montanari, e A. Rengakos (edd.), *The Brill's Companion to Ancient Scholarship* (in corso di stampa).
5 La documentazione è stata recentemente raccolta da A. Anastassiou e D. Irmer, *Testimonien zum Corpus Hippocraticum*, voll. 1, 2.1, 2.2, 3 (Göttingen: Vandenhoeck & Ruprecht, 1997–2012), nella sezione A relativa a ognuno dei trattati ippocratici. Particolare interesse riveste per la questione dell'autenticità il proemio al quarto libro del commento ad *acut. (sp.)* (*in Hp. acut. comm.* 15.732–34 K. [= 271–72 Helmreich]).
6 Un'idea che dipende dalle pratiche di rielaborazione di materiali altrui contemporanee a Galeno.

οὐλομελίης), Galeno confuta l'autenticità del trattato con lo stesso titolo che era attribuito ad Ippocrate e che ancora oggi possediamo. Dice Galeno:

> οὐ μὴν διασῴζεται γέ τι τοιοῦτο Ἱπποκράτους, ἀλλά τις τῶν νεωτέρων Ἱπποκρατείων ἔγραφε μικρὸν βιβλίδιον ἐπιγράψας Ἱπποκράτους περὶ ἀδένων οὐλομελίης, ὃ <u>καὶ τῇ λέξει καὶ τῇ διανοίᾳ</u> λείπεται πάμπολυ τῶν γνησίων Ἱπποκράτους συγγραμμάτων. Οὐ μὴν οὐδὲ ἐμνημόνευσέ τις περὶ αὐτοῦ τῶν ἔμπροσθεν ἰατρῶν, ἀλλ' οὐδ' οἱ τοὺς πίνακας ποιήσαντες ἴσασιν τὸ βιβλίον. (in Hp. art. comm. 18A.379.6–14 K.)
>
> Non è conservato un tale trattato di Ippocrate ma uno degli ippocratici recenti ha scritto un libriccino con l'intestazione: "Di Ippocrate, sulla natura generale delle ghiandole," un libro che <u>e per lingua e per dottrina</u> è di molto inferiore agli autentici scritti di Ippocrate. E del resto nessuno dei medici precedenti ha fatto menzione di questo libro né lo conoscono coloro che hanno redatto i *pinakes*.

Trattandosi di un libro intero, per quanto breve, Galeno può addurre come prova della non autenticità elementi sia esterni al testo, come l'assenza dai cataloghi (πίνακες) e il fatto che non ne facciano menzione i medici antichi, sia argomenti interni, e cioè la non corrispondenza di lingua e dottrina con la lingua e la dottrina di Ippocrate.[7] Il nesso lingua e dottrina (λέξει καὶ διανοίᾳ) ha una tradizione autorevole che risale alla *Poetica* di Aristotele, cf. 1450a29–30, dove è riferito alle *rheseis* tragiche ben costruite sia per la *lexis* sia per la *dianoia* (ῥήσεις λέξει καὶ διανοίᾳ εὖ πεποιημένας), e 1459b16, dove *lexis* e *dianoia* sono i campi specifici in cui si manifesta l'eccellenza di Omero (λέξει καὶ διανοίᾳ πάντα ὑπερβέβληκεν).[8]

Galeno invoca i criteri relativi a *lexis* e *dianoia* anche per dirimere la questione dell'autenticità nel commento a *Regime nelle malattie acute*, a proposito dell'autenticità di *acut. (sp.)*. In quel caso Galeno distingue anche, in forma piuttosto astratta, le quattro diverse possibilità di inadeguatezza di un testo rispetto a ciascuno dei due criteri: testo degno di Ippocrate per lingua e per

---

[7] Sull'autenticità del *de glandulis* si sono misurati anche i filologi moderni; si vedano i dettagliati resoconti di R. Joly, *Hippocrate, Du système des glandes*, t. XIII (Paris: Les Belles Lettres, 1978), 104–11, che propende per la non autenticità, e di Potter, *Hippocrates*, 103–104, che invece è decisamente convinto dell'antichità dello scritto, proprio per il suo contenuto e per la lingua. Potter non manca di rilevare il carattere spesso strumentale della filologia di Galeno che toglie il crisma dell'autenticità a tutti gli scritti che contraddicono l'immagine di Ippocrate e dell'autentica dottrina ippocratica che egli si è fatta e che intende trasmettere.

[8] Non fa parte di questa ricerca ricostruire la storia della relazione *lexis–dianoia* nella letteratura precedente Galeno, ma vale almeno la pena di ricordare il ruolo che la relazione tra i due termini gioca nel proemio di Erotiano (29.10–16 Nachmanson) proprio per la retta comprensione di Ippocrate.

dottrina, o degno di Ippocrate solo per la lingua, o solo per la dottrina, o per nessuna delle due.⁹

Come è ovvio, nell'ottica di un commento destinato alla comprensione del suo contenuto medico lingua e dottrina non si collocano sullo stesso piano. Nel commento a *Epidemie VI* Galeno osserva più volte, a proposito di alcune varianti attestate nei suoi manoscritti o dai commentatori, che esse toccano la *lexis* ma non modificano la *dianoia* (e in questo caso egli mostra un certo disinteresse per la scelta della variante da preferire);¹⁰ lo stesso accade nel commento al *Prognostico*,¹¹ in quello a *Fratture*,¹² e a *Officina del medico*¹³ e anche, sebbene più sporadicamente, in quello al *Prorretico*.¹⁴ È del tutto evidente che, nonostante l'interesse di Galeno per la lingua di Ippocrate, e in particolare per la lingua attica, la sua attenzione è concentrata sulle varianti che modificano il significato. L'aderenza alla lingua ippocratica è funzionale all'esigenza di stabilire l'autenticità dei testi piuttosto che un obiettivo da raggiungere mediante la restituzione della lezione originale.

Per quanto riguarda la lingua, il criterio principale secondo cui Galeno giudica una parola come non ippocratica è quello della sua rarità e della sua recenziorità,¹⁵ che gli pare provata quando la parola è assente negli altri scritti ippocratici autentici: p. es., i termini οὐρήματα ("urine") e σύνοχος ("febbre continua") che si trovano in *nat. hom.* 14 e 15 – e sono anche per noi due *hapax* ippocratici come lo erano per Galeno –, sarebbero entrati in uso solo recentemente;¹⁶ il criterio ci appare oggi non sempre condivisibile ed è stato criticato.¹⁷

---

9 Cf. Gal. *in Hp. acut. comm.* 15.733.2–4 K. (= 271.9–11 Helmreich) e 15.733.17–734.4 K (= 271.22–273.3 Helmreich).
10 Cf. *in Hp. epid. VI comm.* 17A.800.12–13 K. (= 7.13–14 Wenkebach); 17A.975.17–976.3 K. (= 104.16–17 Wenkebach); 17A.1006.2–3 K. (= 121.29–30 Wenkebach); 17A.1008.11–14 K. (= 123.12–14 Wenkebach); 17B.131.7–9 K. (= 195.11–14 Wenkebach), dove occuparsi della *lexis* è considerato pura sciocchezza (λεπτολογία); 17B.310.5 (= 320.13 Wenkebach).
11 Cf. *in Hp. prog. comm.* 18B.104.2–3 K. (= 255.2–3 Heeg).
12 Cf. *in Hp. fract. comm.* 18B.425.15–17 K.; 18B.438.5 K. (dove però la variante comporterebbe oscurità sui due piani della *lexis* e della *dianoia*).
13 Cf. *in Hp. off. comm.* 18B.684.12–13 K., dove Galeno osserva che la *dianoia* del *technikos* è resa poco perspicua dalla *lexis* non chiara con cui egli si è espresso: οὐ σαφῆ τῇ λέξει τὴν ἑαυτοῦ διάνοιαν ἡρμήνευσεν ὁ τεχνικός.
14 Cf. *in Hp. prorrh. I comm.* 16.736.9–11 K. (= 121.21–22 Diels).
15 Gli autori del II sec. hanno chiara consapevolezza dell'evoluzione della lingua e Galeno, che parla spesso del nuovo vocabolario tecnico dei medici più recenti, ha affrontato il tema in maniera diretta nel proemio del suo *Glossario*.
16 In questo caso, certamente σύνοχος è termine usato dai medici pneumatici per designare un tipo di febbre. Si noti, *per incidens*, che anche il *De glandulis*, che Galeno giudica non autentico,

Un altro criterio per la valutazione lingua ippocratica è fornito dalla chiarezza (σαφήνεια), caratteristica propria del *Prognostico* (cf. *de diff. resp.* 7.905.10 K.) ma anche di altri scritti ippocratici, ma decisamente assente nel *Prorretico*,[18] un criterio che riguarda specialmente la sintassi, e del quale ci occuperemo più avanti. Vedremo anche che, nel caso specifico del commento al *Prorretico*, i difetti maggiori che gli vengono imputati riguardano la *dianoia*.

III Il proemio del *Commento al Prorretico I*

Il *Commento al Prorretico I* si apre con un proemio, ora mutilo all'inizio; non sappiamo quindi se e come Galeno vi avesse trattato la questione dell'autenticità dell'opera che si accinge a commentare.[19] Di esso si è conservata solo la parte finale, una ventina di righi che concludono il discorso sul significato di

---

contiene un termine *hapax* negli scritti ippocratici (λύματα, "liquidi prodotti dalla purificazione"), termine che è stato glossato da Erotiano, cf. Anastassiou-Irmer, *Testimonien*, 1:275 e 489 (il *de glandulis* però non compare nella lista iniziale di Erotiano). Per il sospetto nei confronti di parole rare cf. anche *in Hp. epid. III comm.* 17A.682–83 K. (= 130.3–4 Wenkebach): "non è consuetudine di Ippocrate servirsi di parole inconsuete e rare" (οὐ κατὰ τὴν Ἱπποκράτους εἶναι συνήθειαν ἀήθεcι καὶ cπανίοιc ὀνόμαcι χρῆcθαι).

17 J. Jouanna, *Hippocratis De natura hominis*, CMG I 1,3 (Berlin: Akademie-Verlag, 2002), 292 e 294, osserva giustamente che, per quanto non attestati in altri scritti ippocratici, cύνοχοc e οὐρήματα sono parole perfettamente compatibili con la prosa ionica, e dunque l'assenza di documentazione in età più antica non è prova di recenziorità.

18 Sulla lingua di Ippocrate, le sue caratteristiche di chiarezza, correttezza e brevità, sul vocabolario ippocratico e la sua evoluzione e sul valore della *sunētheia*, cf. Sluiter, "The Embarrassment of Imperfection"; D. Manetti, "Galeno, la lingua di Ippocrate e il tempo," in J. Barnes e J. Jouanna (edd.), *Galien et la philosophie*, Entretiens sur l'antiquité classique XLIX (Genève: Fondation Hardt, 2003), 171–228, che offre una base di partenza anche per comprendere le osservazioni di Galeno relative al *Prorretico*, ed eadem, "Galen and Hippocratic Medicine: Language and Practice," in C. Gill, T. Whitmarsh, e J. Wilkins (edd.), *Galen and the World of Knowledge* (Cambridge: Cambridge University Press, 2009), 157–74.

19 Diels ha per primo ipotizzato che il proemio sia mutilo e che sia andato perduto il primo folio dell'archetipo dei nostri mss. Pur non essendo ancora codificati i *kephalaia* degli *accessus* che saranno tipici dei commenti neoplatonici, già in Galeno alcuni dei temi proemiali sono trattati in alcuni dei proemi ai commenti, si vedano i proemi a *Fratture*, *Officina del medico*, *Natura dell'uomo*. Certo, i proemi dei commenti di Galeno sono di varia estensione e talora del tutto assenti, ma non mi pare in nessun modo possibile che questo proemio sia integro come vorrebbe Smith, *The Hippocratic Tradition*, 156 n. 78. Per una sintesi delle intenzioni di Galeno nell'affrontare questo commento si veda D. Manetti e A. Roselli, "Galeno commentatore di Ippocrate," in W. Haase (ed.), *Aufstieg und Niedergang der römischen Welt*, II, 37.2 (Berlin: De Gruyter, 1994), 1529–635, specialmente 1550–51.

πρόρρηcιc ("predizione"), da cui il titolo Προρρητικός, con la distinzione tra πρόγνωcιc ("prognosi") e πρόρρηcιc ("predizione"). Nella frase di raccordo tra il proemio e il commento dei lemmi, tuttavia, Galeno esprime un giudizio critico sulla scrittura del *Prorretico* e il proposito di porvi rimedio:

> ἀλλὰ γάρ, ὅπως ἂν ἔχῃ τὰ τῆc ἐπιγραφῆc τοῦ βιβλίου καταλιπόντες ἑτέροις ἐπιζητεῖν, ὅcοι τῶν ἔργων τῆc τέχνης οὐ πάνυ τι πεφροντίκαcιν, αὐτοὶ τῶν γεγραμμένων κατὰ τὸ βιβλίον ἕκαστον ἐπισκεψώμεθα, καθότι κἀπὶ τῶν ἄλλων ἐποιήcαμεν ἃ προεξηγηcάμεθα, τοῖc μὲν *διοριcμούc* τινας ἀναγκαίους προcτιθέντες οὐκ εἰρημένους ὑπὸ τοῦ γράψαντος, ἔνια δὲ δεικνύντες ὡς *οὐκ ἔcτι καθόλου,* κἂν τῷ χαρακτῆρι τῆc λέξεως ὡς περὶ τοιούτων ὁ cυγγραφεὺς φαίνηται διαλεγόμενος. (*in Hp. prorrh. I comm.* 16.490.10–18 K. [= 3.20–27 Diels])

e infatti, lasciando ad altri l'indagine su come dovesse essere il titolo, a quanti cioè non si curano della pratica dell'arte medica,[20] noi indagheremo sulle singole cose scritte in questo libro, come abbiamo fatto anche per gli altri libri che abbiamo già spiegato, ad alcune aggiungendo i necessari *diorismoi*, che non sono stati detti da colui che [lo] ha scritto, e dimostrando che alcune cose *non sono "in generale,"* nonostante che, per il carattere della *lexis*, sembri che l'autore parli come se si trattasse di affermazioni "in generale."

La mancanza di elementi che delimitano la portata di certe affermazioni (nel linguaggio tecnico di Galeno i *diorismoi*), non è una peculiarità del *Prorretico*, e Galeno la riscontra anche in altri scritti ippocratici senza mostrare troppo disagio per la loro assenza;[21] la confusione tra affermazioni di portata generale e

---

20 Osservazioni sulla sostanziale indifferenza di alcune scelte di lessico e formali sono frequenti negli scritti di Galeno, e spesso si riferiscono alla terminologia tecnica adottata da medici recenti.
21 Cf. Manetti e Roselli, "Galeno commentatore di Ippocrate," *passim*; spesso Galeno integra le necessarie distinzioni senza neppure notare esplicitamente che sta aggiungendo i dovuti *diorismoi*. In questo commento il problema viene evocato più volte (29 occorrenze complessive del sostantivo e di suoi derivati in forma verbale e avverbiale, positiva e negativa), cf. almeno, a proposito di Hipp. *prorrh*. I 22, 16.558.3–7 K. (= 36.31–37.2 Diels): "Il discorso sui dolori dei fianchi, poiché è stato detto così manchevolmente e senza distinzioni preliminari (*aprosdioristōs*), produce una considerevole oscurità; ma noi affronteremo il commento curandoci di insegnare con un insegnamento chiaro quello che da lui è stato detto in maniera oscura" (Ἑλλιπῶς οὕτως καὶ ἀπροcδιορίcτως ῥηθεὶς ὁ περὶ τῶν ἀλγημάτων τῶν πλευρῶν λόγος ἀcάφειαν οὐ τὴν τυχοῦcαν ἐμποιεῖ. ἀλλ' ἡμεῖς cπουδάζοντες ἐν cαφεῖ διδαcκαλίᾳ τὰ πρὸς αὐτοῦ ἀcαφῶς εἰρημένα διδάξαι τῆc ἐξηγήcεως ἐφαψώμεθα); e il commento al primo lemma del II libro (*prorrh*. I 37), 16.588.17–589.1 K. (= 52.1–2 Diels): "io dunque sempre aggiungo le distinzioni preliminari (*prosdiorizomai*) alle cose che nel libro sono dette senza distinzioni (*adioristōs*) e che, per questo motivo, recheranno danno a coloro che le leggeranno" (ἐγὼ μὲν οὖν ἀεὶ προcδιορίζομαι τὰ κατὰ τὸ βιβλίον ἀδιορίcτως εἰρημένα καὶ διὰ τοῦτο μέλλοντα βλάψαι τοὺς ἀναγνωcκόντας αὐτά). Poco tempo dopo, nel commento a *Epidemie III* (17A.580.5 K. [= 63.16 Wenkebach]), Galeno ripete queste parole quasi alle lettera: "dopo che avevo ormai completato i

affermazioni particolari, invece, è più raramente riscontrabile negli altri scritti ippocratici, e, secondo Galeno, è uno dei difetti peculiari del *Prorretico*. Questo libro, infatti, come Galeno ripeterà altre volte, è stato scritto da qualcuno che si è servito delle osservazioni raccolte nelle *Epidemie*, attribuendo validità generale ad osservazioni particolari, da qualcuno che ha raccolto in una stessa "sindrome" (cυνδρομή), intrecciandoli tra loro, sintomi non omogenei, ed ha così costruito un libro prognostico oscuro e confuso.

L'idea soggiacente a questo giudizio è che il materiale ippocratico sia stato rielaborato da medici che hanno scritto loro testi autonomi, con l'intento di restare sulla linea autenticamente ippocratica; si tratta di un'operazione che a Galeno appare legittima; il *Prorretico* però è opera di un empirico – non necessariamente di un appartenente alla scuola empirica[22] –, che aspira ad essere dogmatico ("si diffonde al modo degli empirici [ἐμπειρικῶc]... non essendo in grado di ricondurre al generale le sue osservazioni al modo dei logici [λογικῶc]"),[23] come per l'appunto era Ippocrate, ma che proprio nella scrittura si è tradito e non è riuscito a sfuggire all'analisi critica di Galeno e di alcuni dei commentatori precedenti.[24]

---

commenti al secondo libro delle *Epidemie*, nel frattempo mi è accaduto di scrivere tre commenti al *Prorretico* perché, come dicevo, alcuni mi invitavano a farlo; in essi ho mostrato che in quel libro la maggior parte delle cose sono dette male, e molte sono dette come generali, pur essendo rare, e molte senza distinzioni [ἀδιόριcτα], pur essendo bisognose di distinzioni [πολλὰ δ' ἀδιόριcτα διοριcμῶν δεόμενα]."

22 Galeno non tocca il problema della datazione del trattato, limitandosi a dire che l'autore dispone di *Epidemie* e *Prognostico*.

23 Cf. *in Hp. prorrh. I comm.* 16.550. 12–14 K. (= 33.7–8 Diels). In questo stesso commento Galeno osserva che il termine cυνδρομή è tipicamente empirico. K. Deichgräber, *Die Griechische Empirikerschule*, um Zusätze vermehrter anastatischer Neudruck der Ausg. von 1930 (Berlin: Weidmann, 1965), 226–28, segnala tuttavia tracce di dottrina autenticamente empirica anche nell'esegesi di Galeno al *Prorretico*, cf. i suoi frr. 329, 330, 331, 332, 332 A.

24 L'indicazione dei commentatori che hanno dubitato dell'autenticità del *Prorretico I* proprio per la presenza di espressioni solecistiche resta vaga, cf. *in Hp. prorrh. I comm.* 16.511.3–5 K. (= 13.27–29 Diels): "non poche espressioni in questo libro sono solecistiche nella *lexis*, perciò anche per questo alcuni lo hanno sospettato con buone ragioni" (Cολοικώδειc κατὰ τὴν λέξιν εἰcὶν οὐκ ὀλίγαι τῶν ἐν τούτῳ τῷ βιβλίῳ ῥήcειc, ὡc καὶ διὰ τοῦθ' ὑποπτεύειν αὐτό τιναc εὐλόγωc Ἱπποκράτουc οὐκ εἶναι). Più avanti Galeno scrive che alcuni commentatori attribuiscono quest'opera a medici della prima generazione post-ippocratica, ma mostra scarsissimo interesse per la precisa definizione dell'autore, cf. 16.625.4–14 K. (= 67.29–68.8 Diels): "Quel che ho già detto molte volte, lo ripeterò anche adesso: colui che ha composto questo libro pare appartenere alla stessa arte del grande Ippocrate ma è molto inferiore a lui; per questo ad alcuni è sembrato che questo scritto sia di Draconte figlio di Ippocrate, ad altri di Tessalo (si ritiene concordemente che Draconte e Tessalo siano i due figli del grande Ippocrate, e che ciascuno di loro abbia avuto un figlio di nome Ippocrate). Ma chiedersi se questo libro sia stato scritto da uno di loro o da un

Va detto che questa posizione critica di Galeno nei confronti del *Prorretico* si esprime solo in questo commento e nel contemporaneo commento a *Epidemie III*, e mai altrove: non nel commento agli *Aforismi* (17B.706.15 K.), né nel *De comate* (7.644.9 K.; 7.662.16 K.), né nel *De crisibus* (9.758.10 K.), né nel *De difficultate respirationis* (7.863.8 K.), e va aggiunto che qualche volta Galeno chiama l'autore Ippocrate anche in questo commento.²⁵

IV *Prorretico I* versus *Epidemie I–III* e *Prognostico*

Il *Prorretico* era evidentemente un testo apprezzato al tempo di Galeno e faceva parte del gruppo di scritti usualmente oggetto di commento; non c'è dunque nulla di strano nel fatto che anche Galeno lo abbia commentato a sua volta, per rispondere, come al solito, alle insistenze degli amici.²⁶ La rilevanza generalmente riconosciuta al *Prorretico* e l'assenza di dubbi sulla sua autenticità, (attribuirlo ai figli di Ippocrate non è una vera contestazione di autenticità) offre una spiegazione sufficiente per quella che ci appare come un'anomalia – scrivere un commento continuo ad un'opera di scarso valore dal punto di vista della dottrina e piuttosto pericolosa per la sua debolezza teorica. Galeno aggiunge però una giustificazione specifica, e cioè la volontà di mettere fine alla pratica invalsa tra i commentatori di spiegare il *Prorretico* copiando interi passi di *Epidemie*²⁷ e viceversa; il commento di Galeno è destinato dunque a stabilire

---

altro, o se l'autore è morto prima di pubblicarlo, è lavoro superfluo; basti quanto si è detto adesso su questo argomento."

25 *In Hp. prorrh. I comm.* 16.556.10 K. (= 36.12 Diels), 16.560.16–17 K. (= 38.12 Diels), 16.595.15 K. (= 55.7 Diels) e 16.835.5 K. (= 175.17 Diels), passi opportunamente segnalati nell'indice dell'edizione Diels (Diels, *Galeni In Hippocratis Prorrheticum*, 383).

26 Si tratta di un topos (Galeno come "reluctant commentator," secondo la definizione di von Staden, "'A Woman Does Not Become Ambidextrous,'" 133), ma questa volta Galeno menziona anche l'occasione in cui la richiesta gli è stata fatta. Nel proemio al secondo libro del commento ad *Epidemie III* (17A.578.12–579.2 K. [= 61.18–62.7 Wenkebach]) Galeno ricorda che, quando si accingeva a scrivere quel commento, durante la visita ad un malato il discorso cadde su cose dette negli *Aforismi* e nel *Prognostico*, e poiché egli aveva detto che chi si fosse attenuto alla dottrina del *Prorretico* avrebbe commesso un grave errore, gli fu richiesto con garbo e insistenza, di commentare il *Prorretico*.

27 Smith, *The Hippocratic Tradition*, 155–60, ritiene che Galeno abbia assunto questa posizione critica nei confronti del *Prorretico* in polemica con Lico che aveva commentato le *Epidemie*; si tratta di un avversario a cui Galeno non evita in altre circostanze di rivolgersi direttamente e in modo molto aggressivo; anche per questo l'ipotesi mi pare difficile da sostenere.

una netta separazione tra *Prorretico* e *Epidemie*. Qui, come mai altrove, la critica di Galeno non è una presa di distanza circoscritta a singole interpretazioni ma costituisce la motivazione dell'esistenza stessa dell'intero commento:

> μοχθηρῶς δὲ πάντων ἐξηγηcαμένων τὸ cύγγραμμα τοῦτο κἀγὼ τοῖc ἀναγκάcαcί με πειcθεὶc ἐπὶ τὴν ἐξήγηcιν ἧκον. (16.547.9–11 K. [= 31.24–26 Diels])
>
> poiché tutti hanno spiegato male questo scritto anche io, convinto da coloro che mi costringevano a farlo, mi sono dedicato alla sua esegesi.

Partendo da queste premesse, Galeno dà del *Prorretico* una lettura contrastiva rispetto alle *Epidemie* (per la materia) e al *Prognostico* (per la finalità),[28] le due opere autentiche alle quali l'autore si sarebbe ispirato.[29]

Il materiale che può essere addotto per illustrare la relazione con questi due trattati è moltissimo; mi limiterò dunque ad un esempio di commento enfatico e retoricamente costruito, relativo a *prorrh.* I 86 (16.676.1–678.11 K. [= 91.22–93.4 Diels]): "Gola dolorante senza gonfiore con sofferenza, soffocamento e morte rapida" (Φάρυγξ ἐπώδυνοc ἰcχνὴ μετὰ δυcφορίηc πνιγώδηc ὀλεθρίη ὀξέωc); qui il confronto è limitato al *Prognostico*.

> Ἀναμνήcθητι τῆc ἐν τῷ Προγνωcτικῷ ῥήcεωc, ἧc ἡ ἀρχή· αἱ δὲ κυνάγχαι δεινόταται μέν εἰcι καὶ τάχιcτα ἀναιροῦcαι, ὅcαι μήτ' ἐν τῇ φάρυγγι μηδὲν ἔκδηλον ἐμποιέουcι μήτ' ἐν τῷ αὐχένι, πλεῖcτόν τε πόνον παρέχουcιν καὶ ὀρθόπνοιαν, ἔνθα cοι καὶ παραβάλλειν ἔξεcτιν ἑρμηνείαν ἐπιcτήμονοc ἀνδρὸc ὧν λέγει πραγμάτων τῇ τοῦ πλανωμένου κατὰ τὰc cυνδρομὰc ἀδυναμίᾳ τοῦ πρὸc τὸ καθόλου ποιεῖcθαι τὴν cυναγωγήν. ὁ μὲν γὰρ Ἱπποκράτηc τό τε τοῦ πάθουc ὄνομα προεῖπεν καὶ τὰc διαφορὰc ἁπάcαc ἐφεξῆc εἶπεν ἀρξάμενοc ἀπὸ τῆc χαλεπωτάτηc αὐτῶν, ἧc νῦν παρέγραψα τὴν ῥῆcιν. ὁ δὲ τοῦτο τὸ βιβλίον γράψαc πρῶτον μὲν ἥμαρτεν εἰπὼν τὴν φάρυγγα πνιγώδη, δέον τὴν κυνάγχην. οὐδὲ γὰρ τοῦτ' ἔcτιν εἰπεῖν ὡc ἀντὶ τοῦ πάθουc τῷ πεπονθότι μορίῳ κατεχρήcατο· τὴν ἀρχὴν γὰρ οὐδὲ πέπονθέ τι τοῖc οὕτωc ἔχουcιν φάρυγξ, ὡc Ἱπποκράτηc ἐδήλωcεν δεινοτάταc εἰπὼν εἶναι κυνάγχαc,

---

28 In questo commento Galeno menziona le *Epidemie* (e gli *Aforismi*) molto più frequentemente di quanto non faccia per altri trattati ippocratici, cf. l'*Index nominum* dell'edizione Diels (Diels, *Galeni In Hippocratis Prorrheticum*, 382–83).

29 Cf. il commento a *epid.* III 1 β′ (17A.532.17 e 534.5–10 K. [= 31.15–17 e 32.18–21 Wenkebach]): "‹ho già spiegato questa frase anche nel commento al *Prorretico*›, biasimando colui che ha scritto quel libro: sbaglia infatti riferendo quel che accade per un certo motivo ad un altro motivo"; e "pare infatti che, a partire da quel che qui è scritto su Ermocrate, colui che ha scritto quel libro (sc. il *Prorretico*) abbia messo insieme un'affermazione generale, ma non coglie affatto né la natura dell'oggetto, né l'opinione di Ippocrate, né le cose osservate nei pazienti." Così, secondo quanto Galeno dice a *in Hp. epid. III comm.* 17A.636.9–14 K. (= 101.22–26 Wenkebach), il caso della donna descritto in *epid.* III 1 ιαʹ (3.60.10 L.) potrebbe contenere il materiale che ha ispirato *prorrh.* I 1 (ma Galeno non esclude che l'autore possa aver visto egli stesso un caso simile).

ὅcαι μήτε ἐν τῇ φάρυγγι ἔκδηλον μηδὲν ποιοῦcι μήτ' ἐν τῷ αὐχένι. ὁ δὲ τὸ Προρρητικὸν τοῦτο γράψας ἔφη· φάρυγξ ἐπώδυνος ἰcχνή. τὸ γὰρ ἰcχνὴ ταὐτὸν τῷ μηδὲν ἔκδηλον ἔχειν δύναται. προcέθηκε δὲ οὗτος μάτην τὸ μετὰ δυcφορίης, ἐξ ἑτέρου γένους ὑπάρχον cυμπτωμάτων. εἶτα καὶ τὸ πνιγώδης ὁ μὲν τοῦτο τὸ βιβλίον γράψας ἀδιορίcτως τε καὶ κοινῶς εἶπεν, ὁ Ἱπποκράτης δὲ οἰκείῳ τῆς διαφορᾶς ὀνόματι προcηγόρευcεν εἰπὼν ὀρθόπνοιαν ἐπιφέρειν αὐτήν [...]. ἀλλὰ καὶ τὰς ἄλλας ‹τῆς› κυνάγχης διαφορὰς ὁ Ἱπποκράτης ἔγραψεν ἐν τῷ Προγνωcτικῷ cαφέcτατά τε καὶ ἀκριβέcτατα κατ' ἐκεῖνο τοῦ βιβλίου τὸ χωρίον, ἐν ᾧ καὶ ἡ ῥηθεῖcά μοι νῦν εἴρηται ῥῆcις. ἐν δὲ τῷ Προρρητικῷ τούτῳ καίτοι πολλάκις φάρυγγος μνημονεύcας ὁ γράψας τὸ βιβλίον οὐδὲν ἡμᾶς cαφὲς οὐδὲ διηρθρωμένον ἐδίδαξε περὶ τῶν κατὰ ταύτην παθῶν.

Ricordati la frase del *Prognostico* che inizia: "le angine più pericolose e che portano molto rapidamente alla morte, sono quelle che non producono nulla di ben visibile né nella gola né nel collo, e provocano moltissima sofferenza e *orthopnoia*" [cap. 23, 2.176.2–5 Littré], nella quale puoi anche confrontare il modo di esprimersi di uno che conosce le cose che dice con quello di uno che va errando tra le sindromi per incapacità di raccogliere i dati in un discorso generale.[30] Ippocrate dunque [sc. nel *Prognostico*] ha detto per prima cosa il nome dell'affezione [sc. le angine] e poi tutte le [sue] varietà, cominciando dalla più difficile tra di loro, quella che ho appena trascritto [sc. l'*orthopnoia*]. Colui che ha scritto questo libro per prima cosa ha sbagliato nel dire che la gola è πνιγώδης, mentre avrebbe dovuto dire [che lo è] l'angina. E non si può neppure dire che, invece dell'affezione, per catacresi ha detto la parte affetta. Per prima cosa, infatti, in coloro che sono in questa condizione la gola non soffre affatto, come ha mostrato Ippocrate dicendo che le angine più pericolose sono quelle che non mostrano sintomi né nella gola né nel collo. Colui invece che ha scritto il nostro *Prorretico* ha detto: "Gola dolorante senza gonfiore" e senza gonfiore equivale a dire senza che nulla sia evidente; egli poi ha aggiunto vanamente anche "con stato di sofferenza," che è un sintomo non omogeneo [di altro genere]. E ancora: colui che ha scritto questo libro ha detto πνιγώδης senza distinzioni e nel senso comune, mentre Ippocrate ha nominato [l'affezione] col nome proprio della varietà dicendo che produce l'*orthopnoia* [...].[31] E anche le varietà dell'angina, Ippocrate nel *Prognostico* le ha scritte nella maniera più chiara e più precisa nella parte del libro in cui si trova la frase che ho detto, nel nostro *Prorretico*, invece, sebbene colui che ha scritto il libro abbia fatto menzione spesso della gola, non ci ha insegnato nulla di chiaro e di articolato sulle affezioni di questa parte.

---

30 Per l'incapacità di concentrare in poche parole le cose essenziali cf. anche 16.599.10–12 K. (= 56.23–24 Diels) *ad prorrh.* I 41: "avendo egli scritto un segno comune a molte malattie come proprio di una sola *per la sua incapacità* di comprendere l'insegnamento in un solo discorso generale" (τὸ κοινὸν πολλῶν ὡς ἴδιον γράφοντος αὐτοῦ διὰ τὸ μὴ δύναcθαι περιλαβεῖν ἑνὶ λόγῳ καθόλου τὴν διδαcκαλίαν); e anche 16.815.6–8 K. (= 163.12–14 Diels) *ad prorrh.* I 152: "Molte sindromi una di seguito all'altra ha scritto in tutto questo discorso, *essendo incapace* di circoscrivere in un discorso breve l'intero metodo relativo a coloro che hanno avuto emorragie" (Πολλὰς ἐν τούτῳ παντὶ τῷ λόγῳ cυνδρομὰς ἐφεξῆς ἀλλήλων ἔγραψεν μὴ δυνάμενος ἐν ὀλίγῳ περιγράψαι τὴν καθόλου μέθοδον ἐπὶ τῶν αἱμορραγηcάντων).
31 Tralascio un breve *excursus* sul significato del termine *orthopnoia*.

Per ben quattro volte Galeno oppone Ippocrate a "colui che ha scritto questo libro";[32] l'uno è un "uomo di scienza" (ἐπιστήμων), l'altro è uno "che va errando tra le sindromi" (πλανώμενος κατὰ τὰς συνδρομάς), uno scrive in maniera appropriata, l'altro senza *diorismoi* e specificazioni, uno scrive in modo chiarissimo e precisissimo, l'altro in maniera oscura e male articolata. Sarà adesso opportuno fornire un'esemplificazione più ampia degli errori logici e di espressione che Galeno individua nel corso del suo commento.

V Errori nella raccolta di segni in una sindrome: il caso di *prorrh.* I 1–2 (16.8–506.11 K. [= 9.9–11.20 Diels])

La prima parte del *Prorretico I* contiene aforismi prognostici che riguardano diversi disturbi mentali.[33] Il primo elenca sintomi che riguardano i pazienti affetti da coma (οἱ κωματώδεες); il secondo contiene un solo sintomo, e cioè una debole evacuazione (κοιλίης περίπλυσις), sintomo cattivo per tutti i malati, ma soprattutto per quelli detti prima, cioè i κωματώδεες.[34]

Legare tra loro due aforismi è inusuale,[35] ma a Galeno interessa piuttosto segnalare la scorrettezza di questo nesso dal punto di vista medico. L'autore,

---

32 Questa l'espressione con cui Galeno lo definisce abitualmente; altre varianti sono: ὁ τοῦ βιβλίου συγγραφεύς (16.566.6 K. [= 40.27–28 Diels]; 16.606.14 K. [= 59.27 Diels]; 16.657.3 K. [= 82.26 Diels]; 16.754.14 K. [= 130.12–13 Diels]; 16.761.9–10 K. [= 133.21 Diels]), ὁ συνθεὶς τὸ βιβλίον τοῦτο (16.625.5 K. [= 67.29–30 Diels] e 16.632.5 K. [= 71.14 Diels]); qualche volta però Galeno lo chiama anche col nome di Ippocrate (cf. supra, n. 25).
33 Osserva Potter, *Hippocrates*, 167, che nel *Prorretico* "the conditions included are weighted towards the neurological (i. e. loss of speech, deafness, delirium, coma, strabismus, headache, paralyses)"; osservazione inserita in un'efficace analisi della struttura complessiva dell'intero trattato (166–69).
34 Cf. *prorrh*. I 1: "Quelli che all'inizio sono comatosi e sono insonni con dolore della testa, del fianco, dell'ipocondrio e del collo forse che sono frenitici? Il naso che cola in costoro è segno mortale, soprattutto se comincia al quarto giorno" (Οἱ κωματώδεες ἐν ἀρχῇσι γινόμενοι μετὰ κεφαλῆς, ὀσφύος, ὑποχονδρίου, τραχήλου ὀδύνης, ἀγρυπνέοντες ἆρά γε φρενιτικοί εἰσιν; μυκτὴρ ἐν τουτέοισιν ἀποστάζων ὀλέθριον, ἄλλως τε καὶ ἢν τεταρταίοισιν ἀρχομένοισιν); *prorrh*. I 2: "Una debole evacuazione del ventre, molto rossa è un segno cattivo in tutti i casi, ma soprattutto <u>nei casi detti prima</u>" (Κοιλίης περίπλυσις ἐξέρυθρος κακὸν μὲν ἐν πᾶσιν, οὐχ ἥκιστα δὲ ἐπὶ τοῖσι προειρημένοισιν).
35 Questo è il solo caso nel *Prorretico*; la stessa formula ἐπὶ τοῖσι προειρημένοισιν ricorre solo in *Coac*. 318, che non è un passo parallelo. È sempre possibile che la formula sia un relitto di riferimenti che si trovavano nelle opere da cui Hipp. *prorrh*. I 2 e *Coac*. 318 sono stati tratti.

ipotizza Galeno (16.501.9-12 K. [= 9.9-12 Diels]), deve essersi ispirato al finale del *Prognostico* (cap. 25, 2.188.10 L.), dove Ippocrate, dopo aver illustrato la forza (δύναμις) di ciascun segno, raccomanda di confrontarli tra di loro, calcolando il loro peso complessivo (ἐκλογιζόμενον, ὅπερ ἐςτὶν ἀλλήλοις παραβάλλοντα) per procedere alla formulazione della prognosi. Ora, dei segni alcuni sono "dello stesso tipo" (ὁμοειδῆ), altri non sono né "dello stesso tipo" (ὁμοειδῆ) né "dello stesso genere," "omogenei" (ὁμογενῆ).[36] Ma, mentre l'intreccio di segni omogenei è utile, quello di segni eterogenei (la ἑτερογενῶν cημείων ἐπιπλοκή) è inutile, sia che lo si scriva in un libro sia che lo si faccia negli esercizi scolastici in forma orale (si noti, fin dall'inizio del commento, la preoccupazione didattica di Galeno). Se si mettono insieme i segni relativi alle urine con i segni che riguardano la testa (come nel caso di *prorrh.* I 1 e 2 sono messi insieme i segni del ventre con quelli della testa) si avrà un "vano intreccio di segni" (ματαία ἐπιπλοκὴ cημείων), buono solo per chi ha carta e tempo da sprecare! E ancora. Mentre nel *Prognostico* sono riuniti sintomi diversi solo nel capitolo iniziale sulla *facies* mortale e, più avanti, a proposito degli empiemi, degli *hyderoi* e di qualche altra malattia, nel *Prorretico* ciò avviene spesso: in un'unica sindrome sono intrecciati molti segni eterogenei, in modo superfluo.[37] Non c'è nessun altro medico – conclude Galeno – che in un suo scritto insegni tali intrecci.[38]

---

36 Commentando il passo (*in Hp. prog. comm.* 18B.298.3 K. [= 367.20 Heeg]), Galeno aveva parlato della necessità di confrontare non solo le *dynameis* dei segni omogenei (ὁμογενῶν), i segni buoni con quelli buoni, i segni cattivi con i cattivi, ma anche di confrontare i segni buoni con quelli cattivi. Il termine ὁμοειδῆ costituisce una variazione rispetto a ὁμογενῆ del commento al *Prognostico*. Galeno usa spesso i due aggettivi insieme, apparentemente senza insistere sul loro diverso significato cf. *de temper.* 1.542.16-543.1 K. (= 21.22-24 Helmreich): ὡς πρὸς τὸ cύμμετρον ὁμογενὲς ἢ ὁμοειδὲς παραβάλλων; 1.545.4-5 K. (= 23.7 Helmreich): πρὸς τὸ cύμμετρον ὁμογενὲς ἢ ὁμοειδές.
37 *In Hp. prorrh. I comm.* 16.505.16-17 K. (= 11.9-11 Diels): "Qui spesso in un'unica sindrome intreccia in modo superfluo molti segni eterogenei" (ἐνταυθοῖ δὲ πολλάκις ἐν μιᾷ cυνδρομῇ τῶν ἑτερογενῶν ἐπιπλέκεται πολλὰ περιττῶς).
38 *In Hp. prorrh. I comm.* 16.506.10-11 K. (= 11.19-20 Diels): "Nessuno in uno scritto insegna tali intrecci" (ἀλλ' οὐδεὶς ἐν cυγγράμματι τοιαύτας ἐπιπλοκὰς διδάσκει). Galeno riprenderà il caso di questi due aforismi del *Prorretico* nel già più volte citato commento a *Epidemie III* (17A.583.7-84.3 K. [= 65.4-17 Wenkebach]): "chi menzionasse un sintomo della futura *phrenitis*, e poi di seguito ne aggiungesse un altro proprio della *peripneumonia* e dell'infiammazione del fegato, o del ventre, o di un'altra parte, e poi dicesse che la *diathesis* è più difficile in quei malati che soffrono in due o più parti, proporrebbe un cattivo insegnamento [μοχθηρὰν διδαςκαλίαν; segue citazione di *prorrh.* I 1-2] [...], non si tratta infatti di segni della stessa *diathesis*."

Questo errore riguarda la corretta presentazione dei dati, ma non tocca la veridicità dei fatti, come Galeno esplicitamente chiarisce più avanti a proposito di *prorrh.* I 32: [39]

> ἡ γὰρ ἐπιπλοκὴ τῶν σημείων οὐχ ἁπάντων ἐςτὶν ὁμογενής, τῆς μὲν κωφώςεως τὴν κεφαλὴν ἐνδεικνυμένης πεπονθέναι, τῶν δ' ἐξερύθρων οὔρων τοὺς νεφρούς, τῶν δὲ ἀκαταςτάτων οὔρων ἐπὶ τοῖς ἐναιωρήμαςι πρῶτον μὲν ἐν ταῖς φλεψὶν ἀπέπτων χυμῶν ταραχὴν ἐνδεικνυμένων, κατὰ ςυμβεβηκὸς δὲ καὶ τῇ τῆς παρακρούςεως γενέςει ςυντελούντων. τὸ δ' ἐφεξῆς τῶν εἰρημένων γεγραμμένον τὸ τοὺς τοιούτους ἰκτεροῦςθαι κακόν, ἀληθὲς μέν ἐςτιν, ἀλλ' ὅτι ςυνδρομὰς ἑτερογενῶν ςημείων οὐ χρὴ ποιεῖςθαι κατὰ τὰς τοιαύτας διδαςκαλίας ἔμπροςθεν δέδεικται. (16.575.4–14 K. [= 44.28–45.4 Diels])

> l'intreccio dei segni non è per tutti omogeneo; la mancanza di voce indica che soffre la testa, le urine rossastre che soffrono i reni, le urine non sedimentate indicano in primo luogo turbamento di umori non concotti nelle vene che, per accidente, concorrono anche all'insorgere della follia. Quel che è scritto dopo le cose dette, che cioè è un segno cattivo che tali soggetti diventino itterici, è vero, ma si è già precedentemente indicato che in tali esposizioni didattiche non si devono fare sindromi di segni eterogenei.[40]

---

39 *Prorrh.* I 32: "Mancanza di voce e urina rossa, con elementi non sedimentati in sospensione, (segno di) turbamento mentale" (κώφωςις καὶ οὖρα ἐξέρυθρα, ἀκατάςτατα ἐναιωρήματα, παρακρουςτικά). Anche in questo caso Galeno nel commento ritorna su *prorrh.* I 2 dove la κοιλίης περίπλυςις è segno di affezione del fegato, mentre il colore ἐξέρυθρος dell'urina è segno di un'affezione dei reni.

40 Cf. anche *In Hp. prorrh. I comm.* 16.718.13–14 K. (= 112.29–30 Diels): "anche in questa frase vi è un mucchio di sintomi non omogenei" (καὶ κατὰ τήνδε τὴν ῥῆςιν ἀνομοιογενῶν ἐςτι ςυμπτωμάτων ἄθροιςμα); e 16.789.1–5 K. (= 147.28–148.3 Diels), dove Galeno critica il comportamento incerto dell'autore nei confronti dell'elaborazione dei dati di osservazione: "ma colui che ha scritto questo *Prorretico* si lascia trasportare nelle cose particolari e per questo talora ammucchia una sindrome di segni non omogenei, talora da uno o due casi trae un'affermazione generale, talora è incerto e dubita, come anche qui, dove dice 'forse che ecc.'" (ἀλλ' ὁ τὸ Προρρητικὸν τοῦτο γεγραφὼς ἐν τοῖς κατὰ μέρος φέρεται καὶ διὰ τοῦτο ποτὲ μὲν ἐξ ἀνομοιογενῶν ςυμπτωμάτων ἀθροίζει ςυνδρομήν, ποτὲ δ' ἐξ ἑνὸς ἢ δυοῖν ἀρρώςτοιν ἀπόφαςιν ποιεῖται καθόλου, ποτὲ δ' ἀπορεῖ καὶ ἀμφιβάλλει, καθάπερ καὶ νῦν εἰπὼν ἆρά γε κτλ.). E in generale, cf. 16.660.2–5 K. (= 84.10–12 Diels): "dunque in questo *Prorretico* in molti punti l'esposizione è confusa, poiché tutto ciò che si è manifestato nei malati viene ricondotto ad un'unica sindrome" (κατὰ μέντοι τὸ Προρρητικὸν τοῦτο ςύγγραμμα πολλαχόθι ςυγκέχυται τὰ τῆς διδαςκαλίας τῶν φανέντων κατὰ τοὺς νοςοῦντας ἁπάντων εἰς μίαν ἀναγομένων ςυνδρομῇ). Il gusto dell'autore del *Prorretico* per costruire sindromi diventa perfino una guida per divinare l'assetto originale del testo; a 76.3–11 Diels, Galeno segnala che molti esegeti avrebbero riunito gli aforismi 58, 59 e 60 del *Prorretico* ritenendo che tutti sintomi presenti in essi siano riferiti ad un unico caso clinico (ὡς ἐφ' ἑνὸς ἀρρώςτου μίαν ςυνδρομὴν ἡγεῖςθαι) e commenta: "ed è possibile che con quest'intenzione questi sintomi che abbiamo detto siano stati congiunti da colui che ha scritto questo libro: risulta infatti che egli talvolta ha costruito delle sindromi a partire da segni non omogenei" (καὶ δυνατόν ἐςτιν κατὰ τὴν διάνοιαν ταύτην ὑπὸ τοῦ γράψαντος τὸ βιβλίον ἐζεῦχθαι ταῦτα τὰ εἰρημένα ςυμπτώματα· φαίνεται γὰρ ἐξ ἀνομοιογενῶν ἐνίοτε ςυνδρομὰς ποιούμενος).

Il bersaglio di queste critiche è l'insegnamento della medicina empirica e in particolare l'uso che gli empirici fanno dei dati di osservazione.[41] Potremmo dire, applicando una categoria grammaticale all'intreccio dei sintomi nella sindrome, che le sindromi scorrette degli empirici sono una sorta di solecismo, un intreccio di parole non adatte le une alle altre (cυμπλοκὴ λέξεων ἀλλήλαιc ἀναρμόcτων), secondo la definizione dei grammatici (cf. Herodianus *soloec.* 295.8 Nauck); come accade per i solecismi, la gravità nella costruzione di sindromi errate è di grado diverso: a volte costituiscono un leggero disturbo, ma possono anche arrivare a precludere del tutto la comprensione dei fatti o la comprensione del testo.[42]

VI.1 Oscurità κατὰ τὴν λέξιν

Dal punto di vista della lingua, il difetto più grave che Galeno individua nel *Prorretico I* è costituito dall'oscurità, che per lo più è dovuta ad incapacità di esprimersi,[43] ma che in qualche caso gli pare intenzionale, cf. p. es. *in Hp. prorrh. I comm.* 16.749.17–750.5 K. (= 128.1–5 Diels) a proposito del nesso οὔρηcιc δυcκολαίνουcα:

> è straordinario come [l'espressione οὔρηcιc δυcκολαίνουcα] sia stata escogitata dall'autore per rendere il testo oscuro, così che un commentatore intende "la minzione che avviene con dolore," un altro "quella che presenta ritenzione," un altro "quella che avviene a fatica" e "quella continua" e "quella a gocce," e un altro "quella molto divorante."

e 16.713.7–10 K. (= 110.9–11 Diels):

---

[41] Sull'empirismo nei testi ippocratici, e l'interpretazione che ne dà Galeno, in particolare nel commento a *Epidemie*, cf. van der Eijk, "Exegesis, Explanation, and Epistemology," in particolare 36–41.

[42] Cf. E. Siebenborn, *Die Lehre von der Sprachrichtigkeit und ihren Kriterien. Studien zur antiken normativen Grammatik*, Studien zur antiken Philosophie V (Amsterdam: B. R. Grüner, 1976).

[43] Cf. 16.676.4–7 K. (= 92.2–5 Diels): "Qui puoi confrontare il modo di esprimersi di uno che conosce le cose di cui parla con quello di uno che va errando tra le sindromi, per la sua incapacità di ricondurre al generale" (ἔνθα coι καὶ παραβάλλειν ἔξεcτιν ἑρμηνείαν ἐπιcτήμονοc ἀνδρὸc ὧν λέγει πραγμάτων τῇ τοῦ πλανωμένου κατὰ τὰc cυνδρομὰc ἀδυναμίᾳ τοῦ πρὸc τὸ καθόλου ποιεῖcθαι τὴν cυναγωγήν); 16.569.7–8 K. (= 42.9–10 Diels): "essendo incapace dunque di dare dimostrazione del generale con metodo dimostrativo a partire dalla natura dell'oggetto" (ἀδυνατῶν οὖν ἐκ τῆc τοῦ πράγματοc φύcεωc ἐνδεικτικῶc ἀποφήναcθαι τὸ καθόλου).

le urine δεδαcυμένα è una di quelle espressioni pensate per ottenere l'oscurità, come anche molte altre in questo libro, perché anche qui abbiano su che dire sciocchezze quelli che amano farlo.

In tutt'e due i casi si tratta di scelte lessicali inconsuete, ma il discorso vale anche per le costruzioni sintattiche, come a 16.693.3–4 K. (= 100.9-10 Diels), dove non è chiaro a cosa si debba riferire il participio ἐκλείπουcαι:

> Anche questa frase è simile a quelle appositamente studiate per conseguire l'oscurità [ταῖc ἐξεπίτηδεc ἐπιτετηδευμέναιc εἰc ἀcάφειαν].

Il *Prorretico*, secondo Galeno, è caratterizzato da solecismi di questo tipo, e di diversa gravità. Ecco allora l'esame più dettagliato di qualche passo per mostrare la tipologia e le conseguenze di questo errore.

a) *Prorrh.* I 4 (16.511.3–512.5 K. [= 13.27–14.13 Diels])

Il primo caso che Galeno individua nel corso del commento riguarda *prorrh*. I 4 (Τὰ ἐπὶ ταραχώδεcιν ἀγρύπνοιcιν οὖρα ἄχροα, μέλανα, ἐναιωρούμενα ἐφιδρῶcι φρενιτικά) e gli offre l'occasione di rilevare che in questo libro il fenomeno è così diffuso da far sorgere il sospetto che si tratti di un'opera spuria.[44] La brevità, continua Galeno, dando alle sue parole un valore generale che travalica il caso specifico, non comporta di necessità la cattiva qualità dell'espressione, come provano gli *Aforismi*, che sono brevi ma senza difetti di lingua.[45] Il *Prorretico* invece presenta di frequente una dizione strana (ἀλλόκοτον ἐν πολλαῖc ῥήcεcιν ἔχει τὴν ἑρμηνείαν) ed impropria (ἄκυροc, ἀκύρωc). Qui, in *prorrh*. I 4, il participio ἐναιωρούμενα è detto impropriamente (ἀκύρωc λέλεκται) delle urine mentre dovrebbe essere riferito alle particelle che vi sono in sospensione.

---

[44] Cf. *in Hp. prorrh. I comm.* 16.511.3–5 K. (= 13.27–29 Diels): "Non poche proposizioni in questo libro sono solecistiche per la *lexis*, sicché anche per questo alcuni hanno ragionevolmente sospettato che non sia di Ippocrate" (Cολοικώδειc κατὰ τὴν λέξιν εἰcὶν οὐκ ὀλίγαι τῶν ἐν τούτῳ τῷ βιβλίῳ ῥήcειc, ὡc καὶ διὰ τοῦθ᾽ ὑποπτεύειν αὐτό τιναc εὐλόγωc Ἱπποκράτουc οὐκ εἶναι); perduto l'inizio del proemio, questo è il primo passo in cui Galeno parla della possibile non autenticità del testo.

[45] Anzi, la brevità è una virtù, ovviamente finché non scivola nell'oscurità per l'omissione di elementi necessari alla comprensione, cf. Manetti "Galeno, la lingua di Ippocrate e il tempo," 182–83. Due soli casi di solecismo vengono registrati da Galeno nel commento ad *Aforismi*, nell'uno e nell'altro senza conseguenze per la comprensione del testo, cf. 18A.167.6–10 K., a proposito di *aph*. 7.56 (ἀλύκη, χάcμη, φρίκη, οἶνοc ἴcοc ἴcῳ πινόμενοc λύει), dove alcuni hanno corretto i tre nominativi iniziali in accusativi e 18A.170.3–7 K., a proposito di *aph*. 7.58.

Correggere il testo, come è stato proposto (pratica frequente, come dimostra le proposta di correzione dell'unico caso di possibile solecismo in *Epidemie III*⁴⁶), e sostituire μέλαcιν a μέλανα, dice Galeno, non elimina τὸ coλοικοφανές;⁴⁷ e ancora, in questo testo (κατὰ τὴν προκειμένην λέξιν) non si articola con chiarezza (διήρθρωται cαφῶc)⁴⁸ se μέλανα si debba riferire all'urina o solo alle parti che in essa sono in sospensione (16.513.1-3 K. [= 14.24-26 Diels]); e ancora, è superfluo aggiungere l'aggettivo ἄχροα quando già μέλανα indica il cattivo colore; e infine, è impossibile stabilire se ἐφιδρῶcι si riferisce alle urine, o ai corpi affetti o addirittura alle affezioni; insomma:

> (ἐφιδρῶcι) ἀκύρωc δ' ἡρμήνευται [...]. ὅπωc δ' ἄν τιc ἀκούῃ, τό γε τῆc ἑρμηνείαc ἄκυρον οὐκ ἔcτιν ἐκφυγεῖν. (16.513.16-514.4 K. [= 15.5-9 Diels])

> ἐφιδρῶcι è detto impropriamente [...]; e comunque lo si intenda, non si può sfuggire all'improprietà dell'espressione.⁴⁹

b) *Prorrh.* I 6 (16.527.3-529.9 [= 21.19-22.26 Diels])

Così, poco più avanti, commentando *prorrh*. I 6 "Espettorazione frequente, se si aggiunge anche qualche altro segno, frenitici" (Ἀνάχρεμψιc πυκνή, ἢν δή τι καὶ ἄλλο cημεῖον προcῇ, φρενιτικά), Galeno osserva:

---

46 In *Epidemie III* Galeno non trova solecismi, sebbene proprio nelle parole iniziali i commentatori abbiano intravisto la possibilità di leggerne, e correggerne, uno (*in Hp. epid. III comm.* 17A.481.1 K. [= 1.9 Wenkebach]); è l'incipit del primo caso clinico: Πυθίων ὃc ᾤκει παρὰ Γῆc ἱερὸν ἤρξατο τρόμου. Per evitare il solecismo alcuni hanno corretto Πυθίων in Πυθίωνι, ma Galeno trova una soluzione che salva il testo tramandato ed insieme evita il solecismo – un bell'esempio del suo metodo filologico –, e cioè propone di fare di Πυθίων ὃc ᾤκει παρὰ Γῆc ἱερόν una sorta di titolo a sé (πρόγραμμα), seguito da una pausa forte.
47 Questa è anche la scelta di Potter. L'aggettivo coλοικοφανές compare per la prima volta in Dionigi di Alicarnasso, e poi ritorna frequentemente nella letteratura grammaticale ed esegetica, per lo più riferito alle figure (cχήματα).
48 Per la nozione di articolazione sintattica cf. anche 16.678.10-11 K. (= 93.4 Diels) a proposito di *prorrh*. I 86.
49 Stessa accusa a proposito dell'espressione ἐν τῷ πλανᾶcθαι ἐκλείποντac (ἐκλιμπάνοντac Potter) di *prorrh*. I 71b, che Galeno ritiene vada riferito a coloro che sono in un stato di debolezza grave pur conservando parzialmente la facoltà di muoversi, aggiungendo una sua nota di disagio per il gusto dell'autore per le espressioni improprie ed oscure: "Come se l'autore di questo libro temesse di dire qualcosa di chiaro, così dappertutto, evitando i termini propri e chiari, fa uso di altri nomi" (ὥcπερ δεδιὼc ὁ τοῦ βιβλίου cυγγραφεὺc εἰπεῖν τι cαφὲc οὕτωc ἐπιτηδεύει πανταχοῦ τὰ κύρια καὶ cαφῆ τῶν ὀνομάτων ὑπερβαίνων ἑτέροιc χρῆcθαι, 16.657.3-5 K. [= 82.26-28 Diels]).

Non è chiaro a cosa si riferisca la parola finale, φρενιτικά al plurale, se ai segni o ai πάθη.

L'una e l'altra interpretazione rendono l'espressione (ἑρμηνεία) strana (ἀλλόκοτον).

c) *Prorrh.* I 97 (16.716.14–718.7 K. [= 112.1–22 Diels])

E ancora, commentando *prorrh.* I 97 "Dolore del fianco in seguito a sputi biliosi che sparisce senza ragione, sono fuori di sé" (Πλευροῦ ἄλγημα ἐπὶ πτύcεcι χολώδεcιν ἀλόγωc ἀφανιcθέν, ἐξίcτανται), Galeno trova difficile comprendere a cosa si riferisca il verbo ἐξίcτανται che chiude l'aforismo:

> col suo solito modo di esprimersi [τῷ μέντοι cυνήθει τῆc ἑρμηνείαc εἴδει] chi ha scritto questo libro ha reso la frase strana [λέξιc ἀλλόκοτοc], sicché non abbiamo a cosa riferire ἐξίcτανται. Per quanto riguarda la frase [λέξιc], infatti, [il verbo si deve riferire] al dolore del fianco, avendo egli scritto: πλευροῦ ἄλγημα ἀλόγωc ἀφανιcθέν, ἐξίcτανται. Ma in modo davvero improprio si direbbe πλευροῦ ἄλγημα ἀλόγωc ἀφανιcθέν, ἐξίcτανται. E anche se, invece di ἐξίcτανται, scrivessimo ἐξίcταται senza il ν della penultima sillaba con solo τ e α,[50] anche così l'espressione [ἑρμηνεία] è strana [ἄτοποc] e di molto peggiore. Da una parte infatti il passaggio dalle affezioni al paziente è cολοικώδηc,[51] dall'altra, neppure con ἐξίcταται è possibile trovare fin dall'inizio cosa significa. In modo naturale [κατὰ φύcιν] si sarebbe dovuto scrivere: Πλευροῦ ἄλγημα ἐπὶ πτύcεcι χολώδεcιν ἀλόγωc ἀφανιcθέν, οἱ κάμνοντεc ἐξίcτανται.[52]

---

50 Questa la soluzione di Potter, che traduce: "When pains in the side associated with bilious expectoration disappear without any reason patients become delirious."

51 Questo stesso errore è rilevato anche a 16.583.2–8 K. (= 48.17–22 Diels), a proposito di *prorrh.* I 35 (Αἱ ἐκ ῥίγεοc νωθρότητεc οὐ πάνυ παρ' ἑωυτοῖcιν), dove Galeno osserva: "Non so come, colui che ha scritto questo libro in molte frasi usa una dizione strana passando talora dalle malattie agli uomini malati, a volte dai malati alle malattie; e ora appunto avendo detto 'i torpori che derivano dal freddo' ha aggiunto 'non molto presso di sé,' mentre avrebbe potuto dire 'coloro che sono torpidi a causa del freddo non molto presso di sé'" (Οὐκ οἶδ' ὅπωc ὃc ἔγραψε τοῦτο τὸ βιβλίον ἐν πολλαῖc ταῖc ῥήcεcιν ἀλλοκότῳ χρῆται τῇ λέξει μεταβαίνων ἐνίοτε μὲν ἀπὸ τῶν παθῶν ἐπὶ τοὺc πάcχονταc ἀνθρώπουc, ἐνίοτε δὲ ἀπὸ τῶν παcχόντων ἐπὶ τὰ πάθη. καὶ νῦν γοῦν προειπὼν αἱ ἐκ ῥίγεοc νωθρότητεc ἐπήνεγκεν οὐ πάνυ παρ' ἑωυτοῖcιν, καίτοι δυνάμενοc εἰπεῖν οἱ ἐκ ῥίγεοc νωθροὶ οὐ πάνυ παρ' ἑωυτοῖcιν).

52 Un'analoga correzione della forma linguistica concentrata all'eccesso si trovava già a 16.606.11–14 K. (= 59.25–27 Diels), nel commento a *prorrh.* I 45 (Φωνὴ ὀξεῖα, ὑποχόνδρια τουτέοιcιν εἴcω εἴρυαται). Dice Galeno: "Κατὰ φύcιν si sarebbe dovuto scrivere οἷc ὀξύνεται νοcοῦcιν ἡ φωνή, τούτων ἕλκεται τὰ ὑποχόνδρια πρὸc τὴν ἔcω χώραν, ma, non so come, l'autore di questo libro gode delle espressioni strane [ἀλλοκότοιc]."

VI.2 La cυνήθεια

I casi esaminati riguardano solecismi dovuti alla sintassi non perspicua della frase (manca il soggetto), oppure all'uso improprio di un termine del lessico tecnico medico (ἐναιωρούμενα riferito all'urina invece che, come usualmente, al sedimento); le due anomalie insieme sono richiamate nel commento a *prorrh.* I 93 (οἷcι κῶμα γίνεται ἐπάφρων περιελθόντων[53] πυρετὸc παροξύνεται ὀξύc), all'inizio del terzo libro, dove Galeno introduce la nozione di cυνήθεια.[54]

Galeno qui si dilunga sull'impossibilità di definire il significato di *coma* (sul quale aveva in precedenza scritto un'operetta ancora conservata), parola che nel *Prorretico* non è mai accompagnata da altri termini che permettano di capire cosa intenda l'autore, poi critica un commentatore che ha riferito l'aggettivo ἐπάφρων a secrezioni diverse dalle evacuazioni del ventre (le sole a cui l'aggettivo sia riferito in *Epidemie* e *Prognostico*) e infine conclude con una certa irritazione e, mi pare, attribuendo all'autore del *Prorretico* anche l'errore del commentatore:

> ho già detto più volte che [...] l'autore di questo libro si è dato il compito di cambiare le parole consuete [τὰ cυνήθη τῶν ὀνομάτων ἐξαλλάττειν] e, se non riesce a farlo, di cambiarne almeno i significati [τὰ cημαινόμενα γοῦν ὑπαλλάττειν] e insieme si è curato di configurare la sintassi [cύνταξιν] [delle parole] in figure solecistiche [cολοικώδεcι ... cχήμαcιν]; e se non vuoi chiamarla sintassi [cύνταξιν], chiamala pure composizione [cύνθεcιν].[55] (16.709.2–8 K. [= 108.7–11 Diels])

---

[53] Potter: προcδιελθόντων.
[54] Da qui in avanti Galeno ricorre spesso alla nozione di uso linguistico che non compare nei due libri precedenti. Il sostantivo cυνήθεια compare solo a 16.791.14 K. (= 149.13 Diels), più frequenti l'aggettivo cυνήθηc (qui, poi a 16.717.8 K. [= 112.8 Diels] e a 16.749.2 K. [= 127.14 Diels] riferito all' εἶδοc τῆc ἑρμηνείαc; a 148.19 e 23 riferito alla grafia di una parola) e l'avverbio cυνήθωc, cf. 16.729.4 K. (= 117.23 Diels) per una metafora consueta (λέγεcθαι cυνήθωc); a essi si aggiungono le forme negative ἀήθηc (16.731.9–10 K. [= 119.6–7 Diels]; 16.736.14 K. [= 121.24 Diels]), ἀcυνήθηc (16.720.10–11 K. [= 113.25 Diels]).
[55] Giudizio che Galeno riprende, anche in difesa dei commentatori, a 16.736.6–7 K. (= 121.18–19 Diels), a proposito di *prorrh.* I 103: "E anche qui, sapendo che la sintassi delle parole è solecistica [τὴν cύνταξιν τῶν ὀνομάτων εἰδότεc cολοικώδη], alcuni hanno scritto diversamente ecc." La correzione tuttavia non modifica il senso e dunque Galeno preferisce indagare sulla verità medica dell'aforismo. "Ma neppure si può indagare sulla verità, poiché nella frase si trova la parola ἐπιφόροιcιν che necessita di spiegazione. Non solo colui che ha scritto [il *Prorretico*] si sforza di trovare in ogni frase una composizione dei nomi insolita, ma vi inserisce anche due o tre parole o del tutto strane o non nello stesso significato in uso tra i Greci" (παρεγκείμενον τῇ ῥήcει τοὔνομα τὸ *ἐπιφόροιcιν* ἐξηγήcεωc δεόμενον. οὐ γὰρ μόνον ἀήθη cύνθεcιν ὀνομάτων ὁ cυγγραφεὺc εὑρίcκειν πειρᾶται καθ᾽ ἑκάcτην ῥῆcιν, ἀλλὰ δύο που καὶ τρία παρεντίθηcιν ἤτοι ξένα παντάπαcιν ἢ μὴ κατ᾽ αὐτὸ τὸ cημαινόμενον, ὅπερ παρὰ τοῖc Ἕλληcιν εἴθιcτο, 16.736.12–

La stessa critica alle scelte di lessico e alla compiacenza dei commentatori si ritrova nel caso di *prorrh*. I 128 (διαλεγόμενοι λαθραίως τελευτῶςιν):[56]

> [secondo i commentatori] costoro, come se non fossero in pericolo, "mentre parlano con i presenti improvvisamente muoiono." Se questo, secondo loro, è il significato di λαθραίως, è chiaro che si danno la massima cura di cambiare la dizione naturale con metafore strane [τὴν κατὰ φύςιν ἑρμηνείαν ἀλλοκότοις μεταφοραῖς]; ritengono infatti che la virtù del discorso [ἀρετὴν λόγου] consista nell'allontanarsi comunque dall'uso [διὰ παντὸς ἀφίςταςθαι τῆς ςυνηθείας], e che non importi se ci se ne allontana verso il peggio. (16.791.6–15 K. [= 149.7–14 Diels])

VI.3 Le difficoltà dei commentatori. *Prorrh*. I 39 (16.595.9–596.10 K. [= 55.1–15 Diels])

Gli ultimi casi esaminati contengono riferimenti alle interpretazioni dei commentatori e ai loro vani tentativi di modificare il testo quando esso presenta anomalie; talora il problema dell'interpretazione sale in primo piano. P. es., all'uso strano (ἀλλόκοτος χρῆςις) dei nomi Galeno imputa, con insofferenza, il moltiplicarsi delle varianti (in questo caso πνεῦμα θολερόν e πνεῦμα θαλερόν), proposte da alcuni commentatori in *prorrh*. I 39 (16.595.5–596.10 K. [= 55.1–15 Diels]):

> ἡ γὰρ τῶν ὀνομάτων ἀλλόκοτος χρῆςις ἀφορμὴν παρέχει τοῖς βουλομένοις νοεῖν, ὡς ἂν ἐπέλθῃ. (16.596.17–18 K. [= 55.21–22 Diels])
>
> l'uso strano dei nomi offre il destro a chi vuole di intendere come gli viene in mente.

Né l'una né l'altra espressione, osserva Galeno, è attestata; l'aggettivo θολερόν nel *Prognostico* e negli *Aforismi* è riferito all'urina, l'aggettivo θαλερόν (nel linguaggio comune) è riferito al volto, ma nessuno dei due aggettivi è mai riferito

---

737.4 K. [= 121.23–28 Diels]). E così anche a 16.720.8–11 K. (= 113.23–25 Diels) (commento a *prorrh*. I 98), con maggiore comprensione: "se la sintassi ti pare strana, bisogna comprendere coloro che hanno spiegato in questo modo: molte cose infatti in questo libro sono espresse con parole insolite" (εἰ δὲ ἀλλόκοτός ςοι ἡ ςύνταξις εἶναι δοκεῖ, χρὴ ςυγγινώςκειν τοῖς οὕτως ἐξηγουμένοις· πάμπολλα γὰρ ἐν τῷ βιβλίῳ τούτῳ τοιαῦτα φαίνεται ἀςυνήθεςι τῶν ὀνομάτων ἑρμηνευόμενα).

56 Il lemma di Galeno coincide con *Coac*. 322 che Potter traduce "these patients die from no apparent cause, in the act of speaking"; nella tradizione diretta del *Prorretico* invece διαλεγόμενοι è preceduto da οὗτοι ςυντόμως (οὗτοι ςυντόμως διαλεγόμενοι λαθραίως τελευτῶςιν) che è così tradotto da Potter: "These patients speak only briefly, before dying without any obvious reason."

a πνεῦμα; la sola cosa certa è che l'autore intendeva un tipo di respirazione e non la respirazione *tout court*; a questo punto l'interprete si deve arrestare.

Galeno attribuisce alla presenza di solecismi o alla disposizione delle parole, la pratica dei commentatori di modificare il testo per eliminarli. Galeno, per parte sua, adotta una soluzione meno rigida, e che a noi pare filologicamente più corretta: nel caso che essi non precludano la comprensione, egli esprime la massima tolleranza, se invece il solecismo è dovuto, come spesso nel *Prorretico*, alla debolezza logica e linguistica dell'autore, allora preferisce rinunciare del tutto all'esegesi. E del resto, come Galeno dice in *de meth. med.* 10.43.8-10 K., non il solecismo nuoce alla scienza, ma la mancanza di chiarezza dell'esposizione.

Non sempre Galeno ha dato prova di coerenza nell'esegesi; a proposito di *prorrh.* I 15 (16.545.15-550.14 K. [= 31.1-33.8 Diels]), quando, per la prima volta, affronta il tema della sintassi di molti aforismi di questo libro,[57] Galeno sceglie tutt'altra strategia esegetica. L'aforismo 15, infatti, che appare nella forma "οἱ ἐκcτάντες ὀξέωc ἐπιπυρέξαντες cὺν ἱδρῶτι φρενιτικοὶ γίνονται," per la sua espressione "dissonante" (ἀcύμφωνοc) offre un caso esemplare di *amphibolia* che coinvolge tutta la frase e rende possibile darne almeno quattro diverse letture.[58] L'avverbio ὀξέωc può essere connesso con οἱ ἐκcτάντες oppure con ἐπιπυρέξαντες; il nesso cὺν ἱδρῶτι può essere letto con ἐπιπυρέξαντες oppure con φρενιτικοί. Dopo aver detto che da espressioni di questo genere non si ricava nessun insegnamento e che ogni tentativo di interpretarle è una perdita di tempo, Galeno si impegna tuttavia a spiegare l'aforismo in ciascuna delle quattro maniere possibili. Si tratta certo di uno sfogo aggressivo nei confronti di colui che lo ha scritto, che ha appena definito un empirico,[59] ma forse in quest'esibizione esegetica vi è anche una ragione in più; ognuna di queste quattro interpretazioni contiene uno stesso errore: l'autore può ben aver visto un caso clinico che presentava quei sintomi, ma, in ogni caso, quel che manca è la validità generale che a quei sintomi egli attribuisce.

---

[57] Soffermarsi sulla gravità di questo difetto dell'espressione, che toglie ogni valore agli enunciati, è così importante che si trasforma in una sorta di dichiarazione proemiale, come lascia intendere Galeno a 16.547.11 K. (= 31.25-26 Diels), dove dopo aver dichiarato le ragioni per cui ha scritto questo commento, parla esplicitamente di un nuovo inizio: ἧc [sc. ἐξηγήcεωc] αὖθιc ἀρξάμενοι λέγωμεν.
[58] Cf. 16.546.1-3 K. (= 31.3-5 Diels): "Chiamiamo ἀcύμφωνον questa espressione. La sua opinione infatti diventa oscura perché tutte le parole possono unirsi tra loro e separarsi."
[59] Cf. 16.550.12-14 K. (= 33.7-8 Diels), cit. a n. 23.

VII Errori di dottrina (κατὰ διάνοιαν)

Galeno inizia il secondo libro del commento introducendo il tipo particolare di oscurità (ἀσάφεια) che riguarda la dottrina (διάνοια).[60] Quest'errore, che rileva di frequente, consiste nel presentare come sempre presenti sintomi che si manifestano solo occasionalmente (mentre in uno scritto sulla prognosi si deve menzionare solo ciò che avviene "sempre" o "nella maggioranza dei casi")[61] e nell'attribuire carattere generale ad affermazioni di carattere particolare, cf., p. es., soltanto il commento a *prorrh*. I 73 (καὶ τοῦτο τῶν κατὰ μέρος ὂν ἔγραψεν ὡς καθόλου, 16.663.5 K. [= 85.26 Diels]) e a *prorrh*. I 66, dove c'è l'aggravante di aver raccolto segni non omogenei (μερικὸς οὖν καὶ αὐτὸς ὁ λόγος καὶ οὐ καθόλου καὶ μέντοι καὶ κατὰ σύνθεσιν οὐχ ὁμοιογενῶν σημείων, 16.650.1-2 K. [= 79.14-15 Diels]). L'errore contrario, con conseguenze meno gravi, si ha quando l'autore riferisce a una singola patologia osservazioni che hanno validità generale; p. es., in *prorrh*. I 71c prescrive di non somministrare farmaci evacuanti ai comatosi (κωματώδεις), o a chi ha febbre ardente (καυματώδεις), quando sono molto deboli; questa prescrizione, osserva Galeno (16.658.3-14 K. [= 83.9-19 Diels]), ha carattere generale ed è del tutto superfluo (ἐκ περιττοῦ) riferirla in particolare ad ognuna delle malattie. Si tratta insomma di non appesantire inutilmente il testo, che deve essere essenziale per essere più facilmente appreso e utilizzato.

Non mi pare necessario aggiungere altri esempi, e sarà sufficiente concludere citando due passi in cui Galeno critica l'educazione ricevuta da colui che ha scritto il *Prorretico I*.

Chi fa errori di questo genere, dice Galeno, è uno che ha compreso male le *Epidemie*, dove accade che Ippocrate scriva una frase di validità generale sui segni cattivi menzionando anche una delle loro specie particolari; ma costui non ha capito, o forse non gli è stato insegnato, che quelle affermazioni hanno funzione esemplificativa:

ὁ γὰρ ἄρτι μανθάνων τὴν τέχνην οἴεται ταῦτα μόνα κακοήθη εἶναι, περὶ ὧν ἤκουσεν. εἰ δὲ καὶ φύσει ταχύς ἐστιν, ὑπονοῆσαι ἐνδέχεταί τινα καὶ ἄλλα τοιαῦτα εἶναι καὶ περιμένειν

---

[60] Cf. 16.587.4-5 K. (= 51.6-7 Diels): "Non poco c'è di oscuro in tutto il libro, vediamo qui di seguito quello che riguarda la dottrina" (Ἔστι μὲν οὐκ ὀλίγον ἐν ὅλῳ τῷ βιβλίῳ τὸ ἀσαφές, τὸ δὲ τῆς διανοίας ἐφεξῆς ἴδωμεν).

[61] Cf. 16.808.16-18 K. (= 159.16-17 Diels): "Queste cose si devono scrivere nell'esposizione relativa all'insegnamento della predizione, quelle cioè che avvengono sempre o nella maggior parte dei casi" (χρὴ δὲ ταῦτα γράφειν ἐν τῇ προρρητικῇ διδασκαλίᾳ, ὅσα διὰ παντὸς ἢ πλειστάκις γίγνεται).

χρόνον ἄλλον, ἐν ᾧ κἀκεῖνα μαθήσεται, κἂν οὕτω τύχῃ μαθὼν ἄλλα, τὰ κατὰ μέρος αὖθις περιμενεῖ μὴ γινώσκων, ὁπόσον ἔτι λείπει τῆς διδασκαλίας αὐτῷ. ὁ δὲ τὸ καθόλου μαθὼν ἅπαντα γινώσκει τὰ κατὰ μέρος. (16.788.12–789.1 K. [= 147.22–27 Diels])

colui che sta ancora apprendendo l'arte crede che i soli segni cattivi siano quelli di cui ha sentito parlare [dal maestro]. Se poi è per natura rapido [nell'apprendere], può arrivare a sospettare che ce ne siano anche altri che sono tali e attende il tempo in cui apprenderà anche quelli. E quando ne avrà appresi ancora altri, aspetterà ancora di apprendere i segni particolari, ignorando quanto ancora gli manca da apprendere. Colui invece che avrà appreso gli insegnamenti in generale conosce anche tutte le cose particolari.

E ancora:

Εἴρηταί μοι καὶ πρόσθεν ὡς ἡ διδασκαλία τῆς ἰατρικῆς θεωρίας ὑφ' Ἱπποκράτους διὰ τῶν καθόλου γέγονεν ἐν ἅπασι τοῖς γνησίοις βιβλίοις προχειριζομένου ποτὲ καὶ τῶν κατὰ μέρος ἔνια παραδείγματος ἕνεκα πρὸς σαφήνειαν τῶν μανθανόντων. ἐπεὶ δ' οὐκ ἀπόχρη τὰ καθόλου θεωρήματα μαθεῖν εἰς τὸ διαγινώσκειν ἑτοίμως ἐπὶ τῶν ἔργων αὐτὰ χωρὶς τοῦ γυμνάζεσθαι, διὰ τοῦτο γυμνασίας ἕνεκεν ἔγραψε τοὺς ἐν τοῖς τῶν Ἐπιδημιῶν ἀρρώστους, ἐφ' ὧν οὐδὲν παρέλιπεν τῶν συμβάντων αὐτοῖς. ἡ δ' ὑπὸ τοῦ τὸ Προρρητικὸν γράψαντος ἀνδρὸς γενομένη διδασκαλία δέδεικται μὲν ἤδη πολλάκις οὐκ ἔχουσα τὸ καθόλου, πλὴν εἴ που σπανίως, ἀπολειπομένη δὲ καὶ τῆς τῶν κατὰ μέρος ἀκριβοῦς διηγήσεως, ὁποῖα γέγραπται κατὰ τοὺς ἐν τοῖς τῶν Ἐπιδημιῶν ἀρρώστους, ἔνθα πρῶτον μέν, ἐὰν ἐπίσημόν τι ᾖ τῶν προκαταρξάντων αἰτίων, δεδήλωκε λέγων (οἷον εἴ τις ἐκ κόπων ἢ γυμνασίων ἀκαίρων ἢ μέθης ἤρξατο νοσεῖν), εἶθ' ὅτι τὴν μὲν πρώτην τῶν ἡμερῶν ἐγένετο ταυτὶ περὶ αὐτόν, ἐν δὲ τῇ δευτέρᾳ ταυτί, καὶ οὕτως κατὰ τὴν τρίτην καὶ τὴν τετάρτην καὶ τὰς ἐφεξῆς ἅπαντα διηγεῖται τὰ γενόμενα τοῖς παρ' αὐτοῦ μεμαθηκόσι τὴν καθόλου θεωρίαν ἐφαρμόττειν προτρέπων τοῖς πεφηνόσι κατὰ τοὺς ἀρρώστους μετὰ τοῦ καὶ αὐτὸν (lege αὐτοὺς) ποδηγεῖν ἐνίοτε πρὸς τὴν γυμνασίαν, ὡς ἐπιδέδεικταί μοι κατὰ τὰς ἐξηγήσεις αὐτῶν. ὁ δὲ τοὺς ἐν τούτῳ τῷ βιβλίῳ γράψας ἀρρώστους ἀπὸ τῆς τοιαύτης διηγήσεως ἐπί τισι τῶν ὀφθέντων ἅπαξ ἤδη βούλεται καθόλου θεωρήματα συνίστασθαι σφαλλόμενος ἐν πολλοῖς οὐ σμικρά, καθάπερ ἐν τῷ τὸ ἕτερον γενόμενον ἑτέρῳ προσνέμειν, ὥσπερ καὶ νῦν. (16.723.9–725.3 K. [= 115.3–25 Diels])

ho detto anche prima che in tutti i suoi libri autentici Ippocrate espone la dottrina medica per mezzo di insegnamenti di validità generale, ma che, in qualche caso, offre anche alcuni dei segni particolari, a titolo di esempio, per chiarezza degli allievi. Ma poiché non è sufficiente apprendere gli insegnamenti generali per riconoscerli prontamente nella pratica se non ci si esercita, per questo, come esercizio [Ippocrate] ha scritto i casi clinici che sono nelle *Epidemie*, nei quali non ha tralasciato nulla di ciò che accadde ai malati. Si è dimostrato già molte volte che l'esposizione di colui che ha scritto il *Prorretico* non contiene il generale, se non raramente, e manca anche di una precisa esposizione delle cose particolari come quelle scritte nei casi delle *Epidemie*, dove [Ippocrate] indica per prima cosa se si manifesta una delle cause procatartiche (per esempio se uno ha cominciato ad essere malato a seguito di fatiche o di esercizi inopportuni, o a seguito di ubriachezza), poi che nel primo giorno gli capitarono queste cose, nel secondo queste altre, e così nel terzo e nel quarto e nei giorni successivi, ed espone ai suoi allievi tutto quel che è accaduto, invitandoli a far coincidere la teoria generale con ciò che si manifesta nei malati e insieme indirizzandoli ad esercitarsi, come ho mostrato nei commenti alle *Epidemie*.[62] Colui invece che ha scritto i casi clinici in questo libro [partendo] da una tale esposizione, vuole mettere insieme, sulla base di alcuni segni visti una volta sola, dottrine di carattere generale, sbagliando in molti casi non poco, come nell'attribuire una cosa accaduta in un caso ad un altro caso, come fa anche ora.

VIII La finalità didattica del *Commento al Prorretico*

La critica ai singoli aforismi del *Prorretico I* della quale ho raccolto qui una serie di esempi ha, come credo sia apparso evidente, l'intento dichiarato di evitare danni[63] a chi li legge e anzi di giovare a chi apprende. Il *Commento al Prorretico* costituisce un caso rarissimo, forse unico, di commento antico ad un'opera ritenuta nella sua integralità inadatta all'insegnamento, che è la destinazione ovvia dei commenti. Come emerge nel corso del commento, tuttavia, Galeno è convinto che, sotto la sua guida, sia possibile salvare e apprezzare quanto di buono c'è nel *Prorretico*[64] e anche apprendere la lezione di metodo che Galeno propone, una volta per tutte, quando affronta l'esegesi di *prorrh*. I 4, una frase detta in forma solecistica (ῥῆϲιϲ coλοικώδηϲ) che raccoglie i sintomi della frenite.[65] Qui Galeno insegna il metodo che deve guidare nell'identificare i sintomi pertinenti per non incorrere negli errori in cui è caduto il suo autore – e con lui i commentatori che si sono affannati a portare a confronto passi delle *Epidemie* che menzionano gli stessi sintomi (16.514.14–524.9 K. [= 15.18–20.9 Diels]).

---

62 Cf. *in Hp. epid. I comm.* 17A.13.11–14 K. (= 10.24–26 Wenkebach): "[nei miei libri] io incoraggio coloro che vogliono imparare bene l'arte medica ad esercitarsi nelle cose particolari che apprendono coi sensi così da riconoscere le cose che hanno precedentemente appreso in generale" (προτρέποντόϲ μου γυμνάζεϲθαι τοὺϲ ἐκμαθεῖν θέλονταϲ τὴν ἰατρικὴν τέχνην ἐν τοῖϲ κατὰ μέροϲ αἰϲθητοῖϲ, ὡϲ διαγινώϲκειν αὐτούϲ, ἃ καθόλου προμεμαθήκαϲιν).
63 Galeno si riferisce al danno per i giovani (che apprendono), o più in generale per coloro che leggono il *Prorretico*, sia in questo commento (cf. 16.588.17–589.3 K. [= 52.1–4 Diels]), sia in quello a *Epidemie III* (cf. *in Hp. epid. III comm.* 17A.578.18–579.2 K. [= 62.6–7 Wenkebach]): "[mostrai che] se uno si atterrà a tutto quello che è scritto nel *Prorretico* come se fosse 'in generale,' egli sbaglierà di molto" (ἐὰν δ' ὡϲ καθολικοῖϲ τιϲ προϲϲχῇ τοῖϲ ἐν τῷ Προρρητικῷ γεγραμμένοιϲ ἅπαϲιν, ‹αὐτὸν› πάμπολυ ϲφαληϲόμενον).
64 In un caso Galeno si dice addirittura d'accordo con l'autore, a cui deve essere capitato un caso simile a quello che una volta è capitato anche a lui, cf. 16.582.9–16 K. [= 48.8–14 Diels].
65 Cf. supra, pp. 544–45.

Per raggiungere questo scopo Galeno per prima cosa raccomanda di esercitarsi sui testi di riferimento, i suoi libri *De locis affectis* e *De crisibus* perché il commentatore sia in grado di distinguere cosa è detto bene e cosa è detto male nel libro che commenta e non gli accada di difendere testi errati rischiando di apparire ignorante e presuntuoso. Come un maestro, Galeno comincia con la definizione della malattia in questione; si dovrà sapere qual è la parte affetta (nel nostro caso il cervello) e quali sono le sue attività (nel nostro caso l'attività volontaria, la facoltà di ragionare, la sensazione e la memoria); poi insegna che per la prognosi si possono usare solo i segni che sono presenti *sempre* e *solo* in quella malattia, considerando che se non si verifica la seconda ipotesi si deve verificare almeno la prima. Tutto questo, inutile dire, non è detto nel *Prorretico* ma può essere integrato dal commentatore.

La celebre massima ippocratica "recare vantaggio o non fare danno" si realizza qui nel campo dell' esegesi e dell'insegnamento, ed è presentata come il fondamento di un'etica dell'insegnamento al cui servizio Galeno mette la sua competenza medica, l'esercizio nella valutazione dei sintomi, la logica, l'autorità del testo ippocratico e la sua competenza linguistica:

> ἐγὼ μὲν οὖν ἀεὶ προσδιορίζομαι τὰ κατὰ τὸ βιβλίον ἀδιορίστως εἰρημένα καὶ διὰ τοῦτο μέλλοντα βλάψαι τοὺς ἀναγινώσκοντας αὐτά· τοῖς δ' ἐξηγηταῖς ἀδιόριστος ἡ διδασκαλίη γινομένη βλάπτει μᾶλλον ἢ ὠφελεῖ τοὺς νέους. (16.588.17–589.3 K. [= 52.1–4 Diels])
>
> io dunque sempre aggiungo i *diorismoi* a quanto, essendo detto senza *diorismoi*, proprio per questo farebbe danno a chi legge; se l'esposizione didattica da parte degli esegeti manca dei *diorismoi*, danneggia i giovani piuttosto di giovare.

L'arte medica, che Galeno fa dipendere così strettamente dagli scritti di Ippocrate, trova così nella filologia uno straordinario strumento per la sua affermazione, e, se serve, lo usa anche capziosamente non essendo certo la filologia uno strumento neutrale.

Bibliografia

Anastassiou, A., e D. Irmer. *Testimonien zum Corpus Hippocraticum*, voll. 1, 2.1, 2.2, 3 (Göttingen: Vandenhoeck & Ruprecht, 1997–2012).

Bröcker, L. O. "Die Methoden Galens in der literarischen Kritik," *Rheinisches Museum* 40 (1885): 426–28.

Deichgräber, K. *Die griechische Empirikerschule*, um Zusätze vermehrter anastatischer Neudruck der Ausg. von 1930 (Berlin: Weidmann, 1965).

Diels, H. *Galeni In Hippocratis Prorrheticum I commentaria III*, CMG V 9,2 (Leipzig: Teubner, 1915).

Heeg, J. *Galeni In Hippocratis Prognosticum commentaria III*, CMG V 9,2 (Leipzig: Teubner, 1915).
Helmreich, G. *Galeni In Hippocratis De victu acutorum commentaria IV*, CMG V 9,1 (Leipzig: Teubner, 1914).
Helmreich, G. *Galeni De temperamentis libri III*, ed. stereo. ed. anni 1904, addenda adiecit S. Besslich (Stuttgart: Teubner, 1969).
Joly, R. *Hippocrate, Du système des glandes*, t. XIII (Paris: Les Belles Lettres, 1978).
Jouanna, J. *Hippocratis De natura hominis*, CMG I 1,3 (Berlin: Akademie-Verlag, 2002).
Jouanna, J. "La lecture du traité hippocratique de la *Nature de l'homme* par Galien," in M. O. Goulet-Cazé (ed.), *Le commentaire entre tradition et innovation: actes du Colloque international de l'Institut des traditions textuelles, Paris-Villejuif 1999* (Paris: Vrin, 2000), 273–92 (= "Galen's Reading of the Hippocratic Treatise *The Nature of Man*," in J. Jouanna, *Greek Medicine from Hippocrates to Galen: Selected Papers*, trans. by N. Allies, ed. with a preface by P. J. van der Eijk [Leiden: Brill, 2012], 313–33).
Littré, É. *Œuvres complètes d'Hippocrate*, vol. 5 (Paris: J. B. Baillière, 1846).
Manetti, D. "Galeno, la lingua di Ippocrate e il tempo," in J. Barnes e J. Jouanna (edd.), *Galien et la philosophie*, Entretiens sur l'antiquité classique XLIX (Genève: Fondation Hardt, 2003), 171–228.
Manetti, D. "Galen and Hippocratic Medicine: Language and Practice," in C. Gill, T. Whitmarsh, e J. Wilkins (edd.), *Galen and the World of Knowledge* (Cambridge: Cambridge University Press, 2009), 157–74.
Manetti, D. "Philology and Medicine: A Strategic Relationship," in S. Matthaios, F. Montanari, e A. Rengakos (edd.), *The Brill's Companion to Ancient Scholarship* (in corso di stampa).
Manetti, D., e A. Roselli. "Galeno commentatore di Ippocrate," in W. Haase (ed.), *Aufstieg und Niedergang der römischen Welt*, II, 37.2 (Berlin: De Gruyter, 1994), 1529–635.
Mewaldt, J. *Galeni In Hippocratis De natura hominis commentaria III*, CMG V 9,1 (Leipzig: Teubner, 1914).
Nutton, V. *Galen, On Prognosis*, CMG V 8,1 (Berlin: Akademie-Verlag, 1979).
Potter, P. *Hippocrates*, vol. 8 (Cambridge, Mass.: Harvard University Press, 1995).
Siebenborn, E. *Die Lehre von der Sprachrichtigkeit und ihren Kriterien. Studien zur antiken normativen Grammatik*, Studien zur antiken Philosophie V (Amsterdam: B. R. Grüner, 1976).
Sluiter, I. "The Embarrassment of Imperfection: Galen's Assessment of Hippocrates' Linguistic Merits," in P. J. van der Eijk, H. F. J. Horstmanshoff, e P. H. Schrijvers (edd.), *Ancient Medicine in its Socio-cultural Context*, 2 voll. (Amsterdam: Rodopi, 1995), 2:519–35.
Smith, W. D. *The Hippocratic Tradition* (Ithaca: Cornell University Press, 1979; electronic edition, revised 2002).
Strohmaier, G. "Galen als Hippokrateskommentator," in W. Geerlings e Ch. Schulze (edd.), *Der Kommentar in Antike und Mittelalter. Beiträge zu seiner Erforschung* (Leiden: Brill, 2002), 253–74.
van der Eijk, P. J. "Exegesis, Explanation and Epistemology in Galen's Commentaries in *Epidemics*, Books One and Two," in P. E. Pormann (ed.), *Epidemics in Context: Greek Commentaries on Hippocrates in the Arabic Tradition* (Berlin: De Gruyter, 2012), 25–48.
von Staden, H. "'A Woman Does Not Become Ambidextrous': Galen and the Culture of Scientific Commentary," in R. Gibson e C. Kraus (edd.), *The Classical Commentary: Histories, Practices, Theory* (Leiden: Brill, 2002), 109–39.

von Staden, H. "Interpreting Hippocrates," in C. W. Müller, C. Brockmann, e W. Brunschön (edd.), *Ärzte und ihre Interpreten. Medizinische Fachtexte der Antike als Forschungsgegenstand der Klassischen Philologie*, Fachkonferenz zu Ehren von Diethard Nickel (München: Saur, 2006), 15–47.

Wenkebach, E. *Galeni In Hippocratis Epidemiarum librum I commentaria III*, CMG V 10,1 (Leipzig: Teubner, 1934).

Wenkebach, E. *Galeni In Hippocratis Epidemiarum librum III commentaria III*, CMG V 10,2,1 (Leipzig: Teubner, 1936).

Wenkebach, E. *Galeni In Hippocratis Epidemiarum librum VI commentaria I–VI*, CMG V 10,2,2 (Berlin: Akademie-Verlag, 1956).

Thomas Rütten
Hippokrateskommentare im 16. Jahrhundert: Peter Memms Eidkommentar als Paradigma eines gegenwartsbezogenen Genres

Abstract: After some general remarks about Hippocrates commentaries authored in the sixteenth and early seventeenth century, my contribution provides a case study of this text genre: a Latin commentary of the Hippocratic Oath published in 1577 in Rostock. At the same time, the article is the first research paper entirely devoted to the commentary's author, the physician Peter Memm (1531–1589) from Herentals near Antwerp. Using methodological approaches developed in book and print history, university history, reception studies, discourse analysis, biographical studies, Reformation and Renaissance studies, political history, and classical scholarship, and based on hitherto untapped archival material, an attempt is made to historically contextualize Memm's commentary and secure its author and his work a place in current scholarship.

Einleitung

Hippokrateskommentare hat es seit dem Hellenismus gegeben, wenn sich auch die vorgalenischen Hippokrateskommentare nur in Form von Testimonien erhalten haben.[1] Die Aufstellung der vorsalernitanischen Hippokrateskommentare von

---

[1] Inwieweit der Kommentar zu *De articulis* von Apollonios von Kition aus dem 1. Jahrhundert v.Chr. zur literarischen Gattung des Kommentars gehört, ist strittig. Vgl. A. Roselli, „Tra pratica medica e filologia ippocratica: il caso di Apollonio di Cizio," in G. Argoud und J.-Y. Guillaumin (Hgg.), *Sciences exactes et sciences appliquées à Alexandrie. IIIe siècle av. J.-C. – Ier siècle ap. J.-C.*, Centre Jean-Palerne, Mémoires Bd. 16 (Saint-Étienne: Publications de l'Université de Saint-Étienne, 1998), 217–32. Vgl. auch C. Brockmann, „Philologische Annäherungen an Chirurgie und Anatomie. Beobachtungen an Galens Kommentar zu Hippokrates, *De articulis*," in C. W. Müller, C. Brockmann und C. W. Brunschön (Hgg.), *Ärzte und ihre Interpreten. Medizinische Fachtexte der Antike als Forschungsgegenstand der Klassischen Philologie* (Berlin: De Gruyter, 2006), 61–80. Einen guten Überblick über den Kommentar von der Antike bis in die Gegenwart bietet A. Grafton, „Commentary," in ders., G. W. Most und S. Settis (Hgg.), *The Classical Tradition* (Cambridge, Mass.: The Belknap Press of Harvard University Press, 2010), 225–33.

Sybille Ihm[2] hat eindrucksvoll gezeigt, wie intensiv Hippokratesexegese via Kommentierung bereits in der Antike betrieben wurde, und sie ist seither niemals wirklich und nachhaltig abgerissen.[3] Und wenn man bedenkt, daß der fleißigste Hippokrateskommentator der Antike, Galen von Pergamon, zugleich der einflußreichste Stifter eines nach ihm „galenisch" genannten Hippokratismus war, der zeitweilig nicht nur der einzig verfügbare Hippokratismus blieb, sondern bis in die frühe Neuzeit hinein nahezu jeden Hippokratismus prägte, wird die Anciennität dieses Genres und seine Wirkmächtigkeit sofort augenfällig.[4] Wie modellbildend Galens Hippokrateskommentare auch im späten 15. und 16. Jahrhundert waren, erkennt man an der Fülle der von Richard Durling schon 1961 gelisteten

---

2 S. Ihm, *Clavis Commentariorum der antiken medizinischen Texte*, Clavis Commentariorum Antiquitatis et Medii Aevi, Bd. 1 (Leiden: Brill, 2002).
3 H. von Staden, „Staging the Past, Staging Oneself: Galen on Hellenistic Exegetical Traditions," in C. Gill, T. Whitmarsh und J. Wilkins (Hgg.), *Galen and the World of Knowledge* (Cambridge: Cambridge University Press, 2009), 132–56; K.-D. Fischer, „Der griechische Arzt Soran als vermeintlicher Befürworter periodischer Rauschzustände. Anmerkungen zum 2. lateinischen Kommentar zu den hippokratischen Aphorismen," in P. Defosse (Hg.), *Hommage à Carl Deroux*. Tome 2: *Prose et linguistique, médecine* (Brüssel: Latomus 2002), 481–96; ders., „'Zu des Hippokrates reich gedeckter Tafel sind alle eingeladen.' Bemerkungen zu den beiden vorsalernitanischen lateinischen Aphorismenkommentaren," in W. Geerlings und C. Schulze (Hgg.), *Der Kommentar in Antike und Mittelalter. Beiträge zu seiner Erforschung*, Clavis Commentariorum Antiquitatis et Medii Aevi, Bd. 2 (Leiden: Brill, 2002), 275–313; G. Strohmaier, „Der arabische Galenkommentar als indirekter Textzeuge zu Hippokrates, De aere aquis locis," in Müller, Brockmann und Brunschön, *Ärzte und ihre Interpreten*, 233–44; A. Abu Aly, „A Few Notes on Hunayn's Translation and Ibn al-Nafis' Commentary on the First Book of the *Aphorisms*," *Arabic Sciences and Philosophy* 10 (2000): 139–50; N. P. Josse und P. Pormann, „'Abd al-Laṭīf al-Baġdādī's *Commentary on Hippocrates' 'Prognostic'*: A Preliminary Exploration," in P. Pormann (Hg.), *Epidemics in Context: Greek Commentaries on Hippocrates in the Arabic Tradition* (Berlin: De Gruyter, 2012), 251–83; P. O. Kristeller, „Bartholomaeus, Musandinus, and Maurus of Salerno and Other Early Commentators of the 'Articella' with a Tentative List of Texts and Manuscripts," *Italia Medioevale e Umanistica* 19 (1976): 57–87; P. Morpurgo, „L'interpretazione della medicina ippocratica in Maimonide e nei maestri Salernitani," in *Dalla medicina greca alla medicina salernitana: percorsi e tramiti; atti del Convegno internazionale Raito di Vietri sul Mare, Auditorium di Villa Guariglia (25–27 giugno 1995)* (Salerno: Centro Studi Salernitani Raffaele Guariglia, 1999), 25–39; C. O'Boyle, *Thirteenth- and Fourteenth-Century Copies of the „Ars Medicinae": A Checklist and Contents Descriptions of the Manuscripts*, Articella Studies, Nr. 1 (Cambridge: Wellcome Unit for the History of Medicine; Barcelona: CSIC, 1998), vor allem 86–165 (*Ars commentata*).
4 H. von Staden, „'A Woman Does Not Become Ambidextrous': Galen and the Culture of Scientific Commentary," in R. K. Gibson und C. Shuttleworth Kraus (Hgg.), *The Classical Commentary: History, Practice, Theory* (Leiden: Brill, 2002), 109–39; ders., „Staging the Past"; D. Manetti und A. Roselli, „Galeno commentatore di Ippocrate," in W. Haase (Hg.), *Aufstieg und Niedergang der römischen Welt*, II 37.2 (Berlin: De Gruyter, 1994), 1529–635.

Drucke und Nachdrucke dieser Texte und ihrer Übersetzungen ins Lateinische[5] sowie an der Tatsache, daß es auch im 16. Jahrhundert Autoren in den Sinn kommen konnte, einen Hippokrateskommentar nicht unter eigenem Namen zu publizieren, sondern unter dem des Galen zu veröffentlichen, mithin einen Galenkommentar zu fälschen.[6] Hinzu kommt, daß Hippokrateskommentare, von denen zwischen 1550 und 1650 etwa 100 *expressis verbis* im Titel als solche ausgewiesen sind, formal in der Tradition antiker und mittelalterlicher Vorbilder stehen, daß sie in Zeiten der Emanzipation der Volkssprachen sämtlich auf Latein abgefaßt sind und – last but not least – die Autorität eines Altvordern und dessen Bücher zu beschwören scheinen, wo doch schon allenthalben ein neues, über die Alten hinausweisendes „Könnensbewußtsein"[7] zu verzeichnen ist und in der Medizin der „Codex naturae" (Paracelsus), das „Buch der Natur" im Sinne von Galileo Galilei, – oder später Francis Bacon und René Descartes – zunehmend den Büchern des Hippokrates Konkurrenz zu machen beginnt.[8] Man hat den Eindruck, daß Hippokrateskommentare zurückblicken, für das Alte und Traditionelle stehen und sich dem Fortschritt der Zeit verweigern. Sie wären demnach als epigonal zu charakterisieren und von den hochdynamischen Entwicklungen in Staat und Kirche, Wirtschaft und Kunst, Forschung und Lehre, die die Geschichte Europas zwischen 1500 und 1650 prägen, abzukoppeln.

Nun haben aber schon Gelehrte wie Wesley D. Smith, Walter Pagel oder J.-P. Pittion[9] festgestellt, daß im 16. Jahrhundert eine zunehmende Dissoziierung von

---

[5] R. J. Durling, „A Chronological Census of Renaissance Editions and Translations of Galen," *Journal of the Warburg and Courtauld Institutes* 24 (1961): 230–305.
[6] P. Demont, „L'édition Vigoreus (1555) du traité hippocratique De humoribus et d'un 'commentaire de Galien' à ce traité (= [Galien], De humoribus, XIX, 485–496 Kühn), avec la traduction du De humoribus galénique," in V. Boudon-Millot und G. Cobolet (Hgg.) unter der Mitarbeit von H. Ferreira-Lopes und A. Guardasole, *Lire les médecins grecs à la Renaissance: aux origines de l'édition médical; actes du Colloque international de Paris (19–20 Septembre 2003)* (Paris: De Boccard, 2004), 43–59.
[7] Zu diesem Terminus, der sich auch auf die Medizin des 16. Jahrhunderts beziehen läßt, vgl. C. Meier, „Ein antikes Äquivalent des Fortschrittsgedankens: das 'Könnensbewußtsein' des 5. Jahrhunderts v.Chr.," *Historische Zeitschrift* 226 (1978): 265–316; in erweiterter und überarbeiteter Form wiederabgedruckt in ders., *Die Entstehung des Politischen bei den Griechen* (Frankfurt a. M.: Suhrkamp, 1990), 435–99, bsd. 469–84.
[8] Vgl. H. Blumenberg, *Die Lesbarkeit der Welt*, 3. Aufl. (Frankfurt a. M.: Suhrkamp, 1996).
[9] J.-P. Pittion, „Scepticism and Medicine in the Renaissance," in R. H. Popkins und C. Schmitt (Hgg.), *Scepticism from the Renaissance to the Enlightenment*, Wolfenbütteler Forschungen, Bd. 35 (Wiesbaden: Harrassowitz, 1987), 103–32, bsd. 104; W. Pagel, *Paracelsus: An Introduction to the Philosophical Medicine in the Era of the Renaissance* (Basel: S. Karger AG, 1982), 218; W. D. Smith, *The Hippocratic Tradition*, Electronic ed., revised 2002 (1. Aufl.: Ithaca, N.Y.: Cornell University Press, 1979), 14–31: http://www.biusante.parisdescartes.fr/medicina/Hippo2.pdf.

Hippokrates und Galen zu verzeichnen ist, daß die Zeiten, da der galenische Hippokratismus konkurrenzlos das Rezeptionsbild des Hippokrates dominiert, spätestens mit dem Paracelsismus, dem der Einfluß Vesals in dieser Hinsicht sekundiert, vorüber waren. Seit 1550 unterscheiden Humanisten und Ärzte, Maler und Stecher, Poeten und Philosophen die beiden Sterne am antiken Ärztehimmel säuberlich, wird ihnen der rezeptionslenkende Eingriff Galens in das hippokratische Erbe stärker bewußt, öffnen sich neue Aktualisierungschancen für hippokratische Schriften und ihren Namenspatron. Der Höhepunkt dieses Dissoziierungsprozesses fällt mit dem Gipfelpunkt der Produktion von Hippokrateskommentaren in den beiden letzten Dezennien des 16. Jahrhunderts zusammen. Nun kommt etwas vor, was zuvor nahezu undenkbar gewesen wäre: Galen wird in einem Hippokrateskommentar kaum einmal erwähnt. Auch treten nun Texte aus dem *Corpus Hippocraticum* in den Vordergrund, die von antiken Autoren, *notabene* Galen, nicht kommentiert wurden, so daß sich dem Kommentator – bar eines konkreten Modells – größere Freiräume eröffnen, da das In-Spuren-Gehen nur für den Primärtext, nicht aber für vorlaufende Kommentare gilt. Und schließlich läßt die Beobachtung aufhorchen, daß es den meisten Kommentatoren anscheinend gar nicht so sehr um einen Beitrag zum Projekt des Renaissancehumanismus geht, die antiken Hippokratestexte auf der Grundlage ihrer handschriftlichen Überlieferung zu rekonstruieren, von Artefakten zu reinigen und dadurch verständlicher und übersetzbarer zu machen. Bei einem solchen Primärinteresse würde man in zahlreichen Kommentaren mehr philologische Kärnerarbeit erwarten, reichere Informationen zu Lesarten, Handschriften, lexikalischen, grammatischen und terminologischen Spezialfragen, stärkere Reflexe der Editions- und Übersetzungsgeschichte hippokratischer Texte und größere Anstrengungen in Richtung einer Rekonstruktion antiker Verhältnisse. Kommentatoren treibt in der Mehrzahl – Ausnahmen bestätigen die Regel – nicht das *redite ad fontes* um, sie betreiben nicht die Restaurierung der Medizin mit den Mitteln einer Text- und Sprachbereinigungskampagne. Das alles deutet darauf hin, daß Hippokrateskommentare nicht etwa anachronistische Phänomene in ihrer Entstehungszeit, sondern vielmehr deren dezidierter Spiegel sind. Sie haben Konjunktur, wo zeitgenössische, nicht antike Verhältnisse auf dem Gesundheitsmarkt, zeitgenössische, nicht antike Standortbestimmungen von Arzt, Patient und Gesellschaft, zeitgenössische, nicht antike Verhältnisbestimmungen von Medizin und Religion, Medizin und Recht, Medizin und Staatlichkeit zur Debatte stehen. Mit anderen Worten: wenn sich bei der Vielzahl an überlieferten Hippokrateskommentaren aus dem Zeitraum zwischen 1550 und 1650 neben der Provenienz der Primärtexte ein zweiter gemeinsamer Nenner abzeichnet, dann ist es der enorme Gegenwartsbezug, Aktualitätsgrad und Reformwille, der sich in ihnen artikuliert. Die formal und inhaltlich disparaten Texte eint auch und vor allem jene andere

Perspektive, die nach vorn gerichtet ist und sich mit der oben skizzierten rückwärtsgewandten zu einer Gesamtperspektive fügt, die etwas Janushaftes hat.[10]

Unter diesem Blickwinkel, der den Fluchtpunkt der Kommentare in ihrer Entstehungszeit und nicht in der Antike ansiedelt, lohnt es sich, nach den spezifischen Raum-Zeit-Koordinaten jedes einzelnen Hippokrateskommentars zu fragen, nach den Lebensverhältnissen, Erfahrungshintergründen und Bildungshorizonten jedes einzelnen Kommentators, nach der je eigenen Entstehungs-, Druck- und Rezeptionsgeschichte ihrer Texte, deren paratextuellen Apparat sowie intendierten und nicht intendierten Leserkreisen. Es erweist sich nämlich bei näherem Hinsehen, daß außer der Tatsache, daß sämtliche Hippokrateskommentare einen Text aus dem *Corpus Hippocraticum* zum Gegenstand haben und in lateinischer Sprache abgefaßt sind, kein einziges Kriterium, sei es formaler oder inhaltlicher Natur, betreffe es die Verfasser oder die Leser, die Verbreitung oder Wirkung, die Form oder den Inhalt, den Anlaß oder die Absicht, verbindlich zu sein scheint. Die von Glenn Most vor fünfzehn Jahren aufgestellten Kriterien antiker Kommentare gelten für die hier in Rede stehenden Texte nicht oder nur mit starken Einschränkungen.[11] Multiversalität ist das Charakteristikum dieser Textsorte, zumal die Biographien der Kommentatoren den Kommentaren als singuläre Subtexte unterlegt sind.

I Der Eidkommentar Peter Memms

I.1 Buchgeschichtliches

Der vorliegende Beitrag wird sich daher auf eine Fallstudie beschränken: den Kommentar zum hippokratischen Eid, den Peter Memm (1531–1589) aus Herentals, einem Ort nahe der flandrischen Hauptstadt Antwerpen[12], verfaßte und 1577 in

---

10 T. Rütten, „Traduzioni e commenti del corpus ippocratico e galenico," in A. Clericuzio und G. Ernst (Hgg.), *Il Rinascimento italiano e l'Europa* (Treviso-Costabissara: A. Colla, 2008), 479–93.
11 G. Most, „Introduction," in ders. (Hg.), *Commentaries* (Göttingen: Vandenhoeck & Ruprecht, 1999), VII–XV, XII–XIV.
12 Peter Memm besuchte die Lateinschule seines Geburtsortes. Zu dieser Schule im fraglichen Zeitraum vgl. J. Cools, *Geschiedenis van het College te Herentals* (Herentals: Oud-Leerlingenbond van het Sint-Jozefscollege, 1984), 37–46. Das Stadsarchief von Herentals verlor, wie uns sein Leiter Herr Dr J. Goris freundlicher Weise mitteilte, bei einem Brand des Stadthauses 1512 und bei Soldatenunruhen im Jahre 1657 ältere Bestände fast vollständig. Im sogenannten Pellenboek, einem Sterberegister, das die Jahre 1557–1580 umfaßt, taucht der Name Memm nicht auf.

Rostock drucken ließ.[13] Das Buch ist in wenigen Exemplaren erhalten. Det Kongelige Bibliotek in Kopenhagen, die Universitätsbibliothek zu Rostock und die Herzog August Bibliothek in Wolfenbüttel zählen es zu ihren Beständen. Memms Eidkommentar ist also ein ausgesprochenes *Rarum*. Das Wolfenbütteler Exemplar gelangte erst Anfang des 18. Jahrhunderts in die Herzog August Bibliothek. Die Signatur 143.9 Medica wurde von Georg Burkhard Lauterbach (1683–1751), der seit 1719 als Registrator, seit 1727 als erster Sekretär in der Bibliothek arbeitete, in den

---

13 P. Memmius, *Hippocratis iusiurandum commentario recenter illvstratvm, cvi accessit altera pars, qua ratione Medicorum vita et ars sanctè conseruetur declarans* (Rostock: A. Ferber, 1577). Der jüngste biographische Lexikonartikel zu Memm, der im Gegensatz zu zahlreichen älteren biographischen Lexikonartikeln wenigstens die Eckdaten seiner Biographie und seine Anstellungen korrekt wiedergibt, findet sich in G. Grewolls, *Wer war wer in Mecklenburg und Vorpommern*, DVD (Rostock: Hinstorff, 2011), 6510. Der ausführlichste Artikel zu Memms Zeit in den Niederlanden ist immer noch J. J. Dodt van Flensburg, „Over den Utrechtschen arts Petrus Memmius, den Delftschen Drukker Hermann Schinckel, en den aartsbisschop Frederik Schenck van Tautenburch, een paar aanteekeningen," *Utrecht voorheen en thans: Tijdschrift voor geschiedenis, Oudheden en Statistick van Utrecht*, 2de Serie, 2 (1845): 75–80. Ich danke Frau Dr. Irmgard Broos für die freundliche Übersendung von Fotokopien. Memms Eidkommentar hat bislang in der Forschungsliteratur nur sporadisch Erwähnung gefunden: J. H. Meibom, *Hippocratis Magni Orkos sive Jusjurandvm* (Leiden: J. Lauwikius, 1643), [Brief an den Leser; o.P.] und 111; [Anon.], „Hippocratis Coi ... ," *Etwas von gelehrten Rostockschen Sachen, für gute Freunde* 3 (1739): 581–82; L. Edelstein, *The Hippocratic Oath: Text, Translation, and Interpretation*, Supplements to the Bulletin of the History of Medicine, Nr. 1 (Baltimore: The Johns Hopkins Press, 1943), Anm. 72; T. Rütten, „Ärztliche Ethik in der Renaissancemedizin: Mechanismen der Neukontextuierung des hippokratischen Eides in der späthumanistischen Kommentarliteratur zwischen 1540 und 1640," in I. Garofalo et al. (Hgg.), *Aspetti della terapia nel Corpus Hippocraticum; atti del IXe Colloque International Hippocratique, Pisa 25–29 settembre 1996* (Florenz: Olschki, 1999), 517–42, bsd. 529–33; N. Siraisi, *History, Medicine, and the Traditions of Renaissance Learning* (Ann Arbor: The University of Michigan Press, 2007), 84; T. Rütten, *Geschichten vom Hippokratischen Eid*, CD-ROM (Wiesbaden: Harrassowitz, 2008) (im Pfad Frühe Neuzeit). Ich danke Herrn Dr. M. Repetzki für seine Mitarbeit an diesem Projekt als studentischer Hilfskraft, dem Wellcome Trust für Forschungsförderung im Rahmen eines University Award, Veranstaltern wie Auditorien in Mainz, Greifswald, Leiden und St. Andrews für Anregungen und Nachfragen zu Vortragsfassungen dieses Aufsatzes und Frau Dr. A. Oommen-Halbach, Herrn Dr. R. Kinsky und Herrn Professor K.-D. Fischer für die kritische Durchsicht des Typoskripts und manchen Korrekturvorschlag. Es ist mir eine große Freude, meine Forschungen zu Memms Eidkommentar in dieser Festschrift, dessen Empfänger sie nicht nur begleitet und gefördert hat, sondern ihnen durch seine substantiellen Beiträge zu Kommentaren medizinischer Texte (vgl. Anm. 3 und 4), zum hippokratischen Eid (vgl. Anm. 43) und zur Methodologie der Erforschung vormoderner Fachtexte den Weg gewiesen hat, erstmalig im Zusammenhang in Schriftform zur Diskussion zu stellen.

Bücherradkatalog eingetragen.[14] Gut 100 Jahre zuvor gehörte das Exemplar indes dem Arzt Augustinus Lollius Adama (c. 1570–1616).[15] Er erwarb es in einer Zeit, als er nach seiner medizinischen und philosophischen Promotion als *lector tertianorum*, Lehrer der dritten Klasse, an der Hieronymus-School in Utrecht tätig war, an der auch Memm etwa 40 Jahre zuvor gearbeitet hatte.[16] Von Utrecht dürfte das Exemplar seinem Besitzer nach Kampen, Leeuwarden und Franeker gefolgt sein, wo Lollius am 10.4.1608 zum *Professor medicinae* ernannt wurde und 1615 das Amt des Rektors bekleidete. Mit seinem Tod im darauf folgenden Jahr verliert sich die Spur des Exemplars bis zu dessen Eintreffen in Wolfenbüttel.[17]

---

14 Der Bibliotheksdirektor Otto Heinemann nannte ihn „ein Muster treuen Fleisses und selbstloser Hingabe," der sich „um die Bibliothek die allergrößten Verdienste erworben" habe, „vor allem durch die stille aber rastlose Weiterführung des von Leibniz begonnenen Nominalkatalogs." O. V. Heinemann, *Die Herzogliche Bibliothek zu Wolfenbüttel. Ein Beitrag zur Geschichte deutscher Büchersammlungen*, 2., völlig neugearb. Aufl. (Wolfenbüttel: Julius Zwissler, 1894), 138. Siehe auch C. P. C. Schönemann, „Dem Andenken Georg Burkhard Lauterbachs. Ein Beitrag zur Geschichte der Wolfenbütteler Bibliothek," *Serapeum. Zeitschrift für Bibliothekswissenschaft* Jg. 3, Nr. 14 (1842): 213–24.
15 Der handschriftliche Besitznachweis am oberen Rand des Titelblattes lautet: „Sum Augustini Lollii ab Adama," und die Jahreszahl 1601 ist ebenfalls zu entziffern. Zu Augustinus Lollius Adama vgl. J. W. Napjus, „Augustinus Lollius Adama: Hoogleeraar in de geneeskunde te Franeker," *Nederlands Tijdschrift voor Geneeskunde* Jg. 80 (1936): 1478–84. Ich danke Herrn Professor Christoph Lüthy für die Übersendung dieses Artikels. Siehe auch W. B. S. Boeles, *Frieslands Hoogeschool en het Rijks Athenaeum te Franeker*, 2 Bde. (Leeuwarden: Kuipers, 1878–1889), 2:84–86; G. A. Lindeboom, *Dutch Medical Biography: A Biographical Dictionary of Dutch Physicians and Surgeons 1475–1975* (Amsterdam: Rodopi, 1984), Sp. 6; F. S. Knipscheer, „Adama (Augustinus Lollius)," in P. C. Molhuysen und P. J. Blok (Hgg.), *Nieuw Nederlandsch Biografisch Woordenboek*, 10 Bde. (Leiden: A. W. Sijthoff, 1911–1937), 1: Sp. 18. Zur 1585 gegründeten Universität von Franeker, vgl. G. T. Jensma, F. R. H. Smit und F. Westra, Hgg., *Universiteit te Franeker 1585–1811: Bijdragen tot de geschiedenis van de Friese Hogeschool* (Leeuwarden: Fryske Akademy, 1985); R. E. O. Ekkart, *Franeker Professorenportretten: Iconografie van de Professoren aan de Academie en het Rijksathenaeum te Franeker 1585–1843* (Franeker: Wever, 1977); S. Zijlstra, *Het geleerde Friesland – een mythe? Universiteit en maatschappij in Friesland en Stad en Lande ca. 1380–1650* (Leeuwarden: Fryske Akademy, 1996).
16 A. Ekker, *De Hieronymusschool te Utrecht*, 2 Bde. (Utrecht: Bosch, 1863–864), 1:1474–636, 30–1 (Memm) bzw. 81 (Lollius Adama).
17 Das Kopenhagener und Rostocker Exemplar harren autoptischer Untersuchung. Beide Städte, wie auch Utrecht, wo Augustinus Lollius Adama sein Exemplar des Memmschen Kommentars erwarb, gehören zu den Wirkungsstätten Peter Memms.

I.2 Die *pars prima* des Memmschen Eidkommentars

Memms Eidkommentar gliedert sich in zwei Teile. Die *pars prima* besteht aus einem Widmungsschreiben (ff. A2–a2) an Ulrich III., den Herzog von Mecklenburg-Güstrow (1527–1603)[18], einer lateinischen Übersetzung des Eides (ff. B1–B2), die stillschweigend von Janus Cornarius (1500–1558) übernommen wird[19], und dem eigentlichen *commentarius*, der sich wiederum in eine Art Biographie (ff. B2ᵛ-C4ᵛ) des Hippokrates, des vermeintlichen Verfassers des Eides, und den eigentlichen Kommentar (C4ᵛ-G2) gliedert.

I.2.1 Memms Widmungsschreiben an Herzog Ulrich

In seinem Widmungsschreiben stilisiert Memm seinen Dienstherrn zu einem *Hippocrates redivivus iurisprudentiae*, lobt in Analogie zur *pax Augusta* die *pax Ulriaca* und deren Schöpfer als *pater patriae* und *numen venerandum*. Er rühmt in Ulrich die Personifizierung der Gerechtigkeit und preist dessen *diuina clementia* (f. a). So wie die Religion auf Liebe zu Gott und dem Nächsten beruhe, die Juris-

---

[18] Ulrich regierte seit dem 17.2.1555 als Nachfolger seines Bruders Johann Albrecht I. in Güstrow, von 1576–1585 als Vormund des Sohnes Johann VII., allerdings auch in Schwerin. Vgl. L. Sellmer, „Ulrich III., Herzog von Mecklenburg-Güstrow," in S. Pettke und A. Röpcke (Hgg.), *Biographisches Lexikon für Mecklenburg*, 6 Bde. (Rostock: Schmidt Römhild, 1995–2011), 1:231–35; Grewolls, *Wer war wer*, 10258–59.

[19] [J. Cornarius], *Hippocratis Coi medicorum omnium longe principis opera quae ad nos extant omnia. Per Ianum Cornarium medicum physicum latina lingua conscripta* (Basel: Froben, H. d. Ä. und Episcopius, N. d. Ä., 1546), 5. Zu Cornarius, der 1526 von Herzog Heinrich IV. von Mecklenburg zum Leibarzt des Prinzen Magnus (1509–1550) in Bützow ernannt wurde und am 1.2.1526 – 42 Jahre vor Memm – in die Rostocker Universität einzog, um für etwa zwei Jahre mit einem Stipendium des Herzogs *extraordinarie* an der medizinischen Fakultät über Hippokrates, z. B. dessen Aphorismen, zu lesen und an der Artistenfakultät Griechisch zu unterrichten, vgl. M.-L. Monfort, *Janus Cornarius et la redécouverte d'Hippocrate à la Renaissance* (Turnhout: Brepols, 2014). Vgl. auch B. Mondrain, „Éditer et traduire les médecins grecs au XVIᵉ siècle: l'exemple de Janus Cornarius," in D. Jacquart (Hg.), *Les voies de la science grecque; études sur la transmission des textes de l'Antiquité au dix-neuvième siècle* (Genf: Droz, 1997), 391–417. Wie Cornarius versieht Memm an zahlreichen Stellen (z. B. f. M6ᵛ, M7) seines Kommentars Hippokrates mit dem Beinamen *princeps medicorum*. Andere, die es ihnen gleichtun, werden erwähnt in Siraisi, *History, Medicine, and the Traditions*, 290, Anm. 69 und H. King, „The Power of Paternity: The Father of Medicine Meets the Prince of Physicians," in D. Cantor (Hg.), *Reinventing Hippocrates* (Aldershot: Ashgate, 2002), 23–36. Zum Lehrbetrieb an der Rostocker Universität in den 1520er Jahren, siehe W. E. Wagner unter Mitarbeit von R. Ackermann, *Observantia lectionum in universitate Rostochiensi (1520). Das älteste gedruckte Vorlesungsprogramm der Universität Rostock* (Hamburg: Verlag Dr. Kovač, 2011), vor allem 83–88 (Medizin).

prudenz auf Liebe zur Gerechtigkeit und die Philosophie auf Liebe zur Weisheit, so beruhe die Medizin auf eben jenen Tugenden, die der hippokratische Eid einklage: *fidelitas, taciturnitas, castitas*. Sie seien der „Tugendschatz" (f. A5ᵛ), der die *medica facultas* zur berühmtesten (f. A4) aller Künste mache. Neben dieser Stellungnahme zur *disputa delle arti*, die Memms Landsmann Andreas Vesal in der Widmung seiner *Fabrica* an Karl V. vom 1. August 1542 auf das Wirkmächtigste wiederbelebt hatte,[20] schlägt Memm in seiner *epistola dedicatoria* zwei Themen an, die sich leitmotivisch durch seinen gesamten Kommentar ziehen: das der Nützlichkeit im Hinblick auf das Gemeinwohl und das der Lehrer-Schüler-Beziehung, die nach dem Muster der Vater-Sohn-Beziehung zu gestalten sei. Ersteres Motiv wird in einer Klimax verdichtet (f. A8ᵛ):

> Hoc modo functionibus *vtiliter* aliquis inuigilat, *vtilius* se in ijs exercet, *vtilißime* administrat. [Auf diese Weise wacht jemand *nutzbringend* über seine Aufgaben, führt sie mit *größerem Nutzen* aus und trifft seine Vorkehrungen *auf das Nützlichste*. (Kursivierung v. Verf.)]

Letzteres Thema, das der Lehrer-Schüler-Beziehung, wird in einer Antithese zwischen dem hippokratischen Dankbarkeitsversprechen der Medizinstudenten gegenüber ihren Lehrern und deren Söhnen auf der einen Seite und der tatsächlichen Undankbarkeit zeitgenössischer Medizinstudenten auf der anderen eingeführt (f. A6).

I.2.2 Der Eidtext in der Übersetzung von Janus Cornarius

Was den nachfolgenden Volltext des Eides in der lateinischen Übersetzung des Cornarius betrifft, so fällt auf, daß dessen lemmatische Wiederholungen zu Beginn der einzelnen Kommentarteile – drucktechnisch durch Majuskeln vom eigentlichen Kommentar Memms abgehoben – nicht nur Verkürzungen und orthographische Veränderungen (*foeliciter/feliciter*; *caeterum/coeterum*; *praeces/ preces*), sondern auch gravierendere Umtextuierungen erfahren: *aegrotos* statt

---

20 A. Vesalius, *De humani corporis fabrica libri septem* (Basel: Oporinus, 1543), f. 2: *Nam, ut caeteras quidem silentio praeteream, et de ea quae sanitati hominum praefecta est, aliquanti per sermonem instituam, profectò in hac tametsi reliquarum omnium quas hominis ingenium adinuenit longè commodissima et imprimis necessaria difficilisque ac operosa sit* ... („I will pass over the other arts in silence and direct my words for a while to that which is responsible for the health of mankind; certainly, of all the arts that human genius has discovered, this is by far the most useful, indispensable, difficult, and laborious," übers. nach D. H. Garrison and M. H. Hast, *Andreas Vesalius, De Humani Corporis Fabrica Libri Septem/The Fabric of the Human Body: An Annotated Translation of the 1543 and 1555 Editions*, 2 Bde. [Basel: Karger, 2013], Bd. 1, 1)].

aegros; *ob aegrotantium utilitatem* statt *ob utilitatem aegrotantium*; *Quaecunque vero audiero, aut inter curandum videro* statt *Quaecunque vero inter curandum videro aut audiero*; *ad medicandum tum non adhibitus* statt *ad medicandum non adhibitus*; *apud homines* statt *apud omnes homines*. Solche Retextuierungen und inversen Wortstellungen legen den Schluß auf Extemporierung nahe. Memm kannte den Eid wohl auswendig, aber eben nicht im exakten Wortlaut der Übersetzung, auf die er sich stützte.[21] Obwohl Memm bei Jacques Dubois (1478–1555), einem der führenden humanistischen Galenisten seiner Zeit, in Paris studierte[22]

---

21 Cornarius' (*Hippocratis Coi medicorum, 5*) Übersetzung lautet folgendermaßen: *Apollinem Medicum, et Aesculapium, Hygeamque, ac Panaceam iuro, deosque omnes itemque deas testes facio, me hoc iusiurandum, et hanc contestationem conscriptam, pro viribus et iudicio meo integrè seruaturum esse. Praeceptorem sanè qui me hanc edocuit artem, parentum loco habiturum, vitam communicaturum, eaque quibus opus habuerit impertiturum: eos item, qui ex eo nati sunt, pro fratribus masculis iudicaturum, artemque hanc si discere voluerint, absque mercede, et pacto edocturum: praeceptionum, ac auditionum, reliquaeque totius disciplinae participes facturum, tum meos, tum praeceptoris mei filios, imò et discipulos, qui mihi scripto cauerint, et medico jureiurando addicti fuerint, alij verò praeter hos nulli. Caeterum quod ad aegros attinet sanandos, diaetam ipsis constituam pro facultate et judicio meo commodam, omneque detrimentum et iniuriam ab eis prohibebo. Neque verò ullius preces apud me adeò validae fuerint, vt cuiquam venenum sim propinaturus, neque etiam ad hanc rem consilium dabo. Similiter autem neque mulieri talum uuluae subdicium, ad corrumpendum conceptum uel foetum, dabo. Porrò praeterea et sanctè uitam et artem meam conseruabo. Nec verò calculo laborantes secabo, sed uiris chirurgiae operarijs eius rei faciendae locum dabo. In quascunque autem domos ingrediar, ob vtilitatem aegrotantium intrabo: ab omnique iniuria uoluntaria inferenda, et corruptione cum alia, tum praesertim operum venereorum abstinebo, siue muliebria, siue uirilia, liberorum ue hominum aut seruorum corpora mihi contigerint curanda. Quaecunque uero inter curandum uidero aut audiero, imò etiam ad medicandum non adhibitus, in communi hominum uita cognouero, ea siquidem efferre non contulerit, tacebo: et tanquam arcana apud me continebo: Hoc igitur iusiurandum mihi integrè seruanti, et non confundenti, contingat et uita et arte feliciter frui, et apud omnes homines in perpetuum gloriam meam celebrari. Transgredienti autem et peieranti, his contraria eueniant.*
22 Im Brief an den Leser, der sich in Memms *De recto medicinae vsu* findet, bezeichnet er Iacobus Sylvius, also Jacques Dubois, als seinen Lehrer (*meus praeceptor*). Vgl. P. Memm, *De recto medicinae vsu liber vnus: non solùm medicis, verùm etiam cuiusuis disciplinae sectatoribus perutilis, nunc primùm in lucem editus* (Delft: H. Schinckel, 1564), f. b6. Geht man davon aus, daß Memm in etwa demselben Alter (18) wie sein Landsmann Jacob Bording nach Paris ging, um bei Dubois Medizin zu studieren, wird er um 1550 dort gewesen sein, als Dubois einen Lehrstuhl am Collège Royal übernahm. Einen sicheren *terminus ante quem* markiert Dubois' Todesjahr 1555. Zu Jacques Dubois vgl. G. Baader, „Jacques Dubois as a Practitioner," in A. Wear, R. K. French und I. M. Lonie (Hgg.), *Medical Renaissance of the Sixteenth Century* (Cambridge: Cambridge University Press, 1985), 146–54 (Anmerkungen: 317–18); A. Drizenko, „Jacques Dubois, dit Sylvius, traducteur et commentateur de Galien," in Boudon-Millot und Cobolet, *Lire les médecins grecs à la Renaissance*, 199–208; H. King, „Engendrer 'la femme': Jacques Dubois et Diane de Poitiers," in C. McClive und N. Pellegrin (Hgg.), *Femmes en fleurs, femmes en corps: sang, santé, sexualités du Moyen Âge au Lumières* (Saint-Étienne: Publications de l'Université de Saint-Éti-

und sich als Advokat der Hellenisierung – und übrigens auch Arabisierung – der Medizin nach dem Pariser Modell vernehmen läßt,[23] scheint er sich für die griechische bzw. arabische Eidüberlieferung nicht interessiert zu haben. Sein Eidverständnis beruht vor allem auf Cornarius' lateinischer Übersetzung, die z. B. das chirurgische Eidversprechen auf den Steinschnitt verengt, weshalb auch Memm in dem betreffenden Kommentarteil (ff. E7–E8^v) ausschließlich den Blasensteinschnitt thematisiert und dem *medicinae studioso* sogar einen Einblick in die zu seiner Zeit angeblich gängigste Operationstechnik (*agendi modum maxime vsitatum*) gewährt (E 8^v) – beschrieben wird der mediane Dammschnitt und das Verfahren mit der kleinen Gerätschaft.[24] Das Beispiel zeigt, in welchem Maße

---

enne, 2010), 125–38. Zu Jacob Bording vgl. Lindeboom, *Dutch Medical Biography*, Sp. 218–19; W. Teichmann, „Bording, Jacob d. Ä.," in Pettke und Röpcke, *Biographisches Lexikon für Mecklenburg*, 1:36–39.

23 *Operae pretium itaque videtur, vt sint ditißimi linguarum plurimarum, non quidem ad superbiam, magnitudinemque animi inanem, verùm vt viam, qua ars accumuletur, consequantur ampliorem. Ars enim primos praecipuosque fontes habet graecos. Horum sermo etsi in latinum idioma satis quidem latinè à multis hodie translatus sit, tamen ipsos graecos latentes energias vsurpare inter se intelligo, quarum interpretes in transpositione variant semperque discordes existunt: ... Gallia nobis exemplo esse potest, in qua florida est doctorum virorum congregatio, qualis nunquam cuiquam visa est Athenis. Graecorum sermo cuique in ore est, adeò vt septennes pueri ipsum et legant et intelligant. Medicinae verò est graeca profeßio, quod tantoperè auditorum animos inflammat, vt pauci ad scholas adeant, qui graecis non sunt comitati auctoribus. Consultò insuper fecerint, qui graeco sermoni Arabum linguam adiungunt, quorum purae translationis hactenus extitit inopia.* („Deshalb scheint es der Mühe wert zu sein, wenn sie [die Ärzte] sehr viele Sprachen beherrschen, freilich nicht aus Übermut und falscher Überheblichkeit, sondern um einen besseren Weg zu verfolgen, auf dem ihre Kunst gesteigert werde. Die[se] Kunst verfügt nämlich als früheste und hervorragendste Quellen über griechische. Selbst wenn deren Sprache heutzutage von vielen hinreichend in lateinische Wendungen übersetzt worden sein mag, so ist mir doch bewußt, daß die Griechen selbst die verborgenen Bedeutungen untereinander geltend machen, bei deren Übersetzung die Übersetzer voneinander abweichen und sich nie einig sind. ... Als Beispiel mag uns Frankreich dienen, wo eine Gemeinschaft gebildeter Männer in Blüte steht, wie sie Athen nie gekannt hat. Jeder beherrscht die griechische Sprache derart, daß selbst siebenjährige Knaben sie lesen und verstehen. Die Ausübung der Medizin ist in der Tat griechisch, was die Hörer derart anspornt, daß nur wenige Schulen besuchen, die mit den griechischen Autoren nicht vertraut sind. Überlegt haben außerdem diejenigen gehandelt, die der griechischen Sprache noch die der Araber hinzugefügt haben, von denen bislang klar verständliche Übersetzungen noch nicht in ausreichender Zahl vorliegen,") Memmius, *De recto medicinae vsu*, 61–63 = P. Memmius, *Hippocratis iusiurandum commentario Pars secunda*, K8^v–L1^v; zu der Identität der beiden Texte, siehe unten unter 1.3.)

24 Zur Chirurgie in den nördlichen Niederlanden zur fraglichen Zeit vgl. M. J. van Lieburg, „De genees- en heelkunde in de noordelijke Nederlanden, gezien vanuit de stedelijke en chirurgijnsgilde-ordonnanties van de 16e eeuw," *Tijdschrift voor geschiedenis van de geneeskunde* 6 (1983): 169–84.

Cornarius mit seiner Eidübertragung Interpretamente in den hippokratischen Text hineinträgt, die unausweichlich die weitere Exegese des Textes durch Kommentatoren präjudiziert.

I.2.3 Der *commentarius* überschriebene Kommentarteil

I.2.3.1 Die Kurzbiographie des Hippokrates

Der auf die Eidübersetzung folgende Textteil stellt den Verfasser des Eides vor, der für Memm wie für seine Zeitgenossen unstreitig Hippokrates heißt. Dessen Abstammung (*genus*), Herkunft (*locus*), Ausbildung/Erziehung (*educatio*) und Taten werden in gebotener Kürze abgehandelt, bevor Memm auf die Folgen der Vernachlässigung des Eides zu sprechen kommt: Ansehensverlust der Medizin und der Ärzte, Mißbrauch der Medizin, Abwanderung der Patienten zu Quacksalbern.[25] Hier lassen sich Elemente einer *declamatio*[26] erkennen, wie sie Philipp Melanchthon (1497–1560) anläßlich der Promotion Jakob Milichs durch Augustin Schurff im November 1536 verfaßte.[27] Dieser Text, der 1544 erstmalig im Druck erschien, könnte ein Modell[28] für den biographischen Teil des Memmschen Kommentars

---

[25] Memm spricht von „*[g]aneonibus, meretriculis, vetulis et in genere leuißimis quibusq[ue] hominibus*" [Wüstlingen, elenden Huren, alten Weibern und überhaupt gerade den charakterlosesten Menschen] (f. B 6ᵛ).

[26] Melanchthon hatte dieses Genre 1523 in den universitären Ausbildungsgang in Wittenberg eingeführt. Als kunstvoll stilisierte Reden gehören sie rhetorisch dem *genus demonstrativum* an, besitzen als solche eminenten Gegenwartsbezug und decken, wenn es sich um Biographien handelt, die *loci* Heimat, Geschlecht, Stand, Begabung, Erziehung, Lehre, Taten, Lebensende und Beurteilung durch die Nachwelt ab. Vgl. H. Scheible, *Melanchthon. Eine Biographie* (München: Beck, 1997), 87.

[27] Zu dieser *declamatio* vgl. T. Rütten und U. Rütten, „Melanchthons Rede 'De Hippocrate,'" *Medizinhistorisches Journal* 33 (1998): 19–55. Nachweise ebenda. Vgl. auch R.-D. Hofheinz, *Philipp Melanchthon und die Medizin im Spiegel seiner akademischen Reden*, Neuere Medizin- und Wissenschaftsgeschichte, Bd. 11 (Herbolzheim: Centaurus, 2001), vor allem 137–44 (Übersetzung der *declamatio de Hippocrate*) und 310–11 (Kurzbiographie von Jakob Milich); ders. und R. Bröer, „Gesundheitspädagogik statt Tröstung. Die theologische Bewältigung von Krankheit bei Philipp Melanchthon und Caspar Peucer," *Sudhoffs Archiv* 85 (2001): 18–44; S. Kusukawa, *The Transformation of Natural Philosophy: The Case of Philipp Melanchthon* (Cambridge: Cambridge University Press, 2003); M. Fink-Jensen, „Medicine, Natural Philosophy, and the Influence of Melanchthon in Reformation Denmark and Norway," *Bulletin of the History of Medicine* 80 (2006): 439–64.

[28] Die ubiquitäre Präsenz Melanchthons an der Universität Rostock erhellt ein Hinweis auf den Melanchthonschüler David Chytraeus (1530–1600), der nach seinem Studium in Wittenberg und seine Wohngemeinschaft mit Melanchthon die Geschicke der Universität über vier Jahrzehnte lenkte. Vgl. R. Keller, „David Chytraeus (1530–1600). Melanchthons Geist im Luthertum," in H.

gewesen sein, da auch er nach dem *exordium* von der Abstammung des Hippokrates, seiner Ausbildung und seinen Taten zu einem Fazit seiner wissenschaftlichen Leistungen und schließlich zu einem höchst schmeichelhaften Charakterportrait fortschreitet, das mit dem Hinweis auf den hippokratischen Eid, dem sich auch Hippokrates unterworfen habe, beglaubigt wird.[29] Beide Autoren, Memm und Melanchthon, wählen einen narrativen Kontext, um ihre religiösen Anliegen, theoretischen Konzepte und moralischen Verhaltenspostulate zu konkretisieren. Beide unterwerfen Überlieferungspartikel der assimilierenden Kraft zeitbedingter Theorieverständnisse, Wertvorstellungen, Weltanschauungen, Diskurslagen und Interessenskonflikte. Beide instrumentalisieren den „göttlichen" Hippokrates und dessen Eid als Panazee gegen die Mißstände in der Medizin ihrer Tage.[30] Beide projizieren auf ihn ein geradezu enzyklopädisches Wissen, wie es höchstens von Galen überliefert ist.[31] Allerdings scheinen Sinn und Bedeutung des Eides in Melanchthons Rede evident, während Memm den Text für erklärungsbedürftig hält.

Scheible (Hg.), *Melanchthon in seinen Schülern*, Wolfenbütteler Forschungen, Bd. 73 (Wiesbaden: Harrassowitz, 1997), 361–71.
29 Siehe auch R. Kinsky und T. Rütten, „Text und Geschichte," in T. Rütten (Hg.) unter Mitarbeit von N. Metzger, *Geschichte der Medizingeschichtsschreibung. Historiographie unter dem Diktat literarischer Gattungen von der Antike bis zur Aufklärung* (Remscheid: gardez!, 2009), 9–56, bsd. 50–52. Damit soll nicht unbedingt gesagt sein, daß dieser Textteil des Memmschen Kommentars zum Vortrag bestimmt war. Es ist aber angesichts der Arbeitsweise Memms nicht auszuschließen, daß er hier eine Vorlesung in den Kommentar eingebaut hat. Die Zielgruppe sind jedenfalls Studenten. Der Schlußsatz seines Kommentars (*pars prima*, f. G2) läßt daran keinen Zweifel: *Haec volui breuiter in gratiam studiosorum de jureiurando Hippocratis delineare, eorum est, qualemcumq[ue] hunc laborem boni consulere.* („Dies wollte ich kurz zum Nutzen der Studenten über den Eid umreißen, und ihre Pflicht ist es nun, diese wie auch immer beschaffene Arbeit gutzuheißen.")
30 Memms Satz (f. C 1ᵛ) *Ita hodie omnia praepostere in medicina fiunt* [„So läuft heutzutage in der Medizin alles verkehrt"] ist hier paradigmatisch. Melanchthons Medizinkritik lautet so: *Haec nos obiter tantum diximus, ad castigandam segniciem huius aetatis, quae ita versatur in artibus omnis generis, non ut solidam doctrinam percipiat, sed ut opinionem quandam auferat, qua se venditet apud imperitos.* („Diese Dinge haben wir nur nebenbei erwähnt, um die Schlaffheit unserer Zeit zu tadeln, die sich mit Künsten aller Art beschäftigt, allerdings nicht so, daß sie ein gediegenes Wissen erlangt, sondern so, daß sie davon einen bestimmten Ruf erwirbt, mit dessen Hilfe sie sich bei denen, die keine Ahnung haben, verkaufen kann.") Vgl. C. G. Bretschneider (Hg.), *Philippi Melanthonis Opera quae supersunt omnia*, 28 Bde., Corpus Reformatorum, Bde. 1–28 (Halle: Schwetschke, 1835–1860), 11:503–509, bsd. 508.
31 Siraisi, *History, Medicine, and the Traditions*, 84.

Die Artaxerxesgeschichte aus den hippokratischen Pseudepigrapha[32] dient Memm als *pars pro toto* für den modellhaften, nachahmungswürdigen Charakter des Hippokrates, illustriere sie doch,

> daß Hippokrates eine echt philosophische Haltung zeigte und ein bemerkenswerter Verächter des Geldes, ja sogar von Ehrenstellen war, weil er, obwohl er von Königen hätte in Ehren gehalten und äußerst glänzend alimentiert werden können, seine Freiheit höher schätzte als Gold und mehr an Griechenland, in dem er lebte, dachte als an sich selbst oder Ehren bei Hofe.[33]

Auch Melanchthon hatte die in Antike und Mittelalter vielbesprochene – gleichwohl fiktive – Anekdote in diesem Sinne verstanden.

I.2.3.2 Der Kommentar *stricto sensu*

Der Kommentar schließt sich lückenlos an die Biographie an, da der Primärautor für den Primärtext steht und *vice versa*. Wer nun aber glaubt, der Unterautor Memm trete ganz hinter dem absoluten Autor Hippokrates zurück, sieht sich getäuscht. Memm scheint im Gegenteil den antiken Text lediglich als Stichwortgeber, als thematische Vorgabe zu eigenen Stellungnahmen zu instrumentalisieren. Bei aller Ehrerbietung dem antiken Modell gegenüber läßt er auch Hippokrates selbst bzw. die hippokratischen Texte herzlich wenig zu Wort kommen.[34]

So stellt er das Lehrer-Schüler-Verhältnis, das der Lehrvertrag zu regeln bestrebt ist, in den Deutungshorizont des Vater-Sohn-Verhältnisses. Um die Lehrer-Schüler-Dyade schlinge sich ein Liebesband (f. C7), das dem zwischen Vater und Sohn zu vergleichen sei und gegenseitige Pflichten schaffe. Während der väterliche Lehrer für eine gediegene Ausbildung zu sorgen habe und den unsterblichen Geist (*animum immortalem*) seiner Schüler mit unsterblicher geistiger Nahrung (*immortali cibo*) speisen solle (f. C8), habe der Adoptiv-Student Gehorsam (*obedientia*), freiwillige Disziplin (*voluntaria disciplina*), Beständigkeit beim Lernen (*aßiduitas in studendo*) und Dankbarkeit (*gratitudo*) an den Tag zu legen (f. C7ᵛ). Memm untermauert seine Forderungen nach Einhaltung der Schüler-/Sohnespflichten

---

32 W. D. Smith, Hg., *Hippocrates, Pseudepigraphic Writings: Letters, Embassy, Speech from the Altar, Decree*, Studies in Ancient Medicine, Bd. 2 (Leiden: Brill, 1992), 120–21.

33 *Ex quo constat Hippocratem animo verè philosophico, et memorabilem fuisse pecuniae, imò etiam honorum contemptorem, quòd dum à Regibus coli potuit et ali splendidissimè, pluris fecerit libertatem suam, quam aurum, et maiorem habuit graeciae, in qua viuebat, rationem, quam aut suipsius, aut honorum aulicorum* (f. B 5ᵛ).

34 Memms Umgang mit dem Primärtext wurde an anderer Stelle einmal „enthistorisierend-aktualisierend" genannt. Vgl. Rütten, „Ärztliche Ethik in der Renaissancemedizin," 533.

mit einer flammenden Rede wider die Undankbarkeit, in deren Verlauf er Xenophons Kyropädie (1.2.7) zitiert (f. D).[35] Diese in der Renaissance als Fürstenspiegel höchst beliebte Schrift[36] dient Memm umso mehr als Schützenhilfe, als sie die einzige nicht-hippokratische Schrift darstellt, auf die er sich in diesem Kommentarteil bezieht. Im Rückgriff auf die Geschichte von der Erziehung des Kyros droht Memm den ihm anvertrauten Medizinstudenten mit den drakonischen Strafen, mit denen die persischen Erzieher Fälle von Undankbarkeit ahndeten. Zudem erinnert er an den Schaden, der dem Gemeinwohl entstehe, wenn sich die Undankbarkeit der Jugend nicht nur gegenüber Lehrern, sondern auch gegenüber Gott, den Eltern, Freunden und dem Vaterland breit mache. Und wenn die Dankespflicht der Schüler schon von heidnischen Persern ohne jedes Pardon eingefordert wurde, dann „soll es bei Christen ein unermeßliches, durch Bisse reißender Hunde zu bestrafendes Verbrechen sein," seine Lehrer zu vergessen, im Stich zu lassen und ihnen dadurch nach dem Leben zu trachten.[37] Denn die praktische Konsequenz gelebter Dankbarkeit seitens der Schüler sieht Memm in der Freistellung des Lehrers von existenzsichernden bzw. -erhaltenden Tätigkeiten, die Freiräume für Forschung und gemeinschaftsdienliche Aufgaben schaffe.[38] Ja, er

---

35 Vgl. C. Mueller-Goldingen, *Untersuchungen zu Xenophons Kyrupädie*, Beiträge zur Altertumskunde, Bd. 42 (Stuttgar: Teubner, 1995).
36 Vgl. D. Marsh, „Xenophon," in V. Brown (Hg.), *Catalogus Translationum et Commentariorum: Medieval and Renaissance Latin Translations and Commentaries; Annotated Lists and Guides*, Bd. 7 (Washington, D.C.: Catholic University of America Press, 1992), 75–196, bsd. 116–37. 1579 gab Memms Kollege an der Universität Rostock Johannes Caselius den 2. Band der Kyropädie bei demselben Drucker heraus, dem Memm seinen Kommentar anvertraute: J. Caselius, *Xenophontis Cyri Paediae liber secundus* ... (Rostock: A. Ferber d. Ä. 1579). Zu Caselius, dem Humanisten, Dichter, Gräzisten, Fürstenerzieher und Ireniker vgl. E. Bonfatti, „Johannes Caselius liest Giovanni de la Casas Galateo (Bologna, 1565)," in A. Buck und M. Bircher (Hgg.), *Republica Guelpherbytana: Wolfenbütteler Beiträge zur Renaissance- und Barockforschung*, Chloe, Beihefte zum Daphnis, Bd. 6 (Amsterdam: Rodopi, 1982), 357–91; M. Scattola, „Gelehrte Philosophie vs. Theologie: Johannes Caselius im Streit mit den Helmstedter Theologen," in H. Jaumann (Hg.), *Die europäische Gelehrtenrepublik im Zeitalter der Konfessionalisierung* (Wiesbaden: Harrassowitz, 2001), 155–81; Grewolls, *Wer war wer*, 1736–37.
37 ... *scelus sit apud homines Christianos immensum, rabidorum canum morsibus castigandum, non solum praeceptorum obliuisci, verumetiam vitam, si possint, adimere velle* (f. D 1ᵛ).
38 *Haec obligatio animos praeceptorum ad artem medicam et philosophiam illustrandam et amplificandam effecit liberrimos. Philosophi enim pecuniae contemptores et necessarijs vitae sustentaculis contenti, non tàm proprijs afficiebantur commodis, quàm publicis: nec voluptas maior vlla ijs occurrebat, quàm si res absconditas communem hominum societatem aliquando adijuturas suis lucubrationibus et labore in vsum adducere potuerunt. Quod quicunq[ue] sibi natura duce et amore vtilitatis publicae perficiendum proponunt, animos ab alienis curis liberos poßidere et volunt et debent; simul enim curae domesticae, et studiorum aßiduitas in ijs rebus, quae nullo cum lucro sunt coniunctae, rarò cum vtilitate concurrunt. Nec ex hac ipsa arte medica,*

fordert von den Medizinstudenten sogar, im Notfall ihr eigenes Leben für das des Lehrers hinzugeben. Der Leser gewinnt den Eindruck, als habe Memm hier das Lehrer-Schüler-Verhältnis nach *Joh* 15,9–17 figuriert, die Schüler zu Märtyrern des Lehrers erklärt und quasi die Jesus-Jünger-Konstellation umbesetzt,[39] dies jedoch nicht im Rückgriff auf das Neue Testament, sondern dessen medizinisches Analogon, das *Corpus Hippocraticum*. Daß die Schülerpflicht gegenüber dem Lehrer, wie sie im Lehrvertrag des Eides konkretisiert wird, nicht nur von Memm in diesem weitreichenden Sinne ausgelegt wurde, läßt sich auch an zeitgenössischen Übersetzungen des hippokratischen Eides ablesen.[40] Memm ist sich jedoch durchaus darüber im klaren, daß solche Verhältnisse sich nicht von selbst einstellen, und so sieht er im Eid ein Druckmittel zu deren Realisierung. Die mit einem Eidbruch verbundene Schmach soll die Undankbarkeit als Vorstufe zu Schamlosigkeit und allgemeiner Sittenverderbnis in Schach halten. Die Entlastung der Lehrer von Alltagssorgen durch die Fürsorgeverpflichtung, die die Schüler im Eidschwur bei Gott gelobten, habe seit altersher medizinischen Fortschritt garantiert und sei auch für die Zukunft eine unabdingbare Voraussetzung für die Freiheit der Forschung. Wer als Schüler Memms mit diesen Ansichten in einer *extraordinarie* gehaltenen Vorlesung oder als privater Leser des Eidkommentars in Berührung kam, dem blieb bei aller Beklemmung der Trost, daß die Zuchtrute zur Dankbarkeit im antiken Perserreich auch über einen späteren König herniedergegangen war. Mit diesem Einblick in Memms Auffassung des Lehrvertrages aus

---

in qua floruerunt illi, diuitias comparandas putarunt. Si igitur vita publici commodi studio tandem consumpta ad benè viuendum praeceptoribus aliquid deesse videatur, id discipulorum liberalitate resarcietur. („Diese Verpflichtung hat die Lehrer zur Vermittlung und Erweiterung der Heilkunst und Philosophie in höchstem Maße freigestellt. Als Philosophen und Verächter des Geldes und Menschen, die mit dem zum Leben Notwendigen zufrieden sind, wurden sie nicht so sehr vom eigenen wie vom öffentlichen Vorteil angerührt: Und sie hatten keinen größeren Wunsch, als Tag und Nacht zu arbeiten und auf diese Weise verborgene Dinge, die irgendwann einmal dem Gemeinwesen der Menschen nutzen würden, nutzbar machen zu können. Wer sich vornimmt, dies unter Führung der Natur und aus Liebe zum allgemeinen Nutzen zu tun, will und muss von anderweitigen Sorgen frei sein; häusliche Sorgen und brotlose Studien vertragen sich nämlich selten. Und sie haben auch nicht geglaubt, daß in jener ärztlichen Kunst selbst, in der jene sich auszeichneten, Reichtum erworben werden müsse. Wenn also ihren Lehrern, die darin aufgehen, ihr Leben dem Streben nach dem allgemeinen Vorteil zu widmen, etwas zu einem guten Leben zu fehlen schiene, wird dies durch die Freigebigkeit ihrer Schüler ausgeglichen werden," ff. D 2ᵛ– D 3.)

39 Ich danke Herrn Dr. Rüdiger Kinsky für diesen Hinweis.

40 Vgl. J. Horstius, *Levini Lemnii Occulta naturae miracula. Wunderbarliche Geheimnisse der Natur in des Menschen leibe und Seel* ... (Leipzig: Steinmann, 1588), 409: „Meinen Praeceptorn/ der mich diese kunst gelehret will ich ehren vnd lieben / als ein kind seine Eltern/ mit ihm leben vnd sterben / was er von mir bedürffen / jederzeit gerne darreichen ..."

dem hippokratischen Eid wollen wir seinen *commentarius* verlassen und uns der *pars secunda* seines Eidkommentars zuwenden.[41]

I.3 Die *pars secunda*

Die *pars secunda* folgt dem soeben besprochenen ersten Teil des Memmschen Kommentars unter eigenem Titel und mit eigenem Titelblatt. Allerdings wird der Leser im ersten Teil (f. E6^v) ausdrücklich auf die *pars secunda* als Kommentar zum zentralen Reinheitsgelöbnis des Eides verwiesen, als sei sie erst im Rahmen des Eidkommentars verfaßt worden, was, wie sich zeigen wird, nicht stimmt.[42] Der zweite Teil, der jene zentrale Gelöbnisformel des Eides (*Porrò praeterea et sanctè vitam et artem meam conseruabo*) im Titel zu hinterfragen scheint (*Quomodo medicorvm vita et ars sancte conservetur*[43]), präsentiert zehn *hypotheses* (f. G4), die

---

41 Näheres zum *commentarius* findet sich in Rütten, *Geschichten vom Hippokratischen Eid*. Das Eidversprechen, nicht zu schneiden, bezieht Memm, wie gehört, auf den Steinschnitt, den Hippokrates chirurgischen Spezialisten überlassen habe, um abermals zu demonstrieren, daß der Arzt nicht aus allem „süßen Gewinn" ziehen müsse, also alles andere als ein Geschäftemacher sei. Das Gelöbnis, die Kranken zu deren Nutzen aufzusuchen, betrachtet Memm als rufsichernde und zugleich vertrauensbildende Maßnahme. Denn zum einen hänge das Ansehen der Ärzte und der Medizin von deren Nutzen ab. Zum anderen schaffe die Absage an jedes Unrecht als Absage an jede Form von ärztlicher Habgier und ärztlichem Geschäfts- und Gewinndenken Vertrauen beim Patienten und in der Öffentlichkeit. Eine adäquate Bezahlung des Arztes werde sich auf diesem Boden schon von selbst und aufgrund der Dankbarkeit der Patienten einstellen. Das Versprechen bezüglich der Nichttötung versteht Memm als Absage an Giftmord bzw. Beihilfe zum Giftmord sowie jede Form der Abtreibung, einschließlich der Verbreitung entsprechenden Wissens. In der Unbescholtenheit des ärztlichen Lebens, in der Aufrichtigkeit des Geistes und in der Unschuld bzw. Keuschheit erkennt Memm das einzige Mittel gegen die schlechte öffentliche Presse über die Ärzte. In seinem Kommentar zum hippokratischen Schweigegebot pocht Memm abermals auf die Würde der ärztlichen Kunst, die der „Geistesmitgift" der ärztlichen Verschwiegenheit dringend bedürfe.
42 *Quia verò si quis, quomodo et in quibus vita et ars sanctè conseruanda sit, enumerare velit, opus prolixa oratione foret, remitto beneuolum lectorem ad tractatum sequentem, in quo haec jurisiurandi particula plenißimè explicatur.* („Da es aber, wenn jemand aufzählen wollte, auf welche Weise und worin das Leben und die Kunst zu bewahren wären, einer langen Rede bedürfte, verweise ich den gewogenen Leser auf den folgenden Traktat, in dem dieser Teil des Eides sehr ausführlich erklärt wird.")
43 Zu diesem zentralen Eidversprechen vgl. H. von Staden, „'In a Pure and Holy Way': Personal and Professional Conduct in the Hippocratic Oath," *Journal of the History of Medicine and Allied Sciences* 51 (1996): 404–37; ders., „Character and Competence: Personal and Professional Conduct in Greek Medicine," in H. Flashar und J. Jouanna (Hgg.), *Médecine et morale dans l'Antiquité; dix exposés suivis de discussions, Vandœuvres-Genève, 19–23 août 1996* (Vandœuvres:

im folgenden der Reihe nach als Kapitelüberschriften wieder aufgegriffen und erläutert werden. Sie lauten:

> Die Medizin ist eine Gabe, die auf Geheiß von Gott zu uns gekommen ist.
>
> Was als dieser hochheiligen Gabe Würdiges von den Hütern der Medizin verlangt wird.
>
> Über die Aufgaben und Verfassungen der Universitäten.
>
> Ob die Untersuchung des Urins, die dem Arzt heutzutage wohlvertraut ist, erlaubt werden darf.
>
> Die richtige Heilmethode bedarf der Anwesenheit des Arztes beim Kranken.
>
> Um seiner Pflicht nachzukommen, ist der Arzt gehalten, die Scharlatane in begründeter Weise vor der Versammlung zu tadeln, anzuklagen und zu widerlegen.
>
> Der Magistrat muß gegen jene einschreiten, die sich den Anschein des rechtschaffenen und die Bezeichnung eines guten Mannes geben, dem Gemeinwesen aber schaden.
>
> Die Ärzte in einem Gemeinwesen müssen sich einträchtig und einer Meinung zeigen.
>
> Der Arzt darf den Irrtümern im Gebrauch der Medizin gegenüber keine Nachsicht zeigen, sondern muß sie beheben.
>
> Verständig handelt, wer sich der Dienste eines erfahrenen und tüchtigen Hausarztes versichert.[44]

An diesen apodiktischen Sätzen läßt sich schon erkennen, wie lose der zweite Teil des Eidkommentars mit dem Primärtext des hippokratischen Eides, in dem ja z. B. von Harnschau mit keinem Wort die Rede ist, verknüpft ist, und in welchem Maße

---

Fondation Hardt, 1996), 157–210; ders., „'The Oath,' the Oaths, and the Hippocratic Corpus," in V. Boudon-Millot, A. Guardasole und C. Magdelaine (Hgg.), *La science médicale antique: nouveaux regards; études réunis en l'honneur de Jacques Jouanna* (Paris: Beauchesne, 2008), 425–66. Diese Trias mit ihren weitreichenden historisch-kritischen Rekonstruktionsleistungen kontrastiert mit Memms vergleichsweise geringem Interesse an historischer Rekonstruktion. Der Titel seiner *pars secunda* fährt fort: *et alias traditiones Medicinae studiosis necessarias complectens* [und weitere für die Studenten der Medizin notwendige Lehren enthaltend], und davon ist in seiner Schrift überwiegend die Rede.

44 Memm, *Hippocratis iusiurandum commentario*, f. G4: *Medicina donum est, diuinitus ad nos profectum; Quid in Medicinae cultoribus dono sanctißimo dignum exigatur; De Academiarum officijs institutisq[ue]; An vrinae inspectio sit admittenda, quae hodiè medicis familiaris est; Recta medendi ratio, aegri praesentiam desiderat; Medicus vt suo satisfaciat officio, tenetur adulteros medicinae cultores ratione reprehendere, accusare, vincere coram senatu; Magistratus debet in eos animaduertere, qui specie recti, et titulo boni viri, reipub[licae] noxij sunt; Medicorum vnius reip[ublicae] concordia et vnanimi consensu opus est; Erroribus in medicinae vsu versantibus medicus non indulgere, sed corrigere debet; Consulto faciunt, qui sibi medicum peritum ac probum conciliat domesticum.*

es Memm um scheinbare Universalien der Medizin bzw. um die Gegenwart und Zukunft der Medizin geht.[45] Den eminenten Gegenwartsbezug seines Kommentars, die Metamorphose des antiken Ennealogs[46] zum hypothetischen Dekalog der *pars secunda*, erkennt der Leser bereits an dem ubiquitären Gebrauch des Präsens und an der spärlichen Nutzung von Quellentexten.[47] Zitate und Nachweise fehlen vollständig.

Die mangelnde Kohärenz dieses Kommentarteils mit dem antiken Ausgangs- und Primärtext hängt mit der merkwürdigen Vorgeschichte der *pars secunda* zusammen. Sie wurde nämlich nicht erst in Rostock im Rahmen des Eidbuches verfaßt, sondern erschien als eigenständige Publikation bereits 1564 unter dem

---

45 In der *pars secunda* gibt es nur eine Stelle, an der Memm eindeutig auf den hippokratischen Eid Bezug nimmt (ebenda, f. L7ᵛ): *Ad quos labores nulla re magis, quàm reipub[licae] prosperitatis desiderio instigor: tum etiam ne iusiurandum, quod Deo praestiti, et omnibus Musis, in publicis peccatis violetur conniuendis.* („Diese Mühen auf mich zu nehmen, sehe ich mich durch nichts mehr veranlaßt als durch mein Verlangen nach dem Wohlergehen des Gemeinwesens, und besonders auch deshalb, damit der Gott und allen Musen geleistete Eid nicht verletzt wird, wenn man den anderen öffentlichen Vergehen gegenüber die Augen verschließt.")
46 Dieser Begriff bringt die Struktur- und Sprachanalyse des hippokratischen Eides (*textus receptus*) von Charles Lichtenthaeler, derzufolge sich der Text in neun Abschnitte gliedern läßt, gesellend auf den Punkt. Vgl. C. Lichtenthaeler, *Der Eid des Hippokrates. Ursprung und Bedeutung* (Köln: Deutscher Ärzteverlag, 1984), 42–232.
47 Die im Corpus Hippocraticum überlieferten Texte *De lege, De victus ratione in morbis acutis* und die Aphorismen sowie Xenophons Kyropädie sind die einzigen explizit genannten Referenztexte in der *pars prima*. Die *pars secunda* zitiert ohne Stellennachweis das Buch *Genesis* I, 28,3 bzw. I, 3,19 (stumme Zitate; f. H 3 bzw. I 2ᵛ), wobei die letztere Stelle mit 2. *Thess.* 3,10 kurz geschlossen wird, *Luk* 17, 37 oder *Matth.* 24,28 (stummes Zitat; f. I 8ᵛ), *Spr.* 19, 7 (stummes Zitat; f. I 8ᵛ), Horaz, *Ep.* I, 17 (stummes Zitat f. L4), Ovid, *Rem.* 91 (stummes Zitat; f. Q4ᵛ), verschiedene hippokratische Aphorismen (f. M6, N3, N5), darunter zwei Mal *Aph.* I, 1 (f. M7 und Q1), spielt auf die Denkfigur der geistigen Beschneidung und das Dankbarkeitsgebot in *Kol.* 2,11 bzw. 3,15 an (f. K3), nennt Celsus (f. Q3ᵛ) und mehrmals Galen (f. H6, N6ᵛ, an der ersten Stelle Nikons Traum, der Galens Medizinstudium besiegelte [Gal., *de meth. med.* 10.4, 10.609 K.]), und wenn man liest (f. O6–O6ᵛ), daß angehende Ärzte „von Kindheit an bis zum dreißigsten Lebensjahr Umgang mit gebildeten Leuten pflegen, die Philosophen hören, das Wesen der Dinge erforschen und sich in der Fremde umschauen" sollen, dann scheint auch hier Galens Biographie das Modell abzugeben. Vgl. V. Nutton, „Biographical Accounts of Galen, 1340–1660," in Rütten, *Geschichte der Medizingeschichtsschreibung*, 201–32. Insgesamt scheint eine paulinische Richtung, wie sie für die reformatorische Bewegung charakteristisch ist, in Memms Auswahl von Bibelstellen unübersehbar. Vgl. R. Ward Holder, *A Companion to Paul in the Reformation*, Brill's Companion to the Christian Tradition, Bd. 15 (Leiden: Brill, 2009). Auch der Rückgriff auf das „Wachset und mehret Euch" aus *Gen.* I, 28,3 ließe sich reformatorisch auslegen: K. Crowther-Heyck, „'Be Fruitful and Multiply': Genesis and Generation in Reformation Germany," *Renaissance Quarterly* 55 (2002): 904–35.

Titel *De recto medicinae vsu*.⁴⁸ Und obwohl der Rostocker Nachdruck dieses Buches – mit verändertem Titel, minimalen, meist orthographischen Veränderungen und ohne ursprüngliche Widmung und Brief an den Leser – nahezu identisch ist, sei diesem Erstling Memms ein eigenes Unterkapitel gewidmet, das dem gänzlich anderen Entstehungskontext der Erstveröffentlichung Rechnung trägt.

I.3.1 Die Erstfassung der *pars secunda: De recto medicinae vsu* von 1564

Dem Text war ein Widmungsschreiben (ff. a2–b1ᵛ) an Erzbischof Friedrich Freiherr Schenck von Tautenburg (1503/4–1580) vorangestellt, der 1561 durch Philipp II. zum ersten Erzbischof des zur Metropole erhobenen Utrecht und damit zu einem der wichtigsten Sachwalter des Katholizismus in den Niederlanden ernannt worden war.⁴⁹ In diesem Schreiben vom März 1563 erklärt Memm den zu seinen Lebzeiten herrschenden Mißbrauch der Medizin und Maßnahmen zu dessen Eindämmung zum Gegenstand seines Buches. Die zitierten Hypothesen sind gewissermaßen die konzeptuellen, berufspraktischen, bildungs-, standes- und gesundheitspolitischen Schlußfolgerungen, die Memm aus seiner Diagnose der zeitgenössischen Medizin zieht: Die Medizin sei eine göttliche Gabe, ein Geschenk, das von der Güte und Gnade des Spenders zeuge, und ihrerseits Erkenntnismittel der göttlichen Gnade. Der Arzt sei Vermittler und Verteiler der göttlichen Gnade auf den Menschen, sei Vorbote und Verteidiger zugleich der Ehre Gottes auf Erden. Diesem göttlichen Amte habe sich der Arzt Tag und Nacht sein ganzes Leben lang

---

48 P. Memmius, *De recto medicinae vsu*. Von diesem Buch sind mehr Exemplare nachweisbar als von Memms Eidkommentar aus dem Jahre 1577. Das Oktavbändchen findet sich z.B. in Amsterdam, Basel, Edinburgh, Erlangen, Fulda, Herzogenburg, London, München, Paderborn, Tübingen, Utrecht, Wien und Wolfenbüttel. Ein Vergleich der beiden Werke zeigt, daß der Nachdruck als *pars secunda* des Eidkommentars das Widmungsschreiben und den Brief an den Leser aus *De recto medicinae vsu* ausläßt und das Prooemium (ff. a1–a4) in das erste Kapitel, in dem die erste Hypothese diskutiert wird, integriert.
49 J. F. von Schulte, „Schenck: Friedrich Freiherr S. von Tautenburg," in *Allgemeine Deutsche Biographie*, Bd. 31: Scheller – Karl Schmidt, Neudruck der 1. Aufl. von 1890 (Berlin: Duncker & Humblot, 1970), 66; M. J. Gasman, „Schenk van Tautenburg (Frederik)," in Molhuysen und Blok, *Nieuw Nederlandsch Biografisch Woordenboek*, 2: Sp. 1280–82. Vgl. auch die Kurzbiographie in H. de Vocht, *Literae virorum eruditorum ad Franciscum Craneveldium 1522–1528: A Collection of Original Letters Ed. from the Manuscripts and Illustrated with Notes and Commentaries* (Louvain: Uystpruyst, 1928), 676–77. Memm soll Leibarzt dieses Bischofs gewesen sein. Vgl. Dodt van Flensburg, „Over den Utrechtschen arts Petrus Memmius," 77–79; Lindeboom, *Dutch Medical Biography*, Sp. 1304, formuliert zu Recht vorsichtig: „Others (Dodt van Flensburg, in *Utrecht voorheen en thans* 2d Ser. II (1845), 77) supposes [sic] him to have been physician in ordinary to the archbishop Schenck van Tautenburg."

zu widmen zur Ehre Gottes und zum Wohl und Nutzen des Gemeinwesens, das so einzurichten sei, daß Gottes Geschenk dankbar und fromm Genüge getan werde: ärztliche Tugenden, Prüfungsrichtlinien und Qualitätsprüfungen an den Universitäten, Stärkung der Rolle des Hausarztes und damit der Gesundheitsvorsorge, kompetente Besetzung von Sittengerichten, hartes Vorgehen gegen Scharlatane, Schulterschluß von Ärzten, Fürsten, Juristen und Theologen zum Wohle des Gemeinwesens und vieles mehr klagt Memm in diesem Zusammenhang ein. Es handelt sich also um einen Reformversuch der zeitgenössischen Medizin, die ganz und gar der Religion unterstellt und aus ihr abgeleitet wird, im Interesse des Allgemeinwohls, aber auch im Interesse der akademischen Ärzte.

Initiiert wird dieser Reformversuch in jener katholischen Enklave,[50] die den Brüsseler und Madrilener Behörden lange treu blieb, von einem Arzt, der in Diensten eines katholischen Bischofs steht, dessen Kirche Memm in seinem Widmungsbrief gute Rektoren und treue Minister zum Schutz wünscht.[51] Am Vorabend des Bildersturms (1566), zumal in Utrecht, kann Memm sein umfassendes Reformprogramm nur im Horizont einer innerkirchlichen Reformbewegung artikulieren. Und es dürfte nicht verfehlt sein, in diesem Zusammenhang an die *devotio moderna* zu denken, die im 16. Jahrhundert sowohl in die altgläubige Kirche wie auch in die Kirchen der Reformation hineinwirkte.[52] Streckenweise liest sich Memms Buch mit seinen Meditationen über biblische Stoffe, Tugenden und Laster, echte Frömmigkeit und Liebesgemeinschaft mit Gott und den Menschen, als wolle sein Verfasser jene im 14. Jahrhundert beginnende Frömmigkeitsbewegung, der es um die Verlebendigung der christozentrischen Innerlichkeit der Urgemeinde in der spätmittelalterlichen und frühneuzeitlichen Gegenwart ging, auch in der Medizin verankern. Daß eine solche Verortung des Memmschen Buches nicht ganz abwegig ist, zeigen Memms Kontakte zu einer Gemeinschaft, in der die *devotio moderna* Gestalt gewann, nämlich zu den Schwestern und Brüdern

---

50 Utrecht wird sich erst 1580 zum Protestantismus bekennen.
51 Memmius, *De recto medicinae vsu*, f. b1–b1ᵛ: *Itaque etiam atque etiam oro S. Maximúmque deum, vt hunc Archiepiscopatum amplissimum, atque vna tuam celsitudinem clementia sua semper incolumes conseruet: faueatque idem, vt catholica & orthodoxa ecclesia aduersus omnes diaboli insultus, per bonos rectores & fideles ministros omni tempore muniatur.* Schon Dodt van Flensburg, „Over den Utrechtschen arts Petrus Memmius," 77 hat aufgrund dieser Passage Zweifel an Memms calvinistischer Gesinnung angemeldet, und Lindeboom, *Dutch Medical Biography*, Sp. 1305, bemerkt zu Recht, es sei nicht sicher „that religious reasons induced him to leave Utrecht." Siehe auch E. E. Gewin, „Memmius (Peter)," in Molhuysen und Blok, *Nieuw Nederlandsch Biografisch Woordenboek*, 2: Sp. 895.
52 Zur *devotio moderna* siehe R. R. Post, *The Modern Devotion: Confrontation with Reformation and Humanism* (Leiden: Brill, 1968); J. H. van Engen (Hg.), *Devotio Moderna: Basic Writings* (New York: Paulist Press, 1988).

vom gemeinsamen Leben.⁵³ Denn Memm arbeitete als Lehrer an der von den Brüdern vom gemeinsamen Leben 1474 gegründeten Hieronymus School in Utrecht.⁵⁴ Einen sicheren Beleg für seine Verbindungen zu dieser Schule bietet ein Vertrag, den Vertreter des Konvents von Sankt Hieronymus mit Memm, Cornelius Lauerman (Laurimanus)⁵⁵ und einem Hubrecht (Huybert) Hastreus am 23. Januar 1565 unterzeichneten und der die Rechte und Pflichten der Hausherren wie auch der Lehrer, die in den Konventsräumen unterrichten sollten, regelte.⁵⁶ Und wenn auch Memm keinesfalls als *frater* in Betracht kommt⁵⁷ und lediglich mit dem

---

53 J. H. van Engen, *Sisters and Brothers of the Common Life: The Devotio Moderna and the World of the Later Middle Ages* (Philadelphia: University of Pennsylvania Press, 2008).
54 Vgl. Ekker, *De Hieronymusschool te Utrecht*, 1:30–31.
55 Zu diesem katholischen Kollegen Memms vgl. J. Bloemendal, „Cornelius Laurimanus als Dramatiker. Theater und Theologie gegen Ketzereien," in R. F. Glei und R. Seidel (Hgg.), *Das lateinische Drama der Frühen Neuzeit. Exemplarische Einsichten in Praxis und Theorie* (Tübingen: Niemeyer, 2008), 101–32. Auf S. 103 deutet Bloemendal auch diesen Vertrag als Krisenzeichen für die Nachwuchssorgen des Konvents, für die rückläufigen Einkünfte und Stipendien, für die im Zuge der Reformation erschwerte Mobilität potentieller Schüler und die Lockerung des Nexus zwischen Sprachkenntnissen, Theologie und kirchlichen Ämtern. Siehe auch ders., *Spiegel van het dagelijks leven? Latijnse school en toneel in de noordelijke Nederlanden in de zestiende en de zeventiende eeuw* (Hilversum: Verloren, 2003); O. J. de Jong und J. T. K. Marcelis, *Uti porta latina stat: Hieronymusschool en Utrechts stedelijk gymnasium van 1474 tot 2000* (Utrecht: USG, 2000), 26.
56 J. J. Dodt van Flensburg, „Hieronymus School 1630–1648," *Archief voor kerkelijke en wereldlijke geschiedenis van Nederland meer bepaaldelijk van Utrecht* 7 (1848): 366–92, bsd. 389–91. Dieser Vertrag, der Ostern 1565 in Kraft treten und auf zwölf Jahre mit der Option auf Verlängerung geschlossen wurde, widerlegt nicht nur, daß Memm, wie vielfach überliefert (z. B. in N. F. J. Eloy, *Dictionnaire historique de la médecine ancienne et moderne …*, 4 Bde. [Mons: H. Hoyois, 1778], 3:262; A. J. van der Aa, *Biographisch Woordenboek der Nederlanden …*, Nieuwe Uitgaaf, 21 Bde. [Haarlem: J. J. van Brederode, 1852–1878], 12.1:586; C. E. Daniëls, „Memmius, Petrus M.," in A. Hirsch (Hg.), *Biographisches Lexikon der hervorragenden Ärzte aller Zeiten und Völker*, 2. Aufl. durchges. und erg. von W. Haberling et al., 6 Bde. [Berlin etc.: Urban & Schwarzenberg, 1929–1935], 4:160; E. M. H. de Seyn, *Dictionnaire biographique des sciences, des lettres et des arts en Belgique*, 2 Bde. [Brüssel: Éditions Avenir, 1935–1936], 1:561; M. Asche, *Von der reichen hansischen Bürgeruniversität zur armen mecklenburgischen Landeshochschule. Das regionale und soziale Besucherprofil der Universitäten Rostock und Bützow in der frühen Neuzeit [1500–1800]*, Contubernium, Bd. 52 [Stuttgart: Steiner, 2000], 57, Anm. 67), schon 1561 Utrecht verlassen haben bzw. 1563 in Rostock gewesen sein soll, sondern er legt zudem die Vermutung nahe, daß er sich 1565 noch gar nicht mit Abwanderungsplänen befaßte, was nicht weiter wunder nimmt, da ja in Holland, zumindest äußerlich, bis zum Mai 1566 Frieden herrschte. Die Tatsache, daß die Lehrer in dem genannten Vertrag auf Zahlung einer Kaution und Bürgen verpflichtet werden, spricht für die Ernsthaftigkeit der Verhandlungen.
57 Das hat Post, *Modern Devotion*, 571 klargestellt, meines Wissens zuerst in ders., „Studien over de broeders van het gemeene leven," *Nederlandsche Historiebladen* 2 (1939): 136–62, bsd. 141.

Erziehungsprogramm der Brüder, das eine humanistisch fundierte sprachlich-wissenschaftliche Ausbildung mit moralischer und religiöser Unterweisung im Sinne eines christlichen Humanismus verband,[58] vergesellschaftet werden kann, so spricht doch auch sein Buch dafür, daß er theologisch der Bewegung nahe stand. Es ist ein Dokument des Einspruchs gegen den *status quo* und verrät Reformwillen, ohne die Kirchenspaltung bereits vorauszusetzen oder gar zu propagieren.

An der Verankerung des Memmschen Buches in der Frömmigkeitsbewegung der *devotio moderna* ändert auch nichts die Tatsache, daß Memm seinen Erstling bei Harman Schinckel (1535/36–1568) drucken ließ, den ein Biograph als Humanisten in erasmianischer Tradition, als Gemäßigten, als Vertreter von Toleranz und individueller Religionsfreiheit charakterisiert hat.[59] Denn auch wenn Schinckel wenige Jahre darauf, 1568, vom Delfter Magistrat wegen der Verbreitung klandestiner Literatur zum Tode verurteilt und am 23. Juli hingerichtet wurde, so ist ernst zu nehmen, daß er in seinem Verhör darauf bestand, kein Calvinist zu sein.[60] Memm war es 1564 sicher nicht, wenn auch manche seiner Reformanstöße nach dem Bildersturm und erst recht nach der Ankunft Fernand Álvarez de Toledo (1508–1582), des Herzogs von Alba, der im August 1567 in den habsburgischen Niederlanden als Generalstatthalter und Nachfolger der Margarethe von Parma eintraf,[61] als häresieverdächtig gegolten haben mögen. Jedenfalls muss Memm um diese Zeit Utrecht mit seiner Frau[62] und seinen beiden Söhnen Abraham (1564–

---

58 I. Crusius, „Brüder und Schwestern vom gemeinsamen Leben," in H. D. Betz et al. (Hgg.), *Religion in Geschichte und Gegenwart: Handwörterbuch für Theologie und Religionswissenschaft*, 4., völlig neu bearb. Aufl., 8 Bde. und Register (Tübingen: Mohr Siebeck, 1998–2007), 1: Sp. 1781–82; siehe auch C. Burger, „Devotio moderna," in Betz et al., *Religion in Geschichte und Gegenwart*, 2: Sp. 776.
59 H. de la Fontaine Verwey, *Meester Harman Schinckel, een Delftse boekdrukker van de 16e eeuw*, Oud Delft, Bd. 3 (Rotterdam: Nijgh en van Ditmar, [1963]), 55–56.
60 Das geht aus seinen letzten erhaltenen Briefen hervor: de la Fontaine Verwey, *Meester Harman Schinckel*, 59–70.
61 H. Kamen, *The Duke of Alba* (New Haven: Yale University Press, 2004), 75–105.
62 Memm war mit Gertrud Longolia (gest. 1604), der Witwe eines Johannes (Janus) Longolius, verheiratet. Vgl. E. J. von Westphalen, *Monumenta inedita Rerum Germanicarum praecipue Cimbricarum et Megapolensium*, 4 Bde. (Leipzig: Martini, 1739–1745), 3:1442: „*Habuit* [gemeint ist P. Memm] *uxorem Gertrudem, Jani Longolii, ex antiqua et nobili Longoliorum der von Langenrechte prosapia oriundi Viduam*"; A. Blanck, A. Wilhelmi und G. Willgeroth (Hgg.), *Die Mecklenburgischen Ärzte von den ältesten Zeiten bis zur Gegenwart* (Schwerin: Verlag der Landesgeschäftsstelle des Meckl. Aerztevereinsbundes, 1929), 228; [Anon.], „Leben D. Gvilielmi Lavrembergii ... ," *Etwas von gelehrten Rostockschen Sachen, für gute Freunde* 2 (1738): 155–60, bsd. 157. Gertrud brachte eine Tochter namens Johanna (Jeanne Longoeil) in ihre zweite Ehe, die 1557 in Utrecht zur Welt gekommen war. Vgl. G. Teichmann, „Lauremberg (Laurenberg), Peter," in Pettke und

1602)[63] und Conrad[64] verlassen haben. Denn sein dritter Sohn Isaak kam in Culemborg im Gelderland zur Welt.[65] Die Wahl dieser ersten Zwischenstation ist

---

Röpcke, *Biographisches Lexikon für Mecklenburg*, 2:153–55, bsd. 153. De Seyn, *Dictionnaire biographique*, 561 behauptet, Memm habe Utrecht „après la mort de sa femme" verlassen. Eloy, Foppen und Lindeboom datieren den Tod seiner Frau sogar auf das Jahr 1551, Desselius auf das Jahr 1561. Vgl. Eloy, *Dictionnaire historique*, 3:262; J. F. Foppens, *Bibliotheca belgica sive virorum in Belgio vita, scriptisque illustrium catalogus, librorumque nomenclatura. Continens scriptores a clarissimis viris Valerio Andrea, Auberto Miraeo, Francisco Sweertio aliisque recensitos, usque ad annum MDCLXXX* (Brüssel: Foppens, 1739), 992; Lindeboom, *Dutch Medical Biography*, Sp. 1305; V. A. Desselius, *Bibliotheca Belgica de Belgis vita scriptisque claris* .... Editio renovata et tertia parte auctior (Löwen: Zegers, 1643), 748. Siehe auch J. Moller, *Cimbria literata. Tomus secundus. Adoptivos sive exteros, in ducatu utroque Slesvicensi et Holsatico, vel officiis functos publicis vel diuitius commoratos, complectens* (Kopenhagen: Orphanotrophium Regium, 1744), 550–51. Aber solche Nachrichten gehören zu den zahlreichen Irrtümern, die sich in der prosopographischen Literatur zu Peter Memm finden. Alles spricht dafür, daß Gertrud ihren zweiten Mann, Peter Memm, nach Rostock und 1581 nach Lübeck begleitete und mit ihrem Mann drei weitere Kinder hatte: Abraham, Konrad und Isaak. In diesem Sinne auch Dodt van Flensburg, „Over den Utrechtschen arts Petrus Memmius," 76–77. Leider blieb die Suche nach Archivalien, die mehr Licht in Memms Utrechter Zeit werfen könnten, bislang ergebnislos. Ich danke Frau Dr. Irmgard Broos vom Rijksarchief in Utrecht für entsprechende Recherchen und hilfreiche Auskünfte.

63 Abraham Memm wurde 1589 nach Aufenthalten in Utrecht, Rostock, Lübeck, Jena und Padua Doktor der Medizin in Basel, 1594 Professor der Medizin in Königsberg und 1601 altstädtischer Pestarzt. Vgl. G. Erler (Hg.), *Die Matrikel der Universität Königsberg i. Pr.*, Bd. 1, H. 1 (Leipzig: Duncker & Humblot, 1908), 120 (Dez. 1593 als *medicinae doctor* immatrikuliert); S. Jaster, „Die medizinische Fakultät der Albertus-Universität und ihre bedeutendsten Vertreter im 16. und 17. Jahrhundert," in H. Marti und M. Komorowski (Hgg.), *Die Universität Königsberg in der Frühen Neuzeit* (Köln: Böhlau, 2008), 42–76, bsd. 39.

64 Conrad Memm immatrikulierte sich im Sommersemester 1586 in Helmstedt und wurde dort 1591 begraben. Wann er in Utrecht zur Welt kam, ist nicht bekannt. Daß er dort zur Welt kam, belegt Peter Memms Immatrikulation seiner Söhne an der Universität Rostock. Vgl. A. Hofmeister, *Die Matrikel der Universität Rostock*. Bd. 2: *Mich. 1499 – Ost. 1611*, Nachdruck der Ausg. Rostock 1891 (Nendeln, Liechtenstein: Kraus, 1976), 197.

65 Hofmeister, *Die Matrikel der Universität Rostock: Isaacus Memmius Culemburgensis*. Im Matrikelverzeichnis der Heidelberger Universität wurde Isaak am 15. Mai 1594 mit der Herkunftsbezeichnung „Belga" eingetragen. Siehe G. Toepke (Hg.), *Die Matrikel der Universität Heidelberg von 1386 bis 1662. 2. Theil von 1554 bis 1602*, (Heidelberg: Selbstverlag, 1886; Nendeln: Kraus Reprint, 1976), 173. Isaak studierte seit 1588 in Wittenberg, immatrikulierte sich am 15. Mai 1594 in Heidelberg, promovierte in Jura und arbeitete seit 1604 als Stiftssyndikus in Hamburg. Auch Gedichte schrieb er. Vgl. das Hochzeitsgedicht des Hildesheimer Pastors von St. Andreas, Nikolaus Siegfrid, anläßlich der Vermählung Isaaks mit Angela Neoburia im Oktober 1600: N. Siegfridus, *Epithalamivm De vera laeticia et Angelica custodia status conjugalis. Clarissimo ... viro, D. Isaaco Memmio, I.V.D. apud inclytos Lubecenses peritissimo, nuptias cum honestissima et plentissima Virgine Angela Neoburia Hamburgensi, ... Anno salutis nostrae 1600. 3. Non. Octobr.* (Hildesiae: Excudebat Andreas Hantzsch in Coenobio D. Pauli, 1600).

insofern interessant, als dort Floris van Pallandt (1537–1598), einer der Grafen von Culemborg, seit dem 2. Juni 1566 als Erster in den Niederlanden einen protestantischen Gottesdienst in seinem Schloß zuließ und Mitte September desselben Jahres half, die Kirche von Culemborg im Rahmen des Bildersturms von Kirchenschmuck und Bildwerken, die mit dem päpstlichen Kirchenstaat assoziiert wurden, zu befreien.[66] Er konnte sich bis April 1567 behaupten, ehe er nach Deutschland floh.[67]

Auch die Familie Memm setzte ihre Wanderung fort und gelangte 1568 nach Rostock.[68] Im Jahr zuvor waren bereits zwei niederländische Ärzte, der Vesalschüler Heinrich van den Brock (Henricus Brucaeus, 1530–1593)[69] als herzoglicher Professor der Medizin und höheren Mathematik[70] und der Paracelsist Levinus

---

66 O. J. de Jong, *De Reformatie in Culemborg* (Assen: Van Gorcum, [1957]), 102, 131–35; siehe auch A. Duke, *Reformation and Revolt in the Low Countries* (London: Hambledon and London, 2003), 129–32.
67 S. P. Haak, „Pallandt (Floris van)," in Molhuysen und Blok, *Nieuw Nederlandsch Biografisch Woordenboek*, 5: Sp. 424–27.
68 Dodt van Flensburg, „Over den Utrechtschen arts Petrus Memmius," 76; C. G. Jöcher, J. C. Adelung und H. W. Rotermund, *Fortsetzung und Ergänzungen zu Christian Gottlieb Jöchers allgemeinem Gelehrten-Lexicon: worin die Schriftsteller aller Stände nach ihren vornehmsten Lebensumständen und Schriften beschrieben werden*, 4. Bd. (1813), Unveränderter Nachdruck (Hildesheim: Olms, 1961), CLXIX; Gewin, „Memmius (Peter)"; Blanck, Wilhelmi und Willgeroth, *Die mecklenburgischen Ärzte*, 228; O. Krabbe, *Die Universität Rostock im fünfzehnten und sechzehnten Jahrhundert* (Rostock: Stillersche Hofbuchhandlung, 1854), 707; [Anon.], „D. Petrus Memmivs, Herenthalio-Belga," *Etwas von gelehrten Rostockschen Sachen, für gute Freunde* 2 (1738): 285; J. B. Krey, *Andenken an die Rostock'schen Gelehrten aus den drei letzten Jahrhunderten*, 5. Stück (Rostock: Adler, 1815), 16; G. Heidorn et al. (Hgg.), *Geschichte der Universität Rostock 1419–1969: Festschrift zur Fünfhundertfünfzig-Jahr-Feier der Universität*. Bd. 1: *Die Universität von 1419–1945* (Berlin: Deutscher Verlag der Wissenschaften, 1969), 41; A. Hulshof, „Rostock und die nördlichen Niederlande vom 15. bis zum 17. Jahrhundert," *Hansische Geschichtsblätter* 16 (1910): 531–53, bsd. 539; Westphalen, *Monumenta Inedita*, 1441; K. E. H. Krause, „Memmius, Peter," in *Allgemeine Deutsche Biographie. Auf Veranlassung und mit Unterstützung des Königs von Bayern Maximilian II. hrsg. durch die Historische Commission bei der Königlichen Akademie der Wissenschaften*, 56 Bde. (Leipzig: Duncker & Humblot, 1875–1912), 21 (1885): 310.
69 Vgl. Grewolls, *Wer war wer*, 1417; Lindeboom, *Dutch Medical Biography*, Sp. 276–77. Brucaeus machte kurz vor seinem Tod seinen Übertritt vom Katholizismus zum Protestantismus öffentlich.
70 1571 wurde er auch Leibarzt des Herzogs Johann Albrecht I. Vgl. F. W. Schirmacher, *Johann Albrecht I., Herzog von Mecklenburg* (Wismar: Hinstorff, 1885); H. Schreiber, *Johann Albrecht I., Herzog zu Mecklenburg* (Halle: Niemeyer, 1899); Heeg, J., *Buch, Macht, Bildung: Die Bibliothek des Herzogs Johann Albrecht I. von Mecklenburg. Ausstellung im Kulturhistorischen Museum Rostock vom 20.4. bis 24.6.2007*, Veröffentlichungen der Universitätsbibliothek Rostock, Bd. 137 (Rostock: Universitätsbibliothek, 2007).

Battus (1545–1591)[71] als Professor der Medizin und Leibarzt Ulrich III., in Rostock angestellt worden.[72] Schon im 15. Jahrhundert waren zahlreiche Studenten aus den Niederlanden nach Rostock gekommen, insbesondere aus der Diözese Utrecht.[73] Auch die Brüder vom gemeinsamen Leben stellten eine Brücke nach Rostock dar, wo sie seit 1462 ein eigenes Fraterhaus besaßen, Schulen errichtet und ihren Schulhumanismus verbreitet hatten.[74] Memm erreichte Rostock, als die *Formula concordiae* vom Mai 1563, mit der die gut 30 Jahre währenden Auseinandersetzungen zwischen Landesherrschaft, Rat und Bürgerschaft der Stadt Rostock um die Reorganisation der Hochschule beigelegt worden waren, erste Früchte trug, ein botanischer Garten bestand[75] und die medizinischen Statuten überarbeitet wurden.[76] Der *Formula concordiae* zufolge sollte die Universität auf die Augsburgische

---

71 Levinus Battus stammte wie Heinrich van den Brock aus dem flandrischen Aalst, und war mit seiner lutherischen Familie als Religionsflüchtling bereits 1557 nach Rostock gelangt, wo er sich zunächst dem Studium der Mathematik widmete. Nach einer Studienzeit in Wittenberg wurde er Professor der Mathematik in Rostock (1560–1565). Von einer *peregrinatio academica* nach Norditalien (Padua, Venedig) war er als *doctor medicinae* zurückgekehrt. Vgl. Grewolls, *Wer war wer*, 589; Lindeboom, *Dutch Medical Biography*, Sp. 85.
72 G.-H. Schumacher und H. Wischhusen, *Anatomia Rostochiensis: Die Geschichte der Anatomie an der 550 Jahre alten Universität Rostock* (Berlin: Akademie-Verlag, 1970), 32–35; R. van Roosbroeck, „Die Beziehungen der Niederländer und der niederländischen Emigranten zur deutschen Gelehrtenwelt im XVI. Jahrhundert: Eine Übersicht," in H. Rössler und G. Franz (Hgg.), *Universität und Gelehrtenstand 1400–1800: Büdinger Vorträge 1966*, Deutsche Führungsschichten in der Neuzeit, Bd. 4 (Limburg a.d. Lahn: Starke, 1970), 107–25, bsd. 113, wo neben Memm Jacob Bording, Heinrich van den Brock und Levinus Battus genannt werden.
73 Hulshof, „Rostock und die nördlichen Niederlande," 531; Asche, *Von der reichen hansischen Bürgeruniversität*, 331–32.
74 G. C. F. Lisch, *Geschichte der Buchdrucker-Kunst in Mecklenburg bis zum Jahre 1540. Mit einem Anhange über die niederdeutsche Bearbeitung des Reineke Voß* (Schwerin: Stiller, 1839).
75 Asche, *Von der reichen hansischen Bürgeruniversität*, 107–108.
76 Die beste Studie zur Vorgeschichte der *Formula concordiae* ist M. A. Pluns, *Die Universität Rostock 1418–1563. Eine Hochschule im Spannungsfeld zwischen Stadt, Landesherren und wendischen Hansestädten* (Köln: Böhler, 2007), vor allem 433–94. Siehe auch Asche, *Von der reichen hansischen Bürgeruniversität*, 56–63. Memm wird im November 1568 unter dem Rektorat von Johannes Borcholt als *Petrus Memmius Herendalius, artis medicae doctor et professor Rostochii* ins Matrikelverzeichnis der Universität eingetragen. Vgl. Hofmeister, *Die Matrikel der Universität Rostock*, 166. „A[nno] 1568 d. 15. Nouembr. begehrt ein Ehrbarer Rath ihrem bestalten Medico horam und lectionem zu assigniren. Weiter heist es unter dem 24 Nouembr. *de D[omino] Memmio quaesitum est in Concilio, quid proponere debeat, et quae illi hora attribuenda, et de lectione ad Medicorum consultationem reiectum est, hora autem 4 illi assignata.*" Am 9. April. 1569 heißt es: „*Litterae lectae de recipiendo in Concilium D[omino] Petro Memmio, Medico, Senatus Professore, qui statim post praestitum iuramentum receptus est et locus ei a D[omino] Henrico Broucaeo proximus assignatus est.*" Vgl. [Anon.], „Von D. Memmio. P. 285," *Etwas von gelehrten Rostockschen Sachen, für gute Freunde* 2 (1738): 833–34.

Konfession festgelegt sein. Die Herzöge (Ulrich und Johann Albrecht) und der Rat der Stadt sollten jeweils neun Dozenten zu Mitgliedern im Universitätskonzil ernennen und besolden (*collegium professorum ducalium* und *collegium professorum senatorium*). Die Rektoren waren reihum aus den Mitgliedern des Konzils zu rekrutieren. Der Rektor sollte schwören, „den Statuten Geltung zu verschaffen, die Rechte und Freiheiten der Hochschule zu bewahren, sowie Nutzen und Ehre von Fürsten und Staat unterschiedslos zu fördern."[77] In der Artistenfakultät sollten vier landesherrliche Professoren (ducales) vier städtischen (senatorii) gegenüber stehen. Memm dürfte unmittelbar von der *Formula* profitiert haben, die nämlich vorsah, die Renten, die Konrad Pegel (1487–1567) und Bernhard Mensing (gest. 1567)[78] 1563 noch bezogen, nach deren Ableben zur Besoldung der städtischen Professoren zu verwenden.[79] Diese Summen standen 1567 zur Verfügung. Kurzum, Memm wurde 1568 als *Professor senatorius* eingestellt und zugleich mit dem Amt eines Rostocker Stadtarztes betraut. Rostocks Ratsherren wurden seine Dienstherren, und wie alle Professoren der Rostocker Universität mußte auch Memm schwören, „Nutzen und Ehre" der städtischen wie der fürstlichen Obrigkeit „zu wahren und zu fördern."[80]

Neben seinem Lehrstuhl und dem Amt eines Stadtphysicus bekleidete Memm seit 1571 auch das Amt eines herzoglichen Leibarztes.[81] Herzog Ulrich berief ihn in diese Stellung gegen ein Jahresgehalt von 100 Talern, einem Ochsen und vier Drömt[82] Roggen.

---

77 Pluns, *Die Universität Rostock 1418–1563*, 488. Zum Sitz von Eiden im universitären Leben siehe J. Miethke, „Der Eid an mittelalterlichen Universitäten. Formen seiner Gebräuche. Funktionen einer Institution," in P. Prodi (Hg.), *Glaube und Eid. Treueformeln, Glaubensbekenntnisse und Sozialdisziplinierung zwischen Mittelalter und Neuzeit*, Schriften des Historischen Kollegs, Bd. 28 (München: Oldenbourg, 1993), 49–67.
78 Zu diesen beiden vgl. Grewolls, *Wer war wer*, 7425 bzw. 6523. Pegel hatte sich übrigens schon 1534 bei dem Magister, Ratssekretär und Rostocker Stadtapotheker Petrus Sasse nach dem Namen jenes flandrischen Arztes erkundigt, der Luther und Melanchthon wohl bekannt sei und inzwischen in Hamburg lebe. Vgl. [Anon.], „Ein ungedruckter Brieff Conr. Pegelii aus dem Original," *Etwas von gelehrten Rostockschen Sachen, für gute Freunde* 4 (1740): 657-58. Pegel scheint eine nicht unerhebliche Rolle bei der Akquirierung niederländischer Ärzte an die Rostocker Universität gespielt zu haben. Levinus Battus war übrigens seit 1563 mit Pegels Tochter Anna verheiratet.
79 Pluns, *Die Universität Rostock 1418–1563*, 480.
80 Pluns, *Die Universität Rostock 1418–1563*, 490.
81 Siehe u. a. [Anon.], „D. Petrus Memmivs," 285; Lindeboom, *Dutch Medical Biography*, 1305; Westphalen, *Monumenta Inedita*, 1441.
82 Das alte mecklenburgische Getreidemaß „Drömt" entspricht 466,68 Liter (= 12 Scheffel). Siehe H.-J. von Alberti, *Maß und Gewicht. Geschichtliche und tabellarische Darstellungen von den Anfängen bis zur Gegenwart* (Berlin: Akademie-Verlag, 1957), 471.

I.3.2 Alter Wein in neuen Schläuchen: die *pars secunda* von 1577

1577, das Erscheinungsjahr des Eidkommentars von Memm, war ein Pestjahr in Rostock.[83] Nathan Chytraeus (1543–1598), neulateinischer Dichter und Professor für lateinische Sprache (1564) bzw. Poetik (1567–1593) an der Universität Rostock, Begründer ihrer *Bibliotheca Philosophica* (1579) und erster Rektor der Lateinschule Rostock (1580–1593), dem Memm 1578 ein Hochzeitsgedicht widmete[84] und unter dessen Rektorat er in demselben Jahr seine drei Söhne an der Universität Rostock immatrikulierte und dem er in diesem Amt nachfolgte,[85] schrieb aus Anlaß der Pest ein Gedicht, das vom 13. November 1577 datiert.[86] Dieses Gedicht steht, wie Wilhelm Kühlmann herausgearbeitet hat, für die Opposition gegen ein blindes, lutherisch legitimiertes Gottvertrauen, verbunden mit einem Aufruf zur *emendatio vitae*, womit sowohl das private wie das öffentliche Leben gemeint ist. Im Gewand

---

83 W. Kühlmann, „Selbstverständigung im Leiden: Zur Bewältigung von Krankheitserfahrungen im versgebundenen Schrifttum der Frühen Neuzeit (P. Lotichius Secundus, Nathan Chytraeus, Andreas Gryphius)," in U. Benzenhöfer und W. Kühlmann (Hgg.), *Heilkunde und Krankheitserfahrung in der frühen Neuzeit. Studien am Grenzrain von Literaturgeschichte und Medizingeschichte* (Tübingen: Max Niemeyer, 1992), 1–29, bsd. 16–19 und 23–29. Zu einer früheren, weit verheerenderen Pestepidemie in Rostock siehe K. Koppmann, „Über die Pest des Jahres 1565 und zur Bevölkerungsstatistik Rostocks im 14., 15. und 16. Jahrhundert," *Hansische Geschichtsblätter* 29 (1901): 45–63. Vgl. auch O. Karrig, „Geschichtliches über das Auftreten der Pest in Mecklenburg," *Archiv für Geschichte der Medizin* 5 (1912): 436–46. Auch in den Jahren 1598, 1603 und 1624 wurde Rostock von der Pest heimgesucht.

84 N. Chytraeus, *Poematum Nathanis Chytraei Praeter Sacra Omnium Libri Septendecim* (Rostock: Mylander, 1579), 264–66. Als Rektor der Universität Rostock schrieb Memm am 29.12.1578 ein *scriptum publice* als Auftakt zu einer Reihe von *carmina amicorum* (von Johannes Caselius, Nathan Chytraeus etc.) anläßlich der Beerdigung von Christoph Gärtner, das im folgenden Jahr gedruckt wurde: P. Memmius, *Scriptvm Pvblice propositvm a rectore academiae Rostochiensis in funere opt. et doctißimi viri Christophori Gertneri Lubecensis, bonarum artium magistri et professoris publici, et decani facultatis philosophicae* (Rostock: [o. V.], 1579), A1ᵛ-A3. Ein Brief Memms – datiert auf den 30. Juni, aber ohne Jahreszahl, aber in jedem Fall aus Rostocker Tagen – an seinen Kollegen und Freund Jakob Horst d. Ä. (1537–1600) wurde vom Empfänger in folgendes Werk aufgenommen: J. Horstius, *Levini Lemnii Occulta naturae miracula. Wunderbarliche Geheimnisse der Natur in des Menschen leibe und Seel ...* (Leipzig: V. Vögelin, M. Lantzenberger, 1593), 848–50. Darin gibt sich Memm als Experte in Empfängnisberatung zu erkennen.

85 Hofmeister, *Die Matrikel der Universität Rostock*, 197. Vgl. auch A. Hartwig und T. Schmidt, Hgg., *Die Rektoren der Universität Rostock, 1419–2000*, Beiträge zur Geschichte der Universität Rostock, H. 23 (Rostock: Univ., 2000), Rektorenreihe, o. S. Aus dieser Aufstellung geht hervor, daß Peter Memm am 15.10.1572 bzw. 8.10.1578 zum Rektor gewählt und am 4.11.1572 bzw. 16.10. 1578 in sein Amt eingeführt wurde. Für Memms erstes Rektorat vgl. auch Hofmeister, *Die Matrikel der Universität Rostock*, 177.

86 N. Chytraeus, *Contra Pestem Epistola Satyrica* (Rostock: Lucius, 1578); siehe auch N. Chytraeus, *Poematum Nathanis Chytraei*, 315–20.

eines satirischen Briefgedichts plädiere Chytraeus für Rationalität und tatkräftige Vorsorge und unterstreiche die „Bedeutung der Kategorie des 'Nützlichen' für die Steuerung menschlichen Handelns und die Synthese von wahrer Frömmigkeit und weltimmanenter 'Klugheit.'"[87] Man gewinnt den Eindruck, als hätten Pest- und Krisenerfahrung den Reformwillen im Zuge der sogenannten „zweiten" Reformation erneut gestärkt und damit auch einer größeren Aufnahmebereitschaft für calvinistisches Gedankengut die Wege geebnet. Das gilt zumindest für Nathan Chytraeus. Er bekannte sich schließlich ausdrücklich zu calvinistischen Positionen, weshalb er 1593 seinen Wohnsitz ins calvinistische Bremen verlegen mußte und dort das Rektorat der Lateinschule übernahm.[88]

In jenem Pestjahr 1577 also beschloß Memm, seine Schrift *De recto medicinae vsu* als *pars secunda* seines Eidkommentars erneut drucken zu lassen. Die *pars prima* mit ihrem starken Leitmotiv der Gottgefälligkeit und Dankbarkeit reflektiert auf zwei Sünden, die Gott zahlreichen christlichen und jüdischen Autoren zufolge mit Seuchen bestrafte: Gottlosigkeit und kindlichen Ungehorsam.[89] Indem Memm die Gehorsamspflicht des Kindes gegenüber den Eltern auf die der Studenten gegenüber ihren Lehrern überträgt und damit die Sünde der Undankbarkeit weiter einzudämmen bestrebt ist, betreibt er also auch Pestprophylaxe. Aber was bestimmte ihn, seinen Erstling als *pars secunda* seines Eidkommentars wieder abdrucken zu lassen? Ein solches Verfahren mutet auf den ersten Blick seltsam an, hatte er doch den Schritt von Utrecht nach Rostock, von einem (überwiegend) katholischen Gemeinwesen in ein lutherisches Gemeinwesen, von einem bischöflichen zu einem fürstlichen Brotherrn und von einer weitgehend katholischen zu einer lutherischen Patientenklientel bewältigt. Daß ein und derselbe Text, imprägniert mit religiösen Reflexionen, in beiden Welten, deren Ränder durch Inquisition und Zensur streng bewacht wurden, Resonanz finden würde, war doch reichlich unwahrscheinlich. Zudem gibt es ein Indiz, das darauf schließen läßt, daß die Reformation Memm selbst spätestens nach seiner Ankunft in Rostock in ihren Bann geschlagen haben mag:

---

87 Kühlmann, „Selbstverständigung im Leiden," 18.
88 T. Elsmann, H. Lietz und S. Pettke (Hgg.), *Nathan Chytraeus, 1543–1598. Ein Humanist in Rostock und Bremen. Quellen und Studien* (Bremen: Edition Timmen, 1991); K.-H. Glaser, H. Lietz und St. Rhein (Hgg.), *David und Nathan Chytraeus. Humanismus im konfessionellen Zeitalter* (Ubstadt-Weiher: Verlag Regionalkultur, 1993).
89 Zur Pest als göttlicher Strafe für kindlichen Ungehorsam s. R. Horrox, *The Black Death* (Manchester: Manchester University Press, 1994), 134–35; S. L. Einbinder, *No Place of Rest: Jewish Literature, Expulsion, and the Memory of Medieval France* (Philadelphia: University of Pennsylvania Press, 2009), 125–27; J. K. Stearns, *Infectious Ideas: Contagion in Premodern Islamic and Christian Thought in the Western Mediterranean* (Baltimore: The Johns Hopkins University, 2011), 163.

Es war im Herbst 1593, als Herzog Heinrich zu Braunschweig und Lüneburg (1533–1598) und seine Gemahlin Ursula, Prinzessin von Sachsen-Lauenburg (1545–1620) beschlossen, ihren jüngsten, bald fünfzehnjährigen Sohn August, den späteren Begründer und Namenspatron der Herzog August Bibliothek, auf die Universität zu schicken.[90] Da dies ratsamerweise in Begleitung eines Präzeptors geschehen sollte, wandten sie sich an Augusts früheren Lehrer Martin Nordtmann, seit 1590 Lehrer am Catharineum in Lübeck, mit der Bitte, einen „feinen gelahrten gesellen" für diese Stellung zu empfehlen. Unter den von Nordtmann in die engere Wahl genommenen Kandidaten, die sowohl wissenschaftliche Qualifikationen als auch pädagogisches Geschick und Lehrerfahrung mitbringen sollten, befand sich auch ein gewisser Isaak Memm, höheres Jura-Semester in Wittenberg. Als die fürstlichen Auftraggeber aus Danneberg jedoch darum baten, Erkundigungen über Bekenntniszugehörigkeit und Lebenswandel der avisierten Personen einzuholen, war es um Isaak Memms Kandidatur geschehen. Denn mit Brief vom 18. Dezember 1593 berichtete Nordtmann, diesem werde zwar „kunst," vor allem im Lateinischen, und „geschicklichkeit" nachgerühmt, drei Indizien ließen jedoch an seinem orthodoxen Luthertum zweifeln: zunächst sei die Konfession des Vaters Peter zweifelhaft, zweitens sei Isaak während seiner Zeit in Wittenberg, wo Kurfürst Christian zwischen 1586 und 1591 ein calvinistisches Kirchenregiment führte, Präzeptor bei den Kindern eines Wesenbecks gewesen – und einer von denen, Petrus Wesenbeck, neigte tatsächlich zum Calvinismus –, und drittens sei Isaak Memm Schüler des Caselius gewesen, an dessen Bekenntniszugehörigkeit ebenfalls Zweifel laut geworden seien. Falls man also „so einen gantzen Lutheraner" haben wolle, rate er, Martin Nordtmann, von Isaak Memm ab. Trotz dessen „furtrefflichen ingenium," trotz einer verheißungsvollen Karriere, die man ihm in Wittenberg nachsage, trotz der Möglichkeit, ihm vielleicht sogar mit Verdächtigungen Unrecht zu tun, sei doch mit seiner Religion als dem „furnemesten puncte" nicht zu spaßen, und es helfe nicht, „wan die ware religio nicht lauter und rein bey einem" sei.[91] Mit anderen Worten, die Episode mobilisierte Gerüchte von

---

90 Ich folge in diesem Absatz im wesentlichen dem Aufsatz von M. von Katte, „Zur Erziehung und Ausbildung Herzog August des Jüngeren zu Braunschweig und Lüneburg. Die Präzeptorwahl von 1594 und die Entstehung seiner Devise 'Expende,'" in P. Raabe (Hg.), *Wolfenbütteler Beiträge. Aus den Schätzen der Herzog August Bibliothek*, Wolfenbütteler Beiträge, 5 (Frankfurt a. M.: Vittorio Klostermann, 1982), 9–52. Die Nachweise finden sich dort.
91 Als Herzog August von Braunschweig 1596 Rektor der Universität Tübingen wurde, schrieb der einige Jahre zuvor als Präzeptor verschmähte Isaak Memm eine Ode auf den späteren Fürsten, trug sich mit dem vor dem Hintergrund der geplatzten Präzeptorwahl beziehungsreichen Horazvers „Was einer zu vermeiden sucht, davor kann er sich nie genug hüten" (*Carm.* II, 13, 13) in dessen Stammbuch ein, schloß Freundschaft mit ihm und stand ihm auch später als Rechtsbeistand zu Diensten. Vgl. von Katte, „Zur Erziehung," 29–30.

einem vermeintlichen Kryptocalvinismus, die nicht nur über Isaak Memm, sondern auch über seinen Vater Peter kursierten. Damit stand die Familie im Verdacht, „den lutherischen Bekenntnisstand durch heimliche Annäherung an ref[ormierte] bzw. calvinistische Positionen absichtsvoll" zu unterminieren, wobei sich der Vorwurf im Falle des Vaters auch auf philippistische Lehrpositionen beziehen könnte.[92]

Wenn solche Gerüchte ein *fundamentum in re* hatten, wäre es doch recht verwunderlich, daß ein vor der Konversion verfaßter, religiös stark aufgeladener Text auch nach dem religiösen Bekenntniswandel noch seine Gültigkeit hätte. Wäre der alte Text im Hinblick auf den neuen Widmungsträger, das neue Zielpublikum, den neuen Werkkontext, das neue institutionelle Entstehungsmilieu, den neuen Drucker,[93] die neue Pestsituation und vor allem den „neuen," weil konvertierten, Autor nicht gänzlich überarbeitungs- und aktualisierungsbedürftig, wenn nicht hoffnungslos anachronistisch und damit überhaupt nicht wiederverwertbar gewesen? Mir scheint viel plausibler, daß der Reformvorschlag, den Memm 1564 einem orthodoxen Katholizismus ins Brevier schrieb, ihm 1577 ebenso geeignet schien, einer lutherischen Orthodoxie unterbreitet zu werden, gerade im Pestjahr 1577. Ob dieser Reformvorschlag im Kern calvinistisch ist – Memm scheint, wie Calvin, der Astrologie kritisch gegenüber zu stehen,[94] propagiert Gottesfurcht, das Studium zum Zwecke der Erkenntnis des göttlichen Willens, ein nach der Bibel ausgerichtetes Leben und die Hingabe an den Beruf – müssen weitere Forschungen klären. Vorerst scheint es, als sei Memm von den gemeinreformatorischen Bewegungen getragen, ohne sich in den innerprotestantischen Querelen definitiv zu positionieren.

---

92 T. Kaufmann, „Kryptocalvinismus," in Betz et al., *Religion in Geschichte und Gegenwart*, 4: Sp. 1793.
93 Das Verlagsprogramm des Rostocker Druckers Augustin Ferber d. Ä., der Memms Eidkommentar als eines seiner ersten Bücher 1577 herausbrachte, bestand zwischen 1575 und 1602, als sein Sohn die Druckerei übernahm, überwiegend aus Leichenpredigten, Gesangbüchern und religiösen Texten von Reformatoren wie Jacob Runge (1527–1595). Die Gunst, die Herzog Ulrich ihm erwies, spricht ebenso dafür, daß Ferber Lutheraner war. Vgl. Grewolls, *Wer war wer*, 2710.
94 S. Kusukawa, „Aspectio divinorum operum: Melanchthon and Astrology for Lutheran Medics," in O. P. Grell und A. Cunningham (Hgg.), *Medicine and the Reformation* (London: Routledge, 1993), 33–56.

I.3.3 Der professionspolitische Kontext beider Fassungen (*De recto medicinae vsu* und *pars secunda*)

Wichtiger aber als die konfessionelle Verortung Memms[95] ist die Beobachtung, daß er die Medizin von der Theologie durchaus als eigenständige Disziplin getrennt wissen will, konfessionelle Grabenkämpfe seinem irenischen Ideal, das ihm auch für die Ärzteschaft vorschwebt, zuwiderlaufen und er seinem Kommentar ein dezidiert professionspolitisches Programm einschreibt. Unter dieser Perspektive gewinnt auch die Tatsache des Nachdruckes seines Erstlings unter den gänzlich anderen Lebens- und Arbeitsbedingungen des Jahres 1577 an Plausibilität. Denn die professionspolitische Dimension stiftet eine Kontinuität, die die hier in Rede stehende Periode von 1564 bis 1577 bei allen konfessionellen Diskontinuitäten umgreift und verständlich werden läßt, warum der 1564 publizierte Text 1577 immer noch als brandaktuell gelten konnte.[96] Drei Beispiele sollen diese Dimension des Memmschen Werkes illustrieren:

a. Als Peter Memm 1571 einen Ruf an den schwedischen Hof ausgeschlagen und die Leibarztstelle bei Herzog Ulrich angenommen hatte, verpflichtete er sich in dem der Bestallungsurkunde vom 29. September 1571 beigegebenen Diensteid, dem lutheranischen Fürsten, seiner Gemahlin, der Prinzessin Elisabeth von Dänemark (1524–1586),[97] und ihrer Tochter Sophie (1557–1631),[98] die im darauffolgenden Jahr mit Friedrich II. (gest. 1588), dem König von Dänemark, verheiratet werden sollte, „treu holt und gewertig" zu sein, deren „nutz und bestes, so vihl meine vocation betrifft, euersten vorstandes und vormugens" vorauszusagen und zu befördern, „unheil und schaden" abzuwenden, verschwiegen bis ins Grab zu sein und sich „treulig auffrichtig und redlich" zu verhalten.[99] Die Parallelen

---

[95] Selbst bei reicher belegten und intensiver erforschten Ärzten aus dem 16. Jahrhundert wie Vesal oder Paracelsus ist ein solches Unterfangen bislang nicht völlig überzeugend gelungen. Es genügt an dieser Stelle festzuhalten, daß Memm von katholischen wie protestantischen Reformbewegungen beeinflußt, inspiriert und getragen wurde und bestrebt war, diese Impulse für eine Reform der Medizin seiner Tage fruchtbar zu machen, ohne die Ärzteschaft insgesamt zu spalten.
[96] Von niederen Motiven für das Recycling im Vorfeld seines Rektorats soll hier abgesehen werden.
[97] Ulrich hatte am 16.2.1556 die Witwe seines Onkels Magnus III. zu Mecklenburg geheiratet. Vgl. Grewolls, *Wer war wer*, 10258.
[98] Vgl. Grewolls, *Wer war wer*, 9515.
[99] Mecklenburgisches Landeshauptarchiv Schwerin, Bestand Hofstaatssachen, Bestallungen: Mediziner (Memmius). Die Bestallungsurkunde lautet: „Wir Ulrich von Gotts gnaden hertzog zu Mecklenburgk etc. mit diesem unserm offenen briefe bekennen und thun kundt, daß wir den hochgelerten unsern lieben doctorem, vor unsern leibarz bestalt und angenommen haben.

zwischen Memms „iuramentum" beliteltem Diensteid zum Schweige-, Redlichkeits- und Schadenbegrenzungsgebot des hippokratischen *iusiurandum* sind nicht zu überhören. Den Eid des Hippokrates zu kommentieren, hieß also auch, Sinn und Bedeutung solcher Diensteide zu ergründen bzw. vorzugeben, modern gesprochen, an seiner eigenen Arbeitsplatzbeschreibung mitzuwirken. Memms Biographie zeigt exemplarisch, in welchem Maße der Eid des Hippokrates ein Echo in Bestallungseiden der Zeit findet und somit keineswegs ausschließlich als altehrwürdiges Dokument rezipiert wurde, das von längst Vergangenem raunte, sondern einen echten, professionspolitisch relevanten Sitz im frühneuzeitlichen Leben hatte. Da spielte es kaum eine Rolle, ob der Arbeitgeber ein katholischer Bischof oder ein lutheranischer Herzog war.

b. Der Fluchtpunkt der medizinischen Reform, wie Memm sie konzipiert, ist letztlich nicht Christus, sondern Hippokrates. Hippokrates firmiert in Memms

---

Nehmen ihnen darzu crafft diß unsers briffs uff und an, alß und dergestalt, daß her uff seine Eide und pflichte so ehr unß in uberantwortung dieser unser bestallung wircklich geleistett, wan und zu welcher zeit, auch so offt wir ihn vor uns, unsere freuntliche herzliebste Gemhal, und tochter Freulin Sophie auch andere unsere diener von Rostog abe fordern würden sich bei uns, uff unsere unkosten bei hofe einstellen, und daselbst so lang wir sein vor uns unser mitbenannten zu gebrauchen, vorharren sol, sich auch mit allem fleiß wie einem ehrlichen frommen und geweltem fürsten Medico eignet und gebuertt unsere und unser mitbenante ... und Cur bevholen lassen sein, und alle seine Consilia und Medicin mit gutem tiefsinnigem nachdencken euersten und besten seins vorstands anordnen, sol und wil. Hir entkegen und zu ergetzligkeit solcher seiner dienste wolle wir Ihm jedes jhars Einhundert thaler, als alle halbe Jhar zo zhalen, durch unsern Rentmeister entrichten und zustellen lassen und jherlich vier drömpt Roggen und einen Ochsen zu seiner haushaltung aus unserm Ampt Dobberan auff Michaelis vorreichen lassen, und sol diese bestallung so wol zu seinem als unserm gefallen stehn, daß wen uns Ihn vor ein diener und leibarzt nicht lenger zu haben gelegen wir Ihm im jhar zuvor sein dienstbestallung widderumb auffkundigen muegen und also herwidder do jhars uns lenger zu dienen bedencklichen alles getreulich und sonder geferde. Zu urkunt haben wir unser fürstlich Secret hi unten gedruckt und uns mit eigener Hand unterschrieben Actum Güstrow am tage Michaelis anno 1571." Der Diensteid lautet: „Ich gelobe und schwere, nachdem der durchleuchtige hochgeborne fürst und her, her ulrich, herzogh zu Mecklenburgk Mgh mich vor sein f.g. leibarz gnediglich bestalt und angenommen hatt, daß ich s.f.g., derselben gemal und Jungen freulin, treu holt und gewertig sein wil, Ihrer f.g. semptlichen und sonderlichen nutz und bestes, so vihl meine vocation betrifft, euerstern vorstandes und vormugens, vorsagen und befordern, unheil und schaden so vihl mir vorstendlich und möglich höchstes fleisses abwenden. Do mir auch ettwas so zu schweigen sich gebuert, ... würde wil ich das bei mir vortraulich bis in meine gruebe behalten und niemanden offenbaren, Mich auch in allem vermuege s.f.g. ... mir sonderbaren auff gerichen bestallung wie eim geweltem fürsten Medico und diener, eignet und gebürt, treulig auffrichtig und redlich vorhalten, Als mir Godt helpfe und sein heylligs worth."
Ich danke Herrn Dr. Rakow für die Übersendung von Kopien und Herrn Dr. Henning von Gadow für seine unschätzbare Hilfe bei der Transkription der in diesem Aufsatz zitierten Archivalien aus dem 16. Jahrhundert.

Buch als Stifter der Medizin,[100] zugleich ihr Vollender[101] und mit Galen als der nützlichste aller medizinischen Autoren.[102] Hippokrates ist Memms Vorbild, das es zu imitieren gilt. Das ist gute Rostocker Tradition. Von Cornarius, der sich nichts sehnlicher wünschte, als mit Hippokrates alt zu werden, war schon die Rede.[103] Im Jahr von Memms Dienstantritt an der Universität lasen Heinrich van den Brock und Levinus Battus über die hippokratischen Aphorismen.[104] Daß dieses vermeintliche In-Spuren-Gehen, das natürlich auf eine Rekreierung des Hippokrates hinausläuft,[105] weit in die ärztliche Praxis hineinwirkte, scheint folgende Anekdote nahezulegen, die Memm aus seinem Berufsleben zum Besten gibt:

> Zu spät zu einer erkrankten alten Frau gerufen, erwog ich alles und prognostizierte endlich, daß sie etwa zwei Wochen nach meiner Visite aus dem Leben scheiden werde, und auch in einem Heilmittel keine Hoffnung auf Leben mehr liege. Als ich keine Anstalten machte, ein Rezept auszustellen, warf mir die Alte, die sich über meine Sparsamkeit beim Verschreiben erregte, vor, die Medizin sei doch nicht für Gesunde, sondern für von Krankheit Geplagte bestimmt. Ganz recht, erwiderte ich, sofern man den Dienst der Medizin rechtzeitig in An-

---

100 Memm, *De recto medicinae vsu*, 67–68: *Hippocrate verò in medica disciplina non extant vetustiores, quorum opera aut certior facta est aut lucidior medicina.* („Es gibt aber in der Medizin keine älteren Autoren als Hippokrates, durch deren Bemühen die Medizin entweder sicherer oder einleuchtender gemacht wurde.")
101 Memm, *De recto medicinae vsu*, 68–69: *Nam quibus Hippocratis mens conciliata est, reliquorum traditiones tantum abest vt sequantur, vt ne semel quidem aspicere dignentur: tanta est nobis ab Hippocrate relicta perfectio. Sed quia pleraque aphoristicos ab illo pronunciata sunt, non cuius contingit adire Corinthum.* („Denn wer sich einmal mit Hippokrates vertraut gemacht hat, dem liegt es derart fern, sich der Überlieferung der anderen Autoren anzuschließen, daß er sie nicht einmal eines Blickes würdigt: Einen so großen Grad der Vollendung hat das Werk des Hippokrates für uns erreicht. Da er sich aber meistens aphoristisch geäußert hat, erschließt sich der Sinn nicht jedermann.") *Non cuivis contingit adire Corinthum* ist ein stilles Zitat aus Horaz, *Ep.* 1.17.36.
102 Memm, *De recto medicinae vsu*, 70: *Sed ab his ad Hippocratem et Galenum recedamus, quorum lectio, vt frugifera, ita omnibus praeponenda.* („Laßt uns aber von diesen (Dingen) wieder zu Hippokrates und Galen zurückkehren, deren Lektüre, fruchtbringend wie sie ist, jeder anderen vorzuziehen ist.")
103 Rembert Gilsheim (ca. 1485–1533), der 1520 allein die medizinische Fakultät in Rostock repräsentierte und 1519 einen *Liber collectionum aphorismorum Hypocratis de unaquaque egritudine* bei Ludwig Dietz in Rostock herausgab, wäre hier ebenfalls zu nennen. Vgl. Wagner, *Observantia Lectionum*, 84–86; Grewolls, *Wer war wer*, 3331; Asche, *Von der reichen hansischen Bürgeruniversität*, 106.
104 Vgl. den Einblattdruck *Series lectionum, quae futuro semestri aestivo anni 1568: in Academia Rostochiensi publice proponentur* (Rostock: Lucius, 1568). Es ist also keinesfalls undenkbar, daß der erste Teil von Memms Eidkommentar aus dem Lehrbetrieb hervorging, zumal Memm die Medizinstudenten als Zielgruppe direkt anspricht.
105 Cantor, *Reinventing Hippocrates*.

spruch nehme: Die zweite Woche überlebte sie nicht. Die dermaßen belehrten Anwesenden lobten mein anständiges Verhalten und schätzten die Medizin fortan höher.[106]

Die Erziehung der Bevölkerung zu rechtzeitiger Konsultation universitär ausgebildeter Ärzte als Vorleistung für erfolgversprechende ärztliche Behandlung, die auf einen Reputationsgewinn der Medizin, eine wirtschaftliche und soziale Stärkung ihrer professionellen, d. h. an Universitäten ausgebildeten, Sachwalter und einen gehobenen Gesundheitsstatus der Bevölkerung hinausläuft, wird hier, zumindest implizit, hippokratisch legitimiert. Denn solche Denkfiguren finden sich auch in der hippokratischen Schrift *de Arte* 3,24 (Jouanna 226.15–227.1 [VI, 4.18–6.1 L.]), und auch wenn Memm an der zitierten Stelle einen Verweis auf diese Schrift oder Hippokrates unterläßt – er ist ja ein Meister des stillen Zitats –, zeigt die Passage, wie sehr Memm hippokratische Lehren verinnerlicht und in sein Verhaltensrepertoire als praktischer Arzt aufgenommen hat.[107]

Eine solche Fixierung auf Hippokrates vertrug sich übrigens durchaus mit einer modernen, gegenwartsorientierten Einstellung, wie Memms Kritik an der Überbewertung der Harnschau[108] sowie seine Empfehlung des Unterrichts am Krankenbett[109] zeigen. Hippokrates stand eben über den Zeitläuften und damit übrigens auch über den konfessionellen Auseinandersetzungen.

---

106 Memm, *De recto medicinae vsu*, 154: *Serò ad aegrotantem mulierculam accersitus cuncta circumspexi, tandemque praedixi, circa decimumquartum à nostro accessu diem, ipsam ex hac vita excessuram, neque in medicamine spem vitae esse reliquam. Manumque à scribendo cohibenti obiecit vetula, meam in scribendo parsimoniam immoderatè ferens, non sanis quidem, sed aegritudine vexatis medicinam deberi. Cuius sententiae annuens, si in tempore postuletur, respondi: decimumque quartum diem non superauit. Vnde edocti astantes meam aequitatem laudabant, plurisque deinceps fecere medicinam.*
107 Zur Frage der Nichtbehandlung im Corpus Hippocraticum insgesamt vgl. die hochdifferenzierte Studie von H. von Staden, „Incurability and Hopelessness: The Hippocratic Corpus," in P. Potter, G. Maloney und J. Desautels (Hgg.), *La maladie et les maladies dans la Collection Hippocratique; actes du VIe Colloque International Hippocratique* (Québec: Les Éditions du Sphinx, 1990), 75–112. Zu De arte s. von Staden, „Incurability and Hopelessness," 106.
108 Memm, *De recto medicinae vsu*, ff. e6^v-f5.
109 Memm, *De recto medicinae vsu*, 74: *Proinde ex lectorum munere erunt a lectione, aut alioqui vbi opportunum fuerit, ad aegros auditores deducendi: ibique cum illis ineunda ratio, quo pacto ex praesenti morborum statu, elicienda sit medendi regula.* („Folglich müssen die Hörer gemäß der Aufgabe der Dozenten aus der Vorlesung, oder von wo immer es passend erscheint, zu den Kranken geführt werden und dort mit ihnen erörtern, wie sich aus dem gegenwärtigen Krankheitszustand eine Regel für die Heilung ableiten lässt.") Auch Memms Einstellung zur Humansektion ließe sich hier anführen. Er findet sie oft nützlicher als Vorlesungen. *Quia singularum corporis partium, quae videlicet earum ratione in aegris corporibus oriuntur, signorum rationem aßequimur: quum ipsa naturam partis oculis objicit.* („Weil wir die Logik der Zeichen der

c. Als Memm in Paris studierte, war sein Professor Dubois in eine Fehde mit seinem ehemaligen Schüler Andreas Vesal verwickelt. 1551 schrieb Dubois eine Invektive gegen den Autor der *Fabrica*,[110] der die gelehrte Welt in seinem *opus magnum* aus dem Jahre 1543 nicht nur daran erinnert hatte, daß Galens Anatomie im wesentlichen auf Zootomie und nicht auf Humansektion fußte, sondern auch auf zahlreiche Irrtümer Galens hingewiesen und damit der Galionsfigur der Galenisten einen empfindlichen Ansehensverlust beschert hatte.[111] In seiner Schmähschrift warf Dubois Vesal Mangel an Frömmigkeit, Unwissenheit, Undankbarkeit und Anmaßung vor. Der Verbreiter falscher Anschuldigungen (*calumniator*) und Wahnsinnige (*vaesanus*), d. h. Vesal, wird den Lesern als Dissident vorgeführt. Dubois ruft den Kaiser auf, Vesal übel zuzurichten und dessen Anhängern nur Amnestie zu gewähren, wenn sie zu den Fahnen ihrer Lehrer zurückkehrten.[112] Interessant ist nun, daß all dies auch im Namen des Hippokrates

---

einzelnen Körperteile, d. h. der [Zeichen], die auf Grund ihrer Beschaffenheit [gemeint: die anatomischen Verhältnisse] in den kranken Körpern entstehen, begreifen, weil sie [die Sektion] die Natur eines Körperteils den Augen preisgibt.") Vgl. Memm, *De recto medicinae vsu*, 71.

110 Der Titel der Invektive lautet: *Vaesani cuiusdam calumniarum in Hippocratis Galenique rem anatomicam depulsio*. Ihr Erstdruck erschien 1551 bei C. Barbé in Paris. Ich zitiere aus der Ausgabe Basel: Jacobus Derbilley, 1556.

111 Die Stimmung gab Renatus Henerus in den Einleitungssätzen zu seiner 1555 bei Scotto erschienenen Schrift *Adversus Jacobi Sylvii depulsionum anatomicarum calumnias, pro Andrea Vesalio*, 1 und 3 wieder: *Annvs jam est tertius, quo Lutetiae cum essem, Syluius suum illud aduersus Andream Vesalium virulentum euulgabat scriptum: scriptum tragicum et contumeliosum, viro vt annis et laboribus jam confecto non dignum, ita homine alienae gloriae inuidente dignissimum. ... dum haec (quod nunc facile conijcio) parturiret, tot conuitiorum plaustra pertuleramus, cùm singulis ferè horis in Vesalium crudo inueheretur stomacho, non sine insigni aurium nostrarum taedio, multorumque indignatione.* („When I was in Paris three years ago Sylvius published that tragic, very abusive, and virulent attack upon Vesalius. While it was unworthy of Sylvius – after all, he was a man already worn out by his age and work – , it displayed his envy of the fame achieved by another to the greatest perfection. ... While Sylvius was in labor with this pamphlet, as I now easily realize, we were forced to endure a constant stream of abuse and virtually incessant and furious invective against Vesalius. It wearied our ears and aroused the indignation of many of us." Die Übersetzung von C. D. O'Malley, *Andreas Vesalius of Brussels 1514–1564* [Berkeley: University of California Press, 1964], 459–60 [Anm. 91] und 247 wurde leicht verändert.)

112 Dubois, *Vaesani cuiusdam*, 116: *Caesarem Maiestatem imploro, vt monstrum hoc ignorantiae, ingratitudinis, arrogantiae, impietatis exemplar perniciosissimum domi suae natum et educatum, vti est commeritum, malè mulctet, ac omnino comprimat, ne suo pestilenti halitu Europam reliquam venenet. Nam Gallos quosdam, et Germanos et Italos sua exitiali spuma iam infecit, sed eos rei anatomicae et reliquarum opinor medicinae partium ignaros ... si quoque falsis dogmatibus relictis huius transfugae, ad praeceptorum suorum vexilla redeant, à quibus liberalissimè, humanissimè, integerrimè hactenus fuerant instituti, et deinceps quoque perpolientur.* („Ich flehe Eure Kaiserliche Hoheit an, dieses Monster, das in Eurem Hause geboren und erzogen wurde,

gesagt wird, wie der Titel der Invektive bereits andeutet. Wer Galen, den Chefexegeten hippokratischer Medizin, angreift, hat auch Hippokrates, dessen Vorbild, angegriffen. Also muß auch eine Verteidigung beiden antiken Autoritäten gelten. Es geht um Traditions- und damit Qualitätssicherung in der Medizin, um Rechtgläubigkeit der Ärzte, um Einheit der Lehre und der Wissenschaft. Und natürlich geht es um die Autorität von Medizinprofessoren wie Dubois, die sich als Statthalter des Hippokrates auf Erden in der Nachfolge Galens verstanden. Dubois wirft Vesal sogar einen unmittelbaren Bruch des hippokratischen Eides vor:

> Hence the loyal reader will now clearly realize that this slanderer wickedly renounced his oath of allegiance to his master Hippocrates, in which he had promised the greatest gratitude to his teachers and to their adopted children ...[113]

Die im Lehrvertrag des hippokratischen Eides genannten Lehrer werden zur *pars pro toto* für die Gesamtheit aller Sukzessionsglieder der medizinischen Wissens- und Traditionssicherung. Vesal habe folglich seine im Eid geschworene Loyalitätspflicht verletzt.[114]

Memm wird dieser Disput als jungem Medizinstudenten in Paris nicht entgangen sein, und vielleicht dachte er ähnlich darüber wie Henerus. Doch eine Schlußfolgerung, die er aus dieser Erfahrung eindeutig zieht, kehrt als 8. Hypothese, sprich These, in seinem Erstling von 1564 wieder: „Die Ärzte in einem Gemeinwesen müssen sich einträchtig und einer Meinung zeigen."[115] Und viel-

---

jenes höchst verderbliche Exemplar an Unwissenheit, Undankbarkeit, Arroganz und Gottlosigkeit, verdientermaßen übel zuzurichten und gänzlich zu zerquetschen, damit es mit seinem Pesthauch nicht das übrige Europa vergifte. Denn er hat schon mit seinem verderblichen Schaum einige Franzosen, Deutsche und Italiener infiziert, aber ich meine, daß diese von der Anatomie und den übrigen Teilen der Medizin keine Ahnung haben [... Diese Männer verdienen Mitleid und Vergebung ...] wenn sie, nachdem sie ... die falschen Lehren dieses Fahnenflüchtigen aufgegeben haben, zu den Fahnen ihrer Lehrer zurückkehren, von denen sie bislang äußerst freigebig, menschlich und lauter unterrichtet und darauf auch feingeschliffen werden.") Vgl. auch O'Malley, *Andreas Vesalius*, 250, dessen Übertragung an dieser Stelle recht frei ist.
113 ... *vt iam planè intellecturus sit auditor pius, calumniatorem hunc malè ab Hippocratis magistri iureiurando desciuisse, qua parte suis praeceptoribus, eorumque liberis antipelargían summamque gratitudinem pollicetur.* ... Vgl. Dubois, *Vaesani cuiusdam*, f. 4. Die Übersetzung stammt von O'Malley, *Andreas Vesalius*, 248.
114 O'Malley, *Andreas Vesalius*, 247–48 kommentiert: „Elsewhere, too, there is the suggestion that this special quality of commanding loyalty was handed down by these ancient teachers and that Sylvius felt, as certainly the most vocal adherent of Galen, that Vesalius had been disloyal to him, too."
115 Memm, *De recto medicinae vsu*, f. b8: *Medicorum vnius reipub[licae] concordia et vnanimi consensu opus est.*

leicht sind wir sogar berechtigt, in dem folgenden Satz aus dem entsprechenden Kapitel von *De recto medicinae vsu* auch eine Anspielung auf Dubois zu erkennen, der bekanntlich den jungen Vesal als Pesthauch perhorresziert hatte:

> An dieser Stelle scheint die Leichtfertigkeit gewisser älterer Leute erwähnenswert, die sich aus unterschwelligem Haß in Verleumdungen der Jüngeren ergehen, von denen ihrer Meinung nach statt Heilung nichts als der Tod zu erwarten ist.[116]

Solche von Habgier und Ehrgeiz bestimmten Ärzte müßten aus der Ärzteschaft entfernt werden, ehe Frieden und Eintracht eintreten könnten. Und vielleicht ist es diese am Familienmodell orientierte Eintracht, die Memm veranlaßte, Fragen der Glaubensspaltung nicht offen anzusprechen und sich auch selbst auf diesem Kampfplatz nicht deutlich zu positionieren. Er hätte es als unprofessionell erachtet und dem Ansehen und der Würde der Medizin abträglich. Dieses Bekenntnis war während der brodelnden Reformationswirren in den Niederlanden ebenso aktuell wie in einer Universität, die ihr Überleben einer *Formula concordiae* verdankte, die die Forderung nach Einheit und Reinheit der Lehre in die Statuten ihrer Theologischen Fakultät aufgenommen hatte, wie auch im lutherischen Rostock insgesamt, in dem zunehmend die Angst vor Kryptocalvinisten umging. Daß die Ärzteschaft nach dem Muster der Familie zu organisieren sei und denselben Friedens- und Eintrachtsgeboten wie diese zu unterwerfen sei, wird an der „Familienuniversität"[117] Rostock besonders gern aufgenommen worden sein. Memm scheint selbst danach gelebt zu haben: Er betreute nicht nur Wilhelm Lauremberg (1547–1612) *disputatio circularis* über ein hippokratisches Thema[118] und beschäf-

---

116 Memm, *De recto medicinae vsu*, 137–38: *Quo loco seniorum quorundam leuitas videtur commemorabilis, qui odio latente sunt in obtrectationibus iuniorum frequentes, à quibus eorum iudicio, pro salute mera mors expectanda est.*
117 Asche, *Von der reichen hansischen Bürgeruniversität*, 478–80. Mit diesem Terminus sind die „exklusiven Heiratskreise zwischen Professoren, Stadtpastoren sowie städtischen und fürstlichen Beamten" gemeint, die seit der Mitte des 16. Jahrhunderts „Strukturen einer protestantischen Familienuniversität mit Protektionsmöglichkeiten für die Besetzung akademischer Lehrstühle sowie geistlicher und weltlicher Ämter in den Territorien" ausbilden. „Die Kontrolle des Stipendienwesens, exklusive Heiratskreise innerhalb der Gelehrtendynastien und die Quasi-Erblichkeit von Lehrstühlen waren konstitutive Bestandteile protestantischer Familienuniversitäten." Ebenda, 478 bzw. 485–86. Vgl. auch ders., „Über den Nutzen von Landesuniversitäten in der Frühen Neuzeit. Leistung und Grenzen der protestantischen 'Familienuniversität,'" in P. Herde und A. Schindling (Hgg.), *Universität Würzburg und Wissenschaft in der Neuzeit. Beiträge zur Bildungsgeschichte. Gewidmet Peter Baumgart anläßlich seines 65. Geburtstages* (Würzburg: Schöningh, 1998), 133–49.
118 W. Lauremberg, *Resp. circularis disputatio continens propositiones de flatibus ex doctrina Hippocratis desumptas. Praes. Petro Memmio* (Rostock: [o. V.], 1578). Zirkulardisputationen waren

tigte ihn als Hauslehrer seiner Söhne, sondern gab ihm auch seine Stieftochter Johanna Longolia (Jeanne Longoeil) zur Frau und half, ihn als seinen Amtsnachfolger, als städtischen Professor und Stadtarzt in Rostock zu installieren, als er selbst 1581 diese Positionen aufgab und als Stadtarzt nach Lübeck ging.[119]

I.3.4 Ein professionspolitischer Kontext der *pars prima* (1577)

Memm hatte als Arzt in Rostock und Lübeck immer wieder mit Apothekern zu schaffen und zu kämpfen.[120] Als zum Beispiel sein Dienstvertrag mit dem Herzoghaus am 19. Januar 1587 erneuert wurde – Memm lebte seit 1581 als Stadtarzt in Lübeck –, wurde er auch zum *generalis inspector* der Apotheke in Schwerin bestellt und erhielt den Auftrag, diese Apotheke zu ordnen.[121] Vor Ablauf des ersten Jahres,

---

halböffentliche Veranstaltungen, zu denen jeder Universitätsangehörige zugelassen, der Kreis der Disputanten jedoch vorher festgelegt war. In ihnen ging es um die regelhafte Reproduktion und Applikation bekannter Wahrheiten, und es darf nach dem Gesagten nicht überraschen, wenn letztere in der Medizin den hippokratischen Schriften entnommen wurden. Zur Zirkulardisputation in Rostock vgl. T. Kaufmann, *Universität und lutherische Konfessionalisierung. Die Rostocker Theologieprofessoren und ihr Beitrag zur theologischen Bildung und kirchlichen Gestaltung im Herzogtum Mecklenburg zwischen 1500 und 1675* (Gütersloh: Gütersloher Verlags-Haus, 1997), 415–16. Siehe auch E. Horn, *Die Disputationen und Promotionen an den deutschen Universitäten, vornehmlich seit dem 16. Jahrhundert. Mit einem Anhang enthaltend ein Verzeichnis aller ehemaligen und gegenwärtigen deutschen Universitäten* (Leipzig: Harrassowitz, 1893). Memm fungierte als Praeses bei zwei weiteren Zirkulardisputationen, der von Gerardus Riswick *de temperamento* (1572) und der von Johannes Colerus *de purgandi ratione* (1575). Er selbst disputierte im August 1569 *de catarrho*.
119 Grewolls, *Wer war wer*, 5722; G. Teichmann, „Wilhelm Lauremberg," in Hartwig und Schmidt, *Die Rektoren der Universität Rostock*, 89. Der Grund für Memms Wohnortwechsel dürfte in seiner Beanspruchung als herzoglicher Leibarzt liegen. Am 29.12.1578 sandte Memm seinem Fürsten ein Schreiben mit der Bitte um Aufwandsentschädigung seiner im Dienstvertrag nicht vorgesehenen Auslandsaufenthalte, während derer er seine Familie und seine Patienten habe vernachlässigen müssen und dadurch empfindliche Gehaltseinbußen erlitten habe. Vgl. Mecklenburgisches Landeshauptarchiv Schwerin, Bestand Hofstaatssachen, Bestallungen: Mediziner (Memmius). Insbesondere seinen Verpflichtungen in Kopenhagen konnte er von Lübeck aus besser nachkommen. Auch dürfte ihn die Dreifachbelastung als Professor, Stadtarzt und Leibarzt auf die Dauer überstrapaziert haben.
120 Als Stadtarzt oblag Memm auch in Rostock die Apothekenaufsicht. Dazu gehörte die mindestens jährliche Visitation der Apotheken, die Kontrolle der *materialia und species* und die Sicherstellung der Versorgung der ganzen Bevölkerung mit „frische[n] und taugliche[n] materialien und artzney." Vgl. H. Schümann, *Von apothecarii, physici und clystierweibern. Apotheker und Apotheken der Stadt Rostock in acht Jahrhunderten* (Rostock: Redieck und Schade, 2003), 13.
121 Herzog Johann von Mecklenburg-Schwerin sicherte Memm mit 100 Talern, Bekleidung für 2 Personen, 20 Gulden, drei Drömpt Roggen, drei Drömpt Malz, einem Ochsen, vier Hammeln, vier

während dessen Memm im Dienste des seit 1585 für mündig erklärten und damit als Regent von Mecklenburg-Schwerin installierten Herzogs Johann VII. (1558–1592)[122] stand, trat er in Verhandlungen mit dem Rat der Stadt Lüneburg, wo die Stelle eines Stadtarztes zu besetzen war. In einem Schreiben vom 16. November 1587 resümierte Memm das Verhandlungsergebnis, das bei seinem Besuch in Lüneburg erzielt worden war.[123] Vor allem zwei Streitpunkte bereiteten Memm nach wie vor Sorgen: die Lüneburger Apothekenordnung und das „gebotne gelt." Die Apothekenordnung gestattete nämlich den Lüneburger Apothekern das freie Praktizieren, ein Umstand, den Memm „zum hogsten beschwerlich" fand, da dies Rezeptkorrekturen und Substitutionen seitens der Apotheker Vorschub leiste. Obschon ein strenges Verbot des *quid pro quo* im 15. und 16. Jahrhundert selten war, verlangten doch zahlreiche Apothekenordnungen und Apotheker-Dienstbriefe dieser Zeit Rücksprache mit dem Arzt bei Substitution und die „getreuliche" Anfertigung der Rezepte.[124] Diese Gewaltenteilung zwischen Arzt und Apotheker scheint Memm bei seinem Einspruch vorzuschweben. Bürgermeister und Rat der Stadt Lüneburg sicherten Memm ein absolutes Praktizierverbot für den ansässigen Apotheker zu.[125]

Schweinen und einer Tonne Butter nicht nur ein wesentlich höheres Jahresgehalt zu, sondern auch Sonderzuwendungen (z. B. ein Ehrenkleid) für Einsätze außerhalb des Herzogtums. Memm wird in diesem Diensteid auf den Fürsten, seine Kinder, seine Mutter, seine Gemahlin und künftige Gemahlin verpflichtet, denen er im Krankheitsfall auf schriftlichen Wunsch von Lübeck aus Medikamente bringen und bei denen er Heilmaßnahmen treffen und Besserung der Beschwerden erzielen soll. Im Notfall solle er sich auch der „Räthe Juncker" und anderer vornehmer Personen am Hofe annehmen, die Apotheke in Schwerin ordnen, einen Verwalter bestimmen und im übrigen deren „generalis inspector" bleiben. Vgl. Mecklenburgisches Landeshauptarchiv Schwerin, Bestand Hofstaatssachen, Bestallungen: Mediziner (Memmius).
122 Grewolls, *Wer war wer*, 4680–81. Er war vom 12. September 1585 bis zum 22.3.1591, als er an den Folgen von Messerstichen, die er sich selbst beigebracht hatte, für die aber zwei Frauen büßen mussten, die unter der Folter starben, Regent von Mecklenburg-Schwerin. Ulrich III. übernahm danach wieder die Regierungsgeschäfte, dieses Mal als Vormund der minderjährigen Söhne Adolf Friedrich I. und Johann Albrecht II.
123 Lüneburg, Stadtarchiv, AA A 7 a Nr. 42: Peter Memm an den Rat der Stadt Lüneburg am 16. 11.1587. Ich danke Herrn Archivamtmann Brauss für die Übersendung von Kopien der Korrespondenz zwischen Memm und dem Rat der Stadt Lüneburg.
124 U. Seidel, *Rezept und Apotheke. Zur Geschichte der Arzneiverordnung vom 13. bis zum 16. Jahrhundert*, Diss. rer. nat., Marburg, 1977; P.-H. Berges, *„Quid pro quo." Zur Geschichte der Arzneimittelsubstitution*, Diss. rer. nat., Marburg, 1975, 125–37. Zur Situation in Lübeck vgl. A. Adlung, „Zur Geschichte des lübeckischen Apothekenwesens," *Deutsche Apotheker-Zeitung* 49 Nr. 86 (1934): 1301ff., 1381–84, 1419ff., 1381. Ich danke Christoph Friedrich für Literaturhinweise und ein erhellendes Gespräch in Greifswald.
125 Lüneburg, Stadtarchiv, AA A 7 a Nr. 42 (Brief von P. Memm an Rat der Stadt Lüneburg vom 16.11.1587; Brief vom Rat der Stadt Lüneburg an P. Memm vom 30.11.1587; Brief von P. Memm an

Diese Episode macht deutlich, welche standespolitischen Auseinandersetzungen in den Eidkommentar hineinwirken, wenn es darin heißt:

> In ihrer Liebe zu Müßiggang und Vergnügungen, die in fast allen, von keinerlei Disziplin
> gezügelt, lodert, verzichten sie auf die gebührende Sorgfalt und Genauigkeit bei der Bereitung
> von Heilmitteln. Soweit ist es gekommen, daß die, die die neue Medizin gewerbsmäßig

den Rat der Stadt Lüneburg vom 4.12.1587; Brief vom Rat der Stadt Lüneburg an P. Memm vom 10.12.1587; Brief von P. Memm an den Rat der Stadt Lüneburg vom 15.12.1587; Brief vom Rat der Stadt Lüneburg an P. Memm vom 8.1.1588). Wie aus dieser Korrespondenz zwischen zwischen dem Bürgermeister Lutke von Dassel und den Ratmannen und Apothekenherren Hieronymus Töbing und Heinrich Witzendorff auf Lüneburger Seite und P. Memm hervorgeht, wurde aus Memms Anstellung in Lüneburg nichts. Denn wenn der Rat der Stadt auch in Sachen Apothekenordnung einlenkte, so ging man auf weitere Bedingungen, die Memm stellte, nicht ein: die Deckung der Umzugskosten, Studienstipendien für Memms Söhne, rückdatierter Dienstantritt sowie die Verfügbarmachung des Physikatshauses. Auf der anderen Seite hatten die Lübecker nachverhandelt, nachdem sich etliche Bürger der Stadt dafür eingesetzt hatten, da man angesichts des betagten Johannes Sager und des jungen Dr Theodor Watermann um die Gesundheitsversorgung der Bürger besorgt war, sollte der erfahrene Memm nach Lüneburg abwandern. So offerierte der Rat eine Gehaltserhöhung von 100 Mk. pro Jahr, was Memm schließlich annahm und also in Lübeck wohnen blieb. Am 6. Mai 1588 wurde übrigens Memms Entlassungsgesuch seitens des Herzogs von Mecklenburg stattgegeben. Das Entlassungsschreiben zeugt von der hohen Wertschätzung, die der Herzog seinem „zugethanen Dienstverwandten," dem „hochgelerte(n) liebe(n) getreue(n)" Arzt entgegenbrachte, wobei er dessen Dienste auch nach der Entlassung nicht gänzlich missen wollte. Memms Motive für das Entlassungsgesuch lassen sich dem Schreiben Johanns nur andeutungsweise entnehmen. Offensichtlich störte ihn das mit der Stelle eines Leibarztes verbundene Reisen am meisten, sei es aus privaten, finanziellen oder gesundheitlichen Gründen. Die Schwierigkeiten, die er mit diversen Stadträten gehabt zu haben scheint, mögen ebenfalls zu seinem Gesuch beigetragen haben. Dem Reich als Leibarzt des Herzogs zu dienen und zugleich dem Gemeinwesen als Stadtarzt, mag zu manchen Interessenskonflikten geführt haben, deren leidtragende Memm selbst, seine Familie, aber auch seine Patientenklientel gewesen sind. Falls Memm die vereinbarte halbjährige Kündigungsfrist hat einhalten müssen, so wird er sein Entlassungsgesuch Anfang November 1587 gestellt haben, zu einem Zeitpunkt, als er mit den Lüneburgern in Verhandlungen trat. Diese Koinzidenz legt die Vermutung nahe, daß auch die Vakanz in Lüneburg Memm zu seinem Gesuch veranlaßt haben könnte, zumal sich bereits bei seinem ersten Besuch günstige Vertragsbedingungen abgezeichnet haben mögen. Doch weder in Lübeck, wo Memm noch eine Zeitlang vertretungsweise praktizierte, noch in Lüneburg kam es zum Vertragsabschluß. Wenige Monate später, am 17. Juli 1589, starb Memm im Alter von 58 Jahren. Vgl. Archiv der Hansestadt Lübeck, Altes Senatsarchiv, Interna, Physikat 7,6. Wie knapp die Verhandlungen scheiterten, zeigt das im Lüneburger Stadtarchiv unter der Signatur AA A 7 a, 43[11] aufbewahrte Konzept vom 23.4.1588 der Bestallungsurkunde, das auf den Namen Memms ausgestellt, dann aber auf Abel Sylvius (gest. 1598), der schließlich die Stelle antrat, verbessert wurde. Ich danke Frau Dr. Reinhardt und Herrn Brauss vom Stadtarchiv Lüneburg sowie Herrn Dr. Ulrich Simon vom Archiv der Hansestadt Lübeck für die freundliche Übersendung von Fotokopien. Herr Dr. Ulrich stellte mir dankenswerter Weise sein handschriftliches Exzerpt aus der fraglichen Akte zur Verfügung.

betreiben,[126] unausgebildeten und der natürlichen Dinge (*res naturales*) unkundigen Menschen das ganze Geschäft der Arzneiherstellung überlassen haben. Die Unerfahrenheit, mannigfaltige List und verfälschte Methode der Heilmittelzubereitung dieser Leute, die Falsches für Wahres und Veraltetes für Frisches um des Geldscheffelns willen öffentlich feilbieten, fällt den Leuten überall auf. Durch dieses Laster ist nämlich ein großer Teil jener Menschen verdorben worden und wie gleichsam die Affen, die sich bei der Krankheitsabwehr selbst untreu werden, sind jene Leute, nachdem die Patienten zu diesem Zwecke [der Krankheitsabwehr] (wie sie sagen) in die Bücher der Ärzte erst einmal aufgenommen sind, fröhlicher als billig und machen den Kranken immer wieder größere Hoffnung auf Gesundheit als irgendjemand von denen, die in den Medizinschulen über Jahre ausgebildet worden sind. So wird die unkundige und mit keinerlei Urteil in den hier in Rede stehenden Dingen versehene Masse getäuscht; so mißbrauchen jene die göttliche Kunst, indem sie nicht einmal das Geringste verstehen, was sogar in den Grundlagen der Kunst das primitivste ist. Die Gelegenheit zum Mißbrauch haben jene geboten, durch deren Unvorsichtigkeit die Geheimnisse – wenn es bis dahin in der Volksmedizin noch welche gab – in die Hände unbesonnener Leute gelangten.[127]

Der Gegenwartsbezug dieser Medizinkritik ist evident, so sehr, daß der Bezug zum hippokratischen Eid weitgehend verloren geht. Der Eid dient hier wieder einmal als Stichwortgeber – das Wort *pharmakon* kommt immerhin vor, und von einem Schweigeversprechen, wenn auch anderer Art, ist die Rede –, um den eigenen

---

126 Die „neue Medizin" dürfte die paracelsische sein. Vgl. P. Elmer, „Chemical Medicine and the Challenge to Galenism," in ders. (Hg.), *The Healing Arts: Health, Disease, and Society in Europe 1500–1800* (Manchester: Manchester University Press, 2004), 108–35. Der pharmazeutische Wildwuchs, den Memm hier anspricht, verbreitete sich auch an Fürstenhöfen, an denen die paracelsische Medizin populär wurde. Kein geringerer als Moritz, der Landgraf von Hessen-Kassel (1572–1632), soll in seinem fürstlichen Labor mit eigenen Händen paracelsische Medikamente zubereitet und an seinen Bediensteten getestet haben. Vgl. B. T. Moran, *The Alchemical World of the German Court: Occult Philosophy and Chemical Medicine in the Circle of Moritz of Hessen (1572–1632)* (Stuttgart: Franz Steiner, 1991), 100. Halsabschneiderei und Scharlatanerie gehörten zu den populären Topoi der Paracelsuskritik.

127 Memm, *Hippocratis iusiurandum commentario*, ff. B8ᵛ-C1ᵛ: ... *ocij, deliciarumque amore, quo plerique omnes aestuant nulla disciplina coërciti, debitam curam et parandi medicamenta sollicitudinem deponunt. Hinc factum est, quod qui modernam medicinam factitant, hominibus indoctis et rerum naturalium imperitis omne conficiendi medicamenti negocium committant. Quorum imperitia, dolus varius, et fucata medicandi ratio in oculos hominum vltro citroque incurrunt, qui falsa pro veris, inueterata pro recentibus, pecuniae augendae causa prostituunt. Nam eo vitio magna pars istorum hominum contaminata est, et veluti simiae suijpsius obliti in propulsandis morbis, Receptis ad hoc (vt vocant) Doctorum in libros congestis, plus aequo lasciuiunt, maioremque aegrotantibus spem valetudinis subinde faciunt, quam quisquam eorum, qui in scholis medicorum ad multos annos educati sunt. Ita decipitur vulgus ignarum, et nullo hisce de rebus iudicio praeditum; Ita abutuntur illi diuina arte, ne hilum quidem, quod vel in fundamentis artis minimum est, intelligentes. Occasionem abusus praebuerunt illi, quorum imprudentia arcana (si quae hactenus in vulgari medicina extiterunt) ad manus temerariorum hominum peruenerunt.*

Standpunkt im Grabenkampf mit Konkurrenten auf dem Gesundheitsmarkt – und hier scheint Memm sich vor allem auf Apotheker kapriziert zu haben – mit dem Namen des Hippokrates zu autorisieren und zu adeln. Diesen Grabenkampf hatte Memm erst nach seinem Weggang aus Utrecht zu bestehen.

Schluß

Der in diesem Aufsatz erstmalig untersuchte Kommentar erweist sich als ein komplexes Produkt von Späthumanismus und vielschichtigen reformatorischen Bewegungen (in Kirche und Staat, Schule und Universität, Bürger- und Ärzteschaft). Sein Autor war sicher kein großer Gelehrter, aber ein geschätzter Arzt, der nicht nur von seinen fürstlichen Arbeitgebern, sondern auch von Kollegen – ärztlichen wie Jakob Horst d. Ä. und nichtärztlichen wie David Chytraeus und Johannes Caselius[128] – und Patienten geachtet wurde, wie zumindest die erhaltenen Belege nahelegen. Der Kommentar, der anscheinend keine Verbreitung über die Wirkungsstätten seines Verfassers (Utrecht, Rostock, Lübeck, wo Johann Heinrich Meibom ihn konsultieren sollte, und Kopenhagen) hinaus gefunden hat, imponiert als Programmschrift, die auf eine Reformierung des Gesundheitswesens, die Stärkung der Universitätsmedizin und derer, die sie erlernen, wie auch deren professionspolitische Organisation und Reputation hinausläuft. Der Kommentar ist untrennbar mit der Vita seines Verfassers verbunden, im christlichen Humanismus verwurzelt und der sittlich-moralischen, aber auch medizintheoretischen und medizinpraktischen Vervollkommnung angehender Ärzte verpflichtet. Der Anlaß zu seiner Drucklegung mag das bevorstehende Rektorat Memms an der Rostocker Universität gewesen sein. Das eigentliche Movens ist jedoch ein religiös motiviertes Erneuerungsstreben, das auf eine theologisch gerechtfertigte Medizin und ein theologisch vergesellschaftetes (oder gespaltenes) Gemeinwesen appliziert wird. Daß all dies im Namen und im Auftrag des Hippokrates geschieht, hat mit dem Umstand zu tun, daß Hippokrates in der zweiten

---

128 In einem Brief an Henning Arnisaeus (1570–1636) vom 28. Juni 1610 charakterisierte Johannes Caselius Peter Memm als „humanissimus"; ihn habe die Religion nach Rostock eingeladen; er habe drei Söhne „in litteris" unterwiesen und habe die Universität um einer Stadtarztstelle in Lübeck willen verlassen, wo er weniger als ein Jahrzehnt gelebt habe. Vgl. [J. Caselius], *Io. Caselii ad Henningvm Arnisaevm philosophvm et medicvm clarissimvm epistola* (Helmstedt: J. Lucius, 1610), f. B3 und C1^v. Zur Wertschätzung Memms seitens seiner Arbeitgeber und Patienten vgl. Anm. 125, zu der Jakob Horsts vgl. Anm. 84. Zu Arnisaeus vgl. H. Jaumann, *Handbuch Gelehrtenkultur der Frühen Neuzeit.* Bd. 1: *Bio-bibliographisches Repertorium* (Berlin: De Gruyter, 2004), 46.

Hälfte des 16. Jahrhunderts für die Einheit der Medizin steht und eben nicht nur deren Alpha ist, sondern auch deren Omega.

Bibliographie

Ungedruckte Quellen

Lübeck, Archiv der Hansestadt, Altes Senatsarchiv, Interna, Physikat 7,6.
Lüneburg, Stadtarchiv, AA A 7 a Nr. 42 und AA A 7 a Nr. 43[11].
Schwerin, Mecklenburgisches Landeshauptarchiv, Bestand Hofstaatssachen, Bestallungen: Mediziner (Memmius).

Gedruckte Primär- und Sekundärliteratur

Abu Aly, A. „A Few Notes on Hunayn's Translation and Ibn al-Nafis' Commentary on the First Book of the *Aphorisms*," *Arabic Sciences and Philosophy* 10 (2000): 139–50.

Adlung, A. „Zur Geschichte des lübeckischen Apothekenwesens," *Deutsche Apotheker-Zeitung* 49 Nr. 86 (1934): 1301 ff., 1381–84, 1419 ff., 1381.

[Anon.], „Leben D. Gvilielmi Lavrembergii …," *Etwas von gelehrten Rostockschen Sachen, für gute Freunde* 2 (1738): 155–60.

[Anon.], „D. Petrus Memmivs, Herenthalio-Belga," *Etwas von gelehrten Rostockschen Sachen, für gute Freunde* 2 (1738): 285.

[Anon.], „Von D. Memmio. P. 285," *Etwas von gelehrten Rostockschen Sachen, für gute Freunde* 2 (1738): 833–34.

[Anon.], „Hippocratis Coi … ," *Etwas von gelehrten Rostockschen Sachen, für gute Freunde* 3 (1739): 581–82.

[Anon.], „Ein ungedruckter Brieff Conr. Pegelii aus dem Original," *Etwas von gelehrten Rostockschen Sachen, für gute Freunde* 4 (1740): 657–58.

Asche, M. „Über den Nutzen von Landesuniversitäten in der Frühen Neuzeit. Leistung und Grenzen der protestantischen 'Familienuniversität,'" in P. Herde und A. Schindling (Hgg.), *Universität Würzburg und Wissenschaft in der Neuzeit. Beiträge zur Bildungsgeschichte. Gewidmet Peter Baumgart anläßlich seines 65. Geburtstages* (Würzburg: Schöningh, 1998), 133–49.

Asche, M. *Von der reichen hansischen Bürgeruniversität zur armen mecklenburgischen Landeshochschule. Das regionale und soziale Besucherprofil der Universitäten Rostock und Bützow in der frühen Neuzeit (1500–1800)*, Contubernium, Bd. 52 (Stuttgart: Steiner, 2000).

Baader, G. „Jacques Dubois as a Practitioner," in A. Wear, R. K. French und I. M. Lonie (Hgg.), *Medical Renaissance of the Sixteenth Century* (Cambridge: Cambridge University Press, 1985), 146–54.

Berges, P.-H. *„Quid pro quo." Zur Geschichte der Arzneimittelsubstitution*, Diss. rer. nat., Marburg, 1975.

Betz H. D., et al., Hgg. *Religion in Geschichte und Gegenwart: Handwörterbuch für Theologie und Religionswissenschaft*, 4., völlig neu bearb. Aufl., 8 Bde. und Register (Tübingen: Mohr Siebeck, 1998–2007).

Blanck, A., A. Wilhelmi und G. Willgeroth, Hgg. *Die mecklenburgischen Ärzte von den ältesten Zeiten bis zur Gegenwart* (Schwerin: Verlag der Landesgeschäftsstelle des Meckl. Aerztevereinsbundes, 1929).

Bloemendal, J. *Spiegel van het dagelijks leven? Latijnse school en toneel in de noordelijke Nederlanden in de zestiende en de zeventiende eeuw* (Hilversum: Verloren, 2003).

Bloemendal, J. "Cornelius Laurimanus als Dramatiker. Theater und Theologie gegen Ketzereien," in R. F. Glei und R. Seidel (Hgg.), *Das lateinische Drama der Frühen Neuzeit. Exemplarische Einsichten in Praxis und Theorie* (Tübingen: Niemeyer, 2008), 101–32.

Blumenberg, H. *Die Lesbarkeit der Welt*, 3. Aufl. (Frankfurt a. M.: Suhrkamp, 1996).

Boeles, W. B. S. *Frieslands Hoogeschool en het Rijks Athenaeum te Franeker*, 2 Bde. (Leeuwarden: Kuipers, 1878–1889).

Bonfatti, E. „Johannes Caselius liest Giovanni de la Casas Galateo (Bologna, 1565)," in A. Buck und M. Bircher (Hgg.), *Respublica Guelpherbytana. Wolfenbütteler Beiträge zur Renaissance- und Barockforschung*, Chloe, Beihefte zum Daphnis, Bd. 6 (Amsterdam: Rodopi, 1982), 357–91.

Boudon-Millot, V., und G. Cobolet, Hgg. (unter der Mitarbeit von H. Ferreira-Lopes und A. Guardasole). *Lire les médecins grecs à la Renaissance: aux origines de l'édition médical; actes du Colloque international de Paris (19–20 Septembre 2003)* (Paris: De Boccard, 2004).

Bretschneider, C. G., Hg. *Philippi Melanthonis Opera quae supersunt omnia*, 28 Bde., Corpus Reformatorum, Bde. 1–28 (Halle: Schetschke, 1835–1860).

Brockmann, C. „Philologische Annäherungen an Chirurgie und Anatomie. Beobachtungen an Galens Kommentar zu Hippokrates, De articulis," in Müller, Brockmann und Brunschön, *Ärzte und ihre Interpreten*, 61–80.

Burger, C. „Devotio moderna," in Betz et al., *Religion in Geschichte und Gegenwart*, 2: Sp. 776.

Cantor, D., Hg. *Reinventing Hippocrates* (Aldershot: Ashgate, 2002).

Caselius, J. *Xenophontis Cyri Paediae liber secundus ...* (Rostock: A. Ferber d. Ä. 1579).

Caselius, J. *Io. Caselii ad Henningvm Arnisaevm philosophvm et medicvm clarissimvm epistola* (Helmstedt: J. Lucius, 1610).

Chytraeus, N. *Contra Pestem Epistola Satyrica* (Rostock: Lucius, 1578).

Chytraeus, N. *Poematum Nathanis Chytraei Praeter Sacra Omnium Libri Septendecim* (Rostock: Mylander, 1579).

Cools, J. *Geschiedenis van het College te Herentals* (Herentals: Oud-Leerlingenbond van het Sint-Jozefscollege, 1984).

[Cornarius, J.] *Hippocratis Coi medicorum omnium longe principis opera quae ad nos extant omnia. Per Ianum Cornarium medicum physicum latina lingua conscripta* (Basel: H. Froben d. Ä. und N. Episcopius d. Ä., 1546).

Crowther-Heyck, K. „'Be Fruitful and Multiply': Genesis and Generation in Reformation Germany," *Renaissance Quarterly* 55 (2002): 904–35.

Crusius, I. „Brüder und Schwestern vom gemeinsamen Leben," in Betz et al., *Religion in Geschichte und Gegenwart*, 1: Sp. 1781–82.

Daniëls, C. E. „Memmius, Petrus M.," in A. Hirsch (Hg.), *Biographisches Lexikon der hervorragenden Ärzte aller Zeiten und Völker*, 2. Aufl. durchges. und erg. von W. Haberling et al., 6 Bde. (Berlin: Urban & Schwarzenberg, 1929–1935), 4:160.

Demont, P. „L'édition Vigoreus (1555) du traité hippocratique De humoribus et d'un 'commentaire de Galien' à ce traité (= [Galien], De humoribus, XIX, 485–496 Kühn), avec la traduction du De humoribus galénique," in Boudon-Millot und Cobolet, *Lire les médecins grecs à la Renaissance*, 43–59.

Desselius, V. A. *Bibliotheca Belgica de Belgis vita scriptisque claris* .... Editio renovata et tertia parte auctior (Löwen: Zegers, 1643).

De Jong, O. J. *De Reformatie in Culemborg* (Assen: Van Gorcum, [1957])

De Jong, O. J., und J. T. K. Marcelis. *Uti porta latina stat: Hieronymusschool en Utrechts stedelijk gymnasium van 1474 tot 2000* (Utrecht: USG, 2000).

De la Fontaine Verwey, H. *Meester Harman Schinckel, een Delftse boekdrukker van de 16e eeuw*, Oud Delft, Bd. 3 (Rotterdam: Nijgh en van Ditmar, [1963]).

De Seyn, E. M. H. *Dictionnaire biographique des sciences, des lettres et des arts en Belgique*, 2 Bde. (Brüssel: Éditions Avenir, 1935–1936).

De Vocht, H. *Literae virorum eruditorum ad Franciscum Craneveldium 1522–1528: A Collection of Original Letters Ed. from the Manuscripts and Illustrated with Notes and Commentaries* (Louvain: Uystpruyst, 1928).

Dodt van Flensburg, J. J. „Over den Utrechtschen arts Petrus Memmius, den Delftschen Drukker Hermann Schinckel, en den aartsbisschop Frederik Schenck van Tautenburch, een paar aanteekeningen," *Utrecht voorheen en thans: Tijdschrift voor Geschiedenis, Oudheden en Statistick van Utrecht*, 2de Serie, 2 (1845): 75–80.

Dodt van Flensburg, J. J. „Hieronymus School 1630–1648," *Archief voor kerkelijke en wereldlijke geschiedenis van Nederland meer bepaaldelijk van Utrecht* 7 (1848): 366–92.

Drizenko, A. „Jacques Dubois, dit Sylvius, traducteur et commentateur de Galien," in Boudon-Millot und Cobolet, *Lire les médecins grecs à la Renaissance*, 199–208.

Dubois, J. *Vaesani cuiusdam calumniarum in Hippocratis Galenique rem anatomicam depulsio* (Basel: Jacobus Derbilley, 1556).

Duke, A. *Reformation and Revolt in the Low Countries* (London: Hambledon and London, 2003).

Durling, R. J. „A Chronological Census of Renaissance Editions and Translations of Galen," *Journal of the Warburg and Courtauld Institutes* 24 (1961): 230–305.

Edelstein, L. *The Hippocratic Oath: Text, Translation, and Interpretation*, Supplements to the Bulletin of the History of Medicine, Nr. 1 (Baltimore: The Johns Hopkins Press, 1943).

Einbinder, S. L. *No Place of Rest: Jewish Literature, Expulsion, and the Memory of Medieval France* (Philadelphia: University of Pennsylvania Press, 2009).

Ekkart, R. E. O. *Franeker Professorenportretten: Iconografie van de Professoren aan de Academie en het Rijksathenaeum te Franeker 1585–1843* (Franeker: Wever, 1977).

Ekker, A. *De Hieronymusschool te Utrecht*, 2 Bde. (Utrecht: Bosch, 1863–1864)

Eloy, N. F. J. *Dictionnaire historique de la médecine ancienne et moderne* ..., 4 Bde. (Mons: H. Hoyois, 1778).

Elsmann, T., H. Lietz und S. Pettke, Hgg. *Nathan Chytraeus, 1543–1598. Ein Humanist in Rostock und Bremen. Quellen und Studien* (Bremen: Edition Timmen, 1991).

Erler, G., Hg. *Die Matrikel der Universität Königsberg i. Pr.*, Bd. 1, H. 1 (Leipzig: Duncker & Humblot, 1908).

Fink-Jensen, M. „Medicine, Natural Philosophy, and the Influence of Melanchthon in Reformation Denmark and Norway," *Bulletin of the History of Medicine* 80 (2006): 439–64.

Fischer, K.-D. „Der griechische Arzt Soran als vermeintlicher Befürworter periodischer Rauschzustände. Anmerkungen zum 2. lateinischen Kommentar zu den hippokratischen Aphorismen," in P. Defosse (Hg.), *Hommage à Carl Deroux. Tome 2: Prose et linguistique, médecine* (Brüssel: Latomus 2002), 481–96.

Fischer, K.-D. „'Zu des Hippokrates reich gedeckter Tafel sind alle eingeladen.' Bemerkungen zu den beiden vorsalernitanischen lateinischen Aphorismenkommentaren," in W. Geerlings und C. Schulze (Hgg.), *Der Kommentar in Antike und Mittelalter. Beiträge zu seiner Erforschung*, Clavis Commentariorum Antiquitatis et Medii Aevi, vol. 2 (Leiden: Brill, 2002), 275–313.

Foppens, J. F. *Bibliotheca belgica sive virorum in Belgio vita, scriptisque illustrium catalogus, librorumque nomenclatura. Continens scriptores a clarissimis viris Valerio Andrea, Auberto Miraeo, Francisco Sweertio aliisque recensitos, usque ad annum MDCLXXX* (Brüssel: Foppens, 1739).

Garrison, D. H. und M. H. Hast, *Andreas Vesalius, De Humani Corporis Fabrica Libri Septem/The Fabric of the Human Body: An Annotated Translation of the 1543 and 1555 Editions*, 2 Bde. (Basel: Karger, 2013).

Gasman, M. J. „Schenk van Tautenburg (Frederik)," in Molhuysen und Blok, *Nieuw Nederlandsch Biografisch Woordenboek*, 2: Sp. 1280–82.

Gewin, E. E. „Memmius (Peter)," in Molhuysen und Blok, *Nieuw Nederlandsch Biografisch Woordenboek*, 2: Sp. 895.

Gilsheim, R. *Liber collectionum aphorismorum Hypocratis de unaquaque egritudine* (Rostock: Ludwig Dietz, 1519).

Glaser, K.-H., H. Lietz und St. Rhein, Hgg. *David und Nathan Chytraeus. Humanismus im konfessionellen Zeitalter* (Ubstadt-Weiher: Verlag Regionalkultur, 1993).

Grafton, A. „Commentary," in ders., G. W. Most und S. Settis (Hgg.), *The Classical Tradition* (Cambridge, Mass.: The Belknap Press of Harvard University Press, 2010), 225–33.

Grewolls, G. *Wer war wer in Mecklenburg und Vorpommern*, DVD (Rostock: Hinstorff, 2011).

Haak, S. P. „Pallandt (Floris van)," in Molhuysen und Blok, *Nieuw Nederlandsch Biografisch Woordenboek*, 5: Sp. 424–27.

Hartwig, A., und T. Schmidt, Hgg. *Die Rektoren der Universität Rostock, 1419–2000*, Beiträge zur Geschichte der Universität Rostock, H. 23 (Rostock: Universität, 2000).

Heeg, J. *Buch, Macht, Bildung: Die Bibliothek des Herzogs Johann Albrecht I. von Mecklenburg. Ausstellung im Kulturhistorischen Museum Rostock vom 20.4. bis 24.6.2007*, Veröffentlichungen der Universitätsbibliothek Rostock, Bd. 137 (Rostock: Universitätsbibliothek, 2007).

Heidorn, G., G. Heitz, J. Kalisch, et al., Hgg. *Geschichte der Universität Rostock 1419–1969: Festschrift zur Fünfhundertfünfzig-Jahr-Feier der Universität*. Bd. 1: *Die Universität von 1419–1945* (Berlin: Deutscher Verlag der Wissenschaften, 1969).

Heinemann, O. V. *Die Herzogliche Bibliothek zu Wolfenbüttel. Ein Beitrag zur Geschichte deutscher Büchersammlungen*, 2., völlig neugearb. Aufl. (Wolfenbüttel: Julius Zwissler, 1894).

Henerus, R. *Adversus Jacobi Sylvii depulsionum anatomicarum calumnias, pro Andrea Vesalio* (Venedig: [Scotto], 1555).

Hofheinz, R.-D. *Philipp Melanchthon und die Medizin im Spiegel seiner akademischen Reden*, Neuere Medizin- und Wissenschaftsgeschichte, Bd. 11 (Herbolzheim: Centaurus, 2001).

Hofheinz, R.-D., und R. Bröer. „Gesundheitspädagogik statt Tröstung. Die theologische Bewältigung von Krankheit bei Philipp Melanchthon und Caspar Peucer," *Sudhoffs Archiv* 85 (2001): 18–44.

Hofmeister, A. *Die Matrikel der Universität Rostock*. Bd. 2: *Mich. 1499 – Ost. 1611*, Nachdruck der Ausg. Rostock 1891 (Nendeln, Liechtenstein: Kraus, 1976).

Holder, R. Ward. *A Companion to Paul in the Reformation*, Brill's Companion to the Christian Tradition, Bd. 15 (Leiden: Brill, 2009).

Horn, E. *Die Disputationen und Promotionen an den deutschen Universitäten, vornehmlich seit dem 16. Jahrhundert. Mit einem Anhang enthaltend ein Verzeichnis aller ehemaligen und gegenwärtigen deutschen Universitäten* (Leipzig: Harrassowitz, 1893).

Horrox, R. *The Black Death* (Manchester: Manchester University Press, 1994).

Horstius, J. *Levini Lemnii Occulta naturae miracula. Wunderbarliche Geheimnisse der Natur in des Menschen leibe und Seel …* (Leipzig: Steinmann, 1588).

Horstius, J. *Levini Lemnii Occulta naturae miracula. Wunderbarliche Geheimnisse der Natur in des Menschen leibe und Seel …* (Leipzig: V. Vögelin, M. Lantzenberger, 1593).

Hulshof, A. „Rostock und die nördlichen Niederlande vom 15. bis zum 17. Jahrhundert," *Hansische Geschichtsblätter* 16 (1910): 531–53.

Ihm, S. *Clavis Commentariorum der antiken medizinischen Texte*, Clavis Commentariorum Antiquitatis et Medii Aevi, Bd. 1 (Leiden: Brill, 2002).

Jaster, S. „Die medizinische Fakultät der Albertus-Universität und ihre bedeutendsten Vertreter im 16. und 17. Jahrhundert," in H. Marti und M. Komorowski (Hgg.), *Die Universität Königsberg in der Frühen Neuzeit* (Köln: Böhlau, 2008), 42–76.

Jaumann, H. *Handbuch Gelehrtenkultur der Frühen Neuzeit*. Bd. 1: *Bio-bibliographisches Repertorium* (Berlin: De Gruyter, 2004).

Jensma, G. T., F. R. H. Smit und F. Westra, Hgg. *Universiteit te Franeker 1585–1811: Bijdragen tot de geschiedenis van de Friese Hogeschool* (Leeuwarden: Fryske Akademy, 1985).

Jöcher, Ch. G., J. C. Adelung und H. W. Rotermund. *Fortsetzung und Ergänzungen zu Christian Gottlieb Jöchers allgemeinem Gelehrten-Lexicon: worin die Schriftsteller aller Stände nach ihren vornehmsten Lebensumständen und Schriften beschrieben werden*, 4. Bd. (1813), Unveränderter Nachdruck (Hildesheim: Olms, 1961).

Josse, N. P., und P. Pormann. „'Abd al-Laṭīf al-Baġdādī's *Commentary on Hippocrates' 'Prognostic'*: A Preliminary Exploration," in P. Pormann (Hg.), *Epidemics in Context: Greek Commentaries on Hippocrates in the Arabic Tradition* (Berlin: De Gruyter, 2012), 251–83.

Kamen, H. *The Duke of Alba* (New Haven: Yale University Press, 2004).

Karrig, O. „Geschichtliches über das Auftreten der Pest in Mecklenburg," *Archiv für Geschichte der Medizin* 5 (1912): 436–46.

Kaufmann, T. *Universität und lutherische Konfessionalisierung: die Rostocker Theologieprofessoren und ihr Beitrag zur theologischen Bildung und kirchlichen Gestaltung im Herzogtum Mecklenburg zwischen 1500 und 1675* (Gütersloh: Gütersloher Verlags-Haus, 1997).

Kaufmann, T. „Kryptocalvinismus," in Betz et al., *Religion in Geschichte und Gegenwart*, 4: Sp. 1793.

Keller, R. „David Chytraeus (1530–1600). Melanchthons Geist im Luthertum," in H. Scheible (Hg.), *Melanchthon in seinen Schülern*, Wolfenbütteler Forschungen, Bd. 73 (Wiesbaden: Harrassowitz, 1997), 361–71.

King, H. „The Power of Paternity: The Father of Medicine Meets the Prince of Physicians," in Cantor, *Reinventing Hippocrates*, 23–36.
King, H. „Engendrer 'la femme': Jacques Dubois et Diane de Poitiers," in C. McClive und N. Pellegrin (Hgg.), *Femmes en fleurs, femmes en corps: sang, santé, sexualités du Moyen Âge au Lumières* (Saint-Étienne: Publications de l'Université de Saint-Étienne, 2010), 125–38.
Knipscheer, F. S. „Adama (Augustinus Lollius)," in Molhuysen und Blok, *Nieuw Nederlandsch Biografisch Woordenboek*, 1: Sp. 18.
Koppmann, K. „Über die Pest des Jahres 1565 und zur Bevölkerungsstatistik Rostocks im 14., 15. und 16. Jahrhundert," *Hansische Geschichtsblätter* 29 (1901): 45–63.
Krabbe, O. *Die Universität Rostock im fünfzehnten und sechzehnten Jahrhundert* (Rostock: Stillersche Hofbuchhandlung, 1854).
Krause, K. E. H., „Memmius, Peter," in *Allgemeine Deutsche Biographie. Auf Veranlassung und mit Unterstützung des Königs von Bayern Maximilian II. hrsg. durch die Historische Commission bei der Königlichen Akademie der Wissenschaften*, 56 Bde. (Leipzig: Duncker & Humblot, 1875–1912), 21 (1885): 310.
Krey, J. B. *Andenken an die Rostock'schen Gelehrten aus den drei letzten Jahrhunderten*, 5. Stück (Rostock: Adler, 1815).
Kristeller, P. O. „Bartholomaeus, Musandinus, and Maurus of Salerno and Other Early Commentators of the 'Articella' with a Tentative List of Texts and Manuscripts," *Italia Medioevale e Umanistica* 19 (1976): 57–87.
Kühlmann, W. „Selbstverständigung im Leiden: Zur Bewältigung von Krankheitserfahrungen im versgebundenen Schrifttum der Frühen Neuzeit (P. Lotichius Secundus, Nathan Chytraeus, Andreas Gryphius)," in U. Benzenhöfer und W. Kühlmann (Hgg.), *Heilkunde und Krankheitserfahrung in der frühen Neuzeit: Studien am Grenzrain von Literaturgeschichte und Medizingeschichte* (Tübingen: Max Niemeyer, 1992), 1–29.
Kusukawa, S. „Aspectio divinorum operum. Melanchthon and Astrology for Lutheran Medics," in O. P. Grell und A. Cunningham (Hgg.), *Medicine and the Reformation* (London: Routledge, 1993), 33–56.
Kusukawa, S. *The Transformation of Natural Philosophy: The Case of Philipp Melanchthon* (Cambridge: Cambridge University Press, 2003).
Lauremberg, W. Resp. *circularis disputatio continens propositiones de flatibus ex doctrina Hippocratis desumptas*. Praes. Petro Memmio (Rostock: [o. V.], 1578)
Lichtenthaeler, C. *Der Eid des Hippokrates. Ursprung und Bedeutung* (Köln: Deutscher Ärzteverlag, 1984).
Lindeboom, G. A. *Dutch Medical Biography: A Biographical Dictionary of Dutch Physicians and Surgeons 1475–1975* (Amsterdam: Rodopi, 1984).
Lisch, G. C. F. *Geschichte der Buchdrucker-Kunst in Mecklenburg bis zum Jahre 1540. Mit einem Anhange über die niederdeutsche Bearbeitung des Reineke Voß* (Schwerin: Stiller, 1839).
Manetti, D., und A. Roselli. „Galeno commentatore di Ippocrate," in W. Haase (Hg.), *Aufstieg und Niedergang der römischen Welt*, II 37.2 (Berlin: De Gruyter, 1994), 1529–635.
Marsh, D. „Xenophon," in V. Brown (Hg.), *Catalogus Translationum et Commentariorum: Medieval and Renaissance Latin Translations and Commentaries. Annotated Lists and Guides*, Bd. 7 (Washington, D.C.: Catholic University of America Press, 1992), 75–196.
Meibom, J. H. *Hippocratis Magni ΟΡΚΟΣ sive Jusjurandvm* (Leiden: J. Lauwikius, 1643).
Meier, C. „Ein antikes Äquivalent des Fortschrittsgedankens: das 'Könnensbewußtsein' des 5. Jahrhunderts v. Chr.," *Historische Zeitschrift* 226 (1978): 265–316; in erweiterter und

überarbeiteter Form wiederabgedruckt in ders., *Die Entstehung des Politischen bei den Griechen* (Frankfurt a. M.: Suhrkamp, 1990), 435–99.

Memmius, P. *De recto medicinae vsu liber vnus: non solùm medicis, verùm etiam cuiusuis disciplinae sectatoribus perutilis, nunc primùm in lucem editus* (Delft: H. Schinckel, 1564).

Memmius, P. *Hippocratis iusiurandum commentario recenter illvstratvm, cvi accessit altera pars, qua ratione Medicorum vita et ars sanctè conseruetur declarans* (Rostock: A. Ferber, 1577).

Memmius, P. *Scriptvm Pvblice propositvm a rectore academiae Rostochiensis in funere opt. et doctißimi viri Christophori Gertneri Lubecensis, bonarum artium magistri et professoris publici, et decani facultatis philosophicae* (Rostock: [o. V.], 1579), A1ᵛ-A3.

Miethke, J. „Der Eid an mittelalterlichen Universitäten. Formen seiner Gebräuche. Funktionen einer Institution," in P. Prodi (Hg.), *Glaube und Eid. Treueformeln, Glaubensbekenntnisse und Sozialdisziplinierung zwischen Mittelalter und Neuzeit*, Schriften des Historischen Kollegs, Bd. 28 (München: Oldenbourg, 1993), 49–67.

Molhuysen, P. C., und P. J. Blok, Hgg. *Nieuw Nederlandsch Biografisch Woordenboek*, 10 Bde. (Leiden: A. W. Sijthoff, 1911–1937).

Moller, J. *Cimbria literata. Tomus secundus. Adoptivos sive exteros, in ducatu utroque Slesvicensi et Holsatico, vel officiis functos publicis vel diuitius commoratos, complectens* (Kopenhagen: Orphanotrophium Regium, 1744).

Mondrain, B. „Éditer et traduire les médecins grecs au XVIᵉ siècle: l'exemple de Janus Cornarius," in D. Jacquart (Hg.), *Les voies de la science grecque: études sur la transmission des textes de l'Antiquité au dix-neuvième siècle* (Genf: Droz, 1997), 391–417.

Monfort, M. *Janus Cornarius et la redécouverte d'Hippocrate à la Renaissance* (Turnhout: Brepols, 2014).

Morpurgo, P. „L'interpretazione della medicina ippocratica in Maimonide e nei maestri Salernitani," in *Dalla medicina greca alla medicina salernitana: percorsi e tramiti; atti del Convegno internazionale Raito di Vietri sul Mare, Auditorium di Villa Guariglia (25–27 giugno 1995)* (Salerno: Centro Studi Salernitani Raffaele Guariglia, 1999), 25–39.

Müller, C. W., C. Brockmann und C. W. Brunschön, Hgg. *Ärzte und ihre Interpreten. Medizinische Fachtexte der Antike als Forschungsgegenstand der Klassischen Philologie* (Berlin: De Gruyter, 2006).

Mueller-Goldingen, C. *Untersuchungen zu Xenophons Kyropädie*, Beiträge zur Altertumskunde, Bd. 42 (Stuttgart: Teubner, 1995).

Napjus, J. W. „Augustinus Lollius Adama: Hoogleeraar in de geneeskunde te Franeker," *Nederlands Tijdschrift voor Geneeskunde* Jg. 80 (1936): 1478–84.

Nutton, V. „Biographical Accounts of Galen, 1340–1660," in Rütten, *Geschichte der Medizingeschichtsschreibung*, 201–32.

O'Boyle, C. *Thirteenth- and Fourteenth-Century Copies of the „Ars Medicinae": A Checklist and Contents Descriptions of the Manuscripts*, Articella Studies, Nr. 1 (Cambridge: Wellcome Unit for the History of Medicine; Barcelona: CSIC, 1998).

O'Malley, C. O. *Andreas Vesalius of Brussels 1514–1564* (Berkeley: University of California Press, 1964).

Pagel, W. *Paracelsus: An Introduction to the Philosophical Medicine in the Era of the Renaissance* (Basel: S. Karger AG, 1982).

Pettke, S., und A. Röpcke, Hgg. *Biographisches Lexikon für Mecklenburg*, 6 Bde. (Rostock: Schmidt Römhild, 1995–2011).

Pittion, J.-P. „Scepticism and Medicine in the Renaissance," in R. H. Popkins und C. Schmitt (Hgg.), *Scepticism from the Renaissance to the Enlightenment*, Wolfenbütteler Forschungen, Bd. 35 (Wiesbaden: Harrassowitz, 1987), 103–32.

Pluns, M. A. *Die Universität Rostock 1418–1563. Eine Hochschule im Spannungsfeld zwischen Stadt, Landesherren und wendischen Hansestädten* (Köln: Böhler, 2007).

Post, R. R. „Studien over de broeders van het gemeene leven," *Nederlandsche Historiebladen* 2 (1939): 136–62.

Post, R. R. *The Modern Devotion: Confrontation with Reformation and Humanism* (Leiden: Brill, 1968).

Roselli, A. „Tra pratica medica e filologia ippocratica: il caso di Apollonio di Cizio," in G. Argoud und J.-Y. Guillaumin (Hgg.), *Sciences exactes et sciences appliquées à Alexandrie. IIIe siècle av. J.-C. – Ier siècle ap. J.-C.*, Centre Jean-Palerne, Mémoires Bd. 16 (Saint-Étienne: Publications de l'Université de Saint-Étienne, 1998), 217–32.

Rütten, T. „Ärztliche Ethik in der Renaissancemedizin: Mechanismen der Neukontextuierung des hippokratischen Eides in der späthumanistischen Kommentarliteratur zwischen 1540 und 1640," in I. Garofalo, A. Lami, D. Manetti und A. Roselli (Hgg.), *Aspetti della terapia nel Corpus Hippocraticum; atti del IXe Colloque International Hippocratique, Pisa 25–29 settembre 1996* (Florenz: Olschki, 1999), 517–42.

Rütten, T. „Traduzioni e commenti del corpus ippocratico e galenico," in A. Clericuzio und G. Ernst (Hgg.), *Il Rinascimento italiano e l'Europa* (Treviso-Costabissara: A. Colla, 2008), 479–93.

Rütten, T. *Geschichten vom Hippokratischen Eid*, CD-ROM (Wiesbaden: Harrassowitz, 2008).

Rütten, T., und R. Kinsky. „Text und Geschichte," in Rütten, *Geschichte der Medizingeschichtsschreibung*, 9–56.

Rütten, T., und U. Rütten. „Melanchthons Rede 'De Hippocrate,'" *Medizinhistorisches Journal* 33 (1998): 19–55.

Rütten, T., Hg., unter Mitarbeit von N. Metzger. *Geschichte der Medizingeschichtsschreibung: Historiographie unter dem Diktat literarischer Gattungen von der Antike bis zur Aufklärung* (Remscheid: gardez!, 2009).

Scattola, M. „Gelehrte Philosophie vs. Theologie: Johannes Caselius im Streit mit den Helmstedter Theologen," in H. Jaumann (Hg.), *Die europäische Gelehrtenrepublik im Zeitalter der Konfessionalisierung* (Wiesbaden: Harrassowitz, 2001), 155–81.

Scheible, H. *Melanchthon. Eine Biographie* (München: Beck, 1997).

Schirmacher, F. W. *Johann Albrecht I., Herzog von Mecklenburg* (Wismar: Hinstorff, 1885).

Schönemann, C. P. C. „Dem Andenken Georg Burkhard Lauterbachs. Ein Beitrag zur Geschichte der Wolfenbütteler Bibliothek," *Serapeum. Zeitschrift für Bibliothekswissenschaft* Jg. 3, Nr. 14 (1842): 213–24.

Schreiber, H. *Johann Albrecht I., Herzog zu Mecklenburg* (Halle: Niemeyer, 1899).

Schümann, H. *Von apothecarii, physici und clystierweibern. Apotheker und Apotheken der Stadt Rostock in acht Jahrhunderten* (Rostock: Redieck und Schade, 2003).

Schumacher, G.-H., und H. Wischhusen. *Anatomia Rostochiensis: Die Geschichte der Anatomie an der 550 Jahre alten Universität Rostock* (Berlin: Akademie-Verlag, 1970).

Seidel, U. *Rezept und Apotheke. Zur Geschichte der Arzneiverordnung vom 13. bis zum 16. Jahrhundert*, Diss. rer. nat., Marburg, 1977.

Sellmer, L. „Ulrich III., Herzog von Mecklenburg-Güstrow," in Pettke und Röpcke, *Biographisches Lexikon für Mecklenburg*, 1:231–35.

Series lectionum, quae futuro semestri aestivo anni 1568. In Academia Rostochiensi publice proponentur (Rostock: Lucius, 1568).

Siegfridus, N. *Epithalamivm De vera laeticia et Angelica custodia status conjugalis. Clarissimo ... viro, D. Isaaco Memmio, I.V.D. apud inclytos Lubecenses peritissimo, nuptias cum honestissima et pientissima Virgine Angela Neoburia Hamburgensi, ... , Anno salutis nostrae 1600. 3. Non. Octobr.* (Hildesiae: Excudebat Andreas Hantzsch in Coenobio D. Pauli, 1600).

Siraisi, N. *History, Medicine, and the Traditions of Renaissance Learning* (Ann Arbor: The University of Michigan Press, 2007).

Smith, W. D., Hg. *Hippocrates, Pseudepigraphic Writings: Letters, Embassy, Speech from the Altar, Decree*, Studies in Ancient Medicine, Bd. 2 (Leiden: Brill, 1992).

Smith, W. D. *The Hippocratic Tradition*, Electronic ed., rev. 2002 (1. Aufl.: Ithaca, N.Y.: Cornell University Press, 1979): http://www.biusante.parisdescartes.fr/medicina/Hippo2.pdf.

Stearns, J. K. *Infectious Ideas: Contagion in Premodern Islamic and Christian Thought in the Western Mediterranean* (Baltimore: The Johns Hopkins University, 2011).

Stolberg, M. „Active Euthanasia in Pre-Modern Society, 1500–1800: Learned Debates and Popular Practices," *Social History of Medicine* 20 (2007): 205–21.

Strohmaier, G. „Der arabische Galenkommentar als indirekter Textzeuge zu Hippokrates, De aere aquis locis," in Müller, Brockmann und Brunschön, *Ärzte und ihre Interpreten*, 233–44.

Sylvius, siehe Dubois, J.

Teichmann, G. „Lauremberg (Laurenberg), Peter," in Pettke und Röpcke, *Biographisches Lexikon für Mecklenburg*, 2:153–55.

Teichmann, G. „Wilhelm Lauremberg," in Hartwig und Schmidt, *Die Rektoren der Universität Rostock, 1419–2000*, 89.

Teichmann, W. „Bording, Jacob d. Ä.," in Pettke und Röpcke, *Biographisches Lexikon für Mecklenburg*, 1:36–39.

Toepke, G., Hg. *Die Matrikel der Universität Heidelberg von 1386 bis 1662. 2. Theil von 1554 bis 1602* (Heidelberg: Selbstverlag, 1886; Nendeln: Kraus Reprint, 1976).

van der Aa, J. *Biographisch Woordenboek der Nederlanden ...* , Nieuwe Uitgaaf, 21 Bde. (Haarlem: J. J. van Brederode, 1852–1878).

van Engen, J. H., Hg. *Devotio Moderna: Basic Writings* (New York: Paulist Press, 1988).

van Engen, J. H. *Sisters and Brothers of the Common Life: The Devotio Moderna and the World of the Later Middle Ages* (Philadelphia: University of Pennsylvania Press, 2008).

van Lieburg, M. J. „De genees- en heelkunde in de noordelijke Nederlanden, gezien vanuit de stedelijke en chirurgijnsgilde-ordonnanties van de 16e eeuw," *Tijdschrift voor geschiedenis van de geneeskunde* 6 (1983): 169–84.

van Roosbroeck, R. „Die Beziehungen der Niederländer und der niederländischen Emigranten zur deutschen Gelehrtenwelt im XVI. Jahrhundert: Eine Übersicht," in H. Rössler und G. Franz (Hgg.), *Universität und Gelehrtenstand 1400–1800: Büdinger Vorträge 1966*, Deutsche Führungsschichten in der Neuzeit, Bd. 4 (Limburg a. d. Lahn: Starke, 1970), 107–25.

Vesalius, A. *De humani corporis fabrica libri septem* (Basel: Oporinus, 1543).

von Alberti, H.-J. *Maß und Gewicht. Geschichtliche und tabellarische Darstellungen von den Anfängen bis zur Gegenwart* (Berlin: Akademie-Verlag, 1957).

von Katte, M. „Zur Erziehung und Ausbildung Herzog August des Jüngeren zu Braunschweig und Lüneburg. Die Präzeptorwahl von 1594 und die Entstehung seiner Devise 'Expende,'" in P. Raabe (Hg.), *Wolfenbütteler Beiträge. Aus den Schätzen der Herzog August Bibliothek*, Wolfenbütteler Beiträge, 5 (Frankfurt a. M.: Vittorio Klostermann, 1982), 9–52.

von Schulte, J. F. „Schenck: Friedrich Freiherr S. von Tautenburg," in *Allgemeine Deutsche Biographie*, Bd. 31: Scheller – Karl Schmidt, Neudruck der 1. Aufl. von 1890 (Berlin: Duncker & Humblot, 1970), 66.

von Staden, H. „Incurability and Hopelessness: The Hippocratic Corpus," in P. Potter, G. Maloney und J. Desautels (Hgg.), *La maladie et les maladies dans la Collection Hippocratique; actes du VIe Colloque International Hippocratique* (Québec: Les Éditions du Sphinx, 1990), 75–112.

von Staden, H. „'In a Pure and Holy Way': Personal and Professional Conduct in the Hippocratic Oath," *Journal of the History of Medicine and Allied Sciences* 51 (1996): 404–37.

von Staden, H. „Character and Competence: Personal and Professional Conduct in Greek Medicine," in H. Flashar und J. Jouanna (Hgg.), *Médecine et morale dans l'Antiquité; dix exposés suivis de discussions, Vandœuvres-Genève, 19–23 août 1996* (Vandœuvres: Fondation Hardt, 1996), 157–210.

von Staden, H. „'A Woman Does Not Become Ambidextrous': Galen and the Culture of Scientific Commentary," in R. K. Gibson und C. Shuttleworth Kraus (Hgg.), *The Classical Commentary: History, Practice, Theory* (Leiden: Brill, 2002), 109–39.

von Staden, H. „'The Oath,' the Oaths, and the Hippocratic Corpus," in V. Boudon-Millot, A. Guardasole und C. Magdelaine (Hgg.), *La science médicale antique: nouveaux regards; études réunis en l'honneur de Jacques Jouanna* (Paris: Beauchesne, 2008), 425–66.

von Staden, H. „Staging the Past, Staging Oneself: Galen on Hellenistic Exegetical Traditions," in C. Gill, T. Whitmarsh und J. Wilkins (Hgg.), *Galen and the World of Knowledge* (Cambridge: Cambridge University Press, 2009), 132–56.

von Westphalen, E. J. *Monumenta inedita Rerum Germanicarum praecipue Cimbricarum et Megapolensium*, 4 Bde. (Leipzig: Martini, 1739–1745).

Wagner, W. E., Hg., unter Mitarbeit von R. Ackermann. *Observantia lectionum in universitate Rostochiensi (1520). Das älteste gedruckte Vorlesungsprogramm der Universität Rostock* (Hamburg: Verlag Dr. Kovač, 2011).

Zijlstra, S. *Het geleerde Friesland – een mythe? Universiteit en maatschappij in Friesland en Stad en Lande a 1380–1650* (Leeuwarden: Fryske Akademy, 1996).

Mark J. Schiefsky
Technē and Method in Ancient Artillery Construction: The *Belopoeica* of Philo of Byzantium

Abstract: In his *Belopoeica*, Philo of Byzantium presents artillery construction (*belopoiikē*) as a kind of expertise or *technē* that possesses a standardized method for attaining success. I describe this method, which consists of a set of procedures and rules that are systematically organized on the basis of general principles, and discuss Philo's claim that its discovery depended crucially on experience (*peira*). In the second part of the *Belopoeica* Philo presents several designs for artillery engines that allegedly improve on the standard method. I discuss these designs, which draw on both natural philosophy and theoretical mechanics, and conclude with a brief attempt to place Philo's picture of artillery construction as a *technē* involving both experience and theory in the context of roughly contemporary views of *technē* in philosophy and medicine.

Introduction

From the fourth century B.C. to the end of Antiquity, the discipline of artillery construction (*belopoiikē*) was one of the most important and highly developed types of professional expertise (*technē*) in the ancient Greco-Roman world.[1] Starting from the traditional bow, Greek engineers devised a wide array of mechanical shooting devices, weapons which had a significant impact on the course of history. The development of this technology was fostered by royal patronage and carried out by communities of practitioners working in major cultural and political centers such as Alexandria and Rhodes. These practitioners had a high sense

---

1 As is well known, the Greek term *technē* has no single English equivalent. Used to refer to disciplines as diverse as carpentry, sculpture, medicine, and geometry, its meanings include – but are by no means limited to – "art," "craft," and "science." My concern in this paper is with *technē* as a form of knowledge, a kind of "result-oriented expertise" in the formulation of Heinrich von Staden. It is a pleasure to acknowledge my debt to Heinrich for this formulation, which is only a small example of the profound impact that his work in ancient medicine and science has had on my own. I am delighted and honored to have the opportunity to dedicate this paper to him as a small token of thanks for his inspiring example and steadfast support over many years. I would also like to thank Brooke Holmes and Klaus-Dietrich Fischer for their patience and comments on an earlier draft of this essay.

of the importance of their calling and gained widespread recognition for their achievements. While much of the technical expertise that they possessed was transmitted orally, a substantial amount of evidence documenting their methods and activities survives in both the archaeological record and written texts. For these reasons, artillery construction is one of the most promising areas for studying the impact of science and technology on ancient society.[2]

In this paper I shall focus on the earliest and in many ways the richest of the literary sources documenting ancient artillery construction, the *Belopoeica* of Philo of Byzantium, probably written around 200 B.C. Of Philo's life almost nothing is known, but it seems clear that he was active in Alexandria in the late third or early second century B.C., a period which saw rapid growth in both mechanical technology and the theory of machines.[3] The *Belopoeica* originally formed part of the *Mechanical Syntaxis*, a comprehensive treatment of mechanics in nine books; in addition to artillery construction, this work covered such topics as the theory of levers, harbor construction, siegecraft, pneumatics, and the building of automata. Though most of the *Mechanical Syntaxis* is lost, several books are extant in Greek, Arabic, and/or Latin versions.[4] Philo's attempt to

---

[2] The standard work on the development of ancient artillery and its place in ancient society is Eric W. Marsden, *Greek and Roman Artillery: Historical Development* (Oxford: Clarendon Press, 1969) and idem, *Greek and Roman Artillery: Technical Treatises* (Oxford: Clarendon Press, 1971). Marsden's reconstructions build on the pioneering work of Erwin Schramm, e.g., his *Die antiken Geschütze der Saalburg* (Berlin: Weidmann, 1918; repr. with intro. by D. Baatz, Bad Homburg: Saalburgmuseum, 1980). Since Marsden wrote, a number of works have shed new light on the archaeological, historical, and technological aspects of ancient artillery; see especially Dietwulf Baatz, *Bauten und Katapulte des römischen Heeres* (Stuttgart: Steiner, 1994); Rubén Sáez Abad, *Artillería y poliorcética en el mundo grecorromano* (Madrid: Consejo Superior de Investigaciones Científicas, 2005); M. C. Bishop and J. C. N. Coulston, *Roman Military Equipment: From the Punic Wars to the Fall of Rome*, 2nd ed. (Oxford: Oxbow Books, 2006); and Tracey E. Rihll, *The Catapult: A History* (Yardley, Pa.: Westholme Publishing, 2007).

[3] From statements in the *Belopoeica* it seems that Philo was at least one or two generations younger than Ctesibius, the great Alexandrian engineer whose *floruit* is generally placed at around 270 B.C. On Philo's date see K. Orinsky, O. Neugebauer, and A. G. Drachmann, "Philon (48)," in W. Kroll and K. Mittelhaus (eds.), *Paulys Realencyclopädie der classischen Altertumswissenschaft: Neue Bearbeitung*, vol. 20.1 (Stuttgart: J. B. Metzlersche Verlagsbuchhandlung, 1941), 53–54; Marsden, *Technical Treatises*, 7. The earliest text that attempts to provide a theoretical account of the working of machines is the *Mechanical Problems* (*Mēchanika problēmata*), a short text of disputed authorship transmitted in the Aristotelian corpus; its date is uncertain but generally thought to be relatively early in the third century. The work of Archimedes on centers of gravity and the equilibrium of plane figures represents a second crucial strand in the growth of theoretical mechanics in the third century B.C.

[4] For the contents of Philo's *Mechanical Syntaxis* see Orinsky, Neugebauer, and Drachmann, "Philon"; Marsden, *Technical Treatises*, 156; Bertrand Gille, *Les mécaniciens grecs: la naissance*

bring all these activities together under the rubric of a single discipline, the "art of mechanics" (*mēchanikē technē*), is an innovation that should be seen, in part, as a response to the proliferation of new technologies in the third century B.C.[5] The *Belopoeica*, which has survived in Greek, is structured in two main parts.[6] The first (49.1–56.8) describes a set of procedures for the construction of various types of artillery engines. Philo presents these procedures as a standard method that was widely diffused in actual practice. In the remainder of the *Belopoeica* (56.8–78.26), Philo makes a number of criticisms of the standard method and goes on to propose four alternative designs, one of which he claims to have developed himself. Throughout the text Philo adopts the authorial stance of an expert in artillery construction. He claims that his account of the standard method is based on personal association with engineers in both Alexandria and Rhodes, and various passages in the *Belopoeica* indicate that he was in close contact with sources familiar with Alexandrian engineering traditions.[7] The *Belopoeica* is evi-

---

de la technologie (Paris: Seuil, 1980); Astrid Schürmann, *Griechische Mechanik und antike Gesellschaft. Studien zur staatlichen Förderung einer technischen Wissenschaft* (Stuttgart: Steiner, 1991), 7–8. The work is dedicated to one Ariston, about whom nothing else is known. Aside from the *Belopoeica*, the only books that survive in Greek are those dealing with siegecraft. The book on pneumatics survives in both Arabic and Latin, though the relationship of these versions (which differ significantly from one another) to the original Greek text is far from clear; see Frank D. Prager, *Philo of Byzantium: Pneumatica* (Wiesbaden: L. Reichert, 1974).

5 Several passages in Aristotle (e.g., *metaph.* M 1078a16; *an. post.* 76a34, 78b37) refer to "mechanics" (*mēchanikē*, sc. *technē* or *epistēmē*) as a discipline that provides a mathematical account of the motion of physical bodies. The introduction to the *Mechanical Problems* (above, n. 3) conceives of mechanics as dealing with phenomena that take place "against" or "beyond" the ordinary course of nature (*para phusin*: mech. 847a1–b1). The paradigm example of a mechanical device is the lever, which enables a large force to be moved by a small weight (847b1–16); though the treatise discusses a wide range of devices used in particular *technai* and in daily life, its subject matter is not limited to technology. There is no parallel in any pre-Hellenistic source for Philo's conception of mechanics as a single *technē* embracing the wide range of subjects that he mentions. Cf. G. A. Ferrari, "Meccanica 'allargata,'" in Gabriele Giannantoni and Mario Vegetti (eds.), *La scienza ellenistica: atti delle tre giornate di studio tenutesi a Pavia dal 14 al 16 aprile 1982* (Naples: Bibliopolis, 1984), 227–96.

6 Text and references to the *Belopoeica* are according to Marsden's edition (Marsden, *Technical Treatises*); for the text see also Richard Schöne, *Philonis Mechanicae Syntaxis libri quartus et quintus* (Berlin: Reimer, 1893); Hermann Diels and Erwin Schramm, *Philons Belopoiika (viertes Buch der Mechanik): Griechisch und deutsch*, Abhandlungen der Preussischen Akademie der Wissenschaften, Philosophisch-Historische Klasse, Jahrg. 1918 no. 16 (Berlin: Verlag der Akademie der Wissenschaften, 1919). My translations are based on Marsden's, though I have sometimes modified them significantly.

7 For the claim of personal association see *Belopoeica* 51.10–14 (introducing the account of the standard method): "We shall recount to you exactly what we discovered in Alexandria through

dently intended for practitioners of artillery construction. Philo makes no attempt to explain the basic components of artillery engines or the terminology used for them; many aspects of his account would hardly be intelligible to a reader who lacked experience in the discipline. In this respect Philo's *Belopoeica* contrasts with the work of the same name by Hero of Alexandria. Near the beginning of his *Belopoeica*, Hero remarks that earlier writers on artillery construction wrote exclusively for experts, and states that he will explain the construction and uses of the various types of artillery engine in terms that a layperson can understand.[8] A further indication of the relatively specialized character of Philo's *Belopoeica* is the large proportion of the text that is devoted to the modified designs; this reflects the fact that Philo's concern is not only to present a standard method for constructing the best artillery engines, but also to impart the ability to improve their design. In these two aims Philo's *Belopoeica* reveals its character as a sophisticated discourse on engineering intended both to offer a canonical account of practitioners' knowledge and to shape that knowledge.[9]

Philo's *Belopoeica* stands out among the ancient literature on artillery construction for its high degree of explicit methodological reflection, much of which relates to the notion of expertise or *technē*. Philo presents artillery construction as a *technē* that possesses both a goal – "to dispatch the missile at long range, to strike with powerful impact" (τὸ μακρὰν ἀποστέλλειν τὸ βέλος εὔτονον τὴν πληγὴν ἔχον, 51.8–9) – and an established method for reaching that goal. The term "method" (*methodos*) occurs some sixteen times in the text, often in emphatic assertions that a certain result is brought about "not haphazardly, but by means of a method" (*vel sim.*).[10] The idea that a *technē* needs both a goal

---

much association with the craftsmen engaged in such matters and through intercourse with many master craftsmen in Rhodes, from whom we understood that the most reputable engines (τὰ μάλιστα τῶν ὀργάνων εὐδοκιμοῦντα) more or less (*sunengus*) conformed to the method we are about to describe." For Philo's reliance on others for information about Ctesibius cf. n. 48 below.

8 Hero Alex. *bel.* 73–74 Wescher. Hero was probably active in the first century A.D.; in any case his *Belopoeica* certainly postdates Philo's, and his criticism of the specialized character of earlier writings may well be directed, in part, at Philo's text.
9 Philo's text is much more than just a description of successful designs such as we find in Biton's *Construction of War Machines* (Κατασκευαὶ πολεμικῶν ὀργάνων καὶ καταπαλτικῶν; Marsden, *Technical Treatises*, 66–77). At the end of the text (67 Wescher) Biton suggests that reflection on such examples is sufficient to acquire the ability for successful design: "Whatever engines we considered especially suitable for you, we have now described. For we are convinced that you will be able to work out similar designs [τὰ ὁμοιοειδῆ] by means of the ones provided."
10 Cf. *Belopoeica* 50.15–17: "it was necessary for this to be grasped not by chance or at random, but by a fixed method" (μὴ ἀπὸ τύχης μηδὲ εἰκῇ λαμβάνεσθαι, μεθόδῳ δέ τινι ἑστηκυίᾳ); 52.21–2: "This too must not be drawn at random, but by a method" (οὐκ εἰκῇ καταγραπτέον, ἀλλὰ καὶ

and a method for attaining it is a widely shared view in Greek thought that goes back to the fifth century B.C.[11] But if there was general agreement on this point, there was much debate about the character that a discipline's method had to have in order for it to qualify as a genuine *technē*. At a minimum, the existence of a method implied the existence of rules of procedure and techniques, i.e., the ability to carry out the rules. Thus a doctor might recognize that a particular case falls under a general rule ("bloodletting is helpful in cases of fever") and proceed accordingly. But there was significant dispute about whether the practitioner of a *technē* also needed to be able to explain his practice in terms of some sort of general theory, and so give an account of the reason or cause (*aitia*) of his actions. For example, a doctor might justify the administration of a particular drug by saying that it is able to purge phlegm, and the patient is suffering from an overabundance of phlegm. According to an influential line of thought represented by Plato, Aristotle, and some of the early Hippocratic treatises, the practitioner of a genuine *technē* needed to be able to give such explanations, which often appealed to the nature (*phusis*) of the subject matter of the discipline in question. Both Plato and Aristotle contrasted *technē* with experience or *empeiria*, understood as a collection of rough generalizations and rules of thumb that were not based on an explanatory theory.[12] In the early Hellenistic period, however,

---

τοῦτο μεθόδῳ τινί); 55.12: "it is necessary that there be a method" (δεῖ δὲ καὶ μέθοδόν τινα ὑπάρχειν); 69.26: "there was need of another method" (προσεδεῖτο δὲ ἄλλης μεθόδου).
11 The association between *technē* and method may go back to the fifth-century Sophists; see Felix Heinimann, "Eine vorplatonische Theorie der τέχνη," *Museum Helveticum* 18 (1961): 105–30. But it is first clearly attested in some of the Hippocratic treatises that can plausibly be dated to the late fifth century; see especially Hipp. *vet. med.* 1–2, 1.570–74 L. (= 118–20 Jouanna) and *art.* 5, 6.8 L. (= 229–30 Jouanna).
12 For the view that medicine needs to be based on a theory of human *phusis* see Hipp. *vet. med.* 20, 1.620–24 L. (= 145–48 Jouanna); the doctrine of the four canonical humors – blood, phlegm, yellow, and black bile – stated in the Hippocratic treatise *On the Nature of Human Beings* is only the most famous such theory. Important Platonic passages on the nature of *technē* include *Gorg.* 464b–466a, *leg.* 720a-e and 857c-e, and *Phdr.* 268a–270e, where Socrates argues that the genuine doctor needs more than just mastery of a set of effective procedures to cure patients effectively; only an understanding of human *phusis* will enable him to know when, to what extent, and to which patients he should apply those procedures. For Aristotle, the genuine doctor is distinguished from the empiric by the possession of explanatory knowledge: for example, it is a matter of *empeiria* to know that heat cures fever, but a matter of *technē* to know that this is so because heat counteracts bile, which is the cause (*aitia*) of fever (cf. *metaph.* A 981a5–12). Yet Aristotle also remarks that as far as practice is concerned, *empeiria* seems equivalent to *technē*, and that practitioners with experience are more successful than those who have theory without experience; cf. *metaph.* A 981a12–24. For a full discussion of these and other relevant Platonic and Aristotelian passages see Mark J. Schiefsky, *Hippocrates "On Ancient*

the notion that a *technē* must be based on explanatory knowledge was called into question by the so-called Empiricist school of medicine, which took the position that *empeiria*, understood as a body of more or less general correlations based solely on observation, was entirely sufficient to account for both the discovery and practice of medicine.[13]

The dispute between the Empiricists and their so-called Rationalist opponents – who argued that explanatory theory was essential – dominated the methodological debate in medicine for several centuries, and raised general issues that pertained to the very nature of expertise. This debate is relevant for assessing Philo's picture of artillery construction as a *technē* in more than one way. In this paper I shall attempt to clarify Philo's position on the two key issues in this debate: the nature of the generalizations on which the *technē* of artillery construction is based, and the roles of theory and experience in it. As we shall see, Philo takes a distinctive and nuanced approach to these questions, one which emphasizes the importance of both explanatory generalizations and the essential role of experience or *peira* in the discovery and practice of the artillery builder's *technē*.

A second important methodological issue raised in the *Belopoeica* concerns the uses of mechanical and physical theory in artillery construction. What sorts of theories should the practitioner bring to bear in designing and building artillery engines, and how should he make use of them? Here Philo had a wide range of sources at his disposal, including the Aristotelian *Mechanical Problems* (above, n. 3) and the works of Archimedes for theoretical mechanics, as well as the works of thinkers such as Aristotle and his Hellenistic successor as head of the Lyceum, Strato of Lampsacus, for physical theory. The relationship between mechanics and physics had already been raised as an issue in the *Mechanical Problems*, which sets out a conception of mechanics as concerned with phenomena that take place "against" or "beyond" the ordinary course of nature

---

Medicine": Translated with Introduction and Commentary, Studies in Ancient Medicine 28 (Leiden: Brill, 2005), 345–59.

[13] The most accessible introduction to Empiricism is Michael Frede and Richard Walzer, *Galen: Three Treatises on the Nature of Science* (Indianapolis: Hackett, 1985); the fundamental collection and study remains Karl Deichgräber, *Die griechische Empirikerschule: Sammlung der Fragmente und Darstellung der Lehre* (Berlin: Weidmann, 1930). The origin of the school can be traced back to Philinus of Cos (*fl.* ca. 250 B.C.), a renegade pupil of the Alexandrian physician Herophilus (Deichgräber, *Die griechische Empirikerschule*, 163–64). The connection between the Empiricists' conception of *empeiria* and the Platonic and Aristotelian discussions referred to in the previous note is explored in Frede's introduction and Schiefsky, *Hippocrates*, 345–59. That the Empiricist sect was named for a methodological position rather than a founder is a further indication of the importance of methodological discussion in the third century B.C.

(*para phusin*: *mech.* 847a11–b10). Such a formulation obviously raises the question of the sense in which mechanics is based on the study of nature; we shall see that Philo takes an interesting approach to this issue as well.

Finally there is the issue of the values that guided the artillery builder in the practice of his *technē*. A number of passages in the *Belopoeica* suggest that these included factors such as economy of construction and aesthetic beauty as well as performance. Given the wide range of ancient Greek *technē* – which included activities such as the fine arts as well as disciplines such as medicine and mathematics — this should come as no surprise. However, it does suggest that Philo's conception of artillery construction as a *technē* with a unitary goal – the achievement of long range and powerful impact – does not fully capture the range of motivations that guided the activities of its practitioners.

Philo's *Belopoeica* thus offers a precious window into the methodological disputes of an age of great scientific creativity and controversy, and a view of the nature of expertise that is informed both by close association with communities of practitioners and by familiarity with broader trends in Greek philosophical and scientific thought. The aim of this paper is to offer a preliminary assessment of Philo's handling of the methodological issues outlined above through a close examination of the text. Part I discusses Philo's account of the standard method, focusing on his conception of the role of experience or testing (*peira*) in its discovery. Part II discusses the modified designs proposed in the second part of the *Belopoeica*, with an emphasis on Philo's conception of the role of physical and mechanical theory in artillery construction. Throughout both parts I shall draw attention to Philo's concern with values such as beauty and economy. In the conclusion I will briefly sum up Philo's conception of expertise and attempt to place it in a wider context.

I Philo on *Peira:* The Standard Method

I begin with a brief outline of the technological background to Philo's account in the first part of the *Belopoeica* (49.1–56.8).[14] The invention of artillery may plausibly be dated to around 399 B.C., when the tyrant Dionysius of Syracuse brought together a large number of craftsmen with the specific goal of developing new military technology.[15] The earliest artillery was based on extending the power

---

14 For fuller accounts of the developments mentioned here see Marsden, *Historical Development*; Rihll, *Catapult*, 26–75.
15 Diodor. Sic. 14.41; Marsden, *Historical Development*, 48–64; Rihll, *Catapult*, 26–45.

of the traditional bow, as in the so-called "belly-bow" or *gastraphetēs* (fig. 1). At some point in the mid-fourth century B.C. it was realized that the resilient properties of animal sinew could provide much more power than the composite bow; a typical example of this type of engine, known as torsion artillery, is shown in figure 2. Long strands of sinew were wound through the frame, and the arms of the engine were placed inside the bundles of strands (see the front elevation "c" in fig. 2). Different types of torsion engines were designed for shooting arrows and stones.

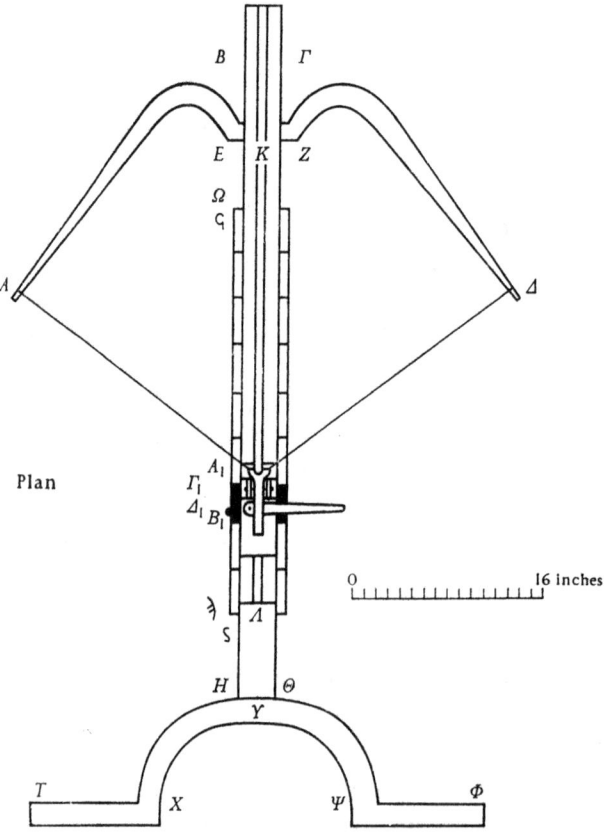

Fig. 1 The belly-bow or *gastraphetēs* (Marsden, *Technical Treatises*, 47)

At a certain point in the early third century B.C., two important advances were made in artillery design. First, lists of dimensions were set out, specifying the size of all components of an artillery engine down to the smallest detail in terms of a single unit: the diameter of the holes in the frame through which

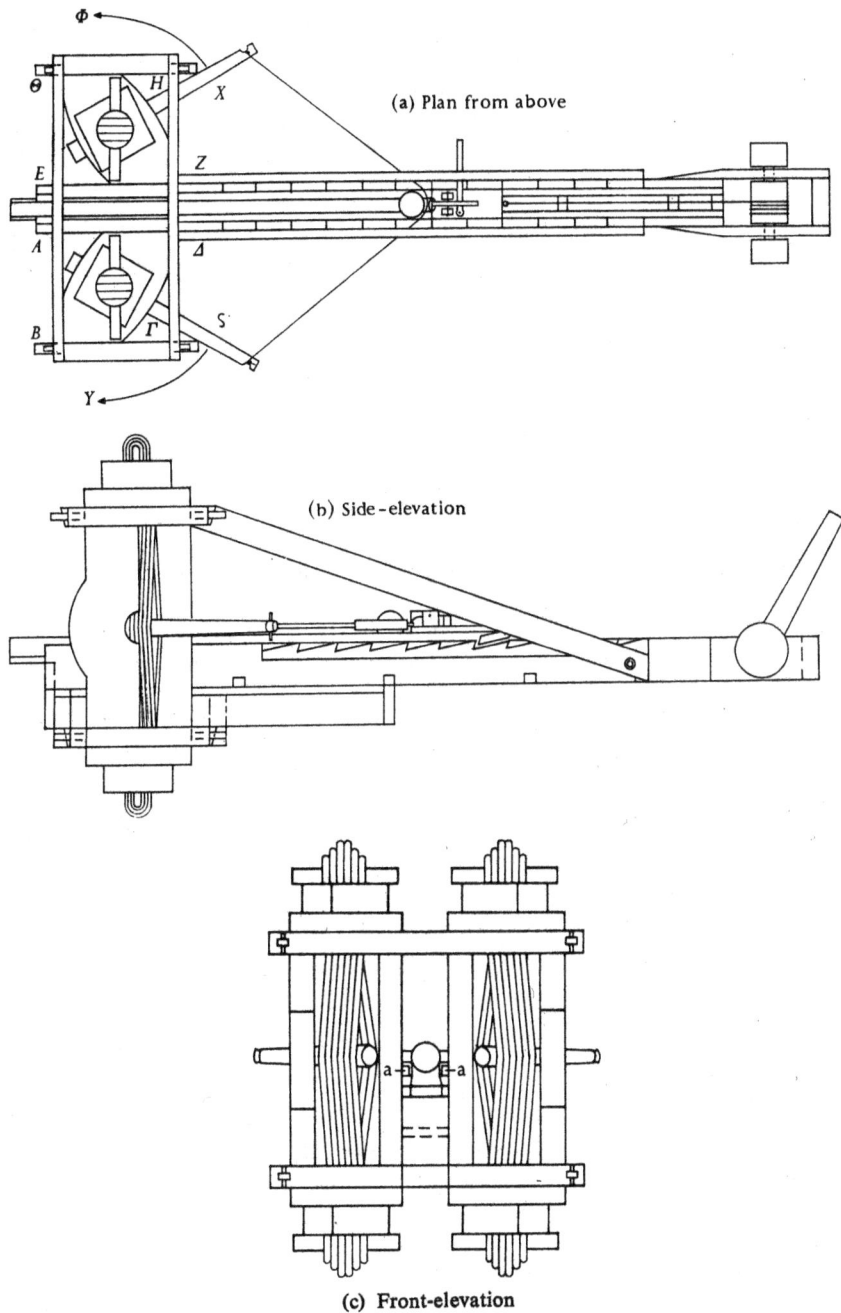

Fig. 2 Torsion artillery (Marsden, *Technical Treatises*, 56)

the spring cords were strung. Thus the size of all the parts of an engine was related to the force it could produce: a larger hole meant a larger bundle of springs and thus a more powerful engine. Second, quantitative relationships were derived, correlating the weight of the stone or the length of the arrow the engine was designed to shoot with the diameter of the spring hole. In the case of arrow-throwing engines, the diameter of the hole was specified as one-ninth the length of the arrow. For stone throwers the diameter of the hole was obtained by taking the cube root of the weight of the shot, then adding one-tenth of that root. Taken together, these spring hole relations and dimensional lists made it possible for a practitioner to construct an artillery engine for a missile of any given weight or length.

Turning to Philo, I shall begin by considering his general characterization of the standard method and what it brings to the artillery builder's activity. Philo contrasts the existence of a method in his own day with the haphazard situation in which earlier engineers found themselves:

> I suppose you are not unaware that the art [technē] contains something unintelligible and baffling to many people; at any rate, many who have undertaken the building of engines of the same size, using the same design, similar wood, and identical metal, without even changing its [i.e., the metal's] weight, have made some with long range and powerful impact and others which fall short of the ones mentioned. Asked why this happened, they could not give the reason [aitia]. Hence the remark made by Polycleitus the sculptor is pertinent to what I am going to say. He maintained that excellence is achieved gradually through many numbers.[16] Likewise, in this art [technē], since products are brought to completion through many numbers, those who deviate slightly in particular parts produce a large total error at the end. Therefore, I maintain that we must pay close attention when adapting the design of successful engines to a distinctive construction, especially when one wishes to do this while either increasing or diminishing the scale.[17] (*Belopoeica* 49.12–50.12)

---

16 The reference is presumably to the numerical proportions between the different components of a Polycleitan sculpture, which Philo takes as analogous to the numerical proportions between the dimensions of the components of the artillery engine.

17 ὅτι μὲν οὖν συμβαίνει δυσθεώρητόν τι τοῖς πολλοῖς καὶ ἀτέκμαρτον ἔχειν τὴν τέχνην, ὑπολαμβάνω μὴ ἀγνοεῖν σε· πολλοὶ γοῦν ἐνστησάμενοι κατασκευὴν ὀργάνων ἰσομεγεθῶν καὶ χρησάμενοι τῇ τε αὐτῇ συντάξει καὶ ξύλοις ὁμοίοις καὶ σιδήρῳ τῷ ἴσῳ οὐδὲ τὸν σταθμὸν αὐτοῦ μεταβάλλοντες, τὰ μὲν μακροβολοῦντα καὶ εὔτονα ταῖς πληγαῖς ἐποίησαν, τὰ δὲ καθυστεροῦντα τῶν εἰρημένων· καὶ ἐρωτηθέντες, διὰ τί τοῦτο συνέβη, τὴν αἰτίαν οὐκ εἶχον εἰπεῖν· ὥστε τὴν ὑπὸ Πολυκλείτου τοῦ ἀνδριαντοποιοῦ ῥηθεῖσαν φωνὴν οἰκείαν εἶναι τῷ μέλλοντι λέγεσθαι· τὸ γὰρ εὖ παρὰ μικρὸν διὰ πολλῶν ἀριθμῶν ἔφη γίνεσθαι. τὸν αὐτὸν δὴ τρόπον καὶ ἐπὶ ταύτης τῆς τέχνης συμβαίνει διὰ πολλῶν ἀριθμῶν συντελουμένων τῶν ἔργων μικρὰν ἐν τοῖς κατὰ μέρος παρέκβασιν ποιησαμένους μέγα συγκεφαλαιοῦν ἐπὶ πέρας ἁμάρτημα· διὸ φημι δεῖν προσέχοντας

Without a method, the artillery designer is at a loss even when attempting to copy a successful design; he is unable to explain (to give a reason, *aitia*) why some engines are more effective than others. Conversely, the existence of a reliable method enables the builder to adapt a successful design to the requirements of a particular situation (i.e., to a particular length of arrow or weight of shot). The reference to the *Canon*, a lost work of the famous Greek sculptor Polycleitus, emphasizes the quantitative precision of the method and its character as a norm or standard.[18] The *Canon* was an attempt to describe the ideal human figure by giving numerical proportions between its various parts.[19] Just as Polycleitus had set out ideal proportions between the parts of the body, so Philo gives an exhaustive list of the dimensions of all components of the engines he describes in terms of the spring hole diameter. Artillery engines thus display harmony and proportion between their various components, and even the slightest deviation from these proportions will result in diminished performance. The reference to Polycleitus further suggests that the *technē* of artillery construction aims at the attainment of an aesthetic ideal comparable to that achieved by a great work of sculpture. Finally, we may note that Philo emphasizes the established character of the method, as at 58.32–5: "though very many years have passed since this design [*suntaxis*] was invented and there have naturally been many machine and artillery makers, no one has dared to depart from the existing method."[20]

In setting out the content of the standard method, Philo begins by giving several alternative procedures for determining the size of the spring hole for a stone-throwing engine (*Belopoeica* 51.15–52.19). (1) The first option is to calculate the size of the hole from the weight of the shot by applying the spring hole relation directly. This requires converting the weight to drachmae, taking the cube root of the result, and adding a tenth of that root; the result of this procedure is the diameter of the hole in "finger-breadths" (dactyls). In case the cube root is not a whole number, Philo recommends a method of approximation: take as the diameter the whole number closest to the root, then diminish it slightly if it is larger

---

μεταφέρειν τὴν ἀπὸ τῶν ἐπιτετευγμένων ὀργάνων σύνταξιν ἐπὶ τὴν ἰδίαν κατασκευήν, μάλιστα δέ, ὅταν τις εἰς μεῖζον μέγεθος αὔξων τοῦτο βούληται ποιεῖν καὶ ὅταν εἰς ἔλασσον συναιρῶν.
18 On *kanōn* as a norm or standard see Herbert Oppel, *Kanōn. Zur Bedeutungsgeschichte des Wortes und seiner lateinischen Entsprechungen (regula-norma)* (Leipzig: Dieterich, 1937).
19 See Andrew Stewart, "The Canon of Polykleitos: A Question of Evidence," *Journal of Hellenic Studies* 98 (1978): 122–31. Cf. Gal. *de temper.* 1.566 K. and *de plac. Hipp. et Plat.* 5.448 K. (where Chrysippus is praised for holding that beauty [*kallos*] lies in the *summetria* of the parts of the body, and Polycleitus' *Canon* is cited as advocating the same view.)
20 πολλῶν σφόδρα ἐτῶν διεληλυθότων, ἀφ' οὗ τὴν σύνταξιν εὑρῆσθαι τήνδε συμβαίνει, καὶ πολλῶν γεγονότων, ὅπερ εἰκός, καὶ μηχανικῶν καὶ βελοποιῶν, μηθένα τετολμηκέναι παρεκβῆναι τὴν ὑποκειμένην μέθοδον.

than the root but increase it slightly if it is smaller. (2) To enable the practitioner to avoid having to extract a cube root, Philo gives a table providing the correspondences between weight of shot and hole size for a number of standard weights (*Belopoeica* 51.21–26); such tables were no doubt much in use among practitioners.[21] (3) Still another alternative (51.28–52.19) would enable a practitioner to find the diameter of the hole for a given weight assuming that the correct diameter for another weight was known; here Philo gives a geometrical procedure for finding two mean proportionals.[22] Once the diameter of the spring hole had been determined, the practitioner would go on to construct the other parts of the engine by referring to the list of dimensions (*Belopoeica* 53.8–54.16). Throughout his account of the standard method Philo notes the importance of symmetry (*summetria*) and precision (*akribeia*), continuing the emphasis on these qualities suggested by the Polycleitus reference.[23] Philo's account is intended for a practitioner who is familiar with the technical terminology of artillery construction and who has mastered certain mathematical techniques; at the same time, his concern to present alternative methods for finding the spring

---

[21] Vitruvius, whose account of artillery in *arch.* 10.10–12 is clearly intended for the practitioner, gives only a list of standard weights with corresponding sizes of shot and does not state the spring hole relation in its full generality. The reason he gives for including the table is "in order that even those ignorant of geometry may have a convenience, so that they will not be held up by calculations in the dangers of war" (*ut etiam, qui geometrice non noverint, habeant expeditum, ne in periculo bellico cogitationibus detineantur*, 10.11.2).

[22] The reasoning behind this is as follows (cf. Hero Alex. *bel.* 113.8–119.2 Wescher). (1) The diameter of the hole is equal to the cube root of the weight times a constant. Thus, for two different weights, the ratio of the weights is equal to the ratio of the cubes of the respective diameters, i.e., $W_1:W_2 = D_1^3:D_2^3$. In a practical situation W_1, W_2, and D_1 will be given, and we will need to find D_2. (2) Suppose for the sake of argument that we wish to construct an engine for a shot of triple the weight, i.e., $W_1:W_2 = 1:3$. We must now find x such that $D_1:D_2 = D_2:x = x:3D_1$; then $D_1^3:D_2^3 = 1:3$, as desired. D_2 and x are the two "mean proportionals" between D_1 and $3D_1$. (3) From line D_1, we construct a line of length $3D_1$, and then construct the two mean proportionals between them. The construction cannot be accomplished by the standard Euclidean uses of ruler and compass; the method Philo gives involves rotating a ruler around a given point until certain conditions are fulfilled, and it is thus inherently approximate (see Marsden, *Technical Treatises*, 59–60, 158–59). The problem is essentially that of "doubling the cube," i.e., given a cube, construct another with double the volume. Greek mathematicians devised a variety of solutions to this problem; it is clear that its importance in artillery construction was a major stimulus to their efforts. Cf. Thomas L. Heath, *A History of Greek Mathematics*, 2 vols. (Oxford: Clarendon Press, 1921), 1:244–70; Philo's construction, which is essentially the same as that given by Hero, is discussed at 262–64.

[23] For *summetria* see *Belopoeica* 53.24–25, 54.15, 54.21; for *akribeia* see 55.13, 55.19, 55.29. Cf. 54.15–16 for the injunction to avoid error in the proportions (*analogiai*) between different parts.

hole diameter suggests an effort to communicate with practitioners at different levels of mathematical competence.[24]

I turn now to Philo's account of the discovery of the standard method:

> Among the ancients, some were on the way to discovering that the element [*stoicheion*], principle [*archē*], and measure [*metron*] for the construction of engines was the diameter of the hole. This[25] had to be obtained not by chance or at random, but by a fixed method which could produce the correct proportion at all sizes. It was impossible to obtain it except by increasing and diminishing the perimeter of the hole on the basis of experience [*ek peiras*]. Now the ancients did not bring this to a conclusion, as I say, nor did they determine the size, since their experience [*peira*] was not based on many trials [*erga*], but they did hit on what to look for. Those who came later, by drawing conclusions from previous mistakes, and by looking for a fixed element on the basis of later tests [ἐκ τῶν μετὰ ταῦτα πειραζομένων], introduced the principle and beginning [ἀρχὴν καὶ ἐπίστασιν] of construction, I mean the diameter of the circle that receives the spring. The craftsmen at Alexandria achieved this first, being heavily subsidized because they had kings who were lovers of reputation and craftsmanship [φιλοδόξων καὶ φιλοτέχνων]. For the fact that it is not possible for everything to be grasped by reasoning and the methods of mechanics [τῷ λόγῳ καὶ ταῖς ἐκ τῶν μηχανικῶν μεθόδοις], but that many things are also discovered through experience [διὰ τῆς πείρας], is clear both from many other things and not least from what is about to be said.[26] (*Belopoeica* 50.14–29)

In this passage, Philo identifies the spring hole diameter as the fundamental parameter in the construction of artillery; it is the "element" (*stoicheion*), "princi-

---

24 Among the other procedures that Philo discusses in his account of the standard method are a technique for drawing the shape of the so-called "hole-carrier" or *peritrēton*, the part of the frame into which the spring holes were drilled (52.20–53.7), and a general method for enlarging and reducing figures in a given ratio (55.12–56.8). The construction of an arrow-throwing engine receives only a brief mention at 54.25–55.11.
25 I.e., the diameter of the spring hole.
26 Ἐπὶ γὰρ τῶν ἀρχαίων τινὲς ηὕρισκον στοιχεῖον ὑπάρχον καὶ ἀρχὴν καὶ μέτρον τῆς τῶν ὀργάνων κατασκευῆς τὴν τοῦ τρήματος διάμετρον· ταύτην δ' ἔδει μὴ ἀπὸ τύχης μηδὲ εἰκῇ λαμβάνεσθαι, μεθόδῳ δέ τινι ἑστηκυίᾳ καὶ ἐπὶ πάντων τῶν μεγεθῶν δυναμένῃ τὸ ἀνάλογον ὁμοίως ποιεῖν. οὐκ ἄλλως δὲ ἦν ταύτην λαβεῖν, ἀλλὰ ἐκ πείρας αὔξοντάς τε καὶ συναιροῦντας τὸν τοῦ τρήματος κύκλον. τοὺς γοῦν ἀρχαίους μὴ ἐπὶ πέρας ἀγαγεῖν, ὡς λέγω, μηδὲ ἐνστήσασθαι τὸ μέγεθος, οὐκ ἐκ πολλῶν ἔργων τῆς πείρας γεγενημένης, ἀκμῇ δὲ ζητουμένου τοῦ πράγματος· τοὺς δὲ ὕστερον ἔκ τε τῶν πρότερον ἡμαρτημένων θεωροῦντας καὶ ἐκ τῶν μετὰ ταῦτα πειραζομένων ἐπιβλέποντας εἰς ἑστηκὸς στοιχεῖον ἀγαγεῖν τὴν ἀρχὴν καὶ ἐπίστασιν τῆς κατασκευῆς, λέγω δὲ τοῦ κύκλου τὴν διάμετρον τοῦ τὸν τόνον δεχομένου. τοῦτο δὲ συμβαίνει ποιῆσαι τοὺς ἐν Ἀλεξανδρείᾳ τεχνίτας πρώτους μεγάλην ἐσχηκότας χορηγίαν διὰ τὸ φιλοδόξων καὶ φιλοτέχνων ἐπειλῆφθαι βασιλέων. ὅτι γὰρ οὐ πάντα δυνατὸν τῷ λόγῳ καὶ ταῖς ἐκ τῶν μηχανικῶν μεθόδοις λαμβάνεσθαι, πολλὰ δὲ καὶ διὰ τῆς πείρας εὑρίσκεται, φανερὸν μὲν καὶ ἐξ ἄλλων πλειόνων ἐστίν, οὐχ ἥκιστα δὲ καὶ ἀπὸ τοῦ μέλλοντος λέγεσθαι.

ple" or "beginning" (*archē, epistasis*), and unit of measure or standard of reference (*metron*). That the diameter of the spring hole has this character was discovered through a long process of empirical investigation, "by increasing and diminishing the perimeter of the hole on the basis of experience [*ek peiras*]." This investigation, furthermore, was a highly directed and systematic one that was carried out with state support. The ancients did not discover the fundamental role of the spring hole diameter "since their experience [*peira*] was not based on many trials [*erga*]"; they succeeded only when the Alexandrian kings provided a subsidized context for research, and by "drawing conclusions from previous mistakes" and "looking for a fixed element on the basis of later tests." Summing up, Philo draws a contrast between what is discovered by "reasoning [*logos*] and the methods of mechanics" and by experience (*peira*), and places the importance of the spring hole diameter in the latter category.

The key term in Philo's account is *peira*, which commonly means "trial" or "attempt." In the *Belopoeica* the term can generally be translated as "test," but it sometimes seems to refer to knowledge acquired on the basis of testing, that is, "experience."[27] In any case, Philo's meaning here is clear: in order to find the optimal size of the spring hole for a given missile, the only possible procedure is to vary its size and evaluate the effect of the variation on the range and force of the missile. If one takes account of the results of one test in setting up the next, eventually one will zero in on the optimal size for the hole. Philo reinforces the systematic character of this procedure by drawing a parallel with architecture. It was not from a single attempt [*peira*] that people discovered how to shape columns in the way that produced the most symmetric and harmonious appearance in buildings, but by evaluating the effects of many slight modifications to the shapes of the individual parts:

> For instance, the correct proportions of buildings could not possibly have been determined right from the start and without the benefit of previous experience [*peira*], as is clear from the fact that the old builders were extremely unskillful, not only in general building, but also in shaping the individual parts. The progress to proper building was not the result of a single or chance experience [*peira*]. Some of the individual parts, which were equally thick and straight, seemed not to be so, because the sight is deceived in such cases by the difference of distance. By experimentally [διὰ τῆς πείρας] adding to the bulk here and subtracting there, by tapering, and by conducting every possible test, they made them appear

---

[27] For the broader sense cf. *Belopoeica* 50.20 (in the passage just quoted in the text): "since their experience was not based on many trials" (οὐκ ἐκ πολλῶν ἔργων τῆς πείρας γεγενημένης) and 53.29–30: "for this reason those who have acquired their experience in action commanded them to make use of the above-mentioned size" (διὸ ἐκέλευσαν ἔργῳ τὴν πεῖραν εἰληφότες τῷ προειρημένῳ μεγέθει χρᾶσθαι).

regular to the sight and quite symmetrical: for this was the aim in that art [*technē*].²⁸ (*Belopoeica* 50.30–51.7)

Several features of this remarkable account of discovery are worthy of comment. First, Philo asserts that the Alexandrian engineers worked out both (1) the relation between the spring hole and the weight of the shot and (2) the specifications of the dimensions of the various parts in terms of the spring hole diameter, in the course of a program of systematic, controlled testing. The process evidently involved (1) keeping the dimensions of the engine fixed and varying the size of the hole, and (2) keeping the size of the hole fixed and varying the dimensions of the different components. The latter procedure is suggested by Philo's account of the dimensional list for stone-throwing engines, where he remarks that those who speak "from experience" (ἐκ τῆς πείρας) gave the optimal dimension for the height of the spring: any larger, and the engine would achieve long range but little force on impact; any smaller, and the range would be impaired.²⁹ Furthermore, the method of discovery that Philo describes does not involve any antecedent commitment to theory; what drives the process is the evaluation of the impact that modifications in construction have on performance.

At the same time, it is crucial to note that the account culminates not just in the discovery of the spring hole relations and dimensional lists, but also in the recognition that the diameter of the spring hole is the fundamental "element" (*stoicheion*), "principle" (*archē*), and "measure" (*metron*) in construction. Each of these terms has a long history in Greek thought in contexts that are relevant to the development of a concern with systematic explanation, such as Presocratic philosophy, Hippocratic medicine, and mathematics. By suggesting that each of them can be used of the spring hole diameter, Philo signals a concern to sys-

---

28 τοὺς γὰρ τῶν οἰκοδομικῶν ἔργων ῥυθμοὺς οὐ δυνατὸν ἦν ἐξ ἀρχῆς συστήσασθαι μὴ πρότερον πείρας προσαχθείσης, καθ' ὅτι καὶ δῆλόν ἐστιν ἐκ τῶν ἀρχαίων καθ' ὑπερβολὴν ἀτεχνῶν οὐ μόνον κατὰ τὴν οἰκοδομίαν, ἀλλὰ καὶ ἐν ταῖς κατὰ μέρος εἰδοποιίαις. μετετέθη οὖν ἐπὶ τὸ δέον οὐ διὰ μιᾶς οὐδὲ τῆς τυχούσης πείρας. τινὰ δὲ τῶν κατὰ μέρος ἐν αὐτοῖς ὑπαρχόντων ἰσοπαχῆ τε ὄντα καὶ ὀρθὰ ἐδόκει μήτε ἰσοπαχῆ μήτε ὀρθὰ εἶναι διὰ τὸ ψεύδεσθαι τὴν ὄψιν ἐπὶ τῶν τοιούτων μὴ τὸ ἴσον ἐχούσαν ἀπόστημα· διὰ τῆς πείρας οὖν προστιθέντες τοῖς ὄγκοις καὶ ἀφαιροῦντες καὶ μύουρα ποιοῦντες καὶ παντὶ τρόπῳ πειράζοντες κατέστησαν ὁμόλογα τῇ ὁράσει καὶ εὔρυθμα φαινόμενα· τοῦτο γὰρ ἦν τὸ προκείμενον ἐν ἐκείνῃ τῇ τέχνῃ.
29 *Belopoeica* 53.17–23; cf. 53.25–30, where Philo makes a similar remark about the length of the arms. As Marsden has noted (*Historical Development*, 25), the Alexandrian engineers were engaged in three closely related investigations: (1) determining the optimal relationship between diameter and height of the spring, (2) determining the optimum size of the spring for a given weight of shot, and (3) demonstrating that all measurements are dependent on the spring hole diameter.

tematize the rules of the standard method in terms of underlying general principles. This in turn suggests that his emphasis on the importance of *peira* in the discovery process should not be understood as a denial of the importance of seeking general explanations. Indeed, the intuition that the power of an artillery engine is connected with the volume of the bundle of spring cords – which in turn depends on its diameter and height – presumably played some role in discovering the cube-root relation between the weight of shot and the diameter of the hole, and Philo's account leaves this possibility open.[30] What he does insist on is the need for constant reference to practical testing. The contrast that he introduces between discovery by "reasoning [*logos*] and the methods of mechanics" and discovery by *peira* should be understood as a contrast between purely theoretical inquiry and research that depends on continued testing and modification.

To be sure, Philo's account of the standard method reflects a certain amount of simplification and idealization. The emphasis on the quantitative precision of the spring hole relations and dimensional lists contrasts with the fact that practitioners would often need to resort to approximation in practice, e.g., when it was necessary to extract a cube root. Philo's conception of the goal of artillery construction as the attainment of both long range and powerful impact ignores the possibility that a practitioner might sometimes need to achieve a very powerful impact at short range.[31] A further problem involves the sense in which the spring hole relations and dimensional lists can be said to yield optimal artillery engines. One might expect that longer range and more powerful impact would be attained by using a lighter shot than that for which the engine had been designed according to the spring hole relations. Evidently these relations are based on consideration of factors other than just performance, i.e., factors such as the size and portability of the engine and the expense of construction.[32] In general, it is unlikely that the practice of artillery construction conformed as closely as Philo suggests to his picture of a standard, established method for attaining a single goal. But that does not diminish the interest of his attempt to

---

30 In his description of the method of finding two mean proportionals at *bel.* 114.8–119.2 Wescher, Hero suggests one way in which theoretical considerations might have influenced the discovery of the spring hole relation for stone-throwers. That a cube root is involved may have been suggested by reflecting on the fact that the force produced by the engine is directly related to the volume of the cylinder made up of the spring cords: a larger weight requires a spring cylinder of proportionally larger volume, and since the ratio of the volumes of two similar cylinders is that of the cubes of their base diameters, we immediately have the relation $W_1:W_2 = D_1^3:D_2^3$ (above, n. 22).
31 Cf. Marsden, *Technical Treatises*, 160–61.
32 Marsden, *Historical Development*, 37–39.

standardize what was presumably a somewhat diverse set of practices. Moreover, Philo is surely correct to emphasize the importance of systematic testing in the development of artillery. The artillery builder's work faced critical tests in battle situations, and the idea of state-sponsored research into military technology is – as modern experience indicates – all too plausible.

To sum up, in the first part of the *Belopoeica* Philo argues that artillery construction possesses a method that enables its practitioners to achieve successful results in a reliable fashion. This method rests on two types of quantitatively precise rules: the spring hole relations and dimensional lists. Philo evidently expects practitioners to know these rules, though he also provides shortcuts (such as the lists correlating the weight of shot and the size of the spring hole) that would make it easier to apply them in practical situations. Yet by identifying the spring hole diameter as the fundamental element, principle, and standard of measure in construction, he asserts that the standard method is based on a principle of greater generality. A grasp of the importance of the spring hole diameter unifies the rules of the artillery builder's *technē* and in a certain sense explains them. It accounts for the different results that were attained in practice by the earlier designers who lacked proper method. And it is the role of the spring hole diameter as a general principle – not just the dimensional lists and spring hole relations – that the Alexandrian engineers are said to have discovered. Philo's stress on the role of *peira* in the discovery process should be understood not as a rejection of the need for explanation, but as a way of emphasizing that the discovery could only have been made by the systematic evaluation of the results of practical tests. In his account of the standard method Philo does not explain *why* the spring hole diameter is fundamental in construction (by appealing to the relationship between the volume of the spring and the power it can generate, for example). Nonetheless, by pointing to its importance, he takes a significant step beyond a conception of *technē* as just a set of practical rules and techniques.

II Philo on Theory: The Modified Designs

I turn now to the second part of the *Belopoeica*, in which Philo presents his criticisms of the standard method and suggests a number of alternative designs. In contrast to the emphasis on experience in the first part of the text, theoretical considerations play a prominent role in Philo's account of these innovations. Philo portrays his own contribution as building on the admirable achievements of his predecessors, a sentiment that finds numerous parallels in the technical literature of other fields such as medicine.³³ In this section I shall consider three of the four modified designs that Philo discusses, each of which illustrates the role of theory in mechanics as he conceives of it.

1 Philo's Wedge Engine

At *Belopoeica* 56.8–67.27 Philo proposes a modified design for an arrow-shooting artillery engine and attempts to show that it avoids various deficiencies in the standard design. The basic technological problem that Philo's modifications were intended to address concerned the need to keep the springs of a piece of torsion artillery at a high state of tension. There was a natural tendency for the springs to loosen after use, and they would then require re-tightening. In an engine of the standard design, this re-tightening would be accomplished by twisting the springs using tightening bars on the top and bottom of the frame. Philo objects to this procedure as weakening the spring cords and proposes that the re-tightening should be done by using wedges, thus increasing the tension of the spring in the vertical direction without any horizontal twist (fig. 3). At 56.18–24 he claims that the excellence of his wedge engine is evident in six respects: (1) it can shoot far; (2) it retains its strength in the heat of battle; (3) it is easy to construct; (4) it is easily assembled, strung, and disassembled; (5) it is "in no way deficient in appearance" (τὴν ὄψιν οὐθὲν καταδεεστέραν) to the standard design; and (6) it is less expensive. This list gives a good idea of the various factors that the ancient practitioner of artillery construction had to keep in mind in his attempt to build a machine that would attain long range and a powerful impact: ease of construction, transportation, and expense, as well as aesthetic considerations.³⁴

---

33 *Belopoeica* 58.26–32: "Yet one must praise those who originally invented the construction of these engines, for they were the authors of the invention and of its characteristic form: they discovered something superior to all other weapons, both in shooting range and in weight of missiles, I mean weapons like the bow, javelin, and sling. To have an original idea and to bring it to completion is the work of a greater nature [μείζονος φύσεως]; to improve or modify something that already exists seems appreciably easier." In the Hippocratic Corpus the closest parallel is *vict.* I 1, 6.466–68 L. (= 122–23 Joly/Byl); cf. also *vet. med.* 2, 1.572–74 L. (= 119–20 Jouanna) on the need to build upon prior discoveries and *art.* 1, 6.2 L. (= 224–25 Jouanna) for the praise of discovery in general.
34 For the importance of aesthetic factors in the wedge engine, see *Belopoeica* 61.29–62.15 and esp. 62.12–14: the frame, since it is smooth and has no protuberances, "presents a fine appearance" (καλὴν τὴν ὅρασιν ἀποδιδόναι). Cf. also 66.18–19, where Philo, having just finished

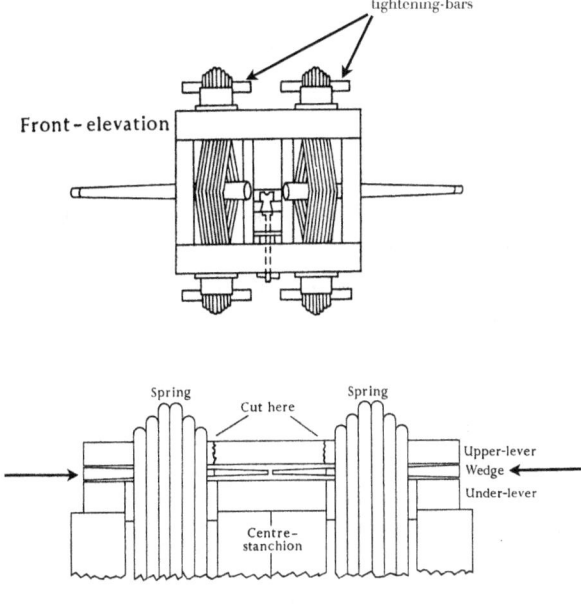

Fig. 3 Tightening bars in the standard design (Marsden, *Technical Treatises*, 57) vs. the wedges in Philo's modified engine (Marsden, *Technical Treatises*, 174)

At *Belopoeica* 59.1–62.21 Philo goes through each of these six points in sequence, arguing systematically for the excellence of his modified design. I shall focus on the argument for claim (1), the ability of the wedge engine to attain a longer range than a standard-design arrow shooter. Philo announces that he will demonstrate this by means of both "mechanical demonstrations" (*mēchanikai apodeixeis*) and "physical arguments" (*phusikoi logoi*, 59.8–10). His appeal to a mechanical demonstration runs as follows:

> Since larger circles overpower smaller ones which lie about the same center, as we have proven in our discussion of levers, and for the same reason [people] also move loads more easily with levers, when they place the fulcrum as near as possible to the load (for the fulcrum has the position of the center; hence when it is brought close to the load it diminishes the circle, through which it happens that ease of motion comes about) – the same thing, indeed, must be imagined [*noēteon*] in the case of the engine. For the arm is an in-

---

explaining how to construct the frame, remarks that "the frame, thus built and strung, is ready for shooting, but it is a little too ugly in appearance [τῇ ὄψει μικρῷ ἀπρεπέστερον], for it seems to have no head, as is the case." He goes on to explain how to build a cover for the frame, which "produces a good appearance [τήν τε ὄψιν ἀποδίδωσι καλήν], conceals the area around the wedges, and protects the spring" (66.30–31). For Philo there is no question of aesthetic factors *conflicting* with considerations of efficiency.

verted lever: the fulcrum is the part of it ⟨in the middle of the string⟩, the [imaginary] load [*baros*] is the bowstring, which is fastened to the end of the arm and sends forth the [actual] load [*baros*]. Now if one arranges the spring cords by spacing them out as much as possible starting from the heel, the fulcrum will clearly be nearer the load and the force [*dunamis*] further from the fulcrum. If this is done, the discharge of the missile will be violent and forceful.³⁵ (*Belopoeica* 59.11–22)

Philo begins by stating the proposition, which he says he has already proved in his work on levers (τὰ μοχλικά, the lost second book of the *Mechanical Syntaxis*), that "larger circles overpower smaller ones which lie about the same center" (*Belopoeica* 59.11–12). He then notes that it is easier to move heavy loads with a lever when the fulcrum is placed close to the load; this corresponds to the common experience that increasing the length of a lever arm makes it easier to move a load. The proposition about the circles explains this well-known fact: if one imagines the fulcrum to be the center of a circle and brings it close to the load, then the effort will move over a greater circle than the load, and thus "overpower" it (cf. fig. 4). Now, just as the fulcrum of the lever can be considered as the center of a circle, so the arm of the arrow-shooter can be imagined as a lever turning about a fulcrum at the center of the point of contact between the spring cords and the arm (fig. 5). The bowstring corresponds to the load; the effort is provided by the spring cords, which exert a rotatory force on the arm. What Philo proposes is to change the arrangement of the spring cords so that the distance between the load (the end of the arm) and the imaginary fulcrum is less (fig. 6b). This, he claims, will make the discharge more violent and forceful.³⁶

---

35 ἐπεὶ γὰρ οἱ μείζονες κύκλοι κρατοῦσιν τῶν ἐλασσόνων τῶν περὶ ⟨τὸ⟩ αὐτὸ κέντρον κειμένων, καθάπερ ἐν τοῖς μοχλικοῖς ἀπεδείξαμεν, διὰ δὲ τὸ ὅμοιον καὶ τοῖς μοχλοῖς ῥᾷον κινοῦσι τὰ βάρη, ὅταν ὡς ἐγγύτατα τοῦ βάρους τὸ ὑπομόχλιον θῶσιν (ἔχει γὰρ τὴν τοῦ κέντρου τάξιν· προσαγόμενον οὖν πρὸς τὸ βάρος [δὲ] ἐλασσοῖ κύκλον, δι' οὗ τὴν εὐκινησίαν συμβαίνει γίνεσθαι)· τὸ αὐτὸ δὴ νοητέον ἐστὶ καὶ περὶ τὸ ὄργανον. ὁ γὰρ ἀγκών ἐστι μοχλὸς ἀντεστραμμένος· ὑπομόχλιον μὲν γὰρ γίνεται τὸ ἐν ⟨μέσῳ τοῦ τόνου⟩ μέρος αὐτοῦ, ἡ δὲ τοξῖτις νευρὰ τὸ βάρος, ἥ τις ἐξ ἄκρου τοῦ ἀγκῶνος ἐχομένη τὸ βάρος ἐξαποστέλλει. ἐὰν οὖν τις τὸν τόνον ὅτι πλεῖστον ἀπ' ἀλλήλων διαστήσας ἀπὸ τῆς πτέρνης θῇ, δηλονότι τὸ μὲν ὑπομόχλιον ἔγγιον ἔσται τοῦ βάρους, ἡ δὲ δύναμις μακρότερον ἀπὸ τοῦ ὑπομοχλίου· τούτου δὲ γενομένου συμβήσεται τὴν ἐξαποστολὴν τοῦ βέλους σφοδρὰν καὶ βίαιον γίνεσθαι.

36 The point of remarking that the arm is an "inverted lever" (μοχλὸς ἀντεστραμμένος) is explained by the fact that the distance between the fulcrum and the effort (which must be imagined to be exerted at the end of the arm closest to the spring cords) is less than the distance between the fulcrum and the load (the bowstring). Marsden is quite right (*Technical Treatises*, 165) to compare the discussion of the unequal-armed balance (i.e., the steelyard) in the Aristotelian *Mechanical Problems* (854a10–11), where the steelyard is said to be a μοχλὸς ἀνεστραμμένος because in it the effort is regarded as exerted by the weight that is closer to the point of suspension.

Technē and Method in Ancient Artillery Construction — 635

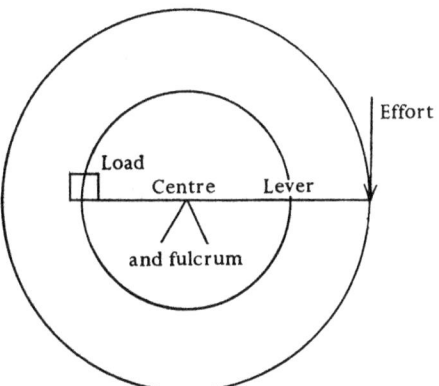

Fig. 4 The lever and concentric circles (Marsden, *Technical Treatises*, 165)

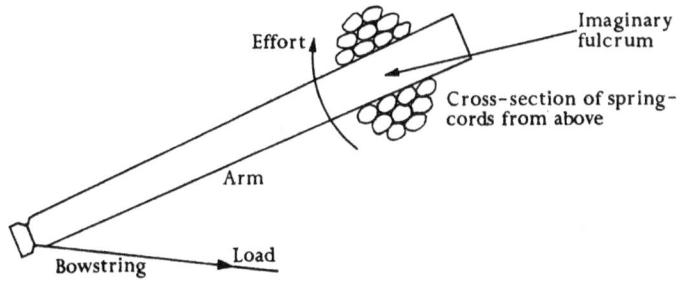

Fig. 5 The arm as a lever (Marsden, *Technical Treatises*, 166)

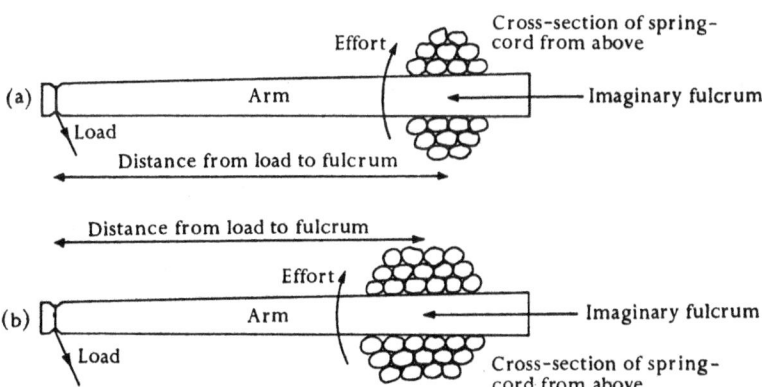

Fig. 6 Arrangement of the spring cords: (a) standard design; (b) Philo's modification (Marsden, *Technical Treatises*, 167)

Several points about this procedure are notable. First, despite the reference to "mechanical demonstrations," the crucial step is not a deductive inference, but the recognition that the arm of the artillery engine can be viewed as a lever for the purpose of the analysis. Philo makes no attempt to prove the proposition about greater circles "overpowering" smaller ones, but refers to the demonstration he has already given elsewhere. The conclusion – that the force will be increased if the spring cords are arranged in the way that Philo suggests – follows immediately once it is realized that the arm can be viewed as a lever. This kind of analysis of the components of complex machines in terms of simpler ones is characteristic of other Greek mechanical texts such as the Aristotelian *Mechanical Problems,* which begins by stating that the circle is the primary cause (the ἀρχὴ τῆς αἰτίης, 847b16–17) of all the wondrous phenomena that take place in mechanics; this amounts to the claim that all mechanical movements can be "reduced" (*anagetai*) to the circle, i.e., explained in terms of the movement of concentric circles.[37] After showing how the balance can be explained in terms of the circle and the lever in terms of the balance, the author goes on to show how a number of puzzling mechanical phenomena can be explained in terms of the lever. These analyses are carried out in a linguistically standardized way, with *baros* used to denote the load moved by the lever, *hupomochlion* the fulcrum, and *dunamis* or *ischus* the effort, i.e., the power or force that causes the movement.[38] In this way, the author of the *Mechanical Problems* employs the lever as a model for understanding the operation of a wide variety of more complex machines; this is exactly the procedure followed by Philo in the present passage.[39]

---

[37] *Mech.* 848a11–14: "Now the things that come about with the balance are reduced to the circle, those that come about with the lever are reduced to the balance, and practically everything else that is concerned with mechanical movements is reduced to the lever" (τὰ μὲν οὖν περὶ τὸν ζυγὸν γινόμενα εἰς τὸν κύκλον ἀνάγεται, τὰ δὲ περὶ τὸν μοχλὸν εἰς τὸν ζυγόν, τὰ δ' ἄλλα πάντα σχεδὸν τὰ περὶ τὰς κινήσεις τὰς μηχανικὰς εἰς τὸν μοχλόν).

[38] Cf. 850b10–16: "Why do the rowers in the middle of the ship contribute most to its movement? Is it because the oar is a lever [*mochlos*]? For the thole-pin becomes the fulcrum [*hupomochlion*], for it is fixed, and the load [*baros*] is the sea, which the oar pushes away; the mover [*ho kinōn*] of the lever is the sailor. And, always, the further the mover of the load [*baros*] is from the fulcrum [*hupomochlion*], the more load it moves; for the line from the center (i.e., the radius) is greater, and the thole-pin, which is the fulcrum, is the center." For further discussion of the use of model-based reasoning in the *Mechanical Problems* see my paper, "Structures of Argument and Concepts of Force in the Aristotelian *Mechanical Problems,*" *Early Science and Medicine* 14 (2009): 43–67.

[39] The linguistic standardization associated with the application of the lever model explains Philo's initially puzzling use of the term *baros* in the passage quoted above, whereby it first

Secondly, Philo's claim that larger circles "overpower" (*kratousi*) smaller ones placed around the same center expresses a principle that in one form or another was fundamental to much of ancient Greek theoretical mechanics. The author of the *Mechanical Problems* bases all his explanations on a proposition about concentric circles: given two such circles, a force applied to the circumference of the larger circle causes a greater movement than the same force applied to the circumference of the smaller circle, because a point on the larger circle covers a greater distance in the same time (assuming the two circles turn at the same angular speed); in this sense, the greater circle may be said to "overpower" the lesser.[40] Pappus of Alexandria, writing in the fourth century A.D., states that "it was proved in Archimedes' *On Balances* and the *Mechanics* of Philo and Hero that larger circles overpower (*katakratousin*) smaller circles when they turn around the same center."[41] How Philo proved the principle of concentric circles is not clear, though his tendency to associate force and movement suggests an approach closer to the *Mechanical Problems* than to that of Archimedes, which is based on an analysis of static equilibrium.[42] At any rate, the type of generalization about the lever that is important for the present passage ("the closer the load to the fulcrum, the more easily it is moved") is much closer to the *Mechanical Problems*, where the law of the lever in its exact quantitative form is only hinted at, than to Archimedes' works, in which the inverse proportionality of weights and distances from the center of the balance beam is stated,

---

refers the bowstring as the "imaginary" load (i.e., the load in terms of the lever model), then to the "actual" load, i.e., the missile that is discharged.
40 See *mech.* 848b1–849b22, esp. 848b3–5 and 849b19–22. At 850b2–6 the principle is applied to the lever. For further discussion of the author's application of the circular motion principle see Schiefsky, "Structures of Argument."
41 Pappus 1068.20 Hultsch: ἀπεδείχθη γὰρ ἐν τῷ περὶ ζυγῶν Ἀρχιμήδους καὶ τοῖς Φίλωνος καὶ Ἥρωνος μηχανικοῖς, ὅτι οἱ μείζονες κύκλοι κατακρατοῦσιν τῶν ἐλασσόνων κύκλων, ὅταν περὶ τὸ αὐτὸ κέντρον ἡ κύλισις αὐτῶν γίνηται. Cf. Hero Alex. *dioptra* 312.20 Schöne.
42 The argument of the *Mechanical Problems* is roughly as follows (for a fuller analysis see Schiefsky, "Structures of Argument"). Given two points on two concentric circles turning at the same angular speed, the point farther from the center will move more quickly, i.e., it will cover a greater distance in the same time. Now if we imagine the two points as lying at the ends of a balance beam or a lever, we can ask what downward force is exerted by the same body at each of the two points. If we consider the force exerted by a body to depend on its speed as well as its weight, then the force exerted by a body at the end of the longer radius will exceed the force exerted by a body of the same weight acting at the end of the shorter radius, for it will cover a greater distance in the same time. Thus a body placed at the end of the longer radius will "overpower" a body of the same weight that is closer to the center. Similarly, Philo closely associates the concepts of force (*dunamis*) and speed (*tachutēs*); cf. *Belopoeica* 69.1–5 and 73.8–13, with the discussion in the text below, pp. 638–42.

proved, and applied.⁴³ Philo makes no attempt to provide a quantitative estimate of the gain in performance that could be attained by the modification that he suggests; he remarks only that his engines shoot farther than standard ones of the same caliber (*Belopoeica* 59.4–8).

It is notable that Philo presents this analysis not only as a justification of his claim that the modified engine will achieve greater range, but also as a consideration which actually suggested that modification. Immediately after the passage just quoted, he goes on as follows:

> I saw that, in existing engines, the spring cords converged on each other, and that most artificers realized that this was what was harming the shooting, but were unable to alter it because this characteristic was naturally [*phusikōs*] inherent in the design and because it could not be removed in any other way. I tried, for this reason ... to change the form and the entire disposition. ... The most important innovation in this design is that the spring cords do not converge, but run parallel, and this, most of all, must produce long range.⁴⁴ (*Belopoeica* 59.23–31)

The rearrangement of the spring cords, suggested by the analysis of the arm as a lever, is the fundamental innovation that improves the range.

What of the "physical arguments" (*phusikoi logoi*) that are supposed to establish the superior range of Philo's wedge engine? Philo's reference to the spring cords converging "naturally" (*phusikōs*) in the passage just quoted suggests that by "physical" considerations he has in mind the inherent properties of materials such as animal sinew that provided the motive power for the engine. In an earlier passage criticizing the standard design, Philo remarks that one reason for its limited range is that the re-tightening necessary after a certain amount of use can only be accomplished by twisting the spring cords in a way that is

---

43 The author of the *Mechanical Problems* states the law of the lever in its exact quantitative form ("as the weight moved is to the moving [weight], so the distance [sc. from the fulcrum] is to the distance, inversely"; 850a39–b2). But he neither proves the inverse proportionality of weights and distances nor makes use of it in his analysis of machines (see Schiefsky, "Structures of Argument"). For Archimedes' proof of the lever principle see his *aequil.* 1.6–7.

44 ὁρῶν οὖν ἐν τοῖς προϋπάρχουσιν ὀργάνοις καταλλήλους πίπτοντας τοὺς τόνους, καὶ νοοῦντας μὲν τοὺς πλείστους τῶν ἀρχιτεκτόνων, ὅτι τὸ λυμαινόμενον τὴν τοξείαν τοῦτό ἐστιν, ἀδυνατοῦντας δὲ μεταθεῖναι διὰ τὸ φυσικῶς ἐν τῇ συντάξει τοῦτον ὑπάρχειν τὸν τρόπον καὶ ἄλλως ἂν μὴ δύνασθαι μεταχθῆναι, ἐπειράθην καὶ διὰ τοῦτο καὶ διὰ τὰ λοιπὰ τὰ προσόντα τῇ συντάξει δύσχρηστα μεταθεῖναι τὸ σχῆμα καὶ τὴν ὅλην διάθεσιν, ὅπως ἂν ἐγὼ βούλομαι τρόπον ἐν πᾶσιν ἀναστραφῶ μηδενὸς ἐμποδίζοντος ἡμῖν. τοῦτο μὲν οὖν μέγιστόν ἐστι τῶν εὑρημένων ἐν τῇδε τῇ συντάξει, τοὺς τόνους μὴ καταλλήλους, ἀλλὰ παραλλήλους πίπτειν, καὶ τοῦτο μάλιστα ἀναγκάζει μακροβολεῖν.

"contrary to nature" (*para phusin*) and which causes them to lose their natural (*kata phusin*) force and tension:

> In the heat of shooting and pulling back, the spring experiences a slackening and needs tightening again. The range of the shooting deteriorates because of the relaxation. But those who wish to tighten it cannot apply the re-stretching vertically and in a straight line, but do it by extra-twisting, imparting an extra-twist unnaturally [*para phusin*], greater than is suitable. ... The engine loses its springiness because the strands are huddled up into a thick spiral and the spring, becoming askew, is robbed of its natural [*kata phusin*] force and liveliness through the excessive extra-twisting.[45] (*Belopoeica* 58.7–16)

In contrast, he claims that his modified design, in which the re-stretching is achieved by driving in the wedges rather than by twisting, makes it possible to

> impart a very strong, natural [*kata phusin*] extra-tension, which will be enduring throughout and can in no way diminish. I maintain that, while there is a tendency in continuous shooting, as we have shown, for relaxation of the spring to occur on account of frequent pullings-back, I can produce additional stretch immediately, not by extra-twisting (for we have shown this to be injurious), but by stretching naturally [*kata phusin*] and vertically all the strands at once, just as they were originally stretched when the machine was being strung.[46] (*Belopoeica* 61.14–21)

It is thus a mark of the superiority of Philo's design that it takes full advantage of the natural properties of the components of the engine, rather than trying to work against them. The clear implication is that a machine functions better when its components are acting "according to nature" (*kata phusin*), even if the resultant effect produced by the machine may be viewed as in some sense "against nature" (*para phusin*; for example, in the sense that it causes a heavy body to move in a way that is contrary to its natural tendencies). However the

---

45 ἐν γὰρ ταῖς τοξείαις καὶ ταῖς πυκναῖς καταγωγαῖς χάλασμα λαβὼν ὁ τόνος ἐπιτάσεως πάλιν προσδεῖται. τὸ γὰρ τῆς τοξείας μῆκος ἀπολήγει διὰ τὴν γεγενημένην ἄνεσιν. συμβαίνει οὖν βουλομένους ἐπιτείνειν αὐτὸν εἰς ὀρθὸν μὲν μὴ δύνασθαι μηδὲ κατ' εὐθεῖαν διδόναι τὴν ἐπέκτασιν, ἐπιστρέφοντας δὲ τοῦτο ποιεῖν διδόντας παρὰ φύσιν ⟨πλείονα⟩ τῆς καθηκούσης ἐπιστροφῆς, ὑπολαμβάνοντας μὲν βοηθεῖν, μέγα δὲ λυμαινομένους τὴν τάσιν καὶ ποιοῦντας, λέγω, τὴν τοξείαν βραδυτέραν καὶ ἀσθενεστέραν ταῖς πληγαῖς, ἀτόνου τοῦ ὀργάνου γινομένου διὰ τὸ τοὺς στήμονας εἰς πυκνὴν ἕλικα ἀνάγεσθαι καὶ πλάγιον γεγονότα ⟨τὸν τόνον⟩ τοῦ βιαίου καὶ εὐτόνου ⟨τοῦ⟩ κατὰ φύσιν ἐστερῆσθαι διὰ τὴν ὑπεράγουσαν ἐπιστροφήν.
46 ἐπιστροφήν τε δώσειν τὴν ὑπάρχουσαν κατὰ φύσιν κρατίστην, μένουσαν δι' ὅλου καὶ μεταπεσεῖν οὐθενὶ τρόπῳ δυναμένην. ἐπεὶ δὲ καὶ ἐν ταῖς συνεχέσι τοξείαις συμβαίνει, καθότι δεδηλώκαμεν, ἀνέσεις γίνεσθαι τοῦ τόνου διὰ τὰς πυκνὰς καταγωγάς, ἐπεντείνειν παραχρῆμα μὴ ἐπιστροφὴν διδούς (τοῦτο μὲν γὰρ ἐδείξαμεν βλαβερὸν ὑπάρχον), ἀλλὰ κατὰ φύσιν εἰς ὀρθὸν ἐντείνων τοὺς στήμονας πάντας ἅμα, καθάπερ ἐξαρτυομένου τὴν ἀρχὴν ἐξετάθησαν.

effect produced by the machine is viewed, it can be understood in terms of the natural behavior of the machine's components.[47] In order to produce the most efficient design, the artillery builder needs knowledge of the natural properties of the materials with which he is working, as well as the principles of theoretical mechanics.

2 The Bronze-Spring Engine (*chalkotonon*)

A second, more far-reaching modification of standard-design artillery that Philo discusses is the bronze-spring engine or *chalkotonon*, in which motive power was supplied by specially manufactured bronze plates rather than animal sinew (fig. 7). When the string is pulled back, the ends of the arms press against the bronze springs; these then recoil and produce the forward thrust when the trigger is pulled. Philo credits Ctesibius, the brilliant Alexandrian engineer of the third century B.C., with the original invention of the bronze-spring engine, but he also claims to have made substantial improvements to Ctesibius' design.[48] He begins his discussion by calling for a general inquiry into the problem of attaining long range:

> As we intend to recount the peculiarity of the springs, we think it a good idea, in this case also, to examine the old engines first and to reckon what is the cause [*aitia*] of their ability to hurl the missile over a long range. We shall not make the inquiry about minor causes [*mikras aitias*], as mentioned above – lengthening or contracting of the springs, extending or shortening of the arms, making them lean further back or further forward, or the merit of sinew or hair. These have been investigated before, as I have said previously; they are public and common to everyone, and have already been tested [*pepeiramena*] frequently and thoroughly. Now we must make a thoroughgoing examination of the problem as a whole

---

[47] On this understanding of the working of machines, the *technē* of mechanics can be understood as completing what nature leaves unfinished, a common Aristotelian view of the relationship between art and nature (e.g., *phys.* 199a15–17). I develop this view in my paper "Art and Nature in Ancient Mechanics," in Bernadette Bensaude-Vincent and William R. Newman (eds.), *The Artificial and the Natural: An Evolving Polarity* (Cambridge, Mass.: The MIT Press, 2007), 67–108.

[48] Ctesibius (*fl.* ca. 270 B.C.) is credited with important inventions in many different areas of technology, including the water pump, water organ, water clocks, and automata, as well as various types of artillery engine. On his achievements see especially Gille, *Les mécaniciens grecs*, and Aage G. Drachmann, *Ktesibios, Philon, and Heron: A Study in Ancient Pneumatics* (Copenhagen: E. Munksgaard, 1948). From the way in which Philo introduces his discussion of the bronze-spring engine, it is clear that he relies on the reports of others for his knowledge of Ctesibius' work; cf. *Belopoeica* 67.28–68.2.

[*peri tōn katholou*], since we intend to introduce a completely revolutionary proposition that is both unique in its design and far different from previous ones.[49] (*Belopoeica* 68.7-17)

The general inquiry described here is contrasted with the investigation of the effects of varying the dimensions of different components: such "minor causes" (*mikrai aitiai*) have been thoroughly investigated and tested (*pepeiramena*) by experience. With this clear reference to the development of the standard method Philo signals that that method too rests on an understanding of causes, in a sense. But if the goal is to make a fundamental improvement to the range of an artillery engine a deeper understanding of the principles that underlie its operation is necessary. Indeed, Philo's emphasis on the need for a general or universal (*katholou*) inquiry into the cause (*aitia*) of long range is strikingly reminiscent of Aristotle's doctrine that that it is the grasp of the universal (*katholou*) and the cause (*aitia*) that marks the transition from experience (*empeiria*) to art (*technē*).[50]

Philo's discussion of the problem of long range is remarkable, both for the dynamical assumptions on which it is based and for the consistency with which he applies them to the analysis of a technological device. He begins by reducing the problem to an analysis of the force provided by the springs. The force with which the arms are propelled determines the speed at which they move and therefore the range: the quicker the movement of the arms, the faster the missile travels, and the longer the range (68.18–29). The springs, in turn, are responsible for the force with which the arms are propelled; the problem thus becomes one of analyzing the force produced by the springs. Now the arm of a piece of torsion artillery, when inserted into the bundle of spring cords, is situated between two "half-springs" or *hēmitonia* (fig. 8). Philo claims that the force with which the arm is moved depends only on one of these half-springs, not on both, because the two half-springs exert equal forces (*dunameis*) and move at the same speed:

---

49 μέλλοντες οὖν περὶ τῆς τῶν τόνων ἰδιότητος ἀπαγγελεῖν καλῶς ἔχειν ὑπελαμβάνομεν καὶ ἐπὶ τούτου πρῶτον ἐπισκέψασθαι περὶ τῶν ἀρχαίων ὀργάνων καὶ συμβάλλειν, τίς ἐστιν ἡ αἰτία τοῦ μακρὰν ἀποστέλλειν δύνασθαι τὸ βέλος, μὴ περὶ μικρὰς αἰτίας τὴν θεωρίαν ποιούμενοι καθάπερ ἀνώτερον δεδηλώκαμεν, περὶ τὸ μακροτονεῖν ἢ συναιρεῖν τὸ μῆκος τῶν τόνων ἢ τοὺς ἀγκῶνας ἐπεκτείνειν ἢ συστέλλειν ἢ προσεστηκότας ἢ ἀναπεπτωκότας μᾶλλον ποιεῖν ἢ τὴν τῶν νεύρων ἢ τριχὸς ἀρετήν· ταῦτα μὲν γὰρ καὶ προεζήτηται, καθάπερ εἶπον ἐν τοῖς πρότερον, καὶ ἐν μέσῳ κείμενα κοινὰ πᾶσιν ὑπάρχει πολλάκις ἤδη καὶ παντοδαπῶς πεπειραμένα· νῦν δὲ ὁλοσχερῆ τινα δεῖ τὴν ἐπίσκεψιν περὶ τῶν καθόλου ποιήσασθαι μέλλοντας δὴ καὶ προσάγειν ὁλοσχερές τι θεώρημα καὶ ἴδιον τῇ διαθέσει καὶ πολὺ παρηλλαγμένον τῶν πρότερον.
50 *Metaph.* A 980a27–981b6; *an. post.* 100a3–b5.

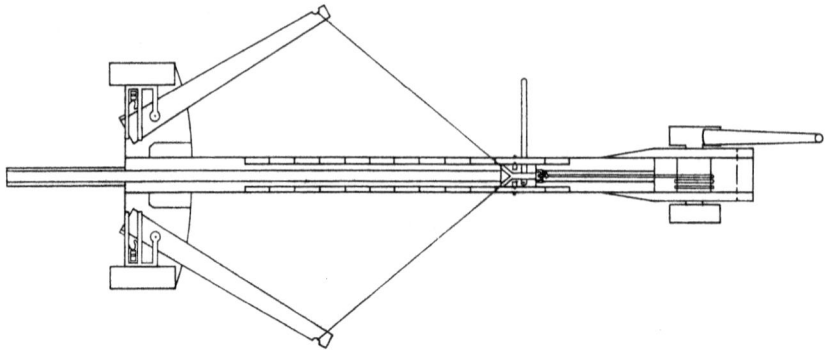

(a) From above, the frame's upper cross-beam removed

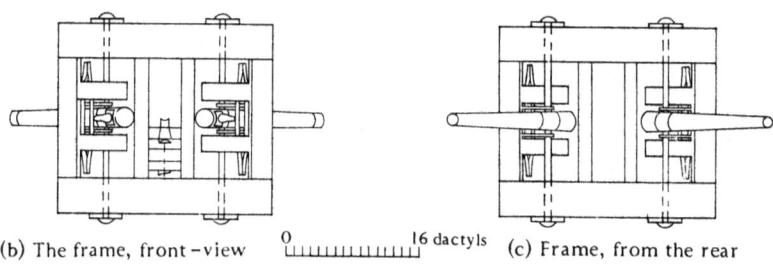

(b) The frame, front-view   0 |⎵⎵⎵⎵⎵⎵⎵⎵⎵⎵⎵⎵⎵⎵⎵| 16 dactyls   (c) Frame, from the rear

Fig. 7 Ctesibius' bronze-spring catapult (Marsden, *Technical Treatises*, 174)

> On the pulling of the trigger, the two forces [*dunameis*] of the half-springs simultaneously move the bowstring, since they have exactly equal speeds [*isotacheis*] because they are composed of equal and like forces [*dunameis*]. Now, the one force would not add to the speed of the arm unless it were greater than the other; for then it would overpower [*katakratoiē*] the lesser and increase the speed.[51] (*Belopoeica* 69.1–5)

To support this claim Philo presents a thought experiment, based on what he takes to be the acknowledged fact that heavier weights fall faster than lighter ones:

---

[51] ἔν τε τῷ σχάσματι τὴν τοξῖτιν ἅμα συμβαίνει ἀμφοτέρας τὰς τῶν ἡμιτονίων δυνάμεις κινεῖν ἰσοταχεῖς αὐταῖς συνυπαρχούσας διὰ τὸ ἐξ ἴσων καὶ ὁμοίων δυνάμεων συνεστάναι. οὐκ ἂν οὖν πρὸς τὸ τάχος τοῦ ἀγκῶνος ἡ μία συμβάλλοιτο δύναμις, εἰ μὴ μείζων εἴη τῆς ἄλλης· οὕτω γὰρ ἂν κατακρατοίη τῆς ἐλάσσονος καὶ ἐπισυνάπτοι τῷ τάχει.

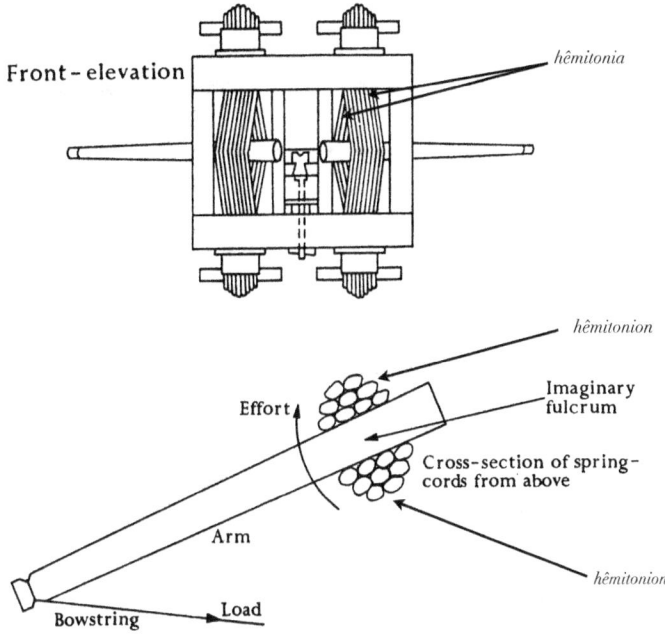

Fig. 8 "Half-springs" (*hēmitonia*), based on Marsden, *Technical Treatises*, 57 and 166

> If one takes two weights, alike in substance and shape, the one of one mina, the other of two minae, and lets them drop simultaneously from a height, I maintain that the two-minae weight will drop far more quickly. In the case of other weights the same account holds: the larger always falls proportionately [*ana logon*] more quickly than the smaller, whether because the greater weight, as some of the natural philosophers [*phusikoi*] say, can displace and disrupt the air more easily, or because greater inclination [*rhopē*] follows upon greater weight [*baros*], and the greater inclination [*rhopē*] increases the motion in the vertical direction.[52] (*Belopoeica* 69.8–14)

A weight of two minae will fall "far more quickly" than a one-mina weight, and the larger falls "proportionately" more quickly than the smaller.[53] Philo indicates

---

[52] ἐὰν γάρ τις λαβὼν δύο βάρη, ὅμοια τῷ γένει καὶ τῷ σχήματι, τὸ μὲν μναιαῖον, τὸ δ' ἄλλο δίμνουν, ἅμα ἀφ' ὕψους ἀφῇ φέρεσθαι, λέγω ὅτι τὸ δίμνουν παρὰ πολὺ τάχιον οἰσθήσεται· καὶ ἐπὶ τῶν ἄλλων δὲ βαρῶν ὁ αὐτὸς ὑπάρχει λόγος, ἀνὰ λόγον ἀεὶ τὸ μεῖζον τοῦ ἐλάσσονος ὡς τάχιον φέρεσθαι, εἴθ' ὅτι τὸ μεῖζον βάρος, καθάπερ φασί τινες τῶν φυσικῶν, μᾶλλον ἐκρομβεῖν δύναται καὶ διαστέλλειν τὸν ἀέρα, εἴθ' ὅτι τῷ μείζονι βάρει καὶ ῥοπὴ πλείων παρέπεται, ⟨ἡ⟩ δὲ πλείων ῥοπὴ μᾶλλον αὔξει τὴν κατὰ κάθετον φοράν.

[53] This of course does not imply that a two-mina weight falls twice as fast as a one-mina weight; much more plausibly, it can be taken to mean that the speed of fall increases in constant ratio to the increase in the weight.

his familiarity with various explanations of this phenomenon given by natural philosophers without committing himself to a single alternative.[54] But then he goes on to develop the example further: a two-mina weight, he argues, will fall more swiftly not only than a one-mina weight, but also more swiftly than two one-mina weights joined together, and indeed more swiftly than several one-mina weights joined together:

> Again, because what has been said comes about, if one takes two weights of one mina and, having connected and fastened them together as well as possible, lets them drop, I affirm that the weight of two minae will once more drop more quickly than the two weights of one mina joined together. Even if three or more are connected together, they will likewise drop more slowly. It becomes clear from this that, when several forces [*dunameis*] equal to each other are connected together, their combined speed will not exceed the natural [*phusikē*] speed belonging to one weight alone.[55] (*Belopoeica* 69.14–20)

On this view, each weight is considered to have a natural force or capacity (*dunamis*) to fall at a certain speed; even if one fastens together two one-mina weights "as well as possible," the combination will not possess the natural *dunamis* of a two-minae weight. This, Philo claims, supports his view that only one of the half-springs contributes to the speed of the arm's motion. The two half-springs correspond to the two equal weights, which when joined together do not augment the natural motion of one weight alone: "Since this is so, it has been clearly shown that the one half-spring does not contribute to the movement of the arm at all, because its speed is equal to that of the other" (69.20–22).[56] Now it is impossible to do away with one half-spring in the case of standard torsion artillery, since the arm is wedged in between the two half-springs. This, Philo suggests, is the motivation for the introduction of the bronze-spring en-

---

54 Cf. Arist. *phys.* 216a11–21, esp. 18–20, on the reason why heavier objects fall more quickly through a medium: "for the object that is moving or thrown divides either by its shape, or by the inclination which it possesses" (ἢ γὰρ σχήματι διαιρεῖ, ἢ ῥοπῇ ἣν ἔχει τὸ φερόμενον ἢ τὸ ἀφεθέν). The problem of free fall was investigated further by Strato of Lampsacus in the early third century B.C.; see fr. 73 Wehrli (= Simpl. *in phys.* 916.4–31) for a report of a test similar to Philo's.

55 πάλιν, ὅτι γίνεται τὸ ῥηθέν, ἐὰν δύο βάρη λαβὼν μναϊαῖα καὶ συνθεὶς ἔπειτα καὶ συναιωρήσας ὡς δυνατόν τις ἀφῇ φέρεσθαι· λέγω δή, ὅτι ταχύτερον οἰσθήσεται πάλιν τὸ δίμνουν βάρος ἢ τὰ δύο αὐτοῖς συγκείμενα μναϊαῖα βάρη· βραδύτερον δέ κἂν τρία καὶ ἔτι πλείονα συντεθῇ, ταὐτὸ ποιήσει. φανερὸν οὖν γίνεται καὶ ἐκ τούτου, διότι πλειόνων δυνάμεων ἐπὶ τὸ αὐτὸ συντεθεισῶν, ἴσων δὲ αὐταῖς οὐσῶν, οὐθὲν ἡ φορὰ κατὰ κοινὸν μᾶλλον αὔξεται τῆς ὑποκειμένης φυσικῆς τῷ ἑνὶ μόνον βάρει.

56 τούτων δὴ τοιούτων ὑπαρχόντων ἐδείχθη σαφῶς τὸ ἓν ἡμιτόνιον μηθὲν συνεργοῦν τῇ τοῦ ἀγκῶνος φορᾷ διὰ τὸ ἰσοταχὲς τῷ ἄλλῳ.

gine, in which a single spring forged from bronze plates provides all the motive power for each arm.

It is easy to dismiss Philo's thought experiment as a piece of purely armchair speculation. To be sure, Philo's claims are unlikely to be based on actual empirical tests with falling bodies. However, the argument is quite Aristotelian in spirit. The basic idea is that two one-mina weights when joined together do not unite into a single substance with the natural motion of a two-mina weight; this would require a genuine fusion of the two weights into a single nature (*phusis*).[57] Similarly, two half-springs acting together will not produce the same motion as a single spring with twice the resiliency. From the point of view of the Aristotelian distinction between forced and natural motion, it is remarkable that Philo draws an analogy with the natural motion of falling bodies to explain the motion of the arms of the artillery engine. There is no suggestion of any fundamental difference between the natural motion of falling bodies and the forced motion of the arms; instead, understanding the former is crucial for understanding the latter. A knowledge of "physical arguments" (*phusikoi logoi*) that includes the behavior of the natural motion of bodies is thus essential to understanding the forced motion of the arms of an artillery engine, and so to improving its design.

Philo applies the analogy with falling bodies quite consistently. At *Belopoeica* 72.24–73.20 he goes so far as to criticize Ctesibius for using more than one bronze spring for each of the arms in his design. This, he says, was due to a failure to grasp the truth revealed by the thought experiment; that is, Ctesibius held the mistaken belief that "more forces [*dunameis*] of equal speed [*isotacheis*] and alike in strength [*ischus*], when joined together, would produce a more violent

---

[57] Similar issues about the additivity of forces are discussed by the Aristotelian commentators; cf. Shmuel Sambursky, *The Physical World of Late Antiquity* (London: Routledge, 1962), 65–68. I note in passing that Philo's analysis shows that a famous thought experiment found in early modern authors such as Galileo was not nearly as conclusive as it is sometimes taken to be. According to this thought experiment, if we imagine two one-pound weights first falling side by side, then coming together as they fall, on the Aristotelian view we would have to conclude that they would suddenly speed up and fall twice as fast – a conclusion so implausible that it would imply the falsity of the Aristotelian view. But Philo's passage shows that an Aristotelian could quite easily accept the absurdity of the conclusion but reject the inference leading to it: indeed, he might say, two one-pound weights joined together do not fall as quickly as a single two-pound weight, but this is because they do not make up a single body with the corresponding natural motion. Only if the two weights really fused into a single two-pound weight would an Aristotelian be committed to the claim that the speed of motion would double; but such a fusion, it could be argued, goes well beyond what was envisioned in the original thought experiment.

thrust [*bia*]" (72.26–28). Just as only one of the half-springs contributes to the movement of the arm in a standard-design engine, so only one bronze spring will contribute to the movement of the arm in the bronze-spring engine; the inclusion of multiple springs just makes the arms harder to pull back, while contributing nothing to the range. Underlying all of Philo's dynamical reflections is a consistent association between force and speed: a greater force is assumed to correspond directly to a greater speed of movement. Now if one thinks of a force simply as a capacity to produce a motion of a certain speed, it might seem reasonable to suppose that a combination of forces, each of which produces a motion of the same speed, will not produce a movement that is any faster than each of the forces taken individually. This at any rate seems to be the idea behind the following passage, from Philo's criticism of Ctesibius:

> Many forces [*dunameis*] of equal speed [*isotacheis*], when joined together, and when all are being compressed, produce a resistance proportional to the sum of their forces, so that the sum total of thrust [*bia*] is considerable. But, in their recoil, as there is no difference in their speed, they all move simultaneously. How then can one of them alone acquire additional speed, when it, too, has the same speed?[58] (*Belopoeica* 73.8–13)

Philo's view is certainly strange from the point of view of modern (i.e., Newtonian) mechanics, and might seem open to obvious criticism even from an ancient perspective. Surely, one might think, if two bronze springs (or half-springs in a torsion engine) acting together dispatch the missile quickly, one will dispatch it with half the speed, or perhaps not provide enough force to move it at all.[59] But this misses the point of Philo's association between force and speed. Philo assumes that both springs have the capacity (*dunamis*) to move the arm at a certain speed when employed individually; if this is so, he might ask, why suppose that the two together would move it more quickly? After all, it is not as though one spring will move more quickly than the other when it recoils.

Philo's discussion of the construction of the bronze springs themselves at *Belopoeica* 70.35–72.4 provides further information about his conception of the

---

[58] αἱ γὰρ πολλαὶ δυνάμεις, ἰσοταχεῖς δ' οὖσαι, ὅταν αὐταῖς συζευχθῶσιν, ἐπισπώμεναι μὲν ἅπασαι τὴν ἀντίβασιν ποιοῦνται κατὰ τὴν ὑπάρχουσαν αὐταῖς δύναμιν, ὥστε πολλὴν τῆς βίας ἄθροισιν γενέσθαι· ἀναπίπτουσαι δὲ καὶ οὐθὲν ἀλλήλων τῷ τάχει διαφέρουσαι πᾶσαι ἅμα φέρονται. πῶς οὖν δυνατόν ἐστι προσλαμβάνειν τάχος τὴν μίαν τούτων μόνην, ἔχουσαν καὶ αὐτὴν τὸ ὅμοιον τάχος;

[59] Cf. Aristotle's famous analysis of forced motion in *phys.* 249b27–250b7: if "force" (*ischus*, *dunamis*) A moves load B over distance D in time T, then 2 A will move B over distance 2D in time T, but it does not follow that A/2 will move B at all.

role of *phusikoi logoi* in artillery design. His chief concern is to respond to the objection that bronze is not naturally resilient in the way that is necessary if it is to be used for powering an artillery engine. As an example of material with the necessary degree of resiliency Philo cites the iron of "so-called Celtic and Spanish swords" (71.9), which can be bent repeatedly but always return to their normal straightness. Philo remarks that the reason (*aitia*, 71.17) why these swords behave in this way has been the subject of special investigation; the answer lies in the way they are made, which involves beating them gently to create hardness on the outside while leaving the middle soft and flexible. The need to justify this procedure leads Philo to appeal once again to the views of natural philosophers:

> Firings soften iron and bronze, because the bodies [*sōmata*] become less dense, as they say; while coolings and beatings harden them, for both processes cause the bodies [*sōmata*] to become tightly packed, because their parts [*moria*][60] move closer to one another and the interstices of void are removed.[61] (*Belopoeica* 71.27–31)

The combination of a corpuscularian theory of matter with the hypothesis of interspersed void is characteristic of Hellenistic matter theories, especially those associated with Strato of Lampsacus.[62] As in the case of the wedge engine, a knowledge of the natural constitution of the components of the machine is necessary for understanding its operation; this is best provided, at least in some cases, by an up-to-date knowledge of the work of natural philosophers.[63]

---

60 These "parts" are the small pieces of matter which, along with the interspersed void space, make up the larger bodies.
61 αἱ μὲν οὖν πυρώσεις τόν τε σίδηρον καὶ χαλκὸν μαλακύνουσιν ἀραιουμένων τῶν σωμάτων, ὥς φασιν, αἱ δὲ ψύξεις καὶ κροτήσεις σκληρύνουσιν· ἀμφότερα γὰρ αἴτια γίνεται τοῦ πυκνοῦσθαι τὰ σώματα συντρεχόντων τῶν μορίων πρὸς ἄλληλα καὶ τῆς τοῦ κενοῦ περιπλοκῆς αἰρομένης.
62 Strato adopted a theory of interspersed void to explain such phenomena as the transmission of light and heat through substances such as air and water; see fr. Wehrli (= Simpl. *in phys.* 693.10–18). Philo's term for "resilience" is *eutonia*, ascribed to the iron of the Celtic and Spanish swords at *Belopoeica* 71.22 and 72.1; it is notable that this concept plays a large role in the theory of matter propounded by Hero of Alexandria in the introduction to the *Pneumatics* (proem lines 76, 249), which itself goes back ultimately to early Hellenistic antecedents (including Strato).
63 That *phusikoi logoi* are in question here is made absolutely clear by the conclusion of the discussion of the resiliency of bronze at 72.1–4: "Let this be enough about bronze-spring engines and the construction entailed therein, lest we inadvertently digress too far and enter deeper into physical arguments [εἰς τοὺς φυσικοὺς ... λόγους]."

3 Ctesibius' Air-Spring Engine (*aerotonos*)

A third innovation in artillery construction that Philo presents as drawing on theoretical knowledge is the air-spring engine or *aerotonos* also invented by Ctesibius (*Belopoeica* 77.9–78.22). In this design, precisely manufactured bronze cylinders and pistons are attached to the arms of a stone-throwing engine. When the arms are pulled back, the pistons are pressed into the cylinders, compressing the air in them; on pulling the trigger, the impulse of the air to return to its natural state pushes the pistons out with great force and propels the arms forward. Philo emphasizes the fact that the air-spring engine has both a "physical" and a "mechanical" character (μηχανικὴν δὲ πάνυ καὶ φυσικὴν εἶχε διάθεσιν, 77.11–12). According to his account, Ctesibius based the invention on his knowledge of pneumatics, the study of the behavior of air in closed vessels, especially under compression; drawing on this knowledge and his experience in mechanics, he realized that the power of compressed air could impart high speed to the arms (77.12–18). The chief technical challenge involved was the construction of the airtight cylinders and pistons. Philo describes the procedure followed in some detail, and responds to those who might doubt the possibility of such construction by citing the example of a well-known mechanical device that also makes use of cylinders and pistons, the water organ (77.27).[64] The "mechanical" excellence of the air-spring engine is reflected in the fact that Ctesibius aimed not only at long range but also at an attractive design.[65] Philo relates how Ctesibius demonstrated the natural compressibility and force of air, as well as the possibility of constructing cylinders with the requisite properties:

> Ctesibius, it was explained to us,[66] demonstrated the nature [*phusis*] of air – namely that it has violent and swift movement – and, at the same time, the fabrication entailed by the cylinders which contain the air; he smeared the cylinder with carpenter's adhesive, set a protective edging over the circular mouth [sc. of the cylinder], and with wedge and mallet drove in the piston with very great force. It was possible to see the piston making gradual progress; but, when the air inside was once compressed, it gave way no more even to the strongest blow on the wedge. On the application of force and the removal of the wedge, the

---

[64] Philo also notes a parallel with medicine: the cylinders were shaped "like doctors' medicine-boxes" (ὅμοια πυξίσιν ἰατρικαῖς, 77.18–19).

[65] "He aimed not only at strength, but also fine appearance so that it should seem to be a [genuine] instrument" (οὐ μόνον τῆς ἰσχύος, ἀλλὰ καὶ τῆς ὄψεως στοχαζόμενος, ὅπως ὀργανικὴ φαίνηται, 78.11–12).

[66] The Greek is peculiar (see next n.), but should be understood in light of Philo's reliance on others for information about Ctesibius; cf. n. 48 above.

role of *phusikoi logoi* in artillery design. His chief concern is to respond to the objection that bronze is not naturally resilient in the way that is necessary if it is to be used for powering an artillery engine. As an example of material with the necessary degree of resiliency Philo cites the iron of "so-called Celtic and Spanish swords" (71.9), which can be bent repeatedly but always return to their normal straightness. Philo remarks that the reason (*aitia*, 71.17) why these swords behave in this way has been the subject of special investigation; the answer lies in the way they are made, which involves beating them gently to create hardness on the outside while leaving the middle soft and flexible. The need to justify this procedure leads Philo to appeal once again to the views of natural philosophers:

> Firings soften iron and bronze, because the bodies [*sōmata*] become less dense, as they say; while coolings and beatings harden them, for both processes cause the bodies [*sōmata*] to become tightly packed, because their parts [*moria*][60] move closer to one another and the interstices of void are removed.[61] (*Belopoeica* 71.27–31)

The combination of a corpuscularian theory of matter with the hypothesis of interspersed void is characteristic of Hellenistic matter theories, especially those associated with Strato of Lampsacus.[62] As in the case of the wedge engine, a knowledge of the natural constitution of the components of the machine is necessary for understanding its operation; this is best provided, at least in some cases, by an up-to-date knowledge of the work of natural philosophers.[63]

---

60 These "parts" are the small pieces of matter which, along with the interspersed void space, make up the larger bodies.
61 αἱ μὲν οὖν πυρώσεις τόν τε σίδηρον καὶ χαλκὸν μαλακύνουσιν ἀραιουμένων τῶν σωμάτων, ὥς φασιν, αἱ δὲ ψύξεις καὶ κροτήσεις σκληρύνουσιν· ἀμφότερα γὰρ αἴτια γίνεται τοῦ πυκνοῦσθαι τὰ σώματα συντρεχόντων τῶν μορίων πρὸς ἄλληλα καὶ τῆς τοῦ κενοῦ περιπλοκῆς αἰρομένης.
62 Strato adopted a theory of interspersed void to explain such phenomena as the transmission of light and heat through substances such as air and water; see fr. Wehrli (= Simpl. *in phys.* 693.10–18). Philo's term for "resilience" is *eutonia*, ascribed to the iron of the Celtic and Spanish swords at *Belopoeica* 71.22 and 72.1; it is notable that this concept plays a large role in the theory of matter propounded by Hero of Alexandria in the introduction to the *Pneumatics* (proem lines 76, 249), which itself goes back ultimately to early Hellenistic antecedents (including Strato).
63 That *phusikoi logoi* are in question here is made absolutely clear by the conclusion of the discussion of the resiliency of bronze at 72.1–4: "Let this be enough about bronze-spring engines and the construction entailed therein, lest we inadvertently digress too far and enter deeper into physical arguments [εἰς τοὺς φυσικοὺς ... λόγους]."

3 Ctesibius' Air-Spring Engine (*aerotonos*)

A third innovation in artillery construction that Philo presents as drawing on theoretical knowledge is the air-spring engine or *aerotonos* also invented by Ctesibius (*Belopoeica* 77.9–78.22). In this design, precisely manufactured bronze cylinders and pistons are attached to the arms of a stone-throwing engine. When the arms are pulled back, the pistons are pressed into the cylinders, compressing the air in them; on pulling the trigger, the impulse of the air to return to its natural state pushes the pistons out with great force and propels the arms forward. Philo emphasizes the fact that the air-spring engine has both a "physical" and a "mechanical" character (μηχανικὴν δὲ πάνυ καὶ φυσικὴν εἶχε διάθεσιν, 77.11–12). According to his account, Ctesibius based the invention on his knowledge of pneumatics, the study of the behavior of air in closed vessels, especially under compression; drawing on this knowledge and his experience in mechanics, he realized that the power of compressed air could impart high speed to the arms (77.12–18). The chief technical challenge involved was the construction of the airtight cylinders and pistons. Philo describes the procedure followed in some detail, and responds to those who might doubt the possibility of such construction by citing the example of a well-known mechanical device that also makes use of cylinders and pistons, the water organ (77.27).[64] The "mechanical" excellence of the air-spring engine is reflected in the fact that Ctesibius aimed not only at long range but also at an attractive design.[65] Philo relates how Ctesibius demonstrated the natural compressibility and force of air, as well as the possibility of constructing cylinders with the requisite properties:

> Ctesibius, it was explained to us,[66] demonstrated the nature [*phusis*] of air – namely that it has violent and swift movement – and, at the same time, the fabrication entailed by the cylinders which contain the air; he smeared the cylinder with carpenter's adhesive, set a protective edging over the circular mouth [sc. of the cylinder], and with wedge and mallet drove in the piston with very great force. It was possible to see the piston making gradual progress; but, when the air inside was once compressed, it gave way no more even to the strongest blow on the wedge. On the application of force and the removal of the wedge, the

---

[64] Philo also notes a parallel with medicine: the cylinders were shaped "like doctors' medicine-boxes" (ὅμοια πυξίσιν ἰατρικαῖς, 77.18–19).
[65] "He aimed not only at strength, but also fine appearance so that it should seem to be a [genuine] instrument" (οὐ μόνον τῆς ἰσχύος, ἀλλὰ καὶ τῆς ὄψεως στοχαζόμενος, ὅπως ὀργανικὴ φαίνηται, 78.11–12).
[66] The Greek is peculiar (see next n.), but should be understood in light of Philo's reliance on others for information about Ctesibius; cf. n. 48 above.

piston shot out with great force from the cylinder. It often happened that fire came out, too, since the air rubbed against the vessel in the speed of its motion.[67] (*Belopoeica* 77.29–78.7)

Once again the construction of a mechanical device is said to depend on knowledge of physics; the speed of the arms is traced back to the natural tendency of air to expand when it has been compressed. The "physical" and "mechanical" character of the air-spring engine thus results from the fact that it makes creative use of natural principles to achieve a beneficial and aesthetically pleasing result.

The three examples[68] we have discussed in this section suggest the following conclusions about Philo's conception of the role of mechanical and physical theory in artillery construction. First, mechanical theory is not applied deductively; rather, a knowledge of theoretical principles guides the practitioner in his attempt to understand the working of a complex piece of artillery and improve its design, by suggesting analogies between the components of complex machines and simple machines such as the lever. Second, the appeal to theory does not result in greater quantitative precision; Philo's appeal to the law of the lever is much closer to the Aristotelian *Mechanical Problems* than to Archimedes' quantitatively precise formulation. Third, a knowledge of "physical arguments" dealing with such phenomena as the resiliency of materials, the natural motion of falling bodies, and the compressibility of air is essential to the process of discovery; this is because the optimal functioning of a machine depends on its components acting in ways that are natural for them. Far from suggesting an opposition between the "mechanical" and the "natural," Philo argues that their proper combination is one hallmark of an excellent machine like Ctesibius' *aerotonos*.

---

67 ἐπεδείκνυτο δὲ ἡμῖν ὁ Κτησίβιος παραδεικνύων τήν τε τοῦ ἀέρος φύσιν, ὡς ἰσχυρὰν ἔχει καὶ ὀξεῖαν κίνησιν, καὶ ἅμα τὴν περὶ τὰ ἀγγεῖα ὑπάρχουσαν χειρουργίαν τὰ τὸν ἀέρα συνέχοντα, περιθεὶς κολλητήριον τεκτονικὸν περὶ τὸ ἀγγεῖον καὶ πρόθεμα ἐπιθεὶς τῷ κυκλίσκῳ, καὶ σφηνὶ καὶ σφύρᾳ εἰσωθῶν τὸ τυμπάνιον μετὰ βίας μεγίστης. ἦν δὲ ὁρᾶν μικρὰν μὲν ἔνδοσιν ποιούμενον τὸ τυμπάνιον, ὅτε δὲ ἅπαξ ὁ ἀπειλημμένος ἀὴρ ἔσω πιληθείη, μηκέτι εἶκον μηδὲ ἐκ τῆς ἰσχυροτάτης πληγῆς πρὸς τὸν σφῆνα· καὶ βίας προσαχθείσης ἐκκρουσθέντος τε τοῦ σφηνός καὶ τὸ τυμπάνιον ἐξήλλετο μετὰ βίας πολλῆς ἐκ τοῦ ἀγγείου. πολλάκις δὲ συνέβαινε καὶ πῦρ συνεκπίπτειν διὰ τὴν ὀξύτητα τῆς φορᾶς παράτριψιν λαβόντος τοῦ ἀέρος πρὸς τὸ τεῦχος.
68 The fourth modified design that Philo discusses is the so-called repeating catapult invented by Dionysius of Alexandria (73.21–77.8). This was an engine that could shoot a large number of arrows in rapid succession. Although he praises this design for its inventiveness (φιλότεχνον, 76.22, μὴ ἀμηχάνως, 77.8), he claims that it is useless in battle: since the operator normally faces a moving target, repeated shots to the same location would only result in a waste of missiles. The account of the repeating catapult does not involve any theoretical considerations.

Conclusion

Let me now sum up by offering a general characterization of Philo's conception of expertise in light of the foregoing analysis and attempting to place it in a wider context. We have seen that a concern with general explanations runs throughout Philo's account of the artillery builder's *technē*. The rules of the standard method are based on recognizing the character of the spring hole diameter as the fundamental principle of construction, while the practice of artillery design, as reflected in the second part of the *Belopoeica*, demands a general inquiry into the cause (*aitia*) of long range. In keeping with this picture, Philo views both theory and experience as crucial to the *technē* of artillery construction, though the emphasis is different in the two parts of his account. The discussion of the standard method stresses the importance of testing or *peira*; on the other hand, the modified designs depend crucially on an understanding of basic principles of theoretical mechanics such as the law of the lever, as well as knowledge of physics, understood broadly to include the properties of materials, natural motion, and the behavior of compressed air. Physical and mechanical theory is applied in a flexible, non-deductive manner, with a view towards qualitative improvements in performance and in order to ensure that the components of the artillery engine behave in ways that are natural for them. Finally, despite Philo's claim that artillery construction is a *technē* with a unitary goal, his many references to aesthetic and other factors suggest that the ancient practitioner of this *technē* was motivated by the need to balance the competing claims of a variety of goals.

In its concern with generalization and explanation Philo's conception of *technē* shows clear affinities with the Platonic and Aristotelian discussions mentioned in the Introduction. Indeed, by insisting on the need for the modified designs to be based on theoretical understanding, Philo arguably goes beyond Aristotle, who sometimes grants that *empeiria* is sufficient for practice (above, n. 12). Philo's emphasis on the importance of *peira* is characteristic of the early Hellenistic period, as exemplified by the rise of the Empiricist school of medicine in the third century B.C. Yet Philo's stress on the systematic, directed nature of the discovery process sharply differentiates his conception of the role of *peira* from that of the Empiricists. As Philo puts it, it was not just one chance observation that led the Alexandrian engineers to their discovery, but a program of directed research that was only possible with state support. The Empiricists did accord an important role in discovery to repeated testing: on their view, a generalization such as "bloodletting cures fever" needed to be confirmed on many occasions if it was to be accepted as part of the medical *technē*.[69] Yet they were quite willing to acknowledge that the ultimate source of the propositions they put to the test was chance observation. The overall impression given by the Empiricist accounts of the discovery of medicine (as far as we can judge from the sources) is that it is a passive process driven by chance factors, not the kind of active, directed investigation that Philo associates with the Alexandrian engineers.[70] Two other features that distinguish Philo's spring hole relations and dimensional lists from the types of generalizations normally associated with *empeiria* in medical and philosophical sources are their universal character and their exactness.[71] Finally, we may note that Philo's emphasis on the need for theory to make innovations in artillery design resonates with one of the charges frequently directed against the Empiricists, viz., that their rejection of theory made it impossible for them to deal with new diseases.[72]

In conclusion, the particular combination of concerns with generalization, systematic testing, and mechanical and physical theory that we have discerned in Philo's conception of the *technē* of artillery construction finds no exact parallel in any ancient philosophical or medical source. Rather than viewing Philo's conception of expertise as the result of influence from a particular thinker or school, we should see it as the result of sophisticated reflection on the methodological situation in the discipline of artillery construction as Philo encountered it, informed by broad familiarity with philosophical and scientific thought.

---

69 This is the so-called "mimetic" (*mimētikon*) type of experience, which arises when cures suggested by chance or some other source are "put to the test" (εἰς πεῖραν ἄγηται, Gal. *de sectis* 2, 1.67 K. [= 3.7–8 Helmreich]).

70 Only the procedure known as "transition to the similar" (ἡ τοῦ ὁμοίου μετάβασις), which involved applying cures effective against certain diseases to other similar diseases, or cures effective on certain parts of the body to other similar parts, could be said to involve the formulation of hypotheses on the basis of specialized experience. This procedure did not result in discovery until confirmed by a test known as πεῖρα τριβική (Gal. *de sect*. 2, 1.67–68 K. [= 3–4 Helmreich]). But the Empiricist attitude to this procedure was highly ambivalent, and many refused to recognize its legitimacy. See Heinrich von Staden, "Experiment and Experience in Hellenistic Medicine," *Bulletin of the Institute of Classical Studies* 22 (1975): 178–99.

71 For quantitative precision as distinguishing *technē* from *empeiria* cf. Plat. *Phlb*. 55e1–56c7. The Empiricists acknowledged that the generalizations on which medicine was based were typically not universally valid, and they developed a fourfold classification: some held in all cases, others in most, others half the time, others only rarely (Gal. *subf. emp*. 45.24–30 Deichgräber; *de exp. med*. 15 [112–13 Walzer]).

72 For a closer medical parallel to Philo's position we might look to the doctor Erasistratus. According to Galen (*de sectis* 5, 1.75 K. [= 9.13–19 Helmreich]), Erasistratus claimed that experience was sufficient for the discovery of "simple" cures, such as antidotes, but not for more complex cures, where theory was needed.

Philo's picture of artillery construction is a response to the distinctive features of a field of technology that had developed, through empirical investigation, a quantitatively precise set of rules and procedures that exerted significant influence on its practice. In attempting to bring together the various strands of this most dynamic of ancient technological traditions under the rubric of a unified conception of *technē*, Philo offers one of the most detailed, sophisticated, and distinctive accounts of expertise that we have from the ancient world.

Bibliography

Baatz, Dietwulf. *Bauten und Katapulte des römischen Heeres* (Stuttgart: Steiner, 1994).
Bishop, M. C., and J. C. N. Coulston. *Roman Military Equipment: From the Punic Wars to the Fall of Rome*, 2nd ed. (Oxford: Oxbow Books, 2006).
Deichgräber, Karl. *Die griechische Empirikerschule: Sammlung der Fragmente und Darstellung der Lehre* (Berlin: Weidmann, 1930).
Diels, Hermann, and Erwin Schramm. *Philons Belopoiika (viertes Buch der Mechanik): Griechisch und deutsch*, Abhandlungen der Preussischen Akademie der Wissenschaften, Philosophisch-Historische Klasse, Jahrg. 1918 no. 16 (Berlin: Verlag der Akademie der Wissenschaften, 1919).
Drachmann, Aage G. *Ktesibios, Philon, and Heron: A Study in Ancient Pneumatics* (Copenhagen: E. Munksgaard, 1948).
Ferrari, G. A. "Meccanica 'allargata,'" in Gabriele Giannantoni and Mario Vegetti (eds.), *La scienza ellenistica: atti delle tre giornate di studio tenutesi a Pavia dal 14 al 16 aprile 1982* (Naples: Bibliopolis, 1984), 227–96.
Frede, Michael, and Richard Walzer. *Galen: Three Treatises on the Nature of Science* (Indianapolis: Hackett, 1985).
Gille, Bertrand. *Les mécaniciens grecs: la naissance de la technologie* (Paris: Seuil, 1980).
Heath, Thomas L. *A History of Greek Mathematics*, 2 vols. (Oxford: Clarendon Press, 1921).
Heinimann, Felix. "Eine vorplatonische Theorie der τέχνη," *Museum Helveticum* 18 (1961): 105–30.
Helmreich, Georg. *Galeni Scripta Minora*, vol. 3 (Leipzig: Teubner, 1893).
Hultsch, Friedrich. *Pappi Alexandrini Collectionis quae supersunt*, 3 vols. (Berlin: Weidmann, 1876–1878; repr., Amsterdam: Hakkert, 1965).
Joly, Robert, and Simon Byl. *Hippocrate, Du régime*, CMG I 2,4 (Berlin: Akademie-Verlag, 1984).
Jouanna, Jacques. *Hippocrate, Des vents; De l'art*, t. VI.1 (Paris: Les Belles Lettres, 1988).
Jouanna, Jacques. *Hippocrate, L'ancienne médecine*, t. II.1 (Paris: Les Belles Lettres, 1990).
Marsden, Eric W. *Greek and Roman Artillery: Historical Development* (Oxford: Clarendon Press, 1969).
Marsden, Eric W. *Greek and Roman Artillery: Technical Treatises* (Oxford: Clarendon Press, 1971).
Oppel, Herbert. *Kanōn. Zur Bedeutungsgeschichte des Wortes und seiner lateinischen Entsprechungen (regula-norma)* (Leipzig: Dieterich, 1937).

Orinsky K., O. Neugebauer, and A. G. Drachmann. "Philon," in W. Kroll and K. Mittelhaus (eds.), *Paulys Realencyclopädie der classischen Altertumswissenschaft: Neue Bearbeitung*, vol. 20.1 (Stuttgart: J. B. Metzlersche Verlagsbuchhandlung, 1941), 53–54.
Prager, Frank D. *Philo of Byzantium, Pneumatica* (Wiesbaden: L. Reichert, 1974).
Rihll, Tracey E. *The Catapult: A History* (Yardley, Pa.: Westholme Publishing, 2007).
Sáez Abad, Rubén. *Artillería y poliorcética en el mundo grecorromano* (Madrid: Consejo Superior de Investigaciones Científicas, 2005).
Sambursky, Shmuel. *The Physical World of Late Antiquity* (London: Routledge, 1962).
Schiefsky, Mark J. *Hippocrates "On Ancient Medicine": Translated with Introduction and Commentary*, Studies in Ancient Medicine 28 (Leiden: Brill, 2005).
Schiefsky, Mark J. "Art and Nature in Ancient Mechanics," in Bernadette Bensaude-Vincent and William R. Newman (eds.), *The Artificial and the Natural: An Evolving Polarity* (Cambridge, Mass.: The MIT Press, 2007), 67–108.
Schiefsky, Mark J. "Structures of Argument and Concepts of Force in the Aristotelian *Mechanical Problems*," *Early Science and Medicine* 14.1–3 (2009): 43–67.
Schöne, Hermann. *Heronis Alexandrini Opera quae supersunt omnia*, vol. 3: *Rationes dimetiendi et commentatio dioptrica* (Leipzig: Teubner, 1903; repr., Stuttgart: Teubner, 1976).
Schöne, Richard. *Philonis Mechanicae Syntaxis libri quartus et quintus* (Berlin: Reimer, 1893).
Schramm, Erwin. *Die antiken Geschütze der Saalburg* (Berlin: Weidmann, 1918; repr. with intro. by D. Baatz, Bad Homburg: Saalburgmuseum, 1980).
Schürmann, Astrid. *Griechische Mechanik und antike Gesellschaft. Studien zur staatlichen Förderung einer technischen Wissenschaft* (Stuttgart: Steiner, 1991).
Stewart, Andrew. "The Canon of Polykleitos: A Question of Evidence," *Journal of Hellenic Studies* 98 (1978): 122–31.
von Staden, Heinrich. "Experiment and Experience in Hellenistic Medicine," *Bulletin of the Institute of Classical Studies* 22 (1975): 178–99.
Walzer, Richard. *Galen on Medical Experience* (London: Oxford University Press, 1944).
Wehrli, Fritz. *Die Schule des Aristoteles. Texte und Kommentare*, Heft V: *Straton von Lampsakos*, 2nd ed. (Basel: B. Schwabe, 1950)
Wescher, Carle. *Poliorcétique des Grecs: Traités théoriques – Récits historiques* (Paris: Imprimerie Impériale, 1867).

Heinrich Schlange-Schöningen
Herrschaftskritik bei Galen

Abstract: While we encounter many statements about Roman society in his works, Galen only rarely offers political comments. This article will look into Galen's attitude towards monarchical rule. It is well known that at times Galen openly criticized the Roman Empire and its leading representatives, including members of the Roman Senate. If we look much further and include all instances found in his œuvre, starting with the mythical figure of Priam and continuing with Croesus, Cyrus, and Polycrates, on to Alexander and some of the latter's successors in Hellenistic times, finishing with the Roman emperors, we notice that Galen more often than not will express a negative opinion of monarchic rule; the reason is that he compares them to his ideal of the physician as a philosopher. Only a few rulers, e.g., Attalus, Mithridates, or Marcus Aurelius, meet with his approval; all of them were either themselves interested in medical questions or at least in healthy living. In all other cases, Galen sees examples of a lack of interest in science and an abuse of power, which applies even to rulers like Cyrus or Hadrian, who are both usually seen as positive figures in ancient sources.

> „Wo könnte heute jemand einen Diogenes finden,
> der dazu in der Lage wäre, die Wahrheit auszusprechen,
> vor allem wenn man einer der reichen Männer ist oder gar ein Herrscher?" [1]

Mit dieser rhetorischen Frage unterstreicht Galen in seiner moralphilosophischen Abhandlung „Über die Diagnose und Behandlung der seelischen Krankheiten," dass es für mächtige Zeitgenossen kaum Aussicht auf Gesundheit der eigenen Psyche geben könne, denn um eine solche zu erlangen, brauche ein jeder einen kritischen Begleiter bzw. Erzieher. Doch kann es diese unabhängigen kritischen Geister, wie einst Diogenes einer war, in Galens Gegenwart nicht mehr geben, da die römische Gesellschaft durch ihr Machtsystem korrumpiert ist.[2]

Trotz der bekannten distanzierten Haltung, die Vertreter der zweiten Sophistik Rom und dem römischen Imperium gegenüber eingenommen haben,[3] bleibt es

---

1 *De an. aff. dign. et cur.* 3, 5.13 K. (= 11 de Boer).
2 Vgl. Tim Whitmarsh, *The Second Sophistic* (Oxford: Oxford University Press, 2005), 60–63 zur Rolle des Diogenes in den Reden περὶ βασιλείας des Dion von Prusa.
3 Vgl. Ewen L. Bowie, „Greeks and their Past in the Second Sophistic," *Past and Present* 46 (1970): 3–41, bsd. 32–41; Graham Anderson, *The Second Sophistic: A Cultural Phenomenon in the*

doch erstaunlich, dass Galen dermaßen kritische Äußerungen niedergeschrieben und verbreitet hat, während er in Rom lebte.⁴ Die hier formulierte Kritik an Herrschaft und Gesellschaft reicht sehr weit, richtet sie sich doch auch gegen das Klientelsystem, an dessen Spitze der Kaiser steht.

Dem unbekannten Adressaten der Schrift erklärt Galen, dass er dem vermeintlich guten Ruf eines Mannes, der als ehrenhaft gilt, nicht trauen dürfe, wenn

> Du feststellen musst, dass er oft in die Häuser der Reichen und Mächtigen oder auch der Herrscher läuft (ἐπὶ τὰς τῶν πλουσίων τε καὶ πολὺ δυναμένων ἢ καὶ τὰς τῶν μονάρχων οἰκίας). Dann wirst Du wissen, dass Du etwas ganz Falsches über diesen Mann gehört hast, denn mit diesem Verhalten gehen die Lügen einher. Und das wird bestätigt, wenn Du weiterhin siehst, dass er diese Leute begrüßt und in der Öffentlichkeit begleitet oder dass er an ihren Gastmählern teilnimmt.⁵

Einen solchen Mann kann man sich nicht zum moralischen Präzeptor wählen. Nur diejenigen, die der *salutatio* fernbleiben,⁶ kommen als Ratgeber in Frage; sie soll man immer wieder auffordern, das eigene Verhalten zu kommentieren. Ihrem Tadel darf man nicht zürnen, man soll ihn vielmehr mit Dank entgegennehmen, bis man eines Tages in der Lage sein wird, sich selbst zu kontrollieren und zu

---

Roman Empire (London: Routledge, 1993), 122; Simon Swain, *Hellenism and Empire: Language, Classicism, and Power in the Greek World, AD 50–250* (Oxford: Oxford University Press, 1996), 78–79, 87–89; Paolo Desideri, „Galeno come intellettuale," in Daniela Manetti (Hg.), *Studi su Galeno. Scienza, Filosofia, Retorica e Filologia. Atti del seminario, Firenze 13 novembre 1998* (Florenz: Università degli studi di Firenze, 2000), 13–29, bsd. 28.

4 Nach Johannes Ilberg, „Über die Schriftstellerei des Klaudios Galenos," *Rheinisches Museum* 52 (1897): 591–623 gehört die Schrift „in den zweiten römischen Aufenthalt und zwar in die ersten Jahre" (611); so auch Vincent Barras, Terpsichore Birchler und Anne-France Morand, *Galien: l'âme et ses passions. Les passions et les erreurs de l'âme – Les facultés de l'âme suivent les tempéraments du corps* (Paris: Les Belles Lettres, 1995), xxx. Das in Kapitel 4, 5.18–19 K. (= 13–14 de Boer) geschilderte Reiseerlebnis – auf dem Weg von Korinth nach Athen – dürfte sich auf dem Rückweg vom ersten Romaufenthalt nach Pergamon abgespielt haben (vgl. Vivian Nutton, „The Chronology of Galen's Early Career," *Classical Quarterly* 23 [1973]: 158–71, bsd. 168 [wieder in ders., *From Democedes to Harvey: Studies in the History of Medicine* (London: Variorum Reprints, 1988)]), so dass auch eine Abfassung kurz nach 166 möglich erscheint. Dass Galen indes keine Scheu gehabt hat, in Rom Kritik an der römischen Gesellschaft zu äußern, zeigen auch andere, später entstandene und somit sicher in Galens zweiten Romaufenthalt zu datierende Schriften, so z.B. *De optimo medico cognoscendo* (ed. A. Z. Iskandar, CMG Suppl. Or. 4, 1988); vgl. dazu Vivian Nutton, „The Patient's Choice: A New Treatise by Galen," *Classical Quarterly* 40 (1990): 236–57, bsd. 249–53.

5 *De an. aff. dign. et cur.* 3, 5.8–9 K. (= 7 de Boer).

6 Galen kritisiert die *salutatio* z.B. auch in *de opt. med. cogn.* 9.16–17 (111 Iskandar); vgl. Nutton, „The Patient's Choice," 250.

verbessern.⁷ Die hier von Galen vorgebrachte Gesellschaftskritik ist radikal, denn die Ausführungen laufen darauf hinaus, dass kein Mitglied der führenden Schichten des Staates, ob nun in den Provinzstädten oder in Rom, tugendhaft lebt und deshalb als Ratgeber für die wichtige moralphilosophische Frage, wie man Seelenruhe jenseits der Leidenschaften finden könne, konsultiert werden kann.

Selbstbeherrschung, so führt Galen in *De propriorum animi cuiuslibet affectuum dignotione et curatione* weiterhin aus, kann zunächst im Umgang mit den eigenen Sklaven erlernt werden. Galen führt etliche Beispiele dafür an, dass Sklaven hemmungslos malträtiert wurden, wenn die Sklavenbesitzer, oftmals aus geringfügigem Anlass, in Wut gerieten. Auch die eigene Mutter, die in ihrer Wut ihre Sklaven biss, oder Freunde des Vaters, die sich beim Schlagen ihrer Sklaven mit der bloßen Hand verletzten, finden sich hier als abschreckende Belege für ein solches Fehlverhalten.⁸ Die höchste Steigerung erfahren diese Beispiele indes mit einer Erzählung über den Kaiser Hadrian, für die es neben Galen kein weiteres Zeugnis in der Überlieferung gibt.

Als der Kaiser, so berichtet Galen, einmal in Wut geraten war, habe er einem seiner Sklaven mit dem Schreibgriffel ein Auge ausgestochen. Sobald Hadrian bewusst geworden sei, welchen Schaden er verursacht hatte, habe er den Sklaven zu sich kommen lassen und ihn aufgefordert, zur Wiedergutmachung einen Wunsch zu äußern. Der Sklave habe anfangs geschwiegen, auf Nachfrage des Kaisers hin dann aber den Wunsch geäußert, sein volles Augenlicht zurückzuerhalten. Was könnte denn auch, so fragt Galen kommentierend, als Geschenk dem Wert eines verlorenen Auges gleichkommen?⁹

Für Fergus Millar zeigt diese Anekdote, „how the conception of an emperor as the source of gifts and benefits gave rise to the typically monarchic custom of specially favoured persons being given an open invitation to request whatever they wished."¹⁰ Die eigentliche Pointe der Anekdote liegt aber darin, dass hier die Grenzen der kaiserlichen Gnadenfähigkeit aufgezeigt werden. Und man kann noch einen Schritt weitergehen und mit David E. Eichholz in Galens Erzählung einen weiteren Beleg für die distanzierte Haltung erkennen, die Galen Rom gegenüber

---

7 *De an. aff. dign. et cur.* 4, 5.14 K. (= 11 de Boer); vgl. William V. Harris, *Restraining Rage: The Ideology of Anger Control in Classical Antiquity* (Cambridge, Mass.: Harvard University Press, 2001), 385–86.
8 *De an. aff. dign. et cur.* 4, 5.17 K. (= 13 de Boer); 7–8, 5.40–41 K. (= 27–28 de Boer). Vgl. Harris, *Restraining Rage*, 329; Heinrich Schlange-Schöningen, „Galen on Slavery," in Arnaldo Marcone (Hg.), *Medicina e società nel mondo antico: atti del convegno di Udine (4–5 ottobre 2005)* (Florenz: Le Monnier, 2006), 180–93, bsd. 190–92.
9 *De an. aff. dign. et cur.* 4, 5.17 K. (= 13 de Boer).
10 Fergus Millar, *The Emperor in the Roman World* (London: Duckworth, 1977), 212.

eingenommen hat.¹¹ Denn man sollte auch berücksichtigen, dass Hadrian in Galens Heimatstadt Pergamon als νέος Ἀσκληπιός verehrt worden ist,¹² und der damit verbundene Anspruch auf göttliche Ehren wird mit der von Galen erzählten Anekdote zurückgewiesen.¹³ Ein Kaiser, der sich von seinem Zorn dazu verleiten lässt, einen seiner Sklaven auf nicht wieder gut zu machende Weise zu verletzen, ist in seinem Machtmissbrauch das krasse Gegenteil eines fürsorglichen, Asklepios-ähnlichen Herrschers.¹⁴

Die Schärfe der Kritik am römischen Gesellschaftssystem und dann auch an einem der Kaiser der jüngeren Vergangenheit ist bemerkenswert für einen Mann, der seinen beruflichen Aufstieg der Förderung durch die Autoritäten des Reiches verdankte,¹⁵ zunächst den Oberpriestern von Pergamon, die ihn zum Gladiatorenarzt machten, später, während des ersten Romaufenthalts, Angehörigen der Reichsaristokratie, die auf ihn aufmerksam wurden, sich von ihm behandeln und belehren ließen und Mark Aurel Bericht über den erstaunlich befähigten jungen griechischen Arzt erstatteten, und schließlich dem Kaiser selbst, der Galen aus Pergamon nach Rom zurückrief, sich von ihm den Theriak zubereiten ließ, ihm die Behandlung seiner Familienangehörigen übertrug und ihm durch seine kaiserliche Anerkennung zu einer herausragenden Position unter den in Rom versammelten griechischen Philosophen und Ärzten verhalf.¹⁶ Galens Stolz auf diese Erfolge, der vor allem in *De praecognitione* zum Ausdruck kommt, ist unüberhörbar, und man könnte vermuten, dass sich bei Galen im Verlauf des eigenen

---

11 David E. Eichholz, „Galen and his Environment," *Greece and Rome* 20 (1951): 60–71, bsd. 64.
12 Marcel Le Glay, „Hadrien et l'Asklépieion de Pergame," *Bulletin de correspondance hellénique* 100 (1976): 347–72, bsd. 353–55, 367–68. Vgl. auch Anthony R. Birley, *Hadrian. Der rastlose Kaiser* (Mainz am Rhein: von Zabern, 2006), 56. Zur Förderung des Pergamenischen Asklepieions durch Hadrian vgl. auch Florian Steger, *Asklepiosmedizin. Medizinischer Alltag in der römischen Kaiserzeit* (Stuttgart: Steiner, 2004), 125.
13 Vgl. Heinrich Schlange-Schöningen, *Die römische Gesellschaft bei Galen. Biographie und Sozialgeschichte* (Berlin: De Gruyter, 2003), 278 mit Anm. 91. Vgl. auch Susan P. Mattern, *Galen and the Rhetoric of Healing* (Baltimore: The Johns Hopkins University Press, 2009), 40.
14 Vgl. Harris, *Restraining Rage*, 256–57, 329; Steger, *Asklepiosmedizin*, 134 zu den Wunderheilungen, die Hadrian in der *Hist. Aug.* (*Hadr.* 25.1-4) zugeschrieben werden; bei den beiden Wunderheilungen handelt es sich um Blinde, die wieder sehend wurden.
15 Zu Galens Biographie vgl. Nutton, „Galen's Early Career," 158–71; Schlange-Schöningen, *Die römische Gesellschaft bei Galen*; Vivian Nutton, *Ancient Medicine*, 2. Aufl. (London: Routledge, 2013), 222–35; Robert J. Hankinson, „The Man and his Work," in R. J. Hankinson (Hg.), *The Cambridge Companion to Galen* (Cambridge: Cambridge University Press, 2008), 1–33; Véronique Boudon-Millot, *Galien de Pergame: un médecin grec à Rome* (Paris: Les Belles Lettres, 2012).
16 Vgl. Glen W. Bowersock, *Greek Sophists in the Roman Empire* (Oxford: Clarendon Press, 1969), 59; Jutta Kollesch, „Galen und die Zweite Sophistik," in Vivian Nutton (Hg.), *Galen: Problems and Prospects* (London: Wellcome Institute for the History of Medicine, 1981), 6–7.

Aufstiegs unter den sozialen Bedingungen der hohen römischen Kaiserzeit auch ein positiveres Bild von monarchischer Herrschaft eingestellt haben könnte. Dass dies aber nicht der Fall war, hat Simon Swain bereits 1996 in seiner Studie über *Hellenism and Empire* gezeigt, in der er auch Galen ein Kapitel gewidmet und u. a. die Frage behandelt hat, welche Haltung Galen in seinen Schriften dem römischen Reich und seinen Kaisern gegenüber zu erkennen gibt.[17] Swains Befund ist eindeutig, sowohl im Hinblick auf Galens Urteil über die zivilisatorischen Leistungen Roms als auch im Hinblick auf seine Bewertung einzelner Kaiser: „There is no warm admiration for Roman achievements such as we might expect from a man who stresses how closely the Roman elite took him to heart."[18] An positiven Urteilen über Rom findet sich in Galens Schriften nur ein Hinweis auf die hauptstädtische Wasserversorgung[19] und ein Lob für den Straßenbau, mit dem Trajan das Verkehrsnetz in Italien verbessert hat;[20] dabei hat Swain betont, dass das Lob Trajans Galen letztlich dazu dient, seine eigene Verdienste um die Fortentwicklung der Medizin herauszustellen.[21] Und was die römischen Kaiser anbelangt, so kommt Swain zu dem Ergebnis: „Galen mentions most of Rome's emperors and in most cases he is entirely neutral towards them."[22] Nur wenige Abweichungen von dieser Linie lassen sich feststellen: im Negativen ist dies vor allem die oben zitierte Anekdote über Hadrian,[23] und im Positiven die lobenden Worte, die Galen in mehreren seiner Schriften über Mark Aurel findet, wobei das positive Urteil über diesen Kaiser in *De praecognitione* auch wieder dazu dient, Galens eigene Führungsstellung in der Medizin seiner Zeit zu unterstreichen (s.u.). Die vor diesem Hintergrund überraschende, geradezu panegyrische Art, in der Galen in *De theriaca ad Pisonem* über Septimius Severus und weitere Mitglieder der severischen Dynastie spricht, hat Swain zu Recht als „abnormal" für Galen bezeichnet und als

---

17 Swain, *Hellenism and Empire*, 363–72. Vgl. Desideri, „Galeno come intellettuale," 14–19.
18 Swain, *Hellenism and Empire*, 365.
19 *In Hp. epid. VI comm.* 4.11, 17B.159 K. (= 211 Wenkebach).
20 *De meth. med.* 9.8, 10.633 K.
21 Swain, *Hellenism and Empire*, 366.
22 Swain, *Hellenism and Empire*, 371. Römische Kaiser werden von Galen vor allem in *de antidotis* erwähnt; hier finden sich Hinweise auf Trajan, Hadrian, Antoninus Pius, Mark Aurel, Commodus, Septimius Severus (alle in *de antidot.* 1.13, 14.64–65 K.), Tiberius (*de antidot.* 2.5, 14.132 K.) und Nero, der im von Galen zitierten Theriak-Gedicht seines Arztes Andromachos als „Spender sicherer Freiheit" (*de antidot.* 1.6, 14.32 K.: ἀδειμάστου δῶτορ ἐλευθερίης) bezeichnet wird.
23 Vgl. dazu auch Swain, *Hellenism and Empire*, 372.

Reaktion auf den bedrohlichen, autokratischen Herrschaftsstil der Severer gedeutet.[24]

Im Folgenden soll nun, ergänzend zu den bisherigen Befunden, der Frage nachgegangen werden, ob Galens reservierte Haltung gegenüber monarchischer Herrschaft auf das römische Kaiserreich beschränkt war. Da sich in seinen Schriften etliche *exempla* finden, die sich auf vorrömische Herrschaftssysteme beziehen, kann man hier vielleicht zu einem Grundzug von Galens politischem Denken gelangen.[25] Denn die Beispiele, die Galen in seinen Schriften aus der Geschichte der antiken Autokratien anführt, um seine medizinischen Ausführungen oder seine moralphilosophischen Argumente zu stützen, sind zwar nicht sehr zahlreich, doch recht weit verstreut, sowohl was die galenischen Schriften selbst anbelangt, als auch im Hinblick auf ihre historische Verortung, die vom frühen Achämenidenreich über die griechische Tyrannis bis zu Herrschern aus der Epoche des Hellenismus führt.[26]

In seiner Aufforderung, sich um die Wissenschaften zu bemühen (*Adhortatio ad artes addiscendas*), beschreibt Galen die Fähigkeit des Menschen zur Wissenschaft und Philosophie als konstituierend für das Menschsein.[27] Deshalb verfehlt derjenige, der sich nicht um die Wissenschaft, sondern um materiellen Wohlstand bemüht, seine eigentliche menschliche Bestimmung. Galens Urteil über Fortuna lautet folgendermaßen:

> Sie ist wankelmütig und unverständig in einem solchen Ausmaß, dass sie die verdienstvollen Männer zumeist übersieht und die ehrlosen reich macht, allerdings nicht auf Dauer; vielmehr entreißt sie zumeist wieder das, was sie zuvor gegeben hat.[28]

Das Gegenbild zu Fortuna ist Hermes, dessen Anhänger nach Wissen streben und auf diese Weise die dem Menschen gegebene Ähnlichkeit mit den Göttern als

---

24 Swain, *Hellenism and Empire*, 372: „The excessive praise of the Severians is abnormal. ... He [sc. Galen] was also living in an increasingly monarchical regime and under emperors with a different style of government from that of the good Marcus. On Theriac To Piso may be taken as a reflection of these less certain times."
25 Vgl. in diesem Zusammenhang Nutton, „The Patient's Choice," 251–52 zu Galens Kritik an der Beschäftigung mit der Vergangenheit in *de opt. med. cogn.* 13.2 (129 Iskandar: „.... devoting their lives [to the study] of useless traditions of the past ...") und zu seinen wenigen Verweisen auf Thukydides, Herodot, Xenophon und Ktesias; dazu auch ders., „Galen's Library," in Christopher Gill, Tim Whitmarsh und John Wilkins (Hgg.), *Galen and the World of Knowledge* (Cambridge: Cambridge University Press, 2009), 19–34, bsd. 25–26.
26 Zu Galens Fallgeschichten und seiner gelegentlichen Verwendung legendenhafter Überlieferungen vgl. Mattern, *Galen and the Rhetoric of Healing*, bsd. 39–41.
27 *Protrept.* 1.2–4, 1.1–3 K. (= 84–85 Boudon; 114–16 Barigazzi).
28 *Protrept.* 2.3, 1.4 K. (= 86 Boudon; 116 Barigazzi).

Bestimmung ernst nehmen. Wie Véronique Boudon-Millot in der Einführung zu ihrer Edition der *Adhortatio ad artes addiscendas* dargelegt hat, hat Galen hier Gedanken aufgenommen bzw. verändert, die der etwas ältere Favorinus von Arles in seiner im Corpus der Reden des Dion Chrysostomos überlieferten Schrift περὶ Τύχης formuliert hatte.²⁹ So haben beide, Favorinus und Galen, eine ähnliche, aber nicht identische Reihe von *exempla* für die unglücklichen Anhänger der Fortuna. Während bei Favorinus der Lyderkönig Kroisos und Polykrates, der Tyrann von Samos, die nicht verstanden haben, dass dem Glück nicht zu vertrauen ist, die Göttin mit Vorwürfen angreifen, der eine, Kroisos, weil er zuletzt doch seinen Sohn verloren hat, der andere, Polykrates, als er in die Gefangenschaft des Oroites geraten ist, lautet der Passus bei Galen im 4. Kapitel seiner *Adhortatio* folgendermaßen:

> Diejenigen, die der Fortuna Schritt für Schritt folgen, die siehst Du ohne irgendeine Beschäftigung und unkundig jeglicher Wissenschaft, dafür beständig von ihrer Hoffnung vorangetrieben. ... Unter diesen erblickst Du jenen Kroisos aus Lydien und Polykrates von Samos, und vielleicht staunst Du auch über den Anblick des Flusses Paktolos, der dem ersten das Gold zuführte, und über die Fische des Meeres, die dem zweiten gehorchten. In ihrer Gesellschaft wirst Du auch Kyros, Priamos und Dionysios erblicken. Aber kurze Zeit nach diesen wunderbaren Dingen wirst Du Polykrates am Kreuze hängen sehen, Kroisos als von Kyros unterworfen, und Kyros seinerseits ebenfalls von anderen. Und Du wirst sehen, wie Priamos überwältigt wird, und den Dionysios in Korinth.³⁰

Zunächst sind die Veränderungen in der Liste der *exempla* bemerkenswert: Während Favorinus neben Kroisos und Polykrates auch Kyros und Alexander anführt, deren Tod von Persern bzw. Makedonen der Fortuna vorgeworfen worden sei,³¹ bezieht Galen in seine Beispiele Priamos und Dionysios mit ein. Galen hat darüber hinaus aber auch, worauf Véronique Boudon-Millot hingewiesen hat, eine höhere Anschaulichkeit und eine höhere Dramatik erzielt, die bei Kroisos im Verlust nicht des Sohnes, sondern des Reiches liegt, und bei Polykrates darin, dass nicht die Gefangenschaft beklagt, sondern die Kreuzigung in Erinnerung gerufen wird: „Ce qui, chez Favorinus, pouvait à dessein n'apparaître que comme de simples revers de fortune, est cette fois dépeint chez Galien comme une chute aussi brutale qu'irrémédiable."³²

---

29 Véronique Boudon, *Galien, Exhortation à l'étude de la médecine; Art médical*, t. II (Paris: Les Belles Lettres, 2000), 12–14.
30 *Protrept.* 4.1–3, 1.5–6 K. (= 87–88 Boudon; 118 Barigazzi).
31 Adelmo Barigazzi, *Favorino di Arelate, Opere. Introduzione, testo critico e commento* (Florenz: Le Monnier, 1966), 254 (de Fortuna 1); vgl. Boudon, *Galien, Exhortation*, 14, Anm. 21.
32 Boudon, *Galien, Exhortation*, 13–14.

Hinzu kommt noch, dass die *exempla* von Galen mit der Bemerkung eingeleitet werden, die genannten Männer hätten sich in keiner Weise um die Künste bzw. Wissenschaften bemüht (τεχνῶν ἀμαθεῖς), was für Polykrates nicht recht zutrifft, wenn man etwa an den Aufenthalt von Demokedes, Anakreon oder auch von Eupalinos auf Samos denkt.³³ Galen lässt sich hier offensichtlich von einem negativen Tyrannenbild leiten, wie dies mehr noch für Dionysios I. von Syrakus verbreitet war, dem z. B. der Verkauf Platons in die Sklaverei nachgesagt wurde,³⁴ sowie für Dionysios II., an den Galen hier gedacht hat.³⁵ Auch Kyros dient in der antiken Literatur eher selten als *exemplum* für ein unglückliches Lebensende, wie dies bei Favorinus und Galen der Fall ist, die sich auf die Darstellung vom Tod des Kyros im Kampf gegen die Massageten bei Herodot beziehen. Stärker wirkte das positive Bild, das Xenophon in seiner *Kyropädie* gezeichnet hat; Kyros erscheint hier als hoch gebildeter König, der über die Kunst der Selbstbeherrschung und der Menschenführung verfügt und sein Leben mit einem friedlichen Abschied von den Seinen beschließt.³⁶ Galen, der einige Werke Xenophons kannte, die *Kyropädie* aber vielleicht nicht gelesen hat,³⁷ hat sich indes in seiner *Adhortatio ad artes addiscendas* für die Verwendung des Kyros als negativem *exemplum* entschieden.

Dafür dass sich Galen der Tyrannentopik bedient, um seine Argumente im medizinischen Meinungskampf zu stärken, gibt es etliche Belege; erinnert sei hier nur an den Vergleich der Ärzte der von Thessalos begründeten methodischen Schule mit Tyrannen in *De methodo medendi*. Hier sagt Galen, dass Ärzte, die ihre Aussagen nicht begründen, sondern nur apodiktisch verkünden, damit wie „Tyrannen Befehle erteilen" würden (ὡς τύραννοι κελεύουσι).³⁸ Und er fährt fort:

---

33 Vgl. Helmut Berve, *Die Tyrannis bei den Griechen*, 2 Bde. (München: Beck, 1967), 1:107–14; Loretana De Libero, *Die archaische Tyrannis* (Stuttgart: Steiner, 1996), 253–97.
34 Vgl. Karl Friedrich Stroheker, *Dionysios I. Gestalt und Geschichte des Tyrannen von Syrakus* (Wiesbaden: Steiner, 1958), 22; Kai Trampedach, *Platon, die Akademie und die zeitgenössische Politik* (Stuttgart: Steiner, 1994), 105–106. Zur hohen Bildung des Dionysios und dem an seinem Hof versammelten Kreis von Gelehrten vgl. Stroheker, *Dionysios I.*, 38, 99–100; Giuseppe Squillace, *Menecrate di Siracusa: un medico del IV secolo a.C. tra Sicilia, Grecia e Macedonia* (Hildesheim: Olms, 2012), 23–24.
35 Zum Exil Dionysios' II. in Korinth vgl. Berve, *Die Tyrannis bei den Griechen*, 1:277–78.
36 Die positive Rezeption von Kyros reicht bis in die Spätantike; vgl. dazu Werner Portmann, *Geschichte in der spätantiken Panegyrik* (Frankfurt am Mein: Lang 1988), 215, 328 (Stellennachweise), und weiterhin z. B. Lib. *or.* 18.295.
37 Vgl. Nutton, „The Patient's Choice," 251; ders., „Galen's Library," 26.
38 *De meth. med.* 2.5.3, 10.105 K.

> Wir aber verachten Dionysios und Phalaris und die übrigen Tyrannen eben deswegen, weil sie Befehle geben und Vorschriften machen, und weil sie sich nicht darum bemühen zu überreden oder zu lehren wie Solon und Drakon und Lykurg.[39]

Während der für seine Grausamkeit berüchtigte Tyrann Phalaris von Akrigent von Galen nur an dieser einen Stelle erwähnt wird,[40] kommt er auf Dionysios I., der hier gemeint sein dürfte und der, ebenso wie sein Sohn Dionysios II., als besonders grausamer Gewaltherrscher galt,[41] mehrfach zu sprechen. Besonders interessant ist ein Passus aus *De usu partium*, weil Galen für die hier erzählte herrscher- bzw. tyrannenkritische Geschichte – wie bei der eingangs behandelten Anekdote über Hadrian – erneut der einzige Zeuge ist.

Im zehnten Buch dieser Schrift setzt sich Galen mit den verschiedenen Theorien zum Sehvermögen auseinander,[42] um dann seine eigene, Platon folgende Anschauung zu beweisen. Mit der bis auf Pythagoras zurückgehenden Theorie der Sehstrahlen ist Galen ebensowenig einverstanden wie mit der sich auf Empedokles berufenden Abbilder-Theorie, deren unzureichenden Erklärungswert er mit der Frage aufzeigt, wie denn allein ein äußerer Eindruck zu einem präzisen Bild führen sollte? Wie sollte man sich etwa ein Abbild eines großen Berges im Auge vorstellen? Galen hält es auch für unsinnig, im Anschluss an Pythagoras anzunehmen, dass das Auge Teile seines Licht-Pneumas aussenden würde, um etwa einen ganzen großen Berg zu umfassen. Folglich müsse es eine dritte Erklärung geben, und sie liegt Galen zufolge darin, dass die den sehenden Menschen umgebende Luft zum Werkzeug des Sehens wird: Vom Sonnenlicht selbst geht ein Pneuma aus, und auch das Auge verströmt Pneuma, so dass zwischen dem Auge und den äußeren Gegenständen ein Kontinuum besteht, das durch das Augen-Pneuma beim Prozess des Sehens modifiziert wird. Damit stellt sich Galen in die Tradition der sogenannten Synaugie-Theorie, die Platon im *Timaios* dargelegt hatte. In ihr wird das Sehen als Ergebnis des Zusammenwirkens zweier Kräfte beschrieben: auf der einen Seite sendet das Auge die Sehstrahlen aus, aber au-

---

39 *De meth. med.* 2.5.3, 10.105–106 K.; vgl. Robert J. Hankinson, *Galen, On the Therapeutic Method, Books I and II, Translated with an Introduction and Commentary* (Oxford: Clarendon Press, 1991), 53. Vgl. die identische Formulierung in *de meth. med.* 1.3, 10.20 K., wo es über Thessalos heißt, er habe seine Lehren in der Art eines Tyrannen verkündet (ὡς τύραννος κελεύει). Zu Galens Kritik an Thessalos und seiner Schule vgl. auch Nutton, *Ancient Medicine*, 189–90; Elisa Romano, „Modelli intellettuali e modelli sociali in Galeno," in Marcone, *Medicina e società nel mondo antico*, 168–79, bsd. 172.
40 Zu Phalaris vgl. Berve, *Die Tyrannis bei den Griechen*, 1:129–31.
41 Vgl. Stroheker, *Dionysios I.*, 18–20.
42 Vgl. Arthur Erich Haas, „Antike Lichttheorien," *Archiv für Geschichte der Philosophie* 20 (1907): 345–86.

ßerhalb des Auges wirkt auch äußeres Licht mit, das von der Sonne oder anderen Lichtquellen ausgeht. Mit dem äußeren Licht vermischt erreicht das innere Licht das Objekt, das dann auf das Auge zurückwirkt.

Welche Rolle das äußere Licht für das Sehen spiele, dass wisse doch jeder, so führt Galen in *De usu partium* weiter aus, der Xenophons Bericht über den Wintermarsch seiner Soldaten durch den Schnee der armenischen Berge gelesen habe.[43] Allerdings wird der imaginäre Adressat von Galens Ausführungen, der sich der Einsicht in das Wirken der Natur verschließt (ὦ δεινότατε κατήγορε τῆς φύσεως),[44] polemisch als so ungebildet bezeichnet, dass er wohl Xenophons Schriften nicht kenne. Und aufgrund seiner fehlenden Bildung sei auch anzunehmen, dass er noch nie davon gehört habe,

> dass Dionysios, der Tyrann von Sizilien, über seinem Gefängnis eine Kammer bauen ließ, die ganz und gar mit hell strahlendem Kalk bestrichen und auch sonst in jeder Hinsicht sehr hell gehalten war, und dass er dann die Gefangenen, nachdem sie lange in ihrem Kerker festgehalten worden waren, in diese Kammer bringen ließ, und die Gefangenen, die nun aus der langandauernden Dunkelheit ins Licht kamen, voll Freude in das Licht blickten, dadurch aber erblindeten, weil sie das so plötzlich auftretende helle Licht nicht ertragen konnten.[45]

Wie erwähnt, gibt es für diese Erzählung keinen weiteren Beleg; sie passt sich indes recht gut in die anekdotenreiche Darstellung ein, die zuerst Timaios von Tauromenion über die grausame Herrschaftspraxis von Dionysios I. gegeben hat.[46]

Wenn Galens Kritik an den griechischen Tyrannen sich somit auch weitgehend in die antike Deutungstradition einpasst und deshalb auf den ersten Blick wenig überraschend erscheint, liegt doch bereits in dem Argument der angeblichen Bildungsfeindlichkeit dieser Herrscher eine spezifische Ausrichtung, die auf ähnliche Weise auch in Galens Urteil über die hellenistischen Herrscher wiederkehrt. Denn kennzeichnend für die hellenistischen Könige, die Galen als *exempla*

---

[43] *De usu part.* 10.3, 3.775 K. (= 2:66 Helmreich). Galen bezieht sich hier auf Xen. *an.* 4.5.13; vgl. Nutton, „The Patient's Choice," 252.
[44] Zu Galens Konzept des Demiurgen, seiner in *De usu partium* verfolgten Auseinandersetzung mit Erasistratos und den Methodikern sowie der Bedeutung von *De usu partium* für Galens Aufstieg in Rom vgl. Rebecca Flemming, „Demiurge and Emperor in Galen's World of Knowledge," in Gill, Whitmarsh und Wilkins, *Galen and the World of Knowledge*, 59–84.
[45] *De usu part.* 10.3, 3.775–76 K. (= 2:66–67 Helmreich).
[46] Vgl. Richard Laqueur, *Timaios von Tauromenion (= Timaios nr. 3)*, in W. Kroll and K. Mittelhaus (Hgg.), *Paulys Realencyclopädie der classischen Altertumswissenschaft: Neue Bearbeitung*, 2. Reihe Bd. 6 A 1 (Stuttgart: J. B. Metzlersche Verlagsbuchhandlung, 1936), 1076–1203, bsd. 1193–94; Stroheker, *Dionysios I.*, 23; vgl. auch Berve, *Die Tyrannis bei den Griechen*, 1:489. Vielleicht kannte Galen die heute nur noch durch Photios belegte Schrift, die sein Zeitgenosse Amyntianos über Dionysios verfasst hat; vgl. zu Amyntianos Bowie, „Greeks and their Past," 16.

anführt, ist die Ambivalenz ihrer Rolle für die Wissenschaft. Dies gilt für die Ptolemäer ebenso wie für Attalos III. von Pergamon oder Mithridates VI. von Pontos.

Galen, der während seiner Studienzeit auch Alexandria besucht hat, bietet mit seinen Ausführungen zum dritten Buch der *Epidemien* des Hippokrates wichtige Details zu den Methoden der Bücherbeschaffung für die in Alexandria neugegründete königliche Bibliothek. In seinem Kommentar beschreibt Galen, auf welche Weise die frühen Ptolemäer ihre Bibliotheksbestände vergrößert und wie sie sich besonders wertvolle Manuskripte verschafft haben sollen. Da Galen ohne nähere Kennzeichnung von Ptolemaios spricht, ist nicht sicher, ob der erste oder zweite ptolemäische König gemeint ist.⁴⁷ Der König habe in seinem Drang, Bücher für seine Bibliothek zu erhalten, die in Alexandria anlegenden Schiffe kontrollieren und alle auf ihnen gefundenen Handschriften kurzerhand einbehalten lassen, was in der Bibliothek zur Einrichtung einer Abteilung „aus den Schiffen" (τῶν ἐκ τῶν πλοίων) geführt habe.⁴⁸ Galen qualifiziert dieses Handeln als Folge der φιλοτιμία des Königs, also mit einem für ihn deutlich negativ konnotierten Begriff. Erscheint aus seiner Perspektive diese Art der Bücherbeschaffung als eine Gewalthandlung tyrannischer Herrscher, so zögerte „Ptolemaios" auch nicht, mit Betrug und Eidbruch vorzugehen, wie Galen im Hinblick auf die Beschaffung der Staatshandschriften der attischen Tragiker aus Athen berichtet. Denn anstatt die gegen den hohen Pfandbetrag von dreißig Talenten ausgeliehenen Originale, von denen in Alexandria angeblich Abschriften hergestellt werden sollten, wieder nach Athen zu senden, seien die Originale einbehalten und nur die Kopien herausgegeben worden.⁴⁹ Das Licht, das hier auf den ägyptischen König fällt, ist schon sehr getrübt, und Galen beklagt dann an anderer Stelle auch noch ausdrücklich, dass mit solchen Aktionen der Wert der Buchrollen in die Höhe getrieben und als Folge davon massenhaft Fälschungen hergestellt worden seien. Die königliche Förderung der Bibliotheken in Alexandria und auch in Pergamon entfaltet also eine denkbar negative Wirkung.⁵⁰

---

47 Vgl. Peter M. Fraser, *Ptolemaic Alexandria*, 3 Bde. (Oxford: Clarendon Press, 1972), 1:325.
48 *In Hp. epid. III comm.* 2.4, 17 A.606 K. (= 79 Wenkebach).
49 *In Hp. epid. III comm.* 2.4, 17 A.607–608 K. (= 79–80 Wenkebach); vgl. Fraser, *Ptolemaic Alexandria*, 1:324.
50 *In Hp. nat. hom. comm.* 1.44, 15.105 K. (= 55 Mewaldt): πρὶν γὰρ τοὺς ἐν Ἀλεξανδρείᾳ τε καὶ Περγάμῳ γενέσθαι βασιλεῖς ἐπὶ κτήσει παλαιῶν βιβλίων φιλοτιμηθέντας, οὐδέπω ψευδῶς ἐπεγέγραπτο σύγγραμμα. Im Vorwort zum 2. Buch seines Kommentars spricht Galen noch einmal von der Konkurrenz zwischen Attaliden und Ptolemäern um den Bücherbesitz und der daraus folgenden Bücherfälschung (15.109 K. [= 57 Mewaldt]). Vgl. Esther V. Hansen, *The Attalids of Pergamon*, 2. Aufl. (Ithaca: Cornell University Press, 1971), 170; Fraser, *Ptolemaic Alexandria*, 1:325; 2:481, Anm. 150.

Zur medizinischen Forschung unter den Ptolemäern berichten Celsus und Tertullian von Sektionen an lebenden Menschen, die in Alexandria vorgenommen worden seien. Herophilos habe diese Vivisektionen an verurteilten Straftätern durchgeführt, die ihm von den Königen zur Verfügung gestellt worden seien.[51] Eine Bestätigung für diese Überlieferung findet sich bei Galen nicht. Fraser hat in diesem Zusammenhang die Überlegung angestellt, dass Galen möglicherweise den Ruf des Herophilos nicht habe schmälern wollen und deshalb die Menschenversuche vielleicht bewusst verschwiegen habe.[52] Ein Beispiel aus *De antidotis*, in dem Mithridates für Menschenversuche verantwortlich gemacht wird (s.u.), zeigt indes, wie Galen die Verantwortung von der ärztlichen Seite auf die Seite des Königs verlagern konnte, und ähnliches könnte auch in Galens verlorener Schrift über Sektionen geschehen sein, in der Galen, wie Heinrich von Staden vermutet hat, über die Vivisektionen in Alexandria gesprochen haben könnte.[53]

Unter den hellenistischen Königen nennt Galen neben den Ptolemäern auch Attalos III. Philometor Euergetes und Mithridates VI. Eupator. Zu Attalos III. drückt Galen in *De antidotis* eine gewisse Nähe aus, wenn er die Theriak-Versuche des Attalos neben die des Mithridates stellt und dabei die folgende Formulierung verwendet:

> Dieser Mithridates hat sich, so wie bei uns Attalos (ὥσπερ καὶ καθ' ἡμᾶς Ἄτταλος), darum bemüht, Kenntnisse über alle einfachen Heilmittel zu gewinnen, soweit sie gegen die Gifte wirken.[54]

Auch in *De simplicium medicamentorum temperamentis ac facultatibus* findet sich ein Hinweis auf Attalos, wenn Galen im Zusammenhang mit Wirkstoffen, die aus geheimnisvollen Tieren wie etwa dem Basilisken oder dem Nilpferd gewonnen werden könnten, über „unseren König Attalos" (ἡμέτερος βασιλεὺς Ἄτταλος) spricht, der an solchen Fragen sehr interessiert gewesen sei.[55] Diese freundliche Stimmlage findet sich dann noch einmal in einem Abschnitt von *De compositione medicamentorum per genera*, in dem Galen auf Attalos als seinen Vorgänger in der Herstellung eines Heilmittels für einen Wundverband verweist[56] und den König –

---

51 Vgl. Heinrich von Staden, *Herophilus: The Art of Medicine in Early Alexandria* (Cambridge: Cambridge University Press, 1989), 139–53; 187 (test. 63 a); 190 (test. 66).
52 Fraser, *Ptolemaic Alexandria*, 1:349.
53 Von Staden, *Herophilus*, 151 mit Anm. 34. Vgl. Gal. de ord. libr. suor. 2, 19.55 K.
54 *De antidot*. 1.1, 14.2 K.
55 *De simpl. med. temp. ac fac*. 10.1, 12.251 K.
56 Vgl. John Scarborough, „Attalus III of Pergamon: Research Toxicologist," in Louise Cilliers (Hg.), *Asklepios: Studies on Ancient Medicine* (Bloemfontein: Classical Association of South Africa, 2008), 138–56, bsd. 147–48, Anm. 41.

er ist wieder „unser König von Pergamon" (ὑπὸ τοῦ βασιλεύσαντος ἡμῶν τῶν Περγαμηνῶν Ἀττάλου) – als einen um die Medikamentenlehre sehr bemühten Mann bezeichnet (σπουδάσαντος ἀνδρὸς περὶ φάρμακα παντοῖα).⁵⁷ Für Galen ist Attalos, dessen Tätigkeit als pharmakologischer Autor u. a. auch von Plinius d. Ä. bezeugt wird, offensichtlich eine positive Figur, in der Herrschertum und Wissenschaftsinteresse zusammenkamen; zugleich verweist Galens Ausdrucksweise auf eine lokalpatriotische Verehrung des letzten rechtmäßigen Attaliden, die durch die Entscheidung des Königs, Pergamon an die Römer zu vererben, nicht beeinträchtigt erscheint.⁵⁸

Dem intensiven Bemühen des Mithridates VI. von Pontos, sich gegen Giftanschläge zu schützen, verdankte die antike Pharmakologie das bis in die Zeit Neros am häufigsten verwendete Antidot, das dann durch Neros Leibarzt Andromachos abgewandelt und als Theriak verbreitet wurde.⁵⁹ Dieser grundlegenden pharmakologischen Forschung wegen geht Galen in *De antidotis* wiederholt auf Mithridates ein, dessen medizinische Experimente indes gegen das ärztliche Ethos der Antike verstießen und deshalb der besonderen Rechtfertigung bedurften, erprobte der König Mittel gegen tödliche Gifte doch

> an schändlichen Menschen, die zum Tode verurteilt worden waren. So fand er heraus, dass manche Mittel besonders gut gegen Giftspinnen wirken, andere gegen Skorpione oder Giftschlangen. Und im Hinblick auf die tödlichen Gifte stellte er fest, dass einige Mittel gegen Eisenhut wirken, andere gegen den Meerhasen, und wieder andere gegen dieses oder jenes Gift. Indem Mithridates alle diese Stoffe miteinander vermischte, machte er ein einziges Heilmittel daraus.⁶⁰

Die bittere Ironie des unärztlichen, wenn auch wissenschaftlich erfolgreichen Experimentierens des Mithridates mit zahlreichen, wehrlosen Versuchspersonen lag dann aber darin, dass es ihm aufgrund der beständigen Einnahme seines Antidots nicht mehr möglich war, seinem Leben durch Gift selbst ein Ende zu setzen, nachdem er die Macht an seinen Sohn Pharnakes II. verloren hatte und fürchten musste, an Pompeius ausgeliefert zu werden. Auf das von Appian,

---

57 *De comp. med. per gen.* 1.13, 13.416 K. Zu Attalos' wissenschaftlichen Interessen vgl. Hansen, *The Attalids of Pergamon*, 144–45; Scarborough, „Attalus III of Pergamon," 144–47.
58 Vgl. Scarborough, „Attalus III of Pergamon," 145, Anm. 41.
59 Vgl. Gilbert Watson, *Theriac and Mithridatium: A Study in Therapeutics* (London: Wellcome Institute for the History of Medicine, 1966), 33–44; Adrienne Mayor, *The Poison King: The Life and Legend of Mithridates, Rome's Deadliest Enemy* (Princeton: Princeton University Press, 2010), 240–47; Laurence Totelin, „Mithridates' Antidote: A Pharmacological Ghost," *Early Science and Medicine* 9 (2004): 1–19.
60 *De antidot.* 1.1, 14.2 K.

Plutarch, Cassius Dio und anderen überlieferte Ende des Königs verweist Galen ein erstes Mal gleich zu Beginn seiner Schrift:

> Wenn aber jemand das Antidot jeden Tag einnimmt, wie das in unserer Zeit der Kaiser Mark Aurel (ὁ αὐτοκράτωρ Αὐρήλιος Ἀντωνῖνος) getan hat oder Mithridates selbst, so erlangt er vollständigen Schutz vor den sogenannten tödlichen und zerstörerischen Giften, so wie über Mithridates erzählt wird, dass er, als er sich lieber vergiften als den Römern in die Hände fallen wollte, kein Gift mehr finden konnte, durch das er hätte sterben können.[61]

Im weiteren Verlauf seiner Abhandlung kommt Galen auf die beiden Aspekte seiner Mithridates-Geschichte noch einmal zurück. Während er den Selbstmord des Königs gegen Ende seiner Abhandlung nur noch einmal kurz erwähnt,[62] wird im Hinblick auf die Antidot-Experimente noch eine andere Überlieferung präsentiert, von der allerdings unklar bleibt, ob sie alternativ oder komplementär zu den eingangs berichteten Versuchen des Mithridates zu verstehen ist. Unter den Rezepten, die Galen zusammengestellt hat, befindet sich auch das Antidot des Alexandrinischen Arztes Zopyros, und dazu heißt es:

> Über dieses Antidot wird berichtet, dass Zopyros Mithridates brieflich dazu aufgefordert habe, es an einem der Verurteilten zu erproben, und zwar indem er diesem ein tödliches Gift verabreichen sollte, wobei er riet, diesem dann danach das Antidot zu geben. Oder es sollte zunächst das Antidot verabreicht und dann das Gift getrunken werden. Und er sprach auch die Empfehlung aus, bei Schlangenbissen oder tödlichen Giften ebenso zu verfahren. Und als diese Proben so gemacht wurden, habe der Betroffene keinen Schaden erlitten.[63]

Vergleicht man die beiden von Galen präsentierten Varianten, so stellt die zweite eine entschärfte Fassung dar. Der Menschenversuch geht in diesem Fall auf eine Aufforderung des Arztes Zopyros zurück, dessen Rezept von Galen nicht nur referiert, sondern auch als gut wirksam bewertet wird. In der Verschiebung der Details scheinen die unterschiedlichen Bewertungen von Herrscher und Arzt aus der Perspektive Galens auf: Das von Zopyros vorgeschlagene Experiment, das auf der Prämisse beruht, bereits ein wirksames Gegenmittel gefunden zu haben, beschränkt sich auf eine einzige Versuchsperson und fügt dieser keinen Schaden zu, während Mithridates zuvor als Herrscher erscheint, der bedenkenlos das Leben etlicher von ihm zum Tode Verurteilter für seine Experimente eingesetzt hat. Gleichwohl steht Galen auch Mithridates eher positiv gegenüber, bezeichnet er ihn doch in *De theriaca ad Pisonem*, wo Galen noch einmal ausführlich die Sterbe-

---

61 *De antidot.* 1.1, 14.3 K. Vgl. Mayor, *Poison King*, 347–52.
62 *De antidot.* 2.9, 14.154 K.
63 *De antidot.* 2.8, 14.150 K.: ... τούτων γὰρ οὕτω γινομένων συνέβαινε τὸν ἄνθρωπον ἀδιάπτωτον εἶναι.

szene von Mithridates und seinen Töchtern beschreibt, als „großen Krieger."⁶⁴ Bemerkenswert ist in diesem Zusammenhang auch, dass Galen die Antidot-Versuche der Herrscher trotz der Lebensgefahr, die für die Versuchspersonen bestand, rechtfertigt. Herrscher, so führte er in *De simplicium medicamentorum temperamentis ac facultatibus* aus, hätten die Macht, solche Versuche an zum Tode Verurteilten durchzuführen, und daran sei nichts Verwerfliches (οὐδὲν ἔπραξαν δεινόν); anders sei es aber, wenn Privatleute Abhandlungen über das Thema verfassten. Denn entweder wüssten diese Autoren von dem, worüber sie schreiben, nichts Genaues, oder aber sie könnten ein solches Wissen nur durch verbrecherische Versuche an – offensichtlich ahnungslosen – Menschen erlangt haben, die nichts Unrechtes getan hätten.⁶⁵

In der Reihe der hier behandelten *exempla*, in denen Galen auf Herrscher der Vergangenheit verweist, kommt kaum einmal eine positive Wertschätzung zum Ausdruck.⁶⁶ Abgesehen von „unserem Attalos" und dem „großen Kriegsmann" Mithridates gibt es nur negative Urteile, die in der Regel im Zusammenhang mit Galens Bildungs- und Forschungsideal zu sehen sind. Zu diesem Ideal gehört auch die Forderung, dass der philosophisch gebildete und hohen moralischen Ansprüchen genügende Arzt unabhängig von seinen Patienten sein muss, und dies gilt im Besonderen gegenüber reichen und mächtigen Patienten bis hin zu den Herrschern. Als Vorbild für ein solchermaßen unabhängiges ärztliches Handeln kann Galen auf Hippokrates verweisen, etwa indem er in seiner Abhandlung *Quod optimus medicus sit quoque philosophus* daran erinnert, dass Hippokrates eine Einladung des Perserkönigs Artaxerxes zurückgewiesen und sich auch nur kurz am Hof des makedonischen Königs Perdikkas aufgehalten habe.⁶⁷ Wenn Galen in der pharmakologischen Arbeit von Attalos und Mithridates einen Grund dafür sah, diese Herrscher positiver zu bewerten, als etwa Kyros und die frühen Ptolemäer, dann sicher vor allem deshalb, weil sie sich persönlich um einen Teilbereich der medizinischen Wissenschaft bemühten.⁶⁸ Grundsätzlich steht Galen

---

64 *De ther. ad Pis.* 16, 14.283–84 K.: τὸν μέγαν πολεμιστήν.
65 *De simpl. med. temp. ac fac.* 10.1, 12.252 K. Vgl. Scarborough, „Attalus III of Pergamon," 146–47; Mayor, *Poison King*, 58. Vgl. auch die Rechtfertigung der Vivisektion in Alexandria durch Cels. prooem. 26 (21 Marx); dazu von Staden, *Herophilus*, 144–45.
66 Vgl. dagegen den idealisierenden Rückgriff auf die fernere, vorhellenistische Vergangenheit bei etlichen Vertretern der zweiten Sophistik; dazu Bowie, „Greeks and their Past," 7–18, 22–24, 27–28.
67 *Quod opt. med.* 3, 1.58 K.; vgl. Schlange-Schöningen, *Die römische Gesellschaft bei Galen*, 176.
68 Vgl. *de opt. med. cogn.* 1.2 (41 Iskandar): „In ancient times, when men were not yet obsessed, as they are (now), with the pursuit of luxury, students of this art: the virtuous kings and the sons of divine kings and their kinsmen, were worthy of this (art) on account of their superior innate virtues." Vgl. dazu Nutton , „The Patient's Choice," 253–56.

jedoch monarchischer Herrschaft kritisch und ablehnend gegenüber, wobei die Beispiele aus der vorrömischen Zeit bzw. aus den außerrömischen Herrschaftsgebieten deutlich machen, dass man es hier nicht allein mit einer für viele Vertreter der zweiten Sophistik typischen Reserviertheit der römischen Macht gegenüber[69] oder gar mit einem völligen Desinteresse an politischen Fragen dieser Art zu tun hat.[70] Denn dort, wo Galen seine Maßstäbe eingehalten sieht, kann auch ein römischer Kaiser ausdrücklich gelobt werden, wie es Mark Aurel in einigen Schriften Galens widerfährt. Die Wertschätzung, die Galen diesem Kaiser gegenüber formuliert, spiegelt zwar auch die Anerkennung, die Galen selbst durch seine Nähe zum Kaiser in Rom gefunden hat, aber sie entspricht zugleich den zentralen Aspekten des Ideals, das Galen vom philosophischen Arzt entwirft: Mark Aurel bestätigt persönlich die Bedeutung der Medizin bzw. einer auf die Erhaltung der Gesundheit ausgerichteten Lebensführung,[71] und er erkennt ausdrücklich die aus Galens Perspektive notwendige Unabhängigkeit eines guten Arztes an. Da Mark Aurel Galen, wie dieser in *De praecognitione* berichtet, nicht nur „als ersten der Ärzte und einzigartig unter den Philosophen" (τῶν μὲν ἰατρῶν πρῶτον εἶναι, τῶν δὲ φιλοσόφων μόνον) bezeichnet, sondern auch dafür gelobt hat, „frei" (ὡς ἰατρὸν ἔχομεν ἕνα καὶ τοῦτον ἐλεύθερον πάνυ) zu sein,[72] konnte Galen von sich selbst

---

69 Vgl. oben, Anm. 3; zu Galens Stellung im Rahmen der zweiten Sophistik vgl. Bowersock, *Greek Sophists*, 59–75; Kollesch, „Galen und die Zweite Sophistik," 1–11; Peter A. Brunt, „The Bubble of the Second Sophistic," *Bulletin of the Institute of Classical Studies* 38 (1994): 25–52, bsd. 43–52; Heinrich von Staden, „Galen and the 'Second Sophistic,'" in Richard Sorabji (Hg.), *Aristotle and After* (London: Institute of Classical Studies; School of Advanced Studies; University of London, 1997), 33–54.
70 So aber, bezogen auf Rom, Desideri, „Galeno come intellettuale," 28: „... sua assoluta mancanza di interesse per tal genere di problematica."
71 Z.B. *de san. tuenda* 6.5, 6.405 K. (= 178 Koch). Vgl. auch das negative Urteil über Commodus, zu dem Galen in *de antidot.* 1.13 vermerkt, er habe sich, anders als Mark Aurels, nicht um die Herstellung des Theriaks bzw. um die Besorgung der Zutaten gekümmert (14.65 K.). Vgl. Patricia Gaillard-Seux, „La santé du prince: une illustration des vertus impériales," in Michel Molin (Hg.), *Images et représentations du pouvoir et de l'ordre social dans l'Antiquité; actes du colloque, Angers, 28–29 mai 1999* (Paris: De Boccard, 2001), 27–35, bsd. 30; zur politischen Bedeutung der Gesundheit des Prinzeps vgl. z. B. Gabriele Ziethen, „Heilung und römischer Kaiserkult," *Archiv für Geschichte der Medizin* 78 (1994): 171–91, bsd. 178–80.
72 *De praecogn.* 11.7–8, 14.660 K. (= 128 Nutton). Vgl. den Kommentar von Vivian Nutton, *Galeni De praecognitione*, CMG V 8,1 (Berlin: Akademie-Verlag, 1979), 219. Dabei ist bemerkenswert, dass Galen, der an anderer Stelle darüber klagt, dass reiche und mächtige Patienten die eigentlich notwendige medizinische Behandlung nicht ertragen wollten (*de meth. med.* 11.15, 10.784 K.; vgl. auch Nutton, „The Patient's Choice," 249–50 zu ähnlichen Klagen Galens in *de opt. med. cogn.*), bei der Behandlung Mark Aurels erklärt, dass „die Ärzte bei der Behandlung von Königen (ἐφ' ὑμῶν δὲ τῶν βασιλέων) die sichersten Mittel zur Anwendung bringen" sollten (*de praecogn.* 11.6, 14.659–60 K. [= 128 Nutton]).

meinen, dem Vorbild des Hippokrates auch im Umgang mit den Mächtigen gerecht zu werden.

Bibliographie

Anderson, Graham. *The Second Sophistic: A Cultural Phenomenon in the Roman Empire* (London: Routledge, 1993).
Barigazzi, Adelmo. *Galeno, Sull'ottima maniera d'insegnare; Esortazione alla medicina*, CMG V 1,1 (Berlin: Akademie-Verlag, 1991).
Barras, Vincent, Terpsichore Birchler, und Anne-France Morand. *Galien; l'âme et ses passions. Les passions et les erreurs de l'âme – Les facultés de l'âme suivent les tempéraments du corps* (Paris: Les Belles Lettres, 1995).
Berve, Helmut. *Die Tyrannis bei den Griechen*, 2 Bde. (München: Beck, 1967).
Birley, Anthony R. *Hadrian. Der rastlose Kaiser* (Mainz am Rhein: von Zabern, 2006).
Boudon, Véronique. *Galien, Exhortation à l'étude de la médecine; Art médical*, t. II (Paris: Les Belles Lettres, 2000).
Boudon-Millot, Véronique. *Galien de Pergame: un médecin grec à Rome* (Paris: Les Belles Lettres, 2012).
Bowersock, Glen W. *Greek Sophists in the Roman Empire* (Oxford: Clarendon Press, 1969).
Bowie, Ewen L. „Greeks and their Past in the Second Sophistic," *Past and Present* 46 (1970): 3–41.
Brunt, Peter A. „The Bubble of the Second Sophistic," *Bulletin of the Institute of Classical Studies* 38 (1994): 25–52.
de Boer, Wilko. *Galeni De propriorum animi cuiuslibet affectuum dignotione et curatione*, CMG V 4,1,1 (Leipzig: Teubner, 1937).
De Libero, Loretana. *Die archaische Tyrannis* (Stuttgart: Steiner, 1996).
Desideri, Paolo. „Galeno come intellettuale," in Daniela Manetti (Hg.), *Studi su Galeno. Scienza, Filosofia, Retorica e Filologia; atti del seminario, Firenze 13 novembre 1998* (Florenz: Università degli studi di Firenze, 2000), 13–29.
Eichholz, David E. „Galen and his Environment," *Greece and Rome* 20 (1951): 60–71.
Flemming, Rebecca. „Demiurge and Emperor in Galen's World of Knowledge," in Gill, Whitmarsh und Wilkins, *Galen and the World of Knowledge*, 59–84.
Fraser, Peter M. *Ptolemaic Alexandria*, 3 Bde. (Oxford: Clarendon Press, 1972).
Gaillard-Seux, Patricia. „La santé du prince: une illustration des vertus impériales," in Michel Molin (Hg.), *Images et représentations du pouvoir et de l'ordre social dans l'Antiquité; actes du colloque, Angers, 28–29 mai 1999* (Paris: De Boccard, 2001), 27–35.
Gill, Christopher, Tim Whitmarsh und John Wilkins, Hgg. *Galen and the World of Knowledge* (Cambridge: Cambridge University Press, 2009).
Haas, Arthur E. „Antike Lichttheorien," *Archiv für Geschichte der Philosophie* 20 (1907): 345–86.
Hankinson, Robert J. *Galen, On the Therapeutic Method, Books I and II, Translated with an Introduction and Commentary* (Oxford: Clarendon Press, 1991).
Hankinson, Robert J. „The Man and his Work," in R. J. Hankinson (Hg.), *The Cambridge Companion to Galen* (Cambridge: Cambridge University Press, 2008), 1–33.
Hansen, Esther V. *The Attalids of Pergamon*, 2. Aufl. (Ithaca: Cornell University Press, 1971).

Harris, William V. *Restraining Rage: The Ideology of Anger Control in Classical Antiquity* (Cambridge, Mass.: Harvard University Press, 2001).

Helmreich, Georg. *Galeni de usu partium libri XVII*, 2 Bde. (Leipzig: Teubner, 1907–1909; repr. Amsterdam: Hakkert 1968).

Ilberg, Johannes. „Über die Schriftstellerei des Klaudios Galenos," *Rheinisches Museum* 52 (1897): 591–623.

Iskandar, Albert Z. *Galeni De optimo medico cognoscendo libelli versio arabica*, CMG Suppl. Or. 4 (Berlin: Akademie-Verlag, 1988).

Koch, Konrad. *Galeni De sanitate tuenda*, CMG V 4,2 (Leipzig: Teubner, 1923).

Kollesch, Jutta. „Galen und die Zweite Sophistik," in Vivian Nutton (Hg.), *Galen: Problems and Prospects* (London: Wellcome Institute for the History of Medicine, 1981), 1–11.

Laqueur, Richard. *Timaios von Tauromenion (= Timaios nr. 3)*, in W. Kroll und K. Mittelhaus (Hgg.), *Paulys Realencyclopädie der classischen Altertumswissenschaft: Neue Bearbeitung*, 2. Reihe Bd. 6 A 1 (Stuttgart: J. B. Metzlersche Verlagsbuchhandlung, 1936), 1076–1203.

Le Glay, Marcel. „Hadrien et l'Asklépieion de Pergame," *Bulletin de correspondance hellénique* 100 (1976): 347–72.

Marcone, Arnaldo, Hg. *Medicina e società nel mondo antico; atti del convegno di Udine (4–5 ottobre 2005)* (Florenz: Le Monnier, 2006).

Marx, Friedrich. *A. Cornelii Celsi quae supersunt*, CML I (Leipzig: Teubner, 1915).

Mattern, Susan P. *Galen and the Rhetoric of Healing* (Baltimore: The Johns Hopkins University Press, 2009).

Mayor, Adrienne. *The Poison King: The Life and Legend of Mithridates, Rome's Deadliest Enemy* (Princeton: Princeton University Press, 2010).

Mewaldt, Johannes. *Galeni In Hippocratis De natura hominis commentaria III*, CMG V 9,1 (Leipzig: Teubner, 1914).

Millar, Fergus. *The Emperor in the Roman World* (London: Duckworth, 1977).

Nutton, Vivian. *Ancient Medicine*, 2. Aufl. (London: Routledge, 2013).

Nutton, Vivian. „The Chronology of Galen's Early Career," *Classical Quarterly* 23 (1973): 158–71 [wieder in ders., *From Democedes to Harvey: Studies in the History of Medicine* (London: Variorum Reprints, 1988)].

Nutton, Vivian. *Galeni De praecognitione*, CMG V 8,1 (Berlin: Akademie-Verlag, 1979).

Nutton, Vivian. „Galen's Library," in Gill, Whitmarsh und Wilkins, *Galen and the World of Knowledge*, 19–34.

Nutton, Vivian. „The Patient's Choice: A New Treatise by Galen," *Classical Quarterly* 40 (1990): 236–57.

Portmann, Werner. *Geschichte in der spätantiken Panegyrik* (Frankfurt am Mein: Lang 1988).

Romano, Elisa. „Modelli intellettuali e modelli sociali in Galeno," in Marcone, *Medicina e società nel mondo antico*, 168–79.

Scarborough, John. „Attalus III of Pergamon: Research Toxicologist," in Louise Cilliers (Hg.), *Asklepios: Studies on Ancient Medicine* (Bloemfontein: Classical Association of South Africa, 2008), 138–56.

Schlange-Schöningen, Heinrich. *Die römische Gesellschaft bei Galen. Biographie und Sozialgeschichte* (Berlin: De Gruyter, 2003).

Schlange-Schöningen, Heinrich. „Galen on Slavery," in Marcone, *Medicina e società nel mondo antico*, 180–93.

Squillace, Giuseppe. *Menecrate di Siracusa: un medico del IV secolo a.C. tra Sicilia, Grecia e Macedonia* (Hildesheim: Olms, 2012).
Steger, Florian. *Asklepiosmedizin. Medizinischer Alltag in der römischen Kaiserzeit* (Stuttgart: Steiner, 2004).
Stroheker, Karl F. *Dionysios I. Gestalt und Geschichte des Tyrannen von Syrakus* (Wiesbaden: Steiner, 1958).
Swain, Simon. *Hellenism and Empire: Language, Classicism, and Power in the Greek World, AD 50–250* (Oxford: Oxford University Press, 1996).
Totelin, Laurence. „Mithridates' Antidote: A Pharmacological Ghost," *Early Science and Medicine* 9 (2004): 1–19.
Trampedach, Kai. *Platon, die Akademie und die zeitgenössische Politik* (Stuttgart: Steiner, 1994).
von Staden, Heinrich. „Galen and the 'Second Sophistic,'" in Richard Sorabji (Hg.), *Aristotle and After* (London: Institute of Classical Studies; School of Advanced Studies; University of London, 1997), 33–54.
von Staden, Heinrich. *Herophilus: The Art of Medicine in Early Alexandria. Edition, Translation, and Essays* (Cambridge: Cambridge University Press, 1989).
Watson, Gilbert. *Theriac and Mithridatium: A Study in Therapeutics* (London: Wellcome Institute for the History of Medicine, 1966).
Wenkebach, Ernst. *Galeni In Hippocratis Epidemiarum librum III commentaria III*, CMG V 10,2,1 (Leipzig: Teubner, 1936).
Wenkebach, Ernst. *Galeni In Hippocratis Epidemiarum librum VI commentaria I–VIII*, CMG V 10,2,2 (Berlin: Akademie-Verlag, 1952)
Whitmarsh, Tim. *The Second Sophistic* (Oxford: Oxford University Press, 2005).
Ziethen, Gabriele. „Heilung und römischer Kaiserkult," *Archiv für Geschichte der Medizin* 78 (1994): 171–91.

Philip van der Eijk
Galen on the Assessment of Bodily Mixtures

Abstract: A central concept in Galen's understanding of the bodies of living beings and their interaction with the environment is the notion of mixture (*krasis*). According to Galen, mixtures of the elementary qualities hot, cold, dry, and wet constitute the nature of individual living beings and are causally responsible for a large number of generic, specific and individual bodily and psychological features. Knowledge of a person's mixture is therefore of great diagnostic and therapeutic significance for the doctor. Yet how can mixtures be known and identified? This paper studies the various methods by which Galen believes bodily mixtures can be assessed: empirical observation by means of a thoroughly trained sense of touch, dissection and venesection, indication on the basis of the body's performance, and various modes of inferential reasoning on the basis of external signs.

I Bodies as Mixtures

A central concept in Galen's theory of bodily health and disease is the notion of *krasis*, usually translated "mixture."[1] What Galen means by this is the propor-

---

I am grateful to the editors, Brooke Holmes and Klaus-Dietrich Fischer, for their patience and helpful comments; to Peter Singer for his translation of *Mixtures* and to Christine Salazar and Piero Tassinari for their comments on the translation and interpretation of the text of *Mixtures*; and to Claudia Mirrione, Matyas Havrda, and other members of the Berlin research group "Medicine of the Mind, Philosophy of the Body. Discourses of Health and Well-Being in the Ancient World," for their comments and suggestions. The research for this paper was supported by the Alexander von Humboldt Foundation.

[1] Other translations sometimes used are "blend," "fusion," "constitution," or, less suitably, "temperament." In Galen, the word *krasis* does not have the psychological connotations (in terms of character types) attached to it in later antique and early medieval temperament theory as testified by some of the texts edited by J. Jouanna, "The Legacy of the Hippocratic Treatise *The Nature of Man*: The Theory of the Four Humours," in idem, *Greek Medicine from Hippocrates to Galen. Selected Papers* (Leiden: Brill, 2012), 335–59, esp. 342–55. For discussions of the distinctions between humoral theory, constitution types, and temperament theory in Greek medical thought, see R. Klibansky, E. Panofsky, and F. Saxl, *Saturn und Melancholie* (Frankfurt: Suhrkamp, 1992), 110–24; E. Schöner, *Das Viererschema in der antiken Humoralpathologie* (Wiesbaden: Steiner, 1964), esp. 54–58 and 93–100; H. Flashar, *Melancholie und Melancholiker in*

tional relationship between the elementary qualities hot, cold, dry, and wet in the bodies of living beings and in the material substances that are used in the preservation, maintenance, and treatment of living bodies, such as foods, drinks, and drugs.[2] In the work dedicated to the topic of mixtures (*Peri kraseōn*, Latin *De temperamentis* or the medieval *De complexionibus*), he says that the nature (*phusis*) of a living being is made up of the *krasis* of the four elementary qualities hot, cold, dry, and wet (1.675 K. [= 104.1–3 Helmreich]).[3] This *krasis* can vary from one genus or species of living things to another, between individual members within a species, and even within one individual at different stages of his or her physical development or simultaneously between the different parts of his or her body. Galen explains how the concept of *krasis* is to be understood, how it is to be applied to the study of the physiological constitution of different kinds of living beings and of the parts of their bodies, and what distinct types of mixture can be found. As is well known, Galen distinguishes nine distinct types of bodily *krasis*, one "good mixture" (*eukrasia*), in which the four elementary qualities are in the right proportion to each other, and eight "bad mixtures" (*duskrasiai*), in which either one or two qualities are in excess.[4] Furthermore, in the

---

den medizinischen Theorien der Antike (Berlin: De Gruyter, 1966), esp. 11–49, 60–73 and 105–18. On constitution types in Greek medicine see H. L. Dittmer, "Konstitutionstypen im Corpus Hippocraticum," Diss., Würzburg, 1940.

2 Thus the mixtures that Galen regards as fundamental to the natural constitution of living bodies are mixtures of elementary qualities, not of the elements themselves nor (as is often believed) of the humors, although Galen occasionally also talks of mixtures in this latter sense, e.g., the *melancholikai kraseis* mentioned in *de temper*. 1.641 K. (= 83.4 Helmreich); see also *de loc. aff.* 8.183 K.; *de alim. facult.* 6.661 K. (= 333.3 Helmreich); *de simpl. med. temp. ac fac.* 11.438 K.; and *in Hp. acut. comm.* 15.843 K. (= 327.20 Helmreich). For a discussion of Galen's concept of *krasis* see V. Boudon-Millot, "La notion de mélange dans la pensée médicale de Galien: mixis ou crasis?," *Revue des études grecques* 124 (2011): 261–79. See also W. J. Den Dulk, *Krasis. Bijdrage tot de Grieksche Lexicographie* (Leiden: Brill, 1934); E. Montanari, Κρᾶσις e μίξις: un itinerario semantico e filosofico (Florence: CLUSF, 1979); W. Schwabe, "Mischung" und "Element" im Griechischen bis Platon (Bonn: Bouvier, 1980); T. J. Tracy, *Physiological Theory and the Doctrine of the Mean in Plato and Aristotle* (The Hague: Mouton, 1969); P. Needham, "Compounds and Mixtures," in A. I. Woody, R. F. Hendry, and P. Needham (eds.), *Handbook of the Philosophy of Science*, vol. 6: *Philosophy of Chemistry* (Oxford: North Holland, 2012), 271–90.

3 References to Galen's *Mixtures* follow the page and line numbers in the critical edition by G. Helmreich, *Claudius Galenus. De temperamentis libri III* (Leipzig: Teubner, 1904; repr. with additions, Stuttgart: Teubner, 1969), as well as the corresponding volume and page number in the Kühn edition (K.). English quotations from *Mixtures* are taken from the forthcoming translation by P. N. Singer and P. J. van der Eijk, to be published in P. J. van der Eijk and P. N. Singer (eds.), *Galen, Works on Human Nature* (Cambridge: Cambridge University Press, Forthcoming).

4 See *de temper*. 1.559 K. (= 31.28–32.4 Helmreich). As Galen points out, *eukrasia* is to be understood either in the absolute sense of all qualities being exactly in the same proportion to

same work, and in his works on dietetics, pharmacology, and therapeutics at large, Galen analyzes the efficacy of medical treatment in terms of the interaction between the *krasis* of the body and the *krasis* of the factors to which the body is exposed. For example, when he discusses the effects of foods and drinks on human bodies, or the effects of treatment by massage or bathing, or when he discusses the influence of climate and the seasons on human health and disease, he uses the notion of *krasis* to explain the mechanisms through which the human body and its environment interact.[5] Thus we may well say that the concept of *krasis* is of central underlying importance to Galen's physiology, pharmacology, dietetics, and therapeutics.

It is not immediately obvious why this should be so, i.e., why it would make sense to look at the human body in terms of a "mixture" of elementary qualities, or to define differences between bodies (and bodily parts) in terms of different mixtures of elementary qualities. For most of us, at any rate, to think of bodies as mixtures of elementary qualities can hardly be regarded as a straightforward empirical given.

In response to this, one could of course say that Galen was not the first to speak of the nature of the body in terms of mixture and that he simply followed an established tradition in Greek thought. The concept of *krasis* and cognate terms such as *sunkrēsis* has a long history, which can be traced to some of the Presocratic philosophers and the medical writings attributed to Hippocrates (especially *On Regimen*).[6] In classical and Hellenistic times, Aristotle and the Stoics developed elaborate theories about mixtures and the various ways in which the term *krasis* can be understood. Galen was well aware of these theories, as shown by his discussion of the views of predecessors (notably Aristotle) in Book 1 of *Mixtures* and by his reference to the concept of "total mixture" (τὸ μὲν οὖν

---

each other, or in the relative sense of the qualities being present in that proportion which is appropriate (*oikeios*) to a specific kind of living beings. Likewise, *duskrasia* is to be understood either in the absolute sense of one (or two) qualities prevailing over the others, or in the relative sense of one (or two) qualities being in excess in relation to what is the right, appropriate proportional relationship for that specific kind of living beings. See *de temper.* 1.543–44 K. (= 22.11–20 Helmreich); 1.546–47 K. (= 24.4–15 Helmreich).

5 See G. Harig, *Bestimmung der Intensität im medizinischen System Galens* (Berlin: Akademie-Verlag, 1984), 77–153; P. J. van der Eijk, "Galen's Use of the Concept of Qualified Experience in his Dietetic and Pharmacological Works," in A. Debru (ed.), *Galen on Pharmacology: Philosophy, History, and Medicine* (Leiden: Brill, 1997), 35–57, repr. in P. J. van der Eijk, *Medicine and Philosophy in Classical Antiquity: Doctors and Philosophers on Nature, Soul, Health and Disease* (Cambridge: Cambridge University Press, 2005), 279–98.

6 See W. D. Smith, "Regimen, *krêsis* and the History of Dietetics," in J. A. López Férez (ed.), *Tratados hipocráticos* (Madrid: Universidad Nacional de Educación a Distancia, 1992), 263–72.

ὅλα δι' ὅλων αὐτὰ κεράσαι, 1.562–63 K. [=34,5–19 Helmreich]). On one occasion, Galen refers to "those who have left writings [*hupomnēmata*] about mixtures" (1.640 K. [= 82.11–12 Helmreich]), and several times he reports the relevant views of other, apparently more contemporaneous thinkers without mentioning their names. It is clear that the topic of mixture occupied the attention of philosophers and medical writers, as testified, *inter alios*, by Galen's younger contemporary Alexander of Aphrodisias, who devoted a specialized treatise to the notion of mixture.[7]

Yet one should not too easily take Galen's view of the body as a mixture for granted, as if he had no choice but to follow these earlier thinkers. The medical and philosophical tradition was by no means unanimous in the views entertained of the human bodily constitution, and there were several other options open to Galen, at least theoretically. There was, for example, the humoralist view, set out for the first time by the author of *On the Nature of a Human Being* preserved among the writings attributed to Hippocrates, according to whom it is blood, phlegm, yellow bile, and black bile that make up the human natural constitution. Yet we should remind ourselves that humoralism was not the all-pervading way of looking at the body that later interpretations of Greek medicine have made it to be. The theory of the four humors was, in the fifth and fourth centuries B.C., just one among many different, rival theories, and even if Galen himself was largely responsible for raising the theory of *On the Nature of a Human Being* to canonical status, he did not himself make much use of it.[8] In Hellenistic times, encouraged by the developments in anatomical research, several "solidist" views were developed, i.e., ways of looking at the body as a composition of solid parts and structures. And of course there were the corpuscular theories, such as those held by the Epicureans and in medicine by Asclepiades of Bithynia. Yet Galen strongly disliked these corpuscular theories, for a variety of well-known reasons. As far as bodily structures and parts are concerned, Galen was, of course, strongly interested in the anatomical architecture of the body and in functional anatomy when it came to bodies *in general*, for which he relied, *inter alia*, on his own extensive practice in the dissection of non-human animal bodies; yet when it came to *individual, human* bodies and their differences in constitution, material substance, and functioning, he preferred to approach these in terms of varying mixtures of hot, cold, dry, and wet

---

7 See R. B. Todd, *Alexander of Aphrodisias on Stoic Physics: A Study of the "De Mixtione" with Preliminary Essays, Text, Translation and Commentary* (Leiden: Brill, 1976).
8 See Jouanna, "The Legacy of the Hippocratic Treatise," 335–40.

rather than in anatomical terms.⁹ And as he pointed out in *Mixtures* and in his work *On the Elements according to Hippocrates*, he liked to think that in doing so he was in good company, for he saw himself as standing in the long and venerable tradition of Hippocrates the doctor and Aristotle the natural philosopher.¹⁰

Furthermore, to think of Galen's theory of mixtures as solely and simply the product of tradition would be only part of the story. Galen's theory stands out in comparison with his predecessors in several respects. First, Galen elevated the notion of mixture to a very high status. For it is one thing to say, as previous thinkers did, that within our bodies, there are mixtures of elementary components that influence, to a considerable extent, our state of health, well-being, and flourishing. But it is another thing to say, as Galen does, that mixtures are *the key* to the understanding, maintenance, and treatment of the human body. As already mentioned, Galen more or less equates the mixture with the nature of a living being (1.675 K. [=104.1–3 Helmreich]); as emerges from Book 2 of *Mixtures*, Galen holds the mixtures causally responsible for a large number of bodily and also psychological and ethical features of human beings and animals, both generic features and individual variations; and as Galen famously and notoriously points out in another treatise, it is the *mixtures* of the body, not the body as a whole, on which the capacities of the soul are said to depend.¹¹ Thus for Galen, it is the body's mixture, rather than, say, the bodily organs or internal substances such as *pneuma*, in which a large part of the individual nature of a living being finds its cause and origin.¹²

---

9 Only on one occasion (1.631 K. [= 76.27–77.2 Helmreich]) does Galen seem to imply that there are anatomical differences between individual members of the same species.
10 On Hippocrates and Aristotle as Galen's leading authorities in the area of elementary physiology see J. Jouanna, "Galen's Concept of Nature," in idem, *Greek Medicine from Hippocrates to Galen*, 308–10; P. J. van der Eijk, "Quelques observations sur la réception d'Aristote dans la médecine gréco-romaine de l'époque impériale," in Y. Lehmann (ed.), *Aristoteles Romanus: la réception de la science aristotélicienne dans l'Empire gréco-romain* (Turnhout: Brepols, 2013), 183–93; idem, "Aristotle! What a Thing for You to Say!: Galen's Engagement with Aristotle and Aristotelians," in C. J. Gill, J. Wilkins, and T. Whitmarsh (eds.), *Galen and the World of Knowledge* (Cambridge: Cambridge University Press, 2009), 261–81.
11 This is the treatise usually referred to by its Latin title *Quod animi mores corporis temperamenta sequantur*, "That the Capacities (*dunameis*) of the Soul Follow the Mixtures (*kraseis*) of the Body."
12 Galen recognizes that some features of the human body are not due to the mixtures but to what he calls nature's "shaping capacity" (*diaplastikē dunamis*, 1.635 K. [= 79.21–22 Helmreich]), or to the "original plan" (*kata prōton logon*, 1.619 K. [= 69.17 Helmreich]). For a discussion see P. J. van der Eijk, "Galen on the Nature of Human Beings," in P. Adamson and J. Wilberding (eds.), *Philosophical Themes in Galen* (London: Institute of Classical Studies, 2014), 89–134.

A second reason why Galen's discussion of mixtures deserves more attention, and which I wish to examine further here, lies precisely in Galen's belief in their accessibility and perceptibility. To Galen, mixtures are not just theoretical constructs: they have a certain phenomenological immediacy. In *Mixtures*, Galen goes out of his way to spell out how the *kraseis* can be recognized, known, and assessed. He is particularly concerned with the methods of discovering, distinguishing, evaluating, and judging (*exeuriskein, diagignōskein, krinein*) the mixtures of bodies by reference to certain criteria, standards, or yardsticks (*gnōmōn, kritērion, kanōn*).[13]

Galen's concern with the cognitive accessibility of mixtures, and with the rules for a correct epistemic determination of mixture, can be understood, first of all, from the polemical setting in which he found himself: as several passages testify, Galen clearly felt under considerable pressure to justify his own theory of the mixtures.[14] Furthermore, there is a strong didactic purpose to the work. Throughout *Mixtures*, Galen gives the impression of writing for an audience of aspiring doctors who, as one passage puts it, "want to acquire the ability to discern mixtures" (*diagnōstikos*, 1.559 K. [= 32.8 Helmreich]). He often addresses his students in the second person singular or plural, and the whole text has a vividness that strongly suggests a teaching setting.[15] Finally, and relatedly, the interest of getting the bodily mixtures right is not just a theoretical or intellectual exercise: the correct assessment of a body's mixture is of the greatest importance to medical practice. It is vitally important, Galen argues, to know what kind of mixture a particular body or bodily part has, for this is an essential consideration for

---

13 In *Mixtures*, Galen uses *diagnōsis* and related terms not just in relation to pathological states but also for the determination of healthy conditions, i.e., the balances between hot and cold, and dry and wet, or the state of a particular bodily part in a particular stage of natural growth or development (e.g., childhood or old age). This makes good sense, for the original meaning of *diagnōsis* is discernment, recognition, distinction, identification, discrimination, usually in contexts where the item to be identified is not straightforward or immediately accessible but needs to be distinguished from other items or to be brought to light through a mediating cognitive process. The "diagnostic" (in our sense of the word) dimension of Galen's theory, i.e., the determination of pathological states, is rarely mentioned in *Mixtures* but figures prominently in *Affected Parts* (see, e.g., 8.6–9 K.) and the writings on the pulse.

14 Thus in 1.610–11 K. (= 64.7–12 Helmreich) Galen addresses a potential objection to his theory: "These points have been stated by me; and it is now evident that the simple bad-mixtures were shown not only theoretically to exist in animal bodies, but also that the indicators of each of them are plain, too, not just in terms of heat, moisture, softness, and hardness, but also in terms of all the other differences in the condition of the whole body."

15 For a discussion of this feature in *Mixtures* and other Galenic writings, see P. J. van der Eijk, "Galen and the Scientific Treatise: A Case Study of *Mixtures*," in M. Asper (ed.), *Writing Science* (Berlin: De Gruyter, 2013), 145–75.

the diagnosis and treatment of such a body; and indeed, as Galen argues, much damage and many errors in medical treatment are caused by erroneous assessments of bodily mixture (1.534 K. [= 16.24–25 Helmreich]). So how do we find out what mixture a particular body has? How can we tell whether a particular patient has a wet mixture or a dry mixture, or an imbalance that is both wet and hot, i.e., wetter and hotter than is proper? And how do we ascertain an assumption about the bodily mixture in the face of disagreement and rival claims?

II Assessing the Mixtures by the Sense of Touch

The most straightforward method by which bodily mixtures, at least those in their actual state,[16] can be recognized, Galen argues, is by means of empirical observation, especially the sense of touch, for especially hot and cold, and also dry and wet, are among the tangible qualities that are the particular object of the sense of touch.[17] Touch (*haphē*) had been regarded as an important diagnostic tool in ancient medicine from the very beginnings in Hippocratic times; and its role had been enhanced by the increasing significance attributed to the pulse as a diagnostic indicator in late Classical and Hellenistic medicine, especially by Praxagoras and Herophilus.[18] Galen inherited and further systematized this major area of medical diagnosis, in which elaborate theories and classifications of different kinds of pulse were developed whose recognition was attributed to an extremely sophisticated, discriminative sense of touch. Yet there was also a philosophical tradition of thinking about touch, which again goes back to Presocratic times and which finds its main representative in Aristotle, who had emphasized the major role of touch in human cognition and had

---

[16] As distinct from states that are hot, cold, dry, or wet "in capacity" (*dunamei*); for this distinction, see *Mixtures* 1.560 K. (= 32.14–23 Helmreich) and 1.646–47 K. (= 86.9–87.7 Helmreich).
[17] See 1.588 K. (= 50.13–17 Helmreich); see also 1.598 K. (= 56.12–14 Helmreich).
[18] See H. von Staden, *Herophilus* (Cambridge: Cambridge University Press, 1989), 262–88; see also the summary of his paper "Touch in Ancient Medicine: From a 'Harvest of Sorrows' to Nature's Music in the Arteries," read at the 143th Annual Meeting of the American Philological Association, 2012, in http://apaclassics.org/index.php/annual_meeting/143rd_annual_meeting_ abstracts/06.5.von_staden/. In *de temper.* 1.538 K. (= 19.12 Helmreich) and 1.540 K. (= 20.12 Helmreich), Galen refers to the second book of his treatise *On the Recognition of Pulses* (Περὶ διαγνώσεως σφυγμῶν), though mainly for purposes of terminology. In *Mixtures* itself pulsation is mentioned only once: "For if the mixture of the body as a whole is an even one, then not only will the whole chest automatically be very broad in these cases, but the veins will be broad, and the arteries both large and endowed with a very strong, vigorous pulse [σφοδρότατον σφύζουσαι]" (1.625 K. [= 73.8 Helmreich]).

posited a close connection between human intelligence and humans' exceptionally accurate sense of touch: humans have the most sophisticated and discriminative sense of touch of all animals, and within the human species, those who have the most subtle sense of touch are also the most intelligent.[19]

Galen follows suit in making remarkable claims about the accuracy of the sense of touch, especially by means of the skin on the inner side of the human hand:

> Human skin ... is at the precise midpoint between all the extremes, hot, cold, hard, and soft. And this is especially true of the skin of the hands. For this part was meant to be the instrument of assessment [*gnōmōn*] of all perceptible objects, crafted as it was by nature as the organ of touch proper to the most intelligent of animals. It therefore had to be equidistant from all extremes, whether of hot, cold, dry, or wet. ... Man has the best mixture, not only of all animals, but in fact of all bodies generally, and, furthermore, the skin on the inside of the hand is free from all the excesses suffered by the other parts. ... The skin which is middle, not just with regard to all the parts of the human being, but with regard to all that exists – all bodies that are subject to generation and decay – is ... that which has preserved its natural state; and it is by virtue of this, we say, that its sense of touch is made most precise. ... If, then, you take skin as the yardstick [*kanōn*] and, as it were, criterion [*kritērion*] against which to examine all other parts of the animal, and compare these with it, you will find the eight distinct types of imbalance [*duskrasia*] within those parts. [1.563–64 K. (= 34.20–27 Helmreich); 1.564–65 K. (= 35.13–16 Helmreich); 1.567–68 K. (= 37.8–12 and 21–24 Helmreich)]

According to Galen, the reason why *human* touch is so perfectly suited to simply and immediately feeling the mixture of the body is that humans, as distinct from other animals and all other existing things, are at the center of the universe in terms of constituting the perfect, precisely equal balance between hot, cold, dry, and wet: humans present the norm, the standard, the point of reference in comparison to which all other beings, and the mixtures that make up their natures, are assessed. And within the human body, it is the skin on the inside of the hand that is the part of the body that is the precise midpoint.[20] According-

---

19 Arist. *de an.* 413a31, 422b17–424a17, 421a7–21; *part. anim.* 686a25–690b20.
20 Cf. also 1.575–76 K. (= 41.24–42.11 Helmreich): "For indeed it was shown above that human beings are the most well-mixed, not just among animals or plants, but also among all other things. Since, however, human beings are composed of many different parts, it is clear that that part whose mixture is in the middle of all of them with regard to mixture will also be well-mixed in the absolute sense. For, of that animal that is in the middle with regard to mixture, the middle part will be the most well-mixed of all, in the absolute sense. And it was shown that within human beings, this part is that known as skin [*derma*], and more especially the inner side of the hand, provided that this has been preserved in the state produced by nature. It was further shown above that not every human being's skin without further qualification is in the middle of

ly, Galen identifies the human hand, and especially its inner part, as the naturally most suitable part for the determination of mixture. The ideal human being has the most balanced mixture in absolute terms; he therefore not only represents a standard of reference for the determination of departures from the absolute mean, but he is also best equipped by nature to *perform* such determination himself by means of his hands.

In simply positing, without much in the way of argument, that humans are at the center of the universe, Galen adheres to an existing tradition in Greek thought. Yet Galen goes further in making extraordinary claims both about the sense of touch and about the ability to remember tactile sensations several years after they were made:

> The recognition [*diagnōsis*] of things that are hot, cold, dry, and wet in activity is something that is accessible and knowable to everyone; for our sense of touch, the same sense that has taught us that fire itself is hot and ice cold, is naturally able to discern [*diakrinein pephuke*] such differences. ... Let us rely primarily and especially on the sense of touch in making our evaluation [*krinōmen*] of the heat that belongs to different ages. And the evaluation [*krisis*] will be best carried out on one single body, that of an infant. For it is not impossible to remember what its heat was like at two years old, in relation to what it is now, after an interval of, say, two or three years. If any change appears to have taken place at all and the infant has become either hotter or colder, it is then no difficult matter to draw a conclusion [*sullogizesthai*] about the further increase that will take place up to the prime of life. [1.588 K. (= 50.13–17 Helmreich); 1.589–90 K. (= 52.4–12 Helmreich)]

Yet things are not so easy as they seem, for in order to obtain this ability of touching, one has to go through a long and arduous process of training and practice (*askēsis*). This is the didactic component of *Mixtures* which I already mentioned, details of which Galen sets out in the following passage:[21]

> One who wishes to have the ability to distinguish [*diagnōstikos*] mixtures ought to begin his schooling with those natures that are well-mixed and middle within each genus. By comparing the others with these, he will easily find out the predominant or deficient in each case. ... We start from the hottest of all things that reach our senses, such as fire, or water at its extreme boiling point, and go down to the coldest of all those we know, such as ice or snow; we conceptualize a line between them; and we divide this line at

---

all that exists, but only that of the one who is most well-mixed, there being a considerable degree of difference between human beings themselves. The most well-mixed human being is whichever has a body precisely in the middle of all the extremes, thinness and thickness, softness and hardness, and also heat and coldness."

21 Elsewhere, too, Galen speaks of the importance of training and practice (*askēsis*) and of the long way the student has to go to acquire and develop this sense: see 1.592–98 K. (= 53.4–56.11 Helmreich).

its precise midpoint. In this way we will find out conceptually the point of good balance, which is equidistant from each of the extremes. But we can also in a way create it physically, by combining an equal volume of ice and boiling water. ... And so it is no difficult matter, either, to touch the product of this mixture and so to hold that which is at the midpoint amongst all existent objects as regards the opposition of hot and cold, and to remember this, and to evaluate all other objects by using this as a standard with which to compare them. ... Here, too, it is no difficult matter to distinguish such a body by both sight and touch, to consign it to the memory [παραθέσθαι τῇ μνήμῃ] and to use the object as the yardstick [*kanōn*] and criterion [*kritērion*] for the distinguishing of objects that are deficiently or excessively wet and dry. [1.559 K. (= 32.8–12 Helmreich); 1.560–61 K. (= 32.27–33.8 Helmreich); 1.561 K. (= 33.10–13 Helmreich); 1.561–62 K. (= 33.16–20 Helmreich)]

The use of expressions like "easily" and "no difficult matter" is clearly meant to reassure and encourage the student in the beginning stages, for when it comes to the determination of the mean in living beings, matters get more complicated. The discipline of practising (*askēsai*) the recognition (*gnōrizein*) of the middle within each genus of animal, and the middle of all, is not for everyone, Galen says, but only for one who is extremely hardworking, and who can rely on long experience and a great deal of learning.[22] Galen compares this procedure with the way in which sculptors and painters achieve the greatest beauty in their painting or sculpting of animal species by aiming for the middle within that particular species, and he mentions as example the sculptor Polyclitus with his famous *kanōn* ("yardstick") which, Galen argues, represents the perfect proportions (*summetria*) of the human body. Further down in Book 2 of *Mixtures*, he gives more detailed instructions about how to develop one's sense of touch, e.g., by going to Roman bath houses and experiencing the extremities of heat and cold, and he spells out further rules about how to avoid mistakes in assessing by means of the sense of touch the mixtures of living bodies and bodily parts. One such rule is the comparative use of bodies that are as similar as possible in the relevant respects.[23] The tone becomes more stern here, and Galen warns that one should not take the training process lightly,[24] insisting that only long and thorough discipline in schooling the sense of touch will lead to correct assessment of the mixture.[25]

Even if Galen concedes that an extensive training program is required for the aspiring doctor to develop his haptic sensibility, his belief in the accuracy and subtlety of the sense of touch is, at least to most of us, amazing. What is partic-

---

22 See 1.566 K. (=36.10–11 Helmreich) and 1.592–93 K. (= 53.4–13 Helmreich).
23 See 1.592 K. (= 52.29–53.4 Helmreich).
24 See 1.592–93 K. (= 53.4–11 Helmreich).
25 See 1.594 K. (= 54.4–7 Helmreich).

ularly remarkable is that he does not seem to be bothered by what we would consider the problem of subjectivity. In this he is not alone, for the absence of an explicit concern with objective measuring is a more general characteristic in which ancient thought strikes us as so remote from our way of thinking, even if we allow for the possibility that Greek and Roman doctors had more refined haptic abilities than we have today (and similar points can be made for the other senses). One relevant point here is the optimistic belief held by ancient thinkers such as Aristotle and Galen in a kind of natural kinship between humans and their environment, which was thought to manifest itself in the inborn ability of the senses to have cognitive access to the natural world. Yet this cannot be the whole story, for in Galen's time, the reliability of sense perception had come under fierce attack from the Sceptics. Galen is aware of this also in *Mixtures*, and he alludes to it in a context in which he is referring to disagreements between unnamed doctors about the question of whether children are hotter or colder than adults (1.583–85 K. [= 47.3–48.22 Helmreich]). After reviewing arguments produced by both sides, he goes on to argue that the issue cannot be settled unless we clarify the starting-points of our assertions. As the context of the passage confirms, Galen is not arguing here against real Sceptics (whom he despises) but against people within his own camp, i.e., those who have opted to follow the "best sect in philosophy" in positing the hot, the cold, the dry, and the wet as principles and elements (1.589–90 K. [= 51.10–12 Helmreich]). But he warns them that their position comes close to Pyrrhonism without them realizing it. The problem with his opponents is not so much that they question the reliability of senses, but rather that they base their assertions on untested and/or unclear presuppositions. For example, they argue about qualities without making it clear how qualities may be recognized and what evident data of experience there are to confirm if their assertions are right or not. Therefore, in Galen's view, they act as if it were impossible to gain any secure knowledge about these things. While it seems very unlikely that this is in fact what these people claimed, Galen insists that their scepticism is not justified and that their failure to accept certain starting points is inappropriate. He dismisses their views as examples of inappropriate Pyrrhonian scepticism and, in a remarkable passage, brushes them aside with blunt rhetorical force.[26]

---

26 "And so they should declare everything that comes from the senses untrustworthy, and refuse to call swans 'white' without first investigating the subject logically; and the same applies to chalk, the day, and even the sun itself. And they should be equally distrustful of their ears, in relation to sounds, of their nostrils, in relation to smells, and of their sense of touch, in relation to all tangible objects. Well, is this not by now a Pyrrhonist kind of confusion – which is to say, an interminable piece of nonsense. It is assuredly wrong for those who have chosen the best sect

To us, Galen's reply appears rather dogmatic and insufficiently appreciative of the concern raised by his critics, viz. that some questions simply cannot be decided by sense perception, and that for some issues one needs to invoke other methods. Yet a little later in the work, Galen concedes that in a number of cases, touch alone is not sufficient and a process of inferential or theoretical reasoning is needed to accompany, guide, or even replace the haptic observation, even though the question of exactly which issues are empirically decidable and which ones can only be settled by reason may continue to be a matter of dispute.

III Inference from Signs, Reasoning about Causes, Parts, and Wholes

This brings us to the second, more indirect way of determining the mixtures that Galen distinguishes. Galen often claims that we can recognize internal physiological and pathological states and processes within the body on the basis of external signs and symptoms (*sēmeia*) that the body displays. Again, Galen stands in a long tradition here, for the idea of the body as a carrier of meaning, or significance, presenting signals (*sēmeia, tekmēria*) intended to be read or interpreted by the observer (the doctor or the patient), can already be found in several of the writings attributed to Hippocrates.[27] The idea was captured in the phrase that "what is manifest offers a view on what is non-evident" (ὄψις ἀδήλων τὰ φαινόμενα), a slogan attributed to early Greek philosophers and doctors such as Anaxagoras (DK 59 B 21), Democritus (DK 68 A 11), and Diocles (fr. 56 van der Eijk). By Galen's time, the concept of the body as a provider of signals that need to be interpreted was well established in Greek and Roman medicine, and elaborate theories about inference from signs, both in medicine and in other fields, had been developed.[28] In this connection, Galen often uses the term *gnōr-*

---

in philosophy, namely that which posits the hot, the cold, the dry, and the wet as principles and elements, to wander so far from the men who have posited these things as to fail to realize, first, that the starting-points in any inquiry are matters that are evident, both to the senses of perception and to the understanding, and secondly, that one who is in confusion in respect of these matters will be wasting his time in any further inquiry. For he has not even left himself a place from which he may start" (1.589–90 K. [= 51.4–17 Helmreich]).

27 See the discussion by B. Holmes, *The Symptom and the Subject* (Princeton: Princeton University Press, 2010).

28 See the discussions by J. V. Allen, *Inference from Signs: Ancient Debates about the Nature of Evidence* (Oxford: Oxford University Press, 2001); D. N. Sedley, "On Signs," in J. Barnes and J.

ismata, the externally perceptible "indicators," on the strength of which one can get to know the mixture of a particular body.[29] Yet Galen insists that this use of indicators has to follow strict epistemological rules. Not all external features can be used as signs or indicators; one needs to know which features are significant and appropriate, whether the correlation between sign and signified is necessary or whether it applies for the most part or only occasionally;[30] and in order to interpret them correctly, one needs to know the causes through which they have come about and the conditions under which they have arisen. A recurrent point that Galen emphasizes in this connection is the question of whether a particular observable bodily feature is present by nature or as a result of habituation, for this makes all the difference for its correct interpretation.[31]

In Book 2 of *Mixtures*, Galen discusses a large number of observable bodily features that may serve as indicators of a body's internal mixture, such as hairiness and baldness,[32] the shape of particular bodily parts (such as the eyes, the

---

Brunschwig (eds.), *Science and Speculation* (Cambridge: Cambridge University Press, 2005), 239–72.

29 See *de temper.* 1.551–52 K. (= 27.7–16 Helmreich). On the meaning of the term *gnôrisma* in Galen, see A. Debru, "Γνώρισμα chez Galien: vers une sémiologie de la santé," in Isabelle Boehm and Nathalie Rousseau (eds.), *L'expressivité du lexique médical en Grèce et à Rome: hommages à Françoise Skoda*, Coll. Hellenica (Paris: Presses de l'Université de Paris-Sorbonne, 2014), 171–77.
30 Failure to observe this distinction is a fault for which Galen criticizes other doctors in 1.622 K. (= 71.6–9 Helmreich). See also 1.643 K. (= 84.14–18 Helmreich): "All these are mistakes that most doctors make by turning away from specific indicators to those which happen to obtain for the most part, but not throughout."
31 This is shown by the following passage, from which it also emerges that the aspiring doctor needs to possess a considerable degree of anatomical background knowledge in order to draw the correct inferences: "It is better if we are able to tell for ourselves, on the basis of certain signs [*sēmeion*], rather than by hearing it from somebody else, whether an individual has these characteristics by nature or has acquired them by habituation. And the teacher of these indicators [*gnôrismatōn*], as indeed of all others, is the wonderful Hippocrates. Those whose veins are wider are hotter by nature; those whose veins are narrower are colder. The function of the hot is to widen and inflate them; so that it seems reasonable that, as a rule, narrowness of veins goes together with the fatty, thicker kind of condition, and thinness of condition with a wideness of veins. If, however, someone combines being fatty and thick with having wide veins, this person has become fatty by habituation, not by nature; and similarly, if he had narrow veins, but is thin, he cannot have been so naturally. And, he says, in the case of those who are subjected to starving, one must conduct the investigation of the correct proportion on the basis of these, namely, the wideness or narrowness of the vessels, and not on the basis of the other features of the bodily condition as a whole. For those with narrow veins have little blood and cannot endure a long abstention from food, while those with wide veins and a large quantity of blood can abstain from food without even suffering harm" (1.604–605 K. [= 60.17–61.6 Helmreich]).
32 See for example 1.611–12 K. (= 64.14–65.4 Helmreich).

ears, the nose), and the quality of the flesh. Among these features we also find what we would call physiognomical characteristics; and indeed at one point Galen refers to the practice of people whom he calls "those who attempt the art of physiognomics" (1.624 K. [= 72.10 Helmreich]). Contrary to their practice, however, Galen interprets these bodily features not as evidence of character and psychological disposition, but as further indicators of the internal physical states and processes within the body, i.e., the physiological causes that bring these features about.[33] Furthermore, while being sympathetic to the physiognomists' consideration of the need to distinguish between different bodily parts, Galen criticizes them for relying on experience only and for failing to investigate the causal relationship between internal bodily states, psychological states, and external bodily features:

> Even those who attempt the art of physiognomics do not make assertions about all features in absolute terms: they, too, are taught by experience. If someone has a considerable amount of hair on the chest, they assert him to be spirited, but if on the thighs, lustful. Yet they do not add the reason. Even when they state that the former type is similar to a lion and the latter to a goat, they have not discovered the first cause. Why the lion is spirited and the goat lustful is a matter for reason [*logos*] to discover; those [who stop] short of this point have simply described what happens and omitted its cause. However, the natural philosopher [*phusikos anēr*] attempts to discover the causes of these facts, as of all others. It is because they are in an uneven state as regards the mixture of their [different] parts that not only the lion and the goat, but also a very large number of other animals, are well suited to different activities in different ways. Now, for the most part this area has been well discussed by Aristotle; but the point that is of value for present purposes is already apparent: why, in investigating the mixtures of human beings, we must examine each part individually, and not think that if someone has a hairy chest, it necessarily follows that his whole body is comparatively dry and hot, but rather that the hot is very great in his heart, which is also why he is spirited. (1.624–25 K. [= 72.10–73.1 Helmreich])

This passage stresses the need, on which Galen insists throughout his discussion, to distinguish between different bodily parts. He warns against the danger of drawing false inferences about the whole body on the basis of features that are indicative only of a part of the body, or inferences about some bodily parts on the basis of features that are indicative of different parts of the body. As he

---

33 The same approach is found in Gal. *ars med.* 11.3–5, 1.335–36 K. (= 306–307 Boudon-Millot); for a discussion see H. von Staden, "The Physiology and Therapy of Anger: Galen on Medicine, the Soul, and Nature," in F. Optas and D. Reisman (eds.), *Islamic Philosophy, Science, Culture, and Religion* (Leiden: Brill, 2012), 63–87, esp. 72–78.

points out time and again, even within the human body, the mixture may vary from one bodily part to another.[34]

In addition to inferential reasoning, empirical observations sometimes need to be accompanied by theoretical considerations about the conditions under which they are made, especially concerning the causes of the phenomena observed:

> The criterion [*kritērion*] by which we recognize [*diagnōseōs*] this [i.e., the fact that phlegm is the coldest and the wettest of all the humors that are present in the body of a living being], too, is the sense of touch, as laid out by Hippocrates in his *Nature of Man*. But while the fact that it is cold is recognized [*diagignōskei*] by the sense of touch alone, the fact that it is also wet is recognized by the sense of touch in conjunction with that of sight, and with the reasoning-faculty [*logismos*]: the senses of touch and sight, in that it appears cold to both of them, the reasoning-faculty, in that it must make the distinction [*diorisamenos*] that it is not by virtue of the quantity of heat that it (i.e., phlegm) has acquired this characteristic, but through its innate moisture. This, then, is the situation as regards the parts and humors within the body. (1.603–604 K. [= 59.24–60.5 Helmreich])

The "distinction" (*diorismos*) to which Galen refers here is concerned with the way in which the high degree of wetness of phlegm has come about. Why is this relevant for the assessment of phlegm's wetness? The answer is that the wetness is *sumphutos*, "natural" or "connate," rather than "acquired" (*epiktētos*) or "accidental" (*kata sumbebēkos*). This is a point that Galen elaborates on elsewhere (1.604 K. [= 60.11 Helmreich]), when he spells out the indicators on the strength of which one can determine "whether an individual was of a certain kind by nature or became so by habit."[35]

---

34 See 1.629–30 K. (= 75.12–23 Helmreich): "All these distinctions must be made [*diorizesthai*] by one who intends to recognize [*diagnōsesthai*] a mixture properly. If the skin is comparatively dark, it does not immediately follow without qualification that the human being as a whole is comparatively hot; this is only the case if all the other parts are in the same state. The difference between skin being relatively dark or light may depend on whether an individual has spent more time in the hot sun or in the shade; this is not relevant for the overall mixture of the body. The skin itself will become drier through exposure to the sun and wetter through time spent in the shade; but the natural mixture of liver, heart, or any other organ will not automatically be altered. It is therefore advisable, as already stated, to furnish oneself with the specific indicators [*gnōrismata*] of each of the parts." See also 1.635 K. (= 79.18–20 Helmreich).

35 See *de temper*. 1.607 K. (= 62.1–3 Helmreich): "... to make the distinction [*diorizesthai*], under each individual head, whether such-and-such a person is colder by nature or as a result of habit"; and 1.628–29 K. (= 75.3–13 Helmreich): "And indeed this distinction [*diōristai*], too, has been very well made by Aristotle, in many contexts. One should pay attention to it above all others, and investigate whether a body is hot by virtue of its own hot, or an acquired one. So, for example, all [bodies] undergoing putrefaction are hot by virtue of an acquired hot, but cold by

The words *diorismos* and *diorizesthai* used here are familiar from other works of Galen. They refer to the specifications of the conditions under which a claim or statement obtains.[36] A generalizing statement such as "phlegm is cold and wet" is in need of specification of the kind of wetness that is being talked about, and a statement such as "poppy is cooling" is in need of specification of the conditions under which this applies. Galen speaks of *diorismoi* also in relation to empirical examination, testing, or trial (*peira* or *empeiria*), where they refer to the specifications or considerations that the physician or the researcher needs to take into account for a correct execution of the examination of a patient's body, or the trial of a particular treatment, or, indeed, for a correct interpretation of the results of such a trial. In the passage just quoted, the point is that one should be clear about the cause of the heat that one feels by means of touch: Is this natural heat, or is it acquired? This is a relevant point for the correct assessment of the tactile observation. To put it in modern terms: in order to make a correct interpretation of empirical observations, further information is needed about the source or cause of the heat. Failure to use this information may lead one to attribute the symptoms to the wrong causes; and such misattribution may, in turn, lead to erroneous interpretation and diagnostic judgement and even, ultimately, to an erroneous decision about the treatment of the patient.

Diorismoi are also needed for the correct execution of empirical trial, as shown in the following passage:

> To perform an examination on the basis of opposite types of nature involves a considerable distortion [*paralogismon*], as the difference is sometimes not due to age, but to the naturally obtaining mixture of the bodies being tested [*dokimazomenōn*]. One should, in the same way, choose bodies that are basically similar in terms of their overall daily regime, as well as in terms of the time at which the examination is conducted. ... Everything except the actual age should be as similar as possible. So too in the case where you make the comparison with one child with itself, obviously you must take care that all external conditions are precisely the same, in order to avoid the false attribution of some difference in heating

---

virtue of their own; and the bodies of the inhabitants of the southern regions are, similarly, hot by virtue of an acquired hot but cold by virtue of their own. In our own part of the world, in winter, the proportion of hot which is there by nature is greater, and that which is acquired smaller; while in the summer the acquired portion is greater, and the connate less. And so all these distinctions [*diorizesthai*] must be made by one who intends to recognize [*diagnōsesthai*] a mixture properly."

[36] On Galen's concept of *diorismos* see H. von Staden, "Inefficacy, Error, and Failure: Galen on δόκιμα φάρμακα ἄπρακτα" in Debru, *Galen on Pharmacology*, 59–83, and P. J. van der Eijk, "Galen's Use of the Concept of Qualified Experience."

and cooling arising from some of these to the change in age. (1.591–92 K. [= 52.18–23 and 52.28–53.4 Helmreich])[37]

IV Indication on the Basis of Activities

A further method of determining bodily mixture that Galen mentions several times is the indication on the basis of the activities (*energeiai*), either the activity of the living being as a whole or that of a specific bodily part. Thus in Book 1 of *Mixtures*, Galen points out that the state of good mixture (*eukrasia*) can be recognized on the basis of the optimal performance of the characteristic activity of the living being, such as intelligence (*phronēsis*) in the case of humans; again, there is an Aristotelian heritage here, which Galen explicitly acknowledges.[38] He continues this perspective in Book 2, where he discusses how the best human being can be recognized:

> The most well-mixed human being is whichever appears to have a body precisely in the middle of all the extremes, thinness and thickness, softness and hardness, and also heat and coldness. ... Such, then, is the most well-mixed man with regard to his body; in respect of his soul, similarly, he will be precisely in the middle between boldness and cowardice, hesitancy and rashness, pity and envy. Such a person will be good-spirited, affectionate, generous, intelligent. *It is from these things, then, that the most well-mixed person is recognized primarily and especially* [*gnōrizetai prōtōs kai malista*], and quite a few others are present in conjunction with them, which belong to those things that depend [on these] necessarily. For such a person also eats and drinks in a well-proportioned way, and digests his food well, not just in his stomach but also in his veins, and throughout the whole condition of his body; and, to speak generally, all his "natural" and "psychological" activities are faultless. For he is in the best state as regards perception and motion of the limbs; he also always has a good color and good breathing; he is midway between somnolence and insomnia, smoothness and hairiness, and the black and white color; as a child his

---

[37] Galen often criticizes others for failing to make the right distinctions, as in 1.641–42 K. (= 83.14–84.1 Helmreich) and 1.594 K. (= 53.25–54.7 Helmreich).
[38] "Those [indicators] that are applicable to animal species, on the other hand, are evaluated on the basis of the perfection of the activity [*energeias teleiotēti*] appropriate to each. It is appropriate for a human being to be very intelligent; for a dog to be both very docile and very brave; for a lion to be very brave only; for a sheep to be very docile. Moreover, that the activities of the body should be appropriate to the character of the soul has been shown by Aristotle in the *Parts of Animals*, and no less by us too. ... [The well-fleshed person] will be completely well-mixed; for good balance with regard to well-fleshedness is a product of good mixture. It will also automatically be the property of his body that it is in the best state as regards its activities [*energeiai*]" (1.565–66 K. [= 35.27–36.6 Helmreich]; 1.567 K. [= 36.24–28 Helmreich]). For the Aristotelian heritage see *part. anim.* 661b28–36.

hair will be blond rather than dark; when he reaches his prime, the reverse. (1.576–77 K. [= 42.8–11 and 42.16–43.9 Helmreich])

The same point applies to the functioning of specific bodily parts in a large variety of creatures, such as the eyes, the lung, the brain, or the stomach. This good functioning cannot be assessed by touch but only by overall consideration of bodily performance.[39]

V Venesection and Dissection

Most of the time, Galen recommends observation of external features and performance as well as inferential reasoning as the most appropriate ways of finding out a bodily mixture. Sometimes, however, he refers to more invasive methods. Thus on one occasion he refers to venesection (*temnein*) in people who are thin and thin-veined.[40] There are also several references to animal dissection and subsequent probing of the internal organs:

> The flesh of spleen, kidneys, and liver is wetter than skin to the same degree that it is also softer; and hotter to the same extent to which it has more blood in it. That of the heart, meanwhile, is drier than all these to the same extent that it is harder; at the same time it is hotter, not only than these, but than any other part of the body, absolutely. You can even learn this, too, clearly by sense perception, in the context of those animal dissections [*anatomais*] which are carried out on the chest area. If you insert [*kathenti*] your fingers into the left ventricle of the heart, you will find [*heurēseis*] that this place is quite considerably hotter than all others. But while the flesh of liver, spleen, kidneys, and lung is simple in nature, these organs having come into existence as a growth surrounding the arteries, veins, and nerves, that of the heart is not simple: just as there are fibers within the muscles, around which the flesh has solidified, so it is too in the case of the heart. They are not, however, the same kind of fibers: those in the muscles are parts of nerves and of ligaments, whereas the type of fiber which is found in the heart is a specific one, as, too, is that of the casings of arteries and veins, and of intestines, stomach, womb, and both bladders. In the case of all these organs, one may observe [*idein*] that the flesh proper to the organ has solidified around that organ's specific fibers. (1.601–602 K. [= 58.8–59.2 Helmreich])

---

39 See 1.624–25 K. (= 72.21–73.1 Helmreich); 1.626–27 K. (= 73.25–74.5 Helmreich); 1.629–30 K. (= 75.22–27 Helmreich).

40 "Some people, certainly, are both thin and thin-veined; but if you cut [*temois*] any one of those veins there emerges fat, which has clearly grown underneath the skin in the membrane within. This kind of phenomenon is rare among men, but found very frequently among women: it is an indicator [*gnōrisma*] of a comparatively cold nature and idle lifestyle" (1.607 K. [= 62.6–11 Helmreich]).

On one occasion, Galen refers to a relevant anatomical insight (*anatomikon theōrēma*) about the differences between the structures of the channels by which the liver discharges bile into the stomach in various species.[41] Yet when it comes to the question of how these differences are assessed, he says:

> How are these cases to be distinguished? After all, I can hardly recommend dissection in the case of living human beings [πῶς οὖν χρὴ διαγιγνώσκειν αὐτούς; οὐ γὰρ δὴ ἀνατεμεῖν γε ζῶντας τοὺς ἀνθρώπους ἀξιῶ]. The answer is: first, from the overall mixture of the body, as was discussed a little earlier; secondly, from the [substances] passed below. (1.632 K. [= 77.13–16 Helmreich])

He discusses the case of Eudemus, who used continually to evacuate bilious substances of poor mixture through his stomach, because a great quantity of bile was gathering there, none of which would reach the upper part of the digestive tract. He also discusses the nature of excretions in other types of bodies, and in addition mentions a number of psychological factors such as worry, rage, distress, exertion, exercise, sleeplessness, fasting, and hardship, which cause the accumulation of more of the humor of yellow bile, because they also generate more of it in the liver. These, Galen says, are the precise indicators (*akribē gnōrismata*, 1.633 K. [= 78.11 Helmreich]). He concludes by making once again an important distinction (*diorismos*):

> In these cases, then, we have distinguished [*diōristai*] what comes about because of something other than the mixture. In the same way, if a flow of phlegm from the head into the stomach becomes the cause of sharp belching, here, too, one must separate this from a complaint specific to the stomach itself, according to the same methods. And so too [one must distinguish] whether pains in the head come about because of its own bad-mixture, or because of the secretions of the stomach. And indeed, it is certainly preferable to consider the type of the brain's mixture in its own right, rather than on the basis of the disposition of the body as a whole. And this consideration of the brain's mixture in its own right

---

[41] "There is also an anatomical insight [*anatomikon theōrēma*] which is relevant here, in ignorance of which some doctors are plunged into a terrible state of confusion on account of the discrepancy between symptoms; for they do not know that the channel by which the liver discharges bile into the stomach is double in some cases, and single in others, something which may be observed also in the dissection [*anatomais*] of quadrupeds. In most cases it is single, and is implanted into the space between the *pylorus* and the *intestinum jejunum*, the place that is known as the outgrowth of the stomach; or, when double, it connects into the outgrowth by the larger of its two mouths, and into the base, a little above the *pylorus*, by the smaller. In very rare cases the upper part is found to be the larger one, and the lower the smaller. When this is so, the stomach becomes filled during the day with a large amount of bile, which has to be vomited before meals and which causes harm if retained. When the channel is entirely single, all the bile flows down into the *intestinum jejunum*" (1.631–32 K. [= 76.24–77.12 Helmreich]).

[consists in]: graying of the hair, catarrh, coughs, mucus, quantity of saliva. All these indicate [*emphainei*] that it is relatively cold and wet; and especially so if it is brought to such dispositions as a result of everyday circumstances. But baldness brought about by dryness, or the production of plentiful, black hair, is an indicator of good-mixture in the brain. (1.634–35 K. [=78.17–79.5 Helmreich])

The distinction here is whether certain observable bodily features and symptoms are caused by the mixtures or by something else, e. g., a pathological cause, and hence, whether they can be interpreted as evidence for the state of the mixtures or not. This consideration is based on the fundamental distinction Galen makes between, on the one hand, bodily features that are due to the mixtures, and which may therefore serve as indicators for the mixtures, insofar as signs point to their causes, and on the other hand features that have a different cause, e. g., nature's "original plan" as Galen calls it (*kata prōton logon*), or its "shaping capacity" (*diaplastikē dunamis*).[42] Galen distinguishes a variety of causes that may be held responsible for the internal and external features of human bodies; and he insists that it is important not to attribute such features to the wrong causes.

VI A Catalogue of Errors

Much can be learned, finally, from the criticisms Galen makes of other thinkers, in particular when he spells out in detail the nature of their mistakes in trying to assess the mixtures. For, as in many Galenic writings, there is fair amount of polemical argument in *Mixtures* too, and there is a rich vocabulary for error (*hamartanein*), failure (*sphallesthai*), fallacious reasoning (*sophismata*), deception (*apatasthai*), or confusion (*sunchusis*). Apart from criticizing doctors for not making the proper distinctions, as we have already seen, Galen castigates others for drawing false inferences because of confusion between parts and wholes or because of misattribution of signs to causes. He criticizes "some well-reputed doctors who maintain that old age is wet" for being "deceived" by the quantity of bodily secretions;[43] he criticizes some physicians for failing to observe the proper rules of experiential examination by ignoring the potentially distorting effect

---

[42] On this distinction see above, n. 12; a third origin of individual features of a human body is habituation and philosophical training, mentioned in *ars med.* 11.8, 1.336–37 K. (= 309.2–7 Boudon-Millot).
[43] 1.580 K. (= 45.9–10 Helmreich).

(*paralogismos*) of incidental variables, which leads them to false attributions,[44] or for drawing hasty and premature (*euthus*) inferences about the supposed correspondence between external and internal features, e. g., hairiness and a melancholic constitution.[45] As further potential causes of error, he mentions confusion about the relative priority of empirical observation and logical argument;[46] stubborn misunderstanding and misrepresentation of earlier authorities (such as Aristotle);[47] starting at the wrong end, and being caught in circular reasoning, because of confusion between cause and effect;[48] premature and ill-founded embracing of logical demonstration without proper consideration of the factual basis of what one is trying to prove;[49] misrepresentation of the factual situation because of ambiguity as to the terms used;[50] sheer ignorance of the underlying anatomical facts;[51] premature inferences on the basis of insufficient indicators;[52] failure to make relevant distinctions between different origins of heat;[53] "wishful thinking"; and over-systematization.[54]

Conclusions

Much more can be said about Galen's methods of assessing internal bodily states than is possible within the constraints of this paper. Yet what we have seen suf-

---

44 1.592 K. (= 52.18–53.4 Helmreich), quoted above, pp. 686–87.
45 1.641–42 K. (= 82.11–83.13 Helmreich).
46 1.534–35 K. (=16.24–17.5 Helmreich).
47 1.523 K. (= 10.1–3 Helmreich); 1.535 K. (= 17.4–21 Helmreich).
48 1.586–87 K. (= 48.27–28; 49.11–13; 49.20–21 Helmreich): "Both sides begin their deductions from far-off points ... and seem almost to deduce first principles from secondary ones [πόρρωθεν γὰρ ἑκάτεροι καὶ σχεδὸν ἀπὸ τῶν δευτέρων τὰ πρότερα συλλογίζονται] ... and so it results that their demonstrations are conducted in a circular manner [δι' ἀλλήλων καὶ ἐξ ἀλλήλων], in which propositions are mutually interdependent. They use the matters currently under inquiry as if they were already known. ... My claim is that in every form of teaching it is better to define the order [*taxis*] of the concepts."
49 1.590–91 K. (= 51.26–52.4 H.).
50 1.613–14 K. (= 66.3–6 H.).
51 1.631 K. (= 76.24–77.2 Helmreich), quoted above, n. 41.
52 1.638–39 K. (= 81.15–18 Helmreich); 1.640 K. (= 82.6–14 Helmreich).
53 1.678–79 K. (= 106.5–9 Helmreich).
54 1.534 K. (= 16.16–21 Helmreich): "How, then, did certain doctors and philosophers come to declare spring wet and hot? Clearly, from a desire to make each of the four combinations of qualities correspond to one of the four seasons [ἐκ τοῦ βουληθῆναι δηλονότι τὰς τέτταρας συζυγίας τῶν κράσεων εἰς τὰς τέτταρας ὥρας διανεῖμαι]. And this desire in turn arose from the omission [*paralipein*] of the first of all mixtures, the well-mixed."

fices to show that Galen's theory of mixtures is not just a theoretical, speculative construct: he is concerned with empirical evaluation of the bodily mixtures, with the spelling out of what he calls the *gnōrismata*, the relevant indicators on the basis of which one can identify and assess a body's mixture, and the criteria on the strength of which these assessments can be validated or refuted. Thus Galen's *Mixtures* presents a carefully elaborated account of what we might call the evidence for the internal states of the body, how this is to be used, and correct interpretation—in other words: how to make sense of the body. Far from dogmatically presenting his theory of mixtures as something one simply has to accept, he goes out of his way to spell out how a body's mixture is to be assessed. The methods he employs are, as is usual with Galen, a sophisticated combination of empirical observation and theoretical reasoning. Neither of these two is sufficient on its own: empirical observation alone is not able to tell us the whole truth, it needs to be informed by theoretical considerations and by "qualifications," *diorismoi*. Likewise, theoretical reasoning needs to be checked by empirical testing. Underlying the whole operation is a strong belief that one needs to know the causes of the phenomena, and that one should not jump to conclusions by confusing parts with wholes; and underlying Galen's theory is the fundamental assumption of the four elementary qualities as the basic constituents of the human body and of the universe at large, and of the centrality of the human species within what he calls "the whole of existence."

Bibliography

Allen, J. V. *Inference from Signs: Ancient Debates about the Nature of Evidence* (Oxford: Oxford University Press, 2001).

Boudon-Millot, V. "La notion de mélange dans la pensée médicale de Galien: mixis ou crasis?," *Revue des études grecques* 124 (2011): 261–79.

Debru, A., ed. *Galen on Pharmacology: Philosophy, History, and Medicine* (Leiden: Brill, 1997).

Debru, A."Γνώρισμα chez Galien: vers une sémiologie de la santé," in Isabelle Boehm and Nathalie Rousseau (eds.), *L'expressivité du lexique médical en Grèce et à Rome: hommages à Françoise Skoda, Coll. Hellenica* (Paris: Presses de l'Université de Paris-Sorbonne, 2014), 171–77.

Den Dulk, W. J. *Krasis. Bijdrage tot de Grieksche Lexicographie* (Leiden: Brill, 1934).

Dittmer, H. L. "Konstitutionstypen im Corpus Hippocraticum," Diss., Würzburg, 1940.

Flashar, H. *Melancholie und Melancholiker in den medizinischen Theorien der Antike* (Berlin: De Gruyter, 1966).

Harig, G. *Bestimmung der Intensität im medizinischen System Galens* (Berlin: Akademie-Verlag, 1984).

Helmreich, G. *Claudius Galenus, De temperamentis libri III* (Leipzig: Teubner, 1904; repr. with additions, Stuttgart: Teubner, 1969).
Holmes, B. *The Symptom and the Subject* (Princeton: Princeton University Press, 2010).
Jouanna, J. "Galen's Concept of Nature," in *Greek Medicine from Hippocrates to Galen*, 287–312.
Jouanna, J. *Greek Medicine from Hippocrates to Galen: Selected Papers* (Leiden: Brill, 2012).
Jouanna, J. "The Legacy of the Hippocratic Treatise *The Nature of Man*: The Theory of the Four Humours," in idem, *Greek Medicine from Hippocrates to Galen*, 335–59.
Klibansky, R., E. Panofsky, and F. Saxl. *Saturn und Melancholie* (Frankfurt: Suhrkamp, 1992).
Montanari, E. *Κρᾶσις e μίξις: un itinerario semantico e filosofico* (Florence: CLUSF, 1979).
Needham, P. "Compounds and Mixtures," in A. I. Woody, R. F. Hendry, and P. Needham (eds.), *Handbook of the Philosophy of Science*, vol. 6: *Philosophy of Chemistry* (Oxford: North Holland, 2012), 271–90.
Schöner, E. *Das Viererschema in der antiken Humoralpathologie* (Wiesbaden: Steiner, 1964).
Schwabe, W. *"Mischung" und "Element" im Griechischen bis Platon: Wort- und begriffsgeschichtliche Untersuchungen, insbesondere zur Bedeutungsentwicklung von "stoicheion"* (Bonn: Bouvier, 1980).
Sedley, D. N. "On Signs," in J. Barnes and J. Brunschwig (eds.), *Science and Speculation* (Cambridge: Cambridge University Press, 2005), 239–72.
Smith, W. D. "Regimen, krêsis and the History of Dietetics," in J. A. López Férez (ed.), *Tratados hipocráticos* (Madrid: Universidad Nacional de Educación a Distancia, 1992), 263–72.
Todd, R. B. *Alexander of Aphrodisias on Stoic Physics: A Study of the "De Mixtione" with Preliminary Essays, Text, Translation and Commentary* (Leiden: Brill, 1976).
Tracy, T. J. *Physiological Theory and the Doctrine of the Mean in Plato and Aristotle* (The Hague: Mouton, 1969).
van der Eijk, P. J. "Galen's Use of the Concept of Qualified Experience in His Dietetic and Pharmacological Works," in Debru, *Galen on Pharmacology*, 35–57 [repr. in P. J. van der Eijk, *Medicine and Philosophy in Classical Antiquity: Doctors and Philosophers on Nature, Soul, Health, and Disease* (Cambridge: Cambridge University Press, 2005), 279–98].
van der Eijk, P. J. "Aristotle! What a Thing for You to Say!: Galen's Engagement with Aristotle and Aristotelians," in C. J. Gill, J. Wilkins, and T. Whitmarsh (eds.), *Galen and the World of Knowledge* (Cambridge: Cambridge University Press, 2009), 261–81.
van der Eijk, P. J. "Galen on the Nature of Human Beings," in P. Adamson and J. Wilberding (eds.), *Philosophical Themes in Galen* (London: Institute of Classical Studies, 2014), 89–134.
van der Eijk, P. J. "Galen and the Scientific Treatise. A Case Study of *Mixtures*," in M. Asper (ed.), *Writing Science* (Berlin: De Gruyter, 2013), 145–75.
van der Eijk, P. J. "Quelques observations sur la réception d'Aristote dans la médecine gréco-romaine de l'époque impériale," in Y. Lehmann (ed.), *Aristoteles Romanus: la réception de la science aristotélicienne dans l'Empire gréco-romain* (Turnhout: Brepols, 2013), 183–93.
van der Eijk, P. J., and P. N. Singer, eds. *Galen, Works on Human Nature* (Cambridge: Cambridge University Press, Forthcoming).
von Staden, H. *Herophilus* (Cambridge: Cambridge University Press, 1989).

von Staden, H. "Inefficacy, Error, and Failure: Galen on δόκιμα φάρμακα ἄπρακτα" in Debru, *Galen on Pharmacology*, 59–83.
von Staden, H. "The Physiology and Therapy of Anger: Galen on Medicine, the Soul, and Nature," in F. Optas and D. Reisman (eds.), *Islamic Philosophy, Science, Culture, and Religion* (Leiden: Brill, 2012), 63–87.
von Staden, H. "Touch in Ancient Medicine: From a 'Harvest of Sorrows' to Nature's Music in the Arteries," read at the 143th Annual Meeting of the American Philological Association, 2012, in http://apaclassics.org/index.php/annual_meeting/143rd_annual_meeting_abstracts/06.5.von_staden/.

Contributors

Isabella Andorlini, a former researcher at the "Girolamo Vitelli" Papyrological Institute in Florence, is currently Associate Professor of Papyrology at the University of Parma. Her research has been principally concerned with the *Corpus* of Greek Medical Papyri project. She is the editor of the *Greek Medical Papyri* (I, Florence 2001; II, Florence 2009) and the author of a range of essays on other aspects of the medical tradition. She has published literary and documentary papyri from various international collections. In the Academic Year 2002/2003 she was awarded a Membership at the Institute for Advanced Study in Princeton.

Markus Asper received his PhD from Freiburg University in 1994. After some time at Konstanz and Mainz, he spent a year at the Institute for Advanced Study at Princeton (2003–2004). From there, he moved on to appointments at Pennsylvania State University and New York University before he came to Humboldt University at Berlin in 2010, where he is currently Professor of Greek. His main areas of research are Hellenistic poetry, especially Callimachus, and ancient literature on science (see most recently the collection *Writing Science* of 2013).

Han Baltussen is the Hughes Professor of Classics at the University of Adelaide (Australia). He was a Member of the School of Historical Studies at the Institute for Advanced Study in Princeton in 2006 (Fall/Winter). He is the author of *Theophrastus Against the Presocratics and Plato* (2000), co-editor of *Philosophy, Science, and Exegesis in Greek, Latin, and Arabic Commentaries* (2 vols, 2004), author of *Philosophy and Exegesis in Simplicius: The Methodology of a Commentator* (2008), editor of *Greek and Roman Consolations: Eight Studies of a Tradition and its Afterlife* (2012), and co-translator of *Simplicius, Commentary on Aristotle's Physics 1.5–9* (2012). His current work is concerned with the Peripatetics after Aristotle, ancient grief management, and forms of censorship in antiquity.

Alan C. Bowen, Director of the Institute for Research in Classical Philosophy and Science (Princeton), is a historian of ancient Greco-Latin science and philosophy. His most recent books are *New Perspectives on Aristotle's "De caelo"* (2009, with Christian Wildberg) and *Simplicius on the Planets and their Motions: In Defense of a Heresy* (2013). He is currently writing a monograph on Hellenistic astronomy. He is also the editor (with Tracy Rihll) of the journal *Aestimatio: Critical Reviews in the History of Science* and Director (with Francesca Rochberg) of EKOH—Early

Knowledge of the Heavens: A Digital Library for Mediterranean, Near Eastern, and South Asian Contexts.

Philippe Charlier is an Assistant Professor in Forensic Medicine and an anthropologist (MD, PhD, LittD). He is the head of the Laboratory of Medical and Forensic Anthropology at West Paris University (UVSQ, AP-HP). His specialization is in retrospective diagnosis and human identification.

Andrea Falcon is Associate Professor in the Department of Philosophy at Concordia University, Montreal. He works on Aristotle and the Aristotelian tradition in antiquity, with a focus on Aristotle's philosophy of nature. He is the author of *Corpi e Movimenti: la fortuna del "De caelo" nel mondo antico* (Bibliopolis 2001); *Aristotle and the Science of Nature: Unity without Uniformity* (Cambridge University Press 2005); *Aristotelianism in the First Century BCE: Xenarchus of Seleucia* (Cambridge University Press 2011).

Christopher Athanasius Faraone is the Frank C. and Gertrude M. Springer Professor of the College and the Humanities at the University of Chicago. His work is primarily concerned with ancient Greek religion and poetry. He is co-editor (with D. Dodd) of *Initiation in Ancient Greek Rituals and Narratives: New Critical Perspectives* (2003), (with L. McClure) of *Prostitutes and Courtesans in the Ancient World* (2005), (with F. Naiden) of *Ancient Victims, Modern Observers: Reflections on Greek and Roman Sacrifice* (2011), and (with D. Obbink) of *The Getty Hexameters: Poetry, Magic, and Mystery in Ancient Selinous* (Oxford 2013). He is the sole author of *Talismans and Trojan Horses: Guardian Statues in Ancient Greek Myth and Ritual* (1992), *Ancient Greek Love Magic* (1999), *The Stanzaic Architecture of Ancient Greek Elegiac Poetry* (2008) and numerous articles on ancient Greek magic, poetry, and religion. He is currently working on a book on ancient Greek amulets.

Educated at the universities of Munich, Edinburgh, and Berlin (Free University), **Klaus-Dietrich Fischer** was awarded a doctorate in Classics for his edition of *Pelagonii ars ueterinaria* in 1980. His professional career was in the History of Medicine, teaching first at Berlin's Free University (from 1976) and later (from 1987) at Johannes Gutenberg University in Mainz, where he has organised, first with Werner F. Kümmel and now with Tanja Pommerening, annual workshops (Arbeitskreis Alte Medizin). If he remembers correctly, he first met Heinrich von Staden at a conference in Saint-Étienne in 1989, and has been very grateful for his encouragement and friendship spanning almost a quarter of a century. It was Heinrich who convinced him to apply to the Institute for Advanced Study, where he spent many happy months in 2009 as a Member of the School of Historical Studies, literally a few steps from his office, and again as a Visitor in Summer 2011.

Allan Gotthelf (1942–2013), one of the foremost experts on Aristotle's philosophy of biology, was at the time of his death Emeritus Professor of Philosophy at The College of New Jersey and Anthem Foundation Distinguished Fellow for Research and Teaching in Philosophy at Rutgers. A collection of his essays, *Teleology, First Principles, and Scientific Method in Aristotle's Biology*, was published by Oxford in 2012. He was a Member at the Institute of Advanced Study in 2001.

Danielle Gourevitch is an Emeritus University Professor, Directeur of Studies at the École pratique des hautes études, a specialist in the history of ancient medicine, and the author of over three hundred articles and fifteen books. Her latest book recently appeared from De Boccard (Paris, 2013): *Limos kai loimos: A Study of the Galenic Plague*. She has collaborated with Philippe Charlier on a series of colloquia on pathography and the publications of their proceedings. The most recent, *4ᵉ colloque international de pathographie. Saint-Jean-de-Cole, mai 2011*, was published by De Boccard in 2013.

Brooke Holmes is Professor in the Department of Classics at Princeton University. She is the author of *The Symptom and the Subject: The Emergence of the Physical Body in Ancient Greece* (Princeton, 2010) and *Gender: Antiquity and its Legacy* (Oxford, 2012), and two co-edited volumes, *Aelius Aristides between Greece, Rome, and the Gods* (Brill, 2008), with W. V. Harris, and *Dynamic Reading: Studies in the Reception of Epicureanism* (Oxford, 2012), with W. H. Shearin, as well as numerous articles on Greek literature, the history of medicine and the body, and ancient philosophy. She was a Member at the Institute for Advanced Study in 2007–2008 on a Mellon Fellowship for Assistant Professors.

Carl Huffman is Senior Research Professor at DePauw University. He specializes in ancient Pythagoreanism and is the author of *Philolaus of Croton: Pythagorean and Presocratic* and *Archytas of Tarentum: Pythagorean, Philosopher and Mathematician King*, both published by Cambridge. He has held two NEH fellowships, a Howard Foundation Fellowship and a Guggenheim Fellowship. He was a visitor in the School for Historical Studies at the Institute for Advanced Study in 2009 while holding a fellowship from the American Council of Learned Societies.

Katerina Ierodiakonou is Professor of Ancient Philosophy at the University of Athens, Associate Professor of Ancient Philosophy at the University of Geneva, and a member of the research program "Representation and Reality" at the University of Gothenburg. She has edited many volumes and published numerous articles on ancient and Byzantine philosophy, especially in the areas of epistemology and logic. She is currently working on a monograph about ancient theories of color, as well as on the edition, translation, and commentary of Theophrastus' *De sensibus* and of Michael Psellos' paraphrase of Aristotle's *De interpretatione*. She was a Member in the School of Historical Studies at the Institute for Advanced Study in Princeton during the autumn of 2003.

Jacques Jouanna is a Member of the Institut de France (Académie des Inscriptions et Belles-Lettres) and Professor Emeritus of Greek at the University of Paris-Sorbonne, where he taught from 1981 until 2004 and where he founded and directed the CNRS Research Group "Médecine Grecque." His publications include *Hippocrate: pour une archéologie de l'école de Cnide* (1975, re-issued 2009), *Hippocrate* (1992, Engl. trans. 1999), *Sophocle* (2007), and numerous critical editions of Hippocratic and Galenic medical texts in two Collections, CMG and CUF (Collection des Universités de France). A collection of his essays was recently translated into English as *Greek Medicine from Hippocrates to Galen: Selected Papers* (Leiden: Brill, 2012).

Joshua T. Katz is Professor of Classics at Princeton University, on whose faculty he has been since 1998. A linguist by training, a classicist by profession, and a comparative philologist at heart, he has broad interests in pre-modern languages, literatures, and cultures and enjoys moving from the India of the Vedas to medieval Ireland by way of the Bronze Age Near East. His many publications tackle such topics as Sanskrit riddles, Tocharian phonology, Hittite kingship, Greek belly-prophecy, Latin acrostics, and Welsh pronouns. Among his areas of expertise are etymology, poetics, the reconstruction of Proto-Indo-European, the history and practice of wordplay, and zoological terminology. He has a special fondness for badgers.

W. R. Laird is Associate Professor of History at Carleton University, Ottawa, Canada, where he teaches medieval history and the history of science. He is the author of *The Unfinished Mechanics of Giuseppe Moletti* (Toronto, 2000) and a number of articles on medieval and renaissance mathematical sciences, and co-editor (with Sophie Roux) of *Mechanics and Natural Philosophy before the Scientific Revolution* (Dordrecht, 2008). In 2007–2008 he was a member at the Institute for Advanced Study, where he undertook a study of the Renaissance traditions of ancient mechanics that eventually will be a book titled *The Renaissance of Mechanics*.

Helen S. Lang received her PhD from the University of Toronto in 1977 in Ancient Philosophy, with extensive work in Classics, including studying Greek Prose Composition with Heinrich von Staden at Yale. She is currently a tenured Professor in the Philosophy Department at Villanova University. Over the years, she has received fellowships from the A.C.L.S., the N.E.H., and the N.S.F. She has also held appointments at the Dibner Institute for the History of Science and Technology (M.I.T. Fall 1999) and the Institute for Advanced Study (Spring 2000). Her work has focused on the history of ancient physics, an area in which she has published three books and numerous articles.

Since September 2010, **Roberto Lo Presti** has been Wissenschaftlicher Mitarbeiter at the Humboldt University in Berlin as part of Philip van der Eijk's research program "Medicine of the Mind, Philosophy of the Body," and coordinator of the PhD-Program "History of Ancient Science" of the Berlin Graduate School of Ancient Studies. During his doctoral and postdoctoral studies he worked at the Universities of Palermo, Sorbonne-Paris IV, Newcastle, Lausanne, and Leiden. He has also received grants and research fellowships from the Scaliger Institute of Leiden University Library, the Descartes Centre for the History and Philosophy of Science of Utrecht University, and the Fondation Hardt pour l'étude de l'antiquité classique. He was appointed as Visitor at the Institute for Advanced Study for the second term of the academic year 2010–2011.

Arnaldo Marcone is Professor of Roman History at the University Roma III. He was a Fellow of the Alexander von Humboldt-Stiftung and a Member of the Institute of Advanced Study in Princeton in 2003. His main fields of research are the economic and social history of the ancient world; the Late Roman Empire; and the modern historiography on the ancient world. He is a corresponding member of the German Archaeological Institute in Berlin.

Stephen Menn studied mathematics at Johns Hopkins University (PhD 1985) and philosophy at the University of Chicago (PhD 1989). He is Professor of Ancient and Contemporary Philosophy at the Humboldt-Universität zu Berlin, and Associate Professor of Philosophy at McGill University. He works on ancient Greek, and medieval Arabic and Latin, philosophy and science. He is the author of *Plato on God as Nous* (1995) and *Descartes and Augustine* (1998) and of the forthcoming *The Aim and the Argument of Aristotle's "Metaphysics"*; with Rachel Barney, he has translated Simplicius' commentary on Aristotle's *Physics* 1.1–2, forthcoming in the series Ancient Commentators on Aristotle.

Ian Moyer completed his PhD at the University of Chicago and is Associate Professor of History at the University of Michigan, Ann Arbor. He was a member of the Institute for Advanced Study in 2007/2008. His first book, *Egypt and the Limits of Hellenism*, explored cultural and intellectual encounters between Greeks and Egyptians from the time of Herodotus to the high Roman empire. He is currently investigating the novel cultural and political practices developed for managing relations between Egyptians and Macedonian-Greek rulers in the Ptolemaic state.

Vivian Nutton is Emeritus Professor of the History of Medicine at UCL and Emeritus Professor of Classics at the University of Warwick. His extensive publications have centered upon Galen and the Galenic medical tradition over the centuries, but he has also written on plague, the social history of medicine in the Renaissance, and on the history of anatomy. His most recent books include an edition of Galen's *On Problematical Movements* (2011), a revised edition of his *Ancient Medicine* (2013), the historical introduction to the Karger translation of Andreas Vesalius, *The Fabric of the Human Body*, and the first English translation and commentary of Galen's *Avoiding Distress* (2014). He is a Fellow of the British Academy, the Deutsche Akademie der Wissenschaften, and the Academia Europaea.

D. T. Potts is Professor of Ancient Near Eastern Archaeology and History at the Institute for the Study of the Ancient World, New York University. He is a specialist in the archaeology and early history of Iran, Mesopotamia, and the Arabian peninsula and has excavated extensively in Iran, Saudi Arabia, and the United Arab Emirates.

Christine Proust is a historian of mathematics and ancient sciences, specializing in the cuneiform sources. She started her research by studying mathematical cuneiform texts from the ancient Mesopotamian city of Nippur and has published two books on these sources: *Tablettes mathématiques de Nippur* (2007) and *Tablettes mathématiques de la collection Hilprecht* (2008). She is Directeur de Recherche at the *Centre National de la Recherche Scientifique* (CNRS), member of the Laboratory SPHERE (CNRS and University Paris Diderot), and co-director of the European project "Mathematical Sciences in the Ancient World" (SAW) led by Karine Chemla. She was a Member of the Institute for Advanced Study from September to December 2009.

Francesca Rochberg is Catherine and William L. Magistretti Distinguished Professor of Near Eastern Studies in the Department of Near Eastern Studies, the Office for the History of Science and Technology, and the Graduate Group in Ancient History and Mediterranean Archaeology at the University of California, Berkeley. She is the author of *Aspects of Babylonian Celestial Divination: The Lunar Eclipse Tablets of Enuma Anu Enlil*, Archiv für Orientforschung Beiheft 22 (Ferdinand Berger und Söhne, 1988), *Babylonian Horoscopes*, Transactions of the American Philosophical Society 88.1 (American Philosophical Society, 1998), *The Heavenly Writing: Divination, Horoscopy and Astronomy in Mesopotamian Culture* (Cambridge University Press, 2004 and 2007), and *In the Path of the Moon: Babylonian Celestial Divination and Its Legacy*, Studies in Ancient Magic and Divination (Brill, 2010).

Amneris Roselli is Professor of Classical Philology at the University of Naples "L'Orientale." She has devoted the greater part of her research to editing Greek medical and scientific texts (Hippocrates, *Epidemics* VI with D. Manetti; Ps.-Aristotle's *De spiritu*; and papyri) and publishing scholarly essays on ancient Greek and Latin medicine. She was a Member of the School for Historical Studies at the Institute for Advanced Study in Winter 2004.

Thomas Rütten is the Newcastle Director of the Northern Centre for the History of Medicine and Reader in the History of Medicine at Newcastle University's School of History, Classics and Archaeology. With a physician's licence (1986), a PhD in the History of Medicine (1991), and a Habilitation in the Theory and History of Medicine (1995) and after employments at the University of Münster (1990–1994), the Herzog August Library in Wolfenbüttel (1994–1997), and as a freelance translator of fiction (1997–1999), he came to the Institute for Advanced Study on a Fulbright Scholarship for the academic year 1999–2000. During this annus mirabilis he had the privilege and the pleasure to work as Professor von Staden's research assistant and pursue some of his own research projects (primarily on Ludwig Edelstein and Thomas Mann).

Mark J. Schiefsky is Professor of the Classics and Chair of the Department of the Classics at Harvard University. His research focuses on the history of science and its relationship to philosophy in Greco-Roman antiquity. Among his publications are a translation and commentary on the Hippocratic treatise *On Ancient Medicine* (Leiden: Brill 2005) as well a number of articles on ancient medicine and mechanics. He spent the academic year 2006–2007 at the Institute for Advanced Study.

Heinrich Schlange-Schöningen, born in 1960, studied historical sciences, history of religions, and philosophy at the Freie Universität Berlin and the University of

Bordeaux. From 1989–1994 and 1996–2002 he was Assistant Professor in the Department for History at the Freie Universität Berlin. In 2004, he was a Member of the School of Historical Studies at the Institute for Advanced Study. Since 2006, he has been Professor of Ancient History at Saarland University, Saarbrücken.

Philip van der Eijk is Alexander von Humboldt Professor of Classics and History of Science at the Humboldt University Berlin. He has published widely on ancient medicine, philosophy and science, patristics, and comparative literature. Among his current research interests are projects on Aristotle, Aristotelianism, and medical history; the rhetoric of ancient scientific discourse; a series of scholarly translations of Galen in English (Cambridge Galen Translations); processes of canon formation in the historiography of ancient science; ancient concepts of mental health, mental illness, and of the bodily location of mental functions; and the development of the medical encyclopedia in late antiquity.

Heinrich von Staden: Bibliography 1975–2012

"Experiment and Experience in Hellenistic Medicine," *Bulletin of the Institute of Classical Studies* 22 (1975): 178–99.
"Greek Art and Literature in Marx's Aesthetics," *Arethusa* 8 (1975): 119–44.
"Die Hippokrateskommentare im Codex Ambrosianus Graecus 473 und Herophilus," *Philologus* 120 (1976): 132–36.
"A New Testimonium about Polybus," *Hermes* 104 (1976): 494–96.
"Nietzsche and Marx on Greek Art and Literature: Case Studies in Reception," *Daedalus* 105 (Winter, 1976): 79–96.
"The Stoic Theory of Perception and its 'Platonic' Critics," in P. K. Machamer and R. G. Turnbull (eds.), *Studies in Perception* (Columbus, Ohio: Ohio State University Press, 1978), 96–137.
"An Interview with Athol Fugard," *The New Journal* 14 (1982): 14–21.
"An Interview with Athol Fugard," *Theatre* 14.1 (Winter, 1982): 41–46.
"Hairesis and Heresy: The Case of the *haireseis iatrikai*," in B. F. Meyer and E. P. Sanders (eds.), *Jewish and Christian Self-Definition*, 3 vols. (London: SCM Press, 1982), 3:76–100, 199–206.
"Athol Fugard," in Lee A. Jacobus (ed.), *The Bedford Introduction to Drama* (New York: St. Martin's Press, 1988; 2nd ed., 1993; 3rd ed., 1995), 1251–52.
"Cardiovascular Puzzles in Erasistratus and Herophilus," in *Atti: XXXI Congresso Internazionale di Storia della Medicina* (Bologna: Monduzzi Editore, 1989), 681–87.
Herophilus: The Art of Medicine in Early Alexandria (Cambridge: Cambridge University Press, 1989; 2nd ed., 1994; paperback ed., 2009).
"Incurability and Hopelessness: The Hippocratic Corpus," in P. Potter, G. Maloney, and J. Desautels (eds.), *La maladie et les maladies dans la Collection hippocratique: actes du VIe Colloque International Hippocratique* (Québec: Éditions du Sphinx, 1990), 75–112.
"Response to Michael McVaugh, Review of *Herophilus: The Art of Medicine in Early Alexandria*," *Journal of the History of the Behavioral Sciences* 26 (1990): 385–86.
"*Apud nos foediora verba*: Celsus' Reluctant Construction of the Female Body," in Guy Sabbah (ed.), *Le Latin médical: la constitution d'un langage scientifique; réalités et langage de la médecine dans le monde romain. Actes du IIIe Colloque international "Textes médicaux latins antiques" (Saint-Étienne, 11–13 septembre 1989)*, Centre Jean-Palerne, Mémoires 10 (Saint-Étienne: Publications de l'Université de Saint-Étienne, 1991), 271–96.
"Galen as Historian: His Use of Sources on the Herophileans," in J. A. López Férez (ed.), *Galeno: obra, pensamiento y influencia* (Madrid: Universidad Nacional de Educación a Distancia, 1991), 205–22.
"Matière et signification: rituel, sexe et pharmacologie dans le Corpus hippocratique," *L'Antiquité Classique* 60 (1991): 42–61.
"Affinities and Elisions: Helen and Hellenocentrism," *Isis* 83 (1992): 578–95. Reprinted in M. H. Shank (ed.), *The Scientific Enterprise in Antiquity and the Middle Ages* (Chicago: University of Chicago Press, 2000), 54–71.

"The Discovery of the Body: Human Dissection and its Cultural Contexts in Ancient Greece," *Yale Journal of Biology and Medicine* 65 (1992): 223–41.

"Jaeger's 'Skandalon der historischen Vernunft': Diocles, Aristotle, and Theophrastus," in W. M. Calder III (ed.), *Werner Jaeger Reconsidered: Proceedings of the Second Oldfather Conference, held on the campus of the University of Illinois at Urbana-Champaign, April 26–28, 1990*, Illinois Classical Studies, Supplement 3 (Atlanta: Scholars Press, 1992), 227–65.

"Lexicography in the Early Third Century B.C.: Bacchius of Tanagra, Erotian, and Hippocrates," in J. A. López Férez (ed.), *Tratados Hipocráticos (Estudios acerca de su contenido, forma e influencia): actas del VIIe Colloque International Hippocratique* (Madrid: Universidad Nacional de Educación a Distancia, 1992), 549–69.

"The Mind and Skin of Heracles: Heroic Diseases," in D. Gourevitch (ed.), *Maladie et maladies: histoire et conceptualisation; mélanges en l'honneur de Mirko Grmek*, Hautes études médiévales et modernes 70 (Geneva: Librairie Droz, 1992), 131–50.

"Spiderwoman and the Chaste Tree: The Semantics of Matter," *Configurations* 1 (1992): 23–56.

"Women and Dirt," *Helios* 19 (1992): 7–30.

"Author and Authority: Celsus and the Construction of a Scientific Self," in Manuel Enrique Vázquez Buján (ed.), *Tradición e Innovación de la Medicina Latina de la Antigüedad y de la Alta Edad Media* (Santiago de Compostela: Universidade de Santiago de Compostela, 1994), 103–17.

"'Un autre dieu sobre': Théophraste, Érasistrate et les médecins hellénistiques à propos de l'eau," in René Ginouvès, Anne-Marie Guimier-Sorbets, Jacques Jouanna, and Laurence Villard (eds.), *L'eau, la santé et la maladie dans le monde grec: actes du colloque organisé à Paris (CNRS et Fondation Singer-Polignac) du 25 au 27 novembre 1992 par le Centre de recherche "Archéologie et systèmes d'information" et par l'URA 1255 "Médecine grecque,"* Bulletin de correspondance hellénique, Supplément 28 (Paris: De Boccard, 1994), 77–94.

"*Media quodammodo diuersas inter sententias*: Celsus, the 'Rationalists,' and Erasistratus," in Guy Sabbah and Philippe Mudry (eds.), *La médecine de Celse: aspects historiques, scientifiques et littéraires* (Saint-Étienne: C.N.R.S., 1994), 77–101.

"Anatomy as Rhetoric: Galen on Dissection and Persuasion," *Journal of the History of Medicine and Allied Sciences* 50 (1995): 47–66.

"Don des dieux ou responsabilité des hommes?," in *La santé et ses metamorphoses*, Supplément au *La Recherche* 281 (Paris: Société d'éditions scientifiques, 1995), 10–11.

"Un émigré réformateur et conservateur: Henry Sigerist et ses Grèces anciennes," in D. Gourevitch (ed.), *Médecins érudits de Coray à Sigerist* (Paris: De Boccard 1995), 172–95.

"Science as Text, Science as History: Galen on Metaphor," in P. J. van der Eijk, H. F. J. Horstmanshoff, and P. H. Schrijvers (ed.), *Ancient Medicine in its Socio-cultural Contexts*, 2 vols. (Amsterdam: Rodopi, 1995), 2:499–518.

"Alexander (15) Philalethes," "Apollonius (8)," "Apollonius (10) Mys," "Bacchius," "Callimachus (4)," "Chrysermus," "Demetrius (21)," "Dioscurides (2)," "Erasistratus," "Hegetor," "Heraclides (4)," "Herophilus," "Mantias," "Philinus (1)," "Serapion (1)," "Zeno (7)," "Zeuxis (2)," and "Zeuxis (3)," in Simon Hornblower and Antony Spawforth (eds.), *Oxford Classical Dictionary*, 3rd ed. (Oxford: The Clarendon Press, 1996).

"Alexandrien als Zentrum der medizinischen Forschung: Herophilos von Chalkedon und die frühe Menschenanatomie," in H. Schott (ed.), *Meilensteine der Medizin* (Dortmund: Harenberg Verlag, 1996), 67–73, 616, 646.

(with Klaus-Dietrich Fischer) "Ein angeblicher Brief des Herophilos an König Antiochos, aus einer Brüsseler Handschrift erstmals herausgegeben," *Sudhoffs Archiv* 80 (1996): 86–98.

"Body and Machine: Interactions between Medicine, Mechanics, and Philosophy in Early Alexandria," in *Alexandria and Alexandrianism* (Malibu, Calif.: The J. Paul Getty Museum, 1996), 85–106.

"L'idéal de tranquillité et la construction du passé dans la Seconde Sophistique: Aelius Aristide," in J.-M. André, J. Dangel, and P. Demont (eds.), *Les loisirs et l'héritage de la culture classique: actes du XIIIe Congrès de l'Association Guillaume Budé, Dijon, 27–31 août 1993*, Collection Latomus 230 (Brussels: Latomus, 1996), 148–61.

"'In a Pure and Holy Way': Personal and Professional Conduct in the Hippocratic *Oath*," *Journal of the History of Medicine and Allied Sciences* 51 (1996): 404–37.

"Liminal Perils: Early Roman Receptions of Greek Medicine," in F. Jamil Ragep and Sally P. Ragep (eds.), *Tradition, Transmission, Transformation*, with the assistance of Steven Livesey, Collection de Travaux de l'Académie internationale d'histoire des sciences 37 (Leiden: Brill, 1996), 369–418.

"Les manuscrits du *De pulsibus* de Marcellinus," in Antonio Garzya (ed.), *Storia e ecdotica dei testi medici greci* (Naples: M. D'Auria Editore, 1996), 407–25.

"Alcmaeon," "Athenaeus," "Celsus," "Erasistratus," "Heraclides of Tarentum," "Herophilus," and "Hippocrates and the Hippocratic Corpus," in Donald J. Zeyl (ed.), *Encyclopedia of Classical Philosophy*, with the assistance of Daniel T. Devereux and Phillip T. Mitsis (Westport, Conn.: Greenwood Press, 1997).

"Character and Competence: Personal and Professional Conduct in Greek Medicine," in H. Flashar and J. Jouanna (eds.), *Médecine et morale dans l'Antiquité*, Entretiens sur l'Antiquité classique 43 (Vandœuvres and Geneva: Fondation Hardt, 1997), 157–210.

"Galen and the Second Sophistic," in R. Sorabji (ed.), *Aristotle and After*, Bulletin of the Institute of Classical Studies, Supplement 68 (London: Institute of Classical Studies; School of Advanced Study; University of London, 1997), 33–54.

"'Hard Realism' and 'A Few Romantic Moves': Henry Sigerist's Versions of Ancient Greece," in E. Fee and T. M. Brown (eds.), *Making Medical History: The Life and Times of Henry E. Sigerist* (Baltimore: The Johns Hopkins University Press, 1997), 136–61.

"Inefficacy, Error, and Failure: Galen on δόκιμα φάρμακα ἄπρακτα," in A. Debru (ed.), *Galen on Pharmacology: Philosophy, History and Medicine* (Leiden: Brill, 1997), 59–83.

"Teleology and Mechanism: Aristotelian Biology and Early Hellenistic Medicine," in Wolfgang Kullmann and Sabine Föllinger (eds.), *Aristotelische Biologie. Intentionen, Methoden, Ergebnisse*, Philosophie der Antike 6 (Stuttgart: Franz Steiner Verlag, 1997), 183–208.

"Was Cassius an Empiricist?: Reflections on Method," in Ugo Criscuolo and Riccardo Maisano (eds.), *Synodia: Studia humanitatis Antonio Garzya septuagenario ab amicis atque discipulis dicata* (Naples: M. D'Auria Editore, 1997), 939–60.

"Andréas de Caryste et Philon de Byzance: médecine et mécanique à Alexandrie," in Gilbert Argoud and Jean-Yves Guillaumin (eds.), *Sciences exactes et sciences appliquées à Alexandrie (III siècle av. J.-C.—I siècle ap. J.-C.); actes du Colloque international de Saint-Étienne* (Saint-Étienne: Centre Jean-Palerne, Université de Saint-Étienne, 1998), 147–72.

"Gattung und Gedächtnis. Galen über Wahrheit und Lehrdichtung," in Wolfgang Kullmann, Jochen Althoff, and Markus Asper (eds.), *Gattungen wissenschaftlicher Literatur in der Antike*, ScriptOralia, Band 95 (Tübingen: Gunter Narr Verlag, 1998), 65–94.

"The Rule and the Exception: Celsus on a Scientific Conundrum," in C. Deroux (ed.), *Maladie et maladies dans les textes latins antiques et médiévaux; actes du V Colloque international "Textes médicaux latins,"* Collection Latomus 242 (Brussels: Latomus, 1998), 105–28.

"Caelius Aurelianus and the Hellenistic Epoch: Erasistratus, the Empiricists, and Herophilus," in P. Mudry (ed.), *Le traité des "Maladies Aiguës" et des "Maladies Chroniques" de Caelius Aurelianus. Nouvelles approches; actes du colloque de Lausanne, 1996*, with the assistance of O. Bianchi and D. Castaldo (Nantes: Institut Universitaire de France, Université de Nantes, 1999), 85–119.

"Celsus as Historian?," in P. J. van der Eijk (ed.), *Ancient Histories of Medicine: Essays in Medical Doxography and Historiography in Classical Antiquity*, Studies in Ancient Medicine 20 (Leiden: Brill, 1999), 251–94.

"*Dynamis*: The Hippocratics and Plato," in Konstantinos J. Boudouris (ed.), *Philosophy and Medicine*, 2 vols., Studies in Greek Philosophy 29 (Athens: International Association for Greek Philosophy, 1999), 2:262–79.

"Metaphor and the Sublime: Longinus," in J. A. López Férez (ed.), *Desde los poemas homéricos hasta la prosa griega del siglo IV d.C. Veintiséis estudios filológicos*, Estudios de Filología Griega 4 (Madrid: Ediciones Clásicas, 1999 [2000]), 359–80.

"Reading the Agonal Body: The Hippocratic Corpus," in Y. Otsuka, S. Sakai, and S. Kuriyama (eds.), *Medicine and the History of the Body: Proceedings of the 20th, 21st and 22nd International Symposium on the Comparative History of Medicine—East and West* (Tokyo: Ishiyaku EuroAmerica, 1999), 287–94.

"Rostovtzeff a Yale," in A. Marcone (ed.), *Rostovtzeff e l'Italia: Incontri Perugini di Storia della Storiografia Antica e sul Mondo Antico IX, Gubbio, Casa di Sant'Ubaldo, 25–27 maggio 1995)* (Perugia: Università degli Studi di Perugia, 1999), 63–95.

"Rupture and Continuity: Hellenistic Reflections on the History of Medicine," in P. J. van der Eijk (ed.), *Ancient Histories of Medicine: Essays in Medical Doxography and Historiography in Classical Antiquity*, Studies in Ancient Medicine 20 (Leiden: Brill, 1999), 143–87.

"Anatomy and Physiology," "Dioscurides," "Galen," "Health," "Herophilus," "Hippocrates," and "Nicander" in G. Speake (ed.), *Encyclopedia of Greece and the Hellenic Tradition*, 2 vols. (London: Fritz Dearborn, 2000).

"Body, Soul, and Nerves: Epicurus, Herophilus, Erasistratus, the Stoics, and Galen," in J. P. Wright and P. Potter (eds.), *Psyche and Soma: Physicians and Metaphysicians on the Mind-Body Problem from Antiquity to the Enlightenment* (Oxford: Clarendon Press, 2000), 79–116.

"The Dangers of Literature and the Need for Literacy: A. Cornelius Celsus on Reading and Writing," in A. Pigeaud and J. Pigeaud (eds.), *Les textes médicaux comme littérature; actes du Ve colloque international sur les textes médicaux latins* (Nantes: Institut Universitaire de France, Université de Nantes, 2000), 355–68.

"Erasistratus" and "Herophilus" in *Encyclopedia of Life Sciences* (London: Macmillan, 2001).

"La medicina nel mondo ellenistico-romano," in S. Petruccioli et al. (eds.), *Storia della Scienza*, vol. 1: *La scienza antica* (Rome: Istituto della Enciclopedia Italiana, 2001), 708–35.

"The Transmission of Science in Greco-Roman Antiquity: Memory and Loss, Agency and Form," in A. Kida (ed.), *Kotengaku no saikôchiku (Reconstitution of Classical Studies)* 8 (Kobe: Hideaki Nakatani, Kobegakuin University, 2001), 31–41.
"Division, Dissection, and Specialization: Galen's *On the Parts of the Medical Techne*," in V. Nutton (ed.), *The Unknown Galen*, Bulletin of the Institute of Classical Studies, Supplement 77 (London: Institute of Classical Studies, 2002), 19–45.
"La lecture comme thérapie dans la médecine gréco-romaine," *Comptes rendus des séances de l'Académie des Inscriptions et Belles-Lettres* 146 (2002): 803–22.
"L'œil d'après Hérophile, Démosthène Philaléthès et Aglaïas de Byzance," in J. Royer, M.-J. Roulière-Lambert, and A.-S. de Cohën (eds.), *Autour de l'œil dans l'Antiquité: approche pluridisciplinaire; table ronde de Lons-le-Saunier – Jura – 11–12 février 1994* (Lons-le-Saunier: Centre Jurassien du Patrimoine, Musée d'Archéologie, 2002), 83–93.
"'A Woman Does Not Become Ambidextrous': Galen and the Culture of Scientific Commentary," in R. Gibson and C. Shuttleworth-Kraus (eds.), *The Classical Commentary: Histories, Practices, Theory* (Leiden: Brill, 2002), 109–39.
"ὡς ἐπὶ τὸ πολύ: 'Hippocrates' between Generalization and Individualization," in A. Thivel and A. Zucker (eds.), *Le normal et le pathologique dans la Collection hippocratique: actes du Xème colloque international hippocratique*, 2 vols. (Nice: Publications de la Faculté des Lettres, Arts et Sciences Humaines de Nice-Sophia Antipolis, 2002), 1:23–43.
"Galen's *daimon*: Reflections on 'Irrational' and 'Rational,'" in N. Palmieri (ed.), *Rationnel et irrationnel dans la médecine ancienne et médiévale: aspects historiques, scientifiques et culturels*, Centre Jean Palerne, Mémoires 26 (Saint-Étienne: Publications de l'Université de Saint-Étienne, 2003), 5–43.
"Galen's Alexandria," in W. V. Harris and G. Ruffini (eds.), *Ancient Alexandria between Egypt and Greece*, Columbia Studies in the Classical Tradition 26 (Leiden: Brill, 2004), 179–215.
"Early European Conceptions of the Brain and the Nerves: Classical Intuitions and Hellenistic Discoveries," in *Sphinx: 2004–2005 Yearbook of the Finnish Society of Sciences and Letters (Societas Scientiarum Fennica/Finska Vetenskaps-Societeten – Suomen Tiedeseura)* (Helsingfors: Multiprint, 2005), 11–32.
"Interpreting 'Hippokrates' in the 3rd and 2nd centuries BC," in C. W. Müller, C. Brockmann, and W. Brunschön (eds.), *Ärzte und ihre Interpreten. Medizinische Fachtexte der Antike als Forschungsgegenstand der Klassischen Philologie. Fachkonferenz zu Ehren von Diethard Nickel* (Munich: K. G. Saur, 2006), 15–47.
"Aulus Cornelius Celsus on *ars* and *natura*," in A. Ferraces Rodríguez (ed.), *Tradición griega y textos médicos latinos en el período presalernitano* (La Coruña: Universidade da Coruña, 2007), 345–71.
"'The *Oath*', the Oaths, and the Hippocratic Corpus," in V. Boudon-Millot, A. Guardasole, and C. Magdelaine (eds.), *La science médicale antique: nouveaux regards; études réunies en l'honneur de Jacques Jouanna* (Paris: Beauchesne, 2007), 425–66.
"Physis and Technē in Greek Medicine," in B. Bensaude-Vincent and W. R. Newman (eds.), *The Artificial and the Natural: An Evolving Polarity* (Cambridge, Mass.: The MIT Press, 2007), 21–49.
"Purity, Purification, and Katharsis in Hippocratic Medicine," in M. Vöhler and B. Seidensticker (eds.), *Katharsis vor Aristoteles* (Berlin: De Gruyter, 2007), 21–51.

"Animals, Women, and *Pharmaka* in the *Hippocratic Corpus*," in V. Boudon-Millet, V. Dasen, and B. Maire (eds.), *Femmes en médecine: actes de la Journée internationale d'étude organisée à l'Université René-Descartes-Paris V en l'honneur de Danielle Gourevitch* (Paris: Bibliothèque Interuniversitaire de Médecine [De Boccard Édition – Diffusion], 2008), 171–204.

"The Discourses of Practitioners in Ancient Europe," in R. B. Baker and L. B. McCullough (eds.), *The Cambridge World History of Medical Ethics* (Cambridge: Cambridge University Press, 2009), 352–58.

"The Hellenistic Character of Celsus, *Artes X-XI* (*Medicina V-VI*): History, Structure, and System," in A. Ferraces Rodríguez (ed.), *Fito-zooterapia antigua y altomedieval: textos y doctrinas* (La Coruña: Universidade da Coruña, 2009), 11–32.

"Staging the Past, Staging Oneself: Galen on Hellenistic Exegetical Traditions," in C. Gill, T. Whitmarsh, and J. Wilkins (eds.), *Galen and the World of Knowledge* (Cambridge: Cambridge University Press, 2009), 132–55.

"How Greek Was the Latin Body?: The Parts and the Whole in Celsus' *Medicina*," in D. Langslow and B. Maire (eds.), *Body, Disease and Treatment in a Changing World: Latin Texts and Contexts in Ancient and Medieval Medicine*; *Proceedings of the Ninth International Conference "Ancient Latin Medical Texts"* (Lausanne: Éditions BHMS, 2010), 3–23.

"The Living Environment: Animals and Humans in Celsus' *Medicina*," in N. Palmieri (ed.), *Conserver la santé ou la rétablir: le rôle de l'environnement dans la médecine antique et médiévale* (Saint Étienne: Publications de l'Université de Saint-Étienne, 2012), 162–92.

"The Physiology and Therapy of Anger: Galen on Medicine, the Soul, and Nature," in F. Opwis and D. Reisman (eds.), *Islamic Philosophy, Science, Culture, and Religion: Studies in Honor of Dimitri Gutas* (Leiden: Brill, 2012), 63–87.

"La théorie de la vision chez Galien: la colonne qui saute et autres énigmes," in A. Laks and M. Narcy (eds.), *Autour de la perception*, Special issue, *Philosophie antique* 12 (2012): 115–55.

Index locorum qui e scriptoribus antiquis et medii quod dicunt aevi Graecis atque Latinis citantur

Aelianus
de natura animalium
   1.25   269
   5.47   97
   7.22   271
   7.47   276

Aeschylus
septem contra Thebas
   710   311

Aesopus
fabulae
   242 Perry (= 341 Chambry)   270
   243 Perry (= 340 Chambry)   270

Aëtius Amidenus
   6.11   358 *(bis)*
   6.50   3
   6.68   3
   9.6   355
   12.53   4
   16.67   134

Aëtius philosophus
   1.15.6   240
   4.15.3   244

Africanus, Iulius
   fr. D27.11–14 (Wallraff)   101.102
   fr. D27.23–24 (Wallraff)   101.103
   fr. D27.27–30 (Wallraff)   101

Albertus Magnus
de anima
   2.2.7   381
   2.3.29   381
ethica
   7.1.12 (VII, 492a Borgnet)   385
de generatione
   1.6.8   381
physica
   1.1, 1.23–24   78
   1.1, 4.58–61   78
   4.3.3–4   381
quaestiones super de animalibus
   91.60–62   381
   106.65   381
   171.55   381
   206.53–56   381
   233.48   381
   237.42–44   381
de somno et vigilia
   I.1 (IX, 123 Borgnet)   381
   I.8 (IX, 134 Borgnet)   382
   I.9 (IX, 135 Borgnet)   382
   II.1 (IX, 138 Borgnet)   384
   II.3 (IX, 142 Borgnet)   383
   II.4 (IX, 144 Borgnet)   384
   II.8 (IX, 150–51 Borgnet)   384–385
super II sententiarum
   lib.2, d. XIII, art. II (XXVII, 247a Borgnet)   381
de vegetabilibus
   6.2.4 (494 Meyer-Jessen)   386
   6.2.17 (562 Meyer-Jessen)   386

Alcmaeon
   DK 24 B 4   194

Alexander Aphrodisiensis
in Aristotelis libros de anima
   3.28–4.4   241
in Aristotelis librum de sensu commentarium
   51.3   228
mantissa
   130.14–17   245
quaestiones
   49.30–33   241
   52.21–23   241

Alexander Trallianus (cited by volume, page, and line of Puschmann's edition)
2.427.17   3

Anaxagoras
DK 59 B 21   686

Anonymus Londinensis (Londiniensis)
IV 9–12   199

Anonymus Parisinus (ed. Garofalo)
de morbis acutis et chroniis
3.2.1   343
7.2   199

Antyllus (cited by page and line in Dietz)
44.10   199

Apollonius Pergaeus
conica
1   197
4.11–12   415

Apuleius
apologia
43.8–10   343

Archigenes (fragments cited according to page and line in Larizza Calabrò)
68.2–4   199
68.13–14   199

Archimedes
de conoidibus et sphaeroidibus
13   197
de planorum aequilibriis
1.6–7   638
de sphaera et cylindro
1   197

Archytas Tarentinus (Huffman)
fr. 1 A   418
fr. 1C   418
fr. 4   433

Aretaeus
de causis et signis chronicorum morborum (SD)
3.3.1   199
3.16.1   199
7.4.3   464

Aristophanes
Acharnenses
879   280
ranae
523   101
943   6

Aristoteles
analytica posteriora
76a34   617
78b37   617
100a3-b5   641
de anima
413a31   682
415b8–416a19   167
416a14   168
416a15–18   168
416a19   148
418a31-b1   242
421a7–21   682
422b17–424a17   682
429a5–9   372
de caelo
268a1–268b5   335
268b2–3   335
285a29   315
categoriae
9b20   196
10a3–4   196
de divinatione per somnum
463b12–18   359
463b12–19   362
463b14   364
ethica Nicomachea
1149b10–22   386
de generatione animalium
715a1–18   141
715a9–11   141
715a14–15   141
715a19   144

716a2–4   81
729b5–6   143.145
730b8–23   144
732a18–19   143
733b23–734b4   140
732b32–33   81
734b8   145
734b11   145
734b17–735a4   145
734b27–28   153
734b30–31   153
734b31–36   155
734b31–735a4   151
735a1–2   150
735a12–26   140
735a17–18   148
735a30–749a7   80
735b33–37   242
735b37–736a16   146.147
736b29–737a1   155
736b29–737a7   147
737a2–5   143
737a18–22   155
737a18–24   147
737b25   80.81
737b25–27   80
738b1   142
738b8–23   144–145
739a11   143
739b29   142
739b23   143
740b12–741a2   148.149–150
740b13   148
740b25–35   155
740b25–741a2   151–152
740b29–32   143
740b35–37   148
741b15–37   140
742a14–16   158
742a16–17   161
742a16-b17   140.142.161.163
742a18–19   162
742a20–23   162
742a26–28   162
742a28–32   162
742a32–33   162
742a37-b3   1   58

742b1   158
742b6–10   164
742b10–11   164
742b32–743a   21
742b33–35   158
742b33–743a1   163
743a1–5   152.155 (bis).156
743a4   152–153
743a6-b5   155.156
743a11–17   158
743a15   158
743a17-b5   155
743a21–23   154
743a21–26   160
743a26–28   143
743a28   158
743a29   159
743a34   159
743a36–37   159
743a37-b5   161
743b18–36   140
743b32–744b12   153
743b34–744a6   153
744a13   154
744a14   194
744a33–35   154
744a36–38   154
744b22   154
744b23   153
744b28–38   158
745a4–9   166
745a32   154
745b21   155
745b21–22   152
749a10–755a37   268
754a10   194
755b1–756a5   268
756a5-b12   268
756b13–757a2   268
757a2–7   267
757a2–13   268.269
757a3   278
757a4   274
757a6   274
757a14-b30   268
762a20   143
770b6   196

779b28–33   242
786a2–7   242
786a12–21   242
historia animalium
    503a19   279
    579b15–29   269
    579b16–26   277
    579b.15–16   271
    620b35   196
de incessu animalium
    705a26–28   84
    705a28–32   85
    705b6–8   85
    708a9–20   167
de insomniis
    458b15   357
    462a8   357
    463b12–18   359
    462b31–32   196
de iuventute
    467b10–13   87
    478b30–31   88
    478b31–32   88
    479a29–32   89
de longitudine et brevitate vitae
    464b19–21   87
    464b21–26   87
    466a9   88
    467a6–9   88.89
    467a10–12   88
    467b6–9   89
meteorologica
    339a5–8   75
    374a1–3   242
    383a26–32   158
    383a32   158
metaphysica
    847a1-b1   617
    847a11-b10   621
    847b1–16   617
    847b16–17   636
    848a11–14   636
    848b1–849b22   637
    848b1–850a1   295
    848b3–5   637
    849b19–22   637
    850a39-b2   638
    850b2–6   637
    850b10–16   636
    854a10–11   634
    1047b35–1048a7   149
de partibus animalium
    639a19–20   77
    640a33-b1   140
    640b1–3   140
    645b33–35   340
    646a4   81
    650a21–23   83
    655b28–29   81
    655b30–32   83
    656a2–4   84
    656a11–13   81
    656a32–35   83
    661b28–36   691
    663b20–24   167
    663b31–35   166
    672b29   346
    683a18–19   167
    685b12–15   167
    686a25–690b20   682
physica
    192b18–19   310
    198b36–199a1   196
    199a4–5   196
    199a15–17   640
    208a29   325
    209a4–6   308
    216a11–21   644
    216a18–20   644
    249b27–250b7   293.646
    253b14–26   293
poetica
    1450a29–30   536
    1459b16   536
politica
    1309b18–35   167
    1316a6   219 *(bis)*
de sensu
    436a1–6   86
    436a8–17   340
    439b11–12   242
de somno et vigilia
    454a20-b4   345
    454a25–30   382

455a31-b13   354
455b3–13   347 (bis)
455b7–8   355
455b16–29   345
455b17–28   382
456b9–11   347
456b9–16   353
456b9–19   347
456b11   356
456b17–19   347
456b18–29   341
456b34–457a4   341
457a4–8   341
457a8   356
457a8–14   341.354
457a27–29   346
457b20–27   380
458a29   355
462a8   357
topica
   126b36   196
   126b39   196

Pseudo-Aristoteles
de coloribus
   792a24   228
   792b28   228
   793a15   228
   793b9   228
mechanica
   847a1-b1   617
   847a11-b10   621
   847b1–16   617
   847b16–17   636
   848a11–14   636
   848b1–849b22   637
   848b1–850a1   295
   848b3–5   637
   849b19–22   637
   850a39-b2   638
   850b2–6   637
   850b10–16   636
   854a10–11   634
problemata
   953a10–955a40   362
   953a16   346
   953b6   346

954a14–15   346
954a21–26   362
954a34–38   362
954b28–32   362
954b30   346

Pseudo-Arnaldus Villanovanus
de epilepsia
   1   355

Artemidorus Daldianus
2.36   400

Averroes
colliget
   1.11   371
   1.19   371
   1.20   366.367
   1.21   371
compendium libri Aristotelis de sensu et sensato
   193rb52–59   366
compendium libri Aristotelis de sompno et vigilia
   201ra17–24   382
   201rb54–65   371
   201va43–51   371
   201vb48–54   367
   201vb55–58   367
   201vb59–63   367
   202ra2–15   367
   202va66–202vb11   364
   202vb 8–11   365

Avicenna
canon medicine (cited after Bakhtiar's translation)
   1.5 § 389–390   373
   1.8 § 559   373
liber de anima
   4.2   364

Biton (ed. Wescher)
67   618

Caelius Aurelianus
celeres (vel acutae) passiones
   3.21.201   132
tardae (vel chronicae) passiones
   1.3.55   359
   1.4.60–61   343
   1.4.63   343
   1.4.64   359
   4.8.117   10
   5.4.74   132
de speciali significatione diaeticarum passionum fragmentum
   57   355

Callimachus
   fr. 75.13 Pfeiffer   36

Cassius Felix
   20.3   10
   71.1   343

Celsus
de medicina
   lib. 1 praef. 26   669
   2.11.1–2   131
   $5.22.2^B$ 10
   5.22.5   10
   $5.28.12^K$ 10
   6.4.3   10
   6.15.1   10
   6.19.2   10
   8.3.1–2   464
   8.3.7–9   464

Chiron
   88   10
   92   10
   96   10

Cicero
academica
   2.19   230.239
   2.79   229
de finibus bonorum et malorum
   3.32   200
Philippicae
   1.13   400

   3.3   401
Tusculanae disputationes
   1.57   313

Columella
   6.6.4   10

Constantinus Africanus
de melancholia (cited according to pages in the Basel edition)
   289–290   377
pantegni, practica
   liber 9   128

Corpus glossariorum Latinorum
   3.595.18   114

Corpus hippiatricorum Graecorum
Hippiatrica Berolinensia
   55.5   3
Hippiatrica Cantabrigiensia
   100.7   3
Hippiatrica Londinensia
   19   3
Hippiatrica Parisina
   290   3

Ctesias
   fr. 76 Lenfant   271

Cyranides (cited according to book, chapter and line in Kaimakis)
   1.10.18–22   97
   2.12.11–12   102
   2.14.8–9   100.103
   2.14.22–31   97–98.102
   2.35.2–3   102
   2.50.18–19   102.104
   3.9.4–5   102
   3.49.6–7   102
   3.67.14   102]
   3.74.6–7   102

Democritus
   DK 68 A 11   686
   DK 68 B 9   235
   DK 68 B 125   235

Pseudo-Democritus
liber medicinalis
   77 tit.   131
   78.1–2   132
   78.2   132.133

Diocles Carystius (van der Eijk)
   fr. 56   686
   fr. 64   32
   fr. 182   342

Diodorus Siculus
   1.80.5–6   6
   14.41   621
   15.48.4   196
   16.46.5   196
   19.11.7   196
   20.21.3   196
   32.12.2   270

Diogenes Laërtius
   7.86   200
   7.134   241
   7.157   245
   9.86   230
   10.44   237
   10.49   234

Dioscurides
de materia medica
   1.4.1   2
   1.8.1   3
   1.86.1   1.8
   2.76.16   3
   4.148 (4.143 Lat.)   126
   4.162   126
de simplicibus
   1.75.2   3
   1.183.1   3.9
   1.197.3   9

Pseudo-Dioscurides
alexipharmaca
   praef.   199
   18   199

Elias
in Aristotelis categorias
   204.2–5   229

Empedocles
   DK 31 A 85   342
   DK 31 B 23   241
   DK 31 B 71   241

Epicurus
epistula ad Herodotum
   40   196
   49–50   238
   68   196
   68–69   234
   71   233

Erasistratus (Garofalo)
   fr. 147   193
   fr. 284   197

Erotianus (cited according to pages and lines in Nachmanson)
   3.3–4   261
   9.25–26   256
   29.10–16   536
   45.9   264
   90.6–7   252
   116.11–16   261
   fr. 33   199

Esculapius (ed. Manzanero Cano)
liber chroniorum
   38.7   129

Euclides
elementa
   3.35   412.431
   6.8   412.420 *(bis)*
   6.13   412.421
   11   417
   12   417

Euripides
Hercules furens
   905   194

Eutocius
  commentaria in libros de sphaera et
    cylindro
    84.12–88.2 Heiberg/Stamatis 409
  commentaria in conica (ed. Heiberg)
    184.21–26 415
    184.26–28 414

Favorinus
  254 661

Festus (ed. Lindsay)
  178.9–10 400

Galenus (cited by volume, page and line in Kühn)
  de alimentorum facultatibus libri tres
    6.661 676
  de cuiuslibet animi peccatorum dignotione atque medela
    5.8 656
    5.13 655
    5.14 657
    5.17 657 (bis)
    5.18–19 656
    5.49 460
  de antidotis
    14.2 666. 667
    14.3 668
    14.32 659
    14.65 670
    14.64–65 659
    14.132 659
    14.150 668
    14.154 668
  ars medica
    1.335–336 688
    1.336–337 694
  de atra bile
    5.138 198
  de comate
    7.644.9 541
    7.662.16 541
  de compositione medicamentorum per genera
    13.416 667
    13.841.7 3
  de compositione medicamentorum secundum locos
    12.465.16 3
    12.881.2 4
    13.295.17 3
    13.339.13–14 4
  de constitutione artis medicae ad Patrophilum
    1.272 202
  de crisibus
    9.614 204
    9.748 204
    9.758.10 541
  de diebus decretoriis
    9.837 204
  de difficultate respirationis
    7.863.8 541
    7.905.10 538
  de experientia medica
    15 651
  in Hippocratis aphorismos commentaria
    17b.390 204
    17b.643 344
    17b.706.15 541
    18a.167.6–10 548
    18a.170.3–7 548
  in Hippocratis de articulis commentaria quattuor
    18a.379.6–14 536
  in Hippocratis epidemiarum librum I commentaria III
    17a.13.11–14 556
    17a.256 204
    17a.261 204
  in Hippocratis epidemiarum librum III commentaria III
    17a.481.1 549
    17a.500 253
    17a.532.17 542
    17a.534.5–10 542
    17a.535 204
    17a.539 204
    17a.574 253
    17a.575 253
    17a.578.12–579.2 541
    17a.578.18–579.2 556
    17a.579 253
    17a.580 253

17a.580.5   539
17a.583.7–584.3   545
17a.606   665
17a.607–608   665
17a.636.9–14   542
17a.638   204
17a.754   204
17a.682–683   538
in Hippocratis epidemiarum librum VI *commentaria I-VIII*
  17a.794.17–795.1–2   12
  17a.800.12–13   537
  17a.975.17–973.6   537
  17a.1006.2–3   537
  17a.1008.11–14   537
  17b.131.7–9   537
  17b.159   659
  17b.310.5   537
in Hippocratis de fracturis commentaria tria
  18b.425.15–17   537
  18b.438.5   537
in Hippocratis de natura hominis commentaria tria
  15.9.15–16   535
  15.105   665
  15.109   665
in Hippocratis de officina medici librum commentaria tria
  18b.684.12–13   537
in Hippocratis prognosticum commentaria tria
  18b.104.2–3   537
  18b.298.3   545
  18b.312–313   204
in Hippocratis prorrheticum I *commentaria tria*
  16.490.10–18   539
  16.501.9–12   545
  16.505.16–17   545
  16.506.10–11   545
  16.511.3–5   540.548
  16.513.1–3   549
  16.513.16–514.4   549
  16.514   204
  16.514.14–524.9   556
  16.545.15–550.14   553
  16.546.1–3   553

16.547.9–11   542
16.547.11   553
16.550.12–14   540.553
16.556.10   541
16.558.3–7   539
16.560.16–17   541
16.566.6   544
16.569.7–8   547
16.575.4–14   546
16.582.9–16   556
16.583.2–8   550
16.587.4–5   554
16.588.17–589.3   556.557
16.588.17–589..1   539
16.590   204
16.595.5–596.10   552
16.596.17–18   552
16.599.10–12   543
16.599.15   541
16.606.11–14   550
16.606.14   544
16.606.11–14   550
16.625.4–14   540
16.625.5   544
16.632.5   544
16.650.1–2   554
16.657.3   544
16.657.3–5   549
16.658.2–8   550
16.658.3–14   554
16.660.2–5   546
16.663.5   554
16.676.1–678.11   542
16.676.4–7   547
16.678.10–11   549
16.693.3–4   548
16.709.2–8   551
16.713.7–10   547
16.717.8   551
16.718.13–14   546
16.720.10–11   551
16.720.8–11   552
16.723.9–725.3   555
16.729.4   551
16.731.9   551
16.731.9–10   551
16.736.6–7   551

16.736.9–11   537
16.736.12–737.4   551
16.736.14   551
16.749.17–750.5   547
16.749.2   551
16.754.14   544
16.761.9–10   544
16.788.12–789.1   555
16.789.1–5   546
16.791.6–15   552
16.791.14   551
16.80816–18   554
16.815.6–8   543
16.835.5   541
in Hippocratis de acutorum victu commentaria quattuor
15.732–734   535
15.733.2–4   537
15.733.17–734.4   537
15.843   676
de indolentia
7–8   16
59   459
de libris propriis
19.35   455
linguarum seu dictionum exoletarum Hippocratis explicatio
19.69.10   253
19.81.13   253
de locis affectis
8.6–9   680
8.20   202
8.42–43   205
8.173   369
8.174   369
8.183   676
de simplicium medicamentorum temperamentis ac facultatibus
11.438   676
12.94.12–14   3.9
12.251   666
12.252   669
de methodo medendi
10.20   663
10.43.8–10   553
10.64   201
10.65–66   202
10.68   202
10.70   202
10.71   202
10.75   200
10.81–85   198
10.86   200
10.86   200
10.89   201
10.90   202
10.105   662
10.105–106   663
10.382.5–6   3
10.445   464
10.506   205
10.633   659
10.764–765   205
10.784   670
10.811–814   205
10.849–850   205
10.1000.12–13   4
10.1000.14–16   12
10.1001.7–8   4
ad Glauconem de methodo medendi
11.125.8   3
de nominibus medicis (Meyerhof and Schacht)
23.13sqq.   253
de optimo medico cognoscendo (ed. Iskandar)
1.2   669
9.16–17   656
13.2   660
quod optimus medicus sit quoque philosophus
1.58   669
de ordine librorum suorum
19.55   666
de placitis Hippocratis et Platonis
5.448–449   625
5.642   245
de praecognitione
14.659–660   670
14.660   670
protrepticus
1.1–3   660
1.4   660
1.5–6   661

de causis pulsuum
   9.131–142   345
quod qualitates incorporeae sint
   19.467   243
   19.473   243
   19.483   243
de sanitate tuenda
   6.405   670
de venae sectione adversus Erasistrateos Romae degentes
   11.239   197
de sectis
   1.67   651
   1.67–68   651
   1.75   651
subfiguratio empirica (ed. Deichgräber)
   45.24–30   651
de bonis malisque sucis
   6.756   458
de symptomatum causis
   7.139–144   345
   7.141.17–142.12   344
   7.142   343
   7.144   344
de symptomatum differentiis
   7.43   200
   7.44–45   201
   7.50   202 *(ter)*
   7.51   201 *(bis)*
   7.53   201
de temperamentis libri tres
   1.523   695
   1.534   681.695
   1.534–535   695
   1.535   695
   1.538   681
   1.540   681
   1.542.16–543.1   545
   1.543–544   677
   1.545.4–5   545
   1.546–547   677
   1.551–552   687
   1.559   676.680.684
   1.560   681
   1.560–561   684
   1.561   684
   1.561–562   684

   1.562–563   678
   1.563–564   682
   1.564–565   682
   1.565–566   691
   1.566   625.684
   1.567   691
   1.567–568   682
   1.575–576   682
   1.576–577   692
   1.580   694
   1.583–585   685
   1.586–587   695
   1.588   681.683
   1.589–590   683.685.686
   1.590–591   695
   1.591–592   691
   1.592   684.695
   1.592–593   684
   1.592–598   683
   1.594   683.691
   1.598   681
   1.601–602   692
   1.603–604   689
   1.604   689
   1.604–605   687
   1.607   689.692
   1.610–611   680
   1.611–612   687
   1.613–614   695
   1.619   679
   1.622   687
   1.624   688
   1.624–625   692
   1.625   681
   1.624–625   687
   1.626–627   692
   1.628–629   689
   1.629–630   689.692
   1.631   679.695
   1.631–632   693
   1.632   693
   1.633   693
   1.634–635   694
   1.635   679.689
   1.638–639   695
   1.640   678
   1.641   676

1.641–642   691.695
1.643   686
1.646–647   681
1.675   676.679
1.678–679   695
de theriaca ad Pisonem
   14.283–284   669
de usu partium
   3.775   664
   3.775–776   664

Pseudo-Galenus
definitiones medicae
   19.395   200
   19.445   200
de historia philosopha
   19.343   200
   616.2–3 (Diels, *Doxographi Graeci*)   240
introductio siue medicus (Lat.)
   14.783   479
   14.784.14   133
de remediis parabilibus
   14.324.10   3
   14.358.1   4
de succedaneis
   19.728.8   3
   19.729.5   3
   19.739.18   3

Gellius, Aulus
   16.3.10   197

Hero
belopoeica (ed. Wescher)
   73–74   618
   113.8–119.2   626
   114.8–119.2   630
commentatio dioptrica (ed. Schöne)
   312.20   637
pneumatica
   prooem. l. 76   647
   prooem. l. 249   647

Herodianus
de soloecismo et barbarismo (ed. Nauck)
   295.8   547

Herodotus Halicarnasseus
   2.84   28
   2.92.5   6
   2.93   268
   3.129–130   29
   5.62   320

Herophilus (cited according to the testimonies [T] ed. by von Staden)
   T61   15
   T87–89   15
   T193–196   15

Hippocrates (cited according to volume, page, and line in Littré's edition)
de affectionibus
   6.240   193
   6.244   206
de internis affectionibus
   7.252   193
aphorismi
   4.458   57
   4.460   194
de arte
   6.2   632
   6.8   619
de decenti habitu
   9.234   194
   9.238–240   52
epidemiarum I
   2.606   193
   2.654   194
   2.658   194
epidemiarum II
   5.76   194
   5.84–98   52
epidemiarum IV
   5.164   193
   5.178   194
epidemiarum V
   5.210   193
   5.222   342
epidemiarum VI
   5.292   193.194
   5.310   342
   5.356   362

epidemiarum VII
   5.452   194
epistulae
   9.346   197
de foetus exsectione
   8.512   193
de fistulis
   6.452   193
de flatibus
   6.94   192
   6.104   193
   6.110–112   342
   6.110–114   342
ius iurandum
   2–3   54
   3–7   54
   5   54
de locis in homine
   6.324   464
de medico
   9.206–208   52
de morbis II
   7.24–26   26
   7.26–28   24
   7.66   23
   7.70–72   22
   7.72   26
   7.96   23
   7.102   23
   7.106   23
   7.108–110   23
de morbis IV
   7.602   193
de morbis mulierum
   8.10   193
   8.16   194
   8.70   193
   8.172   24
   8.228.20–23   8
   8.282   193
de morbo sacro
   6.354   34
   6.360   35
   6.360–362   35
   6.362   36
   6.354   34
   6.396   34

Coacae praenotiones
   5.610   194
   5.634   200
   5.724.3–5   254
   5.728.7–13   254
prognosticum
   2.110   193
   2.110.6sqq.   264
   2.136.10–138.6   254
   2.138.1–6   257
   2.138.6   257
   2.176   542
   2.188.10   545
prorrheticus I
   5.516.2–4   252
   5.524.1–3   252
de superfetatione
   8.480   193
de ulceribus
   6.408.12   264
de vetere medicina
   1.570–574   619
   1.572–574   632
   1.620–624   619
de victu IV (= *De insomniis*)
   6.640   33
   6.642   33 *(bis)*
   6.646–648   34. 632
   6.656–658   33
   6.652   33
   6.658   34
   6.662   33. 34
de victu acutorum
   2.224   32
   2.226   32
de victu acutorum (spuria).
   2.404   194
de virginum morbis
   8.466.4   32
   8.468.17–20   32
de vulneribus capitis
   3.236–242   464

Historia Augusta
Hadrianus
   25.1–4   658

Homerus
Odyssea
   4.228–232 29
   17.382–385 37

Horatius
carmina
   4.5.34–36 401
epistulae
   2.1.16 400

Isidorus Hispalensis
origines (etymologiae)
   4.7.9 375
   17.7.1–6 128
   17.8–11 128
   20.3–21 128

Iosephus
  antiquitates Iudaicae
   15.144.3 196

Kilwardby, Robert
de ortu scientiarum (ed. Judy)
   48–52 89

Leo medicus
de natura hominum synopsis
   35 13.14

Lucretius
   1.430–432 196
   1.445–448 196.234
   2.150–156 239
   2.757–767 236
   2.795–809 232
   2.795–798 239
   4.185–190 239
   4.332–336 238
   4.337–352 239
   4.353–363 238

Marcellinus
de pulsibus (ed. Schöne)
   14 199

Marcellus
de medicamentis
   8.49 99.102
   8.50 100.101.103
   8.51 102.103
   34.101 10

Meletius
De hominis fabrica (ed. Cramer)
   2 pp. 61–72 15
   2 p. 63.7–10 13
   2 p. 63.7–11 14
   2 pp. 68.3–70.3 14

Menander
   fr. 201.3 Kassel-Austin 400

Michael Ephesius (cited by page and line in Wendland)
in Aristotelis parva naturalia commentaria
   48.29–49.15 356
   49.8–15 354
   49.14 355
   52.19–20 351
   53, 3–22 353
   53.6 355.356
   54–55 355
   54.25–55.3 352
   55.3–12 357
   62.10–15 357
   67.21 351
   75.11–17 357
   77.20–78.2 359 *(bis)*

Nemesius Emesenus
de natura hominis
   6 (p. 56.2 Morani) 368
   8 (p. 64.2 Morani) 368
   12 (p. 68.12 Morani) 368
   13 (p. 69.18 Morani) 368
   13 (p. 69.21–70.11 Morani) 368

Olympiodorus
in Aristotelis categorias
   98.25 228

Oppianus (Apamensis)
cynegetica
  3.262–292 271
  3.288–292 269

Oribasius (Graecus ed. Raeder)
collectiones medicae
  7.1.5 3
  7.16.1 132
  7.16.3 131
  7.18 131
  7.19 131
  7.21.9 4
  8.24.55 3
  12 v 2 3
  44.21.7 4
  46.30.3 4
  50.5.7–8 4
eclogae medicamentorum
  54.6–10 3
  63.7 3
ad Eunapium (euporista)
  2.5.3 3
  4.12.11 3
  4.117.1 358
synopsis ad Eustathium
  1.14 130.132
  1.19.18 3
  3.113 3
  8.2.1–2 358

Oribasius (Latinus) (ed. Molinier)
synopsis
  1.13 131
  liber 3 126
  liber 3 add. p. 894 Mol. 126
  liber 3 add. p. 897 Mol. 126
  7.14.3 133
  liber 9 add. p. 360–362 Mol. 134

Ovidius
metamorphoses
  15.408–410 269

Panegyrici Latini
  6.1.4 405
  11.13.5 405
  12.1.1 405
  12.4.1 405
  12.13.2 405
  12.19.1 405

Pappus Alexandrinus
collectio
  3, 54.7–56.17 415
  4, 258.20–262.2 414
  4, 262.13 418
  4, 262.18–19 418
  4, 270.1–272.14 415
  4, 270.17–24 414
  4, 270.21–22 418
  7, 636.23 413
  7, 660.18–662.10 413
  7, 662.10–15 415
  7, 672.11–13 413
  7, 672.20–21 413
  7, 676.19–678.12 413
  7, 1004.16–1014.24 414
  1068.20 637

Parmenides
  DK 28 A 46b 342

Paulus Aegineta
  3.3.4 3
  3.13.1 359
  3.15 358
  6.3 484
  7.3.16 s. v. πάπυρος 3 *(bis)*
  7.12.1 3

Paulus Nicaeensis (Nicaenus) (cited by chapter and line in Ieraci Bio's edition)
  65.18 3

Pelagonius
  134 10

Philo Alexandrinus
de ebrietate
  173 230

Philo Byzantinus
belopoeica (ed. Marsden)
  49.1–56.8   617.621
  49.12–50.12   624
  50.14–29   627
  50.15–17   618
  50.20   628
  50.30–51.7   629
  51.8–9   618
  51.10–14   617
  51.15–52.19   625
  51.21–26   626
  51.28–52.19   626
  52.20–53.7   627
  52.21–22   618
  53.8–54.16   626
  53.17–23   629
  53.24–25   626
  53.25–30   628.629
  53.29   628
  54.15   626
  54.15–16   626
  54.21   626
  54.25–55.11   627
  55.12   619
  55.12–56.8   627
  55.13   626
  55.19   626
  55.29   626
  56.8–78.26   617
  56.8–67.27   632
  56.18–24   632
  58.7–16   639
  58.8–78.26   617
  58.26–32   632
  58.32–35   625
  59.1–62.21   633
  59.4–8   638
  59.8–10   633
  59.11–12   634
  59.11–22   634
  59.23–31   638
  61.14–21   639
  61.29–62.15   632
  62.12–14   632
  66.18–19   632
  66.30–31   633
  67.28–68.2   640
  68.7–17   641
  68.18–29   641
  69.1–5   637.642
  69.8–14   643
  69.14–20   644
  69.20–22   644
  69.26   619
  70.35–72.4   646
  71.9   647
  71.17   647
  71.22   647
  71.27–31   647
  72.1   647
  72.1–4   647
  72.24–73.20   645
  72.26–28   646
  73.8–13   637.646
  73.21–77.8   649
  76.22   649
  77.8   649
  77.9–78.22   648
  77.11–12   648 *(bis)*
  77.12–18   648
  77.18–19   648
  77.27   648
  77.29–78.7   649
  78.11–12   648

Philodemus
de signis
  17.37–18.10   238
  18.1   234

Philoponus
in Aristotelis libros de anima
  315.28   228
in Aristotelis libros de generatione et corruptione commentaria
  23.9   228

Philumenus
de venenatis animalibus eorumque remediis
  4.11   199
  22.6   199
  25.3   199
  34.1   199

Index locorum — 729

Plato
de legibus
 635a2–5 332
 720a-e 619
 857c-e 619
 884a1–2 316
 885b4–6 316
 885e1–6 316
 889b-c 326
 889b1–890a1 316
 889c5–6 315
 890b3-d8 316
 893a6-b4 316
 893c2 316. 325
 893c3-e1 316
 893e1-894a1 316
 894a2 316
 894a3 434
 894a1–5 317
 894b8-c1 317
 894c10–895a3 317
 896c9-d3 317
 896e1–3 317 *(bis)*
 898b5–8 327
Phaedo
 65c3 312
 65a9-d2 312
 65d4–7 312
 80c7–8 194
 81b5 321
 81c4 321
 81e1 321
 83d5 321
 86a2 321
 96a6–100a7 316
 98b7–99d2 316
Phaedrus
 245c5–246a2 317
 268a-270e 619
Philebus
 20d1–10 310
 55e1–56c7 651
politicus
 273b4 321
 283b1–3 314
 283b4-c1 314
 283c11-d2 314

 283d7–9 314
 284a1-c3 314
 284e4–285a4 314
 285a1–4 314
res publica
 42e6–11 320
 426b1 38
 486b10-d11 310
 511a 224
 514a1–2 223
 518c 223
 518c7 223
 518c8 223
 518d5 223
 521c5–6 223
 521d 215
 521d4 223
 522a 216
 522c 215. 216
 522c-531d 212
 523a2 223
 524a 224
 524d2–3 224
 526b 213. 224
 526d6-e1 212
 526e2 223
 527b4 224
 527c 213. 215
 527c3 213
 527d-e 215
 528a9-b1 434
 528a9-b3 308
 528b 212
 528b1–2 434
 528b2–3 434
 528b-d 217
 528d8 434
 530c 217
 531c 217
 531c5 212
 531c6–7 224
 531d 212. 225
 531d-535a 212
 532b 224
 532c 224
 532c7 321
 533a8 223

533b 212
533c 225
534e-535a 212
536d 224
536d-537a 212
537b-d 221
537c 225
537c2–3 212
540a 212. 221
545c 218
545e1–3 218
546a 220
546b1 221
546b1–3 220
546b3–4 220
546b5-c7 218
546c7–8 220
546d-547a 218
546d1 220
546d1–2 220
547a 218. 220
587c-d 221
587e 217. 221
588a3–4 222
588a3–5 222
597b5-d8 313

sophista
216a2–5 309
216c5–6 309
217a3 310
218d1-e3 310
218d1–230e 310
222a2 310
228c4–5 310
228c5 326
228c6-d1 310
228d4 311
231b7 309
233c10–11 311
234e5–235a9 311
235c5 309
235d1 311
235d6 311
235d6-e2 307. 311
235e5–236c6 311
236a4–6 311
236a8 311

236b1 311
236b7 311
236c4 311
246a4–248e6 330
265c7–9 316
265e5–266d3 313
266b2–4 313
267a1 313
267d4–8 313
268b7–9 313
268b11-d5 313

Timaeus
27a2-b6 319
27a3–4 319
27a3–6 323
27a4 327
27c1–3 319
27d5–28a1 319
28a4–5 319
28a6–29b1 319
28c3 320
29a6-b2 319
29b1-d3 319
29b2 319
30a2-c1 319
30b3 312
30b6-c1 320
31a4 320
31b3 320
31b4 321 *(bis)*
31b5–6 321
31c1–4 321
32a7-b8 321
32a8 321
32b8 321
32c2–4 321
32c5–34a7 323
34b8 323
34b10-c5 323
34c4–35a1 320
35a8 321
36d9 321
38b6–7 323
38c2–39e2 323
39e2–47e2 323
41a7 320
41a7-d3 335

46c7-d3   323
47e2–48a2   324
48a2–3   335
48a2–7   324
48a7-c2   324
48d4-e2   324
49a1–51b7   325
49d4-e4   325
50c7-d2   325
50d2–51a3   325
51b4–6   325
52a8-b2   325
52a8-d1   325
52d2–53a8   326
52e1   326
52e1–5   326
53a7–8   326
53b1   327
53b1–2   326
53b3–4   326
53b4–5   326
53b5–7   327
53c1   328
53c4–6   328
53c6-d1   328
53c7–69a5   328
53d6–7   328
53d7-e2   328
53e6–54a7   328
54e3–55a8   328
55a8-c7   328
55b3–4   328
55d8–57d6   329
57d7–58c6   329
57e1   329
57e3–6   329
58a7   329
58c2–4   329
67e4–68d2   329

Pseudo-Plato
Axiochus
   364c8   195

Plinius
naturalis historia
   8.72   279

8.105   270
8.107   279
13.72   1.6
24.88   4
28.29   101
28.61   4
28.168   4
28.214   5
28.261–262   273
29.106   5
29.129   96.102
29.130   96.102
29.131   101.103
30.70   39
34.170   5

Plutarchus
adversus Colotem
   1109C-E   238
   1110C-D   237. 239
   1111 A-B   235
de communibus notitiis
   1085E   243
de primo frigido
   948D-E   246
   952C   241
de Stoicorum repugnantiis
   1047 A   240
   1053E   241
   1053F   244
   1054B   240
quomodo quis suos in virtute sentiat profectus
   11.81F   52

Polybius
   1.22.1   196
   1.35.2   196
   2.7.11   196
   3.81.7   196
   5.88.2   196
   6.53.3   196
   9.6.5   196
   9.10.9   196
   21.22.6   196

Posidonius (Edelstein-Kidd)
   fr. 92   241

Praecepta diaetetica (cited according to lines in Groenke)
 1 l. 59   129
 2a l. 46   134
 5 l. 53   134

Proclus
in primum Euclidis elementorum librum commentarii
 67.9 – 10   415
 111.19 – 112.8   414
 111.21 – 22   415
 394 – 395   413

Res gestae divi Augusti
 34   401

Rhazes (Lat.)
liber continens
 fol. 326 verso a 26 – 33   (cited from Anastassiou/Irmer)   258

Rufus Ephesius
de nominibus partium corporis humani
 153   15
quaestiones medicinales
 21 – 22   204
de satyriasi et gonorrhoea
 32   199

Pseudo-Rufus
de anatomia partium corporis humani
 15   15

Scribonius Largus
 114   10
 237   10

Seneca
ad Lucilium epistulae morales
 113.2   243
naturales quaestiones
 1.5.6 – 7   246 – 247
 1.5.10   247
 1.5.13   247
 1.7.2   246

Pseudo-Serapion
Practica, fol. 8^v 377

Serenus, Q.
 139   10

Severus iatrosophista (cited by page and line in Dietz's edition)
de clysteribus
 16.2 – 3   199
 32.25   199
 39.6 – 7   3
 41.11 – 13   4

Sextus Empiricus
adversus mathematicos
 10.220 – 223   233
Pyrrhoneae hypotyposes
 1.120   230

Simplicius
in Aristotelis librum de caelo commentaria (cited by page and line in Heiberg's edition)
 492.25 – 504.32   72
 492.25 – 510.35   70
 504.33 – 510.35   72
 510.31 – 35   71
in Aristotelis categorias commentaria
 66.32 – 67.2   242
 166.23 – 26   243
 212.26   243
 214.34 – 35   243
 217.32 – 33   243
in Aristotelis physicorum libros commentaria (cited by page and line in Diels' edition)
 290.3 – 293.15   71

Sophonias
in libros Aristotelis de anima paraphrasis
 121.5   228
in Aristotelis parva naturalia commentarium
 24.6 – 18   354

Soranus
 1.4   199
 1.10   15

1.13.1   2.12
1.16   15
1.28   199
1.33   193
1.35   15
1.40   15
1.46   205
1.48   199
1.54   199
1.58   193
2.6   15
2.6.4   13
2.9   199
3.1   199
3.2–3   15
3.5   199
3.24   205
3.26   198
3.40   199
3.41.8   3.11
3.47   199
3.50   199

Speusippus
fr. 28 Tarán   219

Stobaeus
2.31.115   415

Strato Lampsacenus (Wehrli)
fr. 65a   647
fr. 73   644

Suda
Σ 851   11
Σ 852   11

Suetonius
de vita Caesarum liber II: divus Augustus
31.4   404

Themistius Sophonias → Sophonias

Theodorus Metochita (cited by page and line in Drossaart Lulofs)
commentarium in Aristotelis parva naturalia
18.28–32   358

Theodorus Priscianus
euporista
2.15   343
2.29   132
2.33   132
2.86   132
2.117   132

Theodosius
sphaerica
1 def. 5   427
1.8–10   427
1.11   424
1.12   424
1.13   424. 431
1.14   424. 427
1.15   426

Theon Smyrnaeus
80.15   219

Theophrastus
historia plantarum
1.1.3   82
1.1.4   82
4.8.2–4   6
4.8.4   13
de odoribus
28 (87 Eigler-Wöhrle-Herzhoff)   2
de sensibus
59   241

Thomas Aquinas
summa theologica
q. 14.1   359

Thucydides
4.29.3   191
4.36.3   192

Tractatus de chirurgia (Bambergensis)
(cited according to lines in Sudhoff)
l. 1575   133

Varro
de lingua latina
7.85   400

Vegetius
digesta artis mulomedicinalis
   2.13.5   10
   2.23.2   10
   2.27.3   10
   2.57.1   10
   4.4.4   10

Vergilius
georgica
   1.29–30   401

Vita Hippocratis codice Bruxellensi 1342 tradita (quoted from Pinault; see also *FrGrHist* part IV fasc. 7)
   fol. 52v 34–37, 38–43   41

Vitruvius
de architectura
   lib. 1 praef. 1   401
   10.10–12   626
   10.11.2   626

Xenophon Atheniensis
anabasis
   4.5.13   664
historia Graeca
   7.2.6   219

Inscriptiones Graecae Latinae

Corpus inscriptionum Latinarum
   VII 1308   187
   XI 3303   401
   XII 4333   402
   XIII 2867   179

Inscriptiones Graecae ad res Romanas pertinentes
   502   457
   502–506   453. 457
   503   457
   504   457. 458
   505   457. 458

Inscriptiones Italiae
   XIII 2,115, nr. 17   401

Inscriptions of Roman Tripolitania (IRT)
   342a   402
   467   405

Papyri Graecae Latinae

P. Ant.
   III 127, fr. 2   10
   III 127, fr. 5   10

P. Chassinat
   121   7

Papyri Graecae Magicae
   V 70–95   104
   V 172–212   105

P. Heidelberg
   inv. 1716 verso   400

P. Oxy.
   4148   443. 445. 446. 447. 448

PSI
   II 117   15
   X 1180   9 *(bis)*
   X 1180, fr. A, col. II, 11–12   10
   X 1180, fr. A, col. III, 5–7   10

P. Tebt.
  II 273   9
  II 273, col. VI, 9   10

Supplementum Magicum
  78 ii 3–6 (Daniel and Maltomini)   103
  86 col. ii 1–8 (Daniel and Maltomini)   106

Papyri Aegyptia lingua scriptae

P. Ebers
  295   25
  340   7
  482   7
  669   7

Cuneiform Tablets

ABL 1277 rev. 9   521
Ass. 13955bu   25
BAM V 480 III 57–64   479
BM 42282+42294 obv. 1   523
BRM 4 19   527
BRM 4 20   527
CT 27 38.43   523   522
Diagnostic Handbook (Babylonian), tablets
  26–30   35
DT 72   525
DT 78   525
Enūma Anu Enlil Tablet 14   525
Erm 15639:15′   515
Heeßel, *Babylonisch-assyrische Diagnostik*,
  tablet 28, 35′   33

Heeßel, *Babylonisch-assyrische Diagnostik*,
  tablet 18, 12′-14′   34
K.2486+ 1–22   523
Labat, *Traité akkadien*, 1951, p. 114, tablet 13,
  42′-44′   23
LAS I, text 318   521
Louvre, AO 9071   503.504–510
SAA 10 no. 30   521
SpTU 14, 159   527
STT 300   527
YBC 4657   500.502
YBC 4662   501.502
YBC 4663   498.499.501.502

Other

Charaka Saṃhitā
  4.52–54   53
  4.55–62   53

General Index

Abdera 468
accessus 538
accident, accidental 154, 169, 191, 194–96, 198, 233–34, 242–44, 326, 689
acculturation 19–21, 28–30, 40, 42
Achaea Klauss 467
Actium 398, 401
Adam, James 217–18, 220, 222
Aelian (Aelianus, Claudius) 97, 100–102, 276, 452
Aelius Aristides 452
Aelius Isidotus 457–60
Aelius Nicon 453, 457–60
– hymn to the Sun 457
Aenesidemus 230
Aeschylus 311
Aesculapius (medical author). *See* Esculapius
aether 69
Aëtius of Amida 3, 8, 358
agramen (agriamen) 115–16, 118–20, 123–24
Agrippa, Marcus Vipsanius 403
aitía 202, 619, 624–25, 640–41, 647, 650. *See also* Aristotle: cause(s) in
Albert the Great 78, 89, 339, 349, 377, 380–87
Alcmaeon of Croton 194
Alexander of Aphrodisias 69, 452, 678
Alexander of Tralles 3, 8, 127
Alexander of Tralles (Pseudo-)
– *de pulsis* [sic] *et urinis* 127, 129
Alexandria (Egypt) 1–2, 11–12, 15, 170, 437, 439, 442, 460, 615–17, 620, 627–29, 631, 640, 650–51, 665–66, 669
Alexandru, Stefan 451–52, 454–60
Alfarabi 363, 381
Algazali 381
almanacs 526. *See also* diaries, Babylonian astronomical
Alpinus, Prosper 5–6
Álvarez de Toledo, Fernando 583

amulet 24, 93, 95–98, 100–104, 107–109, 477, 485
– inscription on 97
– with pebbles 96–97, 100, 102, 104
– in rings (*see also* nails [metallic]) 96–102, 104, 108
analogy
– between art and nature (*see* art/nature analogy)
– between epilepsy and sleep (*see* epilepsy: analogy with sleep)
– "persuasive" 96, 102, 105
Anastassiou, Anargyros 253, 255, 257–59, 264, 535
Anatolia 468, 472, 475, 478–79
anatomy. *See* hyena: anatomy of; Soranus of Ephesus: anatomical knowledge in
Anaxagoras 268, 316, 686
Andromachus (Nero's *archiatros*) 251, 261
angler 310
Anna Comnena 350
Annas, Julia 211, 213–17, 220, 223–24
Antiochus of Ascalon 230
anus 25, 270, 272, 275, 277–78
Apollo 35, 63, 177, 282–83
Apollonius of Citium 561
Apollonius of Perga 40, 197, 413, 415, 417
Apulia 470
Arad 476
archē. See Aristotle: cause(s) in
Archimedes 40, 290, 294–95, 304, 407, 417, 616, 620, 637–38, 649
– *On the Equilibrium of Planes* 294
architecture 42, 628
Archytas of Tarentum 219, 407–34
Argyropoulos, Johannes 456
Aristaeus 413, 415, 417
Aristides, Aelius. *See* Aelius Aristides
Ariston 617
Aristophanes 6, 280
Aristoteles Arabus 274. *See also* Aristotelianism: Arabic

Aristotelianism
- Arabic   347–48, 351, 360–77, 382, 387–88
- Byzantine   347–48, 350–60, 358, 387–88
- Latin   339, 347–48, 350, 360–63, 377–88
Aristotle   40, 42, 67–73, 75–90, 139–73, 196, 216, 219, 228–29, 241, 267–84, 293, 299, 307–308, 315–16, 321, 329, 335–36, 339–88, 408, 519–20, 529, 536, 617, 619–20, 641, 646, 650, 670, 677, 679, 681, 685, 688–89, 691, 695
- automata in   169
- blooded and bloodless animals in   80–82
- cause(s) in
  - efficient   141, 144, 163–65
  - efficient and final   141–42, 148, 161
  - formal-efficient   144, 164–65, 172
  - intrinsic   165
  - material-efficient   144, 150–51, 155, 164–65, 169, 172
- chemical facts in   158
- definition (*logos*) in   145, 149–51, 159–60, 168
- "double accounting" in   171
- *dynamis* in   142–46, 165, 168
- epilepsy in   339–88 *passim* (*see also* epilepsy: analogy with sleep)
- formal nature in   144, 151, 153–55, 157–61, 165, 168, 172–73
- generation in
  - contribution of female to   142, 144, 147–50, 158–59
  - contribution of male to   142, 144, 147–50, 158–59
  - heat and cold (elemental) in   143, 145–46, 148–49, 151, 153–55, 159–61
  - order of formation in   140–42, 160–63
  - spontaneous   80–81, 144, 156, 473
  - *telos* of   142, 164
- *Generation of Animals*   80–81, 86, 139–73 *passim*, 274, 278
  - aims of   139–41, 164
  - Arabic translation of   274
  - as bearing on the interpretation of Aristotle's teleology of   139–42, 161–65
  - explanations in   139–41

- unity of   165
- *History of Animals*   82
- necessity in
  - conditional   140–42, 164
  - material   144, 152–54, 161, 166
- order of discovery in   76
- order of explanation in   76, 78, 81, 84–86, 90
- *Parts of Animals*   77, 79, 81, 83, 86, 140–41, 167, 691
- *On Prophecy in Sleep*   87, 359, 361, 363–64, 384
- *Physics*   79, 293, 299, 316, 408
- plants in   75–90 *passim*
- sleep in   77, 196, 340–47 (*see also* epilepsy: analogy with sleep)
- soul in
  - as efficient cause   144–45, 164, 167–68
  - as final cause   151, 159–60
  - as irreducible potential for form   165
  - nutritive, as cause of development   143, 148–49, 151–52, 158–60, 165–66, 172
- substance (*ousia*) and essence (*to ti ēn einai*) in   142, 163
- *symptōma* in   196
- uniform and non-uniform parts in   81, 83–84, 145, 152–55, 157–61, 163–65, 167
Aristotle (Pseudo-)   346, 362
- *Mechanica*   169, 290–91, 294–95, 299, 301–302, 617, 619, 634, 636–38, 649
arito   177, 181, 186
arkomus (Greek)   268, 280
artillery
- construction of (*belopoiikē*)   615–52
- torsion artillery   622–23, 632, 641, 644, 646
art/nature analogy   149, 158, 173
asbos (Greek)   275, 280–81
Asclepiades of Bithynia   678
Asclepius   63, 282, 438, 459
Aşıklı Höyük   468, 478
Asine   466, 483
āšipu   38, 524, 528
aspritudo   181–83, 186–87
Assurbanipal   520–22

astrology 437–38, 444–45, 516, 520, 524, 526–29, 591
astronomical dating 437–48
astronomical hypothesis 68–70, 71–73
astronomy 19–20, 40, 42, 67–73 passim, 212, 215, 217, 332, 418, 434, 437–48 passim
– Babylonian 515–30
– Hellenistic Greek 519
atheism 315–16, 318–19, 322, 324, 326–27, 333, 335–36
Athenaeus of Naucratis 455
atomism, atomist 231, 233, 237, 308
atoms 231–39, 327
Attalus III of Pergamum 655, 665–67, 669
Augustine of Hippo 375, 519–20
August the Younger, Duke of Brunswick-Wolfenbüttel 590
Augustus (Imperator Caesar Divi Filius Augustus) 397–404
Autolycus of Pitane 418–19, 434
Averroes (Ibn Rushd) 78–79, 89–90, 348, 363–68, 370–72, 376–77, 380–82, 384
Avicenna (Ibn Sīnā) 78–79, 89, 348, 363–66, 372–76, 381–82, 384, 483
Azor 476

Babylonian medicine. See medicine: Babylonian
Babylonian, Old (language) 493, 497, 521
Bacchius of Tanagra 251
Bacon, Roger 349, 520
badger (*Meles meles*) 267–68, 275–83
– scent-marking behavior of (musking) 276–78
– subcaudal gland of 277–78, 282
balance (in ancient mechanics) 290, 294–99, 301, 304, 634, 636–37
bandage 2–4, 7, 9–10
Barnes, Jonathan 308
Baron, Richard 206
Bartolotti, Gian Giacomo 452
bārû 524
Basilicata 470
Basque (language) 281
Battus, Levinus 585–87, 594
Berryman, Sylvia 169

Bewick, Thomas 270, 278
Bible 516–17, 529
biology 267, 284, 343
– Aristotelian 170, 339
birds 150, 247, 268, 274–75
birth
– abnormal 516, 524
– "good" 218–21
– oral. See mouth: birth from
blindness, blind 96–109, 350
bloodletting 32, 484, 619, 650
blood-vessels 152–53, 155–158, 464
Bluck, Richard S. 310
Bobonich, Christopher 321, 323, 332
Boethius 520
bone(s)
– and sinews 156, 157–59, 160–61
– size and shape of 165–67
Bowen, Alan C. 156, 163
Boys-Stones, George 332
Brack-Bernsen, Lis 523, 525
brain 340–88 passim
– as center of cognition 367, 383–84
– mixture in 693–94
– vapors in 344, 357
– ventricles of 368–70, 384 (see also ventricular localization of the cognitive faculties, theory of)
breadth. See Plato: dimension in
Bremen
– Latin School at 589
Bröcker, Ludwig O. 534
Buck, Carl D. 320–21
bulla 100, 103–104
Burkert, Walter 20, 27, 29–30, 35, 37, 40
Burnyeat, Myles 213–16, 225
Burzahom (place in Kashmir) 472, 483
Byblos 3
byblos 2–3

Caelius Aurelianus 4, 11, 132, 359
Cairns, Huntington 307–308
Calvinism 581, 583, 589–90
– Crypto-Calvinism 591, 598
Campania 470
Carneades 230
Caselius, Johannes 575, 588, 590, 603

Cassius Dio   452, 455, 668
Cassius Felix   4
Çatal Höyük   468, 478
catalog (as textual artifact)   497, 500–503, 511–12
Cavalli, Francesco   79–80, 82–83, 89–90
Çayönü   468, 478
Celsus, Aulus Cornelius   2, 4, 31, 131–32, 483, 579, 666
chance   156, 158, 160–61, 191, 196, 315–16, 318, 326, 355, 618, 627–28, 650–51
Charax, Aulus Claudius   452
Charles, David   139, 143, 151
Charles V, Holy Roman Emperor   569
charta combusta   10–11
chartēs (*chartion, chartarion*)   2–3, 10–11, 13
Chiron   4, 10
choiropithēkos (Greek)   279
Chrysippus   244–45, 625
Chytraeus, David   572, 603
Chytraeus, Nathan   588–89
Cicero, Marcus Tullius   229–30, 239, 307, 361
circular movement, principle of   290, 295, 304. *See also* concentric circles, principle of
Clavius, Christophorus   425–26, 432
Cleary, John J.   316
clitoris   270, 277
Cnidian medicine. *See* medicine: Cnidian
coat. *See* tunic
cognition. *See* brain: as center of cognition; ventricular localization of the cognitive faculties, theory of
cogwheel   300
cold. *See* Aristotle: generation in: heat and cold (elemental) in
collyrium stamp   181–82, 185–87
colophon (in Ancient Near Eastern texts)   40, 493, 497, 501–503, 522, 526, 528
color(s)
– mixture of   228, 241–42, 244, 246
– in the production of likeness   311, 313–14, 320, 329–30, 333–34
– as sign of good mixture   691–92

– theories of   227–48
Columella   4
commentary (as genre)   388, 528, 533–34, 561–65
concentric circles, principle of   290–91, 293–97, 299–301, 304, 635–37. *See also* circular movement, principle of
cone, conics   408, 411–18, 423, 427–30, 433–34
confidentiality (between doctor and patient)   48, 50, 52, 54, 58–60
conjunctiva   14. *See also* tunic: ocular
consecratio   398
Constantine the Great   404–405
Constantinople   350, 451, 454–55
Constantinus Africanus   126, 128–29, 376–77
– *Pantegni*, Practica book 9 (surgery)   128
Cooper, John   332, 336
Copenhagen   566–67, 599, 603
Coptic (language)   105, 107
Corinth   484
Cornarius, Janus   568–72, 594
cornea   14. *See also* tunic: ocular
Cornford, Francis M.   307, 310–11, 336
crab (animal)   102–104
critical signs. *See* signs: critical
Crönert, Wilhelm   453–54, 456
Croesus (king of Lydia)   655
Ctesias of Cnidus   29, 271, 660
Ctesibius of Alexandria   616, 618, 640, 642, 645–46, 648–49
cube
– as composed of "flat things"   309
– doubling of   407–34 *passim*
cucurbita   131–32
cucurbitatio, onis   132
cucurbito, are   132
cucurbitula   131–32, 134
Culemborg (place)   584–85
cult, public, of Roman emperors   397–405
cuneiform texts   479, 494–512 *passim*, 515–30 *passim*
Cunningham, Andrew   528
Cuomo, Serafina   29, 33, 38
cylinder   298, 300, 407–15, 417–18, 421–23, 428–30, 630, 648–49

Cyperus papyrus   2, 5, 7, 9, 16
Cyranides   97–104
Cyrus the Great   575, 661–62, 669

daemonicus, daimonios   361, 363–64
Daiber, Hans   363
Daremberg, Charles   251, 260–63
Darwin, Charles   139
De Chiara-Quenze, Deborah   314
declamatio   572
dedicant   179, 181, 186–87
Degli Aromatarii, Giuseppe   457
De Groot, Jean   169
Deichgräber, Karl   540, 620
demagogue   313–15
Democedes of Croton   29, 41, 662
Democritus of Abdera   235–37, 686
Democritus of Abdera (Pseudo-)   132–33
Demotic
– handbooks   105
– magical papyri (see *Papyri Demoticae Magicae*)
Denyer, Nicholas   214
depth. *See* Plato: dimension in
devotio moderna   581–83
diagnosis   23, 34, 36, 253, 680–81, 683
diagnostic texts   22–24, 34–35, 40
dialectic. *See* Plato: dialectic in
dianoia   536–38
diaries, Babylonian astronomical   526. *See also* almanacs
Diels, Hermann   533, 538
dies Aegyptiaci   113–14
dies natalis   402
Dilkaya   469, 478
dimensional lists. *See* lists: of the dimensions of all components of an engine
Dimona   476, 484
Dio, Cassius. *See* Cassius Dio
Diodorus Siculus   196
Diogenes of Sinope   655
Dionysius I of Syracuse   621, 662
Dionysius II of Syracuse   661–62
Dionysius of Alexandria   649
diorismos, diorismoi. *See* Galen: application of *diorismoi* by
Dioscorides   1–3, 6, 8–9, 100, 126

dissection   12, 15, 253, 270, 369, 675, 678, 692–93
divination   34, 361, 363, 372, 376–77, 495, 503, 516–17, 520–25, 527–28
diviner's manual, Babylonian   526
doctor/patient relationship. *See* confidentiality (between doctor and patient)
doxa. *See* opinion (*doxa*)
dreams   33–34, 37–38, 196, 313, 318, 342, 359, 361–67, 372, 380, 384, 516
– distinction from visions   352–54, 357, 379
– states engendering divinatory dreams   359, 361–63
Dubois, Jacques   570, 596–98
dynamis. *See* Aristotle: *dynamis* in

East, vs. West   19–42 *passim*
Ebers papyrus   2, 5, 7, 25, 37, 39
Ebla   472, 482
eel   245, 280, 282
Egypt, Egyptian culture   1–16, 20–22, 24–29, 31–32, 37, 39–42, 50, 93–96, 105–109, 187, 274, 437–38, 442, 446, 471
Elea   309, 314
elements (air, earth, fire, and water)   147, 167, 227, 241–42, 244, 246–47, 313, 315, 318–19, 321–22, 324, 326–27, 329–31
Elisabeth, Duchess of Mecklenburg-Schwerin (later Mecklenburg-Güstrow)   592
Elmalı-Karataş   468, 483
embryo   142, 148–49, 151, 153, 160–65, 169
– constitution of (*kyēsis*)   148
– differentiation of (*diakrisis*)   148–50
Empedocles of Agrigentum   37, 168, 241, 342, 663
empeiria. *See* experience
empirical observation/investigation   378, 387, 628, 645, 652, 675, 681, 689–90, 695–96. *See also* experience
Empiricists   206, 540, 547, 553, 620, 650–51
energeia   145–46, 202, 691
engine   615–52 *passim*
– air-spring (*aerotonos*)   648–49

– bronze-spring (*chalkotonon*) 640, 642, 644, 646–47
Enmeduranki 523
Epicurus 196, 230–39, 520
epigennēma 200–202
epilepsy 32, 34–36, 102, 339–88 *passim*, 484
– analogy with sleep 339–88 *passim*
– causes of 36, 376
equilibrium (in ancient mechanics) 294–98, 304, 616, 637
Erasistratus 12, 170–71, 173, 193, 197–98, 461, 651, 664
Erotian 251–65, 536, 538
– *Glossary* 251–65 *passim*
  – as an epitome 256
  – order of glosses 256, 258
  – preface of 256, 261
Esarhaddon 520–21
Esculapius 129, 343, 355
Estienne, Henri (Stephanus) 258, 260
"ethical code," Chinese 55–58
ethics
– four-principles approach to 60
– and mathematics (*see* mathematics: and ethics)
– medical 47–60 *passim*
– and technology 59
Euclid 40, 42, 408, 413–14, 417–19, 421
Eudemus of Rhodes 408–409
Eudoxus of Cnidus 41, 415, 417, 419
euergetism 458, 460
Euripides 6, 41, 457
Eutocius of Ascalon 407–409, 415–16, 423, 427, 429, 431, 433
experience (*empeiria, peira*) 11, 13, 15, 57, 161, 204, 221, 290–93, 303–304, 522, 618–21, 627–31, 634, 641, 648, 650–51, 684–85, 688, 690
expertise (*technē*) 33–34, 63–64, 197, 310, 480, 522, 615–21, 624–25, 629, 631, 640–41, 650–52
ex-voto 175–87 *passim*
– anepigraphic 177–79
– deity's name on 179
– inscribed 175–87 *passim*

eye 14–16, 93–109, 153–54, 161–62, 175–187, 215, 223, 238–39, 244–46, 342–43, 687, 692
– anatomy of 15
– diseases and disorders of 6–7, 96, 103, 108 (see also *lippitudo*; ophthalmia; ophthalmology)
– four-tunic system of (*see* tunic: ocular)
– of Horus 93–96, 104–105, 107–109
– salve for 10, 104

faith
– Christian 69, 71–73, 575, 590
– Late Platonist 69–70
Falcon, Andrea 139, 165
family resemblance 144
Fasti Praenestini 398, 402
Ferrari, G. R. F. 216
Fidenae 484
Fishwick, Duncan 397–99, 402
flax 107
Fleischer, Ulrich 194–95
fly (*musca*) 101, 103–104
Florence (Italy) 93, 452
Floris of Pallandt, Duke 585
Foës, Anuce 194, 255, 259
Forum Clodii 402
Frahm, Eckart 528
Frede, Michael 620
Frederick II, King of Denmark 592
frog 101, 103–104
fulcrum 295–96, 633–34, 636–38

Gadd, C. J. 525
Gätje, Helmut 366, 370
Galatian 281
Galen 2–3, 6, 8–9, 12, 16, 22–23, 48, 76, 111, 170, 193, 198–204, 206, 253, 264–65, 343–45, 351, 360, 367–71, 373–74, 376, 381–82, 451–60, 478, 483, 533–57, 562–564, 573, 579, 594, 596–97, 655–71, 675–96
– *On the Affected Parts* 11, 204
– application of *diorismoi* by 539, 544, 557, 689–90, 693, 696
– *On the Causes of Symptoms* 204

– *Commentary on Prorrhetic I* 533–57 passim
  – didactic function of 556–57
– concept of mixture in 675–96 *passim*
– conceptualization of error in 544–57 passim, 694–95
– critique of monarchy in 655–71
– *On the Differentiae of Symptoms* 200–202, 204
  – definitional work in 202
– identity of the father of 451, 453, 457–60
– *gentilicium* of 451–60
– *On the Method of Healing* 662
– *On My Own Books* 456
– as Roman citizen 453, 459–60
– Vlatadon manuscript of (*see* manuscripts: Greek: Thessaloniki, Monē Blatadōn, 14)
Galen (Pseudo-)
– *de diæta quatuor anni tempestatum & duodecim mensium* 112
– *praeceptum de humani corporis constitutione* 112
– *de succedaneis* 128
Galenus, Claudius. *See* Galen: *gentilicium* of
Galenus, Iulius. *See* Galen: *gentilicium* of
Galfridus de Aspale 349
Galilei, Galileo 562
Gallo-Roman culture 175–87 *passim*
garsa 111, 129–30, 132, 134
gastraphetēs 622
Gaul, Gaulish 187
Geller, Mark 21–29, 31–32, 35–38, 463, 479
Gellius, Aulus 197
gem, gemstone 93–101, 104, 108–109
– inscriptions on 99, 108
gender 51, 283–84
generation. *See* Aristotle: generation in
genius 400, 403–404
geometry, geometrical figures 77, 212–13, 215, 220, 223–24, 240, 322, 329, 408, 417–19, 430, 433–34, 457, 615, 626
– plane 307–308, 434
– solid 212, 307–308, 328–30, 408, 417, 419
Gerard of Cremona 377

geresis 111, 129, 132–33
gersa 111, 129–30
Gill, Christopher 332
gizzard-stones 101–102, 104
goal-year
– methods 525
– periods 525
– texts 525
gold 96–97, 99–104
Goldin, Owen 319
Goltz, Dietlinde 7, 22–24, 27, 29–30, 479
Goodenough, Erwin 94–95, 97, 104, 107–109
Gordon, Robert Jacob 270, 273
Gorgias of Leontini 309
Gotthelf, Alan 75, 82
Gradel, Ittai 397–400
Grafton, Anthony 516, 561
Grosseteste, Robert 71, 351
Grube, George M. A. 219, 221
Guetter, David L. 332
Gundissalinus, Dominicus 374–75
gynecology 1–2, 7, 11, 24, 100, 199

Habicht, Christian 178, 458–59
Hadrian (Roman emperor) 459, 657–59, 663
Hagios Charalambos 466
hand(s) 97–98, 102, 104, 107, 145, 152–53, 169, 240, 245, 657, 682–83
Hansberger, Rotraud E. 363, 372
Harappa 472
Harlfinger, Dieter 455
harmonics 212, 217, 224
Harrison, Peter 516–17, 519, 528
Hastreus, Hubrecht (Huybert) 582
heart
– as causal principle 371
– as center of cognition 344, 384–87
– flesh and fibers of 692
– function of 89, 170–71
– mixture in 688–89
– place in generation of 142, 152, 155, 157–58, 163–64
heat, vital (*psychikē* or *zōtikē thermotēs*) 143, 148, 158. *See also* Aristotle: generation in: heat and cold (elemental) in

Hebrew, Biblical   281–82
Henry, Devin   139, 156, 169
Henry, Duke of Brunswick-Lüneburg   590
Herentals (place)   561, 565
Heringa, Adriaan   259–64
hermaphroditism   267–70, 283–84
Hero of Alexandria
- *Belopoeica*   618, 626, 630
- *Dioptra*   290, 637
- *Mechanica*   169, 289–304, 637
- *Pneumatica*   647
Herodorus of Heraclea   267, 274, 277, 279
Herodotus   6, 38, 192, 234, 268
Herophilus   12, 15, 111–12, 251, 463, 620, 666, 681
- Hierophilus [sic]   111
Hesychius of Alexandria   276, 281
Himera   465, 470
Hipparchus of Nicaea   440–41, 446
Hippocrates of Chios   407, 420
Hippocrates of Cos. *See* Hippocratic corpus, and "Hippocrates"
Hippocrates of Cos (Pseudo-)
- *capsula eburnea*   129
- *dynamidia*   128
- *Epistles*   400
- *epistula ad Antigonum*   128
- *epistula ad Maecenatem*   128–29
Hippocratic corpus, and "Hippocrates"   3, 7, 19, 21–42, 47–60, 63, 192–94, 196, 200, 204–205, 251–65, 342, 360, 362, 451–52, 455, 465, 469–70, 480, 483–85, 533–57, 564–66, 569, 572–79, 592–99, 602, 619, 629, 632, 677–79, 681, 686, 689
- *On Ancient Medicine*   262–63
- *Aphorisms*   126, 262–63, 452, 534–35, 541–42, 548, 552
- *On the Art*   51, 169, 262–63
- authenticity of   534–37
- *Coan Prenotions*   253–58, 264
- *Decorum*   194, 197
- *Epidemics* I   194, 532, 541–45
- *Epidemics* II   535, 541–45
- *Epidemics* III   253, 534, 541–45, 549, 665
- *Law*   262–63
- *Oath*   37, 47–60, 63–64, 260–63, 561–604 *passim*
- polemics in   31–39, 41
- *Prognostic*   24, 204, 251–60, 262–65, 534–35, 537–38, 541–45, 551–52
- *Prorrhetic* I   251–52, 255, 262–64
  - authenticity of   533–57 *passim*
  - Galen's commentary on (*see* Galen: Commentary on Prorrhetic I)
- *Pseudepigrapha* (*see also* Hippocrates of Cos [Pseudo-])   574
- *On Regimen*   263, 677
- *On Regimen in Acute Diseases*   454–55, 535–36
hippopede   414
historicism   518
historiography of science   41, 518
Hittite (language)   275–76, 281
Horace (Quintus Horatius Flaccus)   400–401, 579, 590, 594
horns   166–67, 172
- and nails, hoofs, and bills   156–58
Horst, Jakob (the Elder)   588, 603
Horus. *See* eye: of Horus
Ḥubaiš ibn al-Ḥasan al-Aʿsam   369
Huffman, Carl A.   409, 418
Humanism, Christian   583, 603
humor(s)   24, 36, 116, 119, 121–22, 124–25, 129, 346, 369, 383, 385–86, 619, 675–76, 678, 689, 693
Ḥunain ibn Isḥāq   265, 372
Hungary   484
Hunger, Hermann   522–23, 525
Hunt, D. P.   324
Hussey, Edward   308
Hu-szû-hui   57
hyena   267–73, 277–79, 283
- anatomy of   270, 272, 277
- scent-marking behavior of (pasting)   278
- spotted (*Crocuta crocuta*)   269–73, 277–79
- striped (*Hyaena hyaena*)   271–73, 277–79

"I," authorial   31, 37
iatromathematics   519–20
Ibn Rušd. *See* Averroes
Ibn Sīnā. *See* Avicenna

iconoclasm (1566)   581, 583, 585
Ilberg, Johannes   258–59, 262–64, 656
illness vs. disease   206
imagines   399, 404
incipits
– *Oportet ut intelligas quia (h)ypostasis lenis*   128
– *Pulsus senis et iuuenis*   127
– *Quod ex(s)puerint mitte in carbonibus*   129
– *Sanguis naturaliter est calidus et humidus*   126
– *Species et uultus infirmi*   127
– *Urina naturalis qu(a)e significat corporis sanitatem*   129
– *Urina subalba et uernitia*   127
indicators (*gnōrismata*)   680–81, 687–89, 691–96. See also sign(s)
Indo-European (language)   275–76, 283. See also Proto-Indo-European (language)
initiation oath, Indian. See oath: Indian initiation
interpretation, science of   517, 519, 528
Irmer, Dieter   253, 255, 257–59, 264, 535
iron   96–97, 99–100, 102
Isidore of Seville   374–75, 380
– excerpts from the *Origines* (*Etymologiae*)   128
Isidotus, Aelius. See Aelius Isidotus
Isin (place)   482

Jabal (al-) Buhais (place)   479, 481–84
James of Venice (translator)   351
jarce (Old French)   131
jasper   97–98, 100, 102, 104
je scientifique   31. See also "I," authorial
jealousy   319, 322, 333, 335
Jericho   475–77, 483
Jewish
– angel names   100 (see also Ouriēl; Souriēl)
– exorcism formula   105, 108
– magical rites   106
Johann VII, Duke of Mecklenburg   568, 600
Johansen, T. K.   312
John Philoponus   69–70, 368

Joly, Robert   536
Jones, Alexander   408, 414, 417–18, 446
Jones, W. H. S.   50, 194–95
Jordan, Mark D.   360, 371, 381–82
Josephus, Titus Flavius   196
Jouanna, Jacques   25, 342, 538
Julius Africanus   101

Kaguru (ethnic and linguistic group of east-central Tanzania)   270–71, 283
Kahn, Charles H.   332
Kalbfleisch, Karl   453–54, 456
kalû   524, 528
Kamtekar, Rachana   323
Karagündüz   469, 478
Kashmir   472, 483
katamēnia   142, 148, 158–59. See also Aristotle: generation in: contribution of female to
Karataş. See Elmalı-Karataş
katholou   75, 77, 641
Kavousi (place in Crete)   465, 468
Kilwardby, Robert   89, 349
Kiš   512
Klebs, Ernst   453, 456
Knorr, Wilbur R.   408, 414, 416–17, 419, 421, 423
knowledge, "secret"   523–24, 528
Kral hospital   454, 456
Kurban Höyük   468, 478

la mūdû   522
Laenatius Romulus   404
Lao Tze   56
Lares Augustales   403–04
Lascaris, Constantine   455–57, 460
Lauerman (Laurimanus), Cornelius   580
Lauremberg, Wilhelm   598–99
Lauterbach, Georg Burkhard   566–67
legislator   316, 335–36
length. See Plato: dimension in
Lennox, James G.   77, 85, 156, 167
Leoniceno, Niccolò   452
Leo the Physician   13
leprosy   9–10
Leptis Magna   402, 405
Leunisse, Mariska   139, 141, 158, 167, 172

le'ūtu 521
lever (*mochlos*) 290, 293-96, 298-301, 304, 616-17, 633-38, 649-50
lexis 536-37, 539-40, 548
Liber karastonis 291
lichen 10
life-source (*zōtikē archē*) 147. See also heat: vital (*psychikē* or *zōtikē thermotēs*)
lignite 97, 104
Linacre, Thomas 452
lippitudo 96, 99, 102-103, 182, 185. See also ophthalmia
lists
- of the dimensions of all components of an engine 622, 624-26, 629-31, 651
- in Mesopotamian mathematics 503-12
Littré, Émile 194-95, 259-62
Livia Drusilla 404
lizard (green lizard) 93-104, 107-109
load, dividing and sharing the, principle of 289, 291-93, 296, 299-304
loanwords 29-30
loci
- solid 413-16
- surface 413-14, 418
Lollius Adama, Augustinus 567
Longolia, Johanna (= Longoeil, Jeanne) 583, 599
Lorenzi, Lorenzo 452
Lorsch (monastery, diocese of Mainz) 114, 117
Lothal (place in Pakistan) 484
Lucretius (Titus Lucretius Carus) 231-34, 236, 239
Lübeck 584, 590, 599-601, 603
lupinus aureus 100, 103, 104
Lycus (commentator of Hippocrates) 541

machines, five simple. See powers, five simple
Magee, John 520
magic 93-109 *passim*, 438, 516, 524, 527
- love 272-73, 523
- "sympathetic" 95-96
- vs. medicine 34, 36, 38
Manetti, Daniela 535, 538-39, 548

manuscripts
- Greek
  - Berlin, Staatsbibliothek zu Berlin – Preußischer Kulturbesitz, Phillipps 1524 457, 460
  - Città del Vaticano, Biblioteca Apostolica Vaticana, Vat. gr. 277 261
  - Città del Vaticano, Biblioteca Apostolica Vaticana, Vat. Urb. 68 261
  - Hagion Oros, Monē Ibērōn, 184 454-56
  - Madrid, Biblioteca nacional, 4631 439
  - Paris, Bibliothèque Nationale de France, gr. 2255 261
  - Thessaloniki, Monē Blatadōn, 14 451-60 *passim*
- Latin
  - Amiens, fonds Escalopier, 2 113
  - Bethesda, Md., National Library of Medicine, E 8 111, 113, 118-19, 121-29, 132
  - Città del Vaticano, Biblioteca Apostolica Vaticana, Palat. lat. 485 114-15, 118-19, 121, 123-25
  - Città del Vaticano, Biblioteca Apostolica Vaticana, Palat. lat. 1449 135
  - Città del Vaticano, Biblioteca Apostolica Vaticana, Reg. lat. 1143 118, 125
  - Glasgow, University Library, Hunterian Collection 96 114, 120-25, 127
  - Glasgow, University Library, Hunterian Collection 414 113
  - Laon, Bibliothèque municipale, 426*bis* 114, 130
  - Leiden, Bibliotheek der Rijksuniversiteit, Voss. lat. F. 96 A 113
  - London, British Library, Harley 4977 114
  - London, British Library, Sloane 2839 130
  - Madrid, Biblioteca nacional, 19 113, 135
  - München, Bayerische Staatsbibliothek, clm 7999 113
  - München, Bayerische Staatsbibliothek, clm 11343 127
  - München, Bayerische Staatsbibliothek, clm 13002 114

- München, Bayerische Staatsbibliothek, clm 17403   113
- Paris, Bibliothèque Nationale de France, lat. 11219   128
- Paris, Bibliothèque Nationale de France, lat. 14025   113
- Praha, Knihova metropolitní kapituly, 1358   122, 130
- Salzburg, Museum Carolino-Augusteum, 2169   113
- Trier, Stadtbibliothek, 40   127
- Vendôme, Bibliothèque municipale, 109   126-27
- Wien, Österreichische Nationalbibliothek, 3124   439

Marcellus Empiricus   4, 98-104
Marcus Aurelius   655
Margarethe of Parma   583
Martelli, Matteo   373
materialism, materialist   316, 318, 322, 324, 327, 330, 333, 336
mathematics   19-20, 39-42, 211-25, 407-408, 412-13, 417-19, 434, 494-512, 518-19, 527-29, 585, 621, 626-27, 629
- and ethics   211-25 passim
- military uses of   215-16
- in Plato (see Plato: mathematics in)
Mattern, Susan   458-59
matter, theory of   647
mean proportional(s)   407-34 passim, 626, 630
mechanical challenge (to move a given weight with a given power)   290, 293, 301-302
mechanics   169, 289-304 passim, 616-52 passim
"mechanistic" account(s)   170-73
medicine
- Babylonian   22-24, 27-29, 31-36, 38, 479, 482
- Cnidian   21-23, 26, 29, 32, 37
- Egyptian   7
- empirical school of (see Empiricists)
- folk   5, 28
- palace   27-28, 39
- practitioners of   39, 64
- rational   32, 37, 38, 41-42, 485
- vs. magic (see magic: vs. medicine)
Medinaceli, arch   403
Meibom, Johann Heinrich   566, 603
Melanchthon, Philipp   572-74, 587
Meletius the Monk   13
membrane. See tunic
Memmius, Abraham   583-84
Memmius, Conrad   584
Memmius, Isaac   584, 590-91
Memmius, Petrus   561-604 passim
- De recto medicinae usu   570-71, 580-81, 589, 592, 594-98
- Hippocratis iusiurandum commentario recenter illustratum   565-66, 568-69, 571, 577, 602
memoria   397, 399, 404
Menaechmus (pupil of Eudoxus)   415-17
Menelaus of Alexandria   418
Menn, Stephen   312
Mensing, Bernhard   587
Mesopotamia   19-42 passim, 273, 479, 493-512, 515-30
Methodists   202, 206, 662, 664. See also Soranus of Ephesus; Thessalus of Tralles
Michael Scot   274, 377
midwife   13
Minet el Beida   476
misfortune   192-93, 202-203, 205-206
mission statements   48, 50, 53, 59
Mithridates VI of Pontus   665-69
mixture (krasis)   675-96
Mommsen, Theodor   398, 401
mongoose   276, 278, 282
moon, crescent   93, 98, 100, 109
Morocco   471, 474
mouse   280, 283
mouth
- birth from   268
- copulation by   268
- foaming (as a symptom of epilepsy)   35
- ulcers of   8
mudû   522
Müller, Iwan von   453
Munatius Apsyrtus   403
Muses   218-20

musking. *See* badger (*Meles meles*): scent-marking behavior of
Mycenae, Mycenaean   21, 27–28, 466, 470, 483

Nabû   522
Nachmanson, Ernst   252, 255, 259, 263–64
nails (metallic)   97, 100–102, 104, 106, 108. *See also* amulets: in rings
nails (of fingers and toes). *See* horns: and nails, hoofs, and bills
namburbi ritual   36
natural phenomena   516, 518, 528–29
nature
– "according to nature"   200–201, 317, 639
– and art   310, 640 (*see also* art/nature analogy)
– "contrary to/against nature"   200–202, 204–205, 617, 620, 639
– formal (*see* Aristotle: formal nature in)
– as making nothing superfluous   154
– "shaping capacity" of   679, 694
Near East, Ancient   20–21, 24, 25, 27, 29, 34, 37, 40, 42, 471–75, 479, 494, 515–16, 529
necessity. *See* Aristotle: necessity in; Plato: necessity in
Nechepso   437–38
nēmequ   522
Nemesius of Emesa   368, 375
Neoplatonism   69–71, 367–68, 373–74, 387, 538
Newton, Isaac   67–68
Niccolò da Reggio   454
Nicodemus, Iulius   453, 457–59
Nicomachus of Gerasa   418
Nicon, Aelius. *See* Aelius Nicon
Nicon, P. Aelius   459
nişirtu   523
Nippur   35, 496, 523
nomenclature   451–60 *passim*
Nordtmann, Martin   590
Notomi, Noburu   310
numen   397–405, 568
– *Augusti*   397–99, 401, 403
– *Augustum*   402–403

Nutton, Vivian   12, 16, 29, 47, 112, 191, 656, 658, 660, 663, 670

oath
– Hippocratic (*see* Hippocratic corpus, "Hippocrates": *Oath*)
– Chinese (*see* "ethical code," Chinese)
– Indian initiation   53–55, 58, 60, 64–65
– medical   47–60 *passim*
obscurity (*asapheia*)   547–50, 554
obstetrics   13, 15
Octavian. *See* Augustus
omina   20, 28, 34, 515–16, 519–21, 524–27
"openness" (of the Greek medical market)   39
ophthalmia   93–109 *passim*, 175–87 *passim*. *See also* eye; *lippitudo*
ophthalmology   93–109 *passim*, 175–87 *passim*. *See also* eye
opinion (*doxa*)   317, 327
Oribasius   3, 8, 126–28, 130–34, 358–59, 480, 483
Oribasius (Latin translation of)
– *euporista*   127
– *synopsis*   128, 130–34
Osborne, Catherine   332–33
Ossetic (language)   276
Ouriēl   94, 100, 108
Overwien, Oliver   265, 372–73
Ovid (Publius Ovidius Naso)   400–401, 579

palace medicine. *See* medicine: palace
Pappus of Alexandria   290, 292, 413–16, 418, 637
– *Collection*   292
Papyri Demoticae Magicae   106–107
Papyri Graecae Magicae   104, 106–108
papyrus
– ashes   3–5, 7–9
– industry   1, 15
– manufacture   1, 5, 10, 16
– medicinal use   1, 3–5
– mineral components   9
– plant (root, stem)   1–11, 13
– roll, sheet   1–2, 4, 8, 12, 16
– strips   9, 12–14
Paracelsianism   563–64, 585, 602

parentalia 400
Paris 67, 73, 570–71, 596–97
– University of 73
Parma 403, 583
Parmenides 309, 342
Parpola, Simo 521
Parry, R. D. 315, 326, 480
Parry, T. W. 480
pasting. *See* hyena: scent-marking behavior of
pathēma 192, 200
pathos 195, 197, 199–202
Paul of Aegina 2–3, 8, 358–59, 480, 483
Pegel, Konrad 587
peira. *See* experience
Pellegrin, Pierre 310, 361, 570
Peloponnesus 467
penis 270, 272, 279, 283
perception
– of colors 227–48
– Epicurean theory of 231–39
– Stoic theory of 239–47
– *See also* sensation (*aisthēsis*); touch, sense of
peregrinus 452, 454
Pergamum 451–53, 457–60
– *Asclepieion* of 459
– inscriptions from 453, 457–60
persuasion 335–36
Pestavecchia 470
Petersen, Walter 320–21
Petosiris 438. *See also* Nechepso
phallus. *See* penis
Philaretus 118, 129
– *de pulsibus* (Latin) 118, 129
Philodemus 520
Philolaus 222
Philo of Byzantium 615–52
– analogy with fallen bodies in 645
– concern with beauty in 621, 625
– methodological reflection in 617–21
Philoponus, John. *See* John Philoponus
physics 240–41, 308, 312, 316, 329, 332–33, 336, 620, 649–50
physikoi logoi ("physical arguments") 633, 638, 645, 647
physiognomy 516, 688

physis. *See* nature
Pietrobelli, Antoine 454–55, 460
pirištu 523, 528
Piso, Cnaeus (pater) 397–99, 404, 660
plague 588–89, 591
plane, inclined 290, 298–300, 304
planetary motion 69–71
plants
– in Aristotle (*see* Aristotle: plants in)
– medicinal use of 1–16, 126, 129, 437
– as upside-down animals 84
plaster 2, 4, 11
Plato 37, 41, 69–70, 201, 211–25, 245, 307–36, 367, 434, 519, 529, 619–20, 650, 662–63
– allegory of the cave in 223–25
– the beautiful and the good in 224–25, 311–12, 319, 328
– body as governed by the soul in 307–36
– construction of the cosmos in 309, 312–13, 315, 318–24, 327–31, 333–35
– dialectic in 212–14, 224–25, 310
– depiction of the sophist by 307–16, 318–20, 322, 326, 330–35
– dimension in 307–309, 311, 313, 315, 331, 333–36, 434
– distinction among "kinds" in 309–10, 312–13
– the Good in
  – and the Beautiful 224–25, 311–12, 319, 328
  – knowledge of 211, 214, 216, 223
  – as paradigm for the Demiurge 320, 333
  – as unity 214
– guest/host relationship in 309, 318, 332–34
– idea of justice in 225, 312, 332, 335
– image-making in
  – human 309–15
  – divine 313–31
– imitation in 309, 311–13, 319, 333
– impulse (*hormē*) toward knowledge in 310–11
– intelligible realm in 224–25
– *Laws* 220, 307–309, 312, 315–16, 318, 320–21, 323–36

- likeness and semblance in 309, 311–15, 318–23, 325–26, 329, 332–35 (*see also* color[s]: in the production of likeness)
- mathematics in
  - constitutive role in education of the Guardians 211–16, 221, 223–24
  - instrumental role in education of the Guardians 211–12, 214, 216, 221, 223, 225
- necessity in 312, 324, 326–28, 331, 334–35
- non-uniformity in 329
- "nuptial number" in 218–22
- philosopher in
  - as hunter 310
  - king 211–16, 221–22, 225
- predecessors of 308
- proper measure in 309–14, 318–20, 322, 326–27, 330, 333–34, 336
- proportion as producing unity in 321–22
- receptacle (*hypodochē*) in 325–27, 330, 333–34
- *Republic* 37, 211–22, 314–15, 319, 332–33, 335, 434
  - calculation (*logismos*) in 220–22, 224
  - role of the Guardians in 211, 213–18, 220–21, 223, 225
- role of the Demiurge in 225, 315, 319–21, 331, 333, 335
- soul in
  - eye of 215, 223
  - as mover of the body 317, 323, 329
  - as self-moving 225
- *Sophist* 307–16, 318–20, 322, 326, 330–34
  - character of the Stranger in 309–10, 312–15, 319
  - method of division in 318, 329
- *Statesman* 307–310, 313–16, 318–19, 323, 331, 334
- *Timaeus* 69, 216, 222, 245, 307–309, 312–13, 315–16, 318–35
Platonism, Middle 364. *See also* Neoplatonism
plebs Narbonensium 401
Pliny the Elder 2, 4–6, 11, 38, 96, 100–104, 273

Plotinus 520
Plutarch 235–37, 240–41, 364, 452, 668
pneumatics 170, 616–17, 647–48
pnigmos = *suffocatio matricis* 134
Poggiardo (place in Apulia) 470
Pohlenz, Max 195
Polybius 195–96
Polyclitus 624–26, 684
Polycrates of Samus 661–62
polyspaston 289, 294
Pontano, Giovanni Gioviano 516
Pontecagnano 470
Posidonius of Apamea 68
Potter, Paul 206, 536, 544, 549–52
poultice 7–9, 11
powers, five simple 289–90, 293–94, 296, 300–301
Praxagoras of Cos 681
"prayers for justice" 106
presentism 518, 521
projectile 299
proportion 87, 156, 160, 168, 214, 216, 311, 321–24, 327, 331, 333–35, 411, 433, 624–28, 676–77, 684, 687, 690
proportional(s), mean. *See* mean proportional(s)
Prosper Alpinus. *See* Alpinus, Prosper
Protagoras 229, 309
Proto-Indo-European (language) 275–76, 281–82. *See also* Indo-European (language)
psychology 323, 363, 373
Ptolemy I Soter 665–66, 669
Ptolemy II Philadelphus 665–66, 669
Ptolemy (Claudius Ptolemaeus) 68–69, 72, 440, 443, 445–48, 453, 456
pulley 289–304 *passim*. *See also polyspaston*
pulse (in the arteries) 680–81
pyktidas (acc. plur., Greek) 280
Pythagoras 41, 128, 663
Pythagoreans 72

quadratrix 415
qualities, elementary (hot, cold, dry, wet) 676–77, 681, 685, 695–96. *See also*

Aristotle: generation in: heat and cold (elemental) in
quid pro quo   600
Quṣṭā ibn Lūqā   292

Ras Shamra   476
Rationalists   206, 620. *See also* Empiricists
rational medicine. *See* medicine: rational
rational-practice texts   20–21
reasoning
– deductive   636, 649–50
– illegitimate   325, 327
receptarium   9–10
recipe   2, 7, 9–10, 24–25, 31, 93, 95–108, 126–27
Reeve, C. D. C.   219
regeneration, regenerative power   95–97, 100–104, 108
Regimen of Health   111–35 *passim*
Regimen XII mensium. See *Regimen of Health*
reproduction
– methods of   80–81, 268
– organs of (*see* penis; vagina)
resiliency   622, 645, 647, 649
rest (and motion)   234, 293, 316, 325, 329, 420, 435
retardation   290–91, 297, 300–301, 304
retina   14–15. *See also* tunic: ocular
Reynolds, J. J.   333
Rhazes   257–58, 264–65
– *Continens*   257–58
  – Book 16 (on prognostic signs)   258
Rheinberger, Hans-Jörg   518
Riddle of the Sphinx   283
Ritter, Jim   20
Robinson, D. B.   336
Rosen, Stanley   310
Rostock   566–68, 572, 575, 579–80, 582, 584–89, 591, 594, 598–99, 603
Rougiers   175, 177
Rufus of Ephesus   15, 203, 483
Rufus of Ephesus (Pseudo-)   15
Rushd, Ibn. *See* Averroes

sacred disease. *See* epilepsy
Sanskrit   53, 275–76, 279, 281

saphēneia   538, 555
sarmina   114
sarminium   114
scent-marking   277. *See also* musking; pasting
Sceptics   229–31, 239, 247–48, 685
Schenck of Tautenburg, Friedrich   580
Schinckel, Harman   583
Schöne, Hermann   453, 458
Schoos, D. J.   332
Schurff, Augustin   572
Schwerin (place in Mecklenburg-Vorpommern)   568, 592, 599–600
screw   290–94, 296, 300–301
scribes
– cuneiform scholar-scribes   515–30 *passim*
– school for   496–97
Scribonius Largus   4, 11, 48
Scultetus, Johannes   483
Second Sophistic   655, 670
Sedley, David   214, 216
seer (*mantis*)   32, 37
semen (*sperma, gonē*)
– heat in   143, 148, 156, 158
– material constitution of   146–47
– motions/movements in   143, 146–49, 156
Seneca, Lucius Annaeus   246–47
sensation (*aisthēsis*)   196, 201, 220, 233–34, 312, 327, 329, 357, 365, 368, 372–73, 379, 683. *See also* perception
senses, internal, theory of   365–70, 372–73, 376, 382–83
Serenus, Quintus   4
series texts   494–512 *passim*, 516, 526
Seth (Egyptian god)   107–108
Severus Iatrosophista   3
Sextus Empiricus   233–34, 520
sexuality   272–73
sexual organs. *See* penis; vagina
Shapin, Steven   29, 38
Shields, Emily L.   348
shiphaulers' argument   293, 299
Shorey, Paul   219
shrew (mouselike insectivorous mammal)   283
Siberia   472
Sicily   470, 475

Siebenborn, Elmar   547
Siger of Brabant   73
sign(s)
- critical   204
- favorable   33, 253
- inference from   520, 686–91
- and symptom(s)   192–94
- unfavorable   33, 253
Simplicius of Cilicia   67, 69–73, 243, 408
Sīnā, Ibn. *See* Avicenna
sinew   153, 156–61, 622, 638, 640. *See also* bone(s)
Sippar   512, 523
Sisters and Brothers of the Common Life   582
skin
- of animals   100–101, 103–104
- of human beings   4, 11, 228, 682, 689, 692
slave/master relationship   54, 64
Slavic (language)   280–81
sleep
- analogy with epilepsy   339–88 *passim*
- in Aristotle   77, 339–88 *passim*
- physiology of   340, 360, 371, 382–83
"slowing up." *See* retardation
Sluiter, Ineke   535, 538
Smith, Wesley D.   197, 534, 538, 541, 563
snake   101, 103–104, 282–83
Socrates   212–13, 215, 218–24, 243, 312, 314–16, 318, 320, 619
solecism   540, 547–49, 551, 553, 556
solid geometry. *See* geometry: solid
solid loci. *See* loci: solid
solid problems   415–16, 433
Sophie of Mecklenburg-Güstrow, Queen of Denmark and Norway   592–93
Soranus of Ephesus   2, 11–16, 41, 198, 205, 453
- anatomical knowledge in   11–16
- use of analogy by   11, 13
Soranus of Ephesus (Pseudo-)
- *Isagoge*   121–22, 125, 130
Souriēl   94, 100, 108
soul. *See* Aristotle: soul in; Plato: soul in
spermatic residue   143, 156, 158–60. *See also* semen

sphere   413–14, 417–19, 423–27, 429, 431–34
spherics   418–19, 424–26, 432–34
Sphinx   283. *See also* Riddle of the Sphinx
spontaneous generation. *See* generation: spontaneous
spring-hole
- diameter as *stoicheion*, *archē*, and *metron* in construction   627–29
- relation   624–31, 650–51
Statilii, family of the (from Heraclea in Caria)   460
Steel, Carlos   332
steelyard   634
Stephanus. *See* Estienne, Henri
Stoicism, Stoics   200, 227–31, 239–48, 520, 529, 677
Stol, Marten   22–24, 29, 35–36, 479
stools (feces)   252–53, 258, 264–65
Strange, Steven K.   312
Strato of Lampsacus   620, 644, 647
Strohmaier, Gotthard   360, 376, 534, 562
Sudhoff, Karl   454
Suetonius   404
Sun Ssu-miao   56
Supplementum magicum   100, 103, 105–106
supplicationes   400
surface
- loci (*see* loci: surface)
- plectoid   415, 418
surgery, surgical incision   11, 38, 52, 58, 463, 465, 470, 479, 481–83, 571
Swain, Simon   659–60
Sylvius, Jacobus. *See* Dubois, Jacques
symmetry   313, 625–26
sympiptō   192–93
symptom   22–23, 25–27, 35, 40, 191–206 *passim*, 358, 385, 686, 690, 693–94
- subjective   203, 206
symptōma   191–206 *passim*
syndrome   540, 543–47
synētheia (customary usage of language)   551–52
Szeged-Kiskundorozsma-Kettőshatár   484

Taforalt (place in Morocco)   471
Tantillo, Ignazio   402, 404–405
taxo (Late Latin)   276
Tebtunis   9–10
technē. *See* expertise
technology
– and ethics (*see* ethics)
– mechanical (*see* mechanics)
– metallurgical   474
– military   215–16, 621, 631
– of writing   30
teleology   139–73 *passim*
– and necessity   156, 158–59
– "secondary"   172
Tell Beit Mirsim   476
Tell Duweir   477
Tell Mozan   482
Temkin, Oswei   48, 199, 355, 361, 376–77
Tempier, Étienne   73
testicle   275, 284
Thebes (Egyptian)   438
theology, political (of Diocletian)   405
Theophrastus of Eresus   5–6, 82, 237
theory, vs. practice   42
theriac   658–60, 666–67, 670
Thessalus of Tralles   437–48, 662–63
thief-detection spell   95, 104–108
Thomas Aquinas   67, 71, 349, 359
Thompson, D'Arcy W.   166, 272, 274–75
Thoth   108
Thucydides   38, 191–92, 203
Tiberius (Roman emperor)   659
Timaeus of Locri Epizephyrii   319–35
– as "the best astronomer"   319, 327
Timaeus of Tauromenium   664
Timna   477
Tiresias   283
Tomasini, Giacomo Filippo   456–57, 460
Tornikes, Georgios   350
Tortelli, Giovanni   452
torus   408, 412–14, 417, 421–13, 428–30
touch, sense of   245–46, 330–31, 479, 675, 681–86, 689–90, 692
Trasano   470
trepanation   24, 28, 463–85
triangle   77–78, 94, 328–30, 411–12, 417–20, 429–30, 468

trochilos (Greek)   274–75
trochos (Greek)   267–68, 273–75, 278–79, 281–83
Trophonius   282
Troy   469
truth, true account   64, 215, 219, 224, 310–11, 320, 333
tunic
– ocular   13–16
– uterine   11–14, 16
ṭupšar Enūma Anu Enlil   524, 528
ṭupšarrūtu   521
tylos (Greek)   300
tyrant
– happiness of   217, 221–22
– negative portrayal of   660–65
– three times removed from the oligarch   221

udjat-eye. *See* eye: of Horus
uentosa   121, 124, 130–32, 134
ulcer. *See* wound
Ulf, Christoph   20, 30
Ulrich III, Duke of Mecklenburg-Güstrow   568, 586, 600
ummânu   523–24
Urartu   478
uroscopy   578
Ursula of Saxe-Lauenburg   590
uterus   2, 11–16, 142, 147, 193, 199. *See also* tunic: uterine
Utrecht
– Hieronymus School in   567, 582
uvea   14. *See also* tunic: ocular

Vagelpohl, Uwe   265
vagina   270, 277, 279
van den Brock, Heinrich (= Brucaeus, Henricus)   585–86, 594
van der Eijk, Philip J.   22, 32–33, 40, 49, 146, 200, 340, 345–46, 364, 533–34
van der Waerden, Bartel L.   408, 410, 419, 421–23
van Riet, Simone   348, 363–64, 372–74
Vatican Library   260. *See also* manuscripts
Vedic myth   279
Vegetius   4

venesection 675, 692
ventricular localization of the cognitive faculties, theory of 367–70, 375
Vergil (Publius Vergilius Maro) 400–401
Vesalius, Andreas 569, 596–97
Villela-Petit, M. 311
violence 316, 318
virtual velocity, principle of 291
virtual work, principle of 289, 291, 304
visions 23, 34, 352–54, 357–59, 361, 364–65, 370, 372, 376, 379, 384. *See also* dreams: distinction from visions
vivisection 666, 669
void 196, 231, 233–35, 329–30, 647
von Staden, Heinrich 12, 15, 24, 31, 48, 54, 63, 111–12, 170–71, 173, 251, 284, 463, 518, 595, 615, 666, 681
Vronda 468, 480

Wanderwort 275, 281–82
Watson, Morrison 270–71, 273, 278
weasel 101–103, 268
wedge 290–94, 296, 299–301, 303–304, 632–33, 638–39, 647–48
Wendland, Paul 348
West. *See* East, vs. West
wheel 281, 294–97, 300–301
– and axle 290, 294, 301 (*see also* windlass)

William of Moerbeke 71–72, 152, 348, 351
William of Ockham 520
Williams, Raymond 529
windlass 290, 293–95, 300–301, 304. *See also* wheel: and axle
worm gear 300
wound 5–6, 8, 11, 22, 466, 469, 475, 479, 483
– of war 469, 479, 483

Xenophon 219, 575, 579, 660, 664
– *Cyropaedia* 575, 579, 662
Xenophon, Stertinius 460

year
– great (argued for by Philolaus) 222
– length of (according to fifth-century thinkers) 222
– *See also* goal-year

Zabarella, Jacopo 89–90
Zeno of Citium 240–41
Zeno of Elea 309
Zeyl, Donald J. 308, 321, 328
Zhmud, Leonid 332
zodiac 438–39, 441–42, 524, 526–27

www.ingramcontent.com/pod-product-compliance
Lightning Source LLC
Chambersburg PA
CBHW052007290426
44112CB00014B/2154